Geology and the Environment

SEVENTH EDITION

Dr. Bernard W. Pipkin is Professor Emeritus in Earth Sciences at the University of Southern California. He received his doctorate from the University of Arizona. After graduation, he worked for the U.S. Army Corps of Engineers on a variety of military and civil projects, from large dams to pioneer roads in rough terrain for microwave sites. He has been a consulting engineering geologist as well as a university teacher since 1965. Dr. Pipkin is past president of the National Association of Geoscience Teachers and is a Fellow in the Geological Society of America. He hosted the PBS 30-part program *Oceanus* that won a local Emmy for Best Educational Television Series. He recently shared the Clare Holdredge Award (1995) with Richard Proctor from the Association of Engineering Geologists for their book *Engineering Geology Practice in Southern California*. Dr. Pipkin is a private pilot with a flight instructor's rating; he took many of the aerial photos in this book. He and his wife, Faye, have three grown children and live south of Los Angeles in Palos Verdes.

Dr. D. D. "Dee" Trent has been working at, or teaching, geology since 1955. After graduating from college, he worked in the petroleum industry, where his geological skills were sharpened with projects in Utah, Arizona, California, and Alaska. When the company decided to send him to Libya, he decided it was time to become a college geology teacher. He taught for 28 years at Citrus Community College in Glendora, California, and along the way, worked for the National Park Service, did field research on glaciers in Alaska and California, visited numerous mines in the United States and Germany, and picked up a Ph.D. from the University of Arizona. He appeared in several episodes of the PBS telecourse *The Earth Revealed* and in *California's Glaciers*, an episode in the PBS series *California's Gold*. He served as an adjunct faculty member at the University of Southern California, where he taught field geology, and he continues to teach at the Desert Institute in Joshua Tree National Park. He and his wife raised two children in Claremont, California, and when not involved with geology, he's playing banjo in a Dixieland band.

Richard W. Hazlett teaches environmental geology, introduction to environmental studies, and a course in agroecology at Pomona College in southern California. He is a physical volcanologist by training, with research experience in the Pacific Northwest, Central America, Italy, New Zealand, the Aleutian Islands, Iceland, and Mongolia. He is a four-time winner of the Pomona College Wig Award for outstanding teaching, was recently selected as one of America's "Top 300" professors by the Princeton Review, and is a National Park Service media award winner for his publications *Pu'uhonua O Hōnaunau Discovery Book* and *Joshua Tree National Park Geology* (the latter co-written with Dee Trent). Other books he has written or co-authored include *The American West at Risk, Volcanoes: A Global Perspective, Roadside Geology of Hawaii,* and *Field Geology of Kilauea Volcano*. He presently holds the Steven M. Pauley Chair in Environmental Studies at Pomona.

Dr. Paul Bierman is a professor of Geology at the University of Vermont, where he engages people of all ages in the study of how Earth's surface works. For more than 25 years, he's done research in Vermont and many other places around the world, including far northern Canada, Greenland, central Australia, southern Africa, Brazil, Israel, and the American Southwest. His latest efforts use historic imagery to document the impact of people on the Vermont landscapes and the impact of landscape events on people and societies. Paul earned his B.A. from Williams College in 1985 and his M.S. and Ph.D. from the University of Washington, the latter in 1993. He has been at the University of Vermont ever since then, with appointments in Geology and the School of Natural Resources. In 1996, Paul was awarded the Donath Medal as the outstanding young scientist of the year by the Geological Society of America; he has since received a CAREER award from the National Science Foundation specifically for integrating scientific education and research. In 2005, Paul was awarded the National Science Foundation's Distinguished Teaching Scholar award in recognition of his ongoing attempts to integrate these two strands of his academic life. Together, Paul, his graduate and undergraduate students, and collaborators have authored 50 publications in refereed journals and books.

Geology and the Environment

SEVENTH EDITION

Bernard W. Pipkin
Emeritus, University of Southern California

D. D. Trent
Citrus College

Richard Hazlett
Pomona College

Paul Bierman
University of Vermont

Australia • Brazil • Japan • Korea • Mexico • Singapore • Spain • United Kingdom • United States

Geology and the Environment,
Seventh Edition, International Edition
Bernard W. Pipkin, D. D. Trent,
Richard Hazlett, Paul Bierman

Acquisitions Editor: Aileen Berg
Publisher: Yolanda Cossio
Developmental Editor: Jake Warde
Assistant Editor: Alexis Glubka
Editorial Assistant: Margaux Cameron
Media Editor: Katie Walton
Market Development Manager: Janet del Mundo
Content Project Manager: Carol Samet
Creative Director: Bruce Bond
Art Director: Pam Galbreath
Manufacturing Planner: Becky Cross
Rights Acquisitions Specialist: Tom McDonough
Production and Composition Service: Graphic World Inc.
Photo Researcher: PreMedia Global
Text Researcher: PreMedia Global
Copy Editor: Graphic World Inc.
Text and Cover Designer: Jeanne Calabrese
Cover Image: Steve and Donna O'Meara/National Geographic Stock

International Edition:

ISBN-13: 978-1-133-95820-8

ISBN-10: 1-133-95820-6

Cengage Learning International Offices

Asia
www.cengageasia.com
tel: (65) 6410 1200

Australia/New Zealand
www.cengage.com.au
tel: (61) 3 9685 4111

Brazil
www.cengage.com.br
tel: (55) 11 3665 9900

India
www.cengage.co.in
tel: (91) 11 4364 1111

Latin America
www.cengage.com.mx
tel: (52) 55 1500 6000

UK/Europe/Middle East/Africa
www.cengage.co.uk
tel: (44) 0 1264 332 424

Represented in Canada by
Nelson Education, Ltd.
www.nelson.com
tel: (416) 752 9100 / (800) 668 0671

Cengage Learning is a leading provider of customized learning solutions with office locations around the globe, including Singapore, the United Kingdom, Australia, Mexico, Brazil, and Japan. Locate your local office at: **www.cengage.com/global**

For product information and free companion resources: **www.cengage.com/international**

Visit your local office: **www.cengage.com/global**

Visit our corporate website: **www.cengage.com**

Printed in Canada
1 2 3 4 5 6 7 16 15 14 13 12

BRIEF CONTENTS

© 1996/MEDFORD TAYLOR/National Geographic Image Collection

© 2003/MICHAEL FAY/National Geographic Image Collection

© 2003/MICHAEL FAY/National Geographic Image Collection

© 2009/JASON EDWARDS/National Geographic Image Collection

© 2010/STEVE AND DONNA O'MEARA/National Geographic Image Collection

© 2010/PETE RYAN/National Geographic Image Collection

© 2010/PATRICK MCFEELEY/National Geographic Image Collection

© 2006/PETE RYAN/National Geographic Image Collection

© 2009/MICHAEL MELFORD/National Geographic Image Collection

Our Changing Planet

Environmental geology is the study of the relationship between humans and their geological environment. An underlying assumption is that this relationship is interactive. Not only do naturally occurring geological phenomena affect the lives of people each day, but also human activities affect geological processes, sometimes with tragic consequences. Given the fact of growing global populations, the potential risk for experiencing abrupt, if not catastrophic, geological and climatic changes has never been greater for more people. During your lifetime there is a strong possibility that you will either indirectly or directly experience an earthquake, flood, landslide, volcanic activity, runaway release of pollutants, or some other significant environmental event that impacts your local community. A basic, practical understanding of geology is essential for understanding such events and for accepting and moving beyond them proactively.

This edition of *Geology and the Environment* benefits greatly from collaborative association with the National Geographic Society (NGS) in Washington, DC, which has made its resources available, including map and image archives, Society explorer materials, and online links supporting this textbook. In this way, the NGS enriches and extends our ability to teach the very important topic of environmental geology. The authors, publishers, and NGS all share the same educational goals. In fact, proceeds from the sale of this product support NGS's mission to increase global understanding and promote conservation of our planet through exploration, research, and education.

Overview of Organization

We must be specific in speaking about geology—the study of the Earth. Unlike many introductory geology texts, this book does not deal extensively with the distant past, but rather concentrates on the here and now. Its contents may be divided into several general sections. The first three chapters "set the stage," providing a context in terms of time, space, and process, for all of the chapter material to follow. This broadly includes an introduction to environmental issues, the climate and oceans of Earth, the schematic of geological time, and the raw materials that make up the bedrock and soil beneath your feet.

Following the introductory section are two chapters strictly related to two formidable products of plate tectonics: earthquakes and volcanoes. As you'll learn throughout the book, threats to human welfare come not so much from the unleashed forces of nature, but rather from the placement of people, quite often unintentionally or through no personal choice, in the path of generally avoidable threats. Sometimes disaster is simply a matter of poor building and home construction and planning. Earthquakes, volcanism, flooding, and landslides have always been a part of the Earth scene and always will be. We can ease our experience of them by learning as much as we can about their triggers and potential impacts.

The section on hazards precedes a general treatment of more ordinary, but also equally important, natural processes: the formation and development of soils, and the dynamics of the hydrological cycle, including the behavior of streams, rivers, waves acting on coasts, groundwaters, and glaciers. Deserts—distinguished by a low level of hydrological activity—also warrant our attention as an important example of Earth's "marginal lands," regions in which it is difficult for people to survive sustainably without careful adaptation.

The final section of this book takes us home, in a sense, examining human extraction of nonbiological resources, that is, mineral and energy supplies. How do we obtain from the Earth what we need to facilitate our lifestyles? Are we extracting resources too rapidly, and are there ancillary environmental issues we must address? How can we be more responsible in our resource-consuming ways?

We wrap up by looking at how we deal with all the wastes that we generate. As you'll see, effective waste management requires a keen awareness of how pollutants can move through the environment when something goes wrong. Indeed, we must prepare for things occasionally not to work out as well as we would like. This brings us back to the basic premise of this book—that a keen understanding of geology and how the natural world "works" is our best hope for living well today and building for a better future.

In a culture where science and technology are interwoven with economics and political action, an understanding of the sciences is increasingly important. The National Science Foundation, the National Center for Earth Science Education, and several other prestigious earth science organizations are promoting earth science literacy in the United States for students from elementary school through college. The hope is that, through education, today's students will become better stewards of our planet than their parents have been. How much responsibility should we as individuals take for maintaining the planet that we share, and how much can we accomplish as individuals? Earth science literacy opens our minds to such questions. It helps us to appreciate Earth's beauty while also recognizing its limitations. One thing is certain: The planet can sustain just so much life, and it is now being pushed to the limit. The fragility of Earth was eloquently described by *Apollo 14* astronaut Edgar Mitchell:

> It is so incredibly impressive when you look back at our planet from out there in space and you realize so forcibly

that it's a closed system—that we don't have unlimited resources, that there's only so much air and so much water. You get out there in space and you say to yourself, "That's home. That's the only home we have, and the only one we're going to have for a long time." We had better take care of it. We don't get a second chance.

For You the Reader—Don't Be Scared! This book is meant to inform (we'd like to say "enlighten!") readers who are taking a one-semester undergraduate introductory-level geology course. As authors, each of us has taught undergraduate students extensively during our careers (see author profiles). Each of us is passionate about our efforts to reach out and provide insights that we have found exciting, even emotionally engaging. We attempt, we hope with reasonable success, to reduce the complex, beautiful mechanism of the geological world to terms that stimulate your interest and desire to learn more. Wherever feasible, we also try to make the topical material directly relevant. Not all is "gloom and doom," nor should it be. Some great, unsung success stories in adaptation, engineering, and shear human ingenuity also shine through. But be prepared to face some possibly strange new facts and perspectives about Planet Earth—all of which, we hope you'll agree, can be potentially useful to know.

This book is also a "work in progress." Science is an iterative, continually evolving pursuit. New areas of importance continually appear, while people lose interest in others and let them recede into the past—never fully lost, but dormant or meaningless in the present day. Simply compare this textbook with similar ones written 10, 20, or even 40 years ago and perhaps you'll be amazed, both by what hasn't changed and what has in terms of emphasis and basic understanding.

Science is also an intensely self-correcting process, an effort to arrive at the "truth" about the raw, natural state of the world (and cosmos) in as impartial and nonjudgmental a way as possible. Scientific progress is driven both by new technology and by ordinary, very human curiosity. Although it is possible to tell a lie as a scientist, one can't get away with doing that for long, because science *requires* intensive peer review and public correction. Science is all about the sharing of carefully nurtured insights with others.

Recently, one of us was confronted by a student who said, matter-of-factly, "I'm a humanities major; I don't *do* science." We must acknowledge that (however inadvertently) teachers sometimes fail to make science as clear and easy as it can and should be. The ongoing student refrain in a well-run science class should be, "Wow, that makes sense!" Unfamiliar mathematical shorthand—wonderful in its utility—all too readily obfuscates, and the timeline imposed by a syllabus is often the enemy in trying to impart a lot of information and new training. On the other hand, to close one's mind to possibly learning about how science works in the most basic way, and to blind oneself to the awesome insights it provides about simply being alive in the world, is a sad thing, a form of self-deprivation we hope you'll strive to avoid as best you can—even if you empathize with the earlier comment. We expect that you will have to work hard to grasp some ideas and concepts, especially in the first few "orientation" chapters. We believe that there is simply no other way to bootstrap to a higher level of understanding without some modicum of hard thinking. Learning is a dance with two partners: the instructor to provide the information as skillfully as possible and the student to aggressively seek out its meaning, irrespective of personal background. It cannot be merely a passive data dump—a simple matter of memorization and regurgitation. If we sometimes fail to make things as clear as we might in our writing, we offer our apologies. Know at least that we have tried to place ourselves in your shoes, well aware too of just how deep and broad this field is.

We attempt to present a portrait of environmental geology as honestly as we can. We certainly take full responsibility for any scientific errors in our text (and indeed expect there will be a few that crop up in the normal course of evolving research). May your class instructor be quick to jump in with fresh, critical insights as you learn! In the end, this is all about *you*.

Distinctive Features of the Book

At most schools, the course for which this book is used is a component of the general education curriculum. General education broadens and enriches students' lives and minds beyond the specialization of their major interest. To this end, **Galleries** of photos at the end of each chapter illustrate many geological wonders and the ways in which humans and Earth interact. The Galleries are intended to stimulate students' curiosity about and appreciation for natural geological wonders, and the ever dynamic engagement of people with the environment.

Have You Ever Wondered? questions open each chapter, stimulating thought about the material that follows. In addition, **Questions to Ponder** accompany the chapter opener section and selected Case Studies. These questions help students develop critical thinking skills and apply the scientific method. **Case Studies** at the end of each chapter highlight the relevance of the text discussion. These cover a broad spectrum of subjects and geographical areas, but many of them focus on the causes and aftereffects of bad environmental and geological planning. Also within each chapter are several provocative **Consider This** questions.

These questions require students to apply the information just presented in the text and, thus, reinforce their learning. The questions will stimulate classroom discussion as well.

At the end of each chapter is a list of **Key Terms** introduced in the chapter, a **Summary** of the chapter in outline form, **Study Questions** geared to test understanding of the chapter's key concepts, and a list of related books and articles (**References and Further Readings**).

The Seventh Edition

- In this seventh edition of *Geology and the Environment*, we have attempted to incorporate as many of the reviewers' suggestions as possible. Recognizing that Case Studies are a natural focal point in many courses, we have incorporated more critical information into many Case Studies. We continue to strive to present the most effective figures and photos to support student learning. This includes the careful placement of the figures near their references in the text.
- With the robust support of the NGS, we have replaced and upgraded many photographs and added some key maps and diagrams to improve learning (Spotlight Figures). We have also added a new section in some chapters highlighting the achievements of select NGS explorers—exemplary scientists, entrepreneurs, and activists who provide students with models for where they might, with the best type of ambition, direct their own lives.
- This edition continues to include the interactive media program called CourseMate, which has been seamlessly integrated with the text, enhancing students' understanding of important geological processes. It brings geology alive with animated figures based directly on figures in the text (Active Figures), media-enhanced activities, tutorials, and more. Like other features in our new edition, it encourages students to be curious, to think about geology in new ways, and to connect with their newfound knowledge of the world around them.
- Chapter 2: This chapter has been expanded to include a section dealing with potential human modification of the global climate system, introducing students to the concept of the Anthropocene—a proposed new epoch in the geological time scale. Much public controversy in recent years has swirled around these topics, and we have had to navigate them with one directive in mind—to present the view of the world representing the balance (in most cases overwhelming) of *scientific* consensus.
- Chapter 4: We provide important relevant updates in our discussion of earthquakes, highlighting the important events and lessons learned from the 2011 Christchurch and Tohoku-Oki disasters. We also have added new sections dealing with some of the peculiar but nonetheless characteristic observations people make about quakes, including seismo-luminosity and observable surface waves. The sections dealing with seismic wavelength, wave period, and building design have also been strengthened and *linked* more coherently. The chapter retains its overall nuts-and-bolts practicality. We wrote it with special regard for students who live in earthquake country and may play important future roles in community development.
- Chapter 5: This chapter has been almost entirely rewritten with several objectives in mind: to provide a practical, descriptive framework for classifying and measuring volcanic eruptions; to give students a modern insight into how volcanoes "work"; and to clarify both the dangers and benefits of volcanic activity. Like Chapter 4, most of the deeply technical information is provided up front, with human interest appearing further along. Instructors can pick and choose specific sections in each chapter without need for students to read through them entirely, though we certainly encourage reading the "whole story."
- Chapters 6 through 10 and 12 have all new Galleries with exciting and informative images and captions that illustrate some of the most germane material in the chapters. In Chapter 10, we've added information about the Fukushima earthquake and the damaging tsunami that followed.
- Chapter 11: The section on global climate change has been brought up to date with the most recent information on changes in Arctic ice cover between 1979 and 2011, the Arctic Oscillation, and the impact of these changes on European weather in the past few winters. New are the effects of El Niños and La Niñas on changing global sea level and up-to-date data on the rate of sea-level rise.
- Chapter 13: The genetic classification of mineral resources is expanded, and data on U.S. import reliance on nonfuel materials and the value of nonfuel minerals produced in the United States have been brought up to date. Mining of base metals and precious metals in Europe has been in limbo for decades, but information from recent exploration is presented that illustrates the potential for resumption of mining base and precious metals in the United Kingdom and mainland Europe. Health issues associated with mountaintop removal mining in Appalachia have been added, and a new Case Study, Porphyry Copper and Plate Tectonics, details the massive size and enormous global reserves of porphyry copper deposits, their global economic impact, and their geological origin that make these deposits worthy of a close look. In addition, the Case Study covering the ongoing saga of the Zortman-Landusky mines has been updated.
- Chapter 14: This chapter has been updated extensively. Included is an evaluation of the light, tight-oil deposits of

the Bakken Shale of North Dakota, and the Barnett and Eagle Ford shales of Texas, accompanied by a discussion of the pros and cons of hydraulic fracturing necessary for recovering tight oil. We have included a recalculation of world and domestic exponential expiration dates of coal reserves based on minimal increases in production. We also cover the new Environmental Protection Agency regulations on air quality and mercury emissions that threaten closure of more than 100 older fossil-fueled power plants. In addition, the number of new nuclear plants under construction worldwide and the specter of danger from nuclear meltdown spawned by the 2011 Fukushima, Japan, disaster are discussed.

- Chapter 15: This chapter has also been extensively rewritten and updated to include a detailed look at plastics, the sociology and patterns of littering, current modes of interim nuclear waste storage, the history of radwaste disposal at sea, and groundwater contaminant plumes. We have also provided a tighter description of landfill design and function and added a new Case Study dealing with the historical roots of the environmental justice movement originating in Warren County, North Carolina. The text continues to emphasize remediation and prevention, an outgrowth of the authors' professional geological experiences. All of the chapters on geological hazards have dedicated sections on mitigation options, and resource and pollution issues are considered in terms of the problems we face and the potential ways to help forestall or lessen the impacts of these problems.
- We also continue to use the systems approach in this edition—the idea that all of Earth's reservoirs (atmosphere, hydrosphere, solid earth, biosphere, and extraterrestrial) and the processes acting within them are interconnected.
- Environmental legal issues are discussed in the text where they are applicable, rather than being placed in a separate chapter near the end of the text.
- All chapters have been updated in terms of data (where available), art, and photos.

Supplements

Instructor Resources

Global Geoscience Watch

Updated several times a day, Global Geoscience Watch is an ideal one-stop site for classroom discussion and research projects for all things geoscience. The site is broken into the four key course areas (Geography, Geology, Meteorology, and Oceanography), so you can easily get to the most relevant content available for your course. You and your students will have access to the latest information from trusted academic sources, news outlets, and magazines. You will also receive access to statistics, primary sources, case studies, podcasts, and much more!

Geology CourseMate

Cengage Learning's Geology CourseMate brings course concepts to life with interactive learning, study, and examination preparation tools that support the printed textbook. Watch student comprehension soar as your class uses the text-specific flashcards, videos, animations, quizzes, and other interactive tools to enhance their learning. CourseMate goes beyond the book to deliver what your students need! Learn more online at www.cengage.com/coursemate.

WebTutor Toolbox for WebCT or Blackboard

Jump-start your course with customizable, rich, text-specific content within your Course Management System. WebTutor offers a wide array of Web quizzes, activities, exercises, and Web links. Robust communication tools—such as a course calendar, asynchronous discussion, real-time chat, a whiteboard, and an integrated e-mail system—make it easy to stay connected to the course.

PowerLecture with JoinIn™ Student Response

A complete, all-in-one reference for instructors, the PowerLecture DVD contains customizable PowerPoint presentation slides with images from the text, including stepped art and zoomable art figures, a video library, and Active Figures that interactively demonstrate key geological concepts. In addition to providing you with fantastic course presentation material, the PowerLecture DVD also contains electronic files of the Test Bank and Instructor's Manual. This resource is available at no additional charge to qualified adopters.

ExamView

Create, deliver, and customize tests and study guides (both print and online) in minutes with this easy-to-use assessment and tutorial system. Also available on the PowerLecture DVD.

Online Instructor's Manual with Test Bank

This comprehensive resource provides chapter summaries and lecture suggestions along with video and Web site resource references. The Test Bank provides multiple-choice, true/false, short answer, and essay questions.

Student Resources

Global Geoscience Watch

Use Global Geoscience Watch to stay current in your course. Updated several times a day, this resource gives you access to the latest information from trusted academic sources, news outlets, and magazines. You will also receive access to statistics, primary sources, case studies, podcasts, and much more. The Global Geoscience Watch is an ideal one-stop site for all your research needs.

Geology CourseMate
Make the most of your study time by accessing everything you need to succeed in one place. Read your textbook; take notes; review flashcards; watch videos, animations, and Active Figures; and take practice quizzes online with CourseMate. Log in or purchase access at www.cengagebrain.com.

Acknowledgments

We gratefully acknowledge the thoughtful and helpful reviews by a great number of individuals, some of whom also reviewed earlier editions. We also acknowledge the contributions of other reviewers and those who have generously contributed published and unpublished materials and photographs as this book evolved into its seventh edition. These have truly helped us build this book. The list is long and we offer our thanks to all:

Herbert G. Adams, *California State University, Northridge*
Jürg Alean, *Kantonsschule Zücher Unterland, Bülach, Switzerland*
Richard M. Allen, *University of Wisconsin–Madison*
Thomas B. Anderson, *Sonoma State University*
Kenneth Ashton, *West Virginia Geological Survey*
James L. Baer, *Brigham Young University*
Ed Belcher, *Wellington, New Zealand*
William B. N. Berry, *University of California, Berkeley*
Robert Boutilier, *Bridgewater State College*
Tom Boving, *University of Rhode Island*
Kathleen M. Bower, *Eastern Illinois University*
David Bowers, *Montana Department of Environmental Quality*
Lynn A. Brant, *University of Northern Iowa*
T. K. Buntzen, *Alaska Department of Natural Resources*
Don W. Byerly, *University of Tennessee, Knoxville*
Susan M. Cashman, *Humboldt State University*
Elizabeth Catlos, *Oklahoma State University*
Dan Cayan, *U.S.G.S.*
Ward Chesworth, *University of Guelph*
Christopher Cirmo, *State University of New York, Cortland*
Robert D. Cody, *Iowa State University*
Kevin Cornwell, *California State University–Sacramento*
Jim Cotter, *University of Minnesota, Morris*
Rachael Craig, *Kent State University*
John Dassinger, *Maricopa Community College*
Charles DeMets, *University of Wisconsin, Madison*
Terry DeVoe, *Hecla Mining Company*
Lisa DuBois, *San Diego State University*
Greg Erickson, *Sullivan County Community College*
Mark W. Evans, *Emory University*
Edward B. Evenson, *Lehigh University*
Larry Fegel, *Grand Valley State University*
John Field, *Western Washington University*
Lydia K. Fox, *University of the Pacific*
Earl Francis
Robert B. Furlong, *Wayne State University*
Marion M. Gallant, *Colorado Department of Public Health and Environment*
John Gamble, *Wellington, New Zealand*
Josef Garvin, *Eden Foundation, Falkenberg, Sweden*
Rick Giardino, *Texas A&M University*
Gayle Gleason, *SUNY, Cortland*
Raymond W. Grant, *Mesa Community College*
John E. Gray, *U.S.G.S.*
Bryan Gregor, *Wright State University*
Tark S. Hamilton, Consultant
Gilbert Hanson, *State University of New York, Stony Brook*
Edwin Harp, *U.S.G.S.*
Raymond C. Harris, *Arizona Geological Survey*
Douglas W. Haywick, *University of South Alabama*
Eric Henry, *University of North Carolina, Wilmington*
Lynn Highland, *U.S.G.S.*
Barbara Hill, *Onondaga Community College*
Roger D. Hoggan, *Ricks College*
Bryce Hoppie, *Mankato State University*
Alan C. Hurt, *San Bernardino Valley College*
Pam Irvine, *California Division of Mines and Geology*
David D. Jackson, *UCLA*
Gaoming Jiang, *China Academy of Sciences*
Randall Jibson, *U.S.G.S.*
Bill Kane, *University of the Pacific*
Steve Kenaga, *Grand Valley State University*
Marcie Kerner, *Arco, Anaconda, Montana*
Hobart King, *Mansfeld University*
Joe Kirschvink, *California Institute of Technology*
Peter Kresan, *University of Arizona*
Robert Kuhlman, *Montgomery County Community College*
Kevin Lamb

Kenneth A. LaSota, *Robert Morris College*
Douglas J. Lathwell, *Cornell University*
Rita Leafgren, *University of Northern Colorado*
Joan Licari, *Cerritos College*
Rick Lozinski, *Fullerton College*
Lawrence Lundgren, *University of Rochester*
Michael Lyle, *Tidewater Community College*
Berry Lyons, *University of Alabama*
Harmon Maher, *University of Nebraska, Omaha*
Alex K. Manda, *East Carolina University*
Peter Martini, *University of Guelph*
John Maurer, *CIRES, University of Colorado*
Larry Mayer, *Miami University*
David McConnell, *University of Akron*
Garry McKenzie, *Ohio State University*
Lisa McKeon, *U.S.G.S.*
Matthew L. McKinney, *University of Tennessee*
Robert Meade, *California State University, Los Angeles*
Chuck Meyers, *U.S. Department of the Interior, Surface Mining Reclamation and Enforcement*
Siddhartha Mitra, *East Carolina University*
William Mode, *University of Wisconsin, Oshkosh*
Marie Morisawa, *State University of New York, Binghamton*
Jack A. Muncy, *Tennessee Valley Authority*
George H. Myer, *Temple University*
William J. Neal, *Grand Valley State University*
James Neiheisel, *George Mason University*
Jennifer Nelson, *Indiana University–Purdue University Indianapolis*
Michael J. Nelson, *University of Alabama, Birmingham*
Ed Nuhfer, *University of Colorado, Denver*
June A. Oberdorfer, *San Jose State University*
Lloyd Olson
Alberto E. Patiño Douce, *University of Georgia*
Darryll Pederson, *University of Nebraska–Lincoln*
Libby Pruher, *University of Northern Colorado*
Kristin Riker-Coleman, *University of Wisconsin Superior*
Albert J. Robb III, *Mobil Exploration and Producing U.S., Inc., Liberal, Kansas*
Charles Rovey, *Missouri State University*
Robert Sanford, *University of Southern Maine*
Steven Schafersman, *Miami University*
James L. Schrack, *Arco, Anaconda, Montana*
Feride Schroeder, *Cuesta College*
Robert Schuster, *U.S.G.S.*
Geoffrey Seltzer, *Syracuse University*
Conrad Shiba, *Centre College*
Jennifer Shosa, *Colby College*
Edward Shuster, *Rensselaer Polytechnic Institute*
Kerry Sieh, *Earth Observatory of Singapore*
Susan C. Slaymaker, *California State University, Sacramento*
Joe Snowden, *University of Southeastern Missouri*
Frederick M. Soster, *DePauw University*
Neptune Srimal, *Florida International University*
Konrad Steffen, *CIRES, University of Colorado*
Dean Stiffarm, *Environmental Control Officer, Fort Belknap Reservation*
Hongbing Sun, *Rider University*
Terry Swanson, *University of Washington*
Siang Tan, *California Division of Mines and Geology*
Glenn D. Thackray, *Idaho State University*
Peter J. Thompson, *University of New Hampshire*
Joan Van Velsor, *California Department of Transportation*
Adil M. Wadia, *University of Akron, Wayne College*
Peter W. Weigand, *California State University, Northridge*
Todd Wilkinson, *Bozeman, Montana*
Nancy S. Williams, *Missouri State University*
Nathaniel W. Yale, *Pomona College*
Simon Young, *Montserrat Volcano Observatory*
Ning Zing, *University of Maryland*

Special thanks are due to Jake Warde, Developmental Editor, who managed the production of this edition, and to some critically important people at Cengage Learning: Aileen Berg, Sr. Acquisitions Sponsoring Editor; Margaux Cameron, Earth Science Editorial Assistant; and Carol Samet, Senior Content Project Manager. Many thanks to Leila Hishmeh and Jen Shook for help in acquiring the many photos and maps from the National Geographic Society. At Graphic World Publishing Services, we would like to thank Jeff Somers and Rhoda Bontrager, who skillfully managed all aspects of the production process. At PreMedia Global, we also thank Christina Ciaramella for her tireless work tracking down photos and photo permissions, and Melissa Tomaselli for acquiring the text permissions.

Geology and the Environment

SEVENTH EDITION

Natural gas flames from oil-drilling pipes set in the rain forest of the Amazon Basin, Ecuador.

© 2003 STEVE WINTER/National Geographic Image Collection

Have you ever *wondered...*

1

Humans, Geology, and the Environment

The science of the environment is just the classic natural science of the past century in the context of modern environmental issues. It is Thoreau with a computer rather than a pen.

—Professor Richard Turco, University of California at Los Angeles

1. If and why we really need to be concerned about "overpopulation"?
2. Why many environmentalists are so concerned about high levels of consumption in wealthy countries such as the United States?
3. Why many environmentalists are so concerned about the way we produce our food?
4. What a scientist is and how scientists work?
5. What GPS and GIS are?

Ever-Present Environmental Questions

During the past few years, people have become increasingly aware of human vulnerability to nature's power: A giant tsunami drowned nearly a quarter of a million people in the Indian Ocean Basin in just a few terrible hours in late December 2004; eight months later, a major hurricane, Katrina, slammed into the city of New Orleans, causing the greatest urban disaster in American history since the shocking San Francisco earthquake and fire of 1906; a few weeks later, hurricane Wilma, the strongest Atlantic storm in history, forced evacuations in Mexico's Yucatan Peninsula. Suddenly, the topic of global warming became a part of everyday conversation, together with concerns about Peak Oil, organic versus industrial farming, water depletion in the American Midwest, and a possible "supervolcanic" eruption in Yellowstone National Park. To borrow a phrase from well-known environmentalist David Orr, we now have "Earth in mind," and many of us have grown uneasy about the environmental future.

Disasters and environmental changes cannot always be prevented, of course, although some *are* avoidable, given foresight and planning. In fact, the perils posed by unleashed natural forces and seemingly inexorable environmental change are largely the fault of people making bad choices, failure to heed warning signs, and inadvertent or deliberate placement in harm's way (**Figure 1**).

questions to ponder

1. To what extent does society make things harder on itself through lack of understanding about how and where potentially harmful natural phenomena take place?
2. In what ways does this lack of understanding become evident?
3. What important environmental questions, if any at all, potentially impact your own life?

© Richard Hazlett

Figure 1 Home versus the ocean (Malibu, California)—Hmmm, what a view! What's it like in a storm, though?

Roger Tomlinson

CourseMate Connection
Start off this chapter by listening to a podcast with Roger Tomlinson, the inventor of Geographic Information System (GIS), which made it possible to digitize maps. **Log in to the CourseMate at www.cengagebrain.com to listen now.**

EACH OF THESE CONCERNS PERTAINS to our relationship with planet Earth on a grand scale. However, even at a modest level, you are likely to face questions in your own life that deal with issues of humans in nature: "What do I do if someday I have to buy a house in an area where there might be landslides?" "What becomes of the waste I'm washing down the sink, and is my tap water safe to drink?" "Is it really bad to be opening a mine in my local forest, as the Sierra Club says? Or are they exaggerating the problems?" This textbook will take you on a journey to build your awareness of the ultimate environment—the physical Earth—helping put great events in a more practical perspective. It will also help you evaluate such questions as these in ways that someday might prove to be personally meaningful.

What Is Environmental Geology?

In his play *As You Like It*, William Shakespeare (1564–1616) writes, "All the world's a stage, and all the men and women merely players. They have their exits and their entrances, and one man in his time plays many parts." This is one of the Bard's best-loved and most-quoted lines. Starting with that metaphor, this book takes a close look at Shakespeare's "stage" and how its changing scenery can affect the many exits, entrances, and parts that we play on it. The stage, of course, is meant to be Earth itself, together with its retinue of ever-changing societies and civilizations—the "background" for our individual human stories. The technical term for the study of the physical Earth is *geology*, a word unknown in Shakespeare's day.

An earlier British author, Geoffrey Chaucer (1342–1400), first introduced to the English literature a word used in this book's title: **environment**. The word originally appeared in early medieval France as *environement*, meaning "surroundings." Chaucer found *environment* a clever choice for building rhymes, and he composed many lines of poetic verse describing people in nature.

Today, the news is abuzz with references to "environmental problems" and "environmentalism" in general. Although we can clearly appreciate an environmental problem when we see it—smog, for instance, or a startlingly red trickle of fish-killing mine water—the meaning of the term *environmentalist* is less clear. Just *who* or *what* is an environmentalist?

If such a person is someone who believes that it's important that we strive to keep our physical environment from deteriorating to the point of harming people and wiping out wildlife, then who among us is *not*, on at least rare occasions, an environmentalist? No one (rationally speaking) wakes up in the morning thinking it would be a fine day to destroy Earth. Unfortunately, years of hard politicking to bring about solutions to environmental problems have given the term a somewhat pejorative overtone. And many of us have become dependent on jobs that *do* destroy at least small parts of the natural world each day, creating conflicting values and awkward intellectual compromises within our own minds. Some opponents of environmentalism regard green activists not only as protectors of nature but also as attackers of livelihoods and even of an entire national way of life.

However, when we begin to face problems on a global scale that we all agree need to be tackled—the deterioration of the ozone layer, the threat of PCB pollutants, the need to protect marine fisheries—then political divisiveness and dissonance almost magically disappear. People are often great at finding solutions in a mutually agreeable way once they clearly recognize the urgency of an issue (**Figures 1.1** and **1.2**).

Environmental geology is the study of how the natural workings of the physical Earth apply to the problems we face in dealing with our environment. Virtually all of these problems have arisen because people have not appreciated

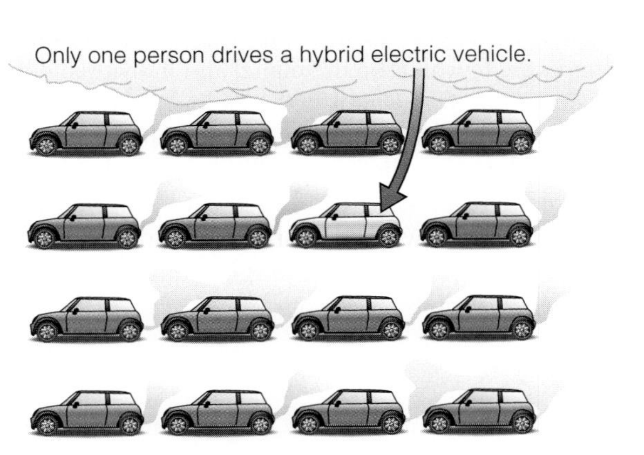

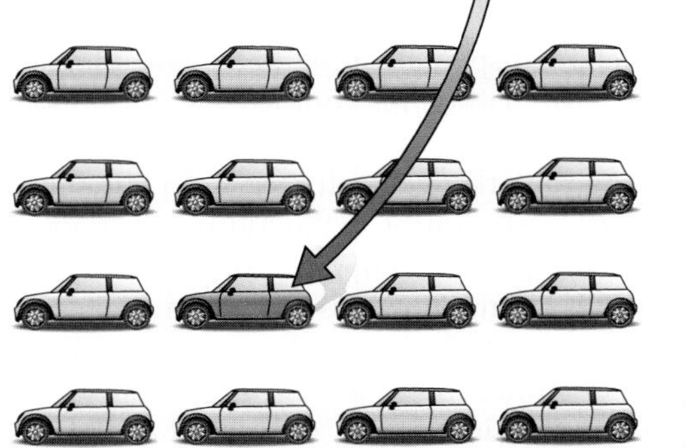

© Cengage Learning

Figure 1.1 Collective behavior can change the global-scale environment, but lone behavior usually counts for little.

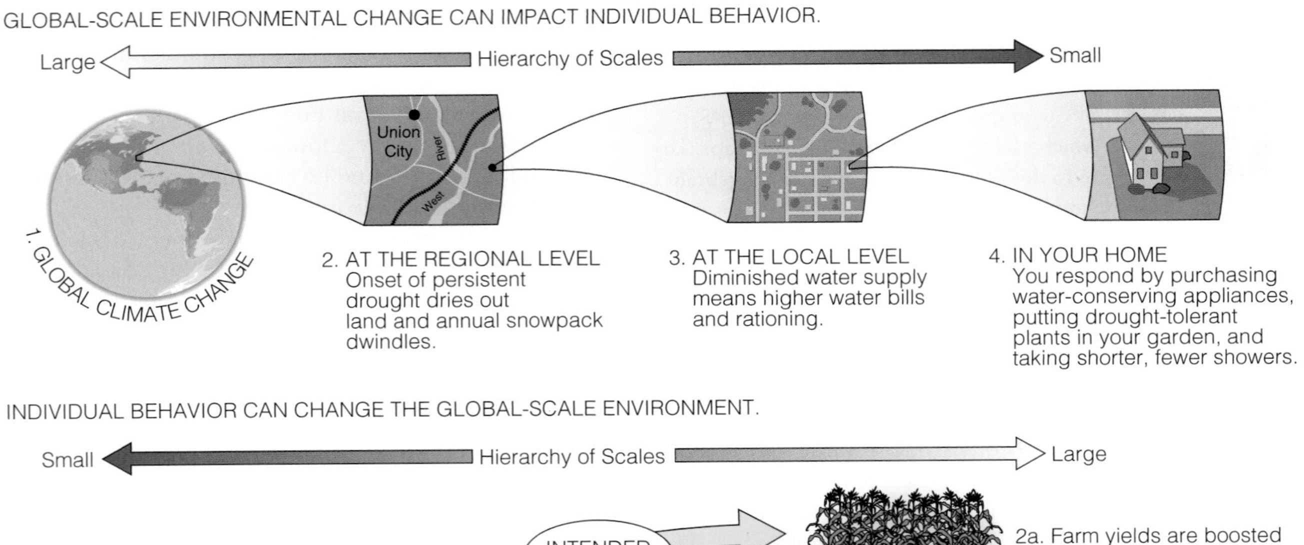

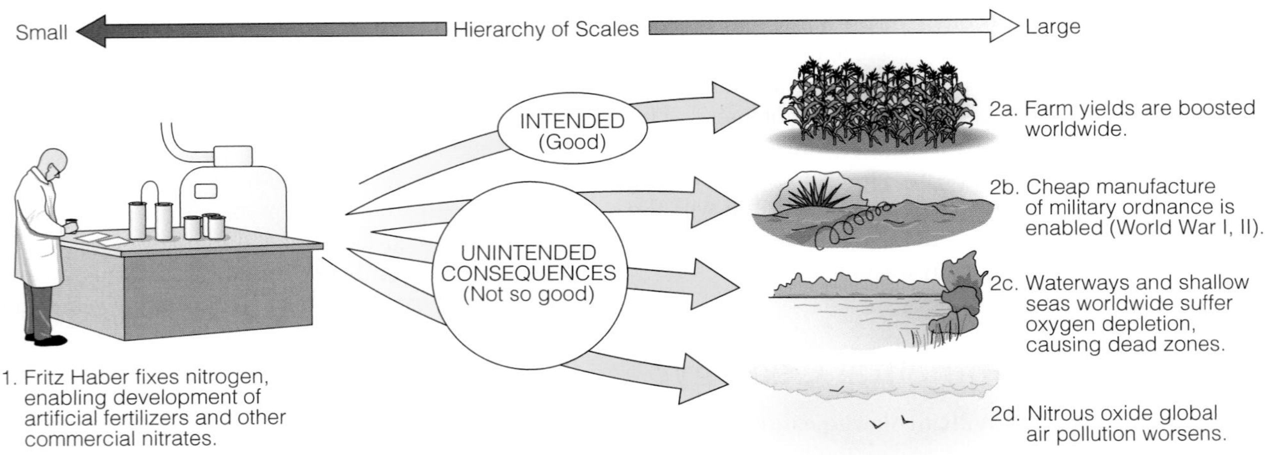

Figure 1.2 Human–Earth interaction is a two-way street.

how the natural world works. We are inventive creatures, to say the least, and introducing a new technology, such as synthetic fertilizer or the internal combustion engine, brings with it an almost childlike excitement—the thrill of playing with a new toy. But history shows that we tend to "leap" before we "look," unable or unwilling to assess the potential impacts of new inventions on the natural world, and ultimately on ourselves. Those consequences, ranging from water pollution to air pollution and worse, often come back to haunt us, because we live in a finite world where everything is physically and chemically interlinked. Only so many resources and so much land exist, and everything is connected in a remarkably self-sustaining natural system—one that has functioned spontaneously to support life, according to radiometric dating, for approximately 4 billion years. The more you learn about how Earth functions, the more wonderful and interlinked it all seems. To eliminate further unintended consequences in the future, it is important to understand our planet's behavior well.

That is the spirit of this book; it is not only a vehicle to give some of you general elective college credit, but we hope a stimulant for building new lifelong insight, appreciation, and sensibility for interacting with Earth. Harking back to Shakespeare, it is our time to appear onstage. Let's make our act a good one!

Humans and the Geological Environment

Throughout most of our history, we have liked to think of ourselves as somehow apart from nature. This, it turns out, is a painful delusion, reinforced in the past by ignorance and religious traditions. Moreover, we have liked to think of Earth as a comfortably big place. We can throw something away—say, by flushing it down the toilet or by tossing it out the car window—and it will simply disappear into the eternal vastness of the environment—out of sight, out of mind. When there were few of us on the planet and individual possessions were minor, throw-away behavior had no significant impact on the world. But now there are more than 7 billion people, and within your lifetime you'll likely see the world become even more pressingly crowded. According to the United Nations, there will be nearly 9 billion people by 2050, even factoring in the declining birthrates that have appeared since the 1960s.

IPAT

The **IPAT equation** is a measure of human environmental impact: $I = P \times A \times T$, where I = impact, P = population, A = affluence (standard of living), and T = technology. It is a qualitative expression; no one can put exact numbers to its factors, but it is important to bear in mind as a conceptual framework when we talk about how we relate to the geological environment.

The IPAT equation tells us that wealthy developed states with a lot of available technology (e.g., computers per household) and affluence (disposable incomes) have enormous environmental impacts. In a typical lifetime, for instance, an American will consume more than 35 times the resources of an average citizen of India, whose country is much less economically developed. One of the conveniences of globalization and export-import trade economies is that we in the developed world *rarely see our personal environmental impacts;* they take place indirectly thousands of kilometers away, often on different continents.

A case in point concerns wood products. Prior to World War I (1914–1918), the wood we used in the United States was harvested domestically, but by 1900, the nation's forests had been so severely cut back that people began to worry about an American timber famine, rather like we worry about the energy crisis today. President Theodore Roosevelt and his chief forester, Gifford Pinchot, established the National Forest Service to help address this concern. They believed that, with the loss of its forests, the economy of the United States would falter, and, with that, the country's national power. Although conservation imposed by the federal government helped save America's forests, the real salvation came in the forms of introducing new technologies, developing new resources, and building a global economy. The first petrochemical plant in the world opened in Bayonne, New Jersey, in 1923, and soon we were mass producing new construction materials, such as plastics, which could substitute for wood in many manufactured products. Likewise, after World War II, we no longer needed to satisfy our demand for wood from an exclusively domestic market. Lumber could be imported in tremendous quantity from tropical countries, such as Brazil and Indonesia, thanks to the low cost of fuel. By the year 2000, more land was under forest cover in the United States than a century before. Denuded landscapes had recovered, with secondary forest growth in many parts of the country, especially in the South and East. But over the same interval of time, the American per capita demand for wood actually *rose*. In fact, world demand for forest products grew threefold between 1975 and 2005, after more than doubling in the previous 70 years.

At present, the average minimum annual consumption of paper required per person for basic communication and literacy is about 30 to 40 kilos (81–120 pounds). In the United States, each of us uses a staggering 330 kilos (890 pounds) of paper averaged per annum—much of it in the form of throwaway commercial packaging. Contrast this with sub-Saharan Africa, a large region where poverty and hunger are almost overwhelming; there the average consumption of paper is only about 15 kilos (47 pounds) per person each year.

The clear-cutting of forests for paper, pulp for plywood, and roundwood for construction creates many geological

Table 1.1 Estimated World Soil Degradation, 1945–1990

	Degraded Land Area as a Percentage of Vegetated Land		
Region	Total	Light Erosion*	Moderate to Extreme Erosion†
World	17	7	10
Asia	20	7	13
South America	14	6	8
Europe	23	6	17
Africa	22	8	14
North America (United States, Canada)	5	1	4
Central America, Mexico	25	1	24

Sources: Various sources compiled by World Resources Institute and the United Nations. 1992. *World resources 1992-93*. New York: Oxford University Press.

**Light:* crop yields reduced less than 10%.

†*Moderate:* crop yields reduced 10% to 50%. *Extreme:* no crop growth possible. About 9 million hectares worldwide exhibit extreme erosion, less than 0.5% of all degraded lands.

problems, including landslides, soil erosion, and the silting up of streams and waterways. By transferring our demand for wood products overseas, we transfer environmental problems to other countries as well, where standards and regulations may be much less strict and the ensuing environmental damage much greater.

In less-developed countries, the $I = P \times A \times T$ equation must be reformulated, given that there is often little technology or affluence to take into consideration. $I = P$ means not only that numbers of people correlate more closely with actual environmental impacts but also that the problems that develop tend to be closer to home, because resources usually must be collected and used locally. There is much less "out of sight, out of mind" warping personal perspectives. Farmers can see their land eroding away as a result of windstorms blowing across their freshly plowed fields (**Table 1.1**), and household refuse is not hauled away to a sanitary landfill kilometers away, but accumulates in the backyard or in the streambed behind it. In fact, at least one sixth of the world's population—more than a billion people—suffer malnutrition on a daily basis, most in what we would regard as environmentally squalid conditions. When we speak about the world suffering from *overpopulation*, we must also speak of *overconsumption* and *resource inequities*. The characteristics and remoteness of the stresses we put on Earth are related directly to the society and economy in which we live.

Carrying Capacity

One of the most famous passages from all of environmental literature comes from Aldo Leopold (**Figure 1.3**), an ecologist who was a forest ranger in New Mexico assigned to shoot wolves for the then-new National Forest Service. The

© CORBIS

Figure 1.3 Aldo Leopold (1887–1948) is widely regarded as the father of ecology. His book *A Sand County Almanac* is a classic in the environmental literature.

wolves threatened the livestock that ranchers were allowed to let graze on public forestland by permit from the federal government. Leopold wrote an essay, "Thinking Like a Mountain" (1948), about what his experience taught him. Read this excerpt carefully and reflect about how it pertains to geology and the natural environment:

> Only the mountain has lived long enough to listen objectively to the howl of a wolf.…My own conviction on this score dates from the day I saw a wolf die. We were eating lunch on high rimrock, at the foot of which a turbulent river elbowed its way. We saw what we thought was a doe fording the torrent, her breast awash in white water. When she climbed the bank toward us and shook her tail, we realized our error: it was a wolf. A half-dozen others, evidently grown pups, sprang from the willows and all joined in a welcoming melee of wagging tails and playful maulings. What was literally a pile of wolves writhed and tumbled in the center of an open flat at the foot of our rimrock. In those days we had never heard of passing up a chance to kill a wolf. In a second we were pumping lead into the pack, but with more excitement than accuracy. When our rifles were empty the old wolf was down, and the pup was dragging a leg into impassible slide-rocks. We reached the old wolf in time to watch a fierce green fire dying in her eyes. I was young then, and full of trigger-itch; I thought that because fewer wolves meant more deer, that no wolves would mean a hunters' paradise. But after seeing the green fire die, I sensed that neither the wolf nor the mountain agreed with such a view.
>
> Since then I have lived to see state after state extirpate its wolves. I have watched the face of many a newly wolfless mountain and then seen the south-facing slopes wrinkle with a maze of new deer trails. I have seen every edible tree defoliated to the height of a saddlehorn. Such a mountain looks as if someone had given God a new pruning shears, and forbidden Him all other exercise. In the end the starved bones of the deer herd, dead of its own "too much," bleach [in the sun] with…the dead sage, or molder under the high-lined junipers. I now suspect that just as a deer herd lives in mortal fear of its wolves, so does a mountain live in mortal fear of its deer. And perhaps with better cause, for while a buck pulled down by wolves can be replaced in two or three years, a range pulled down by too many deer may fail of replacement in as many decades. So also with cows. The cowman who cleans his range of wolves does not realize that he is taking over the wolf's job of trimming the herd to fit the range. He has not learned to think like a mountain. Hence we have dustbowls, and rivers washing the future into the sea [through soil erosion].

© Eastcott Momatiuk/National Geographic/Getty Images

Figure 1.4 Populations of predator and prey are dependent on natural carrying-capacity limits.

Leopold's essay illustrates the reason "out of sight, out of mind" can be such a destructive attitude. It also highlights the meaning of **carrying capacity**—the maximum size to which a population can grow and be maintained indefinitely, given an existing resource base. In the case of the wolves, the carrying capacity under natural conditions is determined by the size of the deer herd (**Figure 1.4**). In the case of the deer, it is the amount of edible forest vegetation. The wolves actually hold the deer in check by preying on them, so that the actual deer population in a forest with predators is less than the carrying capacity would allow. Conversely, as Leopold saw, the removal of the predators allowed the deer population to grow without bounds, and it soon exceeded the capacity of the forest to support it, consuming the food supply base so intensively that the herd eventually starved

Figure 1.5 Two hundred years ago, Thomas Malthus used a detailed economic argument to explore the concept of human carrying capacity.

© Bettmann/CORBIS

© Chuck Keeler/Stone/Getty Images

Figure 1.7 Agricultural pests and diseases reduce human carrying capacity. Pesticides, herbicides, fungicides, and other chemical applications help boost yields from farmlands, expanding carrying capacity for people but often causing serious environmental problems.

and collapsed. The exceeding of carrying capacity with destructive consequence is termed **ecological overshoot**.

If it is true that we are not apart from nature, that we are beholden (like every other species) to a specific carrying-capacity resource base, then what is *our* resource base? For wolves and deer, it is defined by food supply, and the same is true for us. The size of our population is limited by the productiveness of our fisheries and agricultural lands—by the amount of food that we can grow.

As a case in point, consider this thumbnail history of modern farming: In the late 19th century, scientists grew worried as the global population exceeded 1 billion people; they were concerned that the world could not feed all the newborns coming into it, coupled with the fact that improved health care was keeping more and more people alive into old age. Surely we would soon enter a time of global famine and societal breakdown, as Thomas Malthus thus (**Figure 1.5**) warned in 1798, in his influential *An Essay on the Principle of Population as It Affects the Future Improvements of Society*. In fact, if the scale and method of growing food had not improved, then this "Malthusian" prediction might well have come to pass; by 1950, with 2.5 billion people in the world, we would have exceeded our **human carrying capacity** and would be suffering from global ecological overshoot. But several key developments kept this catastrophe from taking place.

© Hulton-Deutsch Collection/CORBIS

Figure 1.6 Fritz Haber provided the technological basis for modern farming through his research on artificial nitrogen fixation. He greatly expanded the limit of human carrying capacity.

In 1912, German chemist Fritz Haber learned how to fix nitrogen from the atmosphere in the form of ammonia compounds to make synthetic fertilizer (**Figure 1.6**). Suddenly, we could fertilize farmlands to greatly increase crop yields. Haber had no way of understanding at the time that this would eventually lead to almost apocalyptic global water-pollution problems, about which you'll learn more later in the book. At the time, he was hailed (and rightly so) as a savior of the human condition. It is hardly hyperbole to say that Fritz Haber is one of the least known and most important people who ever lived. Artificial fertilizer had suddenly boosted the planet's carrying capacity to sustain people—by billions.

Haber's success was followed soon after World War II with the intensive genetic modification of grains and other crops, and the regular application of new and powerful strains of pesticides and herbicides, including DDT and other compounds related to nerve gas (orthophosphates) and nicotine (neonicotinoids). These innovations led to the industrial-scale farming we see today (**Figure 1.7**). Coupled with further improvements in modern medicine and sanitation, the human population surged. Since 1950, for the first time ever, some people have lived to see the population more than double in a lifetime, and some have lived long enough to see it triple. This is economically challenging and socially disorienting, but even more staggering is the general level of improved well-being for most of the world's population. The lesson seems to be clear: Human ingenuity has rendered irrelevant the warnings of Thomas Malthus. We do not have to worry about carrying capacity like the species found in Leopold's quaintly wild nature.

Or is this dangerous and delusional thinking? The production of the world's most important commercial fertilizers depends on fossil fuels, especially natural gas and petroleum. These are finite resources, and we have good reason to believe that we are approaching the peak of our ability to exploit them at a time when the population is continuing to grow and become more prosperous, especially in the less-developed world. Likewise, intensive cultivation has led to further extensive deforestation, especially in tropical countries such as Brazil, where soybeans now cover thousands of square kilometers, much of which was virginal forest until just a few years ago. Soil erosion continues unabated. Since 1950, about a third of all U.S. cropland has been abandoned because of erosion, and about 90% is losing soil faster than it can naturally replenish. If there is no soil, there is no farming, and certainly then *no food.* Many experts believe a more urgent concern is the depletion of vital groundwater supplies needed for irrigation, which may become even more important as the world experiences possibly severe climate shocks from the buildup of industrial carbon dioxide in the atmosphere in coming decades. Thus, there is reason to believe that the technologically supported human carrying capacity we presently enjoy may not be *sustainable*; that, in the end, it will be an unstable demographic frontier.

For the purposes of this book, it is worth keeping in mind how the topical material of some of the chapters relates to human carrying capacity. Any good stewardship of carrying-capacity resources will require an understanding of related geological processes. It may well be, after all, that the famous baseball catcher Yogi Berra was right when he said, "Nature bats last, and Nature owns the stadium."

Putting Ourselves at Risk

The upswing in human population over the past 50 years is wholly without precedent in human history (see Case Study 1.1). During World War II, there were about 15 persons/square kilometer (40 persons per square mile), on average, across the world's land surface. At present, mean population density stands around 45 persons/square kilometer, (117 persons per square mile) and by 2050, even with declining birthrates, this will increase to 65 persons/square kilometer (170 persons per square mile) Because only about 10% of Earth's land surface can be farmed, most of this population is crowded close to or directly on precious agricultural areas, and nearly half the world's population is living in cities (75% in developed countries, 40% in less-developed countries). Most of the world's urban population lives under arguably wretched conditions—ghettos, barrios, slums, and shantytowns (**Figure 1.8**). (The names vary according to the part of the world, but each term evokes a level of human misery to which most readers of this book are unaccustomed.)

A greater number of people means that there are more people at risk from geological processes that otherwise might not inflict much damage and loss of life. These include earthquakes, volcanic eruptions, landslides, and floods—all vital components of the Earth System responsible for creating new lands, revitalizing soils, and developing some of the most strikingly beautiful scenery. A part of environmental geology, therefore, deals with how we can learn to reduce losses while living in greater numbers with these powerful, necessary, and inevitable forces of nature. Especially important is the recognition of how we can actually worsen conditions and increase risk through foolish land-use decisions (e.g., building atop a fault could add new meaning to the term *split-level home*).

Figure 1.8 The fastest growing segment of human population in recent decades has been poor and urban. Vulnerability to natural disasters and environmental problems is high for this part of humanity.

The Environmental Geologist

Many famous detectives from fiction, such as Sherlock Holmes, are like scientists, using the *scientific method* in their investigations (**Figure 1.9**). Environmental geologists also use this multistep, straightforward approach to learn more about situations involving people and the physical environment, and to achieve conclusions that will be useful, either to a client who has hired the geologist or to the scientific community at large. Using the scientific method often leads to fascinating discoveries:

1. At the beginning of the process, a geologist poses a question—perhaps simply indulging some personal curiosity about nature or addressing a problem presented by someone who needs authoritative assistance. One typical question is, How did the water in this stream come to have so much dissolved copper, iron, lead, and zinc? To answer the question, the geologist might take successive water samples upstream, leading to an abandoned mine that is emitting acid mine drainage loaded with the toxic metals (**Figure 1.10**).
2. Next, the geologist goes into the field to gather preliminary information through direct observation or the collection of samples to be analyzed later in a laboratory

© Alessandro Della Bella/epa/CORBIS

Figure 1.9 Detectives are scientists, and scientists are detectives.

D. D Trent

Figure 1.10 Acidic mine water draining from the abandoned Friday-Loudon Mine, Shasta County, California. A U.S. Forest Service geologist is measuring the pH of the mine drainage. The water has a pH of 3.9, and it carries dissolved copper, iron, zinc, and other heavy metals into the stream, which eventually flow into the sport fishing waters of Shasta Lake.

P. Bierman

Figure 1.11 Environmental geologists 50 kilometers (30 miles) from Upernavik, a town of 900 people on an island in northwestern Greenland. They and their field vehicle, a helicopter, are on a *nunatak,* an island of rock surrounded by the ice sheet. The geologists are organizing rock samples to take to the laboratory for analysis. These samples will be dated to determine when and how much the Greenland ice sheet has grown and shrunk in response to climate change.

(**Figure 1.11**). The integrity of an entire scientific study depends largely on the effectiveness of this step. A good geologist would like to leave "no stone unturned" (a useful pun), but limited time, stamina, and budget require that only so much data can be collected. The investigator must make a judgment call in the field about whether sufficient information has been gathered. In other words, thoughtful selectivity and a certain amount of serendipity enter the early stage of every scientific investigation.

3. The geologist then considers the data carefully, formulating a **hypothesis**—an idea about what is going on—which then must be tested with an additional visit to the field. The process of formulating a hypothesis gives the geologist a better idea of what data need to be collected next. Like a good detective homing in on the culprit, the geologist begins to feel in control of the situation. An answer is shaping up, even if several hypotheses must be tested and discarded in succession. An example of a geological hypothesis is the following statement: "The groundwater has a large amount of perchlorate in it because there was once a fireworks company located upslope. The company must have been leaking or dumping wastewater." Note that this statement is not yet a *conclusion*.
4. Finally, once a hypothesis has been verified, the geologist is in a position to develop and announce conclusions. These need not be definite; sometimes scientists are able to say only that something is "probably this" or "probably that"—and in the legal cases in which environmental geologists frequently provide testimony, it may be up to a jury to decide whether the scientific argument is persuasive. In the case of the sample hypothesis in step 3, the geologist might conclude, "The *most likely source* of pollution was a fireworks company located upslope." The public and the political system often find routine scientific uncertainties such as this frustrating, but science is commonly an iterative process in which many people are engaged over long periods of time; as different scientists assemble their own, independently collected data to confirm the conclusions of their peers, certainties of vast importance do emerge (e.g., Earth revolves around the Sun, the tectonic plates move). In fact, the act of independent scientific verification is a necessary part of scientific culture. A scientist can feel honored to have his or her work under scrutiny by other scientists; this generally means that the work is considered "relevant and important."

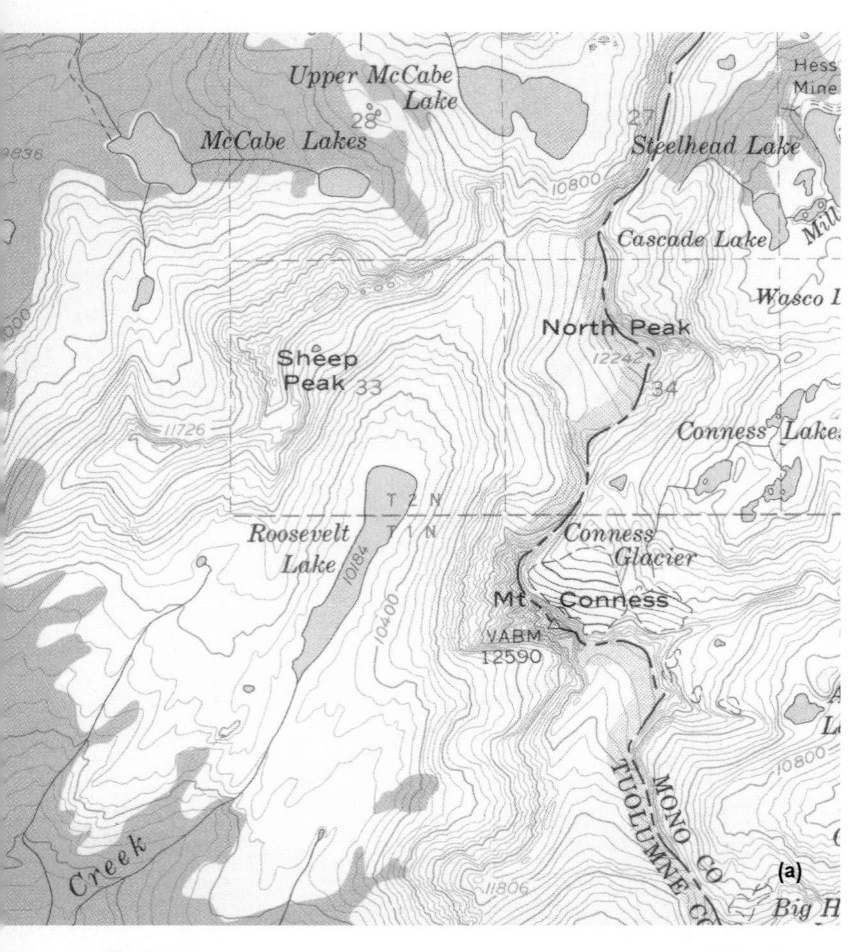

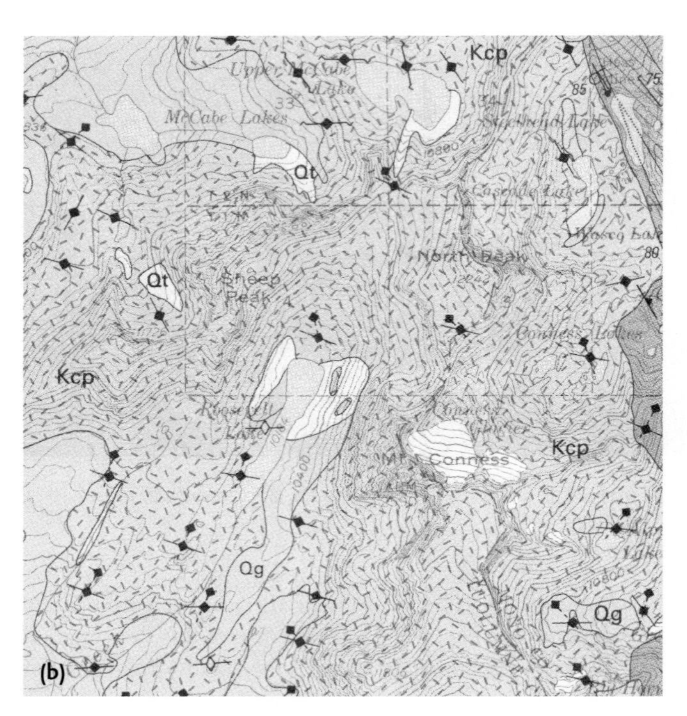

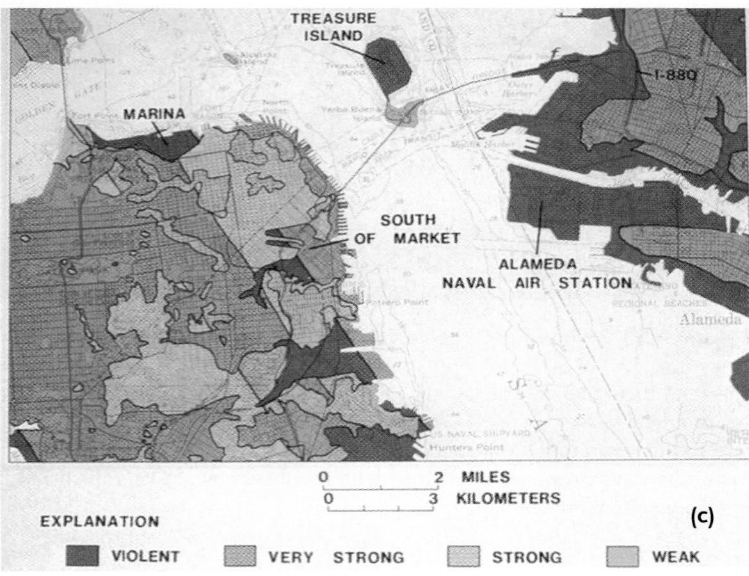

Figure 1.12 (a) A topographic map shows the roughness and vegetative cover of a landscape, using lines of equal elevation called *contours* and colors to represent vegetation, clear areas, and water bodies. (b) A geological map uses a topographic contour base to show where different kinds of bedrock are distributed in a landscape. (c) A derivative map takes data collected from various research studies to show information of use for a particular purpose—in this case, seismic hazards in the central San Francisco Bay area.

A theory is a cause-and-effect relationship regarded by scientists as certain or nearly certain. It is a hard and fast, widely accepted scientific conclusion. Hypotheses are far more readily modified and overturned than theories. In Chapter 3, you'll see how a number of hypotheses, some of which were erroneous, eventually led to the revolutionary theory of plate tectonics.

Tools of the Trade

Environmental geologists work with and prepare a wide range of maps, tables, and diagrams. *Topographic maps* (also known as "contour" or "elevation" maps) illustrate the roughness of Earth's land surface (**Figure 1.12a**). The topographic maps used by wilderness backpackers might be useful, but quite often maps have to be made of much smaller areas than the ones represented in standard topographic formats.

Using a topographic base, the geologist may draw in, through accurate and precise surveying, the locations of special features, such as the boundaries between different rock bodies (known as "geological contacts"), and the locations of faults, springs, seeps, and landslide scarps. The angle of rock layers in various locations may be depicted using "strike and dip" map symbols. Different soil types and soil thicknesses or depths to water tables beneath the surface as measured in wells might also be shown. In any event, these *geological maps* are pictures of reality—readily understood visual representations of the vital data (**Figure 1.12b**).

The geologist may use compasses, clinometers, measuring tapes, and other surveying tools, such as laser theodo-

Kerry Sieh, Earth Observatory of Singapore

Figure 1.13 A typical GPS station in the field. The GPS instrument proper is housed beneath the gray dome in the background, while its electronics are contained in the opened cabinet beneath the solar panel.

lites, in the field to gather additional information. Quite often, environmental geologists use highly sensitive **Global Positioning Systems (GPSs)** to pinpoint locations using satellites (**Figure 1.13**).

In some cases, the data may already have been collected by someone else and entered into a computer-based format that can be tapped and analyzed numerically to make *derivative maps*—maps derived for a special application, such as illustrating for an insurance company the susceptibility of the ground in a particular region to intensive shaking during earthquakes. The online data sets and their processing by high-speed computers constitute a **geographic information system (GIS)**.

GIS technology has refined and improved the power of traditional mapping by facilitating the production of cross sections and three-dimensional images of buried geological features. Being able to ascertain what lies underground is of vital importance to environmental geologists. The practice of environmental geology is very visual, requiring a keen understanding of the structure of the shallow crust and of how fluids move through it. An ability to work three-dimensionally is one reason that many students with artistic inclinations often make very fine geologists. One important arena of three-dimensional space visualization that environmental geologists must work is in areas of high geological hazard potential, such as sites subject to earthquakes or landslides (**Figure 1.14**). Their efforts have significantly reduced injuries and economic losses.

P. Bierman

D. D. Trent

Figure 1.14 (a) Geologists with an auger drilling rig installing a groundwater monitoring well through floodplain deposits along the Hoosic River near a landfill at Williamstown, Massachusetts. Their goal is to determine whether the groundwater has been contaminated by seepage from the landfill waste. (b) Geologists carefully log the sidewalls of a trench across the San Andreas Fault near Wrightwood, California, to find and age date the youngest rupture (an 1857 earthquake) and then earlier breaks. The data will allow the determination of earthquake recurrence intervals for the fault.

explorer's
perspective

National Geographic Image Collection, http://www.nationalgeographic.com/explorers/bios/jennifer-burney/

JENNIFER BURNEY, an environmental scientist, uses modern technology to improve the lives of African farmers.

Jennifer Burney began her academic career studying physics at Harvard University, with a special interest in the impact of science on American history. She moved on to Stanford, earning a Ph.D. in physics. She retained her Stanford affiliation while also becoming a President's Postdoctoral Fellow at the Scripps Institute of Oceanography at the University of California San Diego. The Center on Food Security and the Environment (FSE) at Stanford supports her current research about the potential impacts of climate change upon agriculture, and vice versa. She has worked in Benin and Niger, Africa, to study the potential for solar panels to power irrigation pumps for village farms. Coupled with crop diversification, this application promises to relieve hunger and poverty by increasing consumption and developing tradable farm surpluses, especially where shallow groundwater supplies are readily accessible. Jennifer's research has also demonstrated that general increases in global crop yields (food produced per hectare or acre of land) over the past half century have stored huge amounts of carbon dioxide that otherwise might have entered the atmosphere—on the order of 317–590 billion tons. CO_2 levels might be 18 to 34% higher today had farm yields not improved significantly since the 1950s. Her solar-irrigation work points in a sustainable direction to keep this trend going, while keeping a growing world population adequately fed.

Jennifer says, "I love my job. I'm a scientist, so I still 'dork out' with the data...but I also get to go into the field." The most demanding part of her work, she says, is "dealing with the climate skeptics."

Source: http://education.nationalgeographic.com/education/news/real-world-geography-dr-jennifer-burney

Marginal land such as this near the Cederberg in the Western Cape, South Africa, may only be intensively farmed if a supply of fresh water can be found and careful soil stewardship practiced. Inability to reach groundwater through pumping greatly limits development of such terrain for agricultural purposes.
Courtesy author Rick Hazlett

Typical Projects in Environmental Geology

Some examples of the kinds of projects environmental geologists undertake are hinted at in the preceding sections. More specifically, an environmental geologist might be called on to do the following:

- Investigate the pollution caused by mining operations, golf courses, industries, culverts, or waste-storage ponds.
- Assess the risk of land destabilization from the extraction of groundwater, oil, or gas, or from the construction of structures on slopes.
- Evaluate the risk from natural disasters, such as flooding, earthquake activity, or volcanic eruptions, and help governments formulate zoning or regulatory restrictions to reduce costly impacts.
- Locate future dam sites or investigate the reasons dams have failed.
- Study the quality and erosive potential of soils being considered for agricultural use or suffering from faulty farming practices.
- Deal with coastal erosion and storm-related issues.
- Track the movement of contamination plumes underground.
- Determine the best way to store solid and liquid wastes coming from an urban area or a radiation source, such as a hospital, college laboratory, or government establishment.

Environmental geologists must work extensively with engineers, ecologists, microbiologists, chemists, legal and civil officials, and other specialists. There is rarely a dull moment in such work, although data collection can sometimes seem tedious before answers begin to emerge. And no two environmental projects are alike.

Regardless of whether you seek a career path in environmental geology, however, it is worth grasping the perspectives that guide this profession. To a certain extent, each of us may find reason to apply practical lessons from this field at some time in our lives.

case study

1.1 Exponential Growth, Wealth, Poverty, and Population

The growth of any population of anything is easy to quantify using a **compound growth equation**. The equation simply states:

$$\text{Future quantity of something} = \text{Present quantity of something} \times (e)^{kt}$$

where e is the exponential growth rate constant or *natural log* 2.71828, k is the rate of change in amount of whatever is being studied (usually measured as an *annual* change and expressed as a decimal, so a growth rate of 10% equates to 0.1), and t is the number of years during which the growth takes place.

The symbol N is commonly associated with "future quantity" and N_o with "present quantity"; therefore, the formal expression of the equation is

$$N = N_o \times e^{kt}.$$

Consider the following example, which you may want to crank out with a pocket calculator for confirmation: In June 2000, the world's total population was 6.1 billion. Given a global population growth rate that *holds steady* at 1.25%, this means that 7.8 billion people will live in the world 20 years hence, or

$$7.8 \text{ billion} = 6.1 \text{ billion} \times (2.71828)^{1.25(20)}.$$

The vital point to note is that the rate of increase seems almost trivial to most people (1.25%), but this small number applied to a big quantity (6.1 billion) means an enormous growth in population. In other words, we tend to be lulled into a false sense of security by the exponential aspect of the compound growth equation and to underestimate its often staggering consequences or sometimes unrealistic implications. For example, the huge boom in global population that has altered the world's environment so significantly over the past 50 years was fueled by a growth rate that peaked at a "mere" 1.99% in 1965 (**Table 1.2**).

Since 1965, the global population growth rate has declined to 1.22%, but this does not mean that the world is losing total population—merely that the total is not growing quite so fast. A rough estimate of how long it takes for a population to double can be made by dividing the number 70 by the percentage of rate increase. In other words, the present population would double in (70/1.22) years, or 57 years, if the growth rate remained at 1.22%. This is called the **doubling time** (**Table 1.3**).

The rate of growth varies according to the wealth of countries; it is correlated directly with the IPAT equation. With few exceptions, countries that have the greatest environmental impacts have the lowest growth rates in population. Immigration from poorer to richer countries can radically offset this trend, as in the case of the United States, which has the sixth-fastest-growing population on Earth, exceeded only by India, China, Pakistan, Bangladesh, and Nigeria.

Table 1.2 World Population and Growth Rate, 1950–2002

Year	Population, Billions	Growth Rate, %*
1950	2.52	
1955	2.75	1.77
1960	3.03	1.95
1965	3.34	1.99
1970	3.77	1.90
1975	4.08	1.84
1980	4.45	1.81
1985	4.85	1.75
1990	5.30	1.70
1995	5.76	1.68
2002	6.26	1.56
2009	6.8	1.19

Source: United Nations, Department of Economic and Social Affairs, Population Division, Population Estimates and Projections Section, http://www.un.org/esa/population/unpop.html

*Average annual rate for the previous 5-year period.

The current annual increase in population is only 0.25% in wealthier regions, whereas in poor countries it is 6 times faster, 1.46%; in the 49 poorest nations, home to nearly 700 million people, it is 2.41%.

What accounts for these differences? Many social and economic factors are involved. In poor countries that lack welfare, health care, and institutional care for the elderly, having many children ensures that parents will be cared for in old age or times of unemployment. Contraceptives and family planning are lacking. Children are cheap to have and to rear because expensive educational systems, clothes, toys, and other items we take for granted as part of an ordinary childhood in the United States are simply unavailable to most people. Women, who might take an active role in earning family incomes and reducing family size if professionally trained and employed, have little incentive not to stay at home and bear more children. Indeed, the United Nations regards the education and professional empowerment of women as one of the most powerful tools we have to combat overpopulation and, with it, many of the related environmental problems we'll explore further in this book.

Table 1.3 How Populations Grow

Growth Rate, %	Doubling Time, Years*
1.0	70.0
2.0	35.0
3.0	23.3
4.0	17.5
5.0	14.0
6.0	11.7
7.0	10.0

*Calculated by using the formula 70 ÷ growth rate (%), which yields a close approximation up to a growth rate of 10%.

Town at Ushuaia Bay near the southern tip of Patagonia in Argentina. Reduction of life-protecting ozone in Earth's upper atmosphere from industrial manufacture of chlorofluorocarbons and other chemicals is a special concern for local residents, although they had virtually nothing to do with generating this problem. Ushuaia Bay illustrates Aldo Leopold's dictum that there is "no away" when it comes to human pollution, and that the path to trouble is often paved with good intentions. Richard Hazlett

As emphasized in this chapter, overpopulation is responsible for environmental damage, societal problems, and human suffering. In some countries, population growth is utterly out of control. In other countries—China, for instance—measures and incentives are in place for curbing the growth and achieving a sustainable population.

1 Coney Island beach in the 1930s, when the U.S. population was a fraction of what it is today. Imagine this beach on a summer weekend today. B. Pipkin

2 Deforestation near the city of Manaus in Amazonia, Brazil's largest state. (a) A false-color satellite photo of the confluence of the Rio Negro (black) and the sediment-laden Amazon River (light green). Healthy rain forest is shown in red, and deforested areas appear in light colors. Manaus is on the point of land that extends into the Rio Negro. The waters of the Rio Negro are recognizable for more than 80 kilometers (48 miles) downstream in the Amazon. (b) Adjacent rain forests have been removed as the population of the area has grown. Manaus's population is now more than a million people, and the city is rapidly expanding into areas that were once sparsely populated. © Owen Franken/Sygma/CORBIS

3 A model Chinese family on a billboard encouraging the one-child family. Although the People's Republic of China is pushing population control, its population of more than a billion people makes continued growth inevitable. © Alain Le Garsmeur/CORBIS

SUMMARY Humans, Geology, and the Environment

Environmental Geology

Defined

The study of how humans interact with the geological part of the natural environment.

Utilizes

Most of the traditional geological specialties, including petrology, engineering geology, geophysics, geochemistry, structural geology, and hydrology. Tools used by environmental geologists include precision surveying instruments, maps, and computerized databases.

Areas of Interest

Geological hazards, such as volcanism, earthquakes, and flooding; mineral and water resources; and land-use planning. Environmental geologists are concerned not only with what is happening at Earth's surface but with what is happening underground as well. They are "three-dimensional" thinkers and interact intensively with many other kinds of professionals.

Goals

To predict, anticipate, and solve geological problems by collecting data in the field and analyzing it in the laboratory. Data may be portrayed on special-use (derivative) maps and used by planners and public officials.

Methods

Environmental geologists follow the scientific method:

1. Pose a question.
2. Collect initial data and frame a hypothesis.
3. Test the hypothesis.
4. Reach a conclusion based on the results of their testing, which might include the development of a new theory.

Humans and the Geological Environment

The IPAT Equation

I = PAT defines the human factors that cause environmental impacts (I). Those factors include population (P), affluence (A), and technology (T). In very poor countries, the equation reduces to I = P.

Out of Sight, Out of Mind

In wealthy countries, environmental impacts tend to be "out of sight, out of mind," whereas impacts are more direct and local in less-developed countries.

Carrying Capacity

Carrying capacity is the size of the population that can be supported in a stable, long-term manner by the natural environment. The most important factor in carrying capacity is food supply. Human ability to grow food and support carrying capacity has increased in recent decades, because of the introduction of commercial fertilizers, pest and disease control, and the genetic modification of crops.

Ecological Overshoot

Ecological overshoot occurs when a population exceeds its carrying capacity. This can cause great environmental problems, including many of a geological nature (soil erosion, land subsidence, etc.).

Rural to Urban Transformation

The greatest transformation in the human population over the past century has been from rural to urban growth. Larger concentrations and numbers of people mean greater vulnerability to natural disasters, such as earthquakes and floods, topics of great importance in environmental geology.

KEY TERMS

carrying capacity
compound growth equation
doubling time
ecological overshoot
environment
environmental geology
geographic information system (GIS)
Global Positioning System (GPS)
human carrying capacity
hypothesis
IPAT equation

STUDY QUESTIONS

1. Do you consider yourself an environmentalist? Why or why not? How has reading this chapter changed your view of environmental issues, if at all?
2. Think of one good example of an "out of sight, out of mind" environmental impact that you have caused or are presently causing on a routine basis. Was (is) this impact avoidable? Why or why not?
3. Aldo Leopold asks that we "think like a mountain." This is a metaphor for changing our attitudes and broadening our perspectives, but in what ways? Do you think that it is practical to "think like a mountain"?
4. Do you agree with the assertion that "we do not have to worry about carrying capacity like...wild species" of animals? Are you much concerned with the possibility of "human ecological overshoot"? What is the evidence for your answer to these questions?
5. Given the importance of farmlands, do you think that the continuing growth of huge cities in or upon agricultural areas is such a bad thing? Why or why not? What other information would you like to know to answer this question?
6. Distinguish between a *hypothesis* and a *theory*. Cite an example of a hypothesis you have formulated in the past, even to solve a simple problem at home or at school. How would a society in which hypotheses were not expected to be tested differ from one in which they were routinely tested?
7. Can you think of ways we can reconcile the reality of probability and much uncertainty in science with the absolute certainties required by our political-economic and legal systems? Can you think of a great scientific question today that has caused heated debate because it has not been definitely, concretely answered?
8. The United States has a growth rate of 0.7%. At this rate, how much time is required for the national population to double? Discuss the impact of a doubling of your community's population in that length of time.
9. What will the population of the world be in the year 2100 if a steady growth rate of 1.22% is maintained?
10. How does the education and training of women in less-developed countries relate to geological problems, such as soil erosion and flooding?
11. The IPAT formulation implies that technology and affluence always have negative environmental impacts. Can you think, though, of situations in which impacts have been reduced, or actually reversed, by technology and affluence? Do you think that the IPAT relationship is too simple to be useful?

REFERENCES AND FURTHER READINGS

Brown, L. 2001. *Eco-economy—Building the economy for the Earth.* New York: W. W. Norton & Company, 356 p.

Easton, T. 2008. *Taking sides: Clashing views on controversial environmental issues,* 13th ed. New York: McGraw-Hill, 400 p.

Goudie, A. 2005. *The human impact on the natural environment: Past, present, and future,* 6th ed. Oxford: Wiley-Blackwell, 376 p.

Leopold, A. 1989. *Sand county almanac; And sketches here and there.* New York: Oxford University Press, 256 p.

McNeill, J. R., and P. Kennedy. 2001. *Something new under the Sun: An environmental history of the twentieth-century world.* New York: W. W. Norton & Company, 416 p.

Access an interactive eBook and chapter-specific interactive learning tools, including flashcards, quizzes, videos and more in your Geology CourseMate, accessed through CengageBrain.com.

Melting water creates interesting patterns along the Northwest Passage.

Have you ever
wondered...

2

The Earth System and Climate Change

A doctor can examine our symptoms, try to diagnose our condition, and suggest treatments if the prognosis is not favorable. The success of modern medicine shows clearly that, even when medical knowledge is not perfect, it can still be useful. This is also true for climate scientists studying the Earth—the science is imperfect, but still useful.

—Gavin Schmidt and Joshua Wolfe

1. Why there is so much concern about greenhouse gases and global warming?

2. Why many scientists believe that humans are responsible, at least in part, for global warming?

3. Why the weather is often much more pleasant in coastal areas?

4. How and why the wind blows?

5. Why there have been past ice ages?

Changing Climate and the Fate of Societies: Past and Present

Figure 1 Hohokam farmers caring for their crops with an irrigation canal in the foreground.

In the mid-19th century, when European American settlers first began settling and cultivating the arid, irrigated plain of south-central Arizona, they quickly discovered the ruins of an ancient society that had faded and fallen 300 years earlier—the Native American Hohokam culture. The Hohokam lacked a written language, so their society couldn't be regarded as a true civilization. But they were master canal builders and engineers, tapping the sparse water supply of the Salt River to support widespread cultivation of squash, beans, agave, amaranth, and other crops in an arid landscape (**Figure 1**). The Hohokam dug more than 1,600 kilometers (1,000 miles) of durable irrigation channels, and many of these were simply cleaned out and repaired by the early American farmers, who could find no better way to bring water to their fields. Hohokam towns were well organized socially, with governance and religion centered upon the acquisition and distribution of precious resources, much as in early Mesopotamia. Communities were built of durable adobe, and many included ball courts and ceremonial platform mounds, like those found in the cities of the Aztecs, who came somewhat later. Throughout their primacy, the Hohokam maintained an extensive trade in parrots, bells, and other commerce with Mexico. They prospered for more than four centuries.

But then their society mysteriously collapsed, leaving no direct testimony of what had happened. The metropolis of Phoenix has since risen in their place, building atop their ancient town sites and irrigation works (**Figure 2**). The name *Hohokam* is from the Pima language and means "all used up," perhaps indicative of an environmental catastrophe, although not necessarily one that occurred all at once. In fact, a gradual aggregation of insurmountable environmental problems may have been more effective than any single event in bringing their downfall. Research by anthropologist Donald Graybill of the University of Arizona suggests that climate change was a key factor. An epoch of steady, reliable, annual rainfalls characterized the ascendancy of the Hohokam, who with great efficiency learned how to exploit this natural pattern to their benefit. But then the weather became erratic, with intensive flooding and erosion destroying irrigation channels. There is some indication that the people began protecting themselves, migrating to a few large, multistory "fortress" commu-

Figure 2 A modern view of the City of Phoenix showing its urban sprawl.

© 2012/MICHAEL MELFORD/ National Geographic Image Collection

CourseMate Connection

Start off this chapter by listening to a podcast with National Geographic Waitt grantee Kim Cobb as she discusses climate change and its uncertain consequences. **Log in to the CourseMate at www.cengagebrain.com to listen now.**

nities. Sophisticated traditional ceramic art ended, and new canal works were built farther upslope in the Salt River watershed, depriving users downstream of a potential supply. All of these changes signaled a society growing increasingly desperate, and one that was suffering from periodic resource scarcity. Perhaps in the end, disease, malnutrition, and emigration undid the Hohokam, their great towns reduced to hovels hinted at by the archaeological record.

Could modern-day Phoenix one day share a similar fate? With an exploding population, water supplies continue to be a major concern for the city. And the onsets of drought in recent years, exacerbated by forecasts of worse to come by climate change scientists, suggest a repeat of history in southern Arizona. Unlike the Hohokam, who depended on water from the Salt River, aqueducts have been constructed to support the much more populous modern city from watersheds as far as 600 kilometers (375 miles) away. Nevertheless, the differences with Native American predecessors are merely ones of technology and scale. Phoenix archaeologist Tom Wright reminds us, "The Hohokam were a hydraulic society, and so are we....We depend on the presence of water, the storage of water, the transport of water, and as long as the water is there and can serve our needs, we're fine. But if the water is not falling on the watershed, or if our needs start to outstrip what is available, that's a problem."[1]

questions to ponder

1. How might urban growth, development, or governance be improved to provide for a flexible, successful social response in the face of increasing prospects for drought and a scarce water supply in the southwestern United States?
2. How should the "limit of growth" for a city such as Phoenix be defined and enforced?

[1]Tom Wright quotation from Childs, C. 2007, April 16. *Phoenix falling?* High Country News, p. 9.

THE PRECEDING EXAMPLE SHOWS that climate change, irrespective of its cause, holds the fate of cities and states in balance. Our societies depend on reliable climatic conditions. But climate changes have occurred repeatedly throughout Earth's long history. In this chapter, we explore the workings of Earth's climate system and see how the human race in its own right has become a factor influencing world climate. Earth's environment functions as a finely tuned machine—a natural, self-maintaining spaceship for its inhabitants. Just as the elements of human physiological systems are connected and interdependent, so on Earth are there linkages and interactions between its surface of rock and soil (**lithosphere**), the air (**atmosphere**), bodies of water and ice (**hydrosphere**), and its living realm (**biosphere**). Just as a healthy human body depends on the proper circulation of blood and other fluids and solids within, so does a "healthy Earth"—one hospitable for life, at least—depend on a proper circulation of matter and energy between its major components: the lithosphere, atmosphere, hydrosphere, and biosphere.

A system is simply a set of components, or parts, that work together to perform a particular function. And all systems, including the Earth System, require energy to operate. The human body requires energy from eating food. An automobile requires fuel or electricity to function. A transportation system—the network of streets and interchanges required to move automotive traffic around a city—requires routine street maintenance and the power for traffic signals to work.

The Earth System needs energy to work as well, and this comes from two sources: one internal and one external (**Figure 2.1**). Inside the planet's deep interior, uranium, thorium, and, to a lesser extent, potassium atoms are continuously decaying into more stable, lighter-weight elements. A by-product of this *radiogenic decay* is heat, which gradually escapes to Earth's surface. We see manifestations of this escaping internal heat directly expressed as volcanoes, geysers, and hot springs. Some of the heat converts to mechanical energy, causing earthquakes, mountain building, and the slow shifting of massive tectonic plates (see Chapter 3).

More noticeable to us at the surface, however, is the energy coming from Earth's external heat source—the Sun. Without the daily input of solar radiation, our planet would be a frozen, stony ice ball, despite all the heat slowly escaping from its interior. There would be no wind, rainfall, or circulating surface water (see Chapter 8). Because the biosphere depends on sunlight for photosynthesis and various other vital processes, most life as we know it would be nonexistent.

The interface between the external and internal sources of energy is Earth's surface (Figure 2.1). Soil—an essential factor in human carrying capacity—is a result of conflicting processes—one set internally driven, the other set externally driven—which come together at the surface. Wind, rain, ice, snow, and chemical corrosion break down the land that earthquakes, volcanic eruptions, sedimentation, and mountain building gradually build up. The shape of the landscape on which we live may be thought of as an outcome of these opposing forces. Biological activity modifies the ways in which they operate. For example, a dense cover of vegetation inhibits erosion and sedimentation, whereas a landscape denuded by logging and overgrazing quickly wears away, as Aldo Leopold so clearly observed (see Chapter 1).

Homeostasis and Feedback

The concept of **homeostasis** generally applies to living organisms and is defined as the ability of or tendency for a being to maintain internal equilibrium (e.g., steady body temperature) by adjusting its physiological processes. When you are ill, your personal

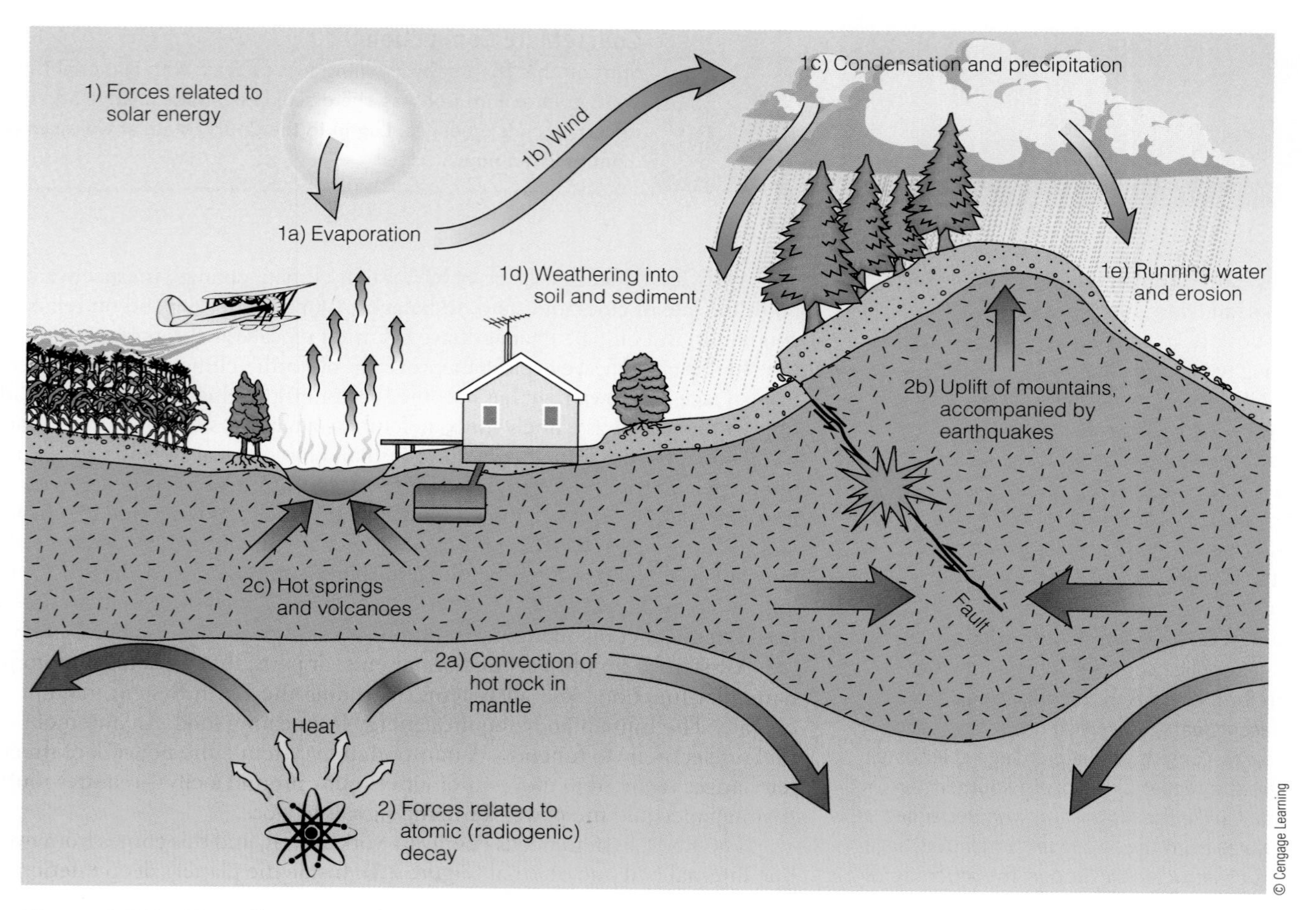

Figure 2.1 Earth's surface is a resolution of two sets of forces, modified by the activity of life.

homeostatic system is strained. Recovery from illness is a good example of homeostasis at work. Your homeostatic system will stay in operation for the rest of your life.

The Earth System shows analogous homeostatic behavior by means of natural processes called **feedbacks**. Geoscientists identify two kinds of feedback: one *positive* and one *negative*. A positive feedback is a self-reinforcing set of reactions, increasing the intensity of the condition that set it into motion in the first place. (A nongeological example of a positive feedback is the simultaneous growth of human agriculture and population; both "feed" on one other, providing the facility for the other to increase).

A negative feedback operates in the opposite sense, *decreasing* over time the intensity of the condition that originally set it into motion. (A nongeological example might be a swimmer; the faster a swimmer goes, the faster the swimmer becomes tired, more quickly slowing down to catch breath and recharge energy).

In a perfectly balanced system, the effects of positive and negative feedbacks cancel out, and the net result is stability. But sometimes a feedback is so powerful that it overwhelms others opposing it and thus becomes a **runaway feedback.** A runaway feedback may continue being expressed indefinitely but quite often terminates in a new stable set of conditions radically different from the ones with which we began. **Tipping points,** or thresholds, are conditions in which even small perturbations in a system—not necessarily runaway feedbacks—will cause it to change significantly. (An example of a threshold is the boiling temperature of water—just a small increase in heat added to the water will cause it to transform suddenly from a liquid to a vapor state.)

In the remainder of this chapter, we'll consider some important feedbacks and tipping points that influence the various "spheres" of the Earth System, focusing on climate. Climate may be regarded as a generally balanced by-product of these feedbacks, and as you'll learn throughout the rest of this book, in many direct and subtle ways, climate influences the geological processes that pertain to environmental geology.

Consider the world population growing at a fixed rate (around 1.1% in 2009). Is this an example of a *positive* or *negative* feedback? What about the example of turning on music in your home or car too loud?

Atmosphere, Hydrosphere

Makeup of Earth's Atmosphere

Earth's atmosphere consists mostly of a colorless gas we can neither smell nor taste: nitrogen, which makes up 78.08% by volume of all the air we breathe. Saturn's largest moon, Titan, also has a nitrogen-rich atmosphere, and nitrogen snow and ice cover the surface of some colder, outer worlds of our solar system. Oxygen is the next most abundant gas in Earth's atmosphere, making up 20.95% of its volume, followed by argon (0.93%), carbon dioxide (CO_2; 0.038%), and trace amounts of other gases.

Among many essential services the atmosphere provides for our natural environment, two other constituents are especially noteworthy: ozone (O_3), which screens out deadly ultraviolet radiation from the Sun in a thin layer at high elevation, and the **greenhouse gases,** including water vapor (H_2O), CO_2, and methane (CH_4), which keep Earth's atmosphere and surface warm enough for water to stay liquid and life to flourish. The explanation for this warmth is a phenomenon that has made much news in recent years: the **greenhouse effect.** Much media coverage of greenhouse gases has been negative, but, in fact, we *need* these gases and the greenhouse effect for our survival. The problem is that there can be "too much of a good thing."

To understand how the greenhouse effect works, consider that the energy reaching Earth's surface from the Sun arrives in a range of wavelengths and frequencies. Visible light is but one band of frequencies of several making up the overall spectrum, which also includes the ultraviolet radiation previously mentioned. On average, about a third of the solar radiation reaching Earth is reflected directly back into space, with the oceans reflecting much less (no more than about 15% reflectivity) than fresh snow and ice (90%–100%). **Albedo** is the scientific term that quantifies this reflectivity. Generally, the darker an object, the less its reflectivity and lower its albedo (**Table 2.1**). What energy is not reflected is absorbed, even by you standing under a bright Sun; then it is reradiated later on—but mostly in a different wavelength, the infrared, which we feel as heat. That heat would escape back into space, except for the fact that it is intercepted in the atmosphere by greenhouse gases, which reabsorb it and reradiate it—**greenhouse warming.** Hence we receive heat not only from the sky above but also from the *ground below and the air all around us.* The extra boost of greenhouse warming makes a huge difference in overall air temperature. Without it, Earth's average air temperature would drop to between −22°C and −36°C (−9°F and −34°F)—well below freezing. With it, Earth's average air temperature is a much more comfortable 14.5°C (58°F).

The most important greenhouse substances in the atmosphere are not the most abundant. Water vapor contributes most to greenhouse heating (60%–70%), followed by CO_2 (15%–20%), a family of industrial gases called the **halocarbons** (7%–10%), methane (5%–7%), ozone, and nitrous oxide (1%–3%). The human contribution to warming is indeed significant, but most greenhouse warming remains natural. The uncertainties in the range of relative greenhouse contributions reflect the difficult task of measuring a complex, planet-wide atmosphere constantly in motion, as well as ambiguity about such specific phenomena as cloud formation. Satellite observations and oceanographic monitoring have greatly improved our ability to study these complex relationships.

Table 2.1 Some Representative Albedos

Types Of Surface	Albedo*
Forests (coniferous and deciduous)	0.05–0.18
Grasslands	0.25
Sandy deserts	0.40
Fresh concrete	0.55
Earth, overall average	0.30
Moon	0.12

Sources: D. Budikova, *Albedo,* the Encyclopedia of The Earth, available at http://www.eoearth/prg/article; A. McEvoy, T. Markvart, T. Castanzer, 2011, *Physical Handbook of Photovoltaics: Fundamentals and Applications,* Elsevier Publishing, 1268 p.; G. Tezlaff, 1983, *Albedo of the Sahara,* in Cologne University, eds., *Satellite Measurements of Radiation Budget Parameters,* pp. 60-63; and D.R. Williams, *Moon Fact Sheet,* National Space Sciences Data Center Moon Fact Sheet, available on line at http://nssdc.gsfc.nasa.gov/planetary/factsheet/moonfact/html

http://en.wikipedia.org/wiki/Albedo.

The reflectivity of water ranges from very low (<0.1) to very high (>0.9) depending on sun angle. Under most conditions of incident sunlight, water is relatively dark and has an albedo of less than 0.1. The Fresnel equation is used to determine the albedo of water bodies.

*Multiplied by 100, the values give you "% of reflectivity" of solar radiation of the surfaces indicated.

Evidence from the fossil record strongly suggests that the atmosphere co-evolved with life, gaining free oxygen and losing a significant amount of CO_2 in response to biological activity and geological burial over hundreds of millions of years. Several negative feedbacks help explain the current homeostasis in temperature. For example, heating of the atmosphere—up to a certain temperature—promotes more frequent and heavier rainfalls, and the growth of vegetation at higher latitudes and elevations in many parts of the world. As fresh plant growth incorporates CO_2 into plant tissues, CO_2 is taken out of the atmosphere, suppressing its role in global warming. If the atmosphere undergoes chilling, this feedback can be thrown into reverse. In fact, an annual cycle of slight worldwide CO_2 buildup and reduction exists, corresponding to the change of seasons from winter to summer and back again. Because most vegetative biomass exists in the Northern Hemisphere, Southern Hemisphere effects do not cancel out those of the Northern Hemisphere in this cycle.

Another important negative feedback is the **cloud-albedo effect.** As the atmosphere warms, more clouds develop across the world, because of increased evaporation at higher temperatures. Increased cloudiness reflects a greater percentage of solar radiation back into space, meaning less heating in the atmosphere. In this respect, the cloud-albedo effect puts the brakes on global warming. But examined more closely, the overall impact of cloudiness is not straight-

Figure 2.2 Two contrasting cloud types with very different thermal impacts on the atmosphere: (a) Nimbostratus clouds, responsible for steady rainfalls, tend to cool the Earth's air below; (b) Cirrostratus clouds, found at high altitudes (as high as 10 km), tend to warm it.

forward, because the moisture in clouds *also traps and reradiates some incident sunlight.* To a certain extent, this offsets cloud-albedo cooling.

We now know that the net effect of all the world's clouds depends on their individual structures and altitudes. Thick, low clouds, such as those associated with steady rains, reradiate more energy into space than they absorb, cooling the surface beneath, whereas high, thin clouds, such as those seen before many approaching storm fronts, absorb more incident sunlight and tend to warm the atmosphere below. The atmosphere contains a complex, ever-changing mixture of both cloud types at any given time. Nevertheless, most earth scientists view the overall impact of the world's clouds as one of atmospheric warming. The cloud-albedo effect reduces that warming significantly (**Figure 2.2**).

Atmospheric Circulation and Climate

As mentioned previously, radiant solar energy warms Earth's surface and atmosphere, with the equatorial latitudes receiving much more sunlight than the polar regions (**Figure 2.3**). Under most circumstances, warm air rises and spreads, whereas cold air sinks, because the various molecules of gas making up the air become more widely spaced as the air heats up. In other words, *warm air is less dense than cold air.* Throughout the atmosphere, air masses are moving and mixing, thus generating wind, in a never-ending quest to *equalize* atmospheric temperature and pressure. The latitude-dependent differences are terrific. For example, the average annual temperature difference between the Philippines, lying in the sultry tropics, and the bitterly cold East Antarctic Plateau is a staggering 75.5°C (138°F). Because the Earth continues to rotate and the Sun continues to shine, such extreme differences persist—and perfect homogeneity of air temperature and density has never existed in our planet's history. Nevertheless, the Earth System, obeying the ordinary laws of physics, keeps trying to achieve it!

Given the natural tendency for the atmosphere to smooth out its physical differences, you might imagine that winds blow in just two directions to be most efficient: hot air moving *directly* from the equator toward the poles and freezing air moving in the *opposite* direction to mix uniformly somewhere in between. However, it doesn't quite happen that way, because of an important fact we haven't yet taken into consideration: The solid Earth and everything attached to it rotate beneath our atmosphere. Consequently, with respect to the surface, what are the actual paths taken by winds? They curve, in much the same way that a ball tossed to a bystander by someone standing on a rotating merry-go-round appears to curve (**Figure 2.4**). We term this illusionary curvature the **Coriolis effect,** and it applies not only to wind paths but also to ballistic missiles, ocean currents, passenger planes, and even rifle bullets traveling through the air.

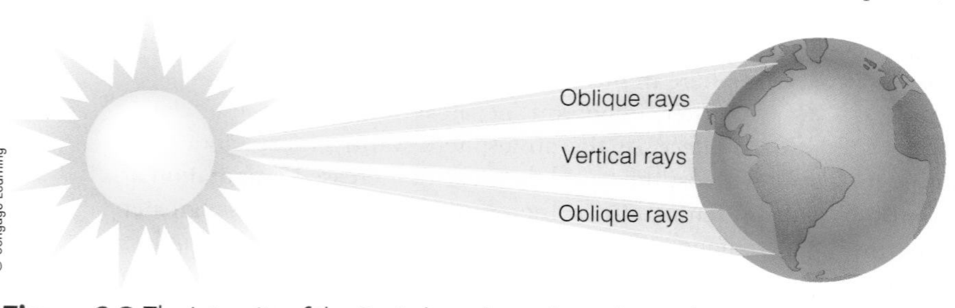

Figure 2.3 The intensity of the Sun's heat depends on the angle at which the Sun's rays hit Earth. The intensity and heating are greatest where the rays hit vertically.

The Coriolis effect is one of the more complicated concepts to explain in earth science, but it is simple when grasped. You just have to think of several things happening at once. Specifically, Earth rotates from west to east, with fixed locations at the equator having to cover a greater distance to get all the way around the planet during the course of a day than locations closer to the poles, through which Earth's rotation axis passes. If you were standing at the North or South Pole, you would find yourself

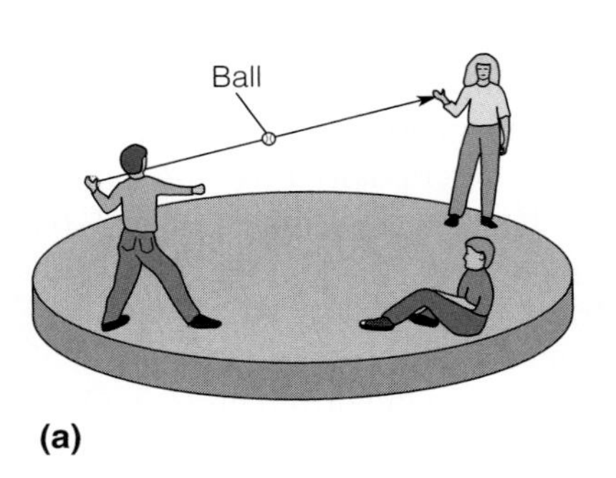

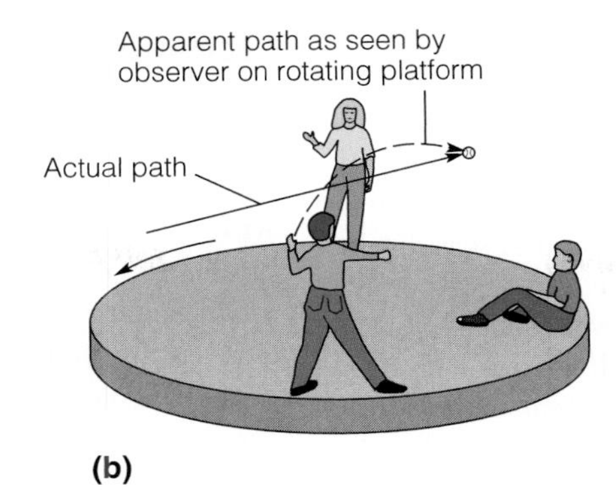

Figure 2.4 The Coriolis effect in a game of catch. (a) The path of the ball is straight when the game is played on a stationary platform, and an observer perceives it as straight. (b) On a rotating platform, the ball still follows a straight path, but to an observer on the platform, the path appears to be curved. This is the Coriolis effect.

slowly turning around in place over the course of 24 hours. But at the equator, your locality would be moving through space at 1,670 km (1,000 mi)/hr to return to your starting position 24 hours later, given Earth's 40,000-kilometer (24,000-mi) equatorial circumference. Because of the latitude-dependent change in rotational speeds, the winds bound for the pole tend to "race ahead" of the surface rotating beneath them, whereas equator-bound winds tend to "lag behind." To observers on the ground, the difference in relative motions of air and surface is apparent as wind paths tend to *bend toward the right in the Northern Hemisphere and toward the left in the Southern Hemisphere.* Coriolis deflection of the winds converging on, and escaping from, cyclones, hurricanes, and typhoons also helps explain the spin of these whirlpool-like storms: *Cyclones in the Northern Hemisphere turn counterclockwise, whereas those in the Southern Hemisphere rotate clockwise.* The take-away point is this: Using Earth's surface as a frame of reference, because of Coriolis deflections, few, if any, winds on Earth blow directly from the equator to the poles. In fact, quite often, winds tend to blow more east-west than north-south. The same is true for the atmospheres of all other rotating planets (**Figure 2.5**).

Quite apart from the impact of Coriolis deflection, our atmosphere's circulation is not so simple because of the fact that winds moving toward the poles converge (just as Earth's longitudinal lines geometrically converge). Let's follow one hypothetical packet of air at low elevation near the equator after sunlight heats it. Our air mass ascends, drawing in fresh air to replace it at low elevation from both north and south of the equator. The air is humid, thanks to evaporation in the torrid tropical heat, but as the air rises, it cools, and the moisture it contains quickly condenses into clouds, from which heavy rain may fall. By the time the air has moved away from the equator, at elevations of 10 to 15 kilometers (6–9 miles), it has become much drier. Moving toward the poles, the air converges with other high-altitude air masses, all headed in the same direction. It grows denser because it cools off and is squeezed by adjacent northward-moving air masses. As a result, all this converging air sinks back to Earth roughly at latitudes 30° north and south of the equator. The air is compressed as it descends, and compression causes it to become drier and warmer, even hot, a process called **adiabatic heating.** The parched descending air soaks up any moisture it can, desiccating the land and helping create the band of temperate-latitude deserts that girdle the planet. If

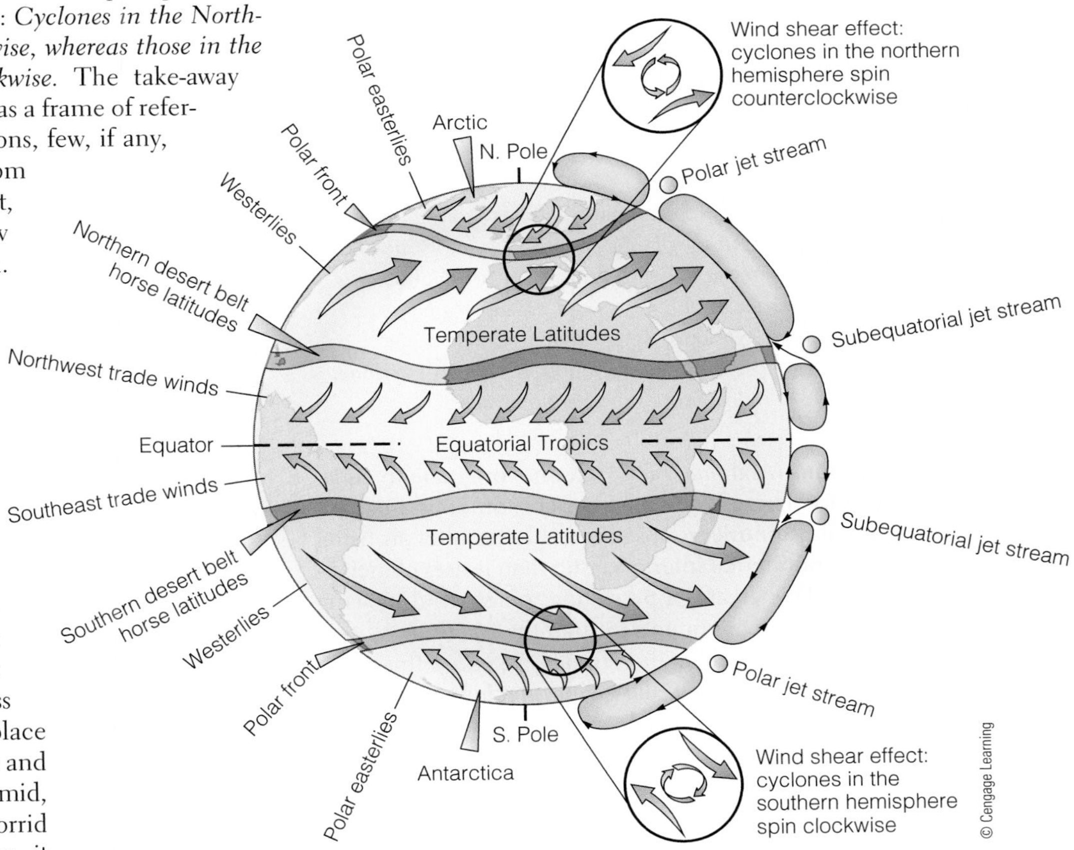

Figure 2.5 Pattern of global surface wind circulation. Note how Coriolis forces play a role in the circulation of surface winds in both hemispheres.

the air packet descends to the ocean surface instead of land, the result is an "ocean desert," creating broad stretches of hot, still water—perilous to early mariners, who referred to them as **horse latitudes,** perhaps because of the death of horses (not to mention the sailors themselves), which could not survive the long, listless journey across these seas. From the desert horse latitudes our air packet divides with some of it racing back toward the equator as westward-blowing trade winds and the rest continuing to move poleward as eastward-blowing westerlies (Figure 2.5). As the wind nears the poles, the warm air encounters a barrier of dense, frigid **polar air**. The colliding air masses partly mix, but some of the poleward-bound air, being warmer, simply avoids the barrier of polar air by rising and turning back toward the equator at high altitude. The zone of collision is called the **polar front**. The result of all this restlessness is an atmosphere divided into sets of overturning wind circulations, called **Hadley cells**, as shown in Figure 2.5.

Hadley cell circulation alone doesn't account for the overall mixing of air in Earth's atmosphere. Also vital are cyclones and monsoons. Cyclones can originate both in the equatorial zone and from the polar front. Fierce tropical cyclones (>119 km/hr winds) are called **hurricanes** in the Atlantic Ocean and eastern Pacific, and **typhoons** in the western Pacific and Indian Ocean. They develop where sea surface temperatures exceed 26.5°C (104.5°F). Cyclones associated with the polar front spin off to generate the great winter cold and warm frontal systems that bring the rain and snow so critical to life at temperate latitudes. As much as 2% of all of Earth's atmospheric energy may be embodied in cyclones. A single Atlantic hurricane may release as much energy in 20 minutes as a 10-megaton nuclear bomb.

consider this . . . The word **hurricane** derives from a Native American (Carib peoples) term, "Hurican." Hurican was the Carib god of evil. **Typhoon** is a modern adaptation of the Chinese word **dàfēng** ("great wind") and the Greek word **typhôn** ("violent wind"). British meteorologist Henry Piddington (1797–1858) introduced the term **cyclone,** meant to derive from the Greek word for "to revolve."

Monsoons (Arab *maivsim*, "season") originate in equatorial regions. There are both summer monsoons and winter monsoons. Summer monsoons result from hot air rising off warmer continental interiors faster than it rises off neighboring seas. This creates a prevailing, moist wind blowing off the oceans into continents for months at a time. Where mountain chains, such as the Himalayas, cause this air to rise and condense, steady, life-giving precipitation results. Hundreds of millions of people are dependent on summer monsoons to sustain their agriculture and water supplies. In the less-developed world, the loss of monsoonal rains in one or two seasons can make all the difference between life and death. Wind direction in winter monsoonal circulation is just the reverse, with air flowing off the continents toward the ocean.

Oceanic Circulation and Climate

Our atmosphere and oceans are *coupled*, meaning that they constantly exchange energy and matter with one another in the form of heat, gases, and moisture. Furthermore, the wind circulation (see earlier description) drives the shallow circulation of Earth's seas. If you have ever sailed a boat, you are familiar with the appearance of a fresh gust of wind across still water. The water roughens up and becomes less glassy. Ripples develop and shift downwind, and soon all of the surface water is moving as a wind-driven current. On a much larger scale, the trade winds and westerlies drive global-scale ocean currents that travel thousands of kilometers and loop around to form **oceanic gyres** (**Figure 2.6**). Gyre patterns are shaped by three factors: (1) the placement of the continents, (2) the Coriolis effect on winds blowing across the waters, and (3) the Coriolis effect on water masses moving across latitudes. At the centers of the gyres, making up some 40% of Earth's total surface, are large areas of calmer water, stirred only by occasional frontal systems or cyclones. Into these parts of the ocean much of the flotsam and jetsam drifts that people discard either accidentally or intentionally from ships and coastal communities, inspiring environmentalists to refer to them as **garbage patches.** In recent years, biologists have discovered that the garbage patches are severely harmful to marine life, especially birds.

Oceanic gyres are important in transferring heat from lower to higher latitudes. Warm-water currents, such as the Gulf Stream in the North Atlantic and the Japanese Current in the North Pacific, heat the air above to make climatic conditions more favorable for people living in such northerly locales as Alaska, British Columbia, and Western Europe. At the same time, as these currents cool and head back toward the equator, they moderate coastal climates at lower latitudes.

Another important climate service provided by the ocean is marine upwelling. Along some shorelines, persistent winds coupled with a Coriolis phenomenon in sea currents known as the Ekman spiral drive shallow waters away from the shore. To replace those waters, cold, deeper water rises ("upwells"), bringing with it important nutrients for marine life and chilling the moisture-laden offshore breezes to create coastal fogs. Seasonal fogs are important for ecosystems such as that of the Atacama Desert of northern Chile, South Africa's *fynbos* (from the Afrikaans, meaning "fine bush," the biologically important shrubland occurring in the Cape region), and California's redwood forests. Marine upwellings occur over only about 0.2% of Earth's ocean surface, but the nutrients they provide sustain more than 20% of the total fishery catch.

Hence the oceans are certainly as important as the atmosphere for determining Earth's climate. But they have an even greater, more subtle impact on keeping the planet hospitable for life, and this requires looking much deeper into their waters.

Wind-driven ocean surface currents involve no more than about 10% of all ocean water at any given time, and they stir the sea to depths of only about 400 meters (1500 feet), whereas the average depth to the seafloor worldwide is around

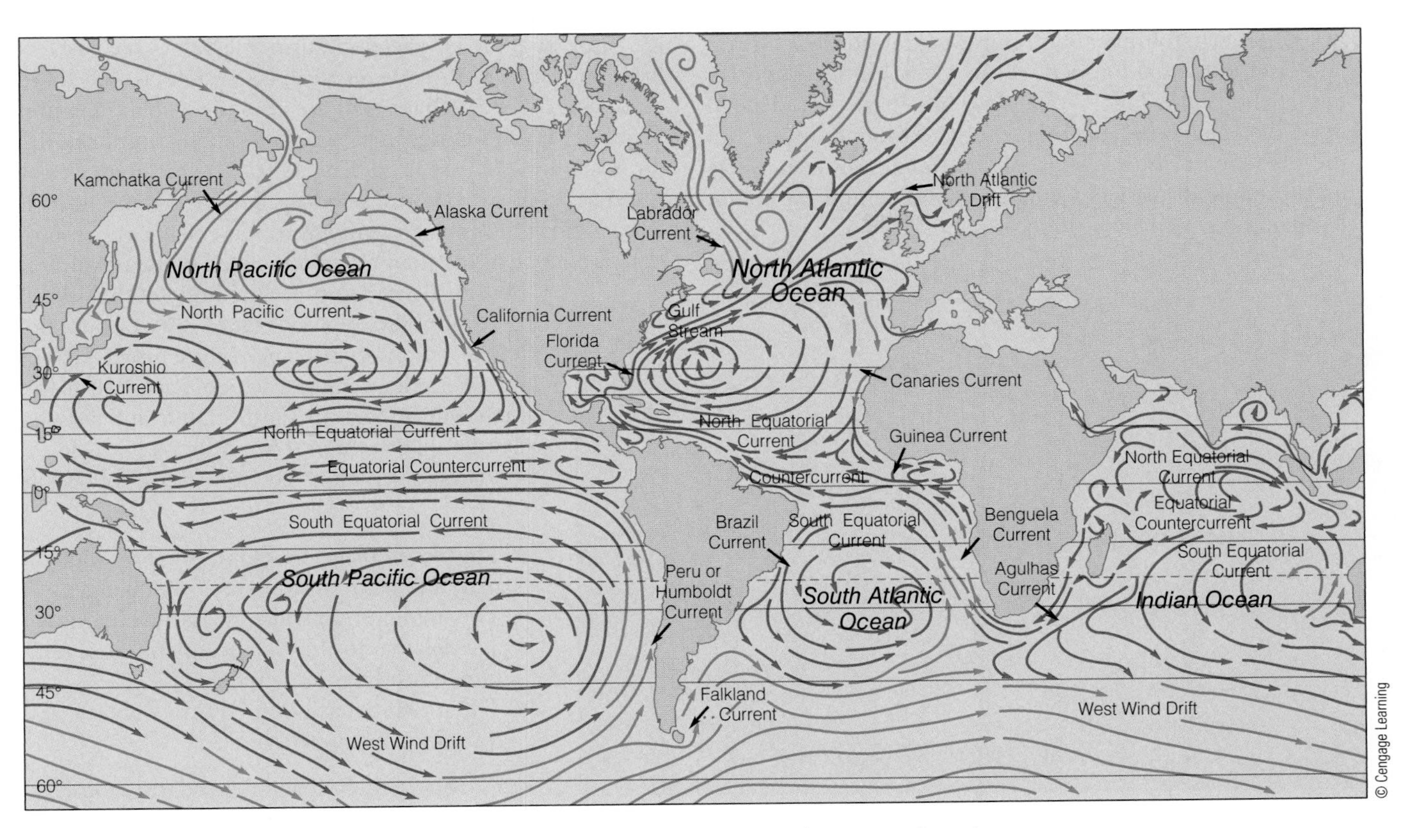

Figure 2.6 Major wind-driven oceanic surface currents. Warm currents are shown in red, cool currents in blue. Currents rotate clockwise in the Northern Hemisphere and counterclockwise in the Southern Hemisphere.

4,000 meters (13,000 feet). That deep, dark, nearly freezing deep sea is far from stagnant, however. It also circulates, albeit sluggishly. What drives this water that few of us will ever get a chance to see? The answer is differences in density caused by salinity (dissolved salt content) and water temperature. Just as dense air sinks, so does dense water. Cold, salty water is significantly heavier than warm freshwater. As it sinks, replacement water rushes in and a current is set in motion.

The major driver of deep marine circulation is the chilling of seawater along the coasts and under the ice shelves of Antarctica. An almost constant cascade of nearly freezing water slips into the deep along the margins of the Ice Continent, then spreads out northward across the seafloor, even crossing the equator into the Northern Hemisphere. This water is also very salty, because of the **brine exclusion effect,** which forces the concentration of dissolved salts in unfrozen water wherever ice forms—sea ice is salt-free. The sinking water must be replaced, and it draws in generally warmer surface water from lower latitudes. In this way, deep marine and shallow marine circulations become connected. In the northern Pacific, the Antarctic deep water wells up as it warms, becoming incorporated into the shallow pattern of global wind-driven currents, in time mixing into the Gulf Stream, where, in the Norwegian Sea east of Greenland, cold temperatures prompt it to sink back to the seafloor as so-called North Atlantic deep water. The complete circulatory system has been named the **global conveyor** by Columbia University paleoclimatologist Wallace Broeker (**Figure 2.7**).

The global conveyor turns out to be of paramount importance to global climate because of a simple chemical property of CO_2, one of the most important greenhouse gases: CO_2 dissolves more readily in cold water than in warm water. Thanks to deepwater formation around Antarctica and near Greenland, billions of tons of CO_2 are transported by solution into storage in the cold deep ocean, from which it takes, on average, more than 500 years for any given molecule of CO_2 to escape back to the atmosphere (**Figure 2.8**). In fact, not all of this dissolved gas ever gets back into the air; much of it over the long span of geological time has ended up precipitating as carbonate minerals in sands and limestone-forming shellfish, and even incorporated into oil and natural gas (see Chapter 14). Were it not for this capacity of the oceans to dissolve and store CO_2, Earth might be as hellish a world as its sister planet, Venus, with atmospheric temperatures hot enough to melt lead.

Natural Climate Change

If Earth's natural environmental system is such a remarkable set of self-regulating checks and balances, why is the geological record full of evidence of variation in the steady states of this system over time? Why have we had fabulously warm times, such as the Cretaceous period (66–145 million years ago), when tropical forests grew as far north as Greenland and dinosaurs romped within a few hundred kilometers of the South Pole? Why, also, has Earth experienced episodic ice ages until quite recently? Throughout most of the existence of our species, *Homo sapiens*, our ancestors had to contend with transitions in climate and far colder temperatures than civilization presently enjoys.

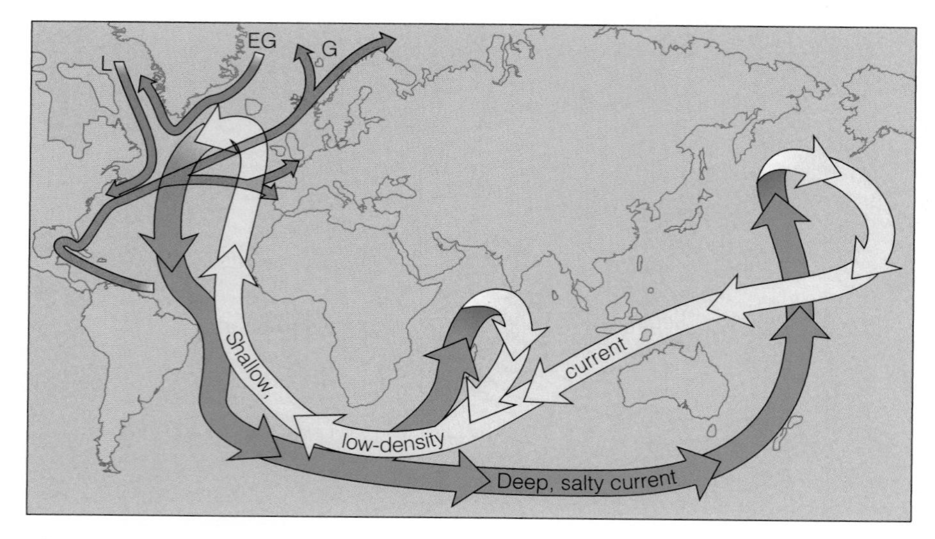

Figure 2.7 The "global conveyor" that snakes through the world's oceans. Two surface currents of major importance in determining the modern climates in Greenland, eastern North America, and Western Europe are shown. The cold Labrador (**L**) and East Greenland (**EG**) currents (both shown in blue) cool the adjacent coasts. The Gulf Stream (**G**, red) transfers heat northward, influencing Arctic air currents and warming the coasts of Iceland, Greenland, and Western Europe. This current was deflected southward toward North Africa during the coldest part of the last maximum glaciation. The importance of these currents, in combination with the atmospheric wind system, is illustrated by comparing the modern equitable temperature of northern Scandinavia, which is essentially unglaciated, with that of Greenland, which, at the same latitude, is covered by an enormous ice sheet. (Data from Broecker, W. S. 1995, November. Chaotic climate. *Scientific American* 273[5].)

Figure 2.8 The near-freezing waters around Antarctica absorb large amounts of atmospheric CO_2, which then descends to the seafloor in sinking, salt-rich water, where it may be stored for many centuries. This process is an important part of the overall global marine circulatory system and climate regulation.

Insofar as the ice ages and warm intervals of the past few million years go, the answer to that question is that there have been cyclical changes in the amount of solar radiation reaching Earth's atmosphere. Output of radiation from the Sun is known to vary from 0.1% to 0.2% over timescales as short as decades. But most scientists believe this has little, if any, significant impact on our atmosphere. Far more important is change in incident radiation related to the revolution of Earth around the Sun and to the tilt of Earth's rotational axis (**Figure 2.9**). Orbital behavior has *externally forced* the climate system into periodic changes, which have had a tremendous impact on Earth's lands and seas—a matter of keen interest to geologists.

Three variations occur as Earth follows its orbit:

- The **eccentricity**, or shape, of the elliptical path it takes in orbiting the Sun changes.
- The angle, or **obliquity**, of Earth's rotational axis relative to the planet's plane of revolution around the Sun changes.
- The rotational axis of the Earth **wobbles**, as a spinning top can wobble.

Let's explore the impact of these variations.

Feeling the gravitational tug of the giant outer planets, Jupiter and Saturn, as these worlds orbit the Sun, Earth's own orbit oscillates from a nearly circular path to a drawn-out ellipse once every 100,000 years. Seasonal variation in temperature is greater when the orbit is more elliptical (or eccentric). Earth's present eccentricity is not great, resulting in only a 6% difference in the heat energy received between January and July. However, at Earth's most eccentric orbit, this difference is 20% to 30%.

The obliquity of Earth's rotational axis has been found to vary between 21.8° and 24.5° over a 41,000-year cycle. The Earth's present tilt is in the middle of this range. This slight change causes variations in the temperature range between summer and winter. When Earth is at the minimum tilt, sunlight hits the polar regions at a higher angle, and the global seasonal temperature variation decreases worldwide.

Earth's axis presently points toward Polaris, the North Star, as generations of campers and wilderness scouts in the Northern Hemisphere have learned. But this is a coincidence that won't last long, geologically speaking. As Earth's axis wobbles, it will point toward other positions in space in years to come, gradually tracing out a circle, not to return to Polaris again for 26,000 years. The wobbling of the planetary axis means that the months in which summer and winter occur are constantly, slowly changing. About 13,000 years from now, the Northern Hemisphere summer will take place in December–February—the times northerners presently associate with freezing snow and short daylight hours, whereas conditions will be wintry in the Southern Hemisphere. Because the two hemispheres have different land–ocean configurations, they have very different net albedos, meaning that changes in seasonal timing greatly influence overall planetary climate.

Taken together, these three variations help explain the waxing and waning of ice ages. But they are not the complete story. It's no surprise that feedbacks come into play also, exaggerating the temperature changes wrought by these external forcings.

In addition to the direct evidence of past glaciations observed in many landscapes (see Chapter 11), geochemical evidence of their occurrence remains preserved in ancient ice records, which can be recovered by drilling into glaciers and ice caps. As snow accumulates and is packed down into ice, it contains information about the atmosphere in which it formed, which remains preserved in glaciers. The relative abundance of two slightly different oxygen atoms found in snow and ice—the oxygen-18 to oxygen-16 isotope ratio, symbolized $^{18}O/^{16}O$—is especially important. ^{18}O contains two more neutrons than does ^{16}O, and when a body of water evaporates, the lighter type of oxygen (^{16}O) becomes proportionally more concentrated in evaporating H_2O molecules than the heavier type. Conversely, when raindrops or snowflakes form from humid air, they concentrate the remaining $H_2{}^{18}O$ from the water vapor, enriching the air even further with $H_2{}^{16}O$. As a mass of warm, moist air moves into colder environments, condenses, and repeatedly precipitates, the ratio of $^{18}O/^{16}O$ decreases—ever less ^{18}O is present in the remaining moisture relative to ^{16}O. In fact, a general correlation can be made between the $^{18}O/^{16}O$ preserved in ice and snow, and *the temperatures at which the initial precipitation occurred.* This allows paleoclimatologists—the scientists who study past climates—to infer trends in regional atmospheric temperature and precipitation. Note that they are not measuring those past temperatures directly, but are using the oxygen isotope ratio as a **proxy** (substitute) for doing so. Matters are complicated by the fact that the $^{18}O/^{16}O$ ratio is also sensitive to the total accumulated volume of ice in the world, because each net addition of ice reduces the total amount of ^{16}O available for additional evaporation and ice formation. But that factor can be untangled from the temperature influence, and measurements of $^{18}O/^{16}O$ have proved very useful in reconstructing the average air temperature of Earth's atmosphere at higher latitudes and elevations for the past several hundred thousand years, facilitated by the fact that each year's snowfall forms an easily recognized surface recorded in glacial ice (**Figure 2.10**). Like the growth rings in trees, one can count these surfaces in an ice core year by year back in time to determine precisely when certain climate conditions prevailed.

The bubbles of ancient atmosphere trapped in glacial ice obtained from 3,348-meter (10,700-feet)-deep cores drilled at the Russian Vostok Station in Antarctica and at the Greenland Ice Sheet Project 2 (GISP2) provide a 420,000-year record of variations in the atmosphere's CO_2 content—one of the longest known. The ancient trapped CO_2, like oxygen isotopes, may be measured as a rough proxy for atmospheric temperature and, in fact, is an easier working substitute when it comes to looking at trends in global atmospheric temperature, given that CO_2 is nearly uniform in its atmosphere distribution, whereas ^{18}O and ^{16}O are not. The Vostok record catalogs four full glacial–interglacial cycles—that is, the atmospheric conditions during four ice ages and intervening warm periods. In 2004, a consortium of Euro-

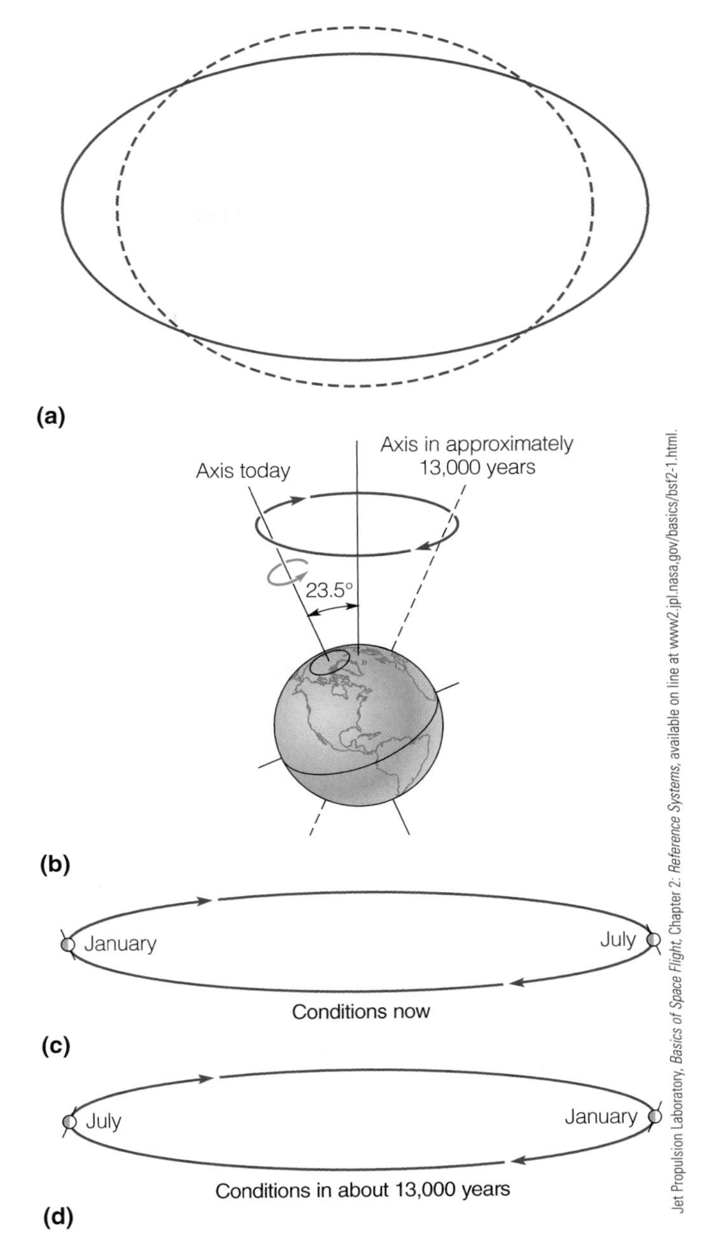

Figure 2.9 Geometry of variation in Earth's orbital eccentricity, axial precession, and axial obliquity. (a) The orbital eccentricity varies from nearly circular (dashed line) to elliptical (solid line) to nearly circular and back again over a period of 100,000 years. (b) Earth's axis slowly processes, like a spinning top tracing out a cone in space, taking about 26,000 years for a complete cycle. (c) Conditions today. Earth is closest to the Sun in January while the Northern Hemisphere is experiencing winter. (d) Conditions in about 13,000 years. Earth will be closest to the Sun in July while the Northern Hemisphere is experiencing summer. Precession cycle = ~26,000 years. (Source: www2.jpl.nasa.gov/basics/bsf2-1.html)

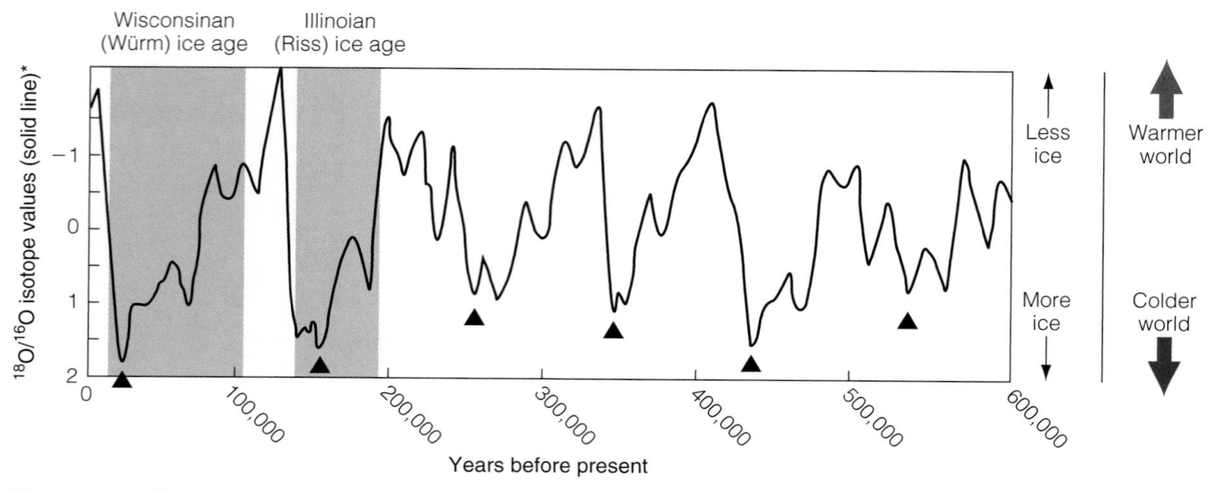

Figure 2.10 Changing oxygen isotope ratios correlated with global ice volume and mean atmospheric temperature. The shaded bands show the timing of the past two ice ages, and the black triangles indicate the coldest times of the past half-dozen ice ages. *The isotope ratios are measured as so-called "del-values": [($^{18}O/^{16}O_{samples}$)/($^{18}O/^{16}O_{modern\ seawater}$) − 1]1000. (Data from Clements, A. C., L.C. Peterson. 2008. Mechanisms of abrupt climate change in the last glacial period. *Reviews of Geophysics,* vol. 46, RG4002: 39p; and SPECMAP Paleoclimate Archives, National Oceanographic and Atmospheric Agency/NCDC World Data Center for Climatology, available at http://gcma.nasa.gov/records/GCMD_EARTH_LAND_NGDC_PALEOCL_SPECIMEN.html)

pean scientists, the European Project for Ice Coring in Antarctica (EPICA), almost doubled the earlier Vostok climate record by pulling a new core from Dome C, about 500 kilometers (300 miles) from the Vostok Station. The ice layers there go back 740,000 years and record eight past ice ages.

Further information and corroboration of past climate changes comes from the study of oxygen isotope ratios preserved in microscopic marine fossils and speleothems—formations such as stalactites and stalagmites found in limestone caves. In these instances, evaporative distillation of ^{18}O relative to ^{16}O still plays an important role; but the isotope proxies indicate temperatures of the residual (post-evaporative) ocean or groundwater at the moment of calcium carbonate ($CaCO_3$) precipitation. Calcium carbonate is the main, oxygen-bearing ingredient making up corals, shells, and cave decorations.

Putting together the data from glaciers, caves, and the shallow sea gives us a general global record of change, with cold and warm atmospheric intervals confirmed independently from sample sites scattered worldwide. Resulting reconstructions clearly show the climate influence of external orbital forcings (**Figure 2.11**). But notice that they are asymmetrical in shape; each ice age starts out as a gradual, erratic descent over tens of thousands of years into frigid condi-

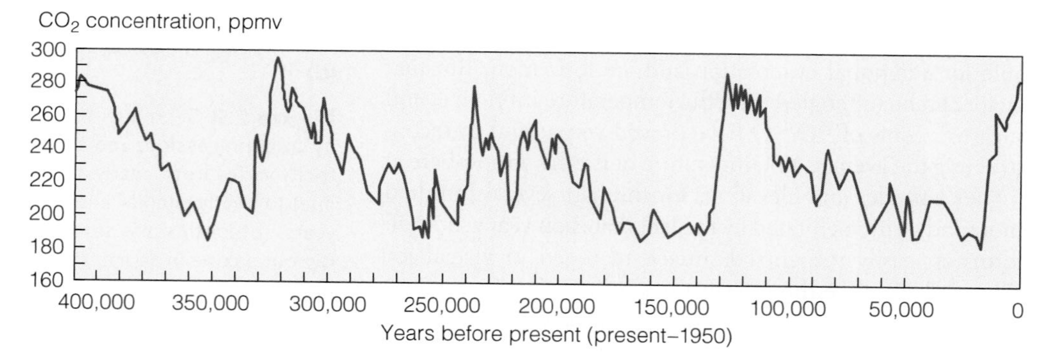

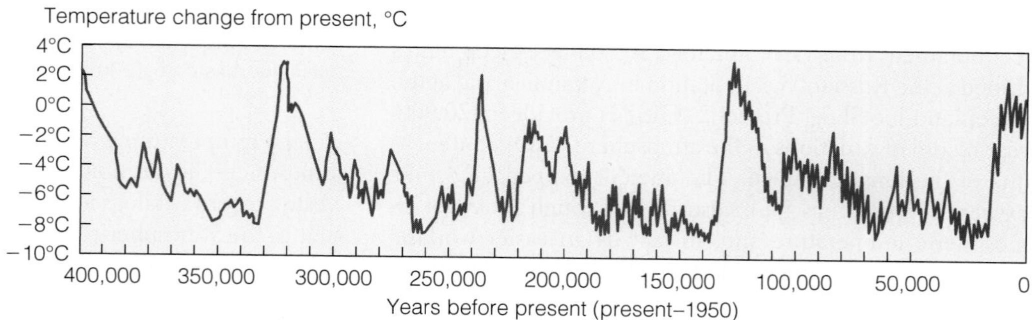

Figure 2.11 Vostok, Antarctica, ice core record of inferred temperature change from oxygen isotopes trapped in bubbles (blue lines) and CO_2 concentration found in the same bubbles. Ice ages are shown by the deep, jagged temperature "troughs." The peaks are relatively brief warm intervals, such as the Holocene, in which we are presently living, which is the temperature/CO_2 peak at the extreme right margin of the diagram. ppmv, parts per million volume. (Reprinted by permission from Macmillan Publishers Ltd: Petit, J. R., D. Jouzel, et al. 1999. *Climate and atmospheric history of the past 400,000 years from the Vostok ice core in Antarctica.* Nature 399.)

tions, the area of ice cover worldwide tripling to almost a third of Earth's total land surface (from the present 10%). Then, just as conditions seem to have become their coldest, the ice age ends abruptly—in just a few thousand years. Nature seems to throw a switch to return to warmth. What explains this strange pattern?

Important clues again lie within the ice cores described earlier. In general, greenhouse gas concentrations in ice-sealed air bubbles vary positively with the temperatures inferred from oxygen isotope measurements, as one might expect. The fit is not perfect, however, and this has led to some interesting chicken-and-egg questions in current debates about climate change. For example, did increasing greenhouse gas concentrations warm the atmosphere to end the last ice age, or did warming of the atmosphere precede increased greenhouse gas concentrations? In fact, *both* cases appear to be true (Chapter 11).

A feedback-driven scenario of how a typical ice age begins and culminates ensues something like this: A small degree of externally forced atmospheric cooling permits small patches of snow to remain on the ground through several summer seasons in certain localities in the Northern Hemisphere (other than in Antarctica, which is already mostly covered with snow and ice, there is not enough land area in the Southern Hemisphere for this to happen). The year-round presence of the snow jump-starts the **snow-albedo positive feedback,** in which the snow chills the air around it and also reflects incoming solar radiation back to space, encouraging additional snowfall at an accelerating pace. The patches grow bigger and thicker, and regional summers become cooler and cooler, until great stretches of land are covered with snow and ice, and eventually a new ice age is in full swing. The feedback is so effective that it sustains frigid conditions around the great continental ice sheets even after external forcing begins to warm the atmosphere again elsewhere. Increasing latitudinal temperature differences as glaciers spread into lower latitudes cause greater windiness worldwide, which increases overall dustiness, especially near the ice fronts. Dust has a low albedo in contrast with ice, and accumulations of dust on ice sheets encourage melting, although the extent to which this negative feedback impedes a deepening ice age is uncertain.

The abrupt end of an ice age has much more to do with atmospheric and oceanic circulation. During the latest ice age, which ended about 10,000 years ago, the global conveyor took a very different form. Cold, CO_2-rich water, especially from North America, formed in much greater quantity. This helped refrigerate the ice age even further. At the same time, the marine density–salinity contrasts, which are enhanced when the climate is warm and evaporation rates are high, were not as pronounced during the ice age, and the CO_2-rich water did not always flow as deeply as it does now. Some geoscientists have linked the erratic, sharp swings in temperature so characteristic of the latest ice age to sudden drainages of giant meltwater lakes at the edges of continental ice sheets—especially in modern-day Canada—and to periodic, rapid degassing of CO_2 from the ocean caused by wind-driven upwellings in water that had lost much of its density layering. As the oceans and atmosphere grew incrementally warmer, the released CO_2 could not be redissolved as rapidly as before, and ultimately a tipping point was reached in which increasingly CO_2-driven climate warming simply overwhelmed the snow-albedo effect. As the ice retreated, tremendous amounts of methane also escaped to the atmosphere from vegetation that had been long buried by glaciers and once again saw the light of day, so that biological decay finally began. Methane is an even more potent greenhouse gas than CO_2.

consider this . . . Earth is not the only planet showing evidence of cyclical ice ages. Various satellite studies indicate that Mars, our close neighbor in space, may have experienced as many as 40 ice age advances and retreats over the past 5 million years. At maximum extent, ice on the Red Planet has extended to latitudes as low as 30°, in contrast with Earth where continental glaciations have not approached closer to the equator than 40° to 50° in the same interval of time. The climate of Mars responds more strongly to orbital forcings than Earth's climate.

The Human Contribution to Earth's Climate System

Aquatic ecologist Eugene Stoermer, from the University of Michigan, coined a new term to describe the present time in Earth's history, in which the forces of human activity have begun to exceed or significantly modify those of wild nature—the **Anthropocene** or "Time of Man." The term first came to public attention in 2000, and in 2008, geologists began formally considering it as valid addition to the geological time scale (see Chapter 3). Many scientists who support this notion believe that the Anthropocene began in the 18th century with development of modern fossil-fuel–based industry. Others, notably William Ruddiman, from the University of Virginia, believe that humankind set the stage much earlier, with the start of widespread agriculture around 8,000 years ago. A "compromise" position is that the Anthropocene starts with rise of the first large empires, such as Rome and the Han Dynasty, some 2,000 years ago. At that time, farming reached the point at which people began plowing, tilling, littering, and disturbing soils across much of the Earth's surface. Because long-lasting change of global scale recorded in soil and rock layers is the hallmark of change from one geological time interval to the next, this rough date perhaps makes the most sense.

In any event, unusual global warming must be regarded as an important aspect of the Anthropocene—and it is primarily an artifact of the Industrial Age (**Figure 2.12**). That CO_2 emissions related to gas, coal, and oil are principally to blame for this is shown by the **Suess effect,** an anomalous loading (at a geologically unprecedented rate) of carbon isotope-12 into the atmosphere, which we'll explore further in Chapter 11, and to a strong correlation between fossil fuel use and rising global temperature since the end of the Second World War (**Figure 2.13**; **Table 2.2**). Today's levels of atmospheric CO_2 exceed those characteristic of previous interglacials by a whop-

© EIGHTFISH/Getty Images

Figure 2.12 Air pollution from industrial activity contributes significantly to carbon dioxide loading of the atmosphere, as here in Shanghai, China, where coal is a major source of heating and electrification. China and the United States are the two top CO_2-producing countries in the world.

ping 100+ parts per million, or 40% (**Table 2.3**). Such high levels of CO_2 have not existed in Earth's atmosphere at any other time for the past few million years. The related increase in temperature has not been uniform; regions of land at high northern latitude have shown the strongest change, whereas—as discussed earlier—ocean areas have responded more slowly because of enormous water mass (**Figure 2.14**). And industrial activity is not the sole aspect of human activity contributing to this change. **Land conversion** is the biggest contributing factor after fossil fuel combustion—the clearing of forests and grasslands for farms, pastures, and urban areas, and many other manufacturing or commercial processes (Table 2.2). Land conversion means significant change in landscape albedo: a general reduction where stands of trees are removed, or an increase where farmers plant irrigated crops in desert areas (**Figure 2.15**). It means a transformation in patterns of cloud formation and precipitation, including drying out of landscapes in otherwise moist, tropical environments, where woodland evapotranspiration plays an important role in the cycling of water. It means a reduction in the biological carbon storage capacity of many regions, because crops and grazing soils involve far less biomass than mature forests.

André Berger and Marie-France Loutre from the Institute of Astronomy and Geophysics in Louvain-de-Neuve, Belgium, have closely researched the changing insolation (incident sunlight) reaching Earth's surface given our planet's various orbital permutations. Presuming that this is the major forcing agent for large-scale global climate change, these two scientists show that under natural conditions, Earth should be plunging into its next ice age around 50,000 years from now. But given the large amount of atmospheric CO_2 held over from the industrial era, the greenhouse effect may simply be too strong for this to happen, according to David Archer, James Zachos, and other researchers. As extraordinary as it may sound, the future

Table 2.2 Origins of Anthropogenic Carbon Emissions

A. Producers of Greenhouse Gases by Fuel Source

Petroleum = 43.3 %
Coal = 36.4%
Natural gas = 20%

Source: Data from National Geographic Society. 2007. "Greenhouse Earth."

B. Categories of Emissions

Electricity and waste heat = 24.6%
Land conversion = 18.2%
Agriculture = 13.5%
Transportation = 13.5%
Industry = 10.4%
Other fuel combustion = 9.0%
Equipment leaks, etc. = 3.9%
Landfills, sewage = 3.6%
Industrial processes = 3.4%

Source: Data from National Geographic Society. 2007. "Greenhouse Earth."

C. Descriptions

Category	Description
Agriculture—synthetic fertilizer production	The first step in the Haber-Bosch process, necessary for making synthetic fertilizers (see Chapter 1) is: $CH_4 + 2O_2 \rightarrow 2H_2O + CO_2$.
Agriculture—livestock	Pigs, goats, sheep, and cows emit lots of CH_4.
Agriculture—rice paddies	Rice fields, through decay, are an increasingly major source of human-produced CH_4.
Land conversion and associated deforestation	Removal of forests means reduction of CO_2 sequestration capacity in the Earth System. Most deforestation is a result of land clearing (conversion) for farming, pasturing, and plantation forestry, increasingly so in the tropics.
Soot-producing stoves in the lesser developed world	Dung and biomass-fueled stoves in developing countries create soot, which settles on snow and ice fields, most critically in the Himalayas. The result is significant albedo reduction.
Industrial processes—quicklime production	Quicklime (CaO) is used to make mortar, cement, and plaster by reduction of limestone in furnaces ($CaCO_3 \rightarrow CaO + CO_2$). As much as 4% of U.S.-generated global warming may be related to the cement industry.
Industrial processes—steel production	Coke (pure carbon from charcoal) is used to flux iron ore to pour into steel in foundries. The reaction is: $2Fe_2O_3 + 3C \rightarrow 4Fe + 3CO_2$.
Industrial processes—beer, wine, and cider production	Fermentation yields carbon dioxide: $C_6H_{12}O_6 \rightarrow 2C_2H_5OH + 2CO_2$.
Burning of fossil fuels	Figure 2.13 speaks for itself! Transportation, manufacturing of medicines, clothing, plastics, and other products, cooking, heating and cooking, and general electrification all contribute.

Table 2.3 Carbon Dioxide Contrasts in Recent Earth History

Time Period	Carbon Dioxide (CO_2) Concentration In The Atmosphere (Parts Per Million)	Mean Global Air Temperature
Ice Age (18,000 years ago)	170	~6°C
Post-Ice Age, pre-Industrial Age	280	14°C
Industrial Age, present	387	14.5°C
End of 21st century	500–750	17°C–20.5°C

planetary climate for the next several hundred thousand years *depends strongly on the carbon emissions we put into the air over the coming century.* Atmospheric CO_2 does not immediately return to the hydrosphere, biosphere, and soils, but will linger. As much as a quarter of peak concentration in an emissions spike can persist for tens of millennia. Ultimately, silicate rocks like granite scrub out CO_2 from the air through natural weathering processes (see Chapters 3 and 6), but the timelines are too long to be humanly meaningful. In a more extreme super-greenhouse scenario in which we load the atmosphere with 5,000 gigatons of CO_2, certainly within the realm of industrial possibility, the pollution tail stretches out more than 400,000 years—more than twice the duration of the human species to date! Under these conditions, mean annual temperatures would rise as high as 6 to 7°C (11-13°F) in the tropics, and 10°C (18°F) at higher latitudes. The chain of environmental consequences, in other words, is immensely long—especially for wildlife unable to shift ranges and adapt.

Of course, one could argue that deferring the next ice age or two is not such a bad thing. What would be worse for humanity, after all? A warmer climate that actually extends the range of human habitability across a much larger area of Earth's surface—albeit including lands subject to dangerous heat waves or flooding at times—or the presence of grinding ice sheets that effectively obliterate whole regions, including Canada, northern Europe, and Eurasia? However one chooses to weigh this situation ethically and morally, we nonetheless find ourselves in a position of unprecedented power. Not since the Precambrian has a life-form (in that instance, primitive blue–green algae) had such immense influence over other forms of life as ours does today. Paleoecologist Curt Stager notes (p. 48):

> Whichever path we take into the Anthropocene future, it's now clear that we have already locked ourselves and our world into some uncomfortably large changes. If you're a hard core fatalist, you might use such points to argue for giving up and doing nothing. That, however, would be a mistake...[Our impact will be] much less extreme and long-lasting than what could happen if we don't reduce our carbon consumption as much and as rapidly as possible.

From a purely anthropocentric point of view, what is fundamentally at stake is the stability of global civilization. Anthropologists tell us that civilization arose at the beginning of the Holocene owing primarily to the development of agriculture, which provided reliable amounts of surplus food for sedentary populations. Previously, human existence had consisted of nomadic hunting and gathering with little established village life. Surplus food allowed new crafts and trades to develop, and elites of political and religious authorities to flourish—people who in the final analysis were sophisticated resource managers and overseers. Before the era of modern globalization and food transportation, a community could not survive after a couple of bad harvests (see chapter opener). A dependable, reliable climate was essential. Likewise, stable sea levels were a must for developing worldwide maritime trade. Complex port cities with all their costly infrastructures could not thrive under conditions of rapidly fluctuating seas. Likewise, a stable high stand in sea level was important for the development of flood plains such as the Great Plain of China, where fertile farmlands support hundreds of millions of people. These conditions remain important even today in the Information Age. It is not so much the threat of warming (or cooling) that need concern us, as the threat of *extreme changes in our physical circumstances.* Environmental flux will occur naturally whether we like it or not, but we can invite and accelerate it disastrously.

We'll touch again on themes related to climate and climate change in later chapters. This chapter merely sets a stage by investigating the environmental aspect of our atmosphere. The next chapter will set stage as well by introducing you to another important "sphere" in the Earth system—the lithosphere, the foundation on which all else ultimately depends.

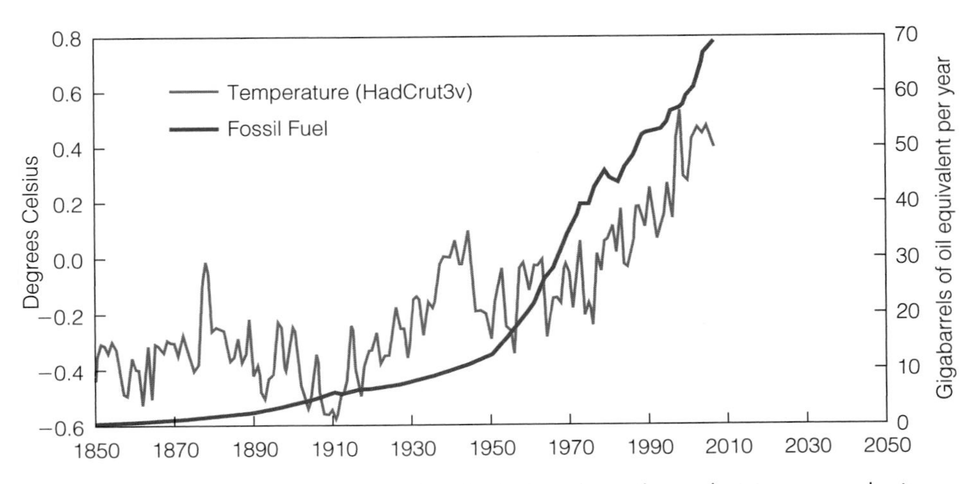

Figure 2.13 Fossil fuel consumption (red line) plotted together with rising atmospheric temperature (blue line). (From Cobb, L. 2007. *The causes of global warming: A graphical approach.* The Quaker Economist, 7(158); and HadCrut3v data series, Climate Prediction Unit, University of East Anglia, Norwich, U.K.)

SPOTLIGHT FIGURE 2.14 The Earth System and Climate Change

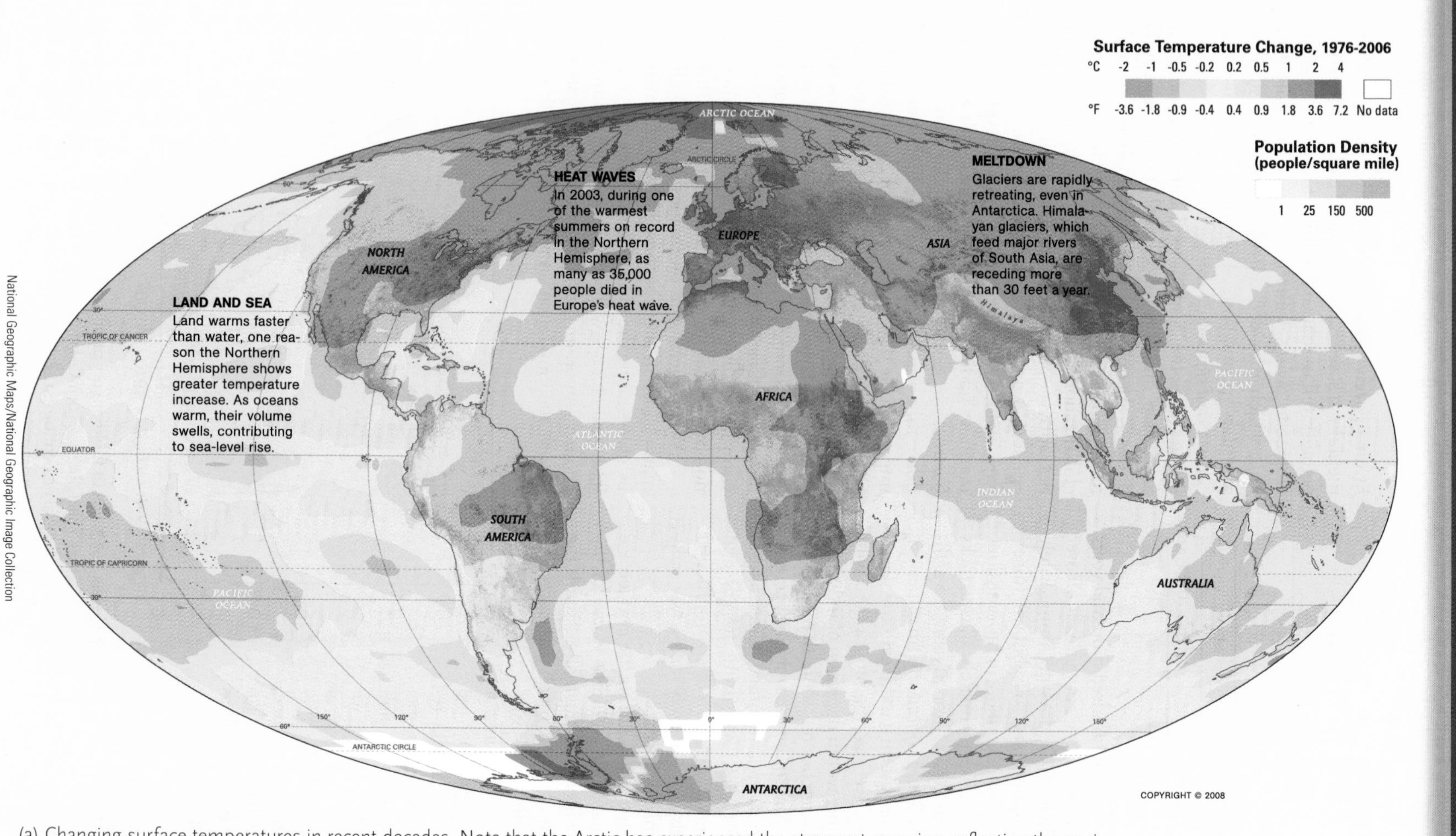

National Geographic Maps/National Geographic Image Collection

(a) Changing surface temperatures in recent decades. Note that the Arctic has experienced the strongest warming, reflecting the positive feedback of melting ice and snow packs, and the release of gases through accelerated organic decay in tundra soils. Biological evidence indicates that temperatures in the Indian and Western Pacific oceans are the highest they've been since the end of the last ice age. Cooling in the Antarctic may be caused by ozone loss, because of industrial inputs of chlorofluorocarbons and bromine-based compounds.

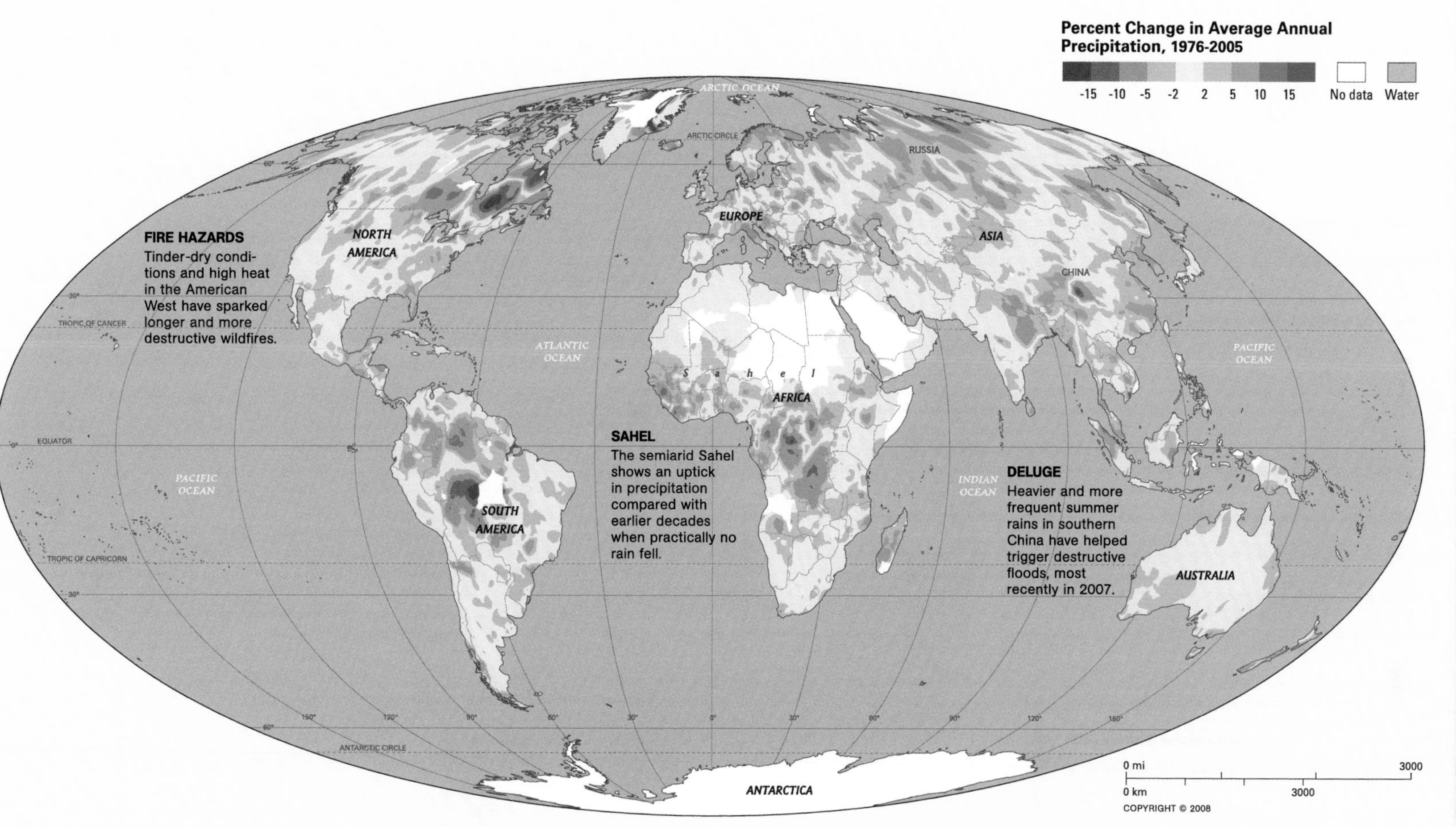

National Geographic Maps/National Geographic Image Collection

(b) A warmer world means a world of more active evaporation and precipitation, leading to increased heavy rainfalls and drought conditions. Changes in recent decades are shown on this map. Of particular concern is the reduction of snowfall in many mountainous areas. Gradual snowmelt provides sustained freshwater for millions of people downslope. Rainwater runs off the land too quickly to support this supply, or to gradually infiltrate and build up groundwater resources for irrigation and drinking.

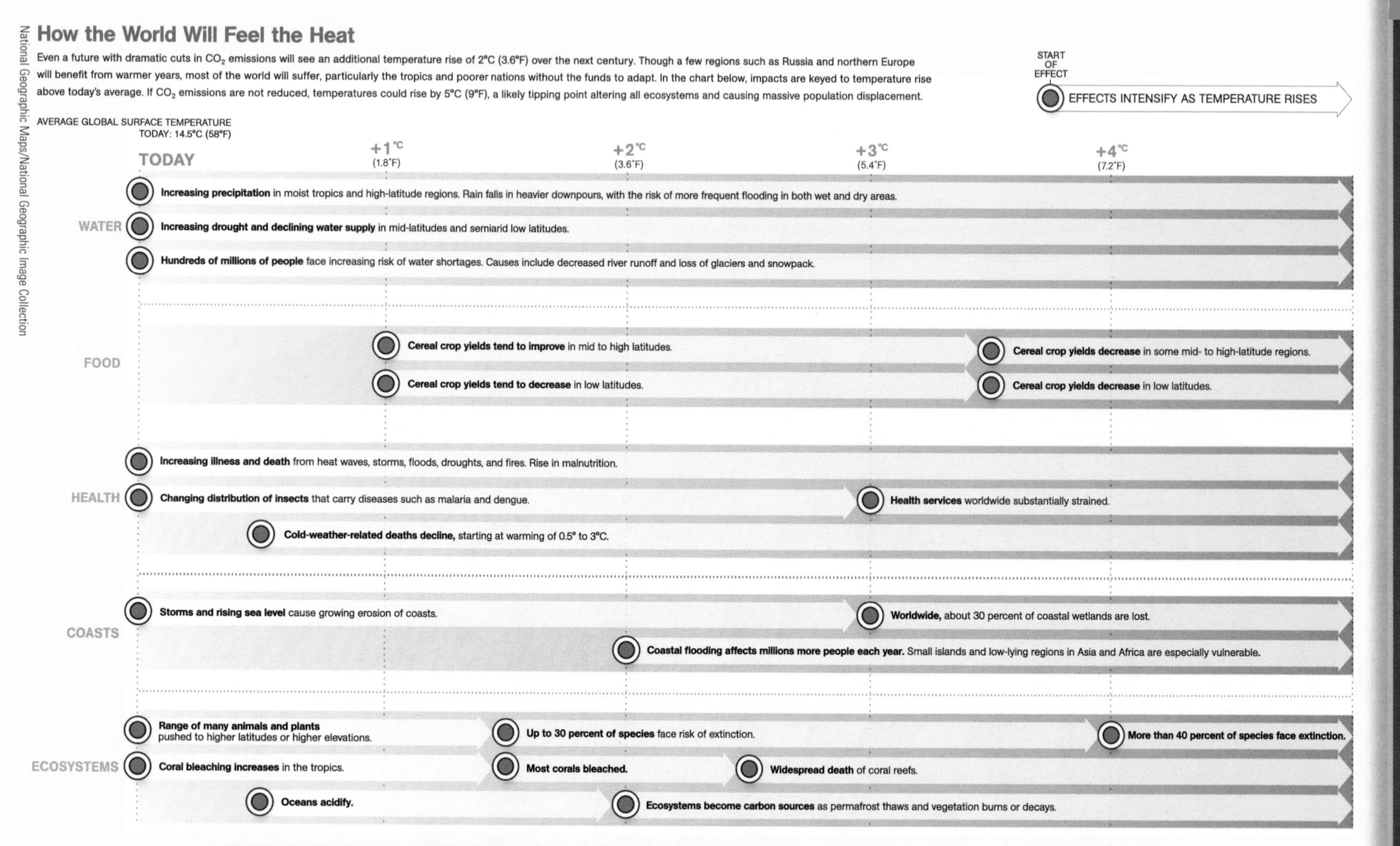

National Geographic Maps/National Geographic Image Collection

(c) Impacts of a warmer world on the human population anticipated by the 2007 Intergovernmental Panel on Climate Change and other researcher bodies. Note particularly the initial rise, then decline in cereal (grain) crop yields with rising temperature. The nutritional value per unit mass of many crops may decline even as yields are boosted. Recall from Chapter 1 that agriculture is the basis for human carrying capacity.

Figure 2.15 A view across the Wellton-Mohawk Valley in southern Arizona, showing irrigated fields amidst the Sonoran desert. Note the albedo change (darkness of color) imposed on this landscape by this modern agriculture. Summed up worldwide, human impact on planetary albedo, and hence atmospheric temperature, has been significant.

case study

2.1 Measuring Atmospheric Temperature

Although best known for his work with one of the first telescopes, in 1593, Galileo Galilei, the Italian scientist, is also credited with inventing the first thermometer, based on water. In 1714, in Germany, Daniel Fahrenheit developed a much more precise thermometer using mercury, which remains in common use, although increasingly displaced by digital instrumentation. By 1850, careful systematic temperature records were being kept at established weather stations over most of the world's surface. But assembling data from these widely scattered and unevenly maintained stations as a basis for establishing a credible "global average" proved impossible, because of a lack of standardization. Warming seemed to take place at many measuring stations located close to growing cities, but this mostly reflected local heating as land was cleared to be replaced by construction materials very effective at releasing thermal energy. Waste heat from intensified energy usage in and around urban development also played a role. This impact, known as the urban heat island effect, has increased over time. But we now know that it contributes very little to overall atmospheric warming.

Since the mid-20th century, to evaluate temperature change in the atmosphere as a whole, scientists have used high-precision thermometers to sample ocean temperatures at depths up to 100 meters (330 feet) worldwide. As it absorbs heat from the air above, the ocean may be regarded as a medium that "smooths" out local thermal differences across broad areas. If the atmosphere is warming, the sea should be getting warmer, too. Since 2000, sampling methods have been further modernized with the release of 3,260 large, automated Argo floats (as of November 2009), which periodically sink to depths of as much as 2,000 meters (6,600 feet), measuring temperatures at selected depths (**Figure 1**). Bobbing back to the surface, their data are reported to satellites for processing. Paralleling this survey on the seas, land-based boreholes have been drilled to measure temperatures underground. Just as in the ocean, a warming trend in the shallow crust has been discerned.

In addition, satellites provide important data by measuring microwave frequencies radiating from oxygen molecules in the air below. Any given frequency range is a function of atmospheric temperature. Combined with the data from weather stations, Argo floats, and boreholes, the information from Earth-orbiting sensors plainly indicates several aspects of temperature change over the past 160 years:

- Local temperature variations from year to year greatly exceed regional or global variations. This characteristic can partially mask longer term, broader changes from the perspective of persons focusing solely on their local environments.
- A net warming of 0.1°C to 0.2°C (0.2–0.35°F) has occurred in sea surface temperature over the past half-century. Given the enormous size of the World Ocean, this is highly significant.
- Whereas the lower atmosphere (below 20 km or 12 mi altitude) has grown warmer by about 0.8°C (1.4°F) at the surface since 1900, the overlying atmosphere has *cooled*—a physical response related to intensification of greenhouse gas activity closer to the surface that is predicted by global warming models.

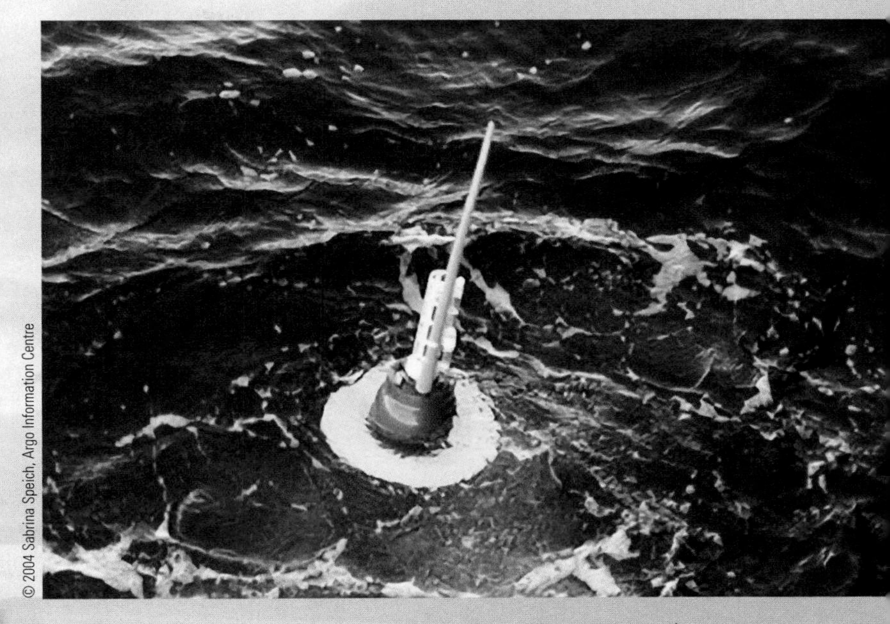

Figure 1 Argo float adrift at sea. This device measures not only sea surface temperature but also salinity and current velocity. Transmitted to satellites, then rebroadcast to Earth, the data become available for the public within hours—a remarkable advance over older methods of sample collection that depended on lowering buckets into the water at various depths, with the publication of the data following weeks or months later.

questions to **ponder**

1. If cited selectively, the data from temperature records since 1850 can be used to refute concerns about the existence of global warming. How so? How does examination of the complete record challenge or overturn this refutation?

When we speak of "climate change," we are speaking of changes in many climate types that occur naturally all over the planet. The first scientist to attempt to classify climates was Russian Wladimir Köppen, who proposed the first global scheme in 1884, applied to Earth's land surfaces. The German climatologist Rudolf Geiger later assisted Köppen to develop the modern *Köppen–Geiger climate classification system.* We recognize major types of climate today, shown on the map and in the following photos.

Cropland-rainforest border at the edge of Iguacu National Park, Brazil. © 2011/FRANS LANTING/National Geographic Image Collection

1 Simplified version of the Köppen–Geiger climate classification map. (From Kotteck, M., J. Griieser, C. Beck, B. Rudolf, and R. Rubel. 2006. World map of the Koppen classification updated. *Meteoroligische Zeitschrift* 15: 259–263.)

2 Tropical climates, as in this image, may not be so wet, humid, and hot that they sustain forests such as this one, but most are. Tropical climates support more than half of the world's known species and store a tremendous amount of its freshwater—more than a third in the Amazon Basin alone. © Hannamariah/Shutterstock.com

3 Dry climates are characterized by deserts, savannas, and grasslands. These mostly lie at low-to-intermediate latitudes. © apdesign/Shutterstock.com

4 Mild to cold winters and warm to hot summers with moderate amounts of annual rainfall take place in *temperate climates,* where most of the population of the developed industrial world lives. mmense mixed conifer/broadleaf hardwood forests thrive in this environment. © S. Borisov/Shutterstock.com

5 *Cold climates* include long, harsh winters and short summers, which may be briefly hot and dry. Cold climates transition to *polar climates*, starting with the edge of the tree line at high latitudes. © TTphoto/Shutterstock.com

6 *Polar climates* feature temperatures near or below freezing most of the year and accommodate the development of immense ice sheets and other glaciers, with pack ice at sea. During the short summers on land, tundra ecosystems thrive. Current warming in Earth's atmosphere appears to impact Northern Hemisphere and high-altitude polar climates more significantly than other climates around the world. © Patrick Robert/CORBIS

SUMMARY The Earth System and Climate Change

The Earth System

Equilibrium Conditions

Earth's surface environment is a balance of forces, some imposed externally (sunlight and related winds, rainfall) and some imposed internally (tectonic forces; see Chapter 6). Life modifies the interaction of these forces, helping create soils and the landscapes to which we are accustomed.

Homeostasis and Feedback

Biological, geological, and chemical processes in Earth's environment respond to externally generated changes in the environment (such as changes in seasonal sunlight) by means of feedback processes. Positive feedbacks in many instances counteract negative ones to help maintain equilibrium (homeostasis) in Earth's natural environment.

Atmosphere, Hydrosphere

Composition of the Atmosphere and the Greenhouse Effect

Earth's atmosphere consists mostly of nitrogen (about 78% by volume) and oxygen (about 21%), with smaller amounts of argon, carbon dioxide (CO_2), and other gases. The atmosphere also contains greenhouse gases, which trap and reradiate heat energy. These include water vapor, CO_2, and methane (CH_4). The living world as we know it could not exist without greenhouse gases. It would be too cold.

Albedo

Reflectivity of sunlight (albedo) plays a critical role in maintaining atmospheric temperature and the global climate. Changes in regional albedos (e.g., seasonal cloudiness, ice cover) set in motion important feedback processes.

Circulation

The atmosphere continuously tries to equalize its pressure and temperature, a process we experience as winds. Warm air generally moves to higher elevations and poleward, whereas cold air generally moves to lower elevations and latitudes. Rotation of the Earth deflects winds—a phenomenon called the Coriolis effect. Wind directions bend toward the right in the Northern Hemisphere and toward the left in the Southern Hemisphere. Atmospheric circulation also drives the major shallow ocean currents, which flow in great loops called gyres. Deeper marine currents move in response to differences in water temperature and salinity. The overall pattern of deep marine circulation is called the global conveyor.

Natural Climate Change

Changes in Earth's orbit and the amount of sunlight received, together with feedbacks in the Earth System, account for the astonishing changes recorded in natural climate over the past few hundred thousand years by oxygen isotope ratio data and other scientific observations.

Human-Induced Climate Change

Human contributions of greenhouse gases, especially carbon dioxide through the burning of fossil fuels, could pre-empt the next ice age and produce an anomalously warm global climate for hundreds of thousands of years.

KEY TERMS

adiabatic heating
Albedo
Anthropocene
atmosphere
biosphere
brine exclusion effect
cloud-albedo effect
Coriolis effect
cyclone
eccentricity
feedback
garbage patches
global conveyor
greenhouse effect
greenhouse gases
greenhouse warming
Hadley cells
halocarbons
homeostasis
horse latitudes
hurricanes
hydrosphere
land conversion
lithosphere
obliquity
oceanic gyres
polar air
polar front
proxy
runaway feedback
snow-albedo positive feedback
Suess effect
temperate climate
tipping point
tropical climate
typhoons
wobble (Earth's rotational wobble)

STUDY QUESTIONS

1. Describe the feedback responses that could take place if Earth's atmosphere suddenly got much cooler (say, for example, the planet's position in space away from the Sun increased). Indicate whether the feedbacks you describe would be positive or negative. Then attempt the same thought experiment considered in reverse: Which feedback responses would be stimulated if Earth's atmosphere suddenly got much warmer?
2. Describe how the circulation in the oceans is indirectly tied to solar radiation.
3. How do the shapes of the continents influence the patterns of marine circulation?
4. Why does deep marine circulation differ from that of the shallow oceans?
5. Why is albedo such a critical factor in climate-related feedbacks?
6. Why do Earth's ice ages begin slowly but end abruptly?
7. Why is the Coriolis effect so important?
8. (a) What suggests that the human consumption of fossil fuels is a key factor in driving atmospheric temperature change? (b) To what extent should we be concerned about that change?
9. What is land conversion, and why is it climatically important?
10. What is the tipping point concept, and why is it environmentally significant?
11. Why is environmental homeostasis critical to the stability and well-being of life on Earth?

REFERENCES AND FURTHER READINGS

Ahrens, C. D. 2003. *Meteorology today: An introduction to weather, climate, and the environment.* Belmont, CA: Brooks/Cole. 544 p.

Archer, D. 2005. The fate of fossil fuel CO_2 in geologic time. *Journal of Geophysical Research* 110: C09805. doi:10.1029/2004/C002625

Berger, A., and M. F. Loutre. 2002. An exceptionally long interglacial ahead? *Science* 297: 1287–1288.

Dow, K., and T. Downing. 2006. *The atlas of climate change: Mapping the world's greatest challenge.* Berkeley, CA: University of California Press. 128 p.

Head, J. W., J. F. Mustard, M. A. Kreslavsky, R. E. Millikin, and D. R. Marchant. 2003. Recent ice ages on Mars. *Nature* 426: 297–802.

Kolbert, E. 2006. *Field notes from a catastrophe: Man, nature, and climate change.* London: Bloomsbury Press, 224 p.

Marland, G., T. A. Boden, and R. J. Andres. 2005. Global, regional, and national fossil fuel CO_2 emissions. In *Trends: A compendium of data on global change.* Oak Ridge, TN: Carbon Dioxide Information Analysis Center, Oak Ridge National Laboratory, U.S. Department of Energy. Available online at: cdiac.ornl.gov.

Ruddiman, W. 2005. *Plagues, plowshares, and petroleum.* Princeton, NJ: Princeton University Press. 224 p.

Schmidt, G., and J. Wolfe. 2008. *Climate change: Picturing the science.* New York: W. W. Norton. 305 p.

Stager, C. 2011. *Deep future: The next 100,000 years of life on earth.* New York: St. Martin's Press. 284 p.

Weart, S. 2000. *The discovery of global warming.* Cambridge, MA: Harvard University Press. 200 p.

Worldwatch Institute. 2009. *State of the world report: Into a warming world.* New York: W.W. Norton. 262 p.

Natural calcium carbonate mineral deposits line a pool called Lake Castrovalva deep inside Lechuguilla Cave in Carlsbad Caverns National Park, New Mexico—one of the most fragile and spectacular geochemical systems in the world.

Have you ever

wondered...

With such wisdom has nature ordered things in the economy of this world, that the destruction of one continent is not brought about without the renovation of the earth in the production of another.

—*James Hutton (1795)*

3 The Solid Earth

1. What the difference is between a rock and a mineral?

2. Why there are different kinds of rocks, and how we can tell them apart?

3. Why some minerals are sparkly and beautiful, and others are dull and not physically attractive?

4. Why are there continents, and why are there many of them rather than a single big one?

5. How we know that the dinosaurs lived more than 70 million years ago, or that Earth is 4.5 billion years old?

Earth's Refrigerator Continent

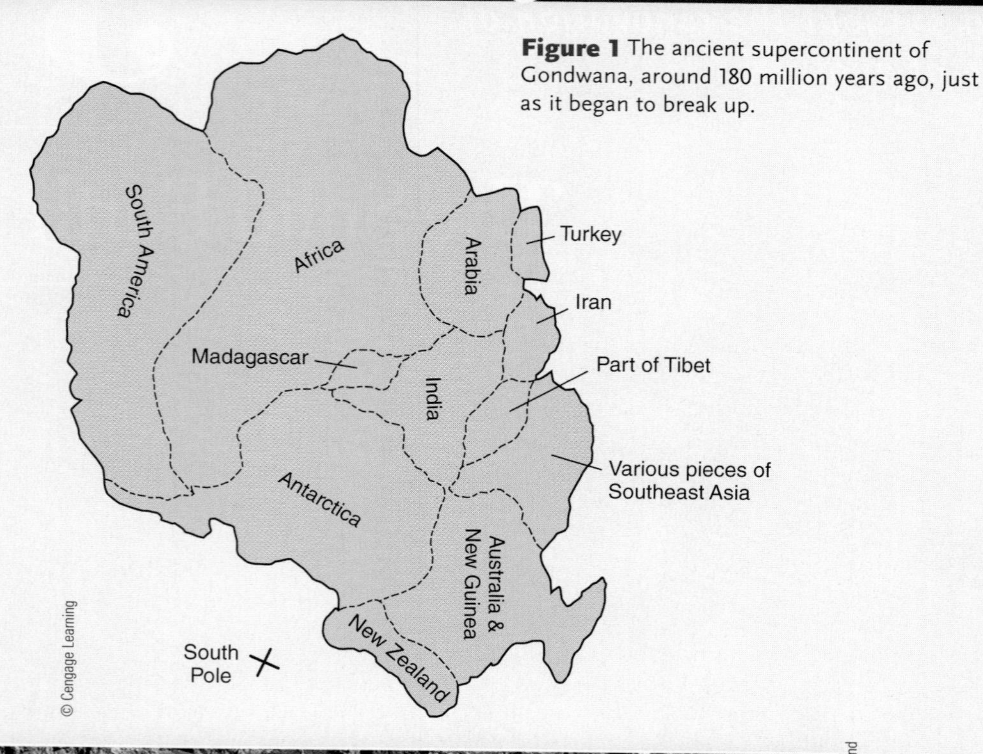

Figure 1 The ancient supercontinent of Gondwana, around 180 million years ago, just as it began to break up.

From outer space, Earth may indeed appear to be a mostly blue planet, largely covered with water and speckled with broad patches of green vegetation, brown deserts, white icecaps, and clouds. Earth is far more active on a daily basis than its planetary neighbors, but the restlessness of Earth's environment occurs on time scales stretching well beyond the day to day. Over millions of years, the very shape and location of Earth's landmasses and oceans change dramatically, responding to our planet's internal heat energy (see Chapter 2). This has profound impacts on planetary climate, oceans, and ecology.

Antarctica did not exist at the South Pole 180 million years ago. Only ocean existed there, with polar water kept much warmer than it is today, because of very different marine circulation. Instead, this continent was positioned several thousand kilometers closer to the equator, near the center of a much larger supercontinent called Gondwana, which has since split into the fragments we now know—in addition to Antarctica—as Africa, South America, India,

Molly Miller, Department of Earth and Environmental Sciences, Vanderbilt University

Figure 2 (a) Fossil leaf of the late Permian (230-million-year-old) tree *Glossopteris,* which at that time formed lush forests, with trees as tall as 15 meters (50 feet) in this part of Antarctica, then at 70° S latitude and ice-free. These were "subpolar forests" of a kind that do not exist in the much cooler world of today. (b) Lamping Peak on the Bowden Neve, near Beardmore Glacier in the Transantarctic Mountains. The fossil leaf shown in (a) was collected in strata just below the summit of this peak.

Molly Miller, Department of Earth and Environmental Sciences, Vanderbilt University

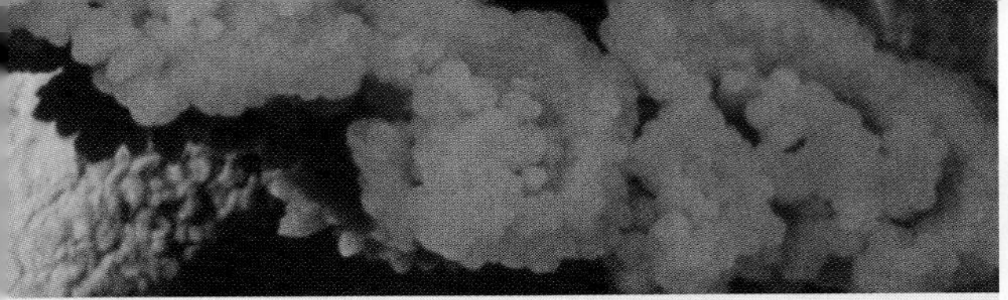

and Australia (**Figure 1**). As Gondwana broke up, Antarctica began drifting south. But 100 million years ago, it was still covered with luxuriant vegetation, fossil leaves of which can be found in sedimentary formations of the Transantarctic Mountains. Around 35 million years ago, Antarctica began to get colder as its ponderous movement carried it closer to the pole. More importantly, it broke free of South America, allowing a circumpolar marine current to develop that isolated the continent from the climate-moderating influence of warmer ocean waters. The ice-albedo feedback activated with a vengeance, and Antarctica's dramatic refrigeration soon chilled Earth's climate worldwide, bringing us to modern climatic conditions. Even as recently as 1.8 million years ago, however, trees resembling the mountain beech *(Nothofagus)* of New Zealand flourished around certain coastal fringes of Antarctica (**Figure 2**). Today, about 7 million cubic miles of glacial ice almost completely cover this continent, in some places weighing down the underlying bedrock so much that it is pushed below sea level. Antarctica hosts 90% of Earth's present ice supply.

But *why* did Antarctica drift to the South Pole in the first place? To answer this question, we must first take a closer look at the solid Earth, beginning with its basic material constituents—the naturally occurring elements, minerals, and rocks. Later in this chapter, we will consider the internal structure of the planet and how this can interact with the crust to move giant continents thousands of kilometers from their places of origin. Finally, we'll see how long these processes have been going on by considering the overall age of Earth.

questions to ponder

1. There is far more ice at the South Pole than at the North Pole. Why?
2. If you were an astronaut who traveled to a newly discovered planet, but had only a couple of days to learn all you could about the geological development of that world to report back to base, what data would you collect and why?

AP Photo/Jon Pall Vilhelmsson

CourseMate Connection

Start off this chapter by watching a video that models the plate movement that caused the tsunami of March 2011 in Japan. **Log in to the CourseMate at www.cengagebrain.com to watch now.**

Earth Materials

Elements, Atoms, and Atomic Structure

ELEMENTS ARE SUBSTANCES THAT CANNOT be changed into other substances by normal chemical methods. They are composed of infinitesimal particles called **atoms**. In 1870, Lothar Meyer published a table of the 57 then-known elements arranged in the order of their atomic weights. Meyer left blank spaces in the table wherever elements of particular weights were not known. About the same time, Dmitry Mendeleyev developed a similar table, but his table was based on the similar chemical properties of the particular known elements. Where an element's placement based on weight did not group it with elements of similar properties, Mendeleyev did not hesitate to suggest that its weight had been measured incorrectly. He also predicted the properties of the "missing" elements, and shortly his predictions proved correct with the discovery of the elements gallium, scandium, and germanium (see Appendix 1).

The weight of an atom of an element is contained almost entirely in its **nucleus**, which contains protons (atomic weight = 1, electrical charge = 1^+) and neutrons (atomic weight = 1, electrical charge = 0). An element's **atomic number** is the number of protons in its nucleus, and this number is unique to that element. The weight of an atom—its **atomic mass**—is the sum of its nuclear protons and neutrons. For example, ^4_2He denotes the element helium, which has 2 protons (denoted by the subscript) and 2 neutrons, for an atomic mass of 4 (superscript). Electrical neutrality of the atom is provided by balancing the positive proton charges by an equal number of negatively charged electrons orbiting in shells around the nucleus (**Figure 3.1**). **Ions** are atoms that are positively or negatively charged because of a loss or gain of electrons (e^-) in the outer electron shell. Thus, sodium (Na) may lose an electron in its outer shell to become sodium ion (Na^+, a *cation*), and chlorine (Cl) may gain an electron to become chloride ion (Cl^-, an *anion*). *Valence* refers to ionic charge, and sodium and chlorine are said to have valences of 1^+ and 1^-, respectively. Elemental sodium is an unstable metallic solid, and chlorine is a poisonous gas. When ionized, they may combine in an orderly fashion to form the mineral halite (NaCl), common table salt, a substance necessary for human existence (**Figure 3.2**).

Isotopes are forms (or species) of an element that have different atomic masses. The element uranium, for instance, always has 92 protons in its nucleus, but varying numbers of neutrons define its isotopes. For example, uranium-238 (^{238}U) is the common naturally occurring isotope of uranium. It weighs 238 atomic mass units and contains 92 protons and 146 (238 *minus* 92) neutrons. Uranium-235 is the rare isotope (0.7% of all uranium) that has 143 neutrons. Similarly, ^{12}C is the common isotope of carbon, but ^{13}C and ^{14}C also exist. A modern periodic table of the elements that shows their atomic number and mass appears in Appendix 1.

Minerals

Earth's crust is composed of rocks, which are made up of one or more minerals. Although most people seldom think about minerals that do not have nutritional importance or great beauty and value, such as gold and silver, the mineral kingdom encompasses a broad spectrum of chemical compositions. **Minerals** are defined as *naturally occurring, inorganic, crystalline* substances, each with a narrow range of *chemical compositions* and characteristic *physical properties.* Thus, neither artificial gems, such as zirconia, nor organic deposits, such as coal and oil, are minerals by definition. The

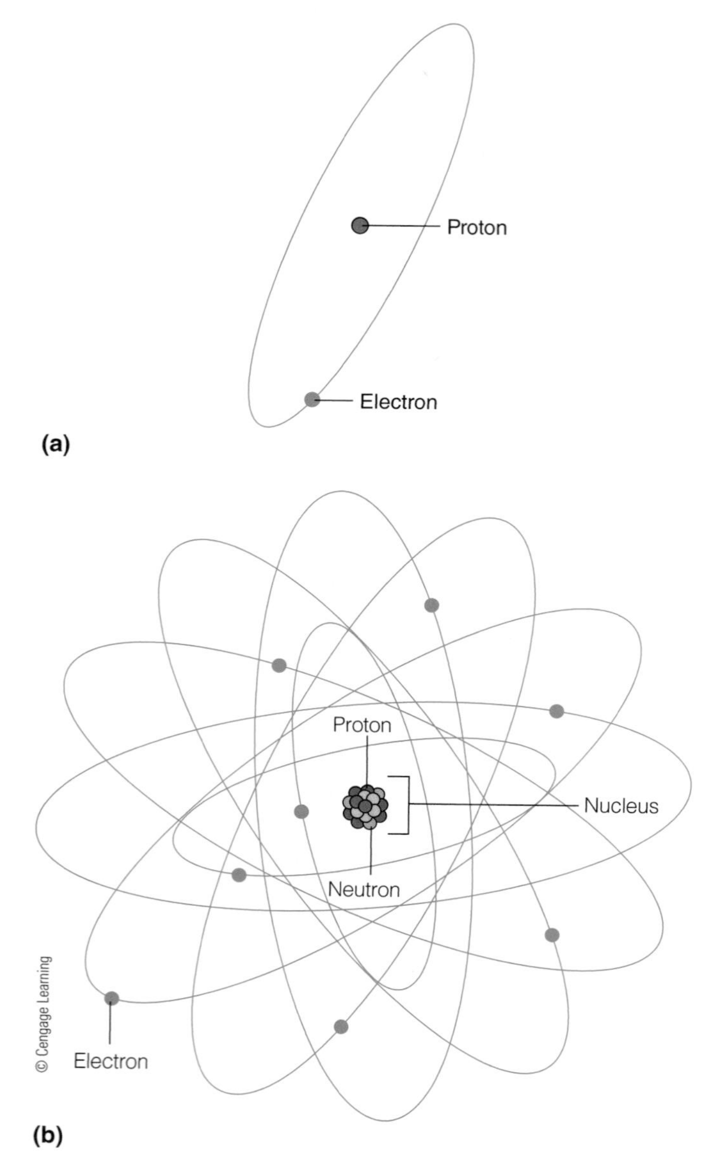

Figure 3.1 Structures of (a) a hydrogen atom, in which an electron orbits a single proton, and (b) an oxygen atom, in which eight electrons orbit a dense nucleus containing eight protons and eight neutrons.

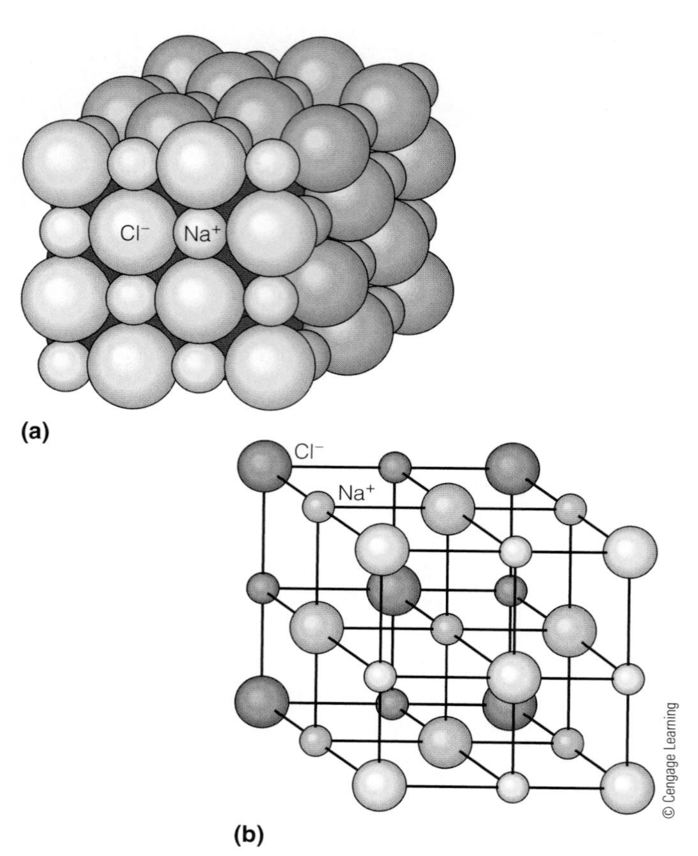

Figure 3.2 Two models of the atomic structure of the mineral halite (NaCl), common table salt: (a) a packing model that shows the location and relative sizes of the sodium and chloride ions; (b) a ball-and-stick model that shows the mineral's cubic crystal structure.

crystal form of a mineral reflects an orderly atomic structure, which, in turn, determines the crystal shape. Solids that have random or noncrystalline atomic structures—glass and opal, for example—are described as **amorphous**. Minerals composed of a single element, such as copper and carbon (as in graphite and diamonds), are known as *native elements*. Most minerals, however, are composed of combinations of two or more elements. Although more than 4,500 minerals have been named and described, only about 20 make up the bulk of Earth's crustal rocks. The chemistry of minerals is the basis of their classification, and the most common minerals are composed primarily of the elements oxygen (O), silicon (Si), aluminum (Al), and iron (Fe). This is not surprising, because oxygen and silicon compose 75% of crustal rocks by weight. A memory aid for these elements in order is "OSAF."

Geologists classify minerals into groups that share similar negatively charged ions (anions) or ion groups (radicals). Oxygen (O^{-2}) may combine with iron to form hematite (Fe_2O_3), a member of the *oxide group* of minerals. The *sulfide minerals* typically are combinations of a metal with sulfur. Examples include galena (lead sulfide, PbS), chalcopyrite (copper-iron sulfide, $CuFeS_2$), and pyrite (iron sulfide, FeS_2), the "fool's gold" of inexperienced prospectors. *Carbonate minerals* contain the negatively charged $(CO_3)^{-2}$ ion. Calcite ($CaCO_3$), the principal mineral of limestone and marble, is an example. **Table 3.1** summarizes some of the mineral groups most important to geologists.

The most common and important minerals are the *silicates*, which are composed of combinations of oxygen and silicon, with or without metallic elements. The basic building block of a silicate mineral is the silica tetrahedron (**Figure 3.3a**). It consists of one silicon atom surrounded by four oxygen atoms at the corners of a four-faced tetrahedron. The manner in which the tetrahedra are packed or arranged in the mineral structure is the basis for classifying silicates. The tetrahedra may be arranged in layers or sheets, as in the clay

Table 3.1 Important Mineral Groups

Mineral Group	Negatively Charged Ion Or Ion Group	Examples	Composition
Carbonate	$(CO_3)^{-2}$	Calcite Dolomite	$CaCO_3$ $CaMg(CO_3)_2$
Halide	Cl^{-1}, F^{-1}	Halite	$NaCl$
Hydroxide	$(OH)^{-1}$	Limonite	$FeO(OH){\cdot}H_2O$
Native element	—	Gold Diamond	Au C
Oxide	O^{-2}	Hematite	Fe_2O_3
Silicate	$(SiO_4)^{-4}$	Quartz Olivine	SiO_2 $(MgFe)_2SiO_4$
Sulfate	$(SO_4)^{-2}$	Gypsum	$CaSO_4{\cdot}2H_2O$
Sulfide	S^{-2}	Galena	PbS

See Appendix 1 for a table of the elements and their symbols.

minerals and mica; in long chains, as in the amphibole and pyroxene groups; and in three-dimensional networks, as in quartz and feldspar (**Figure 3.3b**; see Case Study 3.1).

Rock-forming silicates—Most rocks are aggregates of minerals, and feldspar is the most abundant of the silicate minerals in Earth's crust (**Figure 3.4**). Feldspar is actually a mineral *group* that ranges in chemical composition from potassium-rich orthoclase found in granite (**Figure 3.5**) to calcium-rich plagioclase found in basalt and gabbro. The ferromagnesian silicates are rich in iron and magnesium in addition to silicon and oxygen. Most of them are dark colored—black, brown, or green (**Figure 3.6**). Important among these are hornblende (amphibole), augite (pyroxene), biotite mica, and olivine, which is believed to be the most abundant mineral just below the outer crust of the planet (**Table 3.2**).

Mineral identification—Serious students of mineralogy are able to identify several hundred minerals without destructive testing. They are able to do this because minerals have distinctive physical properties, most of which are easily determined and associated with particular mineral species (Appendix 2). With practice, one can build up a mental catalog of minerals, just like building a vocabulary in a foreign language. The most useful physical properties are hardness, cleavage, crystal form, and to lesser degrees, color (which is variable) and luster. Mineralogists use **Mohs' hardness scale**, developed by Friedrich Mohs in 1812 (**Table 3.3**). Mohs assigned hardness values (*H*) of 10 to diamond and 1 to talc, a very soft mineral. Diamond scratches everything; quartz ($H = 7$) and feldspar ($H = 6$) can scratch glass ($H = 5\frac{1}{2}$–6), and calcite ($H = 3$) and your fingernail can scratch gypsum ($H = 2$).

Cleavage refers to the characteristic way particular minerals split along definite planes as determined by their crystal

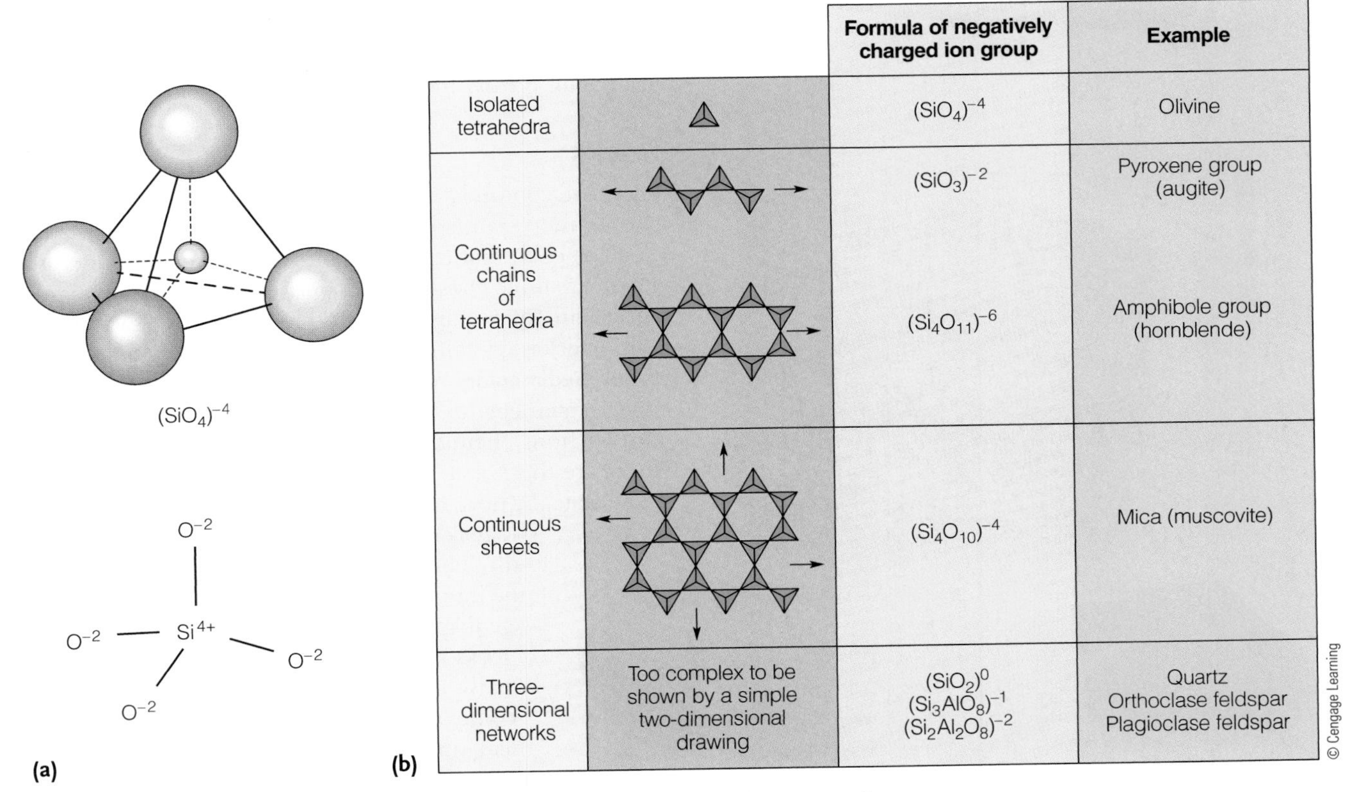

Figure 3.3 (a) Model of the silica tetrahedron, showing the unsatisfied negative charge at each oxygen that allows it to form (b) chains, sheets, and networks.

Figure 3.4 Some common rock-forming silicate minerals: (a) feldspar, (b) quartz, (c) muscovite, (d) hornblende.
© Dr. Parvinder S. Sethi

Figure 3.6 Some common ferromagnesian silicates: (a) olivine; (b) augite, a pyroxene-group mineral; (c) hornblende, an amphibole-group mineral; (d) biotite mica.
© Sue Monroe

structure. Mica has perfect cleavage in one direction; this is called *basal* cleavage, because it is parallel to the basal plane of the crystal structure. Feldspars split in two directions; halite, which has *cubic* structure, splits in three directions; and so on, as shown in **Figure 3.7**. Minerals that have perfect cleavage can be split readily by a tap with a rock hammer or even peeled apart, as in the case of mica. Some minerals do not cleave but have distinctive **fracture** patterns, which help one to identify them. The common crystal forms, shown in **Figure 3.8**, can be useful in identifying a particular mineral. Color tends to vary within a mineral species, so it is not a reliable identifying property. **Luster**—how a mineral reflects light—is useful, however. We recognize metallic and nonmetallic lusters, the latter being divided into types, such as glassy, oily, greasy, and earthy.

Figure 3.5 This granite is an aggregate of the minerals quartz, biotite, and feldspar.
(Top left: © Paul Silverman/Fundamental Photographs; top right and bottom left: © Sue Monroe; bottom right: © Paul Silverman/Fundamental Photographs)

Rocks

Rocks are consolidated or poorly consolidated aggregates of one or more minerals, glass, or solidified organic matter (such as coal) that cover a significant part of Earth's crust. There are three classes of rocks, based on their origin: igneous, sedimentary, and metamorphic. **Igneous** (Latin *ignis*, "fire") **rock** is crystallized from molten or partly molten material. **Sedimentary rocks** include both *lithified* (turned to stone) fragments of preexisting rock and rocks that were formed from chemical or biological action. **Metamorphic rocks** are those that have been changed, essentially in the solid state, by heat, fluids, or pressure within Earth.

The **rock cycle** is one of many natural cycles on Earth. The illustration of it in **Figure 3.9** shows the interactions of energy, earth materials, and geological processes that form and destroy rocks and minerals. The rock cycle is essentially a closed system: "what goes around comes around," so to speak. In a simple cycle, there might be a sequence of formation, destruction, and alteration of rocks by earth processes. For example, an igneous rock may be eroded to form sediment, which subsequently

Table 3.2 Some Common Rock-Forming Minerals

Mineral	Abundance in Crust, %	Rock in Which Found
Plagioclase*	39	Mostly igneous rocks
Quartz	12	Detrital sedimentary rocks, granites
Orthoclase†	12	Granites, detrital sedimentary rocks
Pyroxenes	11	Dark-colored igneous rocks
Micas	5	All rock types as accessory minerals
Amphiboles	5	Granites, other igneous rocks
Clay minerals	5	Shales, slates, decomposed granites
Olivine	3	Iron-rich igneous rocks, basalt
Others	11	Rock salt, gypsum, limestone, etc.

*A series of six minerals within the plagioclase group from albite ($NaAlSi_3O_8$) to anorthite ($CaAl_2Si_2O_8$).

†Feldspar group of minerals.

Table 3.3 Mohs' Hardness Scale

Hardness	Mineral	Common Example
1	Talc	Pencil lead, 1–2
2	Gypsum	Fingernail, 2½
3	Calcite	Copper penny, 3 Brass
4	Fluorite	Iron
5	Apatite	Tooth enamel Knife blade Glass, 5½–6
6	Orthoclase (potassium feldspar)	Steel file, 6½
7	Quartz	
8	Topaz	
9	Corundum	Sapphire, ruby
10	Diamond	Synthetic diamond

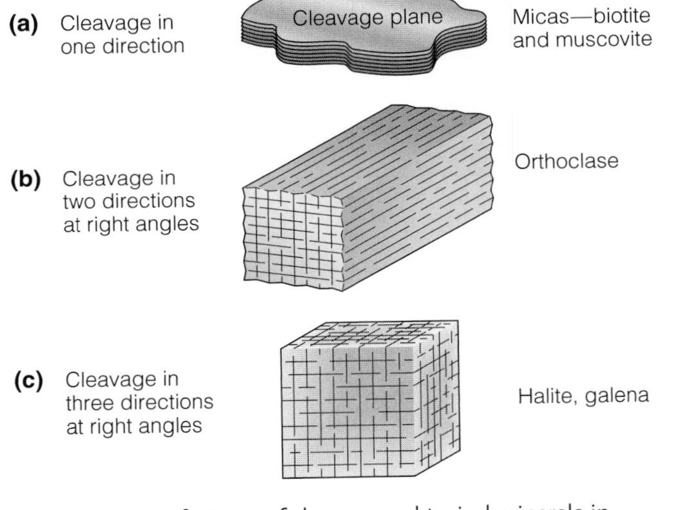

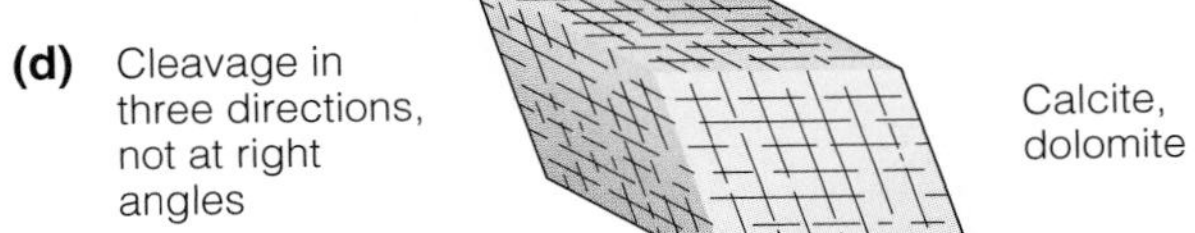

(e) Cleavage in four directions — Fluorite, diamond

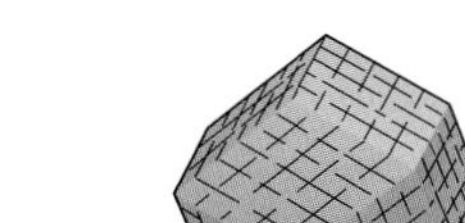

Figure 3.7 (a–f) Types of cleavage and typical minerals in which they occur.

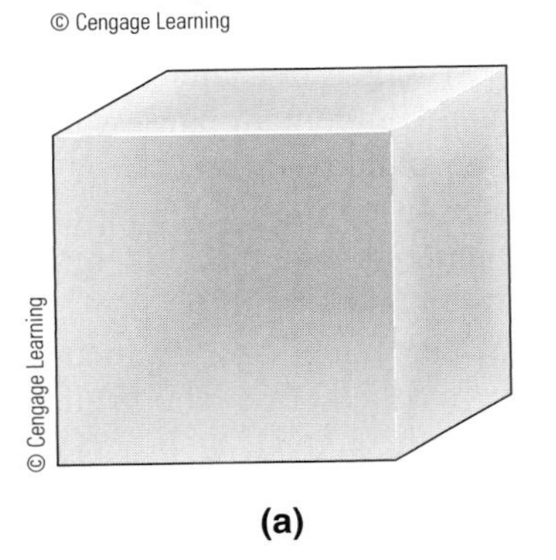

(b)

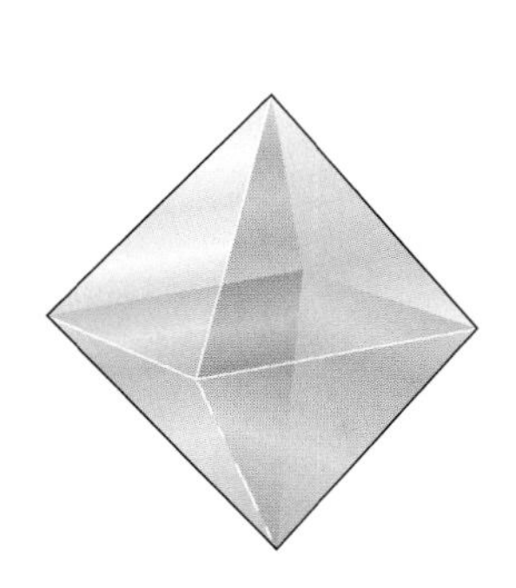

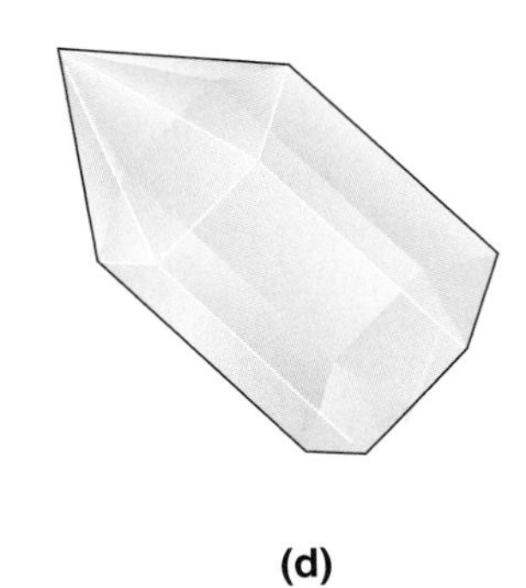

Figure 3.8 Mineral crystal shapes: (a) cube (halite), (b) 12-sided dodecahedron (garnet), (c) 8-sided octahedron (diamond, fluorite), and (d) 6-sided hexagonal prism (quartz).

Figure 3.9 The rock cycle. The three rock types are interrelated by internal and external processes involving the atmosphere, ocean, biosphere, crust, and upper mantle.

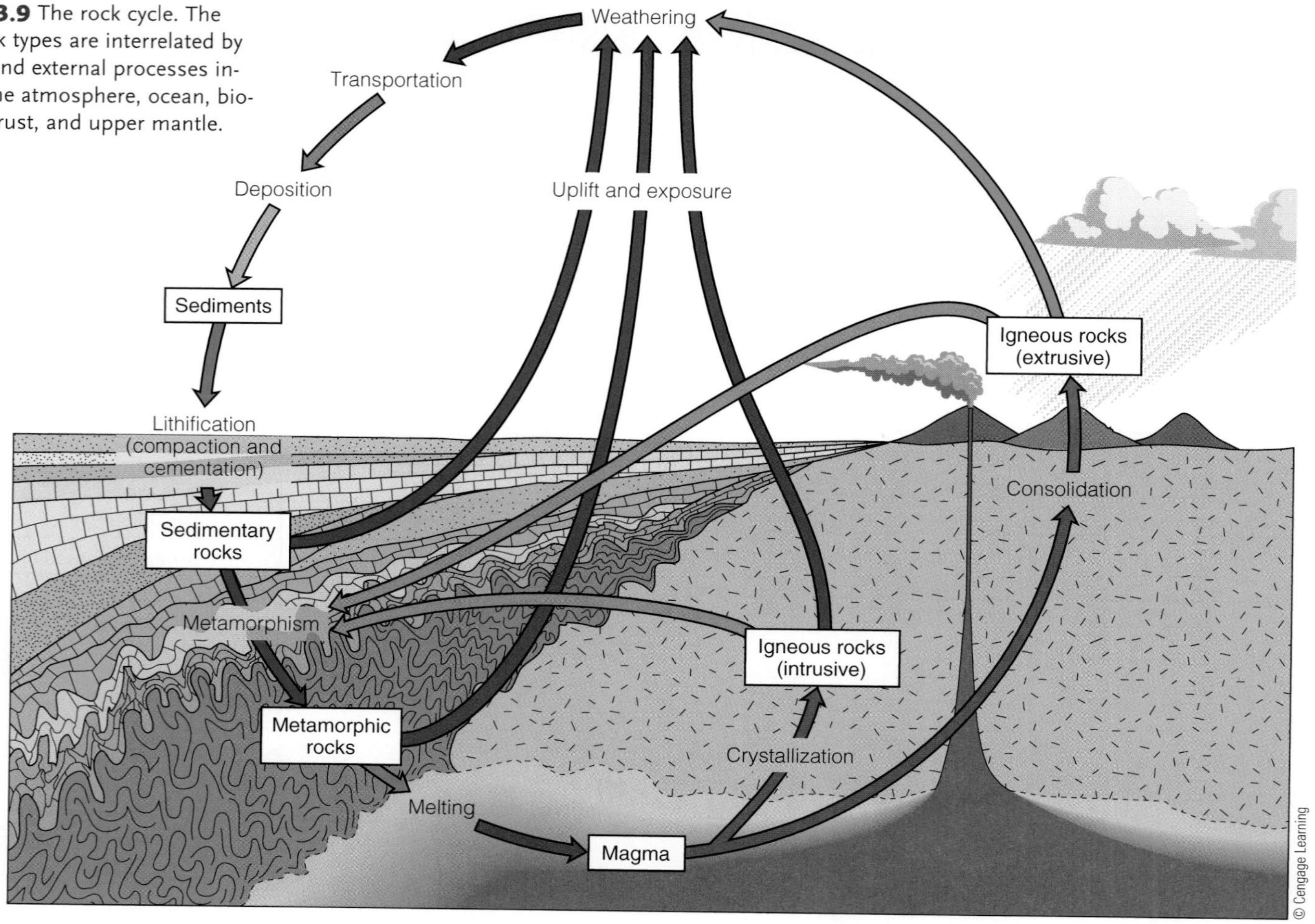

© Cengage Learning

becomes a sedimentary rock, which may then be metamorphosed by heat and pressure to become a metamorphic rock, and then melted to become an igneous rock.

Throughout this book, we emphasize those rocks that, because of their composition or structure, are involved in geological events that endanger human life or well-being, or that are important as resources.

Your school district is told it must remove all asbestos insulation from its older buildings at considerable cost to the taxpayers and at the expense of teachers' salaries and educational materials. What questions should be asked of the government agency requiring this action, and what position will you take if you learn that the asbestos is the long-chain, fibrous type, rather than the sheet-structure, mica type (see Case Study 3.1)?

Igneous rocks—Igneous rocks are classified according to their texture and mineral composition. A rock's texture is a function of the size and shape of its mineral grains, and for igneous rocks, this is determined by how fast or slowly a melted mass cools. If **magma**—molten rock within Earth—cools slowly, large crystals develop, and a rock with a coarse-grained, **phaneritic texture** (such as granite) is formed (see Figure 3.5). The resulting large rock mass formed within the planet is known as a **batholith**. Greater than 100 square kilometers (39 square miles) in area by definition, batholiths are mostly granitic in composition and show evidence of having invaded and pushed aside the *country rock* into which they intruded. Also, many batholiths have formed in the cores of mountain ranges during mountain-building episodes. Batholiths are a type of igneous mass referred to as *plutons*, and rocks of batholiths are described as **plutonic** (after Pluto, the Greek god of the underworld), because they were formed at great depth. The Sierra Nevada mountains of California are an example of an uplifted and eroded mountain range with an exposed core composed of many plutons (**Figure 3.10**).

Most igneous rocks—whether on Earth, the Moon, or Mars—are composed of silicate mineral-containing silicon, oxygen, and aluminum. Silicon combined with oxygen is the compound *silica* (SiO_2). The percentage of silica in an igneous rock is a measure of how "distilled," or fractionated, its original magma was. The higher the percentage of silica in an igneous rock, the more steps of fractionation it has gone through. Minerals that are rich in magnesium and iron and low in silica (minerals such as pyroxene, amphibole, and olivine) will crystallize and settle out of the

Figure 3.10 The Sierra Nevada batholith and Mount Whitney, looking west from Owens Valley, east-central California.

tween these extremes, according to their percentage of silica.

Lava is molten material at Earth's surface produced by volcanic activity. Lava cools rapidly, resulting in restricted crystal growth and a fine-grained, **aphanitic texture.** An example of a rock with aphanitic texture is rhyolite, which has about the same composition as granite but a much finer texture. The grain size of an igneous rock can tell us whether it is an **intrusive** (cooled within Earth) or an **extrusive** (cooled at the surface of Earth) rock. The classification by texture and composition of igneous rocks is shown in **Figure 3.11**. Note that for each composition, pairs of rocks are distinguished by their texture. Granite and rhyolite, diorite and andesite, and gabbro and basalt are the most common pairs found in nature. The end member shown in Figure 3.11, peridotite, is rarely found because it originates deep within

magma first. The resulting rocks, relatively rich in magnesium and iron and poor in silica, are thus designated **mafic** (*ma-* for magnesium, *-fic* for iron). The remaining magma has less magnesium and iron, and a greater percentage of silica. When the next batch of minerals settles out, perhaps with quartz crystals, rocks of *intermediate* composition are formed. During the fractionation process, the magma becomes richer and richer in silica and more impoverished in magnesium and iron. This late-stage magma is largely silica, alumina, and other elements that combine at lower temperatures. The resulting rocks are rich in feldspar and silica, and are thus described as **felsic** (*fel-* for feldspar, *sic* for silica). "Primitive" magmas (<50% SiO_2) produce mafic gabbro and its volcanic equivalent, basalt, the rock common to Hawaii volcanoes. More "evolved" magmas (>50% SiO_2) produce rocks of intermediate composition, such as diorite and its volcanic equivalent, andesite. Highly evolved magmas produce felsic rocks, such as granite and its volcanic equivalent, rhyolite. Mafic rocks are black or dark colored, and felsic rocks are whitish and pinkish. Intermediate igneous rocks vary in color be-

Figure 3.11 Classification of common igneous rocks by their mineralogy and texture. Relative proportions of mineral components are shown for each rock type. Fine-grained *aphanitic* texture results from rapid cooling of molten material. Coarse-grained *phaneritic* texture results from slow cooling.

(Rock photos © Sue Monroe; Gabbro: Copyright and photography by Dr. Parvinder S. Sethi)

Figure 3.12 Extrusive igneous rocks: (a) glassy obsidian; (b) pumice, which is formed from gas-charged magmas; and (c) tuff. The lavas that form obsidian and pumice cool so rapidly that crystals do not grow to any size. The reference scale in each photo is 5 centimeters (2 inches) long.
B. Pipkin

Sediment	Process		Rock
Gravel >2 mm	Compaction/ cementation	Rounded clasts	Conglomerate
		Angular clasts	Breccia
Sand 2 mm–.062 mm	Compaction/ cementation		Sandstone
Silt .062 mm–.004 mm	Compaction/ cementation		Siltstone
Clay <.004 mm	Compaction		Shale

Figure 3.13 How sediments are transformed into clastic sedimentary rocks. The sequence of lithification is *deposition* to *compaction* to *cementation* to hard rock.

Earth. Three important and relatively common glassy (composed of amorphous or finely crystalline SiO_2) volcanic rocks are obsidian, which looks like black glass; pumice, which has the composition of glass but is not really glassy looking; and tuff, consolidated volcanic ash or cinders (**Figure 3.12**). Tuff has a **pyroclastic** (literally "fire-broken") texture, resulting from fragmentation during violent volcanic eruptions.

Sedimentary rocks—Sediment is particulate matter derived from the physical or chemical weathering of the materials of Earth's crust and by certain organic processes. It may be transported and redeposited by streams, glaciers, wind, or waves. Sedimentary rock is sediment that has become **lithified**—turned to stone—by pressure from deep burial, by cementation, or by both of these processes. **Clastic sedimentary rocks** are composed of *clasts*, fragments of preexisting rocks and minerals. Clastic sedimentary rocks are classified according to their grain size. Thus, sand-size clay lithifies to shale, sediment to sandstone, and gravel to conglomerate (**Figure 3.13** and **3.14a–c**). Other sediments result from chemical and biological activity. **Chemical sedimentary rocks**, which may be clastic or nonclastic, include chemically precipitated limestone, rock salt, and gypsum. **Biogenic sedimentary rocks** are produced directly by biological activity and include coal (lithified plant debris), some limestone and chalk ($CaCO_3$, shell material), and chert (siliceous shells) (**Figure 3.14d, e**). **Table 3.4** presents the classification of sedimentary rocks.

A distinguishing characteristic of most sedimentary rock is bedding, or **stratification**, into layers. Because shale is very thinly stratified, or *laminated*, it splits into thin sheets. Some sedimentary rocks, such as sandstone and limestone, occur in beds that are several feet thick. Other structures in these rocks give hints as to how the rock formed. *Cross-bedding*—stratification that is inclined at an angle to the

Table 3.4 Classification Of Sedimentary Rocks

Detrital Sedimentary Rocks (Clastic Texture)		
Sediment	**Description**	**Rock Name**
Gravel (>2.0 mm)	Rounded rock fragments Angular rock fragments	Conglomerate Breccia
Sand (0.062–2.0 mm)	Quartz predominant > 25% feldspars	Quartz sandstone Arkose
Silt (0.004–0.062 mm)	Quartz predominant, gritty feel	Siltstone
Clay, mud (<0.004 mm)	Laminated, splits into thin sheets Thick beds, blocky	Shale Mudstone
Chemical Sedimentary Rocks		
Texture	**Composition**	**Rock Name**
Clastic	Calcite ($CaCO_3$) Dolomite [$CaMg(CO_3)_2$]	Limestone Dolostone
Crystalline	Halite (NaCl) Gypsum ($CaSO_4 \cdot 2H_2O$)	Rock salt Rock gypsum
Biogenic Sedimentary Rocks		
Texture	**Composition**	**Rock Name**
Clastic	Shell calcite, skeletons, broken shells Microscopic shells ($CaCO_3$)	Limestone, coquina Chalk
Nonclastic (altered)	Microscopic shells (SiO_2), recrystallized silica Consolidated plant remains (largely carbon)	Chert Coal

Figure 3.14 Common *clastic* sedimentary rocks: (a) shale, (b) sandstone, and (c) conglomerate. *Biogenic* sedimentary rocks: (d) limestone composed entirely of shell materials (called *coquina*), and (e) coal.
B. Pipkin

D. D. Trent

Figure 3.15 Windblown sandstone exhibiting large cross-beds; Zion National Park, Utah.

D. D. Trent

Figure 3.16 Mud cracks in an old clay mine; Ione, California.

main stratification—indicates the influence of wind or water currents. Thick cross-beds and frosted quartz grains in sandstone indicate an ancient desert sand-dune environment (**Figure 3.15**). Some shale and claystone show polygonal *mud cracks* on bedding planes, similar to those found on the surface of modern dry lakes; these indicate desiccation in a subaerial environment (**Figure 3.16**).

Ripple marks, low, parallel ridges in deposits of fine sand and silt, may be asymmetrical or symmetrical (**Figure 3.17**). Asymmetrical ripples indicate a unidirectional current; symmetrical ones, the back-and-forth motion pro-

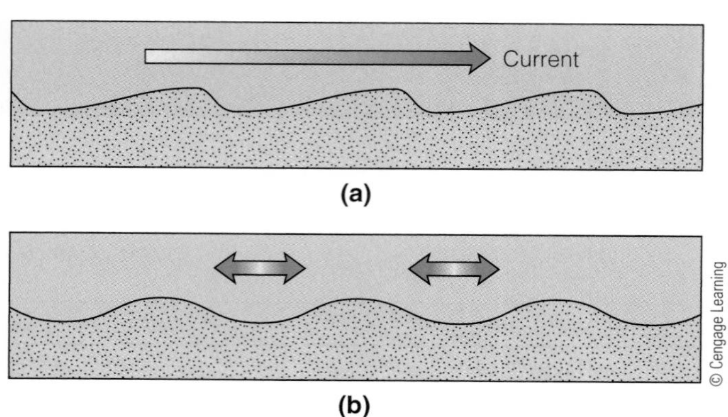

© Cengage Learning

Figure 3.17 Ripple marks found in sedimentary rocks. (a) Asymmetrical ripple marks are common in streambeds and are found on bedding planes in sedimentary rocks. (b) Symmetrical ripple marks are due to oscillating water motion—that is, wave action.

duced by waves in shallow water. Only in sedimentary rock do we find abundant fossils, the remains or traces of life.

Metamorphic rocks—Rocks that have been changed from preexisting rocks by heat, pressure, or chemical processes are classified as metamorphic rocks. The process of metamorphism results in new structures, textures, and minerals. **Foliation** (Latin *folium*, "leaf) is the flattening and layering of minerals by nonuniform stresses. Foliated metamorphic rocks are classified by the development of this structure. Slate, schist, and gneiss are foliated, for example, and they are identified by the thickness or crudeness of their foliation (**Figure 3.18**). Metamorphic rocks may also form by **recrystallization**. This occurs when a rock is heated and strained by uniform stresses, so that larger, more perfect grains result or new minerals form. In this manner, a limestone may recrystallize to marble or a quartz sandstone to quartzite, one of the most resistant of all rocks (**Figure 3.19**). **Table 3.5** summarizes the characteristics of common metamorphic rocks.

Forensic geologists are specialists who are sometimes called on to help solve crimes. They can, for example, compare the mineralogy of soil on a suspect's shoes with the soil at the scene of the crime. They also use geological clues to reconstruct events that happened millions of years ago. For instance, huge, round scars (craters) on the surface of Earth can indicate ancient meteorite impacts. What are some of the "clues" in sedimentary rocks that tell us about the ancient environments in which they formed?

Rock defects—Some rocks have structures that geologists view as "defects"—that is, surfaces along which landslides or rockfalls may occur. Almost any planar structure, such as a bedding plane in sedimentary rock or a foliation plane in metamorphic rock, holds potential for rock slides or falls. The orientation of a plane in space, such as a stratification

Figure 3.18 Metamorphic rocks: (a) coarsely foliated gneiss, (b) mica schist showing wavy or crinkly foliation, and (c) finely foliated slate showing slaty cleavage.
B. Pipkin

plane, a fault, or a joint, may be defined by its **dip** and **strike** (see Appendix 3). **Joints** are rock fractures without displacement and they occur in all rock types. They are commonly found in parallel sets spaced several feet apart. **Faults** are also fractures in crustal rocks, but they differ from joints in that some movement or displacement has occurred along the fault surface. Faults are also found in all rock types and are potential surfaces of "failure." We will examine the relationships of rock defects to geological hazards when we discuss landslides, subsidence, and earthquakes.

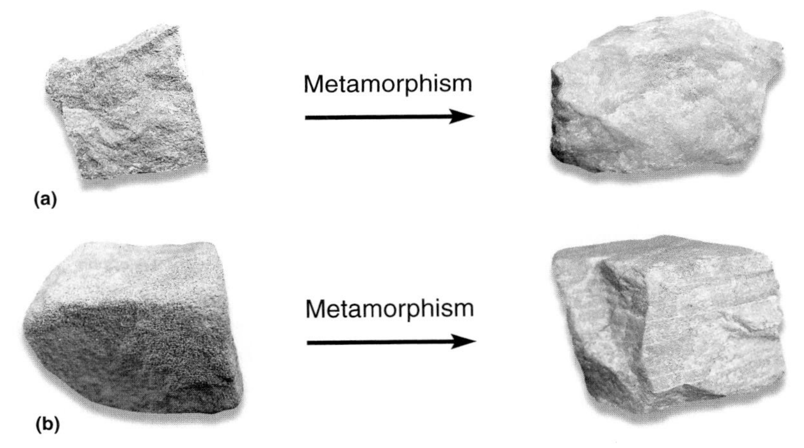

Figure 3.19 (a) Limestone recrystallizes to form white marble. (b) Quartzite, the hardest and most durable common rock, is metamorphosed sandstone.
© Sue Monroe

Earth's Deep Interior and Plate Tectonics

Over the past 75 years, earthquake scientists (seismologists) have learned much about the layering, density, and internal structure of our planet by studying the travel times and patterns of seismic waves passing through its interior. Geologists who study the composition and origin of rocks (petrologists) have been able to simulate conditions deep in Earth's interior through laboratory experiments, learning much about what kinds of rocks must exist to account for the seismic observations. Astronomers have also shed light on how Earth probably originated by the study of stellar nebula and young solar systems elsewhere in the galaxy.

When Earth first formed, the densest material accumulated at the center of the planet, forming a metallic iron-nickel–rich **core**. Around this ball a thick shell of magnesium-silicate matter developed that we call the **mantle**. This is by

Table 3.5 Common Metamorphic Rocks

Rock	Parent Rock	Characteristics
Foliated or layered		
Slate	Shale and mudstone	Splits into thin sheets
Schist	Fine-grained rocks, siltstone, shale, tuff	Mica minerals, often crinkled
Gneiss	Coarse-grained rocks	Dark and light layers of aligned minerals
Nonfoliated or Recrystallized		
Marble	Limestone	Interlocking crystals
Quartzite	Sandstone	Interlocking, almost fused quartz grains

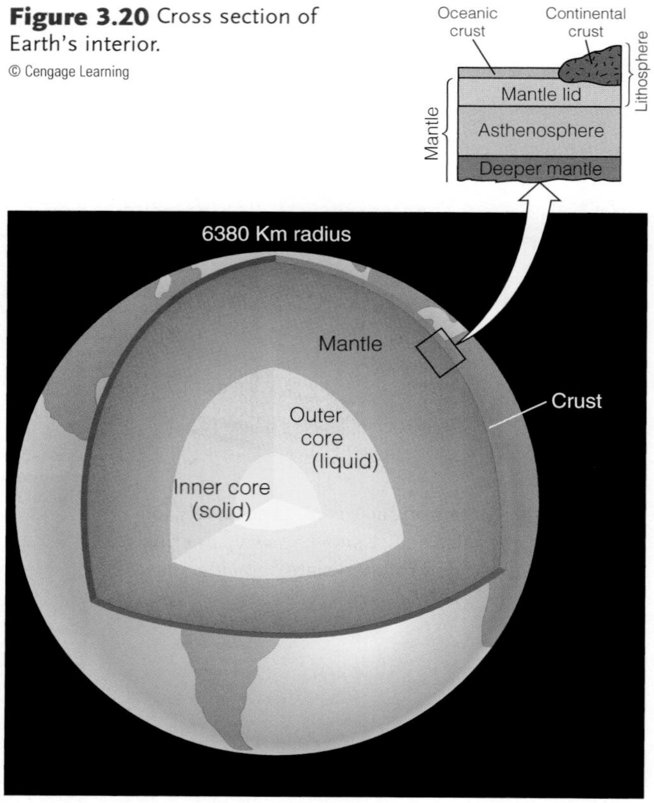

Figure 3.20 Cross section of Earth's interior.

far the thickest layer making up the planet (**Figure 3.20**). A thin "skin" of relatively lightweight aluminum and alkali (sodium and potassium)-rich silicate rocks, termed the **crust**, formed atop the mantle. The crust is still evolving through processes of igneous activity, weathering, sedimentation, and metamorphism. There are two basic kinds of crust: *oceanic* and *continental*. Oceanic crust consists largely of basaltic rocks and is typically 5 to 10 kilometers (3–6 miles) thick. The continental crust, made up largely of silica-rich igneous and sedimentary rocks (such as granite and sandstone) and related metamorphic rocks, ranges from 20 to 90 kilometers (12–56 miles) thick, with the greatest thicknesses found beneath mountain ranges. Oceanic rocks are typically darker and somewhat heavier (3.0 g/cm^3 or 0.108 lbs/in^3) than continental rocks (2.7 g/cm^3 or 0.097 lbs/in^3) because of a higher content of magnesium and iron. Because continental crust is thicker as well as lighter weight, it also stands topographically higher, mostly rising above the sea.

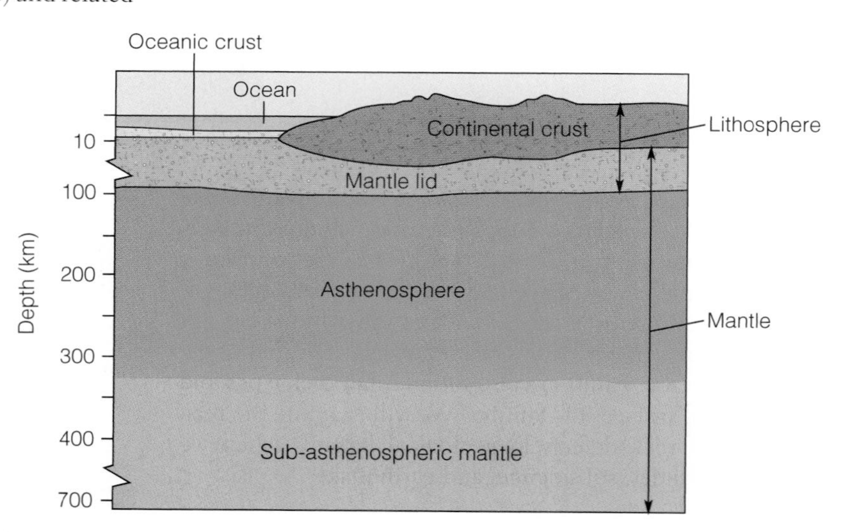

Figure 3.21 Earth's outer layers are divided into *crust* and *mantle* based on rock type, and into *lithosphere* and *asthenosphere* based on rigidity.

Rock climbing is a popular sport in many parts of the world. A favorite spot in the U.S. West is Yosemite Valley, a picturesque area carved by glaciers into a uniform granite body known as the Cathedral Peak granite (see Gallery Figure 3). Why do you suppose climbers would prefer to climb near-vertical cliffs of granite as opposed to similar cliffs composed of schist, gneiss, or shale?

In addition to the compositional layers above, seismologists have also discovered other kinds of boundaries inside Earth. Within the upper mantle, typically between 100 and 350 kilometers (60–210 miles) down, is a seismic zone known as the **asthenosphere** (Greek, "weak sphere"). In the upper part of the asthenosphere is a zone of low seismic wave speeds, indicating that the mantle there is close to its melting point and may contain melted portions. It is a zone of mobility that detaches the overlying, rigid, outer shell of the planet from the hotter, more plastic interior. The rigid, outer shell is called the **lithosphere** (Greek, "rocky sphere"), and it consists not only of the crust but also of **mantle lid**, a narrow layer of rigid, upper mantle rock (**Figure 3.21**).

The asthenosphere owes its existence to the release of heat from Earth's interior. Most of this heat originates from the slow decay of radioactive elements, such as uranium, lying in the core and lower mantle. The pressure in the lower mantle is so great that the rock there remains solid, despite temperatures in the range of 3500°C to 4000°C (6330°F to 7230°F)—hot enough to melt steel at Earth's surface. In the upper mantle, however, pressure is reduced just enough that the hot rock can become partly molten, even though its temperature is a cooler 1500°C (2730°F) at that depth.

The most efficient way for Earth's heat to escape is through the slow, convective churning of mantle rock, much like the churning in a pot of rolled oats cooking on a stove (**Figure 3.22**). Hot rock wells up from the depths of the mantle; becomes partially molten and weakens in the asthenosphere; loses heat to the overlying lithosphere; and being denser because it is now cooler, sinks back down into the depths. The heat imparted to the lithosphere causes magma to form, contributes to earthquakes and volcanic eruptions, and slowly leaks from the ground into the air. (The percentage of this "Earth heat" is trivial in most places, however, compared with the warmth of ordinary sunlight).

Mantle convection has also broken Earth's lithosphere into a set of rigid plates, which shift in response to the currents of sluggishly moving rock underneath. Individual lithospheric plates range in size from just a few hundred kilometers across (microplates) to many thousands of kilometers and may include large expanses of both oceanic and continental crust. Seven major plates make up the surface of Earth (**Figure 3.23**). The plates interact with one another along their margins. As you will learn in later chapters, all of human civilization and its environmental conditions are vitally linked to what happens along these margins. Essentially, three types of plate interactions take place (**Figure 3.24**):

- At **divergent boundaries**, there is tension. Forces from Earth's interior move plates apart along these boundaries. Divergence within the continental crust creates features such as the East African Rift valley. In the oceans, rifting occurs along the axis of a great mountain range, the **mid-ocean ridge**, with molten lava erupting through the fissures opened by the separating plates. The lava cools to form new oceanic crust as the plates move away from one another at rates from a few centimeters to as much as 17 centimeters (6.9 inches)/year—a process called **seafloor spreading**. The mid-ocean ridge winds around the surface of Earth like the seam on a giant baseball and links up with landward rift valleys. Its total length is more than 60,000 kilometers (35,000 mi)—by far the longest (and one of the most circuitous) mountain ranges on Earth, although mostly submerged and little noticed.
- At **convergent boundaries**, there is *compression*. Forces converge to drive plates together. Where a plate capped with oceanic crust collides with a plate made up of continental materials, as around much of the Pacific Rim, the continental material thrusts over the oceanic, which is somewhat denser and sinks back into the mantle, where it may be reassimilated in time. This process is called **subduction** (meaning "under-moving"). The world's deadliest earthquakes and very explosive volcanic eruptions are by-products of subduction expressed at the surface. Volcanic mountain ranges, such as the Andes, mark the tectonic disturbance at the edge of the overriding plate along this type of convergent plate boundary.
- In some instances, two plates made up of oceanic material at their edges collide. Subduction occurs in this situation, too, with the plate having the larger expanse of oceanic crust commonly subducting beneath the plate with the smaller expanse. A result of this is chains of explosively volcanic islands, **island arcs**, of which the Caribbean Islands are a good example. In the West Indies, the Atlantic oceanic lithosphere is subducting beneath the Caribbean Basin at a rate of about 2 centimeters (0.8 inches)/year.

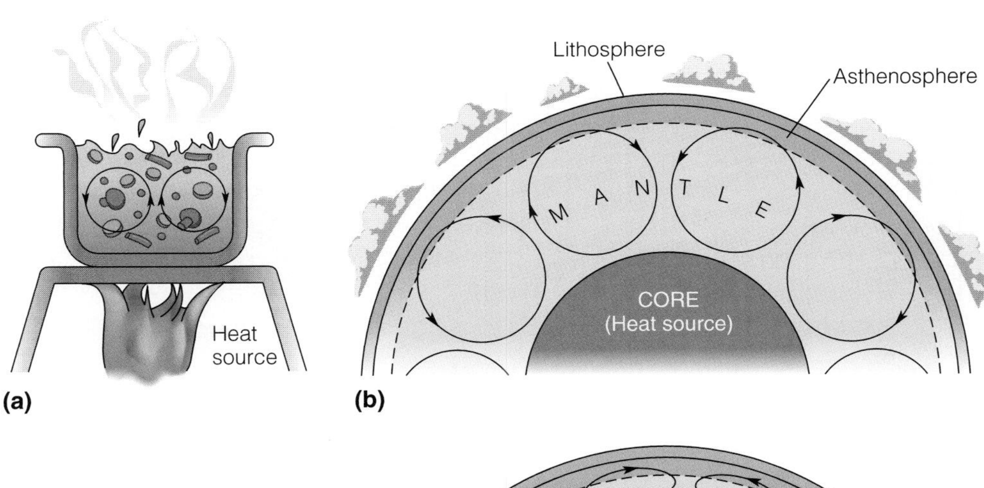

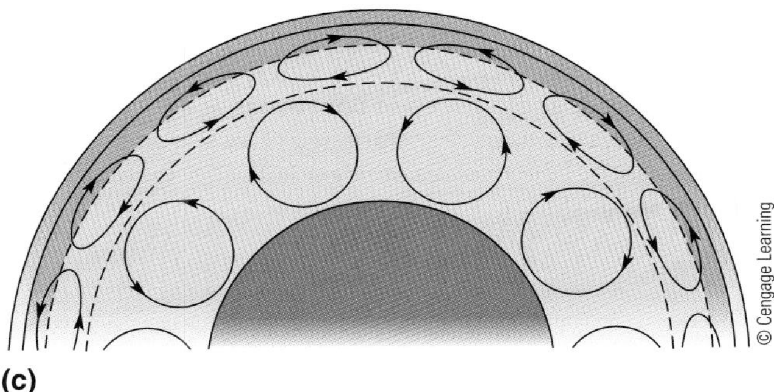

Figure 3.22 Convection (a) in a pot of soup or rolled oats on a stove; (b) in the mantle, involving "whole mantle" overturn. (c) An alternate model of mantle convection involving multiple layers, which is probably closer to reality than (b).

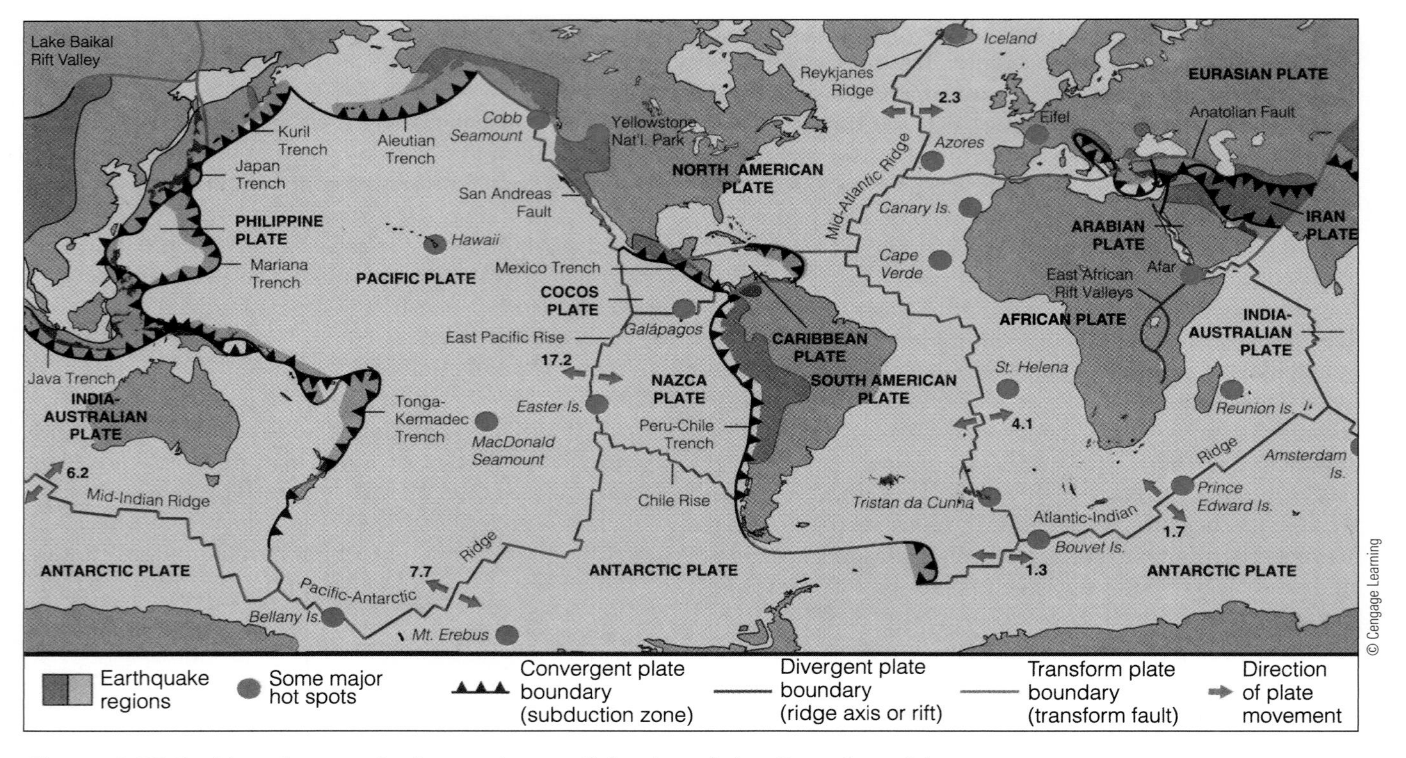

Figure 3.23 Earth's major tectonic plates and types of plate boundaries. Regardless of the type of plate boundary, each one is delineated by earthquake epicenters (not shown). Arrows indicate the present direction of plate motion; numbers indicate the rate of spreading in centimeters per year. The rate of spreading is determined by dividing the distance to a magnetic anomaly (or rock) of known age in kilometers ($\times 10^5$ = centimeters) by the age of that anomaly in years (= centimeters/year).

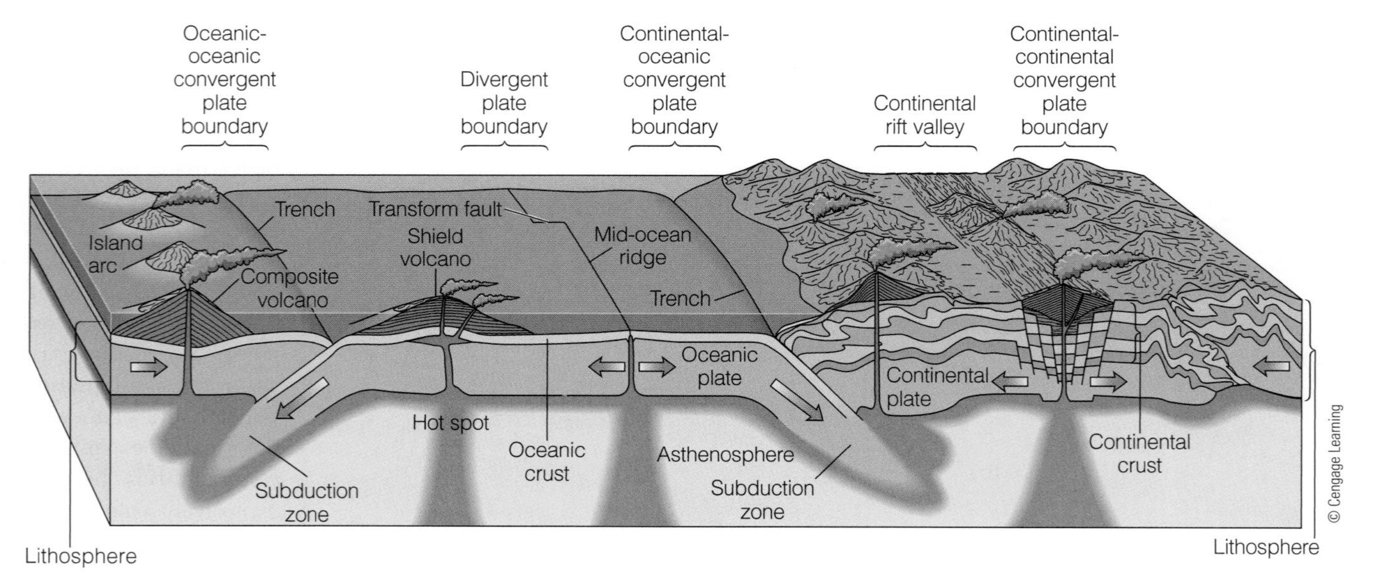

Figure 3.24 The plate tectonic model. Divergent and convergent boundaries and their relationships to seafloor and continental volcanoes are shown. Transform faults on the seafloor are transform boundaries linking different segments of the mid-ocean ridge. Rift-valley formation as a result of divergence under the continent is also shown.

- In some cases, two plates made up of continental material collide, creating a great range of mountains, because continental lithosphere is generally too thick and buoyant to subduct (e.g., Case Study 3.2). The Himalayas and the Alps formed in this way through the collisions of India and Africa, respectively, with Eurasia. Very powerful earthquakes often occur as these intracontinental mountain chains grow, but volcanic activity is essentially nonexistent. The explanation for this is that magma generation by plate convergence *requires the involvement of the upper mantle.* When subduction occurs, a thin wedge of upper mantle is caught between the sinking and the overriding plates. Loss of volatiles (dissolved gases and fluids) from a sinking plate causes the overlying mantle wedge to melt partly. Magmas ascending from this zone (typically, 40–60 kilometers, or 25–40 miles deep) enter the overlying crust and, as heat builds up, can bring about melting there as well (see also Chapter 5). These processes are not possible in Himalayan-style plate convergence.
- Continental margins that correspond to convergent plate boundaries, such as the west coast of South America, are said to be *active,* whereas those that do not correspond to ongoing convergence are said to be *passive.* Passive margins are commonly the by-product of past intracontinental rifting and the formation of divergent plate boundaries. For example, the east coast of the United States is a passive margin originating with the rifting of the ancient supercontinent of Pangaea and opening of the Atlantic Ocean through seafloor spreading beginning 130 million years ago. Prior to rifting, North America formed a continuous landmass with northwestern Africa and Western Europe, and the North Atlantic ocean did not exist (see Case Study 3.3).
- At **transform boundaries**, shear forces cause plates to slide horizontally past one another. Long, linear faults mark transform boundaries in many places (**Figure 3.25**). Outstanding examples include the Alpine Fault in New Zealand and the San Andreas Fault in California. The strong earthquakes along both faults help the world's biggest plate, the Pacific, shift past two other giant plates, the India-Australian and North American plates. Whereas destructive earthquakes are a strong risk from these faults, virtually no volcanic activity takes place along transform boundaries.

Plate tectonics is the general name for these dynamic interactions. *Tectonics* derives from the Greek word for "builder," *tektonikos.* Most, but not all, of the world's great geographical relief may be related to present or past plate tectonic activity. There are, however, some notable exceptions. Within some oceanic basins stretch chains of aligned volcanic islands that become progressively older in one direction. The Hawaiian Islands, Line Islands, and Tuamotus are examples in the Pacific Ocean. Although the Galapagos Islands off the coast of Ecuador—made famous by the writings of Charles Darwin—are only roughly aligned, they are of similar origin. These lines-of-islands are not associated with plate boundaries, but have formed over areas of unusually hot rock, or **hot spots**, fixed (or nearly fixed) in the asthenosphere (**Figure 3.26**). Many geologists believe that hot spots are fed by plumes of hot material welling up from the deep mantle, like the hot air that wafts smoke above a chimney on a still, cold day. These rising intraplate plumes penetrate the general convective circulation that drives plate motion, although some are plainly related to the development and continuation of divergent plate margins—the Iceland hot spot being a good example, lying at the crest of the mid-ocean ridge in the northern Atlantic. The drift of lithospheric plates over hot spots explains the generation of archipelagos such as Hawaii, which includes the biggest volcanoes on Earth. But hot spots can also be found in continental interiors, one of the most famous examples being Yellowstone, Wyoming, the "track" of which is the young, stark volcanic plain of the Snake River in southern Idaho.

Richard Hazlett

Figure 3.25 Geology students standing next to a fresh rupture of the Motagua Fault, the transform boundary between the North American and Caribbean plates in Guatemala. The land on which the students are standing is shifting to the left through time relative to the land in the foreground.

The initial development of the largest, longest-lived hot spots on Earth is signaled by massive outpourings of high-temperature molten lava and the discharge of large amounts of poisonous, climate-cooling volcanic gases. Great extinction events in the fossil record, including the demise of the dinosaurs, may be explained, at least in part, by the intensive volcanic convulsions marking the births of new hot spots.

Plate tectonics also provides an excellent model for explaining the rock cycle, described earlier in this chapter. Igneous rocks form from cooling magma bodies at divergent and convergent plate boundaries. Uplift at convergent boundaries exposes this rock to the harsh surface agents of weathering and erosion, which yields up the loose sediment needed to make sedimentary rocks. Further plate convergence metamorphoses the sedimentary

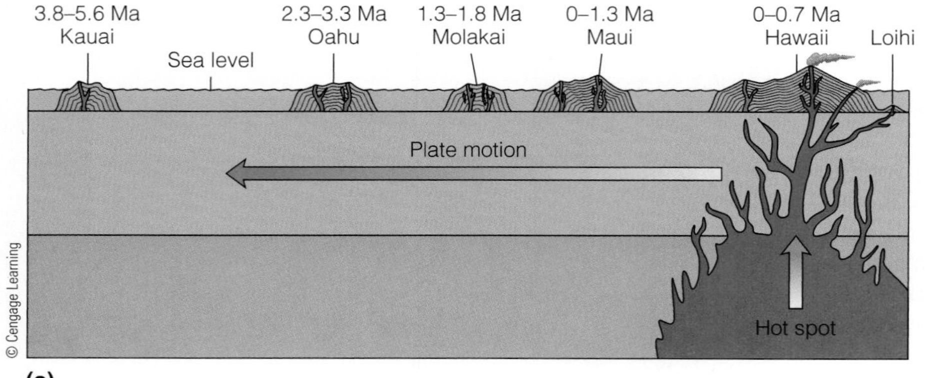

Figure 3.26 (a) Cross section of six islands formed and forming over the Hawaiian hot spot. The Hawaiian Islands are sequentially older in a northwest direction. The new volcano, Loihi, is forming underwater off the southeast end of the Hawaiian chain. (b) Kinked hot-spot trajectories in the Pacific. The kink in each island chain occurred about 40 million years ago because of a change in the Pacific plate's direction of motion. (From Murphy/Nance. *Earth Science Today*, Media Edition (with Earth Systems Today CD-ROM and InfoTrac®), 1E. © 2001 Cengage Learning. Reproduced by permission.)

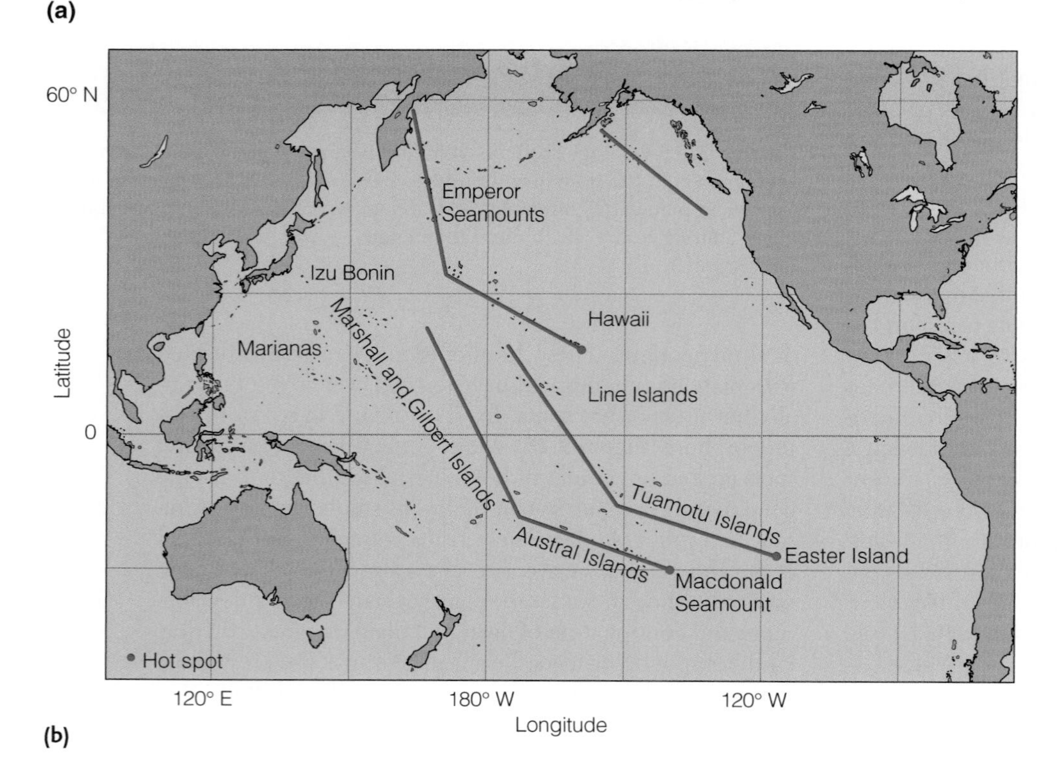

rock in an episode of mountain building, which under intense conditions of heat and pressure causes it to melt partially. This renews the material transfer of the rock cycle.

One tectonic consequence of the rock cycle is that a submerged continental shelf of soft muds and sands in one era can become a towering volcanic mountain range in the next, with the net addition of new, hard crust to a continent. As a result, the total land area of the continents has enlarged over the past few billion years, with new growth concentrated along convergent plate boundaries. Earth is gradually becoming more of a land planet and less of a water world, although the growth in continents may be compensated by an overall deepening trend in the seas as well.

Now we can return to the original question that launched this chapter: *Why did Antarctica drift to the South Pole beginning 180 million years ago.* The answer is that, at that time, a large mantle plume ascended beneath Gondwana, causing it to rift into fragments. Remnants of the many lava flows that erupted during the destruction of this giant ancient landmass may still be found in parts of Patagonia (Chon Aike volcanic province), southern Africa (the Karoo volcanics), and Antarctica itself (as the Ferrar volcanic series). As rifting proceeded, a wide ocean basin opened between Antarctica and the continental pieces to which it originally attached—the southern Atlantic and Indian oceans, or what some geographers simply refer to as the *Southern Ocean*. Will Antarctica eventually drift off the South Pole and return to more benign environmental conditions in the future? It is certainly possible! Might it eventually collide with other landmasses to produce another Gondwana-scale supercontinent? This, too, could happen many millions of years from now.

Geological Time

Interest in extremely long periods of time sets geology and astronomy apart from other sciences. Geologists think in terms of billions of years for the age of Earth and its oldest rocks—numbers that, like the national debt, are not easily comprehended. Nevertheless, the time scales of geological activity are important for environmental geologists because they provide a way to measure human impacts on the natural world. For example, we would like to know the rate of natural soil formation from solid rock to determine whether topsoil erosion from agriculture is too great. Likewise, understanding how climate has changed over millions of years is vital to properly assess current global warming trends. Clues to past environmental change are well preserved in many different kinds of rocks.

Geologists evaluate the age of rocks and geological events using two different approaches. **Relative-age dating** is the technique of determining *a sequence of geological events* based on the structural relations of rocks. **Absolute-age dating**

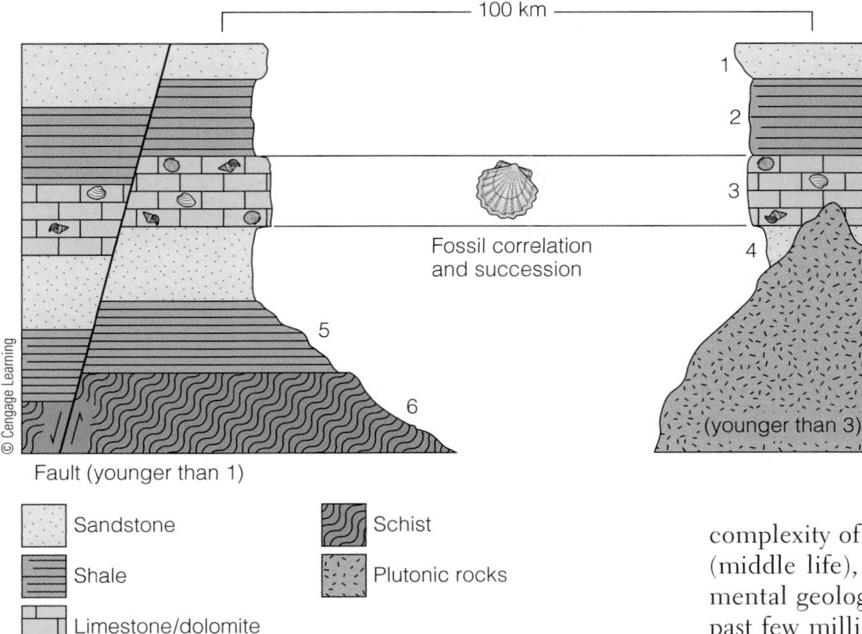

Figure 3.27 Geological cross sections illustrating the laws of superposition, cross-cutting relationships (intrusion and faulting), and fossil succession. The limestone beds are correlative, because they contain the same fossils. Numerals indicate relative ages, with 1 being the youngest.

provides the *actual ages* for rocks in years before the present. Relative age is determined by applying geological laws based on the structural relations of rocks. For instance, the law of superposition tells us that, in a stack of undeformed sedimentary rocks, the stratum (layer) at the top is the youngest. The law of cross-cutting relationships tells us that a fault is younger than the youngest rocks it displaces or cuts. Similarly, we know that a pluton is younger than the rocks it intrudes (**Figure 3.27**).

Using these laws, geologists arranged a great thickness of sedimentary rocks and their contained fossils representing an immense span of geological time. The geological age of a particular sequence of rock was then determined by applying the law of fossil succession, the observed chronological sequence of life-forms through geological time. This allows fossiliferous rocks from two widely separated areas to be correlated by matching key fossils or groups of fossils found in the rocks of the two areas (Figure 3.27). Using such indicator fossils and radioactive dating methods, geologists have developed the geological time scale to chronicle the documented events of Earth's history (**Figure 3.28**). Note that the scale is divided into units of time during which rocks were deposited, life evolved, and significant geological events, such as mountain building, occurred. Eons are the longest time intervals, followed, respectively, by eras, periods, and epochs. The Phanerozoic ("revealed life") eon began 570 million years ago with the Cambrian period, the rocks of which contain the first extensive fossils of organisms with hard skeletons. Because of the significance of the Cambrian period, the informal term *Precambrian* is widely used to denote the time before it, which extends back to the formation of Earth 4.6 billion years ago. Note that the Precambrian is divided into the Archean and Proterozoic eons, with the Archean eon extending back to the formation of the oldest known in-place rocks, about 3.9 billion years ago. Precambrian time accounts for 88% of geological time, and the Phanerozoic for a mere 12%. The eras of geological time correspond to the relative complexity of life-forms: Paleozoic (oldest life), Mesozoic (middle life), and Cenozoic (most recent life). Environmental geologists are most interested in the events of the past few million years, a mere heartbeat in the history of Earth.

Absolute-age dating requires some kind of natural clock. The ticks of the clock may be the annual growth rings of trees or established rates of disintegration of radioactive elements to form other elements. At the turn of the 20th century, American chemist and physicist Bertram Borden Boltwood (1870–1927) discovered that the ratio of lead to uranium in uranium-bearing rocks increases as the rocks' ages increase. He developed a process for determining the age of ancient geological events that is unaffected by heat or pressure—**radiometric dating**. The "ticks" of the radioactive clocks are radioactive decay processes—spontaneous disintegrations of the nuclei of heavier elements, such as uranium and thorium to lead. A radioactive element may decay to another element or to an isotope of the same element. This decay occurs at a precise rate that can be determined experimentally. The most common emissions are alpha particles $^4_2\alpha$, which are helium atoms, and beta particles (β^-), which are nuclear electrons. New *radiogenic* "daughter" elements, or isotopes, result from this **alpha decay** and **beta decay** (**Figure 3.29**).

For example, of the three isotopes of carbon, ^{12}C, ^{B}C, and ^{14}C, only ^{14}C is radioactive, and this radioactivity can be used to date events between a few hundred and a few tens of thousands of years ago. Carbon-14 is formed continually in the upper atmosphere by neutron bombardment of nitrogen, and it exists in a fixed ratio to the common isotope, ^{12}C. All plants and animals contain radioactive ^{14}C in equilibrium with the atmospheric abundance until they die, at which time ^{14}C begins to decrease in abundance and, along with it, the object's radioactivity. Thus, by measuring the radioactivity of an ancient parchment, a log, or a piece of charcoal and comparing the measurement with the activity of a modern standard, the age of archaeological materials and geological events can be determined (**Figure 3.30**). Carbon-14 is formed by the collision of cosmic neutrons

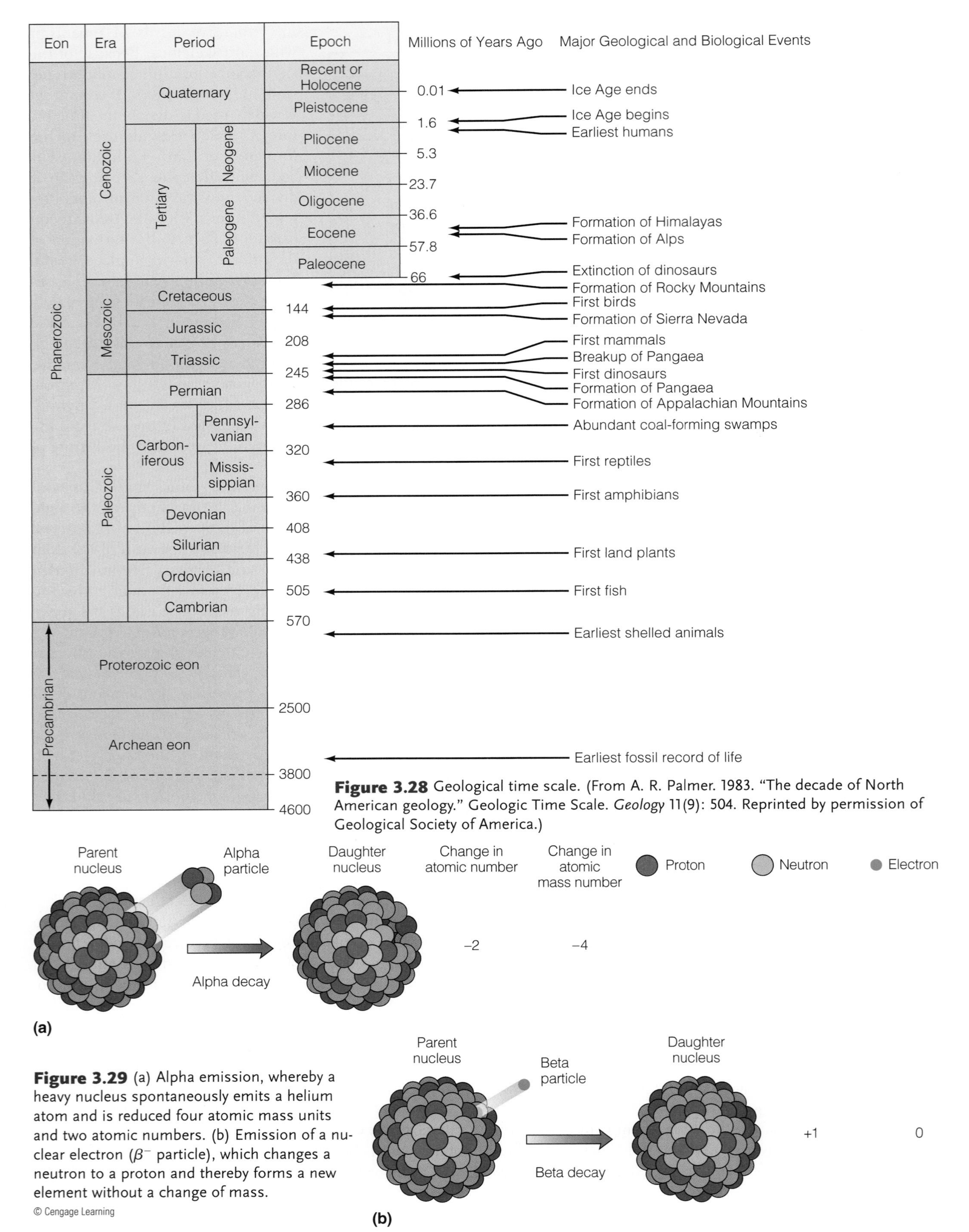

Figure 3.28 Geological time scale. (From A. R. Palmer. 1983. "The decade of North American geology." Geologic Time Scale. *Geology* 11(9): 504. Reprinted by permission of Geological Society of America.)

Figure 3.29 (a) Alpha emission, whereby a heavy nucleus spontaneously emits a helium atom and is reduced four atomic mass units and two atomic numbers. (b) Emission of a nuclear electron (β^- particle), which changes a neutron to a proton and thereby forms a new element without a change of mass.

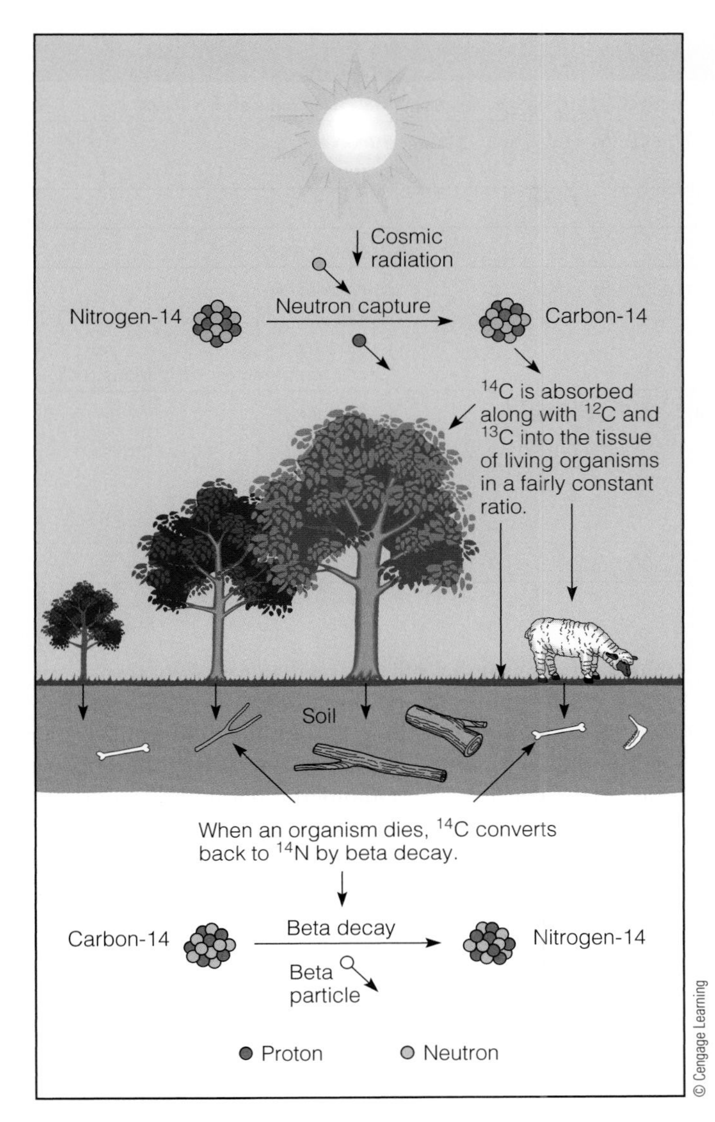

Figure 3.30 Carbon-14 is formed from nitrogen-14 by neutron capture and subsequent proton emission. Carbon-14 decays back to nitrogen-14 by emission of a nuclear electron (β^-).

with ^{14}N in the atmosphere; then it decays back to ^{14}N by emitting a nuclear electron (β^-).

$$^{14}_{7}N + \text{neutron} \rightarrow \, ^{14}_{6}C + \text{proton}$$
$$^{14}_{6}C \rightarrow \, ^{14}_{7}N + \beta^-$$

Both carbon—the radioactive "parent" element—and nitrogen—the radiogenic "daughter"—have 14 atomic mass units. However, one of ^{14}C's neutrons is converted to a proton by the emission of a beta particle, and the carbon changes (or *transmutes*) to nitrogen. This process proceeds at a set rate, which can be expressed as a **half-life**, the time required for half of a population of radioactive atoms to decay. For ^{14}C, this is about 5,730 years.

Radioactive elements decay exponentially; that is, in two half-lives, one fourth of the original number of atoms remain; in three half-lives, one-eighth remain; and so on (**Figure 3.31**). So few parent atoms remain after seven or eight half-lives (<1%), that experimental uncertainty creates limits for the various radiometric dating methods.

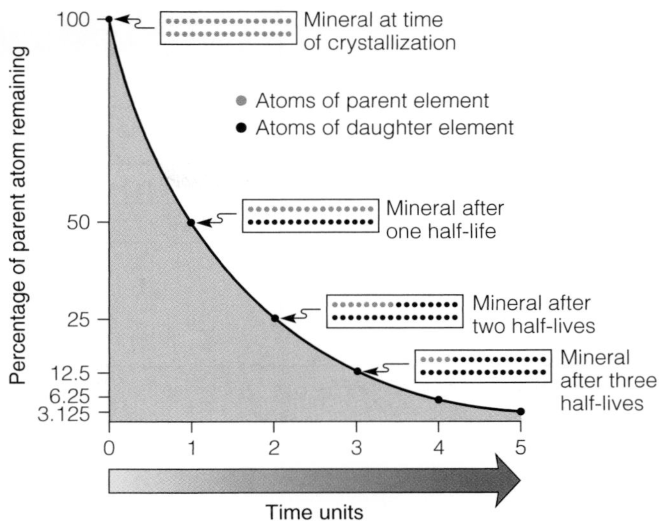

Figure 3.31 Decay of a radioactive parent element with time. Each time unit is one half-life. Note that after two half-lives, one fourth of the parent element remains, and that after three half-lives, one-eighth remains.

Whereas the practical age limit for dating carbon-bearing materials, such as wood, paper, and cloth, is about 40,000 to 50,000 years, ^{238}U disintegrates to ^{206}Pb and has a half-life of 4.5×10^9 years. Uranium-238 emits alpha particles (helium atoms). Because both alpha and beta disintegrations can be measured with a Geiger counter, we have a means of determining the half-lives of a geological or archaeological sample. **Table 3.6** shows radioactive parents, daughters, and half-lives commonly used in age dating.

Age of Earth

Before the advent of radiometric dating, determining the age of Earth was a source of controversy between established religious interpretations and early scientists. Archbishop Ussher (1581–1656), the Archbishop of Armagh and a professor at Trinity College in Dublin, declared that Earth was formed in the year 4004 B.C.E. Ussher provided not only the year but also the day 23 October, and the time, 9:00 A.M. Ussher has many detractors, but Stephen J. Gould of Harvard University, though not proposing acceptance of Ussher's date, was not one of them. Gould pointed out that Ussher's work was good scholarship for its time, because other religious scholars had extrapolated from Greek and Hebrew scriptures that Earth was formed in 5500 B.C.E. and 3761 B.C.E., respectively. Ussher based his age determination on the verse in the Bible that says, "one day is with the Lord as a thousand years" (2 Peter 3:8). Because the Bible also says that God created heaven and Earth in 6 days, Ussher arrived at 4004 years B.C.E.—the extra 4 years because he believed Christ's birth year was wrong by that amount of time. Ussher's age for Earth was accepted by many as "gospel" for almost 200 years.

Table 3.6 Isotopes Used in Age Dating

Parent	Isotope's Daughter	Parent's Half-Life, Years	Effective Dating Range, Years	Material That Can Be Dated
Uranium-238	Lead-206	4.5 billion	10 million to 4.6 billion	Zircon Uraninite
Uranium-235	Lead-207	704 million		
Thorium-232	Lead-208	14 billion		
Rubidium-87	Strontium-87	48.8 billion	10 million to 4.6 billion	Muscovite Biotite Orthoclase Whole metamorphic or igneous rock
Potassium-40	Argon-40	1.3 billion	100,000 to 4.6 billion	Glauconite Hornblende Muscovite Whole volcanic rock Biotite
Carbon-14	Nitrogen-14	5730	<100,000	Shell, bones, charcoal

Table 3.7 Earth's Oldest Known Materials

Material	Location	Age, Billions of Years
Crust	Zircon minerals in rocks of western Australia	4.0–4.3
Rock	Zircon minerals in the Acasta Gneiss, Northwest Territories, Canada	3.96
Sedimentary rock	Isua Greenstone Belt, Greenland	3.7–3.75
Fossils	Algae and bacteria	3.5

By the late 1800s, geologists believed that Earth was on the order of 100 million years old. They reached their estimates by dividing the total thickness of sedimentary rocks (tens of kilometers) by an assumed annual rate of deposition (mm/year). Evolutionists, such as Charles Darwin, thought that geological time must be almost limitless in order that minute changes in organisms could eventually produce the present diversity of species. Both geologists and evolutionists were embarrassed when the British physicist William Thomson (later Lord Kelvin) demonstrated with elegant mathematics how Earth could be no older than 400 million years, and maybe as young as 20 million years. Thomson based this on the rate of cooling of an initially molten Earth and the assumption that the material composing Earth was incapable of creating new heat through time. He did not know about radioactivity, which adds heat to rocks in the crust and mantle.

Bertram Boltwood postulated that older uranium-bearing minerals should carry a higher proportion of lead than younger samples. He analyzed a number of specimens of known relative age, and the absolute ages he came up with ranged from 410 million to 2.2 billion years old. These ages put Lord Kelvin's dates based on cooling rates to rest and ushered in the new radiometric dating technique. By extrapolating backward to the time when no radiogenic lead had been produced on Earth, we arrive at an age of 4.6 billion years for Earth. This corresponds to the dates obtained from meteorites and lunar rocks, which are part of our solar system. The oldest known intact terrestrial rocks are found in the Acasta Gneiss of the Slave Geological Province of Canada's Northwest Territories. Analyses of uranium-to-lead ratios on the gneiss's zircon minerals indicate that the rocks are 3.96 billion years old. However, older detrital zircons on the order of 4.0 to 4.3 billion years old have been found in western Australia, indicating that some stable continental crust was present as early as 4.3 billion years ago (**Table 3.7**). Suffice it to say, Earth is very old and there has been abundant time to produce the features we see today (**Figure 3.32**).

What is most impressive about geological time is how short the period of human life on Earth has been. If we could compress the 4.6 billion years since Earth formed into one calendar year, *Homo sapiens* would appear about 30 minutes before midnight on 31 December. The last ice-age glaciers would begin wasting away a bit more than 2 minutes before midnight, and written history would exist for only the last 30 seconds of the year. Perhaps we should keep this calendar in mind when we hear that the dinosaurs were an unsuccessful group of reptiles. After all, they endured 50 times longer than hominids have existed on the planet to date. At the present rate of population growth, there is some doubt that humankind as we know it will be able to survive anywhere near that long.

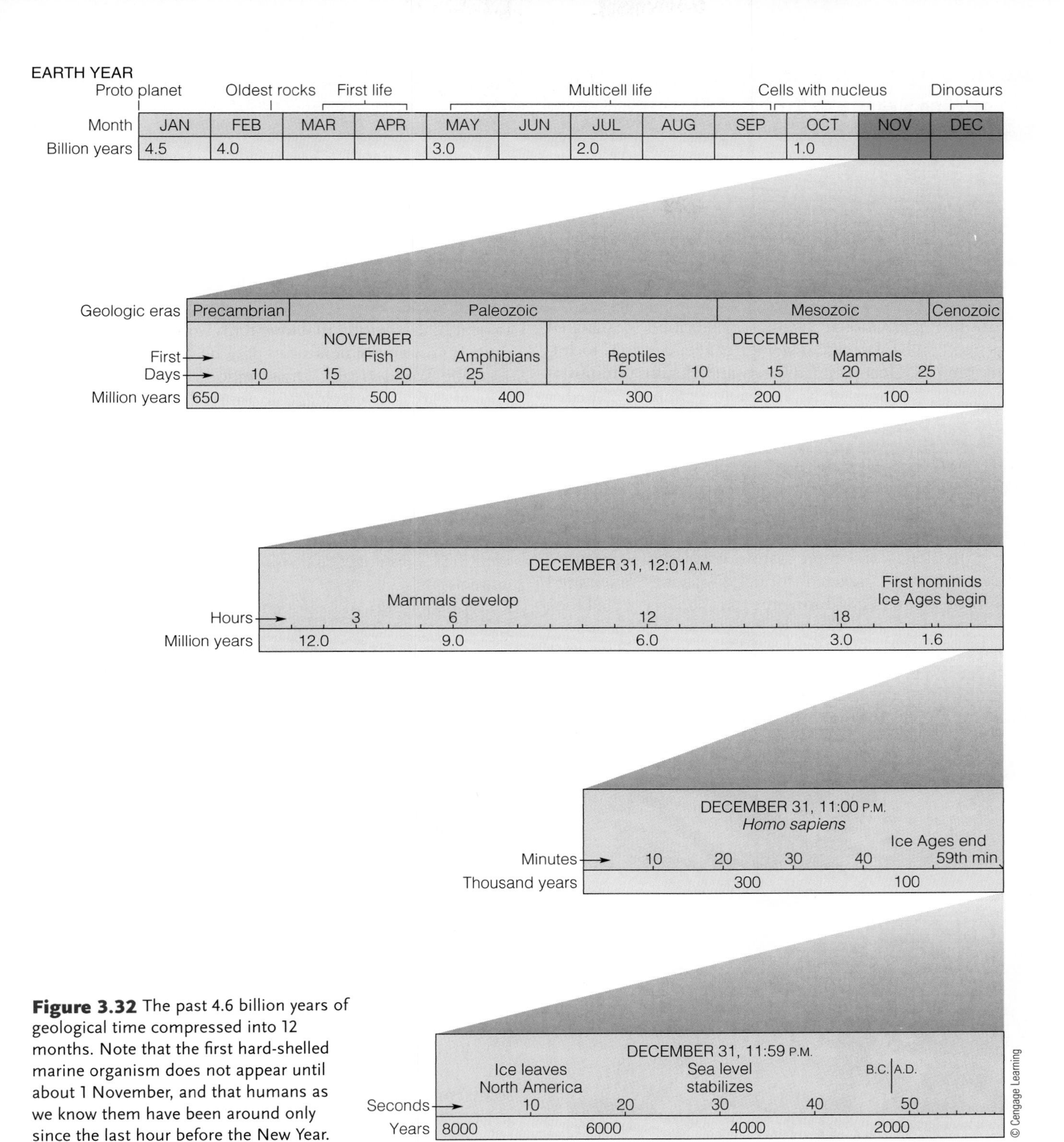

Figure 3.32 The past 4.6 billion years of geological time compressed into 12 months. Note that the first hard-shelled marine organism does not appear until about 1 November, and that humans as we know them have been around only since the last hour before the New Year.

3.1 Minerals, Cancer, and OSHA: Fact and Fiction

When an international health organization designated quartz as "probably" carcinogenic (cancer-causing) to humans, the U.S. Occupational Safety and Health Administration (OSHA) immediately went into action. It issued a mandate requiring products containing more than 0.1% of quartz, or "free silica" (SiO_2), to display warning signs. Because quartz is chemically inert and because it is the most common mineral species, questions arose whether this requirement extended to sandpaper, beach sand, and unpaved roads that may produce clouds of fine-grained quartz dust. Questions—and more—arose when some Delaware truck drivers were "ticketed" for not displaying the required "quartz on board" signs during the transport of crushed rock to construction sites. Such well-meaning OSHA actions have been perceived as ridiculous. They illustrate what happens when a government authority establishes health regulations without understanding the substances involved. In this case, OSHA did not understand the where, how, and chemistry of the mineral—its mineralogy. Unless one has had a long-term craving to inhale beach sand or crushed rock, there are no health hazards in the examples cited. Nonetheless, because exposure to very large quantities of any substance, even table salt or vitamins, can be harmful to health, it is certain that long-term workplace exposure to dusts from some minerals can pose health problems.

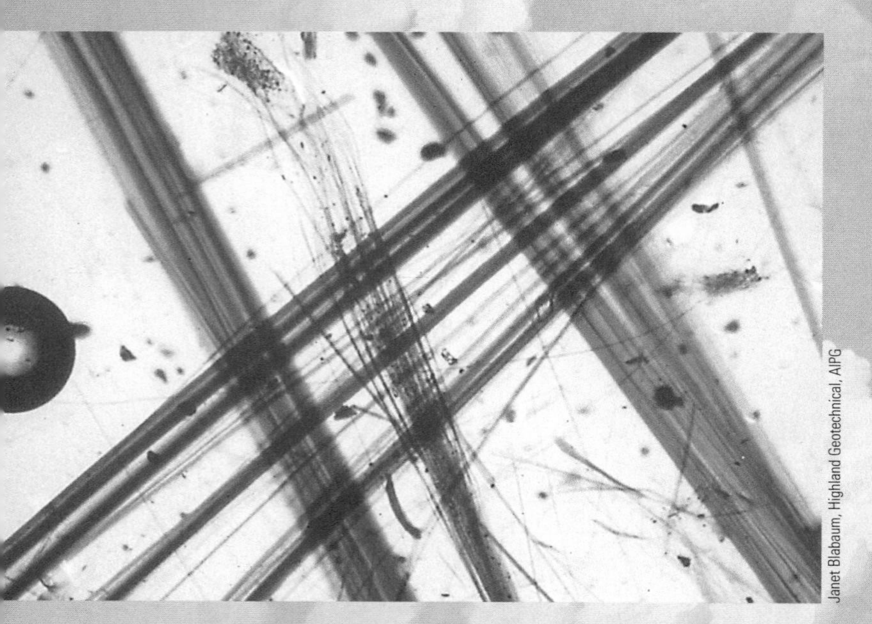

Janet Blabaum, Highland Geotechnical, AIPG

Figure 1 Deadly crocidolite, "blue asbestos," magnified about 100 times. The mineral appears in its characteristic blue and green colors when viewed through a petrographic microscope. It was used in only a small percentage of commercial products.

The World Health Organization's Group 1 classification of "known" carcinogenic minerals includes asbestos minerals (there are six), erionite (a zeolite), and minerals that contain arsenic, chromium, and nickel. Its Group 2 of "probable" carcinogenic minerals includes all radioactive minerals and minerals that contain lead, beryllium, and silica. The classification is based on suspected relationships between specific diseases and workplace exposure to the identified substances and on laboratory experiments with animals.

A major human health concern is the relationship between lung cancer and inhaling the fine fibers of asbestos minerals. The long, fibrous crystal forms of asbestos belonging to the amphibole group (see hornblende structure, Figure 3.3) are considered the most carcinogenic. Of these, blue asbestos, crocidolite (pronounced krō-sid'-əl-īt; **Figure 1**), is thought to be most dangerous and is believed to cause mesothelioma, a relatively rare cancer of the lining of the heart and lungs. Although crocidolite constitutes only 5% of all industrial asbestos, its morbidity statistics are grim. For example, a 1990 survey of 33 men who, in 1953, worked in a factory where crocidolite was used in manufacturing cigarette filters found that 19 of them had died of asbestos-related diseases.

Common white asbestos, chrysotile (pronounced kris'-ə-tīl; **Figure 2**), is a sheet-structure mineral similar to mica. Under high magnification, it appears as long fibers, which are, in reality, rolled sheets, much like long rolls of gift-wrapping paper. Chrysotile constitutes 95% of all industrial asbestos and has never been demonstrated to cause cancer at the levels found in most schools and public buildings. A study conducted in Thetford, Quebec, Canada, evaluated persons who had been exposed to high concentrations of white asbestos in their workplace and found no deaths attributable to it. Many of the diagnosed chrysotile-related lung problems have been traced back to improper handling of the material and inadequate respiratory protection during asbestos removal procedures. Experimental studies at Virginia Tech have shown that respirable-sized 1-nanometer-diameter chrysotile fibers dissolve in about 9 months under conditions expected in lung tissue. This very short lifetime is difficult to reconcile with the general observations that asbestos-related diseases appear in asbestos workers only after many years on the job.

Inhaling finely divided airborne asbestos fibers in large quantities is definitely a health hazard. Under the Toxic Substance Control Act of 1976, the U.S. Environmental Protection Agency (EPA) mandated in 1986 that *all* asbestos minerals be treated as identical health hazards, regardless of their mineralogy, and that they be removed. By the mid-1990s, the cost of removal had amounted to hundreds of millions of dollars, and the total cumulative cost for the remediation of rental and commercial buildings, litigation, and enforcement was estimated at $100 billion. The EPA requirements were sweeping. They applied to buildings containing the relatively benign white asbestos and to those where airborne concentrations were so low that they could not be measured. The fear of asbestos and the EPA regulations are based on the idea that there is no safe "threshold" concentration—that even one fiber can kill. No doctor or public health official would agree with this. A New York school district's administration was besieged by irate parents when it admitted that it could not *guarantee* that *all* asbestos had been removed from its buildings. The district had spent millions of dollars to remove it.

Industry and government can ill afford to squander capital and tax dollars on poorly conceived hazard-mitigation requirements based on the "no threshold" theory of health risk—that is, that one molecule of pesticide or one asbestos fiber can cause health problems. Many government regulations regarding "toxic substances" address risks to humans that are no greater than those incurred by routinely drinking several cups of coffee each day or eating a peanut butter sandwich for lunch. The "no threshold" criterion forces "carcinogen" classification upon otherwise benign minerals and many other useful substances, and needlessly increases public anxiety. (Remember when synthetic sweeteners and cell phones were labeled carcinogenic?) The costs of removal and liability protection in such cases are horrendous, and a fiscal crisis in the environmental field is acknowledged. Liability extends to building owners, real-estate agents who sell the buildings, banks that lend money for property purchases, and purchasers of land on which any material classified as hazardous is found.

A person's chances of being struck by a lightning bolt are about 3.5 per 1 million lifetimes, and the risk of a nonsmoker's dying of asbestos exposure is about 1 in 100,000—about three times the chances of being struck by lightning. Incidentally, the risk of dying from cigarette-related diseases is about 1 in 5 for smokers. Certainly, some substances in the environment pose dangers, and the key to risk reduction and longevity is awareness. Just as a reasonable person would not play golf during a violent thunderstorm because of the risk for being struck by lightning on the course, one should not handle hazardous substances without wearing protective gear.

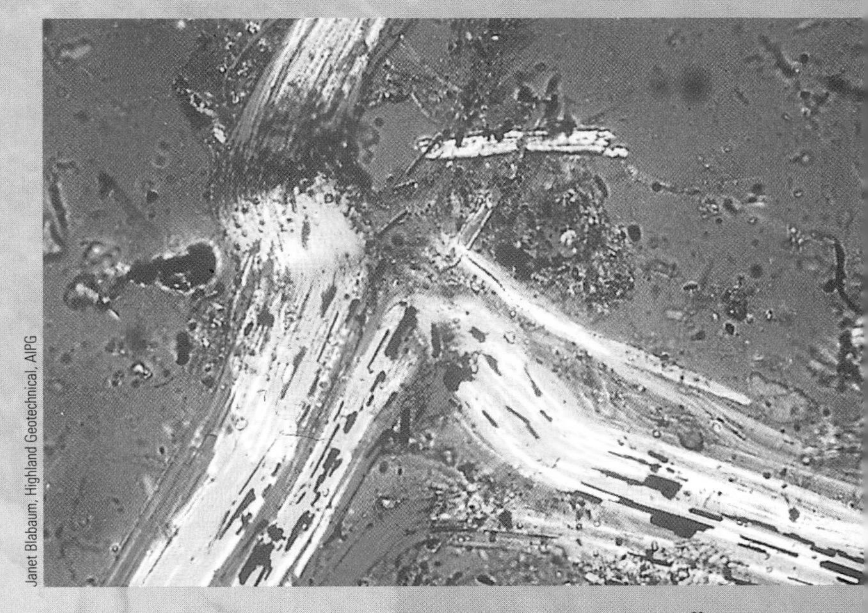
Janet Blabaum, Highland Geotechnical, AIPG

Figure 2 Chrysotile asbestos (×100) has many uses in homes, offices, and public buildings. Although it is the target of removal in thousands of structures, at low levels it poses no documented risk. Note the difference between chrysotile and crocidolite's crystal structures.

questions to ponder

1. How can scientists work better with policy makers to make environmental regulations that avoid the sort of problems highlighted in the earlier OSHA case study?
2. A cavern you are exploring contains large, bladelike, frosty white crystals, some of which are translucent to transparent. What criteria would you use to identify these minerals?

case study

3.2 Exotic Terranes: A Continental Mosaic

Because continental plates divide, collide, and slip along faults, it is not surprising that bits and pieces of continental crust existing within the oceans ultimately collide with continents at subduction zones and become stuck there. Because of their low density, they stand high above oceanic crust, and these microplates or microcontinents become plastered, *accreted,* to larger continental plates when they collide with them. The accreted plates are known as *terranes*, fault-bounded blocks of rock with histories quite different from those of adjacent rocks or terranes.[1] In size, they may be several thousand square kilometers (or miles) or just a few tens of square kilometers (or miles), and they may become part of a continent composed of many terranes. Oceanic crust and the sediment resting on it may be scraped up onto the continents. Terranes may consist of almost any type of rock, but each one is fault bounded, has a paleomagnetic signature indicating a distant origin, and has little geology in common with adjacent terranes or with the continental **craton**—the continent's stable core.

The terrane concept developed in Alaska when geological mapping for land-use planning revealed that the usually predictable pattern of rocks and structures was not valid for any distance. In fact, the rocks the geologists found a few kilometers away were almost always of a "wrong" composition and age. Further studies revealed that Alaska is a collection of microplates (terranes)—tectonic flotsam and jetsam—that have mashed together over the past 160 million years and that are still arriving from the south (**Figure 1**). One block, the Wrangellia terrane, was an island during Triassic time, and it has a paleomagnetic signature of rocks that formed 16° from the equator. It is not known whether Wrangellia formed north or south of the equator, because it is not known whether the magnetic field at the time was normal or reversed. In either case, the terrane traveled a long distance to become part of present-day Alaska (**Figure 2**). It now appears that about 25% of the western edge of North America, from Alaska to Baja California, formed in this way—that is, by bits and pieces being grafted onto the core of the continent, and thus enlarging it. In the distant future, part of California may become an exotic terrane of Alaska.

[1]*Terrane,* as noted, is a geological term describing the area or surface over which a particular rock or group of rocks is prevalent. *Terrain* is a geographical term referring to the topography or physical features of a tract of land.

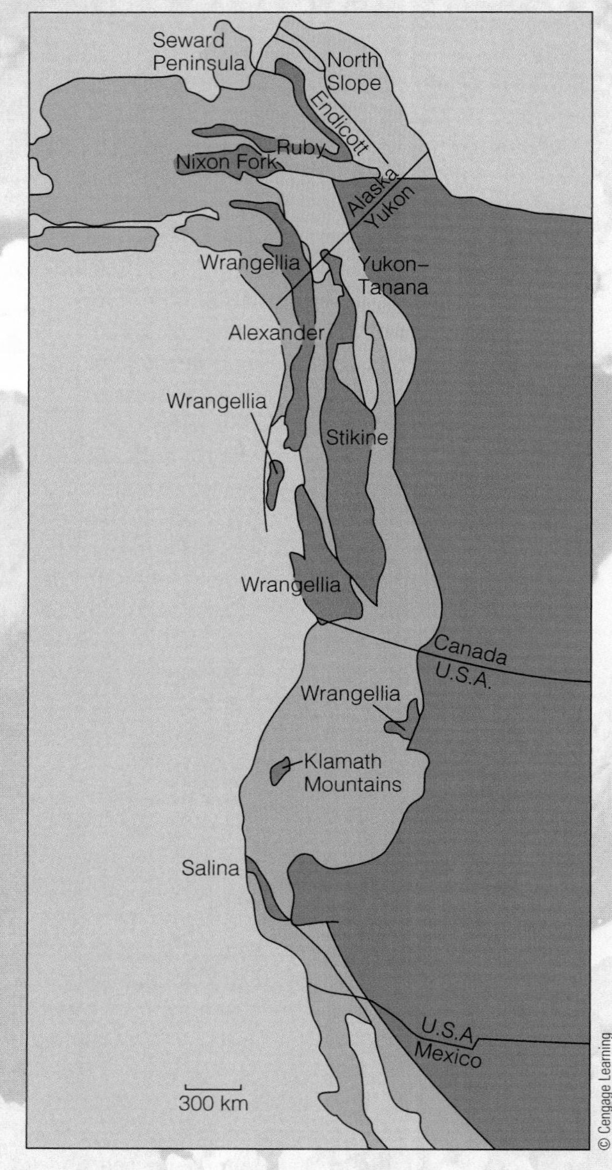

Figure 1 Accreted terranes of the west coast of the United States and Canada. The Salina, Wrangellia, Alexander, Stikine, Klamath Mountains, Ruby, and Nixon Fork blocks (dark brown) were probably once parts of other continents and have been displaced long distances. The terranes shown as light green areas are probably displaced parts of North America, the stable North American craton (darker green). The lighter brown areas represent rocks that have not traveled great distances from their places of origin.

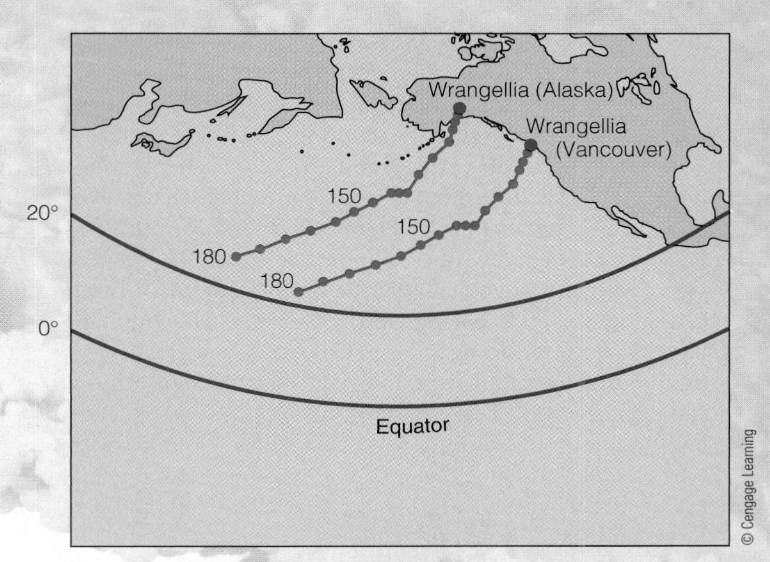

Figure 2 An analysis of Wrangellia terrane trajectories for the period 180 to 100 million years ago, assuming initial position in the Northern Hemisphere. In this interpretation, both Vancouver Island and the Alaskan terranes (see Figure 1) rode with the oceanic plate until they collided (docked) at their present positions on the North American continent about 100 million years ago. The data points (the dots) represent paleomagnetic evidence at the respective locations. (Numbers indicate positions of Wrangellia millions of years ago.)

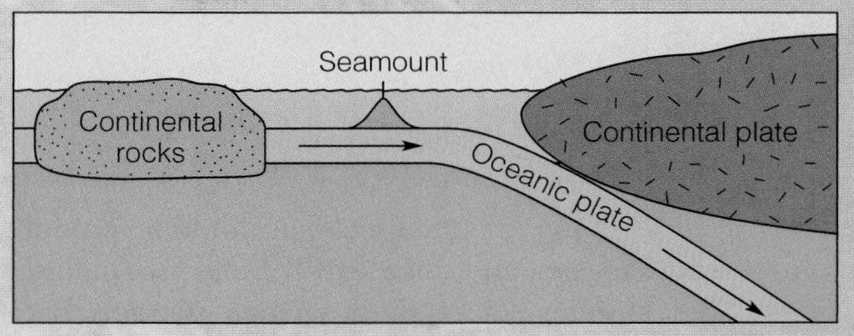

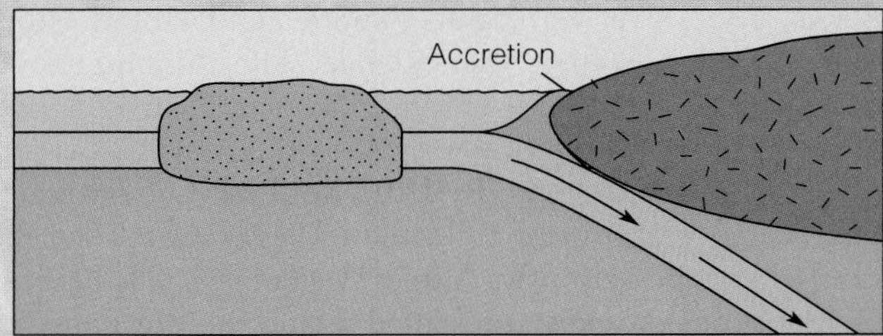

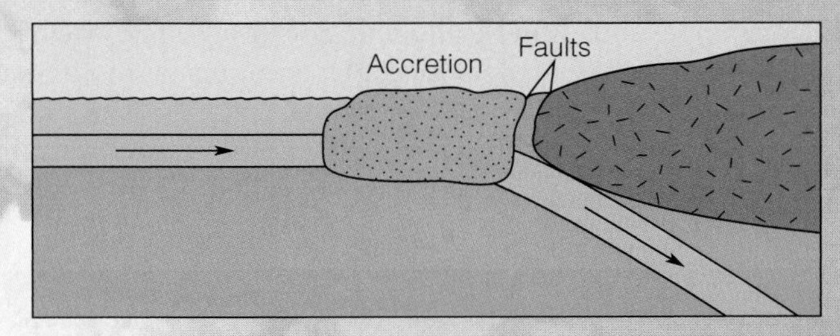

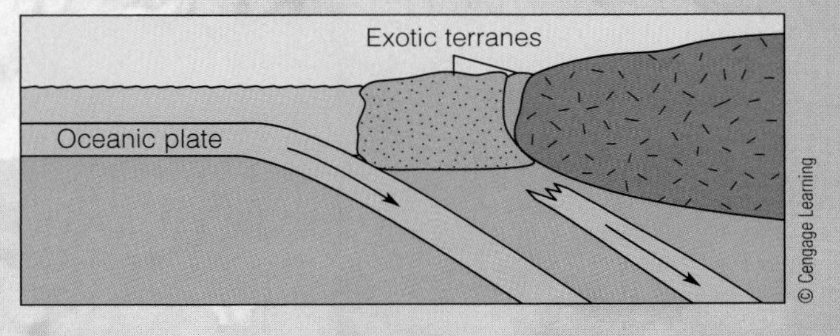

Figure 3 Formation of exotic terranes by a small continental mass and a submarine volcano being scraped off from the subducting plate and onto the continental mass. Many portions of the west coasts of the United States and Canada have grown in this manner.

If you've deduced that terrane accretion onto the continents (also known as *docking*) is a characteristic of active continental margins, you are correct (**Figure 3**). The east coast of North America is a rifted (pulled-apart), or *passive*, margin. Material is added to it by river sediment forming flat-lying sedimentary rocks that become part of the continent. Such rocks are not subject to the mountain-building forces of active margins, and they accumulate in thick, undisturbed sequences.

questions to ponder

1. Why are accretionary terranes much less likely to develop around the margins of the Atlantic Ocean than the Pacific?
2. For many decades in the early history of geology, an influential group of scientists (the Neptunists) thought that rocks like granite must form by crystallization out of water on the floor of an ancient "universal sea." Today, we believe that granite forms from crystallization of magma instead. What evidence would convince you that this must be true? Where would you go to look to find this evidence?

3.3 Discovering Plate Tectonics

The first maps of the Atlantic Ocean began to be produced almost 500 years ago, and some early observers—including one of the founders of modern science, Francis Bacon (1561–1626)—noted that the strong similarity in shape of the coastlines of eastern South America and western Africa was probably not mere coincidence. The thinking of the times suggested that the Atlantic might have carved a path between these two continents, like a giant river—perhaps the work of Noah's flood! By 1910, however, enough scientific evidence had been collected to begin examining this similarity more seriously. Alfred Wegener (1880–1930), a German meteorologist, presented a paper to the Frankfurt Geological Society in which he expressed his opinion that Africa and South America, together with other continents, had once been joined in a single, gigantic landmass—a supercontinent—which somehow split apart. The fragments then separated thousands of kilometers to produce the world map that we see today. He called his hypothesis *die Verschiebung der Kontinente,* meaning "continental displacement," or *continental drift.* Powerful geological evidence existed to support his idea, including fossils, mineralized zones, and glacial scourings, which matched "hand-in-glove" when the Southern Hemisphere continents and India, in particular, were reassembled, like pieces of a gigantic jigsaw puzzle (**Figures 1, 2,** and **3**). But what force of nature drove these landmasses apart in the first place? Wegener suggested that the simple rotation of Earth caused them to break up and separate, each continent plowing its way through the ocean floor, rumpling up a mountain range along its leading edge like the folds in a rug slid against a wall. Unfortunately, no geophysical evidence could be found to support this explanation. Coupled with the facts that evidence of geological matches between continents in the Northern Hemisphere was less clear, and that Wegener lacked professional credibility because his area of formal training was not in geology, the continental drift hypothesis foundered for decades.

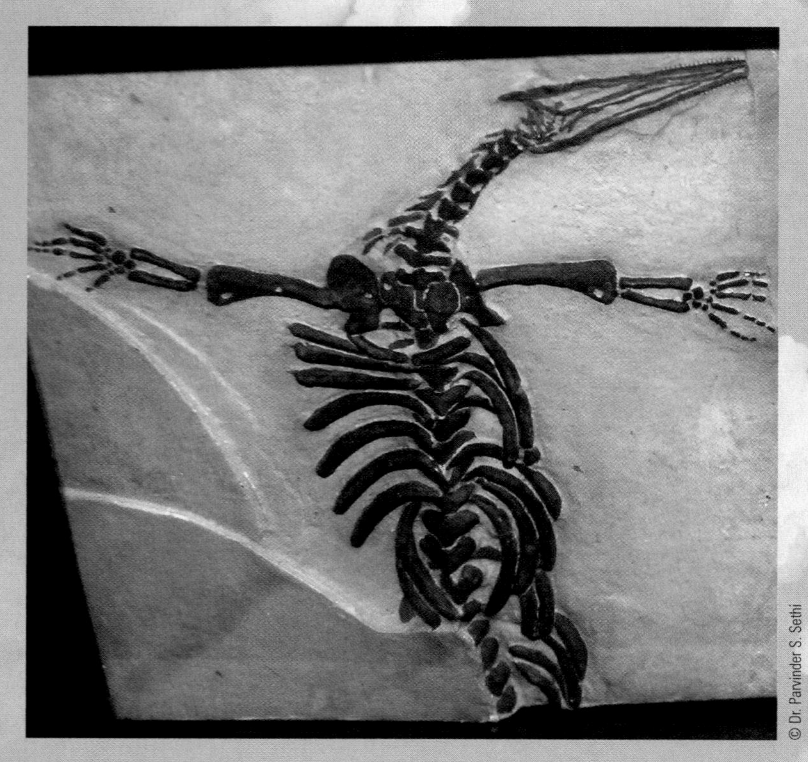

Figure 1 Fossil of the freshwater reptile Mesosaurus, a Gondwanan predator that ranged from a few tens of centimeters to 2 meters (6.6 feet) in length at adulthood.

But Wegener would be vindicated, thanks to the development of new technologies and further mapping of the ocean floor, especially during the early Cold War (1945–1968), when oceanographers revealed in great detail the mid-ocean ridge system and **marine trenches** bordering volcanic and seismically restless island chains and continental margins (see Chapters 4 and 5). Simultaneously, the study of lava flows using the magnetometer led to several important breakthroughs in our understanding of how the ocean floor forms. These studies focused on a property inherent in volcanic rocks called **magnetic polarity**—the alignment of tiny, iron-bearing mineral grains within Earth's magnetic field as molten lava cools into solid stone. The minerals act as natural compass needles, recording the direction toward the north magnetic pole existing at the time that they harden into place. Because Earth as a whole acts like a giant bar magnet, with a north (positive) magnetic pole presently situated in northern Canada and a south (negative) pole near Antarctica, any change in the past orientation of the planetary magnetic field will be recorded in the polarity of ancient rocks formed just before and after a field change.

Geophysicists correctly surmised that the positions of Earth's magnetic poles must have drifted almost continuously within just a few degrees latitude of the planet's geographic (rotational) poles since the planet first formed. However, to their surprise, researchers discovered that the magnetic polarity of different equivalent-aged ancient rocks on opposite sides of the Atlantic mysteriously seemed to give *different locations* for each of the magnetic poles, many localities quite far from Earth's rotational axis. In other words, there appeared to be *multiple* north and south magnetic poles, *each paired to a separate continent* and each

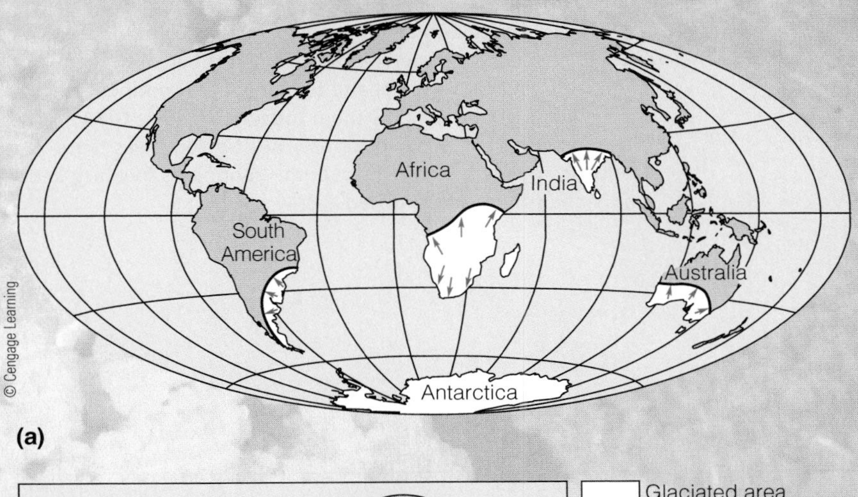

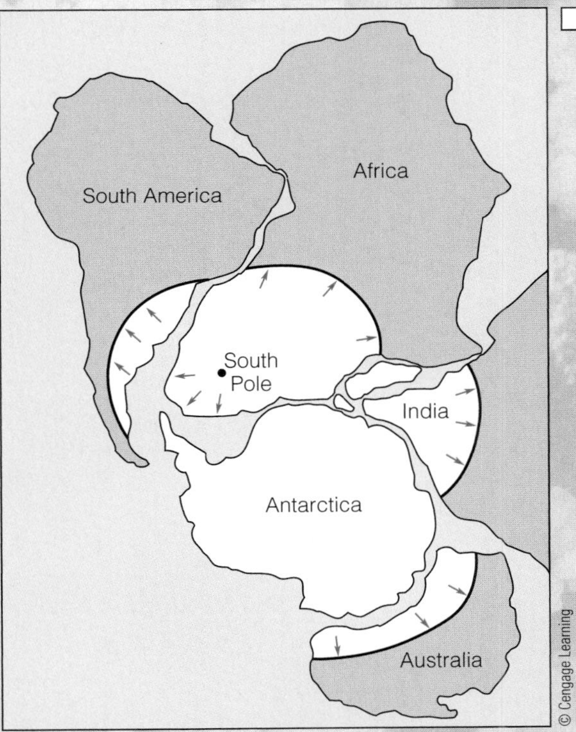

Figure 2 (a) The directions of glacial grooves preserved in bedrock on continents as they are now positioned show the direction of glacier movement during Permian (Gondwana succession) time. For a larger glacier or a number of glaciers to produce the directions of the observed grooves would otherwise require a source area in the Indian and South Atlantic oceans. (b) With Gondwana continents reunited and an ice sheet over a south pole in present-day South Africa, the directions of glacier motion are resolved. (c) Glacial grooves in Hallett Cove, Australia. Formed in Permian time, these grooves are more than 200 million years old. The width of the area shown in the photo is approximately 2 meters (6.6 feet).

with its own history of wandering through many degrees of latitude and longitude across the face of the globe. The farther back in time one looked, the farther from the present magnetic pole positions the inferred paleomagnetic poles seemed to be, and the more widely they were separated from one another. Try as people might, this interpretation of magnetic **polar wandering** made no sense at all—the laws of physics stated that there could be only two magnetic poles at any given time in a planetary magnetic field and, if the continents had remained fixed in position relative to one another as they assembled, the rocks of the same age anywhere in the world should yield identical magnetic polarity data concentrating at the two roughly stable polar positions seen today. But, if the Atlantic was made smaller and the continents rotated back closer to one another *at specific increments of time,* then the polar wandering discrepancies vanished. This could only mean one thing: Wegener was correct—the continents *had* drifted apart (**Figure 4**). Still, few scientists could see how drift could be mechanically possible, and Wegener's ideas remained doubtful throughout the heyday of polar wandering studies in the 1950s.

In fact, a highly speculative and untested explanation already existed. Professor Arthur Holmes, a British geologist and possibly the finest geology teacher of his day, suggested in 1926–1927 that the continents do not plow through seafloor crust like ships moving through sea ice, as Wegener said, but instead move apart as new ocean floor grows *in between* them, driven by convective upwelling in Earth's

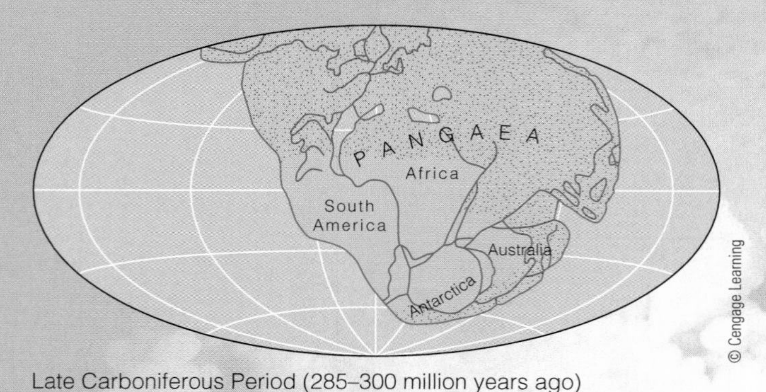

Figure 3 Earth 285 to 300 million years ago according to Alfred Wegener. The supercontinent Pangaea began to break up about 100 million years ago, and the individual continents, roughly outlined as we know them today, have since then "drifted" to their present positions. The stippled areas represent widespread shallow seas over the continents in which sediments and fossils accumulated.

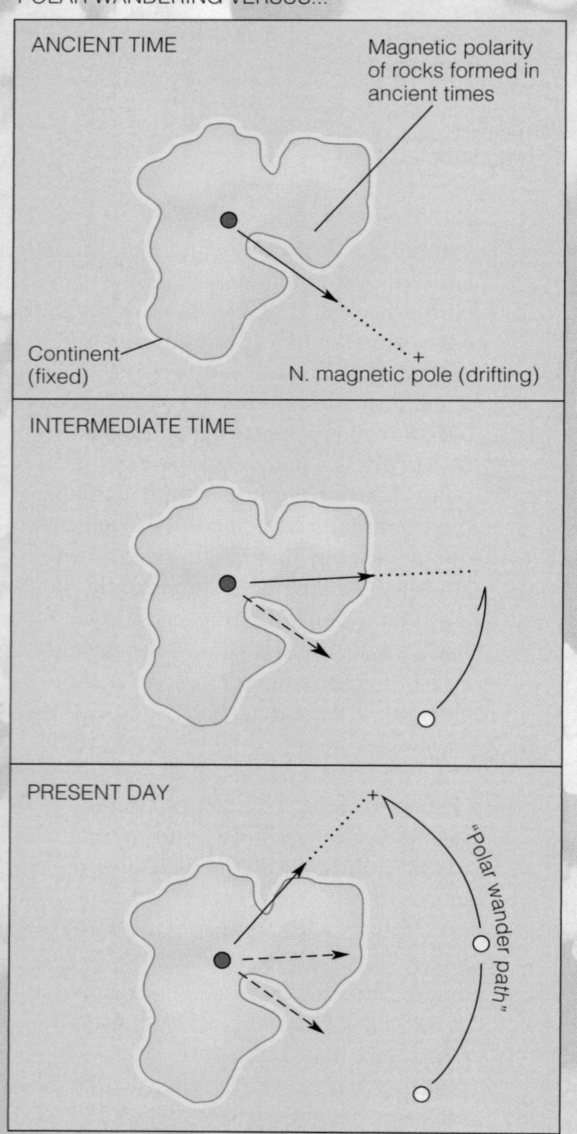

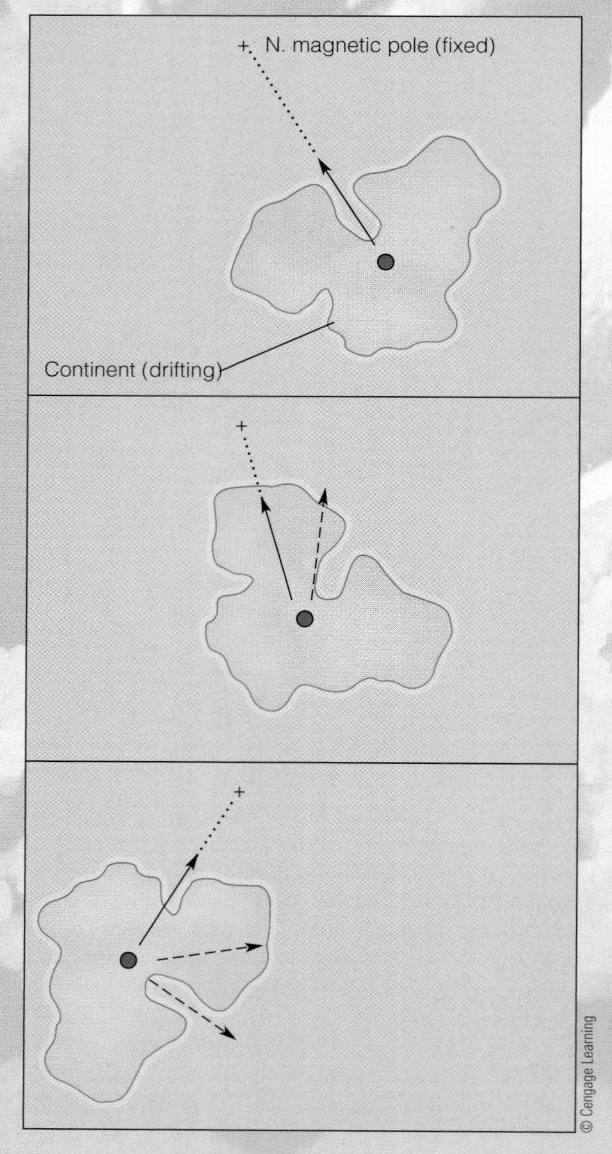

Figure 4 Two interpretations from the 1950s of magnetic polarities found in ancient rocks. The column to the left presumes that the continents remained fixed as the magnetic poles drifted to account for the variation seen in rock magnetization over time. The column to the right presumes just the opposite—that the continents drift through a magnetic field with fixed poles. Both models lead to the same present-day magnetic pattern for any given continent.

deep interior. It is not the continents that are moving so much as the ocean floor, *in which the continents are just passively embedded.* The Atlantic and Indian oceans began as weak seams fracturing Wegener's supercontinent of Pangaea. The seams evolved into narrow seaways as basaltic lavas erupted to fill the gaps between the separating continents, the line of volcanic activity being the axial rifts at the crest of the mid-ocean ridge. Conversely, at the marine trenches, aging oceanic crust sank back into the planet, ultimately to be recycled inside Earth's mantle. This kept Earth at constant volume, so that it wouldn't grow infinitely larger (a physical impossibility) as new oceanic crust continuously formed.

Seizing on the controversial implications of apparent magnetic polar wandering and his own bathymetric studies in 1960, American geologist Harry Hess revived and advanced Holmes's speculations, calling the growth of new oceanic crust seafloor spreading. Perhaps wary of the heavy professional criticism heaped on Alfred Wegener, Hess referred to his own ideas as "geopoetry," but there was growing evidence that he—and Professor Holmes—were correct. In addition to polar wandering studies, seismology began providing support. In 1927, Japanese seismologist Kiyoo Wadati discovered that a zone of frequently occurring earthquakes, including some very destructive and powerful tremors, slants under Japan from the trench along the archipelago's eastern coast. California seismologist Hugo Benioff (1899–1968) later mapped similar earthquake zones worldwide, which are now called *Benioff–Wadati zones.* They mark the paths of subducted oceanic material extending hundreds of kilometers inside Earth, mostly around the edge of the Pacific Basin—clear evidence for Holmes's mantle recycling notion. To lend further support, radiometric dating of oceanic crust showed that the oldest intact seafloor rocks are less than 200 million years old. Where did all of the older oceanic crust go, if it wasn't swallowed back up by the planet? (On the continents, after all, rocks as much as 4 billion years old could be found.)

Researchers were close to verifying continental drift by means of seafloor spreading, but they still needed more evidence. They found it again as they continued studying patterns of natural rock magnetization. Geophysicists discovered that every few hundred thousand to million years, something truly spectacular happens; Earth's magnetic poles swap positions entirely—in other words, the entire magnetic field suddenly flips over, and the poles become *reversed* (**Figure 5**). As a reversal approaches, the strength of the field simply weakens to the point of temporarily disappearing. When the field returns, each pole reappears in the opposite hemisphere from its former position. We regard the present time as having a "normal" magnetic field just because of our place in time, with the field position "reversed" in other epochs. (One can only speculate about what happens, during a reversal, to birds and marine creatures that may depend on the magnetic field to assist their migrations.)

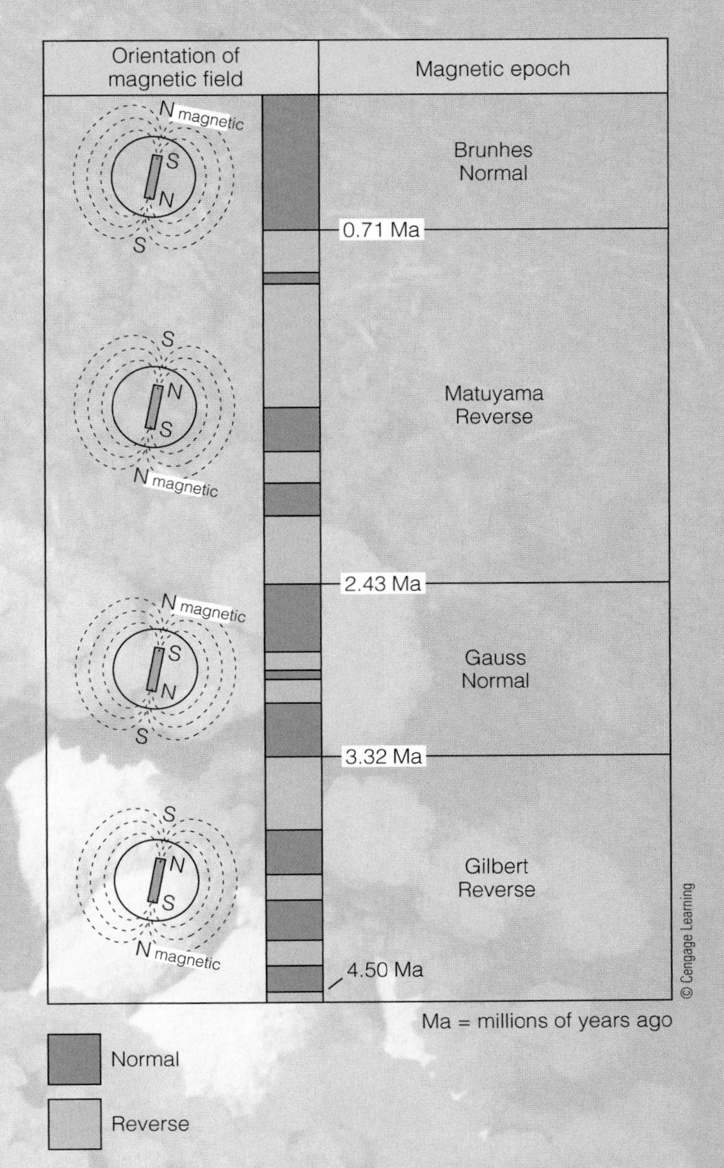

Figure 5 Earth's magnetic field is much like the field that would be generated by a bar magnet inclined at 11° from Earth's rotational axis. Reversals of the field have occurred periodically, leaving "fossil" magnetism in rocks that can be dated. Each epoch is named after an important researcher of earth magnetism.

In 1963, British researchers Fred Vine and Drummond Matthews and Canadian Lawrence Morley proved that all

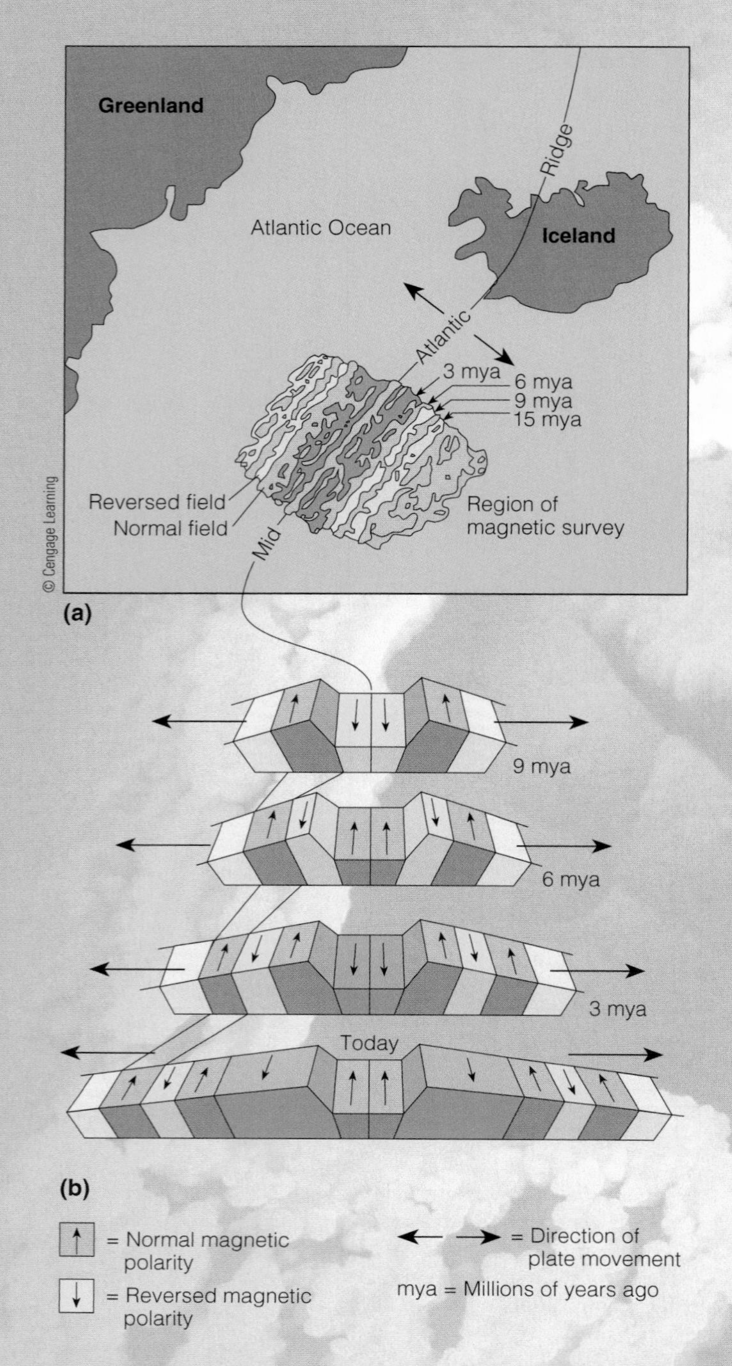

Figure 6 (a) Symmetric pattern of magnetic variation across part of the Mid-Atlantic Ridge southwest of Iceland. Colors indicate seafloor rocks with matching polarity. (b) The origin of the magnetic anomaly pattern across mid-ocean ridges. Lava erupting along the crest of the ridge records Earth's prevailing magnetic polarity as it cools and solidifies. Seafloor spreading then carries the rock away from the ridge in both directions as new lava erupts on the ridge axis.

the ancient basaltic lavas that formed when Earth's magnetic field was the same as today's yield a similar, strong magnetic signal—or *positive anomaly*—whereas those that formed when the field was the opposite of today's field yield a much weaker signal—a *negative anomaly.* Vine, Matthews, and Morley reasoned that a *striped anomaly pattern of seafloor magnetization* would develop if seafloor spreading really happened and oceanic crust grew progressively older with distance from the mid-ocean ridge. The pattern should be geometrically symmetrical about the ridge, assuming that seafloor spreading rates are the same on both sides of its crest. Oceanographers leapt to the challenge and, in 1966, confirmed the presence of this pattern in the bed of the North Atlantic near Iceland (**Figure 6**). Because more than 170 reversals are known to have occurred during the past 76 million years, the striped magnetic record of seafloor spreading is strikingly detailed, and geologists the world over seemed to gasp all at once with excitement. In 1968, on the basis of these discoveries, the theory of plate tectonics was launched. A revolution in scientific thinking was taking place.

At the end of the 1960s, as human beings first reached the Moon, they also attained their first holistic understanding of the geological world. Coupled with the first pictures of Earth from space, scientists began to look at our planet in terms of a single system of integrated processes, rather than an assortment of seemingly unrelated phenomena. Of course, many mysteries and fascinating details remain to be examined. But after centuries of groping in myth, ignorance, and speculation, at last we have a handle on which to build new understanding of our remarkable planet.

questions to ponder

1. What evidence would exist to support the existence of plate tectonics if Earth lacked a magnetic field? Would that evidence be convincing?
2. How can we be reasonably sure that rates of radioactive decay on which we rely to calculate the ages of ancient rocks and fossils have remained nearly constant through geological time?

PHOTO GALLERY **a rock collection**

Specific rocks have had historical, aesthetic, and practical significance to humans; Plymouth Rock, the Rock of Gibraltar, the biblical Rock of Ages, and the Rosetta Stone are examples. Various geological agents of erosion have shaped the rocks of the Grand Canyon, Bryce Canyon, Yosemite, and other national parks into areas that we admire for their great beauty. Early humans lived in rock caves and threw rocks to defend themselves against their enemies and to kill animals for food. Over time, humans came to build crude rock shelters, and then rock fences, and eventually beautiful castles and abbeys. Today, rock and rock products are important mineral resources, used mostly as building materials. The beauty and usefulness of rocks are celebrated in these photographs.

1 Ayers Rock, the world's largest sandstone monolith, is one of Australia's leading tourist attractions. An erosional remnant of a previous highland, it rises abruptly above the surrounding flat land. In geological terminology, it is called an *inselberg* (German, "island mountain") because of its resemblance to an island rising from the sea. Aborigines (native Australians) call Ayers Rock *Uluru*. In 1958, Ayers Rock and the rocks of the Olgas nearby were incorporated into Uluru National Park. (a) Ayers Rock rises 348 meters (1,141 feet) above the surrounding plain and is scaled by 80% of the visitors to the park. The Olgas, a circular grouping of more than 30 red domes, is a subdued but still spectacular sandstone inselberg in Uluru National Park. Its highest point, Mount Olga, rises 460 meters (1,500 feet) above the plain. The Olgas' Aboriginal name is *katajuta*, "many heads." Ayers Rock is in the background. © Reed Wicander

Strangely eroded rock formations, called "tents," dominate the landscape in Kasha-Katuwe Tent Rocks National Monument in New Mexico.
© 2007/ROUL TOUZON/National Geographic Image Collection

2 Sixty-foot-high representations of the heads of four presidents were carved into Mount Rushmore's Harney Peak Granite in southwestern South Dakota between 1927 and 1941. The great work was possible because the Precambrian granite is relatively homogeneous (uniform) and not riddled with defects. Why would such a project not have been considered if Mount Rushmore were composed of slate, shale, or schist? Can you identify the presidents honored here? © Reed Wicander

3 President Theodore Roosevelt (left) and naturalist John Muir at Yosemite Valley in 1903. The rock on which they were photographed is Cathedral Peak granite, one of the many granite intrusions that make up the Sierra Nevada batholith. Note the rounded exfoliation domes in the background (see Chapter 6). Theodore Roosevelt Collection, Harvard College Library

4 A skillfully assembled sandstone wall in Yorkshire, England. Yorkshire is the largest county (shire) in England, and literally thousands of kilometers of these walls are found there. This commonly occurring sandstone breaks readily into platy and blocky pieces, which makes it perfect for wall building. B. Pipkin

GALLERY visions of how the earth works

The evolution of modern plate tectonic theory started with Alfred Wegener's 1912 hypothesis of continental drift. Almost 50 years later, the idea of seafloor spreading was proposed. Although scientists were skeptical at first, this far-reaching concept led to the unifying theory of plate tectonics.

…ting out a long Arctic winter in …lfred Wegener formulated his hy… …t continents move about on the face … ideas were considered so outra… …nore than 50 years passed before …alized he was on the right track. …urce, NY

6 Evidence of colliding and sliding tectonic plates. Collision of the Indian and Asian plates produced the great Himalaya Mountains. The rocks of Nepal's Dhaulagiri I (8,172 meters [26,810 feet]), one of the highest peaks in the world, were uplifted and contorted by the continent–continent collision. The steep-sided south wall is 15,000 feet high. This view shows folded strata in the mountain's lower slope. Rodney Supple

7 (a) “Black smokers” of mid-ocean ridge. Hot brine loaded with minerals leached from basalt emanates from vents on ridges and builds tall “chimneys.” (b) Giant tube worms thrive around the “smokers” and live on bacteria in the Galapagos Rift. (c) Clams more than 25 centimeters (10 inches) long inhabit the warm waters near the vents in the Galapagos Rift. © Woods Hole Oceanographic Institution

SUMMARY The Solid Earth

Earth Materials

Crust
Outermost rocky layer of Earth composed of rocks that are aggregates of minerals.

Minerals
Naturally occurring inorganic substances with a definite set of physical properties and a narrow range of chemical compositions.

Classification
Chemistry of anions (negative ions or radicals), such as oxides, sulfides, and carbonates. Silicates are the most common and important mineral group and are composed of silica tetrahedra $(SiO_4)^{-4}$ units that are joined to form chains, sheets, and networks.

Identification
Physical properties such as hardness, cleavage, crystal structure, fracture pattern, and luster.

Rocks
Consolidated or poorly consolidated aggregates of one or more minerals or organic matter.

Classes
Igneous—rock formed by crystallization of molten or partially molten material.
Sedimentary—layered rock resulting from consolidation and lithification of sediment.
Metamorphic—preexisting rock that has been changed by heat, pressure, or chemically active fluids.

Classification
Classified by texture and composition.

Structures
Many planar rock structures may be viewed as defects along which landslides, rockfalls, or other potentially hazardous events may occur. Stratification in sedimentary rocks and foliation in metamorphic rocks are common planar features. Faults and joints occur in all classes of rock.

Rock Cycle
A sequence of events by which rocks are formed, altered, destroyed, and reformed as a result of internal and external earth processes.

Earth's Deep Structure and Plate Tectonics

The Deep Interior
Earth is a layered planet, with a solid metallic inner core and molten outer core, enclosed in an iron- and magnesium-rich mantle, which is capped by a thin, silicate-rich crust. The uppermost mantle and crust also constitute the lithosphere, which sits atop a partly molten mantle layer called the *asthenosphere*.

Plate Interactions
Earth's rocky surface is divisible into seven major plates of lithosphere, which move independently of one another. That movement is enabled by convection in Earth's underlying mantle, especially in the asthenosphere. Three types of plate interaction occur in response to convection. Plates shift at (1) divergent boundaries, where newer oceanic crust grows; and (2) convergent boundaries, where oceanic crust is consumed, or continental plates can come together to create huge mountain chains. Along (3) transform boundaries, the plates simply scrape past one another. The locations of volcanoes and earthquakes are largely fixed by plate boundaries.

Hot Spots
Hot spots are areas where heat concentrates in a single area beneath the lithosphere, generating magmas that erupt to form chains of volcanoes and volcanic plains as plates drift over them. Unlike the plates, hot spots are fixed, or nearly fixed, with respect to Earth's deep interior.

Geological Time

Relative Time
The sequential order of geological events established by using basic geological principles or laws.

Geological Time Scale
The division of geological history into eons, eras, periods, and epochs. The earliest 88% of geological time is known informally as Precambrian time. *Precambrian* rocks are generally not fossiliferous. Rocks of the Cambrian period and younger contain good fossil records of shelled and skeletonized organisms. Environmental geology is concerned mainly with geological events of the present epoch, the Holocene, and the one that preceded it, the Pleistocene.

Absolute Age
Absolute dating methods use a natural "clock." For long periods of time, such clocks are known rates of radioactive decay. A parent radioactive isotope decays to a daughter isotope at a rate that can be determined experimentally. This established rate, expressed as a half-life, enables us to determine the age of a sample of material.

Age of Earth
Comparing present uranium-to-lead ratios with those in iron meteorites, we can extrapolate backward and obtain an Earth age of 4.6 billion years.

Rocks and Crust
The oldest known in-place rocks are 3.96 billion years old. Dating of minerals derived from older crust yields an age of 4.0 to 4.3 billion years for stable continental crust.

KEY TERMS

absolute-age dating
alpha decay
amorphous
aphanitic texture
asthenosphere
atom
atomic mass
atomic number
batholith
beta decay
biogenic sedimentary rock
chemical sedimentary rock
clastic sedimentary rock
cleavage
convergent boundaries
core
craton
crust
dip
divergent boundaries
element
extrusive rock
fault
felsic
foliation
fracture
half-life
hot spots
igneous rock
intrusive
ion
island arcs
isotope
joint
lava
lithified
lithosphere
luster
mafic
magma
magnetic polarity
mantle
mantle lid
marine trenches
metamorphic rock
mid-ocean ridge
mineral
Mohs' hardness scale
nucleus
phaneritic texture
plate tectonics
plutonic
polar wandering
pyroclastic
radiometric dating
recrystallization
relative-age dating
rock
rock cycle
seafloor spreading
sedimentary rock
stratification
strike
subduction
transform boundaries

STUDY QUESTIONS

1. What physical and chemical factors are the bases of the rock classification system? What are the three classes of rocks, and how does each form? Identify one characteristic of each class that usually makes it readily distinguishable from the others.
2. How and where do batholiths form? What type of rock most commonly forms in batholiths?
3. What are some of the planar surfaces in rocks that may be weak and thus lead to various types of slope failure?
4. How may structures in sedimentary rocks be used to reconstruct past environments?
5. What is the most common mineral species? Name several rocks in which it is a prominent constituent.
6. What is mineral cleavage? How can it serve as an aid in identifying minerals?
7. The most common intrusive igneous rock is composed mostly of: (1) the most common mineral, and (2) a mineral of the most common mineral group. Name the rock and its constituent minerals.
8. What mineral is found in both limestone and marble?
9. Explain how Earth's interior is layered.
10. How and why does plate tectonics take place?
11. Hot spots cannot be described as "plate interaction," yet they provide some evidence for plate motions. Explain.
12. How may absolute-age dating techniques be used to the betterment of human existence?
13. Earth is 4.6 billion years old, and humans have been on Earth for less than two hundred thousand years. What changes of a global nature have humans invoked on Earth's natural systems (water, air, ice, solid Earth, and biology) in this short length of time? Which impacts are reversible, and which ones cannot be reversed or mitigated?

REFERENCES AND FURTHER READINGS

Albritton, C. C. 1984. Geologic time. *Journal of Geological Education* 32(1): 29–47.

Bowring, S. A., I. S. Williams, and W. Compston. 1989. 3.96 Ga gneisses from the Slave province, Northwest Territories, Canada. *Geology* 17: 971–975.

Brown, V. M., and J. A. Harrell. 1991. Megascopic classification of rocks. *Journal of Geological Education* 39: 379.

Dietrich, R. V., and Brian Skinner. 1979. *Rocks and rock minerals.* New York: John Wiley & Sons, 336 p.

Eicher, D. L. 1976. *Geologic time,* 2nd ed. Englewood Cliffs, NJ: Prentice-Hall, 150 p.

Gunter, Mickey Eugene. 1994. Asbestos as a metaphor for teaching risk perception. *Journal of Geological Education* 42: 17.

Harvey, Carolyn, and Mark Rollinson. 1987. *Asbestos in the schools.* New York: Praeger, 133 p.

Hurley, Patrick. 1959. *How old is the Earth?* New York: Anchor Books, 160 p.

Libby, W. E. 1955. *Radiocarbon dating.* Chicago: University of Chicago Press, 491 p.

Mackenzie, Fred T. G., and Judith A. Mackenzie. 1995. *Our changing planet: An introduction to Earth System science and global environmental change.* Englewood Cliffs, NJ: Prentice-Hall, 200 p.

National Research Council, National Academy of Sciences. 1993. *Solid-earth sciences and society.* Washington, DC: National Research Council Commission on Geosciences, Environment, and Resources, National Academy of Sciences Press, 346 p.

Newcott, William R. 1998. Return to Mars. *National Geographic* 194(2): 2–29.

Nuhfer, E. B., R. J. Proctor, and Paul H. Moser. 1993. *The citizen's guide to geologic hazards.* Arvada, CO: American Institute of Professional Geologists, 134 p.

Oreskes, N. 2003. Plate tectonics: An insiders history of the modern theory of the Earth—Seventeen original essays by scientists who made Earth history. Boulder, Colorado: Westview Press, 424 p.

Parker-Pope, T. 1997, April 1. Cat litter breathes new life into region of bentonite mines. *Wall Street Journal.*

Skinner, H., W. Catherine, and Malcom Ross. 1994, January. Geology and health. *Geotimes* 11–12.

________. 1994, January. Minerals and cancer. *Geotimes* 13–15.

Snow, T. P. 1993. *Essentials of the dynamic universe.* St. Paul, MN: West, 592 p.

Access an interactive eBook and chapter-specific interactive learning tools, including flashcards, quizzes, videos and more in your Geology CourseMate, accessed through CengageBrain.com.

Earthquake damage from the magnitude 7.2 earthquake that struck Duzce, Turkey. The epicenter was located 100 kilometers (60 miles) from the city, but inadequate construction and underlying geologic conditions still contributed to serious destruction.

Have you ever
wondered...

4

Earthquakes and Human Activities

A bad earthquake at once destroys the oldest associations; the world, the very emblem of all that is solid, had moved beneath our feet like crust over a fluid.

—Charles Darwin, following the 1835 Concepción, Chile, earthquake

1. What causes earthquakes?

2. Why some parts of the world suffer so much greater damage and loss of life from earthquakes than others?

3. How people *measure* earthquakes and know where they originated?

4. What you should do during an earthquake?

5. What you should look out for after an earthquake?

A Caribbean Catastrophe

If there is one country in the Western Hemisphere that can ill afford to suffer a natural disaster, it is politically unstable and dirt-poor Haiti. The catastrophic magnitude-7 earthquake that hit Haiti on 12 January 2010 was a cruel blow, the astounding devastation resulting from a lethal combination of factors: The country is the poorest in the Western Hemisphere; it has limited, poorly developed infrastructure; and it is near the boundary of two major tectonic plates. The earthquake focus originating at the shallow depth of about 10 kilometers (6 miles) increased the intensity of the ground shaking, which caused the damage to be more localized and catastrophic along the fault. The earthquake was felt throughout Haiti and the Dominican Republic, and as far away as eastern Jamaica, parts of Puerto Rico, the Bahamas, and even Caracas, Venezuela, and Tampa, Florida. The catastrophic main quake was followed by many aftershocks, and on the morning of 20 January, 8 days later, stunned Haitians were jolted awake by another powerful (magnitude-6) tremor, which reignited their fears and sent panic-stricken citizens running into the streets.

Figure 1 Struggling for survival, a girl scours rubble from the January 2010 earthquake in Haiti.

The epicenters of the major shock and most aftershocks were only 15 kilometers (10 miles) from densely populated Port-au-Prince, a city with virtually no building codes and shabbily constructed buildings. Much of Port-au-Prince's 2 million residents lived in shacks that cling precariously to steep hillsides. Entire ravines that had been packed with residential shanties were swept bare by quake-triggered landslides. Even the parliament building and part of the magnificent Presidential Palace collapsed. The precise death toll may never be known, but it has been estimated at 200,000 or more, with thousands of others injured. Houses, schools, the national cathedral, hospitals, and the regional headquarters of the United Nations collapsed within a few murderous seconds (**Figure 1**). Damage to the city's main jail allowed convicted criminals to escape and contribute to the pain and anxiety that stalked the streets.

Geologists have long worried about the region's seismic potential because of its complex geology, which includes a variety of plate boundary interactions. The Island of Hispaniola (shared by the Dominican Republic and Haiti) lies on a small sliver of Earth's crust, the *Gonave microplate,*

Figure 2 Generalized tectonic map of the northern Caribbean region.

squeezed between the Caribbean and North American plates. Fault movement on this plate boundary is dominantly left-lateral with some compression. Relative motion is about 20 millimeters (0.8 inch) per year. The earthquake resulted from tectonic processes acting along this border. The strain between the two plates is partitioned between two major east-west trending strike-slip faults that slice through Haiti, the Enriquillo-Plantain Garden fault zone (EPGFZ) in the south, and the Septentrional fault system in the north. The seismic record reveals that the 12 January 2010 event occurred on the EPGFZ, which accommodates about 7 millimeters (0.25 inch) per year (about a third) of the relative motion between the Caribbean and North American plates (**Figure 2**).

In the popular imagination, Haiti is not known as a region subject to large earthquakes, like, say, Indonesia, or Italy. But memories are short, and in fact, the country does have a history of large shocks. Haiti experienced disastrous earthquakes in 1618, 1673, 1684, 1751, 1761, 1770, and 1860, although none of these events has been confirmed by field studies as due to movement on the EPGFZ. Worldwide, magnitude-7 earthquakes are not uncommon, with about 20 earthquakes of this magnitude somewhere in the world each year.

This chapter explores the relation of faults and plate tectonics to seismicity and examines the causes of building collapse and the remedial measures that may be taken to avoid the costly devastation that accompanies large earthquakes. Please note that we still have much to learn, and that knowledge we gain and usefully apply (often beginning with abstract academic theory) can literally mean life or death for many people.

questions to ponder

1. What construction techniques can be used to avoid the structural collapse of buildings in an earthquake?
2. What *preparations* can be undertaken to limit the devastation in areas of earthquake risk?
3. Is it possible to predict earthquakes?

CourseMate Connection

Start off this chapter by listening to a podcast with National Geographic Grantee Lisa Ely as she discusses what studying past quakes can tell us about future tremblers. **Log in to the CourseMate at www.cengagebrain.com to listen now.**

The Nature of Earthquakes

FAULTS ARE FRACTURES IN EARTH'S CRUST along which great masses of the rock slide past one another either so gradually as to be imperceptible (**fault creep**) or, more often, in violent, sudden jerks we call *earthquakes*. The types of faults and the Earth forces that cause them are shown in **Figure 4.1**.

Most earthquake activity is a result of plate tectonic interactions (see Chapter 3). Because the buildup of **stress** (force per unit area) that produces **strain** (deformation)

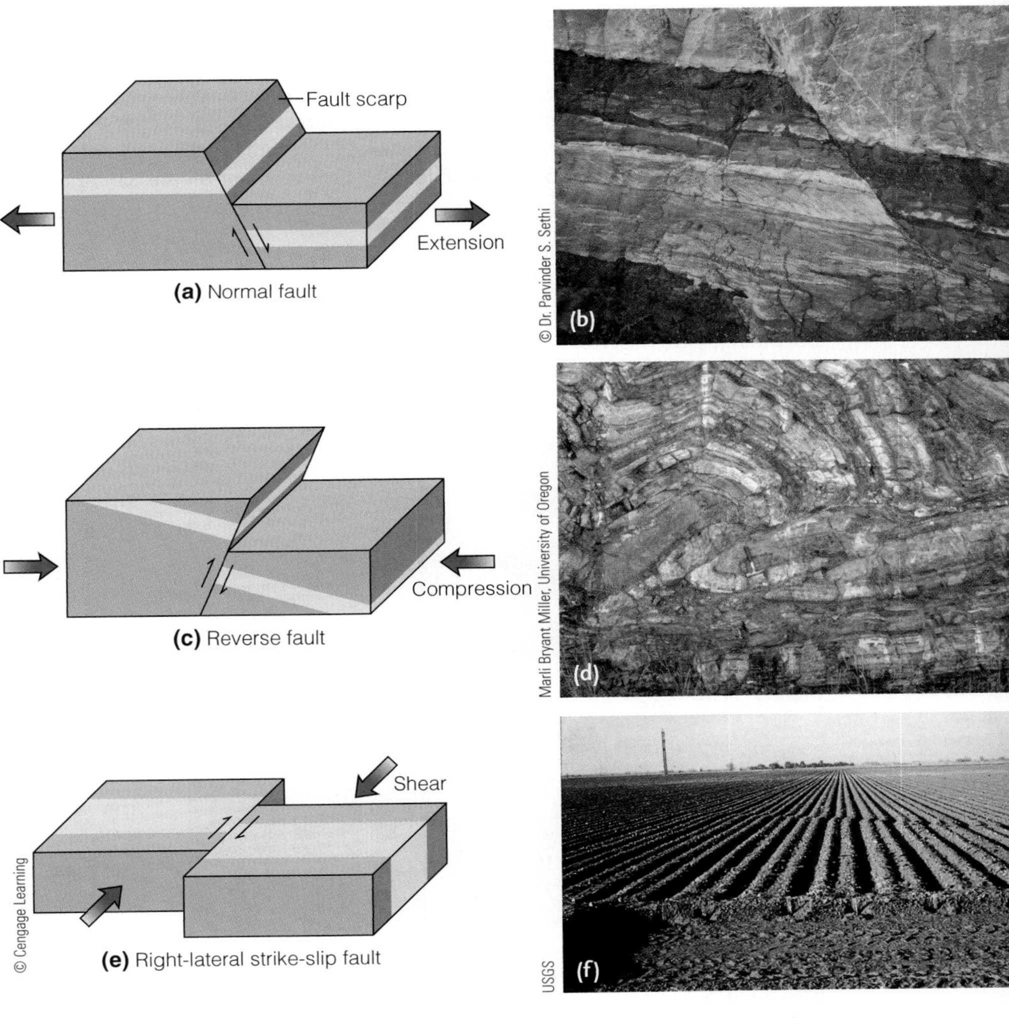

Figure 4.1 (a) Normal fault geometry and (b) example in a road cut on I-40 near Kingman, Arizona. (c) Reverse fault geometry and (d) a low-angle reverse (or thrust) fault, the upper block having moved up and over the lower block, exposed in the road cut on the Coquihalla Highway (Canada Highway 5) near Kamloops, British Columbia. (e) Right-lateral strike-slip fault geometry and (f) a plowed field displaced by a strike-slip fault in the Imperial Valley, California, in 1979.

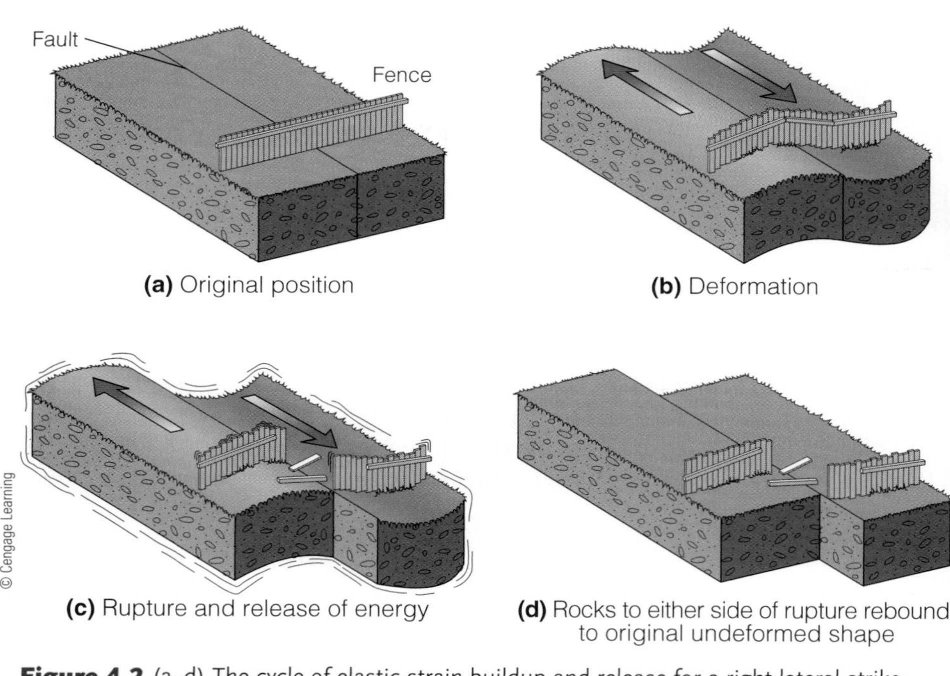

Figure 4.2 (a–d) The cycle of elastic-strain buildup and release for a right-lateral strike-slip fault according to Reid's elastic rebound theory of earthquakes. At the instant of rupture (c), energy is released in the form of earthquake waves that radiate out in all directions, (e) Right-lateral offset of a fence by 2.5 meters (8 feet) by displacement on the San Andreas Fault in 1906; Marin County, California. (www.cengage.com/permissions)

in rocks can be transmitted long distances through the crust, an active fault does not ordinarily occur *exactly* at a plate boundary, but it generally occurs in the vicinity (within a few hundred kilometers) of one. The mechanism by which

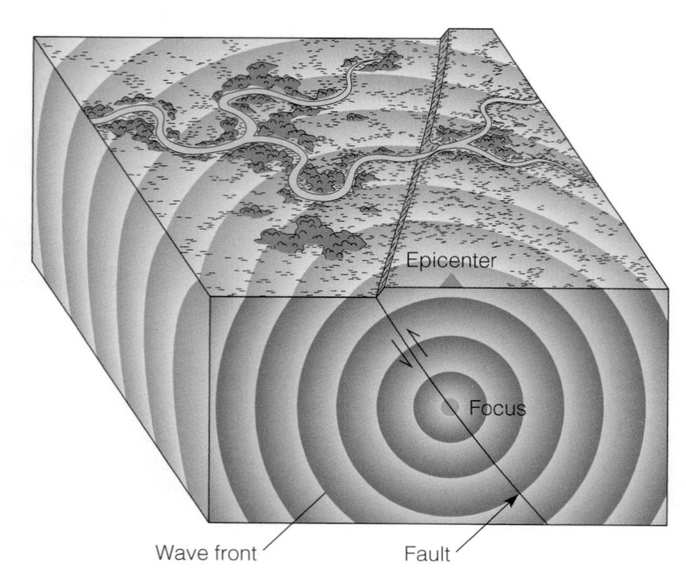

Figure 4.3 The focus of an earthquake is the point in the Earth where fault rupture begins, and the epicenter is directly above the focus at the ground surface. Seismic-wave energy moves out in all directions from the focus. (Source: U.S. Census Bureau. 2010. Historical Income Tables, Table POV-02, People in Families, by Family Structure, Age, and Sex, Iterated by Income-to-Poverty Ratio and Race: 2009. www.census.gov)

stressed rocks store up strain energy along a fault to produce an earthquake was explained by Harold E. Reid in a report to the State of California after the great San Francisco temblor of 1906, which together with an ensuing fire destroyed about 20% of the City by the Bay. Reid proposed the **elastic rebound theory** (**Figure 4.2**), which states that when sufficient strain energy has accumulated in rocks, they may rupture rapidly—just as a rubber band breaks when it is stretched too far—and the stored strain energy is released as vibrations that radiate outward in all directions (**Figure 4.3**). (Think of the shock that passes through your fingers when you release that rubber band!) Most earthquakes are generated by movements on faults within the crust that do not produce ruptures at the ground surface. A fault rarely slips entirely during an earthquake, but the area of interface that does can cover an area of many square kilometers underground, and is called the *rupture zone*. By studying the seismic energy released by an earthquake, we can pinpoint within the Earth where the rupture starts, a position called the **focus**, or *hypocenter*. Even easier to locate is the **epicenter**, or the location on the surface where the seismic energy first reaches Earth's surface and begins spreading outward across the landscape (see Figure 4.3). **Seismology**, the branch of geology dedicated to studying seismic waves and energy release, even allows us to determine *how* faults slip during earthquakes. The "how" of fault rupture is termed an earthquake's *focal mechanism*.

Because rock tends to flow plastically rather than accumulate elastic strain energy under the high-temperature and -pressure conditions of Earth's deep interior, most earthquake foci lie within the crust at depths of less than 15 kilometers (9 miles). The deepest foci in the world occur within subducted oceanic slabs, which rupture as they heat up and sink into the mantle. Subduction zone earthquakes may originate at depths of as great as 700 kilometers (450 miles).

The vibration produced by an earthquake is complex, but it can be described as three distinctly different types of waves (**Figure 4.4**). Primary waves, **P-waves**, and secondary waves, **S-waves**, are generated at the focus and travel through the interior of Earth; thus, they are known as **body waves**. They are designated P- and S-waves because they are the first (primary) and second (secondary) waves to arrive from distant earthquakes. As these body waves strike the planet's surface, they generate **surface waves** that are analogous to water ripples on a pond.

P-waves (see Figure 4.4a) are **longitudinal waves**; the solids, liquids, and gases through which they travel are alternately compressed and expanded in the same direction the waves move. Their speed depends on the resistance to change in volume (in compressibility) and shape of the material through which they travel. *Sound* is a nonseismic equivalent of P-waves that people have evolved to sense with their ears. P-waves travel about 300 meters (1,000 feet) per second in air, 300 to 1,000 meters (1,000–3,000 feet) per second in soil, and faster than 5 kilometers (3 miles) per second in solid rock at the surface.

P-wave velocity increases with depth, because the materials composing the mantle and core become less compressible with distance inside the Earth. This increase causes the waves' travel paths to be bowed downward as they move through the planet. P-waves travel underground at a speed of about 10 kilometers (6 miles) per second. Breaching the ground surface, they have very small amplitudes (motion) and cause little property damage. Although they are physically identical to sound waves, they vibrate at frequencies below what the human ear can detect. The earthquake noise that has been reported could be P-waves of a slightly higher frequency or some other, related vibrations whose frequencies are in the audible range—at least for humans. One of the authors of this book (Richard Hazlett) has felt many earthquakes in Hawaii, and on multiple occasions from the dead of night has heard birds calling and dogs barking moments before actually feeling the ground tremble. These animals plainly sensed "something wicked this way coming" seconds before he did.

S-waves (see Figure 4.4b) are **transverse (shear) waves**. They produce ground motion perpendicular to their direction of travel. This causes the rocks through which they travel to be twisted and sheared. These waves travel much as energy does when a person takes one end of a garden hose or a loosely hanging rope and gives it a vigorous flip. They can travel only through a medium that resists shearing—that is, *the rupturing of a given mass into two parts that then slide past one another, each moving in the opposite direction.* Thus, S-waves travel only through solids, because liquids and gases can't be "ruptured" and have no shear strength (just try piling water in a mound). S-wave ground motion may be largely in the horizontal plane and can result in considerable property damage. S-wave speed is approximately 5.2 kilometers (3.2 miles) per second from distant earthquakes; hence they *arrive* after the P-waves, with rare exception (see later).

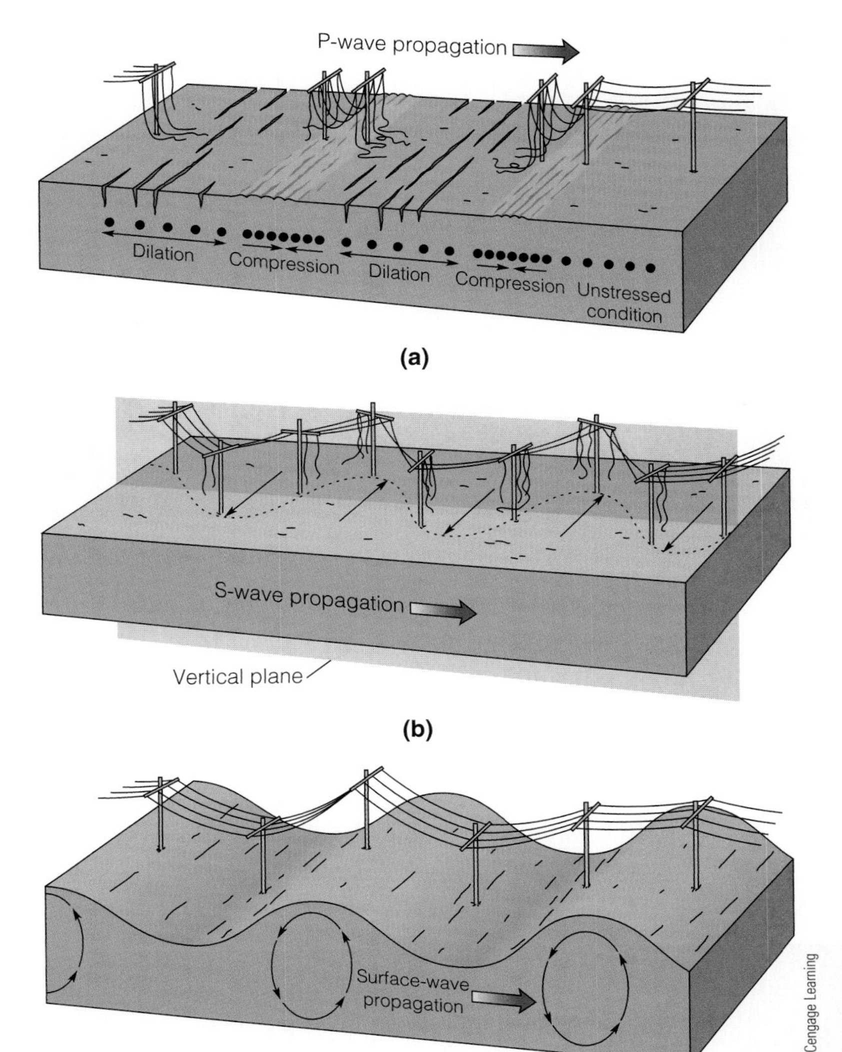

Figure 4.4 Ground motion during passage of earthquake waves. (a) P-waves compress and expand the Earth. (b) S-waves move in all directions perpendicular to the wave advance, but only the horizontal motion is shown in this diagram. (c) Surface waves create surface undulations that result from a combination of the retrograde elliptical motions of Rayleigh waves (shown) and Love waves, which move from side to side at right angles to the direction of wave propagation. (The S-waves shown in (b) are actually behaving as Love waves.)

Unexpended P- and S-wave energy bouncing off Earth's surface generates the complex surface waves (see Figure 4.4c) that are so important for us surface dwellers. These waves produce a rolling motion at the ground surface. In fact, slight motion sickness is a common response to surface waves of long duration. The waves are the result of a complex interaction of several wave types, the most important of which are *Love waves*, which exhibit horizontal motion nor-

B. Pipkin

Figure 4.5 A seismometer (left) and seismogram (right) used to demonstrate a method of detecting and recording earthquakes. The record on the seismogram is not an earthquake; it represents the seismometer's detection of vibrations from students' footsteps amplified about 2,000 times.

mal to the direction of travel, and *Rayleigh waves*, which exhibit retrograde (opposite to the direction of travel) elliptical motion in a plane perpendicular to the ground surface. Surface waves have amplitudes that are greatest in near-surface unconsolidated layers and are the most destructive of the earthquake waves.

James L. Stanfield/National Geographic/Getty Images

Figure 4.6 The first earthquake detector, invented about 130 C.E. by Chinese scholar Chang Heng. Balls held in the dragons' mouths were aligned with a pendulum inside the vase. When an earthquake occurred, the pendulum swung and pushed balls into the mouths of the frogs aligned with the pendulum's swing. If the frogs facing north and south contained balls, for example, Chang Heng could say that the earthquake epicenter was north or south of the instrument.

Love and Rayleigh waves also travel quite quickly through bedrock—in excess of 2 kilometers/second (1.2 mi/s), although not as fast as P and S waves. (The acronym to remember in terms of relative seismic wave speeds, from fastest to slowest, is "PSLR.") Their speeds, like those of body waves, vary according to the composition of the material through which they are moving, with some components traveling slowly enough in poorly consolidated materials to be easily visible during earthquakes—below the threshold of seismometer detection. Many observers (including your author Hazlett) have witnessed unforgettably creepy, wavelike motions rolling down streets and sidewalks once earthquakes are under way. During the 1964 Alaska earthquake, observers reported visible waves roughly a meter (yard) high that traveled through the town of Valdez, causing cracks to open and close with the passage of each wave. Jets of compressed groundwater squirted as high as 6 meters (20 feet) into the air as some fissures shut tight.

Seismic Wave Periods

Wave period as applied to the ocean is *the time it takes for two successive wave crests (or troughs) to pass a particular point*, like a buoy floating in the water. **Wave frequency** is a related concept referring to *the number of wave crests, or troughs, that pass a particular point per unit interval of time*. Just as these physical terms apply to moving waves at sea, so too do they have their seismic equivalents—with each "wave crest," in this case, represented by a fresh pulse of seismic energy, and each "particular point" being a seismometer station. Seismic waves may be classified as **short** and **long period**. Near the source of an earthquake, short-period S- and P-waves predominate. At greater distances from a seismic focus, wave energy diminishes and period increases. Earthquake waves traveling all the way through Earth's core lose their S-waves entirely (because the outer core is liquid magma). The very-long-period seismic waves that may pass all the way through the planet from very big earthquakes to seismometers on the other side are called **teleseisms**. Arrival of teleseismic waves at a station tells seismologists that a potentially disastrous earthquake has occurred somewhere far away—at least hundreds and possibly thousands of kilometers distant.

Locating the Epicenter

A **seismograph** is an instrument designed specifically to detect, measure, and record vibrations in the Earth's crust (**Figure 4.5**). It is relatively simple in concept (**Figure 4.6**), but some very sophisticated electronic systems have been developed. Seismic data are recorded onto a **seismogram** (**Figure 4.7**). Because seismographs are extremely sensitive to vibrations of any kind, they are installed in quiet areas, such as abandoned oil and water wells, cemeteries, and parks.

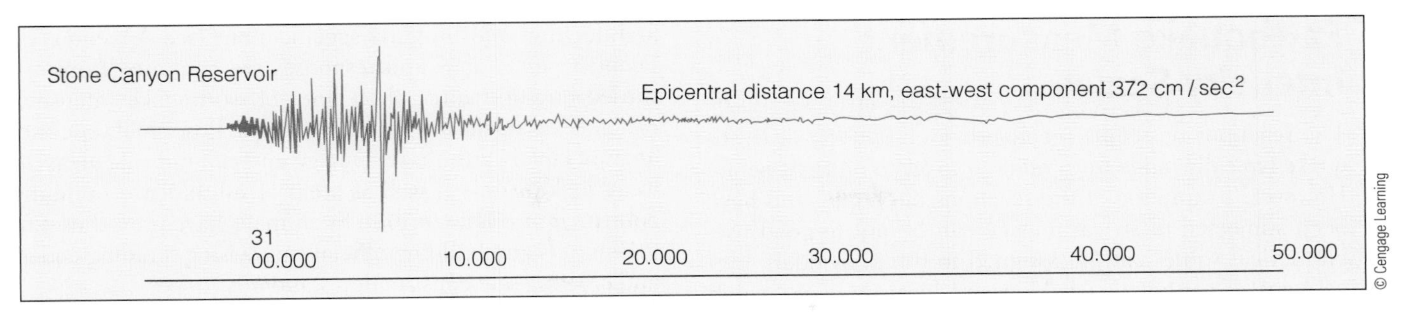

Figure 4.7 A seismogram of the main shock of the Northridge earthquake, 17 January 1994. The time in minutes and seconds after 4:00 a.m. appears at the bottom. Distance from the epicenter, and direction and ground acceleration at the recording site are also indicated. (Acceleration is explained in the next section.)

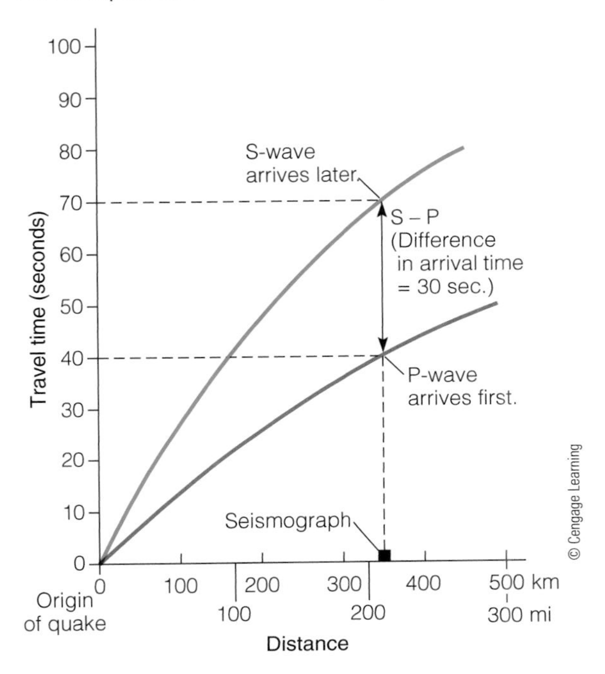

Figure 4.8 Generalized graph of distance versus travel time for P- and S-waves. Note that the P-wave has traveled farther from the origin of the earthquake than the S-wave at any given elapsed time. The curvature of both wave paths is due to increases in velocity with depth (distance from epicenter).

Figure 4.9 An earthquake epicenter can be closely approximated by triangulation from three seismograph stations.

When an earthquake occurs, the distance to its epicenter can be approximated by computing the difference in P- and S-wave arrival times at various seismograph stations. Although the method actually used by seismologists is more accurate and determines the depth and location of the quake's focus and its epicenter, what is described here shows how seismic-wave arrival times can be used to determine the distance to an epicenter. Because it is known that the two kinds of waves are generated simultaneously at the earthquake focus and that P-waves travel faster than S-waves, it is possible to calculate where the waves started.

Imagine that two trains leave a station at the same time, one traveling at 60 kilometers (36 miles)/hour and the other at 30 kilometers (18 miles)/hour, and that the second train passes your house one hour after the first train goes by. If you know their speeds, you can readily calculate your distance from the train station of origin as 60 kilometers (36 miles). The relationship between distance to an epicenter and arrival times of P- and S-waves is illustrated in **Figure 4.8**. The epicenter will lie somewhere on a circle whose center is the seismograph and whose radius is the distance from the seismograph to the epicenter. The problem is then to determine where on the circumference of the circle the epicenter is. More data are needed to learn this—specifically, the distances to the epicenter from two other seismograph stations. Then the intersection of the circles drawn around each of the three stations—a method of map location called *triangulation*—specifies the epicenter of the earthquake (**Figure 4.9**).

Earthquake Measurement Intensity Scales

The reactions of people (geologists included) to an earthquake typically range from mild curiosity to outright panic. However, a sampling of the reactions of people who have been subjected to an earthquake can be put to good use. Numerical values can be assigned to the individuals' perceptions of earthquake shaking and local damage, which can then be contoured on a map. One **intensity scale** developed for measuring these perceptions is the 1931 **modified Mercalli scale** (*MM*). The scale's values range from *MM* = I (denoting not felt at all) to *MM* = XII (denoting widespread destruction), and they are keyed to specific U.S. architectural and building specifications (see Appendix 4) People's perceptions and responses are compiled from returned questionnaires; then lines of earthquake intensity, called **isoseismals**, are plotted on maps. Isoseismals enclose areas of equal earthquake damage and can indicate areas of weak rock or soil, as well as areas of substandard building construction (**Figure 4.10a**). Such maps have proved useful to planners and building officials in revising building codes and creating safe construction standards.

consider this . . . You are shopping for a house in earthquake country. How could an isoseismal map of an earthquake whose epicenter was near a house you think you would like to buy be useful? What patterns would you look for when examining the isoseismal map?

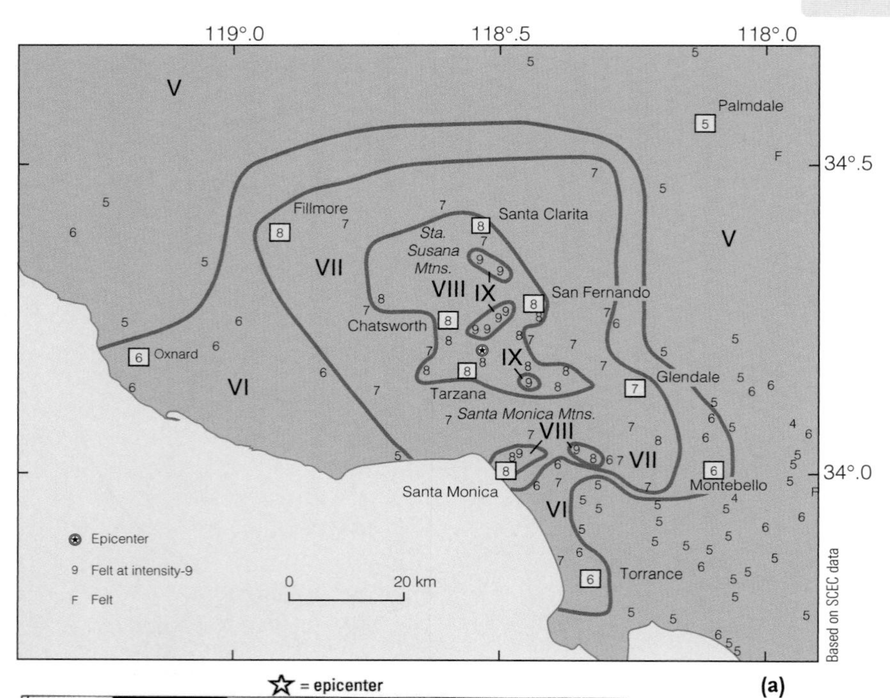

Recently, a method of plotting shaking intensities in the United States has evolved that is faster and compares favorably with the questionnaire method. It is the Community Internet Intensity Maps (CIIM), an example of which is shown in Figure 4.10b. The CIIM takes advantage of the Internet, and the time to generate intensity maps drops from months to minutes. The responses, which can take place within 3 minutes of the event, are summarized by computer, and an intensity number is assigned to each postal ZIP code. This method is particularly useful in areas with sparse seismograph coverage. Although perhaps less simple than the map generated by the traditional *MM* intensities obtained by using postal questionnaires, the CIIM values agree well and have actually proved more reliable in areas of low shaking. However, neither method allows comparison of earthquakes with widely spaced epicenters because of local differences in construction practices and local geology.

☆ = epicenter

PALMDALE, CASTAIC, FILLMORE, NEWHALL, NORTHRIDGE, THOUSAND OAKS, PASADENA, MALIBU, LOS ANGELES, LAKEWOOD

km 0 10 20 30

1837 responses in 452 ZIP-code areas. Max intensity: IX

INTENSITY	I	II-III	IV	V	VI	VII	VIII	IX	X+
SHAKING	Not felt	Weak	Light	Moderate	Strong	Very strong	Severe	Violent	Extreme
DAMAGE	none	none	none	Very light	Light	Moderate	Moderate/Heavy	Heavy	Very Heavy

ZIP-code areas with no data shown in gray.

USGS

(b)

Richter Magnitude Scale: The Best-Known Scale

The best-known measure of earthquake strength is the **Richter magnitude scale**, which was introduced in 1935 by Charles Richter and Beno Gutenberg at the California Institute of Technology. It is a scale of the energy released by an earthquake; thus, in contrast with the intensity scale, it may be used to compare earthquakes in widely separated geographical areas. The Richter value is calculated by measuring the maxi-

Figure 4.10 (a) Isoseismal map of the near-field (close to the epicenter) modified Mercalli scale intensities of the 1994 Northridge, California, earthquake. Intensity-IX values are assigned where there are spectacular partial collapses of modern buildings, destroyed wood-frame buildings, and collapses of elevated freeways (see Appendix 4). (b) Community Internet Intensity Map (CIIM) for the same earthquake. The main difference in appearance is due to the fact that the intensities were not contoured but assigned to ZIP code areas of the respondents.

mum amplitude of the ground motion as shown on the seismogram using a specified seismic wave, usually the surface wave. Next, the seismologist "corrects" the measured amplitude (in microns) to what a "standard" seismograph would record at the station. After an additional correction for distance from the epicenter, the Richter **magnitude** is the common logarithm of that ground motion in microns. For example, a magnitude-4 earthquake is specified as having a corrected ground motion of 10,000 microns ($\log_{10}$ of 10,000 = 4), and thus can be compared with any other earthquake for which the same corrections have been made. It should be noted that, because the scale is logarithmic, each whole number represents a ground shaking (at the seismograph site) 10 times greater than the next-lower number. Thus, a magnitude 7 produces 10 times greater shaking than a magnitude 6, 100 times that of a magnitude 5, and 1,000 times that of a magnitude 4. Total energy released, in contrast, varies logarithmically as some exponent of 30. Compared with the energy released by a magnitude-5 earthquake, an $M = 6$ releases 30 times (30^1) more energy; an $M = 7$ releases 900 times (30^2) more; and an $M = 8$ releases 27,000 (30^3) times more energy (**Figure 4.11a**).

The Richter scale is open-ended; that is, theoretically it has no upper limit. However, rocks in nature do have a limited ability to store strain energy without rupturing, and no earthquake has yet been observed with a Richter magnitude greater than 8.9—although some geophysicists estimate that large meteorite or asteroid strikes on Earth's surface early in its Precambrian history, such as the Vredefort impact in South Africa, could have released energies scaling up to magnitude 14.

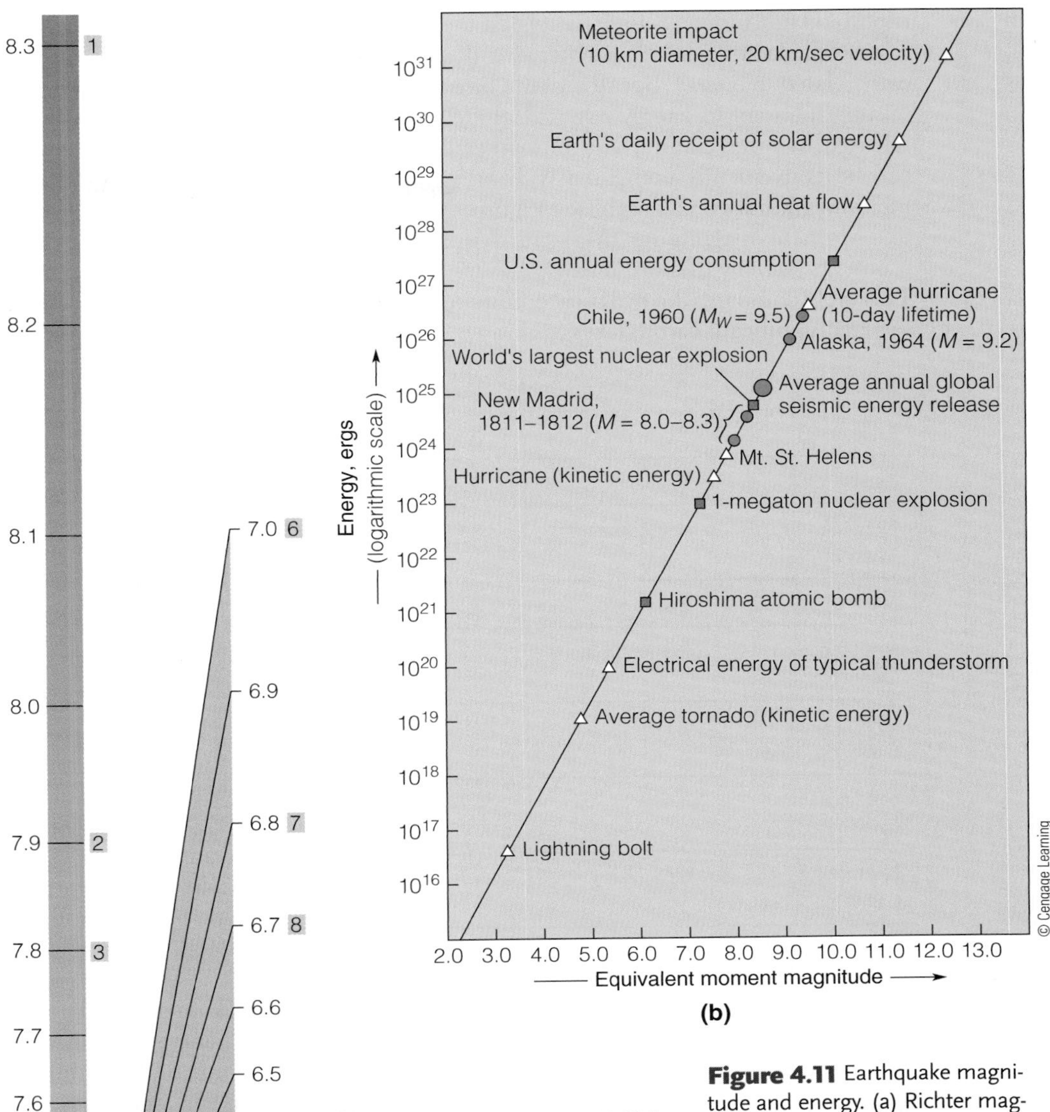

Figure 4.11 Earthquake magnitude and energy. (a) Richter magnitudes of nine selected earthquakes, 1906–1994. (b) Equivalent moment magnitudes of energetic human-caused events (squares), natural events (triangles), and large earthquakes (circles). The erg is a unit of energy, or work, in the metric (cgs) system. To lift a 1-pound weight 1 foot requires 1.4×10^7 ergs.

Moment Magnitude: The Most Widely Used Scale

Seismologists have abandoned Richter magnitudes in favor of **moment magnitudes (M_w or M)** for describing earthquakes. The reason is that Richter magnitudes do not accurately portray the *energy* released by large earthquakes on faults with great rupture lengths. The seismic waves used to determine the Richter magnitude come from only a small part of the fault rupture, and hence cannot provide an accurate measure of the total seismic energy released by a very large event.

Moment magnitude is derived from seismic moment, M_0 (in dyne centimeters), which is proportional to the average displacement (slip) on the fault *times* the rupture area on the fault surface *times* the rigidity of the faulted rock. The amount of seismic energy (in ergs) released from the ruptured fault surface is linearly related to seismic moment by a simple factor, whereas Richter magnitude is logarithmically related to energy. Because of the linear relationship of seismic moment to energy released, the equivalent energy released by other natural and human-caused phenomena can be conveniently compared with an earthquakes' moment magnitudes on a graph (see Figure 4.11b).

Moment magnitudes (M_w) are derived from seismic moments (M_0) by the formula: $M_w = (2/3 \log M_0 - 10.7)$. This table compares the two most commonly used scales for selected significant earthquakes:

Earthquake, Year	Richter Magnitude	Moment Magnitude
Chile, 1960	8.3	9.5
Alaska, 1964	8.4	9.2
New Madrid, Missouri, 1812	8.7 (estimated)	8.1 (estimated)
Mexico City, 1985	8.1	8.1
San Francisco, California, 1906	8.3 (estimated)	7.7
Loma Prieta, California, 1989	7.1	7.0
San Fernando, California, 1971	6.4	6.7
Northridge, California, 1994	6.4	6.7
Kobe, Japan, 1995	7.2 JMA*	6.9

*The Richter scale is not used in Japan. The official earthquake scale in Japan is the scale developed by the Japanese Meteorological Agency (JMA).

Confusion arises because some of the earthquake magnitudes are reported as Richter magnitude (M_L), teleseismic body-wave magnitude (m_b), duration magnitude (m_d), surface-wave magnitude (M_s), or moment magnitude (M_w or M). This is simplified somewhat, however, because most newspapers and the Internet report earthquakes as either Richter or moment magnitude.

As of 2012, the Chilean earthquake of 1960 ($M_w = 9.5$) still holds the record for the greatest seismic moment and energy release ever measured; that is, it had the longest fault rupture and the greatest displacement. The Alaskan earthquake of 1964 had the highest Richter magnitude of the 20th century and theoretically more severe ground motion than the event in Chile.

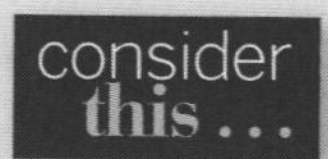

"Then the Lord rained upon Sodom and Gomorrah brimstone and fire...And he overthrew those cities, and all the plain about..." (Genesis 19: 24–25). Sodom and Gomorrah are believed to have been located near the modern industrial city of Sedom in Israel at the south end of the Dead Sea. Could some geological event have caused the destruction of these towns? If so, what events might have been possible?

Fault Creep, the "Nonearthquake"

Fault creep is a well-known but poorly understood form of fault movement. The Hayward Fault, which is part of the San Andreas Fault system, is an example of a creeping fault. It is just east of the San Andreas Fault itself and runs through the cities of Hayward and Berkeley, California. The steady slippage along this fault causes displacements of millimeters per year and is one way that this plate boundary accommodates motion between the Pacific and North American plates. The San Andreas Fault is famously known as the "boundary" between these two plates—the Earth's largest—but, in fact, the fault accounts for only about *half* the net relative plate motion. The rest is allocated piecemeal to parallel or subparallel faults like the Hayward mostly to the east, some of them hundreds of kilometers away.

The obvious benefit of a creeping fault—lack of earthquakes—does not come without a price, as the residents of Hollister, California, can testify. The creeping Calaveras Fault runs through their town and, as in Hayward, it gradually displaces curbs, sidewalks, residences, and at least one winery.

A good strategy in areas subject to fault creep is to map the surface trace of the fault so that it can be avoided in future construction. The University of California-Berkeley Memorial Stadium was built on the Hayward fault before the hazard of fault creep was recognized. The fault creep caused damage to the drainage system, which requires periodic maintenance. It is ironic that creep damage occurs at an institution that installed the first seismometer in the United States.

Why do some faults continuously, silently creep, whereas others exhibit more normal stick-slip, earthquake-generating behavior? The answer seems to lie in the condition of the bedrock that they cross. The Hayward and Calaveras faults slice through large masses of dark gray–green serpentinite, a type of low-grade metamorphic rock that is notoriously weak and deformable, whereas the San Andreas Fault generally cuts through masses of much stronger granite, gneiss, and sedimentary rock—prone to "catch" before breaking.

Sometimes knowledge of the bedrock geology can provide important clues to expected long-term fault behavior.

Seismo-luminosity

One of the strangest phenomena associated with earthquakes is the appearance of lights resembling aurora high in the night sky before, during, or after large earthquakes. The lights reportedly have a bluish or white coloration, and can last for minutes even tens of kilometers (miles) from an epicenter. Not all large nighttime earthquakes report them, or at least generate reports of their occurrence. Nor do earth scientists know why they occur. In fact, scientists tended to dismiss them as folklore before their being first photographed during the 1965–1967 Matsushiro earthquake swarm in Japan. Possibly rapid changes in tectonic stress associated with large earthquakes also impact Earth's magnetic field, which, in turn, causes light-generating flux in the planet's ionosphere. This is not a hypothesis that can be easily tested, however!

Forensic Use of Seismic Records

Seismic records can be used in conjunction with other methods to distinguish between earthquakes and nuclear explosions. The basis for the seismic method is that nuclear explosions are more concentrated in space and time and, therefore, excite more short-period seismic waves. This method requires extensive analysis of seismograms and is one of the tools that will be used to measure compliance by signers of the Comprehensive Test Ban Treaty. The following are select anomalous seismic events from 1995 to 2003, some of which were believed to be nuclear explosions. Note that the 12 August 2000 event marks the tragic explosion and sinking of the Russian submarine *Kursk*, and the 1 February 2003 event marks the *Columbia* space shuttle disintegration. Although not on this list of special studies compiled by Lynn Sykes of Lamont-Doherty Earth Observatory of Columbia University, the collapse of the Twin Towers on 11 September 2001 was also recorded on seismographs at Lamont-Doherty.

Location	Date	Type Of Event
Novaya Zemlya	13 January 1996	Earthquake
Kola Peninsula	19 September 1996	Chemical explosion
Germany	11 September 1996	Mine collapse
East Kazakhstan	3 August 1997	Chemical explosion
India	13 May 1998	Alleged nuclear explosions
Novaya Zemlya	23 September 1999	Alleged nuclear explosion
Kursk, Barents	12 August 2000	Chemical explosion
Novaya Zemlya	23 February 2002	Earthquake
U.S. space shuttle *Columbia*	1 February 2003	Explosion 100 seconds long

Seismic Design Considerations

Ground Shaking

Ground movements, particularly rapid horizontal displacements, are most damaging during an earthquake. Shear and short-period surface waves are the culprits, and the potential for them must be evaluated and then design specifications established. The design objective for earthquake-resistant buildings is relatively straightforward: Structures should be designed to withstand the maximum potential horizontal ground acceleration expected in the region. Engineers call this acceleration **base shear**, and it is usually expressed as a percentage of the *acceleration of gravity* (g). On Earth, g is the acceleration of a falling object in a vacuum (9.8 meters/second^2, or 32 feet/second^2). In your car, it is equivalent to accelerating from a dead stop through 100 meters in 4.5 seconds. An acceleration of 1 g downward produces weightlessness, and a fraction of 1 g in the horizontal direction can cause buildings to separate from their foundations or to collapse completely. An analogy is to imagine rapidly pulling a carpet on which a person is standing; most assuredly, the person will topple.

The effect of high horizontal acceleration on poorly constructed buildings is twofold. Flexible-frame structures may be deformed from cube-shaped to rhomb-shaped, or they may be knocked off their foundations (**Figure 4.12**). More rigid, multistory buildings may suffer "story shift" if floors and walls are not adequately tied together (**Figure 4.13**). The result is a shifting of floor levels and the collapse of one floor on another like a stack of pancakes. Such structural failures are not survivable by inhabitants, and they clearly illustrate the adage that *"earthquakes don't kill people; buildings do."* Damage caused by shearing forces can be mitigated by bolting frame houses to their foundations and by *shear walls*. An example of a shear wall is plywood sheeting nailed in place over a wood frame, which makes the structure highly resistant to deformation. Wall framing, usually two-by-fours, should be nailed very securely to a wooden sill that is bolted to the foundation. Diagonal bracing and blocking also provide shear resistance (**Figure 4.14**). L-shaped structures may suffer damage where they join, as each wing of the structure vibrates independently. Such damage can be minimized by designing *seismic joints* between the building wings or between adjacent buildings of different heights. These joints are filled with a compressible substance that will accommodate movement between the structures.

Wave period is an important consideration when assessing a structure's potential for seismic damage, because if a building's natural period of vibration is equal to that of seismic waves, a condition of resonance exists. **Resonance** occurs when a building sways in step with an oscillatory seismic wave. As a structure sways back and forth under resonant conditions, it gets a push in its direction of sway with the passage of each seismic wave. This causes the sway to increase, just as pushing a child's swing at the proper moments makes it go higher with each push. Resonance also may cause a wineglass to shatter when an operatic soprano sings just the right note or frequency.

consider this . . . Over time, tectonic stresses strain the rocks along a fault, and increase the chance of sudden rupture and a potentially damaging earthquake. For this reason, it is generally believed that a region that has had many small earthquakes (magnitude < 4.0) has a low probability for a large earthquake. For a magnitude-7.0 earthquake with a 50-year recurrence interval, how many magnitude-4.0 earthquakes would be required to "balance the energy books"? How many small earthquakes a day is this? Is this belief realistic in a seismically active region, based on your calculations?

In some instances, buildings of a certain size suffer serious damage, whereas smaller or larger structures in the surrounding area ride through an earthquake intact because of differences in resonance. Low-rise buildings have short natural wave periods (0.05–0.1 second), and high-rise buildings have long natural periods (1–2 seconds). Therefore, short-period (high-frequency) waves affect single-family dwellings and low-rise buildings, and long-period (low-frequency) waves affect tall structures. Close to an earthquake epicenter, high-frequency waves dominate; thus, more extensive home and low-rise damage can be expected. With distance from the epicenter, the short-period wave energy is absorbed or dissipated, resulting in the domination of longer-period waves—much to the chagrin of people who happen to be inside skyscrapers at the time.

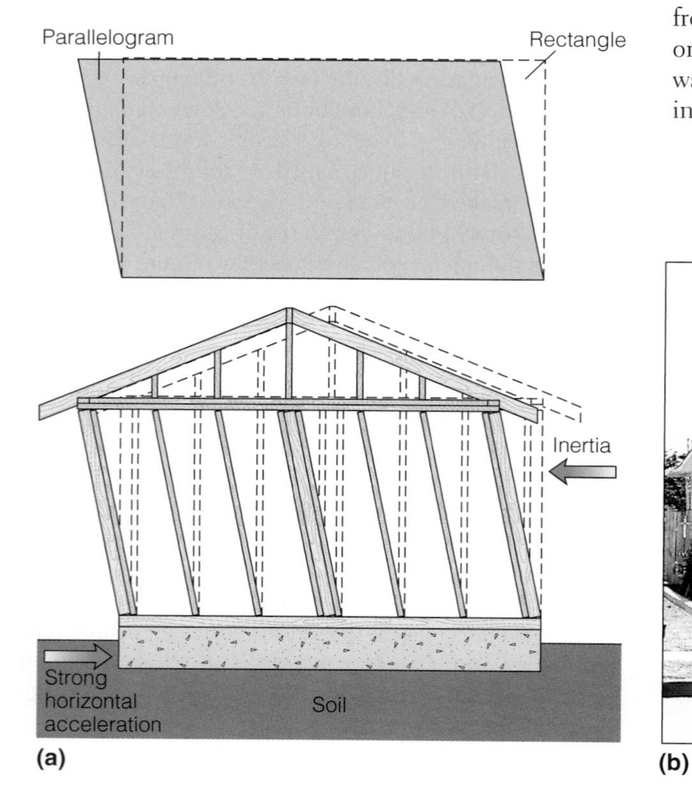

Dorothy Start, Cypress College

Major damage here

Cripple

(c)

Dennis Fox

Figure 4.12 (a) Strong horizontal motion can deform a house from a cube to a rhomboid or knock it from its foundation completely. (b) A Coalinga, California, frame house that was deformed by the magnitude-6.3 earthquake in 1983. (c) A cripple-wall consists of short, vertical members that connect the floor of the house to the foundation. Cripple-walls are common in older construction. (d) A Watsonville, California, house that was knocked off its cripple-wall base during the Loma Prieta earthquake of 1989. Without exception, cripple-walls bent or folded over to the north relative to their foundations.

Figure 4.13 Examples of structural collapse in earthquakes because of poor design or poor construction. (a) Total vertical collapse as a result of "story shift"; Mexico City, 1985. Such structural failures are not survivable. (b) The collapsed Justice Department in Port-au-Prince, Haiti, on 12 January 2010. Men gather helplessly outside the government building where countless victims lie buried. Many major government structures like this one, the national palace, the national cathedral, and many hotels in the capital and the neighboring towns, were constructed with weak concrete cut with sand. Furthermore, most of the collapsed buildings lacked steel reinforcing rods ("rebar"), which would have supported them during earthquakes.

Landslides

Thousands of landslides are triggered by earthquakes in mountainous or hilly terrain; there were an estimated 17,000 during the Northridge earthquake alone (see Case Study 4.1). The 1989 Loma Prieta earthquake caused landslides in the Santa Cruz Mountains and adjacent parts of the California Coast Ranges. Such areas are slide prone under the best of conditions, and even a small earthquake will trigger many slope failures. In the greater San Francisco Bay area, there was an estimated $10 million damage to homes, utilities, and transportation systems because of landslides and surficial ground failures resulting from the Loma Prieta earthquake.

One of the worst earthquake-triggered tragedies in the United States occurred in August 1959, a short distance from Yellowstone Park. A landslide—in reality, a massive rockslide (see Chapter 7)—originated on Red Mountain above the Madison River in Montana. The rockslide was triggered by a moderate earthquake that caused the metamorphic rock (schist) composing the mountain to slide down its foliation planes, which were inclined parallel to the hill slopes (**Figure 4.15**). However, the rockslide occurred above a popular campground, and 26 people were killed by the falling rock. The slide generated a terrific shock wave of air that lifted cars, trees, and campers off the ground like a giant invisible cushion. In addition, the Madison River was dammed by the slide and a lake formed, later named "Quake Lake."

By far the worst quake-related landslide internationally took place on 31 May 1970, when a subduction-related M_w 7.9 to 8.0 earthquake struck Peru. Strong shaking was felt over a vast area (the size of Belgium and the Netherlands combined) and detached from the north face of 6,768-meter (22,200-foot)-tall Nevado Huascarán, an ice-covered Andean peak, a great avalanche of rock, snow, and ice measuring 900 meters (3,000 feet) wide and 1.6 kilometers (1 mile) long. This broken mass roared down the steep flank of the mountain at speeds as high as 355 kilometers/hour (220 miles/hour), mobilized, in part, by a cushion of trapped air. A canyon funneled it toward the towns of Yungay and Ranrahirca, 11 kilometers (7 miles) away, which were buried just a few minutes later as it exited the base of the mountain and spread across the plain below. Of the 20,000 people in Yungay, only 400 survived, mostly children who hap-

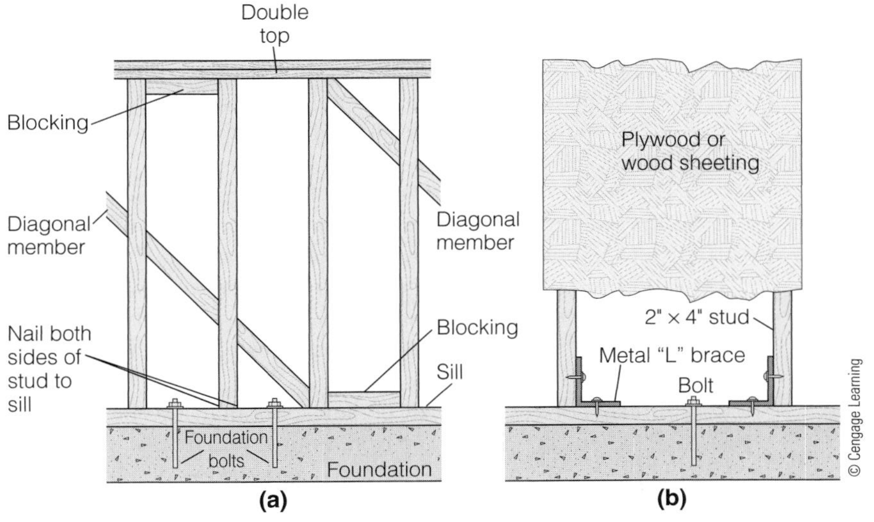

Figure 4.14 Methods of reinforcing structures against base shear. (a) Diagonal cross-members and blocks resist horizontal earthquake motion (shear). (b) Plywood sheeting forms a competent shear wall, and metal L braces and bolts tie the structure to the foundation.

Figure 4.15 The site of the earthquake-induced rockslide at Red Mountain. The Madison River is just off the picture in the foreground. The rockslide was triggered by an earthquake and took place in schist with foliation planes inclined parallel to the mountain slope (see Chapter 7).

pened to occupy a high level in a stadium at the time, and so weren't smothered. The Andes region has the greatest seismic-related landslide risk in the world.

Ground or Foundation Failure

Liquefaction is the sudden loss of strength of water-saturated, sandy soils resulting from shaking during an earthquake. Sometimes called **spontaneous liquefaction**, it can cause large ground cracks to open, lending support to the ancient myth that the Earth opens up to swallow people and animals during earthquakes. Shaking can cause saturated sands to consolidate, and thus to occupy a smaller volume. If the water is slow in draining from the consolidated material, the overlying soil comes to be supported only by pore water, which has no resistance to shearing. This may cause buildings to settle, earth dams to fail, and sand below the surface to blow out through openings at the ground surface (**Figure 4.16a**). Liquefaction at shallow depth may result in extensive lateral movement or spreading of the ground, leaving large cracks and openings.

The ground areas most susceptible to liquefaction are those that are underlain at shallow depth—usually less than 10 meters (about 30 feet)—by layers of water-saturated fine sand. With subsurface geological data obtained from water wells and foundation borings, liquefaction-susceptibility maps have been prepared for many seismically active areas in the United States.

Similar failures occur in certain clays that lose their strength when they are shaken or remolded. Such clays are called *quick clays* and are natural aggregations of fine-grained clays and water. They have the peculiar property of turning from a solid (actually, a gel-like state) to a liquid when they are agitated by an earthquake, an explosion, or even vibrations from pile driving. They occur in deposits of glacial-marine or glacial-lake origin, and are therefore found mostly in northern latitudes, particularly in Scandinavia, Canada, and the New England states. The failure of quick clays underlying Anchorage, Alaska, produced extensive lateral spreading throughout the city in the 1964 earthquake (see Figure 4.16b).

The physics of failure in spontaneous liquefaction and in quick clays is similar. When the earth materials are water-saturated and the Earth shakes, the loosely packed sand consolidates or the clay collapses like a house of cards. The pore-water pressure pushing the grains apart becomes greater than the grain-to-grain friction, and the material becomes "quick," or "liquifies" (**Figure 4.17**). The potential for such geological conditions is not easily recognized. In many

Figure 4.16 (a) These apartment buildings tilted as a result of soil liquefaction in Niigata, Japan, in 1964. Many residents of the building in the center exited by walking down the side of the structure. (b) "House of cards" collapse of quick clay structure in Turnagain Heights, Anchorage, Alaska, in 1964. Total destruction occurred within the slide area, which is now called Earthquake Park.

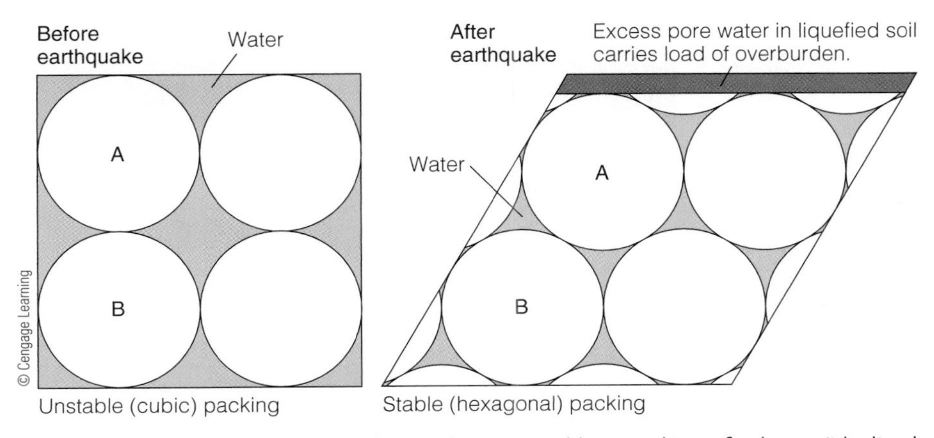

Figure 4.17 Liquefaction (lateral spreading) caused by repacking of spheres (idealized grains of sand) during an earthquake. The earthquake's shaking causes the solids to become packed more efficiently, and thus to occupy less volume. A part of the overburden load is supported by water, which has no resistance to lateral motion.

Figure 4.18 The 1992 Landers earthquake (M_w 5 7.3) in the Colorado Desert of Southern California was felt from Phoenix, Arizona, to Reno, Nevada. Right-lateral offset of 4.27 meters (14 feet) on the Johnson Valley fault created a 2-meter (6.5 foot) vertical scarp caused by lateral offset of a ridge. The geologist is 185 centimeters (6 feet 2 inches) tall.

cases, it can be determined only by information gained from bore holes drilled deeper than 30 meters (100 feet). Because of this expense, many site investigations do not include deep drilling, and the condition can be unsuspected until an earthquake occurs.

Ground Rupture and Changes in Ground Level

Structures that straddle an active fault may be destroyed by actual ground shifting and the formation of a fault scarp (**Figure 4.18**). By excavating trenches across fault zones, geologists can usually locate past rupture surfaces that may reactivate. In 1970, the California legislature enacted the Alquist-Priolo Special Studies Zones Act, which was renamed the Earthquake Fault Zones Act in 1995. It mandates that all known active faults in the state be accurately mapped and zoned for seismic safety. The act provides funds for state and private geologists to locate the youngest fault ruptures within a zone, and requires city and county governments to limit land use adjacent to identified faults within their jurisdictions. Ironically, the faults responsible for the 1992 Landers earthquake ($M_w = 7.3$) near Los Angeles were designated as Earthquake Fault Zones just before the 28 June event.

Changes in ground level as a result of faulting may have an impact, particularly in coastal areas that are uplifted or down-dropped. For instance, during the 1964 Alaskan event, parts of the Gulf of Alaska thrust upward 11 meters (36 feet), exposing vast tracts of former tidelands on island and mainland coasts.

Fires

Fires caused by ruptured gas mains or fallen electric power lines can add considerably to the damage caused by an earthquake. In fact, most damage attributed to the San Francisco earthquake of 1906 and much of that in Kobe, Japan, in 1995 was due to the uncontrolled fires that followed the earthquakes. The Kobe quake hit at breakfast time. In neighborhoods crowded with wooden structures, fires erupted when natural gas lines broke and falling debris tipped over kerosene stoves. Broken water mains made firefighting efforts futile. One of the principal "do's" for citizens immediately after a quake is to shut off the gas supply to homes and other buildings to prevent gas leaks into the structure. This in itself will save many lives.

Tsunamis

The most myth-ridden hazard associated with earthquakes (and submarine volcanic eruptions and landslides) is a tsunami (pronounced "soo-nah-mee"), or seismic sea wave. **Tsunami** is a Japanese word meaning "great wave in harbor," and it is appropriate, because these waves are impulsively generated and most commonly wreak death and destruction inside bays and harbors. They have nothing to do with the tides, even though the term *tidal wave* is commonly used in the English-speaking world. The Japanese written record of tsunamis goes back 200 years, and their power is dramatically displayed in the well-known print by Hokusai (**Figure 4.19**). Tsunamis, their causes, movement, and effects are treated in detail in Chapter 10.

© Metropolitan Museum of Art, H. O. Havemeyer Collection (JP 1847)/Art Resource, NY

Figure 4.19 "The Breaking Wave off Kanagawa," wood-block color print by Katsushika Hokusai (1760–1849) from the series "Thirty-six Views of Mount Fuji," 1826–1833.

Table 4.1 Size Classification and Average Annual Frequency of Earthquakes Worldwide (U.S. Geological Survey)

Size Class	M_w Range	Average Annual Number, Since 1900
Great	>8	1
Major	7–7.9	17
Strong	6–6.9	134
Moderate	5–5.9	1,319
Light	4–4.9	~13,000
Minor	3–3.9	~130,000
Very minor	2–2.9	~1,300,000

U.S. Geological Survey

Five Earthquakes That Make Important Points

More than a million detectable earthquakes occur worldwide every year (**Table 4.1**), and well over a thousand, on average, are strong enough to do significant damage. But few actually prove destructive because of their remote locations or great focal depths. Like dodging the proverbial bullet, however, it's simply a matter of chance before a strong earthquake strikes an area with intensive human development. As the world becomes more and more crowded with people and human infrastructure, statistics suggest that the destructiveness of earthquakes will increase—at least in monetary terms—despite greater understanding and sophisticated monitoring technologies (**Table 4.2**). Geologist Roger Bilham from the University of Colorado notes (1995, p. 8):

> The number of earthquakes resulting in fatalities has increased approximately in proportion to global populations, and although a decreasing fraction of the global population has been killed by earthquakes in this century compared to former centuries, seismic risk in certain regions has increased substantially. The cause of the apparent paradox lies in the growth of urban agglomerations where most of the world's growing population will live, and the location of many of these cities near plate boundaries where earthquakes occur quasi-periodically. There are now dozens of cities near seismic belts with populations exceeding 2 million.

In the literal "School of Hard Knocks," earthquakes are fearsome teachers—each large quake carries important messages about how Earth works, and about how to prepare for the next one. By no means do we know as much as we should, and in many major earthquakes over the past half century, we've learned new lessons painfully. Yet tremendous improvements have been made in construction and emergency services as a result of these disasters, saving countless lives. This accounts substantially for Bilham's observation that the total fraction of earthquake fatalities has gone down over the past hundred years, despite rising global population. Unfortunately, most of these efforts have come about only in developed regions where people can afford to invest in public safety (see chapter opener).

This section describes five earthquakes that occurred between 1994 and 2011. All five had very strong magnitudes but had quite different results because of differences in local foundation conditions, building standards, population density and societal factors (see Table 4.2).

Northridge, California, 1994

The largest earthquake in Los Angeles's short history occurred at 4:30 a.m. Monday, 17 January 1994, on a hidden fault below the San Fernando Valley in the sprawling Northridge district. The $M_w = 6.7$ earthquake started at a depth of 18 kilometers (11 miles) and propagated upward in a matter of seconds to a depth of 5 to 8 kilometers (3–5 miles; **Figure 4.20**). There were thousands of aftershocks, and their clustered pattern indicated the causative fault to be a reverse fault with a shallow dip of 3° to 5° to the south (see Case Study 4.2 and Appendix 3). Such a low-angle reverse fault is called a *thrust fault*, and because the thrust did not rupture the ground surface, it is described as "blind" (**Figure 4.21**). Blind thrusts were first recognized after the 1987 Whittier earthquake 40 kilometers (25 miles) southeast of Northridge, but they were not fully appreciated until after the Northridge event.

Blind thrust faults were a newly identified seismic hazard, and there is evidence that a belt of them underlies the northern Los Angeles basin. They are especially dangerous because they cannot be detected by traditional

Table 4.2 Selected Significant* World Earthquakes, In Chronological Order

Location	Year	Richter Magnitude†	Impact
San Francisco	1906	8.3	700 killed, $7 million damage, fire
Messina (Sicily)	1908	7.5	160,000 killed
Tokyo, Japan	1923	8.3	140,000+ killed, fire
Assam, India	1950	8.4	30,000 killed
Chile	1960	M_w 9.5	5,700 killed, 58,000 homes destroyed, tsunami
Alaska	1964	M_w 9.2	131 killed, tsunami
T'ang-shan, China	1976	7.9	330,000 killed
Mexico City	1985	8.1	10,000+ reported killed
Armenia	1988	6.8	55,000 killed
Loma Prieta (California)	1989	M_w 6.9	67 killed, $8 billion damage
Northridge, California	1994	M_w 6.7	60 killed, $25 billion damage
Kobe, Japan	1995	M_w 6.9	5,378 killed, $100 billion damage
Izmit (Kocaeli), Turkey	1999	M_w 7.1	15,370 killed, $10 to $20 billion damage
Chi-Chi, Taiwan	1999	M_w 7.6	2,300+ killed
Gujarat, India	2001	M_w 7.7	20,000+ reported killed
Alaska	2002	M_w 7.9	Little impact, no casualties
Colima, Mexico	2003	M_w 7.8	Widespread destruction, 27 killed
Bam, Iran	2003	M_w 6.5	30,000 believed killed, 30,000 injured
Sumatra-Andaman Islands	2004	M_w 9.2	230,000 killed, tsunami
Kashmir, Pakistan	2005	M_w 7.6	73,000 killed
Sichuan Province, China	2008	7.8	70,000+ killed
Abruzzo, Italy	2008	6.3	292+ killed
Samoan Islands	2009	8.3	186 killed, tsunami
Pandang, Sumatra	2009	7.6	1,100 killed
Port-au-Prince, Haiti	2010	7.0	150,000–250,000 killed
Maule region, Chile	2010	8.8	At least 700 killed
Christchurch, New Zealand	2011	6.3	185 killed, much liquefaction
Sendai coast, Japan	2011	9.0	19,000+ killed, tsunami

Source: Historic World Earthquakes, U.S. Geological Survey, Earthquake Hazards Program, http://earthquake.usgs.gov

*A "significant" earthquake is defined as one that registers a moment magnitude (M_w) of at least 6.5 or a lesser one that causes considerable damage or loss of life.

†Estimated prior to 1935.

techniques, such as trenching and field mapping (thus, they do not fall within the Earthquake Fault Zones Act as it is written), yet they can generate significant earthquakes. Just as the residents of Southern California were beginning to feel confident that geologists knew the location and behavior of most active faults, blind thrusts made their presence known.

Field inspection of the epicentral region was a depressing experience. Thirteen thousand buildings were found to be severely damaged; 21,000 dwelling units had to be ordered evacuated; 240 mobile homes had been destroyed by fire; and 11 major freeway overpasses were damaged at 8 locations (see Case Study 4.3). The reason for the extensive damage was the high horizontal and vertical accelerations generated by the earthquake. Accelerations of more than 0.30 g are considered dangerous, and the vertical acceleration of 1.8 g measured by an instrument bolted to bedrock in nearby Tarzana probably set a world record (at least until the Christchurch, New Zealand, quake nearly 20 years later). The high ground acceleration explains why many people were literally thrown out of their beds, and objects as heavy as television sets were projected several meters from their stands. Although the damage was not as visually spectacular as that of Gujarat, India,

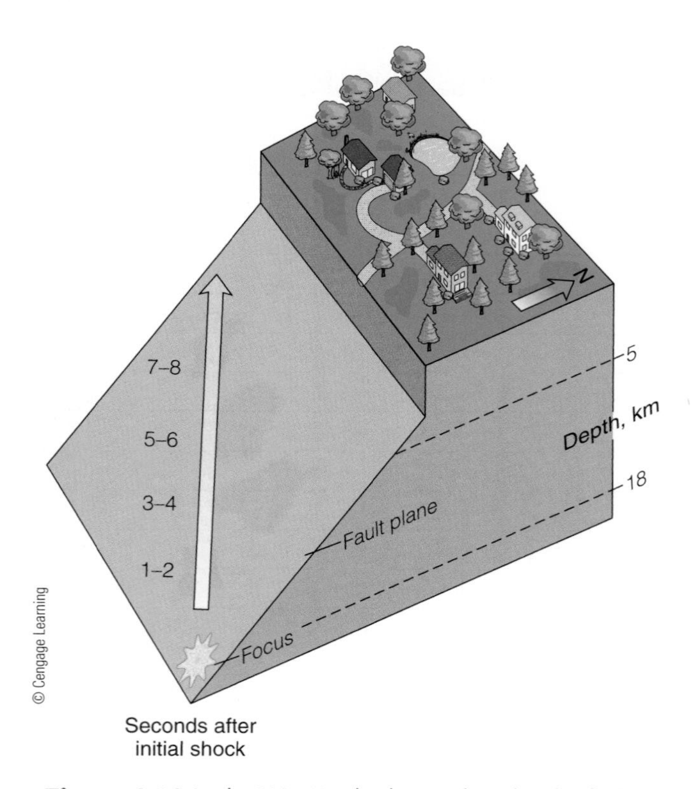

Figure 4.20 In the 1994 Northridge earthquake, the fault rupture progressed up the fault plane from the focus at the lower right of the figure to the upper left in 8 seconds. Rather than rupturing smoothly, like a zipper opening, it moved in jerks along the fault plane, as shown by the pink patches. Total displacement was about 4 meters (12 feet).

described in the following section, it was heartbreaking to many residents—brick chimneys, both those reinforced with steel bars and unreinforced, came down; houses moved off their foundations; concrete-block walls crumbled; gaps broke open in plaster walls; and the number of shattered storefront windows was beyond counting (**Figure 4.22**). The horizontal ground motion was directional; that is, it was strongest in the north-south direction. With few exceptions, block walls oriented east-west tipped over or fell apart, whereas those oriented north-south remained standing. Many two- and three-story apartment buildings built over open first-floor garages collapsed onto the residents' cars. The lack of shear resistance of these open parking areas led to failure of the vertical supports, and everything above came down.

California State University at Northridge sustained almost "textbook" earthquake damage. In its library is "Leviathan II," recognized as one of the most advanced automated book-withdrawal systems in the country. On that day, it recorded a record withdrawal of about 500,000 books—all of them onto the floor of the library. Bottled chemicals fell off their storage shelves (a major concern at any university) and caused a large fire in the chemistry building. In addition, one wall of a large, open parking structure collapsed inward to produce the stunning art deco architecture shown in Figure 4.22.

Gujarat, 2001

A headline in a well-known American newspaper stated, "Bad Quake, Worse Buildings." This just about sums up the impact of this strong event in the State of Gujarat, India

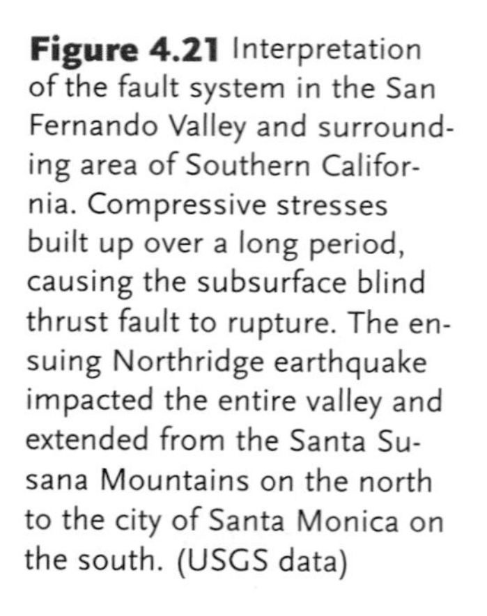

Figure 4.21 Interpretation of the fault system in the San Fernando Valley and surrounding area of Southern California. Compressive stresses built up over a long period, causing the subsurface blind thrust fault to rupture. The ensuing Northridge earthquake impacted the entire valley and extended from the Santa Susana Mountains on the north to the city of Santa Monica on the south. (USGS data)

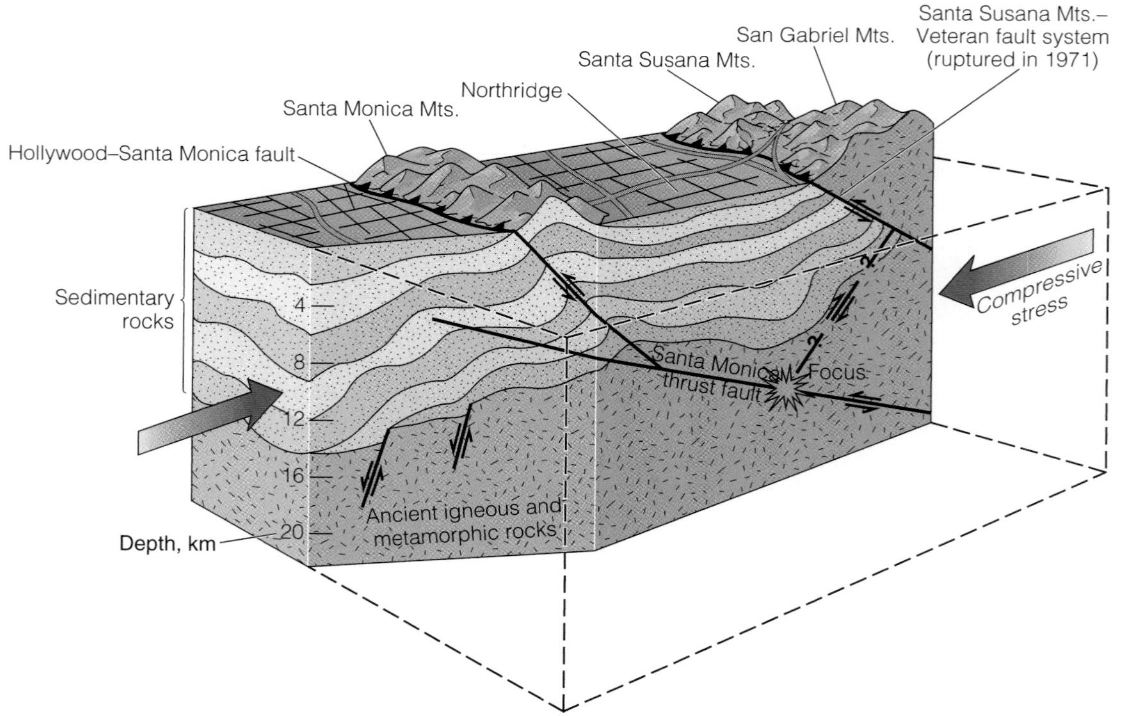

Figure 4.22 (a) Two wings of an apartment building that collapsed toward one another. (b) A staircase that leads to nowhere. In reality, it was the second-floor breezeway that collapsed onto the ground floor. (c) Part of the collapsed parking structure of Fashion Center, Northridge. Such collapses instigated new design criteria for open structures. Note the outside staircase that separated from the main structure.
Greg Pavis, USC

(**Figure 4.23**). The earthquake originated on a reverse fault, and the closest plate boundary lies many hundreds of kilometers away. It was felt 2,000 kilometers (1,200 miles) away and over an area 16 times that of the $M = 7.8$ San Francisco earthquake of 1906. Field investigation in Gujarat revealed no ground rupture, which is unusual for an earthquake of this magnitude. Liquefaction was widespread because of the high water table and thickness of unconsolidated sediment in the region. The quake occurred in an old rift system, which is a well-known seismic zone very similar to one in the central United States (see later). You don't have to be a seismologist to know that an earthquake of this magnitude has major potential to cause devastation. In Gujarat, the lack of strict building codes governing construction methods, some of which date back to the British colonial period, exacerbated the destruction that led to the loss of more than 20,000 lives. Also, in Gujarat, one can become an architect or a building contractor without any special education or license. The damage covered an area more than 500 kilometers (300 miles) wide from the large city of Ahmedabad in the east to the Arabian Sea in the west. Severe damage was reported within an area 50 × 70 kilometers (30 by 40 miles), in which most high-rises (buildings of more than three stories) and low-rise cobblestone structures collapsed because of resonance and engineering factors.

What is surprising to geologists is that this earthquake occurred in an area of relatively flat topography, without the usual tectonic landforms suggestive of faulting and seismic activity. Strong earthquakes in India usually occur as the Indian plate moves northward against the Himalayas. However, a similar event occurred in 1819 in the Rann of Kutch, also in the state of Gujarat, just to the southeast of the 2002 earth-

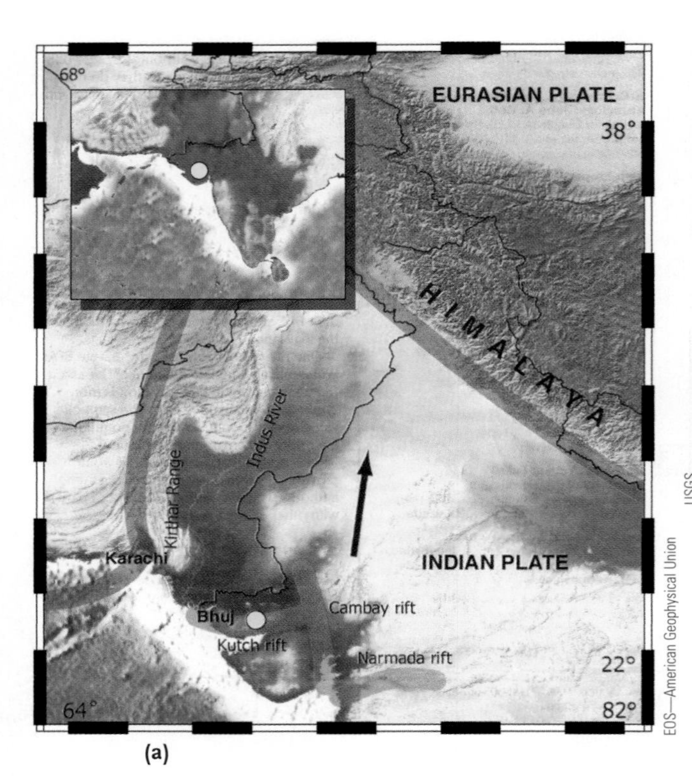

(a)

EOS—American Geophysical Union

USGS (b)

Figure 4.23 (a) The regional geography and plate tectonic setting of the Gujarat (Bhuj) earthquake. The epicenter is shown by the yellow dot. Bhuj lies 400 to 500 kilometers from a plate boundary and a greater distance from the Himalayan range. Also shown are buried rift basins that are seismically active and very similar to those of the central United States. (b) Collapsed houses in the town of Ratnal, in the epicentral region of the Bhuj earthquake.

quake epicenter. This suggests a greater rate of tectonic (earthquake) activity than is indicated by the landforms there.

Sumatra-Andaman Islands, 2004

The third deadliest and second most powerful earthquake in recorded history struck the northwest coast of Sumatra around breakfast time on December 26, 2004. The quake had an M_w of 9.2. Because each whole number increase in magnitude equals 30 times more energy released, the energy released in this earthquake was about 1,000 times greater than in the destructive Haiti 2010 event. In a mere 6 to 8 minutes, the edge of the Indo-Pacific plate boundary flexed and lifted the coral-rimmed coasts of offshore islands closest to the 600-kilometer (350 mile)-long fault rupture as much as 90 centimeters (35 inches), while the shoreline of the main island of Sumatra sank below sea level, inundating important farmlands, roads, and built-up areas (**Figure 4.24**). An enormous tsunami radiated outward from the focal area and battered coasts as far away as eastern Africa, Sri Lanka, Myanmar, and Thailand. Because there was no early warning system in place to alert people of the approaching waves, nearly a quarter of a million people drowned. In one instance, the sea waves derailed and washed away a crowded passenger train that happened to be traveling in the wrong place at the wrong time.

For geologists, the Great Sumatran earthquake was geophysically interesting because it set the whole Earth ringing like a giant bell struck with a hammer. Every location on the planet, including the remote poles and the flat plains of Kansas, rose and fell at least a centimeter in slow, gentle oscillations, with periods ranging from every 20.5 to 54 minutes, too slow for people to notice without instrumental assistance. This wiggling continued with diminishing strength for months. The quake was also strong enough to trigger aftershocks as far away as Alaska. Only a few months later, in March 2005, another great earthquake with a magnitude of 8.5 to 8.7—the biggest aftershock of all—struck the same area of Sumatra. Although it did not generate a large tsunami, 1,300 people lost their lives, mostly on a single island.

Tohoku-Oki, Japan

The M_w 9.0 Tohoku-Oki earthquake (sometimes called the "311 Earthquake" because of the date of its occurrence in 2011) was especially instructive. A set of M_w 6.0 to 7.7 foreshocks preceded the quake by a couple of days with an apparent pause ensuing. Then a modest magnitude-4.9 quake struck the epicentral region, followed by the gigantic rupture nobody anticipated based on the historical record. Only in retrospect do we appreciate that that record was *too short*, although other geological clues existed to suggest we hadn't learned enough about seismicity in the area of northeastern Japan. These included the natural debris left by tsunamis occurring centuries ago that could only have resulted from similarly powerful events. An enormous segment of plate interface, measuring 400 kilometers (240 miles) horizontally and extending 20 to 200 kilometers (12–120 miles) into the Earth, ruptured along the trench in the sea east of Honshu—Japan's main island. The shaking lasted a terrifying 7.5 minutes, far longer than most earthquakes, and the fault displacement was by far the largest ever recorded—an estimated 40 to 80 meters (130–260 feet)! Enough natural energy was released in that 7 minutes to power the city of Los Angeles for a year. Because the rupture reached so close to Earth's surface, the overthrust plate experienced tremendous vertical uplift—as high as 3 meters (12 feet). A giant tsunami resulted (see Chapter 10).

Kerry Seih, Earth Observatory of Singapore

Figure 4.24 *Pontes* coral heads at the Global Positioning System (GPS) survey station NGNG on one of the islands off the west coast of Sumatra. The coral heads were uplifted about 90 centimeters (35 inches) by the 26 December 2004, Great Sumatran earthquake.

Like in the 2004 Sumatra-Andaman earthquake, most fatalities resulting from the earthquake were associated with these powerful sea waves. The quake also threw Earth off its rotational axis by 10 to 25 centimeters (4–10 inches), shortening the length of a day by 1 microsecond. (It will take the planet several years to recover its ordinary polar rotation).

The good news is that only about 5% of the destruction caused by the earthquake was due to the physical shaking itself (**Figure 4.25**). The rest was due to the tsunami, for which inadequate coastal seawall protection existed—even though some walls loomed as high as 10 meters (more than 30 feet). (The tsunami topped 40 meters [>120 feet] in some places). Thanks to modern construction, outstanding engineering, and emergency response technology, structures held up remarkably well. For instance, two dozen of Japan's famous bullet trains were operating on tracks in the region racing at speeds as high as 200 kilometers/hour (120 miles/hour) when the ground beneath began rumbling. Within seconds, automatic braking systems slowed the trains before they were endangered with derailment, and they stopped with no significant harm done. The national earthquake warning system, operated by the Japan Meteorological Agency (JMA), also provided good warning that strong shaking would soon reach people living hundreds of kilometers (miles) away, perhaps saving hundreds, if not thousands, of lives. The JMA warning system includes not only a national network of more than 1,000 seismometers, but undersea sensors between the coast and the trench. In the crowded Tokyo area, people had a minute to prepare themselves. The alarms went off about 30 seconds after the earthquake started far to the north.

On the other hand, the size of the earthquake-in-progress was woefully underestimated by observers in charge of the early warning system—and with it the size of the associated tsunami—because of incompletely processed information about the seismic energy being released. The fact that the earthquake lasted so long, mixing S- and P-wave signals together, further confused data processors. The result was likely a heavier loss of life than would have occurred with more accurate immediate information. Citizens feeling that they were safely protected by their local seawalls because of initial estimates of tsunami height were not protected. They certainly would have had 5 to 30 minutes to flee, depending on location, if they'd known what was truly coming. Better methodology in interpreting incoming seismic data is needed. In addition, geodetic information—that

© MPAK/Alamy

Figure 4.25 Grim panorama of general destruction resulting from earthquake and tsunami, 11 March 2011, northeastern Japan.

is, data about sudden changes in elevation of various points on Earth's surface, as provided by satellites through the Global Positioning System (GPS)—could have provided a better grasp on the tsunami threat. The key relationships we are striving to understand are how fault size relates to earthquake magnitude and tsunami height. Improvements in early warning are now under way. Some geologists have called for expansion of facilities, such as the U.S. Geological Survey National Earthquake Information Center, into a well-integrated international network. The redundancy provided by such a facility might be especially important when a strong earthquake destroys or swamps local systems for seismic monitoring and emergency response.

Christchurch, New Zealand, 2011

In recent years, ordinary people have commented frequently that the number of earthquake-related disasters in the world seems to have really increased. And, in fact, this is largely true. In the time between 1970 and 2000, no mega-earthquakes (greater than magnitude 8.5) took place anywhere in the world. But then on 26 December 2004, the horrific magnitude-9.2 Sumatra-Andaman earthquake took place. Five earthquakes exceeding magnitude 8.5 have occurred since then, the latest on 11 March 2011 in northeastern Japan.

Is this just a statistical quirk, or is the world really falling apart? Doubtlessly the former: If Earth's crust were really degenerating, then one would expect to see a similar increase in earthquakes of *all* magnitude ranges since late 2004 (see Table 4.1). Yet no such pattern is seen for tremors of less than magnitude 8.5 size worldwide. Furthermore, sets of this sort have occurred before in world history. Between 1950 and 1965, seven earthquakes of magnitude 8.5 or greater took place, including the largest shake ever recorded: the magnitude-9.5 Chilean earthquake of 1960. Each of these seismic events originated at a subduction zone.

The latest series of big quakes has taught us that we've underestimated the extent to which one major earthquake can trigger another on adjacent fault surfaces. This helps explain some clustering in frequency. For instance, the 2004 Sumatran earthquake very likely triggered three other great quakes on the same plate boundary in 2005, 2007, and 2012. Strain is accumulating ominously on another portion of the boundary that has not ruptured since 1797—the site of future disaster, *if we are unprepared*.

Another spectacular example of seismic triggering occurred on 28 September 2009, when a magnitude-8 earthquake occurred in the Tonga subduction zone in the southwest Pacific. The shaking initiated a second magnitude-8 earthquake 50 kilometers (30 miles) to the south, even as the fracture from the first event was *still opening*. The resulting tsunami was certainly larger than had each earthquake taken place in isolation.

How wide-reaching can seismic triggering be? Is it conceivable that a major earthquake on one side of the planet could trigger one on the other? Existing data do not support this fear, although as Sumatra demonstrated, many much smaller earthquakes might be stimulated on faults thousands of kilometers from a major epicenter. Possibly, however, a major earthquake might occur *sooner* in a far-removed region than would otherwise take place.

Another instructive example of probable earthquake triggering occurred in central New Zealand in 2010–2011: On the early morning of 4 September 2010, a magnitude-7.1 earthquake rocked the central Canterbury Plain with an epicenter about 40 kilometers (25 miles) west of Christchurch, the second largest city in New Zealand. Casualties were light, despite the strong shaking, because of low population density, the modern construction of most buildings, and the fact that so many people were home asleep in timber-framed homes when the quake struck. Apart from some foundation issues, wood-framed dwellings generally respond well during earthquakes because of their flexibility. At least they do not collapse in a heap, as does stone and masonry. Many commentators compared the effects of this tremor with the one that occurred in Haiti shortly before, where loss of life was orders of magnitude greater despite lower magnitude shaking, because of poor construction and engineering design. Following the quake, various sociological responses became apparent: Although the numbers of heart attacks greatly increased, overall domestic crimes in Christchurch greatly decreased, whereas incidence of domestic violence increased by more than 50%.

But the worst was yet to come. An unusually strong set of aftershocks developed. "Unusual" in this case relates to several fundamental observations in seismology, including the **Gutenberg–Richter law**, which states that *the total number of aftershocks of a particular magnitude range following a big quake diminishes exponentially over time, with large aftershocks much less common than smaller ones*. Aftershocks represent the gradual readjustment of the crust in the area of a fault rupture, rather like the rebalancing you do after making a big jump. The bigger the main triggering event, the greater the length of time before resulting aftershocks fade away completely. Statistically speaking, for instance, aftershocks are *still* taking place following the great New Madrid earthquake of 1812, whereas for most less powerful major earthquakes, aftershocks die out over a period of a few years or decades. In the case of the 2010 Canterbury quake, aftershocks spread eastward along buried faults into the Christchurch area, showing a transfer of strain in that direction from the original rupture. Their magnitudes and frequency exceeded what was expected based on observations of similar seismic events elsewhere in the world.

Many buildings in the region had unnoticed or underestimated pre-weakening from the September 2010 earthquake. When over the lunch hour on 22 February 2011, a magnitude-6.3 quake struck 10 kilometers (6 miles) southeast of central Christchurch, many structures were damaged that might have withstood the shaking had it occurred without a precursor (**Figure 4.26**). This quake was so strong in comparison with ordinary aftershock statistics that some seismologists characterized it as an independent event—not truly an aftershock, but a *triggered quake* similar to the ones described ear-

gmwnz/Shutterstock.com

Figure 4.26 Emergency responder examines a building front that has shed masonry during heavy shaking in Christchurch. Exiting a building once a quake has started risks serious injury.

Figure 4.27 Dazed residents examine fissures and boils (cratered mounds formed by erupting groundwater) during liquefaction in a suburban field, Christchurch, New Zealand, 2011.
REUTERS/KYODO Kyodo

lier. None of this mattered to the victims, of course. The combination of pre-weakening and localization at very shallow depth—only about 5 kilometers (3 miles) down—meant that the city suffered terribly. To compound the problem, the peak ground acceleration in downtown Christchurch reached 1.8 g, and at one local school 2.2 g. What preconditioned the ground to shake so violently?

The shallowness of the earthquake focus was certainly partly responsible. But perhaps the presence of abundant shallow groundwater played a role as well. A water-saturated mass of sediment will shake like a giant bowl of gelatin during an earthquake, whereas solid granite bedrock can transmit seismic energy readily without amplified shaking. The presence of the large amounts of groundwater right beneath metro Christchurch could be easily seen even during the September 2010 earthquake. Sandy, silty liquid—expelled groundwater—erupted from the earth in many districts during the shaking, causing roads, foundations, and fields to crack and settle beneath the flood (**Figure 4.27**).

Nearly 200 people died in the February earthquake, victims not only of pre-weakened structures and strong seismic accelerations, but of a time of day that engaged many people in activities that made them vulnerable, including street activity exposed to falling masonry and office work in non–wood-frame structures that could not stand up. Important to note, however, is that most fatalities occurred at only a few building sites. The Canterbury TV building alone, a fairly modern structure, took 115 lives when it collapsed. This is typical of many earthquakes—human loss is concentrated in relatively few key structures. Such underscores once again the importance of building well in earthquake country.

What We Can Learn

A number of important lessons derive from the earlier examples. For instance, in the Greater Los Angeles area, the real earthquake threat is not from the San Andreas Fault, but rather from much shallower reverse faults that slant unseen beneath the city, which is the second largest in the United States. Not only shallow, such faults tend to generate unusually strong ground accelerations as well. The damage also illustrated some weak links in modern architecture, including chimneys and apartment complexes. (Pity the poor librarian at California State, too!)

The Gujarat earthquake further underscored the importance of architects, engineers, and buildings in seismic safety. But it also showed that destructive earthquakes can sometimes occur where least expected—even by geologists! Where we can count on experiencing them, such as along subduction zones, damage is largely a function of economic development, type of earthquake (e.g., thrust versus strike-slip), and magnitude.

Both the Sumatra-Andaman and Tohoku-Oki earthquakes set the whole earth ringing like a bell and wobbling on its rotational axis for months (a phenomenon called the *Chandler wobble*). Although impressive and maybe even unnerving, this displacement is not threatening, because planetary rotation is self-correcting. More importantly, the Sumatra-Andaman earthquake illustrated the potential for earthquake triggering—something reiterated in the Christchurch example and possibly to be expected with the Tohoku-Oki event. Both Sumatra-Andaman and Tohoku-Oki also make the point that although it's nice to live next to the beach, residents must be ready to leave home *fast* after a sharp tremor.

The most important lesson of Tohoku-Oki perhaps was that our human history *is too short to understand the true seismic risk and patterns of particular regions*—maybe anywhere in the world. Single earthquakes are also capable of far greater displacements than previously thought possible. On a positive note, the disaster showed that modern earthquake warning systems and technologies *do* work, and can save many lives! We know now how to make these systems even better.

A number of key messages stream from the Christchurch example: (1) the pre-weakening of structures by

Dennis Fox

Figure 4.28 The I-880 freeway failed because it was built on bay muds that reacted to seismic waves like a bowl of gelatin. Pictured is the Cypress Street viaduct, with the upper deck collapsed on the lower one. There was loss of life here.

one earthquake will greatly increase the damage caused by another large earthquake that shortly follows; (2) cities built on soft sediment with shallow groundwater are in special danger; (3) *time of day matters* in terms of seismic safety; and (4) life and property loss, at least in developed countries, tends to be concentrated in pockets—even restricted to particularly weak buildings—during large earthquakes. You may be able to think of reasons why this should be so.

Other factors besides construction, population density, and kind of fault movement can cause damage. Along strike-slip faults, seismic shaking tends to be more intense in a zone straddling the fault than at right angles to it, even at equal distances from an epicenter. Seismic energy tends to build up in hilltops and along ridgelines thanks to internal seismic wave reflection, greatly increasing destructive ground acceleration in such localities. And structures built atop firm bedrock close to an epicenter may suffer little relative to similar structures built on soft, water-saturated ground much farther away. Such localized features as determinants of damage are called *site effects*.

A spectacular example of site effects occurred during the Loma Prieta earthquake, which struck the lower San Francisco Peninsula in 1989. Structures built on firm bedrock held up well in the City of San Francisco. But across the bay farther from the epicenter, the two-story Interstate 880 elevated freeway collapsed, killing 42 drivers on the lower deck—two-thirds of total earthquake victims. Collapse could be blamed on two factors: (1) inadequate reinforcement of support columns, which were in many places not tied with rebar to the lower deck or burst from vertical loading during up-and-down shaking; and (2) the fact that this highway was built atop soft, water-saturated bay-fill. In 1949, when the viaduct was designed and the land allocated for construction, no one took the trouble to test the proposal for a large earthquake. (Although geologists began recording earthquakes with seismographs in the 1880s, the first instruments designed to monitor building response to earthquakes weren't installed until the 1940s [**Figure 4.28**].)

Earthquakes Far from Plate Boundaries

Even though the vast majority of the world's earthquakes occur at plate boundaries, areas hundreds and even thousands of kilometers away are not free of seismic activity—as the example of the Gujarat earthquake illustrates. In the

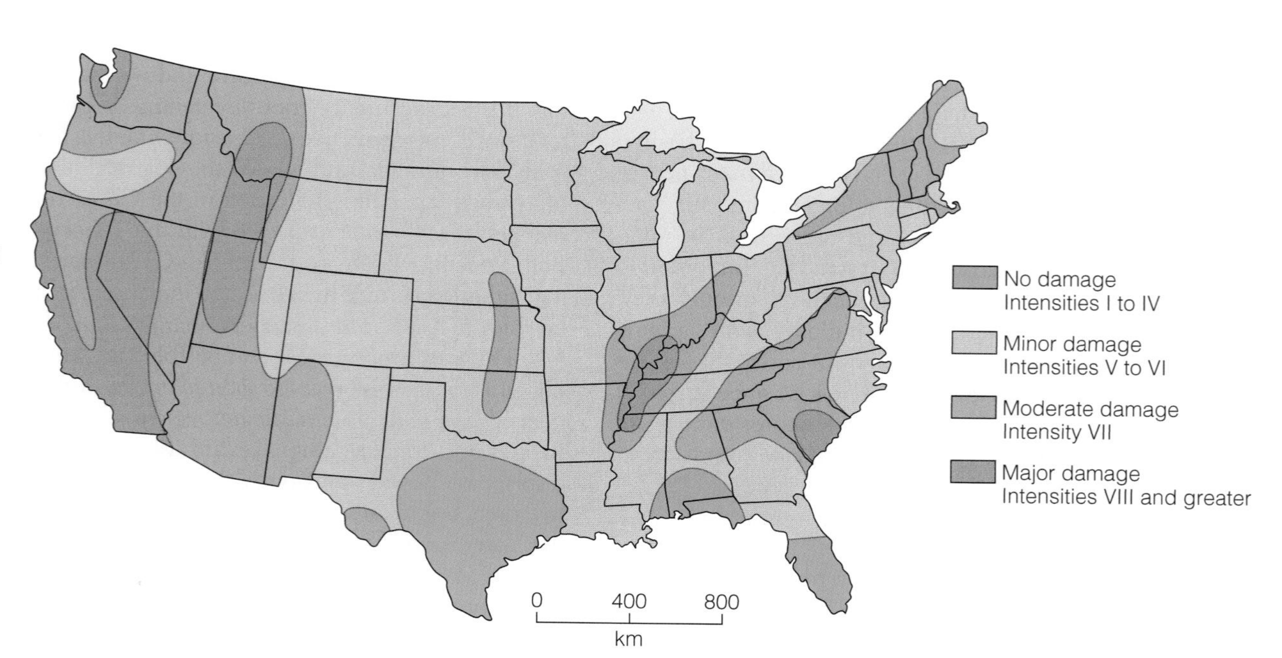

Figure 4.29 Seismic-risk map of the 48 contiguous states based on historical records and intensities collected by the U.S. Coast and Geodetic Survey. The Coast and Geodetic Survey gathers intensity data from questionnaires after earthquakes. (From Blair, M. L., and W. W. Spangle, USGS prof, paper 941-B, 1979.)

Table 4.3 Selected North American Intraplate Earthquakes

Location	Year	Moment Magnitude*	Impact
New Madrid, Missouri	1811	8.2	Reelfoot Lake formed in northwest Tennessee.
New Madrid	1812	8.3	Elevation changes caused Mississippi River to reverse its course locally.
New Madrid	1812	8.1	New Madrid, Missouri destroyed.
Charleston, South Carolina	1886	7.6	It was felt from New York to Chicago; 60 were killed.
Charleston, Missouri	1895	6.8	Damage occurred in six states.
Grand Banks, Newfoundland	1929	7.4	Submarine landslides broke trans-Atlantic cable, disrupting communications.

*Moment magnitude is used in reconstructing the strength of "preinstrument" earthquakes, because we know something of fault length, rupture length, and area felt.

United States, the five most seismically active states between 1980 and 1991 were:

State	Number Recorded	Largest (*M*)
Alaska	10,253	9.2
California	6,732	7.2
Washington	615	5.5
Idaho	536	7.3
Nevada	398	5.6

Although Alaska and California, states smack on plate boundaries, continue to lead the United States in the number of shakers, significant seismic hazards are recognized in 39 *states*. No state is earthquake proof in fact, as the seismic-risk map in **Figure 4.29** shows. It appears that U.S. residents who want to be seismically safest should move to Texas, Florida, or Alabama (all prone to hurricanes and oil spills, by the way).

Why do such strong earthquakes occur *intraplate*—that is, far from a plate boundary (**Table 4.3**)? Intraplate earthquakes have several characteristics in common:

1. The faults that cause them are ancient—even hundreds of millions of years old—and deeply buried with little, if any, evidence of surface rupture before slipping.
2. Because the rocks of the continental interior are structurally more homogeneous than those at plate boundaries, which are laced with faults, they transmit seismic waves better, causing ground motion over a huge area (**Figure 4.30**).
3. Their origins relate to strain that, in most cases, accumulates much more slowly than it does at plate boundaries.

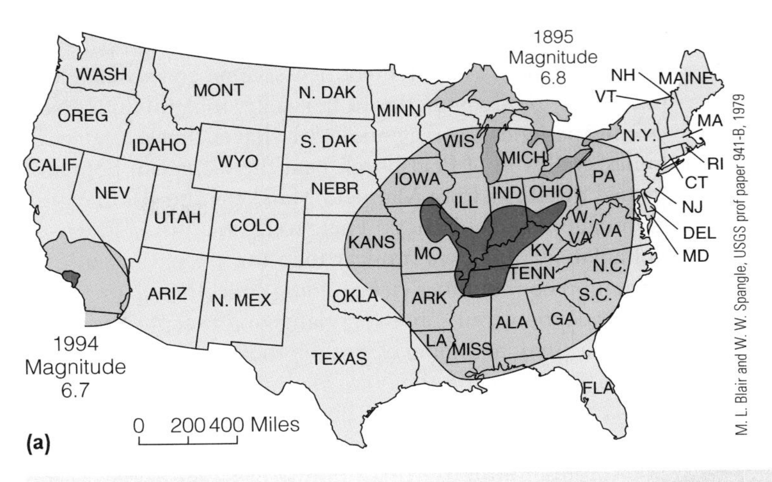

M. L. Blair and W. W. Spangle, USGS prof paper 941-B, 1979

USGS Fact Sheet, 168-95

Figure 4.30 Midcontinent earthquakes. (a) Comparison of the effects of two earthquakes of similar magnitude. The 1895 Charleston, Missouri, earthquake impacted a relatively huge area in the central United States, whereas the 1994 Northridge earthquake's effects were limited to Southern California. Red indicates areas of minor-to-major damage to buildings and their contents. Yellow indicates areas where shaking was felt but there was little or no damage to objects, such as dishes. (b) Damage from the 1886 earthquake along East Bay Street in Charleston, South Carolina.

Principal causes of strain in these settings include the twisting of deep plate interiors by interactions that occur on distant plate boundaries, the loading by sediment over millions of years atop ancient rift-related faults at former divergent margins, and vertical adjustments in the lithosphere caused by deglaciation or heat loss. Ancient faults are fleetingly "reactivated" by these seemingly innocuous, "everyday" processes—even reservoir and river flooding.

Intraplate earthquakes have been reported in "geologically stable" settings as widespread as eastern Australia, South Africa, Quebec (Canada), and the United Kingdom. In the 1980s, an earthquake in Westchester County, New York

(M_w = 4.0) toppled chimneys and caused enough panic for the government to implement new seismic codes that added 2% to 5% to building costs in that state. In April 2002, New York experienced a magnitude-5.1 shaker that centered 15 miles southwest of Plattsburgh near the Canadian border. It rattled dwellings from Maine to Maryland and left cracks in foundations and chimneys. There were no injuries. Rarely do intraplate earthquakes stronger than this occur, and in general, most communities lack building codes to prepare for the rare exception. But some very damaging exceptions can indeed be cited, including one of the most spectacular seismic events ever to occur: the New Madrid (pronounced "MAD-rid") earthquake swarm of 1811–1812 (see Table 4.3).

New Madrid is a small community in southeastern Missouri in the northern Mississippi River Valley. It lies within the Midwest earthquake zone (MEZ), a belt of mostly low-level seismic activity that roughly outlines a down-dropped block of crust largely buried by sedimentary material called the *Reelfoot Rift* (**Figure 4.31**). The Rift probably formed around a half-billion years ago as the young continent of North America began to split apart. The new divergent plate boundary did not fully develop, however, and the continent healed back together before any through-going seaway could form (see Chapter 3). The zone of weakness represented by the ancient rift faults nonetheless remained. Perhaps it is no accident that the Mississippi at least partly follows the path of this failed rift system, too, inherited from a past topographic influence that no longer exists.

Shortly before Christmas in 1811, the MEZ suddenly generated a great earthquake. At that time, few people lived in this frontier region, but the shaking spread so widely across the plate interior that it swayed Washington, D.C., 1,600 kilometers (1,000 miles) away (see Figure 4.30). Church bells rang and scaffolding around the Capitol collapsed. Perhaps no earthquake in recorded history has been more widely felt. The initial shock triggered two additional great tremors in the following 3 weeks. Nobody knows how many people died. An amateur observer in Louisville, Kentucky, recorded almost 2,000 individual quakes with a homemade pendulum. Liquefaction was widespread along the Mississippi River, and the town of New Madrid, Missouri, was completely destroyed as it abruptly subsided 4 meters (12 feet). For a short while during the third main event, near the former Civil War battlefield of Island Number Ten, the Mississippi actually ran backward—even spilling over upstream-facing waterfalls! Subsidence caused some swamps in its floodplain to drain, whereas others became new water bodies—the most famous being Reelfoot Lake in northwest Tennessee, which today is more than 15 meters (50 feet) deep. It seems that the old Reelfoot Rift had suddenly sprung back to life.

Although these events are long forgotten by most Americans, seismic trouble-times remain part of the popular imagination of people in the Mississippi River Valley. Iben Browning, a scientist with a Ph.D. in physiology but who is best known for his work on climate, predicted there would be a repeat of the 1811 New Madrid earthquake on 3 December 1990. His prediction was based on alignment of the planets and consequent gravitational pull, which he believed would be sufficient to trigger an earthquake on that date. Although other scientists discounted the prediction, it created considerable anxiety in Arkansas, Tennessee, Alabama, and Missouri. Months before "the day," earthquake insurance sales boomed, moving companies were booked up, and bottled-water sales rose dramatically—as did sales of quake-related souvenirs. One church sold "Eternity Preparedness Kits," and "Survival Revivals" were held. On the day of the scientifically discredited prediction, nothing earthshaking occurred.

More recently, detailed analysis of the geology shown by trenching across the Reelfoot Fault scarp reveals evidence of three large earthquakes within the past 2,000 years, giving a recurrence-interval estimate of 600 to 900 years.

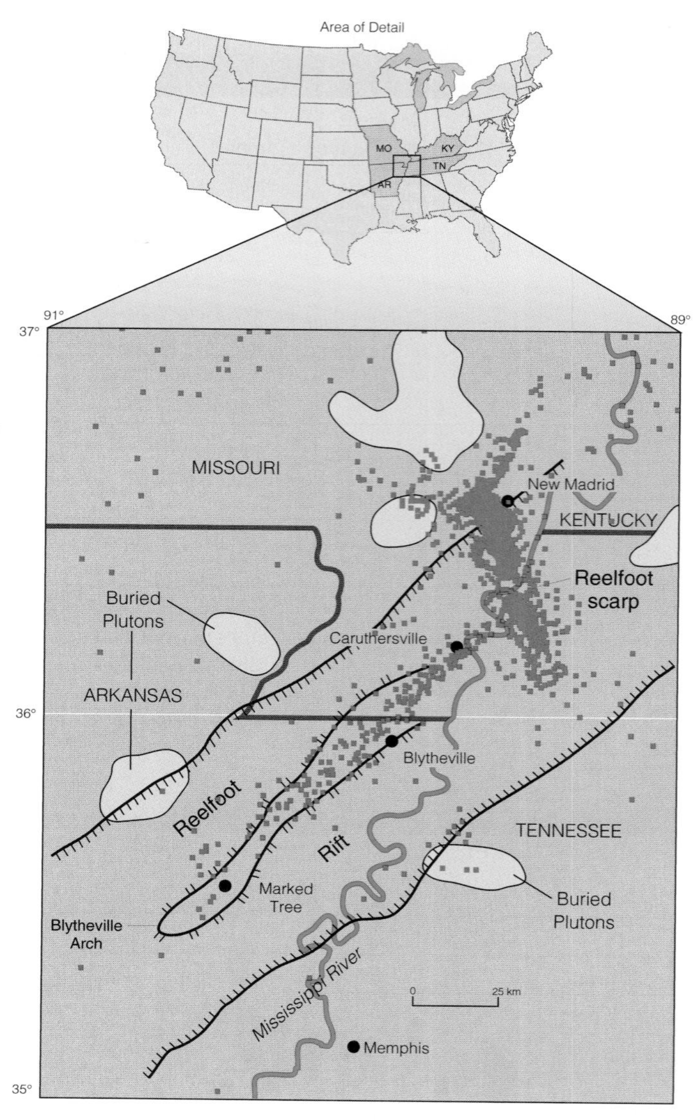

Figure 4.31 The New Madrid seismic zone covers parts of four states and consists of an ancient, fault-bounded depression (rift) buried below 1.5 kilometers (almost a mile) of sedimentary rocks. An area of major seismic hazard, it is studied intensively. (USGS data)

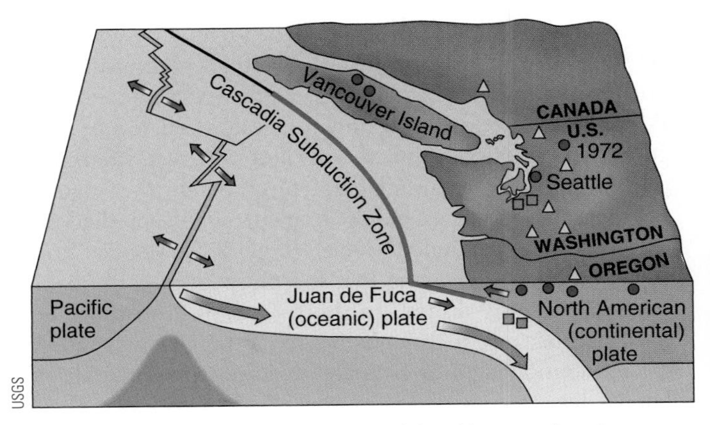

USGS

■ Deep earthquakes (>65 km or 40 mi deep)* occur when the oceanic plate descends beneath the continental plate. The largest deep earthquakes in recent times were in 1949 (M = 7.1) and in 1965 (M = 6.5).

● Shallow earthquakes (<17 km or 10 mi deep)* are caused by faults in the North American continent. Magnitude 7+ earthquakes have occurred along the Seattle fault in 1872, 1918, and 1946.

— Subduction earthquakes are huge quakes that occur when the subduction boundary ruptures. The most recent Cascadia Subduction Zone earthquake was in 1700 and it sent a tsunami as far as Japan.

*Shallow and deep do not refer to focus of quake; rather, depth is subjective and is related to damage caused by faults in the Seattle area.

Figure 4.32 Faults and epicenters in the Seattle, Washington, and Vancouver, British Columbia, area.

The recurring earthquakes were not necessarily of the same magnitude as the 1811–1812 events. Nonetheless, the evidence indicates that the citizens of St. Louis or Memphis should be as concerned about earthquake preparation as those of San Francisco or Los Angeles. Citing field evidence and statistics, some geologists have argued, in fact, that there is a 90% probability of a magnitude 6.7-scale earthquake in the MEZ before the year 2050.

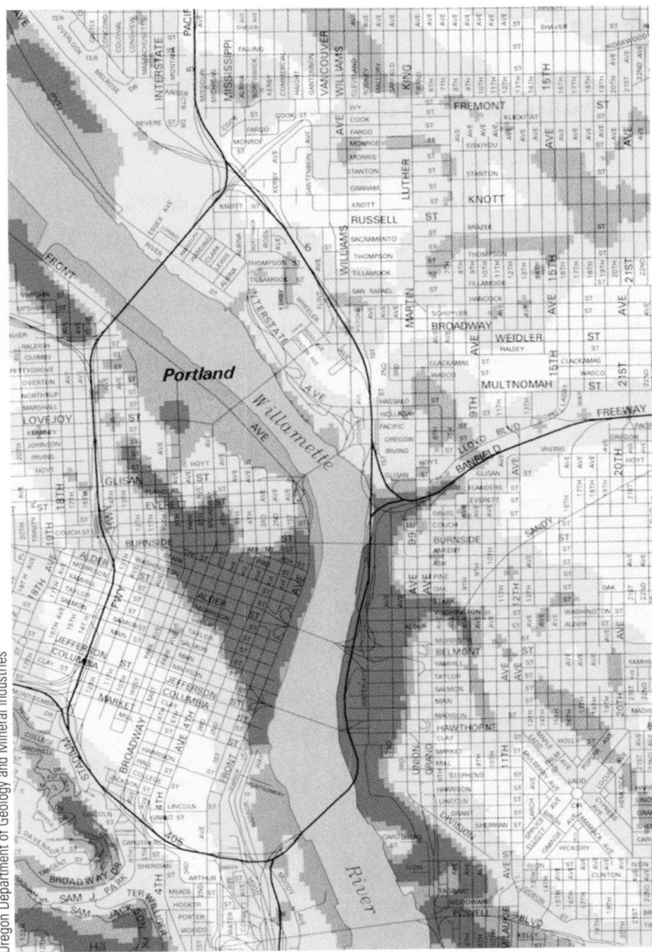

Oregon Department of Geology and Mineral Industries

Figure 4.33 Seismic hazard map of downtown Portland, Oregon. The hazard zones are based on liquefaction potential, landslide potential, and the probable degree of ground shaking. Red denotes greatest hazard; pale yellow denotes lowest hazard.

Pacific Northwest

In years past, earthquake hazards were considered minor in Oregon and Washington—something only people in California and Alaska, sharing the same coastline, had to worry about. During the 1980s, however, research changed this perception, revealing geological evidence that major or great subduction-zone earthquakes occurred repeatedly in the region prior to Euro-American settlement—on average, once every 500 years. Geologists warned on this basis that they will continue striking the region—the next perhaps sooner than we think given, after all, that an "average" implies a "range." The evidence consisted of carbon-dated tsunami deposits and drowned red-cedar forests.

The last large northwestern shake occurred in 1700, approximately 300 years ago. It generated a tsunami that traveled all the way to Japan. The sea-wave deposits are consistent with an earthquake of magnitude 9.0. The source, of course, was the Cascadia subduction zone, which stretches from northern California to Vancouver Island in British Columbia (**Figure 4.32**). Although quakes have rattled Seattle, Eureka, and several other communities in the brief recorded history of the region, nothing catastrophic has taken place. But the Tohoku-Oki earthquake underscores that our recorded histories are, perhaps as a rule, imperfect measures of actual risk when it comes to very large events.

Oregon, Washington, and British Columbia are now preparing for the worst-case seismic scenario at many levels of government and organization. The year 2000 marked the tricentennial of the last great earthquake generated along the Cascadia subduction zone, the source of all the concern. To commemorate this, a hundred earth scientists and public officials gathered in Seaside, Oregon, to review what we know and discuss how we should prepare. Such intense communication between government and science is certainly a prerequisite for riding out the next "big one" in this part of the world (**Figure 4.33**).

Prediction

A guy ought to be careful about making predictions. Particularly about the future.

Yogi Berra, baseball player

It would seem that earthquake prediction has great potential for saving lives and reducing property damage. A good prediction would give the location, time, and magnitude of a future earthquake with acceptable accuracy. Prediction was the hottest area of geophysical and geological research from the 1970s to the early 1990s, in fact, based on the belief that measurable phenomena occurring before large and small earthquakes called **precursors** could be identified. Researchers studied earthquakes and seismograms where presumed precursors were seen. Such things as changes in the ratio of P- and S-wave speeds, ground tilt, water-well levels, and emission of noble gases in groundwater were measured, and piles of data were accumulated. Unfortunately, the hope of finding precursors that would lead to reliable predictions seems to have evaporated. In fact, the U.S. Geological Survey is on record as saying that the prospect of earthquake prediction is very dismal. In short, earthquakes cannot be predicted because the mechanics of earthquake generation are too complicated to evaluate given our present state of knowledge.

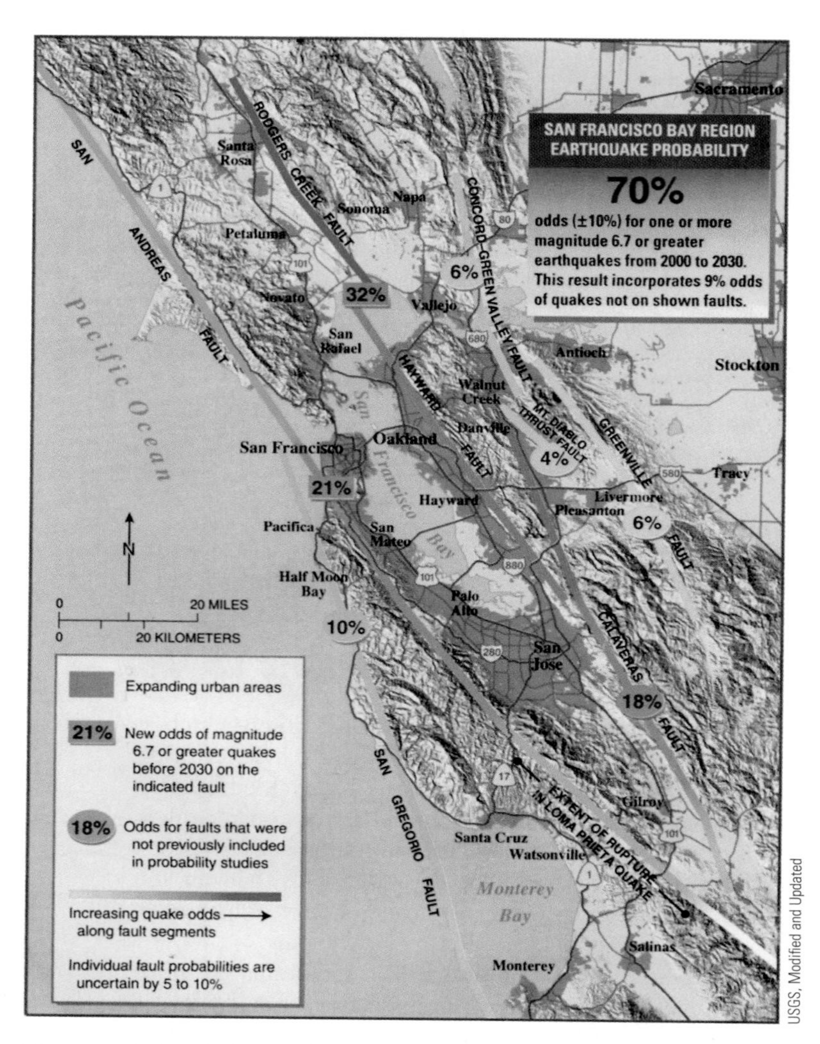

Figure 4.34 Probabilities that one or more magnitude-6.7 or greater earthquakes will strike on specific faults in the San Francisco Bay region between 2000 and 2030. The San Andreas, Rodgers Creek, and Hayward faults have the highest probabilities. Total probability for the region is computed as 70% (±10%). These probabilities were developed by scientists with the U.S. Geological Survey, part of the U.S. Department of the Interior, and thus constitute official long-term *forecasts.* The message is that all communities in the San Francisco Bay region should keep preparing for earthquakes.

Forecasts

We are familiar with weather forecasts and are aware that accuracy declines as the time span of the prediction becomes longer. We can usually rely on one-week forecasts, but a year ahead would be asking too much. The same can be said of earthquakes, only in reverse. There is an old saying in geology that "the longer it has been since the last earthquake, the sooner we can expect the next one."

A new approach is to evaluate the probability of a large earthquake occurring on an active fault during a given time period. This falls under the heading of *long-term forecasting*. An example of this is the U.S. Geological Survey's and other scientists' conclusion that there is a 70% probability of at least one magnitude-6.7 or greater earthquake striking the San Francisco Bay region between 2000 and 2030. Such an event would be capable of causing widespread damage (**Figure 4.34**). Such forecasts are made on the basis of measured plate-tectonic motion and slip on faults. The inexorable movement of the Pacific plate past the North American plate loads strain on the San Andreas network. Periodically, this strain is released on one of the faults in the system and an earthquake occurs.

Statistical Approach

By compiling statistical evidence pertaining to past earthquakes in a region, we acquire basic data for calculating the statistical probability for future events of given magnitudes. These calculations may be done on a worldwide scale or on a local scale, such as the example in **Figure 4.35**. Analysis of the graph indicates that, for the particular area in Southern California, the statistical **recurrence interval**—that is, the length of time that can be expected between events of a given magnitude—is 1,000 years for a magnitude-8 earthquake (0.1/100 years), about 100 years for a magnitude-7 earthquake (1/100 years), and about 10 years for a magnitude-6 earthquake (almost 10/100 years). The probability of a magnitude-7 occurring in any one year is thus 1%, and of a magnitude-6, 10%. On an annual basis worldwide, we can expect, on average (with roughly a century of data), at least one magnitude-8 earthquake, 17

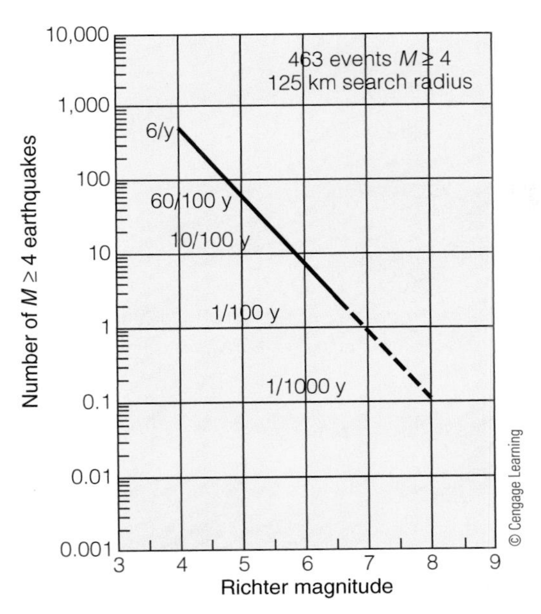

Figure 4.35 Graph of 463 earthquakes of magnitude 4 or greater in a small area near a nuclear reactor in Southern California over a period of 44 years. Statistically, the graph shows that 600 magnitude-4 earthquakes can be expected in 100 years, or 6 per year, on average. The probability of a magnitude-8 earthquake in 100 years is 0.1, which is 1 in 1,000 years. Such plots can be constructed for a region or for the world. The graphs are all similar in shape; only the numbers differ.

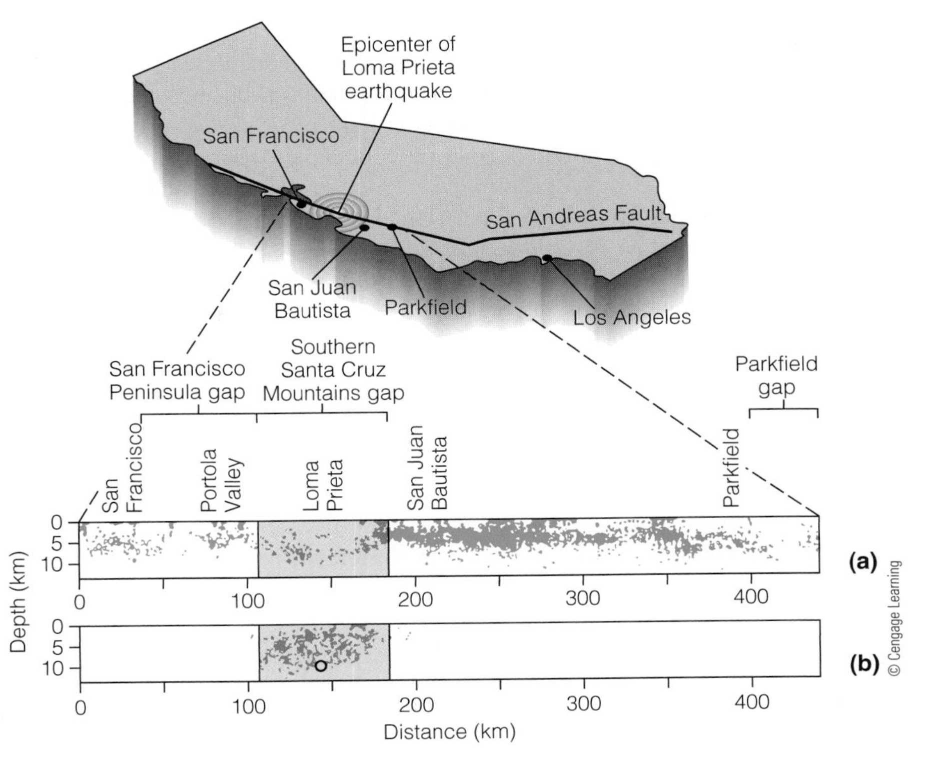

Figure 4.36 (a) Historical seismic activity of about 450 kilometers (275 miles) of the San Andreas Fault south of San Francisco before the 1989 Loma Prieta earthquake. The southern Santa Cruz Mountains seismic gap, where Loma Prieta is located, is highlighted. Other seismic gaps apparent in this historical record at the time were one between San Francisco and the Portola Valley, and one southeast of Parkfield. (b) Seismicity record of the Southern Santa Cruz Mountains portion of the fault after the 1989 earthquake. The former seismic gap had been "filled" by the earthquake (open circle) and its aftershocks.

magnitude-7 earthquakes, and no fewer than 134 earthquakes of magnitude 6 (see Table 4.1). Thus, for seismically active regions, historical seismicity data, possible deficiency notwithstanding, can be used to calculate the *probability* of damaging earthquakes. Although these numbers are not really of predictive value as we defined it, they can be used by planners for making zoning recommendations, by architects and engineers for designing earthquake-resistant structures, and by others for formulating other life- and property-saving measures.

consider this . . . Humans have caused earthquakes by injecting fluids, such as oil-field waters and chemical warfare wastes, into deep wells. How does this open the possibility of controlling earthquake activity, and what might be a major deterrent to the effort?

Geological Methods

Research suggests that active faults and segments of long, active faults tend to have recurring earthquakes of characteristic magnitude, rupture length, and displacement. For example, six earthquakes have occurred along the San Andreas Fault at Parkfield since 1857 (**Figure 4.36**). They had Richter magnitudes of around 5.6, rupture lengths of 13 to 19 kilometers (8–12 miles), and displacements averaging 0.5 meter (1.6 feet). These earthquakes have recurred about every 22 years since the earliest recorded earthquake. A 95% probability of an earthquake was predicted for this segment between 1988 and 1993, the first prediction in the United States to be endorsed by scientists and subsequently issued by the federal government. The prediction was a failure.

Geophysical and seismological precursors, as noted previously, have not proved to be reliable predictors of earthquakes. *Seismic gaps*, however, can indicate stretches of future fault activity. Seismic gaps are stretches along known active fault zones within which no significant earthquakes have been recorded. It is not always clear whether these fault sections are "locked," and thus building up strain energy, or if motion (creep) is taking place there that is relieving strain. Such gaps existed and were "filled," so to speak, during the Mexico City (1985), Loma Prieta (1989; see Figure 4.36), and Izmit (1999) earthquakes. Seismic gaps serve as alerts, or warnings, of possible future events and can be used as forecasting tools.

Because recorded history in North America is short, geologists need other means of collecting frequency data of large prehistoric earthquakes. One method is to dig trenches into marsh or river sediments that have been disrupted by faulting in an effort to decipher a region's **paleoseismicity**, its rock record of past earthquake events. Kerry Sieh of the California Institute of Technology has done this across the San Andreas in Southern California (**Figure 4.37**). Sieh found an intriguing history of seismicity recorded in disrupted marsh deposits at Pallett Creek and liquefaction effects extending from the seventh century to a great earthquake in 1857. Ten large events, extending from 650 to 1857 C.E., were dated using ^{14}C. The average recurrence interval for these ancient earthquakes is 132 years, but they are clustered in four groups. Within each cluster, the recurrence interval is less than 100 years, and the intervals between the clusters are two to three centuries in length. The last big one, also the last one of a cluster, was in 1857. Thus, it appears that this section of the San Andreas may remain dormant until late in the 21st century or beyond.

In summary, we remain uncertain that a quake will follow well-defined precursory phenomena, such as wave-velocity changes or anomalous animal behavior (see Case Study 4.4). It seems that no one phenomenon is a predictor, and many changes will have to be monitored over a long period before our knowledge of fault behavior is refined enough for reliable prediction. Is it worth the price to try?

Kerry E. Sieh, Earth Observatory of Singapore

Figure 4.37 Disrupted marsh and lake deposits along the San Andreas Fault at Pallett Creek near Palmdale, California. Sediments range in age from about 200 C.E. at the lower left to 1910 C.E. at the ground surface. Several large earthquakes are represented by broken layers and buried fault scarps.

Mitigation

Reducing earthquake risks is an admirable goal of scientists and lawmakers. Building codes provide the first line of defense against earthquake damage and help ensure the public safety. Strict building codes in seismic regions, such as the Pacific Rim, reduce damage and loss of life. Laws passed after the 1933 (Long Beach) and 1971 (Sylmar) California earthquakes have proved the effectiveness of strict earthquake-resistant design. A good example is the magnitude-6.2 Morgan Hill (California) earthquake that shook West Valley College 30 kilometers (20 miles) from the epicenter. Seismic instruments on the gymnasium showed that the roof was so flexible that in a strong seismic event it could collapse (**Figure 4.38**). Flexible roofs were permitted by the code at that time, and many gyms and industrial buildings were built that way. As a result of the experience at West Valley College, the Uniform Building Code was revised—this is the code used by hundreds, if not thousands, of municipalities across the country. The revision requires that the roofs be constructed to be less flexible and thus able to withstand nearby or distant strong earthquakes. Most large cities in strong earthquake zones have their own building codes, patterned after the Uniform Building Code, that require construction to modern seismic standards. For instance, ground response due to differing soil types is now more appreciated as contributing to quake damage. **Figure 4.39** shows how the code has evolved from 1955 as more information was obtained. For example, soft clays react more violently with earthquake waves than do granites, and this difference in response is now taken into consideration.

The primary consideration in earthquake design is to incorporate resistance to horizontal ground acceleration, or *base shear.* Strong horizontal motion tends to topple poorly built structures and to deform more flexible ones. In California, high-rise structures are built to withstand about 40% of the acceleration of gravity in the horizontal direction (0.4 g), and single-family dwellings are built to withstand about 15% (0.15 *g*). **Base isolation** is now a popular design option. The structure, low- or high-rise, is placed on Teflon plates, rubber blocks, seismic-energy dissipaters that are similar to auto shock absorbers, or even springs, which allows the ground to move but minimizes building vibration and sway.

The ultimate earthquake resistance for large buildings is provided by base isolation. The century-old Salt Lake

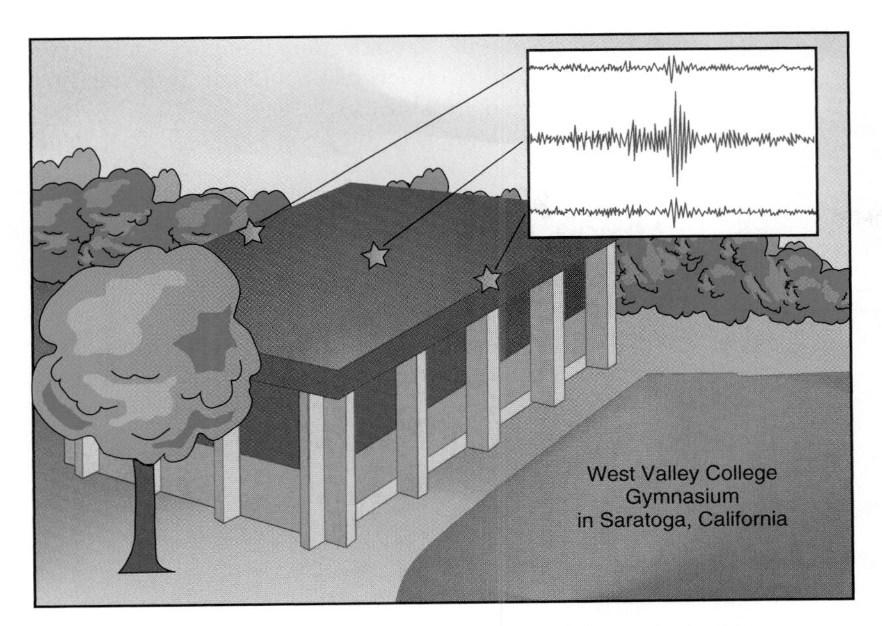

Figure 4.38 Seismic records (upper right) obtained during the 1984 Morgan Hill, California, earthquake led to an improvement in the Uniform Building Code (a set of standards used in many states). The center of the gym roof shook sideways three to four times as much as the edges. The code has since been revised to reduce the flexibility of such large-span roof systems, and thereby improve their seismic resistance. (USGS)

Figure 4.39 Earthquake requirements in building codes have increased over time as scientists and engineers have obtained new information. Note that recent codes specify separate criteria for different ground types. (USGS data)

City and County Building (**Figure 4.40**), scheduled at the time for demolition, was cut from its foundation and retrofitted with 440 rubber base isolators. These permit the structure to move as a single unit during an earthquake while the ground shakes in all directions (**Figure 4.41**). About 75% of Utah's population live near the Wasatch Range and the fault responsible for the uplift. The retrofit of older public buildings is an example of the awareness of earthquake hazards by government officials and the public. In this case, it saved an architectural treasure for future generations.

D. D. Trent

Figure 4.40 Salt Lake City and County Building, Utah. Constructed in 1890, largely of brick, it is now seismically retrofitted to meet the modern building code.

Survival Tips

Knowing what to do before, during, and after an earthquake is of utmost importance.

Before an Earthquake

Figure 4.42 provides the United States Federal Emergency Management Agency's (FEMA) suggestions for minimizing the possibility of damage, fire, and injuries in the home.

During an Earthquake

- Remain calm and consider the consequences of your actions.
- If you are indoors, stay indoors and get under a desk, a bed, or a strong doorway.

D. D. Trent

Figure 4.41 Basement of the Salt Lake City and County Building. The building has been seismically retrofitted by cutting the massive structure from its foundation pillars (one shown on the far left, another in the center of the photograph) and lifting the building onto 440 large rubber isolators (one left of center, another at the extreme right of the photograph).

- If you are outside, stay away from buildings, walls, power poles, and other objects that could fall. If driving, stop your car in an open area.
- Do not use elevators; if you are in a crowded area, do not rush for a door.

After an Earthquake

- Turn off the gas at the meter.
- Use portable radios for information.
- Check water supplies, remembering that there is water in water heaters, melted ice, and toilet tanks—all generally potable in developed countries. Do not drink waterbed or pool water.
- Check your home for damage.
- Do not drive.

Earthquakes and Human Activities

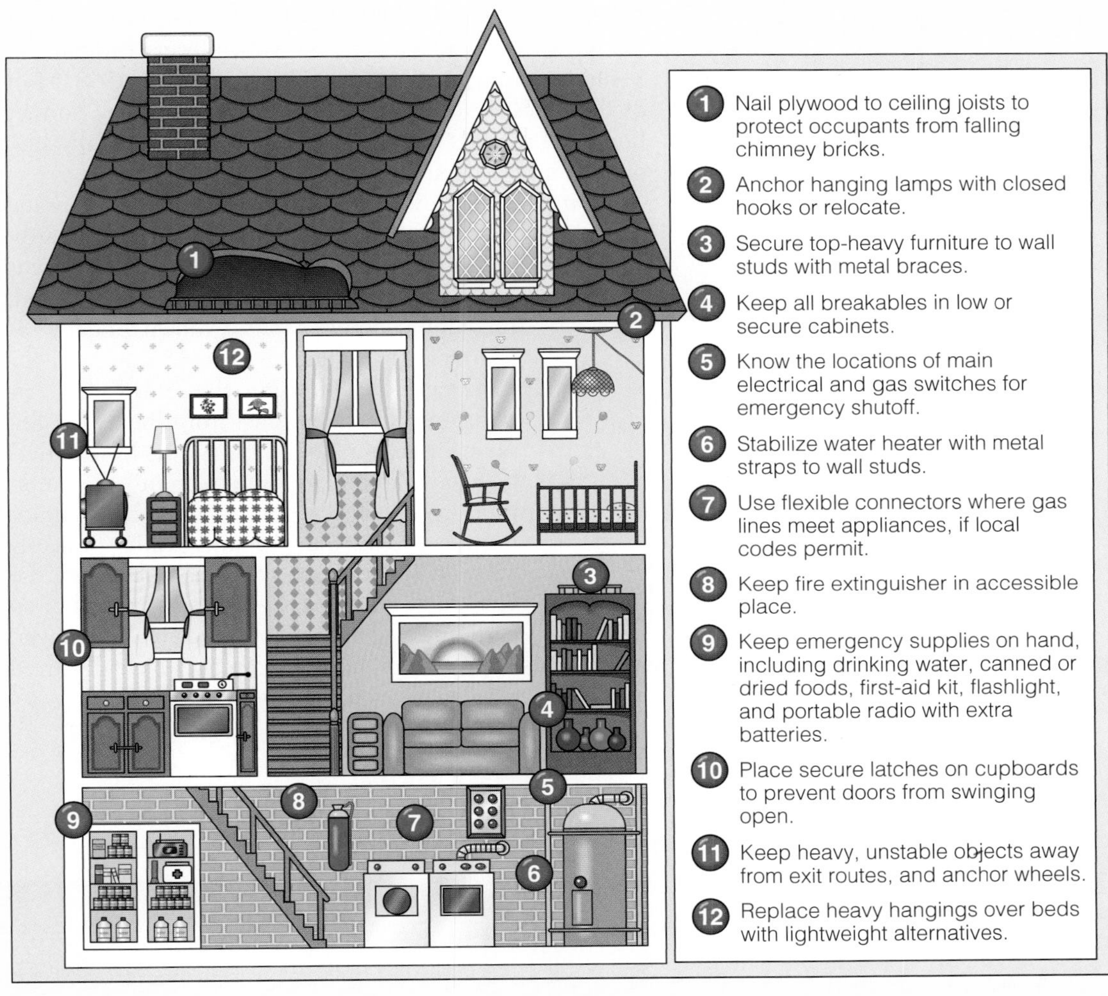

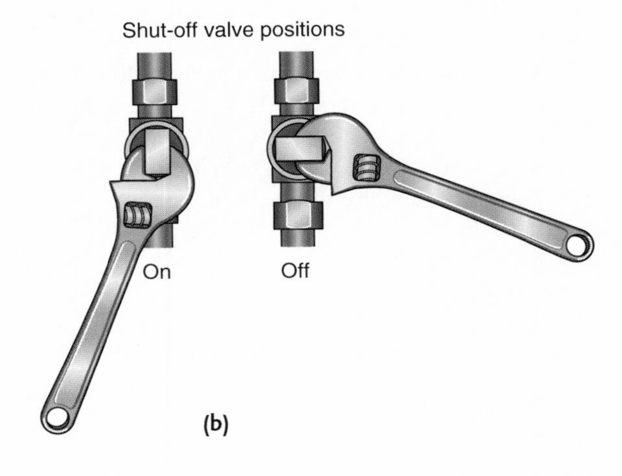

(a) How to minimize earthquake damage in the home in advance. (b) Keep a small crescent wrench at the gas meter. Turn off the gas by turning the valve end 90°. (Source: FEMA)

4.1 Earthquakes, Landslides, and Disease

Dynamically induced landslides are not particularly newsworthy, even though the 1994 Northridge earthquake caused 17,000 of them. However, the soil dislodged during sliding caused an outbreak of *coccidioidomycosis,* commonly known as "valley fever." Endemic to the southwestern United States, the disease causes persistent flu-like symptoms and in extreme cases can be fatal. Exposed to soil dust raised during sliding, the victims breathe in airborne *Coccidioides immitis* spores. From 24 January to 15 March, 166 people were diagnosed with valley fever symptoms in Ventura County, up from only 53 cases in all of 1993. Most of the cases were reported from the Simi Valley, former home of President Ronald Reagan, an area where only 14% of the county's population resides (**Figure 1**).

Large clouds of dust hung over the Santa Susana Mountains for several days after the earthquake, promoted by the lack of winter rains preceding the quake. During this period, pressure-gradient winds, known locally as "Santa Ana" winds, of 10 to 15 knots (11–17 miles per hour) blew into the Simi Valley, carrying in spore-laden dust from the Santa Susana Mountains to the northeast. Researchers believe that the more "coherent" metamorphic rocks of the San Gabriel Mountains northeast of the San Fernando Valley probably account for the lack of cases reported in the epicentral region. This is the first report of an earthquake-associated outbreak of valley fever, even though many earthquakes have occurred in coccidioidomycosis endemic areas.

Figure 1 Histogram of the almost-epidemic outbreak of coccidioidomycosis (valley fever) in Ventura County in early 1994 following the Northridge earthquake.

4.2 Predictable "Future Shocks"

A large earthquake is normally followed by thousands of smaller-magnitude earthquakes, known as aftershocks (**Figure 1**). If the main shock is small—say, in the magnitude-4.0 to -5.0 range—aftershocks are small and nonintimidating. Following Northridge-size and bigger earthquakes, however, the strongest aftershocks (M = 5.0–6.0 at Northridge) can cause buildings damaged by the main earthquake to collapse and, even worse, can greatly increase anxiety in the already damaged psyche of the local citizenry. Aftershocks are caused by small adjustments (slips) on the causative fault or on other faults close to the causative one. For example, try this: Push the eraser on the end of a lead pencil across a desktop. You'll find that it does not slide smoothly; it moves in jerky jumps and starts. This is called "stick-slip" and it is what happens along faults that are adjusting after a big earthquake.

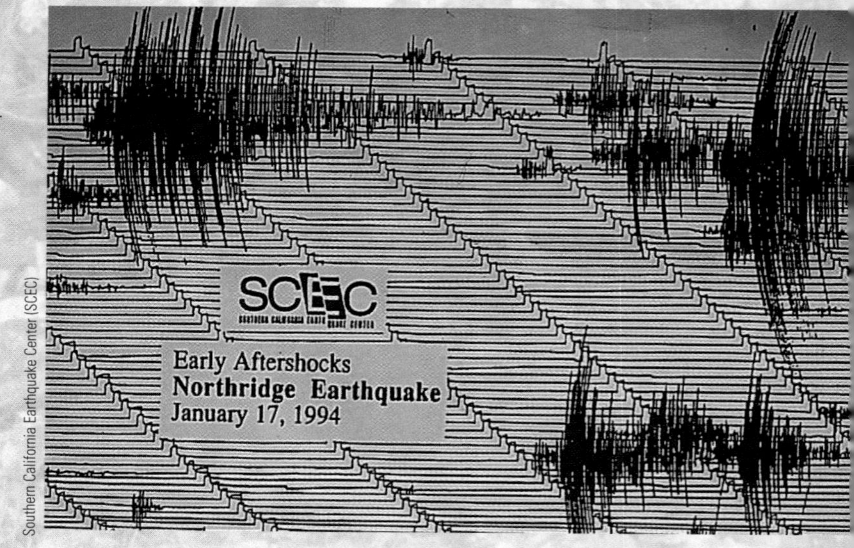

Figure 1 Seismogram of early aftershocks following the Northridge earthquake, one of them a magnitude 5.6.

Los Angeles experienced 2,500 aftershocks in the week following the Northridge earthquake. Three strong ones, magnitude 5 or greater, occurred the first day. The largest (M = 5.6) occurred 11 hours after the main shock, causing concern among rescuers digging for victims beneath the rubble and the already traumatized citizens. Aftershocks follow statistically predictable patterns, as exemplified by the Northridge sequence and described by the earlier mentioned Gutenberg–Richter law. On the first day, there were 188 aftershocks of magnitude 3 or greater; on the second day, only 56 were recorded (**Figure 2**). By fitting an equation to the distribution of aftershocks during the first few weeks, seismologists were able to estimate the number of shocks to be expected in the future. Statistically, there was a 25% chance of another magnitude-5 or greater aftershock occurring within the following year, but it did not happen.

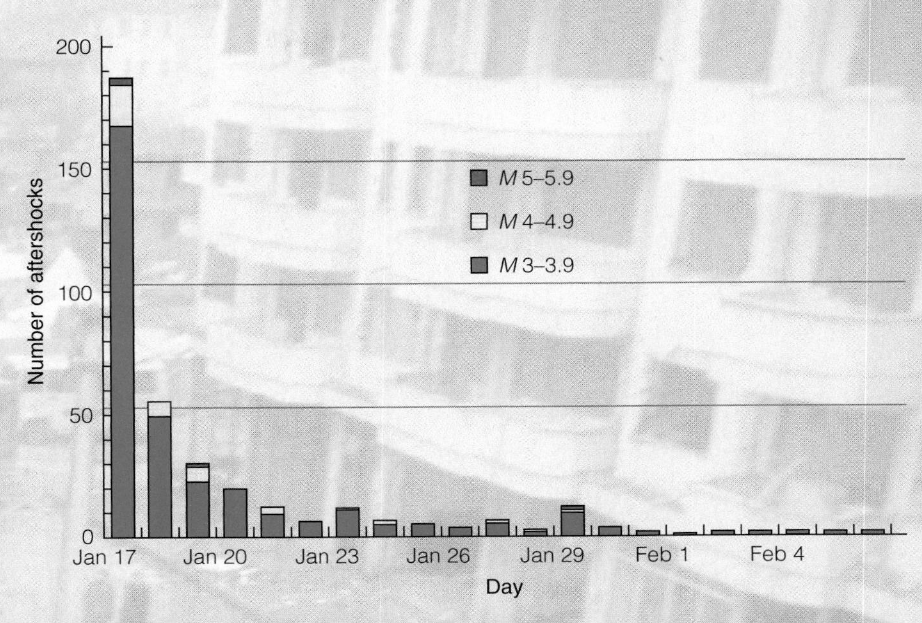

Figure 2 Daily record of aftershocks of magnitude 3.0 to 5.9 during the 3 weeks after the main shock at Northridge. Note the sharp drop in aftershock frequency in the first 4 days. (Redrawn from F. Harp and R. Jibson, USGS)

4.3 Rx for Failed Freeways

Extensive damage to freeway bridges and overpasses typically accompanies earthquakes in large urban areas. Such damage has occurred in Alaska, California, and Japan. Overpass damage commonly is due to failure of the shorter columns, which lack the flexibility of longer ones. During an earthquake, the tall columns supporting a bridge or an overpass system bend and sway with the horizontal forces of the quake. Because the parts of the overpass system are tied together, the stresses are transferred through the structure to the short columns, which are designed to bend only a few centimeters (**Figure 1**). Failure causes the short columns to bulge just above ground level, which breaks and pops off the exterior concrete, exposing the warped, "birdcaged" steel in the interior. In addition, high vertical accelerations can cause some columns to punch holes through the platform deck. With excessive horizontal motion, some deck spans may slip off their column caps at one end and fall to the ground, like tipped dominoes.

In 1971, the California Department of Transportation (CalTrans) decided to retrofit 122 overpasses to alleviate these problems. One aspect of the retrofitting was jacketing the short columns with steel or a composite substance and filling the space between the jacket and the original column with concrete (**Figure 2a**). This allows the columns to bend 12.5 centimeters (5 inches), instead of 2.5 centimeters (1 inch), without shattering. Another solution is to increase columns' horizontal strength by wrapping heavy steel rods around their vertical support bars, particularly on short columns. This allows the columns to bend but prevents "bird-

Figure 1 The difference in flexibility of long and short columns results in failure of the short ones.

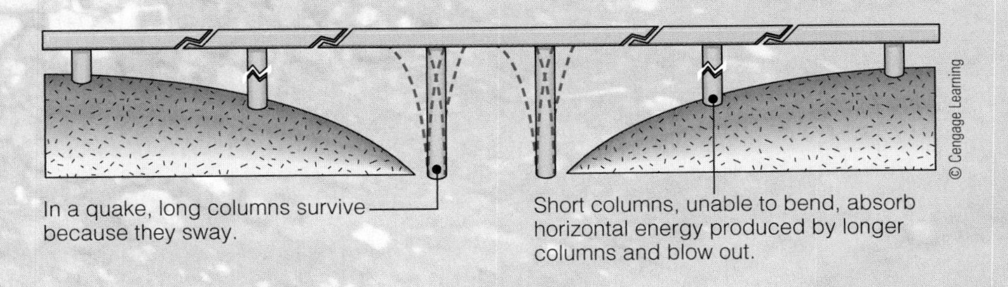

caging" or permanent bending (see Figure 2b). To prevent the decks from slipping off their supports and dropping to the ground, steel straps or cables are installed at the joints (see Figure 2c). Of the 122 overpasses that CalTrans retrofitted, not one collapsed in the 1994 Northridge earthquake; 10 of the 11 that collapsed were slated for future retrofitting.

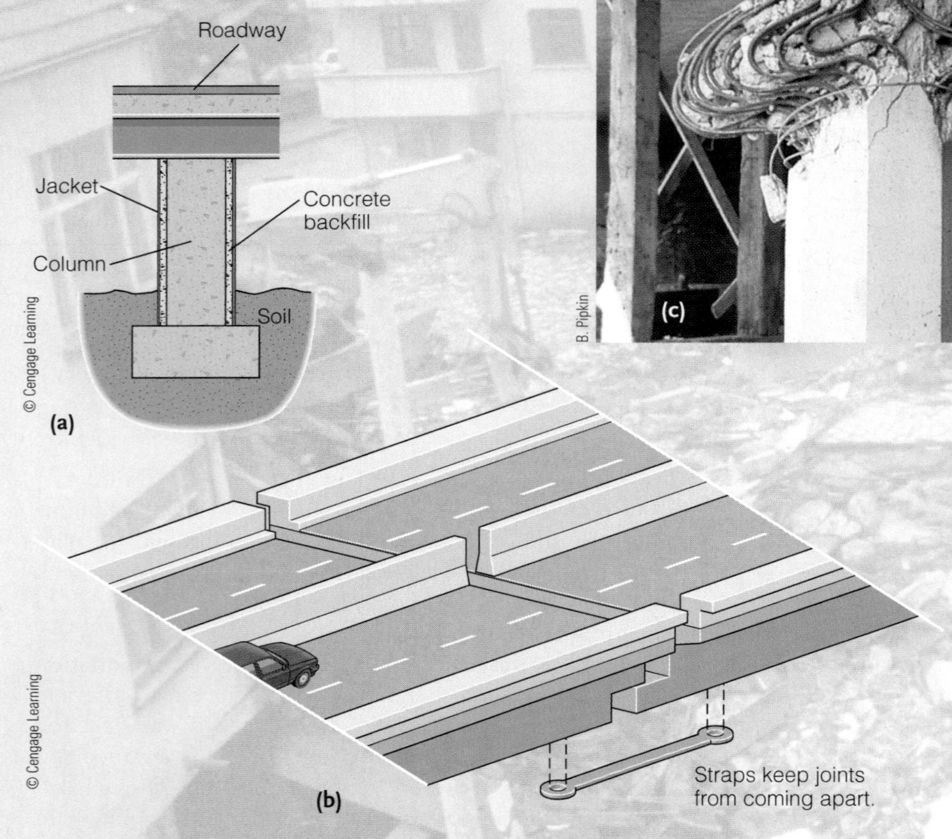

Figure 2 Earthquake-resistant design for bridges and overpasses. (a) Short columns are retrofitted with steel or composite jackets that allow them to bend five times as much. (Source: *The Los Angeles Times*) (b) Steel straps are used to hold deck sections together, preventing them from falling off their support columns. (c) Old method of constructing highway support columns. The vertical steel supports have bent and "birdcaged."

© Aldo Castellani/Alamy

4.4 Depressed Tigers, Restless Turtles, and Earthquakes!

Anomalous animal behavior preceding earthquakes is well documented. Domesticated animals, such as barnyard fowl, horses, cats, and dogs, have been known to behave so peculiarly before big events that they attracted the notice of people not knowledgeable in what is normal or abnormal animal behavior (**Figure 1**). Remember that, although anomalous animal behavior precedes many earthquakes, it does not appear to precede every earthquake, and abnormal animal behavior is not always followed by an earthquake. Other natural phenomena, such as atmospheric disturbances, can also cause animals to behave strangely. Following are some reported incidents of unusual animal behavior noted before earthquakes:

- Tientsin Zoo, China, 1969: 2 hours before a magnitude-7.4 earthquake, the tiger appeared depressed, pandas screamed, turtles were restless, and the yak would not eat.
- Haicheng, China, 1975: 1½ months before a magnitude-7.3 earthquake, snakes came out of hibernation; 1 to 2 days before, pigs would not eat and they climbed walls; 20 minutes before, turtles jumped out of the water and cried.
- Tokyo, Japan, 1855: 1 day before a magnitude-6.9 earthquake, wild cats cried and rats disappeared.
- Concepción, Chile, 1835: 1 hour and 40 minutes before an earthquake, flocks of seabirds flew inland and dogs left the city.
- San Francisco, 1906: dogs barked all night before a magnitude-8.3 earthquake.
- Friuli, Italy, 1976: 2 to 3 hours before a magnitude-6.7 earthquake, cats left their houses, the village mice and rats left their hiding places, and fowl refused to roost.

© B. Pipkin

Figure 1 Anomalous animal behavior. An earthquake may be in the offing if your dog dons a hard hat.

questions to ponder

1. What do you think might explain the strange animal behavior observed before earthquakes?
2. Do you think that field observation of animal behavior could or should be a part of any earthquake forecasting program? If your response to this question is "yes," what would be the practical pitfalls of attempting this, other than what is already described in this case study?

PHOTO GALLERY

earthquakes are hard on cars

The sight of cars buried beneath a pile of bricks and rubble after an earthquake seems to be standard fare for the press. We captured a few of these scenes just by coming upon them and hope you feel, as we do, that there is a light side to most bad natural disasters. The other photos were just too unusual or amusing to pass up.

1 This luxury car seems to be pursuing the garage in which it is normally kept. The house and garage slid down the hillside, but the car was parked with its rear wheels on the driveway, which was on stable ground. Nobody was hurt; Northridge, 1994. B. Pipkin

2 A row of cars in need of body-and-fender work because their owners were asleep in the apartment building at the time of the earthquake; Northridge, 1994. © Reuters/Corbis

Tensional rifting associated with the divergent boundary between the North American and Eurasian plates, in central Iceland.
© 2011/Lee Frost/Robert Harding Picture Library/Getty Images

3 Earthquakes have no respect for "No Parking" signs; Boys Market Sylmar, California, 1971. James Williams

5 The Hector Mine earthquake of 16 October 1999 was a real bomb! The magnitude-7.0 event occurred in California's remote Mojave Desert. The fault scarp, shown here, forms a gash across the desert for 40 kilometers (25 miles), but the earthquake did little damage. This area happens to be a U.S. Marine Corps bombing range, as you can see, which made geologists' fieldwork interesting. Ted Reeves, Chaffey High School, Upland, California

4 One-hundred-yard dash? The 21 September 1999 earthquake (M_w 7.8) in Taiwan was caused by movement on the Chelungpu fault. This running track at Kuang Fu High School was built across the fault, which here had a vertical offset of 2.5 meters (8.1 feet). It is obvious that no 100-meter-dash records will be broken here. Yi-Ben Tsai, National Central Univ. Taiwan

SUMMARY Earthquakes and Human Activities

Earthquakes

Cause

Movements on fractures in the Earth's crust, known as *faults*, result in three types of wave motion: P- and S-waves, which are generated at the focus of the earthquake and travel through the Earth, and surface waves.

Distribution

Most (but not all) large earthquakes occur near plate boundaries, such as the San Andreas Fault, and represent the release of stored elastic strain energy as plates slip past, over, or under each other. Intraplate earthquakes can occur at locations far from plate boundaries where deep crust has been faulted, probably at "failed" continental margins.

Measurement Scales

Some of the scales used to measure earthquakes are the modified Mercalli intensity scale (based on damage), the Richter magnitude scale (based on energy released as measured by maximum wave amplitude on a seismograph), and the moment magnitude (based on the total seismic energy released as measured by the rigidity of the faulted rock, the area of rupture on the fault plane, and displacement).

Earthquake-Related Hazards and Mitigation

Ground Shaking

Damaging motion caused by shear and surface waves.

Ways to reduce effects: seismic zoning; building codes; construction techniques, such as shear walls, seismic joints, and frames bolted to foundations

Landslides

Hundreds of landslides may be triggered by an earthquake in a slide-prone area.

Ways to reduce effects: proper zoning in high-risk areas

Ground Failure (Spontaneous Liquefaction)

Horizontal (lateral) movements caused by loss of strength of water-saturated sandy soils during shaking and by liquefaction of quick clays.

Ways to reduce effects: building codes that require deep drilling to locate liquefiable soils or layers

Ground Rupture/Changes in Ground Level

Fault rupture and uplift or subsidence of land as a result of fault displacement.

Ways to reduce effects: geological mapping to locate fault zones, trenching across fault zones, implementation of effective seismic zonation, such as the Earthquake Fault Zones Act in California

Fire

In some large earthquakes, fire has been the biggest source of damage.

Ways to reduce effects: public education on what to do after a quake, such as shutting off gas and other utilities

Tsunamis

Multidirectional sea waves generated by disruption of the underlying seafloor (see Chapter 10).

Earthquake Prediction

It is not feasible at this time to predict within useful limits the time, magnitude, and location of an earthquake. However, seismologists continue to search for clues to accurate prediction, which is what scientists do and what science is.

Upside/downside: No proven large earthquakes predicted. Inaccuracies in short-term prediction have made the public wary.

Earthquake Forecasts

Based on plate movement and known fault movement, there is a 70% chance there will be a magnitude-6.7 earthquake in the San Francisco Bay area between 2000 and 2030. This kind of forecast is the research thrust today and forecasts will vary for each seismic region.

Upside/downside: Good for planning purposes but not for short-term warnings.

Statistical Methods

Statistics reveal that worldwide there will be 2 $M = 8.0+$ and at least 20 $M = 7.0+$ earthquakes each year. Statistics can be applied for a smaller area in earthquake country.

Upside/downside: Good for planning purposes but not for short-term warnings.

Geological Methods

Active faults are studied to determine the characteristic earthquake magnitudes and recurrence intervals of particular fault segments; sediments exposed in trenches may disclose historic large fault displacements (earthquakes) and, if they contain datable carbon-14 material, their recurrence intervals.

Upside/downside: Useful for long-range forecasting along fault segments and for identifying seismic gaps; not useful for short-term warnings.

KEY TERMS

base isolation
base shear
body wave
elastic rebound theory
epicenter
fault
fault creep
focus
Gutenberg–Richter law
intensity scale
isoseismal
liquefaction
long-period seismic waves
longitudinal wave
magnitude
modified Mercalli scale (*MM*)
moment magnitude (M_w or M)
paleoseismicity
precursor
P-wave
recurrence interval
resonance
Richter magnitude scale
seismogram
seismograph
seismology
short-period seismic waves
spontaneous liquefaction
strain
stress
surface wave
S-wave
teleseisms
transverse (shear) wave
tsunami
wave frequency
wave period

STUDY QUESTIONS

1. What is "elastic rebound," and how does it relate to earthquake motion?
2. Distinguish among earthquake intensity, Richter magnitude, and moment magnitude. Which magnitude scale is most favored by seismologists today? Why?
3. Why should one be more concerned about the likelihood of an earthquake in Alaska than about one in Texas?
4. In light of plate-tectonic theory, explain why devastating shallow-focus earthquakes occur in some areas and only moderate shallow-focus activity takes place in other areas.
5. How do scientists locate the epicenter of an earthquake?
6. What should people who live in earthquake country do before, during, and after an earthquake (the minimum)?
7. If you were going to purchase a home, or go to work in a particular building, and were concerned about seismic safety, what information would you like to know before making a commitment?
8. Describe the motion of the three types of earthquake waves discussed in this chapter and their effects on structures.
9. What is the distinction between earthquake prediction versus forecasting? Which is more effective, and why?
10. Why do wood-frame structures suffer less damage than unreinforced brick buildings in an earthquake?

REFERENCES AND FURTHER READINGS

Anonymous. 2010. Catastrophe in the Caribbean. *The Economist* 394(8554): 38–39.

Atwater, Bryan, et al. 1999. *Surviving a tsunami: Lessons from Chile, Hawaii, and Japan.* U.S. Geological Survey Circular 1187, Reston, Virginia.

Avouac, J.-P. 2011. Earthquakes: The lessons of Tohoku-Oki. *Nature* 475: 300.

Bilham, R. 1995. Global fatalities from earthquakes in the past 2000 years: Prognosis for the next 30. In Rundle, J., J. F. Klein, and D. Turcotte (eds.), *Reduction and Predictability of Natural Disasters, Santa Fe Institute Studies in the Sciences of Complexity, Vol. XXV* (pp. 19–31). Addison Wesley Publishing Boston.

Bolt, Bruce. 1998. *Earthquakes, newly revised and expanded.* New York: W H. Freeman, 331 p.

Brocher, T. M., T. L. Pratt, C. S. Weaver, A. D. Frankel, A. M. Trehu, et al. 2000. Urban seismic experiments investigate the Seattle fault and basin. *Eos Transactions of the American Geophysical Union* 81(46): 545, 551, 552.

Celebi, M., P. Spudich, Robert Page, and Peter Stauffer. 1995. *Saving lives through better design standards.* U.S. Geological Survey Fact Sheet 176-95, Reston, Virginia.[1]

Drake, D. 1815. *Natural and statistical view, or picture of Cincinnati and the Miami County, illustrated by maps.* Cincinnati, OH: Looker and Wallace, 251 p.

Ellis, M., J. Gomberg, and E. Schweig. 2001. Indian earthquake may serve as an analog for New Madrid earthquakes. *EOS, Transactions of the American Geophysical Union,* 82(32): 345.

Field, Edward H. 2000. Accounting for site effects in probabilistic seismic hazard analysis of Southern California. *Bulletin of the Seismological Society of America* (6b): 90.

Gomberg, Joan, and Eugene Schweig. 2002. *East meets Midwest: An earthquake in India helps hazard assessment in the central United States.* U.S. Geological Survey Fact Sheet 007-02, Reston, Virginia.

González, Frank I. 1999. Tsunami. *Scientific American* (May): 56–65.

Gore, Rick. 1995. Living with California's faults. *National Geographic* (April): 2–34.

Hickman, Steve, and John Langbein. 2000. *The Parkfield experiment—Capturing what happens in an earthquake.* U.S. Geological Survey feet sheet 049-02, Reston, Virginia.

Hough, Susan E. 2002. *Earthshaking science: What we know (and don't know) about earthquakes.* Princeton, NJ: Princeton University Press, 272 p.

[1]Fact sheets are one- or two-page condensations of a geological problem and are free of charge. Some are found on the Web at http://quake.usgs.gov. Hard copies may be obtained from the U.S. Geological Survey, Mail Stop 977, 345 Middlefield Road, Menlo Park, CA 94025.

______ 2010. *Predicting the unpredictable: The tumultuous science of earthquake prediction.* Princeton, NJ: Princeton University Press, 272 p.

Hough, Susan E., and R. G. Bilham. 2006. *After the earthquakes: Elastic rebound on an urban planet.* New York: Oxford University Press, 336 p.

Iacopi, Robert. 1996. *Earthquake country.* Menlo Park, CA: Lane Books, 146 p.

Kanamori, H. 2012. Earthquake hazards: Putting seismic research to most effective use. *Nature* 483: 147–148.

Knopoff, L. 1996. Earthquake prediction: The scientific challenge. In *Earthquake prediction: Proceedings of the National Academy of Sciences* (pp. 3719–3720). Washington, DC: National Academy of Sciences.

Kockelman, William J. 1984. *Reducing losses from earthquakes through personal preparedness.* U.S. Geological Survey open file report 84-765. Reston, Virginia.

Koper, Keith D., et al. 2001. Forensic seismology and the sinking of the *Kursk. EOS, Transactions of the American Geophysical Union* 82(4): 37.

Kovachs, Robert. 1995. *Earth's fury: An introduction to natural hazards and disasters.* Englewood Cliffs, NJ: Prentice-Hall, 214 p.

Lay, T. 2012. Seismology: Why giant earthquakes keep catching us out. *Nature* 483: 149–150.

Michael, Andrew J., et al. 1995. *Major quake likely to strike between 2000 and 2030.* U.S. Geological Survey Fact Sheet 151-99. Reston, Virginia.

Mori, James J. 1994. Overview: The Northridge earthquake: Damage to an urban environment. *Earthquakes and Volcanoes* 25 (1) (special issue).

Renwald, Marie, Tammy Baldwin, and Terry C. Wallace. 2003. Seismic analysis of the space shuttle *Columbia* disaster (abstract). Geodaze Geoscience Symposium, Tucson, AZ: University of Arizona, 75 p.

Richter, C. 1958. *Elementary seismology.* New York: W. H. Freeman, 768 p.

Wuethrich, Bernice. 1995. Cascadia countdown. *Earth: The Science of Our Planet* (October): 24–31.

Yeats, Robert S. 1998. *Living with earthquakes in the Pacific Northwest.* Corvallis, OR: Oregon State University Press, 400 p.

A tropical volcanic landscape. Tavurvur volcano erupts across the waters of Rabaul caldera on the island of New Britain in Papua New Guinea.

Ross Armstrong/Alamy

5

Volcanoes

"Civilization exists by geologic consent, subject to change without notice."

—Will Durant, historian (1885–1981)

1. How volcanoes erupt?

2. How pumice forms?

3. How scientists can forecast volcanic eruptions?

4. Whether there are any benefits of eruptions?

5. How you can protect yourself during an eruption?

Ashes, Ashes, All Fall Down

Figure 1 Pitting and abrasion by volcanic ash on part of a jet engine from the British Airways flight that encountered ash over Java in 1982.
Eric Moody

Figure 3 Mount Spurr volcano, a potential hazard to aviation. (a) Scientists from the Alaskan Volcano Observatory and the University of Alaska install seismometers at Mount Spurr to detect magma activity below the volcano. (b) The ash cloud from Mount Spurr's 1992 eruption traveled across Canada and the United States on prevailing westerly winds.
© USGS

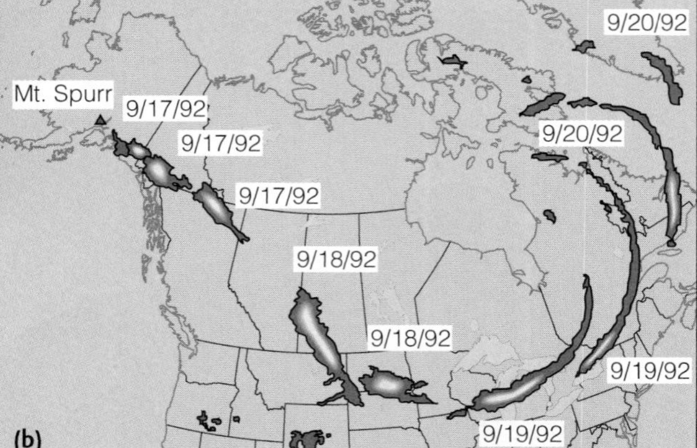

One night in 1982, a British Airways Boeing 747 on a routine flight from Kuala Lumpur, Malaysia, to Australia cruised at an altitude of 12,000 meters (37,000 feet). Just before midnight, sleeping passengers were awakened by a pungent odor. Through the windows, they saw the huge plane's wings lit by an eerie blue glow. Suddenly, engine number 4 flamed out, followed almost immediately by the other three. The plane glided silently for an agonizing 13 minutes. At 4500 meters (14,500 feet), engine number 4 was restarted, then numbers 2, 1, and 3. Nonetheless, an emergency was declared, and the plane landed in Jakarta, Indonesia, with only three engines operating.

This near-death experience gained the attention of the world's airline passengers and pilots. Before this, such a failure seemed virtually impossible in modern aircraft, which have backup systems for almost every contingency. Flying air-gulping jet engines through clouds of volcanic ash, however, turned out not to be one of them (**Figure 1**).

In December 1989, a KLM Boeing 747 encountered airborne ash from Redoubt volcano at about 8,500 meters (28,000 feet) during a descent to land at Anchorage, Alaska. All four engines flamed out, and like the unfortunate British Air flight a few years earlier, the large aircraft suddenly became a glider. After a 4,100-meter fall (13,300 feet, or more than 2 miles), the pilots managed to get the engines restarted, and the plane went on to make what was described as an "uneventful"

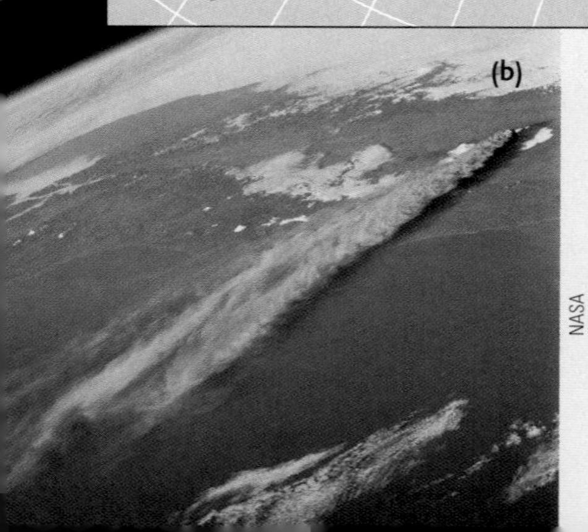

Figure 2 More than 10,000 passengers and millions of dollars in cargo are flown daily on the flight routes shown in red across the northwestern Pacific Ocean and Bering Sea. (a) The flight routes pass over or close by more than 100 active volcanoes (red triangles). (b) Air traffic to the Far East and Asia was disrupted in September 1994 when Russia's Mount Kliuchevskoi, the largest composite volcano in the world, erupted, spewing ash into high-altitude transportation routes.
NASA

landing. All four engines required replacement, as did the windshield and the leading edges of the wings, flaps, and vertical stabilizer, which had all been "sandblasted." The plane's interior was filled with ash, and the seats and avionics equipment had to be removed and cleaned. It cost $80 million to return the aircraft to service. One may wonder what it takes to make a landing "eventful"!

Ash clouds are difficult to distinguish from rain clouds, both visually and with radar. The ash cloud from Mount Pinatubo's 1991 eruption traveled westward more than 8,000 kilometers (5,000 miles) in 3 days from the Philippines to the east coast of Africa. Twenty aircraft were damaged by the cloud, most of them while flying more than 1,000 kilometers (600 miles) from the eruption.

North Pacific air routes are some of the busiest in the world. Since 1980, flying through ash clouds on those routes has caused damage to at least 15 aircraft (including the KLM flight), and worldwide there have been more than 100 inadvertent ash-cloud entries. The Alaskan Peninsula and the Aleutian Islands have 40 historically active volcanoes, and the Kamchatka Peninsula has even more (**Figure 2**). In 1992, Mount Spurr volcano on the Alaskan Peninsula disrupted air traffic several times in the United States and Canada (**Figure 3**).

Historically, there are, on average, five eruptions per year along the 4,400-kilometer (2,750-miles) great circle from Alaska to northern Japan. That means that volcanic ash is present 4 days each year above 19,000 meters (30,000 feet), where passenger jets fly. A massive effort in place to accurately forecast eruptions—the best way to address this potential hazard—involves the Alaskan Volcano Observatory (AVO), U.S. Geological Survey (USGS), University of Alaska, Federal Aviation Administration (FAA), National Oceanic and Atmospheric Administration (NOAA) Weather Service, Michigan Technological University, and Russian Institute of Volcanic Geology and Geochemistry in Petropavlovsk-Kamchatsky. Forecasting data are collected by monitoring seismic activity in the volcanic arc, reviewing satellite imagery to detect ash plumes, and making overflights of the 40 active volcanoes to measure gas emissions, which precede many eruptions—all to help make "flying the friendly skies" of the North Pacific and Far East safe for the traveler. (For more information, see the AVO Web site at: http://www.avo.alaska.edu/.)

questions to ponder

1. How would volcanic ash sucked into a jet engine cause it to seize up? What would allow the engine to start running again as an aircraft glides out from underneath an ash cloud?
2. Why could flying by night be more hazardous than flying by day, with regard to ash-cloud jet encounters?

Courtesy of Federico de la Mano

CourseMate Connection

Start off this chapter by listening to a podcast with National Geographic Young Explorers Grantee Anand Varma as he talks about his experiences photographing the Puyehue-Cordon volcano in Chile. **Log in to the CourseMate at www.cengagebrain.com to listen now.**

VOLCANOES ARE THE MOST DRAMATIC expression of the great heat escaping Earth's interior. They also are conduits for pent-up gases that played an important role in the evolution of Earth's atmosphere, and continue to impact the composition of the air and behavior of our climate today. Volcanoes are not merely terrifying agents of destruction, they also provide surprising benefits to people, as we'll explore later in this chapter. Some of the world's most beautiful landscapes have been created by volcanic activity. Think of the famous national parks and tourist attractions that feature volcanoes prominently. Examples include Yellowstone and Mount Rainier in the United States, Kilimanjaro in Tanzania, Tongariro in New Zealand, and Fujiyama in Japan.

Volcanoes and Plate Tectonics

Plate tectonics plays a role in the origin of the majority of—although certainly not all—volcanoes (**Figures 5.1**, **5.2**), as described in Chapter 3. The mafic material making up Earth's mantle would certainly be hot enough to melt if magically scooped out and brought to Earth's surface all at once. But ordinarily it can't melt where it's positioned deep underground, simply because it is under too much pressure from the overlying rock. The tightly confined individual atoms and compounds making up the mantle and lower crust do not ordinarily have the leeway to move around in a liquid state. It takes special circumstances for this to happen, and plate tectonics helps provide these. At particular depths, primarily less than a few tens of kilometers, natural reduction in pressure ("decompression") provided by the separation of plates at divergent boundaries and continental rifts provides just enough relief to allow partial melting. This is why the largest amount of magma erupted on Earth does so at places like the mid-ocean ridge system (**Figure 5.3**)

But volcanoes also are abundant at convergent plate boundaries, where, if anything, tectonic pressures are much greater. Why should magma form there, too? Recall from Chapter 3 that where oceanic and continental plates collide,

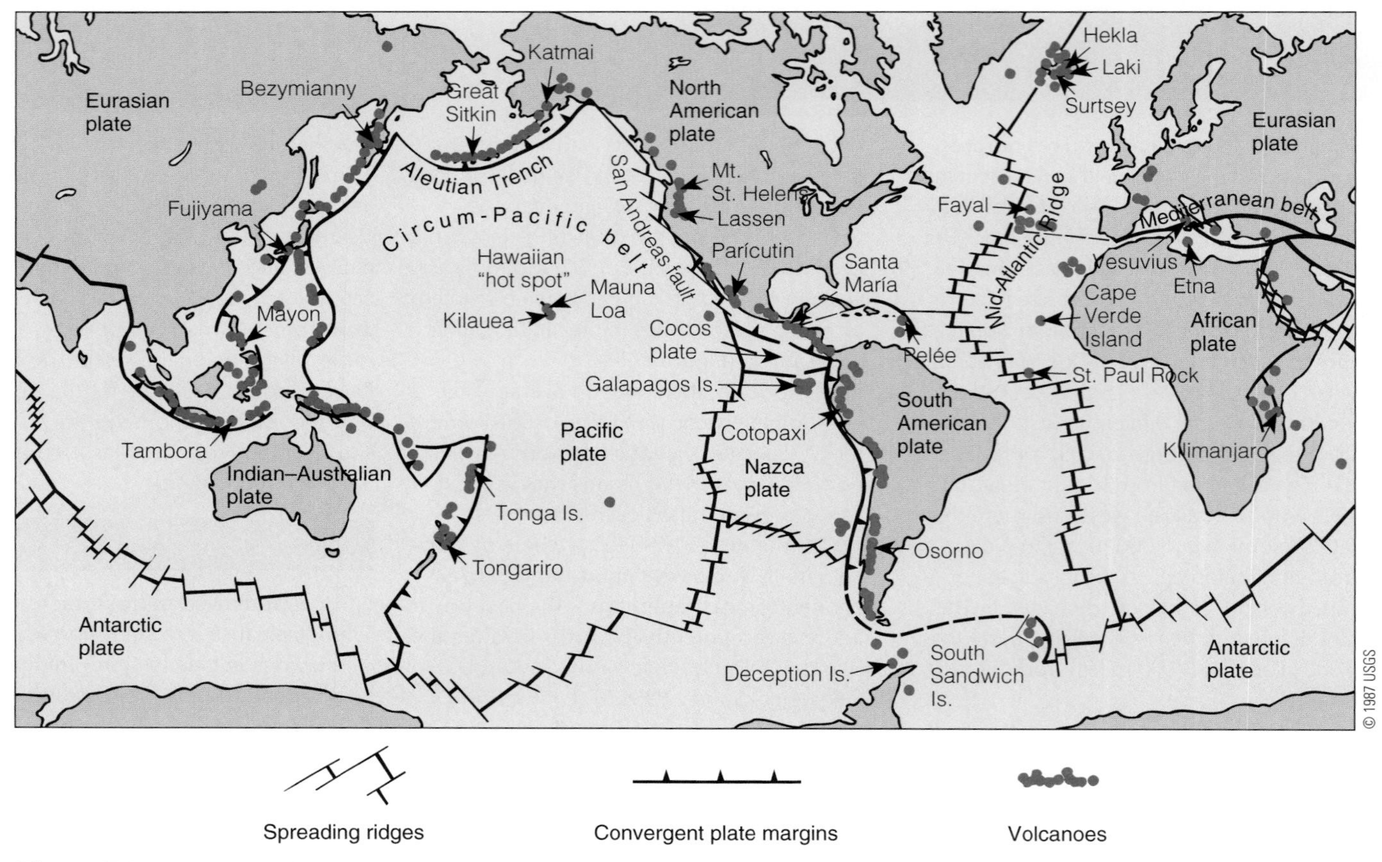

Figure 5.1 Distribution of Earth's active volcanoes at plate boundaries and hot spots. Note the prominent "Ring of Fire" around the Pacific Ocean, which contains 900 (66%) of the world's potentially active volcanoes. The remaining 450 are in the Mediterranean belt (subduction zones) and at spreading centers (divergent boundaries). A few important volcanic centers are related to hot spots, such as those in the Hawaiian and Galápagos islands. (After Tilling, R. I., C. Heiker, and T. L. Wright. 1987. Eruptions of Hawaiian volcanoes: Past, present, and future. Reston, VA: U.S. Geological Survey.)

the dense oceanic lithosphere dives beneath the lighter continental lithosphere. It sinks into the mantle—the process of subduction ("under movement"). At depths typically on the order of 100 to 110 kilometers (60–65 miles), the diving plate is hot enough to rapidly release large quantities of trapped seawater and volatiles because of the metamorphism of certain minerals. The volatiles (including the boiled seawater) play an important role in generating magma in the wedge of mantle that lies between the subducting plate and continental crust above.

Volatiles are mobile substances under very high pressure that form gases and liquids at or near Earth's surface. Examples include H_2O, CO_2, SO_2, and other common vapors issuing from volcanoes (**Figure 5.4**). Because

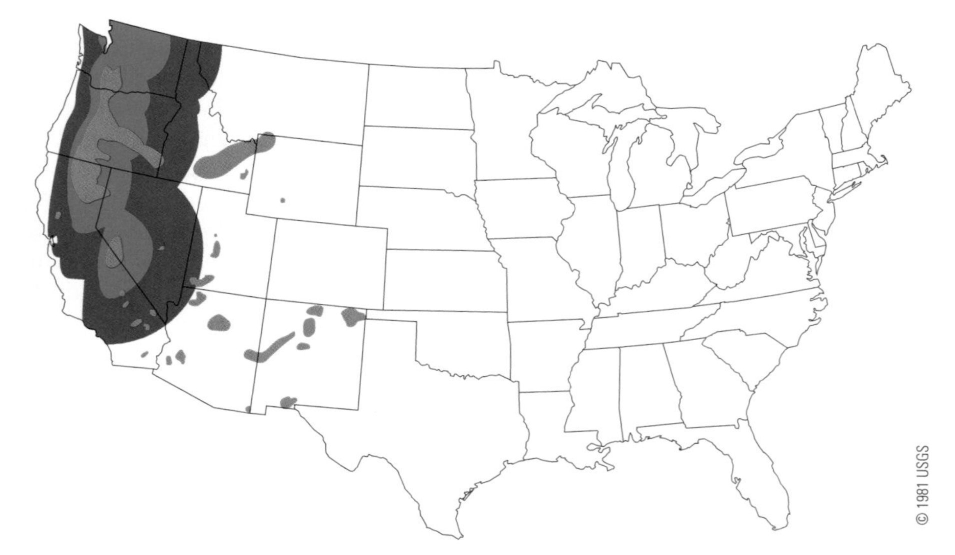

Figure 5.2 Active volcanoes and volcano clusters in the continental United States. The volcanoes of the Cascades are geologically related to the Cascadia subduction zone. (After Hays, W. W. 1981. Facing geologic and hydrologic hazards; Earth-Science considerations. Reston, VA: U.S. Geological Survey Professional Paper 1240-B, 108 p.)

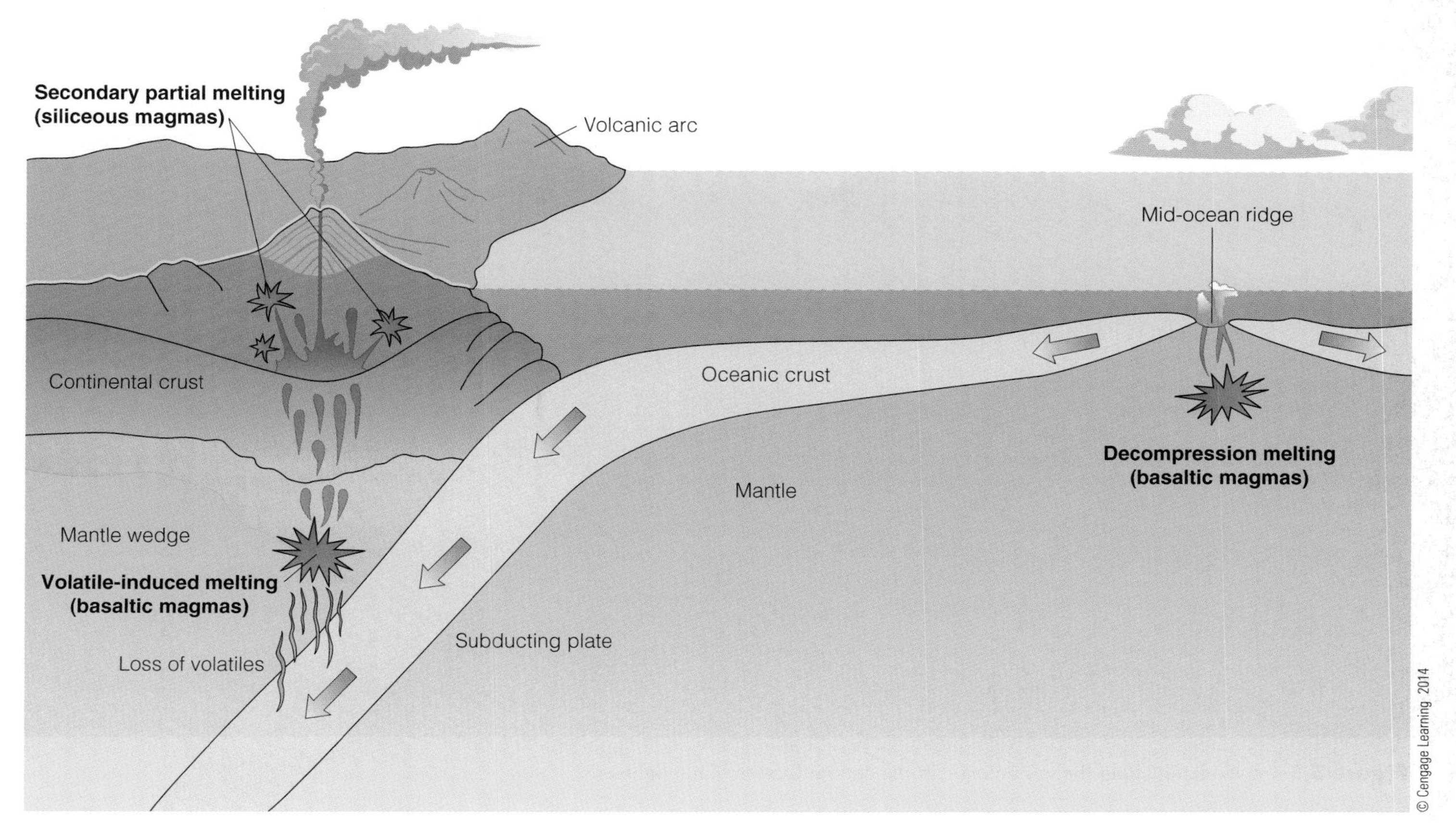

Figure 5.3 Environments of magma formation leading to volcanic activity.

of very high pressures inside the Earth, deep-seated volatiles form "fluids" that are intermediate in physical properties between what we ordinarily call a gas and a liquid. The volatiles may be completely dissolved within magma, much like CO_2 dissolved in a bottle of champagne. Shake the bottle and rapidly pop its cork, and you'll watch CO_2 come out of solution in an uncontrollable swarm of bubbles—an artificial geyser!

As volatiles escape into the mantle wedge, they enable the rock to melt more easily than it would if left completely volatile-free ("dry"). In other words, the melting temperature of the rock just above a subducting plate declines. Furthermore, the introduction of volatiles reduces the overall density of the mantle rock, which responds by shouldering aside the drier, overlying material and ascend-

Figure 5.4 A thick crust of sulfur crystals lines the rock surrounding the mouth of a fumarole in Hawaii. Such sulfur-producing fumaroles are called solfataras.

© 2011/KAREN KASMAUSKI/National Geographic Image Collection

Figure 5.5 View looking along the volcanic arc in the central Guatemalan Highlands.

ing buoyantly. The combination of introduced volatiles and less pressure at a shallower level causes the rising "wet" rock to melt. The resulting magma is compositionally similar to that formed at divergent plate boundaries—basaltic.

Once formed, the new basaltic magma ascends into the lower crust—anywhere from 40 to 70 kilometers (25–45 miles) down. It brings with it a large amount of heat. The process of heat transfer from one part of the Earth's interior to another by moving magma is called **thermal advection**. As discussed in Chapter 3, typical continental crust consists predominantly of silica-rich minerals that also contain a lot of trapped volatiles in the form of H_2O bonded to micas and amphiboles. The melting point of granitic and silicic metamorphic rocks is considerably lower than that of mantle rocks—by as much as 300°C (570° F). The advected heat that rapidly enters these rocks is sufficient to induce secondary partial melting within the crust (see Figure 5.3).

The resulting siliceous magma is not only cooler than the nearby, generally deeper lying mantle magma, but it does not flow as easily (i.e., it's more viscous). It also has a much greater volatile content. Although a typical basaltic magma may contain 1 to 2 weight percent dissolved volatiles, a silicic magma may contain as much as 5% to 10%. Hot basaltic magma can ascend all the way to the surface in a matter of a few weeks' time, erupting with relatively little volcanic violence. Stiffer siliceous magmas, in contrast, tend to erupt less often, and when they do, it is far more explosively than their mafic counterparts.

Eventually, magmas from both the lower crust and mantle above a subduction zone build an arc of volcanoes parallel to the marine trench (**Figure 5.5**; see Chapter 3). Most of these **arc volcanoes** erupt explosively, and many erupt lavas or pumice showing a mix of two source magma compositions—one silicic and the other mafic. This group of volcanoes matter very much to people, because most of us who live close to an active volcano do so at convergent plate boundaries.

Not all volcanoes can be related to plate boundary interactions. Perhaps a quarter of all geologically young volcanoes can be related to hot spots or to the extended effects of plate interactions taking place far from plate boundaries (Figure 5.3; see Chapter 3).

In some regions, such as central Mongolia and western North America, faults develop that cause local stretching and dropping of the crust. Beneath the resulting grabens, decompression melting occurs, and clusters of small, basaltic volcanoes form at the surface. The Mongolian volcanoes can be related to crustal stresses caused by the collision of India with the rest of Eurasia 3,500 kilometers (2,100 miles) away!

Hawaii, Iceland, and Yellowstone are the best known examples of hot-spot–related volcanoes (see Chapter 3). They are each volcanically very active and have been so for millions of years. But many volcanoes, mostly extinct now, have formed from what might best be described as "short-lived" hot spots, especially on the ocean floor, reflecting the restless churnings of the underlying mantle beneath areas of crust that are quite thin. A good example of widespread vol-

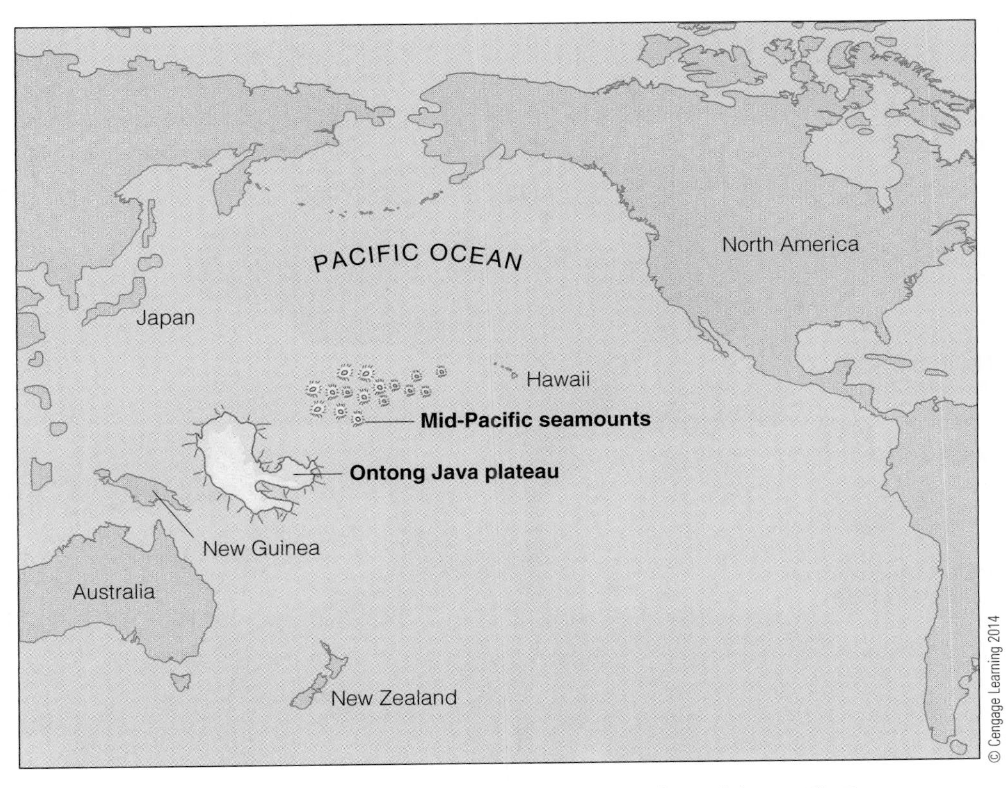

Figure 5.6 Locations of some major volcanic features on the floor of the Pacific Ocean.

canic activity of this sort is the Mid-Pacific Seamounts, which number in the thousands and lie scattered across most of the seafloor between Hawaii and Micronesia (**Figure 5.6**)

The very largest volcanic events ever recorded are also probably related to hot-spot activity. Vast outpourings (or "floods") of basalt lava feeding individual flows extending hundreds of kilometers have built up plains and flat-topped plateaus of lava in central India, Ethiopia, the Columbia River region in the Pacific Northwest of the United States, South Africa, the Paraná Basin of South America, and elsewhere. The largest such **flood basalt** occurrence is submerged in the southwest Pacific—the vast Ontong Java Plateau, which grew over 8 million years during the Cretaceous period (see Figure 5.6). The stack of lava flows making up the plateau is as much as 2,000 meters (roughly 6,000 feet) thick. Some geologists believe that individual flood basalt "provinces" develop where a large, newly rising hot spot first breaches Earth's surface. The unusually great concentration of heat from the deep mantle induces widespread decompression melting at shallow depth, which may be sustained as long as material in the hot spot continues to well up to replace the erupted magma.

Magma Ascent and Eruption

Molten rock is much less dense than its solid equivalent rock, and to equalize its density with the surrounding material, it rises toward the surface. As it does so, however, the density contrast with the host rock steadily lessens, and the magma finally stalls when it reaches a depth at which its density matches that of the enclosing lithosphere, a level called the **neutral buoyancy zone** (NBZ). Gradually, a substantial amount of molten material may accumulate at the NBZ, forming a single large underground body, or magma chamber. Alternatively, a complex network of linked molten chambers develops, which might more appropriately be called a magma reservoir or reservoir system. Most magma chambers lie underground at depths of several kilometers (**Figure 5.7**)

By far the greatest percentage of magma stagnating at the NBZ never erupts at all. Crystalline rock develops as the magma slowly cools and forms minerals (see Chapter 3). But occasionally a magma body becomes critically unstable and the molten rock breaks through to the surface. The result is a volcanic eruption. Why an eruption occurs depends on factors related to what is happening beneath the chamber, conditions within it and what is taking place above it—in the chamber "roof."

The magma chamber may experience overpressurization as a fresh supply of molten rock rises up into it from below (see Figure 5.7). The buildup of pressure, heat, and overall buoyancy cracks the chamber wall, most easily at the ceiling. A ruptured ceiling locally reduces pressure on the underlying magma, and volatiles rush out of solution, infiltrate the new fractures, and help wedge them open farther—and faster. The infiltrating volatiles also weaken the overlying rock through acidic corrosion, part of the general process of **hydrothermal alteration** in which rocks change both compositionally and physically as a result of interaction with heated natural solutions. Some magmatic volatiles such as hydrofluoric acid (HF) are among the strongest natural acids known. A positive feedback process involving volatilization and fracturing begins that may accelerate explosiveness as fracturing continues all the way to the surface. The cork is taken off the champagne bottle and an eruption begins.

The combination of intrusion of fresh melt into magma chambers and ensuing gas buildup, as described earlier, is probably responsible for most eruptions. But seemingly minor external influences can also act to trigger an outburst, especially if a magma chamber is about ready to erupt anyway. For instance, the orbit of the Moon around the Earth not only raises substantial tides in the ocean, but also causes

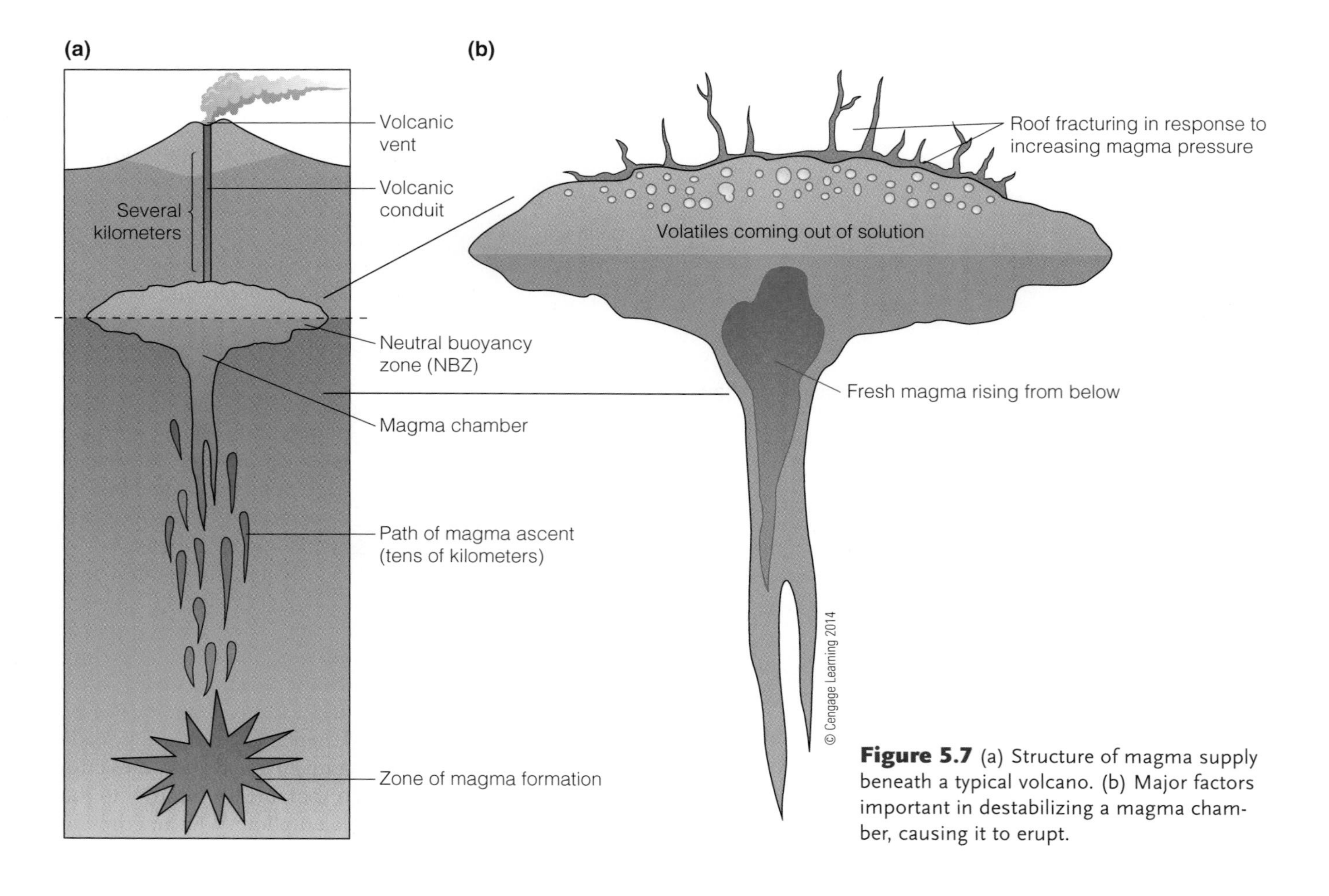

Figure 5.7 (a) Structure of magma supply beneath a typical volcano. (b) Major factors important in destabilizing a magma chamber, causing it to erupt.

the solid crust to rise and fall, too, as much as 40 centimeters (16 inches). We do not notice this solid earth tide given its slow daily pace and planet-wide scale, but for rocks already stressed close to their breaking point, the tidal strain might be the "straw that breaks the camel's back." Sets of eruptions have been recorded at Kīlauea in Hawaii, Ilopango in El Salvador, and at Gamalama volcano in Indonesia correlated with times of maximum or minimum tidal strain. The most important components of the solid earth tide known to act as volcano triggers occur on a once every 12-hour ("semidiurnal") or 14-day ("fortnightly") basis—when the Moon is either full or in new moon status (**Figure 5.8**). Once every 2 weeks (a "fortnight") the Sun and Moon are in alignment with the Earth and the gravitational pull on Earth's crust from these other bodies in space is strongest.

Even sudden changes in atmospheric pressure can be significant enough to trigger an eruption as the crust responds to the changing weight of the overlying air. In mid-June 1991, Typhoon Yunya passed within 75 kilometers (45 miles) of Pinatubo volcano in the Philippines, beneath which fresh magma had intruded only a few months before. Within 3 hours of the greatest barometric pressure drop (6.3 mb) associated with the passing storm, Pinatubo produced one of the biggest eruptions of the 20th century. Was this mere coincidence? Some volcanologists think not.

consider this . . . Not only volcanic eruptions, but swarms of small- to moderate-sized earthquakes along some faults can be correlated with the tidal attraction of the Sun and Moon. The scientific idea that the solid earth tide can cause earthquakes is an old one going back to the 19th century. But inadequate data about the timings and locations of earthquakes existed to test this hypothesis until 2008, when geophysicists at Johns Hopkins, Institut de Physique du Globe de Paris, and other universities found a slight but significant statistical correlation after examining a set of more than 440,000 seismic events occurring worldwide.

Types of Eruptions

Volcanic eruptions in continental settings are mostly explosive, producing great amounts of gas and fragmental (pyroclastic) debris. In extreme instances, they can destroy virtually all life (save for soil microbes) for tens of kilometers around a vent. Others, termed **effusive eruptions**, involve the outpouring of highly fluid molten lava unaccompanied by strong blasts, and may be safely viewed from close by. Still others, called **extrusive eruptions**, involve the slow, steady upwelling of stiff, nearly hardened lava from the ground, punctuated by periodic explosions and collapses. It is possible for

(see Figure 5.7a). The resulting rocks, including rhyolite and dacite, tend to be quite glassy and brittle. Great heaps of sharp rubble form as late-stage extrusion and gravity act on the surfaces and margins of flows.

In contrast, low-silica basaltic eruptions are almost entirely effusive, producing far traveled lava flows issuing, at least initially, from long cracks called eruptive fissures that can stretch for kilometers. Although volcanic explosions generally last only a few days at most, single extrusive and effusive eruptions can persist for weeks or years at a time. Lava shields, spatter ramparts, and **cinder cones** are typical products of basaltic eruptions (**Figure 5.11**). The lava flows themselves may contain extensive volcanic caves called lava tubes, or pyroducts.

Although the three component Geze diagram ("explosive," "extrusive," "effusive") provides a good general framework for looking at volcanic eruptions, the French mineralogist François-Antoine Alfred Lacroix (1863–1948) developed a more specific, observational approach that has proved equally popular with scientists. It has since been modified with the addition of several new eruption types through the studies of English volcanologist George Walker and colleagues in 1972–1973. The best known categories are summarized as follows, going roughly from most effusive, lowest silica, least gaseous events to the most gaseous, explosive, and siliceous. Note that each category is named for a specific type-locality or volcano where such eruptions are typical. Don't be misled, though—they also take place at many other volcanoes around the world.

Figure 5.8 Correlations have been made between the orbit of the Moon and the timing of a statistically significant number of volcanic eruptions at Kīlauea, here on the Island of Hawaii.

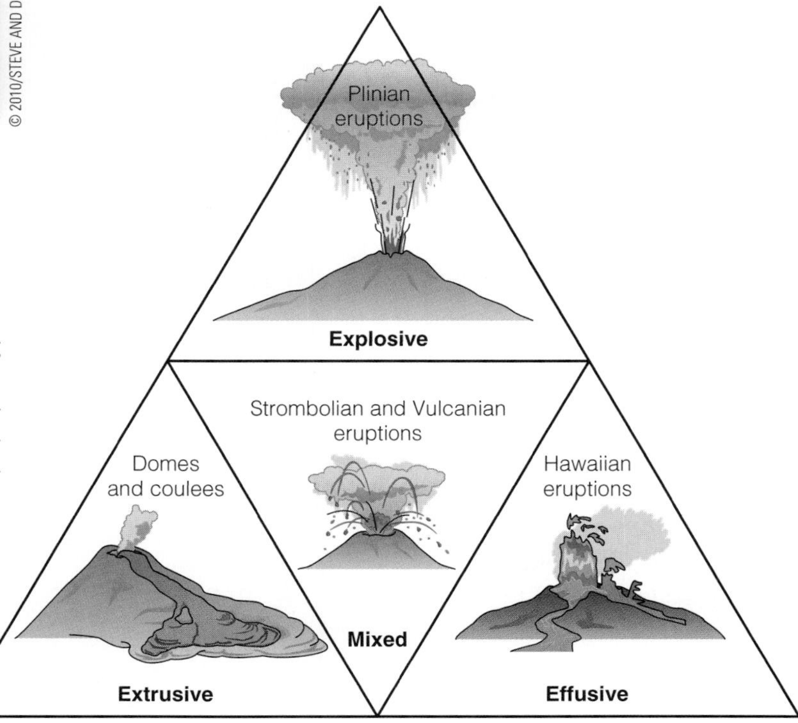

Figure 5.9 Slightly modified version of the Geze eruption classification diagram. (Based on Gèze, B. 1964. Sur la classification des dynamismes volcaniques. Bulletin Volcanologique 27: 237–257.)

a single eruption to exhibit one type of behavior early during an eruption, then transition to another behavior later. Such was the case, for instance, at Mount St. Helens in Washington State, when on 18 May 1980, it exploded with tremendous violence, to be followed almost a month later by an extrusive phase of activity. The Geze diagram developed by a French volcanologist in 1964 summarizes these variations in eruption behaviors (**Figure 5.9**).

The type of eruption depends entirely on the condition of the magma feeding it. As mentioned earlier, high silica magmas erupt explosively, or once they've released their pent-up gases, extrude from the surface as thick, pasty mounds, called **lava domes**, and stubby flows with jagged, steep borders, called *coulees* (**Figure 5.10**). Most such eruptions originate from pipelike **volcanic conduits**

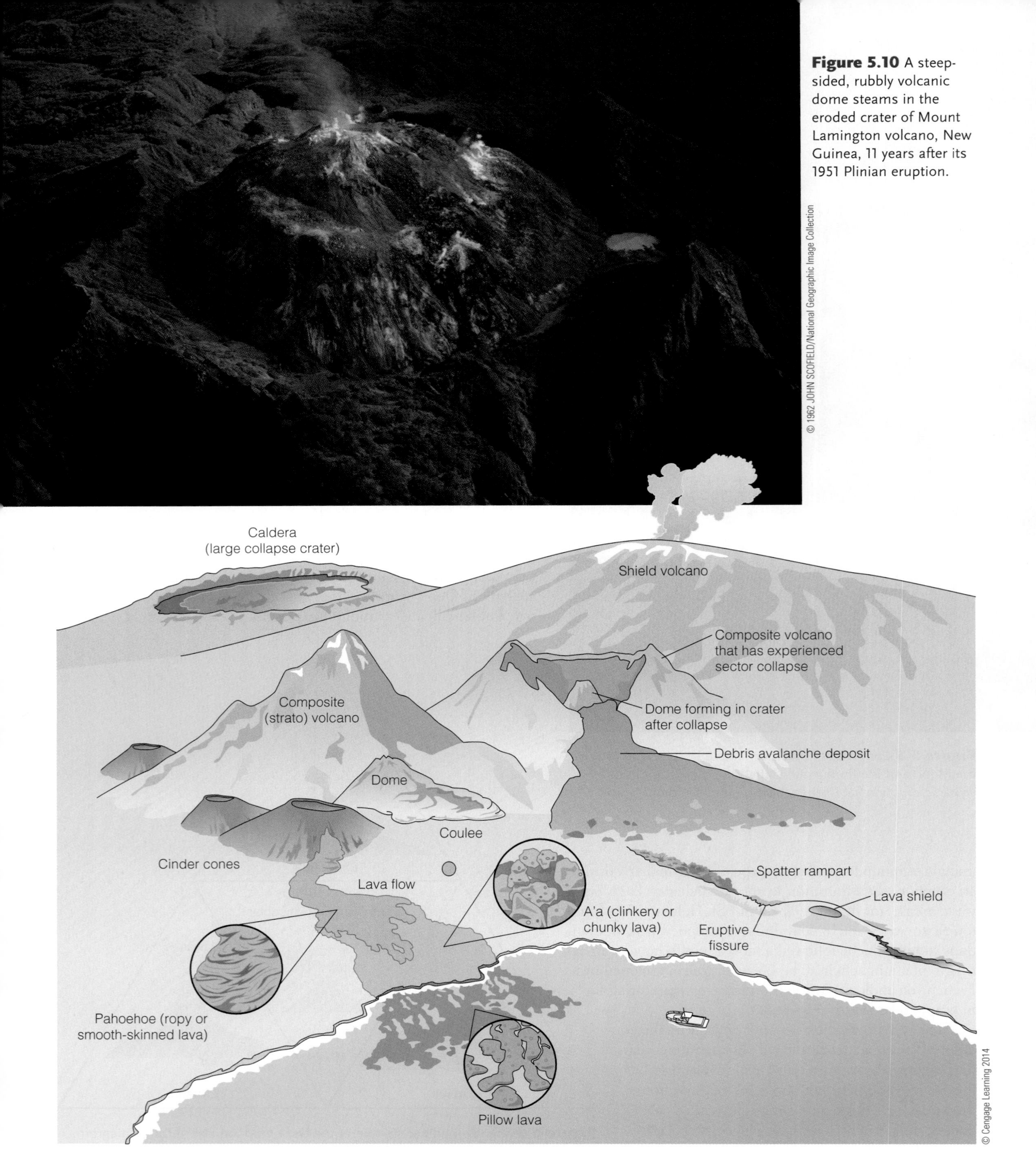

Figure 5.10 A steep-sided, rubbly volcanic dome steams in the eroded crater of Mount Lamington volcano, New Guinea, 11 years after its 1951 Plinian eruption.

Figure 5.11 Common types of volcanic landforms.

J.D. Griggs/U.S. Geological Survey

Figure 5.12 A Hawaiian-style fissure eruption breaks out in the east rift zone of Kīlauea volcano, Hawaii. Blobs of lava pile up around the fissure, building up a spatter rampart on both sides. Note the geologist in light-colored shirt for scale.

Hawaiian-style Eruptions

Hawaiian-style eruptions are typical of oceanic hot spots and divergent plate boundaries. They usually begin with the fountaining of basaltic lava from eruptive fissures (**Figure 5.12**). The lava travels away from the fissures as thin flows that usually move only a few hundred feet an hour—although near a vent the lava can move much faster. The gushing lava sounds like a multitude of fire hoses turned on full blast as it issues from a vent, and its color on a sunny day matches that of tomato soup or blood. Very little pyroclastic material usually issues with the lava—only gases, primarily steam. The steam is "superheated," well above the boiling point of water, so it can only be seen as a shimmering air until it cools off some distance above the vent. As the molten lava cools, it crusts over and consolidates either as smooth or ropy-surfaced **pāhoehoe** (pah-HOY-hoy), jagged, clinkery **a'ā** (AH-ah), or some mix of these two surface textures (**Figure 5.13**). These terms are from the Hawaiian language, and it's easy to distinguish them: The first two letters of pāhoehoe are the same as in "pavement," indicating smooth ground for walking; whereas a'ā roughly translated refers to any rocky surface, which is really difficult to cross. You can imagine exclaiming "ah-ah!" (or worse) as you try to navigate such terrain. Native Hawaiians very sensibly placed flat, water-worn stepping stones to enable easy transit across a'ā flows wherever needed.

Hawaiian eruptions on the seafloor can produce pāhoehoe and a'ā, just as on land. But they commonly also produce a third type of basaltic lava—particularly on gentle slopes submerged by more than a few hundred meters. This is **pillow lava**, a lumpy surface that resembles nothing so much as a great stack of sandbags. The "pillows" of solid basalt form because of the buoyancy and chilling effect of the surrounding water as the lava extrudes. Only rarely have observers been able to observe actively erupting pillow lava. Nonetheless, pillow lava is probably the most common bedrock surface on the ocean floor.

Figure 5.13 (a) An active pāhoehoe flow in Hawaii showing both smooth and ropy textural surfaces. The ropy skin in the lower right forms where molten lava drains from a ruptured lobe, piling up and cooling at the base of a slope. (b) A fresh a'ā flow from Mount Etna covers a railroad track near Randazzo, Sicily, Italy.

a: © 1990 MARC MORITSCH/National Geographic Image Collection; b: © 1982 JONATHAN BLAIR/National Geographic Image Collection

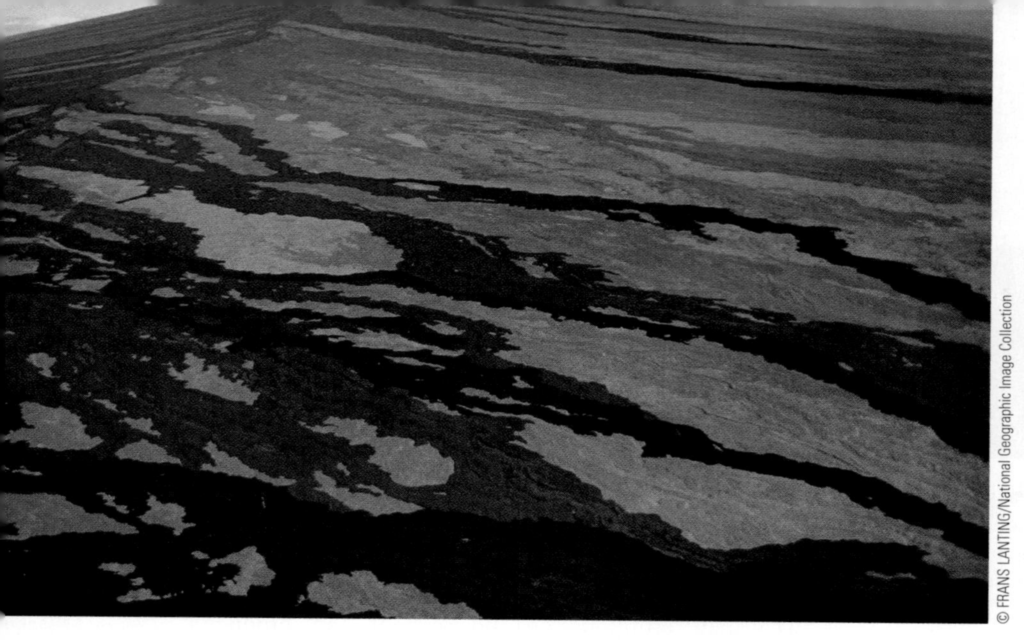

Figure 5.14 Mauna Loa, one of the world's largest shield volcanoes. The irregular strips and fingers of color on its flanks are all lava flows, mostly erupted from a rift zone in the volcano's flank, visible in the upper left corner. The darker and grayer the color, the younger the flow because of unweathered surface rock.

A volcano that grows from Hawaiian-style eruptions is called a **shield volcano**. Slopes generally do not exceed 5-degree inclination. The term shield derives from Iceland, where early European volcanologists suggested these broad, gentle mountains resemble the shields of Vikings lying flat on the ground, concave side up (**Figure 5.14**).

The huge lava plateaus mentioned earlier are also products of Hawaiian-style eruptions, but at a mega-scale. So fast does flood lava erupt and so extensive are the source fissures that shield volcanoes simply cannot grow. The whole landscape is repeatedly "repaved" with fresh lava, forming a level terrain tens to hundreds of kilometers across that may slope no more than a few degrees away from fissure vents. Although modern Hawaiian eruptions produce flows that range from less than 1 to a few cubic kilometers in volume, individual flood basalt lava flows have been measured up to 2,000 to 3,000 cubic kilometers (450–720 miles^3), with source fissures up to 70 to 200 kilometers (40–125 miles) long! Flows can be tens of meters thick—an order of magnitude thicker than ordinary basalt flows. Volcanic features of this size do not often form on Earth (fortunately) but are common on a few of our planetary neighbors, notably Venus and Io, the volcanically active satellite of Jupiter. Flood basalt plains make up the dark spots (maria) on the Moon.

consider this . . . Mark Twain visited Hawaii in 1866. He wrote the following description in his Letters from the Sandwich Islands after watching lava erupt and form a large, crusted pool at nighttime on the crater of Kīlauea volcano:

> The greater part of the vast floor of the desert under us was as black as ink, and apparently smooth and level; but over a mile square of it was ringed and streaked and striped with a thousand branching streams of liquid and gorgeously brilliant fire! Imagine it—imagine a coal-black sky shivered into a tangled network of angry fire!
>
> Here and there were gleaming holes twenty feet in diameter, broken in the dark crust, and in them the melted fire—the color a dazzling white just tinged with yellow—was boiling and surging furiously...Occasionally the molten lava flowing under the superincumbent crust broke through...like a sudden flash of lightning, and then acre after acre of the cold lava parted into fragments, turned up edgewise like cakes of ice when a great river breaks up, plunged downward and were swallowed in the crimson cauldron.

Strombolian Eruptions

Strombolian eruptions derive their name from the island volcano of Stromboli in the Mediterranean, just west of the "boot" of Italy. Strombolian eruptions are the gentlest of the explosive eruption types, although "gentle" hardly seems the case if you witness one of these blasts at close quarters, as countless sightseers do who make the trek to the rim of Stromboli's crater, wine and cheese in packs, to enjoy the spectacle after dark. (This is not recommended, by the way, because of hazardous trail conditions, irrespective of active eruptions).

Strombolian eruptions consist of pulsating bursts that heave molten lava high above the crater rim together with billowing clouds of dark, churning volcanic ash (**Figure 5.15**). Each explosion sounds like a booming cannon, but definitely louder. Although typically basaltic, the lava is generally somewhat more viscous than that of Hawaiian eruptions, and larger clots of it will fall back to the crater rim as aerodynamically streamlined blobs called *volcanic bombs* (**Figure 5.16**). Bombs typically range from a few centimeters to as much as a meter across (**Figure 5.17**). Finer clots of lava called cinder (or scoria) make up most of the volcanic ejecta. Individual pieces almost always contain numerous stretched and flattened vesicles (gas bubbles). Cinder fragments rarely exceed a few centimeters in diameter.

Strombolian eruptions may consist of blasts that occur once every few seconds—or individual explosions may be separated by as much as 10 or 15 minutes. Most volcanologists believe that each explosion is caused when a giant bubble of gas forms in the magma conduit as much as several hundred meters below the surface, eventually building up enough pressure to hurl the overlying molten rock violently out of the vent. The sudden release of gas temporarily reduces the volatile pressure in the conduit—at least until a new bubble forms to start the process all over again.

Although difference in magma viscosity may be important, the fundamental reason basaltic magma produces a Strombolian rather than a Hawaiian eruption may be sim-

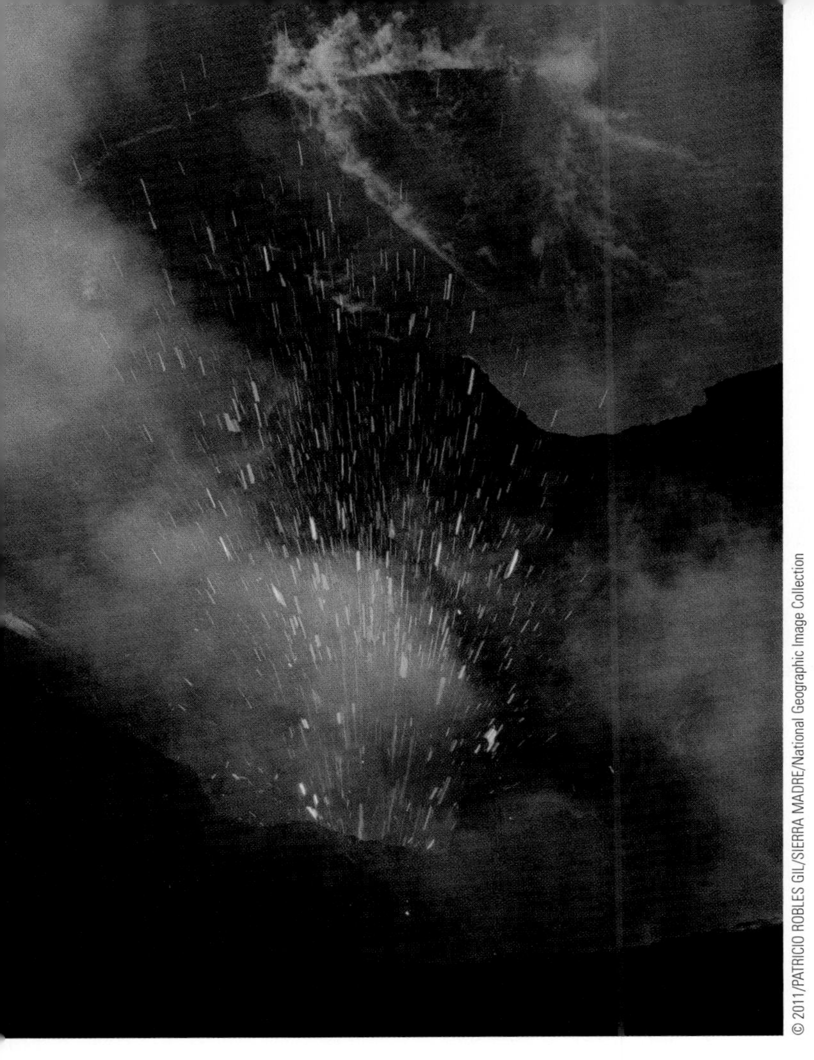

Figure 5.15 Strombolian burst of lava from a vent on Yasur volcano in Vanuatu, southwest Pacific.

Figure 5.17 A volcanic bomb; Galápagos Islands, Ecuador. Note the tapered shape that the bomb acquired as it was flung through the air.

ply a matter of the rate at which magma rises to the surface. A slow-rising magma has the opportunity to build up a great amount of explosive gas pressure episodically within a conduit. In contrast, a fast-rising magma (i.e., several meters a second) will release its gases continuously as it emerges—Hawaiian style.

An isolated, single Strombolian eruption constructs a cone of loose cinders and bombs around a central vent. The cinder cone typically has a cuplike crater and may be oxidized by hot, escaping gases as an eruption wanes to acquire

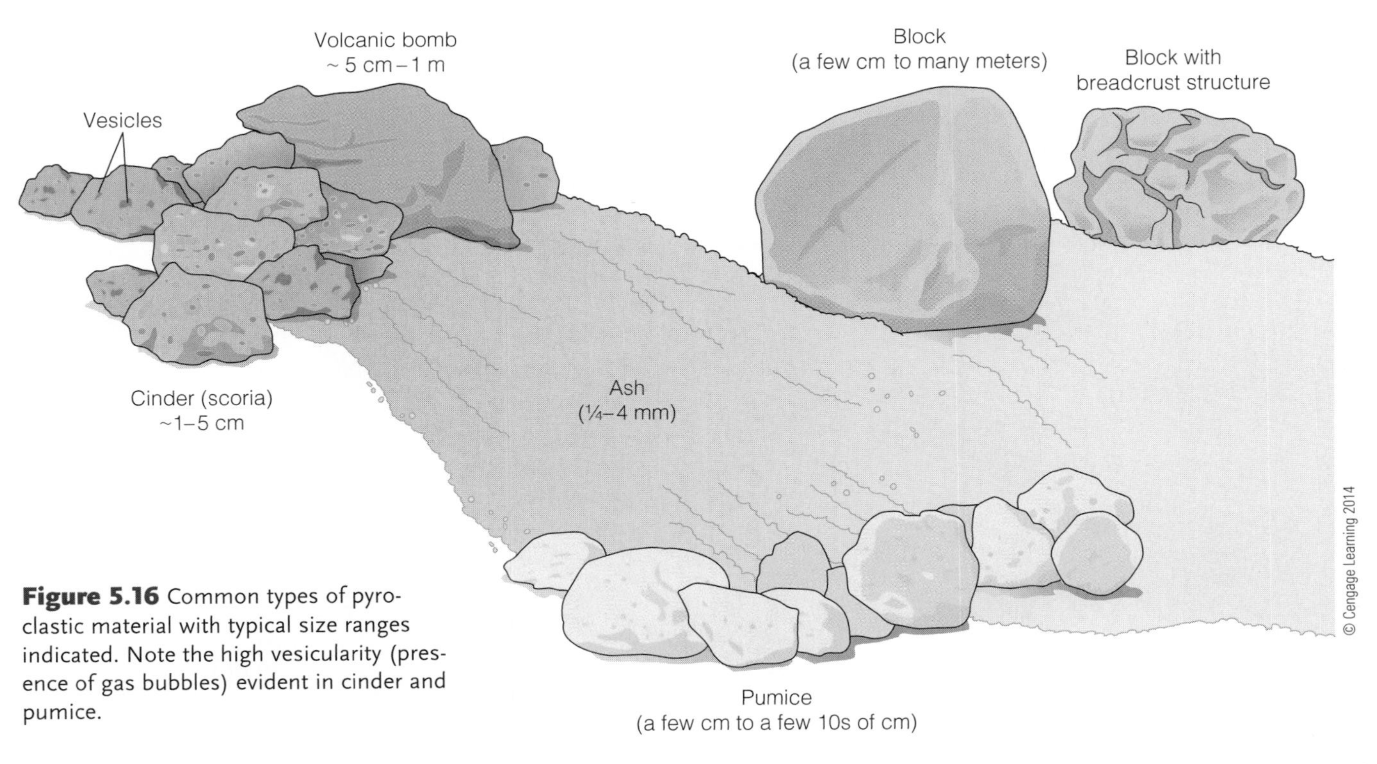

Figure 5.16 Common types of pyroclastic material with typical size ranges indicated. Note the high vesicularity (presence of gas bubbles) evident in cinder and pumice.

a striking, brick-red color. Cones up to several hundred meters tall are typical, and their slopes characteristically lie at the angle of repose—the steepest slope angle possible for loose material without sliding taking place, usually 30 to 35 degrees. Quite often, molten lava escapes from low in the flank of a cone during an eruption, producing a flow that may stretch as long as several kilometers (**Figures 5.11, 5.18**). The lava can pull away the flank of the simultaneously growing cone as it issues, exposing the crater interior if effusion persists long enough. The result is a horseshoe-shaped cinder cone rather than one that is nice and symmetrically circular as seen from the air.

Many cinder cones occur in clusters numbering up to several hundred, with each cone separated by no more than a few miles. In some localities, cities have grown in or adjacent to these clusters, notably Flagstaff, Arizona, and Auckland, the largest city in New Zealand. In both areas, the latest eruptions took place only several centuries ago, and future volcanic activity is virtually certain with a high likelihood of enormous damage. Economic incentives and historical patterns of settlement have trumped geological concerns in both cases, and an understanding of the full risk of eruptions dawned long after each of these cities began growing.

Strombolian activity including lava flows may take place repeatedly from a single vent over centuries or millennia, too, building up a **composite volcano** towering as much as a few thousand meters. The term composite refers to a volcanic cone made up of a mixture of lava flows and pyroclastic materials. Some geologists call composite cones stratovolcanoes (**Figures 5.11, 5.19**).

Strombolian eruptions are the most common type of basaltic volcanic activity in continental settings, and cinder cones are by far more common than composite volcanoes.

Vulcanian Eruptions

Although people can safely view an ongoing Hawaiian or Strombolian eruption from just a few hundred meters or kilometers away, Vulcanian eruptions are generally far larger and more dangerous. Vulcanian eruptions produce only pyroclastic material, and because of the power of Vulcanian explosions, the pyroclastic debris consists of volcanic ash—fine dust to sand-sized particles much smaller than the cinder thrown out by Strombolian activity. Vulcanian blasts are typically continuous rather than pulsating, lasting as long as 12 hours at a time. Rather than periodic cannon bursts, a Vulcanian eruption sounds like a jet engine turned on full blast. The clouds of roiling ash issuing from a crater are a mix of darker material fresh from the magma and the lighter color is pulverized older rock torn from the walls of the volcanic conduit. They can rise kilometers above a vent, and because of the fine size of their constituent particles, they develop strong electrostatic charges. As a result, forks of lightning frequently flash in and out of Vulcanian eruption clouds, adding to the pure terror of their appearance—especially after dark (**Figure 5.20**). Volcanologists studying the eruption of Mount Re-

Figure 5.18 A cinder cone known to locals as "The Stomach of the Elephant," and its associated lava flow spreading to the left, at Lake Turkana in the African Rift Valley, Kenya.

Richard Hazlett

Arturo Aburto Quesada Center for Seismic Investigations, Managua

Figure 5.19 Some examples of stratovolcanoes. (a) Mt. Shasta, California, with Shastina, a parasitic stratocone on Shasta's northwest flank. Shasta, mantled in snow and ice, is deeply eroded, since the volcano's eruption frequency is waning. Shastina grew after the end of the last ice age and preserves a youthful conical shape. (b) Volcan San Cristobal, Nicaragua, erupting in 1976. The dark, cauliflower-shaped cloud is volcanic ash. The white cloud is volcanic gas and steam.

doubt in Alaska in 1992 correlated the energy of lightning bolts with the strength of eruption-related earthquake activity and volatile escape from the vent.

Not all pyroclastic material produced by Vulcanian eruptions consists of ash. Large blocks of rock also frequently toss from a vent. These may be laced with a geometrical network of superficial cracks called bread-crust structure (see Figure 5.11). Bread-crust structure forms from the expansion of hot gases trapped in the cores of blocks during their release, which tears open their brittle, chilled skins. The abundance of blocks results from the initial "throat-clearing" phase of Vulcanian eruptions.

Some Vulcanian eruption clouds are so heavy with ash that they cannot rise far without collapsing back to the surface, running out across the landscape as hot (200–1000°C [390–1830°F]) volcanic dust storms with speeds of up to 200 kilometers (120 miles) per hour. The high temperature and heat of these surface-hugging eruption clouds proves deadly to practically anyone who stands in their paths. The French called them *nuees ardentes* ("glowing clouds") when they were first studied in 1902 erupting on the island of Martinique in the Caribbean. The modern term for this phenomenon is **pyroclastic density current**, or PDC for short. However, we now recognize that there are several kinds of PDCs, including pyroclastic flows and pyroclastic surges. But all share in common extreme destructive force and high temperatures, which blast apart and incinerate any towns in their paths (**Figure 5.21**). Only moving water in the forms of floods or tsunami proves more effective at destroying the infrastructure of a landscape.

Plinian Eruptions

The most fearsome eruptions of all are named for the Roman naturalist and military commander Pliny the Elder, who died trying to help friends escape the eruption of Vesuvius in 79 C.E., which famously destroyed the coastal cities of Pompeii and Herculaneum. Plinian

Figure 5.20 Vulcanian eruption, nighttime, Eyjafjallјökull volcano, Iceland.

Figure 5.21 A pyroclastic flow races down the flank of Mayon volcano on Luzon Island in the Philippines. Meanwhile, a Vulcanian eruption column rises from the vent just left of center.

C. Newhall/U.S. Geological Survey

eruptions are explosive events that characteristically inject large amounts of volcanic gas and pyroclastic material into Earth's upper atmosphere. The eruptions consist of an initial phase of powerful gas release, which can create a pillar of ash and pumice fragments, called an **eruption column**, towering as high as 20 kilometers (12 miles) or more. At the level of stratospheric neutral buoyancy, the eruption cloud spreads out to form a capping **umbrella**, which may extend hundreds of kilometers in the downwind direction. Pliny's nephew, who documented the Vesuvius tragedy, called the combined column and umbrella a pino, or "pine" cloud, because he thought it resembled the shape of the long-trunked Italian pines with their spreading canopy that grew around his estate (**Figures 5.24** [on pages 150–151] and **5.22**). Some powerful Vulcanian deposits can develop pinos, too, but Vulcanian eruptions are generally not as long lasting as Plinian eruptions, which can go on for several days, nor do they impact the upper atmosphere quite so intensely (if at all). An extremely powerful Plinian eruption can produce as much as a hundred billion cubic meters of pyroclastic material—several orders of magnitude greater than the most powerful Vulcanian eruptions.

The most distinctive product of Plinian eruptions is a large amount of pumice, the spongelike, lightweight rock well-known to float on water (see Figure 5.11). Pumice results from the frothing of magma as gases escape a vent rapidly. In much the same way, froth will form at the top of a filling cup of beer. The volume percentage of vesicles in pumice can range from 64% to more than 90%. Plinian siliceous pumice fragments tend to be less vesicular than mafic pumices (called reticulate). Most vesicles do not interconnect in siliceous pumices, so that they will float on water. Mafic pumices tend to sink, however, given their high permeability.

Pumice falling around the base of a Plinian eruption column can accumulate to depths of a few meters or more, creating a thick, moderately well-sorted air-fall deposit. The people who fled Pompeii during the early hours of the Vesuvius eruption carried mattresses and other artifacts on their heads to protect themselves from the falling pumice. Some fragments were heavy enough to kill exposed individuals, as the broken skeletons of several unfortunate victims reveal.

When the strength of the eruption ("vent discharge") wanes, the column cloud collapses, and enormous PDCs may race as far as tens of kilometers across the surrounding landscape. The PDC deposits consist mostly of ash and pumice, but also of blocks of vent-derived fragments and other stony materials derived from the landscape across which the currents travel. Given the great pressures and temperatures associated with PDC deposition, pumice fragments are often flattened and compressed into black obsidian, which the Italians named **fiamme**, because of their resemblance to licks of flame. Pompeii was sealed beneath PDC deposits several meters thick toward the end of the eruption, preserving the ruined city for future excavators to begin discovering 1,700 years later.

Collapse of a volcano's summit, and sometimes the entire volcano, generally occurs during a big Plinian eruption, as the weakening roof of the underlying, rapidly emptying magma chamber suddenly caves in. The term **caldera** (derived from the Spanish word for "cauldron") describes the resulting basin (see Figures 5.4, 5.11). Backfilling by pyroclastic debris may fill in the caldera as an eruption progresses, creating a steaming plain or volcanic tableland. But, in some cases, enough of a physical depression exists to capture water later, forming huge lakes within steep crater rims. Ten-kilometer (6-miles)-wide Crater Lake in Oregon is the most famous example of this. But Lake Toba on the island of Sumatra is far larger. Toba caldera formed 75,000 years ago during the biggest volcanic eruption in the last 2.5 million years. The enclosed lake measures 30 by 100 kilometers (18 by 60 miles). Large calderas may continue to be areas of intense geyser and hot spring activity ("geothermal activity"), and eruption of silicic lavas long after they form (**Figure 5.23**). An excellent case is Yellowstone caldera in Wyoming, preserved by the world's first national park. Yellowstone formed piecemeal in three Plinian eruptions occurring between 1.8 million and 600,000 years ago. Future eruptions in the Yellowstone region are all but certain.

In summary, the distinctive features of Plinian eruptions include thick air-fall pumice and pumiceous PDC deposits, and caldera development. They are so powerful that volcanic cones are destroyed rather than built up by them. Plinian eruptions also can impact the upper atmosphere with significant worldwide effects, as we'll mention later. There is only one logical human response to a Plinian eruption—flee.

consider this . . . How much time do you think would be needed to evacuate the population of a modern, developed city with, say, 500,000 or 1 million residents, if warning developed of an impending volcanic eruption nearby?

McGimsey, Game/Alaska Volcano Observatory/U.S. Geological Survey

Figure 5.22 Plinian eruption column and umbrella cloud above Mount Spurr volcano, Alaska, in 1992. The column is 18 kilometers (11 miles) high.

Volcano Explosivity Index

In addition to the purely descriptive approach toward classifying volcanic eruptions outlined earlier, a new approach was developed in 1982 at Dartmouth College, New Hampshire, to categorize volcanic activity on the basis of relative explosivity. Like the Richter magnitude scale for measuring earthquake strength, the **Volcano Explosivity Index (VEI)** is an open-ended logarithmic scale beginning with 0, running up to 9 to 10 for the most powerful eruptions (**Table 5.1**). The scale contrasts different parameters of volcanic eruptions, including height of eruption clouds (columns), duration of blasts, whether volcanic gases and ash penetrate the stratosphere, and overall volume of explosively ejected materials. **Figure 5.25** presents the scale and relates it to the Lacroix classification scheme. One advantage of the VEI is that a geologist can infer much about the nature of an eruption based on the size of its deposit long after the eruption took place, provided that erosion or burial by younger deposits has not progressed too far. The appearances of materials (e.g., whether pumice is abundant, presence of pāhoehoe and a'ā) factor in to making this inference, too, of course.

The VEI also allows us to study and quantifiably estimate the frequency of different types of volcanic eruptions taking place worldwide. Resembling the example of earthquake magnitudes, we find that eruption frequency tapers off with increasing VEI number in a nonlinear, geometric

© TINO SORIANO/National Geographic Image Collection

Figure 5.23 Eruption of Pohutu Geyser at Te Puia Springs, near the rim of Rotorua caldera in New Zealand.

 Volcanoes

The classical literature of ancient Rome and Greece contains many references to volcanic activity. The Roman god of the forge, Vulcan, who worked beneath Etna's fuming summit, survives in the modern word "volcano." Caldera collapse of the center of the island of Thera (Santorini) in the Aegean probably gave rise to the legend of Atlantis, spoken of by Plato. But the most famous eruption in classical history destroyed the Imperial Roman cities of Herculaneum and Pompeii in 79 C.E. Details of that eruption are profiled here. The map also suggests how vulnerable the modern metro area of Naples is to future volcanic eruptions.

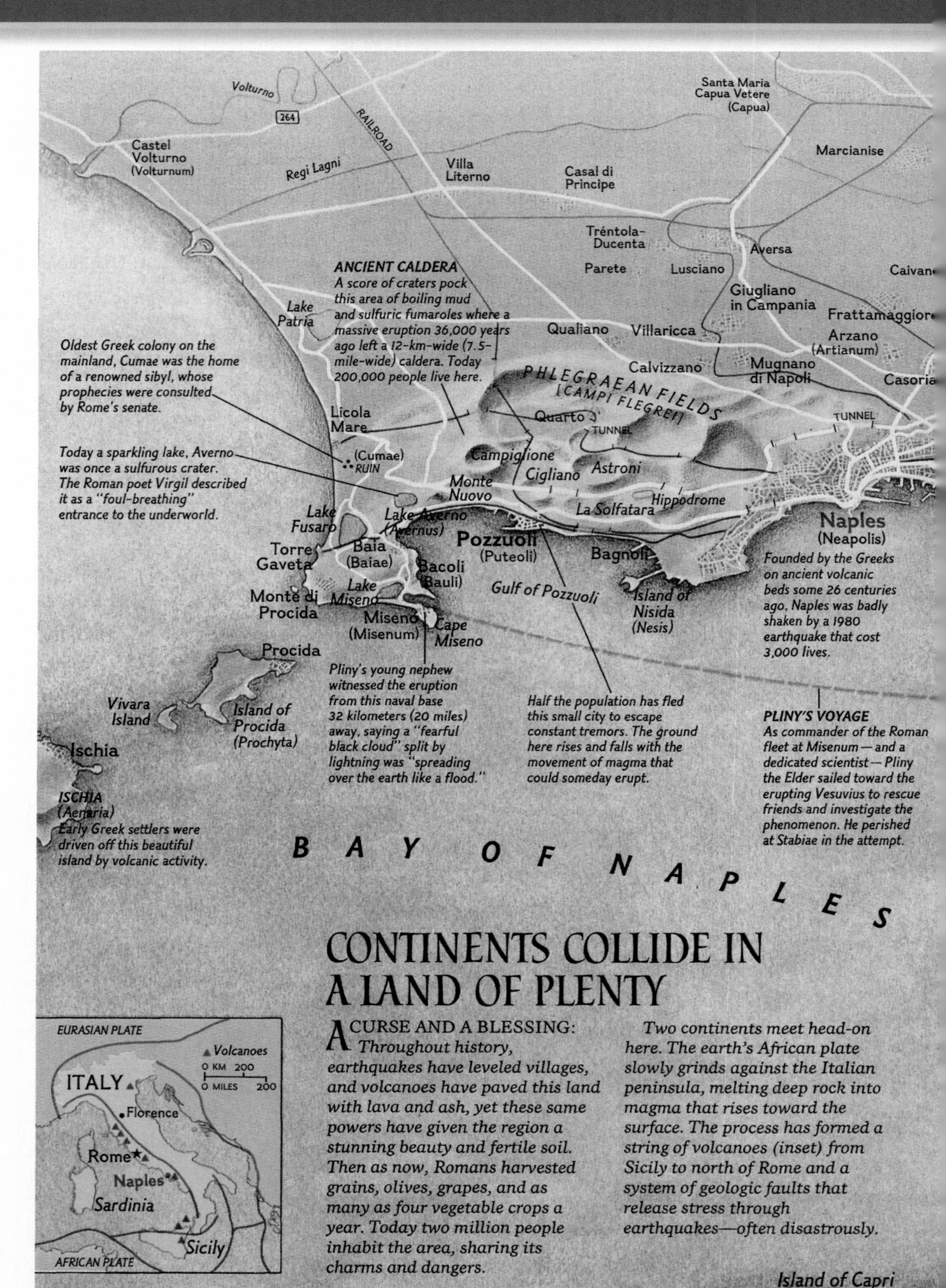

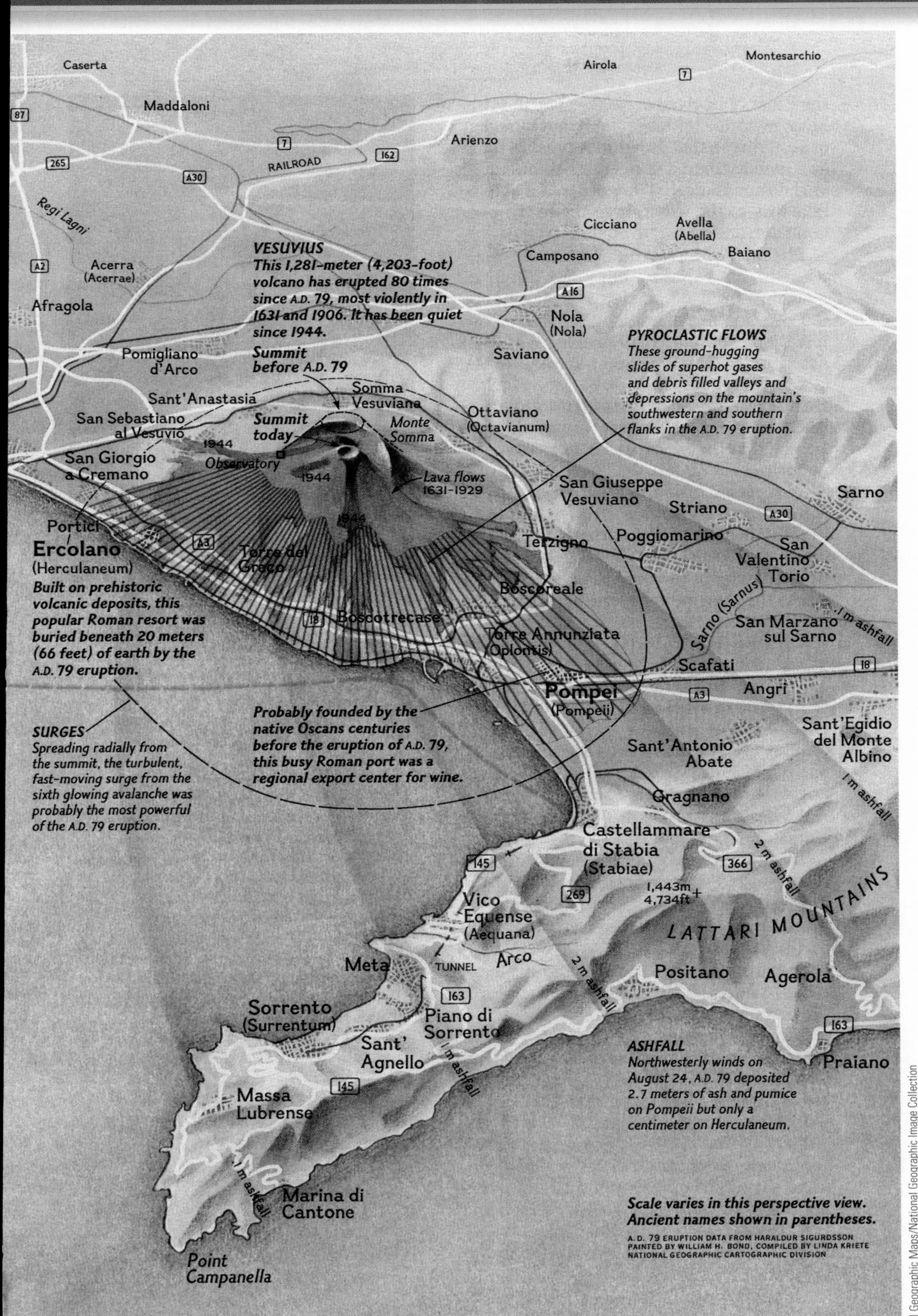

Caserta
Airola
Montesarchio
Maddaloni
Arienzo
RAILROAD
Regi Lagni
Acerra
(Acerrae)
Afragola
Cicciano
Avella
(Abella)
Baiano
Camposano
VESUVIUS
This 1,281-meter (4,203-foot) volcano has erupted 80 times since A.D. 79, most violently in 1631 and 1906. It has been quiet since 1944.
Nola
(Nola)
Saviano
PYROCLASTIC FLOWS
These ground-hugging slides of superhot gases and debris filled valleys and depressions on the mountain's southwestern and southern flanks in the A.D. 79 eruption.
Pomigliano d'Arco
Summit before A.D. 79
Sant'Anastasia
Somma Vesuviana
Monte Somma
Ottaviano
(Octavianum)
San Sebastiano al Vesuvio
Summit today
1944
Observatory
San Giorgio a Cremano
Lava flows 1631-1929
San Giuseppe Vesuviano
Sarno
Striano
Portici
Ercolano
(Herculaneum)
Built on prehistoric volcanic deposits, this popular Roman resort was buried beneath 20 meters (66 feet) of earth by the A.D. 79 eruption.
Torre del Greco
Terzigno
Poggiomarino
San Valentino Torio
Boscoreale
Boscotrecase
Sarno (Sarnus)
San Marzano sul Sarno
1 m ashfall
Torre Annunziata
(Oplontis)
Scafati
Pompei
(Pompeii)
Angri
SURGES
Spreading radially from the summit, the turbulent, fast-moving surge from the sixth glowing avalanche was probably the most powerful of the A.D. 79 eruption.
Probably founded by the native Oscans centuries before the eruption of A.D. 79, this busy Roman port was a regional export center for wine.
Sant'Antonio Abate
Sant'Egidio del Monte Albino
Gragnano
Castellammare di Stabia
(Stabiae)
2 m ashfall
1,443m
4,734ft
Vico Equense
(Aequana)
LATTARI MOUNTAINS
Meta
TUNNEL
Arco
Positano
Agerola
Sorrento
(Surrentum)
Piano di Sorrento
Sant' Agnello
ASHFALL
Northwesterly winds on August 24, A.D. 79 deposited 2.7 meters of ash and pumice on Pompeii but only a centimeter on Herculaneum.
Praiano
Massa Lubrense
Marina di Cantone
Point Campanella
Scale varies in this perspective view.
Ancient names shown in parentheses.
A.D. 79 ERUPTION DATA FROM HARALDUR SIGURDSSON
PAINTED BY WILLIAM H. BOND, COMPILED BY LINDA KRIETE
NATIONAL GEOGRAPHIC CARTOGRAPHIC DIVISION

Table 5.1 Selected Notable Worldwide Volcanic Eruptions*

Year	Volcano Name And Location	VEI	Comments
79	Vesuvius, Italy	5	Pompeii and Herculaneum buried; at least 3,000 killed
1783	Laki, Iceland	4	Largest historic lava flows; 9,350 killed
1792	Unzen, Japan	2	Debris avalanche and tsunami killed 14,500
1815	Tambora, Indonesia	7	Most explosive eruption in history; 92,000 killed; weather changed
1883	Krakatoa, Indonesia	6	Caldera collapse; 36,000 killed, mostly by tsunami
1902	Mount Pelée, Martinique, West Indies	4	Saint-Pierre destroyed; 30,000–40,000 killed; spine extruded from lava dome
1912	Katmai, Alaska	6	Perhaps largest 20th-century eruption; 33 km^3 tephra ejected
1914–1917	Lassen Peak, California	3	California's last historic eruption
1980	Mount St. Helens, Washington	4–5	Ash flow, 600 km^2 devastated
1991	Mount Pinatubo, Philippines	5–6†	Probably the second largest eruption of the 20th century; huge volume of SO_2 emitted
1983–	Kīlauea, Hawaii	2†	Longest continuing eruption, with more than 50 eruption events

*Historic lava or tephra volume ≥2 km^3, Holocene > 100 km^3, fatalities ≥1,500.
†Author's estimate.
VEI, Volcano Explosivity Index.

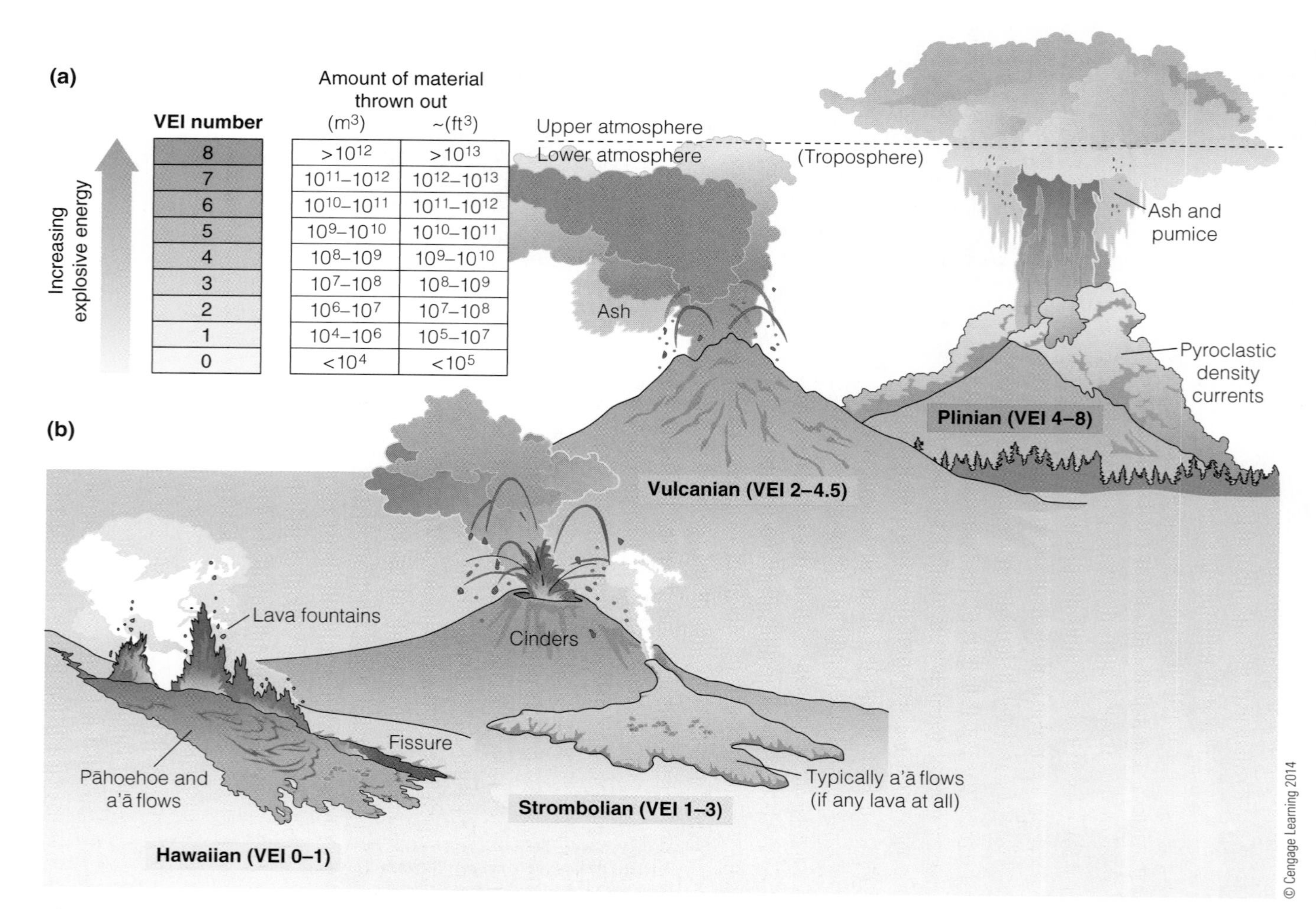

Figure 5.25 (a) The Volcano Explosivity Index of Newhall and Self (1982). See original paper for more complete information. (b) Comparison of qualitative eruption types with VEI numbers.

way (**Figure 5.26**). Although several Strombolian eruptions occur somewhere in the world every year, the incidence of eruptions having a VEI greater than or equal to 5 drops off to 1 every 10 years. Truly terrifying caldera-forming eruptions (VEI > 8), in the meantime, are so infrequent that you shouldn't lose any sleep over their possibility.

Volcanic Hazards

Primary Eruption Hazards

Hazards are potential threats to people. Primary volcanic hazards are those related directly to the exit of magma from an erupting vent and the distribution of volcanic materials across the surrounding countryside. These include hazards related to lava flows, volcanic gases, ash and pumice falls, and PDCs.

In general, lava flows are easy to avoid simply by understanding that they flow downhill. They are generally slow enough to escape by walking. Thick a'ā flows are capable of pushing down building walls and burying smaller structures on slopes. But on flat or gently sloping ground, walls and even barriers made up of loosely piled stones may be sufficient to stop or divert advancing a'ā, provided that the barrier is not directly in the path of travel. Thinner pāhoehoe flows can pour through city streets without destroying the buildings, apart from burning them. They are capable of forming molds around trees, low plants like ferns, even trapped furniture, and in rare, unfortunate instances, trapped animals (**Figure 5.27**). Some fossil molds include very delicate details of the objects they encased, such as the ribbing of charred tree bark and the individual leaves of fern fronds.

Volcanic gases are potentially lethal, although they quickly disperse in the atmosphere. They pose no threat when a person stays upwind of them. If caught in gases by a change in wind, a person can reduce the risk for poisoning by placing a damp cloth such as a handkerchief over the nostrils. Although not foolproof, the moisture in the cloth helps absorb the gases.

Falling (**ashfall**) is a hazard for several notable reasons (**Figure 5.28**). As the chapter opener discloses, drifting ash particles will enter machinery and cause engine lock, as well as expensive abrasive damage to both metal and glass. Ash accumulating on roofs to a thickness of more than about 10 inches (25 cm) can readily collapse roofs. Rooftops need to be periodically observed and swept, then, if people take shelter beneath them. Ash will clog water systems and may leach substances into water supplies such as HF that are hazardous to people using the water—hence it is a good idea for residents to stock fresh water emergency supplies. Fluorosis, including bone and joint pains and deformation, has led to the death of livestock in Chile and Iceland that have consumed grasses contaminated with light ashfall deposits. Ashfalls darken skies, making the movement of people and goods difficult at a time when emergency services may be needed. For most people who don't live so close to a volcano that other, more serious hazards are a consideration, it is best to simply hunker-down during an ashfall. Ashfalls more than a few tens of kilometers from a source vent usually last no more than several hours.

Pyroclastic density currents are far worse, however. Few people have ever survived the burning, battering, and suffocation associated with PDCs unless exposed at the very margins of these clouds. Forensic studies indicate that victims are seared internally as well as externally by taking superheated gases into their lungs (**Figure 5.29**). Of the 30,000 persons in the city of Saint-Pierre, Martinique, during the 1902 eruption of nearby Mount Pelée, only two survived and recovered from the Vulcanian PDC that overwhelmed the town, setting fire to the highly flammable rum distilleries there and tossing objects as large as cannons many meters from their fixed positions. The city burned for days. One survivor was a prisoner, Ludger Sybaris, chained to a basement wall in a jail for assaulting a friend with a dangerous weapon. He was badly injured, but recovered, and for many years after the disaster earned money as a novelty act for Barnum and Bailey circus—an interesting case of "divine justice."

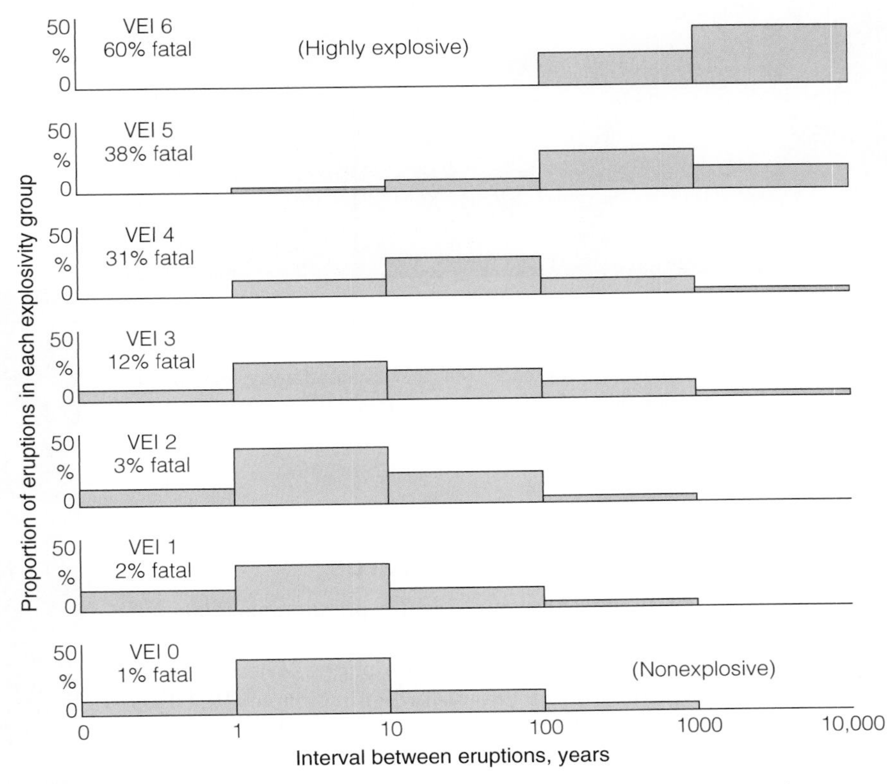

Figure 5.26 The longer the time interval since an eruption, the greater is the next eruption's potential explosivity. These data for 4,320 historic eruptions relate known intervals between eruptions to the Volcano Explosivity Index (see Figure 5.25). Also shown is the percentage of eruptions in each VEI that have caused fatalities. (Data from Siebert, L., T. Simkin, and P. Kimberly. 2010. Volcanoes of the world, 3rd ed. Berkeley, CA: University of California Press, 568 p.)

© FRANS LANTING/National Geographic Image Collection

Figure 5.27 The standing molds of trees ("lava trees") surrounded by a lava flow from Kīlauea volcano, Hawaii. The molten lava chilled around each tree, then drained away, leaving solidified stone pillars behind. Each pillar is hollow inside, preserving the shape of the enclosed trunk, which burned and rotted away.

Secondary Eruption Hazards

Many potentially dangerous volcanic phenomena relate not to direct discharge of material from a vent, but to the rather secondary, peripheral effects of eruptions. These include tsunami, which we will discuss further in Chapter 10, lahars (volcanic debris flows), long-term volcanic degassing, and severe weather conditions.

Tsunami can result from a variety of processes associated with shallow underwater eruptions, or eruptions at coastal volcanoes. In 1883, the collapse of the island of Krakatau during a Plinian eruption created an undersea caldera roughly 8 kilometers (5 miles) wide, simultaneously generating a monstrous sea wave that killed more than 36,000 people. At Unzen volcano in southern Japan in 1792, a giant landslide in the volcano flank created a set of tsunami that destroyed 17 coastal communities and devastated 80 kilometers (50 miles) of shoreline. In some inlets, wave heights reached 60 meters (200 feet)! About 15,000 persons drowned. Pliny the Elder provided the first ever scientific description of a tsunami, formed in the Bay of Naples during intense earthquake activity as nearby Vesuvius roared away. More than 90 volcanogenic tsunami have been reported since Pliny's day, killing more than 60,000 people worldwide.

Lahars have been even more deadly than tsunami—and, in fact, are the most destructive of volcanic phenomena, historically speaking. The Geological Society of America defines a lahar as "a general term for a rapidly flowing mixture

R. Hobblitt/U.S. Geological Survey

Figure 5.28 Fresh ashfall covers cars and landscape at Clark Air Force, Philippines, during the 1991 eruption of Pinatubo volcano.

© 2006 O. LOUIS MAZZATENTA/National Geographic Image Collection

Figure 5.29 Skeletons of victims of the 79 C.E. eruption of Vesuvius, preserved in volcanic ash at the excavated waterfront of the ruined city of Herculaneum. The people were caught by a pyroclastic density current as they scrambled to evacuate by sea.

J. N. Marso/U.S. Geological Survey

Figure 5.30 A lahar pours through a channel near El Palmar, Guatemala, in 1989, originating from the flanks of Santiaguito, an active dome extruding from the side of Santa Maria volcano.

of rock debris and water (other than normal stream flow) from a volcano." A lahar need not take place during an eruption but can occur long afterward. Even heavy rain falling on a slope covered with loose ash miles from a vent active years before can set a lahar in motion (**Figure 5.30**). Quite often lahars are generated when an eruption blasts apart and melts large volumes of snow and ice high on a volcano's slopes. The biggest known lahars in the world originated this way, including multiple lahars from Cotopaxi, Ecuador's 5,900-meter (19,350-feet)-tall volcano, some of which extend at least 100 kilometers (60 miles), all the way to the Pacific Ocean. Another notorious lahar generator is Mt. Rainier in Washington State, which has more than 13 million cubic meters (460 million $feet^3$) of glacial ice present on its upper, hydrothermally weakened flanks. Lahar deposits are by far the largest product of Mt. Rainier produced in the last 10,000 years, including ash, pumice, and lava.

Lahars move fast—up to 8 meters (25 feet) a second on some slopes. Some lahars have discharges that match that of the Amazon, the world's largest river, though confined to much narrower channels! Reaching lowlands, lahars spread out across large distances of surrounding flat terrain. One Mt. Rainier lahar, which erupted 5,700 years ago, widened to 12 kilometers (7 miles) across the Puget Sound plain after exiting the mountains. Many thousands of people live and commute atop this deposit today. Similarly, in 1985, a small eruption atop Nevado del Ruiz volcano in Columbia melted enough snow and ice to generate a nighttime lahar that entered the lowlands only a few hours later, inundating the city of Armero. Having virtually no warning, 23,000 people died, and the city was obliterated. One student who survived the disaster remembered the lahar "was only about two meters deep as it entered town, rushing down streets, and the only lights were from the headlights of cars being tumbled by the flow."[1] In Japan, which has also suffered numerous lahars, a system of engineering works, called sabo, has been established on some streams to reduce potential lahar damage. Sabo works include dams and basins to capture debris, channel building to divert lahar flow, and the installation of large grates to capture boulders washed down with the ashy mud. Such boulders, like battering rams, prove especially destructive when impacting buildings. Electronic pressure pads buried by lahars can provide a warning signal to authorities downslope as danger approaches. Sirens, much like those used by tsunami warning systems in many coastal towns around the Pacific, can warn residents to race for higher ground.

Some of the worst lahars on record have come from the rupturing of the walls around acidic, crater-filling lakes (**Figure 5.31**). Kelud volcano in Indonesia has been a particularly bad source of catastrophic crater-lake discharges. To deal with these, the Dutch colonial government in 1919 began constructing tunnels in the volcano's crater wall to keep the lake level inside at a safer, lower level, following a lahar that killed 5,000 people. Authorities undertook a second tunnel project in 1966 to further maintain drainage. Given the high concentration of population, abundant monsoonal rainfall, and frequent volcanic eruptions, no region in the world suffers lahars more frequently than the island of Java. There, in fact, is where the term lahar originated

Volcanoes release gases ("degas") continuously. Expansion of gases in the shallow magma conduit beneath a vent is the primary propellant of an explosive eruption. But, in fact, most volcanic **degassing** occurs between eruptions or even after a volcano becomes extinct, rather than during its peak eruptive activity. The degassing occurs through **fumaroles**—exhaling cracks near the summit and on the flanks of volcanoes. Steam vents are a class of fumaroles that release solely (or mostly) water vapor.

Such seemingly passive degassing, although it may sound benign, poses the greatest danger of volatiles to people. This is especially true of carbon dioxide, which can collect to poisonous concentration in low spots and cavities where no wind stirs. Such lethal fumaroles are called *mofettes*, or valleys of death. The Swahili speakers who live near the giant mafic volcano Nyiragongo in the East African Rift Valley call them "elephant graveyards."

One of the saddest degassing disasters in world history occurred at Lake Nyos, a 1.8-kilometer (5,900-feet)-wide vol-

[1] J. P. Lockwood and R. W. Hazlett. 2010. *Volcanoes: A global perspective* (p. 357). Oxford, U.K.: Wiley-Blackwell Publishing.

Figure 5.31 Fumaroles steam along the edge of an acidic, green volcanic lake in the crater of Mount Martin volcano, Katmai National Park, Alaska. The pH of such waters can be as low as 0.2.

Pam McPherson/U.S. Geological Survey

canic crater lake in Cameroon, in 1986. Although the eruption that formed Nyos volcano took place 400 years ago, magmatic carbon dioxide has continued to seep into the crater ever since, building up to high concentration in the cooler, deeper waters of the lake mostly below 180 meters (580 feet). Carbon dioxide is more soluble in cold water than warm, and if a situation occurs in which cold water can mix readily with warm, a tremendous amount of CO_2 will suddenly come out of solution as a lethal cloud of gas—a so-called **limnic eruption**. The risk for such eruptions was not unknown to the local people. Bantu folklore is filled with stories of exploding lakes and "rains" of dead fish. According to oral tradition, at least three major degassing catastrophes occurred in the distant past. Many residents accept dead fish found floating on the water surface during minor events as gifts from their ancestors whose spirits live in the lake.

On the night of 21 August 1986, Lake Nyos suddenly discharged its toxic load. Scientists estimate that the gas rushed from the lake as fast as 100 kilometers (60 miles) per hour and rapidly filled the crater basin. It then silently poured over the low northern rim and flowed down a neighboring valley, overtaking livestock and a village of people who had, for the most part, gone to sleep. More than 1,700 villagers died of sudden respiratory failure. Since 2001, observers have used a pipe to allow carbon dioxide to leave the lake without building up deadly concentrations. The gas generates an artificial geyser as it escapes.

consider this... What factors (apart from salary) would you take into consideration if you were presented with an opportunity to live and work near a young-appearing volcano?

Benjamin Franklin is regarded as the first person to recognize the connection between cold weather and volcanic eruptions while he was serving as U.S. Ambassador to France in 1783–1784. Laki Fissure in Iceland erupted in June 1783, putting sufficient sulfate gas aerosol and fine ash into the atmosphere that the following winter was extremely cold in Europe. The high albedo volcanic haze blocked enough sunlight to cool Earth's lower atmosphere by several degrees centigrade. The summer of 1783 was so bleak in Iceland, in fact, that famine ensued, killing a fifth of the island's population and more than half its sheep, cattle, and horses. The eruption of Tambora volcano in Indonesia in April 1815 also hurled a tremendous amount of fine ash and acidic mist into the upper atmosphere, creating spectacular lurid sunsets with "rings around the sun" and other optical phenomena, but also causing temperatures to plummet, especially in Europe and North America. The ensuing famine may have killed as many as several hundred thousand people.

The average length of time that volcanic dust particles (0.0001–0.005 mm) from a large Vulcanian or Plinian eruption can remain in the upper atmosphere, the residence time, is 1 to 2 years. Sulfur dioxide coats on the particles, making them super-reflectors of incident sunlight. An attempt has been made to quantify this phenomenon for individual volcanic eruptions using the so-called *Dust Veil Index* (DVI). Indexes for several important historical examples are shown in **Table 5.2**. As apparent in the table, Tambora and Krakatau have had potentially the most significant effects on climate in recent history, whereas Mount St. Helens, with a DVI of 1, had little effect. Dust veils are obviously sporadic and transient, and many volcanoes would have to erupt in concert to produce significant long-term effects on Earth's climate, which is highly unlikely. Worldwide impact is also seen only after powerful eruptions at low latitudes, where mixing of ash and gas into both the Northern and Southern hemispheres readily takes place.

Volcanoes as Unstable Landforms

The different kinds of volcanic eruptions produce many different kinds of characteristic landforms. As mentioned earlier, Hawaiian volcanoes produce shield volcanoes, flood basalts create plateaus, and Strombolian eruptions produce cinder cones, or with Vulcanian eruption contribute to building beautifully symmetrical, steep-sided peaks called composite volcanoes. Plinian eruptions will create huge calderas—"negative" volcanoes in the sense that these are depressions in Earth's crust rather than hills or mountains.

Some volcanoes erupt only once, then fall extinct as at Lake Nyos. Cinder cones show this behavior and are said to be **monogenetic** ("one-eruption-only formation"). In contrast, composite volcanoes are **polygenetic**, erupting many times or even semicontinuously, and can build cones thousands of meters tall over a period of several hundred thousand years.

Composite volcanoes are also notably unstable landforms. With steep sides made up of strata that in places are loosely consolidated and heavily fractured, these volcanoes can literally fall apart when a high standing mass of magma enters them and/or a strong earthquake takes place. Whole flanks or sectors of a cone will disintegrate and slide away in a debris avalanche at speeds as high as 40 meters (130 feet) per second (**Figure 5.32**). The avalanche can carry blocks of former volcano flank individually up to tens of meters across and

Figure 5.32 Sector collapse in the flank of Kiso Ontake volcano in central Japan. The collapse followed a magnitude 6.8 earthquake and several days of heavy rain in September 1984. The scar measures 1,300 meters (4,250 feet) long by 400 meters (1,300 feet) wide and is 170 meters (560 feet) deep.

U.S. Geological Survey

Table 5.2 Dust Veil of Selected Eruptions

Volcano and Eruption Date	DVI
Tambora, Indonesia, 1815	1,500
Krakatau, Indonesia, 1883	1,000
Mount Pelée, Martinique, 1902	500
Mount St. Helens, Washington, 1980	1

Source: H. H. Lamb. 1970. Volcanic dust in the atmosphere; with a chronology and assessment of its meteorological significance. Philosophical Transactions of the Royal Society, 266 (1178): 425–533, by permission of the Royal Society.

DVI, Dust Veil Index.

can spread out tens of kilometers away from the foot of the mountain. The Pleistocene Nevado del Toluca debris avalanche in Mexico traveled 55 kilometers (30 miles) before finally coming to rest. Given the high concentration of people that live in valleys downslope from active composite volcanoes, this is certainly a concern. But, like high-index VEI eruptions, such events are rare. The only well-documented case of sector collapse in recent history took place at Mount St. Helens, Washington, in May 1980. The avalanche carried a huge amount of snow and ice with it and was soon overtaken by a PDC released from the magma chamber below, which essentially was losing its roof. The PDC melted the snow and ice, creating great lahars and mudflows that reached as far as the Columbia River, closing the shipping channel to Portland, Oregon (**Figure 5.33**). The deposits left by sector collapses form characteristically hummocky landscapes, difficult for farming, transportation routes, and other human land use.

Even larger than the sector collapses on land are the giant avalanches from the slopes of oceanic shield volcanoes. Built on semi-stable platforms of pyroclastic debris, even big shields such as Mauna Loa, made up almost entirely of gently sloping lava flows above sea level, have suffered instability (**Figure 5.34**). A hundred thousand years ago, a substantial section of the western flank of Mauna Loa suddenly slid into the deep ocean, generating a titanic tsunami that reached an elevation of more than 300 meters (1,000 feet) on the neighboring island of Lanai, 160 kilometers (100 miles) away. If such an event were to occur today, the wave would devastate most of the human population of Hawaii, and have serious repercussions for millions of people all around the Pacific Rim. Fortunately, such occurrences take place only once every 100,000 to 200,000 years in the Hawaiian Islands. Of equal interest in the Atlantic Ocean are the unstable flanks of shield volcanoes in the Canary Islands, and in the Indian Ocean, the slowly collapsing eastern flank of Piton de la Fournaise on Réunion Island.

Benefits of Volcanoes

Despite the foregoing litany of nasty hazards associated with volcanoes, we actually owe this phenomenon a tremendous debt of gratitude. "They are a lot more than just Mountains that go 'Boom'!" as French volcanologist Maurice Kraft once said.

Where better to begin than by considering our climate? Russian scientists speak of the Earth as having had three atmospheres during its long history. Atmosphere One consisted of the raw nebular gases left over from the accretion of Earth 4.5 billion years ago. These included hydrogen and helium, lightweight gases that soon escaped into space. Throughout Earth's first few hundred million years, volcanic activity occurred at a scale perhaps greater by an order of magnitude than today; rather than a few thousand active volcanoes and vents, there may have been tens of thousands. Vigorous volcanic degassing created Atmosphere Two—largely water vapor and carbon dioxide, but also sulfurous volatiles and nitrogen. Although we could not breathe this mix of gases, lacking oxygen, it was nonetheless the foundation of Earth's modern atmosphere.

The modern atmosphere, Atmosphere Three, arose as two events modified the primordial volcanic brew: First, volcanogenic steam condensed to build the world's oceans at a very early stage. While they grew, the oceans dissolved a lot of the sulfurous gases and most of the carbon dioxide, leaving nitrogen behind as our planet's predominant atmospheric gas. Secondly, biological activity, in no small part enabled by the presence of an ocean full of dissolved volcanic gases, modified atmospheric composition through algal photosynthesis, which added enough oxygen to enable modern life to arise.

Figure 5.33 Years after the 1980 eruption, National Volcanic Monument visitors view the hollowed out remains of Mt. St. Helens across a plain of pyroclastic density current, lahar, and avalanche debris. The huge new crater of the volcano, resulting from sector collapse, is called an amphitheater caldera.

Figure 5.34 A cliff face more than 300 meters (>1000 feet) high, left by a giant landslide into the ocean from the south flank of Kīlauea volcano, Hawaii. Countless thin lava flows make up the layering visible in the towering sea cliff.

Our foremost debt to volcanoes, then, is simply the air we breathe and take so for granted. The modern atmosphere's composition continues to be largely regulated by the degassing of volcanoes and respiratory activity of life.

We might also thank volcanoes for...food. Agricultural crops need 16 major and minor nutrients to thrive, including potassium, nitrogen, phosphorus, and iron. In many tropical environments, soil leaching rates are high; infiltrating water removes nutrients and washes them out to sea. Soils tend to be nutrient depleted, and the extant nutrient supply is mostly taken up by plants and other organisms, only to be recycled via forest floors or savannah topsoils through death and decay. The Tropics is a tough region to farm, in other words. But with the exception of nitrogen, which can be readily obtained from the air, volcanic ash supplies all the nutrients that crops need. In tropical regions, farmers are quick to turn the soil when an ashfall occurs to fertilize it naturally (**Figure 5.35**). On Borneo, which lacks volcanoes, a farmer will be lucky to have a good harvest once every 3 years. But on neighboring Java, with its 20 recently active volcanoes, good harvests are possible three times a year—thanks to frequent inputs of fresh ash. What this means in terms of human population should be plain: Borneo has a population of 19 million in an area 60% larger than the State of California (population 35 million). Java, on the other hand, is less than a third as large as California and has a population of 135 million. In fact, about 10% of the human population lives close to volcanoes in agricultural belts largely fertilized by pyroclastic eruptions. *Andosol*, the technical name for volcanically derived soil, is among the richest mediums for growing plants on Earth.

Figure 5.35 Mount Rainier looms over Orting, Washington, a lush agricultural area enriched by past volcanic ash deposits.

If air and food aren't sufficient reminders of why we should appreciate volcanoes as a life-giving factor, consider something more mundane—metals, one of the two pillars (in addition to fossil fuels) that make modern industrial civilization possible (see Chapter 13). Most of our principal industrial metals, including copper, nickel, zinc, lead, and silver, derive from shallow magma bodies—the roots of past volcanoes. Gold, chromium, vanadium, platinum, and molybdenum also concentrate in this environment. Former calderas are particularly rich sources of important metallic ores, because of hydrothermal fluids that precipitate many ore-bearing minerals. The Comstock Lode in Nevada, which Mark Twain wrote about in *Roughing It*, is an example of hydrothermal ore mineralization—one of the richest silver deposits ever discovered. In its heyday (1875–1889), miners in the Lode worked shifts in underground passages with floors of boiling mud and air temperatures as high as 57°C (135°F), digging out the very heart of an extinct but still hot volcano. Such heat is a relatively clean source of energy today, exploited as geothermal power in many parts of the world (see Chapter 14).

Volcanic materials make up many other products that we use in ordinary ways. The final polish your teeth receive when you get them cleaned is an example. Dental pumice is refined and flavored volcanic pumice with a hardness just slightly less than tooth enamel. Lava brand hand soap—a rough, abrasive bar soap—contains the same powdered rock. Lightweight bricks, cinder blocks, and many road-foundation and decorative stone products originated deep within the Earth. Where volcanic cinders are mined for road base, they may also be mixed with oil to form asphalt pavement. Pieces of rock pumice are sold in drugstores, supermarkets, and hardware stores for use as a mild abrasive for removing skin calluses and unsightly mineral deposits from sinks and toilets. Powdered, it is used in abrasive cleaners and in furniture finishing.

Glassy volcanic rock, such as obsidian, is easy to chip and shape. Native Americans and Stone Age peoples in many parts of the world used it to make tools and arrowheads. Today, obsidian and many similar volcanic rocks are the raw material of "rock hounds" and artisans for creating polished pieces and decorative art.

Volcanoes provide opportunities for recreational activities, ranging from mountain climbing and skiing to more passive activities, such as photography, painting, and bird watching. Of course, skiing on an active volcano poses some extra measure of risk, as skiers on Mount Etna in Italy and Ruapehu in New Zealand can thrillingly attest.

Volcanoes also have great ecological importance, contributing significantly to the development of biodiversity in many parts of the world. On the island of Hawaii, for instance, lava flows often surround patches of older terrain, called **kīpuka** (meaning "a variation or change in form" in the Hawaiian language; **Figure 5.36**). Biota within kīpuka are isolated by the fresh lava from similar biota elsewhere, and many insects and plants with highly localized reproductive ranges or mobility can evolve into distinctive varieties or whole new species over short periods, biologically speaking. A landscape rich in kīpuka quickly becomes a landscape rich in biodiversity—much richer, in fact, than comparable land areas that lack volcanic activity.

Similarly, great rafts of pumice from Plinian eruptions can float at sea for up to 20 years, traveling hundreds or even thousands of kilometers in marine currents. Many non-planktonic organisms attach themselves to pumice fragments or live sheltered around them. As the pumice washes up on a distant beach, or degrades and finally sinks far from its point of origin, associated organisms become permanently isolated from their parent populations and, in turn, can adapt distinctively to environments they've newly colonized, especially in coastal areas.

The Galilean moon Io is the most volcanically active body in the solar system because of tremendous tidal forces caused by close proximity to Jupiter. Io is about the same size as the Moon, but satellite observations have revealed that it has 74 active volcanoes and dozens of active hot spots. Because of the rarefied atmosphere of Io, eruptions shoot gas and particles as high as 300 kilometers (180 miles) into space, creating a ring of pyroclastic "mist" encircling Jupiter. In contrast, the most powerful Plinian eruptions on Earth project gas and ash no more than about 50 kilometers (30 miles).

Living with Volcanoes

At a handful of volcano observatories around the world, scientists study volcanoes and forecast their eruptions. The oldest is the Vesuvius Volcano Observatory, established by the King of Naples in 1845. Researchers at the Hawaiian Volcano Observatory, established in 1912 on the rim of Kīlauea caldera, developed many of the ground-based technologies used in forecasting eruptions (**Figure 5.37**). More recently, scientists at the Alaska Volcano Observatory, working with similar laboratories in Kamchatka, Japan, and the U.S. Pacific Northwest, have used satellites effectively to study eruptions across thousands of kilometers of sparsely inhabited territory.

Prediction is the ability to identify when and where something will happen. Although volcano scientists have gotten very close to predicting some eruptions within a few days of their occurrence, it is more accurate to say that they are **forecasters**—that is, they state in general terms the probabilities of eruptions taking place. Like hurricane forecasters they do this by looking for certain precursory warning signs. The ability to forecast depends on how well a volcano is monitored with instruments and, to a certain extent, by how well its history of eruptions is known. Almost every volcano shows its unique "personality," or eruptive behavior. Some show patterns of behavior that can be useful in forecasting. Others

Figure 5.36 A kīpuka surrounded by fresh lava, Kīlauea, Hawaii.

are more chaotic, exhibiting great changes in their eruptive styles over time. Scientists at the Hawaiian Volcano Observatory closely monitor the behavior of Kīlauea and Mauna Loa, two of the world's most active volcanoes. Each shield contains a magma reservoir several kilometers below the surface that gradually fills with molten rock from weeks to years before an eruption. The whole mountain swells up, as if air were being blown into a balloon. The uplift causes the slopes of the mountain to steepen gradually, a change that can be measured precisely using a **tiltmeter** (**Figure 5.38**). Gradually increasing earthquake activity typically accompanies the tilting. The locations of earthquake epicenters and foci indicate where the outburst is likely to occur. In the final hours leading up to an eruption, the reservoir suddenly deflates as molten rock works its way toward the surface. Simultaneously, a continuous shuddering of the ground, called *volcanic tremor*, takes place. The scientists alert civil defense authorities that an eruption may be imminent.

Eruptions do not always follow these warnings. Magma may simply intrude at a shallow depth without breaching the surface. Nor do the warnings provide any clue as to the magnitude and duration of an eruption, should one occur. But it is better to be safe than sorry, and thousands of lives have been saved by successful eruption forecasting.

Figure 5.37 Volcanologists in the field. (a) Historical photo of a scientist taking a measurement using a wet-tube tiltmeter, one of the first kinds of tiltmeters invented for volcano studies. As the slope of a volcano changes, so does the water level in three precisely spaced metal water pots connected by hoses that are set on that slope. (b) Hawaiian geologist Gary Puniwai with a proton-precession magnetometer, used to detect shallow magma intrusions. (c) USGS researcher Donald Peterson measures the depth to a lava pool in a vent on the Kīlauea east rift zone using a range-finder. (d) A student researcher collects fumarole gas samples using a gas chromatograph in the crater of Casita volcano, Nicaragua.

Other tools have also proved useful in studying volcanic activity. The ascent of magma into shallow reservoirs usually causes a change in Earth's gravity field, which can be measured using *gravimeters*. A change in the makeup of gases, especially in concentrations of chlorine relative to sulfur issuing from fumaroles, often accompanies the ascent of fresh magma. Telltale swellings may be detected using *satellite interferometry*—a method of comparing images from space at different times to detect even slight changes in the shape of a volcano. Interferometry is an excellent way to study volcanoes that cannot be easily equipped with tiltmeters and seismometers. *Global Positioning System* (GPS) receivers are also placed to help measure landform changes.

Volcano scientists must also be skillful at dealing with the press and government officials, a craft that can only be learned on the job. Scientists must be careful to convey information accurately, so that they do not cause panic or unnecessary and costly evacuations. At the same time, they must judge the precise moment when it is mandatory to urge people to leave their homes. These are not easy calls to make—and throughout a period of crisis, the experts and authorities wrestle with life-threatening probabilities.

The ability to save lives from volcanic activity also depends on the level of public education. People have many troublesome misconceptions about volcanoes. One source of confusion relates to the terms active, dormant, and extinct. Many people think of active volcanoes as presently in a state of eruption. Scientifically, however, an **active volcano** is one that has erupted frequently in historic time and will almost certainly continue to do so in the future, whether or not it is presently erupting. (Just what "frequently" means is a subjective term, however. Also, notice that the length of

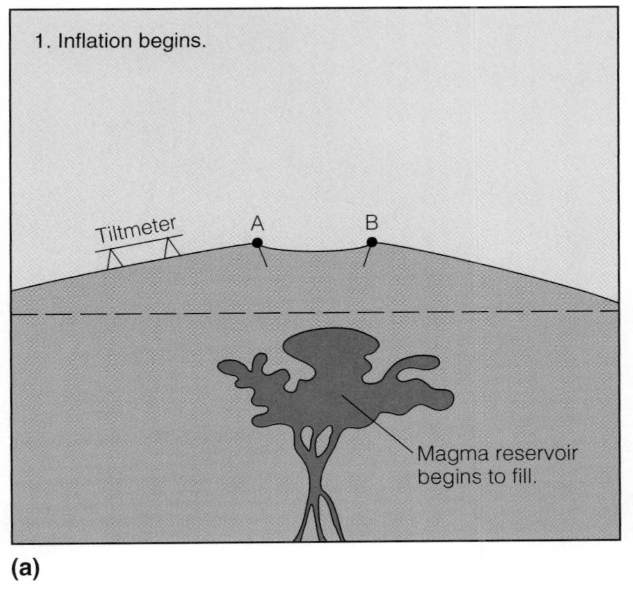

(a)

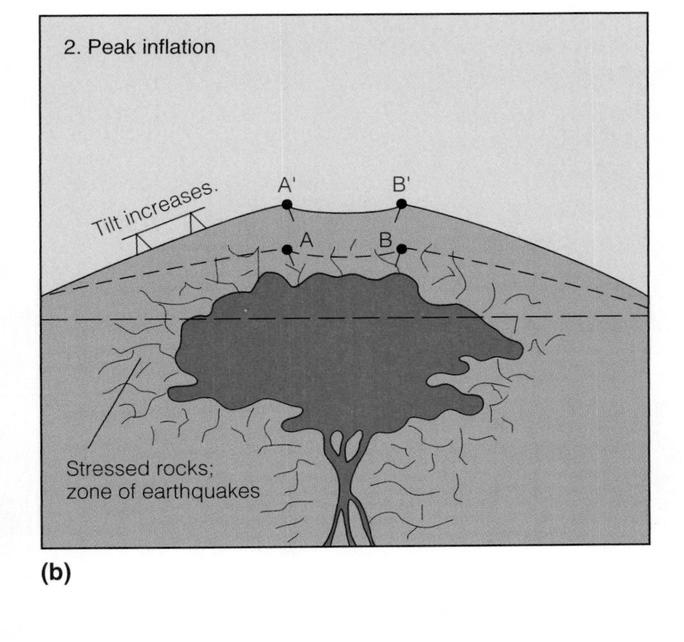

(b)

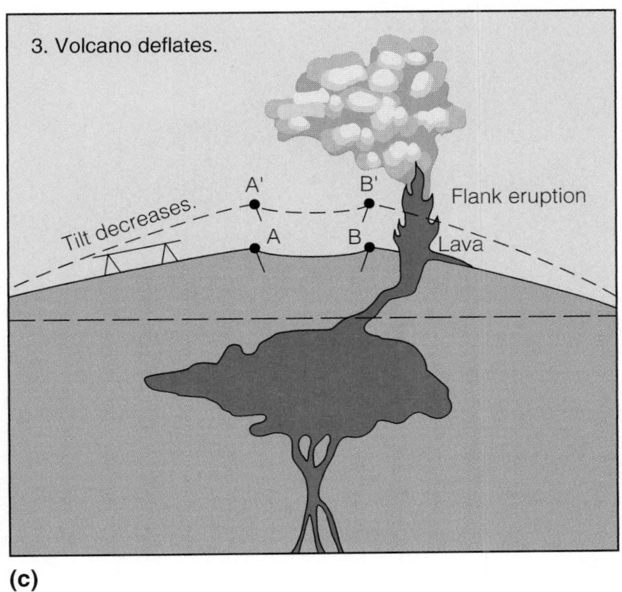

(c)

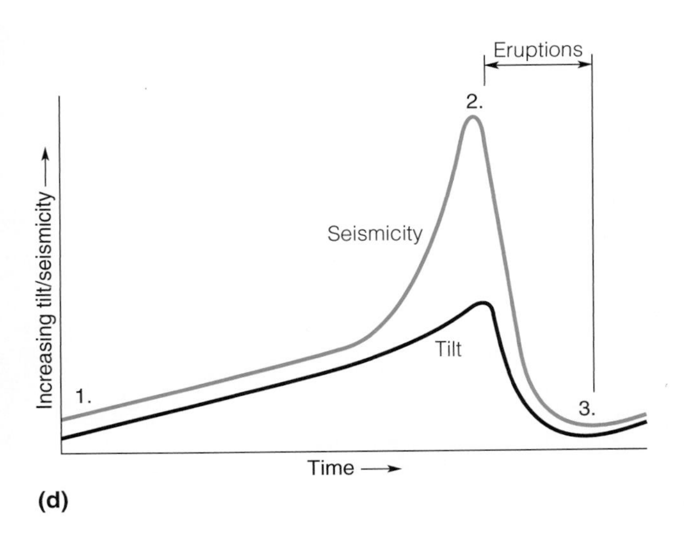

(d)

Figure 5.38 Three stages are apparent.

a–c: © Cengage Learning 2014; d: USGS data

explorer's
perspective

Rich Karstens, http://geology.uwyo.edu/kenwwsims

KENNETH SIMS, a volcanologist, studies isotopes to discover new things about our planet's behavior.

Kenneth Sims graduated with a geology degree from Colorado College at the foot of the Rocky Mountains. He expressed his love of the outdoors through mountain climbing and wilderness guiding and ultimately obtained a Ph.D. from the University of California, Berkeley, specializing in the study of isotopes applied to the chemistry of the Earth—in particular, volcanoes. His research has led him to active vents all over the world, including those of the deep ocean ridges using manned and unmanned submersibles, the frozen wastes of Antarctica, sunny Italy, Nicaragua, and Hawaii—and, more recently, the treacherous summit crater of the African volcano Nyiragongo, study of which is a security need for the nearby Congolese city of Goma (population ~375,000). Sims wants to learn what various radiogenic (unstable) isotopes in rock, water, and gas samples can teach us about future volcanic behaviors, as well as other important geological phenomena—the origins of plate tectonics, mega-scale change in Earth's climate, how groundwater moves, and much more. His climbing skills have enabled him to collect data under often extraordinary physical conditions. And he remains fascinated with Nature, commenting, "I am always left speechless when I look at scenes such as the Trans-Antarctic Mountains and the Polar Plateau, the deserts of the American Southwest, or the world's largest lava lake...." He is currently a professor at the University of Wyoming, and has increasingly turned his attention with students to the mysteries of Yellowstone National Park's huge and scenic caldera.

Source: http://www.nationalgeographic.com/explorers/bios/kenneth-sims/

Beryl Thermal Spring and fumarole, Yellowstone National Park.
Photo by Nathaniel W. Yale (student of Dr. Rick Hazlett)

Figure 5.39 Ship Rock, the 500-m (1,600-foot)-high chilled magmatic conduit of a 27 million-year-old extinct volcano. The landscape has eroded down 700 to 900 meters (2,500–3,000 feet) since volcanic activity ended. The Navajo peoples who live nearby call this feature Tse Bit a'i ("Winged Rock") in reference to the solidified dikes—possibly the deep traces of former eruptive fissures—radiating from the base of the pinnacle.

"history" differs from one region to another around the world.) An active volcano has to be regarded as threatening, whether it is actually erupting or not.

As a rule of thumb, a **dormant volcano** is one that shows signs of having erupted within the past few hundred or few thousand years but has not erupted in living memory. Many dormant volcanoes have only slightly weathered lava flows, appear not to be greatly eroded, and have little vegetation relative to the surrounding area. However, some dormant volcanoes are so heavily vegetated and eroded that people don't even know that they exist without a careful geological study! Even if a volcano is dormant, it does not mean that it is safe to live nearby. We simply don't know if or when it will erupt again. Moreover, if an eruption does occur, it is likely to be dangerously explosive because of the cap of old, hardened rock that plugs the conduit of the volcano. Some dormant volcanoes that have returned to life catastrophically are El Chichon (Mexico), Mount Lamington (New Guinea), Mount Pinatubo, Vesuvius, and Mount St. Helens. Volcanologists Chris Newhall and Steven Self published in 1982 a study of 25 of the world's most violent historical eruptions. They warned that in 17 cases the eruptions were the first ever reported volcanic activity at the volcanoes where they occurred! For others that had a record of previous activity, the mean repose interval was 865 years—long enough for people to forget and lose concern. Geologists do not know exactly how many dormant volcanoes exist.

An **extinct volcano** is typically deeply eroded because it has not erupted in thousands, if not millions, of years. As you might guess, there are many more extinct volcanoes than dormant and active ones in the world today. Ultimately, erosion removes all traces of a volcanic cone, leaving perhaps the riblike feeder dikes and central plug of the former magma conduit standing behind as monumental landforms, such as Ship Rock in New Mexico (**Figure 5.39**).

Volcano scientists working through the United Nations and other organizations are eager to broadcast information about warning signs, potential hazards, and the steps that people can take to protect themselves during an eruption. They also produce detailed maps of the deposits left by past eruptions around active volcanoes. These deposits, as we've seen, can provide valuable information about the characteristic behavior and, therefore, the future anticipated risks from a volcano. Careful, probability-based risk assessments can provide guidelines for developers, zoning planners, insurance companies, and civil authorities (**Figure 5.40**).

Perhaps the best way to save lives is not to put people in harm's way to begin with. But despite immense progress in eruption forecasting and monitoring in recent decades, the threat to human beings just keeps growing because rising populations and pressure for land use around volcanoes has outstripped our ability to protect us from nature's worst blows. The trade-off in terms of general benefit may seem worth it in the long run, but for those people caught in the path of a volcanic eruption, such an argument has no meaning. The need for bright new scientists to carry on the study and observation of volcanoes has never been greater.

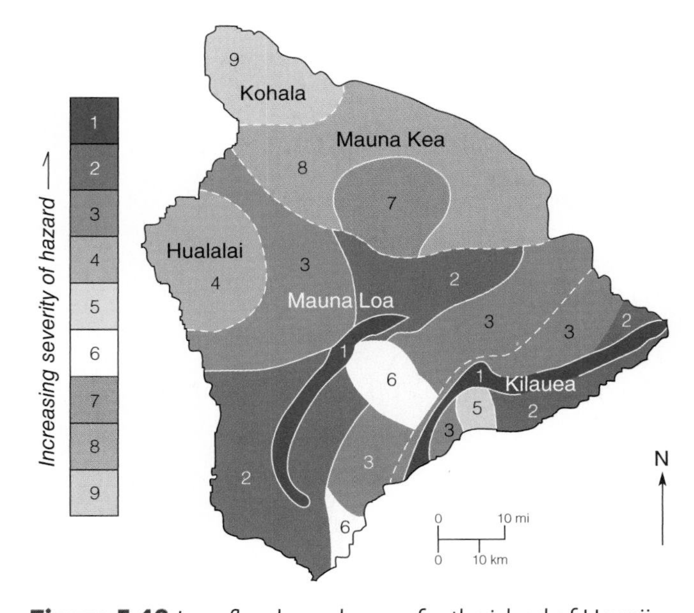

Figure 5.40 Lava-flow hazard zones for the island of Hawaii. The zones range from low risk (9) to high risk (1). Volcano boundaries are denoted by dashed lines. (USGS data)

case study

5.1 New Zealand's Blast in the Past, a Skifield, and Fluidization

The two islands of New Zealand are clearly distinguishable geologically. Whereas North Island exhibits explosive volcanism and some of the most spectacular volcanic landforms in the world, South Island lacks volcanism and is more alpine in nature. North Island's volcanic zone includes Lake Taupo, the site of a cataclysmic Plinian eruption in 186 C.E. that was arguably the largest eruption on Earth in 5,000 years (**Figure 1**). A volcanic source, now beneath Lake Taupo, vented a column of hot gas, ash, and pumice more than 50 kilometers (30 miles) high, which spread downwind, forming light yellow ashfall deposits more than 5 meters (16 feet) thick in some places. At 100 kilometers (60 miles) distance, the deposits are 25 centimeters (10 inches) thick. Toward the end of the eruption sequence, the hot column became so heavy that it collapsed and flowed outward in all directions as a pyroclastic density current. Flow velocity is estimated to have been 600 kilometers (375 miles) per hour, and the material traveled more than 80 kilometers (50 miles) from its source, overwhelming all the topography and vegetation in its path. Fortunately, the native

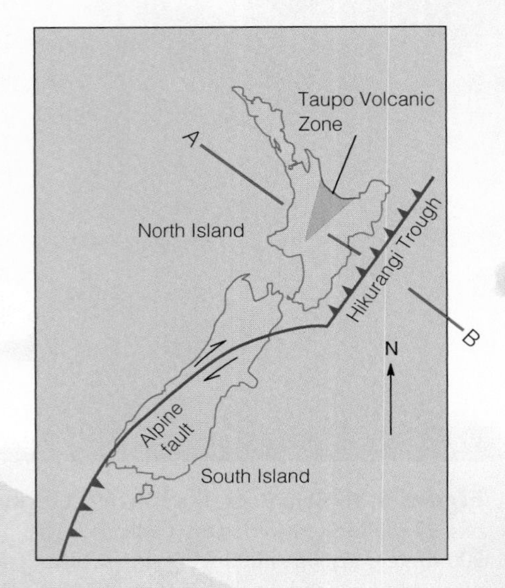

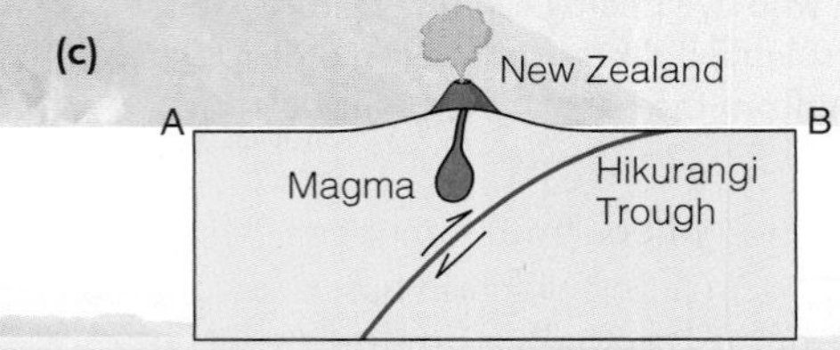

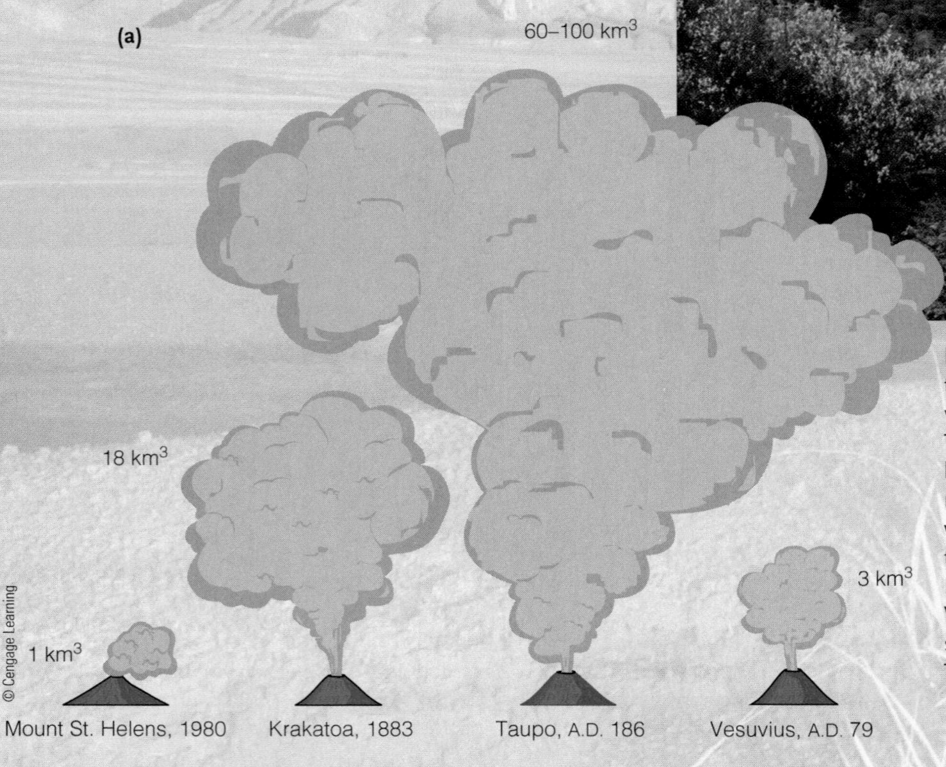

Figure 1 (a) The relative sizes of four large eruptions of the past 2,000 years based on estimated volumes of ejected tephra. (b) The very large Taupo event, together with a series of similar earlier eruptions, contributed to developing 40-kilometer (25-mile) wide Taupo caldera, the floor of which is now submerged by Lake Taupo. (c) Tectonic setting of the Taupo volcanic zone, North Island, New Zealand. (Volcanoes of the South Wind: A Field Guide to the Volcanoes and Landscape of Tongariro National Park. Karen Williams, Tongariro Natural History Society.)

New Zealanders, the Maori, had not yet settled the island when the eruption occurred. When they did go to the island about 1000 C.E., they regarded the Tongariro–Taupo area as sacred. They believed that their gods and demons had brought volcanic fire with them from Hawaiki, the Maori's ancestral homeland.

How can a pyroclastic material move at such a phenomenal speed and travel such a great distance, seemingly defying friction and gravity? Some PDCs are **fluidized**, meaning that the mix of high-pressure gas and fine particles within them moves like a giant liquid flood. Many chemical engineering professors demonstrate fluidization by forcing air through a powder on the bottom of a see-through box. As the air pressure is increased, the powder begins to quiver; then particles start dancing around, and eventually the whole mass is in motion. When the professor has it just right, he or she places a plastic duck (professor's choice) into the container, and it literally floats on the air–solid mixture. If the duck is pushed down, it immediately pops up, proving that it is floating on a denser medium. Part of the weight of the solids is taken by the air, reducing the interparticle friction to such a degree that the particles will not stand in a heap. The particles are not entrained (carried away bodily) like snowflakes in a blizzard; rather, the mass behaves as a fluid substance would. The point of this is that a well-understood process demonstrates how PDCs can fluidize and travel far from their source, destroying everything in their path.

At the south end of the Taupo volcanic zone is a cluster of stratovolcanoes in the Tongariro National Park. Two of the volcanoes are active and are popular playgrounds for New Zealanders, primarily for skiing and tramping (hiking). Mount Ruapehu, with a summit elevation of 2,797 meters (9,100 feet), is the highest point on North Island and has a lake in its summit crater. It is also the location of the Whakapapa Skifield, a very popular, well-equipped ski area. The terrain at Whakapapa (Maori, pronounced "fack-a-papa") is bare, ragged lava and requires a lot of snow to be skiable (**Figure 2**). Throughout the skifield (or lava field, depending on the time of year), signs alert skiers and campers to stay on high ground and out of the canyons, should Ruapehu erupt, which it did in 1969, 1971, 1982, and sensationally in 1995. The eruptions are **phreatic**; that is, superheated water at the bottom of the lake flashes to steam and overflows the crater. The lake's surface-water temperature, which ranges from 20°C to 60°C (68–140°F), is monitored as a potential eruption predictor. When the water heats up, some kind of activity can be expected. A lahar warning system at Whakapapa is based on detection of earthquake swarms, which precede phreatic and magmatic eruptions. Tongariro (Maori tonga-, "south wind," and -tiro, "carried away") National Park is another case of humans both preserving and living wisely within nature.

questions to ponder

1. Can you think of other demonstrations that could be done to illustrate how piles of fine grains can become fluidized? (A rubber duck isn't necessary!)

Figure 2 (a) Whakapapa Skifield is on volcanic Mount Ruapehu, the highest point on New Zealand's North Island and the site of a beautiful summit crater lake. Signs on the slopes warn skiers to stay out of canyons and low spots should a phreatic eruption create a dangerous lahar. (b) On 25 September 1995, a combination of hot magma and cold water produced a violent steam eruption, which hurled boulders over the upper slopes of the Whakapapa Skifield. The eruption did little damage, but projectiles as big as cars were thrown over the rim. The Maori name for Ruapehu means "exploding pit."

case study

© Simon R. Young, Montserrat Volcano Observatory

Figure 1 Pyroclastic flow moving down the slope of Soufrière Hills volcano on the island of Montserrat, British West Indies. Shown is one of many life-endangering flows that occurred during July 1997 and that threatened the city of Plymouth.

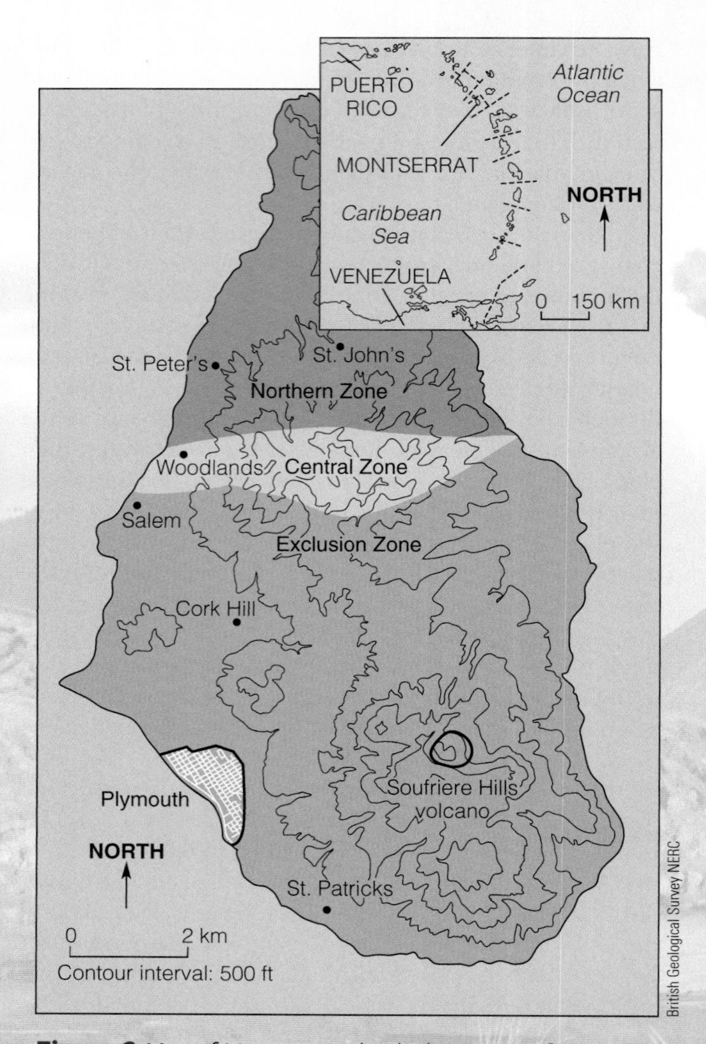

British Geological Survey NERC

Figure 2 Map of Montserrat Island, showing Soufrière Hills volcano and the volcanic hazard zones. Exclusion zone: no admittance except for scientific monitoring and national security matters. Central zone: residents only on a heightened state of alert; all residents to have a rapid means of exit 24 hours a day and to wear hard hats and dust masks when outside. Northern zone: significantly lower risk; suitable for residential and commercial occupation.

5.2 Montserrat, Paradise Lost

The Soufriere Hills volcano on Montserrat, a little island and British crown colony in the eastern Caribbean Sea, began a series of eruptions in 1995 that continued episodically into 2000. The eruptions began with a dome of andesitic lava in the central crater that periodically collapsed, generating pyroclastic flows (**Figure 1**). Ashfalls followed the flows and smothered most of Plymouth, the resort capital, and St. Patrick's, a resort town. The world followed the eruptions on television and on the Internet, particularly the evacuation of half of Montserrat's 12,000 inhabitants away from the shadow of the smoldering peak. Volcanic hazard zones were established in September 1997 by the British Geological Survey's Montserrat Volcano Observatory (**Figure 2**). To add to Montserrat's problems, it was also subjected to the extreme hurricane season of 1999.

Ross Armstrong/Alamy

5.3 Volcanoes and Earth's Climate

In 1815, a typesetter for the Old Farmer's Almanac, a popular American publication, made a mistake and predicted a winter weather forecast for the following July. The error was accepted as such—or as simply being preposterous. But, in fact, it was not far off the mark! The summer of 1816 in northern Europe, the northeastern United States, and eastern Canada was highly abnormal. Temperatures at this time of year in these regions are usually highly stable, averaging from 20°C to 25°C (68–77°F)—too warm for frost and snow. However, that summer, temperature swings between 35°C (95°F) and near freezing took place in some cases within hours, and heavy snowfalls persisted well into June—prime agricultural season—with ice forming on rivers and lakes throughout the summer! Widespread crop failure ensued, and prices for oats, essential for feeding horses—a mainstay of transportation in those days—rose from 12 cents to 92 cents a bushel. In ordinarily peaceful Switzerland, rising prices and food scarcities led to violence, whereas in the United States, thousands of farmers migrated west and south in search of better growing conditions. This climate catastrophe became known as the "Year without a Summer," or the "Poverty Year," one of the worst famines of the 19th century. In time, the cause would be linked primarily to the massive eruption of Mount Tambora on the island of Sumbawa in Indonesia, the largest volcanic outburst to occur in more than 1,600 years—and perhaps the deadliest, killing at least 71,000 people (see Table 5.1 and **Figure 1a**). The blasts from Tambora could be heard as far as 2,600 kilometers (1,550 miles) away, as the volcano explosively ejected some 160 cubic kilometers (38 miles^3) of magma. The eruption cast a dense pall of fine ash and sulfate gas aerosols into the stratosphere, dimming the sunlight reaching the surface below and cooling the global climate by 0.4°C to 0.7°C (0.7–1.3°F). Fine ash remained suspended 10 to 30 kilometers (6 to 18 miles) high worldwide for several years after the eruption. Volcanic blasts large enough to cool Earth's climate temporarily occur roughly once or twice every century. In addition to fine-ash particles and sulfates, volcanoes also release large amounts of water vapor and carbon dioxide (CO_2), gases that contribute to greenhouse warming. So much H_2O and CO_2 already exist in the atmosphere, however, that the warming potentials of big eruptions are trivial compared with the cooling. In a longer term sense, volcanoes have played an important role in replenishing Earth's supply of atmospheric CO_2 throughout geological time.

In human terms, perhaps the severest case of volcano-induced global cooling took place 70,000 to 75,000 years ago, when a far more powerful eruption than that of Tambora created Lake Toba caldera—the world's largest volcanic collapse feature—on the island of Sumatra (see Figure 1b). This super-eruption, which generated on the order of 2,500 to 3,000 cubic kilometers (600–720 miles^3) of ejecta, took place in the midst of the latest ice age, when the world's climate was already very cold compared with today. As much as a hundred million metric tons of sulfuric acid escaped during the eruption, triggering intensive acid rain-

© NASA/JSC

© Wolfgang Kaehler/Corbis

Figure 1 (a) Mt. Tambora volcano showing the immense caldera that remains after the largest volcanic eruption in recorded history. (b) View from the shore of Samosir Island within the giant caldera of Lake Toba, which formed 75,000 years ago.

falls over vast areas. Genetic evidence suggests that the world's human population, already struggling with ordinary ice-age instability, crashed to as low as 10,000 or possibly only 1,000 breeding pairs! The evidence for this catastrophic dieback is not conclusive, but surely volcanic activity 10 to 20 times greater than that causing the Year without a Summer must have had significant global impact.

questions to ponder

1. Do you think that a very powerful volcanic eruption could trigger an ice age? Why or why not?

5.4 Carbon Dioxide, Earthquakes, and the Los Angeles Water Supply

The Eastern Sierra Nevada is a mecca for tourists who want beautiful mountains and outstanding recreation. Much of the area's appeal results from volcanic action. About 760,000 years ago, a huge eruption blew out 617 cubic kilometers (150 miles^3) of magma that spread ash over nine states and as far to the east as present-day Nebraska (**Figure 1**). The Earth surface above the magma chamber sank 2 kilometers, forming the huge, oval caldera (16 × 33 km) that is now called Long Valley. The eruption was 2,000 times the size of the 1980 Mount St. Helens eruption, and it left hard, welded tuff all over today's east-central California. Mammoth Mountain and Mono Craters, a linear chain of glassy **domes** (see Figure 5.16), formed within the last 400,000 years, and the people of Long Valley are living with the caldera's continuing restlessness.

The restlessness is manifest in hot springs (**Figure 2**), carbon dioxide (CO_2) emissions, and swarms of earthquakes, the most intense of which began in 1980 after two decades of quiescence. There were four strong, magnitude-6 earthquakes, three of them occurring on the same day. These quakes prompted studies by the USGS, which found uplift of 60 centimeters (about 2 feet) in the caldera center. Uplift, earthquake swarms, and gas emissions leave little doubt that magma is rising and getting closer to the ground surface. Late in 1997, the volcano acted up, with more than 8,000 earthquakes larger than magnitude 1.2, and a couple of rattlers approached magnitude 5.0. More than 1,000 earthquakes occurred one day, and many scientists believe the volcano was waving a red flag signaling "danger."

Much of the water for the city of Los Angeles comes from Sierra Nevada snowmelt and travels from Mammoth Lakes to the city via open aqueduct. Should the volcano erupt, it would jeopardize or totally cut off this invaluable resource. Furthermore, the area's main highway access is aligned in the direction of the prevailing wind from the mountains. For this reason, an "escape road" was built in the opposite direction, toward the north, to provide an alternate route for the area's residents and visitors.

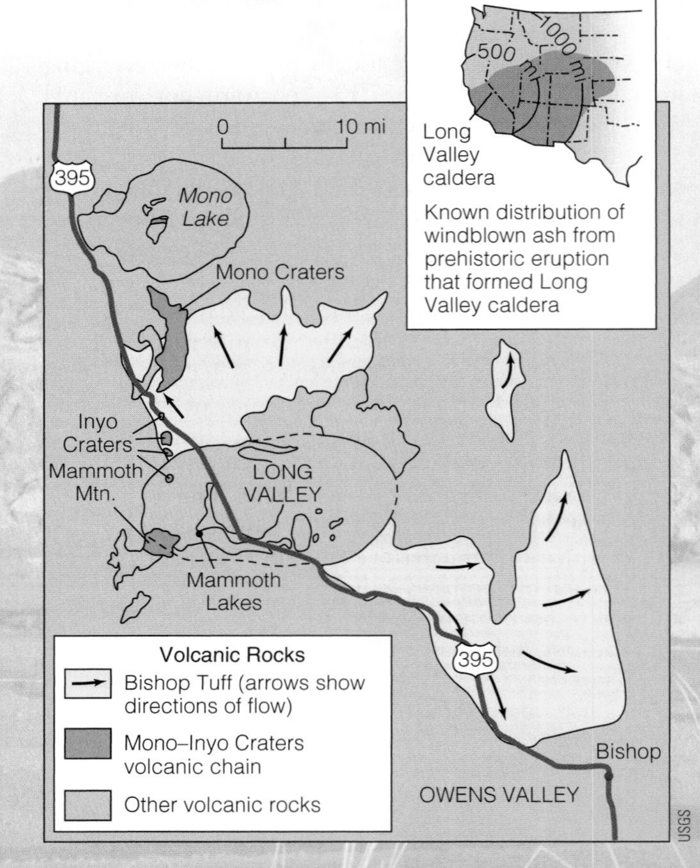

Figure 1 Mammoth Mountain and the Inyo-Mono Craters volcanic chain (red). The cataclysmic eruption 760,000 years ago that formed Long Valley ejected hot, glowing pyroclastic flows, which cooled to form the Bishop Tuff (yellow). Inset map shows the distribution of Bishop Tuff and ash. (Data from U.S. Geological Survey.)

Mammoth Mountain, a large rhyolitic volcano, is still hot. Tree-killing emissions of CO_2 were discovered at the base of the mountain in 1990. This indicated magma activity at shallow depth (**Figure 3**). By 1995, about 30 hectares

Ross Armstrong/Alamy

© 1999 John Rogie

Figure 2 The Horseshoe Lake tree-kill area, Mammoth Lakes, California. (a) Aerial view looking north toward Mammoth Mountain. Inyo-Mono Craters and Mono Lake are shown in the right background. The light-colored area (foreground) at the base of the Mammoth Mountain and on the shore of Horseshoe Lake is the tree-kill area. (b) Close-up of trees killed by leaking volcanogenic carbon dioxide gas.

© 1999 John Rogie

(75 acres) of pine and fir trees had suffocated, and CO_2 concentrations of up to 90% of total gas were found in the trees' soil and root systems. Where CO_2 concentrations exceeded 30%, most of the trees were dead. One large campground had to be closed because park rangers found high levels of CO_2 in the restrooms and cabins after campers exhibited symptoms of asphyxia (**Figure 4**). It appears that CO_2 degasses from the magma and migrates into a deep gas reservoir. When earthquake swarms cause fractures to develop in the reservoir rock, the gas finds its way to the surface. It is estimated that between 200 and 1,000 tons of CO_2 flow into the soil and atmosphere per day. The good news is that gas emissions have been declining, which may indicate that the magma has exhausted its gas supply.

Figure 3 Looking south through the Horseshoe Lake tree-kill area on the south side of Mammoth Mountain. The trees have died from high levels of carbon dioxide in the root zone.

© B. Pipkin

questions to ponder

1. What would a city as large as Los Angeles do if a substantial amount of its water supply were cut off for weeks or months by a volcanic eruption in the Mammoth Lakes-Long Valley area?

Figure 4 A small geyser of scalding water erupts in a formerly popular bathing pool at Hot Creek, not far from Mammoth Mountain. Activity picked up here in the spring of 2006.
© Jason D. Opliger, Inyo National Forest

PHOTO GALLERY volcanic wonders

Volcanoes present some of the strangest, most dramatic, and instructive landscapes on Earth, as can be appreciated with these images.

1 Solfatara at the edge of an acidic volcanic lake, Kawah Ijen crater, Java. Robert and Barbara Decker/U.S. Geological Survey

2 Fossil fumaroles with incrustations of various iron oxides and other gas sublimate minerals, Valley of Ten Thousand Smokes, Katmai National Park, Alaska. Adrienne Kentner/U.S. Geological Survey

Pu'u O'o vent lava lake in the east rift zone of Kīlauea volcano. U.S. Geological Survey

3 Hikers at the rim of Rinjani caldera on the island of Lombok, Indonesia. Volcanic fume and steam rise from a vent at the edge of the caldera lake. © 2011/Aaron Lim Boon Teck/National Geographic Image Collection

4 Call the automobile club! J. D. Griggs, Hawaiian Volcano Observatory, USGS

5 Eroded basaltic lava flow probably less than 100,000 years old, with column-forming cooling fractures; Devil's Postpile National Monument, California. U.S. Geological Survey

SUMMARY Volcanoes

Volcanoes

Defined
A vent or series of vents that issues lava and pyroclastic material.

Distribution
Adjacent to convergent plate boundaries around the Pacific Ocean (the Ring of Fire); in a west-east belt from the Mediterranean region to Asia; along the Mid-Ocean Ridges; and in the interior of tectonic plates above hot spots in the mantle. In the United States, a 1,100-kilometer (680 miles) belt extends northward from northern California through Oregon and Washington.

Measurement
The Volcano Explosivity Index assigns values of 0 to 8 to eruptions of varying size.

Explosivity
Depends on the lava's viscosity and gas content. Viscosity, a function of temperature and composition, determines the type of eruption: felsic (high SiO_2) lavas are viscous and potentially explosive; mafic (low in SiO_2, high in magnesium and iron) lavas are more fluid and less explosive.

Types of Volcanoes and Volcanic Landforms

Shield (Hawaiian or "Effusive" Type)
Gentle outpourings of lava produce a convex-upward edifice resembling a shield.

Continental Flood Basalts
Lava erupts from long cracks, building up broad lava plateaus, such as those found in the Columbia River Plateau (Oregon–Washington–Idaho), Iceland, and India. They are fed by mantle plumes.

Composite Volcanoes (Explosive Type)
Characterized by a concave-upward cone thousands of meters high. Vesuvius (Italy), Fujiyama (Japan), and Mount Hood (Oregon) are examples. Crater Lake (Oregon) is the stump of a stratovolcano that exploded and collapsed inward, leaving a wide crater known as a caldera.

Lava Domes
Bulbous masses of high-silica, glassy lavas, as found at Mono Craters (California). Domes may form in the crater of a composite cone after an eruption, such as at Mount St. Helens (Washington).

Cinder Cones
The smallest and most numerous of volcanic cones, composed almost entirely of tephra.

Benefits of Volcanic Activity

Atmosphere and Hydrosphere
Outgassed from Earth's interior.

Soils
Replenish and fertilize soils for farmers, especially in tropical areas.

Building Materials
Metals and construction stone.

Geothermal Energy
Cooling magma at depth heats water that can be converted to steam to drive electrical generators.

Hazards

Lava Flows
Lava flows destroy or burn everything in their path and can travel a distance of 100 kilometers or more.

Ashfalls
Heavy falls of volcanic ash from Mount Vesuvius buried Pompeii in 79 C.E. There were heavy ashfalls from Mount Pinatubo, the Philippines, in 1991. Vertical plumes of ash may rise many kilometers from a vent, which are then carried by the wind and blanket the terrain. Such plumes rising from a vent are known as Plinian eruptions.

Pyroclastic Density Currents (PDCs)
Hot, fluid masses of steam and pyroclastic material that travel down the flanks of volcanoes at high speeds, blowing down or suffocating everything in their path. Such flows have killed thousands of people.

Lahars
Catastrophic mud flows down the flanks of a volcano. They cause the most volcanic fatalities, more than pyroclastic flows.

Tsunami
Submarine volcanic eruptions (such as that of Krakatoa in 1883) create enormous sea waves, which may travel thousands of miles and still do damage along a shoreline.

Weather
Sulfur dioxide coatings on ash and dust particles increase their reflectivity and cause cooling of the weather, as occurred after the eruption of Mount Pinatubo, 1991–1993. The effects of ash and dust in the stratosphere are short term—a few years.

Gases

Corrosive gases emitted from a volcano can be injurious to health, structures, and crops. Such gases are called vog (volcanic gas) on the island of Hawaii. Carbon dioxide emitted from the bottom of volcanic Lake Nyos in Cameroon formed a cloud that sank to the ground and suffocated 1,700 people in 1986. Tree kills caused by CO_2 have occurred at Mammoth Mountain, California.

Mitigation and Forecasting

Volcanologists (geologists who study volcanoes) forecast eruptions using sensitive high-tech instruments such as seismometers, tiltmeters, and gravitometers. They even use satellites to record changes in the land surface related to magma movement. Volcanologists also provide data to make hazards maps for authorities to use in planning and emergency response. They also provide important information for emergency responders during eruptions.

KEY TERMS

a'ā
active volcano
arc volcano
ashfall
caldera
cinder cone
composite volcano
degassing
dormant volcano
effusive eruption
eruption column
extinct volcano
extrusive eruption
fiamme
flood basalt
fluidized
forecast
fumarole
hydrothermal alteration
kīpuka
lahar
lava dome
limnic eruption
mofettes
monogenetic
neutral buoyancy zone (NBZ)
pāhoehoe
phreatic eruption
pillow lava
polygenetic
prediction
pyroclastic density current (PDC)
shield volcano
thermal advection
tiltmeter
umbrella
volatiles
volcanic column
volcanic conduit
Volcano Explosivity Index (VEI)

STUDY QUESTIONS

1. What are the major kinds of volcanic landforms, and which types of eruptions create each kind?
2. How would you expect the eruption of a volcano that taps mafic (basaltic) magma to differ from the eruption of one that taps felsic (silicic) magma? What kind of rocks will result from each eruption?
3. Explain the distribution of the world's active volcanoes in terms of plate tectonics theory.
4. List the major kinds of volcanic hazards and discuss the steps that should be taken to avoid them or minimize their impacts on people.
5. In what way(s) can dormant volcanoes be especially dangerous?
6. How do calderas form?
7. In what ways does volcanic activity benefit humankind?
8. If you were setting out to establish a program of monitoring a volcano to forecast its eruptions, what would you do?
9. What would you do to make a volcanic hazards map for people living around a volcano?
10. Contrast a volcano such as Mt. Rainier with one like Mauna Loa or Kīlauea. Which, in general, would be the safer volcano to live around, and why?
11. What aspects would you look for to distinguish a Strombolian from a Hawaiian eruption? How about a Strombolian from a Vulcanian eruption?
12. Contrast the different ways of classifying volcanic eruptions introduced in this chapter, and discuss the relative advantages and disadvantages of using each.

REFERENCES AND FURTHER READINGS

Ambrose, S. H. 1998. Late Pleistocene human population bottlenecks, volcanic winter, and differentiation of modern humans. *Journal of Human Evolution* 34: 623–651.

Aspinall, W., S. C. Loughlin, F. V. Michael, A. D. Miller, G. E. Norton, et al. 2002. *The Monserrat Volcano Observatory: Its evolution, organization, role and activities*. Geological Society of London Memoirs 21: 21–91.

Decker, R., and B. Decker. 1997. *Volcanoes*. New York: W. H. Freeman, 320 p.

Duennebier, F. K. (contact person), and the Loihi Science Team. 3 June 1997. Researchers rapidly respond to submarine activity at Loihi Volcano, Hawaii. *EOS* 78(22).

Evans, W. C, L. D. White, M. L. Turtle, G. W. Kling, G. Tanyileke, and R. L. Michel. 1994. Six years of change at Lake Nyos, Cameroon, yield clues to the past and cautions for the future. *Geochemistry* 28: 139–162.

Fierstein, J., et al. 1997. *Can another great volcanic eruption happen in Alaska?* U.S. Geological Survey Fact Sheet 039-97. Reston, VA: U.S. Geological Survey.

Fiske, R. S. 1984. Volcanologists, journalists, and the concerned local public—a tale of two crises in the eastern Caribbean. In *National Research Council Geophysics Study Committee* (eds.), Explosive volcanism: Inception, evolution and hazards (pp. 170–176). Washington, DC: National Academies Press.

Geze, B. 1964. Sur la classification des dynamisms volcaniques. *Bulletin Volcanologique* 27: 237–257.

Hazlett, R. W. 2002. *Geological field guide. Kīlauea Volcano*. Hilo, HI: Hawaiian Natural History Association. 162 p.

Hazlett, R. W., and D. W. Hyndman. 1996. *Roadside geology of Hawaii*. Missoula, MT: Mountain Press, 307 p.

Hill, D., et al. 1996. *Living with a restless caldera, Long Valley, California*. U.S. Geological Survey Fact Sheet 108-96. Reston, VA: U.S. Geological Survey.

________ 1997. *Future eruptions in California's Long Valley area: What's likely*. U.S. Geological Survey Fact Sheet 073-97. Reston, VA: U.S. Geological Survey.

Kling, G. W., et al. 1987. The 1986 Lake Nyos gas disaster in Cameroon, West Africa. *Science* 236:169–174.

Lockwood, J. P., and Hazlett, R. W. 2010. *Volcanoes: A global perspective*. Oxford, U.K.: Wiley-Blackwell Publishing. 624 p.
McCoy, R., and G. Heiken (eds.). 2000. *Volcanic hazards and disasters in human antiquity*. Geological Society of America Special Paper 345. Boulder, CO: Geological Society of America. 99 p.
McGee, K. A., et al. 1997. *Impacts of volcanic gases on climate, the environment, and people*. U.S. Geological Survey open file report 97-262. Reston, VA: U.S. Geological Survey.
Myers, B., P. Stauffee, and J. Hendley II. 1997. *What are volcano hazards?* U.S. Geological Survey Fact Sheet 002-97. Reston, VA: U.S. Geological Survey.
Neal, C. A., T. Casadevall, T. Millar, J. Hendley II, and P. Stauffer. 1997. *Volcanic ash: Danger to aircraft in the North Pacific*. U.S. Geological Survey Fact Sheet 039-97. Reston, VA: U.S. Geological Survey.
Newhall, C. G., and S. Self. 1982. The volcanic explosivity index (VEI)—an estimate of explosive magnitude for historical volcanism. *Journal of Geophysical Research* 87: 1231–1238.
Schuster, R. L., and J. P. Lockwood. April 1991. Geologic hazards at Lake Nyos, Cameroon, West Africa. *Association of Engineering Geologists News* 28–29.
Siggurdsson, H. 1999. *Melting the Earth: The history of ideas on volcanic eruptions*. New York: Oxford University Press. 272 p.
Sigvaldason, G. E. 1989. International conference on Lake Nyos disaster, Yaounde, Cameroon, 16–20 March 1987: Conclusions and recommendations. *Journal of Volcanology and Geothermal Research* 39: 97–109.
Sutton, J., et al. 1997. *Volcanic air pollution: A hazard in Hawaii*. U.S. Geological Survey Fact Sheet 169-97. Reston, VA: U.S. Geological Survey.
Tilling, R. I. 1984. *Monitoring active volcanoes*. Denver, CO: U.S. Geological Survey.
Williams, A. R. July 1997. Montserrat: Under the volcano. *National Geographic* 58–75.
Wright, T. L., and T. C. Pierson. 1992. *Living with volcanoes*. U.S. Geological Survey circular 1073. U.S. Geological Survey, Box 25425, Federal Center, Denver, CO 80225.
See also the Smithsonian National Museum of Natural History Web site (http://www.volcano.si.edu/world/, "Global Volcanism") for updated lists of active and dormant volcanoes and much additional, exciting information about volcanoes.

Access an interactive eBook and chapter-specific interactive learning tools, including flashcards, quizzes, videos and more in your Geology CourseMate, accessed through CengageBrain.com.

Tractors in Kenya tilling soil on farm recently cleared of native vegetation.

Have you ever

wondered...

"A few inches between humanity and starvation..."

—Anonymous

6

Weathering, Soils, and Erosion

1. How rocks weather and soils erode?

2. Why soils around the world look different?

3. How to prevent soil erosion that devastates cropland?

4. What climate change will do to the world's soils?

5. Why polar region soils and those of semiarid areas are so sensitive to a warming climate?

The Wind Blew and the Soil Flew

The Great Depression of the 1930s was exacerbated by a protracted drought in the Great Plains of Oklahoma, Colorado, Texas, New Mexico, and Kansas. This part of the Great Plains was totally dependent on rainfall for crop production, and because the government had guaranteed wheat prices, farmers tilled thousands of acres. A longstanding agricultural practice was to plow the fields after the fall harvest, crop stubble and all, and let the land lie fallow all winter. This was good practice as long as it didn't rain too much—or too little. If too much rain falls, the soil is subject to extreme sheet and rill erosion by water, and if there is drought and the soil dries out, the result is the same: extreme soil erosion, only this time by wind.

In the mid-1930s, there was drought, and the wind did blow in the southern Great Plains. Huge quantities of topsoil were simply blown away (**Figure 1**). Fine particles were lifted 5 kilometers (16,000 feet) in the air and carried eastward as far as Washington, DC, and beyond. Five states that were the source of this airborne dust became known as the "Dust Bowl." Farming was impossible, and farm families, burdened by debt for equipment, seed, and supplies, left their farms to become migrant workers. Many traveled to California and became derisively known as "Okies." These unfortunate victims of severe soil erosion were the subject of John Steinbeck's touching novel *The Grapes of Wrath*. Even today, wind erosion exceeds water erosion in many parts of the Dust Bowl states. With climate changing rapidly and predictions for summer drought in the northeastern United States, could New England be the next dust bowl (**Figure 2**)?

question to ponder

1. What, if anything, do you think the farmers or the government could have done to prevent the Dust Bowl? What are we doing today to increase and decrease the chance that the Dust Bowl could happen again?

Figure 1 A dust storm hits a southwestern Great Plains village (Baca County, Colorado) in October 1935. Storms such as this one destroyed crops and buried pasturelands. Minutes after this photograph was taken, the village was completely engulfed in choking dust.

Dorothea Lange/Corbis

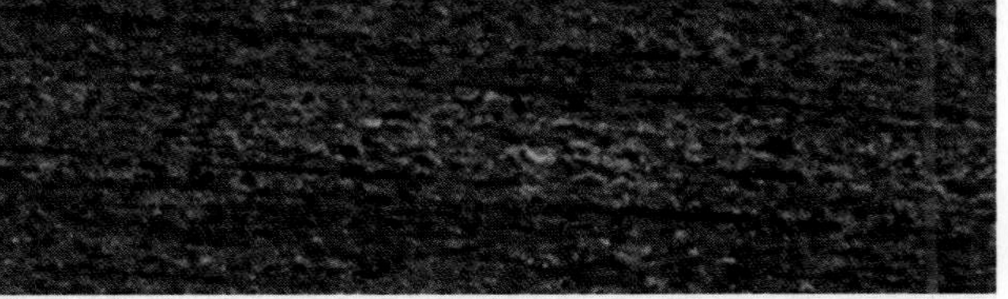

CourseMate Connection

Start off this chapter by watching a video about increased river erosion in Bangladesh that is thought to be caused by global climate change. **Log in to the CourseMate at www.cengagebrain.com to watch now.**

Records 1961–1990

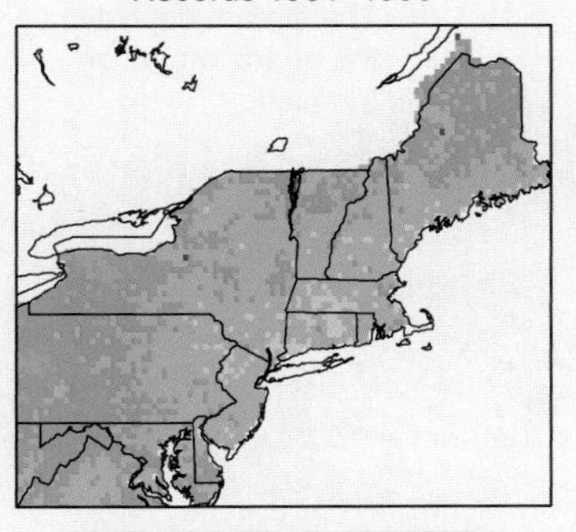

High Emission 2070–2099

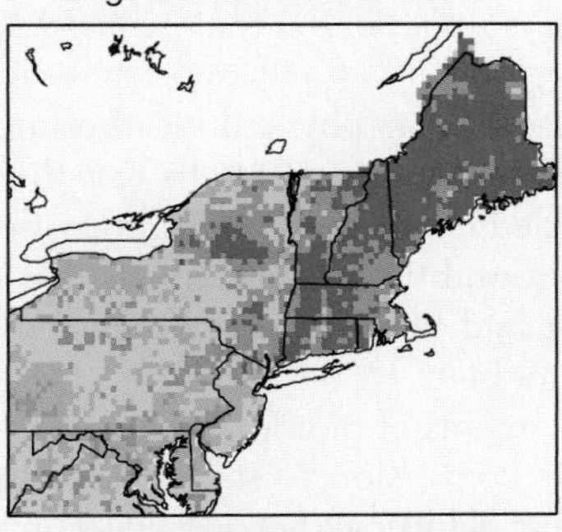

Low Emission 2070–2099

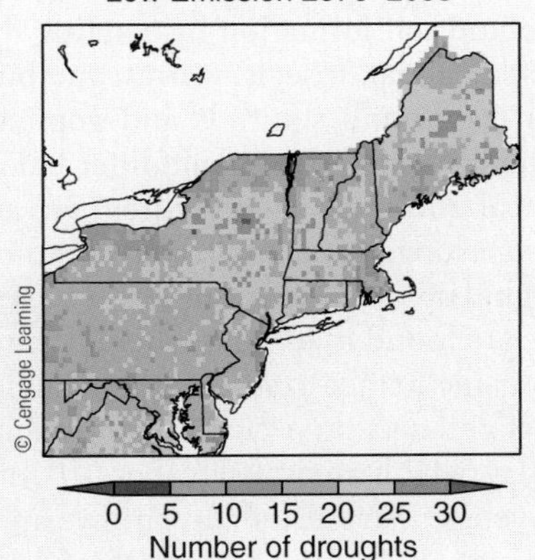

Figure 2 With continued global warming, the northeastern United States could be the next dust bowl. Models show that summers will get hotter and drier; in this image, red indicates an increase in likelihood of summer drought (the middle panel presumes that carbon dioxide emissions increase, and the lower panel suggests that emissions stabilize); with drought common, crops may fail. If the wind blows, the soil will erode unless farming practices change. Scale below the images shows the number of short (1- to 3-month) droughts (soil moisture deficit > 10%) over the 30-year window represented by each map.

SOIL, THE THIN MIXTURE of weathered rock and organic material below our feet, is Earth's most fundamental resource. We live on it, and through it we produce much of our food. Soil supports forest growth, which gives us essential products, including paper and wood. Soil material, organisms within the soil, and vegetation constitute a system critical to life on this planet. It is through soil that the four ingredients needed for plant growth are recycled: water, air, organic matter, and dissolved minerals (**Figure 6.1**).

The multiple uses of soil are reflected in the many definitions and classification schemes that have been developed for soils. To the engineer, a soil is the loose material at Earth's surface—that is, material that can be moved about without first being dynamited and upon which structures can be built. The geologist and the soil scientist see soil as a mixture of weathered rock, mineral grains, and organic material that is capable of supporting plant life. A farmer, in contrast, is mostly interested in what crops a soil can grow and whether the soil is rich or depleted with respect to organic material and minerals.

The carrying capacity of our planet, the number of people Earth can sustain, depends on the availability and productivity of soil. This is why understanding how soil is formed and how it can be best cared for is so important. Soil erosion removes this precious resource, and for all practical purposes, once productive topsoil is removed, it is lost to human use forever. Soil pollution can render areas harmful to enter or unable to grow edible food products. The goal of this chapter is to give you an appreciation of how important soil is to our environment by providing the information you need to understand how soil is formed, how it can be impacted by human actions, and how we as a society can better care for this invaluable resource.

Weathering

Because soil is made up in large part of weathered rock, we start by approaching the question, How does rock weather and erode to produce parent materials on which soils develop?

Weathering is the process by which rocks and minerals are broken down by exposure to atmospheric agents. Specifically, **weathering** is the physical disintegration and chemical decomposition of earth materials at or near Earth's surface. **Erosion**, in contrast, is the *removal and transportation* of weathered or unweathered materials by wind, running water, waves, glaciers, groundwater, and gravity (**Figure 6.2**).

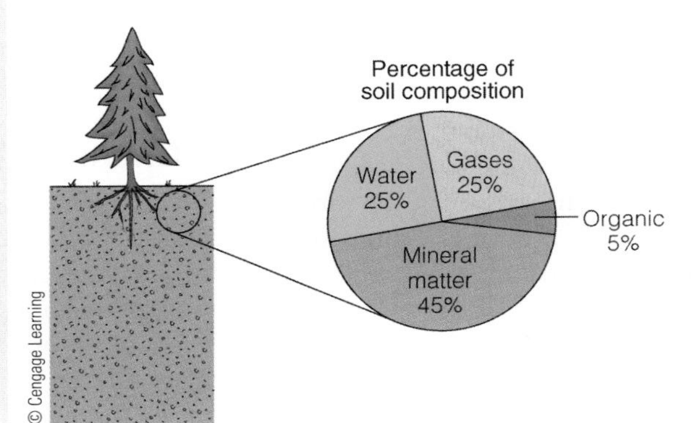

Figure 6.1 Composition of a typical soil. The organic component includes humus (decomposed plant and animal material), partially decomposed plant and animal matter, and bacteria. Mineral matter is what is left of the rock that weathered to form the parent material for the soil.

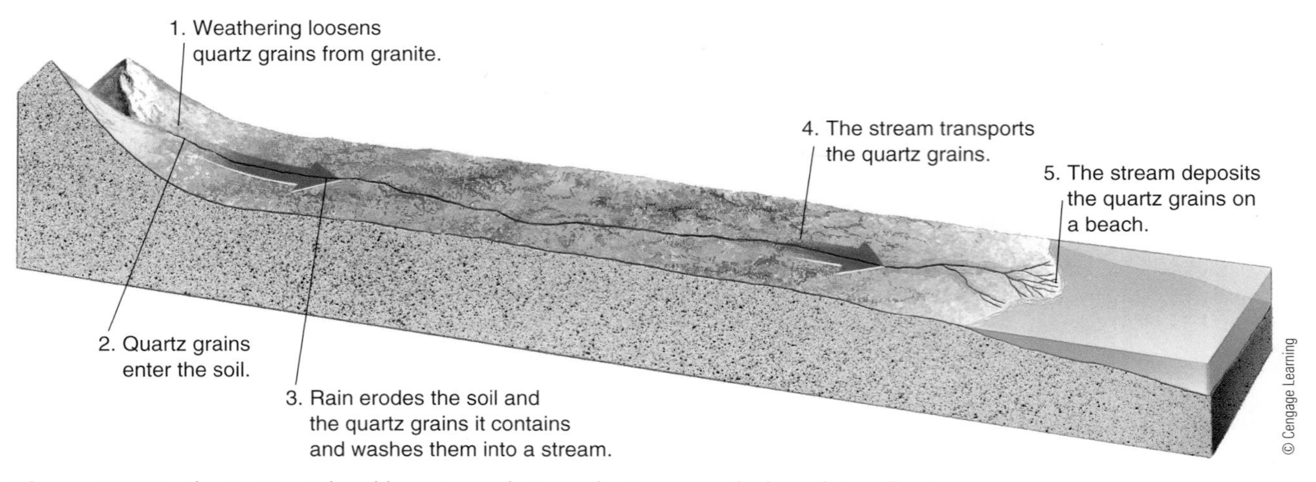

Figure 6.2 Weathering is catalyzed by water and atmospheric gases, which work together to loosen the quartz grains in the host rock, granite. The loosened grains are then incorporated into soil, eroded, and transported (in this example, by a stream) to a beach, which may be at a lake or at the ocean.

Physical Weathering

Physical weathering makes little rocks out of big rocks by many processes. Rocks are wedged apart along planes of weakness by thermal expansion and contraction, and by the activities of organisms. **Frost wedging** occurs when water freezes in joints or other rock openings and expands. This expansion, which amounts to almost a 10% increase in volume, can exert tremendous pressures in irregular openings and joints, but only if the joints are somehow sealed. As one would suspect, this process operates only in temperate or cold regions that have seasonal or daily freeze-and-thaw cycles. Water is not the only material to solidify in rocks and catalyze physical weathering. **Salt-crystal growth** (**Figure 6.3a**) can exert tremendous pressure and is known to pry loose individual minerals in a rock, even granites. It is particularly effective in porous, granular rock such as sandstone and occurs mainly in arid regions. Abrasion of rock by wind-driven sandblasting leaves telltale hints (the smoothing and streamlining of abraded surfaces) and is termed **ventifaction** (see Figure 6.3b). Ventifaction is a common phenomenon in deserts where there is little vegetation to hold soil and sand in place and slow the wind.

Fire is a dramatic way to physically weather rock (see Figure 6.3c). It has long been known that intense forest fires can shatter the surface of rocks, but only recently have people observed similar physical weathering when dry shrubs and thick grasses burn in deserts and semiarid areas. How is it that range fires passing in mere minutes can damage tough rocks, such as granite? Rocks are very poor conductors of heat; therefore, when the fire passes by, the surface of the rock may heat to hundreds of degrees centigrade in a matter of just a few minutes, expanding the rock. Meanwhile, several centimeters below the surface, the rock remains cool. This differential expansion sets up huge stresses, which can shatter even the toughest rock. Some people have speculated that diurnal cycles of heating and cooling in the desert could cause similar physical weathering. Although short-term laboratory experiments to test this hypothesis have never yielded conclusive results, the orientation of cracks in rocks around the world suggests that over millennia, solar heating and the resulting expansion and contraction do damage rocks.

Biota are important agents of physical weathering and help prepare rock for its conversion to soil and sediment. Roots of even the smallest plants can act as wedges in rock in any climate (see Figure 6.3d). The most dramatic example of root wedging is probably the damage to paving and structures caused by tree roots. Animals also contribute significantly to weathering and soil formation by aerating the ground and mixing loose material. Insects, worms, and burrowing mammals, such as ground squirrels and gophers, move and mix soil material, and carry organic litter below the surface. Some tropical ants' nests (see Figure 6.3e) are tunnels hundreds of meters long, for example, and common earthworms can thoroughly mix soils to depths of 1.3 meters (4.25 feet). Earthworms are found in all soils where there is enough moisture and organic matter to sustain them, and it is estimated that they eat and pass their own weight in food and soil minerals each day, which amounts to about 10 tons per acre per year, on average. An Australian earthworm is known that grows to a length of 3.3 meters (11 feet) and is a real earth mover! Charles Darwin was the first person to calculate just how much soil earthworms process.

Chemical Weathering

Chemical reactions alter rock and mineral debris at Earth's surface. These reactions are complex and involve many steps, but the fundamental processes are reactions between earth materials and atmospheric constituents, such as water, oxygen, and carbon dioxide. The products of chemical weathering are new minerals and dissolved elements and compounds. The most important chemical weathering re-

Figure 6.3 (a) Crystallizing salt can help pry rock apart. This granite in the desert of Namibia is spalling in onion skin–like layers as salt crystallizes in small cracks and splits the rock apart. A pen (on the rock) is used for scale. (b) Wind-driven sand can effectively erode rock. In the hyperarid Namibian desert of southern Africa, the hard granite has been physically weathered (polished and shaped) by sandblasting, otherwise known as ventifaction. (c) Range fire is a powerful means by which to physically weather rock. A range fire spreads rapidly through dry sage and grass in Owens Valley, California. Granitic rocks heated by these fires lose centimeter-thick sheets from their surfaces because of rapid heating and thermal expansion. (d) Tree roots and their trunk wedge openings in granite; Yosemite Valley, California. (e) Ants move large amounts of soil as they build their nests underground. Such movement helps stir and aerate soil. This image is from southeastern Australia, as suggested by the eucalyptus leaves near the ant mounds.
a–c: P. Bierman; d: D. D. Trent; e: © 2011/JASON EDWARDS/National Geographic Image Collection

actions are *solution,* the dissolving of minerals; *oxidation,* the "rusting" of minerals; *hydration,* the combining of minerals with water; and *hydrolysis,* the complex reaction that forms clay minerals and the economically important aluminum oxides. **Table 6.1** shows the formulas for the typical chemical-weathering reactions that are explained in the following discussion.

Solution—Carbon dioxide released from decaying organic matter and from the atmosphere combines with water

Table 6.1 Examples of Chemical-Weathering Reactions

Reaction	Equation
Solution	
Step 1: A natural acid forms.	CO_2 (carbon dioxide) + H_2O (water) → H_2CO_3 (carbonic acid)
Step 2: Acid dissolves limestone, placing calcium ions and bicarbonate ions in solution.	$CaCO_3$ (limestone) + H_2CO_3 (carbonic acid) → Ca^{2+} (calcium ions) + $2(HCO_3)^-$ (bicarbonate ions)
Oxidation and Hydration	
Oxygen and water combine with iron, producing hydrated iron oxide ("rust").	$4Fe^{2+}$ (iron minerals) + $3O_2$ (oxygen) + $6H_2O$ (water) → $2(Fe_2O_3 \cdot 3H_2O)$ (limonite (rust))
Hydrolysis	
Orthoclase feldspar combines with acid and water, forming clay minerals and potassium ions.	$2KAlSi_3O_8$ (orthoclase feldspar) + $2H^+$ (acid ions) + $9H_2O$ (water) → $Al_2Si_2O_5(OH)_4$ (clay minerals) + $2K^+$ (potassium ions) + $4H_4SiO_4$ (soluble silica)

to form carbonic acid. This natural, weak acid attacks solid limestone, dissolving it and yielding a watery solution of calcium and bicarbonate ions. Solution in limestone terrain creates caverns, such as the Mammoth Caves in Kentucky and Carlsbad Caverns in New Mexico. (Chapter 7 explains this.) Acid rain, which forms when sulfur dioxide or nitrogen oxides combine with water droplets in the atmosphere to form sulfuric and nitric acids, takes its toll on rock monuments (**Figure 6.4**).

Oxidation and hydration—Oxidation produces iron-oxide minerals (hematite and limonite) in well-aerated soils, usually in the presence of water. Oxidized iron is the rusting commonly seen on metal objects that are left outdoors. Pyroxene, amphibole, magnetite, pyrite, and olivine are the minerals most susceptible to oxidation, because they have high iron contents. It is the oxidation of these and other iron-rich minerals that provides the bright red and yellow colors seen in many of the rocks of the Grand Canyon and the Colorado Plateau.

Hydrolysis—The most complex weathering reaction, hydrolysis, is responsible for the formation of clays, the most important minerals in soil. A typical hydrolytic reaction occurs when orthoclase feldspar, a common mineral in granites and some sedimentary rocks, reacts with slightly acidic carbonated water to form clay minerals, potassium ions, and silica in solution. The ions released from silicate minerals in the weathering process are salts of sodium, potassium, calcium, iron, and magnesium, which become important soil nutrients. Clay minerals are important in soils, because their extremely small grain size—less than 2 microns (0.002 mm)—gives them a large surface area per unit weight. Soil clays can adsorb significant amounts of water on their surfaces, where it stays within

Figure 6.4 Compare the weathering of granite obelisks, both believed to have been shaped 3,500 years ago (1500 B.C.E.). (a) Obelisk amid the ruins of Karnak near Luxor, Egypt, a hyper-arid climate. The hieroglyphs are sharp, and the polished surface is still visible. (b) The red-granite obelisk known as "Cleopatra's needle," which stands in New York's Central Park, was a gift from the government of Egypt in the late 19th century. New York City's humidity and acidic pollution have caused chemical weathering of the incised hieroglyphs, which had endured several thousand years in Egypt's clean air and arid climate.

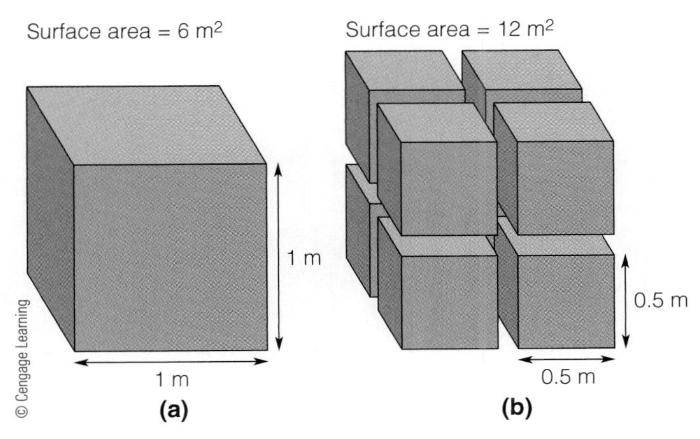

Figure 6.5 Surface area increases with decreasing particle size. A cube that is 1 meter on each side has a surface area of 6 square meters. If the cube is cut in half through each face, eight cubes result, with a total surface area of 12 square meters.

reach of plant roots. Also, because the clay surfaces have a slight negative electrical charge, they attract positively charged ions, such as those of potassium and magnesium, retaining these nutrients and making them available to plants. Other soil components being equal, the amounts of clay and humus in a soil determine its suitability for sustained agriculture. Sandy soils, for example, have much less water-holding capacity, less humus, and fewer available nutrients than do clayey soils. But overly clay-rich soils drain poorly and are difficult to work.

The Rate of Weathering

The rate of chemical weathering is controlled by the surface environment, grain size (surface area), and climate. The stability of the individual minerals in the parent material is determined by the pressure and temperature conditions under which they formed. Quartz is the last mineral to crystallize from a silicate magma, and the temperature at which it does so is lower than the temperatures at which olivine, pyroxene, and the feldspars crystallize. For this reason, quartz has greater chemical stability under the physical conditions at Earth's surface. The relative resistances to chemical weathering of minerals in igneous rocks thus reflect the opposite order of their crystallization sequence in magma. Most stable to least stable, the sequence is quartz (crystallizes at lowest temperature), orthoclase feldspar, amphibole, pyroxene, and olivine (crystallizes at highest temperature).

Micas and plagioclase feldspars form over a range of temperatures, and thus vary in weathering stability. Note that ferromagnesian (iron and magnesium) minerals—olivine, pyroxene, and amphibole—are least stable and most subject to chemical weathering. Igneous rocks, such as basalt and gabbro, are made up of such weathering-susceptible, iron-rich minerals, and thus weather readily in humid climates.

A monolithic mass of rock is less susceptible to weathering than an equal mass of jointed and fractured rock. This is because the monolith has less surface area on which weathering processes can act. A hypothetical cubic meter (about 35 cubic feet) of solid rock would have a surface area of 6 square meters (65 square feet). If we were to cut each face in half, however, we would have eight cubes with a total surface area of 12 square meters (130 square feet) (**Figure 6.5**). Thus, smaller fragments have more area upon which chemical and physical weathering can act. For example, whereas soils have developed on volcanic ash in the West Indies after only 30 years, some nearly 200-year-old solid basalt flows in Hawaii have yet to grow a radish. This reflects differences in particle size and surface area.

The effect of climate has already been discussed. It is important to remember that chemical weathering dominates in humid areas, and physical weathering is most active in arid or dry alpine regions. (Figure 6.4 may help you remember this.) However, humans have accelerated chemical weathering by polluting the atmosphere with acid-forming compounds. Limestone monuments in English cemeteries require 250 to 500 years to weather to a depth of 2.5 centimeters (1 inch), and thus obliterate inscriptions. Buildings in northern England are known to weather more quickly on the side that receives sulfur-laden winds from nearby industrial areas. A 16th-century monastery in Mexico that was built of volcanic tuff has been ravaged by physical and chemical weathering to the extent that some of the carved statues on its exterior are barely recognizable (**Figure 6.6a**). A rock wall protecting Chicago's Lake Shore Drive was built of Indiana limestone. Chemical weathering by water overtopping the wall has dissolved clearly visible pits in a pattern known as "honeycomb" weathering (see Figure 6.6b). You can see examples of weathering in your own community by visiting a cemetery. Compare the legibility of the inscriptions on monuments of various ages and materials. Some rock types hold up better than others, and younger inscriptions are easier to read than those that have been exposed to the elements for a century or two.

In the past 20 years, geologists teaming up with high-energy particle physicists have come up with a new way to measure rates of erosion, the result of physical and chemical weathering of rock (**Figure 6.7**). Measuring erosion is not a simple task, because in most cases, what you are trying to measure, rock, is long gone. The trick to this method is the ability to count rare atoms of several elements, including beryllium, chlorine, and helium. Earth is constantly being bombarded by cosmic radiation (mostly neutrons) that originates outside the solar system. Most of this radiation passes through the rocks that make up our planet, but occasionally there is a direct hit and something like the nucleus of an oxygen atom gets slammed so hard that it breaks apart. One of the pieces left is likely to be a Be atom, with 4 protons and 6 neutrons, giving it a mass of 10 atomic mass units—otherwise known as ^{10}Be. These rare creations are known as **cosmogenic isotopes**, because they are isotopes made (genesis) by cosmic radiation (cosmo).

With expensive and exotic tools, the number of ^{10}Be atoms can be counted (see Figure 6.7a). Because the neutrons that make ^{10}Be penetrate only a meter or two into rock, the number of ^{10}Be atoms reveals how quickly the rock is

B. Pipkin

B. Pipkin

Figure 6.6 (a) A 16th-century monastery in Mexico shows the ravages of weathering, mostly from wind and wind-driven rain. The rock is volcanic tuff. (b) A Lake Michigan seawall built of Indiana limestone is chemically weathered by waves overtopping the structure. Solution of the limestone has caused "honeycomb" weathering.

eroding; for example, lots of ^{10}Be indicates slow erosion, and not much ^{10}Be indicates fast erosion. Using this technique, researchers have shown that parts of the southern continents Australia, Antarctica, South America, and Africa (see Figure 6.7b) are exceptionally stable, eroding only a few tens of centimeters (5–10 inches) over a million years. In contrast, parts of the active mountain ranges, such as the Himalayas, lose a kilometer or more (perhaps up to a mile) of rock over the same length of time.

Geological Features of Weathering

Spheroidal weathering is a combination of chemical and physical weathering in which concentric shells of decayed rock are separated from a block bounded by joints or other fractures (**Figure 6.8a**). The spheroidal form results when a rectangular block is weathered from three sides at the corners and from two sides along its edges, while the block's plane faces are weathered uniformly. It is also called "onion-skin" weathering, because the layers of an onion provide a good analogy. **Exfoliation domes** (Latin *folium*, "leaf," and *ex-*, "off") are the most spectacular examples of physical weathering. These domes result from the unloading of rocks that have been deeply buried. As erosion removes the overlying rock, it reduces pressure on the underlying rocks, which, in turn, begin to expand. As the rocks expand, they

Courtesy Lawrence Livermore National Laboratory

P. Bierman

Figure 6.7 (a) At Lawrence Livermore Laboratory, best known for designing nuclear weapons, is a machine of exquisite sensitivity, an atom sorter and counter known as an accelerator mass spectrometer. (b) In the highlands of Namibia, two geologists collect samples of eroding rock to analyze on the accelerator seen in (a). The data showed that this is a very stable outcropping of rock; it is eroding only about 3 meters every million years.

Figure 6.8 (a) Spheroidal weathering in jointed granite; Alabama Hills, California. (b) Sheet jointing has produced these slabs of granite in Yosemite Valley, California. (c) The generals of the Confederacy are memorialized in a carving on the side of Stone Mountain, one of the largest granite domes in the world. (d) Sometimes, exfoliating slabs of granite can become serious hazards. Here, in Yosemite National Park, California, a slab of granite slid from Turtleback Dome, tumbled down, and blocked the highway.

crack and fracture along **sheet joints** parallel to the erosion surface. Slabs of rock then begin to slip, slide, or spall (break) off the host rock, revealing a large, rounded, domelike feature. Sheeting along joints occurs most commonly in plutonic (intrusive igneous) rocks, such as granite, and in massive sandstone. North Dome in Yosemite National Park in California and Stone Mountain in Georgia are granite exfoliation domes (see Figure 6.8b–d).

Soils

Soil Genesis

Soils form by the weathering of **regolith**, the fragmental rock material at and just below Earth's surface. By studying soils, we can make inferences about their parent materials and the climate in which they formed, and can determine their approximate age. Five environmental factors determine the rate and intensity of soil development: (1) climate, (2) organic activity, (3) relief of the land, (4) parent material, and (5) length of time that soil-forming processes have been acting. These factors make up the "CLORPT equation," a memory aid which is as follows: Soil = $f\{CL, 0, R, P, T\}$, where f is read "a function of" and CL, 0, R, P, and T represent climate, organic activity, relief, parent material, and time, respectively.

Soil Profile

As noted earlier, loose rock and mineral fragments at the surface of the Earth are referred to as *regolith*. Regolith may consist of sediment that has been transported and deposited by rivers or wind, for example, or it may be rock that has decomposed in place. Where weathering has been sufficient, a surface layer that can support plant life—soil—has developed. Climate is the most important factor in soil development; the influence of the parent material is most apparent in young soils, such as decomposed granite, which contains quartz and feldspar grains. Some of these mineral grains having been altered to clay minerals.

A developed **soil profile** has several recognizable **soil horizons**, layers roughly parallel with the ground surface that are products of soil-forming processes (**Figure 6.9**). The uppermost soil horizon, the "A horizon," is the *zone of leaching*. Here, mineral matter is most strongly dissolved by downward-percolating water. It may be capped by a zone of organic matter *(humus)* of variable thickness called the "O horizon," which provides carbon dioxide and organic compounds that make the percolating water slightly acidic. As the dissolved chemicals move downward, some of them are redeposited as new minerals and compounds in the "B horizon," called the *zone of accumulation*. The B horizon is usually redder and harder than the A horizon. Below this is the "C horizon," the weathered transition zone that grades downward into fresh parent material.

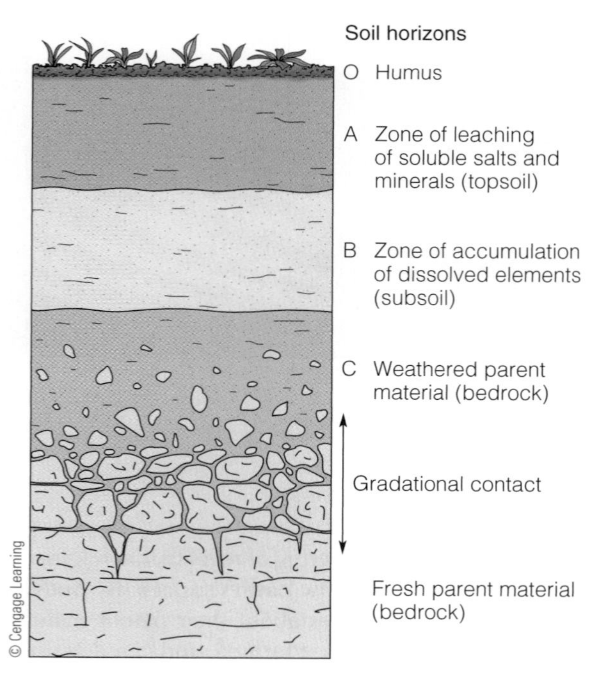

Figure 6.9 An idealized soil profile. Heavily irrigated soils and very young ones exhibit variations of this profile.

Figure 6.10 Caliche exposed in an arroyo (an incised streambed in which water flows only rarely) near Tucson, Arizona.

The thickness and development of soil profiles is profoundly influenced by climate and time. In desert regions of the American Southwest, the A horizon may be thin or nonexistent, and the B horizon may be firmly cemented by white, crusty calcium carbonate known as **caliche**. Caliche forms when downward-percolating water evaporates and deposits calcium carbonate that has been leached from above. Also, if the water table is close to the ground surface, upward-moving waters may evaporate into the layers, leaving deposits of caliche. Some of these deposits are so thick and hard that unusual measures are required to till or excavate through them (**Figure 6.10**).

Residual Soils

Residual soils have developed in place on the underlying bedrock. For example, in humid-temperate climates, rock decomposes into saprolite. If the parent rock were granite, the saprolite would contain quartz, weathered feldspar, and clay minerals formed from the weathering of feldspar. Saprolite can easily be mistaken for its parent rock, except that one can often stick a shovel into it and the density of saprolite can be tens of percent less than the density of the rock from which it was derived (the mass is lost by solution of rock into moving groundwater and the transport away of the dissolved material). Engineers and geologists call slightly weathered granite *decomposed granite*, or simply *d.g.* It is an excellent foundation material for structures and roads. However, heavy clay soils develop from the in situ weathering of shale bedrock. The expansive nature of some of these clays makes them troublesome as they expand and contract. (This is discussed in the next section of this chapter.) Residual soils take tens to hundreds of thousands of years to develop, and they are always subject to erosion and removal by one of many geological processes.

> **consider this . . .** Tennis is a sport that is played throughout the world on a variety of court surfaces. For example, the most prestigious professional tennis tournaments are played on grass (e.g., in England), clay (e.g., in France, Italy, and the eastern United States), and concrete (e.g., in the western United States and Australia). Why do you suppose clay-surface courts are more common in the eastern United States, and in tropical and humid countries, than in the southwestern United States, where concrete is the preferred surface?

Transported Soils

Some soils have formed on transported regolith. These transported soils are designated by the geological agent that is responsible for their transportation and deposition: *alluvial* soils by rivers, *eolian* soils by the wind, *glacial* soils by glaciers, *volcanic* soils by volcanic eruptions, and so on. **Figure 6.11** illustrates a soil that developed on volcanic ash overlying an older basalt on which little or no soil had developed. Known as the *Pahala ash*, soil developed on this parent material supports such exotic and valuable crops as macadamia nuts, sugarcane, and papaya on the Island of Hawaii.

> **consider this . . .** Decomposed granite (referred to as *d.g.* in the building trades) is the preferred sandy-soil material for making durable, attractive garden pathways, trails, and even driveways. What minerals make up this immature soil, and what mineral type specifically makes *d.g.* cohesive and firm when it is rolled?

As Pleistocene glaciers expanded and moved over northern North America, they scraped off soils and regolith, and redeposited them in a band across the United States from Montana to New Jersey. Soils that formed on these widespread deposits—collectively known as *glacial drift*—are extremely variable; they may be thin or thick, bouldery or fine-grained, fertile or relatively barren. Some areas of drift are so bouldery that they are impossible to farm. This is one of the reasons that pastures for grazing dairy cattle, rather than cultivated grain crops, are today the agricultural mainstay of such states as Wisconsin and New Hampshire.

Loess (pronounced to rhyme with *cuss*) is windblown silt composed of very small grains of feldspar, quartz, calcite, and mica. Covering about 20% of the United States and about 10% of Earth's land area (**Figure 6.12**), loess is arguably Earth's most fertile parent material. The sources of loess are glacial deposits and deserts. Strong winds blowing off the ice-age glaciers that covered most of Canada and the northern United States swept up nearby fine-grain meltwater deposits and redeposited the dust far away as loess in valleys and on ridge tops. In the United States, loess has produced especially rich agricultural lands in the Palouse region of eastern Oregon and Washington, and in the central Great Plains, sometimes called the "breadbasket" of the United States. The fertility of the Ukraine, formerly part of

Figure 6.11 Volcanic ash (horizontal layers) overlying weathered basalt; Pahala, Hawaii. This volcanic soil supports rich crops of sugarcane, pineapple, and macadamia nuts.

the Soviet Union but now an independent republic, is due to a favorable climate and its loess-rich soils. During World War II, so highly regarded was this soil that Adolph Hitler had large quantities of it moved to Germany. Loess derives its fertility from its loamy texture, which allows roots to penetrate easily, and retains water well. In addition, many of the minerals in loess are small and not yet weathered. When they do weather in the soil, they release nutrients that plants can use.

The largest loess deposits are in China, where they cover 800,000 square kilometers (310,000 square miles) and are as thick as several hundred meters. These deposits were derived from the Gobi and Takla Makan deserts to the north and west. On windy days, airborne silt from these deserts impacts air quality in Beijing, hundreds of miles away. Similarly, the

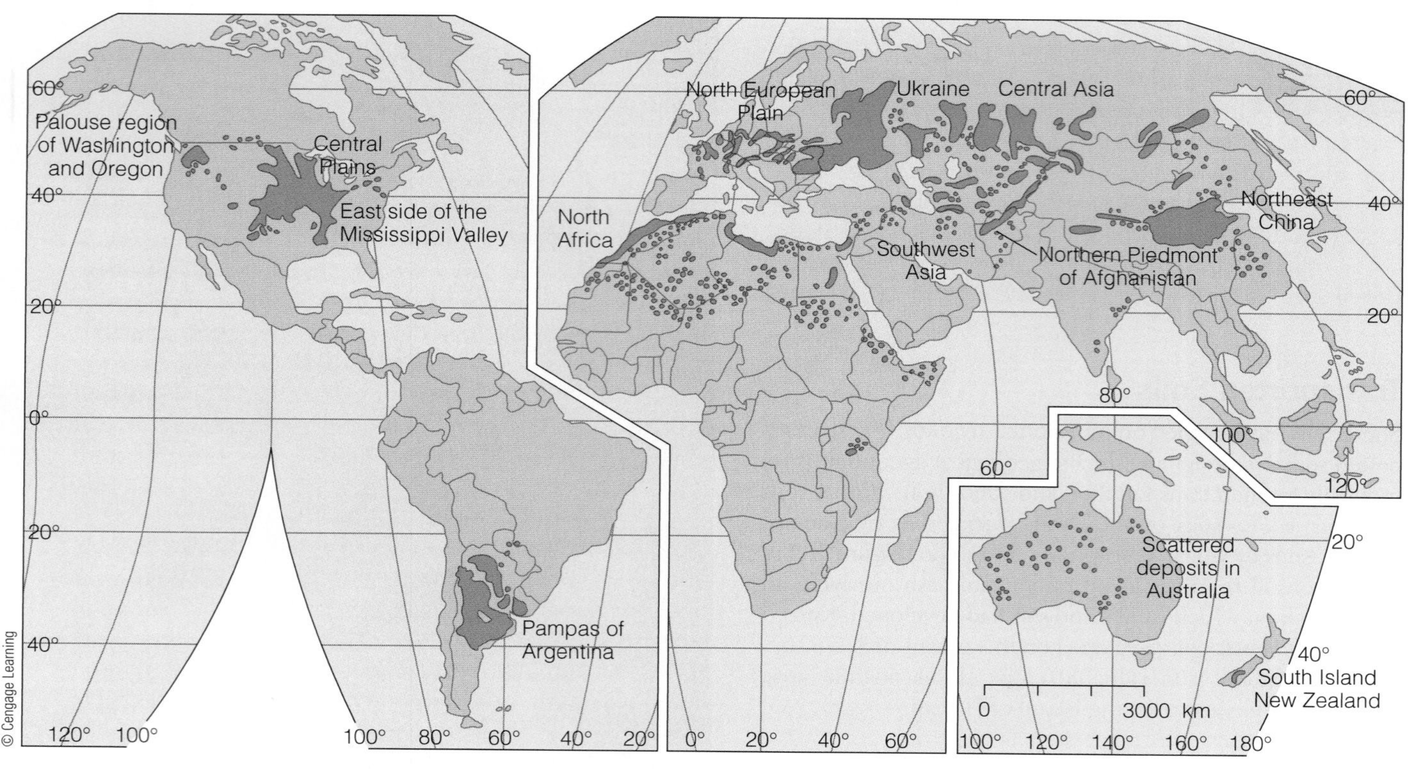

Figure 6.12 Loess-covered regions of the world. Note that the largest areas are in China, central and western Asia, and North America.

loess in Africa's eastern Sudan was carried there by winds from the Sahara Desert. Even more impressive is the wind transport of dust from Asia to Hawaii, where small grains of windblown quartz can be found dropped by wind on the basaltic volcanic rocks and basaltic soils that should contain no quartz at all!

Loess has several noteworthy physical properties. It is tan to almost yellow in color and rather loosely packed (low density), which makes it easily excavated and eroded by running water. Loess has cohesive strength, and thus is capable of standing in vertical slopes without falling (**Figure 6.13**). During the Civil War, these unusual properties served the Union and the Confederacy well in Mississippi. Both sides excavated tunnels, trenches, revetments, and other earthworks in the loess soil before the siege and battle at Vicksburg, which ultimately was won by Ulysses S. Grant's Union forces. In northern China's Shansi Province in the 1500s, elaborate caves scraped out of loess provided refuge for more than a million people. Unfortunately for the cave residents, loess is subject to collapse under dynamic stresses, such as those of an earthquake, and an estimated 800,000 people were buried in their homes during the earthquake of 1556. The load of loess carried by the Huang He River as it flows through this area gave the river its other name, the *Yellow River*—which empties into the *Yellow Sea*.

Soil Color and Texture

Color is a good indicator of a soil's *humus* (organic material) and iron content. As the proportion of humus increases, soil color darkens from brown to nearly black. Oxides of iron impart a reddish or yellowish color and are found mostly in semitropical and tropical areas. Bright red soils occur where there is aeration (oxygen present) and good drainage, whereas yellow hydrated-iron compounds form where there is poor drainage. Gray–green soil colors indicate the absence of oxygen and are most common in water-saturated soils. In arid climates, white B horizons and white-coated soil fragments indicate calcium carbonate or other salts, including gypsum (calcium sulfate) and halite (sodium chloride). Because both gypsum and halite are water soluble, their presence in soils indicates very low amounts of rainfall. Such soils are found in the driest deserts of the world—for example, the southern Negev in Israel.

A soil's texture—the size of the individual grains (particles) composing it—is a major determinant of its utility. A sample of any particular soil, excluding any gravel or boulders in it, can be divided into three grain sizes: sand, silt, and clay. The grain-size range used by the U.S. Department of Agriculture for each texture category is shown in **Figure 6.14a**. Based on the percentages of sand, silt, and clay in a soil sample, the soil can be classified as loam, sandy loam, and so forth (see Figure 6.14b).

The texture of a soil has practical significance, because it determines the soil's properties—that is, its looseness, workability, drainage, and so on. For example, coarse-textured soils are easily worked, readily penetrated by plant roots, and easily eroded. Cemeteries in glaciated regions are almost always dug in sand and gravel. Fine-textured

Figure 6.13 Villagers in Askole, Pakistan, have dug caves, which they use as homes, in these vertical loess cliffs.

(clayey) soils tend to be heavy, harder to work, and sticky when wet. Water retention and infiltration are also functions of soil texture. Loose, sandy soils tend to allow water to pass through them readily, but they have low water retention capacity. The opposite is true of fine-textured soils; they have large water-retention capacity but low infiltration rates. This condition leads to poor drainage and excessive water runoff. Loamy soils are generally considered most favorable for agriculture because they provide a balance between the coarse- and fine-textured soils, as shown in Figure 6.14b.

Soil Classification

Pedologists (Greek *pedon*, "soil," and *logos*, "knowledge of") have long known that soils that form in tropical regions are different from those that form in arid or cold climates. They also know that soils change over time as iron-rich materials oxidize (rust) and grow redder, organic matter accumulates, and clays that form by weathering are transported into the soil by gravity and water, a process called **illuviation**. Where rainfall is heavy, soils are deep, acidic, and dominated by chemical weathering. Soluble salts and minerals are leached (removed) from the soil, iron and aluminum compounds accumulate in the B horizon, and the soil supports abundant plant life. In arid regions, in contrast, soils are usually alkaline and thinner, coarse-textured or rocky, and dominated by physical weathering. Because of infrequent precipitation in arid regions, salts are not leached from soils; when soil moisture evaporates, it leaves additional salts behind (see Case Study 6.1). These salts may form crusts and lenses of caliche ($CaCO_3$ and other salts), most commonly in the B horizon.

These understandings led to the development of a system of **soil orders** used to classify soils. In this system, soils are classified by their physical characteristics; soils that have been modified by human activities are classified along with natural soils; and, most importantly, soil names convey information about the physical characteristics of the soils. There are 13 soil orders (**Figure 6.15a, b**), which can be arranged to follow patterns of age and climate. This type of ordering shows just how useful this soil classification scheme can be to geologists working in the field who want to understand environments and landscape history (**Figure 6.16**). An example is locating a site for a septic system for waste disposal. Such systems require well-aerated soils that are not saturated by water. After meeting with the landowner, the first thing to do is go to the library and consult the county soil survey. If the land is dominated by spodosols (**Figure 6.17**), the soils are probably well drained. However, if there are histosols, they are wetland soils. The next step is to start digging pits. If there are mottled, Cg horizons, it is a histosol, and is not a good

GRAIN SIZE

mm	inch	classification	
2.0	0.08		
		very coarse	sand
1.0	0.04		
		coarse	sand
0.500	0.02		
		medium	sand
0.250	0.01		
		fine	sand
0.100	0.004		
		very fine	sand
0.050	0.002		
		silt	
0.002	0.008		
		clay	

(a)

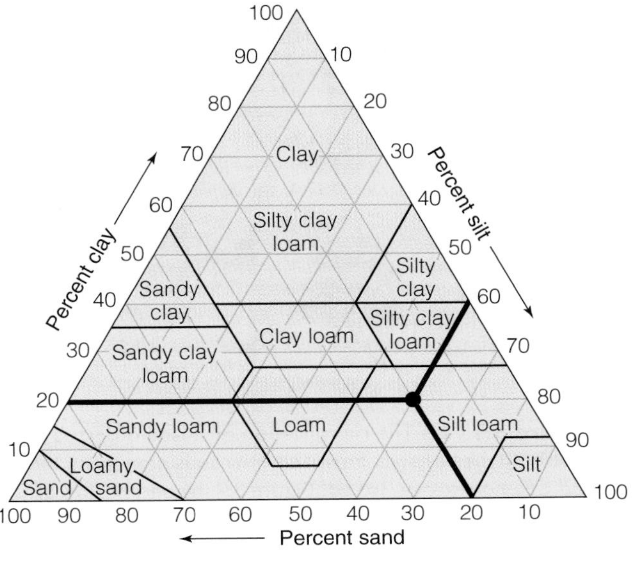

(b)

Figure 6.14 (a) The grain-size scale adopted by the U.S. Department of Agriculture and used by most soil scientists. (b) Soil texture classification triangle. Each side of the triangle is a scale for the percentage content of one of the three textural grades—sand, silt, or clay. A hypothetical soil sample (•) is 60% silt, 20% sand, and 20% clay. Using this triangle, thick black lines are followed inward from the respective percentages to learn the textural class of the soil, "silt loam." (Data from U.S. Department of Agriculture.)

(a)

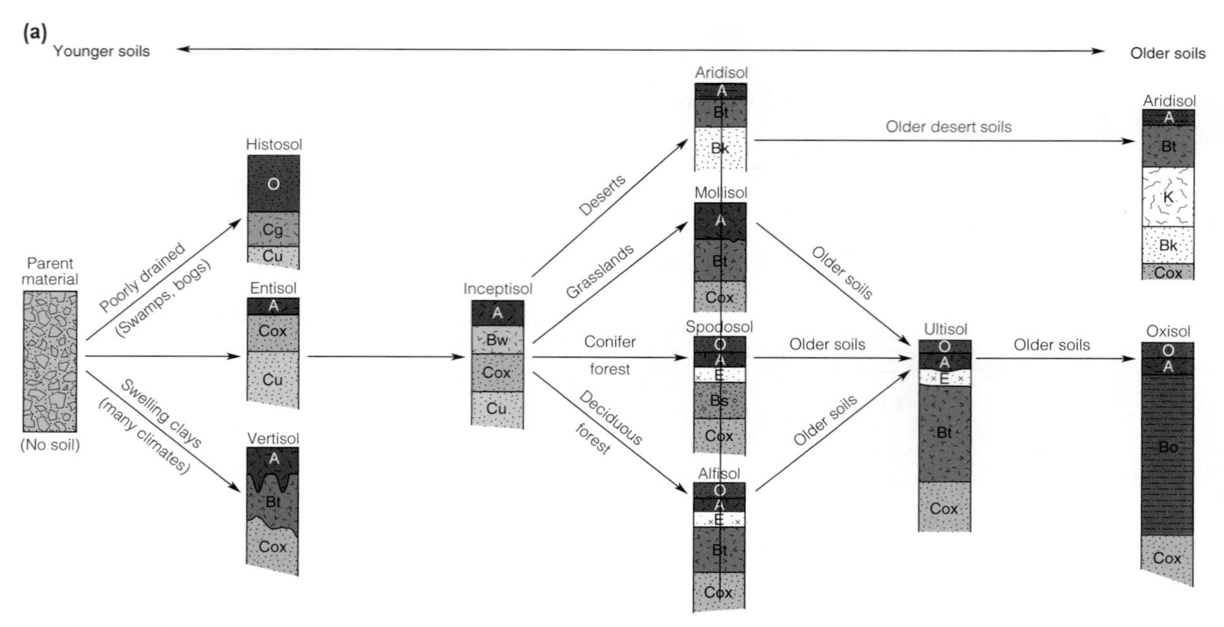

O horizon: Surface horizon dominated by organic material, dark
A horizon: Surface horizon or beneath O horizon, dominated by mineral material but with enough organic material to be dark
E horizon: Subsurface horizon, leached gray by acids from conifer needles
Bw horizon: Subsurface horizon, young and slightly reddened by iron oxide
Bs horizon: Subsurface horizon, reddened where organic matter, aluminum, and iron accumulate
Bk horizon: Subsurface, lightened by the accumulation of calcium carbonate
Bt horizon: Subsurface horizon that is reddened by iron and where clay has accumulated
Bo horizon: Subsurface horizon, deeply weathered and very red with iron, found in very old, tropical soils (laterite)
Cox horizon: Subsurface horizon of oxidized parent material, red from iron
K horizon: Subsurface horizon similar to Bk but so enriched in calcium carbonate that it is white
Cg horizon: Subsurface horizon that has patchy colors of grey and green because it is usually saturated with water
Cu horizon: Unweathered parent material

(b)

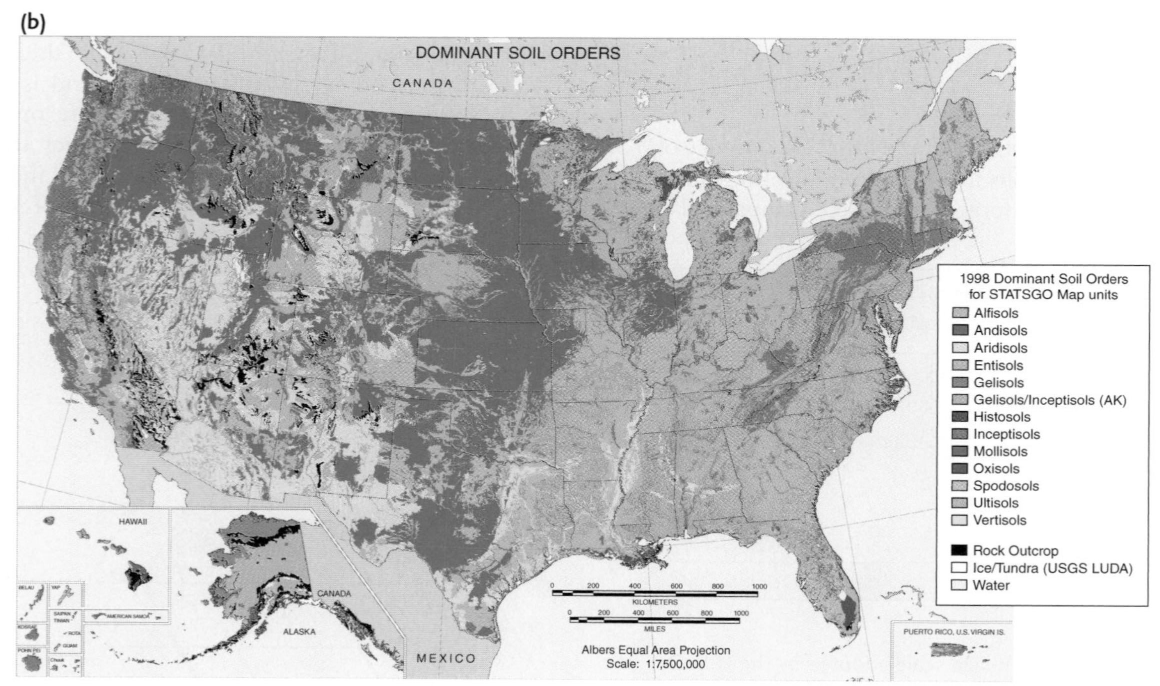

Figure 6.15 (a) The classification of 13 soil orders is an organized way to understand where different types of soils form and how soils change as they age. (From Birkeland, P. W. 1999. *Soils and Geomorphology*, 3rd ed. Figure 2.7, by permission of Oxford University Press, Inc. Courtesy of Dennis Netoff, 1997.) (b) Wonder where to find your favorite soil order? Check the map! If you are in the United States, can you figure out what soil order likely underlies where you are sitting right now, reading this book? (http://soils.ag.uidaho.edu/soilorders/)
b: Paul McDaniel/Univeristy of Idaho

P. Bierman

Figure 6.16 An oxisol, a deep red soil that has formed as a result of extended weathering in a tropical climate, in this case, Brazil. You are looking at the Bo horizon; the A and O horizons have been eroded. This type of soil is also known as a laterite. On the slope above are fields of tomatoes.

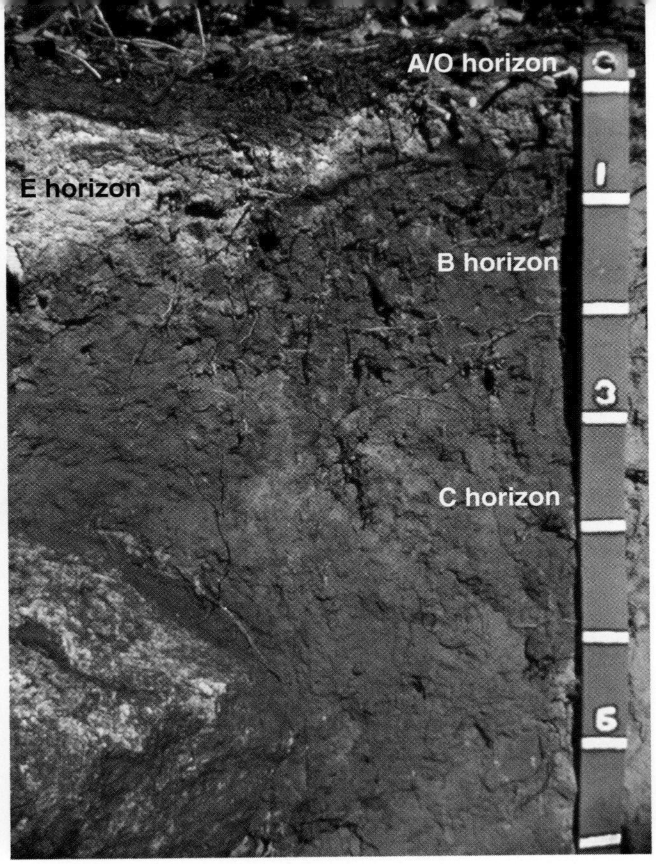

Paul McDaniel, University of Idaho

Figure 6.17 A classic spodosol, with a light gray, leached E horizon and the red B horizon below. At the top is the dark A/0 horizon darkened by decayed vegetation. This soil is formed on parent material of sandy, glacial outwash in Michigan, most likely under the acid needles of conifers.

place to site a septic system, because the ground will be saturated at least part of the year. In actuality, soil orders are the highest level of classification. Below them are suborders, families, and series. The number of soil entries increases astronomically at the lower levels (illustrated by the fact that 14,000 soil series are recognized in the United States).

Soil Problems

Soil Erosion

Although soil is continually being formed, for practical purposes, it is a nonrenewable resource, because hundreds or even thousands of years are necessary for soil to develop. Ample soil is essential for providing food for Earth's burgeoning population, but throughout the world soil is rapidly disappearing. Although the United States has one of the world's most advanced soil-conservation programs, soil erosion remains a problem after 60 years of conservation efforts and expenditures of billions of dollars. Even though rates of soil erosion have been reduced, erosion removed an estimated 2 billion tons of topsoil from U.S. farmlands in 1992—down from slightly more than 3 billion tons a decade earlier. With the average annual soil loss estimated at 10 to 12 tons per hectare (4–4.8 tons/acre) and an assumed annual soil-formation rate of 2 to 4 tons per hectare (0.8–1.6 tons/acre), 3 to 5 times as much soil is being lost as is being formed in the United States. Not only does soil erosion destroy fertility, but tons of soil settle in lakes, and clog waterways and drainages with sediment, pesticides, and nutrients each year. Losses of only 2 to 3 centimeters (1 inch) of topsoil represent about 200 tons per hectare. Such erosion cuts crop yields, and over time renders land less productive.

On a global basis, an area the size of China and India combined has suffered *irreparable degradation* from agricultural activities and overgrazing—mostly in Asia, Africa, and Central and South America (**Figure 6. 18**).

Lightly damaged areas can still be farmed using the modern conservation techniques explained later in this chapter, but "fixing" serious erosion problems is beyond the resources of most developing world farmers and their governments.

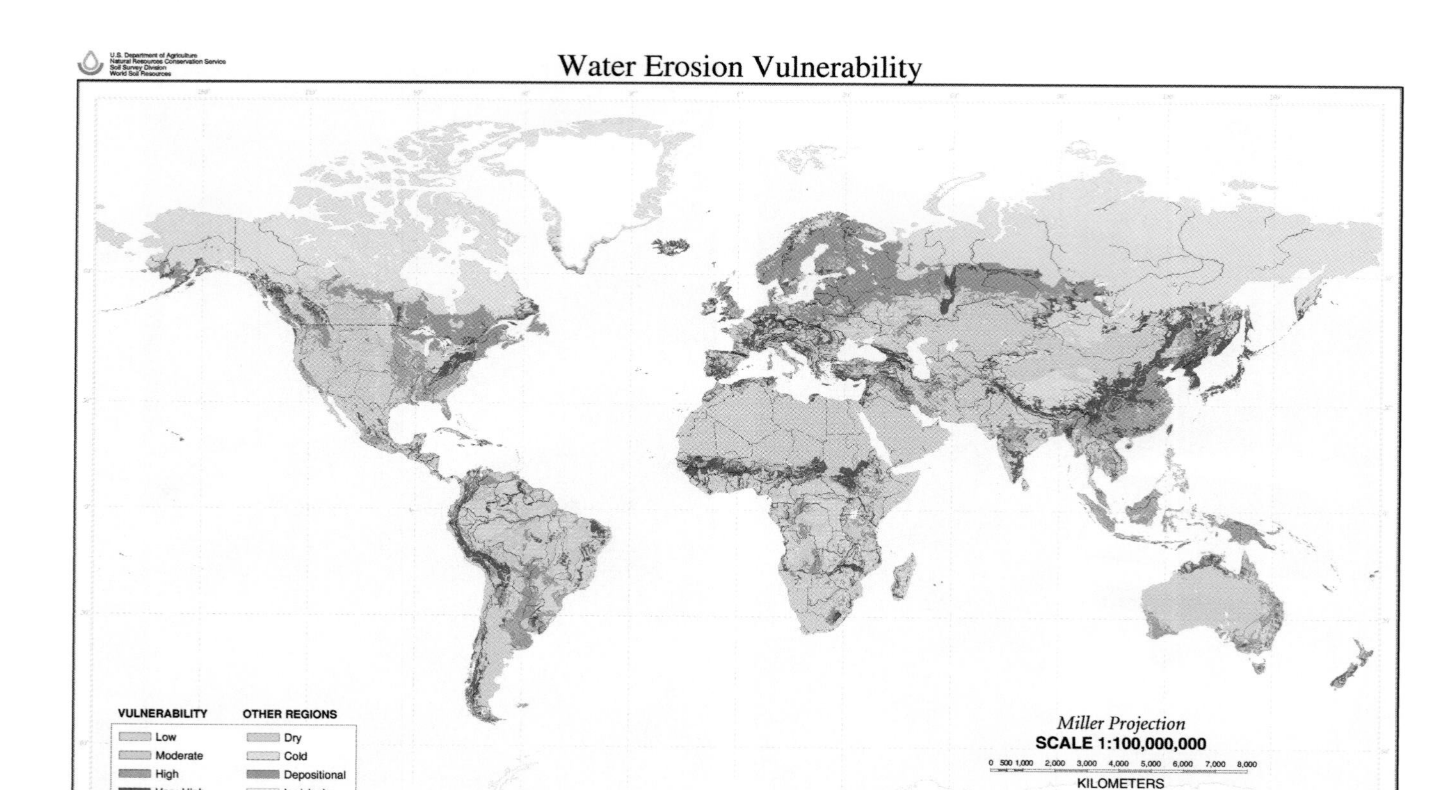

(a)

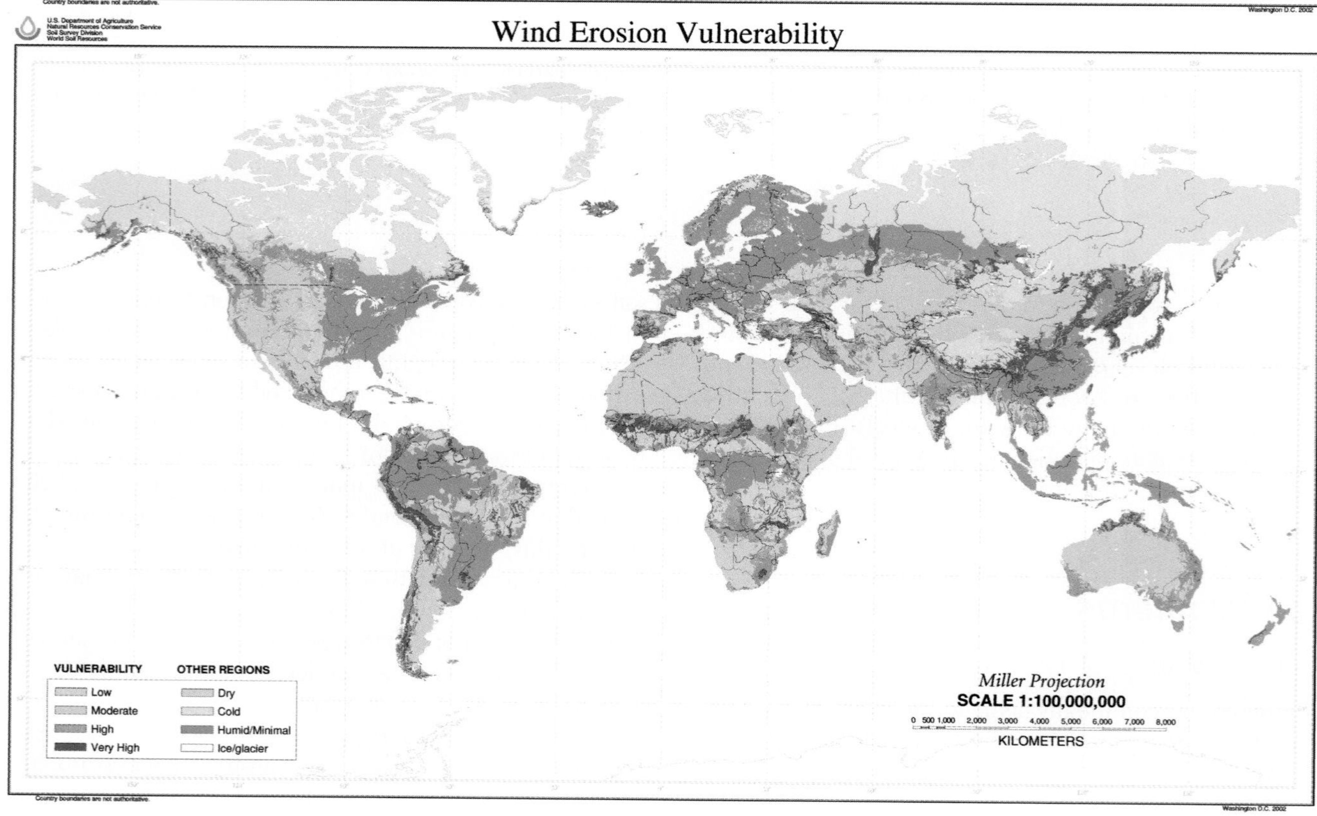

(b)

Figure 6.18 Soil is vulnerable to erosion by both wind and water. These maps show where the high-risk and low-risk areas are in the world for soil erosion. (a) Map of soils vulnerable to water erosion (http://soils.usda.gov/use/worldsoils/mapindex/erosh2o.html). (b) Map of soils vulnerable to wind erosion (http://soils.usda.gov/use/worldsoils/mapindex/eroswind.html). Look hard at the maps. Are the high-risk areas for wind erosion the same as those for water erosion? What do you think is controlling the distribution of erosion by wind and water?
U.S. Dept. of Agriculture

explorer's perspective

ZEPHANIAH PHIRI MASEKO has a knack for saving soil in some very dry places.

Farming in dry lands is not an easy task—getting enough water and keeping the soil intact is always a challenge when months go by without rain and then catastrophic downpours strip fertile soils and leave the land muddied, gullied, and barren.

Zephaniah Phiri Maseko set out to change that cycle and he succeeded in so many ways. Born almost 85 years ago in white-ruled Rhodesia, Zephaniah was repeatedly jailed and blacklisted by the apartheid government for his political activities. Unable to work, he turned to the land to support his family. But the land was no Garden of Eden, his family plot was dry and rocky, and the little rain that did fall ran off quickly and took fertile soil with it, ripping deep gullies.

Rather than giving up, Zephaniah took control. His first task, slow the flow. He built permeable retaining walls along contour that slowed the runoff and held back soil. He planted a variety of trees and rotated his crops; their roots held the soil in place, and over time, their fallen, rotting leaves enriched the soil.

Today, his farm is a well-watered oasis of productivity and Zephaniah is a mentor to many, spreading his gospel of low-tech soil and water conservation throughout much of arid and semi-arid southern Africa. He's a wonderful example of how one man using a simple, low-cost approach can enrich the land and improve people's lives and livelihoods.

Source: http://www.nationalgeographic.com/explorers/bios/zephaniah-maseko/

Gully ripping apart road in the arid highlands of western Namibia.
P. Bierman

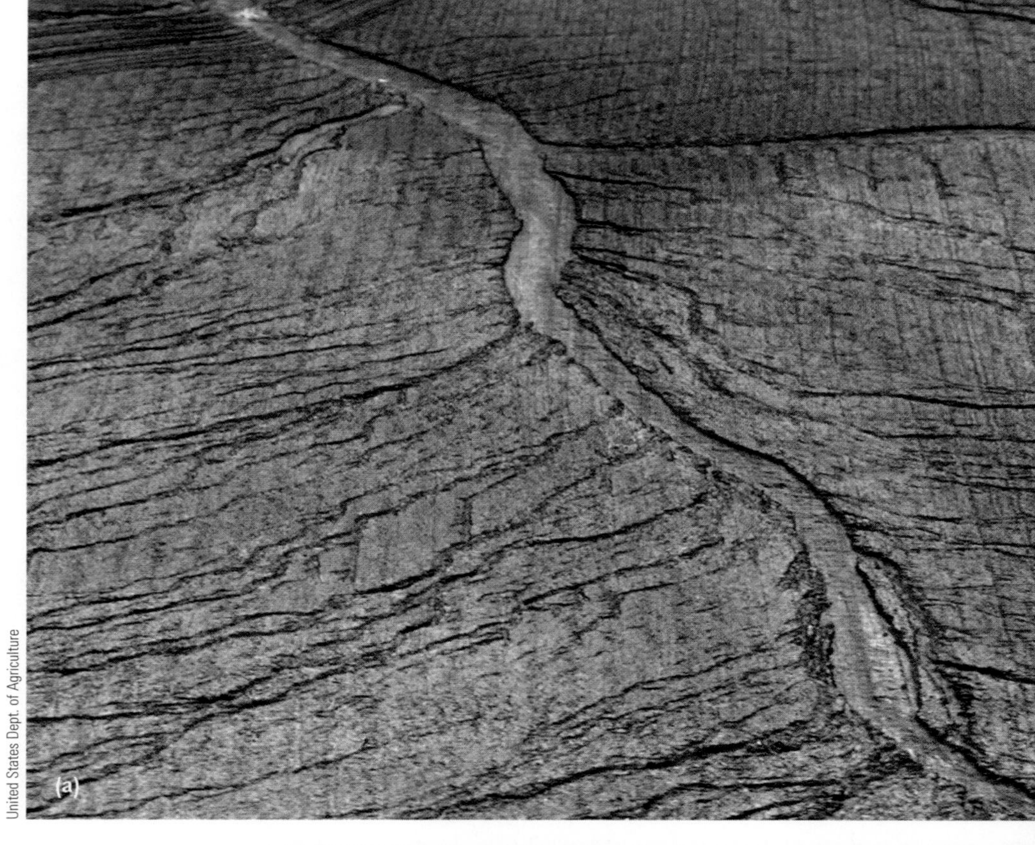

Figure 6.19 (a) Soils in northwest Iowa can be highly erodible. Here, after heavy spring rains fell on bare, plowed soils that had no protection against soil erosion, deep rills were cut into the bare soil surface. (b) Gullies have eaten away at the landscape, dramatically incising the lateritic, highly weathered soil. This erosion was near the Poco das Antas Reserve, Atlantic Forest, Brazil. The trees in the distance provide scale; this is a deep gully!

Erosion processes—The agents of soil erosion are wind and running water. Heavy rainfall and melting snow create running water, which removes soil by sheet, rill, and gully erosion. **Sheet erosion** is the removal of soil particles in thin layers more or less evenly from an area of gently sloping, bare land. It goes almost unnoticed and can easily be stopped by planting a cover crop. The roughness provided by the vegetation and the effective cohesion provided by roots keep soil in place.

Rill erosion, in contrast, is quite visible as discrete streamlets carved into the soil (**Figure 6.19a**). If these rills become deeper than about 25 to 35 centimeters (10–14 inches), they cannot be removed by plowing, and gullies form. Estimated annual U.S. cropland soil loss from water (sheet and rill) erosion is shown in **Figure 6.20a**. In 1980, Tennessee had the greatest rate of soil loss, followed by Hawaii, Missouri, Mississippi, and Iowa (**Table 6.2**). Gullies are created by the widening, deepening, and headward erosion of rills by water both running over the surface and moving through pores in the soil and parent material. **Gullying** is a problem because it removes large areas of land from production or even transit (see Figure 6.19b). The U.S. Army worries about gullies on its training grounds because entire tanks have fallen into gullies during night maneuvers.

Wind erodes soils when tilled land lies fallow and dries out without vegetation and root systems to hold it together. The dry soil breaks apart, and wind removes the lighter particles. If the wind is strong, a dust storm may occur, as it did in the 1930s, creating the Dust Bowl (described at the beginning of this chapter). Much of the reason for the Dust Bowl was the widespread conversion of marginal grasslands, which receive only 25 to 30 centimeters (10–12 inches) of rainfall per year, to cropland. Unfortunately, this is still occurring in some places in the United States and in the rest of the world. Figure 6.20b presents estimates of the amount of soil lost each year to wind erosion in the United States. When the 50 states' estimated wind and water erosion losses are combined (see Table 6.2), the state with the greatest annual loss is Texas, followed by Tennessee, Hawaii, and New

Table 6.2 Significant U.S. Soil Erosion by State, 1980

Rank	State	Tons/ (Acre × Year)
Sheet and Rill Erosion		
1	Tennessee	14.1
2	Hawaii	13.7
3	Missouri	11.3
4	Mississippi	10.9
5	Iowa	9.9
Wind Erosion		
1	Texas	14.9
2	New Mexico	11.5
3	Colorado	8.9
4	Montana	3.8
5	Oklahoma	3.0
Total Soil Erosion, Water and Wind*		
1	Texas	18.4
2	Tennessee	14.1
3	Hawaii	13.7
4	New Mexico	13.5
5	Colorado	11.4

Source: *Environmental trends.* 1981. Washington, DC: Council on Environmental Quality, cited in Sandra Batie. 1983. *Soil erosion: A crisis in America's cropland?* Washington, DC: The Conservation Foundation.

*The U.S. average annual rate of total soil erosion is estimated at 4.0–4.8 tons/acre.

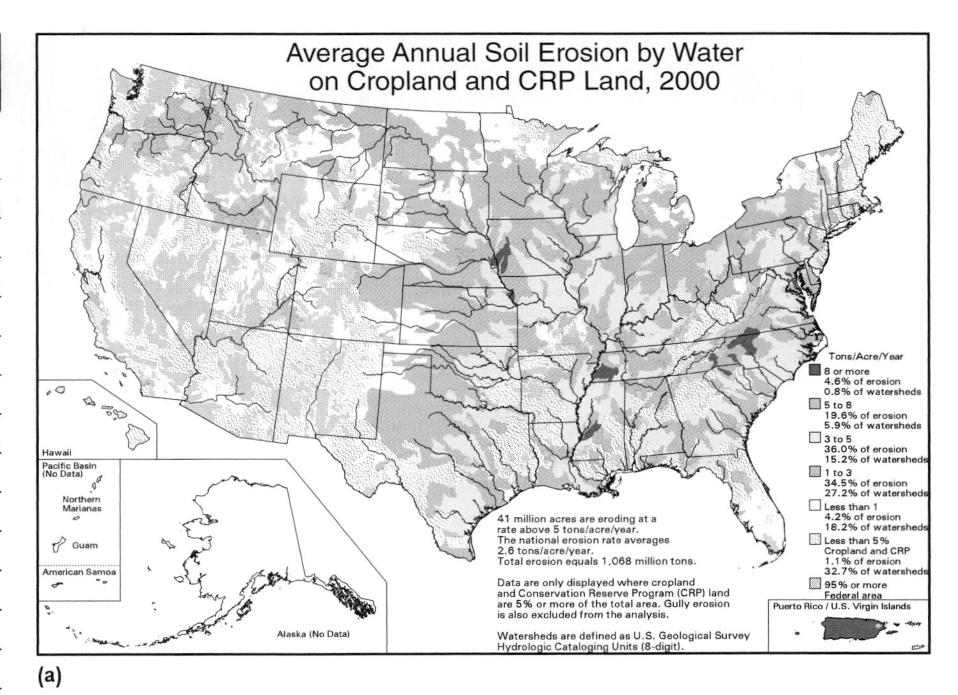

(a)

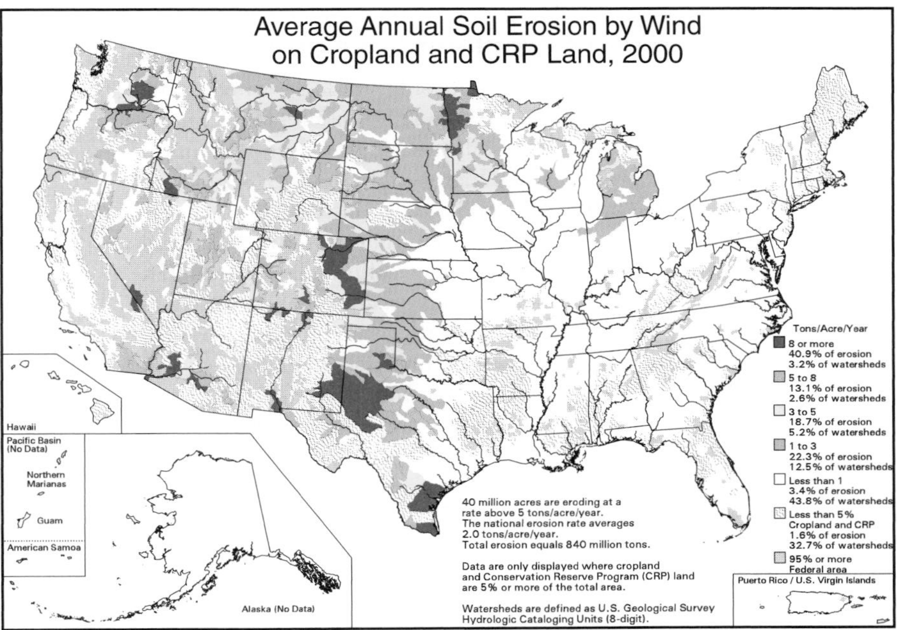

(b)

Figure 6.20 Estimated annual rates of cropland soil loss in the United States. (a) Sheet and rill erosion losses; (b) wind erosion losses in the Great Plains. (From Muckel, G. B. [ed.]. 2004. *Understanding soil risks and hazards, using soil survey to identify areas with risks and hazards to human life and property.* Lincoln, NE: U.S. Department of Agriculture, Natural Resources Conservation Service, National Soil Survey Center.)

Mexico. **Table 6.3** summarizes the recognized causes of accelerated soil erosion worldwide and assesses their relative impact.

In addition to the soil erosion caused by overgrazing, deforestation, and unsustainable agricultural practices, recreational and military pursuits (Case Study 6.2) cause soil erosion, particularly those that involve off-road vehicles (ORVs).

A 1974 study indicated that recreational pressure on the semiarid lands of southern California by ORVs was "almost completely uncontrolled," and that those desert areas had experienced greater degradation than had any other arid region in the United States (**Figure 6.21**). In the three and a half decades since that study, much has been accomplished in controlling off-road traffic in California and other states, but ORVs continue to exert severe impacts on areas established for their use.

Although significant soil erosion has gone on for millennia, starting with the beginning of intensive agriculture in the Middle East, it has not always been clear where the eroded soil goes and how far and fast it travels. More recent work, particularly that of Stan Trimble, a geographer, suggests that most eroded soil does not go far at all. Trimble tracked soil eroded from slopes in the Midwest and found that only a small percentage of it has made the journey all the way to the Mississippi River. Most of the eroded soil is sitting at the bottom of hillslopes or on valley bottoms (**Figure 6.22**). This is also the case in the eastern states. In Pennsylvania, Dorothy Merritts and Bob Walter found that small mill dams, built by

Table 6.3 Causes of World Soil Erosion and Deterioration

Cause	Estimated % of Total
Overgrazing	35
Deforestation	30
Bad agricultural practices	28
Other causes, natural and human	7

Source: World Resources Institute, United Nations Environment and Development Program. 1992. *World resources: A guide to the global environment.* New York: Oxford University Press.

colonial settlers to power the mills that were the backbone of their economy, have trapped much of the soil eroded from the then clear-cut slopes. Today, these dams are an environmental disaster waiting to happen. They are failing or being removed, and all that sediment, eroded more than 200 years ago but in storage ever since, is headed downstream to the Chesapeake Bay, one of the most important estuaries in the country, and a place where millions of dollars are being spent to improve water quality and fisheries.

The Impact of Cropland Loss on Humans

The main concern when addressing soil loss—whether the loss is due to erosion or the conversion of cropland to nonagricultural uses—is that soils are biological and chemical factories that produce food. Increasingly, the U.S. population is moving into urban areas and, to accommodate them, cities are expanding outward onto agricultural land. In 1992, about 20% of the United States was under cultivation (**Figure 6.23**), and each year more of that land is urbanized or converted to forestland, particularly in the northeastern and southeastern states. This trend is also seen in India and other nations with high growth rates. Between 1950 and 1998, the amount of grain-growing land per person worldwide declined. World cropland decreases were not regarded as a major problem until the late 1900s, because increases in productivity had been great. Increases in productivity leveled off in 1990, however, and in the face of a rapidly changing climate, it is doubtful that future increases will keep pace with the increasing population and decreasing cropland soils.

It is also important to consider that much of the increase in agricultural productivity has been leveraged off of the use of fossil fuels to power large-scale industrial agriculture in developed nations and to make synthetic fertilizer. This cheap energy also allows for the shipment of food products long distances, so that in the dead of winter, people in Maine or Wisconsin can eat tropical fruits from Hawaii or lettuce from southern California. If energy prices continue to increase and fossil-fuel supplies tighten, one needs to ask: Is there still enough fertile and available land nearby to support the food needs of the local population? The answer in many intensively developed areas may well be, no. Fossil fuels have other impacts on soils, namely, contamination when those fuels are released into the environment (this problem is considered in Case Study 6.3).

U.S. Geological Survey

Figure 6.21 Off-road vehicles (ORVs), in this case, a quad-bike, erode slopes and release dust in the Calico Mountains in California. Located on land owned by the federal government and administered by the Bureau of Land Management, this area gets heavy use by ORVs and the impact on the soil is clear.

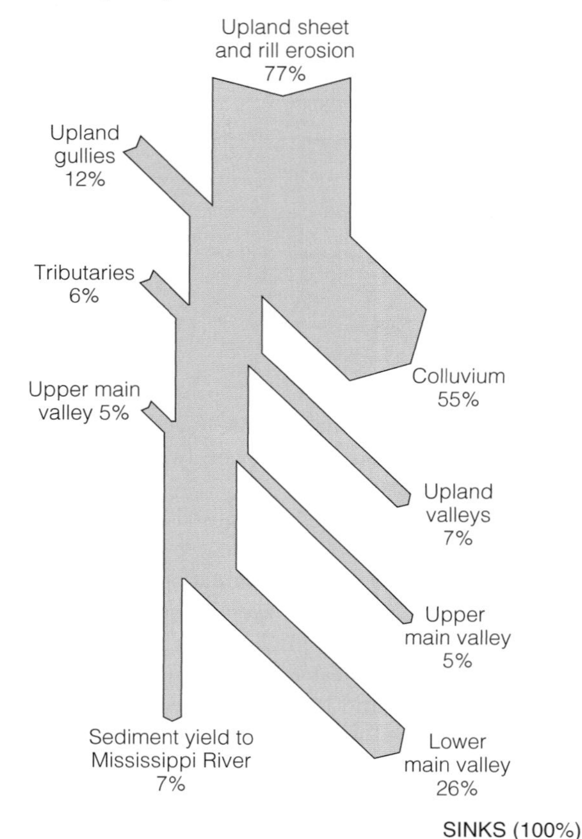

Figure 6.22 A "Trimblegram" for Coon Creek, a stream that is a tributary of the Mississippi River in Wisconsin. The diagram shows sources (top and left side) of eroded soil and sinks (bottom and right side) where the eroded soil has gone between 1938 and 1975. Most of the eroded soil ends up as colluvium, the poorly sorted mass of material at the base of hillslopes. Only 7% of the eroded soil makes it to the Mississippi River. (From Trimble, S. W. 1999. Decreased rates of alluvial sediment storage in the Coon Creek Basin, Wisconsin, 1975-93. *Science* 285(5431): 1244–1246. Reprinted with permission from AAAS.)

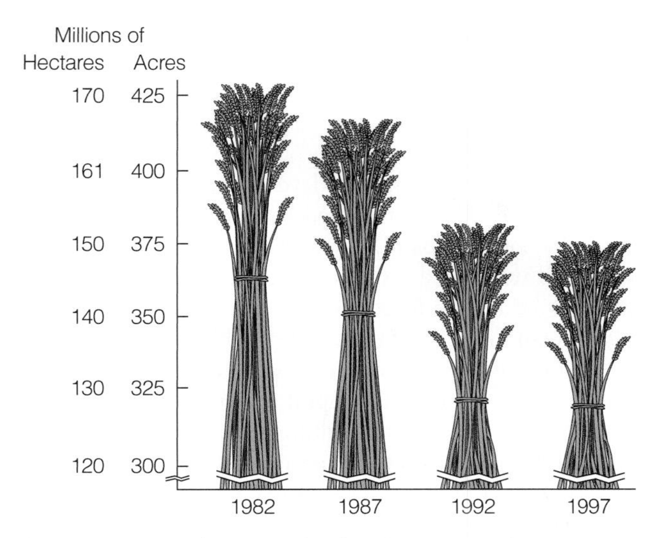

Figure 6.23 Total U.S. cropland, 1982–1997. Reductions were due to urbanization, conversion to other uses, and soil erosion. (American Geological Institute)

Mitigation of Soil Erosion

Thomas Jefferson was among the first people in the United States to comment publicly on the seriousness of soil erosion. He called for such soil-conservation measures as crop rotation, contour plowing, and planting of grasses to provide soil cover during the fallow season. These practices are now common in industrialized nations, but, unfortunately, only minimally used in developing countries. Proper choices of where to plant, what to plant, and how to plant are primary components of soil conservation. The major erosion mitigation practices are described in this section.

- *Terracing*—creating flat areas, terraces, on sloping ground—is one of the oldest and most efficient means of conserving soil and water in steep, easily eroded terrain (**Figure 6.24**).
- *Strip-cropping*, where ground-covering plants are alternated with widely spaced ones, efficiently traps soil that is washed from bare areas and supplies some wind protection. An example of this is alternating strips of corn and strips of alfalfa.
- *Crop rotation* is the yearly alternation of crops. Rotating corn and wheat with clover in Missouri, for example, has been found to reduce soil erosion from 19.7 tons/acre for corn to 2.7 tons/acre. Groundcovers, such as grasses, clover, and alfalfa, tend to be soil-conserving, whereas row crops, such as corn and soybeans, can leave soil vulnerable to runoff losses. Retaining crop residues, such as the stubble of corn or wheat, on the soil surface after harvest has been found to reduce erosion and to increase water retention by more than half.

Figure 6.24 Terracing steep hillsides is an ancient soil-conservation practice in Asia, here illustrated in the loess plateau of China.

Edward B. Nuhfer and AIPG

Figure 6.25 This well-built concrete-block wall in Colorado was uplifted and tilted by stresses exerted by the underlying expansive soil.

- *Conservation-tillage* practices minimize plowing in the fall and leave at least 30% of the soil surface covered with plant residue. In 1990, only 26% of U.S. agricultural lands used conservation tillage. By 2004, more than 40% of U.S. lands used conservation-tillage methods.
- *No-till* and *minimum-till* practices are the most conservative of soil. With no-till farming, seeds are planted into the soil through the previous crop's residue, and weeds are controlled solely with chemicals. Specialized no-till equipment is required. No-till has been especially successful in soybean and corn farming. In 1990, U.S. farmers practiced no-till agriculture on only 6% of their land. By 2004, more than 23% of U.S. fields were managed using no-till. Tennessee leads the nation, with more than 50% of its fields in 2004 using no-till practices. No-till agriculture is most beneficial during times of drought, when plowed fields dry out and topsoil is more vulnerable to wind erosion. No-till practices saved much of the agricultural soil that would have otherwise been eroded by floodwaters of the Great Flood of 1993 in the Mississippi River Valley.

Wind-caused soil losses can also be reduced by planting windbreaks of trees and shrubs near fields, and by planting row crops perpendicular to the prevailing wind direction. Wind losses are almost zero with cover crops in place, such as grasses and clover.

Expansive Soils

Certain clay minerals have a layered structure that allows water molecules to be absorbed between the layers, which causes the soil to expand. (This process differs from the *adsorption* of water to the surface of nonexpansive soil clays.) Soils that are rich in these minerals are said to be **expansive soils** and fall in the soil order *vertisol.* Although this reaction is reversible, as the clays contract when they dry, soil expansion can exert extraordinary uplift pressures on foundations and concrete slabs, with resulting structural and cosmetic damage (**Figure 6.25**). Damage caused by expansive soils costs about $6 billion per year in the United States, mostly in the Rocky Mountain states, the Southwest, and Texas and the other Gulf Coast states (**Figure 6.26**). The swelling potential of a soil can be identified by several standard tests, usually performed in a soil engineering laboratory. A standard test is to compact a soil in a cylindrical container, soak it with water, and measure how much it swells against a certain load. Clays that expand more than 6% are considered highly expansive; those that expand 10% are considered critical. Treatments for expansive soils include: (1) removing them, (2) mixing them with nonexpansive material or with chemicals that change the way the clay reacts with water, (3) keeping the soil moisture constant, and (4) using reinforced foundations that are designed to withstand soil volume changes. Identification and mitigation of damage from expansive soils are standard practice for soil engineers.

Permafrost

The term **permafrost**, a contraction of "permanent" and "frost," was coined by Siemon Muller of the U.S. Geological Survey in 1943 to denote soil or other surficial deposits in which temperatures below freezing are maintained for

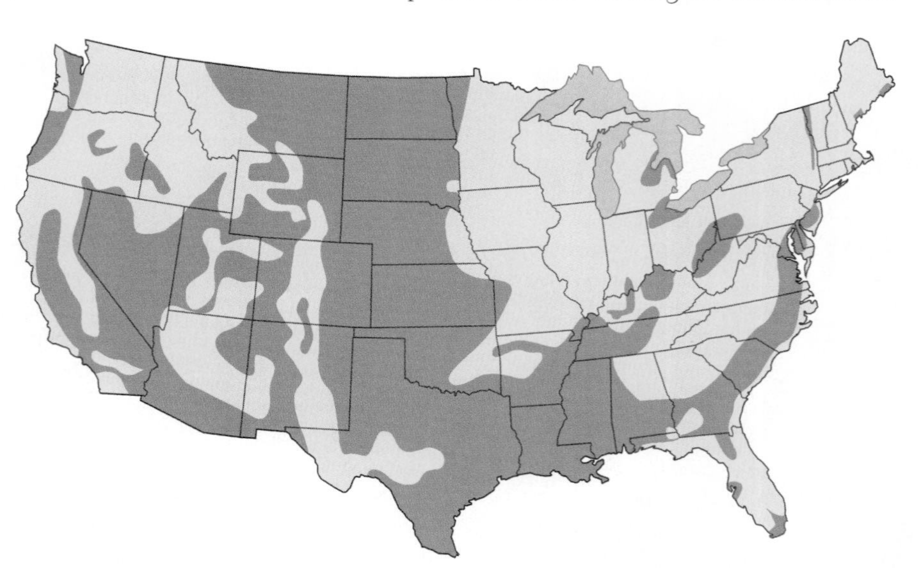

Figure 6.26 Distribution of expansive soils (shown in brown) in the United States.

several years. More than 20% of Earth's land surface is underlain by permafrost, and it is not surprising that most of our knowledge about frozen ground has been derived from construction problems encountered in Siberia, Alaska, and Canada. Permafrost becomes a problem for people when we change the surface environment by our activities in ways that thaw the near-surface soil ice. Such melting results in soil flows, landslides, subsidence, and related phenomena. Permafrost forms where the depth of freezing in the winter exceeds the depth of thawing in the summer. If cold ground temperatures continue for many years, the frozen layer thickens until the penetration of surface cold is balanced by the flow of heat from Earth's interior. This equilibrium between cold and heat determines the thickness of the permafrost layer. Thicknesses of as much as 1,500 and 600 meters (almost 5,000 and 2,000 feet) have been reported in Siberia and Alaska, respectively. Such thick permafrost is very old; it must have formed over hundreds of thousands to perhaps millions of years as Earth cooled during repeated Pleistocene glaciations.

consider this . . . Expansive soil is the soil problem most likely to be encountered in maintaining a home. Assuming that you, as a homeowner, have found that your newly purchased dream house is built on expansive soil, what can you do to minimize its impact on your home?

The top of the permanently frozen layer is called the **permafrost table** (**Figure 6.27**). Above this is the **active layer**, which is subject to seasonal freezing and thawing, and is always unstable during summer. When winter freezing does not penetrate entirely to the permafrost table, water may be trapped between the frozen active layer and the permafrost table. These unfrozen layers and "lenses," called *talik* (a Russian word), are of environmental concern, because the unfrozen groundwater in the talik may be under pressure. Occasionally, pressurized talik water moves toward the surface and freezes, forming a dome-shaped mound 30 to 50 meters (100–160 feet) high, called a **pingo**. The presence of ice-rich permafrost becomes evident when insulating **tundra** vegetation (the hearty, stunted shrubs and trees that cover much of the arctic) is stripped away for a road, runway, or building. Thawing of the underlying soil ice results in subsidence, soil flows, and other gravity-induced mass movements (**Figure 6.28**). Even the casual crossing on foot or in a Jeep can upset the thermal balance and cause thawing. Once melting starts, it is impossible to control, and a permanent scar is left on the landscape.

Structures that are built directly on or into permafrost settle as the frozen layers thaw (**Figure 6.29, 6.30**). Both active and passive construction practices are used to control the problem. Whereas active methods involve the complete removal or thawing of the frozen ground before construction, passive methods build in such manners that the existing thermal regime is not disturbed—for example:

- A house can be built with an open space beneath the floor, so that the ground can remain frozen (**Figure 6.31a**).
- A building can be constructed on foundation piers whose temperature is held constant by heat dissipaters or coolant (see Figure 6.31b).
- Roads are constructed on top of packed coarse-gravel fill to allow cold air to penetrate beneath the roadway surface and keep the ground frozen.

The most ambitious construction project on permafrost terrain to date is the 800-mile oil pipeline from Prudhoe Bay on Alaska's North Slope to Valdez on the Gulf of Alaska. About half the route is across frozen ground, and because oil is hot when it is pumped from depth and during transport, it was decided that the pipe should be aboveground, to avoid the thawing of permafrost soil (**Figure 6.32**). The 4-foot-diameter pipeline also crosses several active faults, three major mountain ranges, and a large river. The pipeline's aboveground location allows for some fault displacement and for easier maintenance if it should be damaged by faulting. Placing the pipeline aboveground also accommodates the longstanding migratory paths of herding animals, such as caribou and elk.

Permafrost is likely to be one of the most troublesome victims of global climate change. All climate models and much recent observational data (such as the dramatic thinning of Arctic sea ice during the last decade; see Case Study 6.4 for an explanation of the impacts of climate change on weathering and erosion) indicate that the warming of Earth as carbon dioxide levels continue to rise will not be a uniform affair. Climate change is predicted to have some of its most extreme impacts in the polar regions, with warming several times greater than at mid-latitudes. The effects will be dramatic, with the disappearance of areas of permafrost that may have existed for hundreds of thousands of years. Once permafrost is gone, traditional building techniques can be used, but the real problems will come during the decades-long transition when once-solid frozen ground melts, and structures, roads, and other engineering works that depend on that solid, frozen base are damaged and destroyed.

Settlement

Settlement occurs when a structure is placed upon a soil, a rock, or other material that lacks sufficient strength to support it. Settlement is an engineering problem, not a geological one. It is treated here, however, because settlement and specific soil types are inextricably associated. All structures settle, but if settlement is uniform—say, 2 or 3 centimeters (about an inch) over the entire foundation—there is usually no problem. When differential settlement occurs, however, cracks appear or the structure tilts.

The Leaning Tower of Pisa in Italy is probably the best known, most-loved example of poor foundation design. It may have scientific importance as well, as Galileo is said to have conducted gravity experiments at the 56-meter (183 feet) structure. In 1174, the tower started to tilt when

Weathering, Soils, and Erosion

(a) Map showing the distribution of permafrost today, and in 2050 and 2100 if global warming continues unabated. (b) Cutaway showing permafrost. Permafrost can be continuous, discontinuous, or sporadic—descriptions of the continuity of frozen ground.

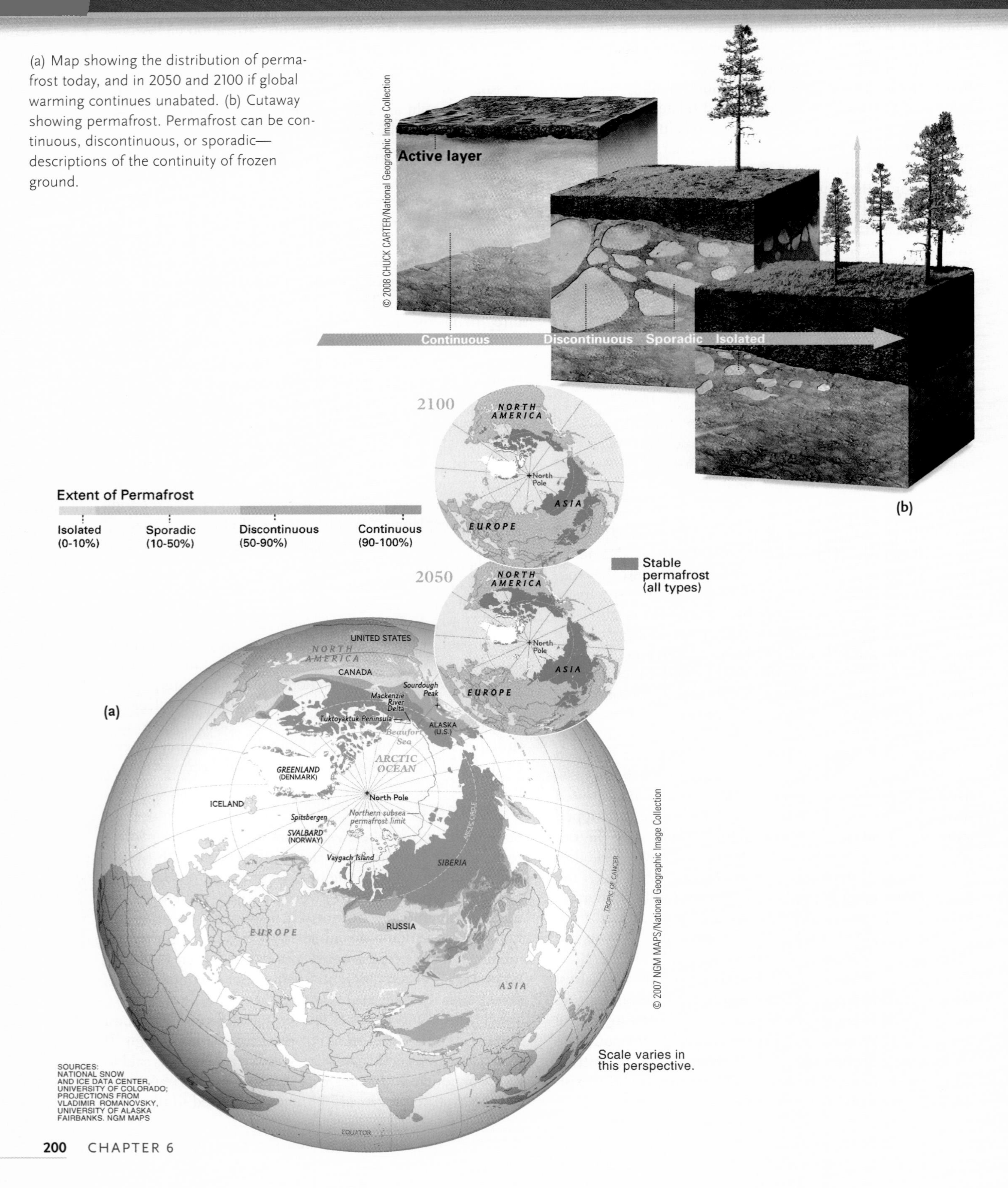

Troy L. Péwé, USGS

Figure 6.28 Utility poles extruded from permafrost; Alaska. One solution has been to support the poles with tripods placed on the ground surface.

Larry Hinzman, University of Alaska

Figure 6.30 Differential melting of permafrost caused this cabin to subside at one end. Note that part of the foundation of the cabin is open to the atmosphere, which prevented the permafrost from melting.

Figure 6.29 Historic wooden buildings in Dawson City, Canada, were built on permafrost. Now, they are collapsing as the underlying soil ice melts and the soil deforms along with the buildings.

ALASKA STOCK IMAGES/National Geographic Stock

Figure 6.31 Two methods of mitigating permafrost damage: (a) Russian workers live in barrel-shaped barracks perched on pilings to isolate them above the permafrost at Bovanenkovo, a settlement near one of the country's northernmost gas fields; and (b) heat-dissipating foundation piers on a modern structure; Kotzebue, Alaska.

Figure 6.32 The 800-mile Alaskan oil pipeline was constructed to accommodate permafrost and wildlife. The meandering pattern allows for expansion and contraction of the pipe, and for potential ground movement along the Denali fault, which it crosses (see Chapter 4). The pipeline supports feature heat-dissipating radiators.

the third of its eight stories was completed, sinking into a 2-meter-thick layer of soft clay just below the ground surface. At first, the tower leaned north, but after the initial settlement and throughout the rest of its history, it has leaned south. To compensate for the lean, the engineer in charge had the fourth through eighth stories made taller on the leaning side. The added weight caused the structure to sink even farther. At its maximum lean, it tilted toward the south about 5.2 meters (17 feet) from the vertical, about 5.5 degrees, which gives one an ominous feeling when standing on the south side of the tower (**Figure 6.33**).

In 1989, a similarly constructed leaning bell tower at the cathedral in Pavia, Italy, collapsed, causing officials to close the tower at Pisa to visitors. In 1990, the Italian government established a commission of structural engineers, soil engineers, and restoration experts to determine new ways to save the monument at Pisa. The commission's goal was to reverse the tilt 10 to 20 centimeters (4–8 inches), which they believed would add 100 years to the lifetime of the 800-year-old structure. After considering many plans, it was decided to remove soft sediment from the north side of the tower, causing it to reverse its lean to the south. The method of soil extraction was proposed by a commission member and professor of soil mechanics, John Burland of London's Imperial College. This method was simple in concept but difficult to implement. It was done with drills 200 mm (about 8 inches) in diameter, which remove cores of soil about 15 to 20 liters (4.8–5.2 gallons) in volume. The weak sediment under the north side foundation quickly filled the void left by the drill and the tower started to straighten—that is, tilt back toward the north. The concept was simple, compared with installing stabilizing cables or constructing enlarged and stronger foundations. After being shut down for 11 years, the tower was reopened, still leaning, but 40 centimeters (16 inches) and 0.5 degree less. It is estimated that the realignment will add 300 years to the life of the tower. The project cost $30 million, but that is dwarfed by the enthusiasm of the 4 million visitors to the tower each year.

Tilting towers are not exclusive to Pisa, however. Few of the 170 campaniles (bell towers) still standing in Venice are vertical, and many other towers throughout Italy and Europe tilt. These graceful towers are certainly testimony to past architects' ability and talent, but also to their failure to understand that a structure can remain truly vertical only if local geology permits.

Figure 6.33 Inadequate foundation investigation and design in the 1100s led to the world-famous leaning tower in Pisa, Italy.

Other Soil Problems

A number of other problems can affect soils. Usually, these problems arise when people take actions that impact either the chemical or the physical properties of soil; many such problems are region specific.

In the tropics, where weathering conditions are extreme due to high rates of rainfall and warm temperatures, soils are very susceptible to damage. These oxisols or **laterites** are bright red and clay rich. They are typically covered by dense canopies of jungle vegetation. However, if the trees are cut and the land used for agriculture, the soils can bake in the sun. When they dry out, they become brick-hard and of little use (**Figure 6.34**). Indeed, the famous Buddhist temple complex at Angkor Wat in Cambodia is partly constructed of laterite bricks that have endured since the 13th century. Deforestation of areas underlain by laterites is an invitation to this hardening, as U.S. forces discovered in Vietnam when they attempted to grade roads and runways on lands that had been defoliated for military reasons. Deforested laterites make for swampy conditions during the rainy season because they lack permeability; there is little infiltration of the water. This can raise havoc with wheeled vehicles and restrict mobility in such regions.

Douglas J. Harthwell, Cornell University

Figure 6.34 Dried and hardened laterite; Brazil. This soil is so hard that soil scientists call it "ironstone."

Desert soils can also be quite fragile, as many military forces have discovered. The dry desert conditions limit plant growth, and what organic matter is produced is rapidly oxidized, leaving soils with only a minimal A horizon. Often, a delicate biological crust of algae, bacteria, and lichen forms on desert soils, along with a pavement (armoring) of stones that allows the soils to resist erosion by both wind and water. When this crust or armoring is disturbed, the soils can erode rapidly. It can take decades for the crust to become reestablished. After disturbance, dust is the biggest issue, and clouds of it can be raised by traffic and wind once the soil's protective armor has been removed. This dust can choke the engines of vehicles operating in the desert, clogging their air filters and bringing them quickly to a halt. It fills the lungs of people living and working nearby. At best, it leads to the "concrete boogers" that plug the nose of every scientist who has ever worked in the desert. At worst, it leads to poisoning when the dust is laden with toxic metals, such as selenium.

Have you ever hiked on a trail through the woods that became a stream after an afternoon thunderstorm passed? If so, you have seen the effect of soil compaction firsthand. In the adjacent woods, there is no water flowing over the soil surface; it has all been absorbed by the soil. People and vehicles moving across soils change soil structure by collapsing the pores that allow water and air to pass between grains. Some of these pores are microscopic, not visible with the naked eye. Others are large enough to put your hand in, perhaps the abandoned burrow of a gopher or the cavity left when an old root rotted away. When these pores are crushed, there is nowhere for the water to go, except to run off the surface.

The pounding of hiking boots is one way to compact soil. The tires of automobiles are an even better way (**Figure 6.35**). Park on a lawn a few times. The soil's ability to soak up rainwater might go down by a factor of 2. Park in the same place until the grass is dead and the soil packed tight, and what was once a lawn now behaves as asphalt and rainwater runs right off rather than soaking in. Repeat this process across a city, and storm sewers that used to be able to handle all the runoff in a major rainstorm no longer can, because so much more water is spilling into the streets every time it rains. However, with a pick axe, some compost, some fencing, and a handful of grass seed, the problem can be fixed. Soil, once aerated and protected from future compaction, will regain its ability to absorb water.

consider this . . . It is jokingly said that the 12th-century architect of the Leaning Tower of Pisa saved money by not having a foundation investigation. Had the 12th-century engineers and geologists had our knowledge and tools, what routine tests would they have performed on borehole soil samples that would show that settlement was inevitable? What recommendations might the consultants have made to the client?

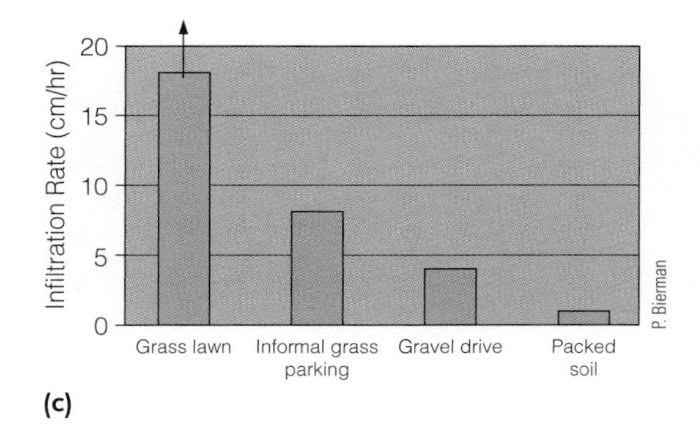

Figure 6.35 Soil compaction is a significant urban and suburban problem caused, in large part, by large objects that we drive. (a) Cars parked on what used to be lawn near college student apartments in Burlington, Vermont. (b) University of Vermont undergraduate research students measure how little water infiltrates on a part of campus where hundreds of people walk over the soil every day. Here, they are simulating a rainstorm and measuring runoff. (c) The results of many such infiltration tests show how well lawns soak up rainfall and how packed soil has almost no ability to infiltrate rainfall.

6.1 Salinization and Waterlogging

Irrigation is an amazing thing, allowing the desert to bloom and many hungry mouths to be fed. But irrigation can adversely impact soils. When irrigated, some soils get so salty that plants cannot grow. Salinization is the oldest soil problem known to humans, dating back at least to the fourth century B.C.E. At that time, a highly civilized culture was dependent on irrigation agriculture in the southern Tigris-Euphrates Valley of Iraq, the "Cradle of Civilization." By the second century B.C.E., the soils there were so saline that the area had to be abandoned. Soils in parts of California, Pakistan, the Ukraine, Australia, and Egypt are now suffering the same fate.

Salinization can be human induced by intensive irrigation, which raises underground water levels very close to the soil surface. Capillary action then causes the underground water, containing dissolved salts, to rise in the soil, just as water is drawn into a paper towel or sugar cube. The soil moisture is then subject to surface evaporation, leaving behind salts that render the soil less productive and eventually barren (**Figure 1**). The productivity of 15% of all U.S. cropland is totally dependent on irrigation, and 30% of U.S. crops are produced on this land.

Figure 1 The white crust is composed of salts, which, as the irrigation water dried, precipitated in the soil near Grand Junction, Colorado.

For millennia, annual flooding of the Nile River added new layers of silt to the already rich soil of the Nile River Delta, and removed the salts by dissolving them and flushing them downstream. After completion of the Aswan Dam in 1970, however, the yearly flooding and flushing action stopped, and salts have now accumulated in the soils of the floodplain.

Salinization is particularly evident in the arid Imperial Valley of California. Soils of the Colorado Desert—or any desert, for that matter—can be made productive if enough water is applied to them. Because deserts have low rainfall, irrigation water must be imported, usually from a nearby river. The All-American Canal brings relatively high-salinity water from the Colorado River to the Imperial Valley, where it is used to irrigate all manner of crops, from cotton to lettuce. In places, the water table has risen to such a degree that the cropland lies barren, its surface covered by white crusts of salt. The Imperial Valley's Salton Sea is one of the largest saltwater lakes in the United States, and it grows saltier and larger every year by the addition of saline irrigation water. This internally drained basin has no outlet, so all the salt that comes in stays in, unless, of course, it dries and blows away. The return of degraded irrigation water in canals to the Colorado River in southern Arizona has required that a desalinization facility be established near Yuma to make the water acceptable for agricultural use in Mexico.

Soil salinization can be reversed by lowering the water table by pumping and then applying heavy irrigation to flush the salts out of the soil. Salinization can also be mitigated, or at least delayed, by installing perforated plastic drainpipes in the soil to intercept excess irrigation water and carry it off-site. Neither remedy is an easy or inexpensive solution to this age-old problem.

question to ponder

1. If you live where it snows, salinization can be a problem. Do you know why and where salinization kills plants in snowy regions?

6.2 General Patton's Soils Decades Later

Most people know General George S. Patton (**Figure 1**) as a legendary military commander who used his tank corps to defeat the Germans in North Africa during World War II; however, few know that he was a great compactor of soil.

In preparation for the North Africa campaign, Patton ordered the building of a dozen training camps in the desert Southwest, camps with such names as Iron Mountain. There, tens of thousands of recruits, most of whom had never seen the desert, went to live and train for months. Imagine the shock, going from the woods of Minnesota or the bayous of Louisiana to the Mojave Desert, where the biggest plant was a sagebrush bush, and there wasn't any-

Figure 1 General George S. Patton, tank commander and desert soil compactor.

Figure 3 University of Vermont geologists lay out lines of numbered and painted pebbles at General Patton's Iron Mountain camp in the Mojave Desert. Several years of surveying showed that most of these pebbles move only slowly or not at all—until massive rain from thunderstorms carries them downstream.

thing green for miles. The camps were spartan places with rows of tents and roads bulldozed onto the gently sloping expanse of the desert. The soldiers, whom we can only assume were more than a bit homesick at times, did what they could to make the place home, building stone walls and intricate stone designs around their tents. Only the roads and little stone walls remain today. When the war was won, the camps were abandoned and never reoccupied. For 60 years, they have been an ongoing experiment in how quickly desert soils recover from the pounding of soldiers' boots and tons of force that tank treads impose on the landscape.

The impacts of Patton's men more than 60 years ago are very much affecting the area today: Soil compaction in deserts is a long-lasting insult. The soil beneath the abandoned roads is still compacted, and water flows along these roads during rainstorms, rather than through natural drainage.

In the late 1990s, geologists from the University of Vermont and Duke University reoccupied one of Patton's camps to determine how the desert soils had recovered over the more than 50 years since Patton and his men had packed up and gone home. They used all sorts of tools to understand how the landscape and soils behaved, including high-precision surveys (**Figure 2**), low-tech soil pits, and measurements of ^{10}Be (see the explanation earlier in this chapter). One of them set outlines of small, painted pebbles, 1,600 in all, to track how pebbles move (**Figure 3**).

Some of the pebbles moved in the several years that they were tracked. The pebbles in channels moved the farthest and fastest as they were swept away by moving water (it rained several times during the three-year experiment). Even the pebbles away from channels moved a bit, probably by the action of gophers or other animals. The rock berms that the soldiers made to outline the walkways might seem insignificant, but they were a major influence on which way the water moved.

What can be done? At present, nothing because this camp is a protected area, but this experiment suggested some

Figure 2 A survey rod equipped with a prism reflects laser light, allowing precise measurement of distance. In the background, a specially designed Global Positioning System (GPS) unit is used to measure locations to within a centimeter or two. Hundreds of locations will be measured in a day, and the data will be used to map the walkways and channels that carry water over this desert surface.

simple ways in which the impact of future camps on desert soils might be minimized. When a camp is abandoned, plow up the roads and disturb the soils to reduce compaction. This will accelerate erosion initially but allow the system to more quickly reestablish natural drainage patterns in the long run.

All of this should make us wonder about the impact people have had on the expansive deserts of the Middle East, where so much of our petroleum exploration and military activity goes on these days.

question to ponder

1. What effects on soils and soil erosion do you think recent wars in Afghanistan and Iraq might be having? Do you think it would be any different than that caused by General Patton and his troops more than 60 years ago?

6.3 Rx for Contaminated Soils

For millennia, soil has been a convenient place to dump waste nobody wanted, as the waste would seem to disappear, and soil was plentiful. Indeed, the use of soil for disposal has been a gold mine for archaeologists who routinely sort through century-old privies to find the remains of long-vanished civilizations. But today the stakes are higher as we manufacture and process many chemicals far more hazardous than what went into colonial outhouses—heavy metals, toxic pesticides, oil wastes. We rely on groundwater for drinking, and much of what is in the soil eventually finds its way into the groundwater. Unfortunately, it took us a while to realize how bad the problem was, so we've been playing catch up as a society for several decades. Fortunately, some soil contamination is reversible, and soils contaminated by organic pollutants, such as oil, can be "cleaned" (i.e., *remediated*) by a variety of methods. A large soil-remediation industry has sprung up in response to U.S. Environmental Protection Agency and state environmental protection requirements.

Various soil-remediation techniques are used, depending on the thickness of the soil to be cleaned and the chemical nature of the contaminant. *Dilution*, mixing contaminated soil with clean soil, is one technique. It works, but it is expensive and it contaminates an even larger volume of soil. The old saying "the solution to pollution is dilution" is not really a solution.

Bioremediation, a natural process, uses microorganisms to degrade and transform organic contaminants including petroleum products like solvents. Bacteria and fungi flourish naturally in soils where oil, sewage sludge, or food-production waste have been discarded. Biodegradation can be stimulated by increasing the amount of air in the soil through cultivation or stockpiling the soil and forcing air into it. Adding easily decomposed organic matter, such as hay, compost, or cattle manure, also speeds up the degradation of organic contaminants. Bioremediation is relatively inexpensive. Some contaminated soils simply require turning (disking or plowing) to speed up oxidation.

A successful bioremediation project at Mobil oil company's natural gas refinery at Liberal, Kansas, is shown in **Figure 1a.** The soil contained up to 26,000 parts per million (2.6%) of hydrocarbons—an oily mess. To help degrade these oils, the contaminated soil was spread in a layer less than 20 centimeters (8 inches) thick; then a nutrient and microbial amendment, cow manure, was applied. The soil was periodically stirred to promote biological activity and was subsequently reclaimed and planted (see Figure 1b).

Al Robb III

Figure 1 A successful bioremediation project. (a) Oil-contaminated soil (2.6% hydrocarbons) at a Kansas refinery is removed for bioremediation by a bulldozer. Note the patchy distribution of the oil contamination. The soil is spread in a layer about 20 centimeters (8 inches) thick. Nutrient amendment (cow manure) is added to stimulate biological degradation of the hydrocarbons. (b) Remediated soil is spread and planted with grass. Note the clods of manure and the white chunks of caliche in the redistributed soil. The structure in the background is a grain silo, in which wheat is stored.

(b)

Al Robb III

6.4 Climate Change: The Effects on Weathering and Erosion

Our climate is changing, and changing rapidly. Everywhere, temperatures are rising, driven by increasing carbon dioxide and methane gas concentrations in the global atmosphere. Some parts of the world are drying, whereas others are getting wetter (**Figure 1**). Let's examine how these changes affect two parts of the Earth that couldn't be more different—the tundra surrounding the Arctic Ocean and the Great Plains of North America.

Within just a few decades, both current trends and computer models indicate that little, if any, permanent sea ice will remain (**Figure 2**). The ice is melting because both the atmosphere and the ocean are growing warmer, the result of climate change driven by greenhouse gas emissions. The loss of sea ice has major consequences for the rest of the planet. By reflecting incoming sunlight back into space during the Northern Hemisphere summer, white sea ice cools the planet. Relatively dark ocean water absorbs more solar energy than the sea ice, further warming the planet in a positive feedback loop. Everyone is familiar with the sad images of starving polar bears stranded on small, remaining bits of ice, but have you thought about what an open ocean in the Arctic means to the survival of the permafrost on which Arctic communities are built? Open seawater in the Arctic allows nearby land to warm, and the results can be disastrous. With permafrost melting, once-frozen soil erodes quickly, and buildings and other types of infrastructure on which people depend (such as gas pipelines) are frequently damaged or destroyed.

Thousands of kilometers to the south, climate change is affecting the breadbasket of North America, the Great Plains. There, where water has always been scarce, atmospheric models predict that, with increasing carbon dioxide levels in the atmosphere, the climate will get warmer and, although there may be more precipitation in the northern plains, the net effect will be drying of the soil. Predictions indicate that, on average, summer temperatures will be between 5° and 10° F (2.8° to 5.2 °C) warmer by the end of this century (**Figure 3**). It was not that long ago that clouds of soil rose from the plains in the Dust Bowl. With climate warming, soils drying, and the water level in aquifers that supply irrigation water to this part of North America dropping rapidly, the stage is set for major soil erosion by wind.

Figure 2 Predicted change in sea ice concentration from 1885 to 2085, showing how little permanent sea ice will be left by the end of this century. Vanishing sea ice will spur the melting and erosion of permafrost around the Arctic Ocean.

question to ponder

1. Given that warming and drying are so likely for the Great Plains, if you were advising the president about policy decisions related to the future of agriculture there, what would you suggest the president do and why? How would you prevent severe soil erosion and groundwater depletion while protecting people's jobs?

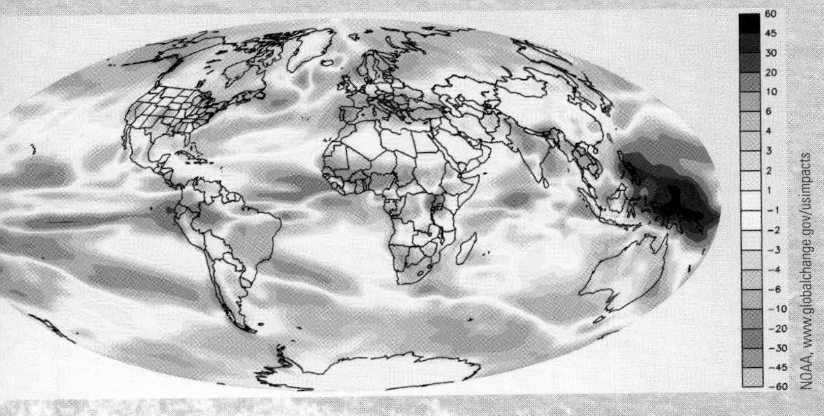

Figure 1 Map shows how the distribution of precipitation around the world is expected to change significantly over the next 100 years. In general, the low latitudes will become drier and the high latitudes wetter. Many semiarid regions that are now marginal areas for farming, such as the southern Great Plains and Australia, will become even drier. Key shows percent difference in mean annual precipitation. Blue = increase; yellow = decrease.

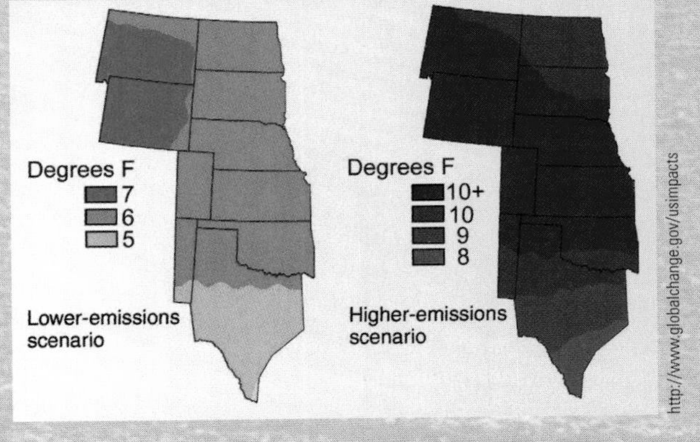

Figure 3 The Great Plains will be getting warmer in the next 100 years, much warmer. The central and northern plains will see the greatest warming. No matter what we do as a society now, the climate will continue to warm, the result of fossil-fuel–derived carbon dioxide released over the last century. However, lower carbon emissions will minimize the warming, whereas higher carbon emissions will exacerbate warming.

PHOTO GALLERY

erosion around the world

Erosion is a natural process, but human actions almost always increase the rate of erosion and focus erosive activities in specific areas. How do we do this? Removal of vegetation for farming or clear-cutting for timber destroys root networks that hold soils together. Mining and construction disturb stable soil, making it much more vulnerable to erosion by running water. Agriculture is probably the worst culprit. Tilled fields are vulnerable to erosion by wind and water, while grazing animals can compact soil and strip vegetation, making the soil more vulnerable to erosion.

Post plowing deposition
Cumulative plow horizon
Post settlement fan deposit
Soil A horizon
Soil B horizon

1 A fish-eye view of a soil that has lots of stories to tell. The bottom two layers are forest soil, with well-developed soil A and soil B horizons. The dark brown soil A horizon is a meter (several feet) below today's surface. The material burying the soil came pouring off the slope above when settlers first cleared this site in Huntington, Vermont, more than 200 years ago. There is a sharp line between the dark and light gray material that marks the bottom of the plow zone (it is called the *cumulative plow horizon*); it is a distinctive way to know that soils have been plowed and planted. The thickness of the plow zone is more than twice the depth to which plows can reach, which means that every time it rained and the slope eroded, it had to be plowed again. The field has not been plowed in several decades, allowing sediment to pour off the hillside and cover the well-mixed plow zone. (Reprinted by permission of Paul Bierman and *GSA Today*.) P. Bierman/University of Vermont

Gold mining at Pantanal, Brazil, is causing massive erosion. The mining clears the land and loosens the rock and sediment. Heavy rainfall then runs off the surface, carving deep rills into the loose mine waste. © 2011/ FRANS LANTING/ National Geographic Image Collection

2 Dirt road cuts through the grassland of Mongolia. Repeated passage by wheeled vehicles has eroded the road deep into the plain. © 2010/FRANS LANTING/National Geographic Image Collection

3 Herding cattle across bare soil in the Serengeti of Tanzania, Africa. The lack of vegetation and the impact of the animals' hooves allow the wind to erode the soil.

4 Cattle in the outback of Western Australia (the Kimberly) forage for grass between termite mounds that dot the landscape. The termites stir the soil, bringing materials from depth to the surface.

5 What remains of a stone wall in an area that was once cleared for farming in Massachusetts. Today, soil erosion rates in these old, abandoned farm fields are very low. The leaves absorb the energy of rainfall, preventing erosion and allowing water to soak into the forest soil. The roots of the trees, which have regrown since the field was abandoned, hold the soil in place. Robert M. Thompson/The Stone Wall Initiative, University of Connecticut

SUMMARY Weathering, Soils, and Erosion

Soil

Defined

Soil is a mixture of weathered rock, mineral grains, and organic material that can support plant life.

Environmental Factors of Soil Formation

Climate, organic activity, relief (topography), parent material, and time. Soils form from the weathering of rock and sediment.

Weathering

Physical and chemical breakdown of bedrock and regolith. They are linked processes.

1. *Physical weathering* is caused by temperature differences and disruption by plants and animals.
2. *Chemical weathering* processes include solution, oxidation, hydration, and hydrolysis. Chemical weathering forms clays, the most important mineral group in soils. Clay minerals are important because they hold moisture and exchange nutrients with plants.

Rate of Weathering

Depends on the rock's or regolith's mineral composition and surface area, as well as climate. Smaller particles weather faster than large masses, and iron-rich minerals weather faster than quartz or feldspar.

Features and Landforms

Spheroidal weathering and exfoliation domes are common features.

Soil Profile

Defined

Distinguishable layers, or horizons, in mature soils composed of the organic-rich O horizon on the top, the dark A horizon or zone of leaching (where minerals are dissolved), and a lower B horizon known as the zone of accumulation, where salts and clays accumulate. Below is the C horizon of weathered parent material. The thickness of the horizons and the developmental rate of soil are profoundly influenced by climate.

Residual Soil

Developed in place on underlying weathered bedrock, also known as saprolite. Oxisols, or laterites, are the brick-red soils of humid (subtropical and tropical) regions that represent intensive weathering of parent rock.

Transported Soil

Developed on regolith, transported and deposited by wind (eolian), glaciers, rivers, or volcanic action. Loess, for example, is an eolian soil associated with ice-age continental glaciation. It is highly productive and is found throughout the central U.S. "breadbasket."

Soil Classification

A classification of soils into 13 orders by their physical characteristics that often reflect soil age and climate.

Soil Problems

Erosion

Erosion is a major problem because soil and the food it grows determine Earth's carrying capacity. Globally, irreparable soil degradation has impacted an area the size of China and India combined. U.S. soil losses vary greatly with location.

Processes of Erosion

Sheet erosion by running water (thin layer is removed) can become rill erosion (defined channels carved in the soil), which leads to gullying (the most destructive). Sheet and rill erosion can be halted, but once gullying starts, it is extremely difficult and costly to mitigate. Wind erodes soil on marginal lands that were once cultivated and then dried out. Without plant root systems to hold the soil, wind erosion can lead to "dust bowl" conditions. Off-road vehicles disturb soils and catalyze soil erosion.

Mitigation

Good agricultural practices include terracing, no-till and minimum-tillage farming, strip-cropping, and crop rotation. Wind losses can be minimized by planting windbreaks, strip crops, and cover crops, such as clover.

Other Soil Problems

Expansive Soils

Certain clay minerals absorb water and expand, exerting high uplift pressures, which can crack or even destroy structures. Effects can be mitigated by removing expansive soil, mixing with other soils, or keeping the moisture content constant.

Permafrost

Permanently frozen ground (soil or rock) underlies more than 20% of Earth's surface. Structures built directly on soil permafrost are likely to settle differentially. Mitigation involves building with an open space below the structure or thermal isolation, so as not to disturb the thermal regime in the soil and melt the permafrost. Climate change will likely cause major issues in polar and alpine regions as permafrost thaws.

Settlement

Settlement occurs when the applied load is greater than the bearing strength of the soil (e.g., the Leaning Tower of Pisa). This is an engineering problem that can be mitigated by proper design or by strengthening of the underlying soil.

Other

Salinization and waterlogging (high water table), hardening of laterites caused by deforestation and desiccation, compaction by vehicles, and dust generation by wind erosion.

Contamination

Brines, petroleum products, crude oil, and chemicals of all sorts can contaminate soils. Bioremediation (bacteria) is useful for reducing the concentration of organic pollutants in soil.

KEY TERMS

active layer
bioremediation
caliche
cosmogenic isotopes
erosion
exfoliation dome
expansive soils
frost wedging
gullying
illuviation
laterite
loess
pedologist
permafrost
permafrost table
pingo
regolith
residual soil
rill erosion
salt-crystal growth
sheet erosion
sheet joint
soil
soil horizon
soil order
soil profile
spheroidal weathering
tundra
ventifaction
weathering

STUDY QUESTIONS

1. Sketch a soil profile and describe each soil horizon.
2. Distinguish among chemical weathering, physical weathering, and erosion.
3. What kinds of soils would you expect to develop on basalt in Hawaii, New Jersey, and Arizona? Explain the differences in terms of the CLORPT equation.
4. What is permafrost? Where is it found?
5. What soil problem does permafrost cause structures, and what techniques are used to prevent damage?
6. What is expansive soil? What problems and challenges does it pose to soil engineers? What can be done to prevent distress caused by this condition?
7. Soil erosion and degradation are two of humankind's oldest problems. How has human-induced soil loss come about?
8. What agricultural practices have led to salinization and waterlogging of soils?
9. An inscribed rock obelisk dating from 1500 B.C.E. was resurrected from the ruins at Karnak, Egypt. Its inscriptions were crystal clear when it was taken to New York's Central Park as a gift of the Egyptian government in 1880, but today they are totally illegible. Explain the reason for this.
10. Laterites, or oxisols, present special problems when they are deforested. Why?
11. Compare the rates of soil erosion and soil formation in the United States.
12. How will climate change affect soil? In particular, what will be the effects of a warming world on rates of wind and water erosion?

REFERENCES AND FURTHER READINGS

Bade, Sandra S. 1983. *Soil erosion: A crisis in America's croplands?* Washington, DC: The Conservation Foundation, 136 p.

Bierman, P. R., and A. R. Gillespie. 1991. Range fires: A significant factor in exposure-age determination and geomorphic surface evolution. *Geology* 19: 641–644.

Birkeland, R W. 1999. *Soils and geomorphology*. New York: Oxford University Press, 372 p.

Brady, N. C. and R. R. Weil. 2007. *The nature and properties of soils*, 14th ed. New York: Prentice-Hall, 295 p.

Brady, Nyle C. 2008. *The nature and properties of soils*, 14th ed. New York: Prentice-Hall, 992 p.

Brevik, Eric. 2002. Problems and suggestions related to soil classification as presented in introduction to physical geology textbooks. *Journal of Geoscience Education* 50(5): 539–543.

Brown, K. W. 1994. New horizons in soil remediation. *Geotimes* 14–17.

Brown, Lester. 1999. *State of the world*. New York: Norton, 259 p.

Ferrians, O. J., et al. 1969. *Permafrost and related engineering problems in Alaska*. U.S. Geological Survey professional paper 678. Washington, D.C: Government Printing Office, 37 p.

Heiniger, Paolo. 1995. The Leaning Tower of Pisa. *Scientific American* 273(6): 62–67.

Loynachan, Thomas E., Kirk W. Brown, Terence H. Cooper, and Murray Milford. 1999. *Sustaining our soils and society*. Alexandria, VA: American Geological Institute, 64 p.

Montgomery, D. R. 2007. *Dirt: The erosion of civilizations*. Berkeley: University of California Press, 295 pp.

Singer, M. J., and D. N Munns. 2002. *Soils: An introduction*, 5th ed. Upper Saddle River, NJ: Prentice Hall, 446 p.

Winkler, Erhard M. 1998, September. The complexity of urban stone decay. *Geotimes* 25–29.

Access an interactive eBook and chapter-specific interactive learning tools, including flashcards, quizzes, videos and more in your Geology CourseMate, accessed through CengageBrain.com.

Two men in Glendora, California, near Los Angeles try in vain to divert a mudflow by shoveling. The mudflow, which buried homes, was caused when heavy rains pounded steep slopes mantled by loose sediment.

Have you ever **wondered...**

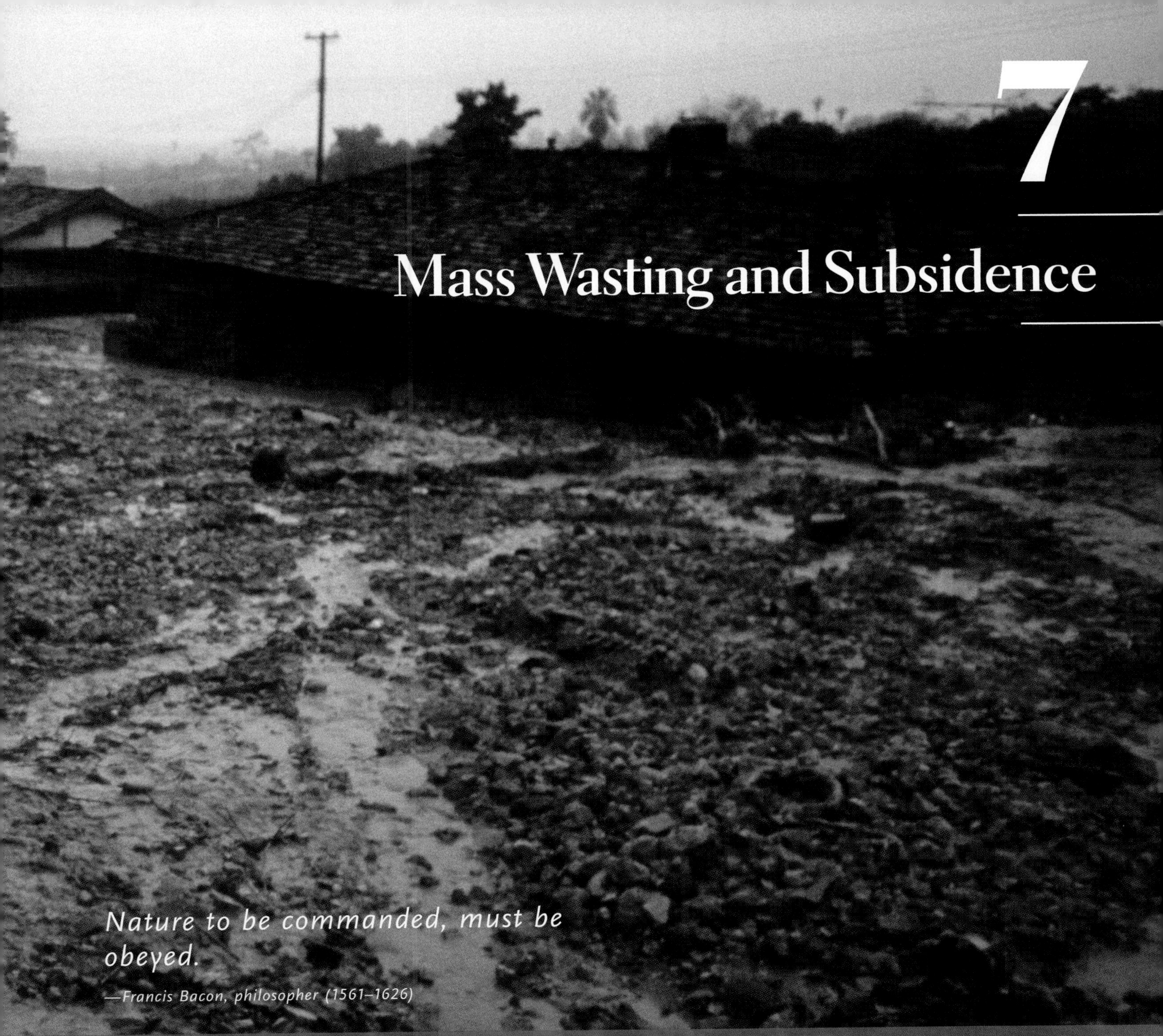

7

Mass Wasting and Subsidence

Nature to be commanded, must be obeyed.

—*Francis Bacon, philosopher (1561–1626)*

1. What processes move soil and rock down slopes?

2. How climate change might affect the frequency of debris flows?

3. How engineers stabilize eroding slopes?

4. What you can do to avoid getting caught in an avalanche?

5. What hazards affect both Venice and New Orleans?

A Bad Slice on the Golf Course

Golf's explosion in popularity in the late 1990s led to many new golf courses. Some are in resort communities, but others are in urban and suburban areas, like the beautiful Ocean Trails Golf Course south of Los Angeles on the Palos Verdes Peninsula, now known as the Trump National Golf Club (**Figure 1**). Geologists know the peninsula well for its many landslides, due in part because it is uplifted 370 meters (1,300 feet) above sea level and in part because of the sea, which is constantly battering the cliffs' base with waves. It is easy to see in 30- to 50-meter cliffs, where the layers are exposed, that the stratified sedimentary rock underlying the course inclines gently to the sea. Just as a slice of bread slides down a breadboard when tilted, some weak layers have slid toward the sea in the past, carrying houses and roads with them. Ocean Trails Golf Course was almost ready for its early summer 1999 opening when, without warning, a fissure opened parallel to the cliff, and 300 meters of the 18th fairway slid into the ocean. Local resident Tony Baker and his dog were temporarily stranded on a precarious 215-meter (700 foot)-long island between fissures. "I heard crumbling earth and started seeing dust rising," Baker said later. "The trail started cracking up. I was doing a little running around, jumping over big cracks. I just found a place and hunkered down. I wasn't sure if I was going to make it back."

After three years of wrangling, the developers and lenders reached an impasse, and the landslide stabilization work stopped. The beautiful golf course, now only 15 holes long, was blemished by the scar of the landslide and stabilization work. Enter Donald Trump, the New York entrepreneur, real-estate magnate, and enthusiastic golfer. He bought the troubled project (at a bargain price) in 2002 and vowed to finish the project by the summer of 2003 (not quite fulfilled). On 20 January 2006, the Trump National Golf Club officially opened with all 18 holes, and the buttress fill designed to stabilize the "slice" is buried under the last two holes.

question to ponder

1. If they had been careful, what geological and landscape clues might the designers of the golf course have noticed that would have alerted them to potential slope stability issues before they built the course?

Figure 1 Ocean Trails Golf Course, Palos Verdes Peninsula, California, shortly after the "slice" occurred, disrupting grand-opening plans.
B. Pipkin

CourseMate Connection

Start off this chapter by watching a video about landslides that are causing residents to demolish houses along the Yorkshire cliffs in Scarborough, England. **Log in to the CourseMate at www.cengagebrain.com to watch now.**

MASS WASTING IS THE GENERAL TERM that denotes any downslope movement of soil and rock under the direct influence of gravity. Mass-wasting processes include landslides, rapidly moving debris flows, slow-moving soil creep, and rockfalls of all kinds. Annual damage from landsliding alone in the United States is estimated at between $1 and $2 billion and is widespread, affecting much of the country (**Figure 7.1**). If we include other ground failures, such as subsidence, expansive soils, and construction-induced slides and flows, total losses are, on average, many times greater than the annual combined losses from earthquakes, volcanic eruptions, floods, hurricanes, and tornadoes—unless, of course, it happens to be the year that hurricane Katrina strikes.

Mass wasting and subsidence are global environmental problems. Earthquake-triggered landslides in Gansu Province in China killed an estimated 200,000 people in 1920, and debris flows left 600 dead and destroyed 100,000 homes near Kobe, Japan, in 1938. The largest loss of life from a single landslide in U.S. history, 129 fatalities, occurred at Mameyes, Puerto Rico, in 1985. Landslides and debris flows are terrifying environmental phenomena that test our very certainty that Earth will always be stable below our feet. However, as climate changes, so probably will the distribution and intensity of rainfall. With an increased frequency and duration of hurricanes and other strong storms, some parts of the world are likely to see more landslides and debris flows—the result of a landscape trying to adapt to a new and different hydrological regimen.

The goal of this chapter is to reveal the variety of ways in which mass movements and subsidence can occur and place people and their property in harm's way. We consider the relevant physical processes and means by which different types of Earth-movement hazards can be effectively mitigated or avoided.

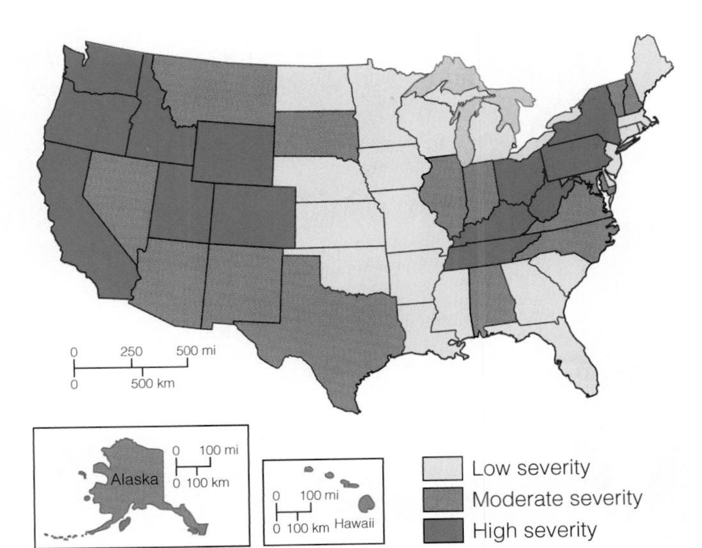

Figure 7.1 Severity of landsliding in the United States. Note that the most serious problems are found in the Appalachian Mountains, the Rocky Mountains, and the coastal mountain ranges of the Pacific Rim. (Based on data from Krohn, J. P., and J. E. Slosson. 1976. Landslide potential in the U.S. *California Geology* 29(10).)

Classifying Mass Movements

The classification of mass movements used in this textbook is similar to the one widely used by engineering geologists and soil engineers (**Figure 7.2**). The bases of the classification are:

- The type of material involved, such as rock or soil
- How the material moves, such as by sliding, flowing, or falling
- How fast the material moves, its velocity

For instance, soil creep occurs at imperceptibly slow rates; debris flows are water saturated and move swiftly; and slides are coherent masses of rock or soil that move along one or more discrete failure surfaces, or **slide planes**. Failure occurs when the force that is pulling the potentially unstable material downward (gravity) exceeds the strength of the earth materials that compose the slope. Because the force of gravity acting on a given slope is constant, either the earth materials' resistance to sliding must decrease or the gradient of the slope must increase for sliding to occur. The presence of water is key because it has the potential to diminish the resisting force that makes slopes stable. The strong shaking of a major earthquake can momentarily change the balance of forces on a just-stable landslide block and send it screaming downslope.

Flows

Types of Flows

Creep is the slow (a few millimeters per year, perhaps a tenth of an inch), downslope movement of soil and rock on steep slopes. It involves any number of specific and different processes that are all typically lumped together in this one term. Physical creep involves deforming or changing the shape of the slope materials. Some deformation of weak materials, such as soft clay, may be driven completely by gravity. For other materials, freezing and thawing or alternate wetting and drying of a hill slope, which causes upward expansion of the ground surface perpendicular to the face of the slope, causes creep. As the slope dries out or thaws, the soil surface drops vertically, resulting in a net downslope

MECHANISM		Rock	Fine-grained Soil	Coarse-grained Soil	VELOCITY
		MATERIAL			
SLIDE		Slump	Earth slump	Debris slump	SLOW
SLIDE		Block glide	Earth slide	Debris slide	RAPID
FLOW		Rock avalanche	Mudflow	Debris flow	VERY RAPID
FLOW		Creep	Creep	Creep	EXTREMELY SLOW
FALL		Rockfall	Earthfall	Debris fall	EXTREMELY RAPID

Figure 7.2 Classification of landslides by mechanism, material, and velocity.

movement of the soil (**Figure 7.3a**). Biota of all sorts contribute to soil creep, and burrowing animals can move soil onto steep slopes. More animal-moved soil ends up moving downslope than upslope; the net result is downslope transport. Trees are an important catalyst for soil creep. Every time one blows over and its root wad tips up (see Figure 7.3b), soil is moved, usually downslope.

How do we know soil creep is active? Bent trees, leaning fence posts and telephone poles, and the bending of rock layers downslope are all evidence of soil creep (see Figure 7.3c). Homes built with conventional foundations 45 to 60 centimeters (18–24 inches) deep may develop cracks caused by soil creep. This is seldom catastrophic; typically, it is a cosmetic and maintenance problem. The influence of soil creep can be overcome by placing foundations through the creeping soil into bedrock.

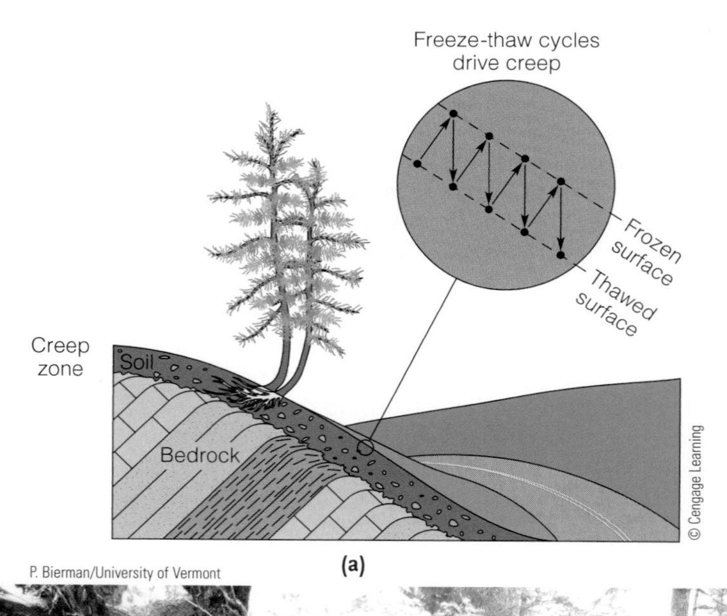

Figure 7.3 (a) Diagram of soil creep. (b) Soil creep on an Oregon Coast Range scale. Here, in the Drift Creek Wilderness Area, geomorphologist Tom Dunne (1.8 meters, 5 foot 10 inches) shows just how large tip ups and root wads can be. This is an old-growth forest; when trees fall on the 25° slopes, soil moves. (c) Soil creep, shown by bent trees; east slope of the Sierra Nevada, California.

P. Bierman/University of Vermont

(b)

B. Pipkin

Debris flows are dense, one-phase fluid mixtures of rock, sand, mud, and water. The one-phase part is important. Add too much water and the sediment settles out. Don't put enough water in the recipe and the debris flow won't flow. Debris flows are often generated quickly during heavy rainfall or snowmelt where there is an abundant supply of loose soil and rock. They often begin as landslides, which mobilize and break up as they move downslope. Moving with the consistency of wet concrete, debris flows are very destructive, with velocities up to many meters or tens of feet per second. Because of their high viscosity and high density, commonly 1.5 to 2.0 times the density of water, debris flows are capable of transporting large boulders, automobiles, and even houses in their mass (**Figure 7.4**).

Areas most subject to debris flows are characterized by sparse vegetation and intense seasonal rainfall, or are in regions

Figure 7.4 (a) Two debris flows in 1.5 hours hit and moved a farmhouse and destroyed several other buildings. The farm structures were on an alluvial fan without a well-defined main channel. The flow occurred in Madison County, Virginia, which shows that debris flows are not limited to the southwestern United States. (b) These are old boulders on a debris flow fan in Madison County. The boulder in the background is the size of a small house; a debris flow dropped these two where they sit today. The soils that partially bury them suggest that repeated debris flows have covered this fan over the past tens of thousands of years. (c) In 2011, very heavy rains triggered widespread landslides that spawned debris flows in the mountains west of Rio de Janeiro, Brazil. Here, Nelson Fernandes, a Brazilian hydrologist, examines a swimming pool filled with boulders in front of vacation homes destroyed by one of the flows.

Kevin Lamb/U.S. Geological Survey

P. Bierman/University of Vermont

P. Bierman

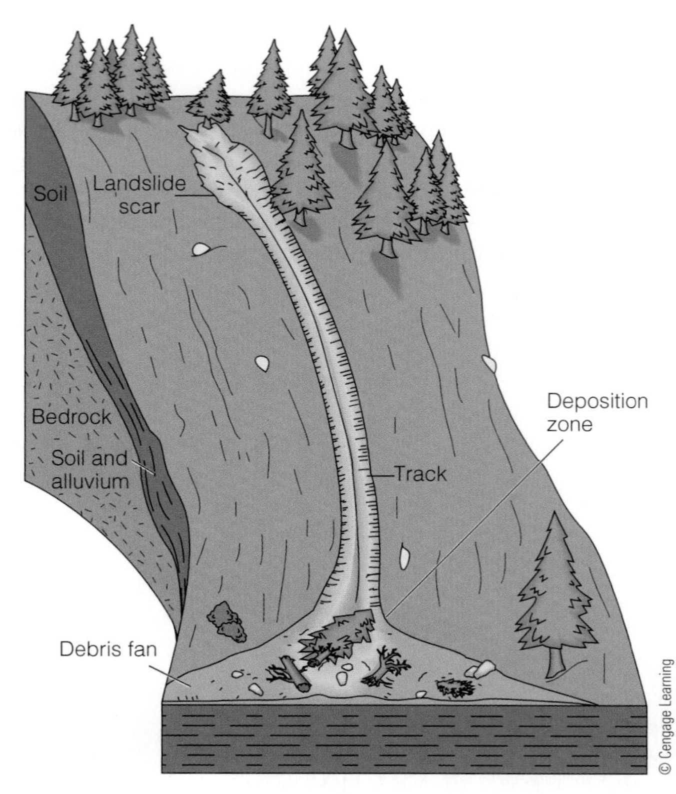

Figure 7.5 Debris flow track and zone of deposition. A debris flow may travel thousands of feet beyond the base of the slope.

that are subject to drenching rains (see Case Study 7.1). Debris flows are relatively common in steep, alpine, tropical, and desert environments worldwide. Debris flows associated with volcanic eruptions (lahars) are potential hazards in all volcanic zones of the world. They pose a particular danger in populated areas near volcanoes and are the most significant threat to life of all the mass-movement hazards (see Chapter 5).

Causes, Behavior, and Prediction of Debris Flows

The recipe for debris flows is pretty short. Take a steep slope, provide plenty of loose material (including some clay and silt), add just enough water to get everything wet but not too wet, mix well, and let the mixture loose down a stream channel or steep canyon. But something needs to start the flow and get things mixed up. Often, that something is a small landslide or sediment collapsing off a steep bank. Once initiated, debris flows usually follow tracks or stream channels, often bulking up as they move by incorporating rocks and water from the stream. Small flows may end at the bottom of their track on a debris fan (**Figure 7.5**).

Depending on the slope and its composition (the amount of water and the type of sediment), debris flows can just creep along or they can thunder down canyons at speeds in excess of 50 kilometers/hour (30 miles/hour). Several flows may merge and travel many miles from their source area. Because debris flows have a finite yield strength (unlike water, which will flow on any slope), at some combination of low slope and critical thickness (or thinness), they will cease to flow and just sit there, slowly drying up and solidifying.

Although predicting the timing and location of specific debris flows is not possible, the conditions that lead to debris-flow initiation are well-known. Researchers have collected the field data required to create what are called *intensity/duration envelope diagrams* (**Figure 7.6**). Simply put, these are graphs that show the combination of how long it needs to rain and at what rainfall intensity before debris flows will begin. Reading the graphs, it is clear that very heavy rainfall need occur for only short periods of time to trigger flows, whereas moderate rain needs to fall for much longer before slopes will give way. **Antecedence**, or the amount of rain that has fallen over the past week or two, is also important. As you might suspect, saturated or nearly saturated soils are more likely to fail than those that have plenty of dry pore space available to soak up the rainfall.

Intensity/duration data have been compiled for flows and slides in the San Francisco Bay area of northern California (see Figure 7.6). These curves show that, for a rainfall intensity of 0.5 inch per hour, the threshold time for the onset of debris flows is 8 hours in Marin County and 14 hours in Contra Costa County. The differing rainfall thresholds in these relatively close areas are due to the variability of geological materials and topography. The lower curve of Figure 7.6 shows that less-intense rainfall will produce debris flows in semiarid and arid areas of California than in more humid areas. This is because these dry areas have little vegetation and abundant loose surface debris; thus, there are few root systems to retain the weak soils. Although these curves are re-

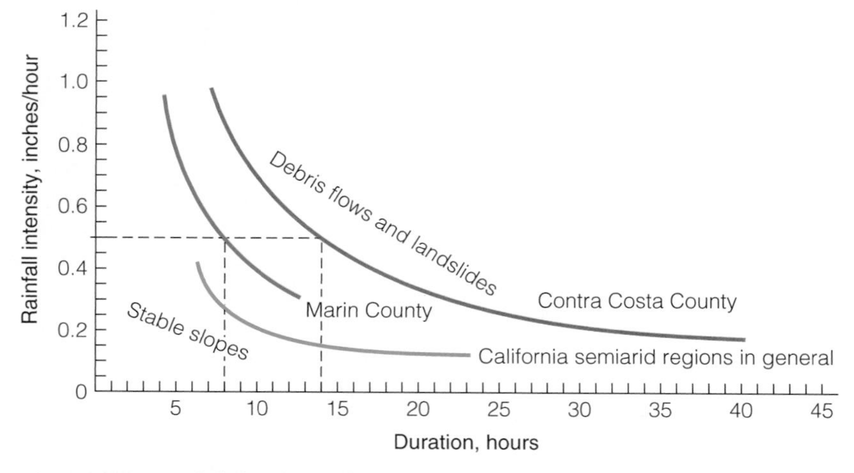

Figure 7.6 Rainfall thresholds for debris flows and landslides for two central California counties and for California semiarid regions in general. The dashed lines show that rainfall intensity of 0.5 inch/hour for 8 hours is sufficient to initiate flows in Marin County. The same rainfall intensity for 14 hours is required to trigger debris flows in Contra Costa County. (California geology data from California Division of Mines & Geology.)

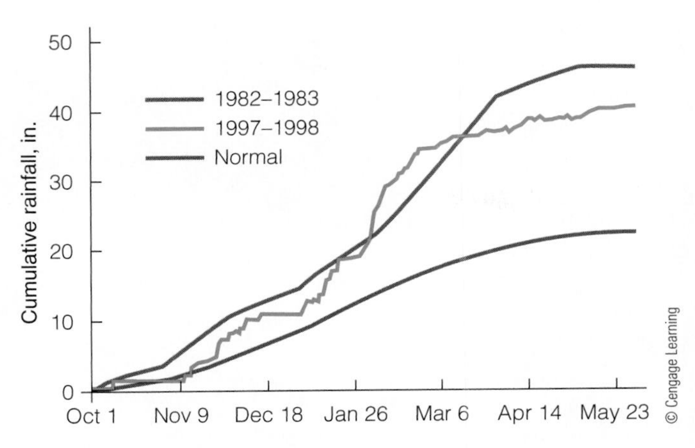

Figure 7.7 Rainfall was plainly in excess of normal in the San Francisco Bay area during El Niño years 1982–1983 and 1997–1998. These heavy rainfalls translated directly into a greater frequency of landslides and debris flows as saturated and nearly saturated soils failed and came roaring down slopes.

gional, they are useful because they allow authorities to alert residents in critical areas once the threshold conditions for debris flows have been attained.

Not all debris-flow seasons are created equal. Heavy rains (**Figure 7.7**) associated with El Niño events (the change in Pacific Ocean water temperature distribution) have resulted in exceptional landslide and debris-flow activity, particularly along the west coasts of North and South America. The major El Niño of 1982–1983 was marked by widespread landsliding (slumps, slides, and flows) in various parts of the Western Hemisphere. The 1997–1998 El Niño, possibly the largest of the 20th century, also had exceptional flows and slides. Hard hit was the San Francisco Bay area, where some of the 85,000 landslides were reactivated. This is an excellent example of interaction among the atmosphere, hydrosphere, and solid-earth systems. Climate change could exacerbate landslide and debris-flow hazards in the Bay area because, with warming climate, some models suggest that El Niño-like conditions will become more common.

Mountain slopes burned by range and forest fires during the dry season are very susceptible to debris flows during the wet season. John McPhee, in his book *The Control of Nature*, does a marvelous job of describing the fire/storm/flow continuum in the hills overlooking the Los Angeles basin, as well as the societal response (or lack thereof). High in those very steep hills, wildfires (**Figure 7.8a**) scorch the scrubby chaparral vegetation (see Figure 7.8b), releasing waxy chemicals that essentially waterproof the ground. When the winter rains come, the soils are impermeable and, with the vegetation burned off, slopes quickly lose strength, letting loose debris flows that sweep through canyons (see Figure 7.8c), with people and their homes and vehicles in the way (see Figure 7d). This cycle plays out over and over again in the southern California mountains (see Figure 7.8e). In October 2003, extreme forest fires burned in Southern California. In December of that year, when the rains came, 13 people lost their lives in a campground at the mouth of one of the burned canyons in the San Gabriel Mountains.

Debris flows are not limited to just the arid and semiarid Southwest and West; they occur anywhere slopes are steep, water is available, and loose sediment is ready to move. In Madison County, Virginia, in 1995, an intense, long-lasting thunderstorm dumped more than 10 inches (25 centimeters) of rain on the steep slopes of the Blue Ridge Mountains, triggering landslides that set off numerous debris flows, which roared down steep stream channels, carrying boulders the size of houses (see Figure 7.4) and devastating the area. Livestock were killed and buildings, bridges, roads, and crops were damaged. There was one loss of life directly attributable to the debris flows. It is estimated that such damaging debris flows may occur every 3 years in the Blue Ridge and southern Appalachian mountains. Often, they are triggered by the intense rains of tropical storms and hurricanes as the warm, moist air is lifted over the steep mountains.

Not all debris flows and avalanches are triggered by intense rainfall. On 30 May 1983, a debris flow was generated on Slide Mountain, Nevada, which killed one person, injured many more, and destroyed a number of homes and vehicles, all in less than 15 minutes. How did this happen without any rainfall? About 720,000 cubic meters (about a million cubic yards) of weathered granite gave way from the mountain's steep flank and slid into Upper Price Lake. This mass displaced the water in the lake, causing it to overflow into a lower lake, which, in turn, overflowed into Ophir Creek gorge as a water flood. Picking up sediment as it went, it became a debris flow that emerged from the gorge, spread out, destroyed homes, and covered a major highway.

Landslides That Move as a Unit

The term **landslide** encompasses all moderately rapid mass movements that have well-defined boundaries, and that move downward and outward from a natural or artificial slope. Landslides can range widely in size from a small planar failure no bigger than a car to one of the largest landslides on record, the sector failure of Mount St. Helens, during which the entire north side of the mountain slid away. Landslides occur in all 50 states and are an economically significant factor in more than 25 of them (see Figure 7.1). Areas with the highest slopes are at greatest risk, because gravity is the force that drives landslides. For example, more than 2 million landslide scars dot the steep slopes of the Appalachian Mountains between New England and Alabama. Almost 10% of the land area of Colorado is landslide terrain, and landslides and debris flows in Utah caused $300 million worth of damage in 1983–1984. In Southern California, eight wet years between 1950 and 1993 averaged $500 million per year in landslide damage. Slope stability problems have such an impact that a national landslide-loss–reduction program was implemented by the U.S. Geological Survey in the mid-1980s. A high priority of the program is to identify and map landslide-prone lands.

Figure 7.8 Fire, flood, and flow. (a) Wildfires rage in the forest and chaparral above Los Angeles. Once scorched and the vegetation removed, the land is primed to generate debris flows. (b) Bare slopes, in Carbon Canyon, denuded by the Yorba Linda–Corona fire. With the coming of heavy winter rains, fire-scorched slopes like this are more likely to produce debris flows than those covered by mature vegetation. (c) A debris flow, caused by heavy rain on the mountains above and carrying rocks and mud, overflows stream channel as emergency responders and homeowners look on; Glendora, California. (d) Mudflows triggered by heavy rains on steep slopes, especially those denuded by fire, are common in the Los Angeles Basin. In December 2010, this car was caught in a mudflow near Highland, California. Damage estimates from this one storm exceeded $10 million. (e) Mudslides in Glendora, California, a town at the base of the San Gabriel Mountains, flowed down a street, burying the homes on each side of the street in mud.

Types

Geologists recognize two kinds of coherent slide masses, distinguished by the shape of the slide surface. They are **slumps**, or **rotational slides**, which move on curved, concave-upward slide surfaces and are self-stabilizing, and **block glides**, or **translational slides**, which move on inclined slide planes (see Figure 7.2). A block glide moves as a unit until it meets an obstacle, the slope of the slide plane changes, or the block disintegrates into a debris flow.

Slumps—Slumps are the most common kind of landslide, and they range in size from small features a few meters (feet) wide to huge failures that can damage structures and transportation systems. They are spoon-shaped, having a slide surface that is curved concave-upward, and exhibit a backward rotation (**Figure 7.9a**). The rotation part is key to the definition. If you cleaned away the landslide and looked just at the failure plane, it would be curving and bowl-shaped. Slumps occur in geological material that is fairly homogeneous, such as soil or intensely weathered or fractured bedrock or clay, and the slide surface cuts across geological boundaries. Slumps move about a center of rotation; that is, as the toe of the slide rotates upward, its mass eventually counterbalances the downward force, causing the slide to stop (see Figure 7.9b). Slumping produces repeated uniform depressions and flat areas on otherwise sloping ground surfaces. Slumps grow headward (upslope) and are encouraged by undercutting at the toe. This is because, as one slump mass forms, it removes support from the slope above it. Thus, a stair-step surface is common in landslide terrain. Slump-type landslides are the scourge of highway builders, and repairing or removing them costs millions of dollars every year.

Block glides—Block glides are coherent masses of rock or soil that move along relatively planar sliding surfaces (failure planes), which may be sedimentary rock bedding planes, metamorphic rock foliation planes, faults, or fracture surfaces. Block glides are translational landslides—imagine the landslide block moving down a plane, just like sliding your cell phone down the cover of your notebook. For a block glide to occur, it is necessary that the failure plane be inclined *less steeply* than the inclination of the natural or manufactured hill slope. Slopes may be stable with respect to block glides until they are steepened by excavation for building subdivisions or roads, thus leading to landsliding (**Figure 7.10**).

A classic block glide of about 10 acres is seen along a sea cliff undercut by wave action at Point Fermin in San Pedro, California (**Figure 7.11**). Movement was first detected there in January 1929, and by 1930, the landslide had moved 2 meters (7 feet) seaward. It has been intermittently active ever since. The rock of the slide mass is coarse sandstone (not the type of earth material usually involved in block glides), but a thin layer of **bentonite** clay dipping 15° seaward forms the slide plane. Bentonite was volcanic ash that has chemically weathered to clay minerals, which become very weak and slippery when wet. Bentonite is commonly involved in slope failures; addition of water is all that is needed to initiate a landslide where the slope of the bentonite layer is as little as 5°.

Some of these translational slides, an alias by which fast-moving block glides are also known, are large and move with devastating speed and tragic results. In 1985, tropical storm Isabel dumped a near-record 24-hour rainfall that averaged almost 470 millimeters (18.5 inches) on a mountainous region near the city of Ponce on the south coast of Puerto Rico. Rainfall intensities peaked in the early morning hours of 7 October, reaching 70 millimeters (2.8 inches) per hour in some places. At 3:30 that Monday morning, much of the Mameyes residential district was destroyed by a fast-moving rock block glide initiated during the most intense period of rainfall. This resulted in the worst loss of life from a landslide in U.S. history—129 deaths. The landslide was in sandstone with stratification that parallels the natural slope of the slide mass, a condition called a *dip slope*. It moved at least 50 meters (165 feet), probably on a clay layer in the sandstone, before breaking up into the large blocks. The scarp at the top of the slide was 10 meters (32 feet) high, and the maximum thickness observed at the toe of the landslide was 15 meters (49 feet).

A similar landslide occurred in the Alps of northeastern Italy in 1963 at Vajont Dam, the highest thin-arch dam in the world at the time (275 meters [900 feet]). The geology at the reservoir consists of a sedimentary-rock structure that is bowed downward into a concave-upward **syncline** with its axis parallel to the Vajont River canyon. Limestones con-

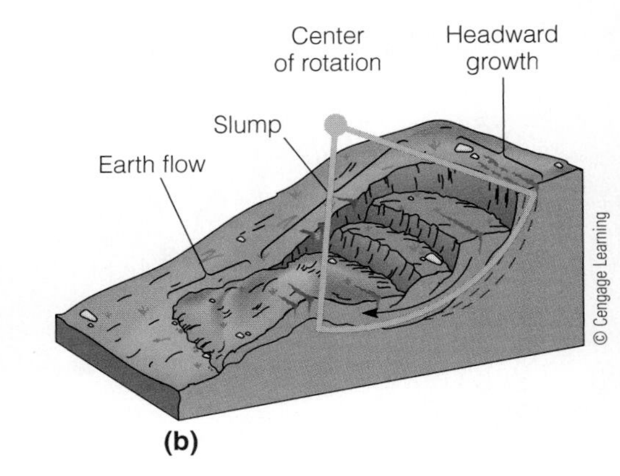

Crown
Toe
(a)

Figure 7.9 A rotational landslide: (a) an oblique stylized view, showing the ideal spoon shape and crown and toe of the landslide; (b) an oblique view of a rotational landslide (slump), showing toe with earth flow and potential headward growth of the landslide.

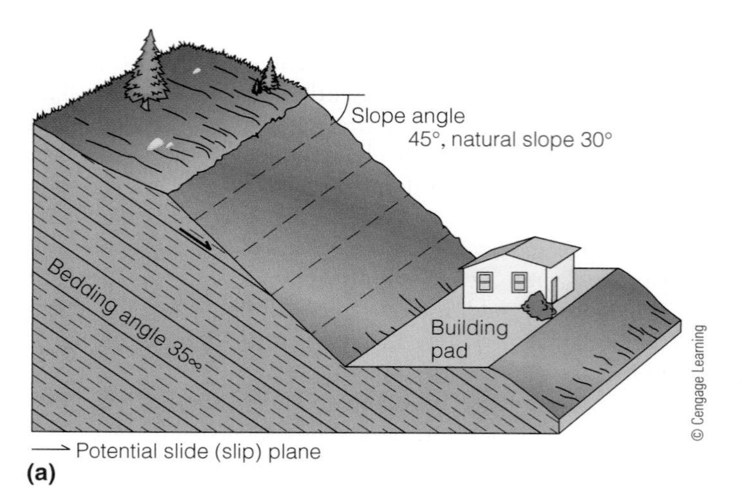

Figure 7.10 Cutting a natural slope for homesites can lead to landslides on bedding or foliation planes. (a) After cutting a natural slope for a building pad, the manufactured slope exposes potential slide planes. (b) A slide plane (smooth surface) and moving mass of shale after toe of the slide was cut away; Santa Monica Mountains, California. (c) Striations gouged in clay on the slide plane by the sliding mass above it, as shown in (b).

B. Pipkin

Figure 7.11 The Point Fermin landslide at San Pedro, California, is a rock block glide that has moved intermittently since 1929. Damaged homes have been removed from the slide area in the foreground. The slide plane emerges just in front of the houses in the background, which are on relatively stable ground.

taining clay layers dip toward the river and reservoir from both sides of the canyon because of this structure (**Figure 7.12**).

Slope movements and slippage along clayey bedding planes above the reservoir had been observed before the dam was constructed. This condition gave engineers and geologists sufficient concern that they placed survey monuments on the slope above the dam for monitoring such movement. Heavy rains fell for two weeks before the disaster, and slope movements as large as 80 centimeters (31 inches) per day were recorded as slow-moving block glides began.

Then, on the night of 9 October, without warning, the speed of the glides dramatically accelerated, and a huge mass of limestone slid into the reservoir so fast it generated a wave 100 meters (330 feet) high. The wave burst over the top of the dam and flowed into the Piave River valley, destroying villages in its path and leaving 2,500 people dead. The landslide velocity into the reservoir was so great and the slide block so large that the slide mass almost emptied the lake. The dam did not fail and is still standing today—a monument to excellent engineering but poor site selection.

Slump (rotation) and block glides (translation) are the most common types of landslides, but there are also **complex landslides**, combinations of the two types. These landslides have elements of both rotational and translational failure, and there are many examples, most of them large landslides, such as the Slumgullion landslide in Colorado (see Case Study 7.2).

Falls—Rockfalls are the simplest type of mass movement. They occur when a mass of rock detaches from a cliff and falls to the bottom. They usually occur very rapidly and are among the most common of mass-wasting processes. Road signs alerting motorists to watch for falling rocks are a frequent sight. Chunks of rock of all sizes fall from vertical or very steep cliffs, having separated from the main mass along joints, faults, foliation planes, and other rock defects. The rocks are loosened by root growth, frost wedging, and heavy precipitation. A pile of rock debris may build up at the base of a cliff, forming a slope called **talus** (**Figure 7.13**). Two huge rockfalls in Yosemite Valley in 1996 killed 1 visitor and injured 14 others. The force of the shock wave from one of these rockfalls toppled nearby trees and stripped bark from more distant ones.

Rockfalls may well increase as a side effect of global warming. Here's how. As the Earth warms, alpine glaciers are retreating ever more rapidly (most alpine glaciers have been in retreat since the Little Ice Age began waning several hundred years ago). As these glaciers melt away, they leave steep bedrock walls in the canyons they used to occupy. Without the buttressing of ice, these walls are often unsta-

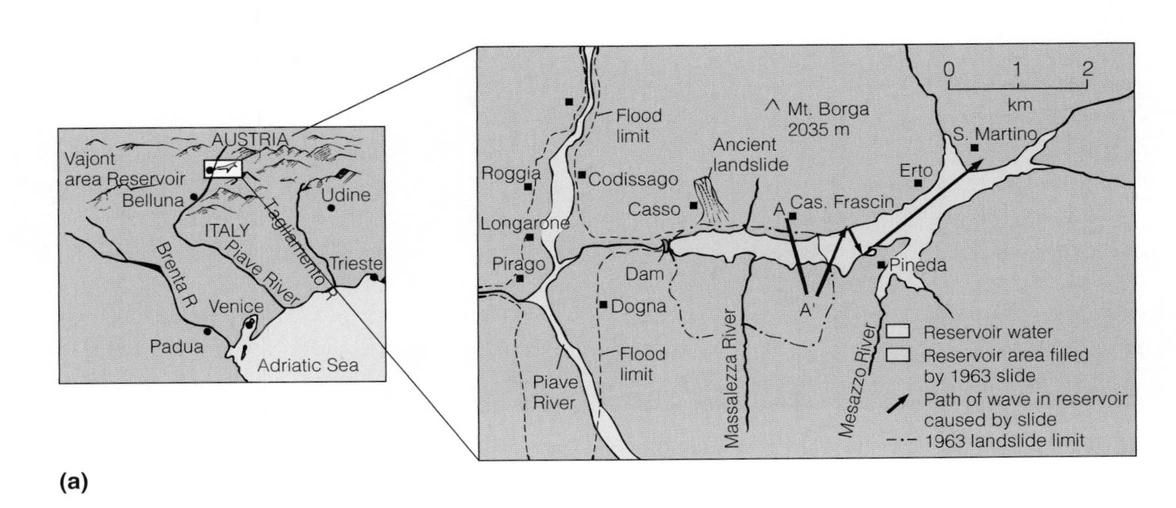

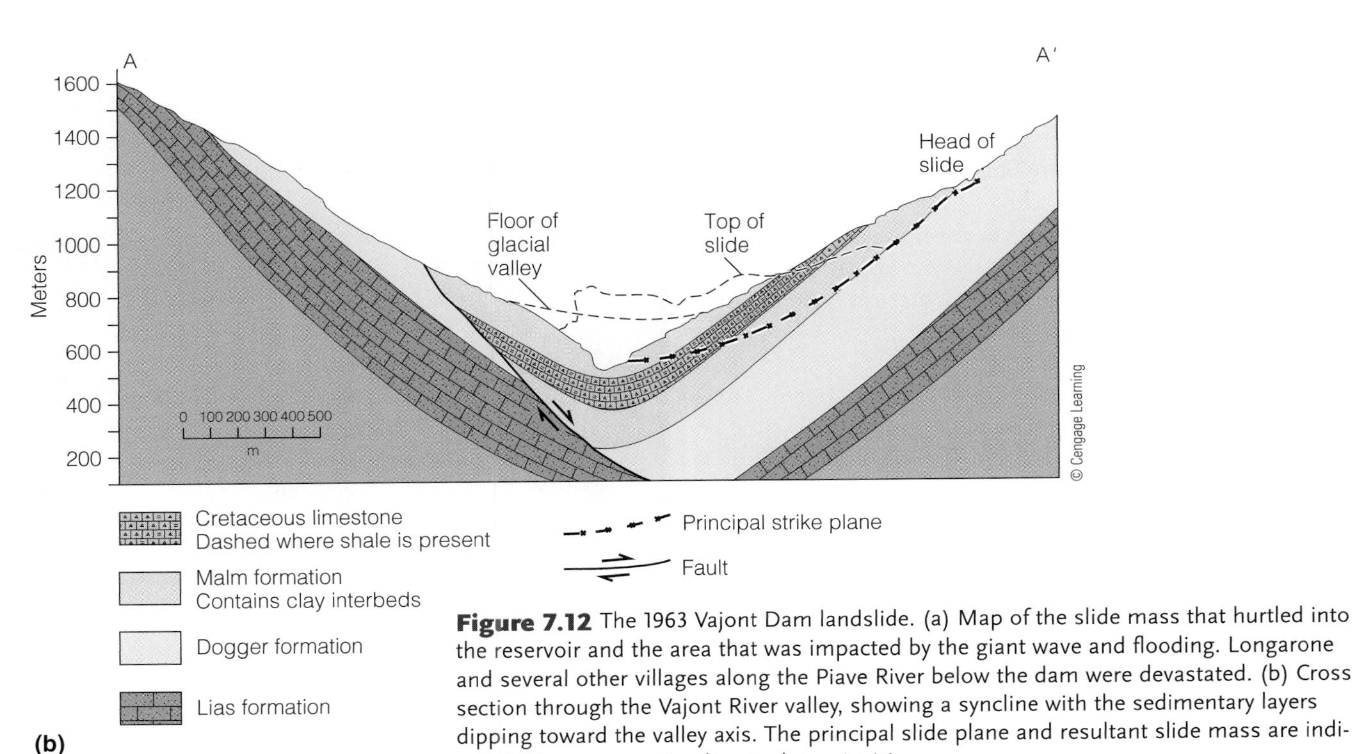

Figure 7.12 The 1963 Vajont Dam landslide. (a) Map of the slide mass that hurtled into the reservoir and the area that was impacted by the giant wave and flooding. Longarone and several other villages along the Piave River below the dam were devastated. (b) Cross section through the Vajont River valley, showing a syncline with the sedimentary layers dipping toward the valley axis. The principal slide plane and resultant slide mass are indicated. The cross section line is shown in (a).

Figure 7.13 Talus slopes caused by frost wedging and other physical weathering processes breaking apart intact rock on the cliffs above; Banff National Park in the Canadian Rocky Mountains.

ble. In summer 2006, the famous Eiger face in the Swiss Alps, a victim of glacial retreat, collapsed, dumping 20 million cubic feet of rock on the glacier below, equal to about half the volume of the Empire State Building (**Figure 7.14**).

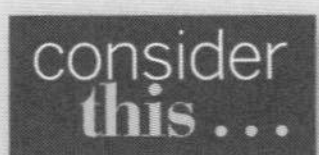

You must choose between two hillside homesites. One site is underlain by bedded sandstone, the other by homogeneous granite. Which site would you choose, and why?

Lateral Spreading

In just the right geological conditions, nearly flat ground can give way and collapse into a morass of sliding blocks. In what are termed **lateral spreads**, coherent blocks move horizontally, overlying a liquefied layer of sediment at depth. Such spreads involve elements of translation, rotation, and flow. Typically triggered by earthquakes, lateral spreading results from the spontaneous liquefaction of water-saturated sand layers or the collapse of **sensitive clays**—also known as **quick clays**—which lose all strength when they are shaken (see Chapter 4). Much of the damage in the Marina district of San Francisco in the 1989 earthquake was due to lateral

Figure 7.14 With climate warming, melting glaciers have left crumbling rock faces unsupported. Soon after the glacial ice melts, the rock gives way. Here, hikers point to the massive fractures on the face of Eiger mountain, where hundreds of tons of rock fell soon after; above Grindelwald, Switzerland, July 2006.

Figure 7.15 Turnagain Heights as it appeared after the quick-clay lateral spread triggered by the 1964 Alaskan earthquake.

spreading caused by the liquefaction of debris piled there after the 1906 earthquake (and then built upon). During a massive 1964 quake in Anchorage, Alaska, the most dramatic damage was due to quick-clay–induced lateral spreads in the Turnagain Heights residential neighborhood and the downtown area. The spreading was so extensive that two houses that had been more than 200 meters (650 feet, more than two football fields) apart collided within the slide mass. The Turnagain Heights area is now a tourist attraction known as "Earthquake Park" (**Figure 7.15**). Such quick clays are usually marine in origin and can also be found in Norway and eastern Canada.

The Mechanics of Slides: A Matter of Balance

Let's briefly examine the physics of landslides, because by understanding how landslides work, you will be better able to understand how they can be predicted and controlled. Landslides occur when the balance of forces shifts (**Figure 7.16**). Actually, there are only two forces to consider. One is the **driving force**, pulling the landslide block downslope. That force is Earth's gravity filtered through the steepness of the slope. Opposing the driving force is the **resisting force**: the strength of whatever material makes up the landslide. So there we have it, a face-off between gravity and the strength of earth materials. If the slide material is strong and the driving force is weak (say, the slope is gentle), nothing happens, the potential landslide is stable, and it will remain so until something changes the balance. If the slide mass is just stronger than the driving force, then the landslide is primed to fail. If things keep changing, when the driving force exceeds the resisting force, the landslide occurs. It's all about balance.

Now, let's understand more about what goes into this balance. First, consider the driving force. As mentioned earlier, gravity is the big player. The gravity force is directed downward, toward the center of Earth. Indeed, it's what keeps your feet planted solidly on the ground (see Figure 7.16a). Now imagine a block-glide type of landslide on a 15° slope (see Figure 7.16b). We can partition that gravity force into two separate forces: One is oriented downslope, trying to get the block to slide, and the other is pushing onto the slope; this one is called the **normal force**, and it helps anchor the slide block. Now, let's crank up the slope to 45° (see Figure 7.16c). The partitioning of forces has now changed. More of

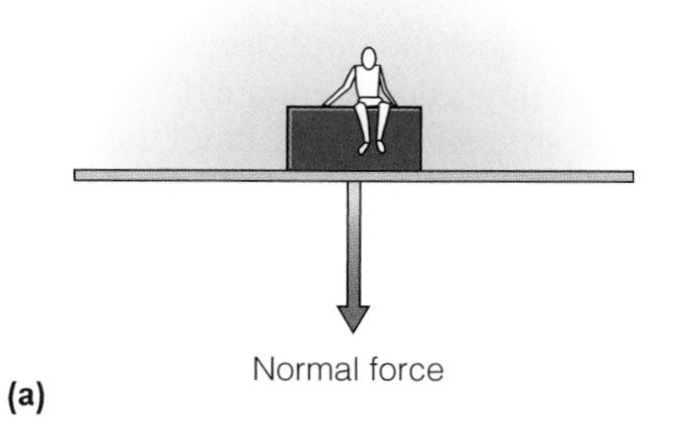

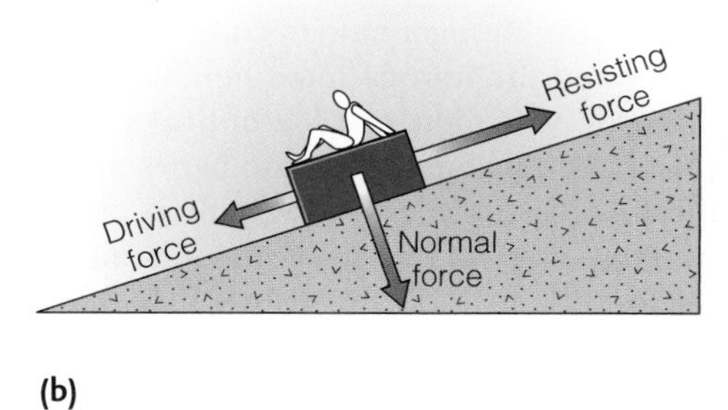

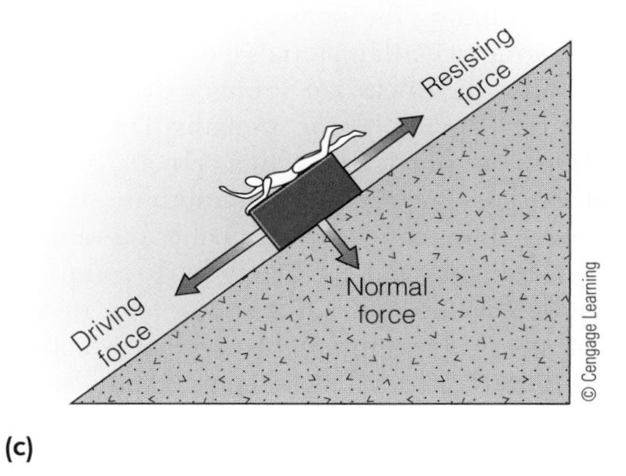

Figure 7.16 Landslide physics is all about balance. In the diagrams, the red block represents the landsliding material. (a) In the middle of the Great Plains, all of the gravity force is directed straight down and there are no landslides. Long arrows mean big forces. Small arrows mean little forces. (b) As slopes steepen, some of the gravity force is directed downslope and makes up the driving force. Some of the gravity force is directed into the slope as the normal force. (c) Here's a slide about to happen on a steep slope. Look carefully at the length of the arrows, compared with (b). What happens to the driving, normal, and resisting forces? How do you know this slope is about to fail?

the gravity force is oriented downslope and less into the slope. Imagine what the partitioning must look like on a 70° slope. The steeper the slope, the more the gravity force is directed downslope, and the more likely the landslide will fail.

The resisting force is all about the physical character of the rock or soil that makes up the landslide. It has two components; both are familiar through life experiences. The first component is cohesion, or the material's tendency to hold together. Consider clay. It has **cohesion** and can hold a vertical face; you can mold and shape it. Now consider sand, a common **granular material** with no cohesion. It has what is termed **frictional strength**, the result of frictional interaction between grains. The harder you push the grains together, the stronger sand becomes. Thus, as the slope gets steeper and the normal force holding sand grains together decreases, you soon reach a critical point. For sand, that critical point where the friction between grains is no longer able to resist the downslope pull is about 35°, which is termed its **angle of repose**. Try as you might, you'll never be able to stack dry sand any steeper than 35°, nor can nature.

Most materials that make up landslides have both cohesion and frictional strength. Even granular materials, such as sand and gravel, have small amounts of cohesive strength in nature where networks of tree roots provide effective cohesion, binding the grains together (see Case Study 7.3). The strongest earth materials have both frictional and cohesive strength. Rocks are stronger than soils, because they have very high cohesive strengths. For sedimentary rocks, cohesion is provided by the cements holding grains together. For igneous and metamorphic rocks, interlocking mineral grains cause the high cohesion values.

Geologists and engineers have come up with many ways to quantify this balance of forces, most of which involve physical measurements and mathematical models. All of this experimentation and computation can be reduced to one number, referred to as the **factor of safety**, which is a ratio between the resisting and driving forces. When the two forces are equal, the factor of safety equals 1.0, and the slide is teetering on the brink of failure. When resisting forces are greater than driving forces, the factor of safety is greater than 1.0 and all is well. The moment the resisting forces are less than the driving forces, the factor of safety is less than 1.0 and the landslide starts. Most modern building codes require that manufactured and natural slopes have a factor of safety of 1.5 or greater to provide a margin of error in the measurements and calculations.

Several times in this chapter, we have mentioned that water has a great influence on slope stability and that, in wet years, landslides are very common. Why is this? Let's go back to the force balance and your personal experience in swimming pools and at the beach. When rainfall saturates a landslide block with water, some of the weight of the grains is buoyed up by that water, just as when you are floating in water. This buoyancy is caused by the pressure of groundwater in the soil and rock pores; it's referred to as **pore pressure**, and it partially offsets the gravity force holding the slide on the slope. Because the water pressure supports a portion of the block's weight, it reduces the normal force, thus reducing the frictional strength of the material and the all-important resisting force. Sometimes this reduction in strength can be just enough to tip the balance, drop the factor of safety below 1.0, and trigger the slide. The greatest effect of heavy rains, raising groundwater tables and increasing pore pressure, is on landslides that are made up of low-cohesion material, such as sand and gravel or weathered rock.

To trigger a landslide, something must change the factor of safety so that it declines to less than 1.0 and driving forces exceed resisting forces. A short-term change in this balance is most often driven by water. Water usually saturates slopes as a result of heavy rain, but occasionally human errors (such as leaky swimming pools and sprinkler systems) are the cause. Other human triggers of landslides include cutting away slide toes (which is equivalent to steepening the slope) and adding weight at the top of a slope, such as by building a house, which increases the driving force. Over the longer term, on geological time scales of millennia or more, minerals in rock and soil will weather, weakening the material, especially if the weathering products are clay. This is the underlying factor in many basalt landslides, such as those that plague parts of the Oregon Coast Range (see Case Study 7.1).

Reducing Losses from Mass Wasting

Landslide Hazard Zonation

The first step in reducing losses caused by mass wasting over the long term is to make a careful inventory or map of known landslide areas. Ancient landslides have a distinct topography characterized by hummocky (bumpy) terrain with knobs and closed depressions formed by backward-rotated slump blocks. On topographic maps, slide terrain has curved, closely spaced, amphitheater-shaped contour lines indicating scarps at the heads of landslides. In the field, one may see scarps, trees, or structures that are tilted uphill; unexpected anomalous flat areas; and water-loving vegetation, such as cattails, where water emerges at the toe of a landslide.

With an adequate database provided by maps and field observations, it is possible to define landslide hazard zones. Maps can be generated that show the *landslide risk* in an area; then these areas can be avoided or work can be done to stabilize them (**Figure 7.17**). Mapping programs do not prevent slides, but they offer a means for minimizing the impact on humans. The human tragedy of landslides in urban areas is that they cause the loss of land, as well as of homes. **Table 7.1** summarizes mitigating measures that can slow or stop landsliding and other mass-wasting processes.

Building Codes and Regulations

Building codes dictate which site investigations geologists and engineers must perform and the way structures must be built. Chapter 70 of the *Uniform Building Code*, which deals

Table 7.1 Summary of Mass-Wasting Processes and Their Mitigation

Landslides		
Cause	**Effect**	**Mitigation**
Excess water	Decreased strength	Horizontal drains, surface sealing
Added weight at top	Increased driving force	Buttress fill, retaining walls, decrease slope angle
Undercut toe of slope	Decreased resistance	Retaining walls, buttress fill
"Daylighted" bedding	Exposure of unsupported bedding planes	Buttress fill, retaining walls, decrease slope angle, bolt

Other Mass-Wasting Processes	
Process	**Mitigation**
Rockfall	Rock bolts and wire mesh on the slope, concrete or wooden cribbing at bottom of the slope, cover with "shotcrete"
Debris flow	Diversion walls or fences and catch basins
Soil creep	Deep foundations
Lateral spreading	Dewater, buttress, abandon

with slopes and alteration of the landscape, has been widely adopted by cities and counties in the United States. It specifies compaction and surface-drainage requirements, and the relationships between such planar elements as bedding, foliation, faults, and slope orientation. Code requirements for manufactured slopes in Los Angeles resemble the specifications found in codes elsewhere (**Figure 7.18**). The maximum slope angle allowed by code in landslide-prone country is 2:1—that is, 2 feet horizontally for each foot vertically (27°). The rationale for the 2:1 slope standard is that granular earth materials (such as sand and gravel) have a natural angle of repose of about 34°; 2:1 slopes allow for a factor of safety.

Losses have been cut dramatically by the enforcement of grading codes, and they will continue to decrease as codes become more restrictive in hillside areas. For instance, before grading codes were established in Los Angeles in 1952, 1,040 building sites were damaged or lost by slope failures out of each 10,000 constructed—a loss rate of 10.4%. With a new code in effect between 1952 and 1962 that required minimal geological and soil-engineering investigation, losses were reduced to 1.3% for new construction. In 1963, the city enacted a revised code requiring extensive geological and soils investigations. Losses were reduced to 0.15% of new construction in the following 6 years, which included 1969, the largest rain year in recent history.

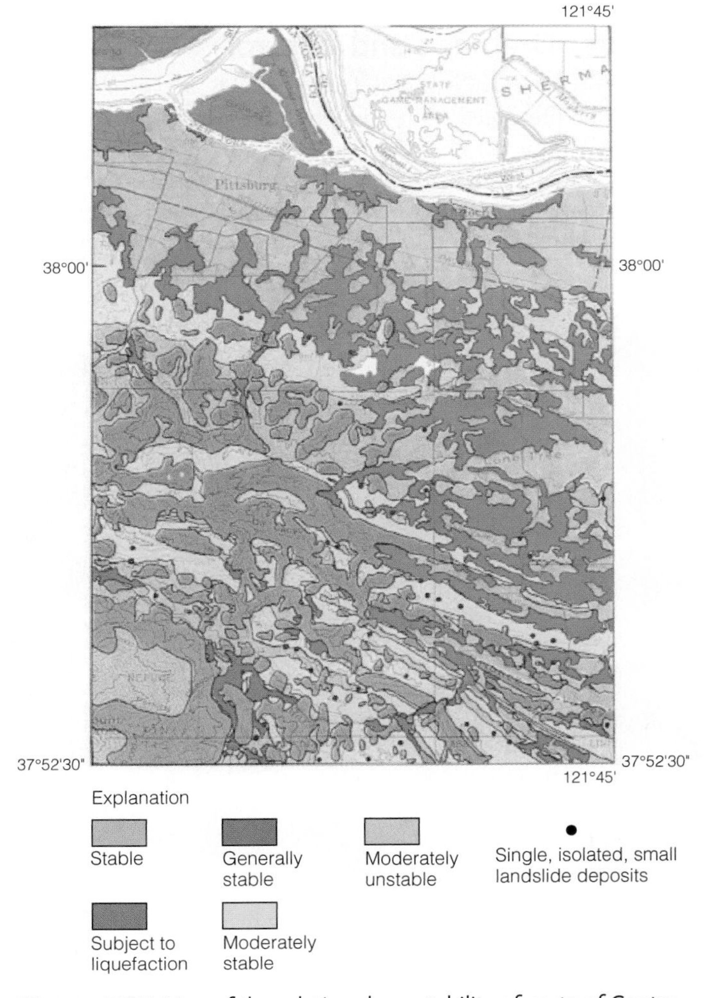

Figure 7.17 Map of the relative slope stability of parts of Contra Costa County and adjacent counties near San Francisco, California, derived by combining a slope map, a landslide-deposit map, and a map of susceptible geological units. The area shown here is 15 kilometers (9 miles) wide. (USGS)

Control and Stabilization

The basic principles behind all methods of preventing or correcting landslides relate to strengthening the earth materials (increasing the resisting forces) and reducing the stresses within the system (decreasing the driving forces). For a given slide mass, this requires halting or reversing the factors that promote instability, which will probably be accomplished by one or more of the following: draining water from the slide area, excavating and redistributing the slide

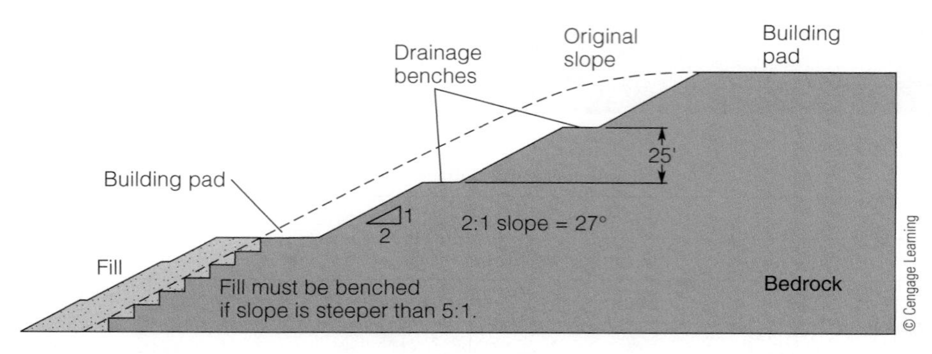

Figure 7.18 Part of a typical grading ordinance (code) that shows the limitation of cut-slope steepness, the required benches on slopes for collecting rainfall and reducing slope-face erosion, and an approved method of constructing a fill slope with bedrock benches for increased stability.

mass, and installing retaining devices. Debris flows require large structures that either divert the path of the flow away from buildings or capture the debris in basins.

Water Drainage and Control

Water is the major culprit in land instability, and surface water must be prevented from infiltrating the potential slide mass. Water within the slide mass can be removed by drilling horizontal drains, called *hydrauger* (water bore) holes, and lining them with perforated plastic drainpipe that weeps water. This is commonly done on hillside terrain where underground water presents a problem. In some areas, the land surface of a building site above a suspect slope is sealed with compacted fill to help keep surface water from percolating underground. Also, plastic sheeting is widely used to cover cracks and prevent water infiltration and erosion of the slide mass (**Figure 7.19a**). The efficacy of drainage control and planting is shown in Figure 7.19b.

Excavation and Redistribution

Recontouring is one method of stabilizing a slide mass. Material is removed from the top of the landslide and placed at the toe. Compacted-earth structures called *buttress fills* are often designed to retain large, active landslides and known inactive ones believed to be vulnerable. They are constructed by removing the toe of a slide and replacing it, layer by layer, with compacted soil. Such fills can "buttress" huge slide masses, thus reclaiming otherwise unusable land. The finished product is required to have concrete terraces (for intercepting rainwater and preventing erosion) and downdrains for conveying the intercepted water off the site (**Figure 7.20**). Buttressing unstable land has aesthetic and economic advantages, in that the resulting slope can be landscaped and built upon. The cost of the additional work needed to stabilize unstable land is usually offset by a relatively low purchase price, which can make it economically "buildable." Of course, the ultimate mitigating measure is total removal of the slide mass and reshaping of the land to buildable contours.

Retaining Devices

Many slopes are oversteepened by cutting back at the toe, usually to obtain more flat building area. The vertical cut can be supported by constructing steel-reinforced concrete-

Figure 7.19 Excess water is the culprit in initiating many landslides. (a) Plastic sheeting placed over the head scarp of a landslide. This is, perhaps, closing the barn door after the horse has escaped but is common practice in landslide-prone areas. (b) Surface-water-erosion-prevention measures on a graded slope; San Clemente, California. Whereas slope planting and terracing have kept the middle slope in good condition, the slope on the left has gullied badly and will need maintenance in the future.

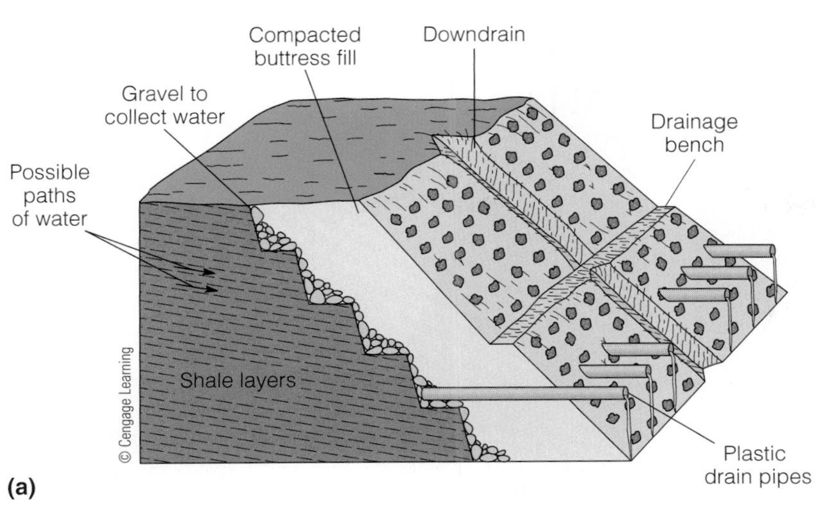

Figure 7.20 (a) Buttress fill with horizontal drains for removing water and surface drains for preventing erosion. The fill acts as a retaining wall to hold up unstable slopes, such as those that result from excavation. (b) A slope with benches and surface drains built according to accepted standards.

block retaining walls with drain (weep) holes for alleviating water-pressure buildup behind the structure (**Figure 7.21a**). Retaining devices are constructed of a variety of materials, including rock, timber, metal, and wire-mesh fencing. An interesting, if not particularly attractive, means of stabilizing weak slopes is the application of a cement cover. This material, known as shotcrete, is sprayed onto rock slopes, where it hardens into a skin. Steel nets can be used to catch falling rocks and protect people and roads below steep cliffs (see Figure 7.21b).

Steep or vertical rock slopes that are jointed or fractured are often strengthened by inserting long rock bolts into holes drilled perpendicular to planes of weakness. This binds the planes of weakness together and prevents slippage. Rock bolts are used extensively to support tunnel and mine openings. They add considerably to the safety of these operations by preventing sudden rock "popouts." They are also installed on steep roadcuts to prevent rockfalls onto high-

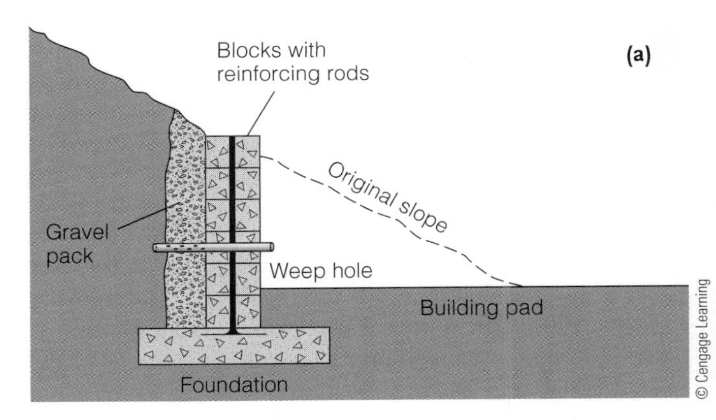

Figure 7.21 (a) Typical retaining wall with drain (weep) holes to prevent water retention and the buildup of hydrostatic pressure behind the wall. The wall is constructed of concrete blocks with steel reinforcing rods and concrete in their hollow centers. (b) Along the steep, rocky sea coast just outside of Cape Town, South Africa, is an amazing road hung along steep, rock sea cliffs. Rockfalls here are common. In addition to rock bolting, rock fall hazard is mitigated by the use of strategically placed steel nets that catch falling rocks before they land on the road or cars below.

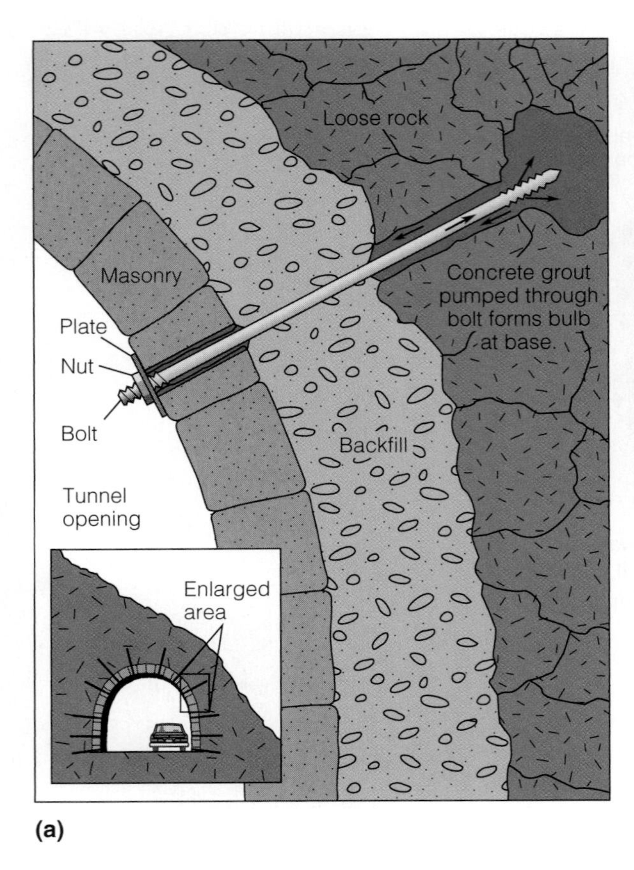

(a)

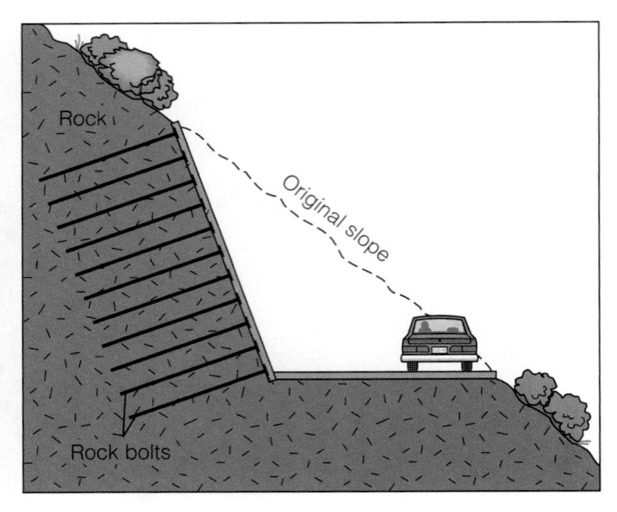

(b)

Figure 7.22 (a) Typical rock bolt installation in a tunnel or mine opening. (b) Rock bolts support jointed rocks above a highway.

ways (**Figure 7.22**). Retaining devices are also effective for mitigating rock and debris falls.

Mitigating Debris Flows

Mitigating debris flows presents a different challenge than landslides, because flows can originate some distance away from the site of interest. How can one avoid debris-flow hazards? The most obvious answer is to live away from steep slopes, and if you choose to live in a mountainous area, stay away from canyon bottoms and the mouths of canyons that act as conduits for debris flows. Home sites by streams seem attractive until that stream is filled with a raging mixture of rock and mud. This is particularly true in areas of high mean annual rainfall, areas that are subject to sudden cloudbursts, and areas that have been burned, such as in the tragic October 1993 firestorms in Southern California. As long as people continue to build houses in the "barrels" of canyons, as is done in the foothills of Southern California, debris flows will continue to take their toll.

Debris-flow insurance is expensive, but the cost can be lowered if *deflecting walls* are designed into house plans (**Figure 7.23a**). These walls have proved to be effective in protecting dwellings, but they can have adverse consequences if they divert debris onto neighbors' structures, in which case neither an architect nor a geologist would be as helpful as a lawyer. Los Angeles has taken a larger scale approach to debris-flow mitigation. It has built a series of large catch basins designed to trap debris flows before they can harm property (see Figure 7.23b). These basins are situated in the upper reaches of deep canyons that drain the mountains and, until they fill, are effective means of capturing flows. Of course, the basins need to be cleaned out with great frequency and the material dumped somewhere else. Some have argued that the debris basins, similar to flood-control levees, provide a false sense of security for those living beneath them and encourage development in what are actually very hazardous places to live. John McPhee's book *The Control of Nature* has wonderful descriptions of these basins, including stories of how they do and do not work.

consider this . . . In many parts of the world, large numbers of people live on or close to potentially unstable hillslopes. How do grading codes help reduce the risk for landslides? Do you think they are effective if they are enforced?

Snow Avalanches

The mechanics of snow avalanches are similar to those of landsliding, the differences being in the material and the velocity. Although snow avalanches have been around as long as there have been mountains and snow, the relatively recent emergence of the recreational skiing industry as big business and the growing interest in backcountry over-snow travel has made avalanche awareness and control all the more critical (**Figure 7.24**). Unfortunately, it's too late for Hannibal. His crossing of the Alps in 218 B.C.E. was plagued by avalanches. Purportedly, 18,000 of his troops and many of his elephants were killed by avalanches. Modern generals may have learned from Hannibal! During World War I, 6,000 troops were buried in one day by avalanches in the Dolomite Mountains of northern Italy. During this war,

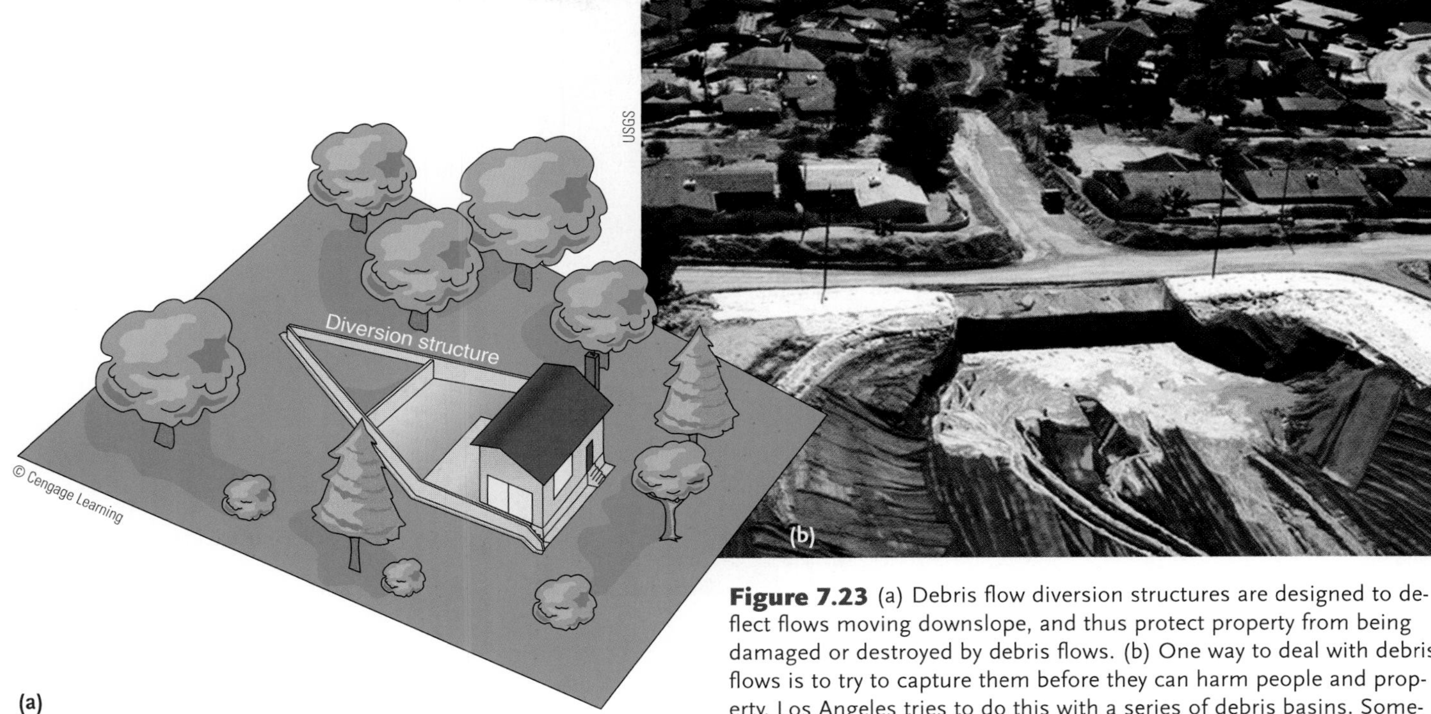

Figure 7.23 (a) Debris flow diversion structures are designed to deflect flows moving downslope, and thus protect property from being damaged or destroyed by debris flows. (b) One way to deal with debris flows is to try to capture them before they can harm people and property. Los Angeles tries to do this with a series of debris basins. Sometimes this approach works well, but other times the flows are so large that they overtop the basins and cover the houses below. Can you find the result of just such an overtopping in this photograph?

armies intentionally triggered avalanches, with the hope of burying their opponents. Avalanches in high mountain regions have wiped out villages, disturbed railroad track alignments, blocked roads, and otherwise made life difficult for millennia. Even eastern North America is affected. In many years in New Hampshire, on Mount Washington, above the tree line, a skier or climber dies in Tuckerman Ravine when an avalanche buries him or her alive.

Snow stability is determined by exactly the same factors that determine the stability of landslides—the driving and resisting forces. In the case of snow, strength changes can happen in just a few hours as snow warms and weakens or wet snow freezes into a strong, solid mass. Many avalanches are caused by an addition of new snow (an increase in driving force). Indeed, the most hazardous time to be in the mountains is when it's snowing hard and the wind is blowing strongly. During such high avalanche-hazard conditions, snow is rapidly loaded onto slopes and avalanches may occur constantly. Avalanches have one resisting force that landslides do not—**terrain anchors**. Trees are probably the most effective terrain anchor, but large rocks can help, too. These roughness elements on the landscape hold snow and minimize avalanche hazard by providing an additional resisting force.

Most avalanches run on weak layers in the snowpack. One such weak layer is an ice crust to which the snow above is not well bonded. **Ice crusts** can form from freezing rainstorms or when bright sun melts the snow, which then refreezes. Other weak layers are **surface hoar**, the delicate crystals of ice that form on top of the snow during cold, moist nights, and depth hoar. **Depth hoar** forms when the ground beneath the snowpack is moist and warm, and the air above the snowpack is very cold. Water vapor moving upward forms a layer of very large and weak crystals, which can fail catastrophically. Depth hoar is common at high altitude, in thin snowpacks, and on north-facing slopes in places such as the Rocky Mountains and the Wasatch Range. In these situations, the gradient in temperature and moisture is extreme between the ground and the air.

There are several kinds of avalanches. A **slab avalanche** is a coherent mass of snow and ice that hurtles down a slope—much like a magazine sliding off a tilted coffee table. This type of avalanche is most frequent on slopes between 35° and 40°, just about black-diamond territory (see Figure 7.24b). Slab avalanches account for the majority of avalanche deaths because of their size and their tendency to solidify when they stop. **Climax avalanches** occur in spring, when the snowpack melts and weakens. They often run over the ground, snapping off trees and burying anything in their run-out zones with thick, wet, heavy snow. **Powder-snow avalanches** are most common during storms. These "dry" avalanches are dangerous because they entrain a great amount of air and behave as a fluid, traveling far and fast. Furthermore, the strong shock wave that precedes them can uproot trees and flatten structures. Victims of dry avalanches tend to be buried in light snow close to the surface, which makes their chances of being rescued alive better than those of wet-avalanche victims.

Avalanches can be triggered by snow loading, rockfalls, skiers, loud noises, or explosives. Mimicking nature, avalanche control is usually achieved by triggering an avalanche with explosives or cannon fire when dangerous snow conditions exist in areas frequented by people, such as ski areas and roads. The explosions collapse the depth hoar, break loose slabs, or detach powder snow, causing a "controlled" avalanche. Various other means have been developed for

Figure 7.24 Avalanches can be devastating. (a) Avalanches pour down steep slopes above the ski town of Telluride, Colorado. (b) Percentage of slab avalanches versus slope angle in degrees. (Data from Williams, Knox. 2000, October. Avalanche formation and release. *Rock Talk* 3(4).) (c) What a great place to have your home, until the avalanches from the bare slope above come roaring down. Look carefully and you will see the rock prow that defends the upslope side of this home in the Swiss Alps above Davos. (d) In Davos, more intensive avalanche-control measures are used to keep the road open. This is an avalanche shed designed to let the traffic go under and the snow go over.

mitigating the destructive might of avalanches. Where buildings already exist, structures can be built to divert avalanches away. In some areas in the Alps, the uphill sides of chalets are shaped similar to a ship's prow, so that they will divert avalanching snow (see Figure 7.24c). Railroads and highways can be protected by building snow sheds along stretches adjacent to steep slopes (see Figure 7.24d). On some slopes high in the Alps, miles of large concrete, metal, and wood barriers have been built to help anchor snow to the slopes.

The best personal defense against avalanche hazard is to stay away from steep, open slopes, especially when conditions are hazardous. Avalanche forecasts and hot lines exist for most mountain areas near population centers. If you spend time in the backcountry or ski out of bounds, take an

Table 7.2 Subsiding Cities, 1986

World		United States	
City	Maximum Subsidence, Centimeters*	City	Maximum Subsidence, Centimeters
Mexico City	850	Long Beach	900
Tokyo	450	San Joaquin Valley	880
Osaka	300	San Jose	390
Shanghai	263	Houston	270
Niigata, Japan	250	Las Vegas	66
Bangkok	100	Denver	30
Taipei, Taiwan	190	New Orleans	22
London	30	Savannah	20

Source: Dolan, R., and H. Grant Goodell. 1986. Sinking cities. *American Scientist* 74(1). Reproduced with permission of SIGNMA XI.

*Since measurement began.

avalanche safety course and learn how to read the snow. It could very well save your life.

Subsidence

Two million people live below the high-tide level in Tokyo. The canals of Venice overflow periodically and flood its beloved tourist attractions (see Case Study 7.4). Earthen dikes prevent the sea from flooding the land in some places in California, and Mexico City's famous Palace of Fine Arts rests in a large depression in the middle of the city (**Table 7.2**). These areas suffer from natural and human-induced **subsidence**, a sinking or downward settling of Earth's surface. Subsidence is not usually catastrophic, and loss of life due to it is rare, but land subsidence is currently observed in 45 states and was estimated to cost $125 million annually in 1991, the equivalent of nearly $200 million in 2009. In the United States, the area of human-induced subsidence is estimated as at least 44,000 square kilometers (17,000 square miles), about the size of Maryland and New Jersey combined, but the actual area is probably even greater. Planners and decision makers need to be aware of the causes and impacts of subsidence to assess the risks and reduce material losses.

Human-Induced Subsidence

Human-induced subsidence occurs when humans extract underground water, when they engage in mining or oil and gas production, and when they cause loose sediments at the ground surface to consolidate or compress. The effects can be local or regional in scale. In the United States and Mexico alone, an area the size of Vermont has slowly subsided 30 centimeters (1 foot) because of withdrawal of underground water. Sinking of the land changes drainage paths, and it is particularly damaging to coastal areas and lands adjacent to rivers, because it increases flood potential. Subsidence can also result from the compaction of sediments rich in organic matter. Parts of the city of New Orleans, built on cypress swamps of the delta of the Mississippi River and now isolated from the sediment-laden floods of that river, are now below river level and sea level (see Chapter 9 and Case Study 7.5).

Annual losses from subsidence caused by fluid withdrawal along the Texas Gulf Coast, for example, were estimated at $109 million between 1943 and 1973. Lowering of the water table in areas of cavernous limestone and gypsum has caused surface subsidence and collapse, costing many millions of dollars each year (see the Chapter 9 discussion of hydrological hazards). Each year, subsidence of the land surface into abandoned coal mines costs the United States $30 million. About a quarter of the 7 million acres that have been mined for coal has subsided, and some of this area underlies cities.

Natural Subsidence

Natural subsidence is caused by earthquakes, volcanic activity, and the solution of limestone, dolomite, and gypsum. Earthquake-related subsidence occurs rapidly. It is best known in Alaska and California but also occurs in other states, including Oregon and Washington. Displacement along large faults can raise or lower the land surface over a large area. For instance, subsidence of 1 meter over 195,000 square kilometers (70,000 square miles) took place during the Alaskan earthquake of 1964. Much of the subsidence was along the coast, and the subsided area is now flooded at high tide. In fact, the land actually tilted, and a large area of the Gulf of Alaska was uplifted several meters.

Severe ground shaking that leads to liquefaction can also lower the ground surface. This happened along the valley of the Mississippi River during the New Madrid earthquakes of 1811 and 1812, and in the Marina district of San Francisco in 1989. A small-scale subsidence problem is the collapse of roofs of shallow lava tunnels and tubes in volcanic areas. Volcanic activity that empties magma chambers can cause collapse and subsidence over much larger areas when it forms calderas (see Chapter 5). Subsidence caused by tectonic (mountain-building) processes occurs so slowly that it is not considered an environmental hazard.

Classification of Subsidence

One method of classifying land subsidence is according to the depth at which the subsidence is initiated. *Deep subsidence* is initiated at considerable depth below the surface when water, oil, or gas are removed. *Shallow subsidence*, in contrast, takes place nearer the ground surface when underground water or solid material are removed by natural processes or by humans. In addition, many poorly consolidated shallow deposits, such as peat, are subject to compaction by overburden pressures, groundwater withdrawal, and, in some soils, simply by water saturation.

Settlement differs from subsidence. It occurs when an applied load, such as that of a structure, is greater than the bearing capacity of the soil onto which it is placed. Settlement is totally human induced, and it is a soils- and foundation-engineering problem. The Leaning Tower of Pisa is a classic example of foundation settlement; its interesting construction history is discussed in Chapter 6.

Deep subsidence—The removal of fluids—water, oil, or gas—confined in the pore spaces in rock or sediment causes deep subsidence. Porewater pressure—that is, the hydrostatic pressure of water in the pores between sediment grains—helps support the overlying material. As pressure is reduced by extraction, the weight of the overburden gradually transfers to mineral-and-rock-grain boundaries. If the sediment was originally deposited with an open structure, the grains reorient into a closer packed arrangement, thus occupying less space, and subsidence ensues (**Figure 7.25**).

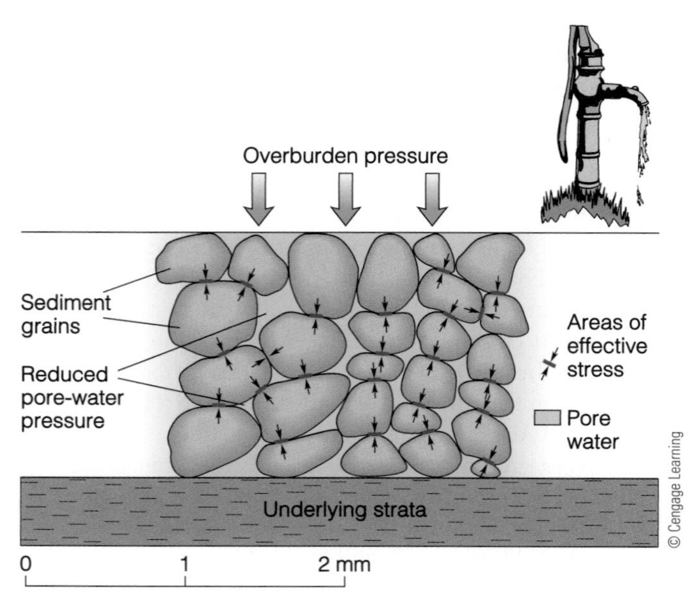

Figure 7.25 A reduction of pore pressure, as groundwater is pumped away, causes increased effective stress and the rearrangement and compaction of sediment particles.

Because clays are more compressible than sands, most compaction takes place in clay layers.

At least 22 oil fields in California have subsidence problems, as do many fields in Texas, Louisiana, and other oil-producing states. A near world record for subsidence is held by the Wilmington Oil Field in Long Beach, California, where the ground has dropped 9 meters (about 30 feet). This field is the largest producer in the state, with more than 2,000 wells tapping oil in an upward-arched geological structure called an *anticline* (see Chapter 14). The arch of the Wilmington field's anticline gradually sagged as oil and water were removed, causing the land surface to sink below sea level. Dikes were constructed to prevent the ocean from flooding the adjacent Port of Los Angeles facilities and the naval shipyard at Terminal Island.

People who have their boats moored on Terminal Island must walk *up* to board them at sea level. When it rains, it is necessary to pump water out of low spots upward to the sea, and oil-well casings and underground pipes have "risen" out of the ground as the land has sunk. To remedy the situation, in 1958, the city of Long Beach initiated a program of injecting water into the ground to replace the oil being withdrawn. Subsidence is no longer a problem, and the surface has even rebounded slightly in places.

Visible effects related to groundwater withdrawals are water-well damage, cracking of long structures (e.g., canals), and the development of tension cracks and fissures. In the Antelope Valley, California, home of Edwards Air Force Base, water levels beneath the dry lake have been lowered by more than 60 meters (200 feet). The dry lakebed at Edwards is the landing strip for the space shuttle and other high-performance aircraft. A ground fissure appeared in the lakebed; initially only a few centimeters wide, the fissure became greatly enlarged by surface-water runoff and erosion. This halted operations on the lakebed until repairs could be made. Similar fissuring has occurred in Arizona, New Mexico, Texas, and Nevada. In Arizona, groundwater withdrawal in some places has caused subsidence of 5 meters (more than 15 feet) and the development of cracks and earth fissures, particularly along the edges of some basins (**Figure 7.26a**). The subsidence has changed natural drainage patterns and caused highways to settle and crack. The magnitude of the subsidence caused by groundwater withdrawal is similar to oil-field subsidence. In the San Joaquin Valley of California, for example, 8.9 meters (29 feet) of subsidence resulted from 50 years of water extraction for agricultural irrigation (see Figure 7.26b). The Houston suburb of Baytown subsided almost 3 meters in the 1900s, and 80 square kilometers (31 square miles) of this coastal region is permanently underwater (see Figure 7.26c).

Mexico City sits in a down-dropped fault valley that has accumulated nearly 2,000 meters (6,600 feet) of lake sediments, mostly of pyroclastic (volcanic) origin. Subsidence occurred there because of the withdrawal of water from these sediments. Total subsidence varies throughout the city, but almost 6 meters (20 feet) of subsidence had occurred by the 1970s in the northeast part of the city, mostly

Figure 7.26 Subsidence wreaks havoc on Earth's surface. (a) An old earth fissure at Queen Creek, Arizona, due to subsidence caused by groundwater removal; the fissure reopened in August 2005 from erosion caused by heavy rain. This is a small segment of a fissure system that is more than a mile (1.5 km) long and locally up to 30 feet (9 meters) wide. (b) Subsidence caused by groundwater withdrawal for agriculture in the San Joaquin Valley, California. The numbers on the pole indicate the ground level for the years indicated. The area is still subsiding, but at a much slower rate. (c) What used to be a suburban lawn is now a pond due to subsidence. The pumping of groundwater in Baytown, Texas, has caused the earth to subside under homes. Look closely and you'll see that it might be a bit difficult to get the cars out of the garage these days.

in the water-bearing top 50 meters (160 feet) of sediment. In 1951, an aqueduct was completed that brought water to the city; as a result, a number of wells were closed. The effect on subsidence rates was dramatic, and by 1970, subsidence had decreased to less than 5 centimeters (2 inches) per year. The famous Our Lady of Guadalupe (the "brown Madonna") Cathedral tilts because its foundation is partly on lava flows and partly on the subsiding lake sediments.

Subsidence can occur for a variety of other reasons as well, related to tectonics and to sediment supply. Great sub-

a: USGS; b: USGS/Brian Atwater; c: Marli Bryant Miller, University of Oregon

duction earthquakes, such as the ones that affected Alaska in 1964 and the Pacific Northwest in the 1700s, changed land levels by meters. In Alaska, whole villages were submerged by quake-induced subsidence. Since 1964, rapid sedimentation has raised land levels somewhat as the abandoned garage shown in **Figure 7.27a** crumbles. In the Pacific Northwest, the last great subduction quake occurred more than a century before anyone built a garage. In this case, prior land level is clearly shown by the submergence of an entire forest into saltwater, killing the trees (see Figure 7.27b). With sediment supply greatly diminished, most of the Mississippi River delta is subsiding, dropping Civil War age forts into the water (see Figure 7.27c).

consider this . . . Subsidence of coastal cities, such as Venice, Houston, and New Orleans, is particularly dangerous because it threatens them with severe flooding. What can be done to prevent or slow such potential catastrophes?

Shallow subsidence—Subsidence that results from the heavy application of irrigation water on certain loose, dry soils is called **hydrocompaction**. The in-place densities of such "collapsible" soils are low, usually less than 1.3 grams/cubic centimeter, about half the density of solid rock and just 30% more dense than fresh water. On initial saturation with water, such as for crop irrigation or the construction of a water canal, the open fabric between the grains collapses and the soil compacts. Subsidence of monumental proportions has occurred because of hydrocompaction. In the United States, it is most common in arid or semiarid regions of the West and Midwest, where soils are dry and moisture seldom penetrates below the root zone. Known areas of hydrocompaction are the Heart Mountain and Riverton areas of Wyoming; Denver, Colorado; the Columbia Basin in Washington; southwest and central Utah; and the San Joaquin Valley of California.

Collapsing soils were a major problem for the designers of the California Aqueduct in the San Joaquin Valley. It was necessary to identify the areas of loose soil that could not be avoided, to maintain a constant ground slope for the flow of water. Sediments were field-tested by saturating large plots and measuring subsidence at various levels beneath the surface. Subsidence was found to average 4.2 meters (close to 14 feet) over a period of 1.3 years, during which thousands of gallons of water had been applied to the test plots. On the basis of the field tests, 180 kilometers (112 miles) of the aqueduct alignment in the valley were "presubsided" by soaking the ground 3 to 6 months before construction. This assured that the gradient would be maintained after construction and water would continue to flow where it was intended.

Mitigation of Subsidence

The best strategy for minimizing losses is to restrict human activities in areas that are most susceptible to subsidence and collapse. Unfortunately, most instances of land subsidence caused by fluid withdrawals were not anticipated. But knowledge of the causes, mechanics, and treatment of deep subsidence has increased with each damaging occurrence, so that we can now implement measures for reducing the incidence and impact of these losses.

Some of the measures taken to minimize subsidence are controlling regional water-table levels by monitoring and limiting water extraction, replacing extracted fluids like oil by water injection, and importing surface water for domestic use to avoid drawing on underground supplies.

Underground-mining operators are required to leave pillars to support mined-out coal beds, and mine openings are being backfilled with compacted mine waste to support the overburden. This latter procedure, however, may introduce chemicals that could contaminate underground water unless the backfill is first stripped of pollutants. Pennsylvania's Mine Subsidence Insurance Act of 1962 allows some property owners to buy subsidence insurance after inspection of their property. Another Pennsylvania bill allows homeowners to buy the bituminous coal beneath their property at a fair price, and therefore exercise control over mining operations.

By far the most expensive and ambitious subsidence mitigation program is that developed and now being implemented for Venice. Massive flood gates have been built to close off the lagoon from the ocean during periods of high water (**Figure 7.28**). Only time will tell whether this multibillion-dollar engineering will pay off.

Figure 7.27 Subsidence can be triggered by earthquakes and by changing sediment supply. (a) The Portage, Alaska, garage was dropped into the drink by the 1964 Alaska earthquake; sediments quickly filled the area and restored the land level even as the garage collapsed and lowered over time. (b) A ghost forest on the coast of Washington State was the victim of submergence when the 1700 Cascadia subduction zone earthquake dropped land levels several meters along the coast on 26 January, submerging the tree roots in saltwater and killing the trees. (c) Fort Proctor, built in the 1850s on the Mississippi Delta, is now surrounded by water as the delta subsides, largely because the sediment supply from the river has been cut off by channelization.

SPOTLIGHT FIGURE 7.28 Mass Wasting and Subsidence

Venice is slowly sinking (subsiding) as the soft lagoonal sediments beneath the city compact under their own weight—the result, flooding. To mitigate the flooding, a series of floodgates are being built to close off the lagoon from the sea during periods of exceptionally high water. Meanwhile, pedestrian walkways keep people above flooded piazzas like the Piazza San Marco (the inset photograph).

The Venetians put down roots—on a cluster of islands in a lagoon at the north end of the Adriatic Sea—by driving alder and oak piles into the sandy ground. Atop these foundations they built homes and palaces and began a battle against the ceaseless rise and fall of the tides. The city's structures, despite reinforcements, have suffered the assault of brackish water, sea-level rise, and subsidence (sinking)—some five inches in the past century. Excessive pumping of groundwater contributed to subsidence.

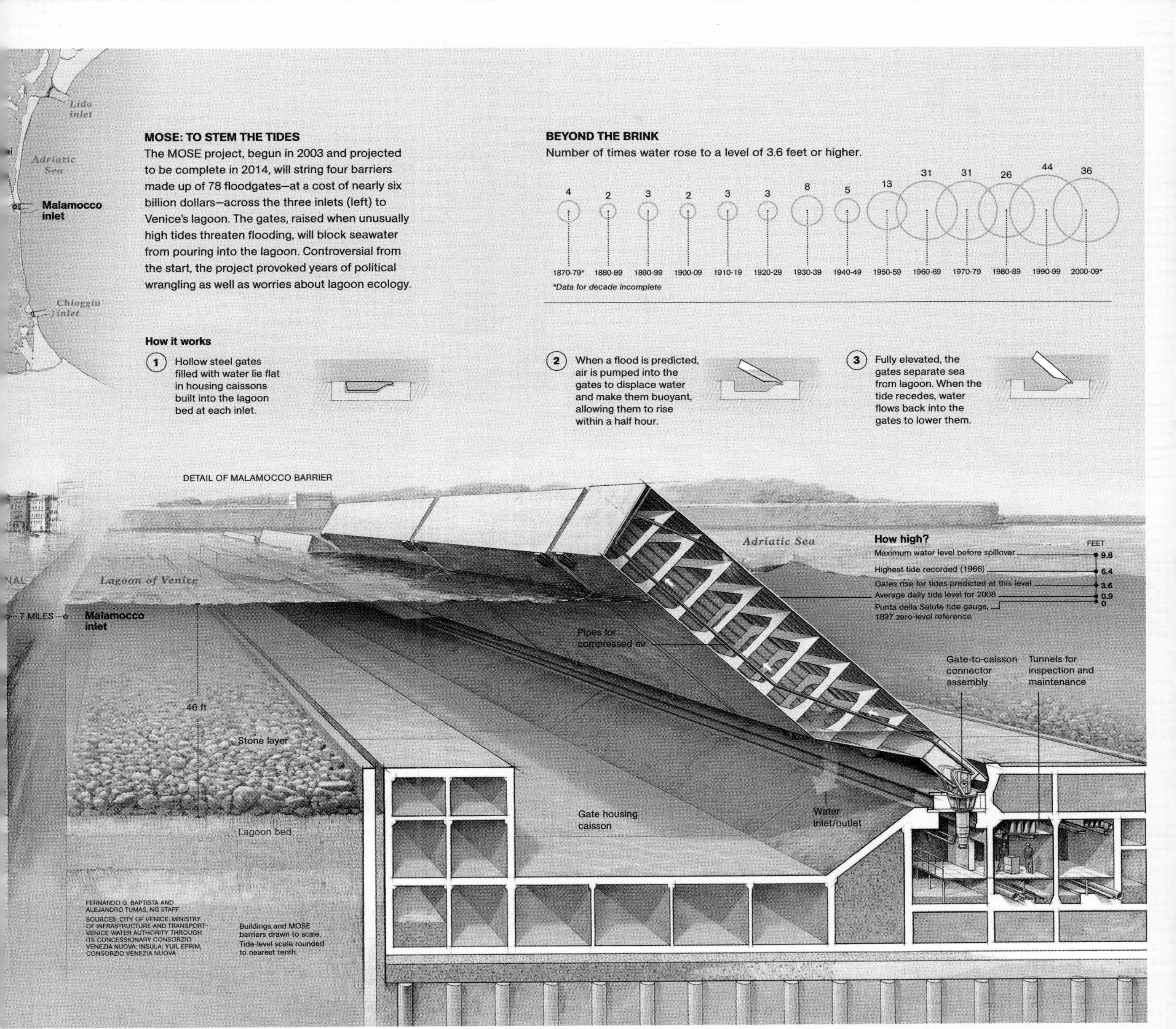
Lido inlet
Adriatic Sea
Malamocco inlet
Chioggia inlet
MOSE: TO STEM THE TIDES
The MOSE project, begun in 2003 and projected to be complete in 2014, will string four barriers made up of 78 floodgates—at a cost of nearly six billion dollars—across the three inlets (left) to Venice's lagoon. The gates, raised when unusually high tides threaten flooding, will block seawater from pouring into the lagoon. Controversial from the start, the project provoked years of political wrangling as well as worries about lagoon ecology.
BEYOND THE BRINK
Number of times water rose to a level of 3.6 feet or higher.
1870-79* 4
1880-89 2
1890-99 3
1900-09 2
1910-19 3
1920-29 3
1930-39 8
1940-49 5
1950-59 13
1960-69 31
1970-79 31
1980-89 26
1990-99 44
2000-09* 36
*Data for decade incomplete
How it works
1 Hollow steel gates filled with water lie flat in housing caissons built into the lagoon bed at each inlet.
2 When a flood is predicted, air is pumped into the gates to displace water and make them buoyant, allowing them to rise within a half hour.
3 Fully elevated, the gates separate sea from lagoon. When the tide recedes, water flows back into the gates to lower them.
DETAIL OF MALAMOCCO BARRIER
Adriatic Sea
Lagoon of Venice
7 MILES
Malamocco inlet
How high?
FEET
Maximum water level before spillover 9.8
Highest tide recorded (1966) 6.4
Gates rise for tides predicted at this level 3.6
Average daily tide level for 2008 0.9
Punta della Salute tide gauge, 1897 zero-level reference 0
Pipes for compressed air
Gate-to-caisson connector assembly
Tunnels for inspection and maintenance
46 ft
Stone layer
Gate housing caisson
Water inlet/outlet
Lagoon bed
FERNANDO G. BAPTISTA AND ALEJANDRO TUMAS, NG STAFF
SOURCES: CITY OF VENICE; MINISTRY OF INFRASTRUCTURE AND TRANSPORT–VENICE WATER AUTHORITY THROUGH ITS CONCESSIONARY CONSORZIO VENEZIA NUOVA; INSULA; YUIL EPRIM, CONSORZIO VENEZIA NUOVA
Buildings and MOSE barriers drawn to scale. Tide-level scale rounded to nearest tenth.

case study

7.1 Dynamic Real Estate in Oregon

By volume, the Columbia River is the largest river flowing into the Pacific Ocean and the second-largest river in the United States. Its narrow gorge is a site of great scenic beauty, with towering basalt cliffs, beautiful river views, and coniferous forests cloaking side slopes. No wonder people want to live there.

There is a problem, however. On a stretch of the gorge between the towns of Warrendale and Dodson, Oregon, massive debris flows have occurred every 50 to 100 years. The record-setting 1996 snowfall of 4 meters (12 feet) and then 32 centimeters (13 inches) of rain on the 2,000- to 4,000-foot cliffs above the towns changed the idyllic environment to an agonizing one. About 3:00 a.m. on 6 February 1996, Dodson resident Mark Chandler went outside to call the family dog. He lived on Tumult Creek and could not help but notice when the creek's rampaging waters suddenly stopped. He stared, unbelievingly, at the creek and then heard an ominous rumbling in the distance, like "two Union Pacific trains colliding" on the nearby tracks. Realizing that something was amiss, he dashed back to the house with the dog, awakened his family, and got them all out immediately.

A giant landslide had dammed Tumult Creek and was mobilized by the creek's water to form a debris flow. With boulders the size of minivans, the flow broke off 20-meter (65 foot) pine trees, filled houses with mud, and blocked I-84 and the railroad. When it was all over, geologists calculated that there was enough debris to cover a football field with mud and boulders 215 meters (700 feet) thick (**Figure 1**).

In all, seven landslides struck the gorge between the two towns. County officials quickly placed restrictions on building along this stretch of the gorge and designated the towns as "hazardous areas." This means residents will have to hire a geological consultant for any building whatsoever. Property values have plummeted, and some owners think the government should buy them out. Others want to rebuild, but with government help. The last major slide in the area had occurred in about 1920, and the lessons learned then apparently were forgotten. Remarkably, no one was killed in 1996, only one person was injured, and a horse and a dog that were caught in the muck were rescued.

B. Pipkin

Figure 1 This home was submerged to its second floor in the Tumult Creek debris flow. The beautiful high cliffs of basalt in the background were the source area of the landslide/rockfall and ensuing debris flow.

question to ponder

1. With more than 70 years between major landslides, do you think it makes sense to restrict building in this area?

7.2 Colorado's Slumgullion Landslide: A Moving Story 300 Years Old

The Rocky Mountain states, particularly Colorado, have some of the nation's highest levels of landslide hazard. Each year, landslides in the Rockies take several lives and cause millions of dollars in damage to forests, roads, pipe- and electrical-transmission lines, and buildings. What may be the largest active landslide in the United States is between Gunnison and Durango, Colorado, and bears the unusual name *Slumgullion,* a name that was probably bestowed upon it by early prospectors (**Figure 1**).

> **slumgullion** (slum-guryn), *n.* 1 *Slang,* a meat stew with vegetables, as potatoes and onion. 2. *Mining,* a muddy red residue in the sluice. (*Webster's New International Dictionary*)

Slumgullion, the stew, was a standard of mining camp cooks, and slumgullion, the red residue, was what miners usually found instead of gold in the sluice box.

C. W. Cross/USGS

Figure 1 The Slumgullion landslide in southwestern Colorado has a tremendous length and a sinewy path. The head scarp is more than 500 meters (1,600 feet) high. This photo shows it flowing from its source in the mountains to the lake it dammed below.

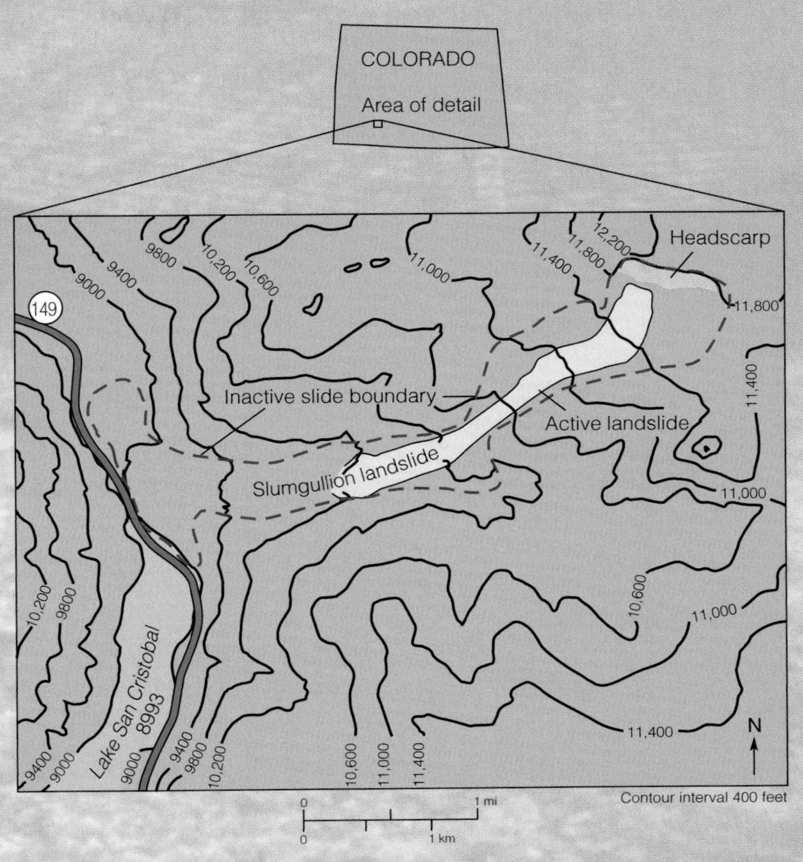

Figure 2 A generalized contour map showing the 300-year-old active landslide (colored yellow) within an inactive older slide mass. Lake San Cristobal formed when the landslide dammed the Lake Fork River. (Redrawn from Highland, Lynn. 1993. *Earthquakes and Volcanoes* 24(3); USGS.)

The Slumgullion landslide originated 700 years ago, when highly weathered and altered Tertiary volcanic rocks gave way on a ridge above the Lake Fork tributary of the Gunnison River. The slide site is a few kilometers upstream from Lake City, a historic mining town in the San Juan Mountains. The landslide dammed Lake Fork River, forming the largest natural lake in Colorado, Lake San Cristobal. The landslide is huge—6 kilometers (3.6 miles) long, 1 kilometer (0.6 mile) wide, and an average of 40 meters (132 feet) thick. As **Figure 2** shows, the Slumgullion is a landslide within a landslide; its 300-year-old active portion is flowing within the larger, 700-year-old mass. The total drop from head scarp to toe is about 762 meters (2,500 feet), and the slide moves at about 6 meters (20 feet) per year. The Slumgullion exhibits three distinct regions of deformation (**Figure 3**):

- The top of the slide is in tension (extension). Normal faulting dominates this region.
- The middle region flows as a rigid block, called *plug flow.* Nearly vertical strike-slip faults are found along the plug's lateral edges.
- The toe is a region of compression. It exhibits thrust (reverse) faults.

Within the active landslide, the terrain is jumbled, and trees are either highly tilted or dead.

The landslide poses several geological hazards. If it overruns Colorado Highway

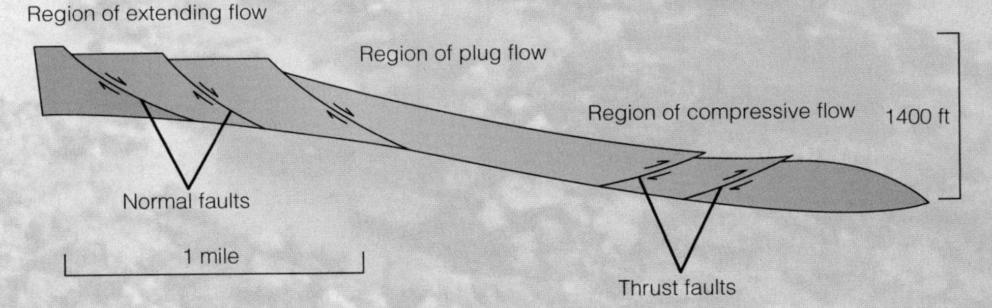

Figure 3 Flow regions within the landslide. Extension (pull apart) at the top of the landslide leads to slumping there. The coherent plug flow in the middle of the landslide moves with lateral slip at the sides of the plug. Thrusting takes place at the toe of the landslide as the coherent plug attempts to override the shattered but stable toe. (Redrawn from Highland, Lynn. 1993. *Earthquakes and Volcanoes* 24(3); USGS.)

149, it will isolate recreational development upstream. Downstream flooding will result if the lake level rises and overtops the natural dam. For these reasons, the U.S. Geological Survey (USGS) is studying the area. In addition to 30 years' worth of observations, the USGS is:

- Making precise measurements of the landslide's rate of movement, using conventional survey techniques, as well as Global Positioning System (GPS) technology.
- Examining old photographs of the area. These enable the researchers to estimate the slide's rate of headward growth. Because headward growth feeds rock debris to the head of the slide, it stimulates movement and growth of the slide.
- Monitoring Lake San Cristobal and its landslide dam. These will provide information that is applicable to other landslide dams in Colorado and the Rocky Mountains.
- Making three-dimensional models of the active and inactive portions of the slide based on the accumulated data. Such models facilitate the prediction of the slide's future behavior, especially in the face of changing climate, including the effects on precipitation, vegetation, snowfall amounts, and temperature.

questions to ponder

1. Can you propose ways in which this landslide could be mitigated?
2. Do you think these mitigations would be accepted by local residents?

7.3 When the Trees Come Down, So Do the Hills

Trees do so much for us. Their wood holds up our homes, and their trunks and leaves store carbon, which would otherwise warm our climate. They give us oxygen to breathe and shade under which to relax on a warm day. But trees do something you may not have given much thought to before. They hold up many of our hillslopes. Trees have a dense network of roots, which binds the soil together. In terms of resisting forces, these roots provide cohesion until the tree dies and the roots rot, or maybe the stumps are pulled, so that sheep can graze on open slopes. On dry days and for soils with high natural cohesion, the loss of roots and their assistance in holding the soil together matters little. But let it rain—let it rain for a long time on sandy soils—and the hillside will fail in landslides, perhaps sending debris flows hurtling down steep slopes to the valley below.

If the trees are grown for wood products, then the slope will soon be replanted for the next rotation. The window is short between when the old roots rot and the new roots are numerous and big enough to hold up the slope—perhaps a decade or less (**Figure 1**). If no big storms hit in this window, everything's okay. But if the slope is kept clear and not replanted, then it's only a matter of time before the mass movements will start. We know this because the experiment has been run over and over again—not by scientists, but by the people who settled North America. First they cleared the Mid-Atlantic states and the slopes failed, then New England, then the Southeast, and finally the Pacific Northwest. Everywhere the loggers went, the slopes responded. The erosion is well documented in writings of the time (see the work of George Perkins Marsh, writing *Man and Nature* in the 1850s, who lamented the clearing of hillslopes and the erosion that followed). More recently, scientists such as Lee Benda, David Montgomery, and Beverley Wemple, geomorphologists who specialize in debris flows and erosion, have directly observed the effects of land clearance.

P. Bierman

Figure 1 A logging road in southwestern Washington goes through a recently clear-cut stand of timber. In the distance are forests of different ages, because each was replanted after it was cut.

Vermont Historical Society

Figure 2 The deforested slopes of Tunbridge, Vermont, in the late 1800s. The sandy river terraces are being ripped apart by gullies and small landslides. All the sand is landing on the alluvial fans below. A horse stands at the apex of one of these fans.

There is another data source, often overlooked, but available just about everywhere—historic images. Chances are, with a little detective work, you can find 100-year-old images of where you live today and, with those images, do your own research about forest clearance and landscape response. Let's look at what the historic images of Vermont can tell us about the results of that clear-cutting experiment our ancestors ran more than a century ago.

Study **Figure 2**. It's an attractive image of a stream in one of the most picturesque towns in Vermont, a town best known for having the longest running agricultural fair in the state. First, find the river terraces, the three fat surfaces. Then, see if you can find the alluvial fans. Here's a hint, find the horse. It's standing at the apex of one of the fans. Then, look above the horse and you'll see the eroding hillslope and the gully that is feeding the sandy sediment to the alluvial fan. Can you find a second fan? Can you find more erosion dumping sediment on the terrace below the alluvial fan? Geologists, graduate students, and faculty have dug up more than a dozen alluvial fans like the one in the photograph. Every one of them was covered with at least a meter (3 feet) of sediment deposited after land clearance; today all the fans and the slopes above them are stable, covered in trees. But in the photograph is the smoking gun: evidence that the cleared slopes eroded and fed sediment to the fans.

Today forest clearance is an issue in developing nations where trees are removed for timber and so that agriculture can feed burgeoning populations (**Figure 3**).

question to ponder

1. Biomass is one form of alternative energy that can help reduce our dependence on fossil fuels. In the northeastern United States, wood-fired power and heating plants are common. What is the trade-off we may face if wood-fired power plants become more common?

© 2011/FRANS LANTING/National Geographic Image Collection

Figure 3 In Madagascar, forests have been cleared and the slash burned so that maize (corn) can be planted on the land. This has increased rates of soil loss and erosion.

case study

7.4 A Rescue Plan for Venice?

Called "the city where streets are paved with water," Venice has special interest because of its historical importance, priceless antiquities, and beauty, as well as its charming canals. The city has a major flooding problem, however. St. Mark's Square and other squares are flooded during periods of the extremely high tides the locals call the *aqua alta* (Italian, "high water"). It is said that the first Venetians settled there in 421 C.E. "to escape the barbarian Goths" sweeping down from the north. Sitting on a low mud bank in a lagoon sheltered by barrier islands (see Chapter 10), the city has subsided about 3 meters (10 feet) relative to sea level since it was founded. The sinking is due, in part, to the existence of 20,000 water wells in the region (now water is brought in by aqueduct), settlement into the soft sediments on which it was built, the diversion of sediment-carrying rivers that starved the lagoon of new sediment, and rising sea level. The traditional building practice was to drive wooden piles into the soft sediment to form level foundations on which to build structures, and, of course, timbers rot and settle. According to records, one cathedral is founded on more than a million timbers.

During periods of aqua alta, the avenues and squares of Venice become flooded, and gondola traffic comes to a screeching halt (**Figure 1**), because the gondolas cannot get under the bridges! The aqua alta occurs when an onshore flow of water (a storm surge) from the Adriatic Sea coincides with high tides—any level more than 80 centimeters (32 inches) or so above average water level leads to significant flooding. There were an average of 70 high waters per year between 1924 and 1933, and by the last decade, the number had increased to more than 250 high water days annually (**Figure 2**). Dan Stanley of the Smithsonian Insti-

Carlos Naya photo circa 1870, public domain

René Seindal/venicekayak.com

Figure 1 When the aqua alta floods Venice, kayaks can go places that people usually walk, such as the Ca' d'Oro, one of the best-preserved Gothic buildings on the Canal Grande. (a) An image from the 1870s shows the portico high and dry. (b) On 9 November 2009, an aqua alta of 100 centimeters (a little over 3 feet) above historic average levels (not enough for the flood barriers to go up) provided an opportunity for kayaking. Can you see how much the water has risen between the two photos? (Source: http://rene.seindal.dk/)

tution in Washington, DC, likens Venice to a man standing in a swimming pool: His feet are embedded in lead, and the water level is rising. It's already to his nose. He'd better not make any waves.

The solution is twofold. The first part is to raise buildings and plazas to reduce flooding during small-to-moderate events, projects known as *rialto*. Second, install 78 movable floodgates that can temporarily cut the lagoon off from the sea, closing the three major inlets. The cost of this, the *Mose* project? Four-and-a-half billion euros (that's nearly $6 billion at 2012 exchange rates). The idea is to raise the floodgates, like dams, in anticipation of high water and then reopen them when the water level drops. There are environmental reasons, as well as financial ones, that make the project controversial. Assuming a sea-level rise of 50 to 100 centimeters over the next century (a conservative range of estimates), the gates would have to be raised more than 70 times a year, and that would interfere with boat traffic and tourism. According to British expert Edmund Penning-Rowsell, "The problem is that as sea level rises, they will have to be closed off so often that they will be quite ridiculous." However, in the face of global warming and rising sea level, is there anything else to do?

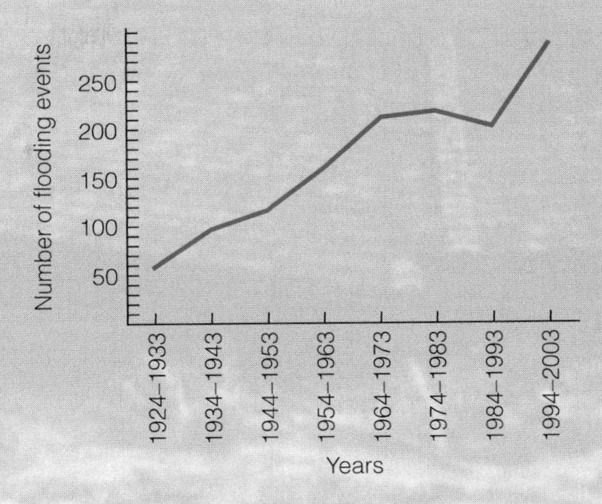

Figure 2 The frequency of aqua alta flooding in Venice since 1924. The increase in frequency is due to subsidence, settlement, and rising sea level.

question to ponder

1. Given the cost of the floodgates and rapidly rising sea level, should we abandon Venice to the floods or invest more time and money to save the city?

7.5 Water, Water Everywhere

New Orleans has the flattest, lowest, and youngest geology of all major U.S. cities. The city's "Alps," with a maximum elevation of about 5 meters (16 feet) above sea level, are natural levees built by the Mississippi River, and the city's average elevation is just 0.4 meter (1.3 feet). No surface deposits in the city are older than about 3,000 years. The city was established in 1717 on the natural levees along the river. The levees' sand and silt provided a dry, firm foundation for the original city, which is now called *Vieux Carre* (French, "old square"), or the "French Quarter." Farther from the river, the land remained undeveloped, because it was mainly water-saturated cypress swamp and marsh formed between distributaries of the Mississippi River's ancient delta. These areas are underlain by as much as 5 meters (16 feet) of peat and organic muck.

When high-volume water pumps became available in the early 1900s (see Chapter 10), drainage canals were excavated into the wetlands to the north. Swamp water was pumped upward into Lake Pontchartrain, a cutoff bay of the Gulf of Mexico. About half the present city is drained wetlands lying well below sea and river level. The lowest parts of the city, about 2 meters (6.5 feet) below sea level, are also the poorest parts of the city and the parts most devastated by the flooding of hurricane Katrina.

Subsidence is a natural process on the Mississippi River delta because of the great volume and sheer weight of the sediment laid down by the river. The sediment compacts, causing the land surface to subside. The region's natural subsidence rate is estimated to have been about 12 centimeters (4 inches) per century for the past 4,400 years. This estimate is based on carbon-14 dating of buried peat deposits and does not take into account any rise in sea level during the period. Urban development in the 1950s added to the subsidence (**Figure 1**). The compaction of peaty soils in reclaimed cypress swamps coincided with the construction of drainage canals and the planting of trees, both of which lowered the water table. Peat shrinks when it is dewatered, and the oxidation of organic matter and compaction contribute to subsidence. Most homes constructed on reclaimed swamp and marsh soils in the 1950s were built on raised-floor foundations. These homes are still standing, but they require periodic leveling. Unfortunately, homes built on concrete-slab foundations, a technique that had just been introduced, sank into the muck and became unusable.

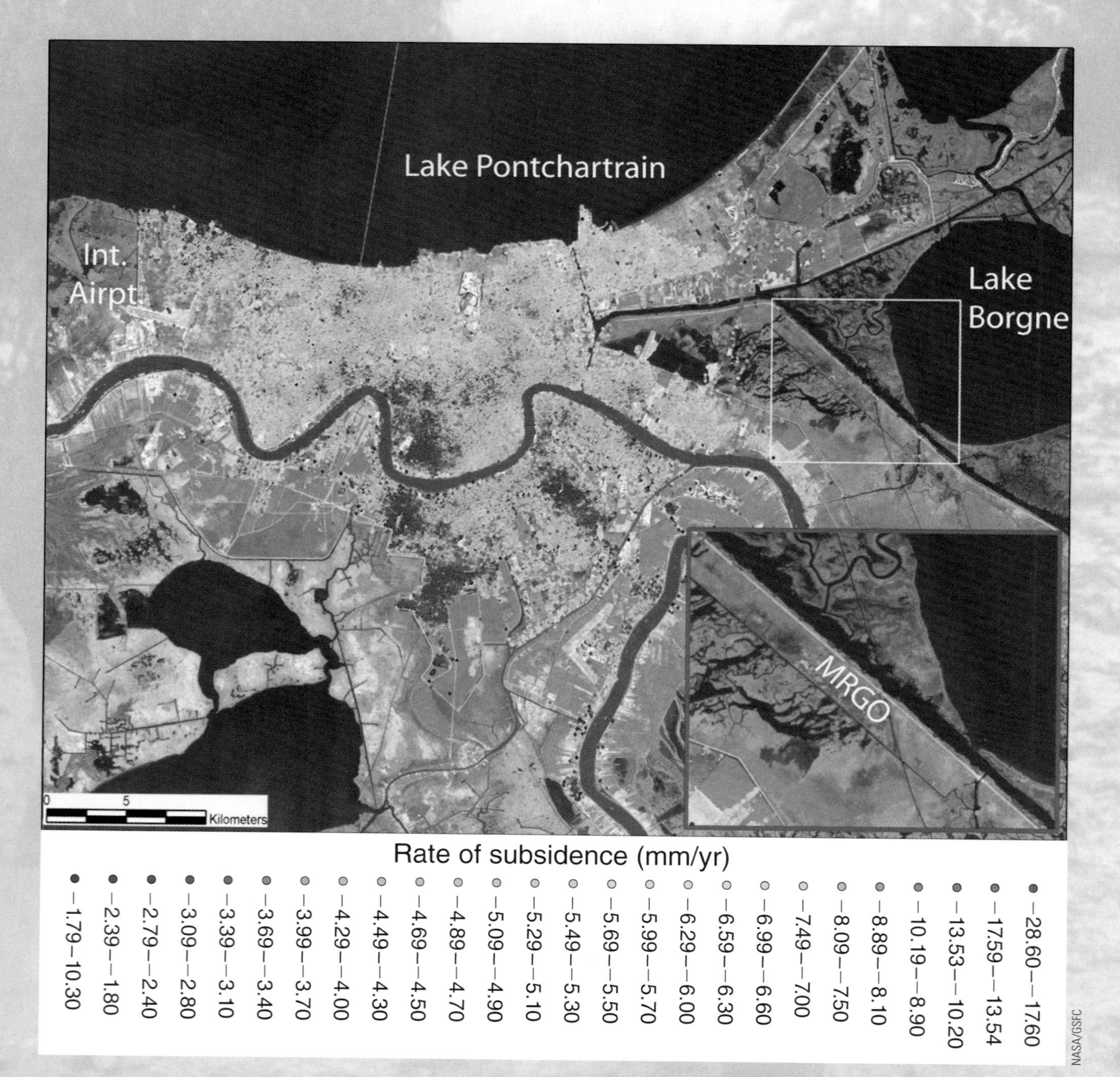

Figure 1 Image of New Orleans with rates of subsidence between 2002 and 2005 color coded on the image. Note that the Mississippi River meanders through New Orleans, Lake Pontchartrain is to the north, and smaller Lake Borgne is to the northeast. The blue areas are subsiding slowly (millimeters per year); the red areas are subsiding quickly (centimeters per year). The fastest subsidence rates are more than an inch (2.5 centimeters) per year. These subsidence estimates were made by analyzing radar data collected from satellites.

Other home foundations were constructed on cypress-log piles sunk to a depth of 10 meters or more (at least 30 feet). These houses have remained at their original level, but the ground around them has subsided. This *differential subsidence* requires that fill dirt be imported to make the sunken ground meet the house level. It is very common to see carports and garages that have been converted to extra rooms when subsidence has cut off driveway access to them (**Figure 2**). One also sees many houses with an inordinate number of porch steps; the owners have added steps as the ground has sunk. In 1979, Jefferson and Orleans parishes (counties are called "parishes" in Louisiana) passed ordinances requiring 10- to 15-meter (33- to 50-foot) deep wooden-pile foundations for all houses built on former marshland and swampland. This has been beneficial, but differential subsidence continues to damage New Orleans' sewer, water, and natural gas lines, as well as its streets and sidewalks. It is this subsidence that will require ever-higher levees to protect New Orleans against storm surges and Mississippi River floods.

Figure 2 Differential subsidence around this pile-supported house has left the carport high and dry. The carport was converted to a family room, and fill was imported to bring the yard surface up to the previous level.

question to ponder

1. Given rising sea level catalyzed by global warming and the melting of ice caps, does it make sense to rebuild New Orleans as it sinks from subsidence?

PHOTO GALLERY slip, slide, and fall

Mass-wasting processes are tragic, devastating, and ongoing. Both the geological and photographic records are clear—this hazard is not going away. Old photographs show clear-cutting and its effects. Take a typical New England scene in the late 1800s (**Figure 1**). Look in the distance and see a perfect landslide scar and then wonder, "Is the man sitting on an old growth stump that is all that's left of the first forest that was cleared at settlement, maybe 100 years earlier?" It's not clear we have learned much as a society despite repeated tragedies and the loss of life and property. Houses are still built on the edges of unstable cliffs and on the deposits of old debris flows. The past is a clue to the future.

1 A man stares out at clear-cut slopes sometime in the late 1800s. Can you find the shallow, translational landslide at which he's looking? The slope failed because the trees were cut and the effective cohesion their roots provided was lost. University of Vermont, Special Collections, Bailey-Howe Library

2 Rockfall is a hazard wherever steep bedrock slopes tower over roads and buildings. In Yellowstone National Park, a ranger examines trees crushed by a fallen boulder that now blocks the road. © 2011/ANDREW H. BROWN/National Geographic Image Collection

A wet, slushy spring avalanche pours down a rock face adjacent to a well crevassed glacier. © 2009/MICHAEL MELFORD/National Geographic Image Collection

3 In 1903, the side of Turtle Mountain in Alberta collapsed in the massive Frank rock slide, killing almost 100 coal miners and causing Canada's worst mining disaster. The slide path and the debris apron of broken rock are clearly visible today, more than 100 years later. © 2010/PETE RYAN/National Geographic Image Collection

4 Landslides along steep seacoasts are common. In San Clemente, California, a large slide not only "pulled the rug out" from under several luxury homes along the top of the bluff, but the debris buried both the Pacific Coast Highway and the Amtrak rail line. Luckily for those buildings on the beach, the slide stopped just short of burying them in broken rock. © 1989 RICK RICKMAN/National Geographic Image Collection

5 The 2011 debris flows in eastern Brazil destroyed these homes. The same flows cut away the stream channel below the homes, exposing older debris flow deposits. These homes were built on bad ground, in a place debris flows have ravaged before and will ravage again. P. Bierman

SUMMARY Mass Wasting and Subsidence

Mass Wasting

Defined

Downslope movement of rock and soil under the direct influence of gravity. The term *mass wasting* includes falls, flows, and slides that have well-defined boundaries.

Classification

1. Type of material—rock or earth
2. Mechanism of movement—flow, slide, or fall
3. Velocity

Falls

Defined

Free fall of earth materials from a steep cliff or slope.

Types

1. Rockfall—involves hard bedrock of any size
2. Debris fall—collapse of weathered rock material and/or soil from a steep slope or cliff

Flows

Defined

Mass movement of unconsolidated material in a fluid state.

Types

1. Creep—slow downslope movement of rock and soil particles
2. Debris flow—fluid mixture of rock, sand, mud, and water that is fast moving and destructive

Causes

Heavy rainfall on steep slopes with loose soil and sparse vegetation. Most common in arid, semiarid, tropical, and alpine climates and where dry-season forest fires expose bare soil.

Slides

Defined

Movement of rock or soil (landslide) as a unit on a discrete failure surface or slide plane.

Mechanics

Driving force (gravity) exceeds resisting forces (friction and cohesion).

Types

1. Rotational slide (slump) on a curved, concave-upward slide surface
2. Translational slide (block glide) on an inclined plane surface
3. Complex landslide, a combination of rotational and translational slides
4. Lateral spreads of water-saturated ground (liquefaction of sand and quick clays)

Causes

1. Weakening of the slope material by saturation with water and weathering of rock or soil minerals
2. Steepening of the slope by artificial or natural undercutting at toe
3. Added weight at the top of the slope such as by building a home or adding fill
4. A change in effective acceleration of gravity such as that provided by the strong shaking of an earthquake

Reduction of Losses

Geological Mapping

To identify active and ancient landslide areas, and thus avoid them.

Building Codes

To limit the steepness of manufactured slopes and specify minimum soil and fill conditions, as well as surface-water drainage from the site.

Control and Stabilization

Insert Drainpipes

To drain subsurface water and lower pore-water pressure.

Excavation

To redistribute soils and rock from the head of the potential slide to the toe. This decreases driving force and increases resisting force.

Buttress Fills and Retaining Devices

To retain active slides and potentially unstable slopes by increasing resistance.

Rock Bolts

To stabilize slopes or jointed rocks.

Deflection and Capture Devices

To divert debris flows around existing structures. Debris basins capture flow material before it impacts buildings and people.

Snow Avalanches

Types

Slab avalanche is old snow and ice that moves as a coherent mass. A powder-snow avalanche is composed of new snow that moves as a fluid at high velocities. Climax avalanches are made up of weak, wet snow and occur in the spring.

Causes

Weak boundary layer between unstable layer above and stable mass below. Loading by new snow from wind or heavy storm.

Mitigation

Controlled avalanching triggered by artillery, avalanches diverted by structures, and snow sheds.

Subsidence

Defined

A sinking of Earth's surface caused by natural geological processes or by human activity, such as mining or the pumping of oil or water from underground.

Types

1. Deep subsidence caused by the removal of water, oil, or gas from depth.
2. Shallow subsidence of ground surface caused by underground mining, groundwater extraction, or by hydrocompaction, the consolidation of rapidly deposited sediment by soaking with water.

3. Collapse of ground surface into shallow underground mines or into natural limestone caverns, forming sinkholes.
4. Settlement occurs when the load applied by a structure exceeds the bearing capacity of its geological foundation.

Causes of Deep Subsidence

Removal of fluids confined in pore spaces at depth transfers the weight of the overburden to the grain boundaries, causing the grains to reorient into a closer-packed arrangement.

Causes of Shallow Subsidence and Collapse

The ground surface may collapse into shallow mines (usually coal) or sinkholes in limestone terrains. Sinkholes form by the collapse of soil and sediment cover into an underground limestone or dolomite cavern (see Chapter 9). Heavy irrigation causes some loose, low-density sediments and soils to consolidate (hydrocompaction). Compaction caused by dewatering of peat and cypress-swamp (delta) clays and oxidation of the highly organic sediments causes subsidence.

Subsidence Problems

Ground cracking, deranged drainage, flooding of coastal areas, and damage to structures, pipelines, sewer systems, and canals. Whole buildings may disappear into sinkholes.

Mitigation

Underground water can be monitored and managed. Extracted crude oil can be replaced with other fluids. Mine operators are now required to leave pillars to support the mine roof, and rock spoil can be returned to the mine to provide support. Collapsible soils along canals can be precollapsed by soaking. Foundations in areas rich in organic deltaic sediments can be built on piles seated on firm geological material. This protects structures from damage when the surrounding land subsides.

KEY TERMS

angle of repose
antecedence
bentonite
block glide (translational slide)
climax avalanche
cohesion
complex landslide
creep
debris flow
depth hoar
driving force
factor of safety
frictional strength
granular material
hydrocompaction
ice crust
landslide
lateral spreading
mass wasting
normal force
pore pressure
powder-snow avalanche
resisting force
sensitive clay (quick clay)
settlement
slab avalanche
slide plane
slump (rotational slide)
subsidence
surface hoar
syncline
talus
terrain anchor

STUDY QUESTIONS

1. What grading and land-use practices can reduce the likelihood of landslides?
2. How can vegetation, topography, and sediment characteristics aid in identifying old or inactive slumps, slides, and debris flows?
3. Name the processes that move earth materials downslope (gravity is the *cause*, not the *process*). Upon what factors does each depend?
4. Modern building codes require a factor of safety of 1.5 or greater to build on or at the top of a cut slope. What is a factor of safety, and why should it be greater than 1.5?
5. What are some common, cost-effective methods of landslide control and prevention?
6. How did geological structure factor into the disaster at Vajont Dam?
7. What is the difference between settlement and subsidence? What can be done in advance to prevent settlement?

STUDY QUESTIONS (*continued*)

8. Distinguish between subsidence and collapse that are due to natural causes and due to human causes.
9. What can be done to slow or stop subsidence that is caused by withdrawal of water or petroleum?
10. What engineering problems make constructing canals and aqueducts in areas of hydrocompaction such an insidious geological hazard?
11. What surface manifestations of subsidence caused by groundwater withdrawal make it a serious problem, particularly in California and Arizona?
12. What particular hazard does subsidence present in coastal areas?

REFERENCES AND FURTHER READINGS

California Department of Conservation. 1998, February. *Hazards from mudslide, debris avalanches and debris flows in hillside and wildfire areas.* California Div. of Mines and Geology, note 33.

Dolan, R., and H. Grant Goodell. 1986. Sinking cities. *American Scientist* 74(1): 38–47.

Fleming, R. W., and T. A. Taylor. 1980. *Estimating the costs of landslides in the United States.* U.S. Geological Survey circular 832.

Gori, Paula L., Carolyn Driedger, and Sharon L. Randall. 1999. *Learning to live with geologic and hydrologic hazards.* U.S. Geological Survey Water Resources Investigations Report 99–4182.

Highland, Lynn. 1993. Slumgullion: Colorado's natural landslide laboratory. *Earthquakes and Volcanoes* 24(5): 208.

Highland, Lynn. 2000, May. *Landslide hazards.* U.S. Geological Survey Fact Sheet FS-071-00. Menlo Park, CA: USGS.

Highland, Lynn, J. Godt, David Howell, and W. Z. Savage. 1998. *El Nino 1997–1998: Damaging landslides in the San Francisco Bay area.* U.S. Geological Survey Fact Sheet 089–98. Menlo Park, CA: USGS.

Highland, Lynn, Ellen Stephenson, Sarah Christian, and William Brown III. 1997. *Debris flow hazards in the United States.* U.S. Geological Survey Fact Sheet FS-176-97. Menlo Park, CA: USGS.

Holzer, T. L. (ed.). 1984. Man-induced land subsidence. *Geological Society of America Reviews in Engineering Geology, Vol. VI.*

Larsen, Matthew C., et al. 2001, October. *Natural hazards on alluvial fans: The Venezuela debris flow and flash flood disaster.* U.S. Geological Survey Fact Sheet FS 103-01. Menlo Park, CA: USGS.

Marsh, G. P. 1874. *The Earth as modified by human action.* New York: Scribner and Sons, 656 p.

McPhee, John. 1989. Los Angeles against the mountains. In *The control of nature*. New York: Farrar, Straus & Giroux.

Monastersky, Richard. 1999. Against the tide: Venice's long war with rising water. *Science News* 156(4): 63.

Montgomery, D. R. 2007. *Dirt: The erosion of civilizations.* Berkeley: University of California Press.

National Academy of Sciences. 1978. *Landslides: Analysis and control.* Transportation Research Board special report 176. Washington, DC: National Academy Press.

National Academy of Sciences. 2004. Partnerships for Reducing Landslide Risk: Assessment of the National Landslide Hazards Mitigation Strategy.

Newman, C. 2009, August. Vanishing Venice. *National Geographic*.

Nilsen, Tor H., et al. 1979. *Relative slope stability and land-use planning in the San Francisco Bay region, California.* U.S. Geological Survey professional paper 944. Washington, DC: U.S. Government Printing Office.

Pearson, Eugene. 1995. *Environmental and engineering geology of central California: Salinian block to Sierra foothills.* Stockton, CA: National Association of Geology Teachers, Far Western Section, Fall Conference, University of the Pacific.

Reid, Mark E., R. G. LaHusen, and William Ellis. 1999. *Real-time monitoring of active landslides.* U.S. Geological Survey Fact Sheet 091–99. Menlo Park, CA: USGS.

Saucier, Robert T., and Jesse O. Snowden. 1995. Engineering geology of the New Orleans area. *Geological Society of America Annual Meeting Field Trip Guidebook* 6a and 6b.

Schultz, Arthur, and Randall W. Jibson (eds.). 1989. *Landslide processes of the eastern United States and Puerto Rico.* Geological Society of America special paper 236. Geological Society of America, Boulder, CO.

Tremper, Bruce. 1993. Life and death in snow country. *Earth: The science of our planet* 2.

U.S. Geological Survey. 1982. *Goals and tasks of the landslide part of a ground-failure hazards reduction program.* U.S. Geological Survey circular 880. Washington, DC: U.S. Government Printing Office.

Varnes, D. J., and W. Z. Savage (eds.). *The Slumgullion earth flow: A large-scale natural laboratory.* U.S. Geological Survey bulletin 2130.

_____. 1995. *Debris-flow hazards in the San Francisco Bay region.* U.S. Geological Survey Fact Sheet 112–95. Menlo Park, CA: USGS.

Varnes, David. 1984. *Landslide hazard zonation: A review of principles and practice.* Paris: UNESCO.

Wieczorek, G. E., L. S. Eaton, T. M. Yanosky, and E. J. Turner. 2006. Hurricane-induced landslide activity on an alluvial fan along Meadow Run, Shenandoah Valley, Virginia (eastern USA). *Landslides* 3(2): 95–106.

Williams, Knox. 2000, October. Avalanche formation and release. *Rock Talk.* Colorado Geological Survey publication no. 4.

Wilshire, Howard, Keith A. Howard, C. M. Wentworth, and Helen Gibbons. 1996. *Geologic processes at the land surface.* U.S. Geological Survey bulletin 2149.

Access an interactive eBook and chapter-specific interactive learning tools, including flashcards, quizzes, videos and more in your Geology CourseMate, accessed through CengageBrain.com.

A child in a Darfur refugee camp carries water to her family from a water hole shared by many.

Have you ever

wondered...

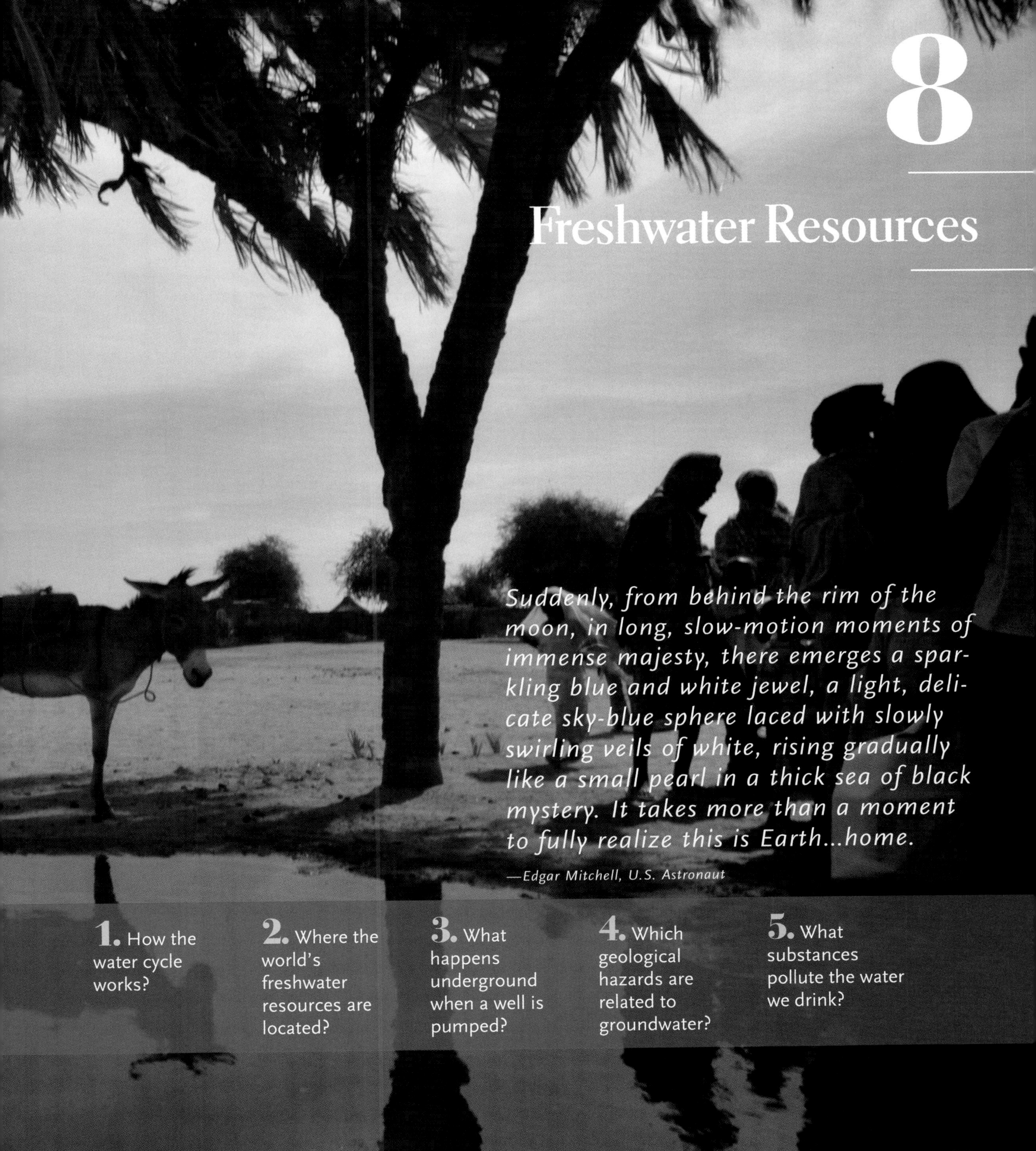

8

Freshwater Resources

Suddenly, from behind the rim of the moon, in long, slow-motion moments of immense majesty, there emerges a sparkling blue and white jewel, a light, delicate sky-blue sphere laced with slowly swirling veils of white, rising gradually like a small pearl in a thick sea of black mystery. It takes more than a moment to fully realize this is Earth...home.

—Edgar Mitchell, U.S. Astronaut

1. How the water cycle works?

2. Where the world's freshwater resources are located?

3. What happens underground when a well is pumped?

4. Which geological hazards are related to groundwater?

5. What substances pollute the water we drink?

The High Plains Aquifer and Water

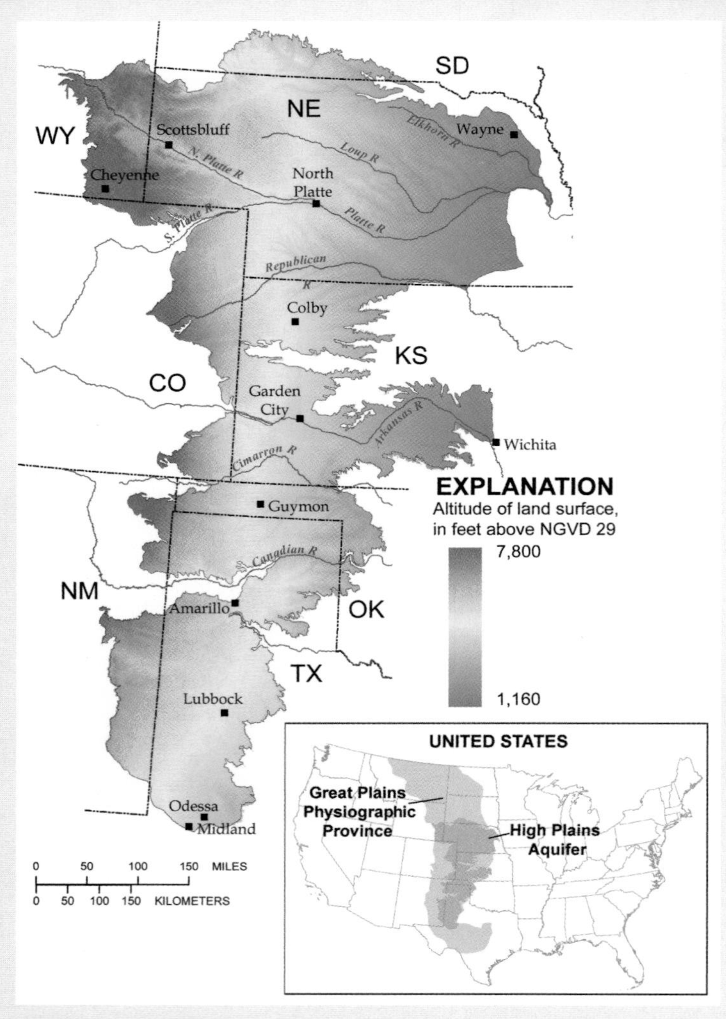

Figure 1 Location map showing the boundary of the High Plains aquifer, major cities and roads, and altitude of land surface. (Sharon Qi, U.S. Geological Survey, written communication, 2010.)

Mark Twain, a writer who could twist a phrase, is credited with saying, "Whiskey is for drinking, water is for fighting." There is much truth in the saying, as demonstrated in the conflict over water in the early settlement of the western United States. History may record that the 21st-century conflict in the Middle East was not over terrorists, territory, or even oil—but water. Most of the water in this region comes from three rivers: the Jordan, the Tigris-Euphrates, and the Nile. By 2025, the populations of water-scarce countries will be between 2.5 and 3.5 billion people, or close to half of the world's population in 2012. Syria, Iraq, Israel, Jordan, and Egypt will be no exception to the population increase, and they will need more water for their people. You can bet that water sharing will be high on the agenda of any future peace talks between the leaders of these countries.

The same might be true, without the threat of military action, in the water wars over the High Plains aquifer. The High Plains aquifer contains as much water as Lake Huron and underlies 480,000 square kilometers (147,000 square miles) of parts of Kansas, Colorado, New Mexico, Wyoming, South Dakota, Nebraska, Oklahoma, and Texas (**Figure 1**). Water distributed by the irrigation system in **Figure 2** is from a well drilled into the aquifer's water-bearing strata. Indeed, 97% of the water drawn from this aquifer is used for irrigation. Congress, worried about the sustainability of such withdrawals, ordered the U.S. Geological Survey to monitor the aquifer and report on its condition every other year. The reports are not promising.

Figure 2 For decades, groundwater from the High Plains aquifer has irrigated our nation's "breadbasket." This irrigation system draws water from a well that taps the aquifer. The well's potential yield was estimated at 4,000 liters per minute (about 1,000 gallons/minute) 24 hours a day. But the real question remains: Is that rate sustainable over the long term when rates of water withdrawal outstrip rates of groundwater recharge?

Sharing

Water levels in the High Plains aquifer have been dropping dramatically, an average of more than 4 meters (14 feet) since pumping began for all the states, with Texas coming in at an amazing 12 meters (37 feet) of average lowering (see Figures 8.21 and 8.22). The latest data suggest that water levels are continuing to fall. When the amount of water withdrawn from an aquifer exceeds the amount replenished to it by rainfall and rivers, an overdraft condition exists, a shortfall; as with shortfalls of all kinds, it will have to be reckoned with, later if not sooner. This overdraft condition is called *groundwater mining,* because the resource is being exploited without adequate replenishment. Models indicating the impact of human-induced climate change on rainfall suggest that the southern Great Plains (including Texas, where much of the water is used) will get much drier; the northern plains will get wetter. We'll see major shifts in where our food is grown.

questions to ponder

1. With water levels in the High Plains aquifer still dropping precipitously, what should be done? Should pumping be limited? Should water be allocated?
2. How would you solve this "tragedy of the commons"?

Courtesy of Virginia Smith

Coursemate Connection

Start off this chapter by listening to a podcast with National Geographic Fellow Sandra Postel as she talks about water issues. **Log in to the CourseMate at www.cengagebrain.com to listen now.**

WHEN PEOPLE GO INTO SPACE and are able to view Earth from afar, they cannot help but be struck by our planet's heavy cloud cover and its expanse of ocean, clear indications that Earth is truly "the water planet." Water is indeed what distinguishes our planet from other bodies in our solar system. Three-quarters of Earth's surface is covered by water, and the human body is 65% water. Water provides humans with means of transportation, and much human recreation is in water—and *on* it, where it exists in solid form as ice and snow. Because water is not equally distributed on Earth's surface, we have droughts, famine, and catastrophic floods. The absence or abundance of water results in such marvelous and diverse features as deserts, rain forests, picturesque canyons, and glaciers. This chapter investigates the reasons for this uneven distribution and its consequences. Water is involved in every process of human life; it is truly our most valuable resource.

The goals of this chapter are to show just how interconnected the hydrologic system truly is and how we, as humans, are dependent on clean, plentiful water for life. As the chapter progresses, we will demonstrate how closely connected groundwater and surface-water systems are, and how fragile the balance of drought and plenty can be.

Water as a Resource

Water is the only common substance that occurs as a solid, a liquid, and a gas over the temperature range found at Earth's surface. More than 97% of Earth's water is in the oceans, and 2% is in ice caps and glaciers; about 0.6% is available to humans as underground or surface freshwater (**Figure 8.1**). Fortunately, water is a renewable resource, which is illustrated by the **hydrologic cycle**, Earth's most important natural cycle (**Figure 8.2**). Water that is evaporated from the ocean or land surface goes back into the atmosphere and forms clouds. Precipitation from the clouds that falls over land areas eventually flows in streams back to the sea, and the cycle is repeated. Remarkably, very little water is in the atmosphere at any one time, about 0.001% of the total. Because of this, we know that water must recycle continuously to supply the precipitation we measure. The average annual rainfall over the surface area of the 48 contiguous United States is about 75 centimeters (30 inches, or 2.5 feet).

Evapotranspiration is the direct return of surface water to the atmosphere by evaporation, as well as its indirect return through transpiration from the leaves of plants. It accounts for the return of about 55 centimeters (22 inches) of the average annual precipitation. River runoff to the sea and infiltration into the ground account for the remaining 20 centimeters (8 inches), but runoff varies from 2.5 centimeters (1 inch) per year in the dry West to more than 50 centimeters (20 inches) per year in the East. Even though runoff varies greatly with location, over time the net water transport from land to sea must balance the net transport from sea to land. The sources and users of public water supplies in 1995 are shown in **Figure 8.3**. The data do not include water used for agriculture or water that is self-supplied, for example, by private wells.

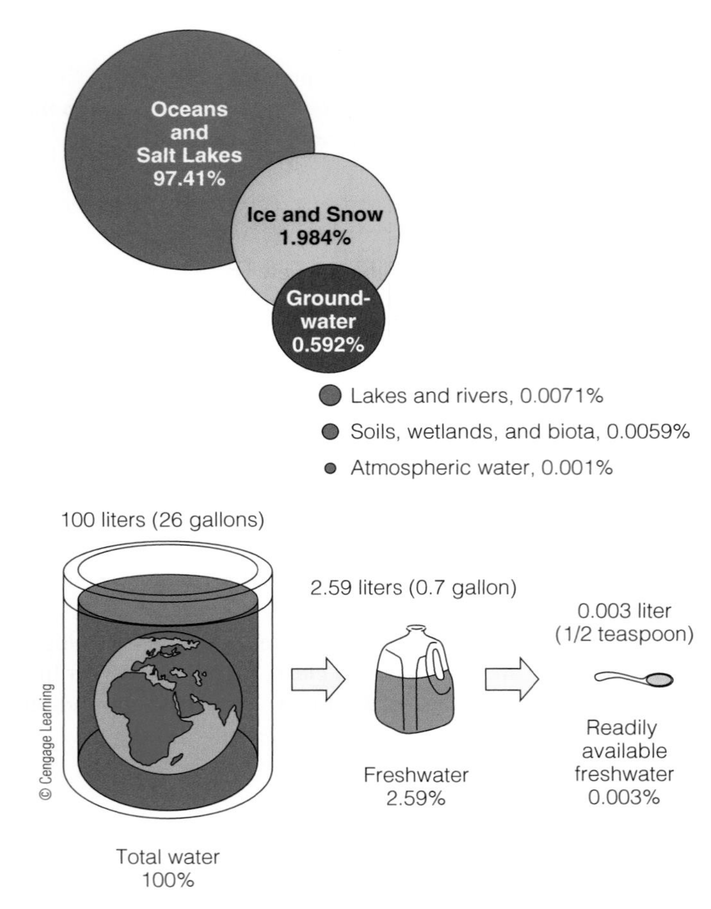

Figure 8.1 The planet's water budget. Only a tiny fraction by volume of the world's water supply is freshwater available for human use.

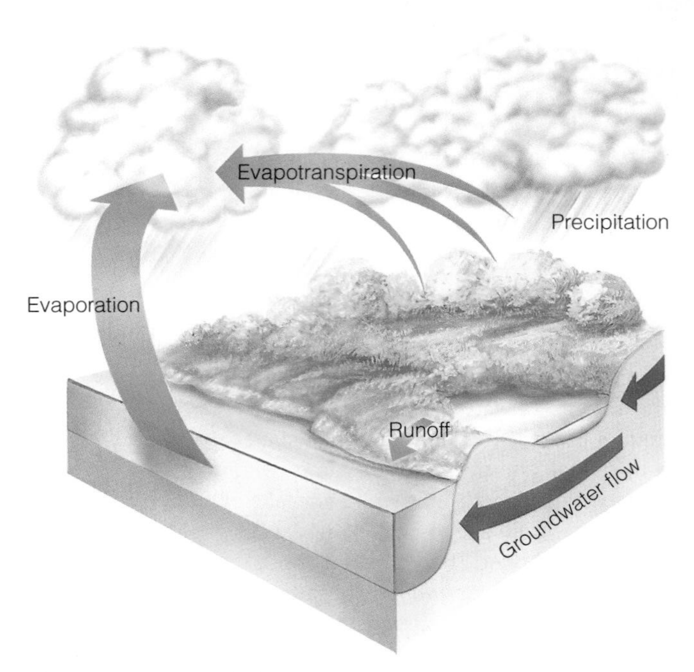

Figure 8.2 The hydrologic cycle. (From Wicander, R., and J. Monroe. 1999. *Essentials of geology,* 2nd ed. Brooks/Cole.)

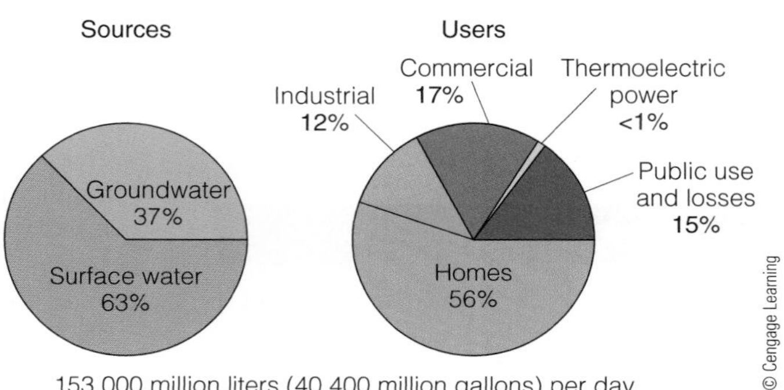

Figure 8.3 The sources and users of the U.S. public water supply in 1995. This does not include freshwater supplied for agricultural use or water that is self-supplied.

Freshwater at Earth's Surface

Rivers

Rivers provide abundant water to cities, some far from the river's location. Usually, as water suppliers, rivers have to be dammed and the waters metered out downstream. Planners recognize several categories of river-water use:

- *Instream use*—all uses that occur in the channel itself, such as hydroelectric power generation and navigation.
- *Offstream use*—the diversion of water from a stream to a place of use outside it, such as to your home. The amount of water available for offstream use along any stretch of a river is the total flow *minus* the amount required for instream purposes and for maintaining water quality.
- *Consumptive use*—water that evaporates, transpires, or infiltrates and cannot be used again immediately. Forty-four percent of all water that is withdrawn is used consumptively, and agriculture accounts for about 90% of that.
- *Nonconsumptive use*—water that is returned to streams with or without treatment so that it can be used again downstream. Domestic (household) water is used nonconsumptively: It is returned to the cycle through sewers and storm drains.

In many areas, particularly the western United States, consumption exceeds local water supplies. This necessitates importing water from other areas or mining groundwater. For example, about 75% of Southern California's water is imported from sources more than 330 kilometers (200 miles) away. Two aqueducts import water from west and east of the Sierra Nevada (**Figure 8.4**), and a third aqueduct brings water from the Colorado River. Irrigation uses about 40% of the total water withdrawn in the United States, a startling 80% in California. It is not surprising, then, that California, with the largest population and a huge irrigation system, is the largest U.S. consumer of water, using almost as much as the two next-largest consumers, Texas and Illinois, combined (**Figure 8.5**).

Total world withdrawals of water for offstream purposes amount to about 2,400 cubic kilometers per year (1.7 trillion gallons/day), of which 82% is for agricultural

© 1910/National Geographic Image Collection

Figure 8.4 The Los Angeles aqueduct supplies water to the city, but at a cost. When construction was complete, in the early 1900s, the aqueduct took the water from Owens Valley, infuriating local residents; some even formed dynamiting parties and breached the channel, spilling water. The aqueduct used open channels, covered channels, and siphons to take water hundreds of miles from the eastern Sierra to the Los Angeles basin.

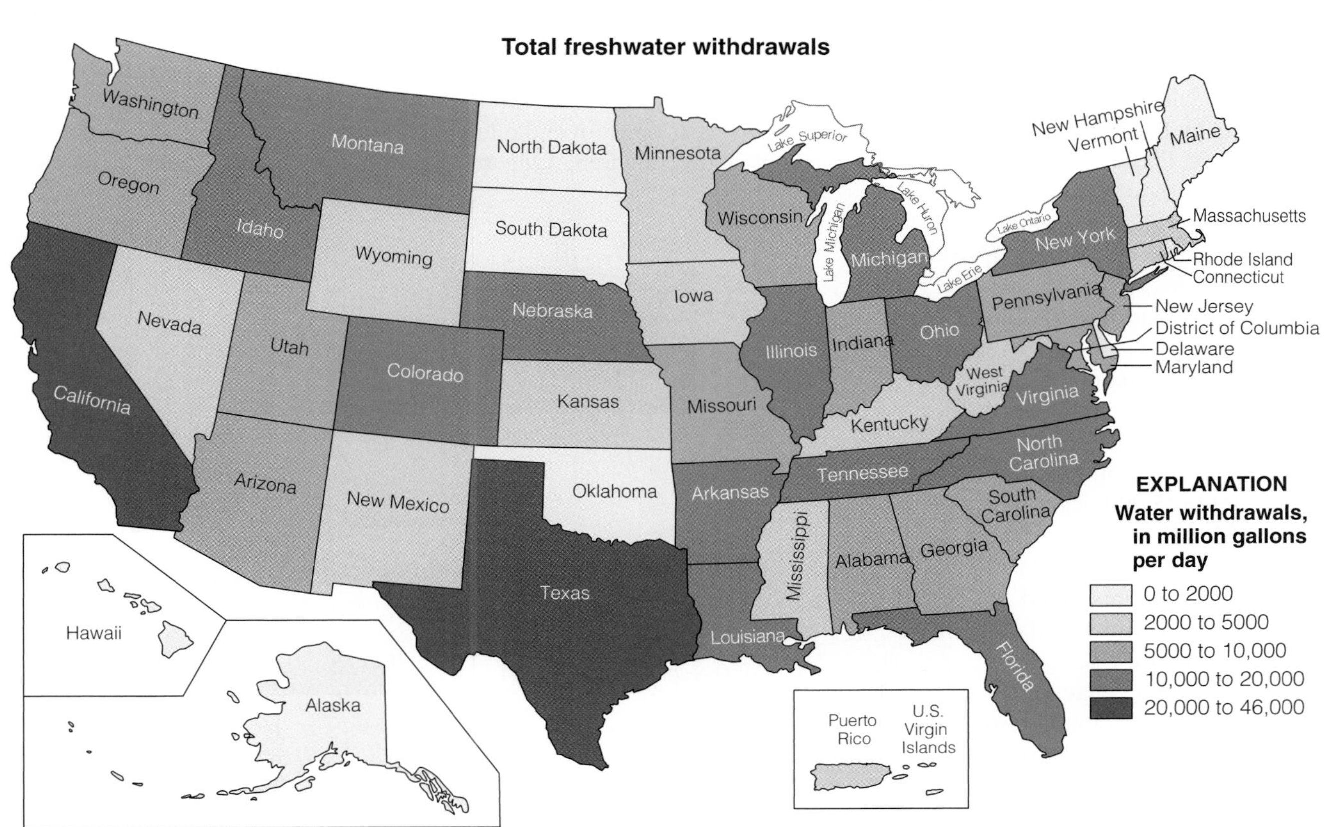

Figure 8.5 Total freshwater withdrawals by state, 2005. (USGS)

irrigation. This amount is expected to increase to 20,000 cubic kilometers (14.5 trillion gallons/day) by the year 2050, which is beyond the dependable flow of the world's rivers. The availability of water will limit population growth in the 21st century, either by statute in developed countries or by natural means, most likely drought and famine, in developing nations. Repetition of the terrible droughts experienced in Africa in the 1970s and 1980s, compounded by overtaxed water supplies, would be disastrous for those countries.

Lakes

Lakes are bodies of water in landlocked basins that are formed by many different geological processes. For example, the geological origin of the Great Lakes (except Lake Superior) is glacial, Crater Lake is volcanic, and the Great Salt Lake is a downfaulted basin. Lakes are not important quantitatively in the global water balance, holding less than 0.4% of all continental freshwater; however, they are of great local importance for agriculture and recreation and as a source of water. It is significant that about 80% of all the water in lakes worldwide is held in fewer than 40 large lakes.

Lakes are important as a source of freshwater, and salt lakes can be useful for transportation and mineral extraction. Whether a lake is fresh, salt, or dry depends on the balance between precipitation (P) and runoff (R) into the lake, and evaporation (E) and outflow (O) from the lake. If a condition exists where $P + R = E + O$, the result is a freshwater lake. If $P + R = E$ and there is no outflow, the result is a salt lake with a balance that is similar to the oceans. However, in the southwestern United States, the condition exists where $P + R < E + O$, producing in time a dry lake bed, also called a *playa*. Dry lake beds, especially the salt flats of the Great Salt Lake, have been used for automobile speed trials, because they are among the most level surfaces above sea level. Rogers Dry Lake in California has the distinction of being used as the landing strip for the space shuttle and is the home of Edwards Air Force Base. Playas are also known for a variety of salts and useful clays, products found in their sediments. Some lakes have their source of water cut off by humans, usually for agriculture, occasionally with catastrophic results (see Case Study 8.1).

Changing climate can dramatically alter the water balance in lakes. Consider this: If you had visited the southwestern United States 15,000 years ago, you would have seen lakes in the Mojave Desert. Those lakes would have been connected by rivers with rapids and waterfalls, places to swim and take a drink of freshwater. The glacial climate was cooler and cloudier, so evaporation was lower and precipitation may have been higher in some places. Even Death Valley, one of the hottest and driest places in the United States today, had a lake in it, Lake Manix. How do we know? Because the shorelines and lake-bottom sediments of this and other now-vanished lakes can easily be found.

A process that keeps the bottom waters in lakes from becoming stagnant is **overturning**. This occurs in lakes in temperate and cool climates where surface waters are subject to near-freezing temperatures. The temperature of maximum density of freshwater is 4°C (about 40°F). Water at this temperature is heavier per unit volume than water that is warmer or colder. In the summer, the lake is stratified and the warmer, less dense surface water is separated from the cooler, denser bottom waters by the *thermocline*, which acts as a floor below the surface water and a ceiling above the bottom waters. As it gets colder in the fall, the surface water temperature decreases to 4°C and the thermocline weakens; that is, the temperature of the lake becomes about equal, or *isothermal*. The surface water sinks and the bottom waters that are oxygen-deficient rise to replace them. As **Figure 8.6** illustrates, the bottom waters stagnate during the winter but are refreshed again in the spring when the surface-water temperature increases to 4°C and sinks, and the bottom waters rise. The net result is the oxygenation of the lake waters and recycling of nutrients, which are transported from the bottom to the surface. This makes for abundant plankton and good fishing.

Another factor in the well-being of lakes and their suitability as a source of water is how well "nourished" they are. Nourishment consists of inorganic compounds (nitrate, phosphate, and silica) that are carried into the lake by streams and released by decaying organic matter on the lake bottom. *Oligotrophic* lakes are usually deep, with clear

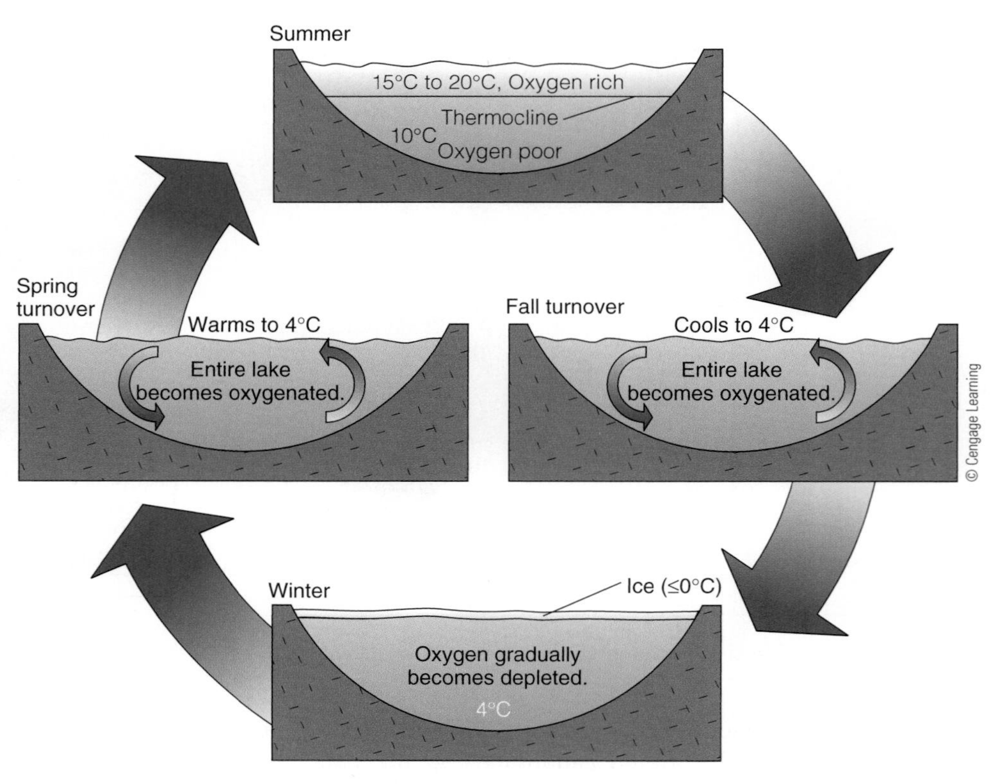

Figure 8.6 Fall and spring overturning occurs in lakes in temperate and cold regions. Overturning is a mechanism whereby deep lake waters are reoxygenated and do not become stagnant.

explorer's perspective

Photograph by Daniel Byers/See: http://www.nationalgeographic.com/explorers/bios/alton-byers/

ALTON BYERS knows glaciers and the water they provide.

Alton Byers is a geographer who has followed his own path to an exciting career that takes him to the world's greatest mountains where he studies water and the people it affects. He's a climber by trade and by hobby and a man who loves to live out of doors, getting up early, doing fieldwork in remote places, and living a full life with his family and friends.

Much of his recent work centers on water and the effect of climate change on both water resources and the hazards that water can cause people.

In Nepal, Byers and his High Mountain Glacial Watershed Program have taken students and highly technical gear into the high country to map out lakes dammed by glaciers—the goal, to understand these potentially hazardous systems and minimize the risk, to the communities below, of glacial lake dam failure and the huge floods that ensue.

What's so ironic is that mountain glaciers are also the lifeblood of mountain peoples. For millennia, their ice has been a reservoir, storing water as snow in the winter and releasing it for use as the glaciers melted in the summer. In arid Peru, where water is scarce and Byers also works, such natural reservoirs are critical for survival, providing a critical ecosystem service at no cost. With climate change causing these glaciers to shrink dramatically, the summer meltwater may soon be gone. Will we be building dams and reservoirs to take the place of what nature has always provided mountain and lowland peoples, a reliable supply of glacier meltwater?

Source: http://www.nationalgeographic.com/explorers/bios/alton-byers/

Tongues of glacial ice, such as this one in Greenland, provide reliable summer water supplies as they melt.
P. Bierman

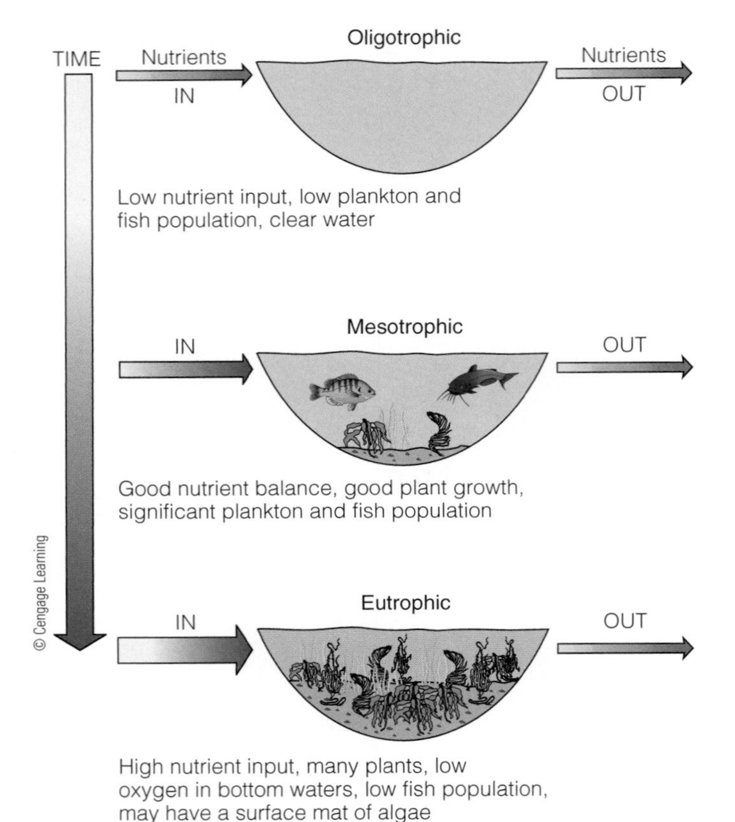

Figure 8.7 Over time, lakes go through a nutrient cycle, which can be accelerated by waste products from humans living around the lake or by agricultural chemicals.

water, and low in nutrients, with a relatively low fish population (**Figure 8.7**). They are a source of good drinking water and are usually visually appealing. Crater Lake is an example. A *eutrophic* lake is "well fed" with nutrients. Such lakes are often shallow, have plants, contain a population of so-called trash fish (such as carp), have a low oxygen content, and usually support a rich growth of algae or other plants at the surface (**Figure 8.8**). Between the two extremes are *mesotrophic* lakes, which have a well-balanced nutrient input and output. The result is a good fish population, a variety of shore and bottom plants, and an abundance of plankton (see Figure 8.7).

Freshwater Underground

Some of the water in the hydrologic cycle infiltrates underground and becomes **groundwater**, one of our most valuable natural resources. In fact, most of Earth's liquid freshwater exists beneath the land surface; its geological occurrence is the subject of many misconceptions. For example, it is commonly believed that groundwater occurs in large lakes or pools beneath the land. The truth is that almost all groundwater is in small pore spaces and fractures in rocks. Another misconception is that deposits of groundwater can be found by people with special skills or powers, people referred to as *water witches* or *dowsers*. They locate underground water using two wires bent into L shapes, forked branches, or other forked devices called *divining rods*. A dowser locates an underground "tube" or lake of water by walking over the ground, holding the divining rod in a prescribed manner until it jerks downward (**Figure 8.9**).

Some dowsers claim they can locate good places to drill water wells using only a map of the land. The divining rod is passed over the map until it jerks downward, where presumably a well should be located. Water dowsers are found throughout the world, particularly in rural areas, and many of them will guarantee finding water. This is one difference between dowsers and groundwater geologists, or **hydrogeologists**. Hydrogeologists will not make this guarantee. However, some people will swear that experienced dowsers can

Figure 8.8 Lake Dianchi, Yunnan Province in the People's Republic of China is eutrophic as indicated by the bright green algae in the water. Such eutrophication is most likely the result of excess nutrients being adding to the lake by agricultural runoff or sewage. Eutrophication caused by human activities is known as "cultural eutrophication."

Figure 8.9 Early dowsers looking for water.

find water reliably. Could it be that, after years of work in the field, dowsers get to know an area and use subtle landscape clues to find places where groundwater is likely to be found?

Groundwater Supply

Groundwater provides about 40% of public water supplies and 30% to 40% of all water used exclusive of power generation in the United States. It is by far the cheapest and most efficient source of municipal water, because obtaining it does not require the construction of expensive aqueducts and reservoirs. Furthermore, groundwater usually requires far less cleaning and purification than surface water. Thirty-four of the largest 100 cities in the United States depend entirely on local groundwater supplies; Miami Beach, San Antonio, Memphis, Honolulu, and Tucson are just a few. Groundwater provides 80% of the water for rural domestic and livestock use, and it is the only source of water in many agricultural areas. It is not surprising that some of the largest agricultural states, such as Texas, Nebraska, Idaho, and California, are the largest consumers of groundwater, accounting for almost half of all the groundwater produced and used. Groundwater is important even to communities that import water from distant areas, because, in almost every case, local groundwater provides a significant, low-cost percentage of their water supply.

Although it is not generally known, groundwater maintains streamflows during periods when there is no rain. This process, by which water from the surrounding landscape finds its way into streams, maintains what is known as baseflow. Baseflow is critical for maintaining the ecological integrity of streams, because it keeps water flowing long after rains have ceased. In arid climates where there are permeable, water-saturated rocks beneath the surface, groundwater seeping into stream channels may be the only source of water for rivers. The fabled oases of the Sahara and Arabian deserts occur where underground water is close to the surface or where it intersects the ground surface, forming a spring or watering hole.

Location and Distribution

Between the land surface and the depth at which we find groundwater is the *zone of aeration,* a zone where voids in soil and rock contain only air and water films. This is also known as the unsaturated zone. The contact between this zone and the *zone of saturation* below it is the **water table** (**Figure 8.10**). Above the water table is a narrow moist zone, the *capillary fringe,* where pores are filled with water, but the water is held by surface tension and cannot freely drain from the soil. In the saturated zone, free (unattached) water fills the openings between grains of clastic sedimentary rock, the fractures or cracks in hard rocks, and the solution channels in limestones and dolomites (**Figure 8.11**).

Water-saturated geological formations whose porosity and permeability are sufficient to yield significant quantities of water to springs and wells are known as **aquifers**. A rock's **porosity** (root word *pore*) is a function of the volume of openings, or void spaces, in it and is expressed as a percentage of the total volume of the rock being considered. Without such voids, a rock cannot contain water or any other fluid. **Permeability** is the ease with which fluids can flow through an aquifer; it is thus a measure of the connections between pore spaces.

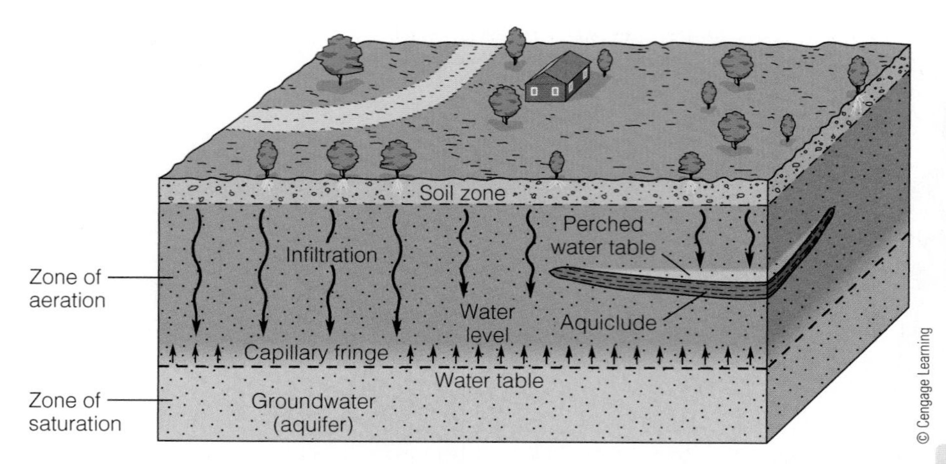

Figure 8.10 A schematic cross section of groundwater zones, showing a static water table with a capillary fringe. A perched water table exists where a body of clay or shale acts as a barrier (otherwise known as an aquiclude) to further infiltration.

Hydraulic conductivity is the term hydrologists prefer, but *permeability* will suffice for our purposes, as it has a long history of use and is easy to remember (see Case Study 8.2). **Aquicludes** are rocks or sediments, such as shale and clay, that lack significant permeability, and hence transmit little water. In some places, aquicludes occur as *lenses* that trap water and prevent it from percolating down to the water table. Such a condition creates a **perched water table** (see Figure 8.10). Aquifers are important because they transmit water from recharge areas—areas where water is added to the zone of saturation—to springs or wells. In addition, they store vast quantities of drinkable water. In terms of the hydrologic cycle, we can view aquifers both as transmission pipes and as storage tanks for groundwater. The Nubian sandstone in the Sahara Desert, for instance, is estimated to contain 600,000 cubic kilometers (130,000 cubic miles) of water, which is just now being tapped. It is the world's largest known aquifer and is a resource of great potential for North Africa. However, like the High Plains aquifer, it contains much fossil water, so that extracting large amounts is unsustainable mining. Started in 1984, the Great Man-Made River Project, a brainchild of Libya's former leader, Muammar Al Qadhafi, now carries 6.5 million cubic meters of water each day from desert wells to the cities and agricultural fields of this arid nation (**Figure 8.12**).

Static Water Table

Unconfined aquifers are formations that are exposed to atmospheric pressure changes and that can provide water to wells by draining adjacent saturated rock or soil (**Figure 8.13a**). As the water is pumped, a **cone of depression** forms around the well, creating a gradient that causes water to flow toward the well (see Figure 8.13b). A low-permeability aquifer will produce a steep cone of depression and substantial lowering of the water table in the well. The opposite is true for an aquifer in highly permeable rock or soil. This lowering, or the difference between the water-table level and the water level in the pumping well, is known as **drawdown**. Stated differently, low permeability produces a large drawdown for a given yield and increases the possibility of a well "running dry." As pointed out, a pumping well in permeable materials creates a small drawdown and a gentle gradient in the cone of depression. Some closely spaced wells, however, have overlapping "cones," and excessive pumping in one well lowers the water level in an adjacent one (**Figure 8.14**). Many lawsuits have arisen for just this reason. Water law is discussed briefly in Case Study 8.3.

consider this . . . As explained in Case Study 8.3, the legal issues surrounding the withdrawal and use of groundwater are complex, and several legal doctrines apply to them. From a philosophical viewpoint, which of the four doctrines seems the fairest for multiple users of an aquifer? Please explain your choice.

A good domestic (single-family) well should yield at least 11 liters (2.6 gallons) per minute, which is equivalent to about 16,000 liters (3,700 gallons) per day, although some families use as little as 4 liters per minute (1,350 gallons/day). Many people get by with domestic wells that yield considerably less water than this. The reason is simple. If the well casing is long, large amounts of water can be stored in the well itself. For a 6-inch well, which is typical

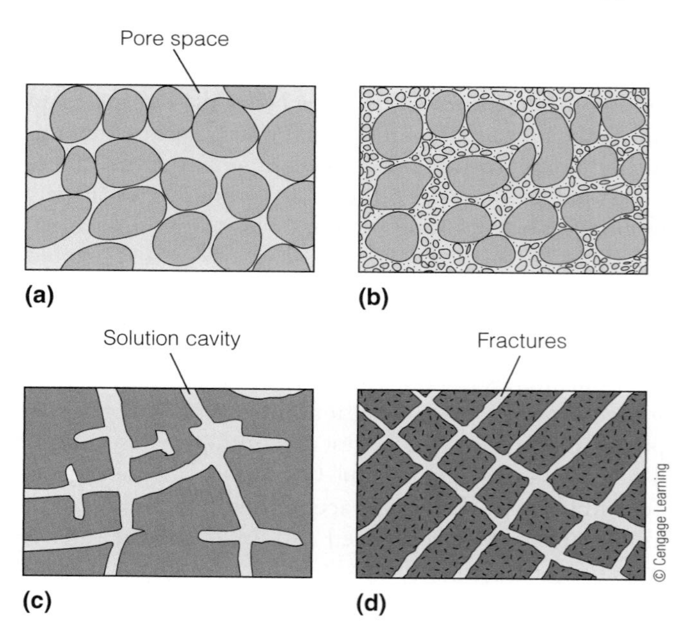

FIGURE 8.11 A rock's porosity is dependent on the size, shape, and arrangement of the material composing the rock. Whereas (a) a well-sorted sedimentary rock has high porosity, (b) a poorly sorted one has low porosity. (c) In soluble rocks, such as limestones, porosity can be increased by solution, whereas (d) crystalline metamorphic and igneous rocks are porous only if they are fractured.

Figure 8.12 Before he was deposed in 2011, Libya's leader, Muammar Al Qadhafi, had overseen an ambitious plan to water his nation with fossil water from the Nubian sandstone aquifer. (a) A map of Libya and the system that takes ancient water (recharged during cooler, moister glacial times) to the cities of today. (b) A billboard showing Qadhafi and the water system. (c) One of the 4-meter-wide (14 foot) pipes that carry the water.

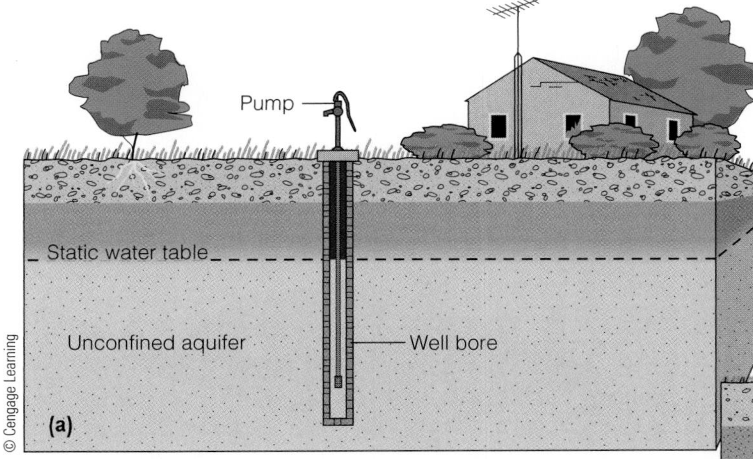

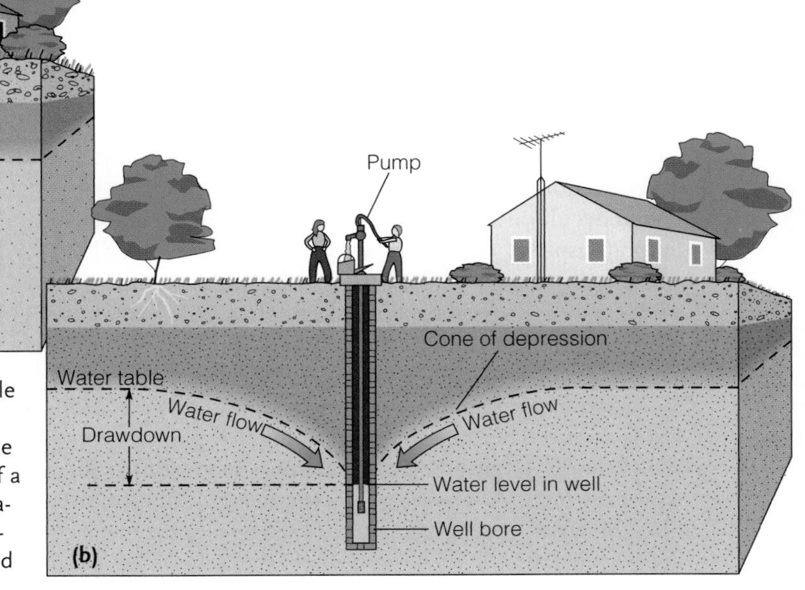

Figure 8.13 (a) A horizontal (static) water table. The water table will remain static until pumping begins. (b) When pumping begins, the water table is drawn down an amount determined by the pumping rate and the aquifer's permeability. The development of a cone of depression creates a hydraulic gradient, which causes water to flow toward the well and enables continuous water production. The amount of water the well yields per unit of time pumped depends on the aquifer's hydrologic properties.

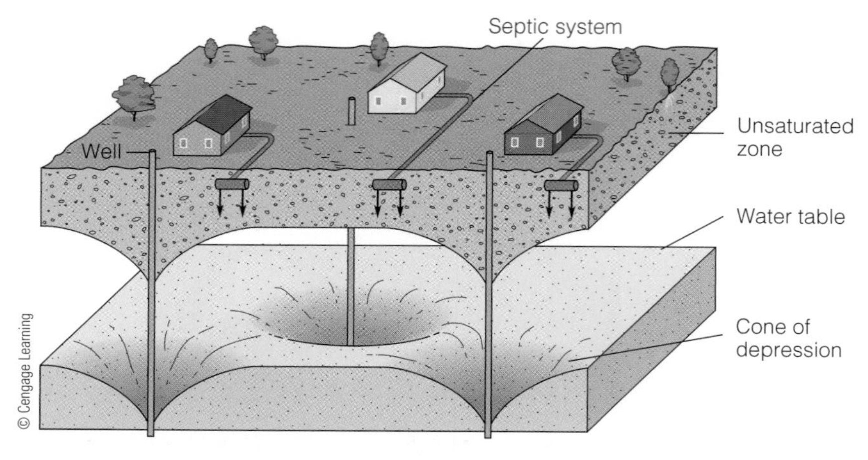

Figure 8.14 Cones of depression overlap in an area of closely spaced wells. Note the contributions to groundwater from septic tanks.

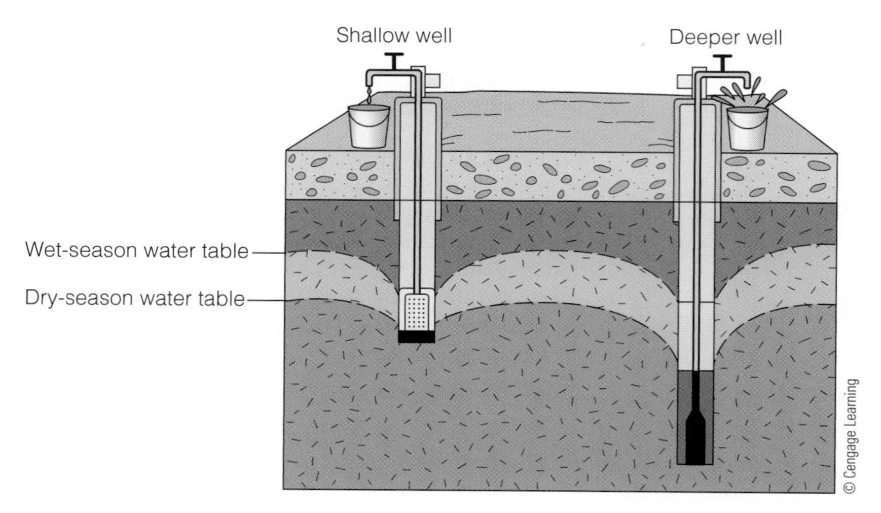

Figure 8.15 The effect of seasonal water-table fluctuations on producing wells. A shallow well may go dry in the summer.

for a residence, each foot of casing stores about 1.5 gallons of water, so a deep well of 200 feet gives people a several-hundred-gallon buffer, good enough for very long showers but not enough to fill the family swimming pool.

The water table rises and falls with consumption and the season of the year. Generally, the water table is lower in late summer and higher during the winter or the wet season. Trees use lots of groundwater, extracting it with their roots. When deciduous trees lose their leaves in the fall and cease evapotranspiring (pumping water), water tables rise. Thus, water wells must be drilled deep enough to accommodate seasonal and longer-term climatic fluctuations of the water table (**Figure 8.15**). If a well "goes dry" during the annual dry season, it needs to be deepened.

Streams located above the local water table are called **losing streams**, because they contribute to the underground water supply (**Figure 8.16a**). Beneath such streams, there may be a mound of water above the local groundwater table, known as a *recharge mound*. Losing streams are most commonly found in the desert, where the water table is usually far below the ground surface and streamwater frequently originates from runoff in high mountains towering above the desert floor. A stream that intersects the water table and is fed by both surface water and groundwater is known as a **gaining stream** (see Figure 8.16b). These streams are most common in humid regions where precipitation is regular and abundant, keeping groundwater tables near the surface. A stream may be both a gaining and a losing stream, depending on the time of year and variations in the elevation of the water table. The gaining of water by streams is what makes up baseflow.

Where water tables are high, they may intersect the ground surface and produce a spring. Springs are also subject to water-table fluctuations. They may "turn on and off," depending on rainfall and the season of the year. **Figure 8.17** illustrates the hydrology of three kinds of springs.

Pressurized Underground Water

Pressurized groundwater systems cause water to rise above aquifer levels and sometimes even to flow out onto the ground. Pressurized systems occur where water-saturated permeable layers are enclosed between aquicludes, and for this reason they are called **confined aquifers** (**Figure 8.18**). A confined aquifer acts as a water conduit, very similar to a garden hose that is filled with water and situated so that one end is lower than the other. In both cases, a water-pressure difference known as a *hydraulic head* is created. Water

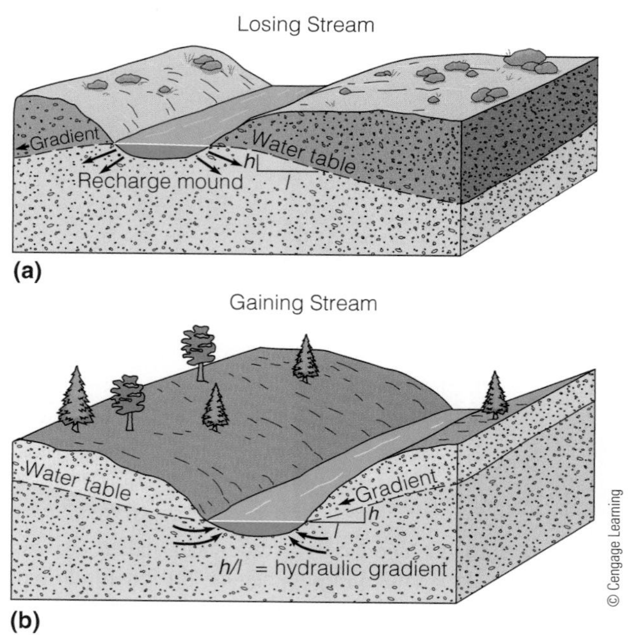

Figure 8.16 Hydraulic gradient (*h*/*l*) in (a) a losing stream, which contributes water to the water table, and (b) a gaining stream, which gains water from the water table.

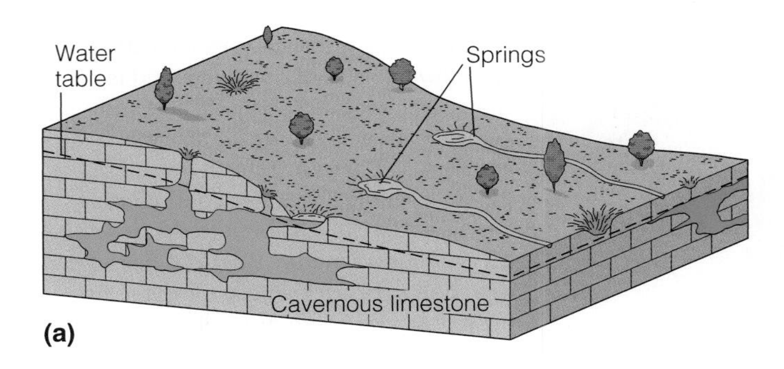

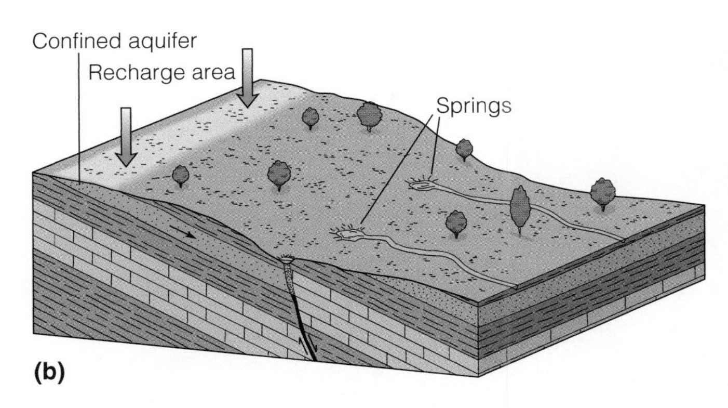

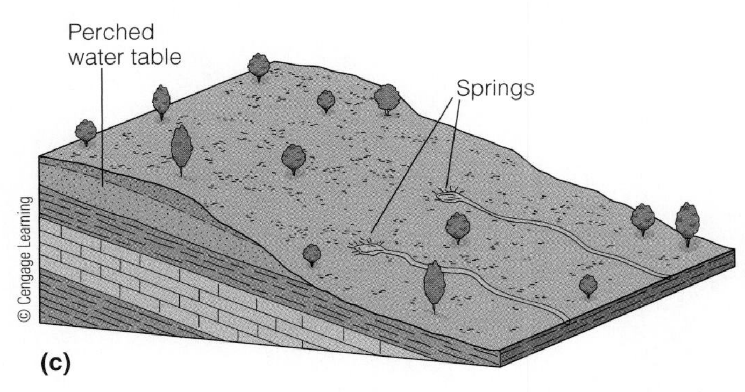

Figure 8.17 Geology and hydrology of three kinds of springs. (a) Springs form where a high water table in cavernous limestone intersects the ground surface. (b) Artesian springs. A fault barrier can cause pressurized water in a confined aquifer to rise as springs along irregularities in the fault. (c) Springs form where a perched water table intersects the ground surface. Such springs are intermittent, because the small volume of perched water is quickly depleted.

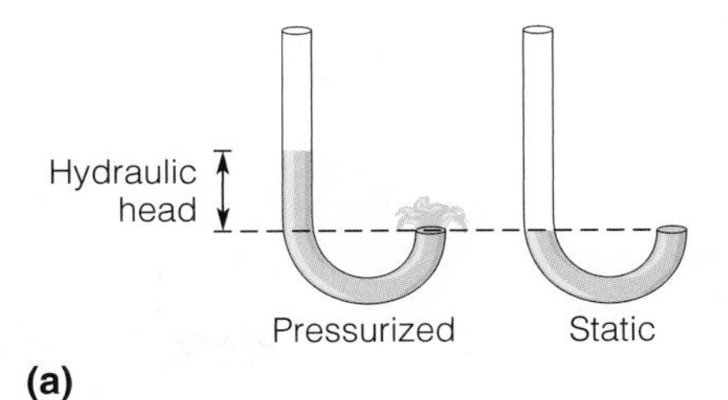

Figure 8.18 Water under pressure. (a) The hydraulic head in a hose lowers as water flows out the low end. Eventually, a static condition prevails. (b) A confined aquifer, commonly called an *artesian* aquifer. The potentiometric surface is the level to which water will rise in wells at specific points above the aquifer. Why is one well flowing, whereas the other is not?

gushes from the low end until there is no longer a pressure difference. At that point the water level stabilizes, becoming *static*.

Wells that penetrate confined aquifers are called *artesian wells*, after the province of Artois, France, where they were first described. Water in some artesian wells is under enough pressure that it flows out onto the ground. In other cases, water simply rises up the well casing above the top of the aquifer. Artesian systems are distinguished by a water-pressure gradient called the **artesian-pressure**, or **potentiometric, surface** (**Figure 8.19**). This "surface" is the level to which water will rise in a well at a given point along the aquifer. It is important to note that the slope of the water table of an unconfined aquifer is also a pressure surface. Underground water moves from areas of high hydraulic head to areas of low hydraulic head (pressure).

Contour maps of the potentiometric surface can be very useful, because they indicate the direction of groundwater flow and the slope of the hydraulic gradient (**Figure 8.20**). If an aquifer's hydraulic conductivity can be estimated or measured, the rate and volume of groundwater flow can be determined using Darcy's law (see Case Study 8.2).

Artesian wells differ from water-table wells in that the latter depend on pumping or topographic differences to create the hydraulic gradient that causes the water to flow. Because the energy required to raise water to the surface is the greatest expense in tapping underground water, artesian systems are much desired and exploited (see Case Study 8.4). Not all artesian wells flow to the ground surface, because the potentiometric surface lowers as water is extracted; water may rise naturally only partway up the well bore. Nevertheless, a natural rise of water any distance reduces the cost of its extraction.

Many bottled-water and beverage producers attempt to convince us that deep-well or artesian water is purer than other kinds of water. *Artesian* refers only to an aquifer's hydrogeology and has no water-quality connotation except that recharge zones for artesian wells are often in high-elevation, mountainous zones, which are less densely settled than valleys and less likely to be impacted by industrial waste dumping and polluting agricultural practices. Because artesian aquifers are under pressure where they are tapped as water supplies, water leaks out of them rather than

Figure 8.19 Artesian wells indicate that groundwater is under pressure and is recharged at a location both distant and higher than the well. (a) In this colorized photograph, taken in Tunisia (northern Africa) in the 1920s, water pouring from an artesian well flows off into the desert. (b) Sometimes, nature does not need a well to release artesian groundwater. In the outback of South Australia, a freshwater mound spring bubbles up through an arid desert plain. The groundwater here is clearly under pressure, but the recharge area is nowhere to be seen. It is hundreds of kilometers (miles) away in the highlands where more precipitation falls.

in, protecting water quality. Do not be misled by advertisements attributing greater purity to artesian or deep-well water. Such waters may or may not have low levels of dissolved contaminants.

Groundwater Storage and Management

To manage underground water supplies in a groundwater basin, the hydrologist or planner must have two measurements: the quantity of water stored in the basin, and the **sustained yield**, the amount of water the aquifers in the basin can yield on a day-to-day basis over a long period.

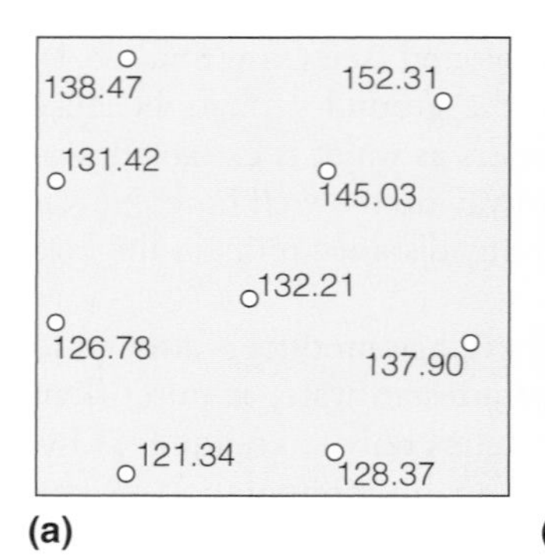

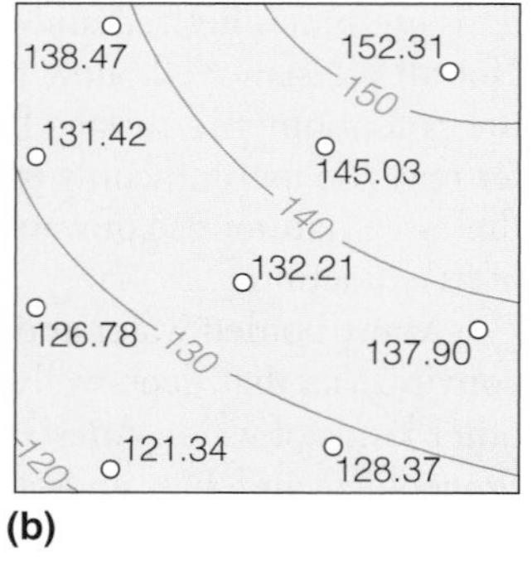

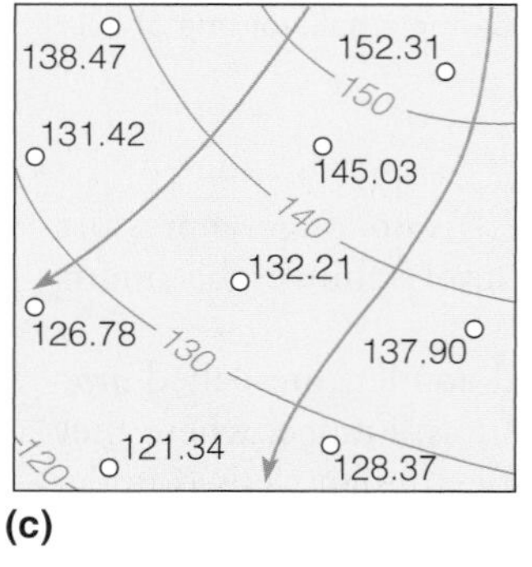

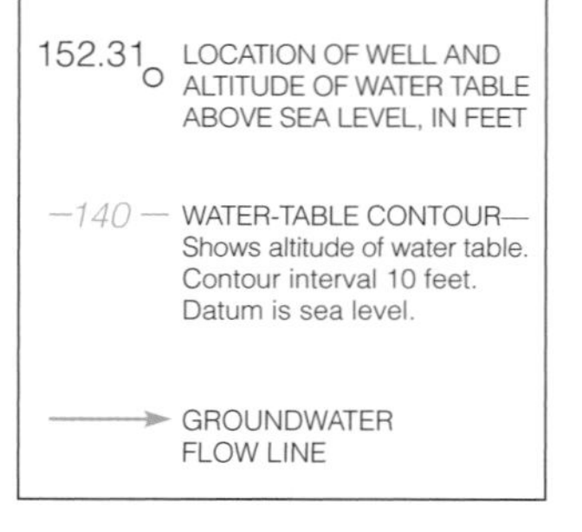

Figure 8.20 Using (a) known altitudes of the water table at individual wells, (b) contour maps of the water-table surface can be drawn and (c) directions of groundwater flow along the water table can be determined, because flow is perpendicular to the contours. Elevations are in feet above a datum, usually sea level, as English units are used almost exclusively on pressure-surface contour maps in the United States. (Source: USGS circular 1139.)

A groundwater basin consists of an aquifer or a number of aquifers that have well-defined geological boundaries and *recharge areas*—places where water seeps into the aquifer (see Figure 8.18b). To determine the amount of usable water in the basin, we rely on the concept of specific yield of water-bearing materials. **Specific yield** is the ratio of the volume of water an aquifer will give up by gravity flow to the total volume of material, expressed as a percentage. For example, a saturated clay with a porosity of 40% may have a specific yield of 3%; that is, only 3% of the clay's volume is able to drain into a well. The remaining water in the pore spaces (37%) is held by the clay-mineral surfaces and between clay layers against the pull of gravity. This "held" water is its *specific retention*. Similarly, sand with a porosity of 35% might have a specific yield of 23% and a specific retention of 12%. Note that the sum of specific yield and specific retention equals the porosity of the aquifer material.

Thus, the total available water in an aquifer can be calculated by multiplying the specific yield by the volume of the aquifer determined from water-well data. The process is a bit more complicated for artesian aquifers, but it follows the same procedure after accounting for the compaction of the aquifer and the expansion of the water when the pressure is reduced. The amount of water available in a groundwater basin can be determined fairly accurately if the basic data are available from numerous wells.

consider this . . . Some hot springs form geysers that are celebrated for their regularity. Old Faithful in Yellowstone National Park is an example. Cold springs, in contrast, do not generate attention-grabbing geysers, and their flow is often far from regular. Indeed, many springs operate on a sort of stop-and-go basis. How can we explain this intermittent flow?

As noted earlier, sustained yield is the amount of water that can be withdrawn on a long-term basis without depleting the resource. Sustained yield is more difficult to assess than aquifer storage capacity, because it is affected by precipitation, runoff, and recharge. One of the challenges facing groundwater resource managers is how to deal with the effect on sustained yield of climate change, because climate change alters rates of rainfall, evaporation, and thus aquifer recharge. Global circulation models predict that low to mid-latitudes will both warm and dry over the next century, reducing aquifer recharge at the same time that water demand will be increasing. Over high latitudes, both precipitation and temperature will increase. In the face of changing climate, and thus aquifer recharge, it's more important than ever to estimate sustained yield, so that groundwater resources can be managed intelligently.

Groundwater Mining

As explained in the chapter opening, when the amount of water withdrawn from an aquifer exceeds the aquifer's sustained yield, an overdraft condition, called *groundwater mining*, exists (**Figure 8.21**).

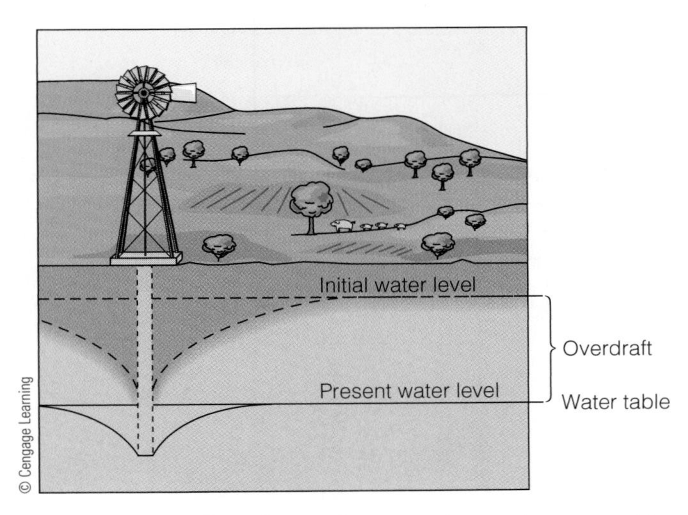

Figure 8.21 Effect of groundwater mining on the water table. The drop from the initial level to the present level is the *overdraft*, the amount of water that has been mined.

The Ogallala sandstone makes up about 80% of what is collectively known as the High Plains aquifer. In general, water levels in this important aquifer are dropping. In Texas, overdrafts have exceeded 65%, and overdrafts of 95% are known in a few places. In the latter case, 20 times more water was extracted from the aquifer each year than was recharged. This magnitude of overdraft is a serious problem, because much of the water in the aquifer infiltrated during wetter glacial climates more than 10,000 years ago. Most water pumped from the aquifer is fossil.

The High Plains aquifer was not massively tapped until the 1950s, and by 1980, the water-saturated zone had decreased an average of 3 meters (10 feet), as much as 30 meters (100 meters) under some parts of Texas and Nebraska (**Figure 8.22**). Government subsidies have encouraged growing water-gulping corn, rather than less-water-intensive wheat, sorghum, and cotton. Over the last decade, increasing amounts of corn have been grown as the feedstock for a popular and government-subsidized biofuel, ethanol. In 2009, about one third of the U.S. corn crop was used for ethanol production. Much of this corn was irrigated by fossil water in an attempt to replace fossil fuel.

In the early 1980s, the rate of water-table decline decreased because of heavy rain and snow, better water management, and new technologies. New low-energy, precision-application (LEPA) irrigation nozzles use less water and less energy, and decrease evaporation by as much as 98% from the amounts lost with spray-type irrigation. Furthermore, treated wastewater is now being recycled on fields.

A 1982 Department of Commerce study found that 6 million hectares (15 million acres) were being irrigated with water drawn from 150,000 wells in the Ogallala and other High Plains aquifers. By the year 2020, according to

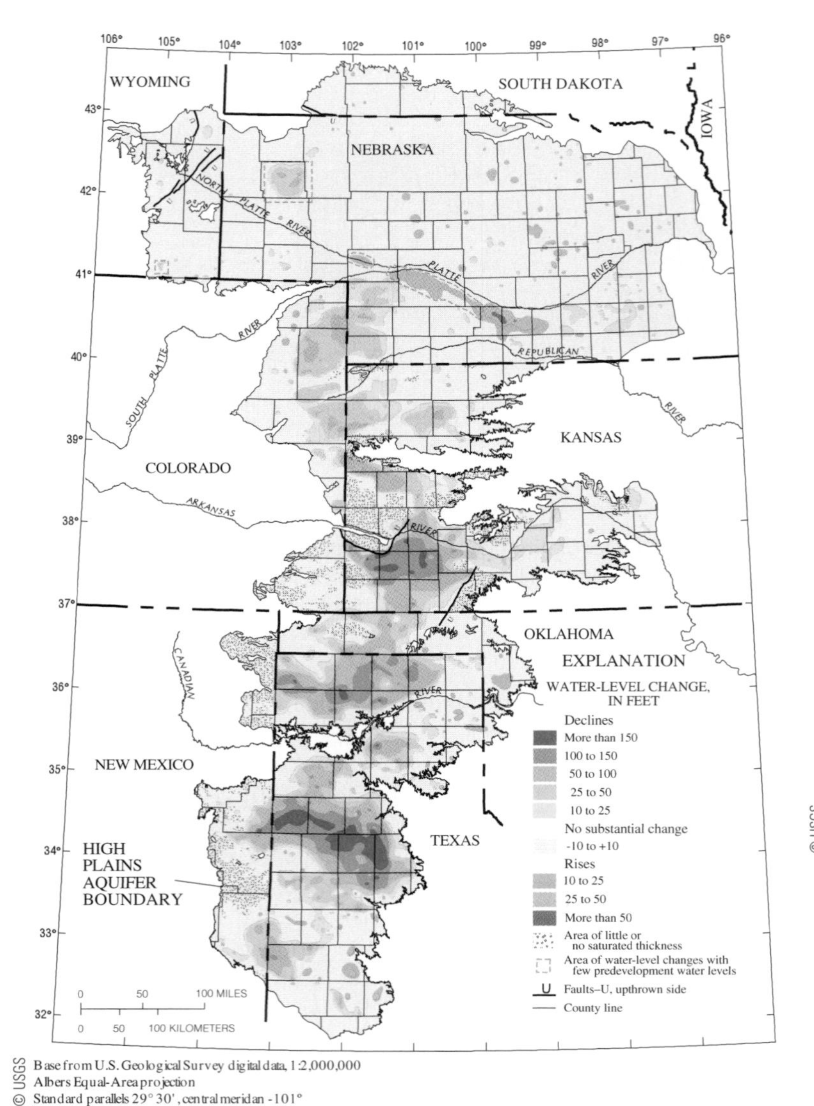

Base from U.S. Geological Survey digital data, 1:2,000,000
Albers Equal-Area projection
Standard parallels 29° 30', central meridian -101°

Figure 8.22 Water-level changes in the High Plains aquifer through 2007. (Source: McGuire, V. L., 2009. Water-level changes in the High Plains aquifer, predevelopment to 2007, 2005–06, and 2006–07: U. S. Geological Survey Scientific Investigations Report 2009–5019, 9 p. Available at: http://pubs.usgs.gov/sir/2009/5019/.)

Figure 8.23 Should the High Plains aquifer be treated as a bathtub or an egg carton? The egg carton analogy assumes that a user is not subject to water depletion by the action of neighboring irrigators because of the local nature of the cone of depression. The bathtub analogy assumes that the aquifer acts as a common pool and the water level responds as if water were being drawn from a lake or bathtub. The actual situation lies somewhere between these two analogies because different parts of the High Plains aquifer are interconnected. (Source: Alley, William M., Thomas Reilly, and O. Lehn Franke. 1999. *Sustainability of ground-water resources.* U.S. Geological Survey circular 1186, p. 47.)

the study report, about a fourth of the aquifer's water will have been mined at the present rate of withdrawal. A 1995 estimate of the total overdraft of the Ogallala is that it equals 1 year's flow of the Colorado River. The Ogallala aquifer is certainly the life-blood of High Plains agriculture, and most users of its water are aware that mining the resource must eventually be curtailed. Strict management that includes conservative usage, monitoring, and utilization of new technologies is required for an attempt at balancing its recharge and withdrawals (**Figure 8.23**). Groundwater mining causes shallow wells to go dry, increases the cost of lifting water to the surface as water levels drop, and may eventually cause ground subsidence (see Chapter 7).

The Santa Clara Valley of central California is a classic case of regional overdraft. Water levels there declined 46 meters (150 feet) between 1912 and 1959, and downtown San Jose dropped 2.76 meters (9 feet). Overpumping in Alabama dropped the potentiometric surface in one aquifer 62 meters (200 feet); it is now below sea level more than 80 miles (130 kilometers) inland from the sea. The solution to such a problem is *artificial recharge*, or "water spreading," which is accomplished by importing or diverting surface water and ponding it where it can percolate into the aquifer. Such a program was initiated in the Santa Clara Valley in 1959, which, in combination with decreased pumping, has restored the water table to 1912 levels. Raising the water table does not bring the land back to its original level, however, because pore spaces are irreversibly lost as subsidence occurs.

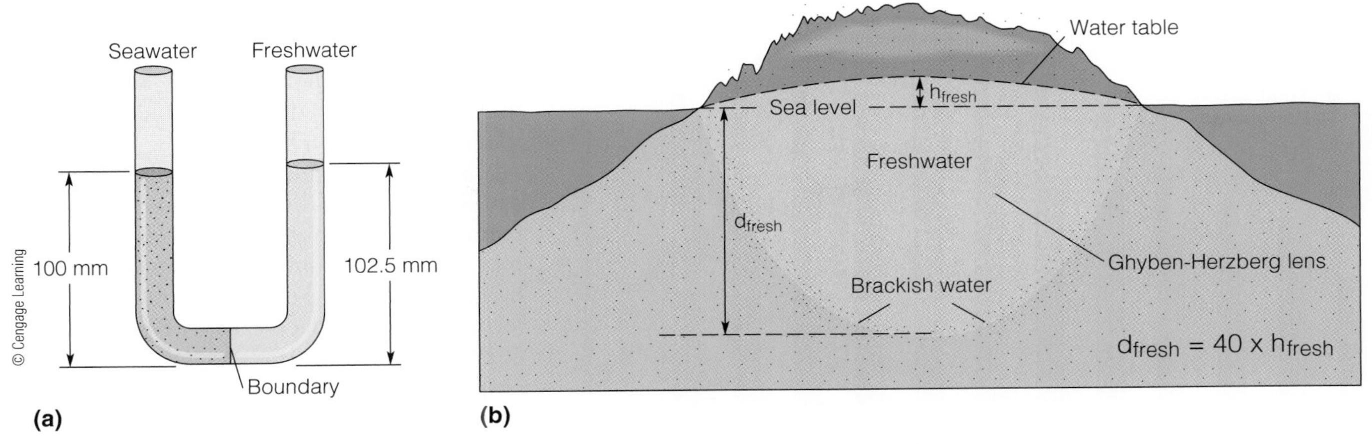

Figure 8.24 (a) The density contrast between freshwater and seawater. Because seawater is 2.5% denser than freshwater, a 102.5 millimeter column of freshwater is required to balance a 100 millimeter column of saltwater. (b) A hypothetical island and Ghyben-Herzberg lens of freshwater floating on seawater. The freshwater lens extends to a depth 40 times the height of the water table above sea level (not to scale).

One inadvertent but very useful product (for hydrologists) of nuclear testing in the 1950s and early 1960s was the introduction into the atmosphere, and thus the hydrosphere, of radioactive tritium. This rare isotope of hydrogen has 2 neutrons and 1 proton (most hydrogen has no neutrons); once incorporated in water (^3H_2O), it labels that water as younger than the mid-1950s. Tritium has a short half-life (little more than a decade) and so, over several decades, it decays radioactively and disappears from the aquifer. Tritium has been used as a tracer of groundwater recharge, because water older than the mid-1950s and modern water both contain very little tritium.

Groundwater–Saltwater Interaction

Aquifers in coastal areas may discharge freshwater into the ocean, creating bodies of diluted seawater. If the water table is lowered by pumping to or near sea-level elevation, however, saltwater invades the freshwater body, causing the water to become saline *(brackish)* and thus undrinkable. The direction of flow between the two water masses is determined by their density differences. Seawater is 2.5% (1/40th) denser than freshwater. This means that a 102.5-foot-high column of freshwater will exactly balance a 100-foot-high column of seawater (**Figure 8.24a**). Freshwater floats on seawater under the Hawaiian Islands, the Outer Banks of North Carolina, Long Island, and many other coastal areas (see Case Study 8.5).

This relationship is known as the *Ghyben-Herzberg lens*—named for the two scientists who independently discovered it and for the lens-like shape of the freshwater body (see Figure 8.24b) that floats on the denser saltwater. In theory, the mass of freshwater extends to a depth below sea level that is 40 times the water-table elevation above sea level. If the water table is 2.5 feet (75 cm) above sea level, the freshwater lens extends to a depth of 100 feet (30 m) below sea level. Any reduction of water-table elevation causes the saltwater to migrate upward into the freshwater lens until a new balance is established. If the water table drops below the level necessary for maintaining a balance with denser seawater, wells will begin to produce brackish water and eventually saltwater. Thus, it is important to maintain high water-table levels in coastal zones, so that the resource is not damaged by **saltwater intrusion**. High water tables are maintained by good management combined, when necessary, with artificial recharge using local or imported water (water spreading), or with the injection of imported water into the aquifer through existing wells.

Groundwater–Surface Water Interaction and Usage Trends

As has been emphasized thus far, surface water and groundwater are not separate entities. Nearly all surface-water features—lakes, streams, rivers, artificial lakes, and wetlands—interact with water underground. Some surface-water bodies gain water and acquire a different chemistry from groundwater, and other surface waters contribute to underground water and may even pollute it. Thus, in water planning, it is important to have a clear understanding of the linkages between groundwater and surface water, and of the geology that controls both.

The greatest volume of freshwater is used for irrigation, and the greatest total volume of water, both fresh and saline, is used for power generation. However, 98% of the water used for power generation is returned to streams or other reservoirs; only 2% is consumed. How water is used affects the reuse potential of return flows. For example, irrigation water may be too salty or contaminated by pesticides and insecticides to have any reuse potential unless it is purified naturally or artificially. In contrast, water used in power generation has great reuse potential, because the principal

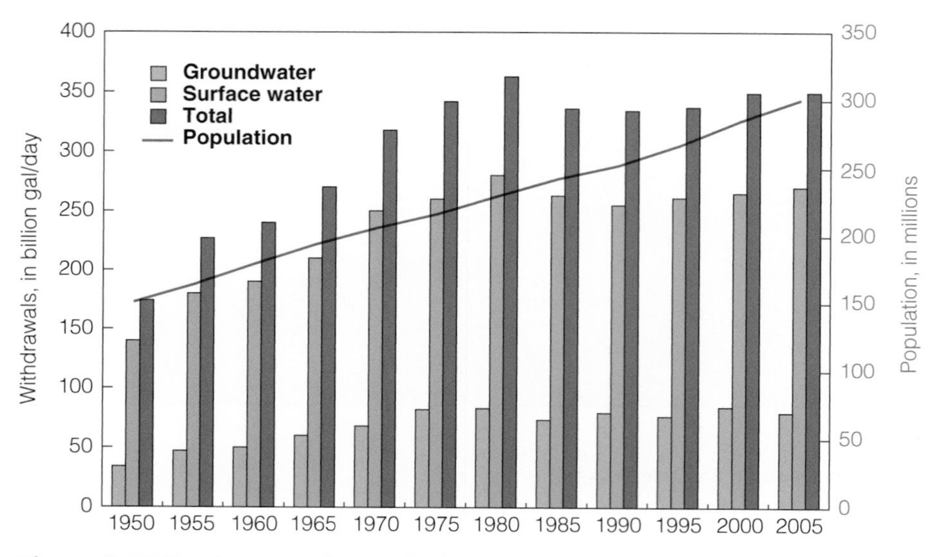

Figure 8.25 Trends in United States fresh groundwater and surface-water withdrawals and population, 1950–2005. (Source: USGS)

change is usually an increase in its temperature and location (or head) as it is released downstream.

Since 1950, U.S. water-use statistics have been compiled every 5 years. There was a steady increase in per-person water usage from 1950 to 1980, the all-time high. Between 1990 and 2005, there was a steady decrease in water use per person (in 2005, the average was ~1,400 gallons a day per person, about the same as 1955), but the U.S. population grew by 33 million people (**Figure 8.25**), so total withdraws were steady at about 340 billion gallons per day. Some of the reasons cited for this decrease were the development and use of more efficient irrigation methods, conservation programs, increased water recycling, and a downturn in farm economies.

Geological Work by Underground Water

The solution (a chemical reaction discussed in Chapter 6) of limestone and gypsum by underground water leads to the formation of caves, caverns, disappearing streams, and high-yielding and extensive groundwater sources, such as the Edwards aquifer (see Case Study 8.4). Land area exhibiting these features is called **karst terrain** (German *karst*, for the Kars limestone plateau of northwest Yugoslavia) (**Figure 8.26**). Urbanization on karst terrain typically results in problems such as the contamination of underground water and collapse of cavern roofs to form sinkholes.

The first sign of impending disaster in Winter Park, Florida, occurred one evening when Rosa Mae Owens heard a "swish" in her backyard and saw, to her amazement, that a large sycamore tree had simply disappeared. The hole into which it had fallen gradually enlarged, and by noon the next day, her home had also fallen into the breach. Six Porsches in a nearby storage yard suffered the same fate, as did a city swimming pool and two residential streets. Deciding to work with nature, the City of Winter Park stabilized and sealed the sinkhole and converted it to a beautiful urban lake. Few people who read of this notorious sinkhole in May 1981 realized the extent of the sinkhole problem in the southeastern United States.

Sinkholes are most commonly found in carbonate terrains (limestone, dolomite, and marble), but they are also known to occur in rock salt and gypsum. They form when carbonic acid (H_2CO_3) dissolves carbonate rock, forming a near-surface cavern (**Figure 8.27**). Continued solution leads to collapse of the cavern's roof and the formation of a sinkhole. About 20% of the United States and 40% of the country east of the Mississippi River are underlain by limestone, dolomite, or carbonate rocks. The states most impacted by surface collapse are Alabama, Florida, Georgia, Tennessee, Missouri, and Pennsylvania.

Tens of thousands of sinkholes exist in the United States, and the formation of new ones can be accelerated by human activities. For example, excessive water withdrawal that lowers groundwater levels reduces or removes the buoyant support of shallow caverns' roofs. In fact, the correlation between water-table lowering and sinkhole formation is so strong that "sinkhole seasons" are designated in the Southeast. These seasons are declared whenever groundwater levels drop naturally because of decreased rainfall and in the summer, when the demand for water is heavy. Vibrations from construction activity or explosive blasting can also trigger sinkhole collapse. Other mechanisms that have been proposed for sinkhole formation are fluctuating water tables (alternate wetting and dry-

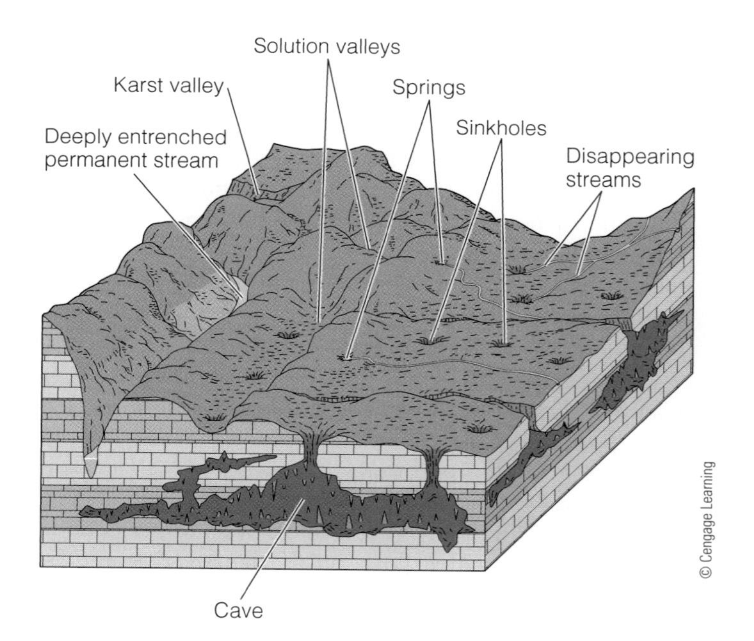

Figure 8.26 Surface and subsurface features, as well as discontinuous surface drainage, are indicative of karst topography.

Table 8.1 Common Water Quality Problems

Problem	Probable Cause	Correction
Mineral scale buildup	Water hardness	Water softener or hardness inhibitor (such as Calgon)
Rusty or black stains	Iron and/or manganese	Aeration/filtration; chlorination; remove source of Fe/Mn
Red-brown slime	Iron bacteria	Chlorination/filtration unit; remove source of slime
Rotten-egg smell or taste	Hydrogen sulfide and/or sulfate-reducing bacteria; low oxygen content	Aeration; chlorination/filtration unit; greensand filters
Gastrointestinal diseases, typhoid fever, dysentery, and diarrhea	Coliform bacteria and other pathogens from septic tanks or livestock yards	Remove source of pollution; disinfect well; boil water; chlorinate; abandon well and relocate new well away from pollution source

Source: Daly, D. 1994. Groundwater quality and pollution. Geological Survey of Ireland Circular 85-1, 1985. In Moore, J. E., et al. *Ground water: A primer.* American Geological Institute.

ing of cover material reduces its strength) and high water tables (which erode the roofs of underground openings).

Although the development of natural sinkholes is not predictable, it is possible to assess a site's potential for collapse resulting from human activities. This requires extensive geological and hydrogeological investigation before site development. Under the best of geological conditions, good geophysical and subsurface (drilling) data can tell *where* collapse is most likely to occur, but they cannot tell exactly *when* a collapse may occur. Armed with this information, planners and civil engineers can implement zoning and building restrictions to minimize the hazards to the public.

You could become a millionaire if you developed a relatively inexpensive method of locating caverns that may collapse and form sinkholes. Got any ideas? If not, what types of more sophisticated (and expensive) methods might be used to locate caverns?

Water Quality

Dissolved Substances

Varying concentrations of all the stable elements known on Earth are dissolved in natural waters. Some of these elements and compounds, such as arsenic, are poisonous. Others are required for sustaining life and health; common table salt (sodium chloride) is one, and it occurs in all natural waters. Some are nuisance pollutants that either make the water taste bad or appear unattractive. **Table 8.1** lists some common water-problem symptoms and suggestions for correcting them.

Years ago, the U.S. Public Health Service established **potable** (drinkable) **water** standards for public water supplies based on purity, defining allowable concentrations of particular dissolved substances in public water supplies. Today, the U.S. Environmental Protection Agency (EPA) is responsible for administering the *National Primary Drinking Water Regulations*. These are legally enforceable standards that apply to public water systems. In fact, if you get your water from such a centralized system, once every year, along with your water bill, you will get a pamphlet letting you know how well the water you drink meets these standards and what your water company is doing to help make the water even cleaner. These primary standards protect your health by limiting the levels of potentially harmful chemicals in your drinking water. These concentrations are expressed in weight per volume (milligrams/liter). The standards can also be expressed in a weight-per-weight system. For example, in the English system, 1 pound of dissolved solid in 999,999 pounds of water is one part per million (ppm). The standards allow only minuscule amounts of pollutants in public water supplies. For example, 1 ppm, or 1 milligram/liter, is equal to about 1 ounce of a dissolved solid in 7,500 gallons of water, or 0.0001%. The standard for "sweet" water—the preferable quality classification for domestic use—is less than 500 ppm of *total dissolved solids*, and the standard for "fresh" water is less than 1,000 ppm of total dissolved solids.

The quality of groundwater can be impaired either by a high total amount of various dissolved salts or by a small amount of a specific toxic element. High salt content may be caused by solution of minerals in local geological formations, seawater intrusion into coastal aquifers, contamination by road salt, or the introduction of industrial wastes into groundwater. Organic contaminants, mainly animal wastes, solvents, and pesticides, also degrade water quality. They are discussed in the next subsection.

Several types of water quality standards have been set by the federal government to protect public health. The most important enforcement standard is the **maximum contaminant level (MCL)**, which is the highest level of a contaminant that is allowed in drinking water. **Maximum contaminant level goals** (MCLGs) are the level of a pollutant at and below which there is no known or expected risk to health. MCLs are set as close to MCLGs as is politically and scientifically feasible, while considering the available treatment technology and taking cost into consideration. **Table 8.2** lists EPA limits for some of the inorganic substances that occur in groundwater and impair its safety.

Sinkholes

Sinkholes are found in areas underlain by limestone, which is soluble in water. Over time, rainwater and groundwater dissolve weaker parts of the rock, opening tunnels and caves. If the roof of a cave collapses, a sinkhole is born. (a) This cutaway view shows the interconnected systems of caverns and water-filled fractures that underlie karst terrain. (b) In Guatemala City, a giant sinkhole swallowed several homes and left the road impassable. (c) The Blue Hole, in Belize, is not only a very popular diving site, but it is a very large sinkhole (note the dive boat for scale). The sinkhole formed when sea level was lower, during a period when the ice sheets were larger and held more of the world's ocean water. Then, when the ice melted, the Blue Hole flooded with sea water. (d) Sinkholes can form in unlikely places as long as there is limestone to dissolve. In the former Soviet Union, a sinkhole nearly swallows an abandoned industrial facility.

(a)

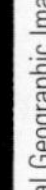
National Geographic Image Collection

(b)

REUTERS/Stringer

(c)
(d)

Table 8.2 Standards (Maximum Contaminant Level or Action Level, 2009) and Health Effects of Significant Concentrations of Common Inorganic Environmental Pollutants

Inorganic Pollutant	Standard, Milligrams/Liter	Health Effects
Arsenic	0.01	Dermal and nervous system toxicity effects, paralysis
Barium	2.0	Gastrointestinal effects, laxative
Boron	1.0	No effect on humans; damaging to plants and trees, particularly citrus trees
Cadmium	0.005	Kidney effects
Copper	1.3	Gastrointestinal irritant, liver damage; toxic to many aquatic organisms
Chromium	0.1	Liver/kidney effects
Fluoride	4.0	Mottled tooth enamel
Lead	0.015	Nervous system/kidney damage; highly toxic to infants and pregnant women
Mercury	0.002	Central nervous system disorders, kidney dysfunction
Nitrate	10.0	Methemoglobinemia ("blue baby syndrome")
Selenium	0.05	Gastrointestinal effects
Silver	0.1	Skin discoloration (argyria)
Sodium	20–170	Hypertension and cardiac difficulties
Sodium	70.0	Renders irrigation water unusable

Source: U.S. Environmental Protection Agency. 2009, May. *National primary drinking water regulations.* Available at: http://www.epa.gov/safewater/consumer/pdf/mcl.pdf.

Water with a high salt content, particularly bicarbonate and sulfate contents, can have a laxative effect on humans. Dissolved iron or manganese can give coffee or tea an odd color, impart a metallic taste, and stain laundry. Concentrations greater than 3 ppm of copper can give the skin a green tinge, a condition that is reversible, and lead is poisonous.

A *persistent pollutant* is one that builds up (or bioaccumulates) in the human system because the body does not metabolize or excrete it. The Romans, for example, used lead plumbing pipes, and skeletal materials exhumed from old Roman cemeteries retain high lead concentrations. Lead concentrations in the water supply greater than about 0.1 ppm can build up over a period of years to debilitating concentrations in the human body. Arsenic is also persistent, and the standards for arsenic and lead are similar, less than 0.015 ppm, preferably absent. Nitrate in amounts greater than 10 ppm is known to inhibit the blood's ability to carry oxygen and to cause "blue baby syndrome" *(infantile methemoglobinemia)* during pregnancy. The element with the greatest impact on plants is boron, which is virtually fatal to citrus and other plants in concentrations as low as 1.0 ppm.

Dissolved calcium (Ca) and magnesium (Mg) ions cause water to be "hard." In hard water, soap does not lather easily, and dirt and soap combine to form scum. Hard water is most prevalent in areas underlain by limestone ($CaCO_3$) and dolomite ($CaMgCO_3$). A water softener exchanges sodium ions (Na) for Ca and Mg ions in the water using a zeolite mineral (a silicate ion exchanger) or mineral sieve. Water with more than 120 ppm of dissolved Ca and Mg is considered hard (**Table 8.3**). Because sodium is known to be bad for persons with heart problems (hypertension) and water softeners increase the sodium content of water, people on low-sodium diets should not drink softened water.

The good news regarding dissolved ions is that fluoride ion (F^-) dissolved in water reduces the incidence of tooth cavities when it is present in amounts of 1.0 to 1.5 ppm. Tooth enamel and bone are composed of the mineral apatite, $Ca_5(PO4)_3(F, Cl, OH)$, whose structure allows the free substitution of fluoride (F^-), chloride (Cl^-), or hydroxyl $(OH)^-$ ions in the mineral structure. In the presence of fluoride, the crystals of apatite in tooth enamel are larger and more perfect, which makes them resist decay. Too much fluoride in drinking water, however—concentrations greater than 4 ppm—can cause children's teeth to become mottled with dark spots. Indeed, it was this cosmetic blemish—among children of Colorado Springs, Colorado, in particular—that led to the discovery of the beneficial effect of fluoride ions and to the establishment of concentration standards for it.

In 1908, Dr. McKay, a Colorado Springs dentist, described a malady he called *Colorado brown stain,* a discoloration of the tooth enamel that we now know results from ingesting too much fluoride. Not only did nearly 90% of the schoolchildren in Colorado Springs have stained teeth, but they also had far fewer cavities than their peers in nearby Boulder, Colorado. It took decades and many studies to prove that fluoride is an effective way to dramatically reduce tooth decay. Now much of the nation drinks fluoridated water and gets far fewer fillings. Dissolved fluoride is also believed to slow osteoporosis, a bone-degeneration process that accompanies aging.

Table 8.3 Water Hardness Scale

Concentration of Calcium, Milligrams/Liter	Classification
0–60	Soft
61–120	Moderately hard
121–180	Hard
>180	Very hard

Note: Dissolved calcium and magnesium in water combine with soap to form an insoluble precipitate that hampers water's cleansing action.

Hardness is expressed here as milligrams of dissolved Ca^{2+} per liter.

Water Pollutants

The term *water pollution* refers to the introduction of chemical, physical, or biological materials into a body of natural water, affecting its future use. Groundwater pollution is widespread, as revealed by studies by state and federal agencies, and by complaints from users. It is a serious problem, because thousands of years may be required to flush contaminated water from an aquifer and replace it with clean water. River water, in contrast, is exchanged in a matter of hours or days.

Residence time, the average length of time a substance remains *(resides)* in a system, is an important concept in water-pollution studies. We may view it as the time that passes before the water in an aquifer, a lake, a river, a glacier, or an ocean is totally replaced and continues on its way through the hydrologic cycle. Residence time can be calculated simply by dividing the volume of the water body (cubic meters) by the flow rate into and out of it (cubic meters/second), assuming everything is in balance. The average residence time of water in rivers is days to weeks; in large lakes, several decades; in shallow gravel aquifers, days to years; and in deeper, low-permeability aquifers, perhaps thousands or even hundreds of thousands of years. This knowledge allows us to estimate the approximate time a particular system will take to clean itself naturally and indicates those cases in which a pollutant must be removed from a water body by human means because it would otherwise cause harm over a long residence time.

Although most pollution is due to the careless disposal of waste at the land surface by humans, some is the result of the leaching of toxic materials in shallow excavations or mines. Groundwater pollution occurs in urban environments because of the improper disposal of industrial waste and leakage from sewer systems, old fuel-storage tanks, wastewater settling ponds, and chemical-waste dumps (**Figure 8.28a**). In suburban and agricultural areas, septic-tank leakage, fecal matter in runoff and seepage from animal feedlots, and the use of inorganic fertilizers including nitrates and phosphates, and the application of weed- and pest-control chemicals can all degrade water quality (see Figure 8.28b).

In the past few decades, leaking gasoline-storage tanks have been recognized as a major cause of soil and groundwater pollution (see Figure 8.28c). By federal law, all underground tanks now must be inspected for leakage, and all single-wall tanks are being replaced with double-wall tanks so fuel leaking from the inner tank is contained rather than being released to the environment; if gasoline is found in the adjacent soil, the tanks must be removed and the soil cleaned. (Although the problem is not amusing, the technology created by the need for hydrocarbon cleanup has come to be lightly referred to as *yank-a-tank*.) If gasoline is found floating on groundwater, it must be removed. Both the gasoline and the decontaminated groundwater are then usable. Soils can

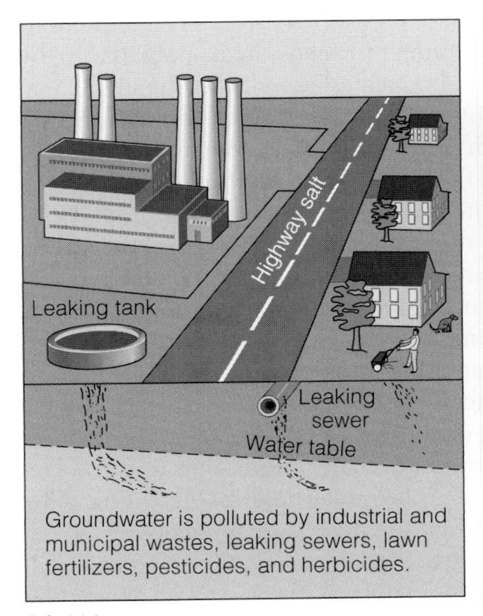

(a) Urban areas

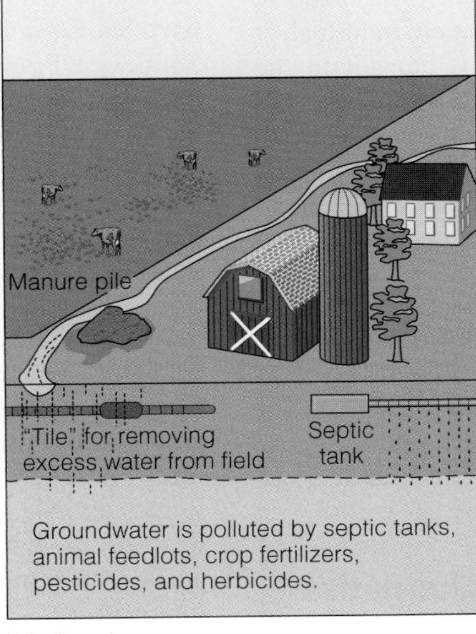

(b) Rural areas

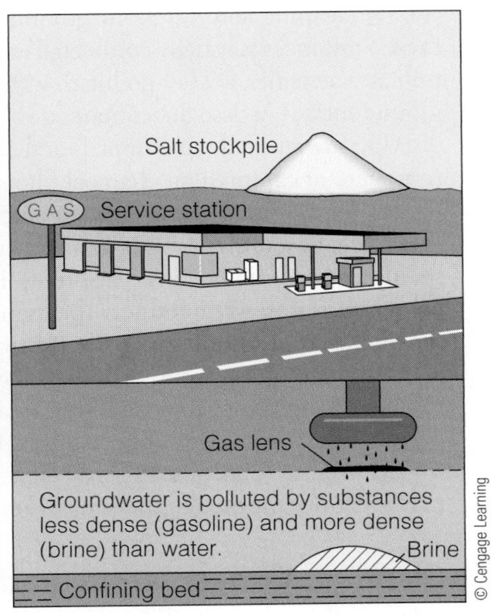

(c) Density effects

Figure 8.28 (a) Urban-area and (b) rural-area pollution sources. (c) Some pollutants float and others sink because of differences in density between the groundwater and the pollutants. Brine, being denser than water, sinks. Gasoline, being less dense than water, floats.

Table 8.4 Health Effects of Common Organic Pollutants

Organic Substance	Human Health Effects	Environmental Effects
DDT	Convulsions, kidney damage	Reproductive failure in animals, eggshell thinning
Polychlorinated biphenyls (PCBs)	Vomiting, abdominal pain, liver damage	Eggshell thinning in birds, liver damage in mammals
Benzene	Anemia, bone marrow damage	Toxic to some fish and aquatic invertebrates
Phenols	Death at high doses	Decreased phytoplankton productivity
Dioxin	Acute skin rashes, systemic damage, mortality	Lethal to aquatic birds and mammals

Sources: U.S. Public Health Service; Environmental Protection Agency.

be cleansed of hydrocarbons biologically with bacteria, a process known as bioremediation, or by aeration and oxidation of the hydrocarbon films (see Case Study 6.3).

Table 8.4 lists some of the detrimental effects of selected organic pollutants. Water standards for common organic pollutants of water bodies are given in **Table 8.5**. The very low limits allowed for the pesticides lindane and endrin indicate their toxicities. Some of these toxic chemicals are also carcinogenic (cancer-causing), which makes them doubly dangerous.

Density differences between groundwater and particular pollutants often govern the remedial measures that can be used; for example, gasoline and oil float on groundwater, whereas salt brines from industry or agriculture and some organic chemicals sink to lower levels (see Figure 8.28c). In some cases, it is possible to skim a layer of gasoline from the top of the water table. Gasoline or fuel oil that is floating on the capillary fringe may yield vapors that can rise and be trapped beneath or in the walls of buildings. Gasoline fumes contain BTX (*b*enzene, *t*oluene, and *x*ylene), and are thus carcinogenic. "Air-stripping" is currently the preferred method of removing volatile organic pollutants such as BTX compounds and trichloroethylene (TCE) and perchloroethylene (PCE) cleaning solvents from groundwater. (The origin of these substances is usually connected to refinery, industrial, or military activities.) The polluted water is brought to the ground surface, where the contained volatile pollutants (TCE or BTX) are gasified (air-stripped) and either vented to the atmosphere or captured on charcoal filters and then destroyed.

In the past decade, increasing attention has been paid to the role of sediment as a pollutant. Increased loads of fine sediment can smother instream biota including plants and fish. Silty, turbid water reduces the amount of light available for photosynthesis and clogs the gills of fish. Many pollutants, such as phosphorus and lead, have affinities for sediment, and thus can be carried into streams attached to solid particles. Sediment pollution is often the result of uncontrolled mining (**Figure 8.29**), as well as rapid land development without proper engineering controls.

Table 8.5 Water Standards for Common Organic Pollutants

Substance	Maximum Contaminant Level (Environmental Protection Agency), Milligrams/Liter
Lindane	0.0002
Endrin	0.002
2,4-D	0.07
Toluene	1.0

Source: U.S. Environmental Protection Agency. 2009, May. *National primary drinking water regulations.* Available at: http://www.epa.gov/safewater/consumer/pdf/mcl.pdf.

Common organic pollutants include insecticides, fuel, and animal-control compounds.

Cleanup, Conservation, and Alternative Water Sources

Water is a precious resource, the value of which is often underappreciated. Increasing scarcity of clean water and a realization that actions taken in one part of the hydrologic system affect other parts of the system have spawned an increasingly active citizen-driven movement to protect the quality of surface and groundwater. For example, many storm drains lead directly to water bodies; used motor oil someone pours down the drain might go directly to the local trout stream (**Figure 8.30**).

Many parts of the United States are experiencing drought conditions. Excessive groundwater withdrawals have led to water mining in many areas, particularly the Southwest. Proposed remedies include desalinizing seawater, recycling wastewater, tugging Antarctic icebergs up the coast of South America to drier areas in the Northern Hemisphere, and floating 35-ton water bags down from the northwestern United States to Southern California. However, in the long run, as population increases, we will be required to conserve water. Voluntary water conservation has been quite successful, reducing consumption by as much as 15%, and even greater reductions have been achieved where rationing is mandatory. The advent of Energy Star appliances, such as water-efficient dishwashers and front-loading clothes washers, lets each of us contribute while often earning rebates from local utilities. Despite conservation, global water use is increasing. The world's more than 7 billion people are intercepting 54% of all freshwater from rivers, lakes, and aquifers. Humans will intercept 70% by 2025 just as the result of population growth. If total consumption increases at the current rate, humans could soon be using 90% of all available freshwater, leaving 10% for the rest of life on Earth. We should think "conservation."

Figure 8.29 Near Ariquemes, Brazil, in the Amazon, the Bon Futuro (good fortune) mining camp may be Brazil's richest tin deposit. Tens of thousands of miners have come to the site to extract the tin ore by washing the sediment with water. The result, water polluted with silt smothers the remaining forest and destroys aquatic ecosystems.

Figure 8.30 Storm drains provide a quick path for pollutants (such as waste oil and antifreeze) to enter streams, rivers, and the groundwater system. In Atlanta, Georgia, during October, which is Georgia Rivers month, people volunteer to stencil storm drains with a "NO DUMPING" message.

8.1 Who Shrank the Aral Sea?

The Aral Sea was once the fourth-largest lake in the world, second only to the Caspian Sea in the former Soviet Union (**Figure 1**). The level of the sea in 1960 was about 53 meters (170 feet) above sea level, and it contained a fish population that supported more than 60,000 persons in fishing and fish processing. In 2003, the level of the Aral Sea was only 30 meters (100 feet) above sea level, and it ranked as the world's sixth-largest lake, having lost more than 74% of its surface area since 1960. The result of this shrinkage is that the fishing industry is no more (**Figure 2**). In addition, the salinity of the sea has increased from 10 grams/liter (about 1%) to more than 100 grams/liter (10%) in some areas, making it inhospitable to most marine life. This story leaves a question that begs an answer: What caused this water body to disappear?

Figure 1 Location map of the Aral Sea. (Redrawn from Philip Micklin, Western Michigan University)

Agricultural demands of the former Soviet Union deprived this great salt lake of water inflow sufficient to overcome evaporation losses. Two major rivers feed the lake: the Amu Darya (*darya*, "river") flowing into the lake from the south, and the Syr Darya, which reaches the sea at its north end. The Soviet leaders wanted exportable crops, and cotton was high on their list. Irrigated agriculture was expanded and the Kara Kum Canal was opened in 1956, which diverted water from the Amu Darya into the deserts of Turkmenistan. Millions of hectares of arid lands came under irrigation. By 1960, the level of the sea had begun to drop, and where in the past it received 59 cubic kilometers (about 13 cubic miles) of water annually, this amount had fallen to zero by the early 1980s. As the lake area receded, the salinity increased,

Figure 2 Fishing boat grounded by shrinkage of the Aral Sea. This spelled doom to a thriving fishing industry and loss of an important source of protein for the local inhabitants.

and by 1977, the fish catch had dropped to one-fourth its level before water diversion. By 1990, the area of the Aral was cut in half, and its water volume was but a mere one-third its historic amount. The lowering of the sea exposed 30,000 square kilometers (about 11,000 square miles) of salt- and chemical-filled lakebeds (**Figure 3**). Blowing dust, particularly south of the lake, has resulted in air pollution, producing a high incidence of throat cancer, eye and respiratory diseases, anemia, and infant mortality. In addition, the shoreline habitat was lost and delta wetlands were desiccated. Animal populations shifted, because conditions favored animals better adapted to drought and high salinity. By any measure, the desiccation of the Aral Sea is an ecological disaster. The central Asian republics that surround the sea—Kazakhstan, Turkmenistan, Uzbekistan, and the little-known Karakalpak Republic (within Uzbekistan)—are the poorest of the former Soviet bloc and have no money to invest in rebuilding irrigation systems. In addition, Afghanistan and Iran, with small areas within the drainage basin, have little interest in the problem.

A sad Uzbek poem goes "You cannot fill the Aral Sea with tears."

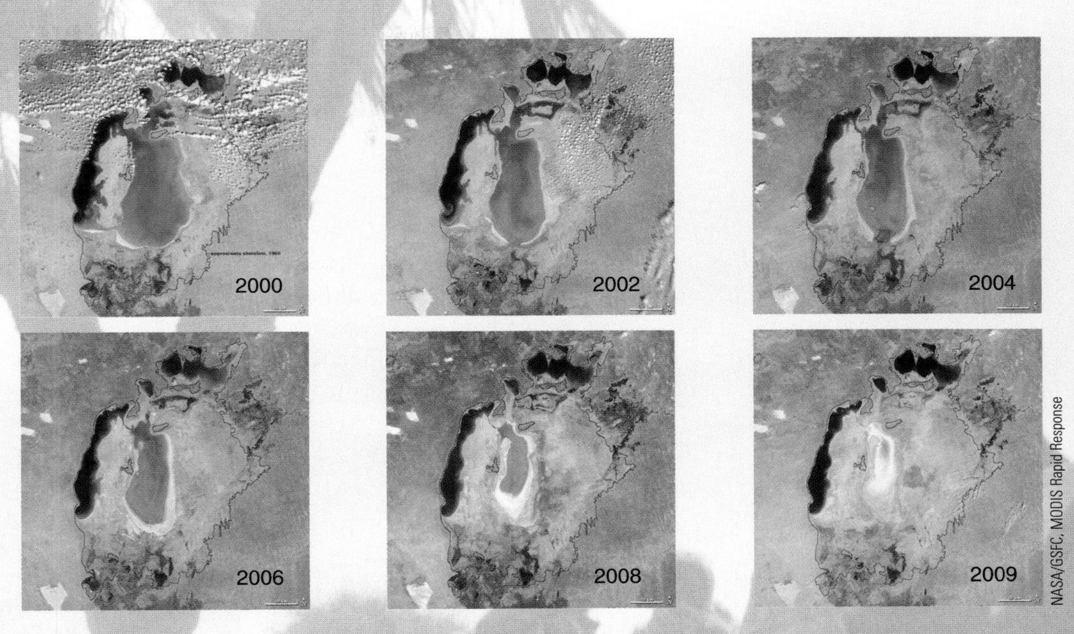

Figure 3 The shrinkage in area of the Aral Sea from 2000 to 2009 is dramatic. By 2003, compared with 1960, the sea had lost 74% of its area and 90% of its volume. In 2005, Kazakhstan built a dam and cut off the southern Aral Sea, keeping all of the water to the north. After the dam was closed, the water levels in the south plummeted. Can you see the changes?

questions to ponder

1. Do you think those diverting water from the rivers entering the Aral Sea could have predicted the environmental impact their actions would cause?
2. Do you think history could repeat itself and a similar scenario could happen again somewhere else in the world?

8.2 Fluid Flow in Porous Rocks

Porosity is the ratio of the volume of pore spaces in a rock or other solid to the material's total volume. Expressed as a percentage, it is

$$P = \frac{V_p}{V_t} \times 100$$

where P = porosity, V_p = pore volume, and V_t = total volume.

If an aquifer has a porosity of 23%, a cubic meter of the aquifer is composed of 0.77 cubic meter of solid material and 0.23 cubic meter of pore space, which may or may not be filled with water. Porosities of 20% to 40% are common in sand and gravel aquifers.

Permeability is a measure of how fast fluids can move through a porous medium and is thus expressed in a unit of velocity, such as meters or feet per day. Knowing the permeability of an aquifer allows us to determine in advance how much water a well will produce and how far apart wells should be spaced. To determine permeability, we use **Darcy's law**, which states that the rate of flow (Q in cubic meters/day) is the product of hydraulic conductivity (K), the

hydraulic gradient, or the slope of the water table (S in meters per meter), and the cross-sectional area (A in square meters). Thus,

$$Q = K * A * S$$

where Q is the rate of flow.

Hydraulic conductivity, K, is a measure of the permeability of a rock or a sediment, its water-transmitting characteristics. It is defined as the quantity of water that will flow in a unit of time under a hydraulic gradient of 1 through a unit of area measured perpendicular to the flow direction. K is expressed in distance of flow with time, usually meters (feet) per day (**Figure 1**).

If we use a unit hydraulic gradient, then Darcy's law reduces to

$$Q = K * A$$

Unit hydraulic gradient is the slope of 1-unit vertical to 1-unit horizontal ($S = 1/1 = 45°$), and it allows us to compare the water-transmitting ability of aquifers composed of materials with different hydraulic conductivities. Flow rates may be less than a millimeter per day for clays, and many meters (feet) per day for well-sorted sands and clean gravels. In general, groundwater moves at a snail's pace.

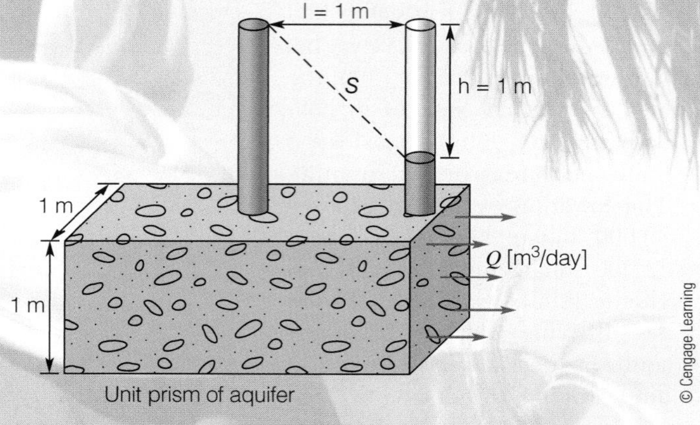

Figure 1 The unit hydraulic gradient (I) and unit prism (1 square meter) of a sandstone aquifer. When the hydraulic conductivity (K) is known, then the discharge (Q) in cubic meters per day per square meter of aquifer can be calculated.

questions to ponder

1. Hydrofracking involves using pressure to shatter rocks near wellbores, and thus increase the amount of water flowing into the well. What parameter in Darcy's law does this process increase?
2. When a well is pumped, a different parameter in Darcy's law changes. What is that parameter and how does it change?

8.3 Groundwater Law

Ownership of groundwater is a property right and has traditionally been regulated as such by individual states. Each state applies one or more of four legal doctrines to groundwater rights, none of which considers the geological relationship between surface-water infiltration and groundwater flow. The doctrines are as follows:

- The *riparian doctrine* (Latin *ripa*, "river or bank") holds that landowners overlying well-defined underground "streams" have absolute right to that water. They may exercise that right whenever they wish without limitation.
- The *reasonable-use doctrine* restricts a landowner's right to groundwater to "reasonable use" on the land above that does not deprive neighboring landowners of their rights to "reasonable use." This doctrine results in a more equal distribution of water when there is a shortage.
- The *prior-appropriation doctrine* holds that the earliest water users have the firmest rights; in other words, "first come, first served."
- The *correlative-rights doctrine* holds that landowners own shares to the water beneath their collective property that are proportional to their shares of the overlying land.

In the San Joaquin Valley of California, for example, surface-water rights are governed by the doctrine of prior appropriation (first come, first served), whereas groundwater rights are governed by the doctrine of correlative rights (proportional shares). In the late 19th century, recharge in the Sierra Nevada to the east of the valley and the lower ranges to the west maintained high water tables, and they both contributed water to the flow of the San Joaquin River (**Figure 1**). By 1966, however, intensive groundwater withdrawals on the west side of the San Joaquin Valley had reversed the regional groundwater flow, so that the river was losing, rather than gaining, water. The San Joaquin had become a losing stream. Imagine a courtroom scenario in which hundreds of people with first-come, first-served rights to *surface* water sued thousands of people holding proportional shares rights to *underground* water for causing their surface water to go underground.

As explained throughout this chapter, it is impossible to treat groundwater and surface water as separate entities. Surface water contributes to underground water supplies, and groundwater contributes to streamflow and springs.

question to ponder

1. Do you know the applicable ground and surface water laws in your state or county? Where would you go to learn more about these laws?

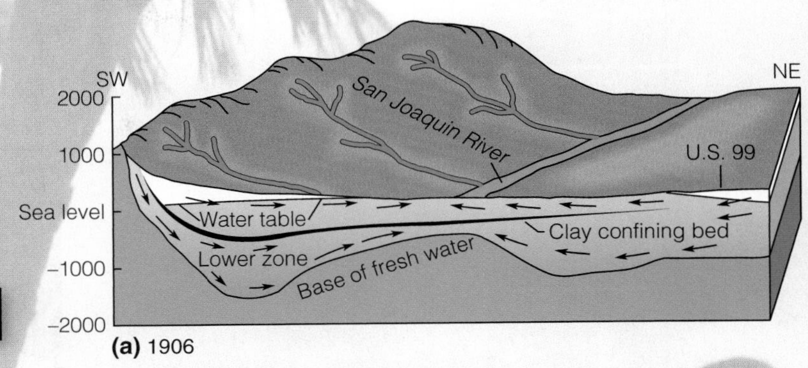

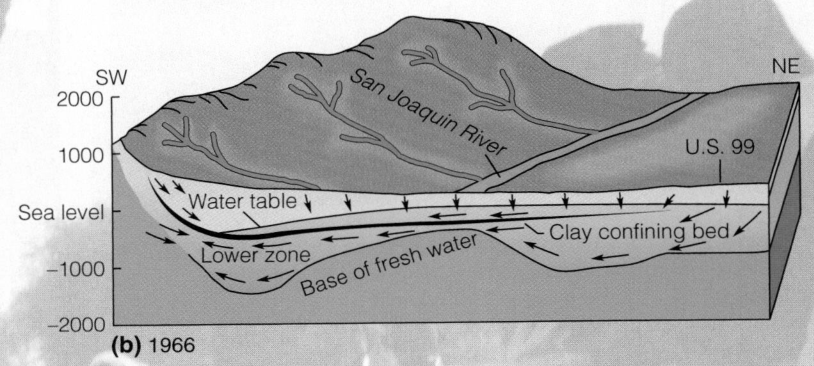

Figure 1 Groundwater flow in the central San Joaquin Valley in (a) 1906 and (b) 1966. Note that the San Joaquin River was a gaining stream in 1906 and a losing stream in 1966. The change reflects intensive development and the extraction of underground water. (Source: American Institute of Professional Geologists. 1985 Ground Water: Issues and Answers. 11 10.07 Ame AIPG, Arvada, CO.)

8.4 The Edwards Aquifer, a Texas Bonanza

The Edwards Aquifer in South-central Texas

The Edwards aquifer in south-central Texas is one of the most prolific aquifers in North America (**Figure 1**). It provides water for 2 million people, including the city of San Antonio. A well drilled for the city in 1991 is reportedly the world's greatest flowing well, yielding 95,000 liters per minute (25,000 gallons/min). The aquifer is composed of limestone of Cretaceous age that is a *karstified* flow system; that is, it is honeycombed with solution cavities, caves, and caverns, and feeds many springs. About 80% of the aquifer's recharge is from losing streams that flow over its exposed outcrops; the rest is from direct infiltration of precipitation. Where the aquifer dips down below the ground surface toward the south, it is under pressure, and water rises in well bores and sometimes to the ground surface (see Figure 8.18). At depth, the water quality and potability deteriorate to a zone marked on maps by the "bad water line." The region served by the aquifer is one of the fastest growing areas in the United States, and water supply is critical to this growth. Recharge areas are environmentally sensitive to

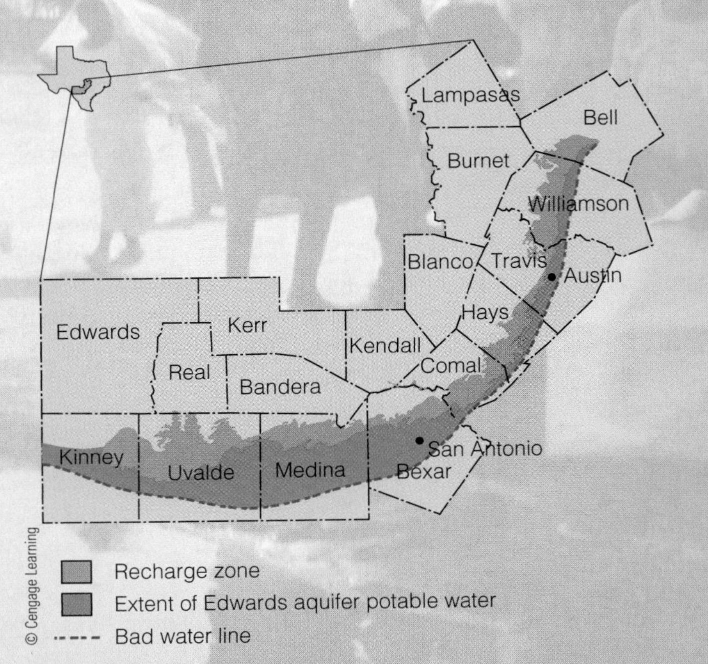

Figure 1 Area of south-central Texas's Edwards aquifer.

pollutants, and the resource is threatened by disregard for this potential problem (**Figure 2**). Threats are from human sewage, highway construction, agricultural chemicals, runoff from paved surfaces, and leaking underground storage

Figure 2 (a) Sign alerting motorists they are entering an aquifer recharge zone. (b) Midnight Cave in the Edwards aquifer before cleanup, 1993; southern Travis County near Austin, Texas. The trash is household garbage, oil filters, pesticide bottles, partially filled turpentine cans, and automobile parts. Trash had floated up to higher ledges of the cave during high-water aquifer conditions. Cleanup efforts involved conservation groups, government agencies, and cavers. An estimated 3,000 cubic feet (80 cubic meters) of trash was removed from the cave.

tanks, to name a few. Government agencies at all levels are committed to protecting the aquifer. The federal government designated the aquifer a Sole Source Aquifer, which makes it eligible for funding for some protection projects. The state declared the Edwards aquifer the most susceptible to pollution in the state and established conservation districts for overseeing its protection.

questions to ponder

1. With climate change predictions suggesting that Texas will get hotter and drier, what do you think the future will be like for the Edwards aquifer?
2. What actions do you think the people of Texas should take to protect this resource over the next century?

8.5 Long Island, New York—Saltwater or Freshwater for the Future?

By far, Long Island's most important source of freshwater is its aquifers, which are estimated to hold 10 to 20 trillion gallons of recoverable water (**Figure 1**). This water infiltrated underground over centuries and the excess was discharged naturally to the sea. Groundwater was exploited as the island was urbanized, causing the water table on the mainland side of the island to drop below sea level. By 1936, saltwater had invaded the freshwater aquifer. Pumping wells were converted to recharge wells using imported water, and by 1965, the water tables had recovered to acceptable elevations above sea level.

To the east in Queens County, pumping had increased and recharge had diminished because of the construction of sewers and the paving of streets. Treated wastewater was thus being carried to the sea, rather than infiltrating underground, as it would with cesspools and septic tanks. With lowered water tables, saltwater invaded the aquifers below Queens County, and in 1970, the State of New York asked the U.S. Geological Survey to conduct groundwater studies.

The basic challenge was to offset groundwater withdrawals with equal amounts of surface infiltration or injection. A line of injection wells was proposed for replacing the decreased natural recharge. These wells would inject treated sewage effluent or imported water and build a freshwater barrier against saltwater intrusion (**Figure 2**). Spreading highly treated sewage effluent and natural runoff into recharge basins could be carried out along with injection into the deeper aquifer. In both the injection-well and the spreading-basin alternatives, the treated sewage is purified to drinking-water standards, so that it does not affect local shallow wells. Recharge, like injection, builds a freshwater ridge that keeps saltwater at bay.

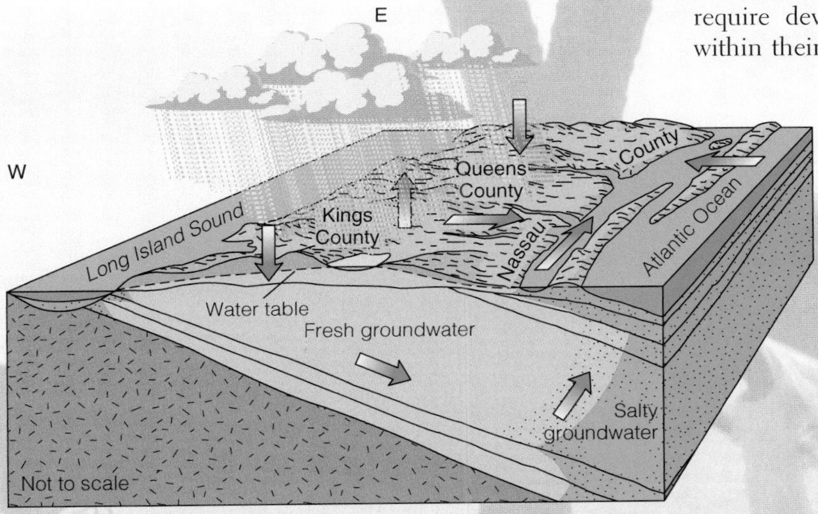

E
W
Queens County
Kings County
Long Island Sound
Nassau County
Atlantic Ocean
Water table
Fresh groundwater
Salty groundwater
Not to scale

Figure 1 The hydrologic relationship between freshwater and salty underground water; Long Island, New York. Arrows indicate the direction of water movement in the hydrologic cycle. (Source: Foxworthy, B. L. 1978. USGS prof. paper 950.)

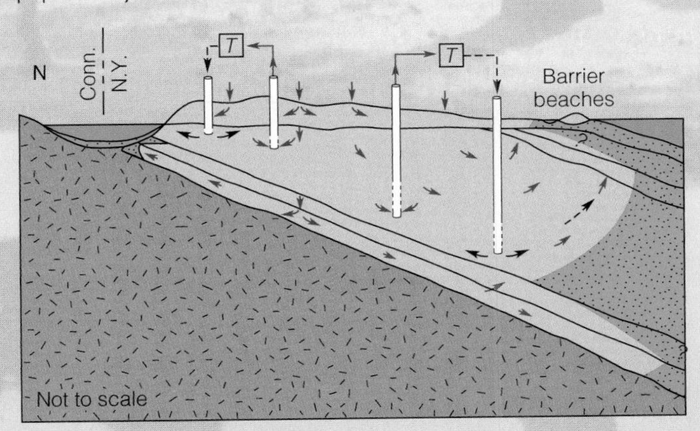

Figure 2 Method of balancing freshwater outflow and seawater inflow; Long Island, New York. The aquifer is recharged with highly treated wastewater (*T*) using injection wells. This method reverses saltwater intrusion and improves the outlook for long-term water yield. (Source: Foxworthy, B. L. 1978. USGS prof. paper 950.)

Each method of balancing outflow and inflow—injection wells, spreading basins, and reduction of groundwater withdrawals—has advantages and disadvantages. As is the case with many geological problems, applying a combination of methods yields the most fruitful and cost-effective solution. Kings and Queens Counties now rely entirely on importation for their freshwater needs. Public water in Nassau and Suffolk Counties is entirely underground water, and only a very small amount of seawater intrusion occurs in the southwest corner of Nassau County. Recharge basins collect surface-water runoff and allow it to percolate underground into the shallow aquifer (**Figure 3**). In Long Island, counties require developers to dedicate land for recharge basins within their housing or commercial projects. This method of recharge has proved effective in maintaining the balance between seawater and freshwater.

question to ponder

1. Look around your community. Can you find recharge basins?

Figure 3 In Florida, these sandy recharge basins recycle treated waste water, letting it infiltrate into the aquifer, raising the water level and preventing saltwater intrusion.

PHOTO GALLERY

everyone needs water

1 On Corfu, in the Greek Ionian Islands, a Prior and his monk stand next to a well at the Europulus monastery. © WILHELM TOBIEN/ National Geographic Image Collection

2 Industrial-scale agriculture is a very thirsty proposition, requiring large amounts of water. In California, the Palo Verde Diversion Dam allows an irrigation district to take water from the Colorado River. © 2011 PETE MCBRIDE/National Geographic Image Collection

Water is a basic human need. Without freshwater, you can live only a few days. Water slakes our thirst and cleans our skin. It cools us in the heat of summer and keeps the plants that feed us from wilting.

The people of the world get the water they need in very different ways. Some gather untreated water from hand-dug wells, whereas others lower buckets down more formal, masonry wells. In developing nations, impromptu water systems are constructed to serve newly prosperous homes, whereas industrialized nations divert entire rivers to feed thirsty agricultural endeavors. Sometimes nature cooperates and water flows freely from the ground, cooling people on a hot day and providing copious amounts of clean drinking water.

A hand-dug well in the Kalahari Desert provides water for the people in this part of Botswana. © FRANS LANTING/National Geographic Image Collection

3 In central Calcutta, India, people get water from a public well. © STEVE RAYMER/National Geographic Image Collection

4 In Shanxi Province of the People's Republic of China, plastic tubing carries water from a well pump to homes on the hillside above, an impromptu water distribution network. © 2008 GREG GIRARD/National Geographic Image Collection

5 Groundwater under pressure soaks two drillers as they try to cap an artesian well in western Massachusetts. The water is rising over 100 feet (30 meters) from a deep glacial gravel aquifer, fed from the highlands, and capped by nearly imperme-able silt and clay. P. Bierman

SUMMARY Freshwater Resources

The Hydrologic Cycle

Less than 1% of Earth's water is readily available to humans. It is accessible because of the hydrologic cycle: continuous evaporation from the oceans, precipitation on land, runoff back to the oceans, and the storage of freshwater in lakes and underground.

Water Use

1. Consumptive use refers to water that cannot be immediately reused. Where consumptive use exceeds supplies, water must be imported.
2. Irrigation and thermoelectric power plants are the largest water consumers. Irrigation water is usually highly degraded, whereas 98% of power-plant water is returned to the system, the only change being a higher temperature and lower elevation.
3. Worldwide, about 82% of water use is for irrigation. Water availability will limit population growth in the 21st century, either by statute in developed countries or by drought and famine in developing nations. The world population projected for 2050 will require water beyond the reliable flow of the world's major rivers. Conservation will then be our only choice.

Lakes

Defined

Lakes are bodies of water in landlocked basins formed by many geological processes.

Budget—depends on precipitation plus inflow equaling outflow plus evaporation. If these do not balance in the long term, the result is an ephemeral or salt lake, which when dry is known as a playa.

Overturn—in temperate regions, overturn, driven by changing lake temperature and density with the seasons, keeps lake water oxygenated.

Nutrients—oligotrophic lakes are relatively sterile and have a small nutrient input. Eutrophic lakes have a high nutrient input and can be choked with plant life. Mesotrophic lakes have a good balance between inflow and outflow of nutrients.

Rivers and Streams

Dams are built to impound freshwater and improve navigation, and for power generation and flood control. In the United States, dams are unpopular because of loss of land by inundation and the subsequent relocation of humans and wildlife.

Groundwater

Defined

Water in the hydrologic cycle that infiltrates underground.

Zones

1. The *zone of aeration*, where the voids in rock or sediment are filled with air and water films.
2. The lower *zone of saturation*, where voids are filled with water.

Water Table

The contact between the zone of aeration and the zone of saturation.

Aquifers

Defined

Bodies of porous and permeable rock or sediment that are saturated with water.

1. *Porosity:* the volume of pore spaces in a geological material, expressed as a percentage of the total volume.
2. *Permeability:* the ease with which fluids flow through a geological material.
3. *Aquiclude:* a geological material of very low permeability that inhibits water flow.

Types

1. *Unconfined:* an aquifer in which the water table is exposed to atmosphere directly. A slope, or gradient, of the water surface is required to induce subsurface flow in a given direction. A well that taps the static water table creates a cone of depression, which provides the necessary gradient for water to flow to the well. Drawdown is the amount of lowering of the water table by a pumping well.
2. *Artesian:* an aquifer that is confined between aquicludes. The hydraulic head causes water to rise in a well when the aquifer is penetrated. The height to which water will rise at any point along the aquifer is known as the artesian pressure or potentiometric surface.

Groundwater Management

Defined

Managing the amount of water contained in a groundwater basin so that it will produce water in the future.

1. *Sustained yield:* the amount of water an aquifer will sustainably yield on a day-to-day basis because it is compensated by recharge.
2. *Specific yield:* the amount of water a body of rock or sediment will yield by gravity drainage alone.
3. *Specific retention:* the amount of water held by a body of rock or sediment against the pull of gravity. Specific yield *plus* specific retention *equals* porosity.

Seawater Encroachment

The invasion of coastal aquifers by saltwater. Management requires maintaining high freshwater tables to balance the incursion of denser saltwater.

Groundwater–Surface Water Interaction

Surface water and groundwater cannot be treated as separate entities; what affects one impacts the other.

Surface-Water Sources

Lakes, rivers, streams, and artificial reservoirs.

U.S. Use Trends

U.S. water use increased from 1950 to 1980. Between 1980 and 1985, it decreased 10% and held steady through 2005, even though the population increased by more than 50 million people. The reasons include more economical application of irrigation water, public conservation programs, recycling, and a downturn in farm economies.

Solution by Groundwater

1. Solution of limestone and gypsum by underground water forms caves, caverns, disappearing streams, sinkholes, and prolific groundwater aquifers. When terrains underlain by carbonate rocks exhibit these features, they are said to have *karst* topography.
2. Water and carbon dioxide form carbonic acid, a weak natural acid, which, over long periods of time, will dissolve carbonate rocks and gypsum.
3. Urbanization on karst terrain results in increased flooding, contamination of underground water, and collapse of the ground surface to form sinkholes.
4. Forty percent of the country east of the Mississippi is underlain by carbonate rocks. The states most impacted by karst and sinkhole formation are Alabama, Florida, Georgia, Tennessee, Missouri, and Pennsylvania.

Water Quality

Defined

Purity or drinkability (potability) of water as established by the U.S. Public Health Service. Dissolved constituents are measured in milligrams per liter (volume) or parts per million (weight).

Standards

Less than 500 ppm total dissolved solids for "sweet" water, and limits on specific ions and organic molecules that impair the quality of water.

Pollution

Defined

Chemical, physical, or biological materials that impair the current and future use of water and that may be a health hazard.

Residence Time

The average length of time a given substance will stay in a system, such as water in a lake or an aquifer. Whereas the residence time of road salt in a stream may be days to weeks, pollutants may reside in some deep aquifers for thousands of years.

Common Pollutants

Hydrocarbons, carcinogenic BTX chemicals that make up gasoline (benzene, toluene, and xylene), fertilizers, pesticides, livestock fecal matter, and gasoline.

Conservation and Alternative Sources

Conservation includes the use of treated wastewater for irrigation and recharging aquifers, water-saving irrigation devices, water spreading during wet years, and voluntary water conservation. Desalinization and importation are the only viable alternative sources of water at present.

KEY TERMS

aquiclude
aquifer
artesian-pressure surface
cone of depression
confined aquifer
Darcy's law
drawdown
evapotranspiration
gaining stream
groundwater
hydraulic conductivity
hydraulic gradient
hydrogeologist
hydrologic cycle
karst terrain
losing stream
maximum contaminant level (MCL)
maximum contaminant level goal (MCLG)
overturn
perched water table
permeability
porosity
potable water
potentiometric surface
residence time
saltwater intrusion
specific yield
sustained yield
unconfined aquifer
water table

STUDY QUESTIONS

1. Sketch a hydrogeological cross section that shows the groundwater zones and a static water table.
2. Sketch and define a confined aquifer. Explain the potentiometric surface and its importance in production of water from an artesian well.
3. How does topography affect the shape of the static water table?
4. How does one determine the amount of water a well that taps the static water table might produce on a long-term basis?
5. Explain the distinction between gaining and losing streams.
6. Explain the relationship between the amount of drawdown in a water well and the aquifer's permeability.
7. What states are the largest consumers of groundwater, and why is their usage so heavy?
8. What is meant by groundwater *mining*? Cite the case history of an aquifer where this has happened.
9. Many beverage manufacturers claim the water in their drinks is purer because it comes from artesian wells or springs. Comment on these claims.
10. What are the Public Health Service standards for total dissolved solids in drinking water?
11. Name three organic compounds sometimes found dissolved in groundwater that are hazardous to human health.

REFERENCES AND FURTHER READINGS

Alley, William M., Thomas Reilly, and O. Lehn Franke. 1999. *Sustainability of ground-water resources.* U.S. Geological Survey circular 1186.

American Institute of Professional Geologists. 1985. *Ground water: Issues and answers.* Arvada, CO: AIPG.

Baldwin, H. L., and C. L. McGuinness. 1963. *A primer on ground water.* Washington, DC: U.S. Geological Survey.

Cohen, P., O. L. Franke, and B. L. Foxworthy. 1970. *Water for the future of Long Island, New York.* New York Division of Water Resources, Department of Environmental Conservation, in cooperation with the U.S. Geological Survey.

Dunne, T., and L. Leopold. 1978. *Water in environmental planning.* New York: Freeman, 818 p.

Fetter, C. W. 2001. *Applied hydrogeology.* New York: Prentice Hall, 598 p.

Galloway, Devin, David Jones, and S. E. Ingebritsen. 1999. *Land subsidence in the United States.* U.S. Geological Survey circular 1182.

Hauwerrt, Nico M., and Shawn Vickers. 1994. *Barton Springs/ Edwards aquifer: Hydrogeology and ground water quality.* Austin, TX: Texas Water Development Board, Barton Springs/Edwards Conservation District.

Kenny, Joan F., Nancy L. Barber, Susan S. Hutson, Kristin S. Linsey, John K. Lovelace, and Molly A. Maupin. 2009. *Estimated use of water in the United States in 2009.* U.S. Geological Survey circular 1344.

Kidd, Mary A. 1996. *Nutrients in the nation's drinking water: Too much of a good thing?* U.S. Geological Survey circular 1136.

McGuire, V. L. 2009. *Water-level changes in the High Plains aquifer, predevelopment to 2007, 2005–06, and 2006–07.* USGS, Ground-Water Resources Program, Scientific Investigations Report 2009–5019.

Micklin, Philip P. 1988. Desiccation of the Aral Sea: A water management disaster in the Soviet Union. *Science* 241 (4870): 1170. [DOI: 10.1126/science.241.4870.1170]

Moore, John E., A. Zaporozed, and James W. Mercer. 1994. *Ground water: A primer.* Environmental awareness series. Alexandria, VA: American Geological Institute.

Sanders, Laura L. 1998. *A manual of field hydrogeology.* Upper Saddle River, NJ: Prentice Hall, 381 p.

Sharp, John M., Jr., and Jay L. Banner. 1997, August. The Edwards aquifer: A resource in conflict. *GSA Today* 1–10.

U.S. Geological Survey. 1999. *The quality of our nation's waters—Nutrients and pesticides.* U.S. Geological Survey circular 1225.

Winter, Thomas C., Judson W. Harvey, O. Lehn Franke, and William M. Alley. 1998. *Ground water and surface water: A single resource.* U.S. Geological Survey circular 1139.

Zwingle, Erla. 1993. Ogallala aquifer: Well spring of the High Plains. *National Geographic* 183 (3): 80–109.

Access an interactive eBook and chapter-specific interactive learning tools, including flashcards, quizzes, videos and more in your Geology CourseMate, accessed through CengageBrain.com.

Indian commuters make their way through a flooded street after a heavy downpour in 2009, just after the arrival of the annual monsoonal rains.
NARINDER NANU/AFP/Getty Images

Have you ever
wondered...

9

Hydrologic Hazards at Earth's Surface

Rain added to a river that is rank
perforce will force it overflow the bank.

—*William Shakespeare (1564–1616)*

1. Why some parts of the world are wetter than others?

2. What causes floods?

3. Whether the 100-year flood actually happens every 100 years?

4. If levees really prevent flooding?

5. What are the impacts of dams on the environment?

The Blue Danube?

Figure 1 The Danube River at Regensburg, Germany, during the great flood of August 2002. Note that the river walk behind the high-water *(Hochwasser)* sign is inundated.

During the summer (August) of 2002, land along rivers in eastern Europe and western Russia was underwater, as these areas experienced their worst flooding in a century (**Figures 1** and **2**). The flooding was driven by exceptionally heavy rains over large areas. The losses have been estimated in excess of $20 billion, which would place it on a par with the Northridge, California, earthquake of 1994 in terms of dollar loss, but greater in terms of loss of life. There were 59 deaths in Northridge versus an estimated 105 for the 2002 flood—62 in Russia alone near the Black Sea coast. Global climate models suggest that such heavy precipitation events are likely to become more common in this area over the next 100 years if greenhouse gas emissions continue at present rates.

- Twelve countries were affected: Russia, Ukraine, Romania, Bulgaria, Croatia, the Czech Republic, Austria, Slovakia, Switzerland, France, Germany, and Italy.
- Along the Elbe River in Germany, a dam broke, threatening a town of 16,500 people.

Figure 2 Diners enjoy a Sunday lunch in a restaurant below the flood level of the Danube River in Regensburg, Germany. The water is kept at bay by the red inflatable flood-fence around the dining area.
B. Pipkin

- More than 400 houses were destroyed and at least 7,000 were damaged in Russia.
- About 220,000 residents living near the Vltava River (German, *Moldau*) in the Czech Republic were displaced.
- More than 20,000 persons were evacuated in Dresden, Germany.
- A 30-meter (100 foot) long steel bridge in Schwertberg, Germany, was torn loose by the flooded Danube River.

Probably the worst flood impact, after the human loss, was the toll on the Prague Zoo. Zoos are a habitat that most humans don't think about during a flood. At the peak of the flood, the lower half of the zoo was completely underwater. A 12-year-old seal named Gaston made headlines by swimming 120 kilometers (72 miles) in the river before being recaptured in Dresden, Germany. An elephant and hippo were lost, but another hippo, which was believed to have died, was saved after the floodwaters began to recede. Ten big animals were lost, including an aging lion and a bear considered too old to evacuate. A big loss was Pong, a popular gorilla and one of five in the zoo, which is believed to have drowned in the evacuation effort. More than 1,000 animals were saved.

questions to ponder

1. With floods like that which happened on the Danube in summer 2002, how do you think society should respond?
2. Given that the ideal response doesn't always happen, what do you think society will actually do about large floods?

Daniel Byers

CourseMate Connection

Start off this chapter by listening to a podcast with National Geographic grantee Alton Byers as he talks about his work studying glacial lakes in Nepal. **Log in to the CourseMate at www.cengagebrain.com to listen now.**

WATER IS LIFE. Without it, Earth would be a dead planet, its surface pock-marked with craters, as the Moon is. There would be no rivers, no mountain streams, no lakes, and no oceans. Forests, grasslands, and swamps wouldn't exist. Most days, water is a good thing. Moving water supports whole ecosystems and provides for recreational and economic activities. Rivers are tapped for drinking water and to irrigate our fields, allowing crops to grow in the desert. But some days, some weeks, some places, water turns deadly. Floodwaters ravage riverside towns, storm surges rip homes from their foundations, flash floods carry people to their deaths, and tsunami waves sweep away entire families and communities. Water and its agents—rivers, streams, and oceans—perform most of the work of erosion and deposition, which shape and modify Earth's landscape.

Hydrologists are scientists who deal with water: its properties, circulation, and distribution on Earth and in the atmosphere. They can be engineers or geologists who make their living solving the very practical problems of water supply, river flooding, and water quality. In this chapter, we will take the view of a practical hydrologist, examining the way in which water moves across the landscape. We focus on hydrologic hazards, which means flooding, the most globally pervasive, environmentally diverse, and continually destructive of all natural hazards. To understand flooding, you need to know a bit about where water comes from and where it goes. We'll start the chapter by examining the types of floods and the forces that drive them. The focus is on rivers and streams, because they carry most of the water moving over Earth's surface. We will look at not only why, where, and how often floods happen, but also how society can deal with water in all the wrong places.

Weather and Climate

Because most floods are driven one way or another by **precipitation**, it makes sense to start by understanding a little about weather and climate. Rain and snow occur when the atmosphere becomes saturated with water, and that water condenses from a diffuse vapor to a localized droplet of liquid or an ice crystal. Then, gravity takes hold and down come the rain, snow, hail, and sleet. If the precipitation falls as rain, it's ready to cause floods. If the precipitation is frozen, it may linger on the ground for days, weeks, months, years, or even millennia before melting and once again moving through the hydrologic cycle.

Precipitation is not equally spread over Earth's surface; rather, there are wet spots and dry spots predictably arranged around the world (**Figure 9.1**). In the tropics, such as the islands of Hawaii or Puerto Rico, moist air evaporated from warm ocean waters generates intense downpours. In contrast, downwind of cold ocean currents, such as in Peru (South America) and Namibia (southern Africa), hyperarid deserts are found where rain may fall only several times a decade. Mountains are almost always wetter than adjacent valleys. Air masses forced up and over high terrain cool, and the moisture in their air condenses, falling as rain or snow (**Figure 9.2**). This "wringing out" of moisture by lifting air over mountaintops is termed the *orographic* effect, and it is why many floods get their start in highlands. Downwind of mountainous terrain, the rung-out air descends. Many valley-bottom deserts of the western United States, such as the Mojave Desert, are the result of mountains to their west having stolen the Pacific moisture before it could move inland.

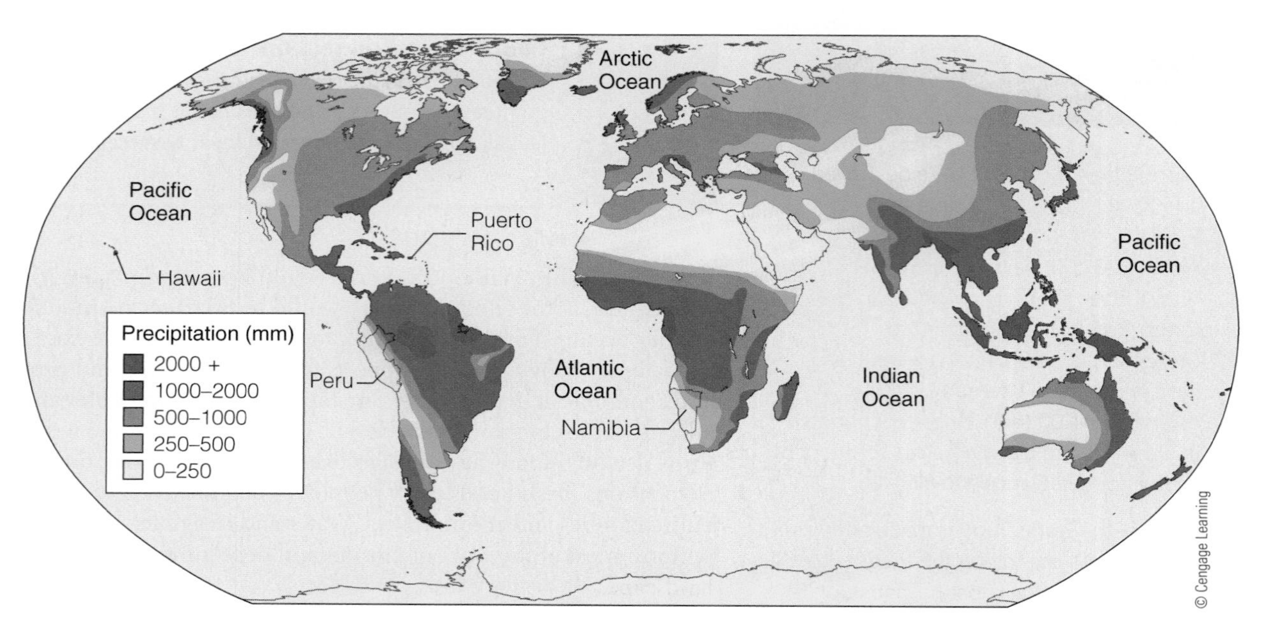

Figure 9.1 The global distribution of precipitation. Can you find areas near tropical oceans where it is very wet? Can you find areas downwind of major mountain ranges where it is very dry?

Precipitation does not fall equally over time either. Storms, or low-pressure systems, are disturbances in Earth's atmosphere that come and go day by day and with the seasons. For example, heavy thunderstorms are a summer phenomenon, the result of an atmosphere made unstable by strong solar heating. They affect mostly smaller drainage basins, with torrential rains capable of generating flash floods. In contrast, extended winter rains commonly cause floods along the west coast of North America as Pacific storms come ashore. On the East Coast, hurricanes can drop a foot of rain in just a few hours during the summer and fall, whereas nor'easters batter the Atlantic Coast with rain, wind, and sometimes very heavy snow, mostly in the winter and spring. In the winter of 1888, much of New England got more than a meter (4 feet) of snow in March from a nor'easter (**Figure 9.3**).

The most dramatic rainfall occurs when storms and topography interact. Let's have a look at southeastern Texas,

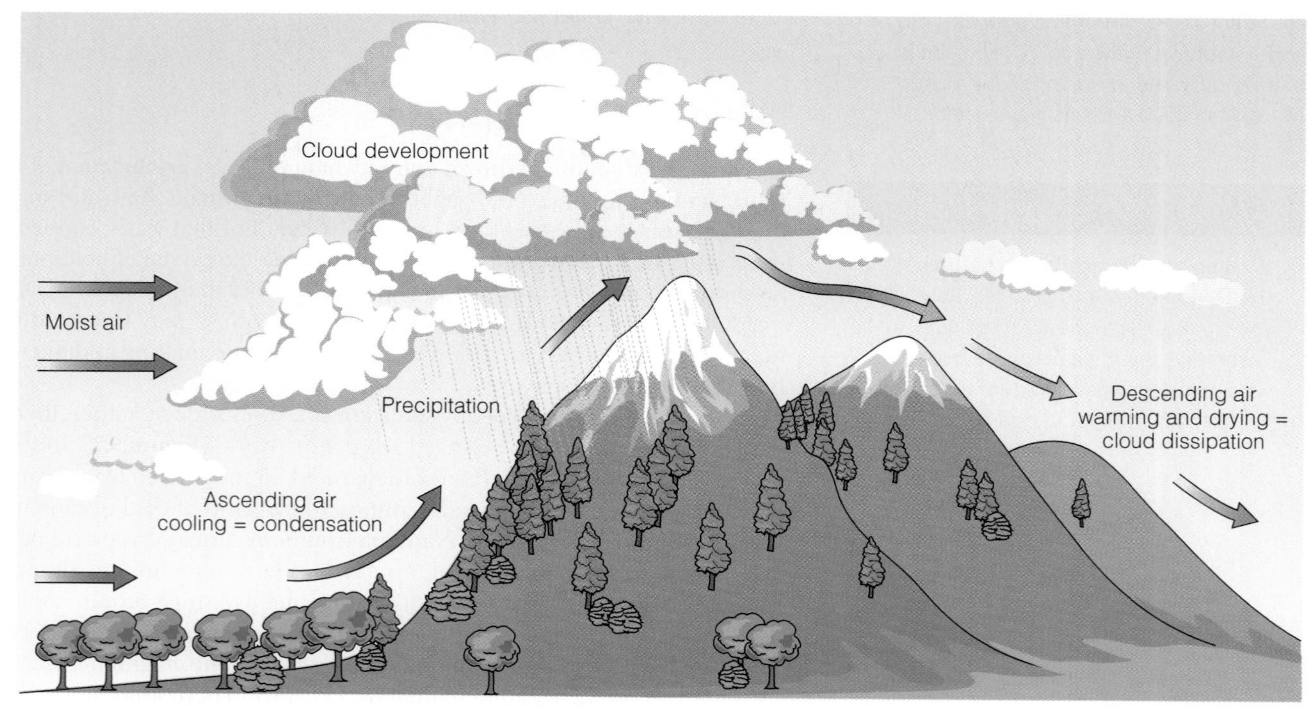

Figure 9.2 Air masses forced up and over mountain ranges cool, allowing the moisture in them to condense and rain out. After passing over the peaks, the air descends and warms, drying the landscape below. Can you predict which side of the mountain range will be cloudier and wetter? (Source: U.S. Geological Survey. 2004. *Large floods in the United States.* USGS circular 1245.)

Vermont Historical Society

© SKIP BROWN/National Geographic Image Collection

Figure 9.3 Nor'easters can be powerful storms. (a) The blizzard of 1888, a classic nor'easter, took the East Coast by surprise in mid-March. Temperatures plummeted from the 50s to a little above zero, and snow fell from Maryland to Maine. Here, in Brattleboro, Vermont, several feet of snow blanketed the city. (b) During a fierce nor'easter, two boys play in the water as wind-driven waves batter the shoreline near Hampton, Virginia, on the southern Chesapeake Bay.

perhaps not what you would think of as a flooding hot spot, but it is. Near Austin, the Balcones Escarpment rises up to 500 feet above the coastal plain. Moisture streaming north from the warm Gulf of Mexico on southerly and southeasterly winds is forced to rise up the escarpment and onto the Edwards Plateau, a rural farming area. Incredible rains can result, swelling normally dry riverbeds into raging torrents in a matter of minutes. During the 1954 flood, the Pecos River, often a dry stream, carried more than a million cubic feet per second, the average discharge of the Mississippi River, yet its drainage is only 0.3% of the area of the Mississippi (**Figure 9.4**). Some of the highest rainfall rates in the continental United States, as well as the greatest 18-hour rainfall total, more than 90 centimeters (36 inches), have been measured on the Balcones Escarpment. There, in 1987, 10 people died when their church bus was swept away by flash flooding. There are dramatic images of students being plucked from trees by helicopter rescue teams as floodwaters swirled around them.

Types of Floods

Floods are dramatic reminders of the power of natural forces. Hydrologic hazards result from a variety of solid Earth, surface, and atmospheric processes. Let's have a look at what causes floods (**Figure 9.5**). Precipitation is the source of most flooding. Whether it is the 1962 Ash Wednesday nor'easter affecting more than 1,500 kilometers (>1,000 miles) of coastline for several days, hurricane Katrina pounding New Orleans with torrential rains for hours, or a thunderstorm parked over Big Thompson Canyon in the Colorado Rockies, water falling from the sky faster than it can be absorbed by the ground or transported downstream sends rivers out of their banks. Warm winds and early spring rains on deep snowpacks are particularly devastating as melting snow adds to the runoff. These rain-on-snow events have caused devastating floods all over the United States from southern Pennsylvania to the Cascades of Washington State. Ice-jam floods occur when river ice breaks up, moves downstream, and wedges against obstacles in the channel. These short-lived ice dams quickly raise water levels upstream. Ice-jam floods are common anywhere rivers freeze over and often occur during rain-on-snow events. Such ice-jam floods may well become more common in the North as midwinter thaws accompany global warming; conversely, in places such as southern New England, the climate will likely become warm enough that the river ice no longer forms (see Case Study 9.1).

During a storm, water need not come from the sky to cause a flood; it might come from the ocean. When hurricanes or nor'easters make landfall, they push before them a wall of water driven by wind and accentuated by low atmospheric pressure near the storm. These surges of water can rapidly raise sea level by meters inundating low-lying coastal areas in an hour or two. The effect of *storm surges* is not

Figure 9.4 Texas is the place for floods. The digital elevation model shows the topography of Texas and the steep Balcones Escarpment as it rises from the coastal plain, ready to intercept storms and warm, moist air from the Gulf of Mexico. Inset (left) is an image of the damage caused by the 13–18 July 1900 flood on the Guadalupe River at Comfort, Texas. The river has flooded many times since, as shown by the modern image (right) taken on 5 July 2002. Star shows location where images were taken.

limited to the coast, however. Storm surges can move inland up coastal rivers. When they do, floodwaters trying to move downstream are blocked, and flooding further intensifies.

Dams are supposed to control floods. Most of the time they do, but some devastating floods have been caused by the failure of dams, both natural and constructed. Natural dams, formed when landslides or glacial moraines block streams, can fail catastrophically. The 1889 failure of an earth-fill dam in Johnstown, Pennsylvania, killed about 3,000 people as they were swept downstream by a wall of water 12 meters (40 feet) high (see Figure 9.5d). The stories are horrific. Riverside homes lifted from their foundations were carried intact and wedged up against a massive stone bridge in such a pile that no one could escape the fires that started when wood-fired cookstoves overturned. Victims were burned alive while their houses floated.

Earthquakes can do more than shake, rattle, and roll: They can cause floods in several ways. The Indian Ocean tsunami of 2004 generated rapid but short-lived flooding, as massive tsunami waves washed ashore. But earthquakes can cause flooding in another way: by changing land level. In the Pacific Northwest, buried marsh soils and standing dead trees, their roots drowned by saltwater, are compelling evidence that giant quakes have yo-yoed the land up and down over the past hundreds to thousands of years, intermittently flooding the coast. In Alaska, after the 1964 magnitude-9+ quake, land level fell far enough that villages were partially submerged, and there was no choice but to abandon them (**Figure 9.6** and Figure 7.27a).

Not all flooding is so obvious or rapid. Consider changing climate and its impact on sea level. At the peak of the last glaciation, 21,000 years ago, wooly mammoths roamed across North America and sea level was about 100 meters (330 feet) lower than today. As the climate warmed and the ice sheets melted, water flowed back into the ocean and sea level rose, rapidly at some times, more slowly at others. The camping sites of the first Americans along the Alaskan coast are now deep underwater, and the Chesapeake Bay, once a river, is now drowned. Today, with the climate warming and most glaciers retreating, the scientific consensus is that sea level will continue to rise, perhaps quite rapidly if melting of the Greenland and Antarctic ice sheets accelerates. It takes only a few meters of sea-level rise to inundate parts of many major and important world cities, such as New York, Miami, Venice, London, Hong Kong, and New Orleans.

Impacts of Floods

Imagine a flood and its effects. You are probably thinking of rising, muddy water covering fields, homes, and cars. Indeed, the most common impact of floods is **inundation**, and because floods tend to carry large amounts of fine sediment, not only do things get wet, but they get very muddy as well. The aftereffects of inundation can be more devastating than the flood itself. After the inundation of New Orleans in 2005, hundreds of thousands of automobiles were damaged, because their engines and electrical

Figure 9.5 There are several ways to trigger devastating floods. (a) In late September 1938, one of the strongest hurricanes to hit New England roared ashore, smashing into Long Island and racing north. Photo shows massive flooding on the Merrimack River, New Hampshire, following hurricane passage. The hurricane was a fast mover, traveling 600 miles in 12 hours. (b) In mid-March 1936, the St. Patrick's Day flood struck Pennsylvania as rain melted the winter's snow. Shown here is a Harrisburg, Pennsylvania, city bus nearly submerged. (c) After Hurricane Dean, a major flash flood filled with mud and debris roars down a river in Rouseau, Dominica Island, Dominica. (d) The Johnstown, Pennsylvania, flood swept huge amounts of debris downstream when a private earth-fill dam failed. Trees and homes were piled high against a stone bridge. A crane had to be used to remove the debris. (e) The failure of the earth-filled Teton Dam in 1976 sent a massive wave of water downstream killing 14 people. Water seeping through the dam weakened it so much that the entire dam failed.

a: © NOAA; b: Susquehanna River Basin Commission PA and USGS; c: © 2009 TYRONE TURNER/ National Geographic Image Collection; d: National Park Service; e: Security Pacific National Bank Historical Collection

(a)

USGS

(b)

Figure 9.6 Inundation is a devastating effect of floods. (a) In Portage, Alaska, the land dropped several meters during the 1964 earthquake, allowing seawater to flood much of the town. (b) In New Orleans, after the floodwaters of hurricane Katrina receded, mold was a devastating problem in the warm, moist Louisiana environment.

systems had filled with floodwater. Many homes needed to be gutted or torn down, not because they were structurally unsound, but because their flood-soaked walls couldn't be dried and had mildewed so badly that inhabiting the buildings was impossible (see Figure 9.6).

Floods do more than get things wet. As floodwaters move down river channels, they pick up sediment, both loose material from the channel bottom and solid material from the banks. At bridge abutments, this tendency of moving water to entrain material from the river bottom can lead to disaster as support is removed and the bridge eventually collapses (**Figure 9.7**). This removal of material from below abutments is termed *bridge scour* and is a hazard worldwide. In the past, not everyone realized that bank erosion and channel migration, processes that are most active during floods, are critical and normal aspects of river function. Human attempts to stabilize rivers using engineering solutions, with concrete and straightened channels at their core, have typically failed over time, because these "solutions" ignore the dynamic nature of flowing water and moving sediment. Today, the emphasis is on river-corridor management and river-channel restoration to accommodate the natural behavior of flowing water.

In the long term, rapidly moving water is a dynamic force, shaping Earth's surface. Large, rare floods can cause river channels to shift dramatically, abandoning their old course and establishing a new channel, sometimes with disastrous results. Some of the most spectacular features on our planet are deep canyons cut into rock. Research suggests that it takes massive and, therefore, rare events—the biggest floods—to generate enough power to rip rock away and carve canyons with moving water. In the Himalayas, raging rivers fed by the Asian monsoon can erode millimeters of rock per year. Surprisingly, the Potomac River, just outside Washington, DC, cut through hard rock at rates almost this fast during the last glaciation, cutting driven presumably by big floods (**Figure 9.8**).

River Systems

The ability of a stream to erode (Latin *erodere*, "to gnaw away") the land and impact humans is related to the speed at which the flowing water moves (its velocity) and the amount of water moving through the stream (its **discharge**). With equal discharge—the same volume of water flowing past a point in a river channel in a given time period—a fast-flowing stream erodes its banks more significantly than does a slow-moving one. In the United States, velocity *(V)* is usually expressed in feet per second (ft/sec), and discharge *(Q)*, the amount of water, in cubic feet per second (cps). A cubic foot contains about 7.4 gallons of water. With increased velocity and discharge, there are increases in the stream's erosion potential and its sediment-carrying capacity (**Figure 9.9**).

Figure 9.7 Bridge scour can be devastating. (a) A bridge on the New York State Thruway collapsed on 5 April 1987, claiming 10 lives as raging floodwaters tore away the support under its footings. (b) The view after the collapse; it's hard to get from here to there anymore.
a: Daily Gazette, Sid Brown; b: USGS

P. Bierman

P. Bierman

Every stream channel has a slope, known as its **gradient**, which can be expressed as the ratio of its vertical drop to the horizontal distance it travels (in meters per kilometer, or feet per mile). A mountain stream may have a steep gradient of several tens of meters per kilometer and an irregular profile with rapids and waterfalls. As the same stream approaches the lowest level to which it can erode, an elevation known as **base level**, its gradient may be almost zero and the water surface is nearly flat. Looked at in a side or cutaway view, the longitudinal profiles of streams are generally concave upward from source to end. Gradients decrease in the lower reaches of a river (**Figure 9.10**), and this is where sedimentation often occurs as streams overflow their banks onto adjacent **floodplains**. Decreased water velocity on the floodplain results in the deposition of fine-grained silt and clay adjacent to the main stream channel. This productive soil, commonly found along low-lying river floodplains, known as *bottomlands*, is much desired in agriculture (see Case Study 9.2).

Figure 9.8 Only big floods can erode rock. At Great Falls on the Potomac River just 12 miles from the U.S. Capitol in Washington, DC, most flows spill gracefully over the rocks. However, when hurricanes or rain-on-snow events strike, the river becomes a raging torrent that can erode rock. (a) A calm November day with flow of several hundred cubic meters per second. (b) After Hurricane Isabel in 2003, the river channel is filled as almost 5,000 cubic meters (160,000 cubic feet) per second of water pour over the falls. The red arrows point to the same rock outcrop at low and at high water.

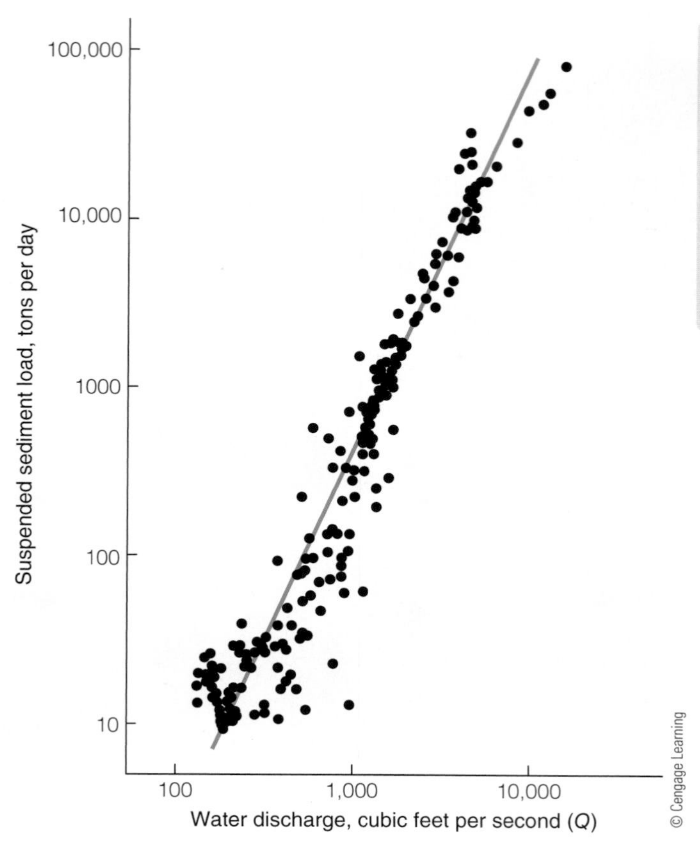

Figure 9.9 The relationship between water discharge *(Q)* and the suspended sediment load for a typical stream. Each point represents a separate measurement. A 10-fold increase in water discharge results in almost a 100-fold increase in sediment load. Look carefully at the axes of this graph. Can you see that the units increase in multiples of 10 because these are logarithmic scales? Consider how variable these measurements are. For any one discharge, suspended sediment load can vary by a factor of more than 50.

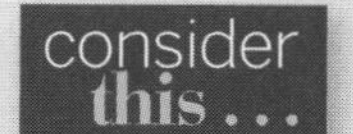

We generally associate high sediment loads with high river discharge, but other variables determine the sediment load transported by a river. With this in mind, would you expect the suspended sediment load per volume of water to be greater in an area dominated by soft rocks, such as the sediments of the Dakota badlands, or hard rocks, such as the granites of the Sierra Nevada Mountains? What role do you think vegetation might play in determining sediment load?

Stream Features

In both humid and arid environments, streams lose the ability to transport sediment as they flow from confined channels in mountainous terrain onto wide-open, flat valley floors. There, where the slope changes, are conspicuous depositional features known as **fans** (**Figure 9.11**). These landforms are, in fact, fan-shaped and build up as the main stream branches, or splays, outward in a series of *distributaries*, each "seeking" the steepest gradient. Fans can provide beautiful, interesting views of surrounding terrain, but they can also be subject to flash floods and debris flows from intense seasonal storms in nearby mountains.

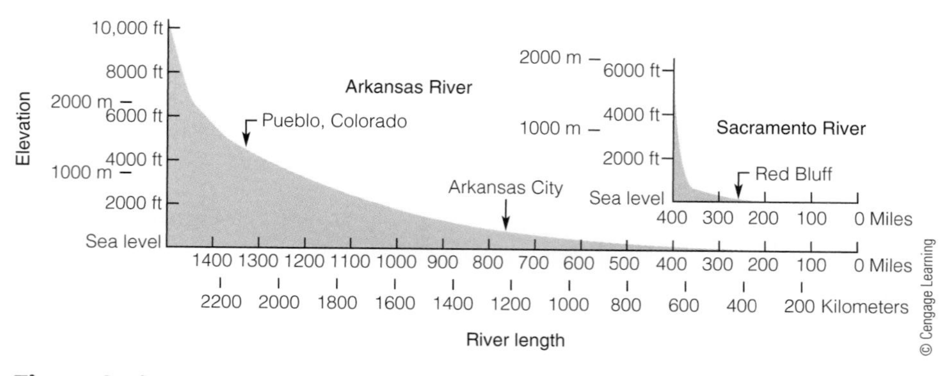

Figure 9.10 Cross sections, or longitudinal profiles, of the Arkansas River in Colorado and the Sacramento River in California. The vertical scale is exaggerated 275 times, meaning that these diagrams make the rivers look far steeper than they really are. Which river is steeper?

Deltas form where running water moves into standing water, causing sediment carried by the water to drop out. The Greek historian Herodotus (484–420 B.C.E.) applied the term to the somewhat triangular shape of the Nile delta in recognition of its resemblance to the Greek letter delta (Δ). Deltas are characterized by low stream gradients, swamps, lakes, and an ever-changing channel system. The "arcuate" (arc-shaped) delta of the Nile and the "bird's-foot" delta of the Mississippi are also described by their appearance on maps (**Figure 9.12**). The tremendous sand output of the Nile has created the huge coastal dunes east of Alexandria that formed the background for the World War II battle at El Alamein. The Ganges-Brahmaputra delta of Bangladesh consists of tidal flats and many low-lying sand islands offshore (see Figure 9.12c). Because deltas are at or near sea level, they are subject to flooding by tropical storms. Coastal flooding is the most deadly of natural disasters, even more deadly than earthquakes or river floods. This type of flooding was particularly disastrous in Bangladesh in 1970, when more than 500,000 people were drowned by a surge of water that moved onshore along with a tropical cyclone.

Streams and rivers with low gradients and floodplains form a number of features that both aggravate and alleviate flood damage. The courses of such rivers follow a series of S-shaped curves called **meanders**, after the ancient name of the winding Menderes River in Turkey. Straight rivers are the exception, rather than the rule, and it ap-

P. Bierman

Figure 9.11 A grassy fan in the Austrian Alps. The fan was created when sediment from the steep mountains above was carried by streams and in debris flows and avalanches to the valley bottom where the slope was less, and thus the sediment could move no more.

pears that running water of all kinds will meander—ocean currents, rivers on glaciers, and even trickles of water on glass. The key to these regularly spaced curves is the river's ability to erode, transport, and deposit sediments. As a stream meanders, it erodes the outside curve of each meander and deposits sediment on each inside curve, producing a deposit known as a **point bar**. **Figure 9.13a** shows the velocity distribution around a stream meander, which explains the erosion-deposition pattern. Erosion occurs where water moves rapidly. Deposition occurs where water moves slowly. Meanders widen stream valleys by lateral cutting of valleyside slopes, causing landslides and eventually creating an alluvium-filled floodplain adjacent to the river. It is across this surface that the river meanders, sometimes becoming so tightly curved that it shortcuts a loop to form a **cutoff** and an **oxbow lake** (**Figure 9.14**). This is quite the hazard if your farm field happens to be on the other side of the cutoff.

River-bottom lands, with wide floodplains, meanders, cutoffs, and oxbow lakes, are prime places for flooding. Every time the river goes out of its banks and the raging floodwaters slow, sand and silt are deposited adjacent to the channel (**Figure 9.15**), helping to build **natural levees**. Following nature's lead, people raise or imitate these natural levees using all sorts of engineering approaches to hold back the floodwaters. **Artificial levees** keep floodwaters channelized, so that the water level in the stream rises above the level of the adjacent floodplain, sometimes far above. This is a tenuous situation, because a breach in the levee causes immediate inundation of the floodplain. Even when the levees hold and protect one area from flooding, they speed the passage of floodwater, exacerbating flooding downstream, particularly in unleveed areas. What is the typical response? Build more levees (see Case Study 9.3).

During the 1993 floods in the Midwestern United States, artificial levees saved some cities in the Mississippi River drainage basin—notably, St. Louis, where the flood crested just several feet (less than a meter) below the top of the levees. Quincy, Illinois, lacking artificial levees, was inundated; but the levees at Hannibal, Missouri, on the opposite side of the river, held and spared the town. Some levees were intentionally breached to reduce flood-crest elevation downstream by allowing water to move onto the floodplain (where it belongs). More than 800 of the 1,400 levees in the 9-state disaster area were breached or overtopped. Most of these levees were simple dirt berms built by local communities, which explains why so many small towns were almost completely submerged by the floodwaters. However, more than 30 levees built by the U.S. Army Corps of Engineers also failed during this record flood event.

River deltas. (a) Bird's-foot delta of the Mississippi River. The main channel and subordinate distributaries are separated by swamps and wetlands. (b) The arc-shaped delta of the Nile River is predominantly wetlands and rice-growing areas. The barrier islands are sand and they contribute to large coastal dunes. (c) The Ganges-Brahmaputra delta of Bangladesh. The small islands rise above the tidal flats. Although subject to catastrophic floods, these islands are nonetheless densely populated. (d) Sediment transported by a river into the Indian Ocean on Mozambique's north coast creates a symmetric, arc-shaped sandy delta.

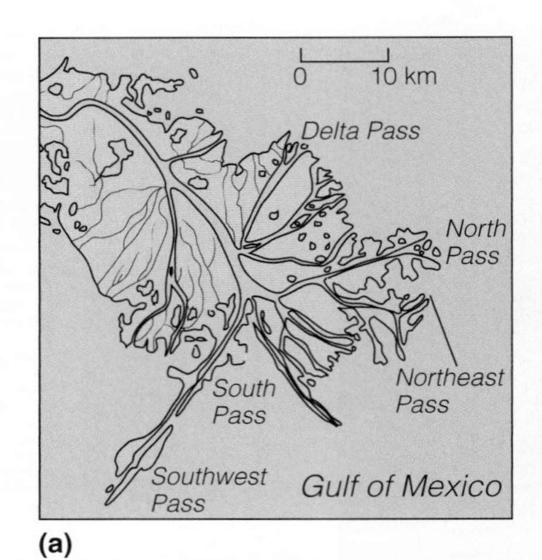

(a)

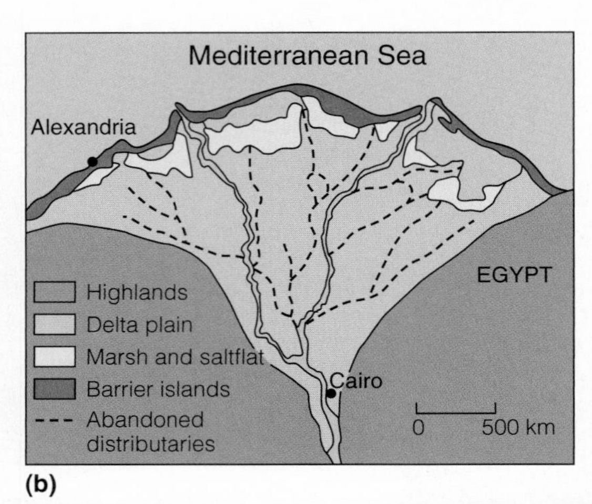

(b)

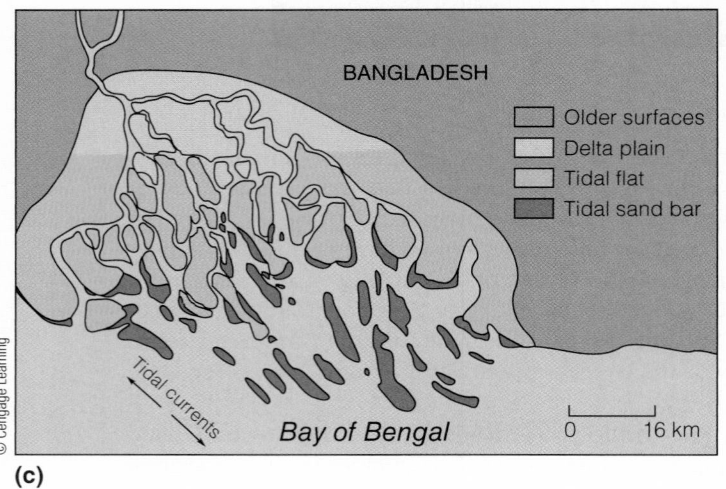

(c)

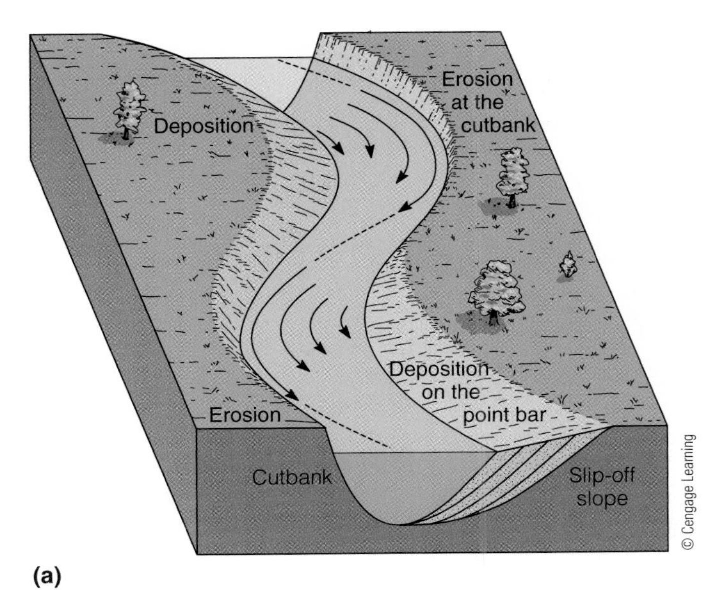

(a)

(b)

Figure 9.13 Erosion and deposition patterns on a meandering stream. (a) Erosion of the cutbank and deposition of a point bar on the gently sloping side of the meander. Longer arrows indicate faster flow. Shorter arrows indicate slower flow. (b) Rillito Creek rampaged through Tucson, Arizona, during the flood of 1983. Damage occurred when the cutbank of the meander migrated west (left) into vacant property immediately upstream from the townhouses (shown by yellow arrow, center left). This allowed erosion to occur behind a concrete bank that was protecting the townhouse property. Note the prominent point bar that is indicated by the red arrow at the bottom of the photograph. Flow direction shown by black arrow.

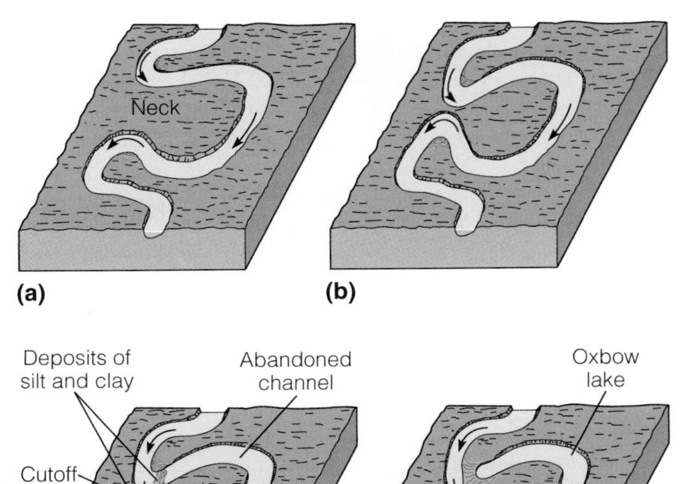

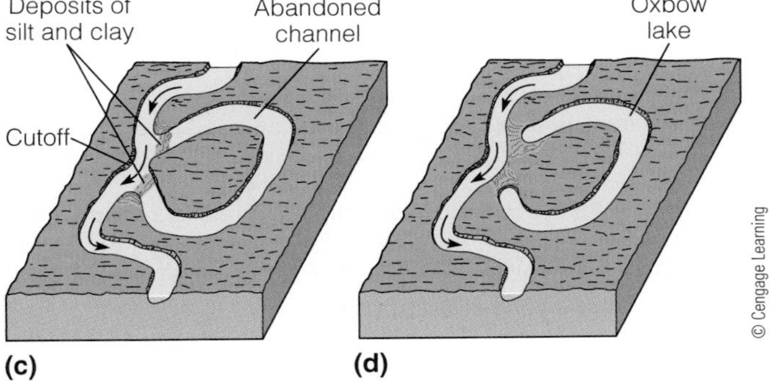

Figure 9.14 (a, b) Over time, cutbanks erode, sometimes bringing meanders closer together. (c) When the two meanders finally join, a cutoff forms, shortening the river channel but leaving (d) an oxbow lake where the river used to flow.

When There Is Too Much Water

Excess water, whether in rivers, in lakes, or along a coast, leads to flooding. The basic hydrologic unit in fluvial (river) systems is the **drainage basin**, the land area that contributes water to a particular stream or stream system. Individual drainage basins are separated by **drainage divides**. A drainage basin may have only one stream, or it may encompass a large number of streams and all their tributaries.

Unless rainfall rates are exceptionally high or the ground has been compacted and there is little vegetation, most of the rain that falls within a particular drainage basin after a

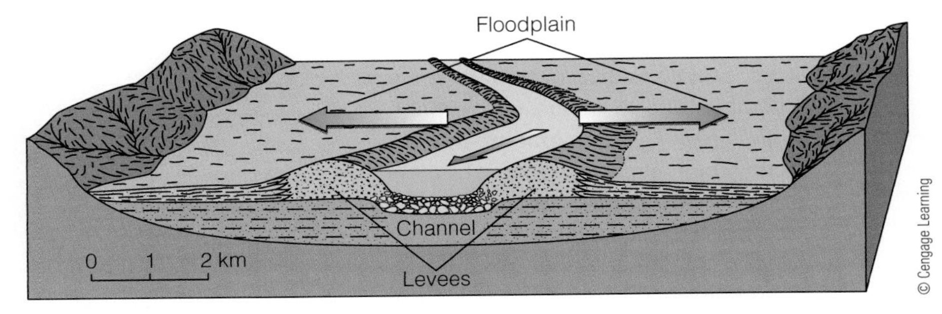

Figure 9.15 Relationship between natural levees and a floodplain. The coarsest sediment is transported and deposited in the stream channel, where the stream's velocity is greatest, and finer silt and clay are deposited on levees and the floodplain as velocity diminishes during overbank flows.

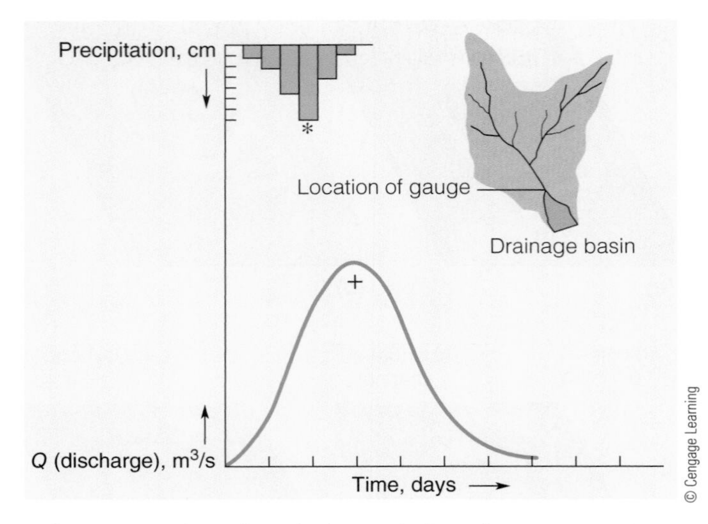

Figure 9.16 A synthetic hydrograph that relates precipitation and discharge to time. Note that peak discharge (+) occurs after the peak rainfall intensity (*). The drainage basin includes the main stream and all its tributaries.

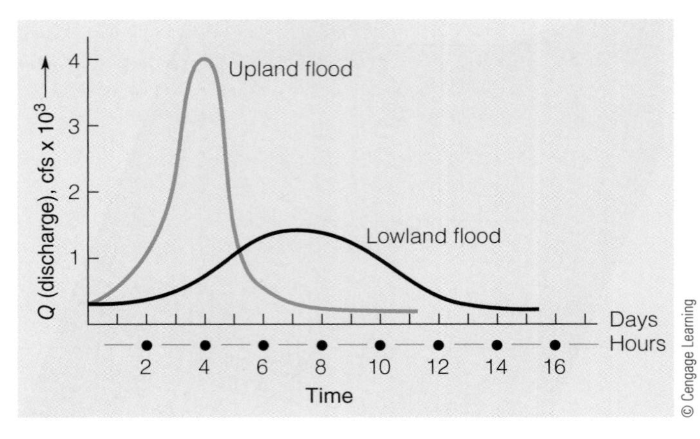

Figure 9.17 Synthetic hydrographs of an upland flash flood and lowland riverine flood, both of which are carrying the same amount of water; therefore, the area under the curves is the same.

dry period **infiltrates** (seeps into) the uppermost layers of soil and rock; little water runs off the land. As the surface materials become saturated or if rainfall rates are extremely high, some rainfall cannot infiltrate, and runoff begins; it begins in saturated areas, flows into small tributary streams, and finally is consolidated into a well-defined channel of the main stream in the watershed. Whether a flood occurs is determined by several factors: the intensity of the rainfall, the amount of antecedent (prior) rainfall, the amount of snowmelt (if any), the topography, and the vegetation.

Upland floods occur in watershed areas of moderate or high topographic relief. They may occur with little or no warning, and they are generally of short duration, hence the name *flash floods*. Such floods result in extensive damage because of the high velocity of the flowing water. Homes and bridges are ripped from their foundations. Lowland, *riverine floods*, in contrast, occur when broad floodplains adjacent to the stream channel are inundated. The water in such lowland floods typically moves quite slowly; thus, this results in damage by soaking and siltation—peoples' homes are wet and full of mud but usually left intact.

The difference in these types of floods is due to watershed morphology (shape). Mountain streams occupy the bottoms of V-shaped valleys and cover most of the narrow valley floor; there is no adjacent floodplain. Heavy rainfall at higher elevations in the watershed can produce a wall of fast-moving water that descends through the canyon, destroying structures and endangering lives. Lowland streams, in contrast, have low gradients, adjacent wide floodplains, and meandering paths. Here, floods tend to move slowly, be less deep, but inundate large areas.

A **hydrograph** is a graph showing a stream's or river's discharge, velocity, or stage (height) over time; it is the basic tool of hydrologists (**Figure 9.16**). The "synthetic" hydrographs in **Figure 9.17** illustrate the basic difference between upland (flash) floods of short duration and high discharge, and lower discharge but longer lasting lowland floods. The timing of flood crests on the main stream—whether they arrive from the tributaries simultaneously or in sequence—in large part determines the severity of lowland flooding.

Floods are normal and, to a degree, predictable natural events. Generally, a stream will overflow its banks every year or two, gently inundating a portion of the adjacent floodplain. Over long periods of time, catastrophic high waters can be expected. Depending on their severity, these events are referred to as 50-, 100-, or 500-year floods, denoting the average period of time between events of similar magnitude. The more infrequent the event, the more widespread and catastrophic is the inundation (**Figure 9.18**).

Floods from all causes are the number-one natural disaster in the world in terms of loss of life. Fifteen disastrous

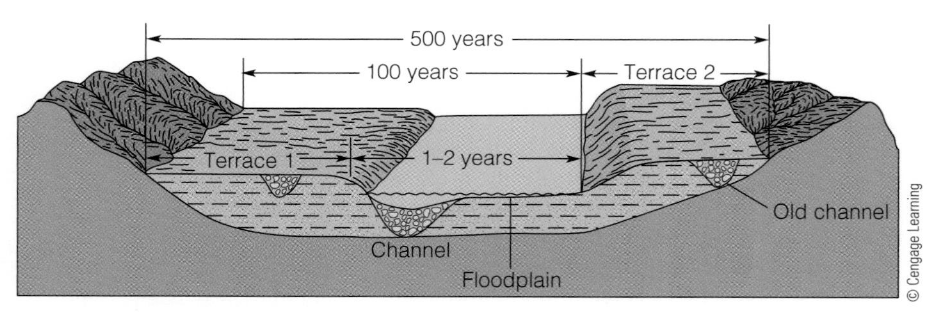

Figure 9.18 The probable extent of flooding over long time intervals on a floodplain. The 1- to 2-year flood just goes over the stream bank onto the adjacent floodplain. The 100-year flood inundates the low terrace (terrace 1), which formed when the river stood at a higher level. The 500-year event inundates an even higher terrace (terrace 2), and all lower terraces and the floodplain.

Table 9.1 Selected U.S. Floods Since May 1889

Type	Date	Location	Lives Lost	Estimated Damage $, Millions
Regional flood	13 September 1928	Lake Okeechobee, Florida	1,836	26
	March 1938	Southern California	79	25
	21 September 1938	New England	600	306
	September/October 1983	Tucson, Arizona	<10	50–100
	Summer 1993	Upper Mississippi River drainage basin	50	12,000
	January/February 1996	Oregon, Washington, Idaho, Montana	1	89
	October 1998	South-central Texas, San Antonio to Houston, Guadalupe River	29	400
Hurricane	8 September 1900	Galveston, Texas	6,000	30
	17–18 August 1969	Mississippi, Louisiana, and Alabama (Hurricane Camille)	256	1,421
	September 1999	North Carolina (Hurricane Floyd and post-hurricane rains)	47	5,000+
	August 2005	Louisiana, Mississippi (Hurricane Katrina)	1,840	125,000+
Flash flood	July 1976	Big Thompson River, Colorado	139	30
Dam failure	May 1889	Johnstown, Pennsylvania	3,000	—
	12–13 March 1928	California, St. Francis Dam	450	14
	February 1972	Buffalo Creek, West Virginia	125	10
	June 1976	Southeast Idaho, Teton Dam	11	1,000

Source: Various sources, including the U.S. Geological Survey.

U.S. floods are listed in **Table 9.1**. Although by no means a complete listing of severe floods, the table gives an idea of the distribution of riverine floods, flash floods, coastal floods, and dam-break floods in the United States. Coastal flooding caused by hurricanes and typhoons is the greatest single killer. The record-holding river floods are due to dam failures (see Figure 9.5). The legendary 1889 Johnstown flood killed about 3,000 people, and the 1928 St. Francis Dam failure in Southern California killed 450 people.

Flood Measurement

Flood studies begin with hydrographs for a particular storm on a river within a given drainage basin. These graphs are generated from data obtained at a stream-gauging station, where stage (the water elevation) is measured and related to discharge. Rapid runoff in a drainage basin is accompanied by high stream discharge and high water velocity. As discharge increases, the water rises to the *bank-full stage* and then spills onto the adjacent floodplain at *flood stage*. **Figure 9.19** shows such a condition. It is a hydrograph superimposed on the cross section of a stream valley; the *stage hydrograph* is most useful for flood-planning purposes, because it relates water height over time. One can see at a glance at

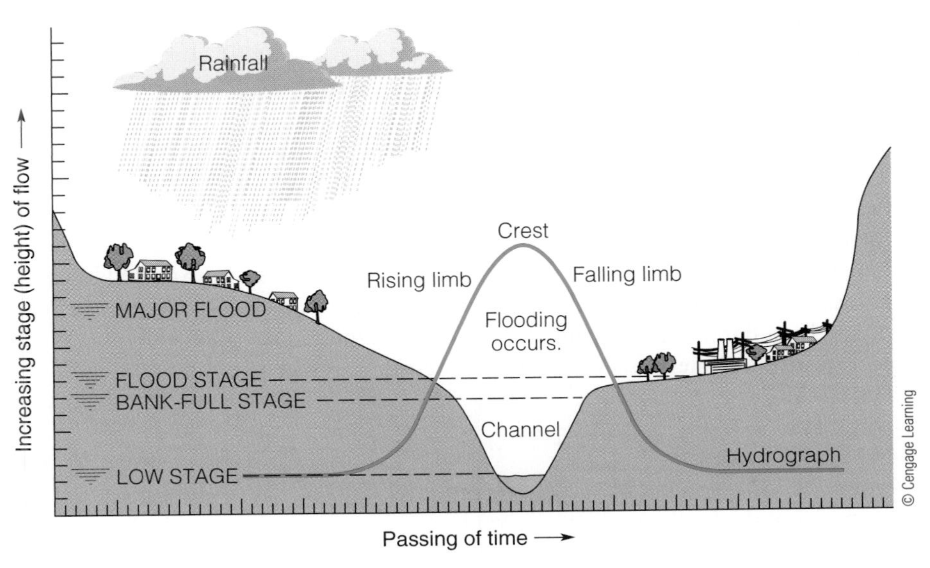

Figure 9.19 A stage hydrograph superimposed on a schematic stream valley illustrates the impact of increasing stage height on the flooded area.

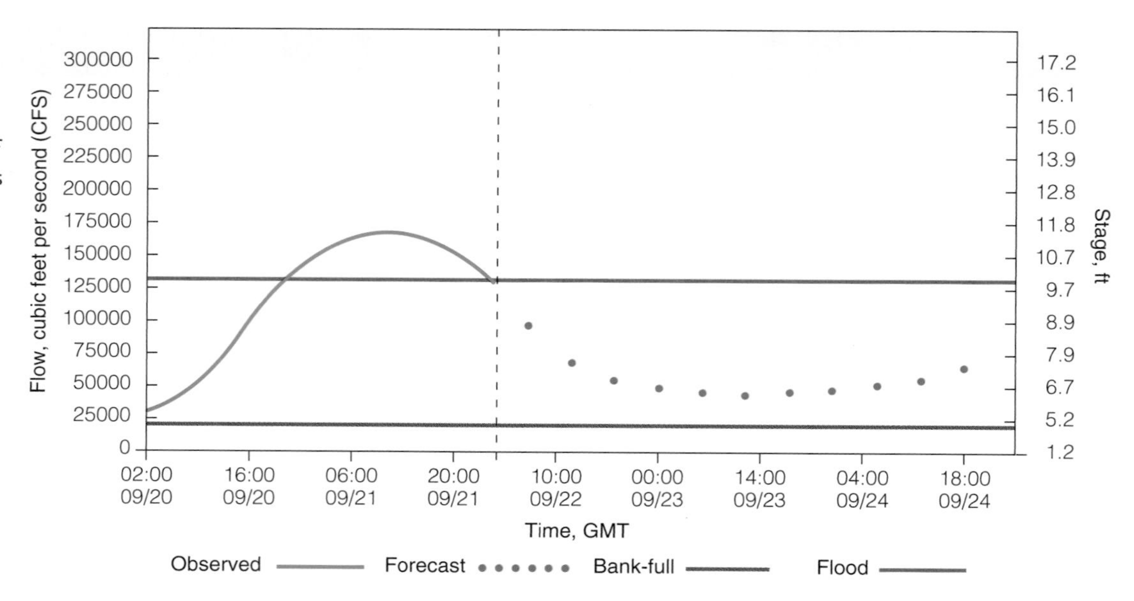

Figure 9.20 A combined hydrograph showing both stage and discharge for the Potomac River at Little Falls after the passage of Hurricane Isabel in fall of 2003. Note that the flood exceeds both the bank-full and flood stages. The blue line represents data collected at the gauging station. The green dots are predicted stage values calculated by the advanced hydrologic prediction service based on rainfall information and gauge data from upstream. Such predictions are available for many North American rivers at the Advanced Hydrologic Prediction Service Web site (http://water.weather.gov/ahps2/). Next time it rains, find a prediction for a river near you.

what point the stream will overflow its banks and at what elevations floodwaters will contact specific structures. Note that both stage and discharge are shown in **Figure 9.20**, which is for the Potomac River, just downstream of where the photographs in Figure 9.8 were taken. This flood, triggered by the torrential rains of Hurricane Isabel several days before, is the type of event that occurs on the Potomac River once or twice every decade.

Flood Frequency

One of the most useful relationships that can be derived from long-term flood records is **flood frequency**, or **recurrence interval**—that is, how often, on average, a flood of a given magnitude can be expected to occur at a particular location. With this kind of information, we are better equipped to design dams, bridge clearances over rivers, the sizes of storm drains, and the like.

The information needed for determining recurrence intervals is a long-term record of annual *peak discharge* (the largest flow for each year) for a particular location on a river. The longer the period of record, the more reliable the statistics. The peak discharges are then ranked according to their relative magnitudes, with the highest discharge being ranked as 1, the second largest as 2, and so forth. To serve as an example, **Table 9.2** lists selected annual peak discharges on the Rio Grande near Lobatos, Colorado, with their ranks and recurrence intervals. A recurrence interval, T, is calculated as follows:

$$T = (N + 1)/M$$

where N is the number of years of record (109 years in our example) and M is the rank of the event.

The analysis predicts how often floods of a given size can be expected to occur in a watershed if the watershed and climate conditions have not changed. In this example, a flood the size of the 1906 flood ($M = 11$) would be expected to occur, on average, about once every 10 years [(109 + 1)/11]. If 100 years of records are available, the largest flood to occur during that period ($M = 1$) has a recurrence interval of about 100 years, on average. This is the so-called 100-year flood. The 100-year flood can also be obtained by projecting stream-flow data beyond the period of record, although such extrapolation is extremely uncertain.

How often can the "20-year flood" occur? The term is misleading, because it implies that a flood of that size can happen only once in 20 years. The 20-year flood is really a statistical statement of the probability that a flood of a given *rank* will occur in any one year—that is, $1/T$. For the 20-year flood, there is a 5% chance (1/20) of its occurring in any one year; for the 50-year flood (1/50), a 2% chance. The U.S. Geological Survey suggests that a better term would be "the 1-in-20 *chance* flood."

There is, however, no meteorologic or statistical prohibition against more than one 100-year flood occurring in a century, or even in one year, for that matter. Recurrence intervals are not exact timetables; they merely indicate the statistical probability of a given-size flood. However, even rough estimates of the scale of flood discharge that can be expected in a river system every 25, 50, or 100 years enable us to plan better against the flooding that is certain to occur eventually. When this concept is viewed as a probability phenomenon, it is more understandable that we might get big floods in successive years or every few years (see Case Study 9.4).

Keep in mind that, the longer the rainfall or stream-flow record, the better and more robust the statistical inference. Look at the stream-flow data collected between 1940 and 2008 on the Chehalis River in Washington State in **Figure 9.21**. Note that the 1-in-20 chance flood calculated for data from 1940 to 1973 is lower than the same 1-in-20

Table 9.2 Annual Peak Discharges on the Rio Grande Near Lobatos, Colorado, 1900–2008

Year	Discharge, Cubic Meters/ Second	Rank *(M)*	Recurrence Interval *(T)*
1900	133	30	3.66
1901	103	43	2.56
1902	16	99	1.11
1903	362	2	55.0
1904	22	93	1.18
1905	371	1	110
1906	234	11	10.0
1907	249	8	13.7
1908	65	60	1.83
1909	216	15	7.33
1974	22	92	1.20
1975	70	56	1.96
1985	177	21	5.24
2007	44	77	1.42
2008	83	51	2.15

Source: U.S. Geological Survey.

Table 9.3 Flow Statistics for the Chehalis River, Doty, Washington

1940–1973	Return Interval	Annual Probability	1975–2008
12,600	5 years	1:5	17,400
15,400	10 years	1:10	21,000
21,600	20 years	1:20	29,000
10,200	Mean annual maximum flow		13,400

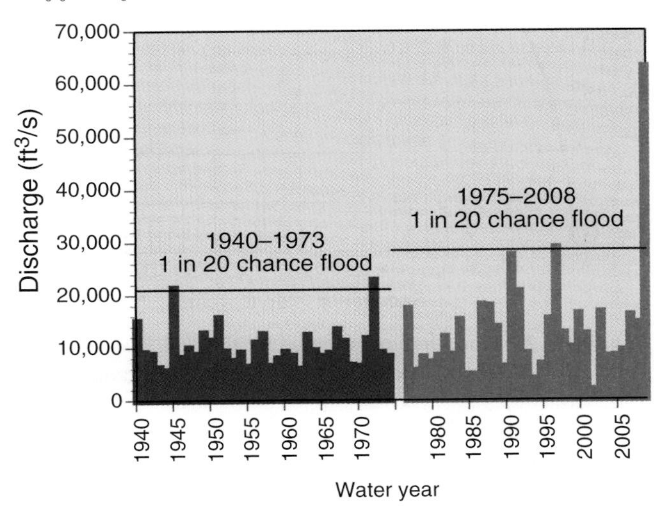

Figure 9.21 Annual maximum discharge data for the Chehalis River near Doty, Washington. Note the difference in the predicted value for the 20-year flood (the horizontal lines on the graphs) based on the first 34-year period and the second 34-year period. (Source: USGS data.)

chance flood based on data for the period 1975 to 2008. A similar relationship occurs for the 1-in-10 and 1-in-5 chance floods (**Table 9.3**). This illustrates the importance of long-term data collection in assessing the probability of floods and illustrates a phenomenon called *nonstationarity*. That is the observation, noted on some rivers, that flow characteristics change over time. Sometimes the driving force is land-use change; other times, it is climate change. Whatever the cause, it is clear that, over the last 35 years, the Chehalis River has produced more frequent, larger floods than it did previously. Most of these large floods occur in December, January, and February, and are presumably related to rain-on-snow events. Might this be a signal of a warming climate, more rain on snow, and/or a more active hydrologic cycle as predicted by global circulation models? Only time will tell.

Once we have determined recurrence intervals from the historical record, it is possible to construct flood-frequency graphs like the ones in **Figure 9.22**. Such graphs allow us to estimate at a glance how often a flood level or discharge of a particular magnitude is likely to occur. For example, if we wish to build a warehouse with a useful life of 50 years on the floodplain of the Red River at Grand Forks, North Dakota, we see in Figure 9.22 that the river reaches a reference elevation of about 49 feet every 50 years, on average. We would probably want to build the warehouse above that elevation, because there is a significant chance of a flood within the 50-year window of interest (and our chances for insurance would be better, too). However, urbanization, artificial levees, and other human-generated changes can alter the amount and timing of runoff and cause flood events to become more common. In urbanizing areas, a 10-year flood can become the 2- or 3-year flood as paved area expands and the ability of water to infiltrate the soil decreases (see Case Study 9.5).

Land-use changes, engineering modifications of the river channel, and long-term climate changes require that recurrence intervals, flood levels (stages), and floodplain management be updated periodically. Indeed, changing climate and climate cycles influence flood frequency, mandating that we continue to collect flow data from as many stream gauges as possible.

Flood Mitigation

Options

Dams: The solution or the problem?—In the contiguous 48 states, there are more than 75,000 dams over 2 meters (6.5 feet) high, and nearly every major river is controlled by at least one dam. Dams provide such societal benefits as cheap electricity, flood control, recreation, and drought mitigation. However, dams destroy habitats, disturb

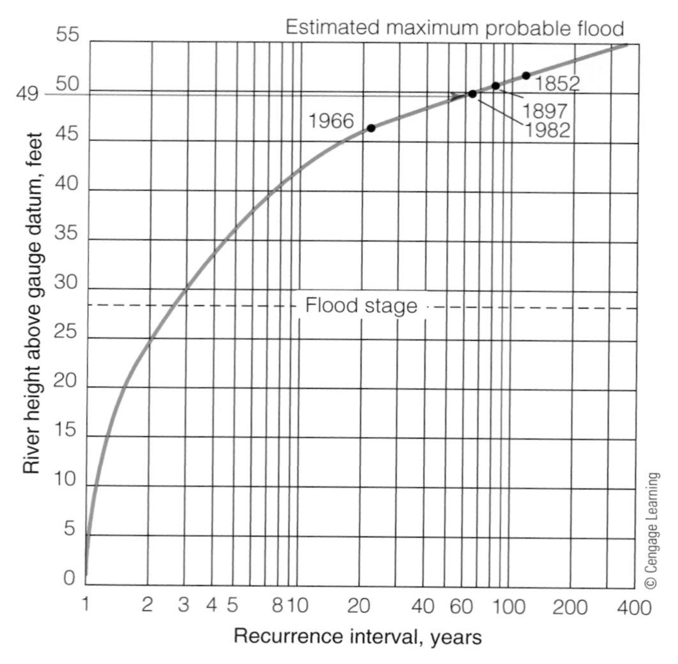

Figure 9.22 Flood-frequency curve for the Red River at Grand Forks, North Dakota, using water elevation as the standard.

natural cycles of flooding and sediment deposition, change aquatic and riverine biology, and destroy important spawning grounds for migratory fish, such as salmon. The days of big dam construction in the United States are over. Although more large dams are being built in developing countries, only a few sites in the United States, Canada, and Europe are seriously being considered.

The water emerging from a dam has different physical properties than the water upstream. Sediment-free outflow water scours the riverbed below the dam; because the temperature is different, native fish are stressed and some die. Even with fish ladders, adult fish have trouble migrating upstream to spawn, and many fry cannot survive the trip back downstream. Different **riparian** aquatic communities arise in the downstream area as the frequency and magnitude of disturbance change (floods are now controlled). In addition, some dams fail, and dam-failure floods have caused loss of life (see Table 9.1).

For all these reasons, some dams are being dismantled and others are being altered to make them more ecologically friendly. The Kennebec River in Maine now flows freely because the 162-year-old Edwards Dam in Augusta, Maine, was breached in 1999. Other dams being breached (to make way for spawning Pacific salmon) are the Elwha Dam in Washington State, which backs up water into Olympic National Park, and 10 hydroelectric dams on the Deerfield River in Vermont and Massachusetts (to allow passage by Atlantic salmon and other migratory species). This dismantling is occurring because a 1996 federal law requires that environmental concerns be addressed before dams can be relicensed. Dam removal is by no means straightforward. Excessive sediment loads after removal must be considered, as decades of trapped sediment are released, sometimes in just a few days. Some of these sediments can be rich in nutrients, as well as toxic metals and organic compounds.

Artificial Levees: The Solution or the Problem?—Dams, retaining basins, floodwalls, and artificial levees are means of flood control, all of which attempt to keep floodwaters within the river channel or store water for slow release at a later time (**Figure 9.23**). Ironically, these flood-control structures may increase flood risk, because floodplain development occurs in response to the

Figure 9.23 Earthen levees are a popular engineering solution to the problem of flooding. (a) Bulldozers building an earthen flood protection levee in Terrebonne Parish, Louisiana. (b) Some important features of earthen levees.

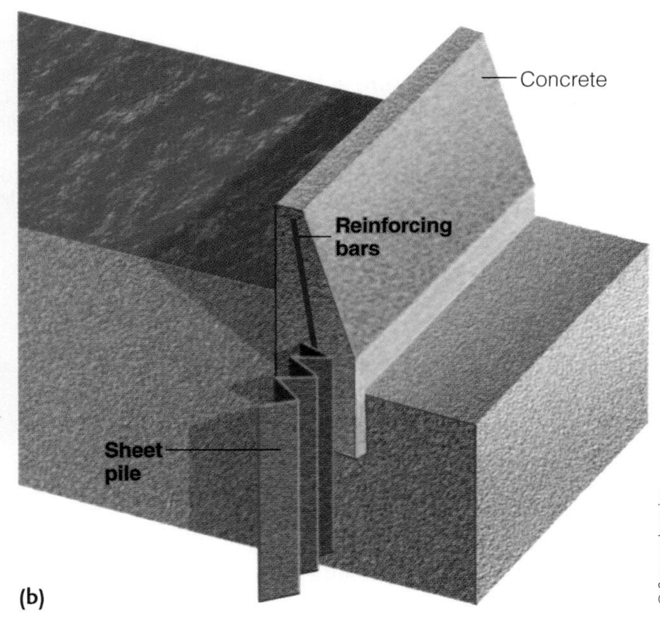

Figure 9.24 (a) The French Quarter as seen from atop the natural levee on the Mississippi River. Note that the levee slopes downward toward the French Quarter. The wall and gate are part of a floodwall built to hold back floodwaters, should the river overtop the levee. (b) A floodwall. The sheet pile helps stabilize the wall and keep water from seeping underneath.

assumptions that floods have been controlled. Population and property values adjacent to and below flood-control structures increase, placing people and their property at risk when the levees fail. For example, many artificial levees were built along the Mississippi River to protect farmland. The floodplain was then urbanized, and the existing levees, although adequate to protect farmland (which can tolerate occasional flooding), were not adequate for protecting homesites.

An artificial levee is usually built upward on a natural levee or riverbank to increase the amount of water the channel can accommodate, and thus prevent the natural periodic flooding on adjacent land. When stream flow is high, the area's would-be floodwater flows downstream to areas that have lower natural levees or no levees. Downstream floodplains then experience greater flooding than they would if the artificial levee had not been built upstream. Some of these downstream communities build their own artificial levees to prevent flooding in the future. Artificial levees are thus self-perpetuating, and the Mississippi River is largely contained within them. Even at nonflood river levels, the river's water surface in some places is above adjacent floodplains, so the danger of flooding is ever present.

Riverside land adjacent to the old French Quarter of New Orleans is extremely valuable tourist property. In the 1700s, the natural levees were about 2.1 meters (7 feet) above sea level, and natural flood stages probably rarely exceeded 3.0 meters above sea level. However, confinement of the river upstream by artificial levees raised the flood level to 6.5 meters (21 feet) in modern times, and artificial levees were constructed. Because of the real-estate's historic and economic value, it was felt that additional protection should be provided as a safety factor; the decision was made to construct a **floodwall** just landward of the existing levee's crest. The floodwall has gates, which close during flooding events (**Figure 9.24**). The 13-kilometer (8-mile) wall looks innocent enough, but it is built of 2-foot-thick reinforced concrete resting on a continuous wall of wooden piles driven 7.6 meters (25 feet) into the ground. The "sheet" piles, in turn, rest on concrete piles, which extend down another 15 meters (almost 50 feet). The sheet piles prevent the underseepage of water during floods. New Orleans' solution to the flooding problem served the city well for some time as it dodged this hurricane and that, but eventually Hurricane Katrina (2005) and its storm surge breached many of the city's levees and floodwalls, and left nearly 80% of the land within the city limits underwater.

Insurance, Flood-proofing, and Floodplain Management

The National Flood Insurance Act of 1968 provides federally subsidized insurance protection to property owners in flood-prone areas. The Federal Insurance Administration chose the 100-year-flood level as the **regulatory floodplain** on which to establish the limits of government management and set insurance rates. Maps delineate regulatory floodplains in thousands of communities in the United States. The subsidized flood insurance is not available on new homes within the regulatory floodway (**Figure 9.25**). Pre-1968 houses in the floodway were "grandfathered" (allowed to stay and to be insured under the government program), but they cannot be rebuilt at that site with flood-insurance money if they are more than 50% damaged by flooding.

Three methods of "flood-proofing" are: (1) raise structures above the 100-year-flood level by artificial filling with soil, (2) build walls and levees to resist flood waters, and (3) use water-resistant building materials. Good floodplain management limits the uses of floodplains in chronically flood-prone areas, the floodway (**Figure 9.26**).

Desert regions provide a particular challenge to flood-hazard management. There, the problem is less inundation and more the migration of channels during large flood events. In some cases, channels have moved laterally hundreds of feet without overflowing as banks were undercut by high flows. The National Academy of Sciences recently came up with a different flood-hazard evaluation strategy more appropriate for arid regions; it explicitly considers the likelihood of channel movement during floods.

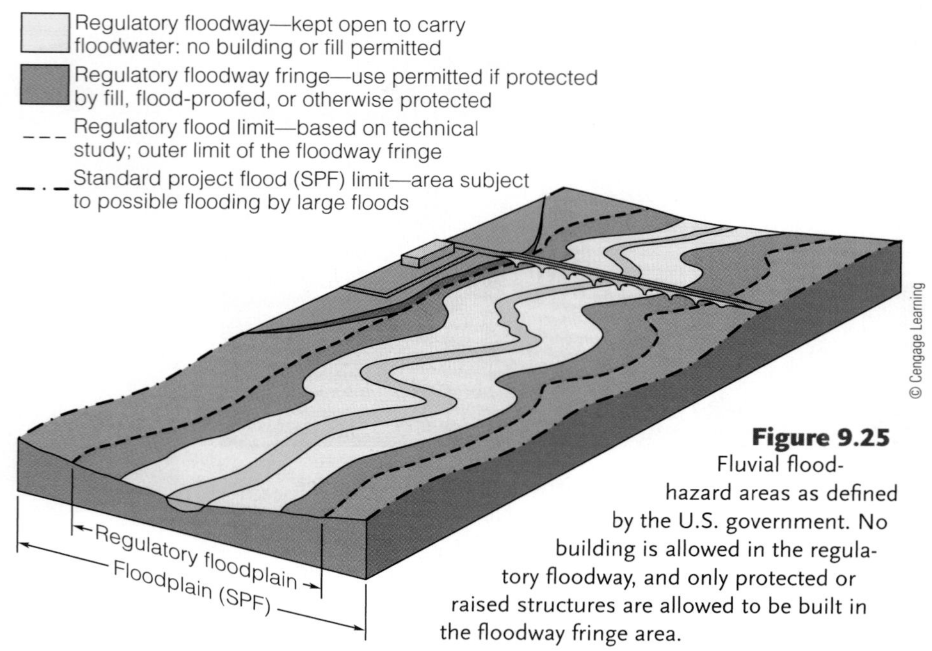

Figure 9.25 Fluvial flood-hazard areas as defined by the U.S. government. No building is allowed in the regulatory floodway, and only protected or raised structures are allowed to be built in the floodway fringe area.

Development has literally poured onto flood-prone areas in the United States. Annual flood damage (in constant, inflation-adjusted dollars) appears to have increased over the last 100 years to an average of several billion dollars per year, but damage amounts are extremely variable on a year-to-year and decade-to-decade basis. For example, a National Oceanographic and Atmospheric Administration (NOAA) compilation of flood damage since 1903 suggests that, between 1998 and 2007, the average loss per year was $9 billion, adjusted for inflation, a figure driven in large part by the $44 billion flood loss attributed to Hurricane Katrina. By comparison, the periods 1905–1914 and 1950–1959 had $3.2 billion and $4.4 billion worth of flood damage (adjusted for inflation).

As a result of the flood of 1993 in the Midwest, the Senate Committee on Environment and Public Works stopped legislation that would have authorized dozens of new flood-control projects. The new belief is that flood policies must be changed, because traditional structural solutions—levees and dams—contribute to the problem.

Floods are natural and inevitable, as demonstrated in this chapter. A congressional report perhaps summed it up best in noting that "floods are an act of God. Flood damages result from acts of people."

SPOTLIGHT FIGURE 9.26 Hydrologic Hazards at Earth's Surface

Monsoonal rains move large amounts of sediment and predictably leave river bottomlands and deltas underwater for part of each year. In areas where high mountains are near the ocean, summer heating of the land draws moist air inland and upward, wringing out moisture and causing summer monsoons—a rainy season lasting many weeks during which daily storms dump large amounts of water on the landscape. (a) Rivers pour over their banks, leaving large amounts of fertile sediment on their floodplains, only to dry as floodwaters recede. (b) For example, Inle Lake is a freshwater lake located in the Shan Hills in Myanmar (Burma). The lake level rises during the monsoons as water pours in. When the floods come, the area around the lake is flooded. (c) Stilt homes like this are one way to mitigate flood hazards.

case study

9.1 When Ice Melts and Moves

One moment, it's a pleasant early spring day by the river; the first robins have appeared, and it's warm enough to pull off your wool hat for the first time since December. The snow around your feet is soggy; it's a good thing you are wearing high rubber boots. It rained hard last night and the river should be rising, but you can't see it through the cover of gray ice. Suddenly, there's a deafening crash, and the ice on the river breaks up before your eyes. Blocks of ice, half a meter (20 inches) thick and the length of small cars, move downstream in a crazy jumble, grinding past one another, ripping trees from the river banks, and jamming against the bridge piers just downstream. You watch in amazement as more and more ice stacks up against the bridge, ignorant of the water rising over your boot tops until it cascades down to your toes, bathing you and your wool socks in an extension of the icy-cold river. The rising water and your soaking feet are telling you that an ice-jam flood has started, and it's time to move to high ground—fast!

Ice-jam floods can happen anywhere that streams freeze in the winter; that means northern latitudes and high elevations. These floods occur when warm winds melt the snowpack, rain pummels icy slopes, and river ice weakens. Rising floodwaters can tear ice cover apart, rapidly adding to the hazard. When ice jamming occurs, water levels rise fast as dams of ice block downstream flow and pond water. Unlike most river floods during which stage rises gradually and predictably, ice-jam floods are fast, devastating, and hard to predict. There's little warning. But it's not quite that grim. Many times, ice jams, and the floods they trigger, happen in the same place where natural or

Figure 1 Ice jams have been damaging human constructs for as long as people have lived near frozen rivers. As shown in this 1863 stereoview, an ice jam diverted the White River right down Main Street in Hartford, Vermont.

Figure 2 A spring ice jam on the Saranac River in New York State floods a city park. The jam is caught in a tight meander bend with a bridge just downstream.

human-made obstacles force ice blocks to hang up (**Figure 1**). Just imagine a bridge abutment or a tight meander bend. Can you see the blocks of ice, stacking one behind the next, the temporary pond behind them growing and spilling over the riverside road (**Figure 2**)? What's the remedy? If you can be patient, time—to melt the ice. If people or their homes are in danger, dynamite and heavy equipment will do the trick, destroying the ice dam and letting the river flow freely again.

question to ponder

1. How do ice-jam floods differ from floods caused by heavy rain?

9.2 The Nile River—Three Cubits between Security and Disaster

Flooding is a natural part of the hydrologic cycle and, as farmers, hydrologists, and geologists know, it can be beneficial.

The annual flood on the Nile River, the longest river in the world (depending on how one measures the Amazon), was a source of life for at least 5,000 years. Eighty percent of the water for these floods came from the faraway Ethiopian Highlands, and it replenished the soil moisture and deposited a layer of nutrient-rich silt on the floodplain (**Figure 1**). The Greek historian and naturalist Herodotus, in 457 B.C.E., recognized the importance of annual flooding when he spoke of Egypt as "the gift of the Nile." The priests of ancient Egypt used principles of astronomy to calculate the time of the river's annual rise and fall. Records of the height of the flood crests go back to 1750 B.C.E. (some say to 3600 B.C.E.), as measured with stepped wells connected to the river and known as Nilometers (**Figure 2**). Water height was measured in cubits, 1 cubit being the distance from one's elbow to the tip of the middle finger, approximately 0.5 meter or 20 inches.

The annual flooding cycle has changed since completion of the Aswan High Dam on the Upper Nile in southern Egypt in 1970 (**Figures 3** and **4**). The dam's purpose was to increase the extent of arable land, generate electricity, improve navigation, and control flooding on the Lower Nile. It is accomplishing all of those goals. Now that the flooding is controlled, crop maintenance depends on irrigation. Severe environmental consequences have resulted, including depletion of the soil nutrients that are beneficial to farming (necessitating the use of artificial fertilizers) and the proliferation of parasite-bearing snails in irrigation ditches and canals. *Schistosomiasis* is a human disease found in communities where people bathe or work in irrigation canals that contain snails serving as temporary hosts to parasitic flatworms (schistosomes).

© Images of Africa Photobank/Alamy

Figure 1 Irrigated fields on the banks of the River Nile near the Temple of Kom Ombo, Egypt.

Gianna Dagli Orti/The Art Archive at Art Resource, NY

Figure 2 The Nile outside of Cairo as it flows past the pyramids. Inset is a schematic view of a Nilometer, a stepped well that allowed the measurement of river stage or height. Water height was measured in cubits, a cubit being the distance between one's elbow and the tip of the middle finger (usually about 18 inches or half a meter). The environmental interpretations shown here were used during the time of Pliny the Elder (23–79 C.E.) of Mount Vesuvius fame.

Satellite image by GeoEye

Figure 3 Aerial view of the Aswan High Dam, showing Lake Nasser at the bottom of the image.

© Barry Kass/imagesofanthropology.com

Figure 4 Lake Nasser is brilliant blue in contrast with the hyperarid desert that surrounds it. The once free-flowing Nile has been stilled by the Aswan High Dam.

These creatures, commonly called *blood flukes,* invade humans' skin when people step on the snails while wading in the canals. The flukes live in their human hosts' blood, where the females release large numbers of eggs daily and cause extensive cell damage. Next to malaria, schistosomiasis is the most serious parasitic infection in humans. Its outbreak in Egypt is a result of controls on the river that prevent the generous floods that formerly flushed the snails from canals and ditches each year.

Another negative consequence is the loss of a huge shrimp fishery off the Nile delta; the dam traps the nutrients that encouraged the growth of plankton, upon which the shrimp larvae fed. Every dam has a finite useful lifetime, which is determined by how fast the lake it impounds fills with sediment. Lake Nasser, behind the Aswan Dam, is rapidly filling with the (precious) sediment that formerly enriched the floodplain of the Lower Nile and the waters off the delta.

question to ponder

1. Do you think that the trade-off between regular water supplies provided by Lake Nasser and the loss of silt to the Nile delta was worth it?

9.3 New Orleans: The Flood Happens

New Orleans has always been a city about water and about risks. Built on the banks of the Mississippi River, scarcely above sea level and flanked by the expansive waters of Lake Pontchartrain to the north, the city has faced flooding since the first building went up 300 years ago in what today is the French Quarter. Thanks to the invention and installation of a huge pumping network in the early 1900s (**Figure 1**), the marshes were drained, the groundwater table lowered, and storm water removed, allowing the city to expand. It took an intricate system of levees and canals to move the water around, but one by one these were built, first of earth and then of steel and concrete. In the last half a century, the Army Corps of Engineers has managed this levee system,

NARINDER NANU/AFP/Getty Images

Figure 1 Massive pumps keep New Orleans from filling with water. In 1913, Albert Baldwin Wood, an engineer with the New Orleans sewer and water board, designed a screw pump that could lift huge amounts of water—a million gallons every 5 minutes. These pumps made the draining, and thus the growth, of New Orleans possible.

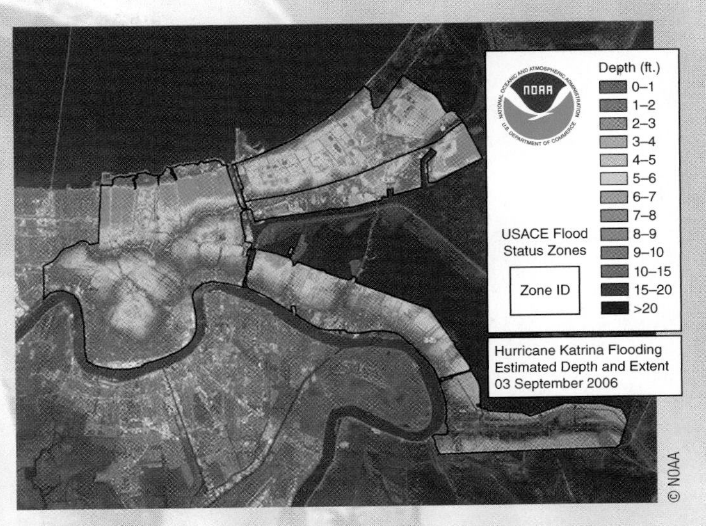

Figure 2 A map of flooding in New Orleans shows just how bad the situation was 5 days after Hurricane Katrina swept ashore. The map is based on combination aerial photography and laser elevation data. The color coding conveys water depth in the flooded areas. Clearly, most of the city is at least partially submerged.

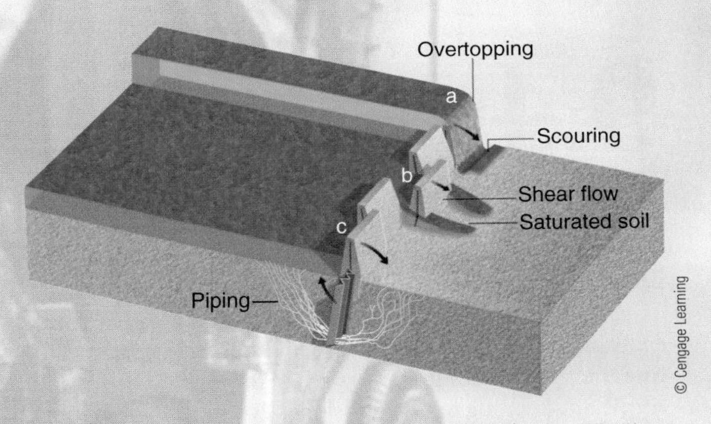

Figure 3 Levees and floodwalls can fail in several ways. (a) Water can pour over the top, scouring away the base. (b) Weak soils beneath can fail, allowing the levee to be pushed outward. (c) Water can move under the levee, called *piping*, a process by which the levee is demolished by erosion from below.

which allows much of New Orleans to exist even though it's below sea level.

Ironically, the combination of pumping and levees has caused the city to sink even farther below sea level as dried, organic-rich soils decay and compact, and the river's floods, which used to deposit sediment and raise the land, are funneled south, beyond the city's reach. Indeed, by walling in a city that now exists below sea level, engineers and politicians made it possible for people to live in a flood-hazard zone and created the potential for disaster. The levees and floodwalls were designed to withstand flooding from a rapidly moving category 3 hurricane, a moderately severe storm.

Several times in the last few decades, the levee system was threatened, but each time the hurricanes changed course at the last minute, sparing New Orleans their full fury. In general, the levees held and flooding of the city was not catastrophic, until Katrina hit in August 2005. The storm hammered the city with wind and rain, but the real flooding came a day later, when the storm surge breached the flood-control system in numerous places, allowing water from Lake Pontchartrain to pour into the city. Nearly 80% of New Orleans was flooded (**Figure 2**), more than 1,800 died, and nearly a million were homeless. This was flooding and destruction on a scale the United States had never seen.

An independent commission issued a detailed report (available online at: www.ce.berkeley.edu/~new_orleans) that examined why the levee and floodwall system failed. It determined that some sections were underdesigned and others poorly installed. The levees failed in several ways (**Figure 3**). Those that were not built high enough were

Vincent Laforet/The New York Times/Redux Pictures

Figure 4 After Hurricane Katrina, water spills over a breached levee along the Inner Harbor Navigational Canal in New Orleans, Louisiana.

overtopped by water. As this water poured over the levees, it eroded and undermined them, leading to failure. Other flood-control features, built on soft soils, were simply pushed aside by the pressure of the water they were meant to contain as the soils beneath them gave way. Sometimes water moved under the levees in what are termed "piping failures," eroding material under the levees until they lost support and collapsed. These engineering failures (**Figure 4**) can be remedied by improved design and construction techniques; however, the commission pointed out that what really needed to be fixed was the administrative system that allowed the design and construction of substandard flood control when millions of people's lives were at stake.

questions to ponder

1. Why do you think the Katrina disaster happened?
2. Was this an engineering failure or a political one?

9.4 A Very Wet Fall

New England is not the kind of place one imagines floods that take out a thousand bridges, destroy towns of 6,000 people, and remove miles of roadway. Instead, people think of Vermont and New Hampshire as pastoral places where snow blankets the landscape in winter and gentle rains nurture the maple trees that provide sap for maple syrup in the spring and flaming orange hillslopes every fall. Not so in the fall of 1927.

October 1927 was dreary and wet in New England. It rained and rained. Many towns got twice their average rainfall. By early November, the ground was sodden and the trees, having lost their leaves, weren't pumping water from the ground. Groundwater levels were on the rise. Then, on 4 November 1927, the rain began, driven by moist, tropical air from the south trying to override cold air from the north. As the warm air rose over the Green Mountains of Vermont, the moisture was squeezed out of the saturated atmosphere. Over the next 48 hours, mountainous parts of Vermont saw more than 20 centimeters (8 inches) of rain fall from the sky; even the lowlands recorded 10–15 centimeters (4 to 6 inches) of rain. Because the ground was already soaked from the October rains, much of this rain ran off into creeks and streams, driving river levels up. Riverside homes that had stood for 150 years were swept away, railroads were twisted into pretzels,

NARINDER NANU/AFP/Getty Images

Fairbanks Museum and Planetarium, St. Johnsbury, VT

Figure 1 The 1927 flood picked up houses and dropped them in some unusual places. This robust post and beam farmhouse survived the half-mile trip from its foundation to the railroad tracks intact. How do you think today's balloon-framed homes might do in a similar situation?

Figure 2 Rampaging flood waters carried away more than 1,000 bridges in the tiny state of Vermont, devastating the transportation network. (a) This photo was taken just before the bridge connecting the cities of Burlington and Winooski washed away. (b) After the flood, a temporary pontoon bridge linked the mills of Winooski (shown in the background) with the city of Burlington until a permanent bridge could be rebuilt the next year.

a: Fletcher Free Library, Burlington, VT; b: Vermont State Archives and Records Administration

and bridges were ripped from their piers and swept downstream (**Figures 1** and **2**).

That flood is now the legend to which every other flood is compared. On the Winooski, the main river draining northern Vermont, the flood of 1927 was the flood of record, the largest to affect the basin since western settlement. At the flood's peak, 3,200 cubic meters (113,000 cubic feet) per second of water was flowing downstream, 2.5 times the second largest flood on record and almost 70 times the average flow. Simple calculations suggest that 1927 was the 100-year flood, but this is misleading, because it does not plot in line with other floods on the flood-frequency curve. This was an odd event, a perfect combination of widespread heavy rain on saturated ground. Miraculously, only 84 Vermonters and 111 people in New England died in the flood, but the region was forever changed. It took years to rebuild the lost bridges, and in many places, rivers found new paths, leaving fields as islands separated from their farms.

How rare was the 1927 flood? Lake mud answers that question: It was not that rare. Geologists from the University of Vermont ventured onto frozen lakes for several winters, bringing back cores of lake sediment that had been secreted away on lake bottoms for the last 14,000 years since the glaciers left (**Figure 3**). When they cut open those cores, they found lots of black, sticky, stinky mud. However, every once in a while, they cut open a core to find a layer of bright gray sand carried into the lake when big storms struck the drainage basin (**Figure 4**). By dating these sand layers from

Anders Noren

Figure 3 Collecting 10,000 years' worth of lake sediment is serious business easily accomplished off the winter ice. Here, one author (Bierman, right) and University of Vermont graduate student Josh Galster (left), working on a warm spring day, struggle to loosen a piece of coring equipment before the ice melts.

lakes all over New England, the geologists, led by graduate student Anders Noren, determined that very big storms hit New England, on average, every few hundred years. What's even more interesting is that the frequency of these storms has changed over time, and if the past is a clue to the future, the forecast for the next 600 years is an increase in big storms.

questions to ponder

1. Do you know what the "flood of record" was like for the river nearest to where you go to school?
2. Can you find out what weather event caused that flood of record?

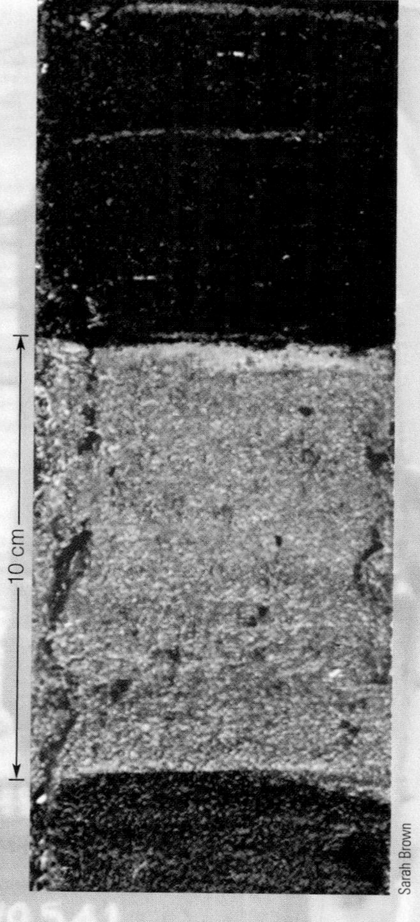

Sarah Brown

Figure 4 Once the cores are cut open, storm layers clearly stand out from the black, organic-rich, and foul-smelling pond sediment. Photo shows a 10-centimeter-thick, light-colored layer of sediment likely washed into Ritterbush Pond by a big storm about 6,800 years ago. At the top of the image are two very thin storm layers.

9.5 The Impermeable Flood

What do you think about when you imagine the cause of floods? Probably heavy rain, storm surges, and ice jams. Do you think about people as a cause of floods? In urban and suburban areas, people and their developments dramatically change the way water moves over and into the ground. How can a new house, a nicely mowed 2-acre lawn, and an SUV parked on the grass make any difference?

When a raindrop falls to Earth, it can take various paths (**Figure 1**). If the ground is not already saturated with water and rain is not falling too heavily, the raindrop will be absorbed and flow underground. Sometimes groundwater moves just below the surface in large and small pores between soil grains or along natural pipes made as roots rot away. Sometimes groundwater moves deeply through the underlying rock or sediment. In forested areas all over the world, almost all rainwater follows these paths. However, if the ground is already saturated or if the raindrop falls on something impermeable—perhaps clay-rich soil, a hiking trail compacted by hundreds of footsteps, or a paved parking lot—then the rainwater runs off, pouring off the slopes and into streams and rivers.

What does development have to do with floods? As forests and grasslands are replaced by parking lots and 6,000-square-foot homes with shingle roofs, water that used to soak into the ground now pours into storm drains and rapidly enters rivers and streams. When forests covered the land, there were deep pits and high mounds where trees had tipped up and roughened the forest floor. These pits held rainwater, allowing it to infiltrate slowly. When the

NARINDER NANU/AFP/Getty Images

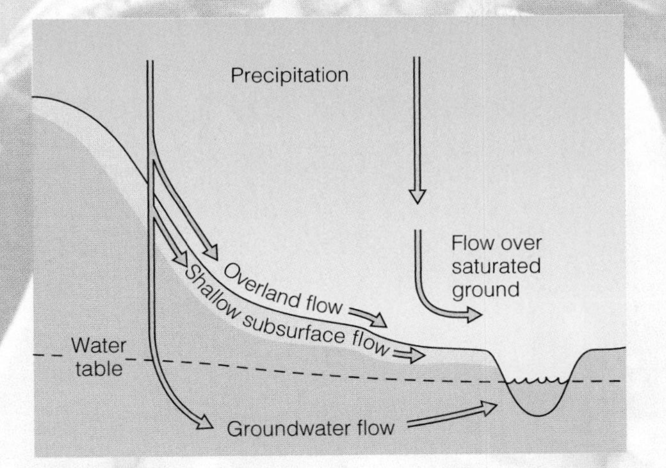

Figure 1 Water, falling from the sky as rain, can move various ways when it hits the ground. Development and other human impacts usually reduce the ground's ability to absorb water, forcing more rainfall downslope as overland flow. (Source: Based on Dunne and Leopold. 1977. *Water in environmental planning*. New York: W. H. Freeman.)

Figure 2 When cars park on lawns, the mud starts quickly. In this photo, several vehicles in Burlington, Vermont, have repeatedly driven over the grass, killing it, compacting the soil beneath, and forming deep ruts. When it rains, that compacted soil can't absorb the rainfall. Its infiltration rate is so low that muddy puddles result. The good news is that, with fencing, rototilling to break up the compaction, and time, the soil will once again be able to absorb rainfall.

trees are cleared for new development and its 5-acre lots, bulldozers smooth the forest floor and uniform, tile-drained grass lawns take the place of rotting logs. Even in the city, the conversion of grassy lawns to parking areas can make a big difference (**Figure 2**). The response is dramatic. Water levels in streams and rivers rise rapidly and attain higher peaks as rain is more quickly and efficiently moved off the landscape. Floods may not last as long, but they can be more devastating. The slug of floodwater, moving at high velocity, tends to erode and deepen channels, causing other unintended consequences.

questions to ponder

1. How has the hydrology of the town in which you go to school changed?
2. Do you think that those changes are affecting runoff?

PHOTO GALLERY

how we cope with floods

For as long as people have been living on Earth, they have been dealing with floods. Archeologists find evidence that ancient peoples moved water out of their way so that they could get on with their lives. In some parts of the world, such as those that experience monsoons or spring snowmelt floods, high water is an important and expected part of life. Those who live on deltas and along rivers know that high water will come year in and year out. Sometimes, though, storms appear and we need to cope immediately—roads get closed, bridges get washed out, and we get wet.

1 Thai Buddist monks carry sandbags to protect the Weeru Wanaram temple in Bangkok, Thailand. In 2011, heavy monsoon rains put a third of Southeast Asia under water. AP Photo/Aaron Favila

2 People have been dealing with flood waters for a long time. Here, in the ancient city of Carthage, archeologists uncovered drains used to guide rainwater from the city, and thus prevent flooding. © Kike Calvo/National Geographic Image Collection

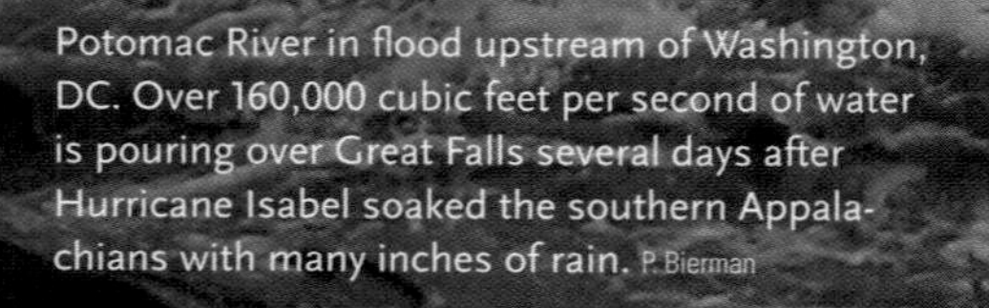

Potomac River in flood upstream of Washington, DC. Over 160,000 cubic feet per second of water is pouring over Great Falls several days after Hurricane Isabel soaked the southern Appalachians with many inches of rain. P.Bierman

3 El Niño–driven storms close a mountain road in Montecito, California, because of flash flooding in the uplands. Why do you think the water is so brown? © 2003 RICH REID/National Geographic Image Collection

4 Sometimes life keeps going right through the flood. In Nairobi, Kenya, a man holding an umbrella rides his bike in traffic during a heavy afternoon rain. © 1969 BRUCE DALE/National Geographic Image Collection

SUMMARY Hydrologic Hazards at Earth's Surface

River Systems

Drainage Basin
The fundamental geographical unit or tract of land that contributes water to a stream or stream system. Drainage basins are separated by drainage divides.

Discharge
The amount of water flowing in a stream channel depends on the amount of runoff and subsurface flow from the land, which, in turn, depends on rainfall or snowmelt, the degree of urbanization, and vegetation.

Erosion
The erosive power of a stream is a function of its velocity—the greater the velocity, the greater the erosion. Stream velocity is determined by discharge, channel shape, and gradient.

Base Level
The lowest level to which a stream or stream system can erode. This is usually sea level, but there are also temporary base levels, such as lakes, dams, and waterfalls.

Stream Features

Alluvium
Sediment deposited by a stream, either in or outside the channel.

Alluvial Fan
A buildup of alluvial sediment at the foot of a mountain where a steep upland stream enters a flat-lying lowland, sediment transport capacity decreases, and sediment drops out.

Delta
A delta forms where a sediment-laden stream flows into standing water and sediment drops out—for example, the Nile River Delta in Egypt.

Floodplain
A low area adjacent to a stream that is subject to periodic flooding and sedimentation; the area covered by water during flood stage.

Meanders, Oxbow Lakes, and Cutoffs
Flowing water will assume a series of S-shaped curves known as meanders. The river may cut off the neck of a tight meander loop and form an oxbow lake.

Flooding and Flood Frequency

Upland floods come on suddenly and move rapidly through narrow valleys. Upland floods are often "flash" floods; the water rises and falls in a matter of a few hours. Lowland floods, in contrast, inundate broad, adjacent floodplains and may take many days, weeks, or even months to complete the flood cycle.

Hydrograph
A graph that plots measured water level (stage) or discharge over a period of time.

Recurrence Interval
The average length of time (T) between flood events of a given magnitude. Mathematically, $T = (N + 1)/M$, where N is the number of years of record and M is the rank of the flood magnitude in comparison with other floods in the record. Recurrence interval calculations enable engineers and planners to anticipate how often floods of a given size are likely to occur.

Flood Probability
The chance that a flood of a particular magnitude will occur within a given year based on historical flood data for the particular location. Using the calculation for T established for the recurrence interval, probability is calculated as $\frac{1}{T}$. Thus, a 100-year flood has a 1% chance of occurring in any one year. It is the 1-in-100 chance flood.

Mitigation
Dams, retaining basins, artificial levees, and structures on artificial fill are common means of flood protection.

Flood Insurance
The National Flood Insurance Act of 1968 provides insurance in flood-prone areas, provided certain building regulations and restrictions are met.

Land-Use Effects
Urbanization causes floods to peak sooner during a storm, results in greater peak runoff and total runoff, and increases the probability of flooding.

Flood Hazards
Floods, including coastal flooding during hurricanes, are the greatest natural hazards facing humankind. Flooding is the most consistently damaging natural disaster in the United States.

KEY TERMS

artificial levee
base level
cutoff
delta
discharge
drainage basin
drainage divide
fan
flood frequency
floodplain
floodwall
gradient
hydrograph
infiltration
inundation
meander
natural levee
oxbow lake
point bar
precipitation
recurrence interval
regulatory floodplain
riparian

STUDY QUESTIONS

1. How does stream erosion shape the landscape?
2. How can we prevent floods or at least minimize the property damage caused by flooding?
3. How is the velocity of a stream related to its erosive power?
4. What has been the impact of urbanization on flood frequency and peak discharge?
5. According to the local newspaper, your town on the Ohio River has just experienced a 50-year flood. What is the probability of a repeat of this event the following year?
6. How does the hydrograph record of an upland flash flood differ from that of a lowland riverine flood?
7. Where is flash flooding most frequent, and what precautions should be observed in such areas?
8. What has the federal government done to alleviate people's suffering because of flooding? Do you believe this is a good thing? Support your answers with examples.
9. Consider a stream the headwaters of which are 1,500 meters (4,900 feet) above sea level and that flows 300 kilometers (180 miles) to the ocean. What is the average gradient of the stream in meters/kilometer and in feet/mile?
10. How do dams benefit society? What harm can those same dams cause people and the environment?

REFERENCES AND FURTHER READINGS

Carlowicz, M. 1996. Controlled flood of Colorado River creates stream of data. *Eos Transactions American Geophysical Union* 77(24): 225.

Carrier, Jim. 1991, June. The Colorado: A river drained dry. *National Geographic* 4–32.

Collier, Michael, et al. 1996. *Dams and rivers: A primer on the downstream effects of dams.* U.S. Geological Survey circular 1126.

Devine, Robert S. 1995, August. The trouble with dams. *Atlantic Monthly* 64–74.

Dinacola, Karen. 1996. The "100-year flood." U.S. Geological Survey Fact Sheet FS-229-96. Menlo Park, CA: USGS.

Federal Emergency Management Agency. 2000. *Guidelines for determining flood hazards on alluvial fans.* Washington, DC.

Fischetti, Mark. 2001, October. Drowning New Orleans. *Scientific American* 78–85.

Gosnold, William, et al. 2000, May. Floods and climate. *Geotimes* 20–23.

Holmes, Robert R., Jr., and Amit Kapadia. 1997. *Floods in northern Illinois, July 1996.* U.S. Geological Survey Fact Sheet FS-097-97. Menlo Park, CA: USGS.

Leopold, L. 1997. *Water, rivers and creeks.* Sausalito, CA: University Science Books, 185 p.

Lucchitta, Ivo, and Luna B. Leopold. 1999. Floods and sandbars in the Grand Canyon. *GSA Today* 9(4): 1–7.

McPhee, John. 1989. *The control of nature.* New York: Farrar, Straus & Giroux.

Milly, R. C., J. Betancourt, M. Falkenmark, R. M. Hirsch, Z. W. Kundzewicz, D. R. Lettenmaier, and R. J. Stouffer. 2008, February. Stationarity is dead: Whither water management? *Science* 319(5863): 573–574.

National Geographic. 1993, October (special edition). Water: The power, promise and turmoil of North America's fresh water. Washington, DC: National Geographic Society, 120 p.

Perry, Charles A. 2000, March. Significant floods in the United States during the 20th century. U.S. Geological Survey Fact Sheet FS-024-00. Menlo Park, CA: USGS.

Rogers, J. David. 1992. Reassessment of the St. Francis dam failure. In B. W. Pipkin and Richard Proctor (eds.), *Engineering geology practice in Southern California.* Belmont, CA: Star.

Saarinen, T. R, V. R. Baker, R. Durrenberger, and T. Maddock, Jr. 1984. *The Tucson, Arizona, flood of October 1983.* Washington, DC: National Research Council, National Academy Press.

Teller, R. W., and M. J. Burr. 1998. *Floods in north-central and eastern South Dakota, spring 1997.* U.S. Geological Survey Fact Sheet FS-021-98. Menlo Park, CA: USGS.

U.S. Geological Survey. 1993. Mississippi River flood of 1993. *U.S. Geological Survey Yearbook, 44-1993,* 37–40.

The 2011 Japanese tsunami comes ashore in Iwate Prefecture, Japan, overtopping the seawall and flowing into the city of Miyako with its load of cars and boats.
JIJI PRESS/AFP/Getty Images

10

Coastal Environments and Humans

For the Earth, it was just a twinge. For a planet where landmasses are in constant motion across geologic time, the event was no great moment. But for people—who make the calendar in days and months rather than eons—a monumental catastrophe had begun, not only the largest earthquake in 40 years but also the displacement of billions of tons of water, unleashing a series of mammoth waves: a tsunami.

—Barry Bearak, "The Day the Sea Came," New York Times Magazine (November 2005), writing about the Indonesian tsunami

1. How water waves work?

2. If eroding beaches can be stabilized?

3. Where tsunami hazards are the greatest?

4. Why some hurricanes are so much more damaging than others?

5. What El Niño is and what geologic hazards it causes?

Reefs in Peril

Figure 1 Healthy hard coral reefs exhibit a variety of colors and diversity, and are host to a large amount of marine life here on the Great Barrier Reef off Port Douglas, Australia.

P. Bierman

Figure 2 A green sea turtle swims over dead coral near Kanton Island in the Phoenix Islands Protected Area, Pacific Ocean.

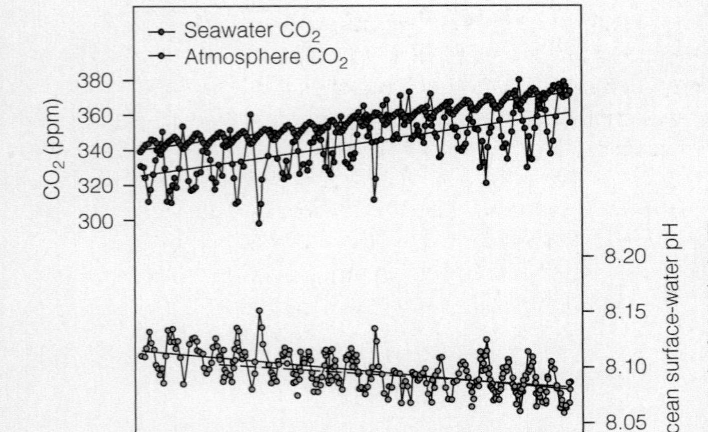

Figure 3 Much of the carbon added to the atmosphere by human burning of fossil fuels ends up dissolved in ocean water. When CO_2 dissolves in water, it forms carbonic acid, lowering the water's pH and making it more acidic. Over the last 20 years, CO_2 levels in the atmosphere and the surface waters of the ocean have risen more than 20 parts per million. As a result, the surface ocean is growing more acidic, making it more difficult for organisms to incorporate calcium carbonate in their shells. (Adapted from Dore, John E., Roger Lukas, Daniel W. Sadler, Matthew J. Church, and David M. Karl. *Physical and biogeochemical modulation of ocean acidification in the central North Pacific.* Proceedings of the National Academy of Science, 2009.

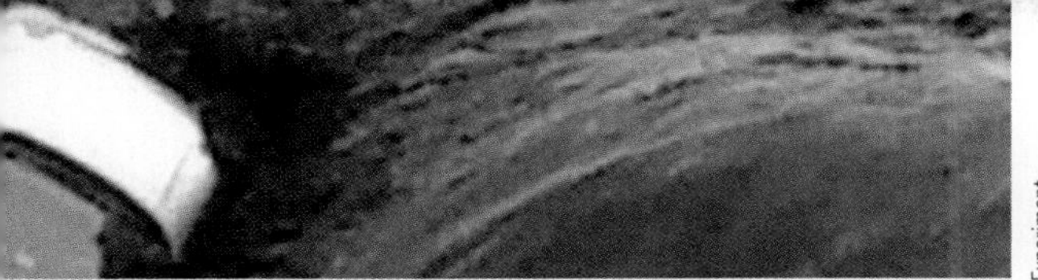

Coral reefs are unique ecosystems. Charles Darwin may have been the first to recognize that they are the largest and most important ecosystems in the marine environment. Corals are animals, related to jellyfish, that secrete a hard skeleton of calcium carbonate (limestone). They grow only in warm water (above 20°C [approximately 68°F]) within the zone of light. Corals require light because they live in symbiosis with the tiny marine algae *zooxanthellae*. These plants live within the coral polyp and give the calcareous skeleton its color (**Figure 1**).

Coral reefs help people in diverse ways. They form barriers to beach erosion, attract tourism, support fisheries, and build atoll islands upon which people live. Unfortunately, reefs are being destroyed worldwide at a phenomenal rate. Mining of reefs for limestone, dumping of mine tailings on reefs, fishing by use of explosives or cyanide, land reclamation, and polluted river runoff all take their toll. Seawater temperature increases, caused by global warming and El Niño events, cause reef destruction. Reefs in Indonesia, North America, Hawaii, and Australia are just a few at risk. At higher temperatures, coral polyps expel *zooxanthellae,* lose their color, and eventually die if the warm water persists. The bleaching turns a beautiful underwater garden into a spiny wasteland (**Figure 2**). The newest threat to coral is the acidification of near-surface ocean waters (**Figure 3**), caused by human-induced increases in atmospheric carbon dioxide (CO_2), the result of burning fossil fuels. The growing acidity of ocean waters as atmospheric CO_2 dissolves in the ocean makes it increasingly difficult for marine creatures to fix $CaCO_3$, which dissolves in acidic waters.

question to ponder

1. Most press coverage around CO_2 is focused on the global warming effects of the gas in the atmosphere. Do you think that if people were better educated about other effects of CO_2 (such as the destruction of coral reefs), they might do more to reduce fossil-fuel use and stabilize the levels of CO_2 in the atmosphere?

CourseMate Connection

Start off this chapter by watching a video about how tsunami waves are formed. **Log in to the CourseMate at www.cengagebrain.com to watch now.**

MANY PEOPLE WANT TO LIVE near a beach. Coastal areas in the United States and around the world experienced enormous population growth in the late 20th century. Fifty percent of the U.S. population can now drive to a coastline in less than an hour. However, as events in the first decade of the 21st century clearly show, the coast can be a dangerous place to be, with tsunamis and hurricanes damaging properties and taking hundreds of thousands of lives. There are more subtle and ongoing coastal hazards, such as the slow erosion of sea cliffs and the retreat of barrier islands. Increasing global sea level ensures that these hazards will not only continue but likely grow worse over the next century. Human actions sometimes contribute to the hazards. For example, in some areas, wetlands are being filled, drained, and made into marinas, whereas in other places, mangrove swamps are cleared—all of this landscape modification reduces the ability of the coastal zone to dampen storm surges and prevent erosion. Shoreline modifications continue to silt in harbors and erode beaches.

In order to understand and approach the environmental issues that are so important in the coastal zone, it is critical that we understand the relevant Earth processes. That is the goal of this chapter. Wind-driven waves that arise in the open ocean eventually strike shorelines, constantly altering beaches. Massive tsunami waves, generated by impulses, including earthquakes and submarine landslides, can change thousands of kilometers of coastline in a matter of minutes. Hurricanes pound smaller stretches of coast for hours, whereas nor'easters batter long stretches of eastern North America, sometimes for days at a time. Armed with the knowledge of how waves erode beaches and of the fate of eroded materials, we are all better equipped to understand and deal with coastal environmental issues.

Wind Waves

Waves in Deep Water and at the Shore

When wind blows over smooth water, surface ripples are created that enlarge with time to form local "chop" (short, steep waves) and, eventually, wind waves. The following factors determine the size of waves at sea and ultimately the size of the surf that strikes the shoreline:

- The wind velocity
- The length of time the wind blows over the water
- The **fetch**, the distance along open water over which the wind blows (**Figure 10.1**)

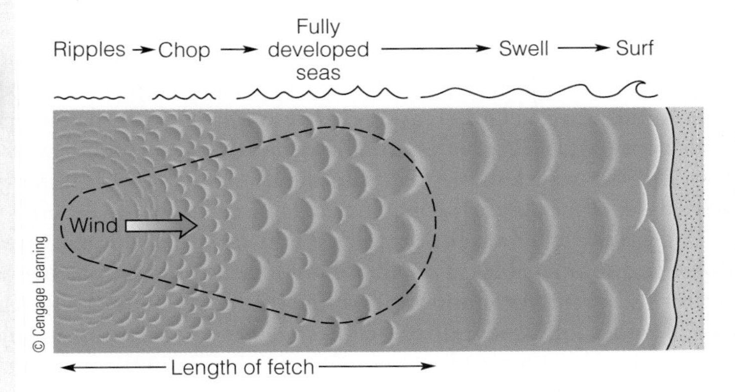

Figure 10.1 The evolution of wind-generated sea waves from a rippled water surface to fully developed seas (storm waves). Long waves with very regular spacing and height that outrun the storm area are called *swell.* Surf only occurs when the waves strike a beach.

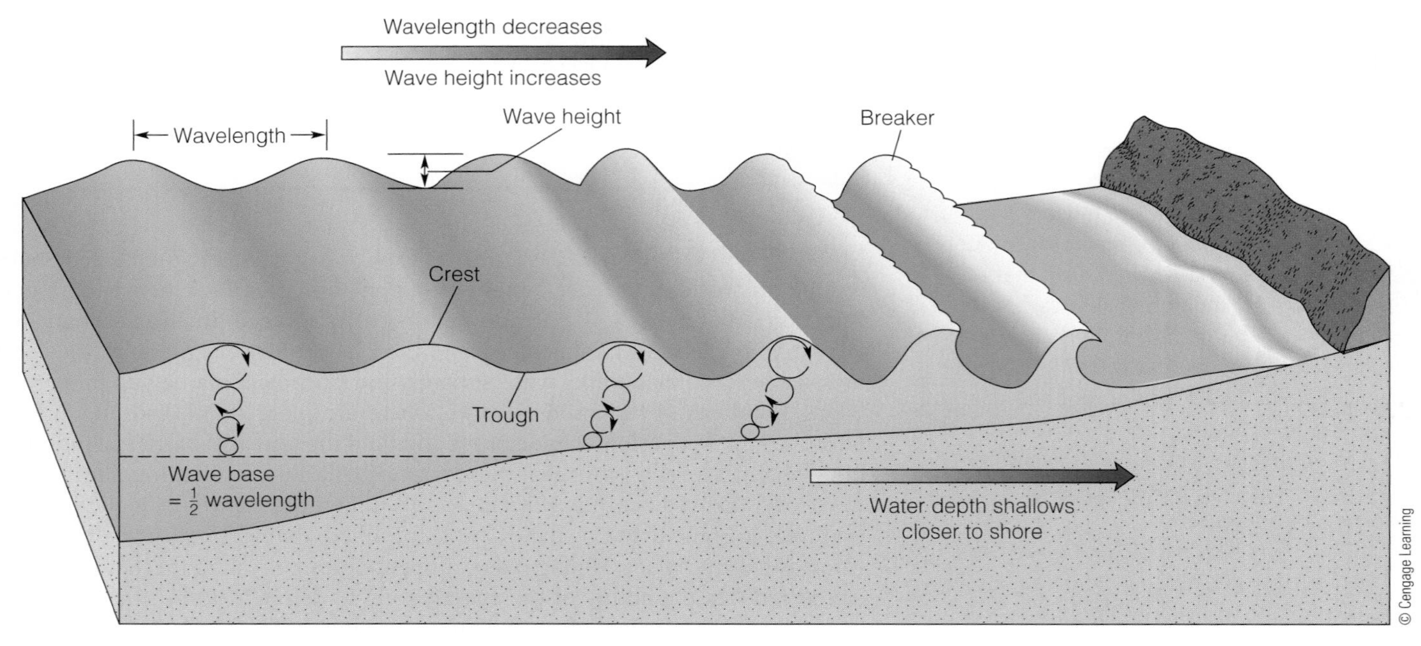

Figure 10.2 Wave motion is complex and changes as the wave approaches shore and begins to "feel" the bottom, or shoal. Small circles with arrows show the orbital motion of water in a wave. The terminology of wave geometry is shown, including wavelength and wave height. Note that the orbital water motion decreases with depth until it is essentially nil at a depth of half the wave length ($L/2$), a depth called *wave base.*

For a given wind velocity, wind duration, and fetch, wave dimensions will grow until they reach the maximum size for the existing storm conditions. This condition, known as a **fully developed sea**, represents the maximum energy that the waves can absorb for wind of a given velocity. For example, wave heights of 32 meters (100 feet) are possible for a 50-knot (93 km/h, 58 mph) wind blowing for 3 days over a fetch of 2,700 kilometers (1600 miles). Wind waves will move out of a stormy area as regularly spaced, long waves known as *swell* and eventually become huge breakers on a distant shoreline.

Important dimensions of wind waves are their length, height, and *period*—the length of time between the passage of equivalent points on the wave. These dimensions are measures of the amount of *energy* the waves contain—that is, their ability to do work (**Figure 10.2**). Although waves with long wavelengths and long periods are the most powerful, both long and short period and wavelength waves are capable of eroding beaches.

The actual water motion in a wind-generated wave describes a circular path, the diameter of which at the sea surface is equal to the wave height. This circular motion diminishes with depth and becomes essentially zero at a depth equal to half the wavelength ($L/2$), the depth referred to as **wave base**. Wave base is the maximum depth to which waves can disturb or erode the seafloor. (This explains why divers and submersibles at depth are not tossed about during storms, as surface ships are.) Although the waveform moves across the water surface with considerable velocity, there is no net forward movement of water. Such waves are called *waves of oscillation*. Floating objects simply bob up and down with a slight back-and-forth motion as **wave crests** and troughs pass.

As waves approach a shoreline and move into shallower water depths, the seafloor begins to interfere with the oscillating water particles. This friction causes wave velocity to decrease. This, in turn, causes the wave to steepen; that is, the wavelength shortens and the wave height increases, a phenomenon easily seen by surf watchers and known formally as **shoaling**. Eventually, the wave becomes so steep that the crest outruns the base of the wave and jets forward as a "breaker." Breakers are *waves of translation*, because the water in them physically moves landward in the **surf zone**. Where the immediate offshore area is steep, we find the classic tube-shaped *plunging breakers* typical of Sunset Beach and Waimea Bay on the north shore of Oahu (**Figure 10.3a**). This is because the wave peaks rapidly on the steeply sloping seafloor. On a flat nearshore slope, in contrast, waves lose energy by friction, so they build up slowly, and the crest simply spills down the face of the wave. This type is called a *spilling breaker* and is typical of wide, flat, nearshore bottoms, such as that at Waikiki Beach on Oahu (see Figure 10.3b). Regardless of the type of wave, *swash*—water that rushes forward onto the foreshore slope of the beach—returns seaward as *backwash* to become part of the next breaker. A common misconception is the existence of a bottom current, called *undertow*, that pulls swimmers out to sea. The strong pull seaward one feels in the surf zone is simply backwash water returning to be recycled in the next breaker.

Figure 10.3 (a) Plunging (tubular) breakers form where there are steep offshore slopes. The front of the wave steepens until the top of the wave "jets" forward and down. This one is near Maui. Such waves only break during the biggest winter storms, which occur a few times each winter. When the conditions are right, big-wave surfers from around the globe descend to ride the monster 15-meter (50 foot) waves. (b) Spilling breakers form where there are gently sloping offshore areas, such as Waikiki Beach on Oahu, Hawaii. The forward face of the wave steepens and the crest spills down the front of the wave.

Wave Refraction

Where waves approach a shoreline at an angle, they are subject to **wave refraction**, or bending, which ends up making the wave crest approach the shore in a nearly parallel fashion. The bending of waves occurs because waves slow first at their shallow-water ends, whereas their deeper water ends move shoreward at a higher velocity. Thus, a fixed point on a wave crest follows a curved path to the shoreline (**Figure 10.4**).

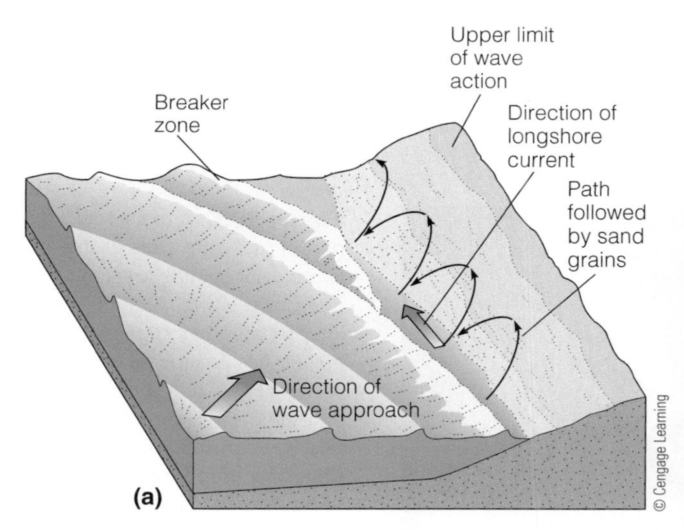

Figure 10.4 (a) Refraction, or bending, of wind waves, takes waves that are approaching the beach at an angle and bends them, so that, by the time they near the beach, the crests are nearly parallel to the shoreline. However, waves approaching at an angle generate a longshore current parallel to the beach. The longshore current transports beach sand, resulting in a dynamic process called *longshore drift.* (b) Waves refracting around two small islands in the Indian Ocean on the north coast of Mozambique.

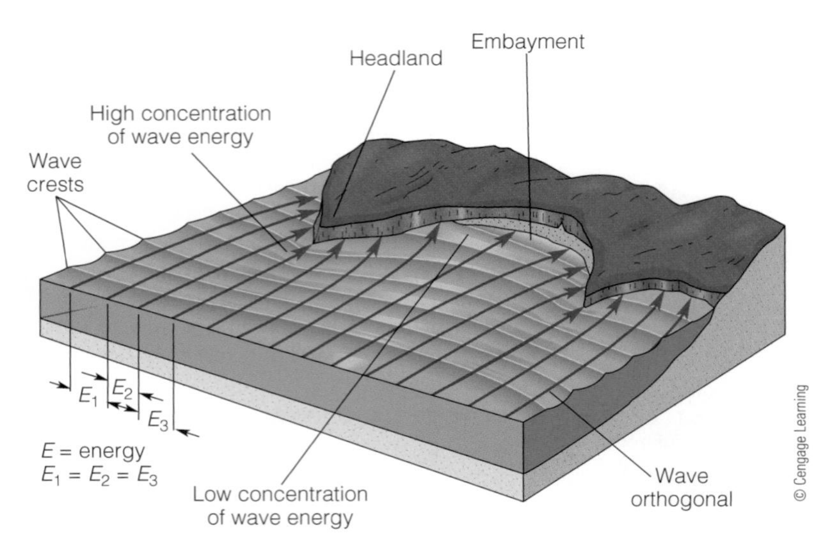

Figure 10.5 Wave refraction leads to a concentration of energy on headlands and dissipation in embayments. The arrowed lines, called *orthogonals,* are constructed perpendicular to the wave crest and show the wave path. E_1, E_2, and so on represent equal parcels of wave energy and, therefore, equal ability to erode. Note that energy is focused toward the headlands and spreads out in the bays. This is not abstract knowledge! Rather, erosion hazards are highest on the headlands and lowest in the embayments.

Because waves seldom approach parallel to the shore, some water moves along the beach as a weak current within the surf zone called a **longshore current**. The greater the angle at which the wave approaches the shoreline, the greater the volume of water that is discharged into the longshore-current stream. This current is capable of moving swimmers, as well as sand grains, down the beach. Sand becomes suspended in the turbulent surf and swash zone and moves in a zigzag pattern down the beach's foreshore slope (see Figure 10.4a). Beaches are dynamic: The sand grains present on the foreshore today will not be the same ones there tomorrow. Beach-sand transport parallel to the shore in the surf zone is known as **longshore drift**, and its rate is usually expressed in thousands of cubic meters or cubic yards of sand per year.

Another consequence of wave refraction is the straightening of irregular shorelines. This occurs because the concentration of wave energy at headlands causes shoreline erosion, and the dissipation of wave energy in embayments allows sand deposition (**Figure 10.5**).

Longshore currents are likened to rivers on land, with the beach and the surf zone constituting the two banks. Just as damming a river causes sedimentation to occur upstream and erosion to occur downstream, so does building structures perpendicular to the beach. Such a structure slows or deflects longshore currents, causing sand accretion up-current from the structure and erosion downcurrent (**Figure 10.6**). **Groins** (French *groyne*, "snout") are constructed of sandbags, wood, stone, or concrete, ostensibly to trap sand and create a wider beach. For groins of equal length and design, the magnitude of the erosion-deposition pattern is a function of the strength of the waves and the angle at which they strike the shoreline. Most groins are short and are intended to preserve or widen a beach in front of a private home or resort. Unfortunately, because this causes neighbors' beaches downcurrent from the groin to narrow, many of those neighbors build groins to preserve and widen *their* beaches, and so on. Like artificial levees on a river, groins tend to generate more groins, and we generally see fields of them along a given stretch of beach instead of just one (**Figure 10.7**).

Just as rivers overflow their banks, longshore currents may flow out through the surf zone, particularly during periods of big surf. **Outflow** to the open ocean occurs because the water level within the surf zone is higher than the water level outside the breakers, making for a hydrau-

Figure 10.6 In this aerial photograph, the ends of the two barrier islands are separated by the thin Ocean City Inlet. Salt marshes separate Virginia's Assateague Island National Seashore (lower left) and the densely populated beach resort of Ocean City, Maryland (upper right), from the mainland. Sand erosion on Assateague, caused by jetties that keep the Ocean City Inlet open to navigation, poses a severe threat to the island. Can you use information in the image to determine which direction the longshore drift is moving the sand?

Figure 10.7 An aerial view of Sandy Hook peninsula in New Jersey clearly shows the effect of jetties on the movement of sand along the shore. Can you use information in the image to determine which direction the longshore drift is moving the sand?

lic imbalance. The higher water inshore flows seaward to lower water offshore as a **rip current** through a narrow gap, or *neck*, in the breakers (**Figure 10.8**). Because the location of rip currents, erroneously called *rip tides*, is controlled by bottom topography, they usually occur at the same places along a beach. Rip currents can become a problem to swimmers under big-surf conditions, when velocities in the neck may reach several miles per hour. Awareness can save your life, so be alert to a gap in the breakers, white water beyond the surf zone, and objects floating seaward—all of which are evidence of a strong rip current flowing out to sea (**Figure 10.9**). Should you find yourself in a rip current, swim parallel to the beach until you are out of the narrow neck region of high velocity.

Table 10.1 Causes of Tsunamis in The Pacific Ocean Region over the Last 2,000 Years

Cause	Number of Events	Percentage of Events	Number of Deaths	Percentage of Deaths
Landslides	65	4.6	14,661	2
Earthquakes	1,171	82.3	>625,000	89
Volcano	65	4.6	51,643	8
Unknown (meteorite?)	121	8.5	5,364	1
Total	1,422	100	>700,000	100

Sources: National Geophysical Data Center and World Data Center A for Solid Earth Geophysics, 1998, and Intergovernmental Oceanographic Commission, 1999, in Bryant, 2001. United Nations (Indian Ocean, 2005 Tsunami estimate).

consider this . . . After a few hours of frolicking in the surf, you find you have drifted along the shore to a spot where the waves aren't breaking. Suddenly, you find yourself being drawn seaward by an invisible hand, no matter how vigorously you swim toward the beach. What is happening? What should you do, and why?

Impulsively Generated Waves

Tsunamis

Tsunamis, the "tidal waves" of the popular press, can be one of the most devastating environmental hazards, causing water levels to rise dramatically in just minutes, sweeping away people, buildings, and entire communities. Tsunami (pronounced "soo-nah-mee") is a Japanese word meaning "harbor" *(tsun)* and "wave" *(ami)* because harbors are where these waves "resonate" and are most observable. In Japanese, the singular and plural are the same, but we will follow the custom of English writers and add the "s" for the plural, *tsunamis*. After the devastating Indonesian tsunami of 26 December 2004, most people now know the devastation these waves can cause—nearly a quarter million people died as water swept onto coasts all around the Indian Ocean (see Case Study 10.1). Only 6 years later, in 2011, another massive quake and tsunami shocked the world. No matter that the disaster struck Japan, probably the best prepared country in the world in terms of natural hazard mitigation; the Fukushima quake and tsunami were soon both well-known, not only because of the size of the waves and the scale of the damage, but because the tsunami overtopped undersized seawalls and devastated a cluster of nuclear reactors along the shoreline. Few other natural phenomena can do so much harm so quickly as a tsunami.

Most tsunamis are produced by underwater earthquakes and less commonly by volcanic eruptions, submarine landslides, and meteorite impacts (**Table 10.1**). These impulses give rise to waves that move out from their source in all directions—analogous to those produced when you

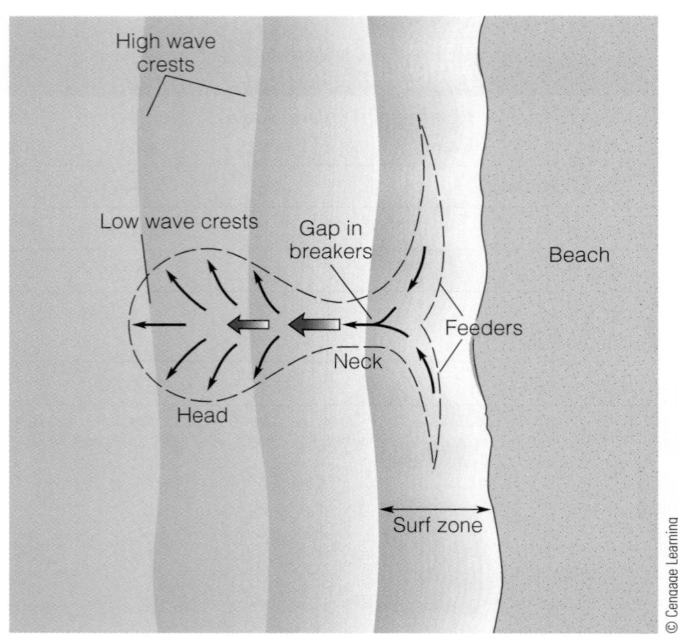

Figure 10.8 A rip current is characterized by relatively weak *feeder currents,* a narrow *neck* of strong currents, and a head outside the surf zone, where the rip dissipates.

Figure 10.9 Dangerous and deadly rip currents along the famous beaches of Rio de Janeiro, Brazil, are clearly identified by the mushroom shapes of white water—these are where water flows rapidly off the beach toward the ocean.

drop a rock into a pond. Tsunamis in deep water may have velocities exceeding 800 kilometers per hour (480 miles/hour) and could easily keep pace with a Boeing 747. With wavelengths in excess of 160 kilometers (100 miles) and wave heights of only a meter or two (3 to 6 feet) in the open ocean, tsunamis have such gentle slopes that they go unnoticed at sea (**Figure 10.10**). As the leading waves move into shallow water, they slow to freeway speeds, and the waves behind begin to overtake them—a process called *shoaling.* Thus, the energy contained in many waves is condensed into a smaller volume of water, and the waves increase in height and steepness. Eventually, the wave runs ashore as a wall of white, foamy water—a breaking, tidelike flood.

Fortunately, most underwater earthquakes do not generate tsunamis. In almost 100 years (1861–1948), only 124 major tsunamis were recorded, despite there being more than 15,000 seafloor earthquakes. The west coast of South America, a seismically active area, experienced 1,100 earthquakes in the same period and only 20 tsunamis. One of the reasons is that many tsunamis have such small amplitudes that they go unnoticed. In addition, earthquake-induced tsunamis require shallow-focus quakes with surface-wave magnitudes greater than 6.5. Earthquake depth is critical. For example, even though the 2010 Chilean subduction earthquake was huge (magnitude > 8.5), it was deep, and thus the tsunami it generated, although devastating in local communities, was smaller than tsunamis generated by some lower magnitude quakes. The Atlantic Ocean experiences only 2% of all tsunamis, whereas the Pacific Ocean experiences 80% of them. Why do you think the Pacific Ocean is so tsunami-rich? (Hint: Consider what you know about plate tectonics.)

Earthquake-induced tsunamis may rise to heights, onshore, of more than 30 meters. In Lituya Bay, Alaska, the 1964 Alaskan earthquake tsunami wave sloshed way up the walls of the fjord, killing trees hundreds of feet above water level and leaving an impressive bathtub ring (**Figure 10.11**).

The time between crests of incoming tsunami waves varies, and the first crest may not be the largest. The *trough* of the initial wave may arrive first; many lives have been lost by curious persons who wandered into tidal regions exposed as water withdrew, some in search of an easy seafood meal. In Lisbon, Portugal, on 1 November 1755, when many people were in church commemorating All Saints' Day, an immense earthquake (magnitude ~ 8.5) struck in three distinct shocks. The worshipers ran outside to escape falling debris and fire, many of them joining a group seeking safety on the waterfront. There was a quiet withdrawal of water followed

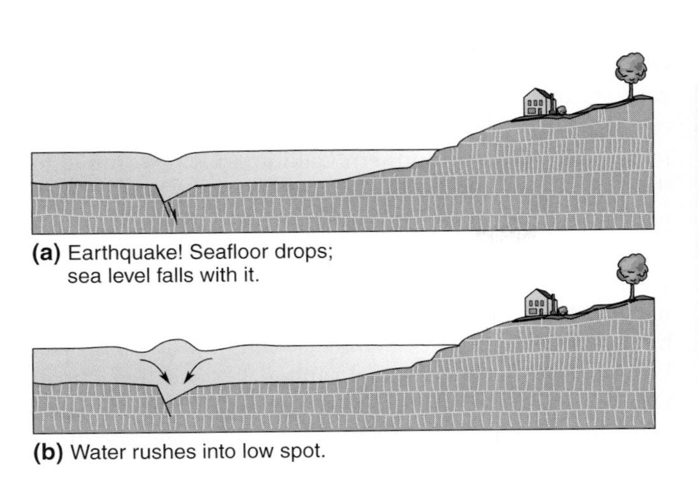

(a) Earthquake! Seafloor drops; sea level falls with it.

(b) Water rushes into low spot.

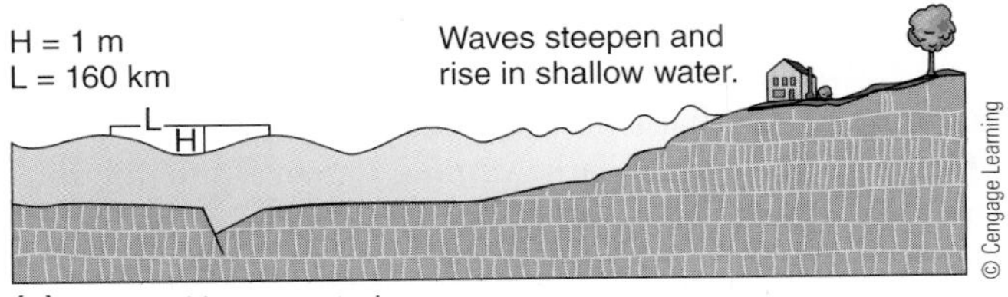

(c) Tsunami is generated.

Figure 10.10 Formation of a tsunami when a portion of the seafloor is down-faulted and the water surface over the fault is lowered. Water rushes in to fill the low spot and creates an impulsively generated wave that moves out in all directions with a very high velocity (hundreds of miles per hour). As the wave approaches shore, it shoals, steepens, breaks, and moves onshore, flooding the land.

Figure 10.11 At Lituya Bay, Alaska, the 1964 Alaskan earthquake generated a wave several hundred feet high. It ran up the forested slopes around the bay, flattening millions of conifer trees and leaving a bathtub ring of dead, yellowed trees around the bay.

by a huge wave minutes later. Sixty thousand people died—about two thirds of them in the quake and the rest in the aftershocks and fires that followed. The quake was felt throughout Europe and northern Africa, and the tsunami is reported to have crossed the Atlantic, raising the level of the ocean 6 meters (20 feet) in Antigua, Martinique, and Barbados. The length of time of strong shaking has been estimated from historical accounts of how many "Ave Maria's" and "Paternosters" the churchgoers recited during their fearful experience.

Water withdrawal before the first high wave also contributed to fatalities in the 1946 Hawaiian tsunami; many people walked offshore to collect exposed mollusks just before the first wave crest arrived. A good rule is to head for high ground at once, should you witness a sudden recession of water along a beach or coastline, or feel an earthquake near the shore.

Most tsunamis are generated by large-scale reverse faulting in subduction zones that ring the Pacific and Indian ocean basins. Japanese studies have shown that strike-slip faulting seldom produces tsunamis. In the late 20th century, earthquake-generated tsunamis around the Pacific Rim damaged Chile (1960), Alaska and the Pacific Northwest (1964), Nicaragua (1992), and New Guinea (1998). The Hawaiian Islands are very tsunami-prone due to multiple tsunami source areas (see Case Study 10.2). Between 1850 and 2000, the Islands averaged a damaging tsunami every 2 years.

A tsunami's damage potential depends on the earthquake mechanism, the distance the waves travel to the shoreline, the offshore topography, and the configuration of the coastline. In 1998, a community on a low, coastal sandbar in Papua New Guinea was obliterated by tsunami waves of up to 15 meters (50 feet) high that killed 2,200 people and left few survivors. Experts were puzzled because the magnitude-7.1 earthquake was considered too small to generate such huge waves. This led to speculation (later confirmed by high-resolution sonar images) that the earthquake had triggered a submarine landslide just offshore. That slide caused the much larger than expected tsunami, which came ashore so soon after the quake that residents had no time to respond.

In 1992, 26 towns along 250 kilometers (150 miles) of Nicaragua's Pacific coast were struck by a 10-meter-high (33 foot) wave that killed 170 people and left 13,000 homeless. Normally, coastal inhabitants run for high ground when they feel rapid earthquake shaking. Only slight tremors were experienced in this case, however, and people did not suspect danger. Most of the earthquake energy was transmitted in much longer waves (long period, up to 200 seconds), which were not noticed or were ignored. This kind of quake, known as a "slow" earthquake, was first described by Hiroo Kanamori of the California Institute of Technology. Kanamori believes these quakes may be 10 times as strong as the magnitude determined from short-period seismometer readings. Slow earthquakes of local origin are particularly dangerous in coastal areas, because they produce little precursory ground shaking to warn people that a tsunami may be imminent. Up to 10% of all tsunamis may be caused by this kind of earthquake.

The entire Pacific Rim is vulnerable to tsunami inundation. Geologic evidence indicates that, within the past several thousand years, many large earthquakes have generated destructive tsunamis on the shores of Washington, Oregon, and northern California. The risk becomes clearer

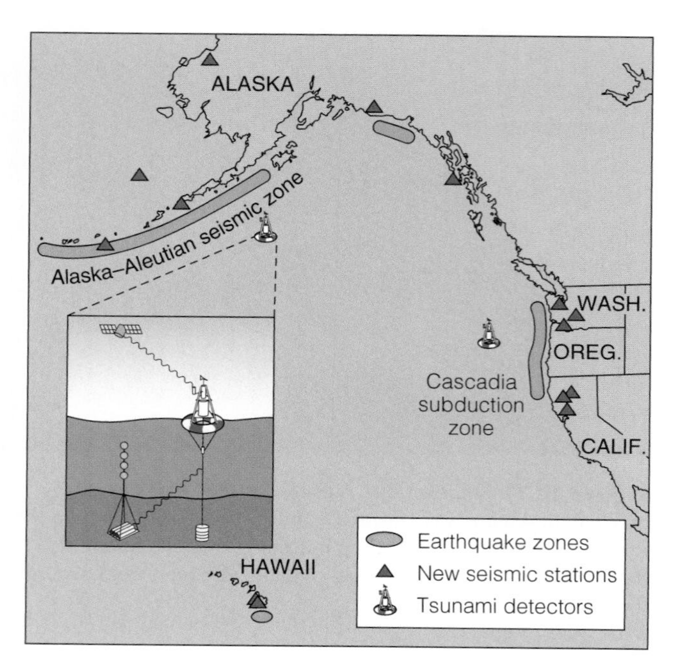

Figure 10.12 U.S. earthquake zones capable of generating tsunamis are the Alaska-Aleutian seismic zone and the Cascadia subduction zone. The five Pacific states are subject to tsunamis. The inset shows a tsunami detector that rests on the seafloor. It transmits acoustic signals of water depth to a surface buoy, which relays signals to a NOAA satellite and then to ground-based warning centers. A slow, steady change in water depth indicates the passage of a long-wavelength, low-amplitude tsunami wave at sea. (From *EOS 79,* no. 2, June 2, 1999.)

when we recognize that 90% of the deaths in the Alaska earthquake of 1964 were tsunami related, and that the geology of the Cascadia subduction zone (off the coasts of Washington, Oregon, and northern California) is very similar to the Alaska coast (**Figure 10.12**)—both are long, linear, coast-parallel, subduction zones.

Indeed, tsunamis have struck the Pacific Northwest quite recently. In 1964, tsunami waves originating from the magnitude-9.2 Alaska subduction-zone quake destroyed much of Crescent City, California, killing at least 10 people. In April 1992, a magnitude-7.1 earthquake generated a small tsunami at Cape Mendocino, California. Estimates suggest that a major Cascadia tsunami could cost the region billions of dollars and endanger more than half a million people who work, live, and play near the Pacific Northwest coast.

The best defenses against tsunami risk are an educated public who knows how to respond, maps showing high-hazard areas, so that land-use decisions can be made wisely (**Figure 10.13**), and an early warning system. Tsunami drills and standardized warning signs are now common in the Pacific states that are most likely to be impacted. Hazard maps have been or are being prepared, and the beginning of a warning system is in place. Useful (life-saving) warning times for the people who live and work in potential tsunami-inundation areas of the Pacific Northwest range from a few minutes in the southern part of the Cascadia subduction zone to 20 to 40 minutes in the northern parts of the zone. The current Pacific-wide warning system is based on rapid estimates of the location and strength of earthquakes, and about an hour is required to issue a warning; that's too long, because it's useful for only coastal areas more than 750 kilometers

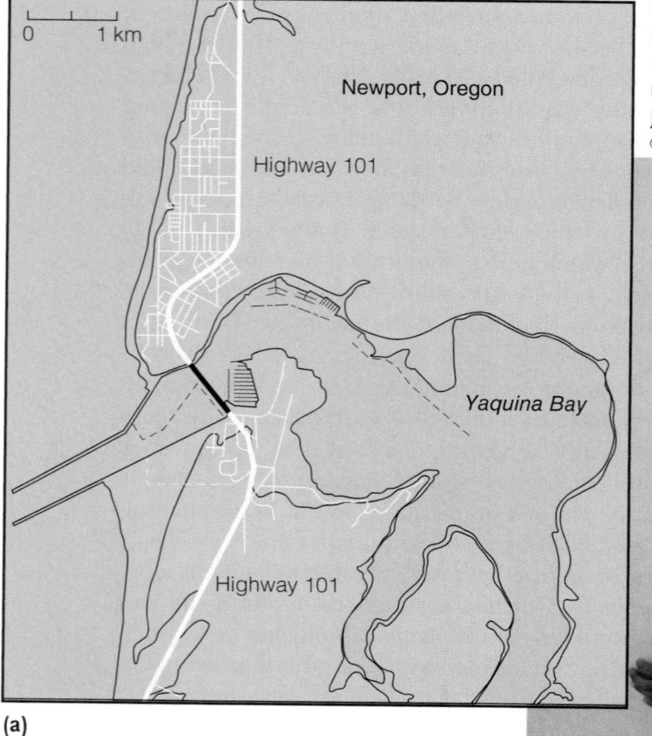

(a)

Figure 10.13 Tsunamis are a hazard in Oregon, with its substantial coastal population. (a) Potential tsunami inundation areas (green) at Newport, Oregon. (From *EOS 79,* no. 2, June 2, 1999.) (b) Tsunami education is a very important mitigation tool in coastal areas where these large waves strike. Here, children in Java, Indonesia, learn about tsunami response from a public education poster.
© ANNIE GRIFFITHS/National Geographic Image Collection

(465 miles) from the tsunami source, the approximate distance a tsunami could travel in that amount of time.

The goal now is to place on the seafloor real-time deep-ocean tsunami detectors that transmit data acoustically to a floating buoy, which relays the information to a satellite; in turn, the satellite transmits the data to a ground station (see Figure 10.12). The sensitive instruments can detect tsunamis as small as 3 centimeters (1.2 inches) high and transmit the data to the ground station in 3 minutes. For nearby tsunami sources, 3 minutes from detection to an alert is a life-saving warning time. Seven instruments were in operation in 2004, and 36 instruments were installed and maintained in the Pacific Ocean by 2012. Lest you think that only the Pacific Northwest, Hawaii, and Alaska are at risk, tsunamis are purported to have struck Santa Barbara, California, in 1812 and Santa Monica, California, in 1930.

Tsunamis initiated by submarine landslides—Recent research now includes landslides as a significant process in tsunami generation. This new awareness may save lives in population centers around both the Pacific and Atlantic margins that are currently under the false impression that tsunamis are not a hazard. San Francisco, Santa Barbara, Los Angeles, San Diego, and New York are some of the cities that could be affected by tsunami-generating landslides on nearby continental slopes. Known landslide-generated tsunamis resulted from seafloor instability after the Alaskan earthquake in 1964, the Papua New Guinea earthquake in 1998, and the Grand Banks earthquake in 1929 (**Figure 10.14**). The Grand Banks quake (magnitude 7.2) is better known for the 12 transatlantic cables severed sequentially by the turbidity currents it triggered than for the tsunami waves (as much as 27 meters or 90 feet high) that struck the Burin Peninsula of Newfoundland about 2.5 hours after the seismic event. Several dozen people perished in that event.

More than just the Grand Banks and Lisbon tsunamis (**Figure 10.15a**) have affected the Atlantic Ocean, and geologic evidence indicates that some Atlantic tsunamis could be massive. Mapping of huge offshore landslide complexes on both sides of the Atlantic Ocean hints of what could happen. On the European shore, there is ample evidence that large underwater landslides have indeed triggered massive tsunami waves. About 8,000 years ago, on the eastern edge of Norway, in a place called Storegga, a massive underwater landslide let loose (see Figure 10.15b). The tsunami it generated ran up more than 25 meters (80 feet) in parts of the Shetland Islands off the coast of Scotland. On the other side of the Atlantic, off the coast of Virginia and North Carolina, there is evidence of a massive submarine landslide that occurred some 18,000 years ago. This slide, which contained more than 33 cubic miles (150 cubic kilometers) of material, came off the edge of the continental shelf and must have generated a substantial tsunami. North of this ancient slide are tension cracks, suggesting another slide may be imminent (see Figure 10.15c).

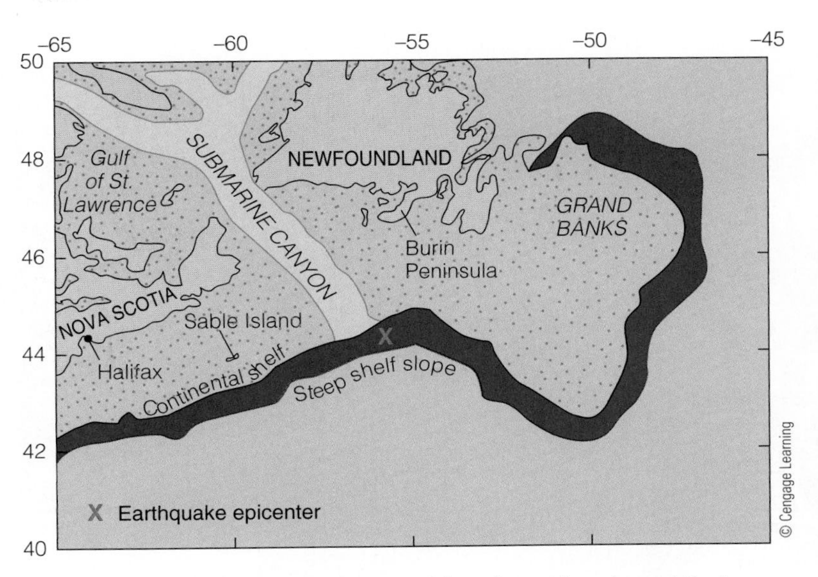

Figure 10.14 The epicenter of the Grand Banks earthquake and Burin Peninsula.

Landslides generated by submarine volcanic action—The explosion of Krakatoa (Indonesia) on 27 August 1883 was one of the largest volcanic eruptions and subsequent tsunamis ever recorded. The volcano collapsed into its own magma chamber and, because it was an island in the Sunda Strait between Java and Sumatra, the ocean flowed in where the volcano had once been. The eruption was disastrous, but the death toll of more than 37,000 from the tsunami was the worst effect. The tsunami had a maximum run up of 42 meters (140 feet) and traveled 5 kilometers (3 miles) inland over low-lying areas. The largest wave struck the town of Merak, because of the funnel shape of its harbor. There the 15-meter (45-foot) wave increased to 40 meters (135 feet) because of concentration of energy, and coral blocks weighing more than 600 tons were transported onshore. About 5,000 boats were destroyed, and for months afterward bloated bodies washed ashore, along with blankets of pumice.

Bolide impacts—**Bolides** are bodies from space—comets, meteorites, and asteroids—that impact Earth. Most meteorites and asteroids come from the asteroid belt between Jupiter and Mars, and smaller bodies incinerate during their trip through the atmosphere before reaching Earth. Comets have a greater variety of orbits, some of which are extremely elongate, taking them away from Earth for decades or even hundreds to thousands of years. Bolides that penetrate the atmosphere and land in a body of water are capable of generating tsunamis. Hollywood has picked up on this particular tsunami-generating process and produced such fine movies as *Deep Impact* and *Armageddon*, both of which feature devastating and massive tsunamis created when bolides landed in the ocean. Indeed, there is good evidence in the geologic record of a major bolide impact and associated tsunami 65 million years ago: the Cretaceous-Tertiary boundary, a time of great extinction.

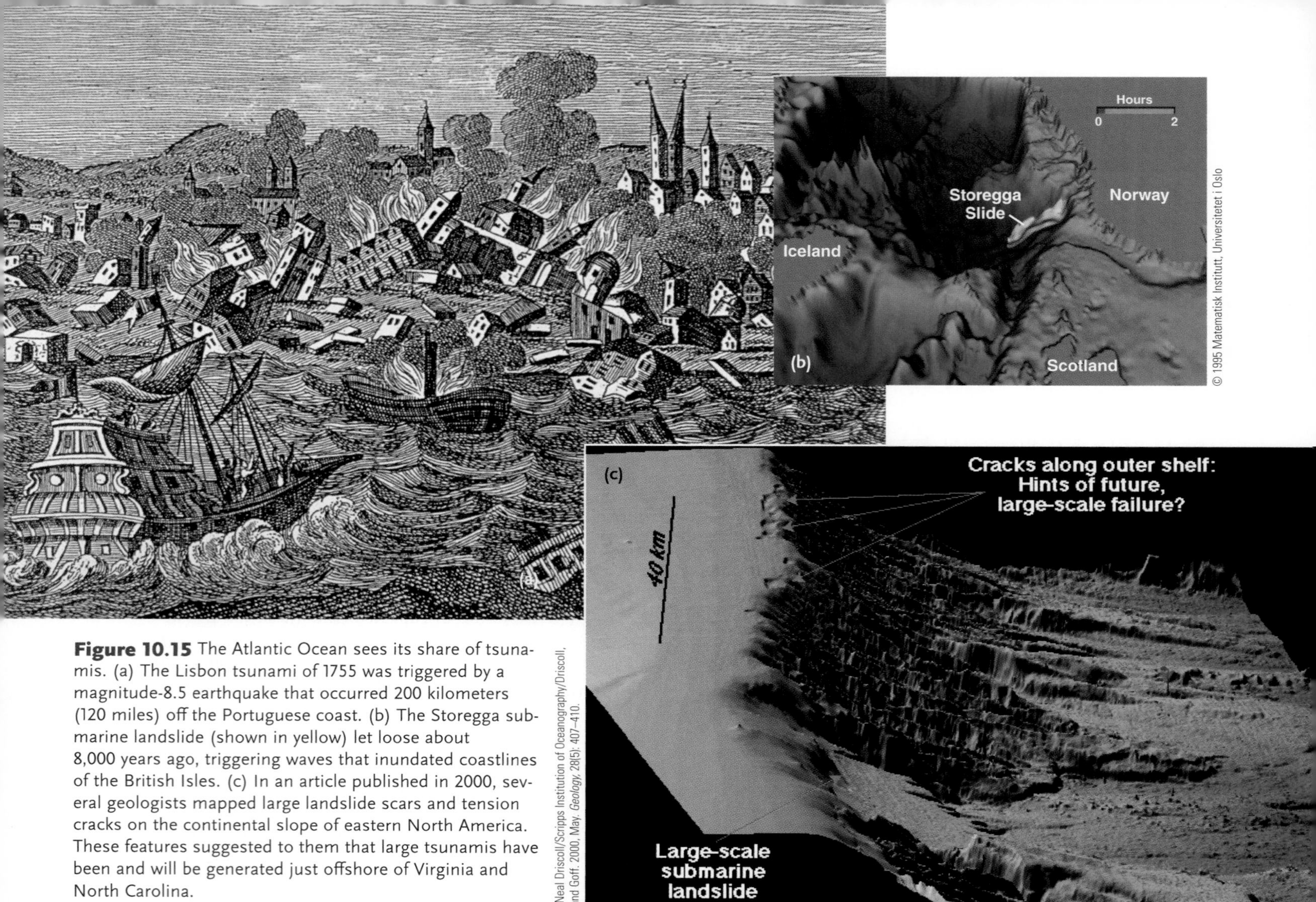

© 1995 Matematisk Institutt, Universitetet i Oslo

Figure 10.15 The Atlantic Ocean sees its share of tsunamis. (a) The Lisbon tsunami of 1755 was triggered by a magnitude-8.5 earthquake that occurred 200 kilometers (120 miles) off the Portuguese coast. (b) The Storegga submarine landslide (shown in yellow) let loose about 8,000 years ago, triggering waves that inundated coastlines of the British Isles. (c) In an article published in 2000, several geologists mapped large landslide scars and tension cracks on the continental slope of eastern North America. These features suggested to them that large tsunamis have been and will be generated just offshore of Virginia and North Carolina.

a: © 2005 the Regents of the University of California. Jan Kozak Collection KZ84

Professor Neal Driscoll/Scripps Institution of Oceanography/Driscoll, Weissel, and Goff. 2000, May. *Geology*, 28(5): 407–410.

Boulder beds are an indication of the great size of the end Cretaceous wave, whereas the extinction of the dinosaurs and about 70% of all other creatures that lived at that time is a testament to the devastation that the impact wrought. Recently, the 180-kilometer (110 miles)-wide crater created by this impact has been found in the northern Yucatan Peninsula, at a place called Chicxulub.

Shorelines

Where the sea and land meet, we find *shorelines*, some of which are shaped primarily by erosional geologic processes and others by depositional geologic processes. Shorelines that are exposed to the open ocean and high-energy wave action are likely to be *erosional* and display such erosional features as sea cliffs or broad, wave-cut platforms. Shorelines that are protected from strong wave action, by offshore islands or by the way they trend relative to the direction of dominant wave attack, are characterized by *depositional* features such as sand beaches. Of course, not all sand beaches are stable, as we'll discover in the next few pages.

The shoreline is the distinct boundary between land and sea that changes with the tides, whereas the *coast* is the area that extends from the shoreline to the landward limit of features related to marine processes. Thus, a coast may extend inland some distance, encompassing estuaries and bays.

Beaches

In addition to their usefulness as recreational areas, beaches also protect the land from the erosive power of the sea. They are where the energy of waves is dissipated. Beaches are composed of whatever sediment rivers and waves deliver to them. For example, in some places in Hawaii, black sand forms from weathered basalt and from hot lava that disintegrated when it flowed into the sea. In contrast, reef-fringed oceanic islands are characterized by beautiful white coral sand and shell beaches. Mainland beach materials are largely quartz and feldspar grains derived from the breakdown of granitic and sedimentary rocks. *Shingle* beaches are composed of gravel; they typically occur where the beach is exposed to high-energy surf. Amusing or not, tin cans from a local dump once formed a beach at Fort Bragg, California.

Beaches change shape with the season. As a rule, winter beaches are narrow, because high-energy winter-storm waves erode sand from the upper part of the beach (the berm) and deposit it in offshore sandbars parallel to the beach (**Figure 10.16**). You may have found just such an offshore sandbar while wading in waist-deep water that abruptly shallowed as you walked onto the bar. Winter beaches are also more likely to be covered in gravel, rather than sand, as the sand is more easily moved offshore. With the onset of long, low-energy summer waves, the sand in the offshore bar gradually moves back onto the beach, and the beach berm widens. Thus, the natural annual beach cycle is from narrower and coarser in the winter to wider and sandier in the summer. On the East Coast, beach narrowing is related to waves generated by long-duration storm events that may occur at any time of the year but are more likely in the winter.

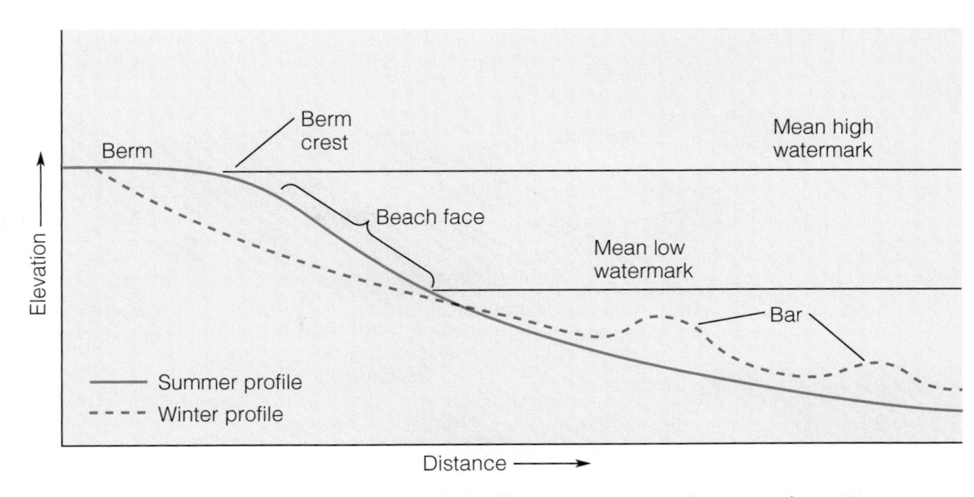

Figure 10.16 Winter and summer beach profiles and some beach terminology. (From Wicander, R., and J. Munroe. 1999. *Essentials of geology* (p. 348), 2nd ed. Brooks/Cole.)

consider this . . . Your favorite beach nearly disappears during a large winter storm. Will you have to find a new beach for next summer? If not, will the summer beach look any different, and if so, how will it be different? Explain your answer.

Barrier Islands

Barrier islands form the most extensive and tenuous beach system in the United States. They extend southward from Long Island and Coney Island, New York, to Atlantic City, continuing as the Outer Banks of North Carolina (see Case Study 10.3) and then down the coast of Florida to Daytona Beach, Cape Canaveral, Palm Beach, and Miami Beach. The barrier system begins again on the west coast of Florida and continues to the Padre Islands in Texas, interrupted only by the Mississippi River delta. These barrier beaches are transient features; formed and reoriented by wave action, they are frequently overtopped by storm surges. Barrier beaches are separated from the mainland by shallow lagoons and marshland over which they migrate as large storms overwash the islands, transporting sand from their seaward sides to their landward sides.

Hog Island, Virginia, is a barrier island that was a popular hunting and fishing spot of the rich and famous in the late 1800s. Its town of Broadwater had 50 houses, a school, a cemetery, and a lighthouse. After a 1933 hurricane inundated the island, killing its pine forest, Broadwater was abandoned. Today, Broadwater is underwater a half-kilometer (1600 feet) offshore. Typical shoreward migration rates for an Atlantic coast barrier island are 2 meters (6.5 feet) per year, but extraordinary displacements have been reported (see Case Study 10.3). The rate of barrier-island overwash, flooding, and migration is likely to increase as sea level rises in response to global warming and the resultant melting of glaciers and the thermal expansion of seawater. Although heavily developed in many parts of the United States, barrier islands are hazardous places to be in the face of any major coastal storm. They are routinely ravaged by hurricanes and nor'easters, and will only become more hazardous places to live over the next century (**Figure 10.17**).

Beach Accumulation and Erosion: The River of Sand

A beach is stable when the sand supplied to it by longshore currents replaces the amount of sand removed by waves. Where the amount of the supplied sand is greater than the available wave energy can remove—typically near the mouth of a river or rapidly eroding cliffs—the beach widens until wind action forms sand dunes. Significant areas of coastal dunes occur in Oregon, on the south shore of Lake Michigan in Indiana, near the Nile River in Egypt, on the tip of Cape Cod near Provincetown, and on the Bay of Biscay on France's west coast, to name but a few (**Figure 10.18**). The more usual case is for wave action and longshore currents to move more sand than is supplied, and here we find eroding beaches or no beaches at all. All 30 of the U.S. coastal states have beach-erosion problems of varying magnitudes resulting from both human activities and natural causes (**Figure 10.19**).

Inasmuch as sand comes to beaches by way of rivers and eroding sea cliffs before leaving by way of waves and

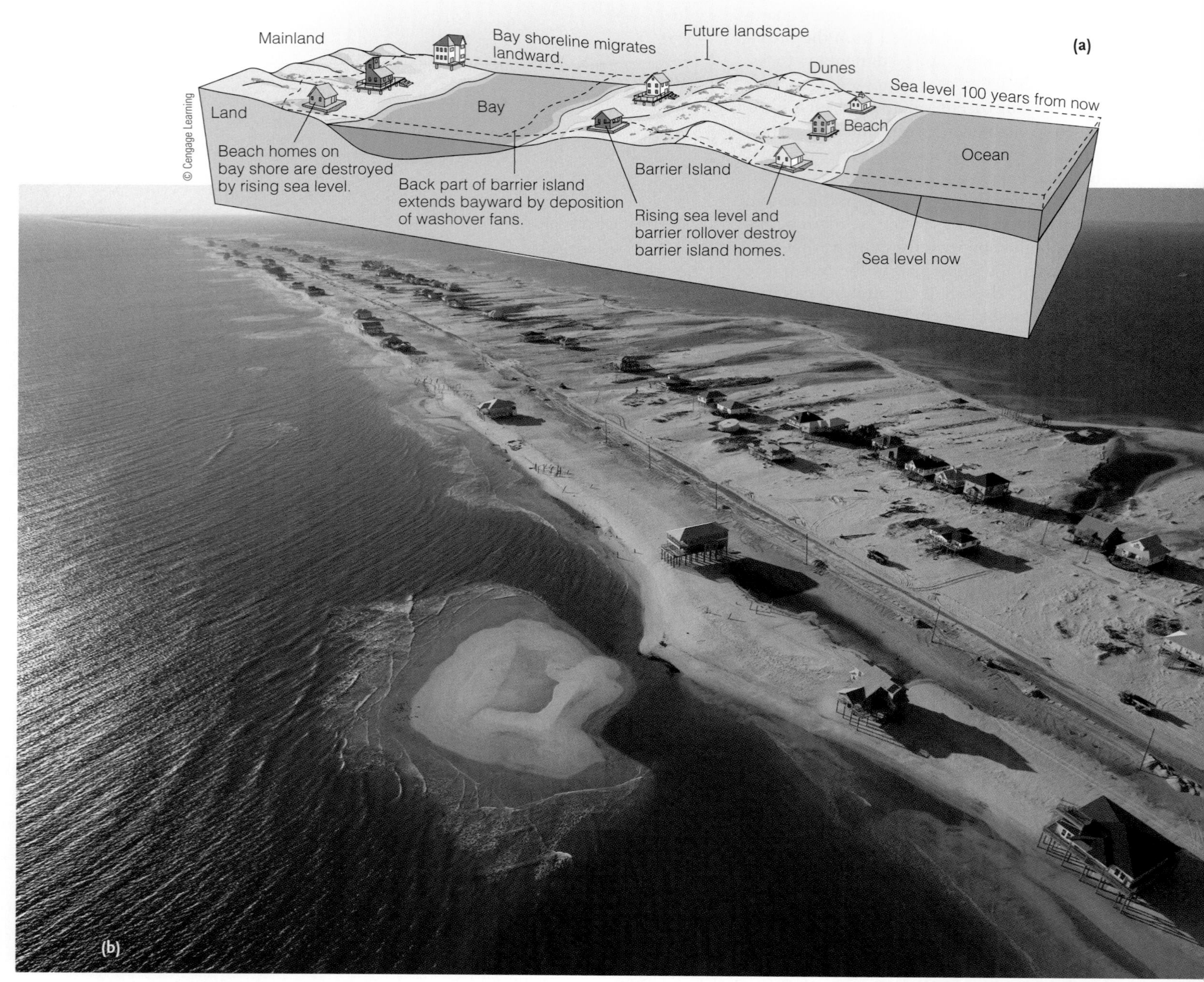

Figure 10.17 Barrier Islands are fragile places where disaster is waiting to strike. (a) This diagram shows the landward movement of barrier islands and the likely effects of storms on homes. (b) Barrier islands are low-lying features and thus at high tides and with high winds, waves and water often wash right over them. Here, at Dauphin Island, Alabama, many homes were destroyed by washover during hurricanes Ivan and Katrina.

currents, we should look to the land for causes of long-term beach erosion. Water-storage and flood-control dams trap sand that would normally be deposited on beaches. The need for upland flood protection is clear, but it can conflict with other needs, including the need for well-nourished beaches. Urban growth also increases the amount of paving, which seals sediment that might otherwise be eroded, find its way into local rivers, ultimately to the beach, and then to the ocean. Significant natural causes of beach erosion are protracted drought and rising

Figure 10.18 Coastal dunes are common and the movement of their sand can become a geologic problem. (a) Oregon has some very large coastal dune fields including those shown here near the Pistol River with sea stacks in the distance. (b) Someone's beach home was buried when sand was blown to the lee side of the 25-meter (80-foot)-high coastal dunes along the southeastern coast of South Africa. This is an example of dune migration (see Chapter 12).

Figure 10.19 Present coastal erosion in the United States. Note that most of the shorelines are eroding, probably because of sea-level rise both in the long term (in response to deglaciation) and in the short term (in response to global warming). Uncolored coastal areas on the map represent shorelines for which data are not available. (After Dolan, R., et al. 1985. Coastal erosion and accretion. *USGS National Atlas.*)

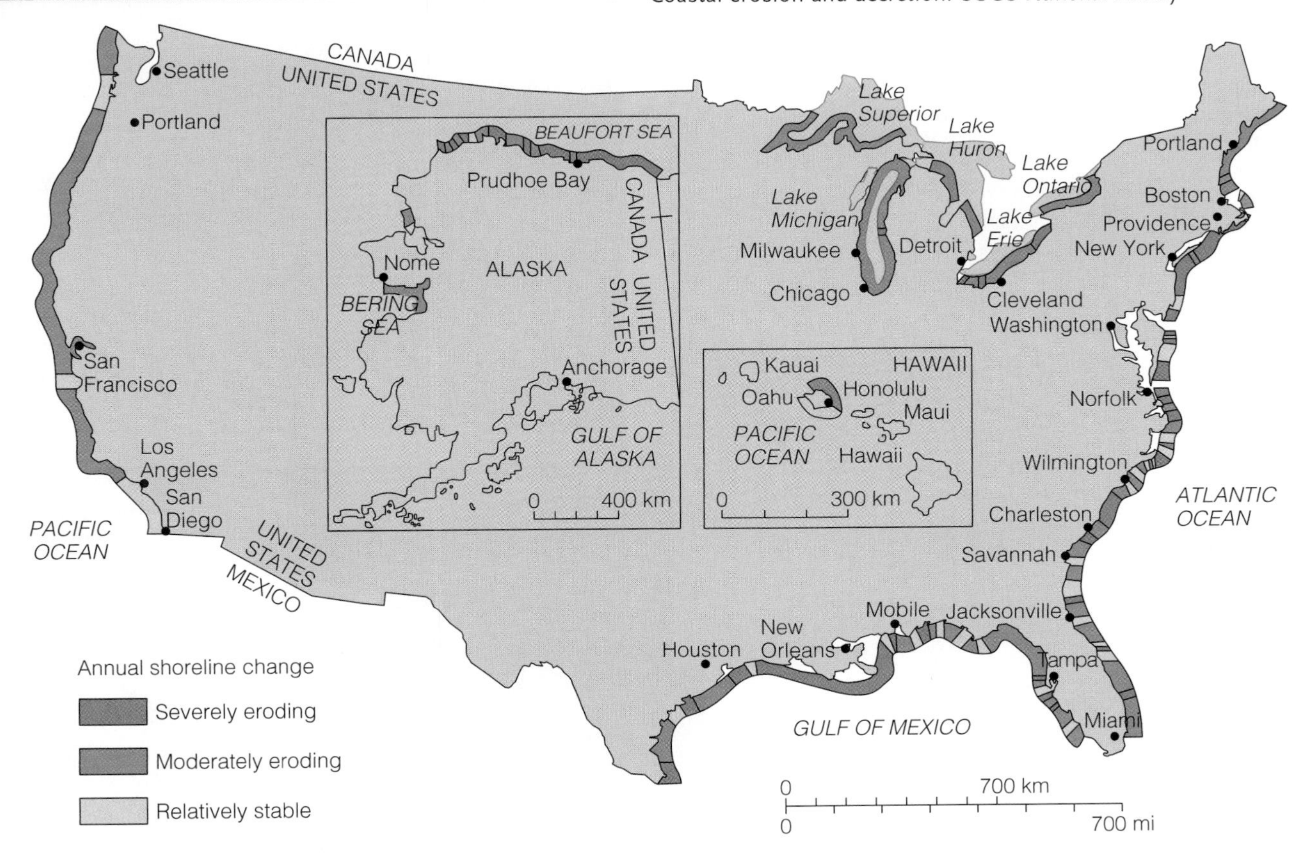

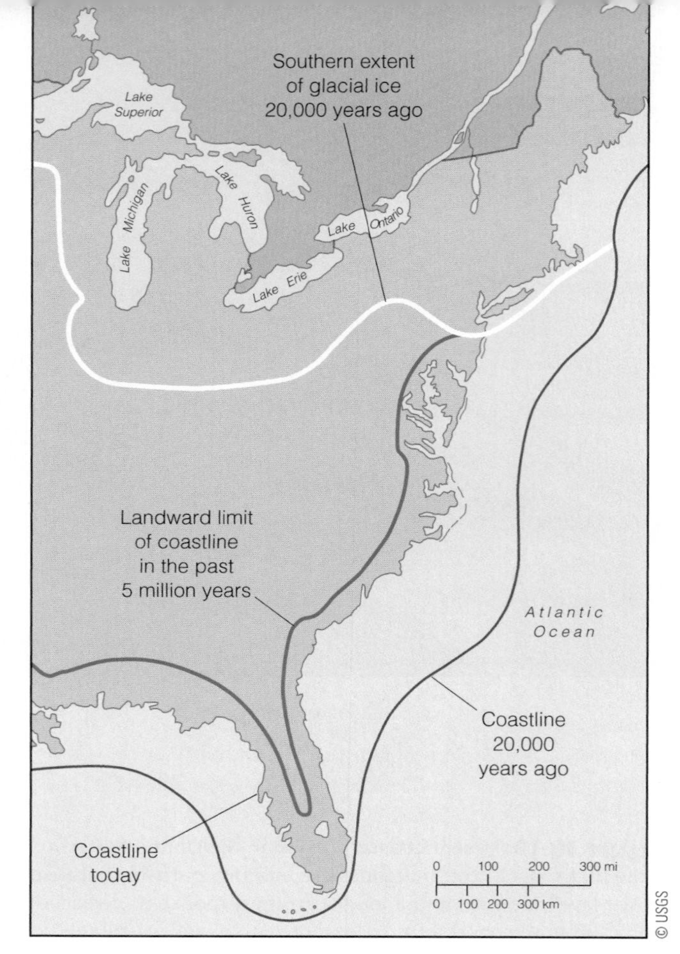

Figure 10.20 Position of the coastline in the eastern United States in late Pleistocene time (20,000 years ago), Miocene time (5 million years ago), and today. (Source: USGS data.)

sea level over geologic time frames—for example, since the last glaciation peaked about 20,000 years ago (**Figure 10.20**). Little runoff occurs during drought conditions; therefore, little sand is delivered to beaches. Rising sea level means that waves are likely to pound frequently at the bottom of sea cliffs, eroding them.

Rising sea level erodes beaches and has catastrophic results. On a human scale, it has been estimated that, as Earth warms through the next century, sea level could rise as much as 1 to 3 meters because of thermal expansion of warming seawater and melting of glacial ice. On flat coastal plains where slopes are as gradual as 0.2 to 0.4 meter per kilometer, a sea-level rise of 2.5 centimeters would cause the shoreline to retreat between 62 and 125 meters. If the coast is subsiding, this shoreline migration landward would be magnified. Careful tidal-gauge measurements at New York City, where the shoreline is tectonically stable, indicated a sea-level rise of 25 centi-

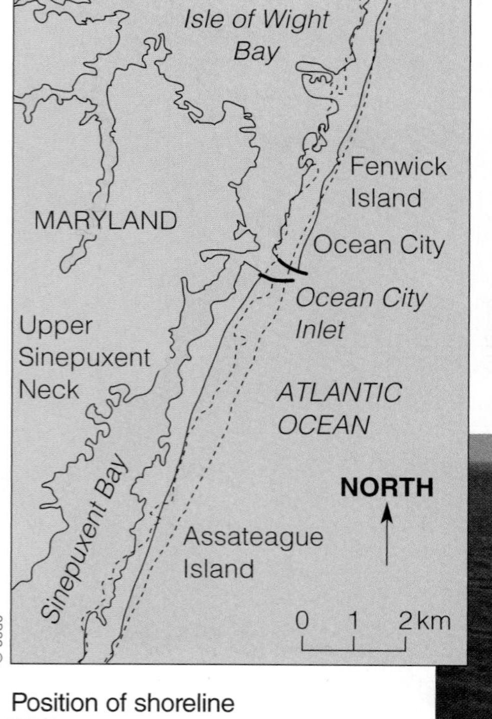

Figure 10.21 (a) Jetties built at Ocean City inlet in the 1930s cut off the supply of sand from the north (see Figure 10.6). The north end of Assateague Island is now 500 meters (1650 feet) shoreward of the south end of Fenwick Island. (Source: Williams, J. S., et al. *Coasts in crisis,* USGS circular 1075.) (b) Using large pipes to pump sand from offshore onto the beach at Ocean City, Maryland. (c) Beach replenishment project at Ocean City, which is doing what it is designed to do: protect the buildings and give beachgoers more room to play.

meters (almost 10 inches) between 1900 and 1970. If that trend continues, plan to take a boat to the Statue of Liberty's big toe and to say good-bye to much of Long Island, Miami Beach, and Texas's Galveston County within the next few centuries.

Erosion examples—Long-term beach erosion is aggravated locally by poorly conceived engineering works, such as *breakwaters* built to create quiet water for safe yacht anchorage and long **jetties** built to prevent sedimentation in harbor or river inlets. Such structures cut off longshore drift, the movement of sand along beaches, causing accretion of sand in the immediate vicinity of the structure and erosion on its downcurrent side. What follows, are a few examples that show ways in which human interaction with coastal processes has not worked out so well.

Ocean City, Maryland, is at the south end of Fenwick Island, one of the chain of barrier islands extending down the U.S. East Coast from New York to Florida. It is a major and highly developed destination beach resort for the millions of people who live in the Baltimore/Washington/Philadelphia corridor. Long jetties were constructed near Ocean City in the 1930s to maintain the inlet between Fenwick and Assateague Island to the south. As a result, the beach has been stabilized at Ocean City—the north jetty blocked the southward longshore transport of sand—and the Assateague Island beach has been displaced 500 meters (1,650 feet) shoreward because of sand starvation (**Figures 10.21** and 10.6). Even though the Ocean City beach is relatively stable, the building of many large hotels and apartments mandates that the barrier island can migrate only at great expense. Indeed, it remains vulnerable despite all the building. A strong nor'easter in March 1962 inundated all but the highest dune areas, causing many millions of dollars of damage. Today, the beach at Ocean City is routinely nourished by sand pumped from offshore. One can argue that this multimillion-dollar investment in sand makes sense, given the number of people who vacation there and the property that the beach protects.

In the 1920s, Santa Barbara's civic leaders decided to mold their community into the "French Riviera" of California. In spite of unfavorable reports by the Corps of Engineers, city officials authorized the construction of a long, L-shaped breakwater for a yacht anchorage in 1929 (**Figure 10.22**). The predominant drift direction here is to the southeast, and as the beach west of the breakwater grew wider, near-record erosion occurred on the beaches east of the structure. Summerland Beach, about 16 kilometers (10 miles) downdrift (east), retreated 75 meters (250 feet) in the decade following construction of the breakwater. Eventually, the southeast-moving littoral drift traveled around the breakwater, forming a sandbar in the harbor entrance. The bar built up, diminishing the size of the harbor and creating a navigational hazard, which remains to this day. When the Santa Barbara city fathers asked Will Rogers, the noted American humorist, what he thought of the harbor in the 1930s, he is alleged to have replied, "It would grow great corn if you could irrigate it." The drift rate here is about 250,000 cubic yards per year, which necessitates dredging sand from the harbor and pumping it to the beach about a mile downdrift. **Table 10.2** compares this drift rate with that at other locations.

Figure 10.22 The Santa Barbara, California, breakwater in 1988. Sand accumulation in the harbor is clearly visible.

Communities along the shores of the Great Lakes have erosion problems also. Early in the 20th century, protective structures were built on the Lake Michigan shoreline. Many of these have now deteriorated, particularly during the high lake level of 1984 (177.2 meters above mean sea level). Lake levels began dropping around 1960, and there was little concern at the time about shoreline erosion. Since 1964 (150-year record low of 175.7 meters above mean sea level), levels have been rising, and there is increasing public pressure to put more protective structures on the shoreline.

Sea cliffs—Most people enjoy ocean views, particularly from atop a sea cliff. Such cliffs are dynamic places, eroded both by wave action at the toe and by mass-wasting processes on the face and the top. An **active sea cliff** is one where erosion is dominated by wave action. The result is a steep cliff face and little rock debris at the base. The erosion of an **inactive sea cliff** is dominated by running water and mass-wasting processes. Its slopes are much gentler, and angular debris accumulates at the toe. The nature of the debris at the foot of a sea cliff can aid in determining whether it is active or inactive. Wave erosion produces rounded cobbles, whereas mass-wasting debris (talus) is angular. Long-term, average rates of cliff erosion vary from a few centimeters to tens of centimeters (many inches) per year, depending on the geology, climate, tectonic setting, and relative sea-level changes. Erosion can be gradual, grain-by-grain detachment or can occur in large, dramatic events, such as rockfalls and landslides.

Location	Drift Rate, Cubic Yards/Year	Direction
West Coast (all California)		
Newport Beach	300,000	South
Port Hueneme	290,000	South
Santa Barbara	250,000	Southeast
Santa Monica	160,000	South
Anaheim Bay	90,000	South
East Coast		
Fire Island Inlet, New York	350,000	West
Rockaway Beach, New York	260,000	West
Sandy Hook, New Jersey	250,000	North
Ocean City, New Jersey	230,000	South
Palm Beach, Florida	130,000	South
Great Lakes		
Waukegan, Illinois	43,000	South
Racine County, Wisconsin	23,000	South

Table 10.2 Rates of Littoral Drift at Selected U.S. Locations

Sources: U.S. Army Corps of Engineers and other sources.

It is standard practice to consider 50 years as the lifetime of a house or other structure. Therefore, a structure should be set back from the top of a cliff at least 50 times the annual rate of cliff retreat. However, when cliff retreat is episodic—say, in large landslides—such annual rates have little meaning. In other cases, the rates are not known or people choose to ignore them, with inevitable results (**Figure 10.23**). Orrin Pilkey of Duke University (see References and Further Readings at the end of this chapter) has developed a checklist for people to use when they are considering buying property on cliffs and bluffs. Pilkey's list includes such items as looking for evidence of failure at the toe of the cliff, determining rock hardness, and learning the history of cliff recession in the area from local land records and photographs.

Figure 10.23 Rapid sea-cliff retreat near Big Lagoon, Humboldt County, California, has left this home in a precarious position. These cliffs near the Oregon border are composed of sandy and gravelly uplifted marine sediments. As shown by the photograph, they are subject to direct wave attack. Drastic undercutting by wave action and the near-vertical cliff indicate that this is an active sea cliff.

Jim Falls, California Department of Mines and Geology

Mitigation of beach erosion—Most coastal engineering works are built to protect the land from the sea. It is perhaps presumptuous to think we can build works that will shelter us from the rise of sea level or withstand enormous storm surges, but that does not keep some people from trying. Below, we provide some examples of what has been done.

Rock revetments and concrete seawalls are built landward of the shoreline and parallel to the beach, not to protect the beach but rather to protect the land and property behind them. Commonly, these "hard" structures reflect wave energy downward, removing sand and undermining the structure. There is a question among coastal engineers and geologists that goes something like, "Do you want a revetment, or do you want a beach?" Because waves reflecting off hard shoreline structures remove the beach material in front of them, the best designed rock revetments have sufficient permeability to absorb some of the wave energy. Another problem is that wave action erodes around the ends of revetments. When that happens, they must be extended to prevent damage to structures located there.

Groins are constructed to trap sand for the purpose of widening beaches. They trap sand on their up-current sides, thus depriving downcurrent beaches of sand. Frequently, another groin is built to widen *that* beach, and the cycle is repeated. Once sand has built out to the end of a groin, it flows around the tip of it and begins filling in the eroded

Figure 10.24 Elevating beach homes on stilts is a short-term mitigation strategy to deal with beach erosion, but as the beach continues to erode, the elevated homes become increasingly vulnerable to wind and water. Here, on the coast of Louisiana, men look at a home on stilts destroyed by waves and wind. Let's hope the oil rig in the distance fared better.

area downcurrent. The result is a beach with groins every few hundred feet and a shoreline with a scalloped or bread knife appearance (see Figure 10.7). The northern New Jersey shore and Miami Beach are textbook examples of multiple-groin shorelines. Permeable groins are often constructed today. They allow some sand to filter through them for nourishing downcurrent beaches. Jetties are similar to groins but longer and often installed in pairs on both sides of a harbor entrance or where a river flows into the sea. They have caused entire beaches and whole communities to disappear. The jetties at Ocean City, Maryland, have already been cited (see Figures 10.6 and 10.21).

consider this . . . Several states have outlawed hard protective structures between beaches and the land. Why did they do this? Should other states, including those on the Great Lakes, enact similar legislation?

At present, the designs of most long structures built perpendicular to the shoreline incorporate permanent *sand-bypassing works*. These features allow for the pumping of sand from the up-current side of the structure to nourish the beaches downcurrent. In addition, downcurrent beaches can be artificially nourished by pumping in sand from offshore areas or by transporting it to them from land sources (see Figure 10.21). Unfortunately, artificially nourished beaches are generally short-lived. One reason is that the imported sand is often finer than the native beach sand and is thus more easily removed by the prevailing waves. The other reason is that artificial beaches usually have a steeper shore face and are therefore more subject to direct wave attack than are gently sloping, natural shorelines, which better dissipate wave energy offshore. Artificial **beach nourishment** is expensive, and it is most commonly done only when large amounts of tourist dollars or infrastructure are at stake.

As a last resort, some homeowners figure if you can't save the beach, then how about rising above the tide? All around the country, on barrier islands from New York to Texas, homes have been jacked up and placed on stilts to keep them above rising sea level (**Figure 10.24**). This approach works until a large enough storm comes along that the pilings are swept away and the house collapses or until the barrier island underneath migrates landward, leaving the house still on stilts but over water instead of land. Raising houses on stilts is at best a short-term approach to coast hazard mitigation.

Although theory and resulting computer models of shoreline behavior are improving, many shoreline-impact studies use scale models to evaluate various remediation schemes before they are built. Physical modeling makes sense, because beaches are complex systems with many feedbacks that might not be obvious at first glance or might be difficult to handle in a mathematical model. Such physical models are an exact replica of the project but require careful scaling of the materials and flows to compensate for differences in size. The U.S. Army Corps of Engineers at its facility in Vicksburg, Mississippi, has been at the forefront of such modeling efforts (**Figure 10.25**).

Engineer Research and Development Center

Figure 10.25 A scale model of Chicago Harbor built by the U.S. Army Corps of Engineers. The physical model, built at a scale of 1:120, depicts more than 5,500 meters (18,000 feet) of the city's shoreline. Along with a wave simulator, the model is useful for assessing potential damage from waves of varying height and source direction. Data from the model will be used to make future design improvements.

Hurricanes, Nor'easters, and Coastal Effects

Andrew, Mitch, Camille, Gilbert, Floyd, and *Katrina* are innocent-sounding names, but onshore surges of water and intense rainfall produced by hurricanes (**Figure 10.26**) with such names are one of the greatest natural threats to humans. Hurricanes (called *typhoons* in the North Pacific Ocean and *cyclones* in the South Pacific and Indian oceans) get their energy from the equatorial oceans in the summer and early fall months, when seawater is warmest, usually above 25°C (about 80°F). Hurricanes begin as tropical depressions when air that has been heated by the Sun and ocean rises, creating reduced atmospheric pressure, clouds, and rain. As atmospheric pressure drops, the warm air moves toward the center of low pressure, what later may become the eye of the impending hurricane. When the moist, rising air reaches higher elevations, the water vapor in it condenses and releases heat, causing the air to rise even faster and creating even lower atmospheric pressure and greater wind velocities. Wind blowing toward the center of low pressure is given a counterclockwise rotation by the Coriolis force in the Northern Hemisphere and a storm is born (see Figure 10.26).

Hurricanes

How hurricanes cause damage—What begins as a localized tropical disturbance, a few thunderstorms in the tropical Atlantic, can grow to be an intense storm, up to 800 kilometers (500 miles) across, with wind velocities far in excess of 120 kilometers per hour (74 miles/hour), the lower velocity limit for hurricane designation. Hurricanes travel as coherent storms with forward velocities of 12–40 miles/hour. Their speed and course, and thus where and if they will strike land, are set by large-scale, atmospheric steering winds, which are controlled by the arrangement of large high- and low-pressure systems. In general, these steering winds guide hurricanes from their birth in the eastern Atlantic westward and then northward to their graves over the cool North Atlantic or the continent of North America. However, hurricane tracks vary as widely as the arrangements of steering winds that guide them (**Figure 10.27**). Hurricanes are storms of the late summer and early fall because they need the energy contained in warm ocean water to grow strong (**Figure 10.28**).

What makes hurricanes so dangerous? Certainly, the heavy rains they spawn cause massive flooding, but the worst actor in the hurricane story is wind. To understand the distribution of wind-induced damage from hurricanes, you must know something about where and how quickly hurricane winds blow. Let's start with the hurricane itself. The winds are wrapping counterclockwise around the eye, as they do in all Northern Hemisphere storms (**Figure 10.29**). But this isn't the entire story, because the hurricane isn't standing still; it's moving along, being carried by the steering winds. Imagine yourself a television reporter. You are stationed to the east of New Orleans as Hurricane Katrina comes ashore. The hurricane has winds of 120 miles per hour, and it's moving toward you from the south at 12 miles per hour. What wind speed do you feel (**Figure 10.30**)? Because both the hurricane and steering winds are blowing in the same direction (from the south), you add them and feel winds of 132 miles per hour. Now your colleague, working for the other network, happens to be stationed west of New Orleans. There, the steering winds are blowing from the south, but the hurricane winds are blowing from the north. As a result, your colleague feels only 108-mile per hour winds (120 − 12 miles per hour)—still fierce, but 24 miles per hour less than what you are feeling and showing your viewers. When a roof is about to fail, the 20- or 30-mile per hour wind-speed difference from one side of the hurricane to the other can be significant.

Although spring can be quite warm, there are rarely spring hurricanes. Why do you think that is the case?

Hurricane winds also drive **storm surges**, a mound of water pushed ahead of the hurricane that can cause sea level to rise many meters above the normal high-tide level (**Figure 10.31**). The storm surge is highest where the winds are moving most quickly. Katrina's storm surge that overtopped levees and flooded New Orleans came from east of the city, where the right side of the hurricane and its strongest winds passed (see Case Study 10.4).

The reduction in atmospheric pressure associated with a hurricane also causes sea level to rise—like fluid rising by suction in a soda straw. Average sea-level atmospheric pressure is 1,013 millibars (29.92 inches of mercury), and the rise in sea level is about 1 centimeter for every millibar of pressure drop. Hurricane Camille in 1969, the strongest storm to hit the U.S. mainland, had a near-record low

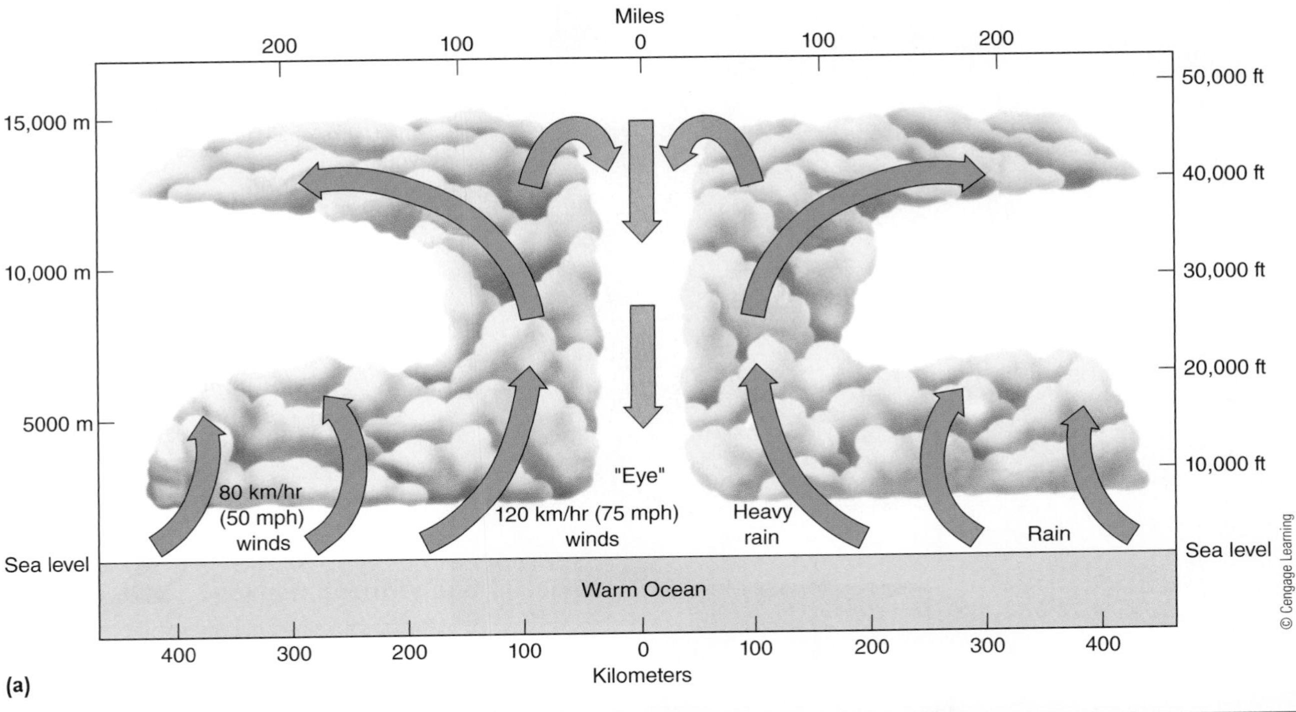

Figure 10.26 (a) Circulation patterns within a hurricane, showing inflow of air in the spiraling arms of the cyclonic system, rising air in the towering circular wall cloud, and outflow in the upper atmosphere. Subsidence of air in the storm's center produces the distinctive calm, cloudless "eye" of the hurricane. (b) A satellite photograph of Hurricane Katrina coming ashore near New Orleans on 29 August 2005. Katrina's rotation, bands of rain-producing clouds, and central eye are clearly visible.

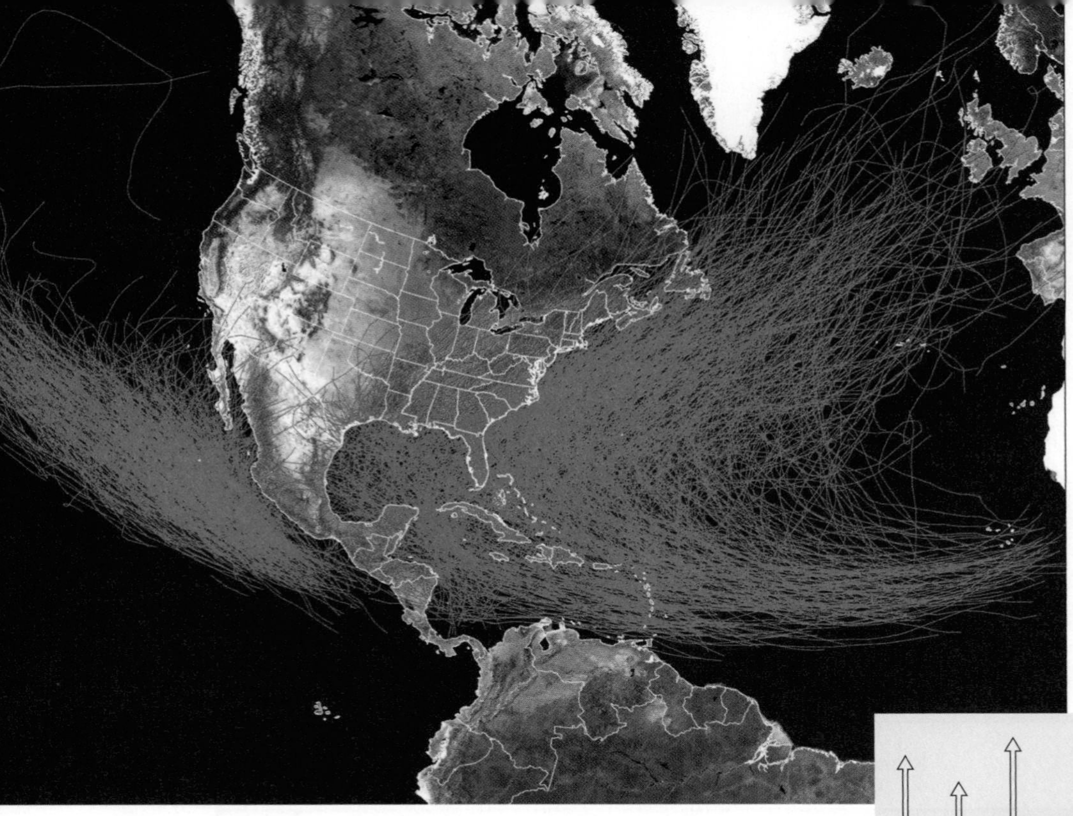

Figure 10.27 Hurricanes follow many different paths from their birthplace in the tropical Atlantic to their demise, usually far to the west and north. Shown are the tracks of all hurricanes from 1851 to 2006.
NOAA/Ethan Gibney

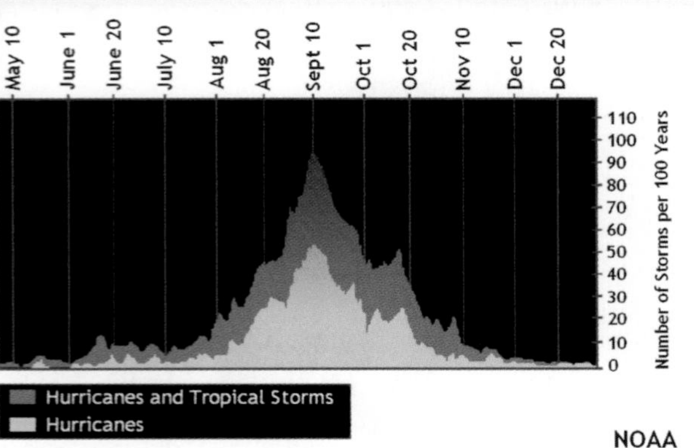

Figure 10.28 Hurricanes and tropical storms are most common in the late summer and early fall, after the ocean has warmed in the strong summer sun. (From http://www.nhc.noaa.gov/climo/)

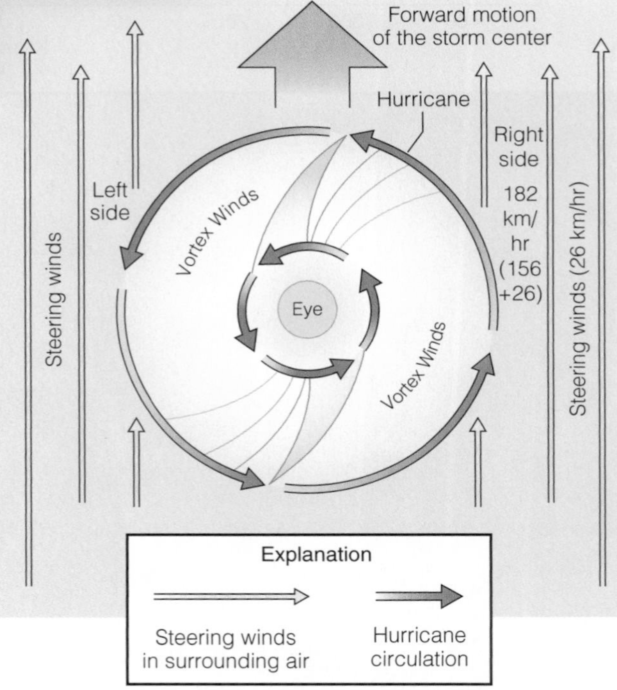

Figure 10.29 This hurricane is being steered by winds of 26 kilometers/hour. The storm winds are 156 kilometers/hour. On the right side of the storm, people on the ground feel 182 kilometers/hour winds (156 + 26). What wind speed do people on the left side of the storm experience? (Reprinted from Coch, N. K. 1994, August. Geological effects of hurricanes. *Geomorphology* 10(1–4): 37–63, with permission of Elsevier.)

barometric pressure of 905 millibars (26.61 inches of mercury). Camille's storm surge and winds devastated the entire Gulf Coast from Florida to Texas.

Hurricane tracks determine where and how severe the damage will be. Consider a hurricane approaching the U.S. East Coast (**Figure 10.32**). If the hurricane moves parallel to the coast, the strongest winds and the storm surge stay offshore. These are referred to as coast-parallel, or CP, hurricane tracks. A hurricane that hits the coast at a right angle has a coast-normal, or CN, track. The full force of the storm surge is pushed ashore and the highest winds (remember where they are and why) slam into the coast. Katrina was a coast-normal storm both when it moved across Florida (going west) and when it moved over Louisiana (going north) and battered New Orleans (**Figure 10.33**). No wonder the damage was so immense.

Katrina was not the only coast-normal storm to wreak havoc on the United States. Almost 70 years earlier, long before weather satellites and hurricane chasers followed the every move of these storms, a September hurricane of great fury snuck up on New England (**Figure 10.34**). The 1938 hurricane, or New England hurricane (this was in the time

Figure 10.30 Wind speed in hurricanes can be so high that it is difficult to stand—unless, of course, you lean into the wind, as this fellow is doing on 25 August 2005, in south Florida, where hurricane Katrina first made landfall.

Figure 10.31 Hurricane Katrina's wind-driven storm surge was massive and severely damaged low-lying areas along the Gulf Coast. Here the surge is coming over Highway 90 near Gulfport, Mississippi, on 29 August 2005.
© Mike Theiss/Ultimatechase.com

before hurricanes were given proper names), slammed into Long Island with no warning. It was a fast-moving but devastating storm with steady winds of 121 miles per hour and gusts reported to 186 miles per hour. Wave heights of 50 feet (15 meters) were recorded in Gloucester, Massachusetts, and the storm surge rose 17 feet (5 meters) above normal high tide along the Rhode Island coast. Nearly 9,000 buildings and about 3,300 boats were destroyed; 63,000 people were left homeless. Even where there was little flooding, the winds on the right side of this fast-moving storm caused terrible damage (**Figure 10.35**), destroying 2 billion trees, many of which were in New Hampshire.

Hurricane examples—In 1969, Hurricane Camille killed 130 people, and 3,000 homes simply disappeared. However, real estate was resold, and new homes were built on the same sites where homes had been removed by the storm. Camille was rated as a category 5 hurricane, the highest category on the Saffir–Simpson Scale (**Table 10.3**).

Some hurricanes veer eastward and follow a track up the U.S. East Coast, causing flooding and beach erosion from the Carolinas to Maine. In 1989, Hurricane Hugo struck the coast just north of Charleston, South Carolina, with sustained winds of more than 130 miles per hour and a storm surge of 6 meters (20 feet). Hurricane Floyd was the 1999 storm in North Carolina that wouldn't go away; it dropped 55 centimeters (22 inches) of rain. Although not a strong hurricane, it created flooding of massive proportions in the eastern part of the state, aided by Hurricane Dennis, which had dropped 25 centimeters (10 inches) of rain 2 weeks earlier. Thousands of square miles were underwater for weeks, and the flood area became a fetid swamp containing the rotting bodies of 10,000 hogs and 2.5 million chickens and turkeys.

In 1970, a tropical cyclone in East Pakistan (now Bangladesh) resulted in perhaps the deadliest coastal flooding of all time, with a death toll of more than 300,000 people. Dissatisfaction with the central government of Pakistan's re-

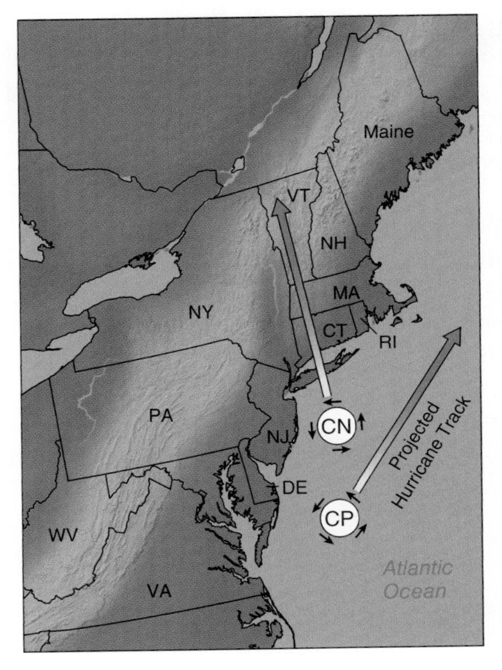

Figure 10.32 The track a hurricane takes makes all the difference. Coast-parallel (CP) hurricanes do little damage to the east coast of North America, because their strong winds and storm surge remain offshore. However, when a hurricane hits the coast straight on—a coast-normal (CN) storm—the damage can be immense. Witness the storm surge damage from both Hurricane Katrina and the 1938 New England hurricane. (Reprinted from Coch, N. K. 1994, August. Geological effects of hurricanes. *Geomorphology* 10(1–4): 37–63, with permission of Elsevier.)

sponse to the tragedy led to the formation of Bangladesh, a new country separated from Pakistan. Another tragedy there in 1991 killed 60,000 people and led to the formation of a tropical-cyclone forecasting center. Experts at the center recommended the construction of raised concrete bunkers above storm-surge level for refuge. A fierce cyclone in 1998 left thousands homeless but killed only 200 people.

One of the key problems in less-developed countries is overpopulation, which forces people to settle on land that is subject to life-threatening natural hazards. Bangladesh is about the size of Wisconsin, with a population of 117 million, more than a third of the population of the United States. Even in areas where people were warned of the 1991 storm, many were reluctant to leave their few possessions to the rising waters.

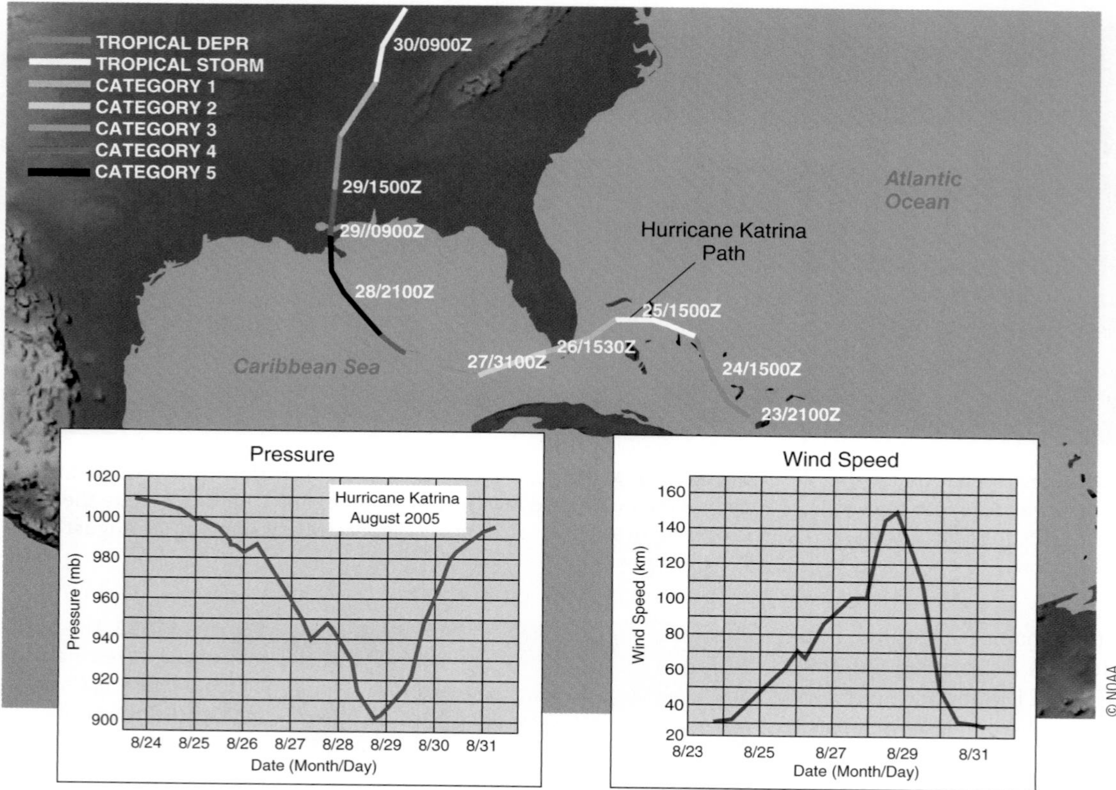

Figure 10.33 When Hurricane Katrina hit land (in Florida and Louisiana), it was a coast normal (CN) storm pushing its highest winds and storm surge (to the right of the eyewall) ashore. Can you see from the graphs how wind speed and barometric pressure are related? What happened to both of these indications of storm strength after Katrina went ashore? (Source: NOAA data from the National Hurricane Center.)

Lest we think such tragedies are limited to less-developed countries, we should look to Galveston, Texas, in September 1900. A hurricane traveled across the Gulf of Mexico and struck the barrier island upon which Galveston is built. The hurricane struck with such force that it smashed 3,600 wooden structures and drowned more than 6,000 people, the largest toll from a natural disaster in American history. Nellie Carey, who lived through the disaster, wrote in a letter: "Thousands of dead in the streets—the gulf and bay strewn with dead bodies. Not a drop of water—food scarce. The dead are not identified at all—they throw them on drays and take them to barges, where they are loaded like cord wood, and taken out to sea to be cast into the waves." The letter was reprinted in its entirety in a book published shortly after the disaster. Today, Galveston is one of the leading cities to deal with evacuation in case of a hurricane.

In September 2008, Hurricane Ike, a 1,000-kilometer (600-mile)-wide category 2 hurricane, pummeled Texas, coming ashore over Galveston, where 20,000 people ignored evacuation orders (but more than 2 million people did leave the Texas and Louisiana coasts). The storm surge was between 3

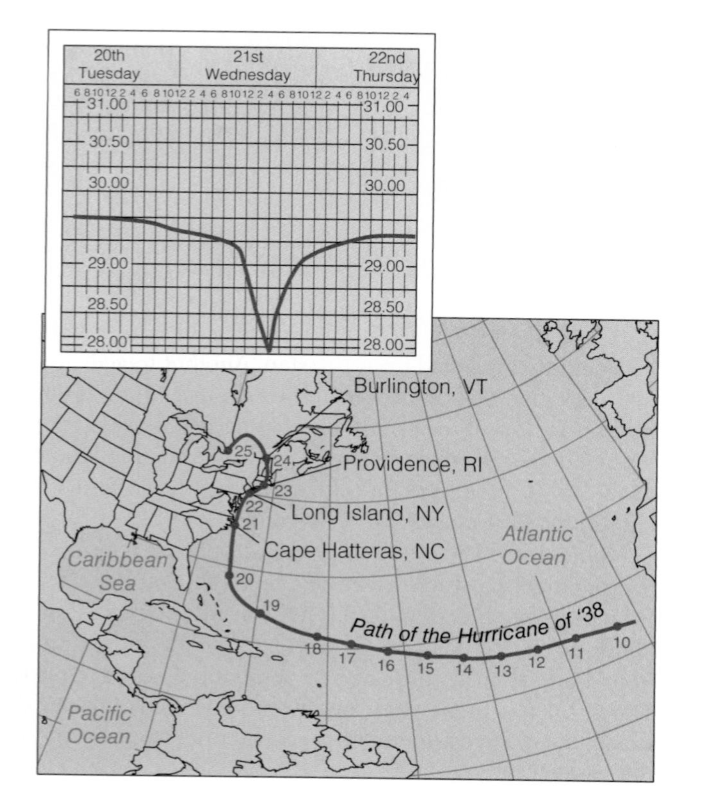

Figure 10.34 The 1938 hurricane roared into New England, slamming into Long Island and taking just about everyone by surprise. The red numbers on the map indicate the hurricane's location on sequential dates in September. It was a fast-moving, coast-normal hurricane when it hit Long Island, which meant high wind speeds and a massive storm surge pounded communities to the right, or east, of the eye. Can you tell from the barometer trace (from a weather station on Long Island) how long the storm lingered and how fast it was moving? (Sources: The Path of the Hurricane of '38 from http://www.pbs.org/wgbh/americanexperience/features/image/hurricane-path/; http://www2.sunysuffolk.edu/mandias/38hurricane/weather_history_38.html; and Hendrickson, Richard G. 1996. *Winds of the Fish's Tail.* Mattituck, NY: Amereon Ltd., pp. 33–62.)

and 5 meters (10 to 15 feet), just overtopping the seawall in some places and deeply flooding low-lying neighborhoods. Quoted in the *New York Times,* one resident, who had tried to ride out the storm at his Galveston home with his family, said, "I know my house was dry at 11 o'clock, and at 12:30 a.m., we were floating on the couch putting lifejackets on." Once the water reached the television, a meter (4 feet) off the floor, he retrieved his boat from the garage and loaded his family into it. "I didn't keep my boat there to plan on evacuating because I didn't plan on the water getting that high, but I sure am glad it was there."

Table 10.3 Saffir–Simpson Hurricane Scale

	Wind Velocity			
Category	Kilometers/Hour	Miles/Hour	Damage	Examples
1	119–153	74–95	Minimal	Juan, 1985
2	154–178	96–110	Moderate	David, 1979; Bob, 1991
3	179–210	111–130	Extensive	Betsy, 1965; Frederic, 1979
4	211–250	131–155	Extreme	Hugo, 1989; Andrew, 1992; Katrina, 2005
5	>250	>155	Catastrophic	Labor Day, 1935; Camille, 1969

Source: Coastal Weather Research Center, University of South Alabama, Mobile.

Hurricane predictors—We know that hurricanes are part of the normal tropical weather pattern and that they will occur seasonally every year. Meteorologist William Gray at Colorado State University has developed an empirical prediction system that can estimate the severity of the upcoming hurricane season. His work is based on understanding the two most important ingredients needed to make strong hurricanes: warm sea-surface temperatures and an atmosphere with little shear, so that hurricanes, once they start to form vertical circulation, are not ripped apart by the steering winds. Gray's model uses the following data:

- Tropospheric winds up to an altitude of 12,000 meters (40,000 feet). Strong winds in the troposphere work against hurricanes by shearing off the tops of their circulation.
- West African climate. Dry years in Africa promote high-altitude winds from the west above the tropical Atlantic, increasing the shear between the upper and lower levels of the atmosphere and diminishing the strength of hurricanes. During wet years, upper atmosphere winds blow from the east, greatly reducing wind shear, and thus promoting hurricane-strength storms.
- El Niño. The warming of the equatorial Pacific Ocean promotes westerly winds that blow into the Caribbean and Atlantic basins, where trade winds blow at lower levels from the east. This promotes wind shear that knocks down hurricane buildups. Between 1991 and 1994, repeated El Niños resulted in few hurricanes. In 1995 and 1999, El Niños gave way to a cooling of the equatorial Pacific Ocean (La Niña), and hurricane activity greatly increased.
- Atmospheric pressure. Low atmospheric pressure in the Atlantic Ocean and Caribbean Sea indicates areas of warm water, which fuel hurricanes, promote the convergence of air masses, and stimulate vertical buildup of moisture-laden clouds.

Although Gray's model has a good record of predicting storm frequency for any one hurricane season, it is up to seasoned forecasters to predict both the track and the strength of

Figure 10.35 Damage from the 1938 New England hurricane differed, depending on location. (a) Near the coast, huge waves and the storm surge ripped apart infrastructure and tossed ships around like toys, as shown here in eastern Connecticut. In New London, Connecticut, the lightship tender TULIP was thrown out of New London harbor and deposited on the New Haven's main rail line. Restoration of passenger service on the Shore Line between New Haven, Connecticut, and Boston, Massachusetts, took nearly 2 weeks. (b) Farther inland, wind was the primary agent of destruction. In Hartford, Vermont, homes were damaged as massive trees were felled by the hurricane's wind.

a: New York, New Haven and Hartford Railroad, courtesy of the New Haven Railroad Historical and Technical Association, Inc.; b: Hartford Historical Society

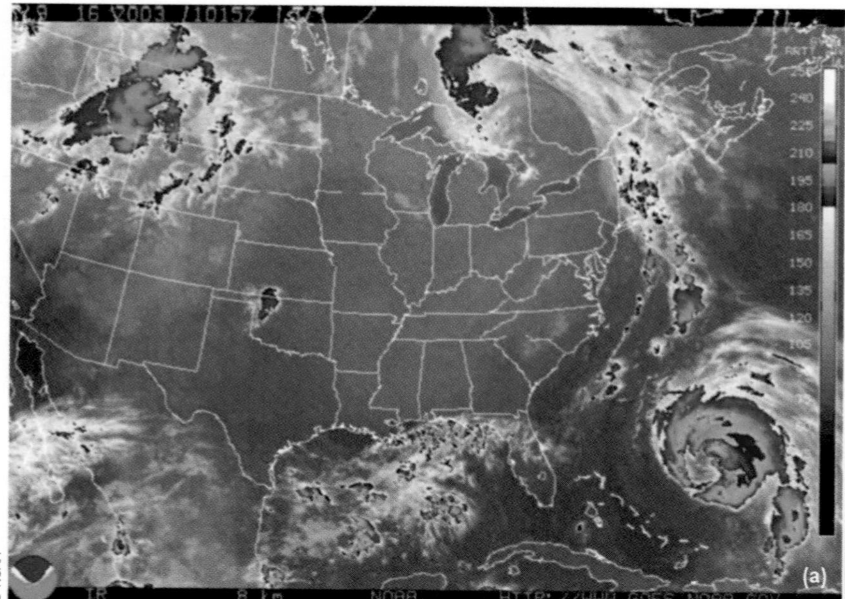

© NOAA

Figure 10.36 Hurricane prediction capability is improving. (a) A satellite image of Hurricane Isabel before it made landfall. (b) The National Weather Service 5-day forecast track for Hurricane Isabel, showing where the storm was likely to go and when. Such data form the basis for informed decisions about whom to evacuate and when those evacuations should begin.

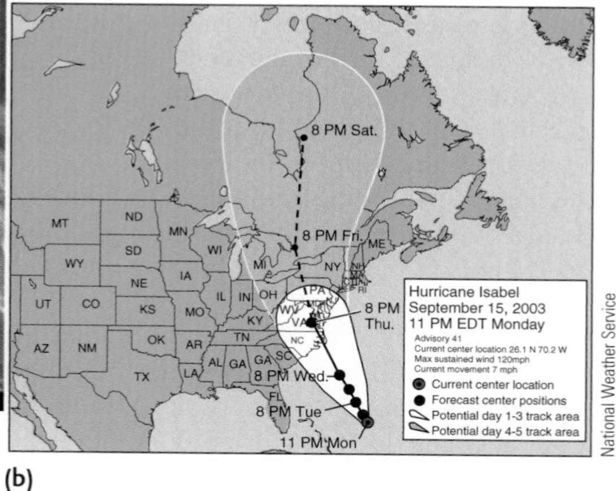

(b)

National Weather Service

any individual hurricane after it is born. The National Weather service is charged with this duty, and it has made great advances in the accuracy of its predictions over the past two decades. Forecasters are aided in their work by several sources of data. First, they use sophisticated atmospheric models to predict the behavior of steering winds that help determine both the track of the hurricane and the degree of shearing that could tear the storm apart. Second, they use data collected in real time by hurricane-hunter planes that fly right through the storms, collecting wind and pressure data. Third, they rely on more than 100 years of hurricane history to understand how such storms behave along different parts of the Gulf and Atlantic coasts. The work of many forecasters at the National Hurricane Center in Miami is condensed into track and intensity prediction maps (**Figure 10.36**). The improvement in hurricane forecasting accuracy is impressive. Today, the accuracy of 5-day hurricane forecasts matches that of the 3-day forecasts only a decade ago.

Nor'easters

The east coast of North America is affected by major coastal storms pretty much all year-round. Although the media focuses heavily on the hurricanes that threaten eastern North America in the summer and fall, during the winter and spring, other powerful storms, termed nor'easters, threaten the Atlantic Coast. Their name comes from the direction that their winds most frequently blow onshore as the storms track north just off the U.S. East Coast. These are winter and spring storms that differ not only in season but also in many other ways from hurricanes. Nor'easters are cold storms, born in the western Atlantic. They are usually triggered by small, low-pressure systems moving from the west and then develop explosively along the coast, often getting their start off Cape Hatteras. Nor'easters are huge and can be slow moving, lingering for days. Their strength and persistence create particular hazards, because they can bring record winter snows and dramatically alter coastlines, as wave heights can build over days and waves relentlessly pound beaches and sea walls (see Figure 9.3).

Perhaps the best-known nor'easter is the Halloween storm of 1991, which pummeled New England with huge seas and is featured in the book and movie *The Perfect Storm*. The Halloween storm was massive and devastating, affecting much of the northeastern United States and Atlantic Canada (**Figure 10.37**). It took the lives of fishermen at sea (yes, George Clooney went down with the ship) and demolished many shorefront homes that had stood for a century or longer. The Bush family compound in Kennebunkport, Maine, suffered serious damage from the pounding surf.

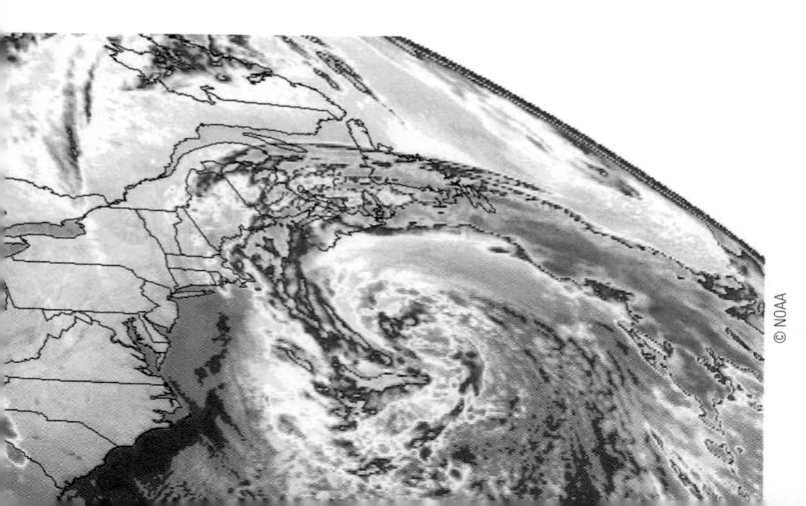

© NOAA

Figure 10.37 The Perfect Storm, the Halloween storm, or the nor'easter of late October 1991 as it battered the northeastern United States and Atlantic Canada. The storm was so large that it extended from Georgia to Labrador!

El Niño and the Coastal Zone

El Niño is a perturbation, or disruption, of the normal cold-water conditions that exist along the west coasts of North and South America. During El Niño years, high-pressure cells replace the normally low atmospheric pressure of the western Pacific Ocean. Because the pressure difference that drives the trade winds between the eastern Pacific and the western Pacific lessens, the trade winds stop, or may even reverse direction. A bulge of warm surface water (as much as a meter or 3 feet high) called a *Kelvin wave* surges eastward toward the coast of the Americas and displaces the cooler water there (**Figure 10.38**). First noticed by Peruvian fishermen around Christmas in 1892, this invasion of warm water was dubbed the *Corriente del Niño* ("Current of the Christ Child"). Today, the atmospheric pressure reversal in the western Pacific region, called the *southern oscillation*, is known to be the trigger for El Niño events. We now call this connection and the ensuing warm-water events the **El Niño–Southern Oscillation**, usually shortened to its easily articulated acronym, **ENSO**.

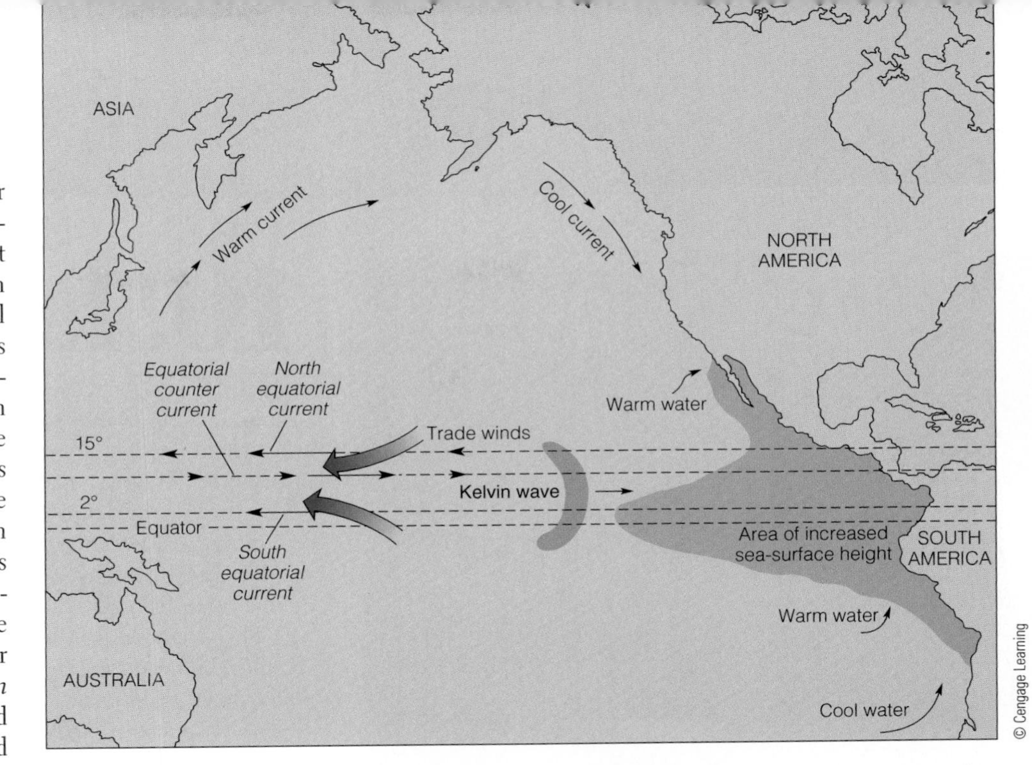

Figure 10.38 Oceanography, meteorology, and geography of an El Niño. At the onset, trade winds maintain warm equatorial currents flowing from the east to the west; these currents cause upwelling of cold, deep ocean water along the west coast of South America. As El Niño begins, the trade winds weaken, and the equatorial countercurrent strengthens. This creates Kelvin waves, surges of warm water moving from west to east. When a Kelvin wave reaches the Americas, it divides northward and southward, traveling along the west coasts.

Significant El Niño events of 1972–1973 and 1976–1977, and the super El Niño of 1982–1983 established it as a global phenomenon. An almost continuous El Niño condition persisted from 1991 to 1994 (**Figure 10.39**), and then another super El Niño occurred in 1997–1998. During El Niño years, because of atmospheric changes globally, there have been droughts in Africa, Australia, and Indonesia; flooding in California, Ecuador, Peru, and Tahiti; and impressive storm waves and beach erosion along both the east and west coasts of the United States. More bad news was the unprecedented number of landslides and debris flows in California, Oregon, and Washington during the 1997–1998 event. According to Jim O'Brien at the Center for Ocean-Atmosphere Prediction Studies (COAPS) at Florida State University, there is also good news about El Niño:

- It decreases Atlantic hurricane activity.
- It reduces forest fires in Arizona, Texas, and New Mexico.
- It reduces tornadoes.
- It produces better citrus and vegetable crops.
- It improves the fishing off California because of warm water.

Interestingly, Hawaiian hurricanes occur only during El Niño years. El Niño is a good example of Earth's functioning as a system—the atmosphere, hydrosphere, and lithosphere interacting to create natural events, some of which are hazardous to humans. El Niño's "sister," La Niña, is a cold-water condition, which most experts call the normal condition for the west coast of the Americas. Global climate models suggest that a warming climate may lead to semipermanent El Niño conditions.

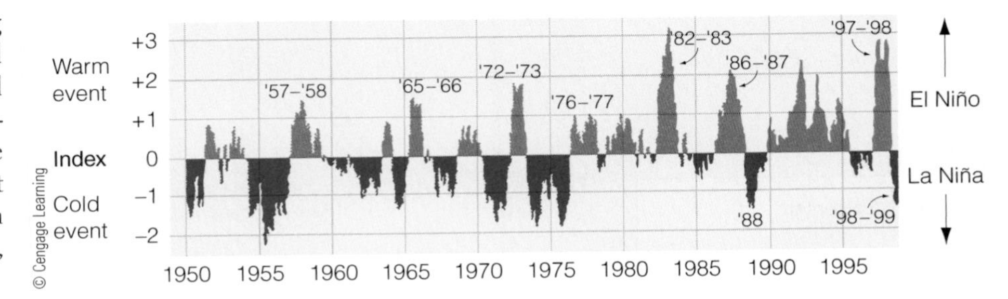

Figure 10.39 The numbers on the left side of the graph are an El Niño–Southern Oscillation (ENSO) index that incorporates these measured parameters: air temperature, atmospheric pressure, cloudiness, and wind speed and direction. El Niño events are shown in red; La Niña events, in blue. (From Ahrens. *Meteorology today* (with Printed Access Card ThomsonNOW™), 8th ed. © 2007 Cengage Learning.)

Figure 1 A week after the 26 December 2004 tsunami, flooding and damage to a village near the coast of Sumatra, Indonesia, were on a scale so massive that it is hard to comprehend. In this image, can you find the limit of inundation that indicates how far the tsunami ran inland?

10.1 Tsunami: A Quarter Million Perish

Indonesia is no stranger to earthquakes, volcanic eruptions, and other spasms of our restless planet. However, within hours after the largest earthquake to strike Earth in 40 years (since the Alaskan and Chilean quakes of 1964) shook the region on 26 December 2004, the shoreline was unrecognizable and hundreds of thousands of people lay dead or dying. What happened that December day?

The magnitude-9.0 quake originated 10 kilometers (6 miles) below the surface, where the Indian and the Burma plates meet in a locked subduction zone. When more than 1,000 kilometers (600 miles) of the fault slipped, the seafloor rose 15 meters (50 feet) in places. Areas close to the quake had little warning as waves, moving at the speed of jet planes, followed the quake within tens of minutes. One coastal city, Banda Ache, on the island of Sumatra, was hit especially hard and fast. Nearly a third of its residents were dead or missing as a result of numerous tsunami waves that surged through the city. The waves, and the debris they carried, smashed everything in their path. Field measurements show that the tsunami waves, when they came ashore, were astonishingly high; in some places, run-ups exceeded 30 meters. Homes, businesses, and people were simply swept away (**Figure 1**).

The damage was not limited to the area near the quake. As the waves spread across the Indian Ocean, so did the tragedy (**Figure 2**). In just a few hours, beach resorts in Thailand were demolished. Just after, the coast of Sri Lanka was assaulted by the waves and finally, hours later, the east coast of Africa, where sufficient warning prevented tragedy. The wave moved across the Indian Ocean for thousands of kilometers and could even be detected in the Atlantic Ocean.

What can be done to prevent this and similar tragedies from reoccurring

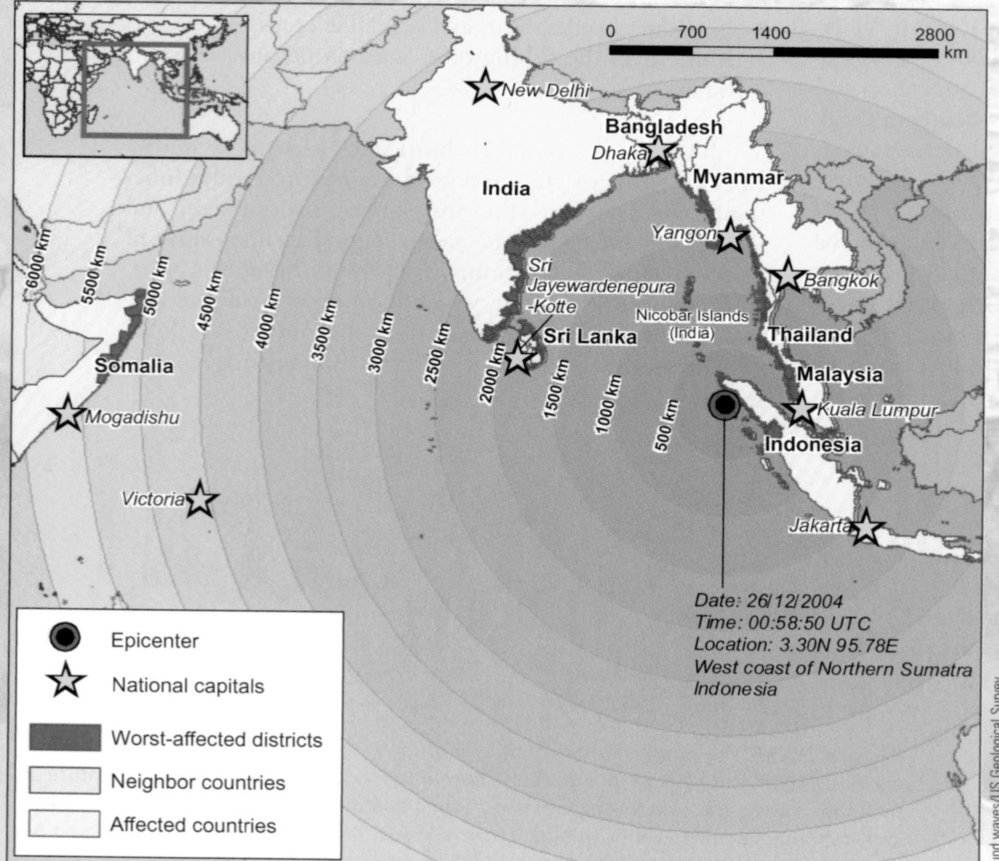

Figure 2 The Indonesian tsunami affected countries all around the Indian Ocean basin. Some of the waves traveled thousands of kilometers before crashing ashore. (Source: USGS Soundwaves, Issue 68, December 2004/January 2005.)

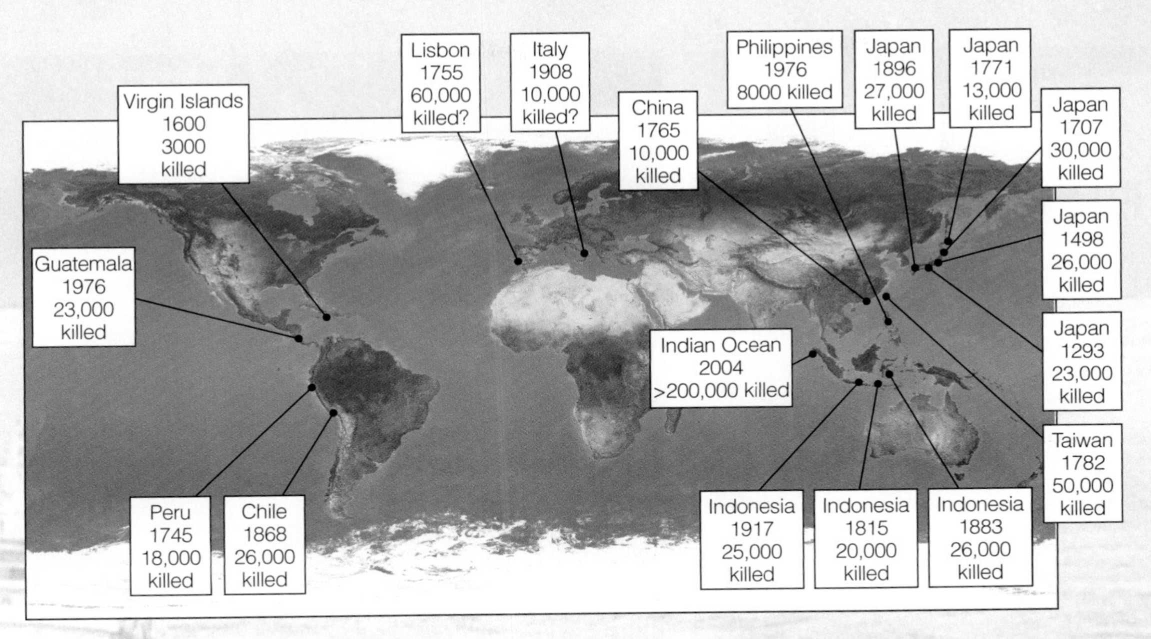

Figure 3 The Indonesian tsunami may be the most deadly we know of, but it is not the first, nor will it be the last tsunami to cause major losses of life and property. (Adapted from Bryant, Edward. 2001. Spacedaily .com/Michael Paine. Data from *Tsunami: The underrated hazard.* Cambridge, UK: Cambridge University Press.)

(**Figure 3**)? Simple measures, such as public education about what to do when the Earth shakes, can save lives. A coordinated, region-wide warning system that relays information (much like that already in place to warn of cyclone hazards) would likely save even more lives. Establishing a direct tsunami monitoring system in the Indian Ocean (much like the one in the Pacific Ocean; see Figure 10.12) is under discussion. In an eerie parallel to post-Katrina analyses, some in India have suggested replanting mangroves to diffuse tsunami energy. The coastal trees and the swamps they create moderate tsunami waves in the same way coastal wetlands tame storm surges. Some people are taking matters into their own hands. If you live in the region and own a Nokia cell phone, you can download a program that uses real-time USGS earthquake data to warn of potential tsunami hazards.

questions to ponder

1. Do you live in an area that could be affected by a tsunami? If so, do you know how to reduce your risk?
2. Do you think the 2004 Indonesian tsunami and the 2011 Japan tsunami will change how the world deals with these hazards in the future?

10.2 Distant Tsunamis: The Silent Threat

Ample warning had been given the residents of Hawaii as the Chilean tsunami raced across the Pacific Ocean at jetliner speeds. A warning had been issued about 6:45 p.m. on 22 May 1960, that large waves were expected to reach Hilo on the island of Hawaii at about midnight. Coastal warning sirens had wailed at 8:30 p.m. and continued for half an hour.

At midnight, a wave arrived, but it was only a few feet high. Hundreds of people had stayed home, and those who had evacuated assumed the danger was over and returned home. Then, another wave appeared, and another. The highest wave struck at 1:04 a.m. (**Figure 1**). Sixty-one people died, and another 282 suffered severe injuries. A restaurant

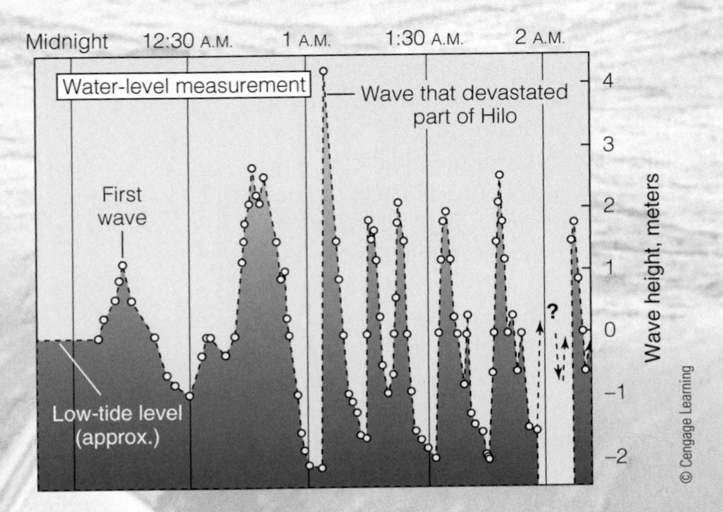

Figure 1 Water-level measurements beneath the Wailuku River bridge during the first hours of the tsunami of 23 May 1960.

Figure 2 A line painted on the window of a rebuilt restaurant commemorates the height to which the tsunami waters rose. Can you explain why the tsunami waves are not tidal waves, as the notation on the window suggests?

B. Pipkin

Figure 3 Parking meters bent by the 1960 tsunami at Hilo, Hawaii. The meters resemble arrows aligned with the direction of wave run-up.

U.S. Department of the Navy

memorialized the event with a waterline on its window (**Figure 2**).

The city of Hilo, with its bay facing northeast toward the Aleutian trench, lies along a section of coast where high waves are expected. These higher than average wave heights are due to the orientation of the coast and the funnel shape of the bay, which focus tsunami energy. An engineering study in the aftermath of the 1960 tsunami resulted in some interesting recommendations for mitigating tsunami damage in that city. For example, parking meters in the run-up area were bent flat in the direction of debris-laden wave travel (**Figure 3**), with the direction varying throughout the affected area. It was recommended that new structures be situated with their narrowest dimension aligned in the direction indicated by nearby bent parking meters. Several waterfront structures had open fronts that allowed water to pass through them; these structures sus-

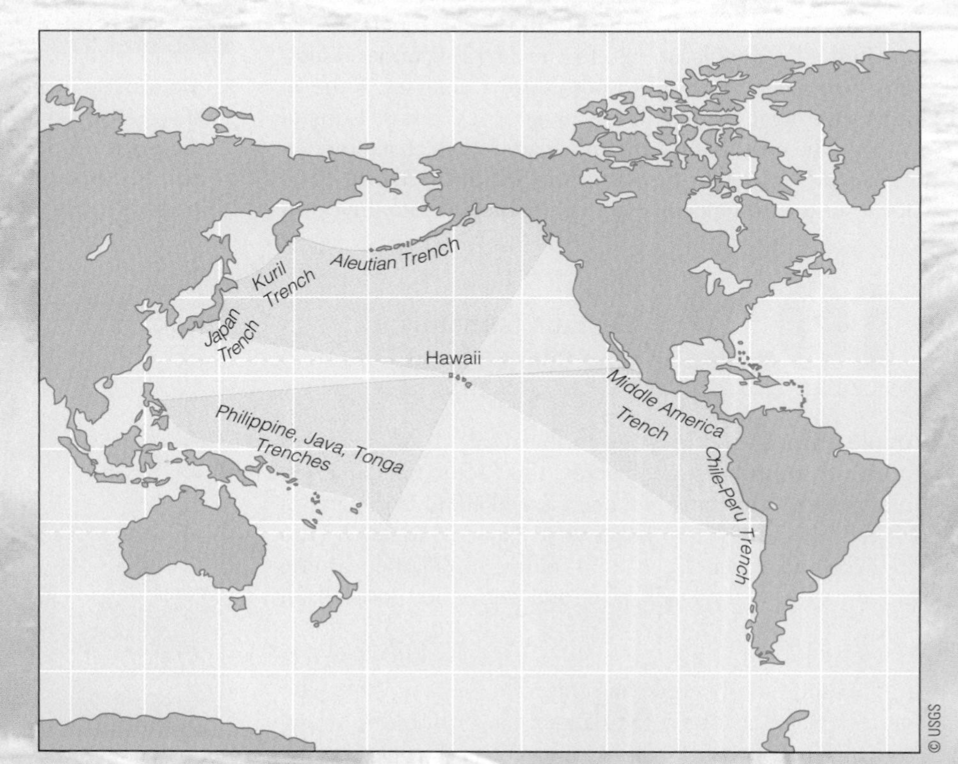

© USGS

Figure 4 Tsunami sources and travel paths to the Hawaiian Islands. (Source: USGS data.)

JIJI PRESS/AFP/Getty Images

tained less damage during the tsunami. Waterfront buildings are now designed with open space on ground level, so that damage from the impact of water and debris can be minimized.

It is no surprise that Hawaiians take tsunami preparation seriously. A glance at a map of the Pacific Ocean basin, ringed by earthquake-generating subduction zones, makes clear the hazard this group of mid-ocean islands face (**Figure 4**).

question to ponder

1. Given what you know about tsunamis, can you think of other ways in which the residents of Hilo can reduce their risk from tsunami hazards?

10.3 A Moving Experience

Cape Hatteras Lighthouse, with its black-and-white, candy-cane paint job, is the tallest brick lighthouse in the world. Built in 1870 with state-of-the-art lenses, the 63-meter (208-foot) structure is so distinctive that it is on the National Register of Historic Places. Besides saving lives, the lighthouse has hosted millions of visitors. It is one of the most enduring landmarks in North Carolina and the nation. This part of the Outer Banks, designated the Cape Hatteras National Seashore by Congress, is a protected environment (**Figure 1**).

Fixed structures built on the shifting sands of barrier islands are inevitably at the mercy of beach erosion and, ultimately, direct wave action. This is because barrier islands "migrate" shoreward. When it was built in 1870, the lighthouse was 460 meters (1,500 feet) from the ocean. By 1919, the sea had come to within 100 meters (310 feet) of its base and, by 1935, to within 30 meters (100 feet). When it comes to a battle of humans against the sea, the sea eventually wins.

Illustrating the futility of shore-protection efforts during times of rising sea level, the National Park Service and the Corps of Engineers spent millions of dollars attempting to save the beloved lighthouse—constructing groins to trap sand, replenishing beaches and artificially depositing a barrier-sand-dune system in front of the structure, installing offshore wave-energy dissipaters, and finally, constructing hard structures (seawalls) to reflect wave energy. Try to find some of these shoreline stabilization features in the photographs of the lighthouse. By the 1980s, the sea was lapping at the base of the lighthouse (**Figure 2**) and more extraordi-

Kitty Hawk
Cape Hatteras National Seashore
Pamlico Sound
Cape Hatteras
Ocracoke Island
Ocracoke Inlet
Atlantic Ocean
Cape Lookout
© NASA

Figure 1 North Carolina's Outer Banks are a chain of islands composed of sand that form a barrier between lagoons and the ocean. Cape Hatteras, near the center of the photo, juts the farthest out into the Atlantic.

Figure 2 Erosion was threatening to destroy the lighthouse; sandbags helped protect it from the waves before the move.
© RAYMOND GEHMAN/National Geographic Image Collection

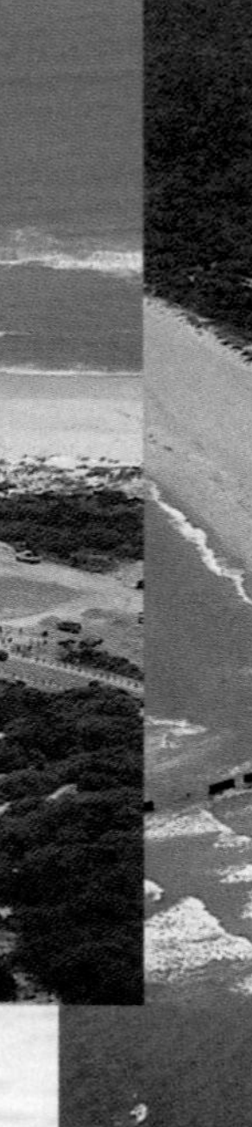

Figure 3 Moving the Cape Hatteras Lighthouse. (a) Ready. (b) Rolling. (c) Home safe.
a-c: Mike Booher/NPS

nary measures were required to save the structure. The National Park Service decided to move the lighthouse in 1990.

Moving structures of great weight has become easier with the development of modern hydraulic systems. The massive Cape Hatteras Lighthouse was moved by using horizontally mounted "pusher" jacks, which pushed it along a track system in 5-foot increments. The move took just 22 days, 19 of which were "pushing" days—a remarkable engineering feat, indeed (**Figure 3**).

The powerful light that had been out for 9 months was relit on Saturday, 13 November 1999, at ceremonies on the original lighthouse site in Buxton, North Carolina.

question to ponder

1. Given rising sea levels driven by climate change, and the likely acceleration of coastal erosion, was it worth the investment to move the lighthouse back?

10.4 Katrina: How Could It Be So Bad?

New Orleans and hurricanes are the stuff of textbooks and movies. Earlier editions of this book, and just about every other geology geography and natural hazards textbook, and hurricane movies have for years presented New Orleans as the city waiting to flood. Why then, with increasingly accurate hurricane track and intensity forecasts (see Figure 10.36) and a scientific consensus that the city was a sitting duck, did more than 1,300 people die, and why were nearly a million more left homeless? The simple answer is that the levees failed, some before their design limits had been met (see Case Study 9.3). Once the levees breached (**Figure 1**), water from the storm surge poured in and inundated the low-lying areas of the city (**Figure 2**). The pumps that keep the city from flooding were overwhelmed and the damage had been done. The more complex answer, stated so clearly by Michael Ignatieff in the *New York Times*, is that the unstated societal contract between the American people and their government was broken—the federal and local governments were simply not there when tens of thousands of

JIJI PRESS/AFP/Getty Images

Figure 1 After Hurricane Katrina, a helicopter dropped a load of sandbags into a New Orleans levee breach.

people were left to fend for themselves in squalid conditions under the roofs of the damaged convention center and Superdome (**Figure 3**).

Katrina will not be the last major hurricane to slam a North American city. Indeed, it may be one of many in the coming years. The year 2005 was the most active hurricane season on record in the Atlantic. There were more named storms than ever before, forcing the weather service to start over with the first letters of the Greek alphabet as they named late-season storms. Why the uptick in storms? Perhaps it's because hurricane frequency is known to vary over decadal cycles, but perhaps it's due to warming of the sea surface. Because hurricanes are heat engines, driven by warm ocean water, this is a plausible mechanism for increasing the energy available for storm formation and maintenance. Two recent studies suggest that the total energy released by hurricanes, as well as the frequency of strong hurricanes, is increasing (**Figure 4**). This is a frightening finding. As global warming raises sea level, storms last longer and their winds blow harder—a triple whammy.

What can be done in the face of rising sea level and increasing storm frequency and energy release? Proposals

Figure 2 In the wake of Hurricane Katrina, two men paddled a boat near the Claiborne Avenue Bridge in the Ninth Ward, New Orleans, Louisiana.

Figure 3 Residents of New Orleans who could not or would not evacuate lined up to enter the Superdome as a shelter of last resort.

Illinoisphotos.com

for New Orleans include raising and strengthening the levee system, raising the land with fill before rebuilding, elevating homes (**Figure 5**), abandoning low-lying neighborhoods, and restoring the marshlands that help dissipate storm energy. A 2009 study, published in *EOS*, a widely read weekly newspaper for geologists, suggests that diverting muddy Mississippi River water into the delta downstream of New Orleans could make a significant difference by rebuilding riverside marshlands, marshlands that effectively temper storm surges and dissipate wave energy (**Figure 6**). Today, extensive levee systems efficiently transport Mississippi River sediment and water directly to the Gulf of Mexico, bypassing the delta. This bypassing is in part responsible for the Mississippi River delta's losing an average of 44 square kilometers (17 square miles) of land to the sea every year since 1940. Using a verified computer model, the authors of the *EOS* article found that, by partially breaching the levees, they could reduce by half the amount of marshland loss predicted for over the next century—even

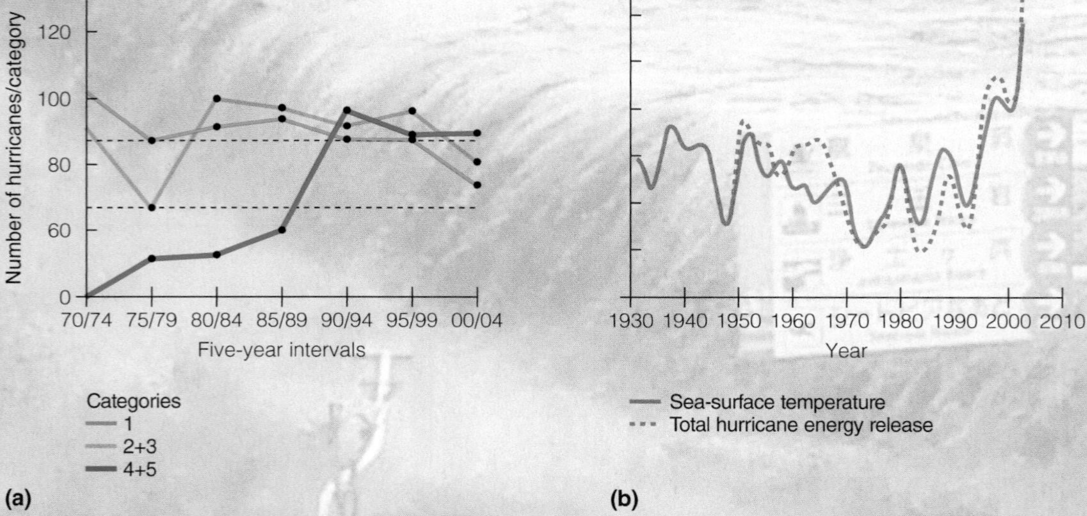

Figure 4 Two research studies that came out in the high-profile journals *Science* and *Nature* within weeks of Hurricane Katrina's striking New Orleans suggest that the potential for hurricane damage has been rising over the past 30 years. Could it be that global warming and the increase in sea-surface temperatures it causes are the driving forces behind changing hurricane dynamics? Time will tell. (a) The number of strong hurricanes (categories 4 and 5) has been increasing globally since 1970. (b) Both sea-surface temperature and total energy release from tropical storms in the Atlantic Ocean have been increasing in lockstep over the past 50 years. Storms are lasting longer and their winds are more intense. Does correlation mean causation in this case? (Sources: a: Based on Webster et al. *Science* 309: 1844; b: Emanuel, Kerry. 2005, August. *Nature* 436(4), and Webster et al. *Science* 309: 1844. 2005.)

JIJI PRESS/AFP/Getty Images

Figure 5 Knowing now that even a category 3 hurricane could flood their city, some New Orleans' residents are taking action. These people who own a lakefront house near the canal are raising it for flood protection.

considering delta subsidence and global sea-level rise. Only time will tell if there is sufficient political will to act responsibly on the scientific data that already exist.

question to ponder

1. In the face of hurricanes, rising sea level, and the subsiding Mississippi River delta, what do you think should be done about New Orleans over the next decade?

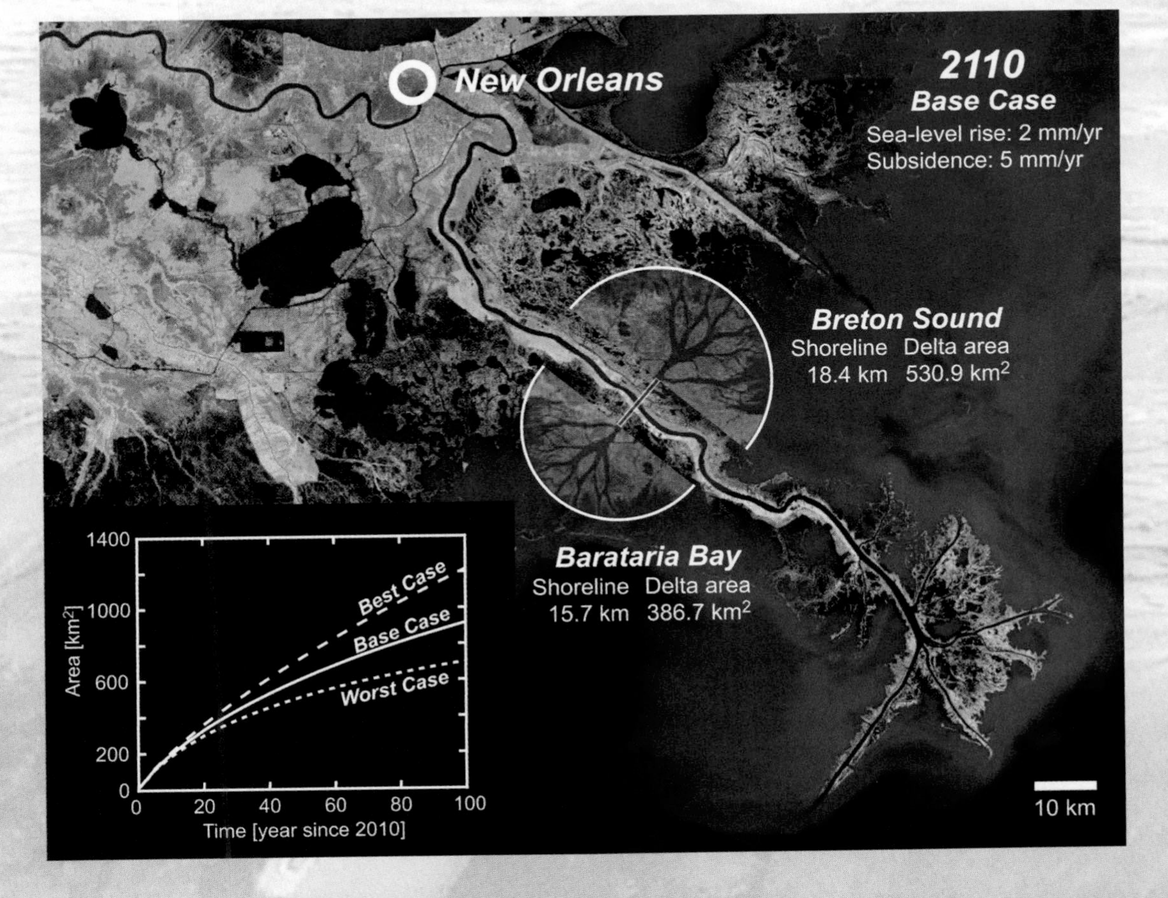

Figure 6 The results of a computer model, showing what is likely to happen if the levees downstream of New Orleans were partially breached, allowing sediment to again spill out onto the Mississippi River delta. Inset shows area of new marshland created under different model scenarios. (From American Geophysical Union. *EOS, Transactions.* October 20, 2009 issue/NASA World Wind.)

PHOTO GALLERY in harm's way

1 Houses in a low-lying area near New Orleans, Louisiana, were flooded by Hurricane Katrina. The city is in the background and the homes are below sea level; no need for a map to determine their elevation. © 2010 TYRONE TURNER/National Geographic Image Collection

Looking at these photographs, one might wonder, "What were they thinking?" Each of these devastating situations could have been avoided. But it's not that simple. Although scientists know why and how coastal environments can be risky places to live, we have not communicated that well to the broader community and there are competing interests. Homes perched on cliff tops have wonderful views, but with the sea eroding the cliffs from below, it is only a matter of time before a landslide carries cliff-top neighborhoods down to the sea. Building on barrier islands gives you quick and easy beach access, perhaps too easy when the island washes over or your stilt house has nothing but waves beneath it. Levees allow cities around the world to persist below sea level, but one day the right storm will hit and the levees will fail. And, of course, fishing villages are built by the sea—why not, except that leaves the villages and their occupants vulnerable to storm surges and tsunami waves. We know enough to move out of harm's way, but will we in the face of other pressures? Can science and scientists effectively communicate the danger?

Beach homes on Dauphin Island, Alabama, were already on stilts to keep them above the waves. That didn't help much with the battering delivered by hurricanes Ivan and Katrina. Such is life on a barrier island, just above sea level. © 2007 TYRONE TURNER/National Geographic Image Collection

2 In Daly City, California, hundreds of homes are perched just above an unstable sea cliff where roads end in midair because landslides have ripped them out from below.

3 At Nags Head on the Outer Banks of North Carolina, a family watches from their stilted house as waves destroy their erosion-control fence. What's next for their barrier island home?

4 The 2011 tsunami obliterates the low-lying, coastal town of Rikuzentakata in Iwate Prefecture, Japan.

SUMMARY Coastal Environments and Humans

Waves

Cause
Wind blowing over the water surface first creates small ripples, then stormy irregular seas, and finally regularly spaced swell that outruns the source area.

Size
Wave size depends on wind velocity, the length of time wind blows over the water, and fetch, the distance the wind blows over the water. A fully developed sea has the highest waves attainable for the three variables.

Motion
Water parcels move in circular paths to a depth of half the wavelength, called *wave base.* Water physically moves shoreward when waves break in the surf zone.

Types of Breakers
Steep offshore slopes develop plunging breakers, and gentle slopes produce spilling breakers.

Wave Refraction
Waves generally approach the shore at an angle. Wave refraction straightens the wave crests as they approach shore and concentrates wave energy on headlands and dissipates it in bays, thus tending to straighten an irregular shoreline.

Longshore Drift
Longshore currents, the result of waves obliquely hitting the shoreline, move sand along the beach. The transported sand is known as *longshore drift.* Groins act as dams to longshore drift, causing sand to be deposited up-current and eroded downcurrent.

Impulsively Generated Waves
Tsunamis (Japanese for "harbor wave") are mostly generated by seismic activity, but they are also generated by volcanic eruptions, submarine landslides, and bolide impacts. They have long wavelengths (hundreds of kilometers) and low amplitudes (about a meter) at sea but transform to an impressive breaker or surge at the shoreline. The 2004 Indonesian tsunami killed nearly a quarter million people. The 2011 Japan tsunami flooded a nuclear power station and caused reactor meltdowns.

Beaches

Defined
Narrow strips of shore that are washed by the waves or tides, usually covered by sand or pebbles.

Beach Erosion

Defined

Less sand is supplied by longshore drift, rivers, and cliff erosion than is removed by wave action.

Natural Causes

Drought and rising sea level.

Human Causes

1. Dams that impound sediment on rivers
2. Groins, jetties, and breakwaters that impound sediment along shore
3. Hard structures, such as seawalls and revetments, that reduce coastal erosion and the sediment it supplies to beaches
4. Human-induced climate warming and the resulting rise in sea level

Mitigation

Zoning to prevent development in the coastal zone is most effective, particularly in the face of rising sea levels. Seawalls, revetments, and artificial beach nourishment are only temporary solutions. Model studies of existing structures and shorelines help us better understand the problem and thus to design and implement remedies.

Sea Cliffs

Defined

Cliffs formed and/or maintained by wave action.

Types

Active and inactive.

Erosion

Average annual rates that vary over place and time between a centimeter and tens of centimeters.

Mitigation

Structures' optimal (safest) setback from the edge of a cliff can be determined when the erosion rate is known.

Coastal Flooding

Hurricanes
Tropical storms in the Atlantic Basin generated at sea and characterized by counterclockwise winds greater than 120 kilometers per hour (74 miles/hour). They travel across the sea at velocities of 20–65 kilometers per hour (12–40 miles/hour).

Nor'easters
Winter and spring storms in the western Atlantic that are massive (can cover the entire U.S. East Coast). They are often slow moving, causing large waves to batter the shoreline for days. Can deposit both heavy snowfall inland (several feet or more than a meter) and heavy rain near the coast and offshore.

Flood Danger

A storm surge is a "mound" of water pushed up by high winds and low atmospheric pressure. It may be many meters above normal high-tide levels. Storm surges on low-lying deltas, such as the Ganges River Delta in Bangladesh, are the greatest natural hazard to human life.

El Niño (ENSO)

Ocean current change associated with floods, landslides, and accelerated beach erosion. Also associated with decreased hurricane activity in the Atlantic, fewer tornadoes in the Midwest, fewer forest fires in the Southwest, and better fruit and vegetable crops. Hawaii experiences hurricanes only in ENSO years.

Hurricane Prediction

Utilizes models of steering winds, West African climate, El Niño, sea-surface temperature, and atmospheric pressure. Heat is the fuel that drives hurricanes.

KEY TERMS

active sea cliff
barrier island
beach nourishment
bolide
El Niño–Southern Oscillation (ENSO)
fetch
fully developed sea
groin
inactive sea cliff
jetty
longshore current
longshore drift
outflow
rip current
shoaling
storm surge
surf zone
tsunami
wave base
wave crest
wave refraction

STUDY QUESTIONS

1. Why do submerged submarines not experience severe storms at sea?
2. What shoreline structures impede longshore drift and cause beach erosion? How can these structures' impact be mitigated to minimize damage?
3. Where and when do hurricanes originate, and how do they obtain their energy? Distinguish a hurricane from a nor'easter.
4. Name two types of breaker shapes and explain what causes one or the other to occur along a given shoreline.
5. How does the width of a beach vary naturally over the year?
6. Where is the sand stored that is removed from the beach in winter and returned to the foreshore in summer?
7. What is a storm surge, and why is it such a danger in low-lying coastal communities?
8. How do longshore currents develop, and how do they impact beaches and swimmers in the surf zone?
9. How can hazardous currents in the surf zone be recognized?
10. Why do warm ocean temperatures stimulate hurricane development?

REFERENCES AND FURTHER READINGS

American Geophysical Union. 2006. Hurricanes and the U.S. Gulf Coast: Science and sustainable rebuilding. Available at: http://www.agu.org/report/hurricanes/.

Bascom, Willard. 1964. *Waves and beaches.* Garden City, NY: Doubleday, 268 p.

Bearak, Barry. 2005, November 27. The day the sea came. *New York Times Magazine.*

Beardsley, Tim. 2000. Dissecting a hurricane. *Scientific American* 282(3): 81–85.

Blum, Michael D., and Harry H. Roberts. 2009. Drowning of the Mississippi delta due to insufficient sediment supply and global sea-level rise. *Nature Geoscience* 2: 488–491. doi:10.1038/ngeo553.

Bryant, Edward. 2001. *Tsunami—The underrated hazard.* Cambridge, UK: Cambridge University Press, 359 p.

Coch, N. K. 1994. Geologic effects of hurricanes. In M. Morisawa (ed.), *Geomorphology and natural hazards* (pp. 37–64). Amsterdam: Elsevier.

Davis, R., and R. Dolan. 1993. Nor'easters. *American Scientist* 81: 428–439.

Dean, Cornelia. 1999. *Against the tide: The battle for America's beaches.* New York: Columbia University Press, 296 p.

Dore, John E., Roger Lukas, Daniel W. Sadler, Matthew J. Church, and David M. Karlb. 2009. Physical and biogeochemical modulation of ocean acidification in the central North Pacific. *Proceedings of the National Academy of Sciences* 106(3): 12235–12240.

Emanuel, K. 2005. *Divine wind: The history and science of hurricanes.* New York: Oxford University Press, 296 p.

Emanuel, K. A. 2005. Increasing destructiveness of tropical cyclones over the past 30 years. *Nature* 436: 686–688.

Emanuel, K. A. 1987. The dependence of hurricane intensity on climate. *Nature* 326: 483–485.

Field, M. E., Susan Cochran, and Kevin R. Evans. 2002. *U.S. coral reefs—Imperiled national treasures.* U.S. Geological Survey Fact Sheet 025-02. Reston, VA: USGS.

Geist, Eric L., Vasily V. Titov, and Costas E. Synolakis. 2006, January. Tsunami: Wave of change. *Scientific American* 56–63.

Gonzalez, Frank. 1999, May. Tsunami. *Scientific American* 56–65.

Griggs, G., and L. Savoy (eds.). 1985. *Living with the California coast.* Durham, NC: Duke University Press.

Ignatieff, Michael. 2005, September 25. The broken contract. *The New York Times Magazine* 15–17.

Junger, Sebastian. 1997. *The perfect storm.* New York: HarperCollins, 240 p.

Kaufman, W., and Orrin Pilkey. 1979. *The beaches are moving.* Garden City, NY: Anchor Press/Doubleday, 336 p.

Kim, W., D. Mohrig, R. Twilley, C. Paola, and G. Parker. 2009. Is it feasible to build new land in the Mississippi River Delta?, EOS, v. 90, p. 373–374.

National Oceanographic and Atmospheric Administration (NOAA). 1996. *Hurricanes—A preparedness guide.* Washington, DC: U.S. Government Printing Office, 16 p.

Pilkey, O. H., and M. E. Fraser. 2003. *A celebration of the world's barrier islands.* New York: Columbia University Press, 309 p.

Pilkey, O. H., and R. Young. 2009. *The rising sea.* Washington, DC: Island Press, 203 p.

Ponton, Mungo. 1870. *Earthquakes and volcanoes.* London: T. Nelson and Sons, 328 p.

Scientific American. 1998. The oceans. *Scientific American Quarterly* (Fall).

Sheets, Bob, and Jack Williams. 2001. *Hurricane watch.* New York: Vintage Books, 331 p.

Shinn, E. A. 1993. Geology and human activity in the Florida Keys. *Public issues in energy and marine geology* (U.S. Geological Survey).

Williams, J. S., K. Dodd, and Kathleen Gohn. 1990. *Coasts in crisis.* U.S. Geological Survey circular 1075, 32 p.

Wolanski, Eric, Robert Richmond, Laurence McCook, and Hugh Swanson. 2003. Mud, marine snow and coral reefs. *American Scientist* 91: 44–51.

Grinnell Glacier, Glacier National Park, as it appears today.
Lisa McKeon, USGS

Have you ever
wondered...

11 Glaciation and Long-Term Climate Change

The ice was here, the ice was there,
The ice was all around;
It cracked and growled, and roared and howled,
Like noises in a swound!

—*Samuel Taylor Coleridge, The Rime of the Ancient Mariner (1798)*

1. Why does glacier ice deform and flow much like silly putty?

2. What causes ice ages?

3. How have scientists been able to learn so much about Earth's ancient climates?

4. How fast do glaciers move?

5. What is the cause of more frequent forest fires in recent years in western North America?

Glaciers and Climate Change

Figure 1 Grinnell Glacier in 1911. More than 120 glaciers within the park have disappeared since 1911.

Stanton Glacier National Park Archives by Kiser

Perhaps the most visually obvious example of current climate change is the worldwide retreat of mountain glaciers. Some 80% of Tanzania's Mount Kilimanjaro's ice disappeared between 1912 and 1970, and since 2000, the summit's Furtwängler glacier has become 50% thinner.[1] At this rate of thinning, it is predicted to become nothing but a damp patch by 2018. Of the 150 glaciers that graced Montana's Glacier National Park in 1910, only 25 remain. Moreover, observations reveal that Grinnell Glacier has shrunk by 90% since 1887, and the rate of ice loss is 3 to 4 times that of the 1960s. Previous computer modeling suggested the park's glaciers would vanish by 2030, but recent studies indicate they could disappear by 2020 (**Figure 1**). Spring in the park starts 45 days earlier than in decades past, and warmer winters bring rain, as well as snow. Comparison of a 1970 aerial survey with measurements by an airborne laser altimeter in 2008 reveals that about 14% of the ice and permanent snow on Washington State's Mount Rainier has melted. Of the mountain's 28 glaciers, all but 2 have thinned and shortened, the exceptions likely because of thickening resulting from large amounts of rock falling on the ice in recent years that insulate the two glaciers from warming temperatures.

Changes in mountain glaciers are significant because they are highly sensitive to temperature fluctuations. Their reaction time

[1]Kilimanjaro lies only 350 kilometers (220 mi) south of the equator. Glaciers in the tropics are especially susceptible to climate change; even slight fluctuations in temperature can have major effects. In this case, more factors than only global warming may be important. Clearing the forest surrounding Kilimanjaro for agriculture causes less moisture to be evaporated and transpired into the atmosphere, leading to reduced cloud cover and precipitation, and increased solar energy and glacier wasting.

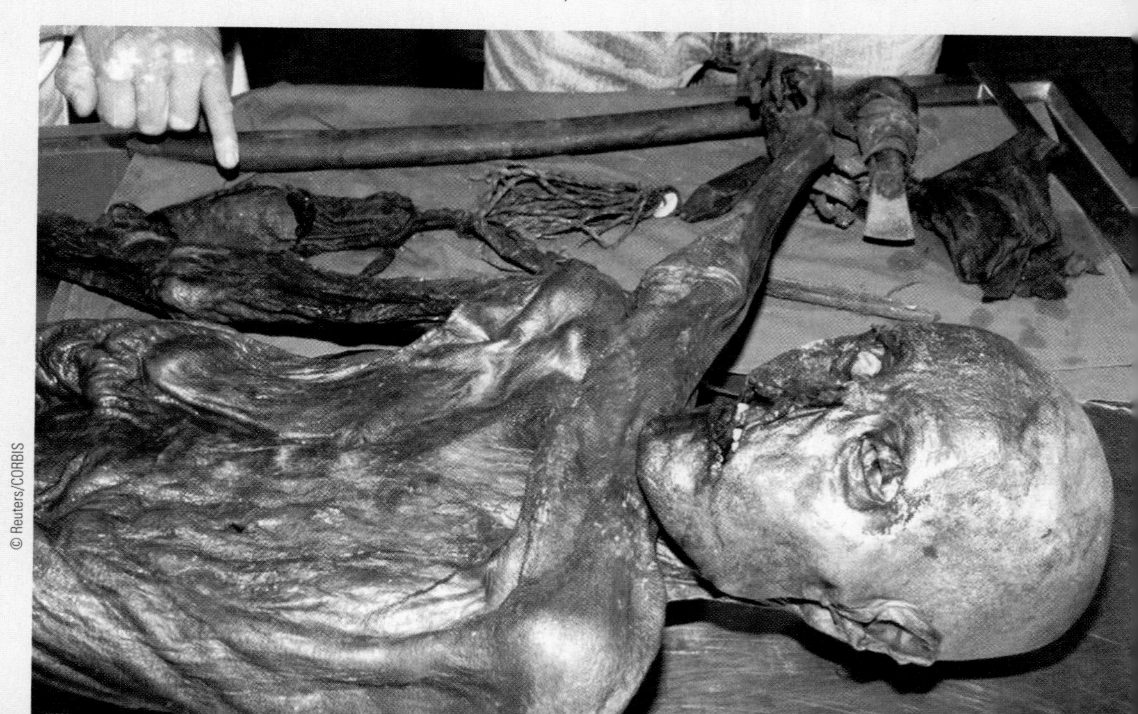

Figure 2 The "ice man," 5,300-year-old remains of a man discovered melting out of high-mountain ice by hikers in northern Italy.

© Reuters/CORBIS

is much shorter than that of the vast ice sheets of Antarctica and Greenland, and they provide significant information about 20th-century global climate change.

The discovery by two hikers in 1991 of the remains of a 5,300-year-old Chalcolithic (Copper Age) "ice man" that had melted out of the summer snowmelt on a glacier high in a mountain pass in the Austrian Ötztal Alps made eye-catching headlines (**Figure 2**). The "ice man," nicknamed Ötzi for the Ötztal (Ötz Valley) region, was remarkably intact, including his clothing, dagger, copper axe, and bow and arrows. The discovery proved to be a treasure trove of information for archaeologists about the life of a Chalcolithic person who lived and died well before the time of the ancient Greek civilization. A glacier typically melts only in its lower part; however, because Ötzi was found high up on an Alpine glacier at 3,200 meters (10,500 ft), the discovery signified to glaciologists and climatologists that the ice was undergoing intensive melting throughout as a consequence of climate warming. Such shrinkage of high-elevation glaciers in temperate regions is currently seen worldwide: in South America, the European Alps, central Asia, tropical Africa, the Himalayas, Irian Jaya, northwest America, California's Sierra Nevada, New Zealand, and Alaska.

questions to ponder

1. Worldwide, most high-elevation mountain glaciers are thinning, retreating, and disappearing. What of importance will be lost with the disappearance of mountain glaciers in temperate regions, especially those such as in Glacier National Park and Tanzania's Kilimanjaro?
2. What effect will the loss of high-altitude mountain glaciers in such places as South America, the Himalayas, and Europe's Alps have on the human scene?
3. Why are mountain glaciers in the temperate latitudes more sensitive to temperature fluctuations than the vast ice sheets in polar regions?
4. What has the discovery of the Ötzi in a high-elevation alpine glacier suggested about climate change?

National Geographic Image Collection

CourseMate Connection

Start off this chapter by listening to a podcast about climate change featuring the late National Geographic Fellow Stephen Schneider. Recorded in December 2009, this podcast includes an engaging discussion between Schneider and podcast host Boyd Matson. **Log in to the CourseMate at www.cengagebrain.com to listen now.**

LESS OBVIOUS CONSEQUENCES of global warming are the decreased snowfalls in New England, 15% since 1953; Australia, 30% since 1960; rising ocean temperatures, up to 2°F since 1985; a global rise in sea level, as much as 20 centimeters (7.8 inches) during the 20th century; and, beginning in the 1960s, the shifting of the ranges of plants, birds, and insects to higher latitudes and to higher altitudes to accommodate warming. No longer is it a question of whether global warming is happening. Instead, it is a question of what we can do about it.

Glaciers advance and retreat in response to the Milankovitch orbital shifts, cyclic changes in Earth's orbital elements (see Chapter 2), and to global climate changes, and consequently may provide evidence of past climates. To better understand global climate changes over time, and the effects of such changes on the landscape, we will first examine glaciers and glaciation.

Glaciers and Society

About 97% of Earth's water is in the oceans, and three fourths of the remainder (2.25%) is in **glaciers**. A tenth of Earth's land area is covered by glaciers, the same amount that is cultivated globally. During the Pleistocene epoch, which radiometric dating places as beginning 1.6 million years ago and lasting until about 10,000 years ago, as much as 30% of Earth was covered by glaciers (**Figure 11.1**). Today's glaciers store enormous amounts of freshwater, more than exists in all of the lakes, ponds, reservoirs, rivers, and streams of North America. Farmers in the Midwest grow corn and soybeans in soils of glacial origin, Bunker Hill in Boston is a glacial feature, and the numerous smooth, polished, and grooved bedrock outcroppings in New York City's Central Park were eroded by a glacier that flowed out of Canada (**Figure 11.2**).

Glaciers in the Himalayas, Norway, Switzerland, Alaska, Washington State, Alberta, and British Columbia have a vital effect on regional water supplies during the dry seasons. They do this by supplying a natural base flow of meltwater to rivers, which helps balance the annual and seasonal variations in precipitation. Meltwater from glaciers is a principal source of domestic water in much of Switzerland. The Arapahoe Glacier is an important water source for Boulder, Colorado, so much so that Boulder residents take great pride in telling visitors that their water comes from a melting glacier. (Actually, the glacier is only part of a drainage basin of several tributaries that together supply the city's water.)

Many lowland areas along the base of the world's glacierized mountains are dependent on glacial meltwater. The desert regions of Pakistan, northwestern India, and parts of China count on water from the ice- and snow-covered Himalayas. Many agricultural regions in South America depend on water from the glacier-covered mountains of the Cordillera (**Figure 11.3**), as do the Swiss in agricultural areas in some of the large valleys that rely on meltwater from Alpine glaciers for irrigation.

Glaciers are equally important when it comes to hydroelectric power generation in mountainous regions such as Norway, Switzerland, and Austria (**Figure 11.4**). Tunnels and pipelines carry meltwater to lakes and reservoirs where it is stored in summer to be released to hydroelectric power stations in winter when the demand for electricity is greatest. The potential impact on the food supply and power generation of the thinning and retreat of glaciers in mountainous regions due to global warming is a matter of concern.

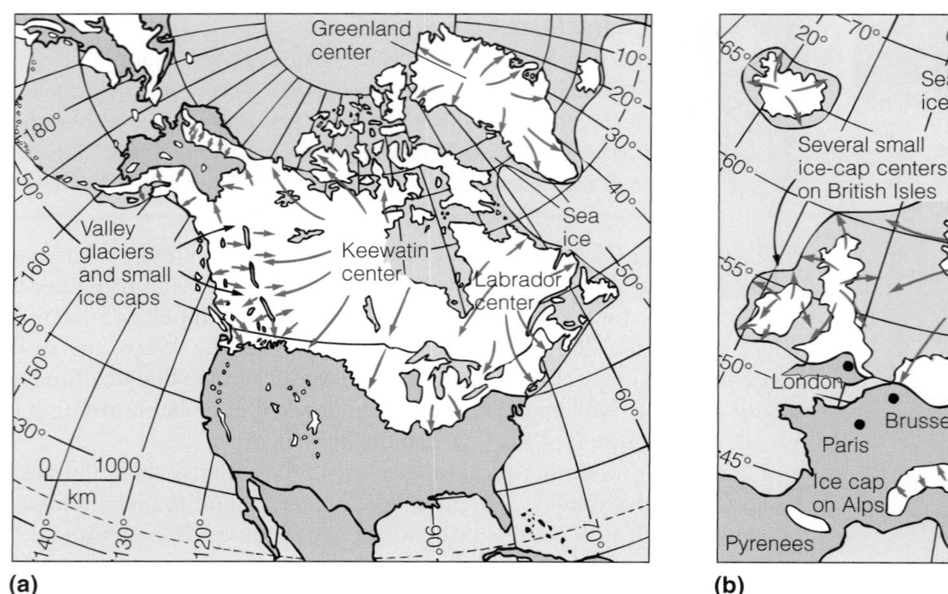

Figure 11.1 Major centers of ice accumulation and the maximum extent of Pleistocene glaciation in (a) North America and (b) Europe.

Origin and Distribution of Glaciers

Glaciers form on land where for a number of years more snow falls in the winter than melts in the summer. Such climatic conditions exist at high latitudes and at high altitudes. The pressure increases that occur as the thickness of the snowpack increases cause the familiar six-sided snowflakes to change into a coarse, granular snow called **firn** ("corn snow" to skiers). Continued accumulation increases the pressure and causes much of the trapped air to be expelled from the firn. Recrystallization into larger crystals occur, and eventually the dense crystals of "blue" glacial ice form.

Once the ice attains a thickness of about 30 meters (100 ft), it begins to deform as a viscous fluid, flowing due to its own weight. It moves downslope if in mountains, or radially outward if on a relatively flat surface. This kind of behavior of an apparently solid material, described as *plastic flow*, takes place in the lower part of the glacier called the **zone of flowage** (**Figure 11.5**). The upper, surficial layer of ice behaves in a more familiar manner as a brittle solid, often breaking into a jagged, chaotic surface of *crevasses*—ominous, deep cracks—and *seracs*—towering prominences of unstable ice. It is the deformation by plastic flow that distinguishes glaciers from the perennial snow fields that persist at higher elevations in many mountain ranges. Glaciers are more than just masses of frozen water. Although they are largely ice, they also contain large quantities of meltwater and vast amounts of rock debris acquired from the underlying bedrock or mountains where they originated and across which they have moved on their journey downslope. There are several types of glaciers, which are illustrated in **Table 11.1**.

Glacier Budget

When the rate of glacial advance equals the rate of wasting, the front (terminus) of the glacier remains stationary, and the glacier is said to be in equilibrium. Glaciers **ab-**

Figure 11.2 Striking evidence of ice-sheet erosion, a bedrock surface showing a grooved and polished rock outcrop in New York City's Central Park.

© D. D. Trent

Figure 11.3 In many parts of the world, people depend on glaciers for their water supply. Melting glaciers on Huascarán Sur, in Peru's Cordillera Blanca, supply most of the water for extensive farmlands along the base of the mountain. At an elevation of 6,748 meters (22,139 ft), Huascarán Sur is the highest peak in Peru and the fourth highest in the Western Hemisphere.

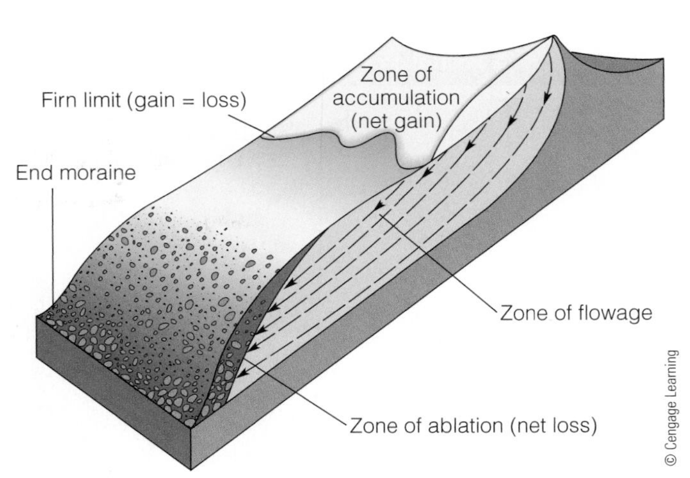

Figure 11.5 Typically, about two thirds of a glacier lies at elevations above the firn limit in the zone of accumulation, where more snow falls in winter than melts in summer. Below the firn limit is the zone of ablation, where glacier ice and the entrained rock debris are exposed by melting in the summer. The ridge of rock debris formed at the edge of the melting ice is an end moraine.

late, or waste away, by melting or, if they terminate in the ocean, by calving. **Calving** is the breaking off of a block of ice from the front of a glacier that produces an iceberg. It is important to realize that glacier ice always flows toward its terminal or lateral margins, irrespective of whether the glacier is advancing (when the net gain of snow and ice in the **accumulation zone** exceeds the net loss in the ablation zone), is in equilibrium (when the accumulation zone covers about two thirds of the glacier's total length), or is retreating (when the net loss in the ablation zone exceeds the net gain in the accumulation zone). Glaciers move at varying speeds. Where the slopes are gentle, they may creep along at rates of a few centimeters to a few meters per day; where slopes are steep, they may move 8 to 10 meters (26–32 ft) a day (**Figure 11.6**). Some glaciers, un-

Figure 11.4 A major benefit of glaciers in developed, mountainous countries is their reliability for irrigation and hydroelectric power generation. A dam at Griesgletscher (Gries Glacier) in Switzerland acquires and stores water in the summer melting season for use in generating power in the winter.

Figure 11.6 Scientists use a jet of hot water to drill a borehole in a glacier. A stake will be placed in the hole that will enable scientists to measure the glacier's movement and summer ablation.

Table 11.1 A Portfolio of Major Glacier Types

© D. D. Trent

(a) A *valley glacier* with lateral and medial moraines; Vaughan Lewis Glacier, Alaska.

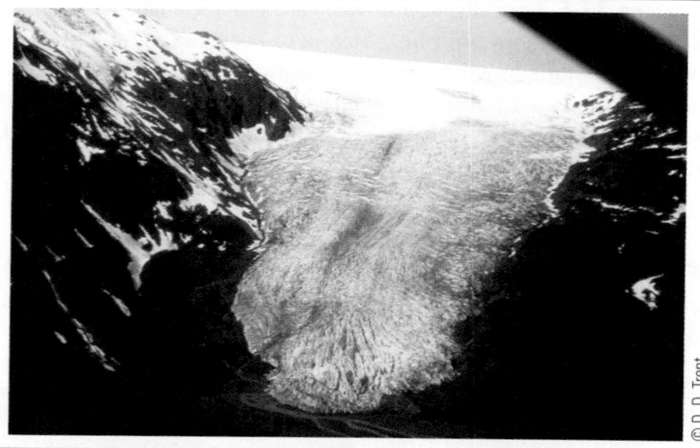

© D. D. Trent

(b) An *outlet glacier*, the Exit Glacier, flowing from the Harding Icefield; near Seward, Alaska. The grey portion is ablating glacier ice; the snow cover at higher elevation is white.

© Jürg Alean

(c) *Piedmont glaciers* spread out as wide lobes upon leaving the confines of narrow valleys and flow onto a plain; Axel Heiberg Island, Canadian Arctic.

© P. Bierman

(d) Two small *ice cap glaciers,* each flat and dome-shaped, in a highland region of west Greenland. Ice-cap glaciers cover areas of less than 50,000 square kilometers (19,300 sq mi). Note that each ice cap nurtures an outlet glacier.

Table 11.1 A Portfolio of Major Glacier Types

© B. Pipkin

(e) A *highland icefield* is a nearly continuous expanse of glacial ice, but with an irregular surface that approximately matches the underlying bedrock contours. Only the highest bedrock peaks and ridges stand above the ice; Juneau Icefield near Juneau, Alaska.

© Michael Hambry/Jürg Alean

(f) A small part of the Antarctic *continental glacier* or *ice sheet*, on which is a research base camp on the Shackleton Glacier in the Transantarctic Mountains. An ice sheet is a mass of ice and snow of great thickness that covers an area of more than 50,000 square kilometers (19,300 sq mi). The base camp, used for one field season, was thoroughly cleaned up after use, and all waste was removed.

der exceptional conditions, may episodically accelerate with speeds up to 100 meters (328 ft) a day, a condition known as *surging*.

Probably the best documented and most dramatic retreat of a glacier is found at Glacier Bay National Park, Alaska. When George Vancouver arrived there in 1794, he found Icy Strait, at its entrance, choked with ice and Glacier Bay only a slight dent in the ice-cliffed shoreline. By 1879, when John Muir visited the area, the glacier, by then named the "Grand Pacific," had retreated nearly 80 kilometers (50 mi) up the bay. The Grand Pacific Glacier had retreated another 24 kilometers (15 mi) by 1916, and today, a 104-kilometer (65 mi)-long **fjord**, an elongate glacial-eroded valley that has been drowned by the ocean, occupies the area that just 200 years ago held a valley glacier that was as much as 1,220 meters (4,000 ft) thick (**Figure 11.7**).

Glacial Features

Glaciers erode and sculpture the underlying bedrock over which they move, leaving tell-tale "footprints" of their former presence called **striations** (see Figure 11.2). Glaciers carry the eroded rock debris that eventually is deposited directly by melting ice as unsorted rock debris, or **glacial till.**

Moraines are landforms composed of till and named for their site of deposition. **End moraines** (also called *terminal moraines*) are deposited at the ends of the melting glaciers. **Recessional moraines** are a series of nested end moraines that record the stepwise retreat, or meltback, at the ends of an ice age (**Figure 11.8**). Also deposited from melting glaciers are *erratic boulders*, very large boulders that can be carried great distances from their point of origin (**Figure 11.9**) and often deposited in what seem like "erratic" places.

After melting, continental and ice-cap glaciers leave a subdued, rounded topography with a definite "grain" that indicates the direction of glacial movement. They may also leave behind numerous **kettle lakes**, water-filled, bowl-shaped depressions without surface drainage (**Figure 11.10**). For example, there are about a dozen kettle lakes in northern Indiana's Valparaiso moraine. These lakes are believed to have formed when large blocks of stagnant ice that were wholly or partly buried in the deposits left behind by retreating glaciers melted (**Figures 11.10**, **11.11**). End moraines now form the hilly areas extending across the Dakotas, Minnesota, Wisconsin, northern Indiana, Ohio, and all the way to the eastern tip of Long Island, New York. Some Chicago suburbs have names that reflect their location on high areas of hilly moraines: Park Ridge, Palos Hills, and Arlington Heights, for example. In addition, poor drainage or

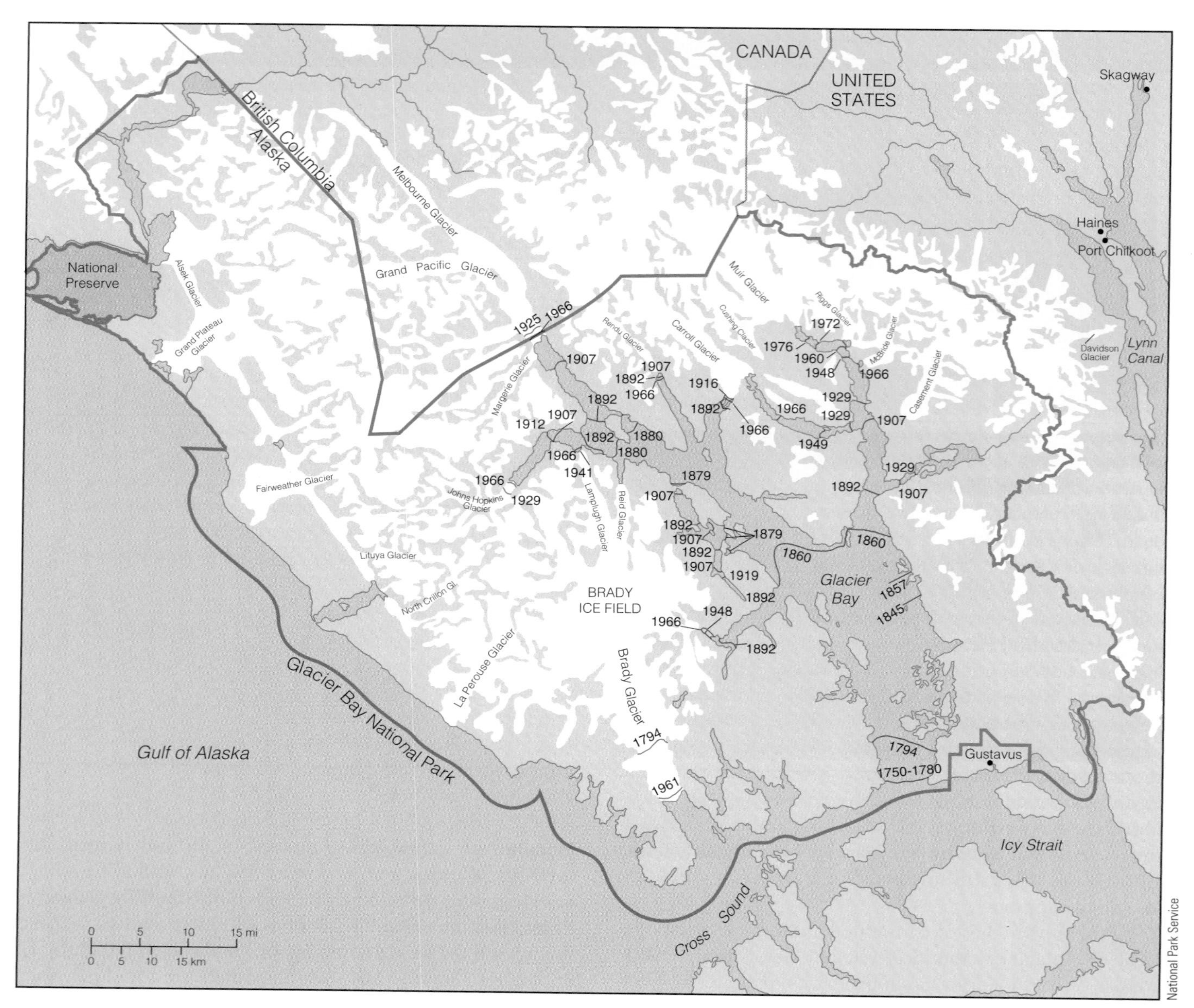

National Park Service

Figure 11.7 Map of Glacier Bay National Park, Alaska, showing the terminal positions of the retreating Grand Pacific and other glaciers from 1750 to the present. (Source: NPS data.)

University of Washington Libraries

Figure 11.8 Recessional moraines; Coteau des Prairies, near Aberdeen, South Dakota.

Figure 11.9 A glacial "erratic" that was transported by the Pleistocene valley glacier that occupied Lee Vining Canyon, California. This erratic boulder's location is more than 25 kilometers (15 mi) from the nearest outcropping of the bedrock type that could have yielded it.

deranged drainage characterizes many areas that have undergone continental glaciation.

Following a period of glaciation, a fjord forms when the seaward end of a coastal U-shaped valley, eroded by a valley glacier, is drowned by an arm of the sea (**Figure 11.12**).

consider this . . . If an Alaskan glacier's ice accumulates at an average rate of 50 centimeters a year and the glacier is 500 meters thick, how many years will be recorded in a complete core drilled to the base of the glacier?

Figure 11.10 Kettle lakes; Kewaskum, Wisconsin.

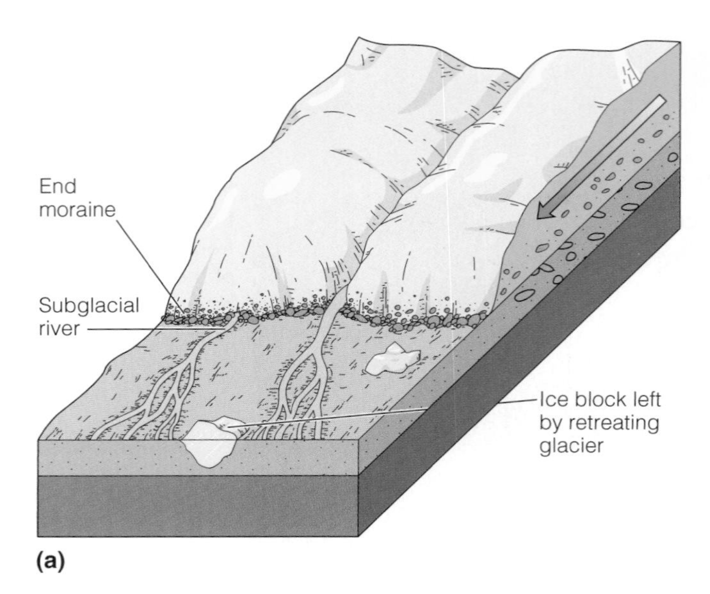

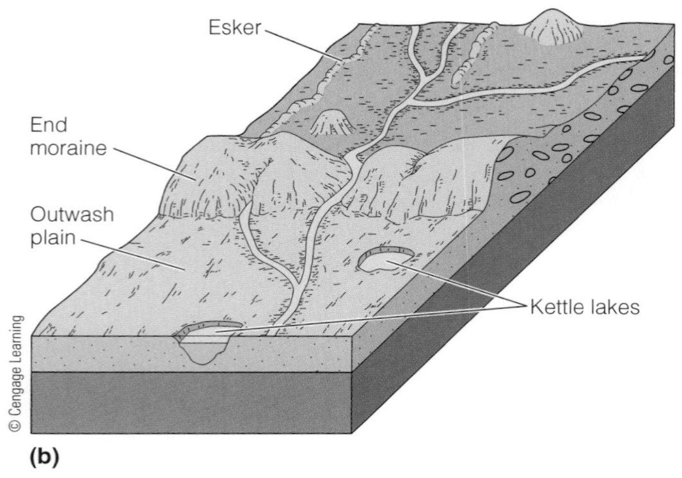

Figure 11.11 Formation of end moraines, kettle lakes, and eskers (a) during glaciation and (b) after glaciation.

Effects of Glaciation

Although the great ice sheets of the Pleistocene epoch seem remote in time and modern glaciers seem remote in distance, both have impacts on modern society that are far from trivial.

Distribution of soils—Farming and forestry generally depend on the distribution and quality of Pleistocene deposits. Some of the best agricultural soils in the world lie on sediments deposited in former glacial lakes and on glacio-eolian (loess) deposits (see Chapter 6). As Pleistocene glaciers expanded and moved over western Europe and North America, they scraped off soils from the underlying land and redeposited them as moraines of glacial till where the ice melted.

Today, some till-covered areas are excellent forestland and excellent-to-poor farmland (**Figure 11.13**). Locally, in some locations—such as south of Buffalo, New York, where the till is clay rich—the clay may become unstable in places such as road cuts along highways and flow following significant wetting.

Groundwater resources—**Glacial outwash** is sediment that was deposited by meltwater rivers originating at the melting edge of a glacier. The great Pleistocene ice sheet that extended out of Canada into the northern Mississippi River valley released copious amounts of sediment-laden meltwater, resulting in vast sheets of outwash sediments. Some of these glacial-outwash deposits serve as important groundwater aquifers in the end-moraine regions of the Midwest. Moraines associated with the outwash are poor aquifers because of the low porosity and permeability of the clays they contain. Thus, geophysical exploration, test-drilling, and careful mapping are necessary to delineate the aquifers of porous and permeable sands within the buried stream channels. Glacial outwash can serve as an important source of sand and gravel (**Figure 11.14**).

Sea-Level changes—The transfer from the oceans to the continents of the millions of cubic miles of water that became the enormous Pleistocene continental ice sheets caused sea level to lower about 120 meters (393 ft). This exposed extensive areas of continental shelves. Consider that a drop of this amount today would increase the area of Florida real estate by more than 30%, would leave New Orleans high and dry, and would cause the Panama Canal to drain. Conversely, if all the polar ice were to melt, sea level would rise on the order of 40 meters (130 ft), and London, New York, Tokyo, and Los Angeles would be mostly under water. As sea level dropped during the ice ages, the rivers in coastal regions began downcutting across the exposed continental shelves as they adjusted to the new conditions. When sea level rose to its present level, about 5,000 years ago, the mouths of the coastal

Figure 11.12 A fjord; Gastineau Channel at Juneau, Alaska.

© D. D. Trent

© D. D. Trent

Figure 11.13 Rocky soil developed on glacial till; Adirondack Mountains, New York.

Figure 11.14 Part of a vast plain, underlain by Pleistocene glacial outwash near Elgin, Illinois. Outwash sediments, deposited along the margins of the melting continental ice sheets, are spread for thousands of square kilometers across the north-central United States. The ponds in the photograph are the excavations remaining from sand and gravel mining operations.

© B. Pipkin

Figure 11.15 Isostatically raised beach ridges; Richmond Gulf, Hudson Bay, Quebec, Canada. The highest ridge in the photo lies about 300 meters (984 ft) above sea level and is about 8,000 years old.

rivers were drowned, forming what are now harbors and estuaries.

The formation of land bridges caused by the lowering of sea level during the ice ages was of major importance in establishing the current biogeography of Earth. Among the land bridges at this time was the Bering bridge. It had existed during much of Cenozoic time and is especially significant because the two-way traffic across the bridge accounts for horses, mammoths, and other large mammals' crossing between North America and Asia. About 20,000 to 30,000 years ago, during the lowered sea level of the last major ice age, humans used the bridge to cross from Asia into North America. Rising sea level beginning about 13,000 years ago drowned the bridge, and the 90-kilometer (54 mi) stretch of ocean between Alaska and Siberia now blocks temperate-climate organisms from migrating.

Isostatic rebound—The Ice Age continental glaciers and ice caps, with thicknesses of 2 miles or more, were so massive that the crust beneath them was depressed by their weight. Since the ice retreated at the end of the last ice age, 8,000 to 11,000 years ago, the land regions exposed by the retreating ice sheets in Scandinavia, Canada, and Great Britain have risen because of the geologically rapid removal of the great load of ice. Uplift of the crust because of unweighting is caused by **isostatic rebound**, the slow transfer by flowage of mantle rock to accommodate uplift of the crust in one area that causes subsidence in another area. Evidence of isostatic rebound in deglaciated areas is seen in the raised beaches around Hudson Bay, the Baltic Sea, and the Great Lakes (**Figure 11.15**). Careful surveying in Europe shows that, as Scandinavia and Great Britain have been rising, a corresponding subsidence of Western Europe has occurred and is especially pronounced in the Netherlands. The uplift and subsidence in these regions remain active. Above the modern shorelines of the Gulf of Bothnia in Finland and Sweden are raised beaches that show a maximum uplift of 275 meters (900 ft). It is estimated that an additional 213 meters (700 ft) of rise will occur before equilibrium is reached. The rise is so rapid in some places in Scandinavia that docks used by ships are literally rising out of the sea. Uplift in Oslo Fjord, Norway, is about 6 millimeters (1/4 in) per year, which adds another 60 centimeters (2 ft) of elevation per century. Sea-level mooring rings the Vikings used for tying up their dragon boats a thousand years ago are now 6 meters (20 ft) above sea level. Not even the tallest Viking would be able to moor his boat to those rings today. Similar uplifting is recorded along the northern shores of Hudson Bay, where the average uplift is about a meter (3 ft) per century (**Figure 11.16**).

Uplifting in the Great Lakes region is slow but continuing, and eventually it may cause problems for the residents

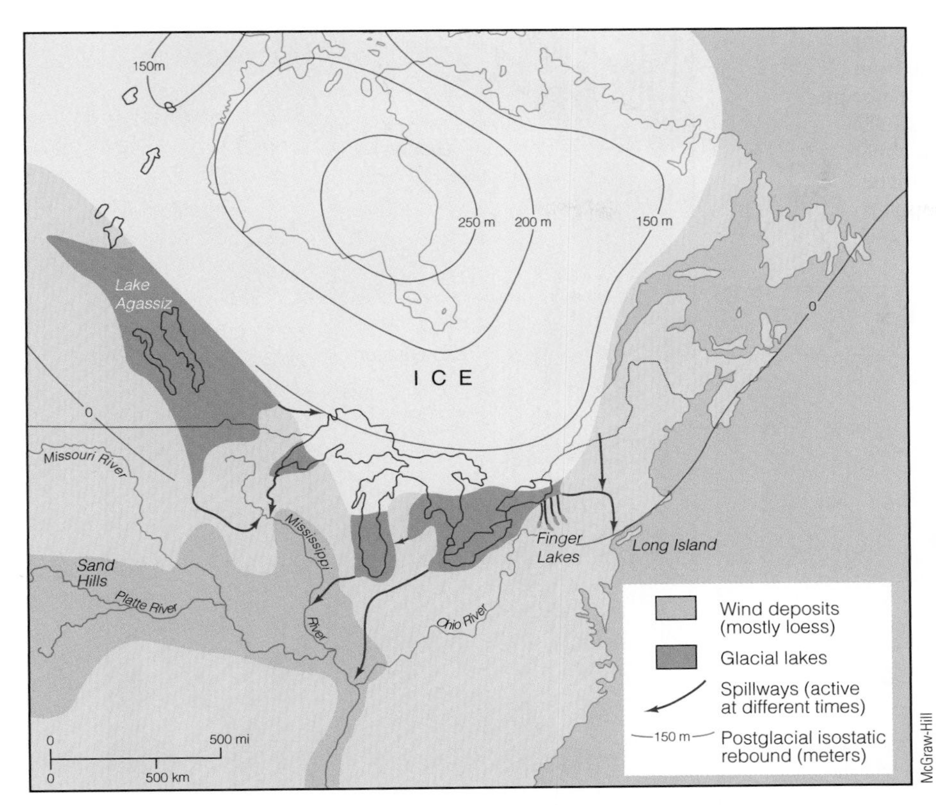

Figure 11.16 Locations of maximum postglacial (after glaciation) isostatic uplift in the Hudson Bay region and of land features resulting from the retreat of the late Pleistocene ice sheet about 10,000 years ago. Proglacial lakes (temporary, ice-marginal lakes) overflowed at different times through various spillways. One of the arrows east of the Finger Lakes represents the spillway pictured in Figure 11.21. The spillway draining Lake Agassiz in South Dakota is pictured in Figure 11.22. Windblown loess was deposited adjacent to major glacial meltwater systems. Waters from Lake Michigan drained southwestward into the Mississippi River at the time, which may occur again as isostatic uplift continues. (Source: Dott and Batten. 1981. *Evolution of the earth*, 3rd ed, New York, McGraw-Hill.)

of Chicago and elsewhere in Illinois. The rebound threatens to cause a drainage change that will force more of Lake Michigan to drain southward into the Mississippi River system.

Glacier Bay, Alaska, is one of the fastest rising places in the world because of the rapid unloading of the crust caused by the recent melting of the area's glaciers. Isostatic adjustment in the last 200 years has resulted in measurable rebound in the lower parts of the bay: the high tide line has moved almost a mile out to sea in only a few decades, small islands are slowly emerging, and existing islands are rising at the rate of about 4 centimeters (1-1/2 in) per year. Near Juneau, Alaska, there's a downside: Water tables lower, channels dry up, salmon have nowhere to spawn, and property owners find they have new and different property lines to argue about.

Human transportation routes—Beginning in 1980, the Columbia Glacier on Prince William Sound, Alaska, began a catastrophic retreat by calving thousands of icebergs each year, some of them as large as houses. The glacier had been in equilibrium at much the same position for at least 200 years—it was first observed in the late 18th century—when it began to show evidence of retreat in the late 1970s. Fortunately, the moraine shoal at its terminus is sufficiently shallow to trap the larger icebergs and prevent them from drifting into Prince William Sound, where they would pose a major threat to the supertankers passing through these waters to and from the oil-loading terminal at Valdez. Because the terminus of the glacier has retreated some 15 kilometers (9 mi) since 1982, it appears that a new fjord is forming.

Juneau, Alaska, is the only state capital in the United States without highway access, even though it is only about 160 miles (by air) west of the nearest highway in Canada. A highway connecting Juneau with Canada would be desirable, but it is not to be because the advancing Taku and Hole-in-Wall glaciers threaten to close off the only possible route along the west shoreline of Taku Inlet (**Figure 11.17**).

Whereas glaciers impede transportation in some areas, in other areas they have served to facilitate it. In wet boggy regions in the early days of New England, the crests of **eskers** provided good travel routes. Eskers are long, winding, steep-sided ridges of stratified sand and gravel deposited by subglacial or englacial streams that flowed in ice tunnels in or beneath a retreating glacier. Hence, especially in Maine, the eskers were called *horsebacks* (see **Figures 11.llb** and **11.18**).

Pleistocene lakes—The cooler climates of the Pleistocene brought increased precipitation, less evaporation, and the runoff of glacial meltwater. Lakes dotted the basins of North America. Accumulations of water, mostly in valleys between the fault-block mountain ranges in the Great Basin, formed hundreds of lakes far from continental glaciers (**Figure 11.19**). Today, many of these valleys contain aquifers that were probably charged by waters of the Ice Age lakes. California had dozens of lakes, now mostly dried up; among them were Owens Lake, Lake Russell (the ancestor of Mono Lake), and China Lake. Lake Manley covered the floor of what is now Death Valley National Park, named after the Manly party, which suffered great hardships there in 1849; and Rogers Dry Lake, near Lancaster, California, is the location of Edwards Air Force Base and a landing site

Figure 11.17 Map recording the behavior of the retreating Norris Glacier and the advancing Taku Glacier, Alaska, from 1890 to 1990. The reason for the different behavior of the neighboring glaciers is that the Norris Glacier has a much smaller and lower-elevation accumulation zone than the Taku. Only 2% of the Norris's accumulation zone is above 1,370 meters (4,500 ft), whereas about 40% of the Taku's is above that elevation.

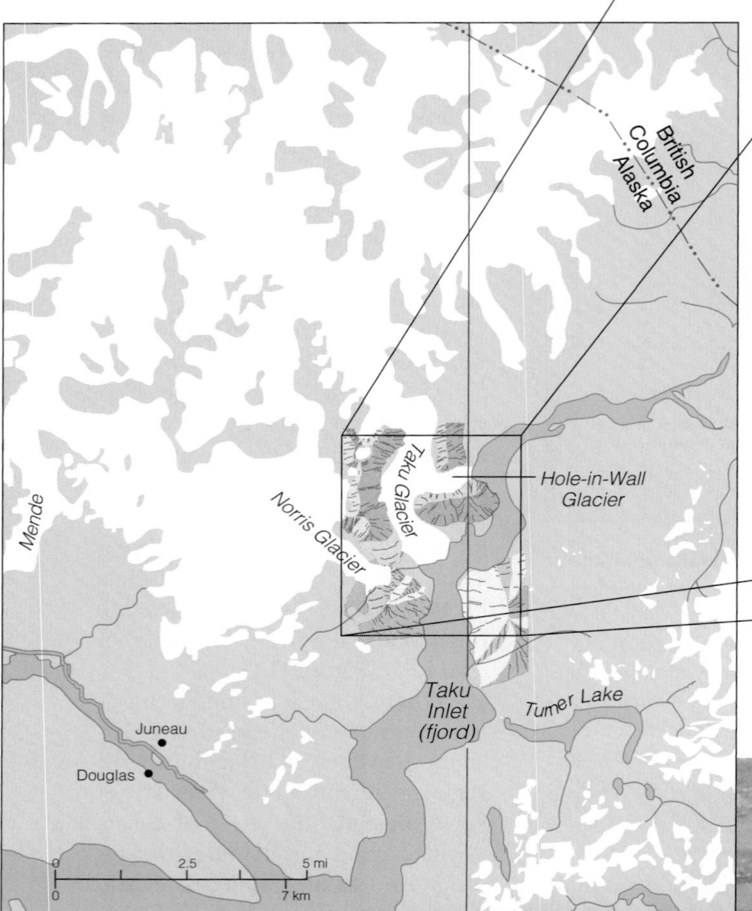

Figure 11.18 Esker with a road on its crest. Malingarna area, southern Sweden.

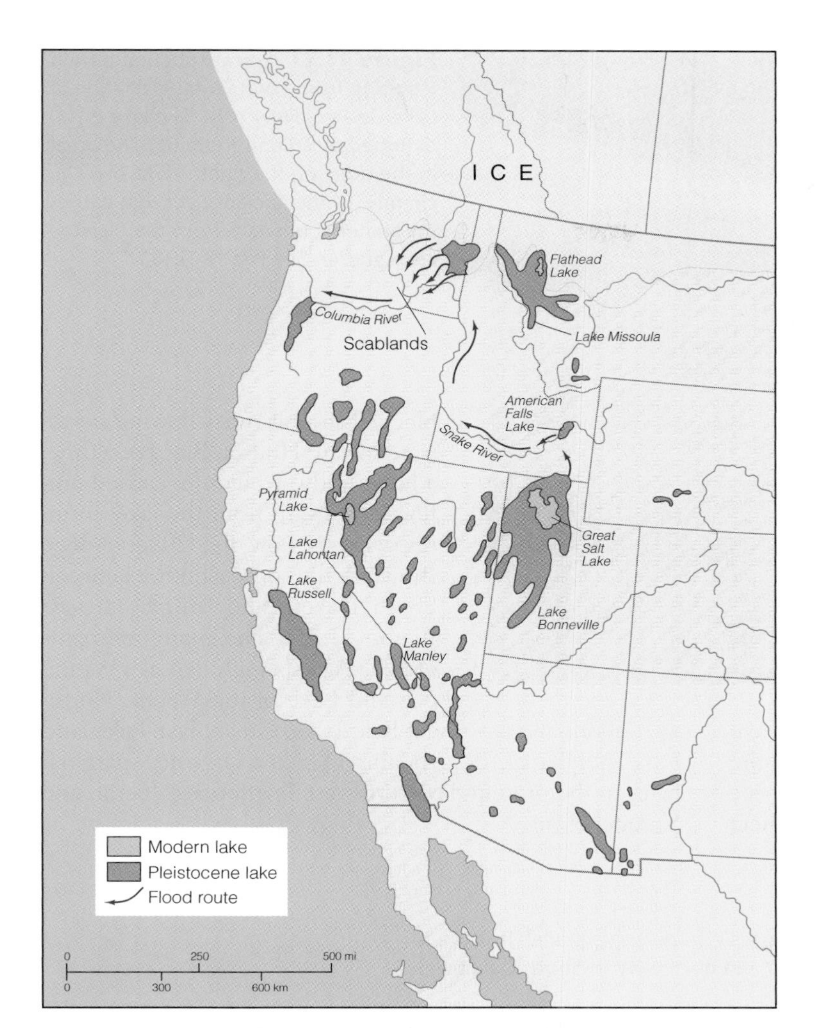

Figure 11.19 Pleistocene pluvial lakes in the western United States. (After P. L. Weiss & W. L. Newman. 1973. "The Channeled Scablands of Eastern Washington," USGS pamphlet.)

for the space shuttle. The Great Salt Lake in Utah is all that remains of the once vast Pleistocene Lake Bonneville (**Figure 11.20**). The Bonneville Salt Flats, an area that is ideal for setting automobile speed records, was formed by evaporation of the ancient Pleistocene lake. Farther east, the Great Lakes and the Finger Lakes of New York State are of glacial origin. In the Pleistocene epoch, when drainage to the north from the Great Lakes and the Finger Lakes was blocked by ice, these lakes drained through various spillways at different times (see **Figures 11.1** and **11.21**). Major disruptions in drainage produced large, temporary ice-marginal lakes, called **proglacial lakes**, which were dammed partly by moraines and partly by glacial ice.

Ice-age lake sediments are the basis of the rich soils of the northern United States and southern Canada. The rich wheat lands of North Dakota and Manitoba, for example, formed as sediments in Pleistocene Lake Agassiz (**Figure 11.22**). Centered in present-day Manitoba, the lake was the largest known North American lake; its area of more than 518,000 square kilometers (200,000

Figure 11.20 Terraces on the flanks of the Wasatch Mountains that mark former shorelines of Pleistocene Lake Bonneville; Brigham City, Utah. See Figure 11.19.

© D. D. Trent

© D. D. Trent

Figure 11.21 Typical subdued terrain resulting from continental glaciation at Lake George, New York. The lake is dammed by an end moraine. The break in the ridge on the right marks the channel of the ancient river that carried meltwater southward from the Pleistocene glacier. See Figure 11.16.

mi^2) was more than twice that of the five present-day Great Lakes combined. (It was named after Louis Agassiz, the 19th-century scientist who fostered the theory of the ice ages.) The lake formed when the great Canadian ice sheet blocked several rivers flowing northward toward Hudson Bay. For a time, a large spillway channel carried outflow southward from the lake, forming what is now the 200-foot-deep Browns Valley on the border between South Dakota and Minnesota (see Figure 11.22). The many remnants of Lake Agassiz include Lake Winnipeg and Lake of the Woods. Northwest of this area are Great Bear Lake, Great Slave Lake, and Lake Athabaska, descendants of other large ice-marginal lakes in the present-day Northwest Territories, Alberta, and Saskatchewan.

Figure 11.22 Big Stone Lake now occupies part of Lake Agassiz's spillway channel. It is in Brown's Valley on the boundary between western Minnesota and northeastern South Dakota. See Figure 11.16.

University of Washington Libraries, Special Collections, John Shelton Collection

If there had been no large-scale Pleistocene continental glaciation in North America, what differences might we find in today's midwestern states? (Hint: Carefully examine Figure 11.1a).

Breaking the Ice: Evidence of Climate Change

Glacial ice provides evidence of cyclic changes in atmospheric chemistry, temperature, and amounts of aerosols (atmospheric dust or fine liquid droplets), which correlate with periods of cooling and warming during the Pleistocene epoch. As snow accumulates, it packs down into glacial ice, preserving atmospheric information that becomes locked into the ice. The relative abundance of two slightly different oxygen atoms—the oxygen-18 and oxygen-16 isotope ratio—symbolized $^{18}O/^{16}O$—in ice molecules indicates the atmospheric temperature when the snow fell from which the ice formed (see Chapter 2). Other atmospheric gases such as carbon dioxide (CO_2), nitrogen oxide (N_2O) and methane (CH_4), dust, volcanic ash, and even traces of pesticides become entrained in glacial ice.

The bubbles of ancient atmosphere trapped in glacial ice obtained from 3,348-meter (10,981-ft)-deep cores drilled at Vostok Station in Antarctica, at the Greenland Ice Sheet Project 2 (GISP2), and at the Greenland Ice-Core Project (GRIP) (**Figure 11.23**) provide a 420,000-year record of variations on the atmosphere's temperature and composition that catalogs four full glacial-interglacial cycles (see Chapter 2). In 2004, a consortium of European scientists, the European Project for Ice Coring in Antarctica (EPICA), almost doubled the earlier Vostok climate record by pulling a new core from Dome C, about 300 miles from Vostok Station (**Figure 11.24**). The bubbles of trapped CO_2 in the Vestok and EPICA ice cores (see Chapter 2 and Figure 11.23) reveal a record of at least eight glacial cycles (**Figure 11.25**).They also clearly show that the amount of **greenhouse gases (GHGs)** in the atmosphere, especially CO_2, and the temperature have gone hand in hand. When CO_2 amounts are high, so is the temperature, and vice versa. This relationship is described by NASA Goddard Institute of Space Studies (GISS) climatologist Gavin Schmidt as a coupled system: "Changes in the climate affect levels of CO_2, and CO_2 levels also change climate. The timing of these cycles is set by variations in the Earth's orbit, but their magnitude is strongly affected by greenhouse gas changes and the waxing and waning of the ice sheets."[2]

In spite of what seem to be large natural variations in CO_2 and CH_4 concentrations, in the approximately 10,000 years from the end of the last glacial stage to the beginning of the Industrial Revolution, the air bubbles in the ice record that concentrations of CO_2 stayed within a narrow range of 260 to 280 ppm, and CH_4 varied from only 0.6 to

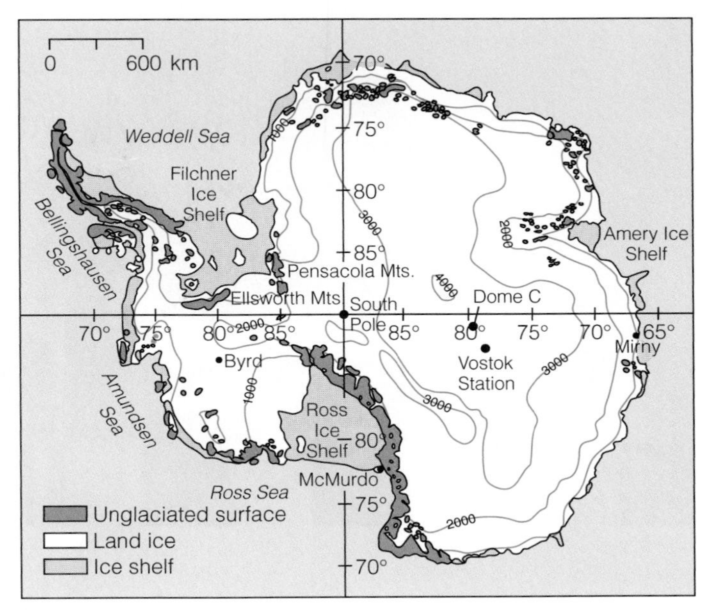

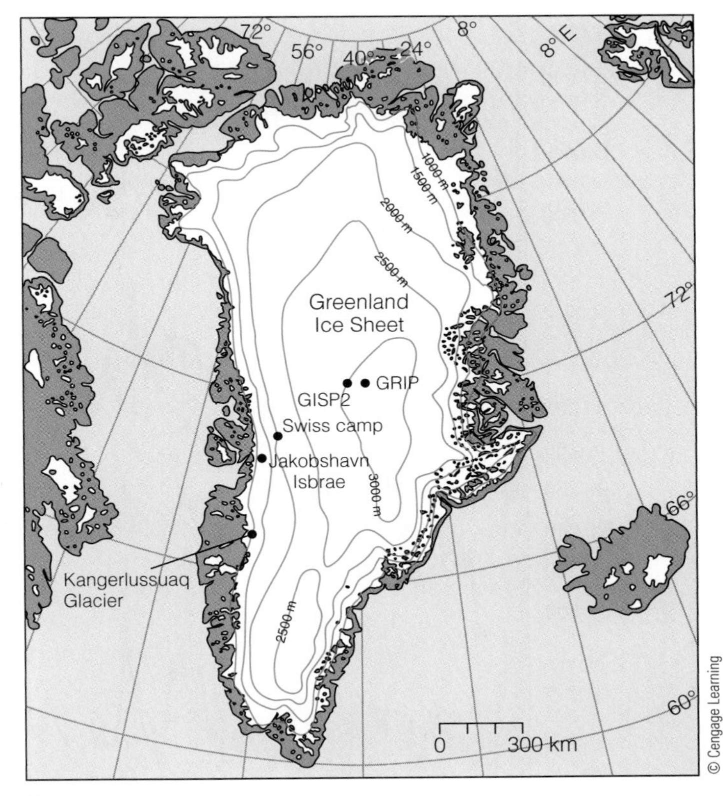

© Cengage Learning

Figure 11.23 Earth's two continental glaciers. (a) The largest glacier complex covers practically all of Antarctica, a continent that is more than half again as large as the 48 states. The ice sheet's thickness averages about 2,160 meters (7,085 ft), with a maximum of 5,000 meters (16,500 ft). (b) Greenland's ice sheet has a maximum thickness of 3,350 meters (10,988 ft). Elevations are contoured in meters. Vostok, Dome C, GRIP, GISP2, Swiss Camp, Kangerlussuaq Glacier, and Jakobshavn Isbrae are localities discussed in the text and Case Study 11.1.

[2]Gavin Schmidt and Joshua Wolfe. 2009. *Climate change: Picturing the science* (p. 144). New York: W. W. Norton.

Figure 11.24 Scientist Michelle Koutnik, of the University of Copenhagen's Center for Ice and Climate, preparing a core of Antarctic ice to be wrapped and put into core tubes for shipment to labs at Brigham Young University in Utah.

0.7 ppm. With the dawn of the industrial age, however, this remarkable stability ended abruptly (Chapter 2 and **Figure 11.26**).

On a human time scale, the ice reveals that, since pre-industrial times, the level of CO_2 has increased about 40% (see Figure 11.26), and CH_4 has increased more than 100%. The warm interglacial period following the termination of one

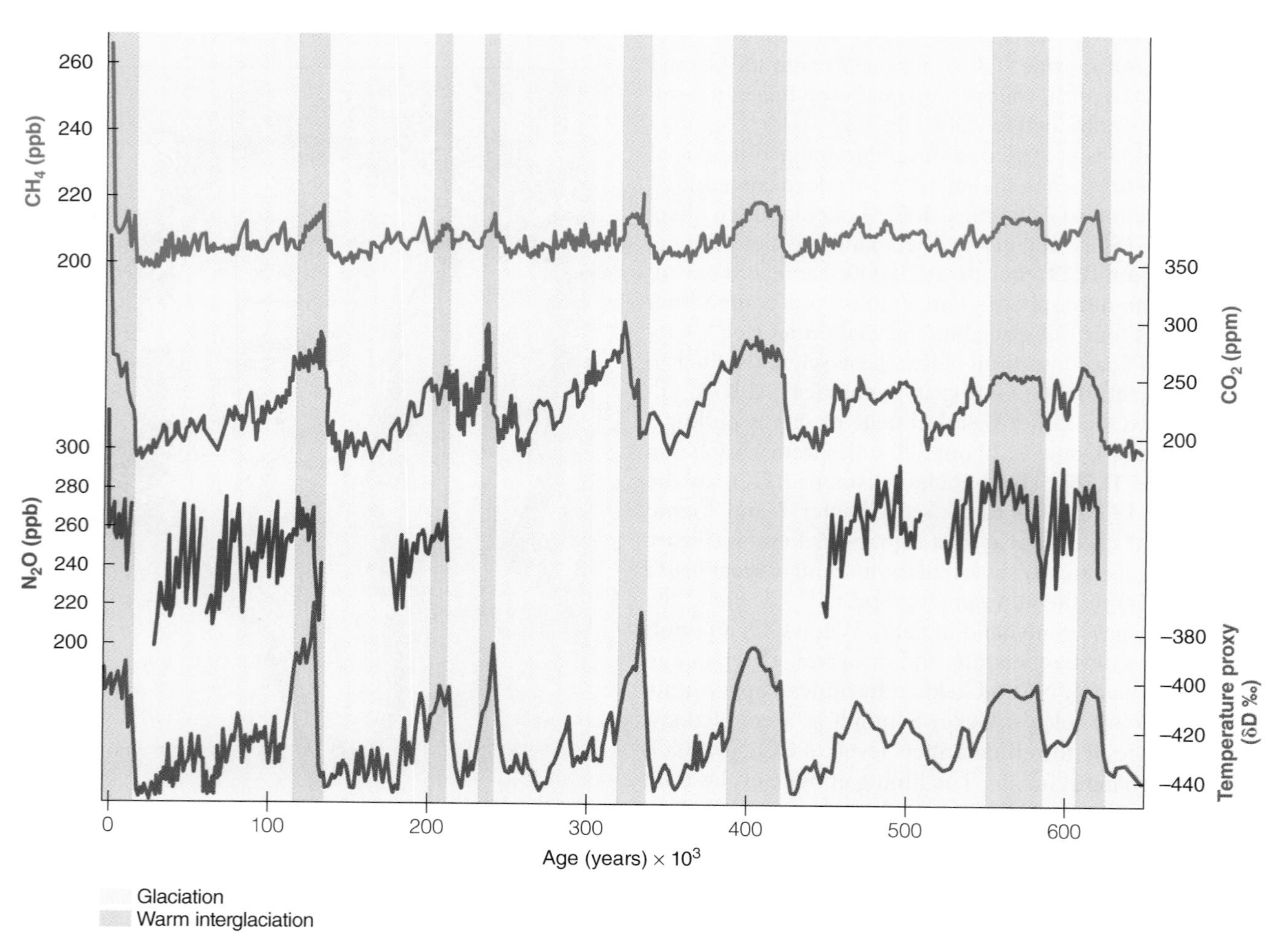

Figure 11.25 The greenhouse gases (CO_2, CH_4, and NO_2) and deuterium (δD) records for the past 650,000 years from the EPICA Dome C and the Vostok ice cores, Antarctica. δD, a proxy for air temperature, is the deuterium/hydrogen ratio of the ice, expressed as a per mil deviation from the value of an isotope standard—the greater negative values indicate colder conditions. ppb = parts per billion; ppm = parts per million. (From Brook, 2005, Nov. 25. "The Long View," *Science* 310: 1285. Reprinted with permission from AAAS.)

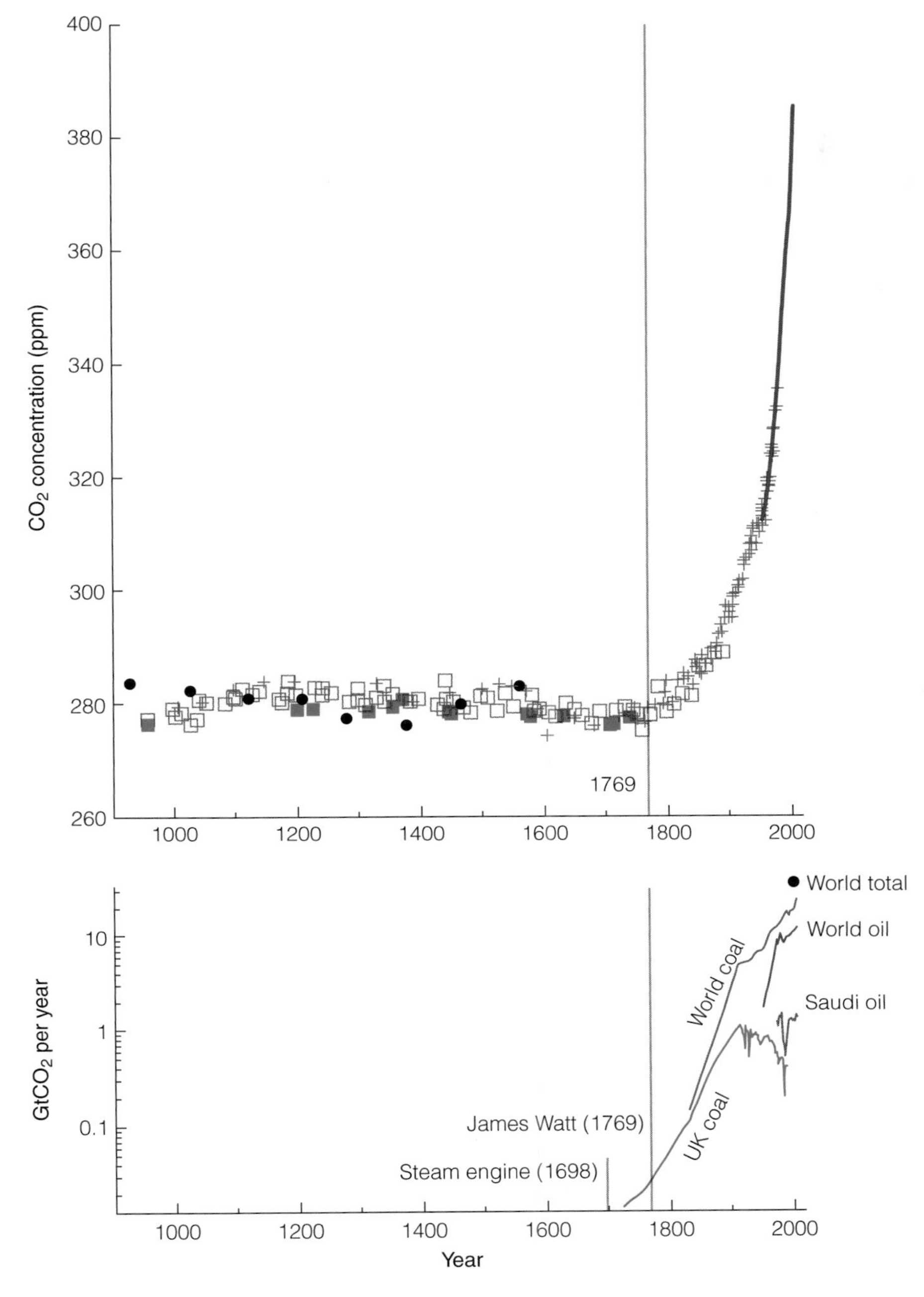

Figure 11.26 The upper graph shows CO_2 concentrations in parts per million (ppm) for the last 1,100 years measured from air bubbles trapped in glacier ice cores (up to 1977) and directly in Hawaii (from 1958). The year 1698 is the date of the first practical steam engine; 1769 marks the year that James Watt patented his more efficient coal-fired steam engine, which marks the start of the Industrial Revolution. The lower graph (on a logarithmic scale) shows the history of world coal production, United Kingdom coal production, Saudi oil production, and the total world greenhouse gas emissions in 2000. All production rates are expressed in billion tons of CO_2 per year. (*Sustainable Energy – without the hot air,* by David JC MacKay, published by UIT: **www.uit.co.uk/** sustainable Also available free to download for personal non-commercial use from **www.withouthotair .com**)

glacial stage was exceptionally long—28,000 years—in contrast with the 12,000 years so far in the present interglacial stage. The similarities discovered in the character of the earlier warm interglacials and that of today suggests that, without human intervention, a climate similar to what we have enjoyed for about the last 200 years might extend well into the future.

Among the discoveries in the ice core record is that increases in GHGs *followed* the onset of warming conditions by several hundred years. Skeptics of climate change, in their attempts to discredit increasing GHGs as the explanation for our current warming, claim the lag proves that climate change is not from increasing CO_2.

This chicken-and-egg situation, discussed previously in Chapter 2, is that factors related to Earth's orbit can set into motion natural processes that release GHGs and significantly boost atmospheric heating. Specifically, because CO_2 is more soluble in cold water, the ocean stored large amounts of CO_2 during the cold periods of the Ice Age. As the ocean warmed, because of the Milankovitch cycle, the CO_2 gradually was emitted to the atmosphere. The release of CO_2 began in the high southern latitudes around Antarctica, where the CO_2 mixed into the atmosphere, eventually spreading to the tropics and the northern latitudes. The record preserved in marine cores confirms this explanation: The increase in tropical temperature lags behind the southern warming by about 1,000 years. Thus, the CO_2 lag amplifies the warming effect and most importantly illustrates the sensitivity of the climate to "tipping points" where warming the ocean leads to a positive feedback to global warming and stimulates an additional warming feedback (Chapter 2).

Glaciation and Climate Change

By early in the 20th century, it was recognized worldwide that Pleistocene—commonly referred to as the Ice Age—glacial deposits of different ages exist. The oldest, weathered and gullied from erosion, have been found beneath younger ones, which are fresher appearing, less weathered, and less eroded. By applying the Laws of Superposition and Cross-

Cutting Relationships to these observations, it is clear that multiple stages of glaciation occurred during the Pleistocene epoch. It is now generally accepted that there have been at least four major glacial stages in North America and Eurasia. Independently, oceanographers studying fossil shells contained in cores of seafloor sediments from three oceans have determined that the oceans experienced 20 or more periods of deep cooling with intervening warm periods during the Pleistocene. Because evidence of glaciations on land is easily destroyed by later ice advances, it is assumed that the continental record is probably incomplete.

Global temperatures during the Pleistocene ice ages averaged only about 4°C to 10°C (7–18°F) cooler than the temperatures at present (Chapter 2). The lowering of the average Ice-Age temperature was not as significant as the range of seasonal extremes; winters were much colder, but summers were moderate. Consequently, climate zones shifted toward the equator, accompanied by similar shifts of plant and animal communities. The Rocky Mountains, the Sierra Nevada, the Alps, the Caucasus, and the Pyrenees were topped with ice fields similar to the modern ice field in the mountains near Juneau, Alaska. Small mountain glaciers existed in high mountains of the midlatitudes much closer to the equator than any similar modern glaciers. Even the summit of Hawaii's Mauna Loa volcano, only 19° north of the equator, shows evidence of glaciation. It also was cooler south of the equator, but because the landmasses in the Southern Hemisphere's middle latitudes are small or narrow, extensive glaciation was limited to the South Island of New Zealand and the southernmost part of the Andes in South America. Antarctica and Greenland were ice covered much the same as they are today.

Changes in modern glaciers—Modern glaciers provide relevant information because they respond to climate changes by either advancing or retreating. Their behavior serves as an indicator of climate change and as a "filter," smoothing out the record of seasonal and annual variations in temperature and precipitation. Because of their small size and ice volume, mountain (alpine and ice-cap) glaciers in the mid- and low latitudes are remarkably sensitive indicators of climate change: They respond quickly even to small perturbations in climatic elements. A worldwide survey of 160,000 mountain glaciers and ice caps reveals that the volume of the world's glacier ice is declining and the rate of loss is accelerating. For instance, in Glacier National Park in Montana's northern Rocky Mountains, glacial fluctuations have been studied for a century, with reports dating back to 1914. By 2006, all but 25 of the park's 150 ice fields had melted since 1850, with the cold slivers that remain estimated to disappear about 2020 at the current rate of shrinking. Similar findings in midlatitudes are reported from Switzerland, where alpine glaciers have lost as much as half their mass since 1850; the Caucasus Mountains of Russia, where glacial ice decreased by about 50% during the 20th century; and New Zealand, where, on average, the 127 glaciers in the Southern Alps have become 38% shorter and 25% smaller in area than when first studied in the 20th century. At low latitudes, the largest glacier on Africa's Mount Kenya has lost 92% of its mass in the last few decades.

At higher latitudes, Alaskan glaciers are in dramatic retreat, having thinned twice as fast in 2000–2005 than during the preceding years. Using a laser altimeter attached to an airplane, a USGS research team flew over 67 glaciers in Alaskan mountains to compare their present surface elevations with those that were mapped in the 1950s. Since the 1950s, most glaciers show several hundred feet of thinning at low elevations and about 18 meters (60 ft) of thinning at higher elevations. Furthermore, the glaciers were thinning twice as fast between 1997 and 2002 as they did from the 1950s to the mid-1990s. This equates to a rise in sea level of 0.1 millimeter (0.004 in) per year from the mid-1950s to the mid-1990s, making Alaska the largest contributor to rising sea level of any ice-bound region on Earth.

Today's Global Warming

Earth's currently retreating glaciers may signal that the climate system has exceeded a critical threshold and that most low-latitude, high-altitude glaciers are likely to disappear in the near future.

Lonnie G. Thompson et al. 2006. *"Abrupt tropical climate change: Past and present."* Proceedings of the National Academy of Sciences 103(28): *10542.*

As I said on the Senate floor on July 28, 2003, "much of the debate over global warming is predicated on fear, rather than science." I called the threat of catastrophic global warming the "greatest hoax ever perpetuated on the American people," a statement that, to put it mildly, was not viewed kindly by environmental extremists and their elitist organizations.

Senator James Inhofe, January 4, 2005

Climate change in modern times is no longer a theoretical idea discussed at international conferences by science nerds. It's happening right now, right here on Earth. Not only is global warming causing major changes in glaciers, but in the United States, in 2011, a dozen weather-related disasters *each* caused more than $1 billion in damage, a record, according to NOAA. The next two sections consider the evidence for, and causes of, global warming.

Evidence of Global Warming

Increasing temperatures of Earth's surface and lower atmosphere—Temperature records from land areas and oceanic shipping lanes reveal that the average Earth temperature increased 0.74 ± 0.18°C (1.33 ± 0.32°F) during the 20th century, with half of that increase after 1970 (**Figure 11.27**). The year 1998 was the warmest year of the

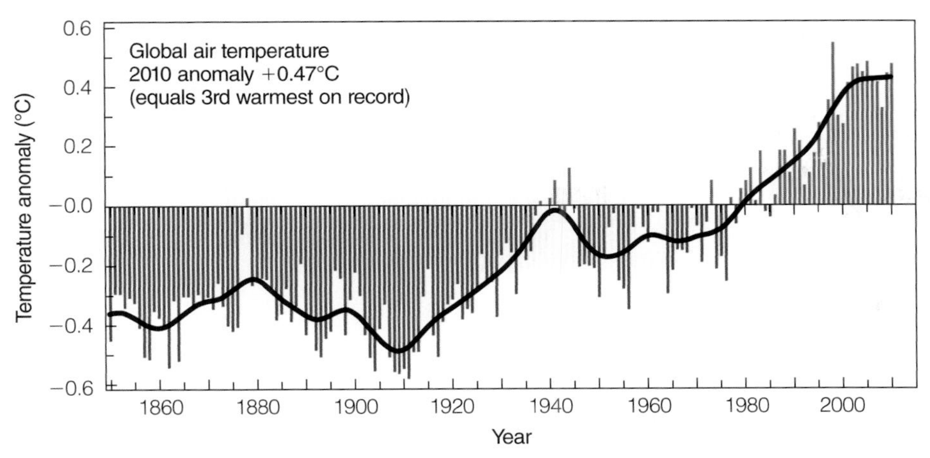

Figure 11.27 The global and ocean surface temperature record from 1850 to 2010. The year 2010 tied with 2003 for the third warmest year on record, exceeded by 1998 and 2005. The years 2003, 2005, and 2010 are distinguishable only by the third decimal place. The error estimate for individual years (two standard errors is about ±0.1°C) is at least 10 times larger than the differences between these 3 years. The warmest 2 years are 1998 and 2005, with the former the warmest at 0.546°C (0.98°F) above the 1961–1990 mean and the warmest year in 341 years. Data are from proxy climate series (ice cores, corals, tree rings), historical and instrumental records. (Courtesy of Phil Jones and Jean Palutikof, Climatic Research Unit, University of East Anglia, and the UK Meteorological Office.)

160-year record, and 1998 and 2005 were the warmest years in the last six centuries. After 1998, the next warmest years in the series were all within the decade 2001–2010 and the years 2003, 2005, and 2010 are distinguishable only by trivial differences in the third decimal place. Although not quite the warmest, 2010 set the record for the warmest 6 months, with 19 nations setting new all-time temperature records; the mercury in Pakistan reached 54°C (129°F), the highest temperature ever recorded in Asia. During this decade, only 2008 is not in the 10 warmest years, but it was still the 12th warmest year of the entire 160-year record. The relative warmth or coolness of individual years is strongly influenced by whether there was an **El Niño** or **La Niña** event occurring in the equatorial Pacific Ocean.

The greatest amount of warming occurred at night in the higher latitudes. Some areas of Canada, Alaska, and Eurasia warmed as much as 5.5°C (10°F) between 1965 and 2011. At high elevations in Glacier National Park, temperatures have risen three times as fast as the global average (see Figure 1 in chapter opener). By 2010, the average midwinter temperature at Palmer Station on the western shore of the Antarctic Peninsula had increased 6°C (10.8°F) since 1950; this is the highest rate of warming anywhere on Earth, about eight times the global average. These differences may seem small, but such changes in the past have had major geological consequences as previously noted.

Earth's present average temperature is the warmest it has been since 1400, which is as far back as we can go with certainty by direct and indirect evidence. Some authorities assert that this may be the warmest period in 1,000 years (see Figure 11.25).

Most temperature reconstructions, such as those shown in Figures 11.25 and 11.26, come from temperature proxies, such as tree rings. A temperature history that is totally independent of instrumental or proxy data was constructed from the records of the lengths of 169 glaciers from different regions of the world. The glaciers of the European Alps, where glaciers have been observed and monitored the longest—beginning in 1534 for one Swiss glacier—provided 93 records, and northwestern America provided 27, with the remaining records from glaciers in Africa, Scandinavia, central Asia, New Zealand, and elsewhere. Unlike biogenic climate proxy data, such as corals and tree rings, glaciers react to climate change in a rather simple manner. By using glacier dynamic theory that treats fluctuations in glacier length as due mainly to precipitation and air temperature, it is possible to calculate the temperature history for any record of glacier length. The temperature record from this reconstruction provides complementary evidence on the timing of the onset of global warming and the magnitude of global warming, and it shows that warming appears to be independent of elevation. The derived temperature record is in general agreement with other temperature reconstructions derived from proxy and instrumental data (**Figure 11.28**).

Alaska's Columbia Glacier moves about 24 meters (80 ft) a day, making it one of North America's fastest moving glaciers. It is the last of Alaska's 51 tidewater glaciers—glaciers that terminate in the ocean and waste away by calving—to exhibit a catastrophic retreat. In 1981, the Columbia Glacier began a drastic retreat and by 2011 had shrunk back more than 20 kilometers (12 miles) from its former terminus (**Figure 11.29**). Accompanying the retreat was about 500 meters (1,600 ft) of thinning at the terminus. Between 1982 and 2000, the glacier's speed increased from 5 to 30 meters (16–98 ft) per day, and the calving rate increased from 3 million to 18 million cubic meters (106–635 ft^3) per day. It is expected to shrink back another 10 kilometers (6 miles) within a few decades, eventually to reach a static equilibrium in shallow water near sea level. The rapid disintegration of this glacier is attributed to a complex set of factors dealing with glacier dynamics, not simply the direct result of global warming, although it is believed that the retreat was triggered by warming. The Columbia Glacier is North America's largest contributor to rising sea level.[3]

But mountain glaciers, such as the Columbia, contain only 6% of the world's ice. Of more concern are the enor-

[3]A time-lapse photo sequence showing the rapid retreat of the Columbia Glacier from October 2007 to November 2010 is available at: www.extremeicesurvey.org/index.php/new_gallery/timelapse_31/

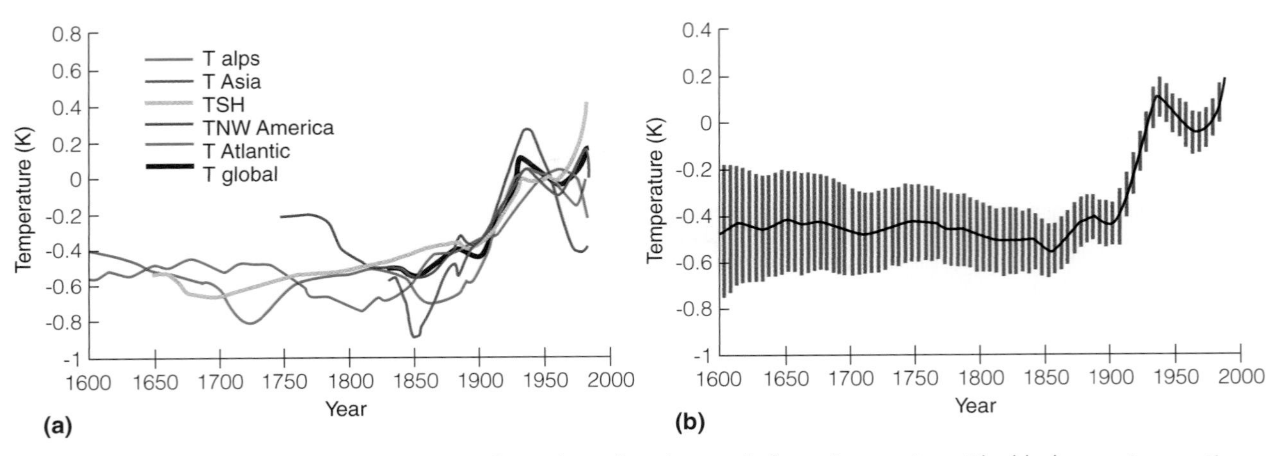

Figure 11.28 (a) Temperature reconstruction from glacier length records for various regions. The black curve is an estimated global average temperature value. (b) The best estimate of the global average temperature from glacier length records. The band indicates the estimated standard deviation, with the changing band width reflecting the increasing number of glacier records beginning in the late 19th century. T = temperature; SH = Southern Hemisphere. (From Extracting a Climate Signal from 169 Glacier Records by J. Oerlemans. 2005, *Science* 29 April 2005: 675–677. Copyright © 2005, The American Association for the Advancement of Science. Reprinted with permission from AAAS.)

mous continental glaciers of Greenland and Antarctica, which together contain about 90% of the world's freshwater. It is estimated that, if only 10% of the water in these enormous ice sheets were to melt and be added to the ocean, sea level would rise by over 6 meters (20 ft) and cause coastal devastation worldwide and force the displacement of hundreds of millions of coastal dwellers. James Hansen, a highly respected senior climate scientist at NASA GISS, states that we are seeing the disintegration of the Greenland and West Antarctica ice sheets, "both losing ice at the rate of about 150 million cubic kilometers per year," which is adding

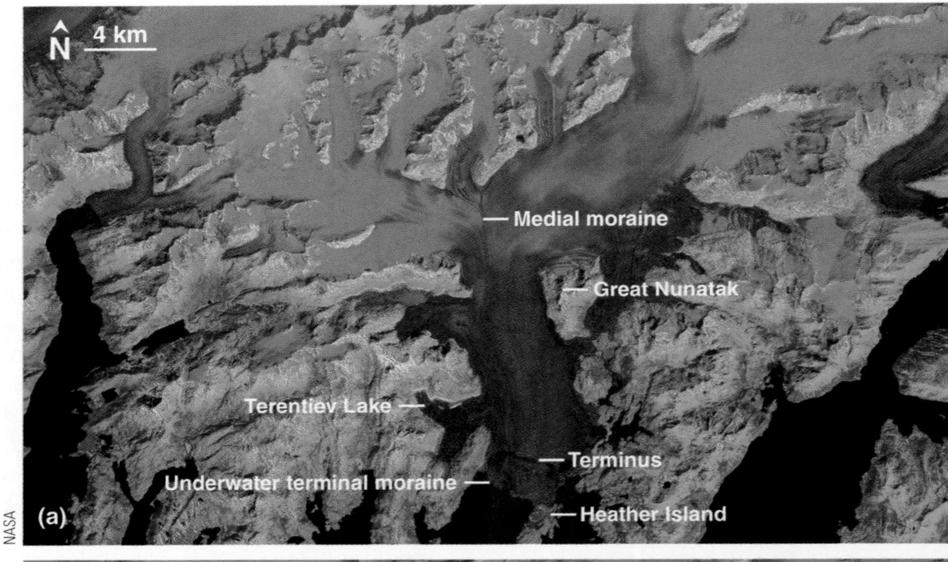

Figure 11.29 (a) The disintegrating Columbia Glacier, Alaska, in July 1986, six years after beginning its catastrophic retreat. When first surveyed in 1794, its terminus was at Heather Island. (b) The glacier in May 2011, when it had retreated more than 20 kilometers (12 miles) and thinned substantially as shown by the expansion of brown bedrock. It is expected to retreat an additional 10 kilometers (6 miles) within a few decades. These are false-color satellite images: vegetation is green, ice and snow are bright cyan, clouds are white or orange, and the ocean is dark blue.

water to the ocean with "sea level rise now going up about 3.5 centimeters (1.4 in) per decade . . . that's more than double what it was 50 years ago." These ice sheets are responding to the higher average global temperature, which is more than 0.5°C (1°F) above the 1940–1976 mean.

The Greenland ice sheet is a true remnant of the vast continental glaciers of the last Ice Age. It contains about 2.5 million cubic kilometers, or 10%, of the world's ice—that's enough water to raise the global sea level about 7 meters (23 ft). In the summer of 2002, along with the Arctic sea ice, Greenland's glaciers shrank by a record 400,000 sq mi, the largest decrease ever recorded. This revelation was followed in 2004 by the discovery that, since 1997, some of Greenland's outlet glaciers—types of valley glaciers that are nurtured by spillover from large ice-cap or continental (ice sheet) glaciers—had thinned as much as 15 meters (49 ft) per year and were melting 10 times faster than previously thought. In addition, tongues of ice from several outlet glaciers, each several hundred meters thick and floating several kilometers out to sea, have disintegrated in recent years, reducing the buttress effect holding back the inland ice. That ice is now sliding down channels toward the sea at a pace never before seen. Satellite radar interferometric measurements of ice speed reveal that the flow rates of several of Greenland's glaciers have about doubled since about 2000 (**Figure 11.30**). For example, in 2001, the Jakobshavn Glacier was already the world's fastest moving glacier, flowing from land to sea at 7 kilometers (4 mi) per year. By July 2010, however, it had accelerated to an unglacial 15 kilometers (9 mi) per year, and its calving terminus had retreated 10 kilometers (6 mi). Combined, the effect has been a greater discharge of meltwater from the Greenland ice sheet as a whole, increasing from about 50 cubic kilometers (12 mi^3) per year to as much as 239 cubic kilometers (57 mi^3) a year, more than 200 times the amount of freshwater used each year by the city of Los Angeles (**Figure 11.31**).

Figure 11.30 The Jakobshavn Isbrae, the largest of Greenland's outlet glaciers, drains 6.5% of the ice sheet. The glacier switched to fast forward in 2004 and now races to the sea at 15 kilometers (9 mi) per year; it is rapidly disintegrating where it enters the ocean.

The conventional wisdom among geoscientists has long been that, in the event of a warming climate, it would take centuries for higher air temperatures to work down through thousands of feet of ice, weaken the ice cap's stability, and turn it to mush and water. In addition, even with a warming climate, a warmer, more humid atmosphere would promote more snowfall at high elevations across the summit of the ice sheet, offsetting the loss caused by melting at lower elevations; the ice sheet could thus maintain a kind of dynamic equilibrium for thousands of years. But this is not happening. The movement of Greenland's glaciers occurs in distinct lurches, which are recorded by seismographs as microearthquakes and even stronger shakes, one reaching a magnitude of 5.2. The incidence of these ice quakes is seasonal, with five times as many in summer as in winter. Such timing suggests that something other than internal glacier dynamics accounts for the increased speeds. The rate of ice quakes has doubled since 1991, beginning about the same time as the increase in surface melting. Evidently, surface meltwater works its way downward through thousands of feet of ice to the rocky base of the Greenland ice sheet. Because of the rough, rocky bed, the basal ice moves in a stick-slip manner, lurching downslope only when enough water builds up to raise the hydraulic pressure enough to momentarily lift the glacier over the irregularities (see Figure 11.31). The microseisms record these lurching events. Greenland's glaciers are feeling the effects of global warming and, depending on the magnitude of warming, the ice may melt in a far shorter time than centuries, raising sea level much faster than was anticipated as recently as even a few years ago. The contribution of meltwater from Greenland's ice sheet to sea-level rise is a matter of considerable scientific and societal significance.

Consequences of Global Warming

Climate, weather, predictions, forecasts, and projections—Before looking at the consequences of global warming and future climate conditions, we must define some terms. Much confusion exists in the media and in the general population over the use and meanings of the terms *climate*, *weather*, *predictions*, *forecasts*, and *projections*. The use of these terms in the scientific community is quite precise. *Climate* encompasses the statistics of temperature, wind, humidity, atmospheric pressure,

SPOTLIGHT FIGURE 11.31 Glaciation and Long-Term

Vanishing Ice

Greenland's ice sheet once seemed too big to melt substantially. Now, as weather warms, the ice is disintegrating faster. Scientists think this could continue for centuries or even millennia, changing the island's geography and the planet's sea level.

Melting Down

SOLAR RADIATION
As average summer temperatures rise in Greenland, more ice melts. Surface water and dark cryoconite absorb solar heat, speeding melting.

CREVASSE
A crack, or crevasse, that forms as the ice moves over uneven bedrock can plunge hundreds of feet. A flood of meltwater can drill it even deeper.

MOULIN
A shaft, or moulin, forms when meltwater weakens ice and suddenly flushes to unseen depths from the bottom of a surface lake or via a crevasse.

HYDRAULICS BELOW
The ice travels on a thin layer of water. When a draining moulin floods the base of the ice sheet, the ice begins to surf on that water and picks up speed.

THE WATER'S JOURNEY
Meltwater that works its way to the bedrock eventually turns into rivers that can carve ice tunnels many feet high as they flow to the sea.

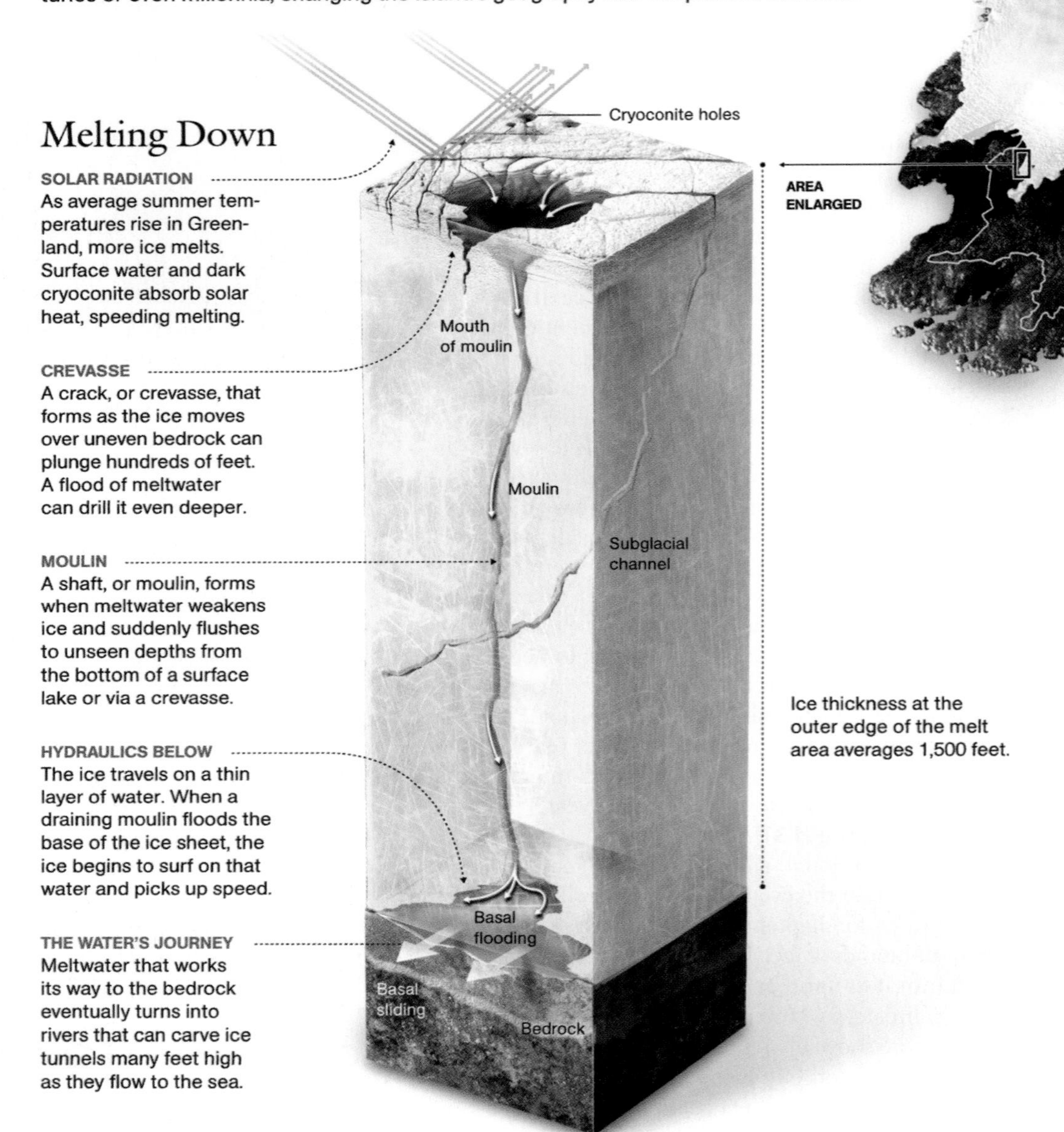

Origins of the Ice Sheet

50-10 million years ago
A warm climate bathed coastal mountains, rivers, and a high central plateau.

3-2 million years ago
As the climate gradually chilled, glaciers began to form in the mountains.

About 2 million years ago
The land continued to grow colder, and glaciers spread across parts of the central plateau.

2 miles
1
0

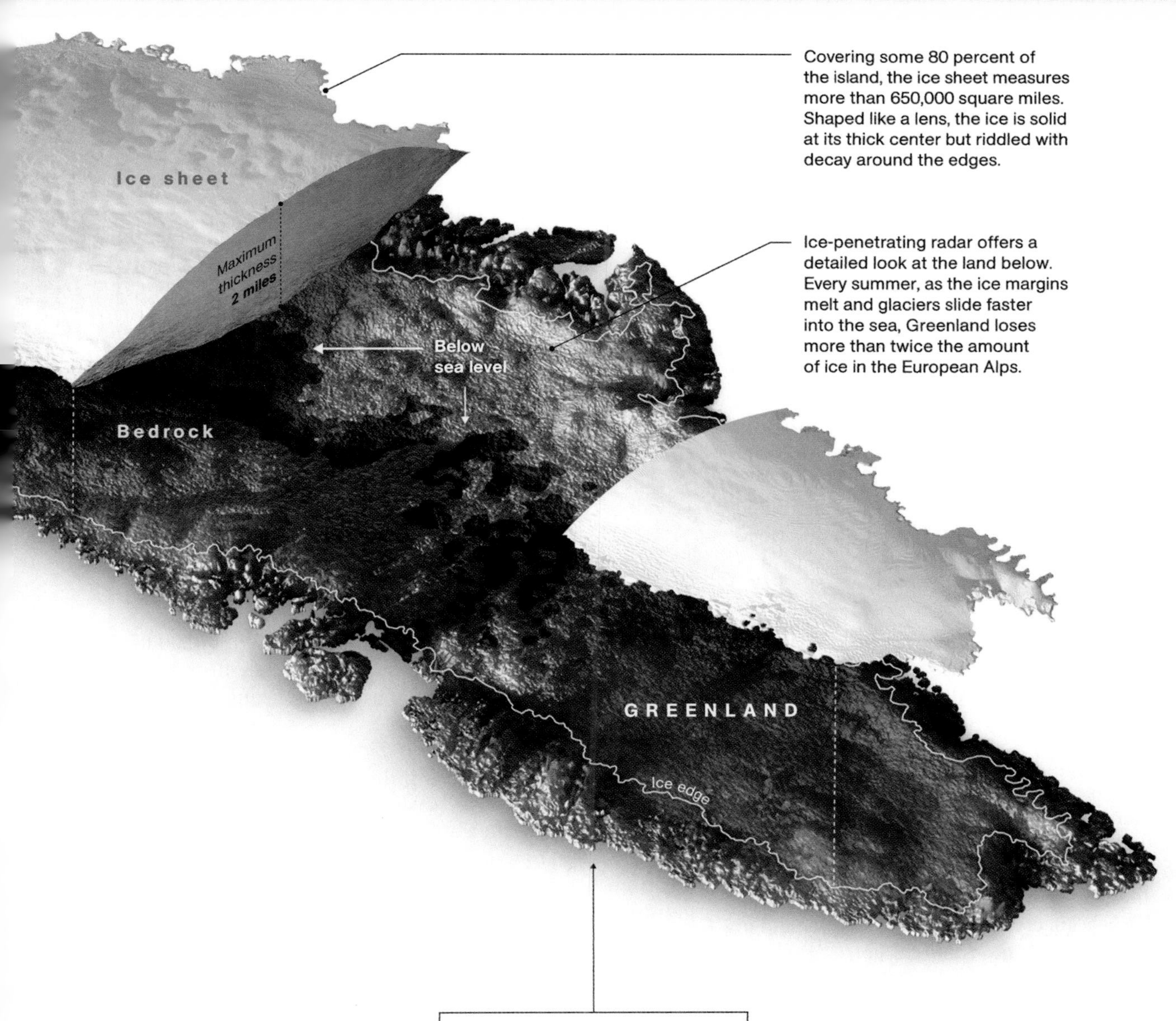

Covering some 80 percent of the island, the ice sheet measures more than 650,000 square miles. Shaped like a lens, the ice is solid at its thick center but riddled with decay around the edges.

Ice-penetrating radar offers a detailed look at the land below. Every summer, as the ice margins melt and glaciers slide faster into the sea, Greenland loses more than twice the amount of ice in the European Alps.

Annual Melt Area

Almost three decades of satellite data reveal an escalating trend in the extent of the ice-melt zone—the part of the ice sheet where melting occurs between April and October. The rest of the year, the entire ice sheet is frozen. Even during the warmer months much of the ice refreezes overnight.

2007

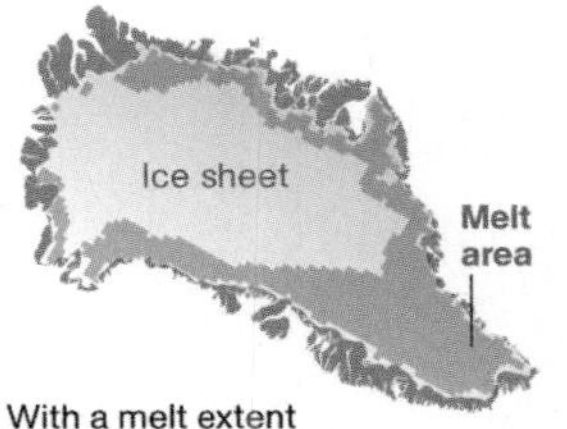

With a melt extent of at least 40 percent, 2007 set a record.

Scenarios for the year 2100
Climate change projections

Medium Estimate:
4.5°F rise in global temperature

Ice sheet

High Estimate:
9°F rise in global temperature

2-1 million years ago
Separate glaciers merged to form a sheet of solid ice that covered most of the island most of the time.

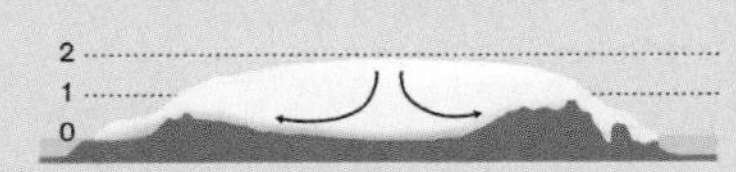

The solid ice sheet has depressed underlying rock by some 3,000 feet. Its weight also squeezes out a thin lip of ice to the sides.

If melting continues at its current rate, the central plateau will slowly begin to rebound as the weight of the ice lifts from it.

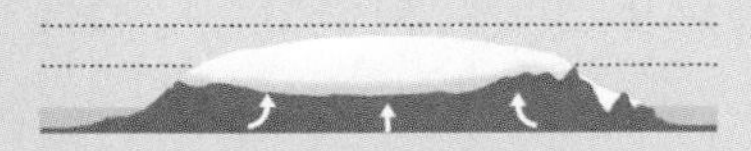

National Geographic Maps/National Geographic Image Collection

rainfall, atmospheric particle count, and other meteorological measurements in a given region over long periods of up to several centuries. In contrast, *weather* is the condition of these elements and their variations over shorter periods of a few days to a few weeks. The bitterly cold winters of 2009–2010, 2010–2011, and February 2012 experienced in some regions of the Northern Hemisphere are good examples of weather. Such was the case in the first 2 weeks of February 2012, when temperatures in parts of Europe nose-dived to –30°C (–22°F), with rooftop-high snow drifts trapping thousands of villagers in their homes, cutting off access to entire towns, and causing nearly 600 deaths. Such cold had not been seen in several decades, and some people claimed the frigid weather marked the end of climate warming. Nonetheless, at the same time as those extremely cold periods, there were less newsworthy, but far more prominent, extreme warm spells elsewhere. A study examining the daily wintertime temperature extremes since 1948 found that the warm extremes were considerably more severe and widespread than the cold extremes during the Northern Hemisphere winters of 2009–2010 and 2010–2011. Moreover, the study concluded that although the extreme cold could be attributed to a natural climate cycle (the Arctic Oscillation described later in this chapter), the extreme warmth was consistent with longer term climatic warming trends.

Thawing in boreal and subarctic regions—Climate changes in Alaska and the Bering Sea region have already had socioeconomic impacts on Alaskans, most of them negative—for example:

- With less sea ice in the Bering Sea, weather events such as severe storm surges, like those accompanying the near-hurricane conditions in November 2011, have increased in frequency and severity, causing coastal erosion, inundation, and destruction of coastal villages and structures.
- Subsistence lifestyles of the native peoples in the region have been adversely affected by such things as changes in sea-ice conditions. Hunting on the ice, for example, is more dangerous.
- Slope instability, landslides, and erosion are increasingly common in thawing permafrost terrain. This threatens roads and bridges, and causes flooding.
- The warmer climate has increased forest-fire frequency and insect outbreaks, reducing timber yields and increasing loss of CO_2 to the atmosphere.
- The speed of permafrost thawing has caused major changes in the tundra, bog, and forest landscape features and wetland ecosystems in subarctic North America and Siberia (**Figure 11.32**). For example, the southern limit of permafrost in the region of James Bay, Canada, has moved 130 kilometers (81 mi) north since about 1960. As Earth warms up in response to millions of tons of CO_2 and other GHGs entering the atmosphere each year, so does the permafrost (see Chapter 6). Once permafrost thaws, bacteria begin eating the organic matter it contains and in so doing release CO_2 and methane, another GHG with 25 times the warming power of CO_2. Also released from the thawing permafrost are phosphates and nitrates, which alter the ecosystem by helping novel plant types become established. Scientists estimate that the world's permafrost contains twice as much CO_2 as the atmosphere. If even a small amount of that were released as CO_2 gas, along with the methane, it would be a strong positive feedback that would strengthen global warming. In recent years, the incidence of tundra fires has increased. The 2007 fire that raged on Alaska's North Slope left a 1,000-square-kilometer (386 mi^2) path of blacked earth. That fire pumped about 1.3 million tons of CO_2 into the atmosphere. Scientists anticipate Arctic fires will become more frequent as tundra and other permafrost plant communities become drier because of global warming (**Figure 11.33**).
- Since about 1960, the number of days that oil company personnel can travel on Alaska's frozen North Slope to search for oil prospects has shrunk in half—from 200 to 100. The tundra, which must be completely frozen before running vehicles on it, has been freezing later and thawing earlier.

Rising sea level—During the last glacial maximum, about 20,000 years ago, sea level was about 120 meters (393 ft) lower than the current level, and its rate of rise has varied. Tide stations around the world reveal that sea level has risen by nearly 20 centimeters (8 in) since 1880. Beginning in 1993, sea level has been accurately measured by satellites, which show sea level rising at the rate of 29.6 centimeters (1.17 in) per decade. (**Figure 11.34**). The latest report of the **Intergovernmental Panel on Climate**

© Vladimir Romanovsky

Figure 11.32 Collapse pit from thawing permafrost can create major changes on the Arctic and subarctic land surface.

Figure 11.33 The 2007 Anaktuvuk River tundra fire on Alaska's North Slope blackened more than 1,000 square kilometers (386 mi^2) and put as much CO_2 into the atmosphere as the entire Arctic tundra normally absorbs in a year. This one fire burned more than all Alaskan tundra fires since 1960. As Arctic warming continues, such tundra wildfires are fully expected to increase, both in frequency and acreage burned, and could lead to a critical positive feedback to global warming.

Change (IPCC) states that sea level has been rising 50% faster since 1961 than computer models predicted. Note the drop in global sea level in 2010–2011, shown in Figure 11.34, results from a switch from a sizable El Niño that began in 2010 that was replaced by one of the strongest La Niñas in recent memory. This sudden shift in the Pacific changed global rainfall patterns that brought devastating floods to the Amazon basin and Australia, and severe drought to the southern United States. The massive rainfall resulted from the evaporation of enormous amounts of seawater to produce severe rainstorms that transferred huge amounts of water from the oceans to extraordinary rainfall on continents. Consequently, global sea level was lowered by about a quarter of an inch. Some may argue that these data reveal us entering a long-term period of lowering sea level, but such sea level drops as this cannot last for long; the overall global trend remains solidly up. After all, water on continents flows downhill and will eventually find its way back to the sea. Inspection of Figure 11.34 reveals similar but lesser drops in sea level in 1994, 1997, and 2007, other years that experienced significant El Niños.

The cause of sea-level rise is two-fold, both as a consequence of global warming: freshwater entering the ocean from the melting of glacial ice and the expansion of warming seawater. (Sea level rises about 24 centimeters [9.5 in] with each 1°C increase in seawater temperature.) It is estimated that 50% to 80% of the rise results from melting of land-based ice, such as the midlatitude mountains and Greenland glaciers, because their meltwater is added directly to the ocean. In contrast, polar and Antarctic ice is floating shelf ice that already displaces its own weight and has no impact on sea level.

It is not possible to accurately predict future sea-level rise. Especially uncertain is how the large Antarctica and Greenland ice sheets will respond to warming temperatures, as their response involves complex flow processes (see Case Study 11.1). For instance, a warming ocean destroys the floating tongues of ice that form where land-based glaciers float out to sea. The land-based tongues are anchored to rock, which holds back the glacier. However, when the ice tongue breaks off, the moving glacier increases its speed. This is already happening to outlet glaciers in Antarctica and some glaciers in Greenland (see Figure 11.30). Furthermore, airborne laser altimetry and modeling show that the rate of loss from Greenland's ice sheets has been increasing. Thermal expansion of the global ocean is predicted to be by far the major contributor to sea-level rise. Climate scientists, working on the fifth climate assessment report (AR5) of the IPCC, due out in 2013, report:

> There is really not any significant difference we can talk about with respect to earlier assessments to the geographic patterns of extremes in precipitation patterns and future warming around the world, but the forecast of sea level rise has been revised downward. Judgments on how warm it will get and how ice sheets will respond result in a new mean estimate of 32 centimeters (12.6 in) of sea level rise by 2100 from melting ice sheets. Sea-level rise from all sources—expansion of seawater and

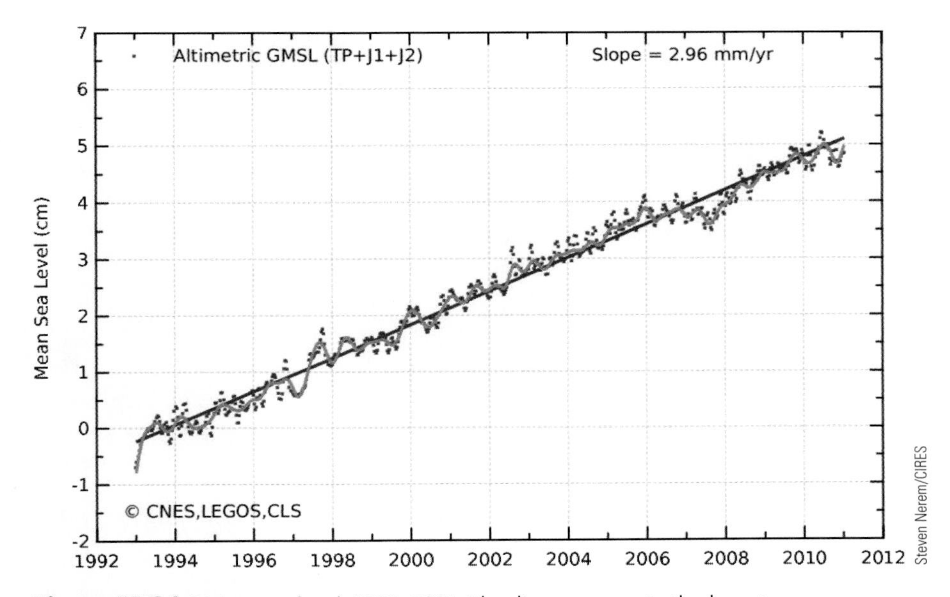

Figure 11.34 Rising sea level, 1993–2011. Blue line represents the long-term average increase in global sea level as determined by satellite altimeters since 1993. Although most years show a worldwide rise in sea level, the drop of nearly a quarter of an inch in 2010–2011 reflects the switch from El Niño to La Niña conditions in the Pacific Ocean. (Source: Sea level data courtesy S. Nerem, University of Colorado)

melting mountain glaciers—will continue and will accelerate until it rises 61 to 73 centimeters (24–29 in) by 2100. The forecast is not as catastrophic as earlier estimates but it is enough to inundate the Maldives, much of Bangladesh and South Florida.[4]

Biological response—Observations in North America and Europe reveal that wildlife in various habitats are responding to warming climates, some in alarming ways. Case Study 11.3 on page 414 provides summaries of some research studies.

The disappearing Arctic sea ice—In the Arctic Ocean, some sea ice persists from year to year, whereas almost all Antarctic and Southern Ocean sea ice is "seasonal"; in other words, it melts away and reforms annually.

[4]A.J. Long. 2009. "Back to the future: Greenland's contribution to sea-level change." *GSA Today* 19(6).

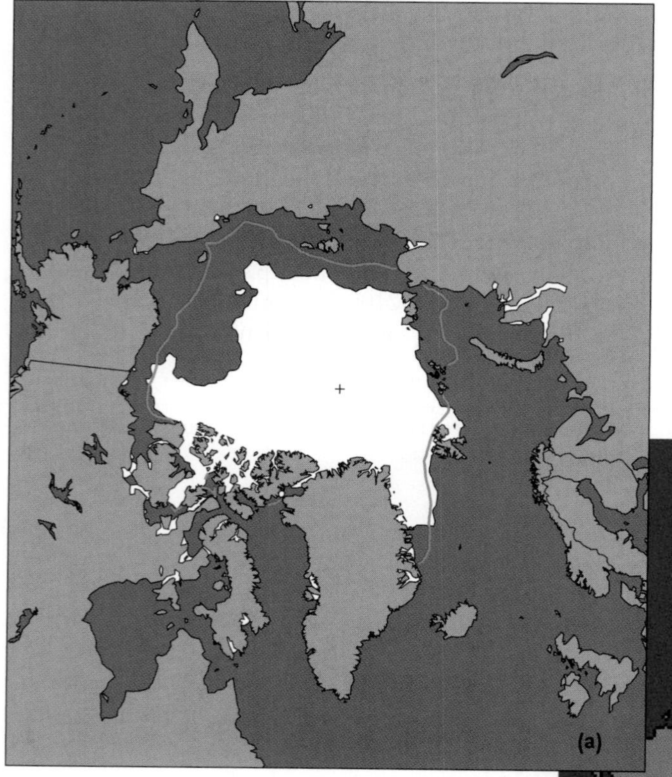

Figure 11.35 A comparison of Arctic sea ice extent (in white) on (a) September 12, 2009, and (b) September 9, 2011. The orange line is the 1979–2000 average ice extent for mid-September. The ice extent for 2011 is the second lowest in the satellite record, only slightly above that of 2007, the record low year, even though the 2011 weather conditions were not as conducive to ice loss as in 2007. Both the Northwest Passage and the Northern Sea Route were open for a period during September. (National Snow and Ice Data Center)

Both Antarctic and Arctic sea ice serve as critical habitats for marine birds and mammals, but Arctic sea ice appears to play another important role: regulating Northern Hemisphere climate.

Comparing ice thicknesses measured by submarines between 1958 and 1976 with measurements made in the 1990s revealed that, on average, Arctic sea ice had thinned by 1.3 meters (4.3 ft) and that it had lost 40% of its volume in that time. Such a major change in the Arctic alerted the scientific community, as it could have a significant impact on Northern Hemisphere weather. Moreover, the summertime minimum ice extent of Arctic ice has been in steady decline since satellite measures began in 1979. The record minimum was set in 2007, with the extent of summer ice in 2011 ranking second during the 32-year record (**Figures 11.35** and **11.36**). Winter Arctic ice extent has also decreased since 1979, about 4.2% per decade. A rapidly warming Arctic encourages more summer ship traffic through Canada's Northwest Passage and the Northern Sea Route. Both routes were open for a period during the summer of 2010 with 18 ships making the voyage, including a 160,000-tonne Russian supertanker aided by two nuclear icebreakers. A Danish shipping company reports it saved a third of its usual fuel costs and half the time in shipping goods to China from Murmansk. A concern is that in the Arctic, soot exerts a double whammy. First, when released into air as a result of incomplete combustion, such as from forest fires or badly tuned diesel engines, the soot particles absorb sunlight and warm the atmosphere. Second, as rain or snow washes the soot onto the ice, it darkens the surface, reducing its albedo, and causes faster melting of the ice. Thus, increased shipping in the Arctic will cause positive forcing according to a joint Canadian–American study that, for the first time, predicts the potential negative impact of sooty engine exhaust particles on the Arctic environment.

The long-term satellite records and earlier records from the 1950s show an earlier onset of the spring melt and an increasing length of the melting season, clear evidence of a warming Arctic. Furthermore, comparison of the recorded sea ice decline with the projections in the 2007 IPCC AR4 shows that the ice loss is more rapid than anticipated in any of the IPCC models.

consider this . . . A recent study of Milankovitch orbital shifts (see Chapter 2) reveals that the Arctic should be cooling at present. Currently, during the Northern Hemisphere's summer solstice, Earth is 600,000 miles farther from the Sun and receives less solar energy than it did 2,000 years ago.

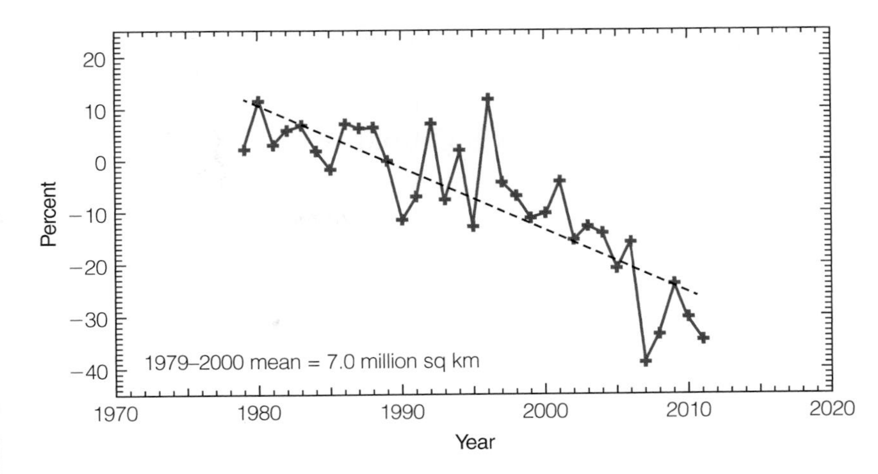

Figure 11.36 Percent of change in Arctic sea ice extent 1979–2011. (National Snow and Ice Data Center)

Combining satellite observations of sea-ice extent with conventional atmospheric observations reveals that decreasing summer Arctic ice impacts large-scale atmospheric phenomena during the following autumn and winter in regions well beyond the Arctic's boundary. Heat rising from the exposed Arctic water in summer warms and destabilizes the lower atmosphere, causing increased cloudiness and weakening of the polar jet stream. This causes less winter precipitation in Scandinavia following the summers of reduced sea ice. This results in less water in reservoirs for hydroelectric power generation. Similarly, reduced sea ice aggravates droughts in the southwestern and southeastern United States.

The Arctic Oscillation and Arctic Sea Ice

It is not entirely certain what is causing the decrease in Arctic ice cover—human-induced global warming, a natural Arctic climatic fluctuation, the Arctic oscillation (AO), or a combination of these factors. The AO is a seesawing pressure difference between the Arctic atmosphere and the atmosphere at lower latitudes. When the pressure difference is high, a vortex of heavy, cold air forms around the Arctic, corralling the cold air and keeping it nearer the pole; this is known as the positive phase. Such was the situation in the early winter of 2011–2012. But as the AO shifted to its negative phase in February 2012, the cold air escaped, sweeping southward across Eastern Europe and Russia, causing blizzards and subzero, bone-chilling cold. It even snowed as far south as North Africa (**Figure 11.37**). Scientists at NASA acknowledge that the breakdown of the vortex may not completely account for the abrupt severe winters that Europe has experienced. Strange as it may seem, some climatologists, among them Judah Cohen of Atmospheric and Environmental Research of Lexington, Massachusetts, attribute the unusual cold episodes of the past few winters to global warming. The basis for his hypothesis is that melting of Arctic sea ice because of warmer temperatures results in more moisture available for the atmosphere to acquire and later fall as snow. Consequently, Siberian snow cover has increased in recent years, which has a cooling effect that extends to Europe and Eastern Asia. But experts caution, "There's always going to be some natural variability. Every episode of high temperatures or extreme cold isn't climate change. Many scientists believe the tendency toward the warm, windy negative phase is due to increasing concentrations of CO_2 in the atmosphere, enhancing the greenhouse effect. This, in turn, forces the decades-long atmospheric swing toward the AO warm, windy phase on which the decadal swings are superposed. A concern expressed by some scientists is that the Arctic sea ice may be unable to recover under the current warm conditions, and a tipping point may have been

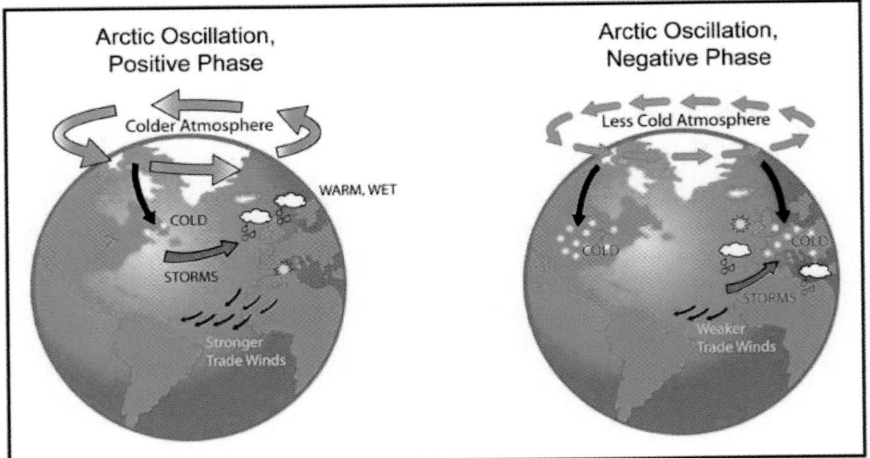

Figure 11.37 The Arctic oscillation (AO). The positive phase (left) has higher atmospheric pressure in the mid-latitudes than in the Arctic, causing a milder winter in the United States and northern Europe. The negative phase (right) has a higher atmospheric pressure over the Arctic, forcing the jet stream south bringing very cold, wet air into the United States and into northern Europe. (Source: NASA)

passed where the Arctic will eventually be totally ice-free during at least part of the summer.

Causes of Today's Global Warming

Of the five factors most likely responsible for the current large-scale temperature changes, three vary naturally:

- Stratospheric volcanic sulfate aerosols
- Climate variability
- Solar radiation

With what you have learned about the increasing concentration of greenhouse gases in the atmosphere, what changes might you expect over the next 50 years where you live if global temperature should rise by 2°C?

Two factors, which have changed decisively, are due to human activity:

- GHGs—mainly CO_2 but also CH_4, NO_3, and CFCs
- Anthropogenic sulfate aerosols

A decisive climatic change caused by a change in one or more of these factors is called a *forcing* (Chapter 2). Some scientists have suggested that solar forcing may have caused warming in the early 1900s, although natural climate variability and an *anthropogenic* (human-caused) factor cannot be ruled out. But warming from about 1880 into the 21st century, irrespective of any natural forcing or increasing solar activity, is considered by the vast majority of scientists as attributable to human activity: emission of the GHG CO_2 from the burning of coal, gasoline, and natural gas (**Figure 11.38**).

What is the evidence that burning coal and oil account for the increase in atmospheric CO_2? The evidence comes by an application of the Suess effect, named for the Austrian chemist, Hans Suess, who first noted the influence of this effect on the accuracy of radiocarbon age dating. Carbon-14 isotopes are produced in the atmosphere by cosmic rays colliding with nitrogen atoms in the atmosphere, and a small amount of carbon-14 is incorporated into CO_2 molecules that are taken up by all living things, including trees. In the course of checking different tree rings from the same tree for radioactive carbon-14, scientists discovered that the amount of carbon-14 in the atmosphere had changed with time. Since the beginning of the 20th century, it had decreased by a measureable percentage. By assuming that the supply of carbon-14 had not changed (that the same number of carbon-14 atoms were being produced in the atmosphere), the scientists reasoned that CO_2 from fossil fuels (which lack carbon-14—they are hundreds of millions of years old and any radioactive carbon-14 has long ago decayed away) must be diluting the atmosphere. This idea led

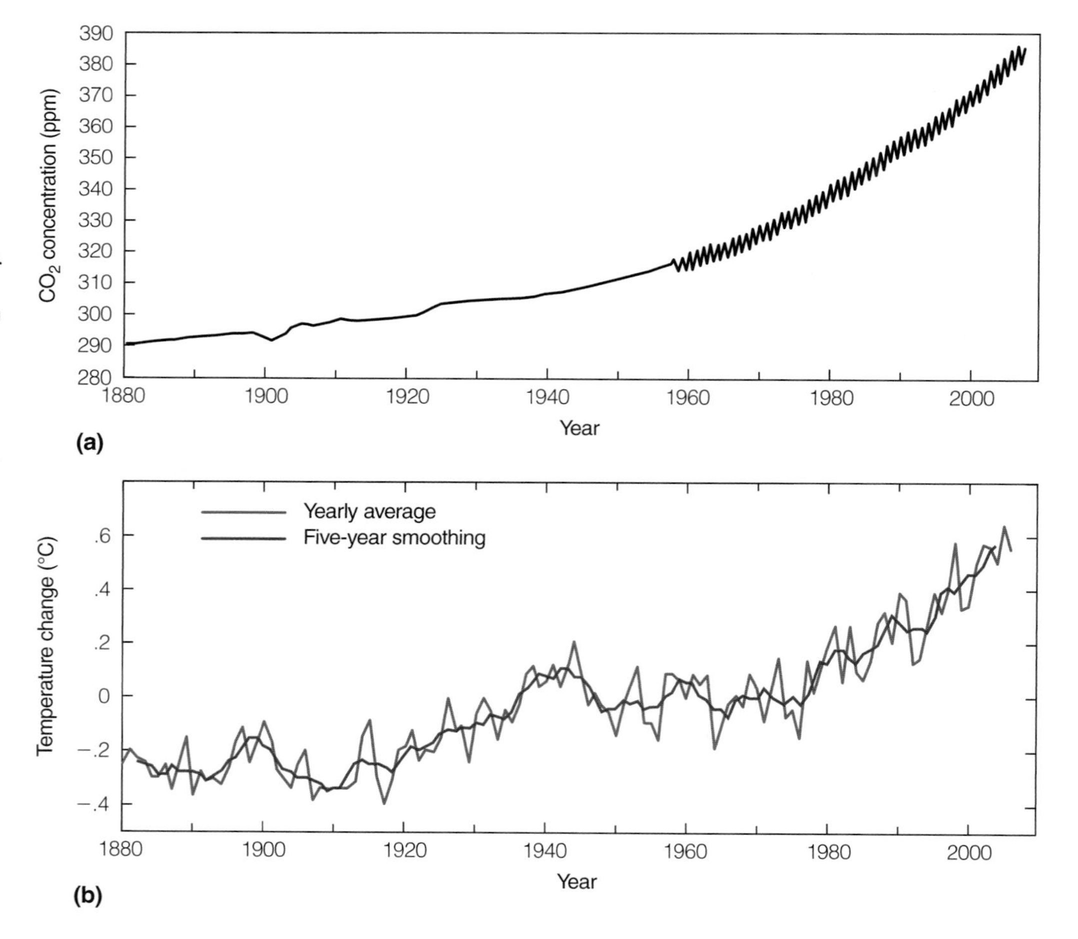

Figure 11.38 Atmospheric CO_2 and temperature changes since 1880. (a) Changes in the concentration of CO_2 obtained from Antarctic ice cores (smooth curve) and measured annual oscillations at Mauna Loa, Hawaii, from 1958 to present. The sawtooth effect results from the seasonal changes in the uptake and release of CO_2 from northern forests. (b) Changes in global mean surface temperature since 1880 relative to 1951–1980 average temperature, denoted as zero (0). Air and ocean temperature data from weather stations, ships, and satellites. (Data from NASA/Goddard Institute for Space Studies.)

to establishing an observatory to measure the amount of CO_2 in the atmosphere. Year after year, the measurements showed a steady increase in atmospheric CO_2. By July 2012, the value was 394.49 ppm, about 25% more than in 1956. The so-called Keeling curve (see Figure 11.38a), named for its discoverer, David Keeling, clearly shows the increase in CO_2, and climate scientists quickly recognized its alarming significance. Even more alarming is the steepening upward curve of the graph illustrating the accelerating rate of CO_2 emissions. Although many expected the severe worldwide economic recession beginning in 2008 to slow the output of CO_2, emissions actually increased significantly, growing 5.9% in 2010, the largest jump since 2003.

It was an analysis of the five factors previously listed that led the IPCC, in its 1995 Second Assessment Report (SAR), to state rather blandly, "the balance of evidence suggests that there is a discernible human influence of global climate," and that the climatic changes caused by human activity add up to just about the amount of expected change. In 2001, the conclusion of the Third Assessment Report (TAR) was that "most of the observed warming over the last 50 years *is likely* to have been due to the increase in greenhouse gas (GHG) concentrations." The most recent IPCC report, AR4 (2007), states the case more strongly, that "the globally averaged temperatures since the mid-20th century *is very likely* due to the increase in anthropogenic GHG concentration." The AR4 report also states that "it is likely that GHG concentrations alone would have caused more warming than observed because volcanic and anthropogenic aerosols have offset some warming that would otherwise have taken place." Of course, the radiation balance can be altered by feedback from factors other than GHGs such as reflection by sulfate particles from volcanic eruptions (negative feedback), absorption by aerosols or various gases (positive feedback), increased output of solar energy (positive feedback), the snow-albedo effect (reflection by snow and ice-covered ocean, negative feedback), reflection by gases or clouds (negative feedback), and emission of heat by various materials (e.g., farmlands, paved surfaces, and roofs radiate more energy to the atmosphere [positive feedback] than woodlands). Since the 1980s, the signal from GHGs has become very strong and is recognized as the dominant forcing over all others.

consider this . . . Global average CO_2 increased from 280 ppm at the start of the Industrial Revolution to 394.49 ppm in 2012, the highest concentration in the last 650,000 years and probably the highest during the last 20 million years. The growth rate of global average atmospheric CO_2 during 2009–2010, about 2.3 ppm per year, is higher than average for both the 1990s (1.5 ppm) and the decade of the 2000s (2.0 ppm).

Earth's surface warming might be explained by an increase in the Sun's energy output as part of the natural cyclic changes in the Sun's irradiance. Those cycles reach a climax about every 11 years and do have a significant effect on year-to-year global temperatures. However, beginning about 1970, despite the cyclic changes in the amounts of solar radiation, the average global surface temperature has continued to increase, even during the unusual warmth of 2007 when the natural effects of the solar cycle and the El Niño-Southern oscillation (ENSO; see Chapter 10) were both in their cooling phases. As shown in **Figure 11.39**, it is obvious that variations in solar irradiance since 1970 cannot account for the increase in global surface temperature.

And What of the Future?

We cannot repeal the laws of nature. If we continue to do nothing, the floods, wildfires, heat waves and extreme weather that have wreaked havoc across our nation—at a cost of tens of billions of dollars annually—will increase and intensify. And future generations will never understand why we squandered our shrinking opportunity to protect the planet.

Rep. Henry Waxman (D-CA), November 13, 2011

About 15 different climate modeling groups around the world all work with the same data but come out with slightly different projections about future global temperatures. For this reason, there is some variation in the range of future temperature projections. Let us examine some of the projected effects of global warming from the 2007 IPCC AR4

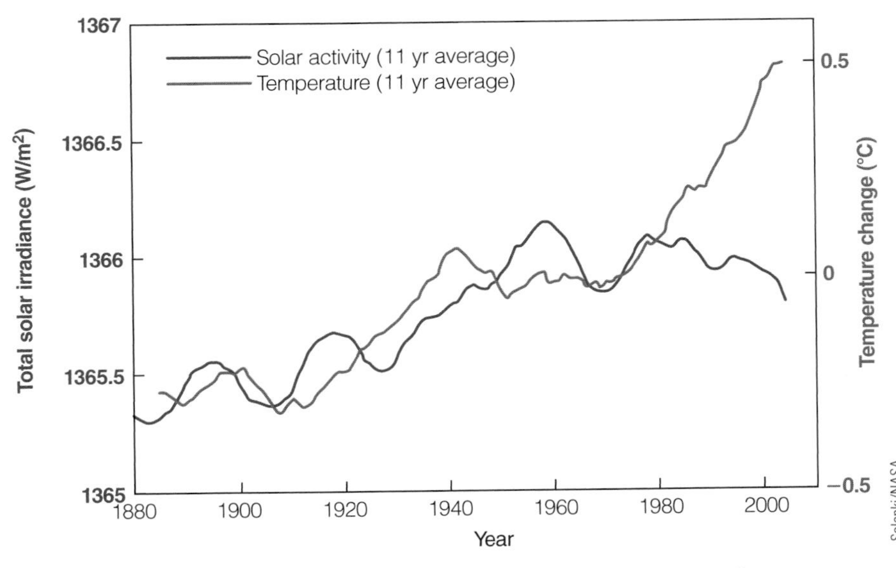

Figure 11.39 Solar radiation received at Earth's surface and global surface temperature. The amount of solar radiation reaching Earth varies in a cycle of about 11 years. Because the average increase in Earth's surface temperature has continued to rise, even during times of minimum solar radiation, the long-term increase in global surface temperature cannot be attributed to variations in solar radiation. (Source: Global temperature [red, from NASA GISS] and total solar irradiance [blue, 1880–1978 from Solanki; 1979–2009 from PMOD] and plot from www.skepticalscience.com.)

and *The Copenhagen Diagnosis: Climate Science Report* of November 2009:

- GHG concentrations have continued to track the upper bounds of IPCC projections.
- CO_2 concentrations, global surface temperature, and sea level are projected to increase under all IPCC AR4 scenarios during the 21st century. Temperatures, especially, will experience major increases. For the business-as-usual scenario (i.e., the worst-case scenario), the 2100 temperature projection in the higher northern latitudes is 1.1°C to 6.4°C (2–11°F) higher than in 2000. The AR4 mid-range scenario gives the land temperature an increase of 3°C to 4°C (5–7°F), although the average global mean change is "only" 2.4°C (4°F). The AR4's lowest emissions scenario is 1.4°C to 3.8°C (3–6°F). Massachusetts Institute of Technology (MIT) climatologists project a median surface warming by 2100 of 5.2°C (9.4°F), doubling its earlier projection of 2.4°C (4.3°F), and the Hadley Center (U.K.) projects a catastrophic rise of 5°C to 7°C (9–12.6°F) by 2100 for the business-as-usual scenario. The MIT group projects a median CO_2 value of 866 ppm (compared with 386 ppm in 2009), with Arctic warming of 20°F by the 2090s. Observed global temperature changes remain entirely in accord with IPCC projections—that is, an anthropogenic warming trend of about 0.2°C per decade with superimposed short-term natural variability. As Gavin Schmidt explains, "To put this into a different context, the global mean temperature during the peak of the last ice age 20,000 years ago was about 5°C (9°F) cooler, a level associated with what can be described as a different planet. Compared with the estimates of temperature variations over the last one thousand years, the projected changes are five to ten times larger."[5] Atmospheric chemist Sherwood Rowland writes that Earth is headed toward an unmitigated catastrophe of 1,000 ppm atmospheric CO_2.
- Arctic sea ice has declined faster than projected by IPCC AR4, and Arctic temperatures will have the largest increase. Modeling indicates a 3°C to 5°C (5–9°F) increase over Arctic lands and 7°C (13°F) over the ocean by 2100, with a corresponding temperature increase in permafrost regions, as discussed previously.
- Globally, average precipitation is expected to increase during the 21st century, although at regional levels both increases and decreases are projected ranging from 5% to 20%. Locally, there will be more intense precipitation events—increased rain in the mid- to high latitudes and at the equator. Drier areas, such as the American Southwest, will become drier. There will be increased variability of Asian summer monsoons, such as the unprecedented record-breaking rainfalls responsible for devastating floods that wracked Pakistan for several weeks in July and August in both 2010 and 2011. For the Sahel, climate models differ greatly: Some projecting a drier Sahel, others wetter, and others little or no change. We are already seeing more intense rainfall in some areas, with rain-on-snow events increasing up to 40% in the Arctic.
- Worldwide, the ice sheets are both losing mass and contributing to sea-level rise. This was not certain in 2007 at the time of the IPCC AR4 report. They are expected to continue shrinking and retreating during the 21st century.
- Worldwide, ecological productivity and biodiversity will be altered by climate change and sea-level rise, with an increased risk for extinction of some vulnerable species (see Case Study 11.3). Sea level has risen more than 5 centimeters (2 in) over the past 15 years, about 80% higher than IPCC projections of 2001.

[5]Schmidt, Gavin, and Joshua Wolfe. 2009. *Climate change: Picturing the science* (p. 201). New York. W.W. Norton.

Combating global warming—James Hansen of NASA GISS has shown that U.S. weather has become more extreme since 1980 and has estimated with 80% certainty that the extremes are due to human-induced greenhouse warming. That is not 100% certain, but, in the case of greenhouse warming, built-in delays in Earth's climate system compound the dangers of waiting for certainty; decades are required to warm or cool the atmosphere and oceans. If the world's policymakers wish to head off the potential of extreme climate changes, they must act before the major effects of GHG pollution are fully developed (some are already apparent). Unfortunately, global warming has become a contentious political topic, because communication between scientists, the public at large, and policymakers is often difficult (**Figure 11.40**).

What individuals can do—What can we do about global warming? We can do nothing about any of the natural factors (the external forcings), to the degree they exist, but there are things that we can do individually to slow the pace of human-generated global warming that will not drastically affect our lifestyles. Because more than 80% of the world's energy comes from burning fossil fuels—oil, coal, and natural gas—which release GHGs, we must first look at ways to cut back or eliminate the use of fossil fuels. Currently, we are getting close to cost-effective alternatives (see Chapter 14), but there is still much to be done. In the United States, there are seven major, simple, inexpensive things that individuals can do:

1. Buy products that reduce energy consumption, such as front-loading washers and dryers, and replace incandescent light bulbs with fluorescents or LEDs.
2. Use energy more efficiently. Add insulation to homes and businesses, seal gaps around windows, and replace old windows with double- or triple-pane windows.
3. Actively recycle.
4. Forego fossil fuels. Use alternatives when possible—biofuels, wind power, and photovoltaic (PV). Invest in companies practicing carbon capture and storage, and divest from oil company stocks.
5. Move closer to work. The transportation sector is a major consumer of fossil fuel and a leading source of GHG

Table 11.2 Reducing Global Warming Begins at Home

Action to Take	Estimated Reduction of CO_2, Pound Per Year
1. Insulate your home, install water-saver shower heads, clean and tune furnace.	2,480
2. Buy a fuel-efficient automobile with a rating of at least 32 mpg to replace your most-used car.	5,600
3. Do not drive your car two days a week.	1,590
4. Recycle all home waste metal, glass, newsprint, packaging, and cardboard.	850
5. Install solar heating to help provide hot water.	720
6. Replace washing machine with low-water-use, low-energy model.	420
7. Buy food and other products with recyclable or reusable packaging.	230
8. Replace refrigerator with a high-efficiency model.	220
9. Use a push lawn mower instead of a power mower.	80
10. Plant two trees.	20

Source: Oregon Energy Office

emissions. Use mass transit, carpool, or ride a bicycle. Cut down on long-distance plane travel, which is one of the fastest growing sources of GHG emissions.

6. Unplug electrical devices when not in use. It seems counterintuitive, but Americans consume more electrical energy to power devices when off than when on. Computers, battery chargers, televisions, stereo sound systems, and other electrical gadgets consume more energy when supposedly switched off than when on.
7. Install smart meters to offer time-of-use incentives to encourage homeowner customers to shift electricity consumption for dishwashers and laundry appliances to off-peak hours.

Table 11.2 provides some more specific suggestions. The consequences of our choices and lifestyles are catching up with us, and even small steps will help. It's important to realize, however, that buying a plug-in hybrid car and replacing a few lightbulbs with compact fluorescents will not be enough to reverse global warming.

What industry and government can do—Major changes in our sources and consumption of energy will be necessary if we are to reduce GHG emissions to a meaningful level. As pointed out in Pulitzer Prize–winning author Thomas Freidman's book *Hot, Flat, and Crowded,* our climate and energy crises are interconnected—we can't solve one without addressing the other. The following are some of the proposals that will help meet the dual challenge of climate warming and the end of cheap oil:

1. Government legislation should be passed encouraging the installation of solar thermal (for heating water) and PV panels (for generating electricity) on roofs. Excess PV energy can be sold to local electric utility companies to be added to the grid. California has been on the forefront, with its Solar Roofs Bill of 2006, with the goal of installing PV panels on 1 million houses.
2. The installation of solar thermal and PV panels in all new home construction should be mandated.
3. Federal incentives for solar and wind energy development should be stabilized. The U.S. government's on-again/off-again incentive policies have proved to be financially disruptive in developing alternative energy.
4. State and federal governments should mandate alternative energy or low-carbon requirements for companies supplying products to state and federal governments. This will promote more efficient and economic manufac-

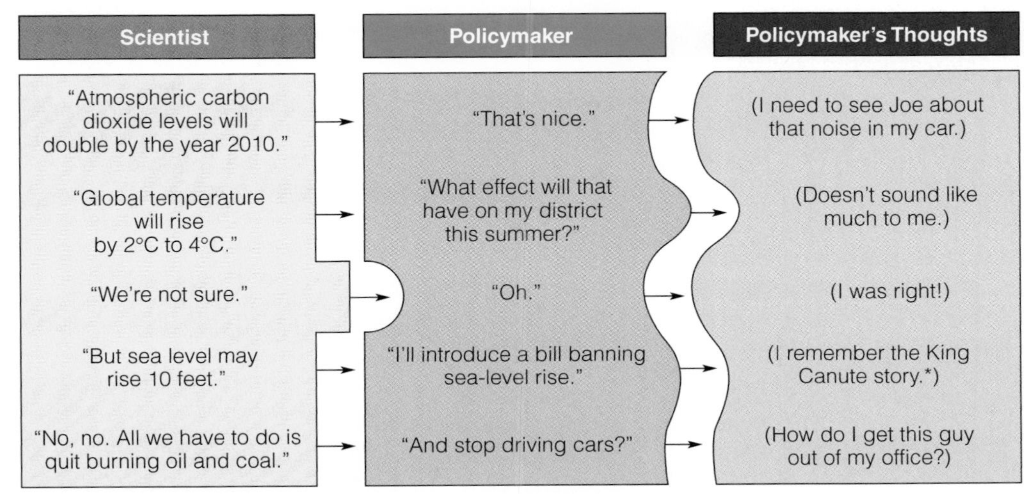

Figure 11.40 A hypothetical scientist briefs a hypothetical policymaker on the dangers of the emission of greenhouse gases. This image was used in a slide presentation at a scientific meeting in 1989. One might ask, "Why has it taken more than 20 years for our policymakers to begin to acknowledge the issue?" (After R. Byerly, Jr., 1989, Spring. "The Policy Dynamics of Global Change." *EarthQuest.*)

explorer's perspective

Photograph by Megan Seldon /See: http://www.nationalgeographic.com/explorers/bios/stephen-schneider/

DR. STEPHEN H. SCHNEIDER, a "climate warrior."

Dr. Stephen H. Schneider obtained his Ph.D. in mechanical engineering and plasma physics from Columbia University but became an internationally recognized biologist and climatologist. He was the Melvin and Joan Lane Professor of Interdisciplinary Environmental Studies in the Department of Biology, a senior fellow at Woods Institute for the Environment at Stanford University, and a research scientist at the National Center for Atmospheric Research from 1973 to 1996, where he co-founded the Climate Project.

Schneider's research focused on climate change science, integrated assessment of ecological and economic impacts of human-induced climate change, and identified viable climate policies and technological solutions. He consulted as an expert advisor for federal agencies and White House staff during six administrations. Since 1988 he was a lead author or convening lead author with the Intergovernmental Panel on Climate Change (IPCC). For example, from 2005 to 2007 he was convening lead author with Working Group II, writing chapter 19 of "Assessing Key Vulnerabilities and the Risk from Climate Change," and was also a synthesis author for the Fourth Assessment Report.

In 2007, Schneider, along with four generations of IPCC authors, received a collective Nobel Peace Prize for their joint efforts. He was elected to the U.S. National Academy of Sciences in 2002, received the American Association for the Advancement of Science/Westinghouse Award for Public Understanding of Science and Technology in 1991, and received a MacArthur Fellowship in 1992 for integrating and interpreting the results of global climate research.

He authored or co-authored over 500 books, scientific papers, proceedings, and legislative testimonies, and edited books, chapters, reviews, and editorials. Schneider counseled policy makers, corporate executives, and non-profit stakeholders in the use of risk management strategies in climate-policy decision-making, given the uncertainties in future projections of global climate change and climate impacts. He was especially active in improving public understanding of science and the environment through extensive media appearances and communications and public outreach. The New York Times referred to Schneider as a "climate warrior" for his longstanding campaign to inspire an aggressive response to warming. Sadly, he died in 2010. He was 65 years old.

Source: http://www.nationalgeographic.com/explorers/bios/stephen-schneider/

Melting ice, Weddell Sea, Antarctica.

turing processes and will lead to reduced costs for the rest of society.

5. The many financial incentives written by the fossil-fuel companies and passed into laws by Congress should be eliminated, or at least reduced. Such incentives include depletion allowances, intangible drilling cost deductions, and enhanced oil recovery credit.[6] This would result in consumer's paying the true cost of fuel upfront and would encourage the automobile industry to manufacture high-mileage vehicles.
6. All new fossil fuel–fired power plants should be required to include carbon-capture and sequestration (CCS) technology or cap-and-trade incentives. (See Case Study 11.4.)

An automobile that gets 26 miles per gallon emits 1 pound of CO_2 per mile.

Most of the scientific community supports global efforts to cut back emissions of GHGs, such as the **Kyoto Protocol**, which set standards for reductions in CO_2. Under conditions of the 1997 Kyoto Protocol, the United States would have reduced emissions to 7% below 1990 levels between 2008 and 2012. The Kyoto Protocol has been ratified by some industrialized nations since 1997, but in 2001, the U.S. government, citing economic concerns, withdrew from the agreement. The United States and some other countries favored giving nations credit for CO_2 reductions by considering the amount of CO_2 that is naturally absorbed by forests, plants, and soil. However, those nations that lacked large areas of farmland and forest objected to this idea. Other issues addressed by scientists that remain unresolved are rates of consumption, energy efficiency, and population growth.

The industrialized countries' GHG emissions may not be the major problem. The United Nations' economic and science advisors on global climate change predict that by 2025, the developing countries will be responsible for nearly 70% of all energy-related CO_2 emissions. For example, China now uses more coal than the United States, Japan, and Europe combined, making it the world's largest emitter of GHGs. (But China has also become the world's leading builder of more efficient coal-fired power plants that use extremely hot steam, and they are completing them at the astonishing rate of about one a month.)

A U.S. trade group representing manufacturers, utilities, mining companies, coal and oil producers, and railroads asserts that reducing industrial emissions by 20% or more below the 1990 levels, as proposed by the Kyoto Protocol, would be disastrous for the U.S. economy. They estimate that cuts of this magnitude could cost an average of 600,000 jobs a year for several years. Others argue, however, that the potential for doing nothing to curb GHG emissions will also hurt the world's economy in agricultural production and human welfare. (A rule of thumb is that with every 1°C increase in average global temperature, we lose 10% of global food production. While some regions stand to benefit from small increases in temperature because of increased photosynthesis, the projected high temperatures elsewhere jeopardize overall global food production. In some places, crops will be damaged by drought; in others, by much higher temperatures. Or, as in Russia's case today, by both.) The nature of these costs is beyond the scope of this book, but they may be greatly in excess of those involved by taking action to address global warming.[7] A small but vocal group claim that global warming is a natural climatic variation such as Earth has experienced in the past and there is nothing we can do to slow or stop it (Case Study 11.5).

Geoengineering and Climate Change

Geoengineering is usually taken to mean technologies that manipulate Earth's climate in an effort to mitigate the effects of global warming from GHG emissions. Geoengineering is defined by the National Academy of Sciences as "options that would involve large-scale engineering of our environment in order to combat or counteract the effects of changes in atmospheric chemistry." A number of wide-ranging technologies have been suggested to counter the effects of increasing atmospheric CO_2 and global warming. These proposals range from massive reforestation to absorb CO_2, to virtual Star Trek technologies to reduce emissions of CO_2, to sucking it out of the air on a massive scale to avert an environmental catastrophe, to "cap and-trade." (See Case Study 11.4.) The most obvious way to cut CO_2, and other GHGs, is to stop burning fossil fuels altogether. However, with coal so cheap, plentiful, and easily mined, and our energy needs so great—coal produces approximately a third of U.S. electricity and 35% of our CO_2 emissions—it is clear that coal will continue to be part of our energy mix for many decades.

Let us first consider the mitigation of point sources of CO_2 emissions. Is there any technology that can remove the CO_2 from carbon-fueled boilers in power plants and other point sources and store it in a suitable place? A technology widely touted by "clean coal" groups for effectively limiting CO_2 emissions is **carbon capture and sequestration** (also referred to as *carbon capture and storage* [CCS]).

Carbon capture and sequestration—Carbon capture and sequestration is a process that traps CO_2 emissions from power plants, refineries, cement plants, and other point sources and securely buries it deep underground. It has the potential to reduce future worldwide emissions by 20%. Several CCS pilot plants are already in trial operation, with 3 megatons of CO_2 (Mt CO_2) per year from natural gas

[6]These are all incentives that allow oil companies to move profits between different business units within the company to reduce their tax burden.

[7]Chapter 25, "Cost, Cost, Cost," in *The Weather Makers* by Tim Flannery (2005) considers the costs of not cutting emissions.

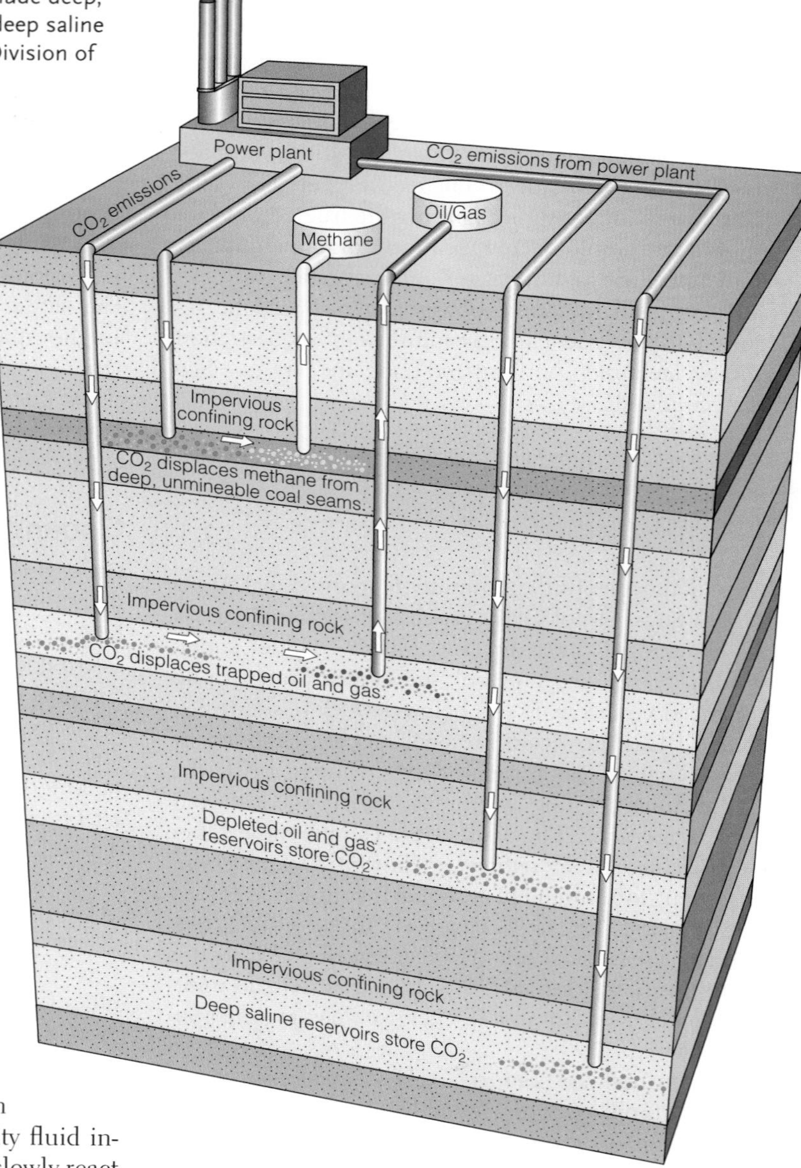

Figure 11.41 Geological storage options by CCS for CO_2 include deep, unmineable coal seams, depleted oil and gas reservoirs, and deep saline reservoirs. (Source: Ohio Department of Natural Resources, Division of Geological Survey)

cleanup or carbon-fired power plants being captured and stored. CCS technology is at the point where it can be upscaled, but there is a lack of incentive for real construction. The design and construction of large-scale demonstration plants must start now if they are to be in operation before 2020. The experience of operating these plants will reveal the feasibility for commercial-scale operation that should carry over into future decades. The question is, will the construction of hundreds of CCS plants be delayed beyond the deadlines needed to meet climate-change forecasts, or is CSS simply an overly ambitious hope that serves as an excuse for electric utilities (and others) to pollute by burning carbon while using "green wash" as a pretext to lock Earth into a future of high CO_2 emissions?

The best sites for potential geological storage (or "sinks") are depleted oil and gas fields, deep saline aquifers, and deep, unmineable coal seams (**Figure 11.41**). Ideally, the sink should be within some 60 kilometers (37 mi) of one or more power plants, have a large reservoir volume, consist of porous and permeable reservoir rock, and be capped by a seal of impermeable shale to prevent leakage. The subsurface storage should be at a depth greater than 795 meters (2,600 ft), so that the overburden pressure will maintain the CO_2 as a high-density fluid instead of a gas. Once underground, the CO_2 will slowly react with mineral material and organic matter to form stable minerals, primarily iron, calcium, and magnesium carbonates. In the United States, more than 5,800 kilometers (3,600 mi) of pipelines already carry and store CO_2 underground, predominantly to oil and gas fields to stimulate additional production, a technology called *enhanced oil recovery* (EOR). The United States is the world leader in EOR and each year injects about 48 million tons of CO_2 to stimulate additional production from abandoned or depleted oil and gas fields.

Worldwide, some 3 to 4 million tons of CO_2 per year are successfully stored in underground reservoirs at several sites. Norway's Statoil has been running an experiment since 1996 that reveals how CCS can play an important role in climate mitigation strategies. The CO_2 comes from the Sleipner natural gas field. The CO_2 is extracted from the natural gas before being piped to the mainland. This is the world's first commercial-scale operation for sequestering CO_2 in a deep saline reservoir. Since 1996, the project has injected some 2,800 tons of CO_2 per day, or about 1 million tons per year, but the demand for CCS is much greater, perhaps more than 10 billion metric tonnes per year. The question has been raised about potential leakage of CO_2, which is a highly reactive compound, from deep reservoirs. Seismic monitoring of the Sleipner Project has shown that the CO_2 is safely stored beneath a thick layer of impermeable shale.

By 2009, there were 16 CCS projects in the United States, more than 40 worldwide, in various stages of planning, development, or operation. The U.S. Department of Energy (DOE) provides financial assistance to seven demonstration projects. For example, two U.S. pilot CCS plants are the 275 Mw FutureGen 2.0 Project, a \$1.3 billion facility near Meredosia, Illinois, which is in the design and construction phase, and the Antelope Valley Power Station Project near Beulah, North Dakota. FutureGen

2.0 will use a technology called **integrated gasification combined cycle (IGCC)**. An IGCC plant only partially burns powdered coal at high pressure in the presence of steam and creates a syngas that is burned in a turbine, with the hot exhaust from that turbine making steam, which is routed to another turbine. Because a double-turbine IGCC system is more efficient than conventional coal-fired plants, the increased efficiency will help pay the cost of carbon capture. The Antelope Valley project is expected to remove 90% of the CO_2 emissions from a 120 Mw lignite-fired boiler. The captured CO_2 will be made available for EOR injection in the nearby Williston Basin oil fields. It is important to realize that scaling up the existing technology for the capture, transport, and storage of CO_2 emitted from a full-scale power plant has yet to be demonstrated (although this is the ultimate goal of the FutureGen 2.0 Project).

Power plant CCS of CO_2 is technically possible now, but with low efficiency and high energy consumption. The cost of IGCC technology is estimated to increase the cost of generating electricity by about 30% above the cost of conventional coal-fired plants, in contrast with the cost of another technology, installing CO_2 scrubbers on conventional coal-fired plants, which might raise the cost of electricity by as much as 60%.

CCS from stationary point sources is technically possible and may prove to be economically possible. But what about dispersed emissions from mobile sources? For example, in the United States, about a third of CO_2 emissions comes from the transport sector—ships, cars, planes, trucks, and railroads. Dispersed emissions also come from heating buildings. Worldwide, about a quarter of the CO_2 emissions comes from mobile sources. Such emissions cannot be captured at the source. Is there any way to capture CO_2 from thin air? There are two possible technologies: using chemical pumps and planting trees.

Scrubbing the skies with chemical pumps—If we could halt anthropomorphic CO_2 emissions today, the risks of the existing CO_2 would be with us for hundreds to thousands of years, assuming that only natural processes dissipated the CO_2. The consequence of CO_2 emissions persists longer than that of nuclear waste. The capture of CO_2 from flue gases at fixed emission sites, such as coal-fired power plants, does nothing to reduce the CO_2 from mobile sources and that which we have already released to the air.

Current air capture research is centered on a handful of small companies. Several techniques are being tested: capture on a solid sorbent system, ion exchange membranes on a porous silica surface, absorption of CO_2 by an alkaline aqueous solution, and capture of CO_2 by alkali hydroxide solutions.

With air capture, the CO_2 would be concentrated, compressed, and then stuffed deep in the crust or pumped deep in the ocean. The energy requirements for such a process are very high, probably about the total energy consumption of the United States. Climate scientist Wallace Broecker has been promoting an unpublished chemical pump process for capturing CO_2 from the atmosphere on a massive scale using "scrubbers." With each American generating on average 20 tons of CO_2 per year, it would take about 17 million scrubbers to remove the CO_2 in the United States, as well as some 60 million scrubbers worldwide. Each air capture unit would be roughly 6 feet in diameter and 50 feet high, and would scrub about a ton of CO_2 from the atmosphere each day. Once captured, the CO_2 needs to be liquefied and pumped through pipes to a storage site. Air capture of CO_2 could be used in making carbon-neutral hydrocarbon fuels (CNHCs) for manufacturing synthetic fuels. CNHCs offer an opportunity for converting wind, solar, or nuclear power into high-energy automobile fuels comparable with the gasoline in current use. An advantage of these "scrubbers" is that they can be placed anywhere but would operate best in low-humidity regions, such as deserts.

The scrubber extraction technology is based on a resin that absorbs CO_2. In pilot studies, it is far cheaper to operate than other chemical pump technology. Nevertheless, it would require about a 20% increase in overall global energy production to operate the scrubbers. And how would all of the CO_2 be disposed of? The most obvious disposal sites are deep saline aquifers, as previously mentioned. Another option is injecting CO_2 dissolved in water into flood basalts, such as those in the Columbia Plateau in the northwestern United States and the Deccan Traps in India. Basalts are rich in calcium silicate minerals, and CO_2, mixed with ground water, will leach out the calcium and form the relatively stable mineral calcite. This chemical reaction, a form of geochemical weathering, goes on naturally all the time, and the technology would simply speed up the natural process. A geothermal plant in Iceland is the site of a pilot study on the sequestration of CO_2 in basalt (**Figure 11.42**). The best rocks for mineral sequestration would be the mantle-derived ultramafic rocks peridotite and serpentinite. These rocks have been brought to the surface at several places on Earth by the collision of Earth's tectonic plates. Because they are more enriched in magnesium silicates than basalts, the CO_2 would react more readily and produce a higher concentration of stable minerals. To date, there is no economically feasible technology to accomplish mineral sequestration.

Disposal in the deep ocean has also been suggested. Below 3,000 meters (9,845 ft), liquid CO_2 is denser than seawater. If injected below that depth, it would sink to the seafloor. Some of the sequestered CO_2 would eventually escape to the atmosphere, but it would take centuries because of the slow speed of the oceanic conveyor belt (see Chapter 2). By that time, we would have exhausted fossil fuels and atmospheric CO_2 would be decreasing. Such disposal would require considerable energy, and there are environmental concerns. Deliberate CO_2 storage in the deep ocean can further acidify the ocean, which is already losing alkalinity as it absorbs atmospheric GHGs. This threatens the ability of many organisms—including the pteropods,

Haraldur Stefansson

Figure 11.42 The Hellisheidi geothermal power plant in Iceland. In a pilot study on the feasibility of CO_2 sequestration in basalt, the CO_2 released from the hot water that powers the facility is dissolved in water and injected into basalts at a depth of 300 to 800 meters (980–2,600 ft), where it reacts with minerals in the basalt to form stable new minerals.

the basis of many marine food chains—to create shells and survive. Scientists have already discovered what may be the most devastating impact yet of human-caused global warming: a 40% decline in marine phytoplankton since 1950 linked to the increase in sea surface temperatures.

Planting trees—Large-scale forestation has been cited as a method to reduce carbon. Trees capture carbon naturally, using sunlight for photosynthesis. With a minimum of energy expended, trees could be grown, harvested, and then safely buried to store the CO_2. However, for trees to undo just the U.S. contribution of CO_2, it would require a land area nearly equal to the total area of the United States.[8]

Fertilizing the ocean—Oceanographers have been cautiously exploring the feasibility of dumping finely ground iron particles into the ocean to stimulate the growth of phytoplankton, which, like all plants, metabolize CO_2 by photosynthesis. From initial experiments, it is estimated that spreading a half-ton of iron across 100 square kilometers (39 mi^2) of a tropical ocean would stimulate enough plant growth to absorb some 350,000 kilograms (771,800 pounds) of CO_2 from seawater. Performed on a much larger scale, the iron fertilization of seawater could absorb billions of tons of CO_2, offsetting perhaps a third of global CO_2 emissions. The environmental side effects of such large-scale "iron fertilization" are difficult to foresee, but the enhanced bloom of plants could enrich the entire oceanic ecosystem.

Solar radiation management—Alternatives to mitigation techniques (e.g., CCS and ocean iron fertilization) are solar radiation management technologies using **sulfate aerosols** and **cloud reflectivity enhancement**. None of these techniques reduces the warming effects of CO_2 or other GHGs, nor would they reduce the ocean acidification, which is already occurring as a result of increasing CO_2 levels.

One way to turn down Earth's thermostat is to inject large amounts of sulfur dioxide aerosols (tiny particles) into the stratosphere, which would act as tiny mirrors and reflect sunlight back to space. Such a process would reduce global temperatures much as happened following the eruption of Mount Pinatubo in 1991. This eruption put millions of tons of sulfate (SO_2) into the air, causing a 0.5°C (1°F) drop in temperature for several months. Artificially injecting SO_2 aerosols into the atmosphere would have the same effect, except that it would require continuous injection. Also, the technology would be excessively expensive and come with the side effects of acid rain and probable damage to the ozone layer.

Cloud reflectivity enhancement, which would create cloud condensation nuclei, would change the albedo of clouds to make them lighter and reflect more sunlight away from Earth. This could be done by spraying a mist of seawater from ships into low-lying clouds. According to two scientists at the University of Texas, the seawater mist could counteract a century's worth of warming for $9 billion. Because the air over the ocean lacks dust and pollution, the technique would be more effective with cloud manipulation over the ocean than over land.

Geoengineering is not a substitute for emissions control, but is merely one of many schemes that need consideration along with emission-reduction efforts. Critics of geoengineering assert that the enormous sums of money required for most geoengineering technologies would be better spent on developing alternative power sources—that is, we need to focus our attention on the source of the problem, not its symptoms. Many critics voice concern that GHG mitigation by geoengineering will be self-defeating, leading to a false sense of security from the consequences of global warming, and reducing the popular and political pressure for emissions reduction and cutbacks on the use of fossil fuels.

[8]For a thorough energy analysis of this topic, see David J. G. Mackay. 2009. *Sustainable energy—Without the hot air* (pp. 245–246). Cambridge, England: UIT Cambridge Ltd.

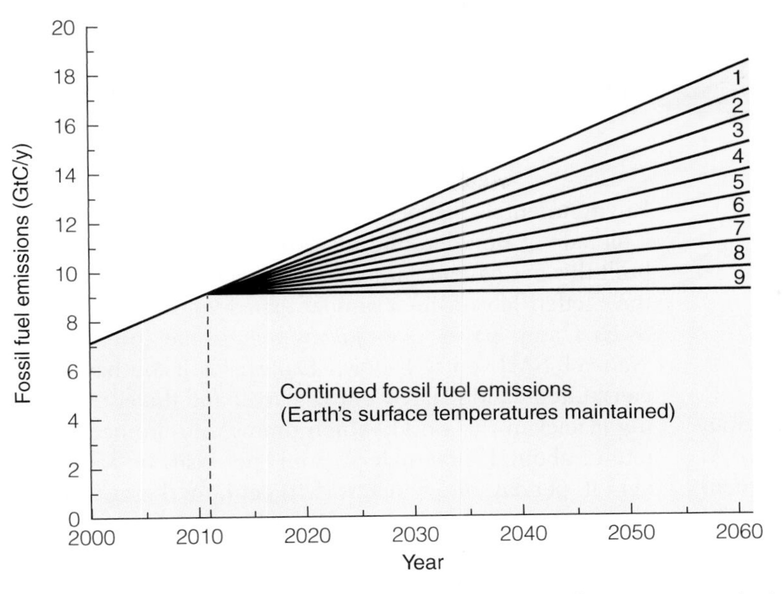

Mitigation strategies to be implemented, assuming implementation in 2011:

- Capture and sequester CO_2 from coal-fired power plants
- 5 million Gwh energy saved by more efficient buildings
- All automobiles average 60 mpg – including diesels and hybrids
- 700 Gw of electricity from major nuclear power plants
- 4,000 Gw of electricity from photovoltaics
- 5 million Gwh energy saved in industry by cogeneration and heat recovery
- Space heating from natural gas (CH_4) captured from unconventional sources such as methane seeps and landfills
- Cellulose-based biofuels from one-sixth of world's cropland
- Termination of all tropical deforestation
- Greater use of rail transportation to replace long-haul trucks

(Gwh = 1 billion watt-hours; Gw = 1 billion watts)

Figure 11.43 Stabilizing CO_2. The uppermost sloping line represents the increase in CO_2 with "business-as-usual" and no effort to reduce burning fossil fuels. Each progressively gentler sloping stabilization "wedge" represents an approximate reduction in the increased rate of CO_2 by 1 GtC/y of the alternatives listed, assuming approximately equal reductions for each wedge, and assuming implementation in 2011. Later implementation steepens the slopes of the wedges. GtC/y = million tons of carbon per year. (Modified from Socolow, Robert, and S. Pacala. 2004, August 13. Stabilization wedges: solving the climate problem for the next 50 years with current technologies. *Science* 305:968–971 and Hyndman, Donald, and David Hyndman. 2009, *National Hazards and Disasters.* Brooks/Cole.)

Multiple Strategies Needed to Combat Climate Change

Avoiding increasing atmospheric GHGs, especially CO_2, will require multiple strategies including improving energy efficiency, reducing gasoline and diesel fuel consumption, increasing use of alternative sources, scrubbing GHGs from the atmosphere, and other strategies listed in **Figure 11.43**. The mitigation goal represented by the figure is to maintain a flat emissions rate for 50 years with each energy-reducing wedge representing a reduction of about 1 gigaton of carbon per year (GtC/y). Proponents for these strategies recognize that achieving such reductions will be difficult, but they are technically achievable with current technology. The problems of implementation are political: Our economy is based on continual growth; people wish to improve their standard of living and continue to enjoy a comfortable lifestyle with its many conveniences; and many of the remediation strategies would come with inconveniences and costs to the public that our decision makers are reluctant to implement. Nevertheless, the eventual consequences of continuing the "business-as-usual" approach will become more serious with time, and the task of remediation will become more formidable and more costly as the atmospheric content of CO_2 increases. To make the energy reduction wedges work will require overcoming the land-use restraints and aesthetic objections in placing wind and solar farms (the not-in-my-backyard syndrome), the fear of nuclear power (and anxiety over possible nuclear weapons proliferation), and the long record of anti-environmental legislation by the fossil-fuel industries. Even if all of these strategies could be implemented today, the risks posed by the existing CO_2 (represented by the space below the horizontal line in Figure 11.43) would be with us for millennia because of the 100-year residence time for a molecule of CO_2 in the atmosphere and the development of positive feedback processes that sustain high levels of CO_2 (see Chapter 2).

case study

11.1 A Summer on Ice: Research on the Greenland Ice Sheet

by John Maurer; member, Greenland Climate Network Expedition

Greenland—Earth's largest island—is mostly covered by an immense sheet of ice as thick as 3,000 meters (9,843 ft). As a result, most of the people and animals residing in Greenland live along a thin strip of nearly barren land along the coasts. Greenland's population in 2006 was about 56,000 people, and the population of the biggest cities runs about 15,000. Greenland is a self-governed Danish territory, so a number of Danish people live there, but the population is 88% Greenlandic Inuit.

Getting to Greenland involves a flight to Kangerlussuaq, a town of about 500 people along the central west coast at latitude 68° north and site of the island's only airport large enough to support jumbo jets. From there, a small utility aircraft, a De Havilland Twin Otter, takes the team and all of their gear and equipment to Swiss Camp (**Figure 1**), a research station on the ice sheet about 300 kilometers (186 mi) north of Kangerlussuaq and 30 kilometers (19 mi) from the ice edge. The camp is strategically located near the *equilibrium line altitude (ELA)* of the ice sheet—the altitude that divides the regions of negative and positive net annual mass balance. Swiss Camp is so named because Dr. Konrad Steffen, a climatologist, a glaciologist, and director of the Cooperative Institute for Research in Environmental Sciences (CIRES) at the University of Colorado at Boulder, and his colleagues from Switzerland built this permanent camp, now serving as a base camp for the Steffen lab on their annual expeditions. The flight to Swiss Camp passes over *Jakobshavn Isbrae* (pronounced "yah-cob-SAH-ven ICE-bray," Danish for Jakobshavn Glacier), Greenland's largest outlet glacier and the fastest moving glacier in the world, which drains into the ocean at a rate of about 13 kilometers (8 mi) per year, or 35 meters (115 ft) per day, or 1.5 meters (5 ft) per hour. Large icebergs calve from its terminus into Disko Bay, to eventually float into Baffin Bay and slowly melt as they travel south into the Labrador Sea. The iceberg that sank the *Titanic* was likely from Jakobshavn.

Although airborne views of Jakobshavn are breathtaking (see Figure 11.30), the interior of the Greenland ice sheet is flat and as boring as a large parking lot painted white. Summer temperature on the ice sheet usually hovers around 0°C (32°F), plus or minus 3°C to 6°C (5–10°F). But strong winds—often greater than 10 meters per second (22 mph)—create a frigid windchill. Although there can be relatively balmy days in the summer when a single layer of clothing will suffice, the weather often requires wearing thick parkas, hats, and gloves. Because of the strong reflection of sunlight from the snow (and its 24-hour persistence at this far-north latitude), one must wear dark sunglasses and use lots of sunscreen.

Swiss Camp is simply three red tents, each the size of a small room. These serve as kitchen, storage, and laboratory. Power is supplied by a generator, and a propane tank fuels

Figure 1 Swiss Camp that serves as base camp for the Steffen lab's annual expeditions to the Greenland ice sheet for servicing the GC-Net and conducting scientific research.

© John Maurer

Lisa McKeon, USGS

both space heaters and the cook-stove. Water for drinking and cooking is from melted snow that is collected in a big bucket. The researchers sleep in their own makeshift tents, held down by bamboo poles, so that strong winds don't blow them away in the middle of the night. Snow, which inevitably is dragged into the tent on boots, melts to form puddles of water on the floor, which one must somehow keep from soaking into clothes and gear. Cargo is organized into designated areas outside, so that it isn't lost and buried by snowstorms.

Travel to nearby automatic weather stations (AWS) is done via snowmobiles. Although these can cruise along at more than 80 kilometers per hour (50 mph), hopping over gentle hummocks of snow and hauling heavy sleds of equipment (propane tanks for steam-drilling, extra AWS instruments, etc.) force them to be operated at slower speeds. Navigation between stations is accomplished using known coordinates and Global Positioning System (GPS) receivers, as there are no useful landmarks to follow and because a bad snowstorm or fog can severely limit visibility. Satellite phones are used for communication (regular cell phones cannot operate that far north). When traveling into the ablation zone on a snowmobile, the team needs to be especially alert for melt ponds, crevasses, moulins (**Figure 2**), and slushy snow. This requires an experienced eye.

Why do scientists subject themselves to such a hostile environment to study and monitor Greenland's climate? Because of the technical and physical difficulties involved, it is not known yet with much certainty whether Greenland's ice sheet is shrinking or growing. Losses of ice associated with global warming, however, are important to monitor because of the potential for rising sea level. Although continued global warming is not expected to melt the ice sheet entirely, or even very quickly, Greenland's contributions to sea-level rise in the 21st century could still be significant. Rising sea level is an ongoing problem for many coastal regions and islands across the planet, and if this trend continues into the future, it could result in large migrations of environmental refugees. Furthermore, meltwater from Greenland could significantly freshen the North Atlantic Ocean (i.e., making it less salty), thereby shutting down or severely weakening oceanic currents that transport heat from the tropics to the Arctic and northern Europe. Should this happen, the Arctic and northern Europe would experience much cooler temperatures, possibly enough to trigger another ice age. For these reasons, it is important to monitor Greenland's climate and mass balance—the changes in its total volume of ice and snow.

To better understand Greenland's climate, a series of 20 AWS have been installed since 1995 to collect data

Figure 2 A moulin (French for "mill"), or glacier mill, a water-worn pothole formed on a glacier where a surface meltwater stream exploits a weakness in the ice. Moulins tend to be cylindrical, up to several meters in diameter, and commonly extend downward to the glacier bed.

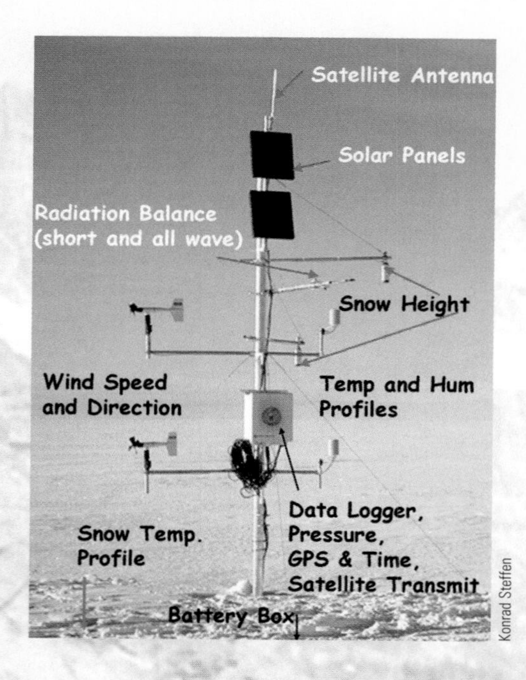

Figure 3 An automatic weather station (AWS) on Greenland's Jakobshavn Glacier. (Konrad Steffen, CIRES)

across Greenland. Sponsored as part of NASA's Program for Arctic Regional Climate Assessment (PARCA), the Greenland Climate Network (GC-Net) provides critical climate information over vast, unpopulated regions of the Greenland ice sheet. Dr. Steffen, who manages this network, returns to the ice sheet every summer with the assistance of other scientists and graduate students to maintain and download data from the instruments at the AWS sites and to conduct research.

Figure 3 is an example of an AWS that is part of the GC-Net. Powered by four large batteries and charged by two solar panels, these stations are 5 to 10-meter (16–33 ft)-tall metal towers that host a variety of meteorological instruments:

- Anemometers, which measure wind speed and direction
- Barometers, which measure atmospheric pressure
- Hygrometers, which measure relative humidity
- Pyranometers, which measure both incoming and outgoing (reflected) solar radiation
- Thermistors, which measure changes in temperature beneath the surface
- Thermometers, which measure air temperature about 2 meters (6.5 ft) above the surface
- Sonic rangers, which measure changes in the height of snow beneath them (due to accumulation, melt, or redeposition) by sending out pulses of ultrasonic sound and timing the echoes

Most AWS have duplicate instruments in case one should fail, and the duplicate records validate the data. Data collected are stored digitally year-round in a *data logger*, and average measurements are transmitted hourly to a computer in Colorado via satellite, thus providing near-real-time monitoring of weather conditions on the Greenland ice sheet. The data are available to anyone via the Internet. Servicing the AWS includes downloading all of the measurements stored in its data logger to keep the logger from becoming filled and unable to record additional data.

The region of the ice sheet nearest the coast is at lower altitudes, where summer warmth can melt the ice sheet's upper layers of snow and ice—this region of the ice sheet is termed the **ablation zone**. Because of the melting, AWS towers in the ablation zone gradually begin to lean as the supporting snow and ice begin to melt. To prevent these towers from melting out of the ice and falling over, the towers must be revisited every few years to steam-drill (see Figure 11.6) new, deeper holes in which to reinstall them. In contrast, the interior of Greenland is at higher altitudes and rarely warm enough for snow to melt; this region is called the *dry snow zone*, where winter temperatures often fall below −50°C (−58°F). Consequently, snow accumulates continually in this zone and eventually will bury weather stations if left unattended. Personnel visit the AWS every few years to increase the station's height. This involves lifting the top portion of the tower with a manual crane and inserting an extension tube to the metal tower. Servicing the necessary AWS towers in this manner is the primary objective of each summer expedition, along with replacing old or broken AWS instruments and downloading all the data.

Digging snow pits is another routine task when visiting an AWS (**Figure 4**). These provide data for analyzing features in the top several meters of snow that can provide clues to climate events of recent years. Visible bands of slightly blue ice provide evidence of refrozen meltwater. Thin layers of relatively dense snow can be proof of previous storms and strong winds that compacted the snow at what was at one time the surface. Thick, dense layers of relatively large ice crystals, called *hoar frost*, provide evidence of prior summers, and the thickness of snow between two summer layers provides a measurement of net annual snow accumulation at the site. In addition to noting the visual *stratigraphy* of the snow pit, other measurements taken at the site include depth profiles of snow density, temperature, and average grain size of snow crystals. Density is measured by weighing a known volume of snow at different depths along the pit wall, whereas grain size is measured with a magnifying glass. Together, snow pit measurements help verify, explain, and give a hands-on sense of what the

Lisa McKeon, USGS

weather instruments at the AWS have been measuring for the previous few years.

Although satellite-borne remote sensing instruments have become important in studying Greenland because of their frequent and complete coverage of the ice sheet, many questions cannot be answered from space. Despite harsh weather conditions and the difficulty of doing field research on the ice sheet, *in situ* observations remain a critical component in our understanding of Greenland's climate and elsewhere in other data-sparse regions of the remote Arctic. The GC-Net will continue to be useful and important in monitoring climate change and assessing the role of the melting of Greenland's ice sheet in our planet's future.

questions to ponder

1. What are the potential hazards of traveling on a glacier?
2. What is the purpose of studying Greenland's Jakobshavn Glacier?
3. For what reasons might the residents of Western Europe be justly concerned about rapid and excessive melting of the Greenland ice sheet?
4. In a snow pit, there are layers of blue ice and layers of white, relatively dense snow. What is the significance of these different ice layers?

Figure 4 A snow pit near one of the automatic weather station (AWS) sites, used to analyze subsurface features in the snow for evidence of recent weather events.

11.2 Climate Data on Ice: The National Ice Core Laboratory

Ice cores from numerous glaciers are stored at the National Ice Core Laboratory (NICL) located on the grounds of the Denver Federal Center in Lakewood, Colorado. The facility currently houses more than 14,000 meters (8.7 mi) of ice cores from 100 different bore holes at 34 sites in Greenland, Antarctica, and high mountain glaciers in the western United States. NICL's archived collection of ice cores is kept in the Main Storage area of the lab, which consists of 1,558 cubic meters (55,000 cu ft) of freezer space held at −36°C (−32.8°F) by a computer-controlled refrigeration system (**Figure 1**). Visiting scientists are able to prepare and measure samples in the 3,050-cubic-meter (12,000 ft^3) Exam Room, which is maintained at –25°C (18°F). A cold, clean room with filtered air is available for the treatment of environmentally sensitive samples. A warm anteroom is provided for suiting up and taking short breaks. The Main Storage collection contains nearly 13,000 tubes of ice samples and contains most of the samples remaining from deep-drilling projects funded by the National Science

© Roger Ressmeyer/CORBIS

Figure 1 The main storage area of the National Ice Core Storage Laboratory in Lakewood, Colorado. More than 14,000 meters (8.7 mi) of ice cores are stored in 1,558 cubic meters (55,000 ft^3) of freezer space held at a temperature of −36°C (−32.8°F).

Foundation (NSE) beginning in 1958, such as Vostock and Greenland Ice Sheet Project 2 (GISP2).

The NICL provides scientists with the capability of studying ice cores without having to travel to remote field sites, and it preserves the integrity of these core samples for the study of past environmental conditions and global climate change. These cores have been recovered for a variety of scientific investigations, most of which focus on the reconstruction of past climates. By investigating past climate fluctuations, scientists hope to be able to understand the mechanisms by which climate change is accomplished; in so doing, they hope to develop predictive capabilities for future climate change. Typical data obtained from the cores include age-depth relationships, chemical analyses, oxygen and hydrogen isotope ratios, snow accumulation rates, and pollen analyses. It is the world's most comprehensive ice core collection available to the scientific community. The NICL is funded by the USGS and the NSF, and overseen by a governing board of research scientists, the Ice Core Working Group.

questions to ponder

1. Of what value is identifying pollen that has been trapped for thousands of years in glacial ice layers?
2. How can the bubbles of trapped atmospheric gases found in glacial cores be analyzed?

11.3 Where Are the Wild Things?

Not where they used to be as Earth warms

With the physical evidence of global warming continuing to accumulate, are there indications of global warming in biological systems? Yes. Studies in the United States and Europe show that some plants and animals are indeed responding to global warming.

As the environment changes, plants and animals adjust in two main ways: by moving to more comfortable living quarters and by changing the calendar dates of important events, such as migration, breeding, and hibernation. The Edith's checkspot (*Euphydryas editha*), a rather plain brown-, white-, and black-spotted butterfly that is a wide-ranging species in western North America, has extended its range 85 kilometers (55 mi) northward, while it is becoming extinct at the southern edge of its range (in Mexico). Furthermore, the rate of extinctions at lower elevations is more than double the rate at elevations above 2,440 meters (8,000 ft). Thus, the butterfly's range is shifting northward and upward. A European study of the ranges of 35 species of butterflies shows similar findings. Two thirds of the species' habitats have shifted northward by 35 to 240 kilometers (21–149 mi) coincidentally with the warming of Europe and only 3% have shifted south. In Britain, a 20-year study of 101 bird species found that, coincidentally with the warming trend, most species had extended their northern range by an average of 19 kilometers (11 mi).

In New England, lilacs are blooming as much as 2 weeks earlier than they did in 1965; April, in one sense, has become May. Maple trees need about 100 days of below-freezing weather. As this becomes more rare, fewer maples will be in the forests and there will be an eventual loss of sugar maples in New England. Cardinals, robins, and cowbirds are all arriving earlier in Vermont than they did a century ago. However, kingfishers, Fox Sparrows, and towhees are not. Why the difference? The former group has

Lisa McKeon, USGS

the ability to respond to climate changes because their arrival is cued to temperature. The latter, apparently, respond to temporal clues that won't change even as the climate does.

The changing of the timing of important life events to accommodate warming is also well documented. Of 677 species studied in North America, two-thirds have advanced important events on their calendar. For example, tree swallows breed about 9 days earlier than in the 1950s, and in the Rocky Mountains, the yellow-bellied marmot now emerges from hibernation, on average, 23 days earlier than in the 1970s. More than 170 species of migrating birds in the lower 48 states have shifted their winter ranges northward (**Figure 1**).

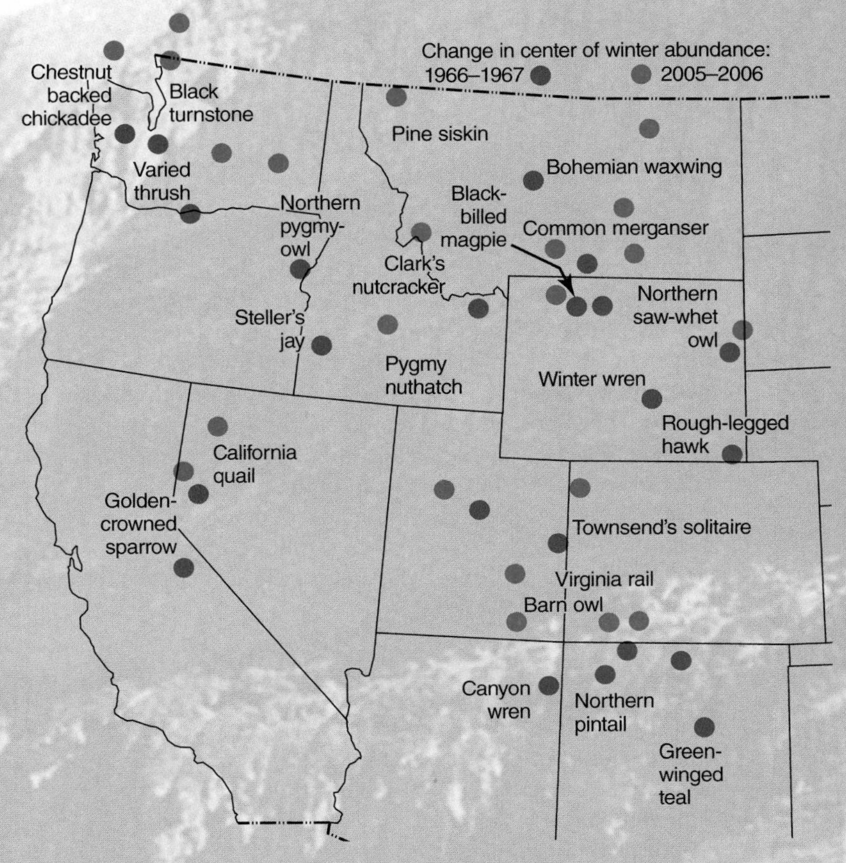

Figure 1 Global warming and migrating birds. Within only 4 decades, more than 170 bird species in the 48 states have shifted their winter ranges northward, apparently because of global warming. Forty years of National Audubon Society's annual Christmas bird counts document those changes for 22 species in the western states. (Courtesy National Audubon Society)

In the mountain forests of Monteverde, Costa Rica, 20 of 50 species of frogs and toads disappeared following population "crashes" in 1987. These crashes probably resulted from ecosystem changes that altered the communities of reptiles, amphibians, and birds because of recent warming. The changes are associated with patterns of dry-season mist frequency, which are related to changes in sea-surface temperatures in the equatorial Pacific since the mid-1970s. The implication is that atmospheric warming has caused the base of the mountaintop cloud bank to raise in elevation and has negatively impacted the mountain forest ecosystem.

Temperature-limited environments in mountains, subarctic, and arctic regions also reveal evidence of warming. In the Swiss and Austrian Alps in the early 1990s, a study compared the plant life on mountain summits exceeding 3,000 meters (9,849 ft) to records collected 70 to 90 years earlier, when the annual temperature averaged 0.7°C (1.26°F) cooler. Altitudinal vegetation belts were found to have shifted to higher elevations by as much as 4 meters (13 ft) per decade, although many belts' shifts were less than 1 meter (3 ft) per decade. A study in Alaska found that boreal forests (forests in mountainous and northern regions of North America in which the mean temperature of the hottest season does not exceed 18°C [64.4°F]) expanded northward at a rate of 97 kilometers (60 mi) per 1°C (2°F) of temperature increase, and that the length of the growing season increased 20% in the last few decades.

At both the North and South poles, climate change is altering ecosystems. Polar bears and penguins may live poles apart, but they both depend on ice and cold for survival. With both poles warming, the temperature increases are the most pronounced at Earth's northern and southern extremes. Penguins and polar bears are what biologists call *indicator species*; their well-being reveals the health of the ecosystem.

Less multiyear ice is surviving in the Arctic, and the remaining ice is thin (see Figure 11.35). This spells trouble for polar bears, which spend spring to midsummer on ice floes, hunting for seals to eat to build energy reserves to get them through the leaner season after the ice has melted and they must move onto land. Once on land, they fast, often for months. The warmer temperatures beginning in the 1990s have made for earlier spring melting, earlier breakup of Arctic sea ice, and later fall freezing, making seal dinners harder to come by. Consequently, polar bears have suffered, their weight dropping on average by 10 kilograms (22 lb) for each week the ice breaks up earlier than normal. Furthermore, cub survival is contingent on a well-fed nursing mother and an adequate food supply after weaning. As a result, the polar bear population has dropped almost 20%

since 1995, and the polar bear is being considered as an addition to the endangered species list. Scientists estimate that the polar bears, currently numbering about 25,000, don't have a very bright future, and they could disappear by the year 2100; the fate of the world's largest carnivore is uncertain, should global warming continue.

At the other end of the world, Antarctica, where the temperature has warmed over 3°C (5°F) in the past 50 years, life is not easy for Adélie penguins, their population dropping by 65% in the last few decades. A change in the abundance of food and more precipitation in the form of rain rather than snow means trouble for the Adélies and their chicks (**Figure 2**). Increased rain causes the chicks to die; the newborn and young chicks lack the insulating plumage to keep them dry, and they die of hypothermia. The emperor penguins, the largest of the penguins, are also in trouble. They depend on stable ice, across which the adult and chicks can march. When the ice breaks up too soon, the chicks fall into the water and die, because their outer coats are not yet waterproof.

© D. D. Trent

Figure 2 Breeding Adélie penguins; Antarctic Peninsula. From 1,000 breeding pairs 30 years ago on Litchfield Island, there are now only 20 pairs. In that time, the temperature there has increased 3°C (5.4°F), 5 times faster than the global average—affecting the krill upon which the Adélie and other penguins feed. Krill feed on plankton and algae stored in pack ice, but with warming temperatures, the ice is no longer forming reliably and, without the ice, the krill die and penguins lose their critical food source.

questions to **ponder**

1. If you live in a region that has cold, snowy winters, how has the timing of the arrival and departure of migrating birds changed in the last 25 years? (Hint: Check with the local chapter of the Audubon Society.)
2. If you live in a region that has cold, snowy winters, how has the blossoming of crocus or lilac changed in the last 25 years?

Sources: Seidl, Amy. 2009. *Early spring*. Guynup, Sharon. 2011, Spring. "A bird's eye view." *Defenders*; Cole, Diane. 2009, April. "The plight of penguins and polar bears." *U.S. News & World Report*. 76–78; Parmesan, Camille, and Gary Yohe. 2003. "A globally coherent fingerprint of climate change impacts across natural systems." *Nature* 421: 37–42; Blaustein, A. R., and Andy Dobson. 2006. "A message from the frogs." *Nature* 439: 143–45; Pounds, J. A., et al. 2006. "Widespread amphibian extinctions from epidemic disease driven by global warming." *Nature* 439: 161–67; Werner, Gian-Reto, et al. 2006. "Ecological responses to recent climate change." *Nature* 416: 389; Parmesan, Camille, et al. 1999. "Poleward shifts in geographical ranges of butterfly species associated with regional warming." *Nature* 399: 579–83; Grabherr, Georg, et al. 1994. "Climate effects on mountain plants." *Nature* 369: 448.

Lisa McKeon, USGS

11.4 Cap-and-Trade

Cap-and-trade is a "clean energy" system that limits the GHG emissions of dirty industries by putting a price on their pollution so that market forces, at least theoretically, provide an incentive for reductions. A complex federal cap-and-trade bill began moving through Congress in 2009. The bill was highly controversial, its supporters asserting that it was essential in fighting climate warming, whereas its opponents claimed its benefits were not worth the price. Proponents of the bill pointed to the already successful cap-and-trade measures that had already proved economically effective in reducing acid rain by lowering nitrogen oxide emissions in the Northeastern United States and reducing SO_4 emissions nationwide. Congress debated the bill, with opponents in Congress attacking it as "cap-and-tax," and many in Congress denying that climate warming was even a problem worth considering. The bill was allowed to die.

Meanwhile, at the state level, a Regional Greenhouse Gas Initiative (RGGI), a modest nine-state cap-and-trade program, was implemented in 2008 in the Northeastern United States to reduce CO_2 emissions by 10% in the utility sector by 2018. Nine states participate in the program: Connecticut, Delaware, Maine, Maryland, Massachusetts, New Hampshire, New York, Rhode Island, and Vermont. One state and three Canadian provinces act as observers: Pennsylvania, Quebec, New Brunswick, and Ontario. Some of the support for the program was weakened by a multimillion-dollar publicity assault on RGGI funded by industry groups who labeled it a tool of economic destruction. Some state legislatures considered withdrawing from RGGI, but only New Jersey actually pulled out of the program claiming the program was a "gimmick." However, a 3-year analysis of RGGI revealed it anything but a gimmick. The program had created 16,000 jobs, helped participating states avoid spending $765 million in out-of-state energy imports, and, all told, electricity customers—households, businesses, and government users—enjoyed a net gain of more than $1 billion in net present value, as their electric bills dropped over time.

A similar California cap-and-trade bill was signed into law in 2011 with its goal to reduce the state's GHG emissions to 1990 levels by 2020. Starting in 2013, California will "cap" the amount of GHGs (mainly CO_2) that industry may emit and gradually lower that cap over time. Permits will be issued to companies for their carbon allowance and companies may sell (trade) their permits. Any firms polluting beyond their cap limits must buy credits.

The European Union has a similar emissions controlling system, with cap-and-trade shares traded in the financial market, just as stocks and bonds are. A similar emissions trading scheme is working its way through the Australian legislature, and even China and India have adopted versions of carbon taxes or emissions trading.

questions to ponder

1. Would a nationwide cap-and-trade of GHG emissions have a significant impact on cutting back CO_2 emissions?
2. Is cap-and-trade all that will be necessary to halt increasing GHG levels in the atmosphere?
3. What impact would cap-and-trade have on U.S. industry, especially the electric utility companies, and the average U.S. consumer?
4. Is it to the long-term advantage of China and India to restrict GHG emissions to meet desired world standards?
5. Cap-and-trade has been used in the United States to control emissions other than CO_2; has it been a proven success?

case study

11.5 Medieval Warming, English Wine, and the Atlantic Flip-flop

Tremendous energy is expended by many people in debating the degree to which the climate is actually warming—despite the obvious evidence of retreating ice masses, rising sea level, and warming oceans—and, if so, to the degree to which humankind is actually responsible (see Chapter 2). Climate change skeptics use a variety of arguments to make their points. Consider the following.

The Early and High Middle Ages, roughly from 800 to 1300, was a time of warmth, with the North Atlantic ice-free most of the year, as shown by the seafaring Vikings, who explored much of northern Europe, colonized Iceland and Greenland, and even reached North America. The northward expansion of vineyards, bountiful harvests, and the building of Gothic cathedrals in Europe, for example, with the great wealth this generated, are other indications that the North Atlantic region was as warm as or warmer than it is today. Across the ocean, great droughts occurred simultaneously in what is now the western United States. Climatologists refer to these times as the Medieval Warm Period (MWP). A gradual cooling trend followed the MWP in the later Middle Ages and beyond (roughly 1300–1750), plunging Europe into long, frigid winters, which threatened food supplies because of dismal harvests. Ice skating was possible for weeks at a time on the River Thames in England, where this cannot be done now. This colder epoch is called the Little Ice Age (LIA). Critics of the idea that people are responsible for the current warming appropriately point out that these climate swings occur naturally, so why shouldn't this be the case at present, too? In fact, they point out, perhaps the world was even warmer 1200 to 1700 years ago than at present—and this certainly did not hurt civilization. In fact, perhaps it made it more pleasant! In 1983, using the distribution of vineyards as a proxy for average annual temperatures, Tkachuck wrote:

> The cultivation of grapes for wine making was extensive throughout the southern portion of England from about 1100 to around 1300. This represents a northward . . . extension of about 500 km from where grapes are presently grown in France and Germany. . . . With the coming of the 1400s, temperatures became too cold for sustained grape production, and the vineyards in these northern latitudes ceased to exist. . . . It is interesting to note that *at the present time the climate is still unfavorable for wine production in these areas* [emphasis added].[9]

We now know that the MWP was a regional, *not* global phenomenon. Coral formations preserve records of past sea temperatures (see Chapter 2), and from these it is clear that, whereas northwestern Europe and parts of North America were warmer in the Middle Ages, conditions were substantially cooler in the central Pacific. Furthermore, it is not necessarily correct to associate vineyard distribution with areas of warmer climate. Many factors besides climate change and the onset of the LIA could account for the decline in English wine making: The Black Death from 1348 into the 1370s reduced the European population by 60% and forced changes in society and agriculture; the feudal working class's dependence on cheaper beer, rather than wine, to avoid drinking contaminated water; the marriage of English king Henry II to French Eleanor of Aquitain in 1152 offered better access to French wines, compared with English wines of lower quality; the dissolution of the monasteries by English King Henry VIII (the English monasteries produced wines in England for religious purposes); and improved techniques for preserving wine, resulting in imported wines arriving in England in good condition. Thus, English wines faced

[9]Tkachuck, R. D. 1983. "The Little Ice Age." *Origins* 10(2): 51–65.

Lisa McKeon, USGS

considerable competition from imported wines, as they do today.

In truth, despite its decline, English and Welsh wine production never completely died out. A resurgence of English and Welsh viticulture began in the 1950s, when 124 vineyards were in production. Currently, there are some 400 commercial vineyards in England (**Figure 1**), significantly more than the 46 recorded in the 11th-century *Doomsday Book*. Today's English vineyards have exceeded the extent of medieval vineyards. Moreover, the best also produce a fine selection of internationally acclaimed white wines. Social preferences have also changed, with a more affluent society now preferring wine. However, the greatly expanded level of production has resulted in European Union wine tariffs, in a manner reflecting French taxes on English wine during the MWP to protect their own trade.

The climate during the last glacial period was notably unstable. Short-term climate changes, the Dansgaard-Oeschger (D-O) events, were rapid swings in temperature once every 1500 ± 500 years, in which a sudden interval of warming over a period of decades was followed by a return to ice-age conditions lasting for centuries. Twenty-five such events occurred between 110,000 and 15,000 years ago. They were first recognized in 1983 in Greenland ice cores. Why couldn't something like this be happening in modern times as well? Perhaps we are experiencing a D-O event today, without a return to "normal" conditions after the current episode of anomalous heating passes.

This line of reasoning ignores the unique conditions prevalent during the past Ice Age and the associated circulation of the global conveyor (see Chapter 2). Moreover, careful study of ice records shows that D-O events did not occur in both the Northern and Southern hemispheres simultaneously, but, rather, "seesawed" between them, unlike current warming, which is worldwide in scope. It is difficult to avoid the conclusion that modern warming is not comparable with the circumstances leading to the MWP. In fact, it has no geological precedent.

Figure 1 Denbies Wine Estate, a 100-hectare vineyard and winery, planted in the 19th century on south-facing chalk slopes of the North Downs, Surrey, England, June 2009. This is the largest single vineyard in the United Kingdom, and it produces an award-winning sparkling white wine.

question to ponder

1. How might a massive surge of freshwater into the Atlantic Ocean from rapid melting of the Greenland ice sheet affect the global conveyor of the marine circulation system? How might this affect the Gulf Stream and the climate of Western Europe? (Hint: See Figures 2.6 and 2.7. Normally, freshwater is less dense than seawater, but the cold glacial meltwater is more dense than seawater.)
2. What might explain the "seesaw" effect of D-O events cooling in the Northern Hemisphere coincident with a warm period in the Southern Hemisphere, and vice versa?

glaciers and climate change

The effects of climate warming and sea-level rise are recognized as a threat to many coastal areas of the world. The Dutch have already taken an expensive major step to minimize the potential threat of rising sea level to the subsiding delta on which much of Holland is built (**Figure 1**). Globally, observations reveal widespread melting and rapid disappearance of glaciers, with the meltwater a major element in the rise of sea level. Consequently, glaciologists are working feverishly at observing and recording glacier behavior. This effort takes them to remote, often nearly inaccessible areas of the world, such as illustrated in **Figures 2, 3,** and **4**.

1 The enormous Maeslant New Waterway Storm Surge Barrier near the Hook of Holland, Netherlands: (a) open, (b) closed. Each gate of the Maeslant Barrier is 300 meters (984 ft) long, and when closed their combined length is almost twice the height of the Eiffel Tower. The largest of Holland's four storm surge barriers, the Maeslant Barrier, protects the Rotterdam region from flooding during storm surge. Storm surge is forecast as more likely in the future because of a combination of rising sea level and subsidence of the delta. This is part of the vast *Delta werken* (Delta Works), an ambitious engineering feat built on the Rhine-Meuse-Schelde river delta, which underlies the southwestern part of the Netherlands. It consists of dikes, surge barrier gates, dams, sluices, and locks that protect the country from storm surge.

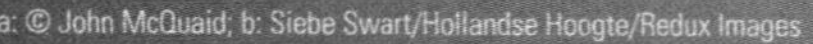

a: © John McQuaid; b: Siebe Swart/Hollandse Hoogte/Redux Images

2 Katmai Caldera, Katmai National Park and Preserve, Alaska. The caldera formed at the summit of the 2,100 meter (7,000 foot)-high Mount Katmai volcano during the 1912 eruption of the volcano centered at nearby Novarupta (see Chapter 5). After volcanic action ended, three glaciers formed on the inside walls of the caldera and flowed down to the lake. Globally, these are the only glaciers for which we know the dates of the origin. In addition, several valley glaciers originate near the rim and descend down the volcano's outer flanks. University of Vermont

3 Geologists sample a glacier meltwater stream near Kulusuk, east Greenland, reachable only by helicopter. The stream is running gray-green with silt. P. Bierman

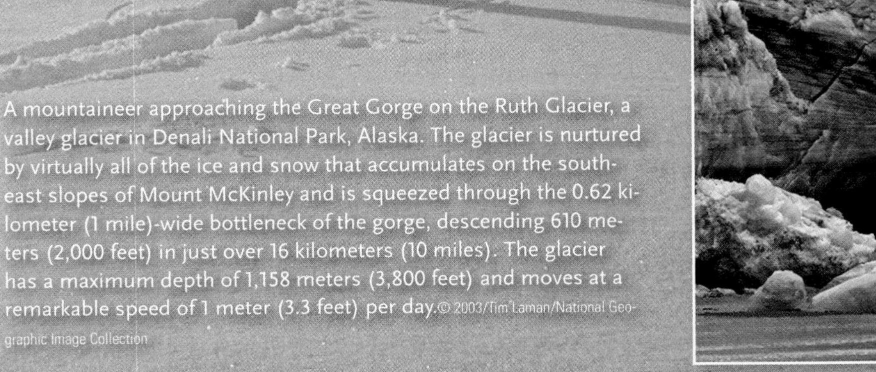

A mountaineer approaching the Great Gorge on the Ruth Glacier, a valley glacier in Denali National Park, Alaska. The glacier is nurtured by virtually all of the ice and snow that accumulates on the southeast slopes of Mount McKinley and is squeezed through the 0.62 kilometer (1 mile)-wide bottleneck of the gorge, descending 610 meters (2,000 feet) in just over 16 kilometers (10 miles). The glacier has a maximum depth of 1,158 meters (3,800 feet) and moves at a remarkable speed of 1 meter (3.3 feet) per day. © 2003/Tim Laman/National Geographic Image Collection

4 Suspended sediment-rich subglacial river gushing from beneath the base of Lamplugh Glacier, Glacier Bay National Park and Preserve, southeast Alaska. Larry Wilson/National Park Services

SUMMARY Glaciation and Long-Term Climate Change

Glaciers

Description
Large masses of ice that form on land where, for a number of years, more snow falls in winter than melts in summer, and that deform and flow due to their own weight because of the force of gravity.

Location
High latitudes and high altitudes.

Classification
Major types are continental (ice sheet), ice field, ice cap, valley (alpine), and piedmont.

Some Important Glacial Features
End moraine—ridge formed at the melting end of a glacier composed of glacially transported rocky debris.
Kettle lakes—water-filled, bowl-shaped depressions formed as glaciers retreated.
U-shaped canyon—remains after a valley glacier melts.
Fjord—U-shaped canyon drowned by the sea.

Effects of Pleistocene Glaciation
Some soil conditions, including loess deposits transported by ice-age winds and some areas of glacial till. Local groundwater conditions. Shoreline configuration due to sea-level changes and isostatic rebound. Human transportation routes.

Ice-Age Climate

Description
Temperatures 4°C to 10°C cooler than at present.

Causes
A number of interrelated factors contribute to climate changes in ways that are not clearly understood. They include obliquity of Earth's axis, eccentricity of Earth's orbit, precession of Earth's axis, variations in content of atmospheric gases, changes in greenhouse gases (GHGs), tectonic changes in the elevation of continents, volume and temperature of the oceans, changes in solar radiation, dust and aerosols, and the oceanic conveyor belt.

Consequences of Global Warming

Rising sea level. Increasing wildfires. Change in subsistence lifestyles for some cultures. Permafrost thawing. A probable ice-free Arctic Ocean in summer. Earlier spring and summer seasons and later fall and winter.

Climate Forcings

Milankovitch cycle. Increasing GHGs. Variation in solar radiation. Energy absorption by aerosols and clouds. Changes in land use. Aerosols from erupting volcanoes.

Some Model-Based Predictions of Climate Change

Globally, average temperatures increase. Globally, average precipitation increases, but by how much and where are unknown. Sea level will rise at an increasing rate, with the drowning of low-level coastal plains and displacement of thousands of coastal dwellers. Atmospheric CO_2 concentration may rise well above the 394.49 ppm of 2012. Arctic temperatures may rise as much as 8°C (13°F) above the 20th-century average, with a corresponding melting of permafrost. Ice sheets and glaciers will continue to shrink and retreat. Decreasing moisture in the Northern Hemisphere may make farming nearly impossible in much of North America and Europe. Plant and animal ranges will expand in some cases, and shrink or disappear altogether in others.

KEY TERMS

ablate
ablation zone
accumulation zone
aerosols (sulfate aerosols)
calving
cap-and-trade
carbon capture and sequestration
cloud reflectivity enhancement
El Niño
end moraine
esker
firn
fjord
geoengineering
glacial outwash
glacial till
glacier
greenhouse gases (GHGs)
integrated gasification combined cycle (IGCC)
Intergovernmental Panel on Climate Change (IPCC)
isostatic rebound
kettle lake
Kyoto Protocol
La Niña
moraine
proglacial lake
recessional moraine
zone of flowage

STUDY QUESTIONS

1. Explain the generalization "glaciers are found at high latitudes and at high altitudes."
2. Glacial ice begins to flow and deform plastically when it reaches a thickness of about 30 meters (100 ft). What is the driving force that causes the ice to flow?

3. Describe the behavior of a glacier in terms of its budget—that is, the relationship between accumulation and ablation. How does a glacier behave when it has a balanced budget, a negative budget, a positive budget? What kind of budget does a surging glacier have?
4. Describe isostatic rebound. What effect has the melting of the great Pleistocene ice sheets had on the crustal regions that were covered by thick masses of ice until about 11,000 years ago? What are the effects of this rebound on society?
5. Describe the greenhouse effect. Is it a bad thing for Earth? What causes greenhouse warming?
6. What can be done to modify (or even control) greenhouse warming?
7. What are some of the potential effects of greenhouse warming on human life?
8. What is the explanation for the speeding up of Greenland's outlet glaciers?
9. What is desirable about signing the Kyoto Protocol? What might be undesirable?
10. What is hoped to be discovered at the FutureGen 2.0 and Antelope Valley projects?
11. Why is ultramafic rock the best for CO_2 sequestration?
12. What is the relationship in ice cores between the concentration of CO_2 and ancient temperatures?
13. How do glacial and interglacial cycles affect sea level?
14. What will be the impact on people living in the northern regions of the Northern Hemisphere if the Arctic oscillation remained in a positive phase for 50 years or more?
15. How do we know that burning fossil fuels is responsible for the increase in atmospheric CO_2 concentrations from 280 ppm in 1900 to 394.49 ppm in 2012?

REFERENCES AND FURTHER READINGS

Books and Periodicals

Alley, Richard B. 2011. *Earth: The operator's manual.* New York: W.W. Norton & Co.

Broecker, Wallace S., and Robert Kunzig. 2008. *Fixing climate: What past climate changes reveal about the current threat—and how to counter it.* New York: Hill & Wang, 253 p.

Brook, E. J. 2005. Tiny bubbles tell all. *Science* 310: 1285–1287.

Economist staff. 2011, September 24. Climate change in the Arctic: Beating a retreat. *Economist* 400 (8752).

Flannery, Tim. 2005. *The weather makers.* New York: Atlantic Monthly Press, 357 p.

Hambrey, Michael, and Jürg Alean. 2004. *Glaciers,* 2nd ed. New York: Cambridge University Press, 376 p.

Hazeldine, Stuart. 2009. Carbon capture and storage: How green can black be? *Science* 325: 1647–1652.

Kaufman, D. S., et al. 2009. Recent warming reverses long-term Arctic cooling. *Science* 325 (5945): 1236–1239.

Keith, D. W. 2009. Why capture CO_2 from the atmosphere. *Science* 325: 1654–1655.

Long, A. J. 2009. Back to the future: Greenland's contribution to sea-level change. *GSA Today* 19(6): 4–10.

MacKay, David J. C. 2009. *Sustainable energy—without the hot air.* Cambridge, England: UIT Cambridge Ltd., 383 p.

Matthews, Vince. 2008. CO_2 sequestration potential for Colorado. *Rock Talk* 11(1): 1–4.

Muller, R. A. 2011, October 21. The case against global-warming skepticism. *The Wall Street Journal.*

New, Mark G., et al. 2011. Four degrees and beyond: The potential for a global temperature increase of four degrees and its implications. *Philosophical Transactions for the Royal Society* 369 (1934): 1–241.

Oleson, Timothy. 2011. Mount Rainier's shrinking glaciers. *Earth* 56(9): 17.

Orr, F. M., Jr. 2009. Onshore geologic storage of CO_2. *Science* 325: 1656–1658.

Romm, Joseph. 2007. *Hell and high water: Global warming—the solution and the politics—and what we should do.* New York: William Morrow. 292 p.

Schmidt, Gavin, and Joshua Wolfe. 2009. *Climate change: Picturing the science.* New York: W. W. Norton, 305 p.

Schrag, D. P. 2009. Storage of carbon dioxide in offshore sediments. *Science* 325: 1658–1659.

Socolow, Robert, and S. Pacala. 2004, August 13. Stabilization wedges: Solving the climate problem for the next 50 years with current technologies. *Science* 305: 968–971.

Weart, S. R. 2008. *The discovery of global warming.* Cambridge, MA: Harvard University Press. 244 p.

Internet

Intergovernmental Panel on Climate Change. 2007. Fourth assessment report: Synthesis report. Available at: http://www.ipcc.ch/publications_and_data/publications_ipcc_fourth_assessment_report_synthesis_report.htm.

Intergovernmental Panel on Climate Change. 2009. Climate science report. Available at: http://www.copenhagendiagnosis.com.

Socolow, Robert. 2011. Wedges reaffirmed. *Bulletin of the Atomic Scientists.* Available at: http://www.thebulletin.org/print/web-edition/features/wedges-reaffirmed.

Steffen Research Group. 2006. Melting glaciers and ice sheet contribute to global sea level rise. Available at: http://cires.colorado.edu/steffen/.

Tedesco, M., et al. 2011. *Year 2011 Greenland melting remains well above the (1979–2010) average: Close-to-record mass loss.* Cryogenic Process Laboratory, City College of New York. Available at: http://greenland2011.cryocity.org/.

Access an interactive eBook and chapter-specific interactive learning tools, including flashcards, quizzes, videos and more in your Geology CourseMate, accessed through CengageBrain.com.

Kolmanskop was a town founded by the wealth of diamonds in 1908, but when the mine gave out, by 1954, the town was abandoned. Here in the Namib Desert, one of the driest places on Earth, blowing sand is slowly reclaiming the town, filling buildings and sandblasting their aging exteriors.

Have you ever

wondered...

12

Arid Lands, Winds, and Desertification

Deserts are natural laboratories in which to study the interactions of wind and sometimes of water on the arid surfaces of planets.

—A. S. Walker, U.S. Geological Survey geologist

1. Do land development and agricultural practices affect soil stability in semiarid and arid regions?

2. Can migrating sand sheets and sand dunes in arid climates be stabilized?

3. We live in a dusty world. Where does the dust originate? How far can it travel?

4. What is the cause of the present drought in parts of Africa where people have lived successfully for thousands of years?

5. Is off-road vehicle use in semiarid and arid regions as damaging as claimed by some environmental groups?

A Dust Bowl in California

Historically, millions of acres of wetlands, including two lakes, flourished in parts of the San Joaquin Valley, California, as reported in the journal of Jedidiah Smith, who, with his company of beaver trappers, traversed the valley in the winter and spring of 1827. Today, only about 5% of the wetlands reported by Smith remain. They exist as publicly owned reserves, parks, and wildlife areas, such as the Kern National Wildlife Refuge west of Delano.

On the morning of 20 December 1977, an extraordinary windstorm generated an enormous dust plume (**Figure 1**), which swept out of the Tehachapi Mountains into the southern end of the San Joaquin Valley, part of California's Great Central Valley, the most productive agricultural region of the United States (**Figure 2**).[1] The dust cloud, rising as high as 1,500 meters (5,000 feet) above the valley floor, vividly illustrates how dry lands can be affected by human activities. The winds, with recorded speeds as much as 300 kilometers/hour (186 miles/hour), mobilized some 25 million tons of loosened soil from grazed lands, resulting in an extensive plume of dust, which extended as far as the northern Sacramento Valley. Wind-stripped agricultural lands added similar amounts of soil to the dust plume. The windstorm damaged vehicles and structures, destroyed orchards, stripped soils and subsoils from intensively farmed and grazed lands, eroded unsurfaced dirt roads, sheared wooden telephone poles about 3 meters (10 feet) above the ground, and toppled steel power-line towers. In some places, soil losses of as much as 60 centimeters (24 inches) were recorded. Mobile homes were blown over, windows in buildings were sandblasted and many were shattered, airplane hangars and barns were blown down, automobiles on one highway were completely destroyed by sandblasting, and semitrailer trucks were blown over by winds channeled through road cuts.

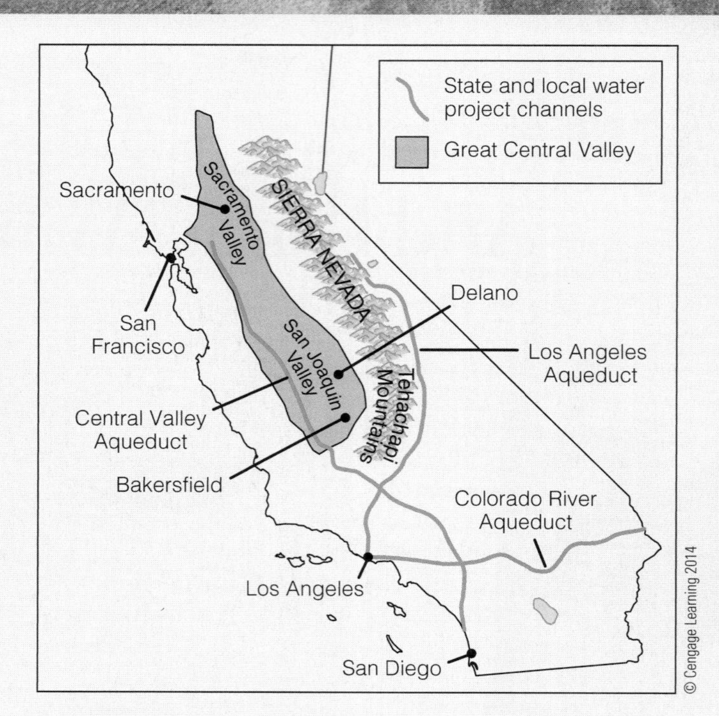

Figure 2 Index map to localities in California.

Nearly 8 kilometers (5 miles) of the Arvin-Edison Canal was filled with sand, orchards of peach and almond trees were uprooted, and immature citrus trees were destroyed. Range cattle in one area drifted downwind until they became trapped in the Arvin-Edison irrigation canal, were buried by the drifting sand, and died of suffocation. More than 2,000 square kilometers (772 square miles) of once-productive lands were affected. Some areas of the valley have not yet recovered, and may not do so for thousands of years.

A soil fungus, *Coccidioides immitis,* endemic to much of the southwestern United States, was carried aloft by the wind and spread valley fever, a respiratory ailment, throughout the region, even causing a widespread increase in reported cases in the San Francisco Bay area and the northern Sacramento Valley. One known fatality from valley fever was a gorilla in the Sacramento Zoo, more than 480 kilometers (300 miles) to the north. Five people were killed in automobile accidents caused by poor visibility. The storm was very unusual because the sizes of sediments carried in traction (dragged along the ground surface), saltation (bouncing), and suspension by the wind greatly exceeded any previously reported for eolian events.

The very high winds were generated by a large and strong high-pressure system positioned over the Great Basin states and a low-

[1]California's agricultural output amounted to $31 billion in 2006, more than that of any other state. For comparison, Hollywood's worldwide ticket sales for films that year amounted to only $25 billion.

Figure 1 Aerial photograph of the enormous dust plume that rose 1,500 meters (5,000 feet) above the southern San Joaquin Valley on 20 December 1977. Winds reached speeds up to 300 kilometers/hour (186 miles/hour). The Tehachapi Mountains are in the lower left.

pressure system over the eastern Pacific Ocean. The high-elevation Sierra Nevada blocked the westward movement of the airflow, but the air mass was able to move across the lower elevation Tehachapi Mountains and sweep into the San Joaquin Valley. The exceptional sediment transport caused by the storm was typical of lands subject to desertification: overgrazing, nearly 2 years of drought, recent plowing of the agricultural land in preparation for planting, recent stripping of natural vegetation in preparation for agricultural uses, and the general absence of windbreaks in the agricultural areas. Less important contributing factors were the stripping of natural vegetation from the numerous oil fields of the region, extensive development near Bakersfield, and the local denudation of land by recreational vehicles.

The San Joaquin Valley windstorm was not unique. In February 1977, a similar storm hit eastern Colorado and New Mexico. The setting was entirely different but involved similar human disturbances (agricultural development on relict sand dunes) and resulted in a dust plume, which was tracked over the Atlantic Ocean by satellite. In the pages that follow, we explore the factors that contribute to desertification.

Sources: Brooks, G. R. 1977. *The Southwest expedition of Jedidiah S. Smith*. Lincoln, NE: University of Nebraska Press, 134–143; Wilshire, H. G., J. E. Nielson, and R. W. Hazlett. 2008. *The American West at risk*. New York: Oxford University Press, 47–48; Wilshire, H. G., J. K. Nakata, and B. Hallett. 1981. Field observations of the December 1977 wind storm, San Joaquin Valley, California. In Pewe, T. L. (ed.), *Desert dust: Origin, characteristics, and effect on man* (GSA Special Paper 186, pp. 233–251), Geological Society of America, Boulder, CO.; Wilshire, H. G. Personal communication, May 2009.

questions to ponder

1. What was the cause of such a drastic change in the wetland character of parts of the San Joaquin Valley?
2. What will be the eventual fate of the San Joaquin Valley?

Kike Calvo via AP Images

CourseMate Connection

Start off this chapter by watching a video about desertification in China.
Log in to the CourseMate at www.cengagebrain.com to watch now.

Wind as a Geological Agent in Deserts

WIND ERODES BY DEFLATION AND ABRASION. **Deflation** occurs when sediment is eroded and blown away by the wind. This produces deflation basins where loose, fine-grained materials are removed, sometimes down to the water table (**Figure 12.1**). **Abrasion** occurs when mineral grains are blown against each other and into other objects. Abrasion by sand can remove paint from vehicles, frost windows, and produce unusual natural features.

Once fine-grained rock material is eroded, it moves, commonly forming migrating **dunes**—hills of windblown, drifting sand, that are common in arid regions and can adversely affect desert highways, railroad lines, and even communities (**Figure 12.2**). Dunes are not limited to arid regions. Strong sea breezes in humid coastal regions coupled with human activities can also exacerbate dune migration (**Figure 12.3**). In some areas, disturbance of dunes and the vegetation that helps anchor them is caused by off-road vehicles. The problems of migrating sand have been mitigated by building fences; planting windbreaks of wind- and drought-tolerant trees, such as tamarisk (**Figure 12.4**); and oiling or paving the areas of migrating sand. Dune stabilization in the humid coastal areas of Europe, the eastern United States, and the California coast has been accomplished by planting such wind-tolerant plants as beach grass *(Leymus)* and ice plant "Hottentot fig" *(Carpobrotus)*.

Figure 12.1 The Devil's Cornfield, a deflation basin that has been lowered by wind to the capillary fringe at the top of the water table; Death Valley National Park, California. The rugged plants growing here—arrow weed, which resemble shocks of corn—are able to tolerate heat and saline soil, and their roots can withstand the effects of exposure to episodic sandblasting.

P. Bierman

Figure 12.2 A windstorm outside of Walvis Bay, Namibia, the country's major port, moves sand across the only paved road leading to the city.

Arid Lands, Dust Storms, and Human Health

Each year, as much as 2 billion tons of dust are lifted into Earth's atmosphere. Most is stirred up by storms, the most important of which are appropriately named **dust storms**. A single large dust storm in North Africa may drop more than 200 tons of sediment in the North Amazon Basin of South America as global atmospheric circulation moves air masses across the Atlantic Ocean (**Figure 12.5**). North African dust storms routinely affect the air quality in the Middle East and Europe; a fine layer of African dust on snow in Western Europe is not uncommon. Also carried aloft with the dust are pollutants such as pesticides, herbicides, and microorganisms—viruses, bacteria, and fungi. About 25% of the microbes detected in Caribbean air during African dust events are plant pathogens; about 10% are opportunistic human pathogens—organisms that infect humans. Evidence suggests that fallout has direct consequences on the health of coral communities and may be responsible for high rates of asthma in the Caribbean—a 17-fold increase in asthma cases in Barbados since 1973 corresponds with the period of increased African dust associated with the drought in the Sahel. Furthermore, African dust events are connected to the meningococcal meningitis pathogen, as outbreaks of meningitis in sub-Saharan Africa often follow Saharan dust storms. The worldwide sources of dust are shown in **Figure 12.6**.

Dust storms originating in the vast steppes of northern China and Mongolia, the region ruled by Genghis Khan in

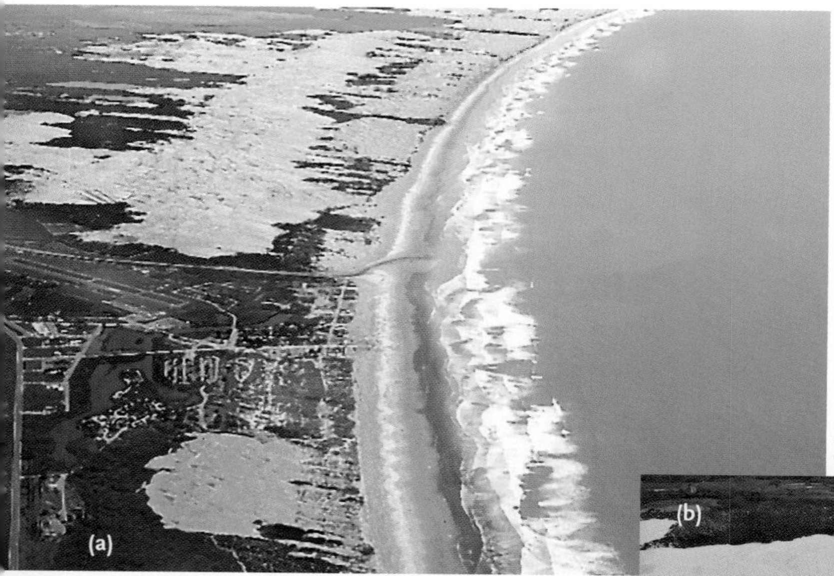

(a)

D. D. Trent

(b)

U.S. Geological Survey

Figure 12.3 (a) Coastal dune field near Pismo Beach, California. In the 1960s, these dunes began migrating onto agricultural lands because of the destruction of stabilizing plant cover by uncontrolled off-road vehicle use in the dune field. (b) The State of California prohibits vehicle use on beaches except for parts of Pismo Beach where many off-road vehicles use the sand. Here, the vegetated areas are fenced to keep vehicles out. The remainder of the barren dune is fair game for vehicular use.

Figure 12.4 Windbreaks of salt-tolerant tamarisk (salt cedar) keep drifting sand from burying railroad tracks; Palm Springs area, California.

D. D. Trent

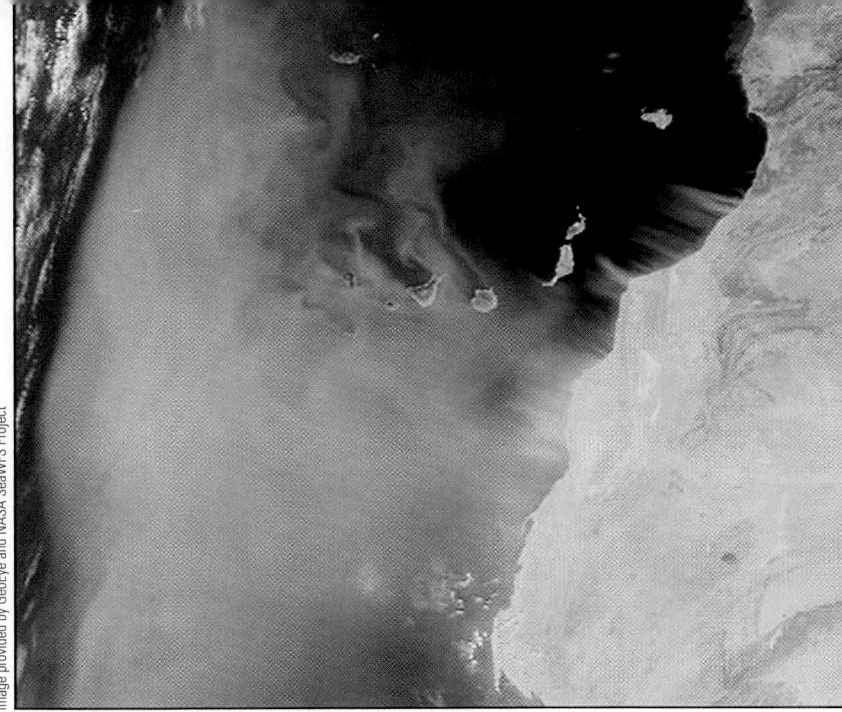

Image provided by GeoEye and NASA SeaWiFS Project

Figure 12.5 Giant dust storm in February 2001, originating in northwestern Africa and blowing over the Canary Islands. In addition to dust, such storms may carry toxic chemicals, viruses, and other infectious microbes, which are believed to pose a significant hazard to the Caribbean and other regions of the Americas. Image was taken by the Sea-viewing Wide Field-of-view Sensor (SeaWiFS) on the OrbView-2 satellite, which has recorded such storms every year since its launch in 1997.

the 13th century, regularly sweep across eastern China, Japan, and Korea; clog jet aircraft engines; cause school closures; and trigger respiratory ailments. In the 1950s, Beijing experienced a dust storm only once every 7 or 8 years, but by the 1990s, China's capital city was being hit by several storms per year, with the number increasing five-fold since the 1980s. Some Chinese scientists believe they will become more frequent and more intense. One particularly severe storm in 1993 resulted in 85 deaths and the loss of

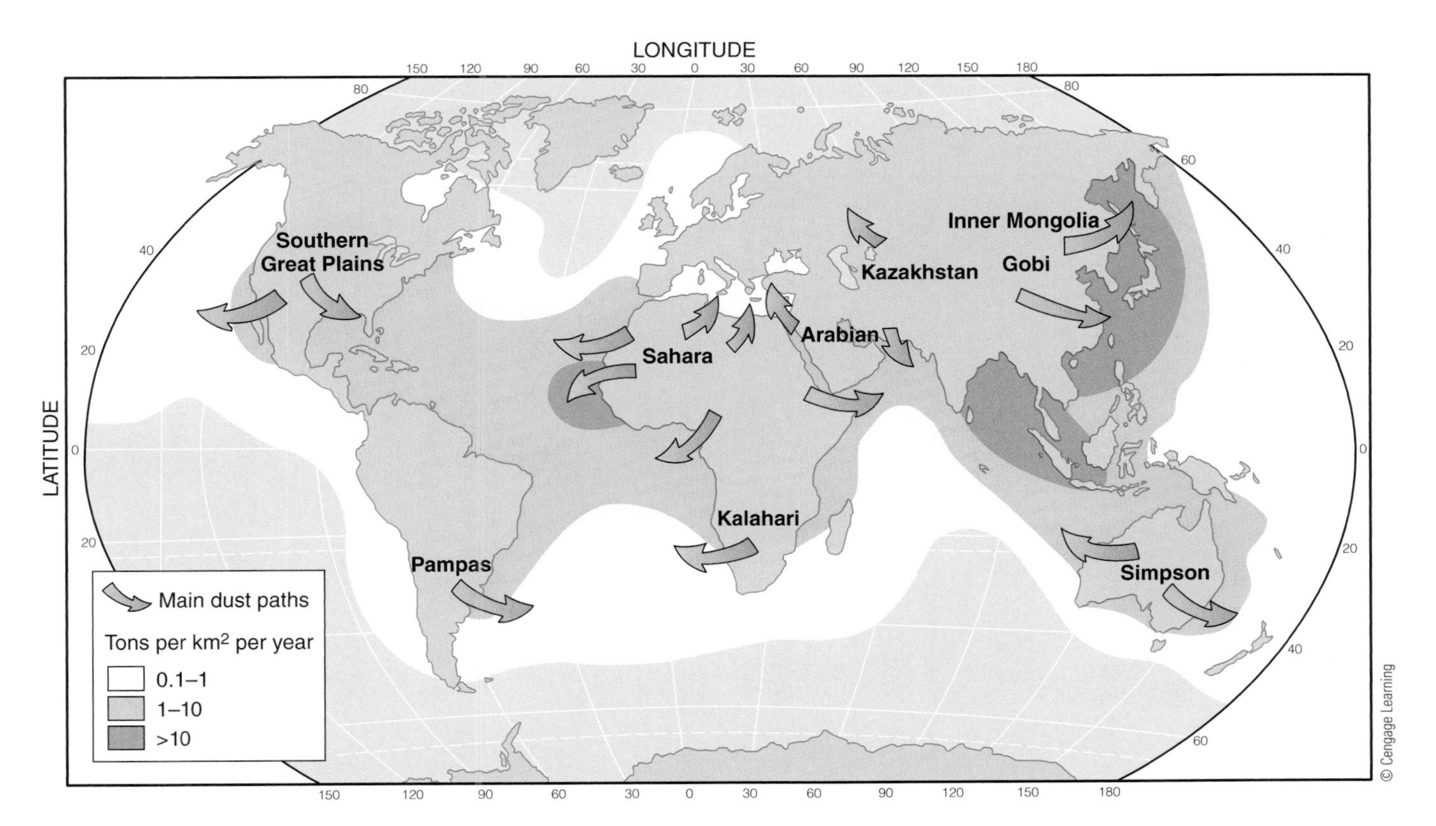

© Cengage Learning

Figure 12.6 The main sources and estimated amounts of atmospheric dust. The rates of dust transport are shown by different colors.

120,000 head of livestock and 2.3 million hectares of crops; it also destroyed more than 4,000 houses. According to the U.S. National Aeronautics and Space Administration (NASA), dust from the March 2001 "Yellow Dragon" storm in China (see Case Study 12.1) traveled across the Pacific Ocean, causing hazy skies in California, and a dust event in April 2002 sullied the skies over Colorado. Some 4,000 tons of sediment per hour can fall in the Arctic during a large dust storm, and the fallout of herbicides and pesticides into the Arctic is common from storms originating in northern China's semiarid farmlands. These chemicals are found in animal tissues, in human breast milk, and among the people living in the Arctic. We examine the underlying causes of severe dust storms in the following section on desertification.

The recognition that airborne sediments from one continent can be carried over great distances and affect air quality thousands of miles from their source raises several important unanswered questions that demonstrate the depth of our ignorance: Can pathogens transported in airborne dust cause outbreaks of disease? Can harmful chemicals transported halfway around the world impact the well-being of ecosystems and human health? What kinds of microorganisms can survive long-range transport in the atmosphere, and can they compete successfully with the indigenous microbial community at a new location?

Deserts

The term **desert** means different things to different people. For some, a desert is a dry, barren region of sand dunes; for others, it is a desolate region of bare, rocky outcrops and cactus; for still others, it is a treasure trove of scenic wonders and unique plants and animals. Regardless of how they are perceived, deserts compose about one third of Earth's surface and are characterized by meager rainfall, scanty vegetation, a distinct landscape, and a limited population of people and animals. In deserts, the effects of running water, tectonic forces, and wind are clearly apparent. Because these factors combine in different ways in different places, the appearances of various desert landscapes differ as well. Deserts may be hot or cold. They may be mountainous and rocky or vast, flat sand sheets, and some are covered with gravel. But all deserts are dry. Although deserts appear harsh and rugged, they are actually quite fragile, and the misuse of arid lands is a serious problem in many parts of the world (**Figure 12.7**).

Kinds of Deserts

A desert (Latin *desertus*, "deserted, barren") is defined as a region with mean annual precipitation of less than 25 centimeters (10 inches), with a potential to evaporate more water than falls as precipitation, and with so little vegetation that it is incapable of supporting abundant life. Note that high temperatures are not necessary for a region to be called a desert—only low precipitation, a dry climate, and limited biological productivity. On this basis, we can identify five kinds of desert regions: polar, subtropical, mid-latitude, rain-shadow, and coastal (**Table 12.1**).

Why are the major deserts of the world located between 20° and 40° of latitude?

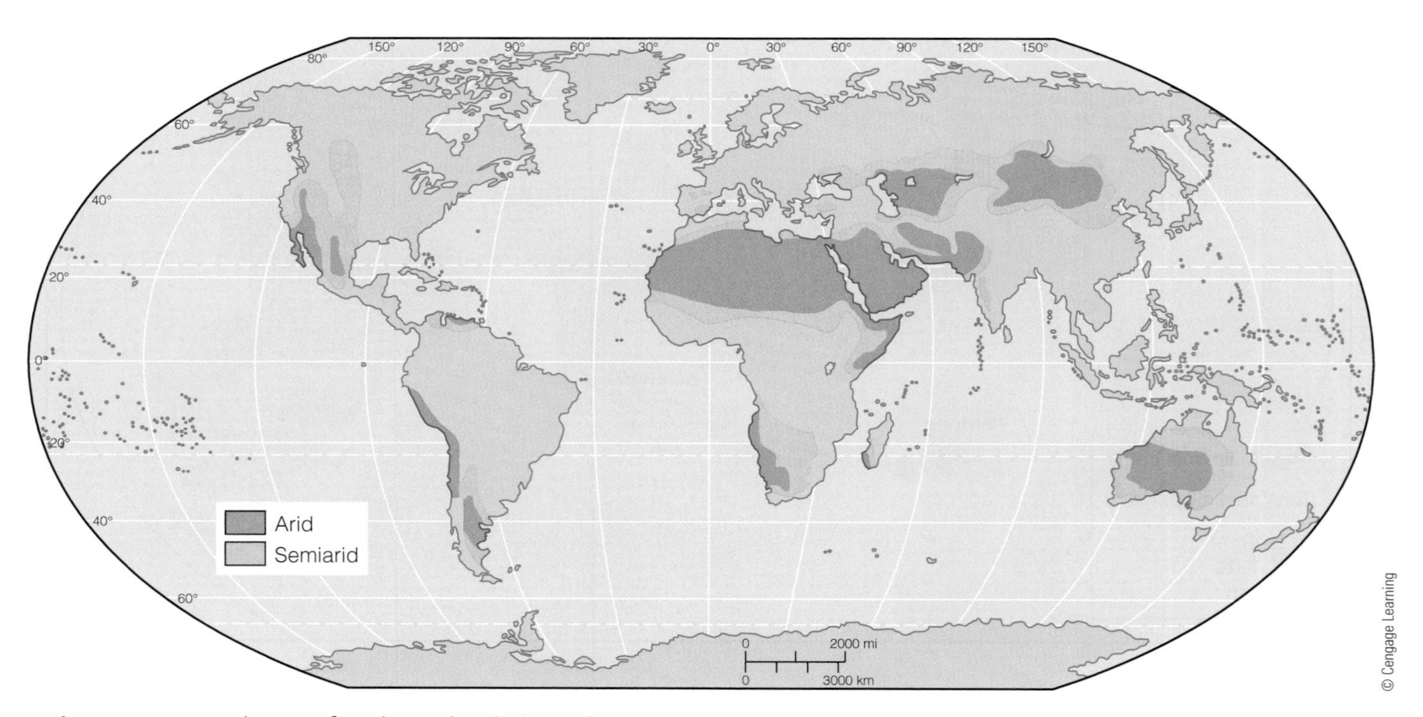

Figure 12.7 Distribution of Earth's arid and semiarid regions.

- **Polar deserts** are characterized by low precipitation and intense cold. The most desert-like polar areas are the ice-free dry valleys of Antarctica. The Antarctic continent and parts of Greenland are deserts. They are very dry, and most of the freshwater there is in the form of ice.
- **Subtropical deserts** (also called *tradewind deserts*) are Earth's largest dry expanses. They lie in the regions of subsiding air beneath semipermanent high-pressure systems between 20° and 40° latitude. They receive almost no precipitation and have large daily temperature variations. The Saharan, Arabian, Kalahari, and Australian deserts are this type.
- **Mid-latitude deserts**, at higher latitudes than the subtropical deserts, exist primarily because they are positioned deep within the interior of continents, separated from the tempering influence of oceans by distance or a topographical barrier. The Gobi Desert of China and Mongolia, characterized by scant rainfall and high temperatures, is an example. Marginal to many mid-latitude deserts are **semiarid grasslands**—extensive, treeless grassland regions. Drier than prairies, they are especially sensitive to human intervention. In North America and Asia, the mid-latitude and subtropical arid regions merge to form nearly unbroken regions of moisture deficiency.
- **Rain-shadow deserts**, also in the mid-latitudes, are due to mountain ranges that act as barriers to the passage of moisture-laden winds from the ocean, a situation leading to the formation of rain-shadow deserts. As water-laden maritime air is drawn across a continent, it is forced to rise over a mountain barrier. This causes the air to cool and to lose its moisture as rain or snow on the windward (upwind) side of the mountains. By the time the air descends on the mountains' leeward side, it has lost most of its moisture, and there we find desert conditions (**Figure 12.8**). The deserts of eastern Washington and Oregon, leeward of the Cascade Mountains, are examples of rain-shadow deserts. Death Valley, the lowest and driest place in North America, is in the rain shadow of the Sierra Nevada. The Sierra Nevada Mountains receive heavy precipitation, and Mount Whitney, their highest point and the highest point in the conterminous 48 states, is a scant 100 kilometers (60 miles) from arid Death Valley.
- **Coastal deserts** lie on the coastal side of a large landmass and are tempered by cold, **upwelling** oceanic currents (**Figure 12.9**). Their seaward margins are perpetually enshrouded in gray coastal fog, but inland, where the air temperatures are higher, the fog evaporates in the warmer, dry air. The Atacama Desert in Chile, which may be the driest area on Earth, is a rain-shadow desert as well as a coastal desert. The Andes on the east block the flow of easterly winds, and air from the Pacific Ocean on the west is chilled and stabilized by the Humboldt Current, the ocean's most prominent cold-water current. A similar geographic setting on the southwestern coast of the African continent keeps the Namib Desert dry.

Collectively, mid-latitude deserts and semiarid grasslands cover about 41% of Earth's land area.

Table 12.1 Classification of Deserts

Type	Characteristic Location	Examples
Polar	Region of cold, dry, descending air with little precipitation	Ice-free, dry valleys of Antarctica
Subtropical	Belt of dry, descending air at 20°–30° north or south latitude	Sahara, Arabian, Kalahari, Australia's Great Sandy and Simpson
Mid-latitude	Deep within continental interior, remote from the influence of an ocean	Gobi, Takla Makan, Turkestan
Coastal	Coastal area in middle latitudes where a cold, upwelling oceanic current chills the shore	Atacama, Namib
Rain-shadow	Leeward of a mountain barrier that traps moist ocean air	Eastern Oregon and Washington State

consider this . . . Why is the Kona coast on the western side of the island of Hawaii arid (25–50 centimeters of rainfall per year, a scant 10-20 inches), yet the eastern side of the island is humid and receives heavy rainfall, as much as 750 centimeters or 300 inches a year near Hilo?

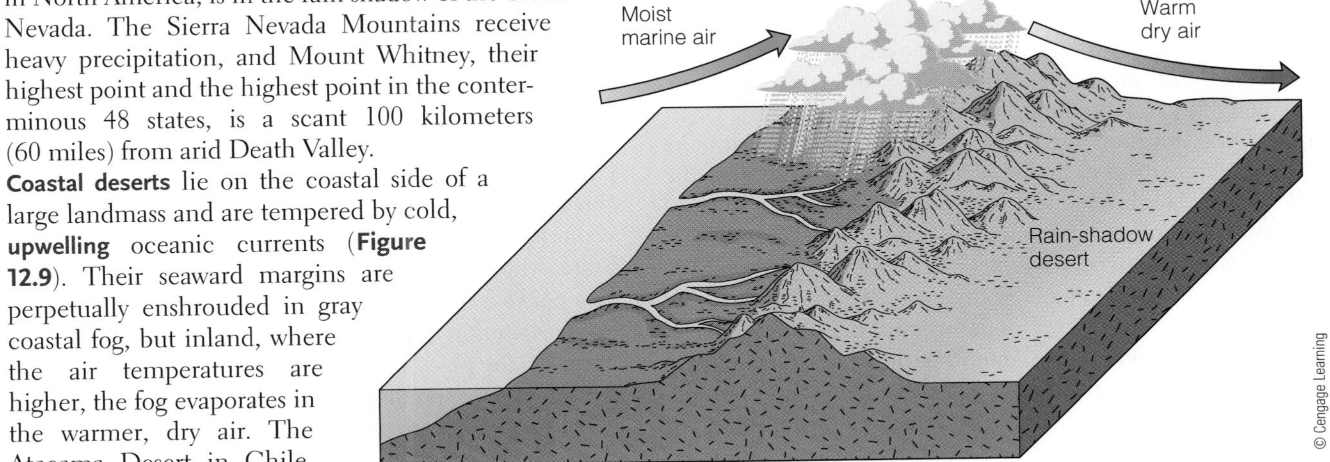

Figure 12.8 The origin of rain-shadow deserts in middle- and high-latitude regions. Moist marine air moving landward is forced upward on the windward (upwind) side of mountains, where cooling causes clouds to form and precipitation to fall. The drier air descending on the leeward (downwind) side warms, producing cloudless skies and a rain-shadow desert.

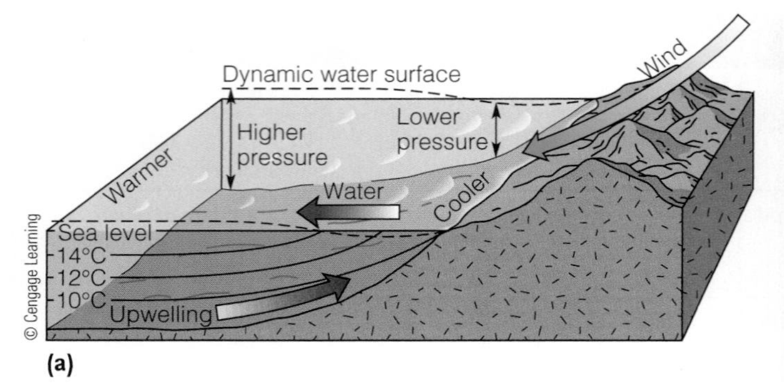

Figure 12.9 (a) Winds blowing offshore drive upwelling of cool water from depth along a coastline. Upwelling not only delivers nutrients, such as phosphates and nitrates, to surface waters, and thus provides for a rich fishery, but it also keeps coastal waters cold, limiting evaporation and thus precipitation. (b) The coastal plain of Namibia is not only flat, but exceptionally dry. What appears to be water in this photograph is a desert mirage. Most years, this area, known as the Skeleton Coast of Namibia, gets less than 10 millimeters of rain (less than half an inch). Many years, it never rains at all.

Desertification

Deluges of blowing sand and dust—such as the one in California's San Joaquin Valley, described in the chapter opener—are common in many arid, semiarid, and subhumid regions. In the past, sandstorms were rare in semiarid grasslands marginal to mid-latitude deserts, but this has changed. Across the globe, we now see deserts expanding into formerly productive semiarid and subhumid areas. Such degradation of these dry lands is called **desertification**, and it is occurring at alarming rates, especially in countries that can least afford to lose productive land (**Figure 12.10**).

The United Nations offers a concise definition of desertification as the "degradation of land in arid, semi-arid, and dry sub-humid areas. It is caused primarily by human activities and climatic variations. Desertification does not refer to the expansion of existing deserts. It occurs because dryland ecosystems, which cover over one third of the world's land area, are extremely vulnerable to overexploitation and inappropriate land use."[2]

Desertification breaks down ecosystems by reducing the productivity of desirable plants, undesirably altering the flora and fauna, accelerating soil loss, and altering the amount of biomass a region can support. Some researchers are modifying the definition of *desertification* to signify a permanent reduction of productivity, when the land becomes desert and can no longer support the plant growth it did in the past. They feel the term is more relevant when the change is permanent on a human scale. The loss of productive land for a few years because of drought or improper use is one thing, but to lose it for perpetuity is far more serious.

Desertification of Earth's arid lands has been occurring for millennia, but until the 20th century it was often seen as a local problem, affecting only a few people, and the problem was ignored because new lands were always available. Currently, with an exploding world population, that's all changing. Desertification is devouring more than 60,000 square kilometers (23,000 square miles) of Earth's arid grasslands each year—an area about the size of Connecticut, Massachusetts, and Vermont combined—and putting another 181,300 square kilometers (70,000 square miles) at risk. More than 250 million people are directly affected by desertification, and perhaps a billion others in more than 100 countries are at risk.

The expansion of deserts is generally attributed more to human abuse than to climate. The overgrazing of sparse vegetation (**Figure 12.11**); trampling and compacting of soil by livestock; clearing of the land of trees and brush for fuel without reforesting; rapid growth of large desert cities, such as Phoenix and Las Vegas in the southwestern United States and other examples in countries such as Egypt and Saudi Arabia; depletion of groundwater for desert irrigation and urban growth; replacement of native vegetation in arid and semiarid regions with cultivated crops; and soil salinization caused by evaporation of irrigation water have all been identified as root causes of desertification. Even inappropriate economic policies—such as those practiced in the United States with agricultural subsidies, which averaged $16.4 billion per year from 1998 to 2007—are cited as a cause. Such policies result in increased production for export, which encourages the placement of more land under cultivation, resulting in corresponding increases in erosion and soil loss.

Drylands occasionally experience drought. When rainfall in these regions is sufficient, the cultivation or grazing of drylands presents no problems. However, once the region is hit by even a modest drought, the crop plants die, few native plants remain that are naturally adapted to survive under drought conditions, and winds carry away what little topsoil exists.

[2] http://www.fao.org/desertification/intro_txt/en/desert.html

SPOTLIGHT FIGURE 12.10 Arid Lands, Winds, and Desertification

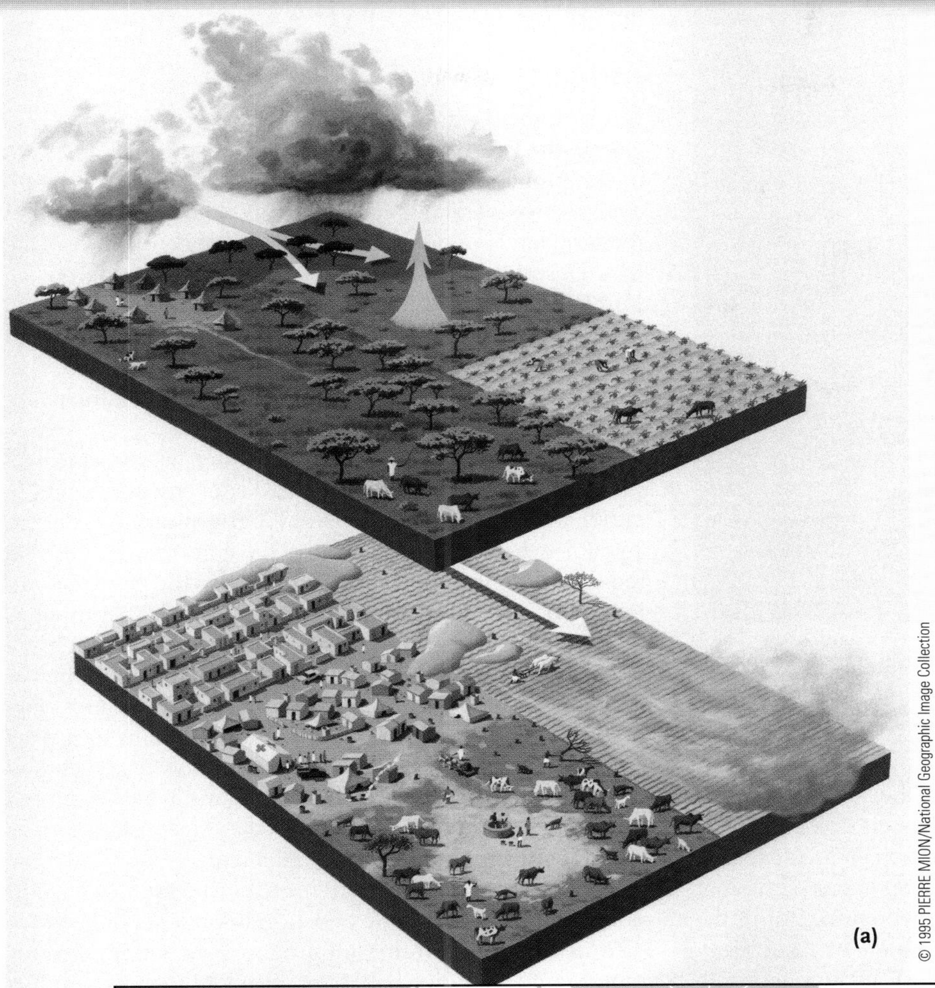

(a) Human impacts on the landscape change both Earth's surface and the resulting water cycle. Clearing land for farming removes vegetation, which can lead to desertification in two ways. First, the trees act as pumps, moving deep groundwater to the atmosphere through their leaves; without the trees pumping water, the atmosphere grows drier and the rains less frequent. Second, tree roots and the roots of other plants keep the soil from eroding. Once the land is plowed and bare, wind storms can erode the land, removing fertile topsoil and burying communities in sand and dust. (b) The map shows that many arid and semiarid regions of the world are currently experiencing desertification.

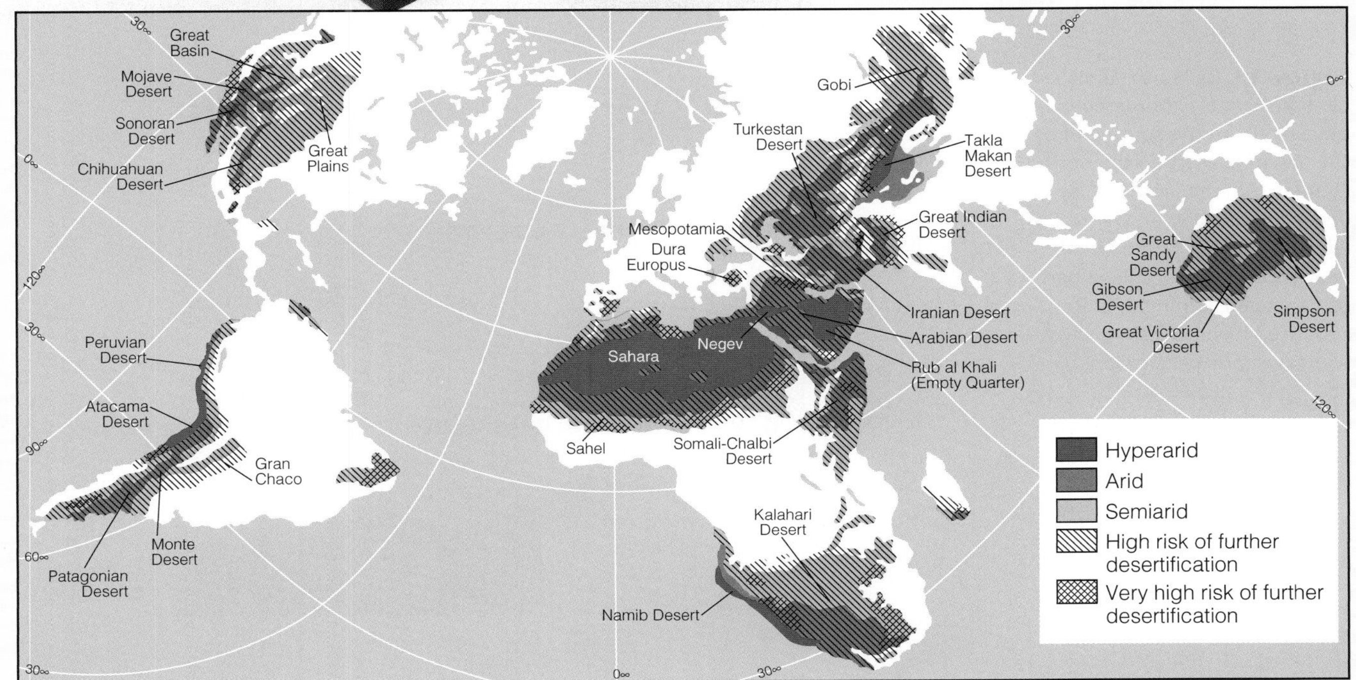

Figure 12.11 Desertification in progress: intensely overgrazed rangeland (left) and lightly grazed (right). Can you describe the impact that grazing has had on this landscape?

Thus, a combination of even wise use of the land and drought can be responsible for the expansion of desert regions. Misuse of the land obviously exacerbates the situation. Generally, droughts have tended to be short-term phenomena of a year or two, and the land recovers after the rains return. But in the past decade, some regions of the world experiencing desertification have not seen the rains return. Instead, they appear to be undergoing a shift to a new, drier climate. Human changes in land use may be an important part driver, but suggesting land use as the sole cause of desertification is no longer adequate. What other factors are involved?

Africa

It was the desertification of Africa's sub-Saharan Sahel region in the late 1960s that first brought the world's attention to the problem of desertification and prompted thorough analyses by several climate specialists. The **Sahel** is a belt of semiarid lands along the southern edge of North Africa's Sahara Desert that stretches from the Atlantic Ocean on the west to the high mountains of Ethiopia on the east—300 to 1,500 kilometers (180–940 miles) wide (**Figure 12.12**)—in which a large and rapidly expanding population resides. In 1980, the population was 30 million, but by 2000, the population had grown to about 50 million.

Typically, the Sahel region has dry winters and wet summers, when the intertropical convergence zone (a rather weak trough of low atmospheric pressure near the equator, where the trade winds converge) moves into the region, bringing rain. Rainfall amounts in the past have varied from year to year, averaging 10 to 20 centimeters (4–8 inches), but in some years it has been as much as 50 centimeters (20 inches) in the southern part where the Sahel blends into the belt of humid savannah grasslands (**Figures 12.13** and **12.14**). In 1968, the summer rains began decreasing, exacerbating the human-induced vegetation changes and initiating drought conditions, which captured the world's attention with grim images of skeletal mothers and starving children with bloated bellies.

Since the 1990s, the rains have returned somewhat, but not before 250,000 people and 12 million head of livestock died in one of the world's most devastating famines. Many more would have died had it not been for the massive influx

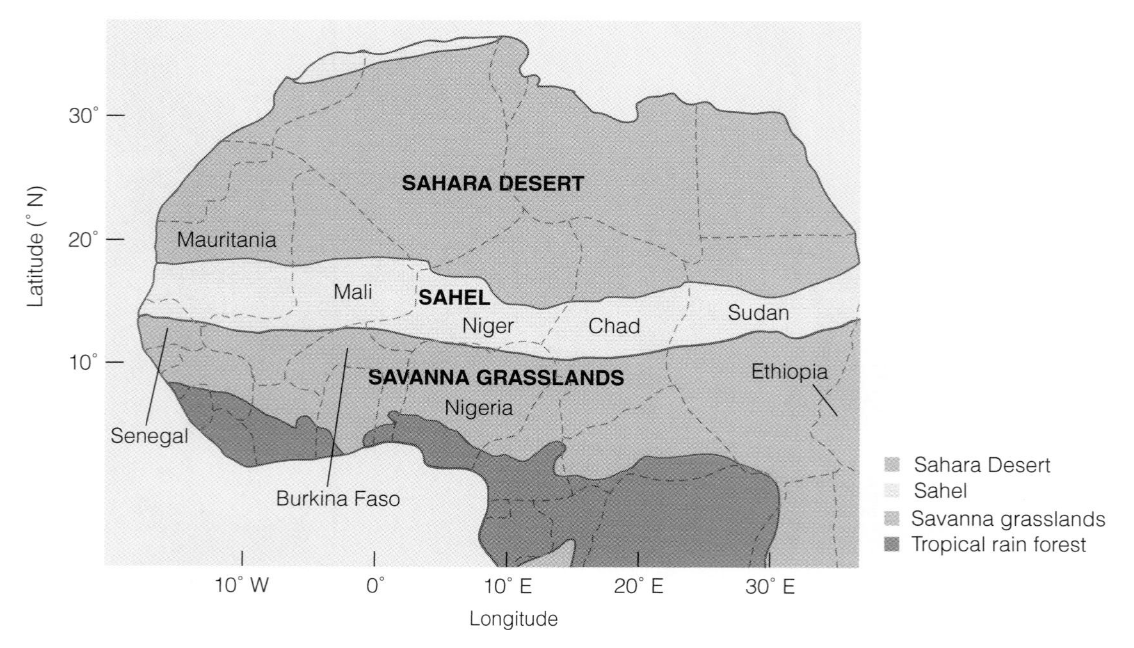

Figure 12.12 North Africa's semiarid Sahel is bordered by grasslands on the south and by the Sahara Desert on the north. (Source: Ahrens. *Meteorology Today Introduction to Weather, Climate, and the Environment,* 4e. © 1991 Brooks/Cole, a part of Cengage Learning, Inc. Reproduced by permission, www.cengage.com/permissions.)

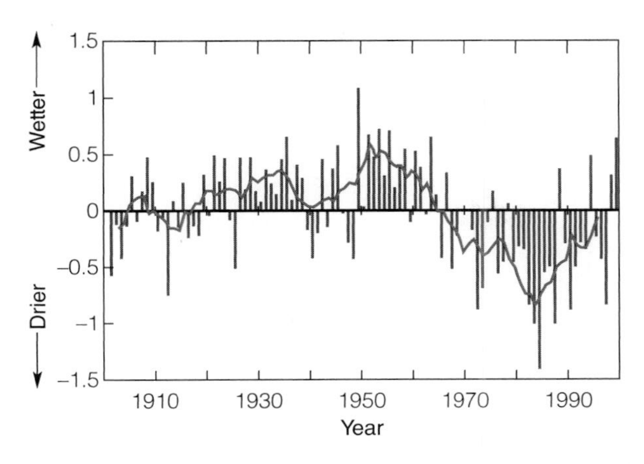

Figure 12.13 The average rainfall in the Sahel varies both yearly and over decade-long periods. The red line shows the Sahel rainfall averaged over a 7-year window; starting in 1952, that average dropped and didn't begin to recover until 30 years later. The black bars indicate yearly rainfall and show great variability superimposed on a longer trend. (Source: Climatic Research Unit, University of East Anglia, England, and Global Historical Climatology Network, Oak Ridge National Laboratory.)

of outside aid. The region is still experiencing a rainfall deficit, and the drought has extended to the Horn of Africa—Somalia, northern Kenya, and the lowland area of eastern Ethiopia (**Figure 12.15**)—where there has been only 1 year of good rain in the region since 1999. Even if the rains return, it is expected that 80% of the cattle in Somalia's Gedo region have died or will die.

The conventional wisdom has been that human use or misuse of dry lands, triggered by a drought, is responsible for desertification. But was that the case for the Sahel disaster? In 2003, climate specialists analyzed the historical record of Indian Ocean sea surface temperatures (SSTs) using computer modeling to simulate rainfall in the Sahel from 1930 to 2000. Surprisingly, the analysis showed that land-use changes in the region were insufficient to cause a climate shift. Instead, the extended drought resulted from rising SSTs in the Indian Ocean, Earth's most rapidly warming ocean. As the ocean transferred heat to the atmosphere, the atmospheric conditions governing the Sahel's monsoon system of wet summers and dry winters were weakened, the consequence of ocean warming likely induced by increasing greenhouse gases. Other potential factors may be variations in ocean circulation and cooling of the North Atlantic Ocean because of dimming from an increase of fine particles in the atmosphere.

The effect of the weakened monsoon was amplified by land–atmosphere interaction, producing a positive feedback to the drying tendency initiated by changes in the global SSTs. The land–atmosphere interaction began with the loss of cultivated and native vegetation cover, because of the rainfall deficit and firewood gathering. Combined, these produced a second feedback mechanism, increased albedo (reflectance of solar energy hitting Earth) and reduced evaporation and transpiration (pumping of water) by trees, factors similar to those attributed to human land-use changes. Once vegetation cover was reduced, little was left to protect the soil from winds, and the region became even further degraded (**Figure 12.16**). Thus, human land-management practices may have played a part, but large-scale external processes were the underlying cause of desertification.

Another aspect of drought in arid and semiarid regions is dust, and the amount of dust originating from the Sahel during wet periods was markedly less than during the drought. Because dusts originating from the Sahel's windstorms are now known to account for about half of

Figure 12.14 A sharp line marks the boundary between natural pasture and a dune encroaching from the Sahara in the West African country of Mali, a graphic example of desertification. The Sahara Desert is creeping southward into the semiarid Sahel region, and the Sahel, in turn, is expanding into the grasslands south of it.

Figure 12.15 The Horn of Africa, an area threatened by desertification. (a) Location map. (b) Drought in Kenya (south of the Horn of Africa) forces women to carry heavy jerry cans filled with water long distances over dry ground to their village.

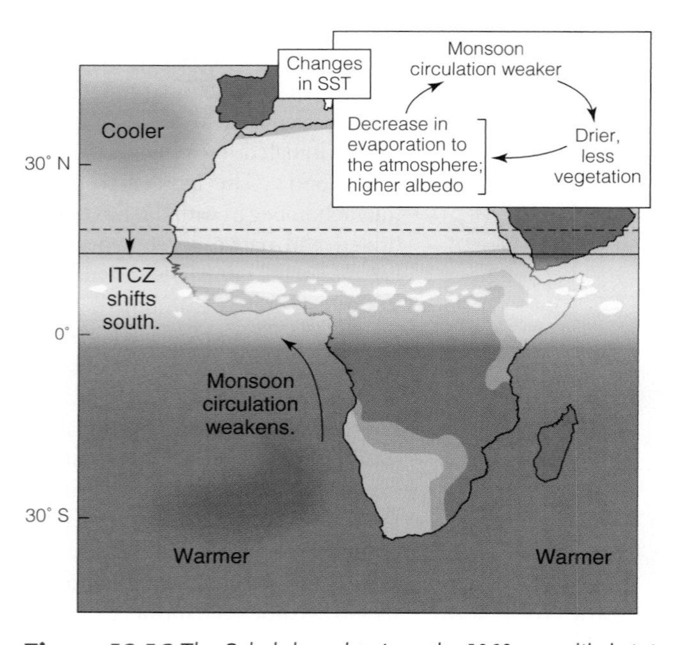

Figure 12.16 The Sahel drought since the 1960s was likely initiated by a change in global ocean temperatures, which reduced the strength of the African monsoon, and exacerbated by land-atmosphere feedbacks through loss of natural vegetation and land cover change. Land use by humans may have played a role. ITCZ = intertropical convergence zone; SST = sea surface temperature. (Adapted from Zeng, N. 2003. Drought in the Sahel. *Science* 302: 999–1000.)

the world's total dust load, and dust causes global dimming, a phenomenon that decreases the amount of solar energy hitting Earth's surface, the drought in the Sahel may have an unexpected and widespread influence on Earth's climate.

United States

In the United States, overgrazing has left an imprint on the arid lands of the West, along with the salinization and water logging that is common in many irrigated valleys. By the mid-1800s, overgrazing in the Southwest was already a problem, but when the railroads arrived in the 1880s, the explosion of cattle grazing that followed resulted in the carrying capacity of the land being exceeded, accompanied by increased erosion and arroyo (gully) deepening. Consequently, the southwestern United States is a textbook example of the conversion of productive semiarid land to desert. In particular, the Rio Puerco Basin of central New Mexico is one of the most severely eroded regions (see Case Study 12.3).

Much of the American West has been in drought conditions for the last decade or more, conditions that have not been experienced for about 700 years. The heavy rains and snowfall in the winter of 2009–2010 in part of the region offered some relief, but 1 wet season in 12 does little to ease the net deficit of so many dry years. Furthermore, the temperature increase throughout the West since the 1950s has reduced the average snowpack (**Figure 12.17a**) and has produced an earlier snowmelt (see Figure 12.17b). Because the years since 1979 have been the warmest three decades in the last 100 years or more (see Figure 11.27), it's likely that the connection between high temperatures and low rainfall is not a coincidence.

As with the Sahel, this deficit seems to be connected to rising ocean temperatures. Between 1998 and 2002, the central western Pacific Ocean was far warmer than normal and the eastern tropical Pacific was cooler. These conditions changed the path of the jet stream, causing storms that would normally move eastward across Oregon and into the central United States to shift northward. Southwest Asia and

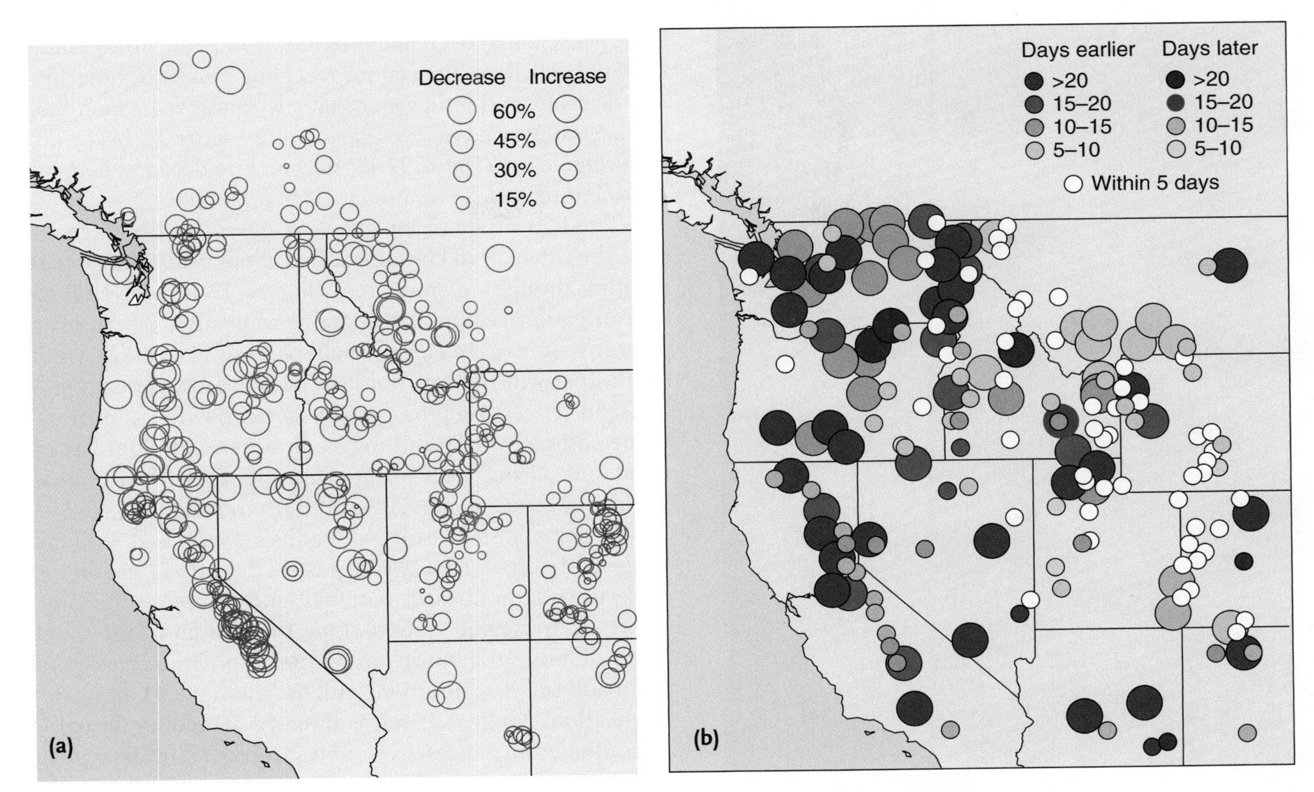

Figure 12.17 A modest temperature increase in the western United States from 1950 to 2000 (a) reduced the snow-water equivalent of the winter snowpacks and (b) shifted the peak snowmelt to earlier in the summer. (Adapted from Robert Service. 2004. *Science* 303: 1124–1127. Reprinted with permission from AAAS.)

southern Europe experienced droughts during the same period, and it appears that the changes in the Pacific Ocean SSTs, coincident with the warming of the Indian Ocean, acted in concert to contribute to widespread mid-latitude drying. The warmth of the west Pacific and Indian oceans was unprecedented and consistent with forcing by increased greenhouse gases in the atmosphere.

Southern Hemisphere

In the Southern Hemisphere, Australia, the desert continent, has experienced unprecedented drought, especially in the continent's southwest. During the first 150 years of European settlement in the southwest, winter rainfall was reliable. Now, however, the region experiences a rainfall deficit that is credited to a combination of global warming and the destruction of the ozone layer. Reduction of ozone has cooled the stratosphere over Antarctica, which has drawn Australia's zone of southern rainfall farther south. This region of former winter wheat farming in southwest Australia is now turning to desert, with the condition exacerbated by increasing summer rainfall. Because of the historically erratic nature of the summer rains, the farmers have never raised summer crops, so the rain now falls on vacant wheat fields, and the water percolates down to the water table, where it encounters salt, which the prevailing westerly winds have been blowing in from the Indian Ocean for millennia.

Under natural conditions, the native plants utilized the summer rains and the salt remained untouched. Currently, however, as the rains fall on bare fields, evaporation draws salty water upward to the surface, poisoning all the plants that it touches (**Figure 12.18**). The former wheat-growing region is the world's worst case of dryland salinization, and it comes at a cost of billions of dollars. In addition to the lost agricultural production, houses, roads, and railroads are being damaged by the salt. There appears to be no solution to the salinization.

The Future?

Scientists have good evidence to suggest that our warming climate will intensify desertification over many parts of the globe, but where desertification is driven by land-use change, it can be reversed or halted—the process can be reversible. An example of reversing desertification occurred in parts of the Great Plains of the United States, which became well-known in the 1930s when a combination of poor farming practices and drought turned much of the region into the Dust Bowl. Vastly improved methods of land and water management and farming methods have prevented the Dust Bowl disaster from recurring. Nevertheless, desertification still affects hundreds of millions of people on nearly every continent.

Desertification is a complex and subtle process of deterioration that may be reversible, but disturbed desert and marginal grassland ecosystems are slow to recover. This is because those ecosystems are characterized by slow plant growth, little bacterial activity in the soil for recycling nutrients, little species diversity, and little water. Already sparse vegetation that has been destroyed by overgrazing (see Figure 12.11) or off-road vehicle use (**Figure 12.19**) may require decades, perhaps even centuries, to be restored to its natural state.

The best program for reversing desertification is to reduce the intensity of land use and to adopt measures such as planting drought-tolerant trees (**Figure 12.20**), developing drip irrigation systems for effective water use, establishing protective screens (e.g., grass belts or strips of trees) as windbreaks that will help reestablish natural ecosystems, building sand fences, keeping grazing stock moving, and in extreme cases, covering drifting sand dunes with boulders or wood and paving areas to interrupt the force of wind near the face of dunes to prevent sand from moving.

In areas just outside Mongolia's Gobi and China's Tengger deserts (see Case Study 12.1), efforts are being made to reclaim abused marginal lands that have become desert. At a research station along the margin of the Gobi Desert, Chinese scientists, in partnership with others from Japan and Israel, experiment with techniques for halting desertification. Their research is directed at finding drought- and salt-tolerant grasses, trees, and shrubs, and utilizing drip irrigation for raising such crops as apples, onions, squash, and watermelons in hope of recovering lands lost to desert dust, turning them once again into productive farmland. On a larger scale, a "Green Wall," which will eventually stretch more than 5,700 kilometers (3,540 miles), much longer than China's famous Great Wall, is being planted in northeastern China to protect sandy deserts that are considered to have been created by human misuse.

A good example of natural recovery is seen in Kuwait in an unexpected consequence of the 1991 Gulf War. Thousands of unexploded land mines and bombs are scattered across the desert in the western part of the country. These have discouraged off-road driving by sport hunters of the region's animals. The native vegetation is reestablishing, and the desert now resembles a U.S. prairie. The number of birds in the area is much greater than before the war.

Figure 12.18 Farming practices raise the water table, bringing with it salt from below. Salt encrusts the bases of dead trees in a lake bed on the property of the Crute family, in Dowerin, Western Australia.

Figure 12.19 Off-road vehicle (ORV) damage to the Utah desert. "ORV play area" at Factory Butte, near Cainveille, Utah, is an example of how human activities contribute to desertification. Because vehicles compact and alter the surface, they cause reductions in vegetation and soil permeability, which lower soil moisture content, and thus affect local biological systems. Desert rains erode the tracks down the slope, transforming them into gullies, which increases runoff and thereby increases soil loss in the delicate arid environment of the United States west of the 100th meridian. Within a few minutes, the simple passage of a few vehicles can disturb soil that required hundreds of years to develop. The area shown in this photograph is underlain by a shale formation that is highly susceptible to direct mechanical erosion by the wheels of ORVs. The lack of plants results from significant amounts of selenium in the shale.

Recognizing the need for an innovative approach in combating desertification, 191 governments joined the United Nations Convention to Combat Desertification in September 2005. The treaty recognizes that the causes of desertification are complex, and that the struggle to reclaim and protect drylands will be difficult—there will be no quick fix.

In the marginal lands in China, the Sahel, and the Horn of Africa that are threatened by desertification, how would you go about convincing local farmers that they should invest money and effort in planting windbreaks and installing drip irrigation?

(b)

(a)

Figure 12.20 Reclaimed farmland lost to desertification at a farm in Niger, on the southern edge of the Sahel. (a) The farm after 1 year's participation in a desert reclamation program. (b) The farmer and his farm after 7 years of participation in the reclamation program. The young trees are drought-tolerant perennials that will eventually provide a market crop and form a windbreak to stabilize the soil in this region of nearly persistent drought. Millet stalks remain following the farm's annual harvest.
Eden Foundation

case study

12.1 Is Desertification Swallowing China?

Is Desertification Swallowing China?

In the People's Republic of China, Longbaoshan, a semi-arid farming area about 65 kilometers (40 miles) northwest of Beijing, is at the leading edge of China's war against desertification. Longbaoshan is on the advancing edge of the desert, where desertification is driven by a combination of drought, poor land management, overpopulation, and overgrazing.

Between 1994 and 2000, desert lands in China grew by nearly 53,000 square kilometers (20,300 square miles)—an area more than the combined area of New Hampshire and Vermont. In the country's driest regions, such as Inner Mongolia (**Figure 1**), drifting sands smother grasslands in scenes reminiscent of the Dust Bowl days of the 1930s in the U.S. Midwest (see Chapter 6). China, a quarter of which is arid or semiarid, has 20% of the world's population but only 7% of the world's arable land. China's current desertification causes financial losses of $6.7 billion a year, affects the lives of 400 million people, and threatens farmers and herders. China will face an unprecedented ecological and human disaster if the marching sands are not halted.

Severe sandstorms, known as Yellow Dragons, frequently turn Beijing's sky yellow, blind its residents by cutting visibility to less than 90 meters (300 feet), and dump as much as 30,000 tons of sand on the city. Yellow Dragon sandstorms sometimes merge with rain clouds to inundate the city with mud. In the last decade, Beijing has been subjected to increasingly frequent and more severe dust storms.

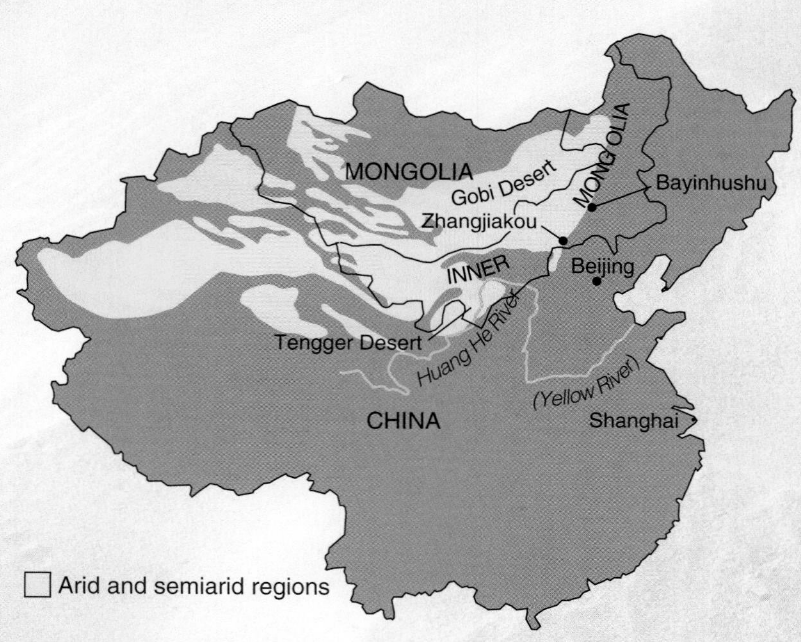

Figure 1 Index map of arid and semiarid regions of China, Inner Mongolia, and Mongolia. (Sources: Jiang, Gaoming. Personal communication, May 2009; Norville, Dennis. 2007, July. Getting to the roots of killer dust storms. *Science* 317(20): 314–316; Ting Ting Zhang, North China desertification critical 2006. Available at: www.china.org.cn/English/2006/Apr/166159.htm.)

Can China Control Desertification?

The Chinese government has been trying to halt the moving sands and dust storms by planting trees, shrubs, and grasses. Green buffer zones of trees have been established in critical regions. A project intended to protect Beijing, and undertaken to prepare the region for the 2008 summer Olympic Games, involved reclaiming a large area recently lost to desertification. In a 10-year project, in the District of Zhangjiakou City, about 150 kilometers (90 miles) northwest of Beijing, 250,000 soldiers tried to stop the advancing desert by planting pine trees and poplars around the Yanghe Reservoir. Officials also encouraged farmers to plant trees and develop oases in the dry fields where they once grew crops, paying the peasants with grain and money if 80% of the trees survived the first year. Raising trees in this barren, mountainous moonscape is arduous, because water must be hauled—in some cases, up to several kilometers (miles)—by oxcart or backpack. Such tree-planting projects in northern China have been ongoing for more than 30 years, but on the whole, they have proved unsuccessful.

The use of grasses rather than trees, however, has been a success story. Grass is much more effective at stopping sandstorms. Once established, it does not even need to be replanted—if protected, it just grows. Trees consume groundwater, but grasses use only rainwater. The dense mat of grass roots binds the soil in place and retains moisture by inhibiting evaporation. Thus, with a substantial grass cover, there is little, if any, dust to be blown away.

Increased grazing pressure from larger herds of livestock can wreck havoc in drylands, tipping the balance between plants and animals, and stripping the land of cover. In an area near Bayinhushu, Inner Mongolia, a village about 180 kilometers (112 miles) north of Beijing, livestock increased from about 1 million in 1940 to more than 24 million in 2000, and villagers reported more dust storms and more pastureland being lost to the desert (**Figure 2a**). In response, a plant ecologist, Dr. Gaoming Jiang from the Chinese Academy of Sciences, worked with the villagers to reduce the size of goat and sheep herds, reserve part of their

Figure 2 (a) Area of a former steppe (semiarid grassland) near Bayinhushu, Hunshandake Sandland, Inner Mongolia, in August 2002 before reclamation. (b) Near Bayinhushu in August 2005, after reclamation, with Dr. Gaoming Jiang, director of the project.

Gaoming Jiang, China Academy of Sciences

(a)

(b)

Gaoming Jiang, China Academy of Sciences

common land for growing animal feed, and encouraged the natural growth of grasses. The result? Within 5 years, the grassland was restored (see Figure 2b), the number of dust storms decreased, milk production doubled, and the villagers' average annual income increased from $145 to $460. Another strategy for halting drifting sands has been successful in stabilizing windblown sand that had regularly buried the tracks of a main railroad line in China's Tengger Desert on the edge of Inner Mongolia. Inexpensive straw mats placed along the railroad right-of-way stabilize the dunes and give desert-adapted plants the time to become established (**Figure 3**).

Water, which is in short supply in China, is the most important weapon against desertification. The demands of rising living standards and heavy industry have created water shortages, leaving many rivers as nothing more than dry, sandy river beds. The lower Huang He (Yellow River), China's "mother" river, was dry for 225 days in 1997, and for most of the year the river failed to reach the Yellow Sea. A similar disappearance of once-raging rivers is repeated elsewhere in the country. For example, in Beijing is a 12th-century stone bridge made famous by Marco Polo, the 13th-century Venetian explorer. It stretches 300 meters (900 feet) across what was once the Yongding River (**Figure 4**). As recently as the 1970s, people fished in the waters and swam in the river beneath the bridge, but today the channel is nothing but gravel, dust, and weeds.

question to ponder

1. Imagine that you live in a rural region of subhumid to semiarid climate that is being affected by desertification. How would you convince local ranchers that reducing the number of grazing livestock would be in their best interest in halting desertification and might even result in an increased income?

Sources: Jiang, Gaoming. Personal communication, May 2009; Norville, Dennis. 2007, July. Getting to the roots of killer dust storms. *Science* 317(20): 314–316; Ting Ting Zhang, North China desertification critical 2006, www.China.org.cn/english/2006/Apr/166159.htm.

© George Steinmetz, National Geographic

Figure 3 Straw barrier mats being placed along the railroad line in the Tengger Desert region of China's Inner Mongolia stabilize the desert's drifting sands and give desert plants time to become established.

Frank Langfitt/SunSource/The Baltimore Sun. Reprinted with permission of the Baltimore Sun Media Group. All Rights Reserved

Figure 4 Southwest of Beijing, an abandoned paddleboat sits on the dry bed over which the Yongding River once flowed. The 12th-century bridge in the background was publicized by Marco Polo after his visit to China in the 13th century.

12.2 Geology, Qanats, and the Water Supply in Arid Regions

The settlement patterns for communities and agricultural lands in Iran and other arid and semiarid regions have been determined by geology. In particular, farm land and associated communities are often found around the toe of **alluvial fans** (**Figure 1**), one of the major landforms of mountainous arid and semiarid regions. Alluvial fans are fan-shaped deposits of gravel, sand, and clay that accumulate at the mouths of canyons close to the base of mountain ranges.

So, why do people choose to settle on alluvial fans in arid regions? It is because of the availability of water that can be extracted by the use of the ancient technology of subsurface tunnels called **qanats** (**Figure 2**). Qanats effectively transport water by gravity without the need for pumping. This ancient technology allows water to be carried long distances underground in arid regions without losses caused by seepage or evaporation.

Although an ancient technology, some 50,000 qanats were still operating in Iran in the middle of the 20th century, and about four fifths of the water used in Iran's plateau region is acquired by qanats. The oldest known qanat in Iran has been in use for 2,700 years and is still the source of drinking and irrigation water for 40,000 people. Qanats have also been in widespread use in other arid regions of the Middle East, as well as Morocco, Spain, Italy, Algeria, Tunisia, Egypt, the Atacama regions of Peru and Chile, and Asia.

© brianafrica/Alamy

Figure 1 Alluvial fan in southern Iran from space. This satellite image shows a dry riverbed, carving through an arid valley in the Zagros Mountains, which ends in a silvery alluvial fan as it enters a valley floor. A broad belt of agricultural land follows the curved outline of the fan. Irrigation water is provided by subsurface water in the fan and is brought to the surface by qanats. The image was captured by the Advanced Spaceborne Thermal Emission and Reflection Radiometer (ASTER) on NASA's Terra satellite.

Communities, gardens, and irrigated fields are located over the qanats slightly above and below their points of emergence along the lower ends of fans. The water is cleanest and coolest upstream of the qanat outlet, where the more prosperous people live. Downstream, below the outlet, the water flows through a series of surface channels to deliver water for irrigation and domestic use.

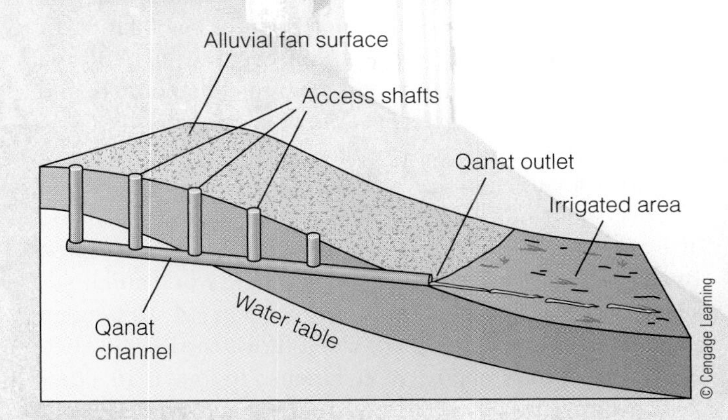

Figure 2 A qanat system for water supply.

For more than 1,000 years, qanat technology has also been used in the Middle East for air conditioning houses. Specifically, air flowing through the qanat is used to cool a basement *shabestan* ("summer room"). A wind tower, a chimney-like structure, is placed above the house and used in conjunction with a qanat to cool the air. In practice, air is pulled from the qanat by atmospheric air flowing across a vertical shaft opening, creating a lower pressure (**Figure 3**) and drawing up cool air. The underground qanat air is cooled by contact with the cool water and tunnel walls, and by giving up latent heat as the water evaporates into the flowing airstream. This technology can reduce air temperatures by more than 15°C (27°F). Furthermore, as long ago as 400 B.C.E., Persian engineers modified this cooled air basement technology into naturally cooled refrigerators in which low temperatures could be maintained, even in the hottest months, for storing winter ice that was brought down from nearby mountains.

In recent years, wars in much of the Middle East have destroyed many of these ancient structures, especially in the southern provinces of Afghanistan. The cost of labor for qanat maintenance has become very high, and maintaining these structures in troubled times has become less and less likely. In many places where qanats have been destroyed or have been abandoned, they have been replaced by diesel-powered groundwater pumps.

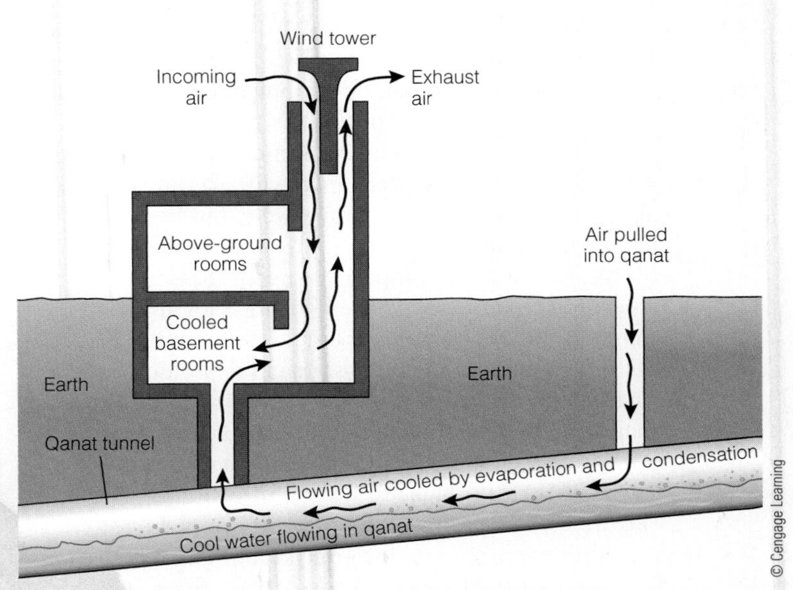

Figure 3 The design of a qanat passive air-conditioning system.

Sources: Bahadori, M. M. 1978. Passive cooling systems in Iranian architecture. *Scientific American* 238(2): 144–154; Beaumont, Peter, Michael Bonine, and Keith McLachlan. 1989. *Qanat, kariz, and kkhattara: Traditional water systems in the Middle East and North Africa* (p. 305). London: Wisbech, Cambridgeshire, England: Middle East & North African Studies Press; Kheirabadi, M. 1991. *Iranian cities: Formation and development.* Austin, TX: University of Texas Press, 89 p.; Motiee, H., et al. 2006. Assessment of the contributions of traditional qanats in sustainable water resources management. *Journal of Water Resources Development* 22(4): 575–588.

12.3 Arroyo Cutting and Environmental Change in the Western United States: Climate or Overgrazing?

In the late 19th and early 20th centuries, the incision of arroyos dramatically changed the landscape of the semiarid southwestern United States. **Arroyos**, common features of the region, are flat-floored, steep-walled stream channels that develop in cohesive, fine-grained material along valley floors. They are complex stream systems and among the most dynamic features of the semiarid landscape, especially in New Mexico and Arizona. Arroyos can be as deep as 20 meters (65 feet) and as wide as 50 meters (165 feet). The deepening and widening of arroyos have been a costly and major nuisance in the West since settlers arrived in the mid-1800s. A striking feature of arroyos is their alternation between periods of aggradation (backfilling) and incision. The most recent period of **arroyo incision**, from about 1865 to 1915, was merely the most recent cycle of cutting; others occurred about 2,000 and about 700 years ago. Areas subject to arroyo formation are illustrated in **Figure 1**, and a record of the arroyo dynamics of the San Pedro River, near Benson, Arizona, from 1900 to 1970, is shown in **Figure 2**.

Arroyo formation can be attributed to three factors, but the relative importance of each is difficult to determine: (1) Climate change producing some years with periods of intense rainfall and other years of virtually no rain is probably a major factor in the arroyo cycle, but the precise role of climate change is unclear; (2) misuse of the land, such as overgrazing and the compaction of soils that leaves the land susceptible to erosion, has been blamed for arroyo deepening during the most recent period of erosion (1865–1915); and (3) a natural cycle of erosion and deposition caused by internal adjustments within the channel system. It is likely that some combination of these three factors explains the arroyo cycle of alternate incision and backfilling.

In the case of the Rio Puerco in central New Mexico (**Figure 3**), the reports of the Spanish explorers in the 17th and 18th centuries cite the river as draining a grassy basin with cottonwood trees lining the banks. Observations of arroyos elsewhere in the Southwest include descriptions of

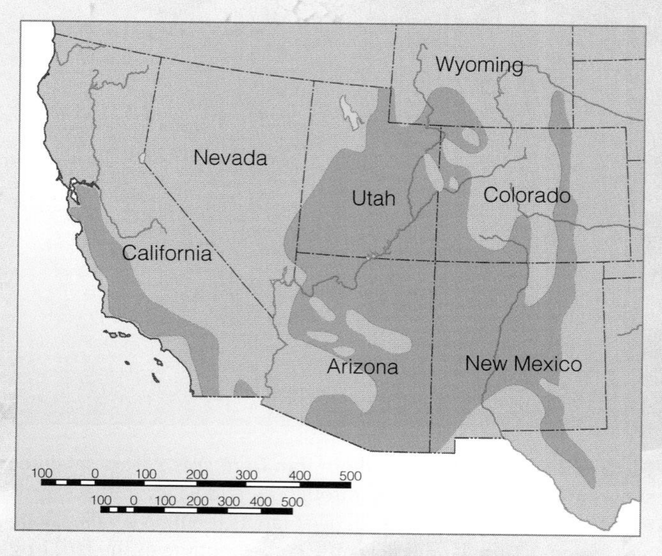

Figure 1 The approximate extent of arroyo development in the southwestern United States. (Modified from Cooke, R. U., and R. W. Reeves. 1976. *Arroyos and environmental change in the American south-west.* London: Oxford University Press.)

oicienegas (riverbed marshes) that contained beaver and fish, as well as tall grasses nourished by high water tables. No cottonwoods now line the Rio Puerco, and elsewhere the marshes have been drained by arroyo incision. Such arroyo cutting can cause water tables to drop as channels intercept the water table and drain water from the area around the river. The result is not good for cows, farms, or people—shallow wells run dry. Consequently, agricultural use of the land ends, farms and towns are abandoned, and the once-productive lands become desert.

Erosion also increases the stream's sediment load, causing downstream flooding and burial of agriculturally productive soils and damage to bridges, roads, railroads, and irrigation works. The Rio Puerco carries one of the highest average loads of suspended sediment in the world and holds a record for the greatest concentration of suspended sediment—an extremely silt-laden flow carried 600,000 parts per million (ppm) of sediment. To put it in other terms, on that day, every 10 pounds of Rio Puerco water contained 6 pounds of dirt.

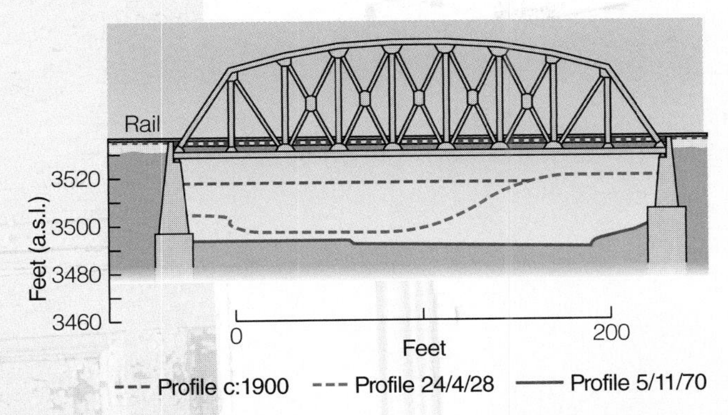

Figure 2 Incision of the San Pedro arroyo at the Benson, Arizona, railroad bridge. The arroyo was estimated to be only 1 meter (3 feet) deep and 49 meters (160 feet) wide in 1900. A flash flood destroyed the bridge in 1926, when the channel was deepened nearly 3 meters (10 feet). By 1930, it had entrenched 9 meters (30 feet) below the base of the bridge. In 1970, the date of the last available measurement, the floor of the arroyo was 13 meters (42 feet) below the bridge, and much of the sediment composing the bank material between the abutments in 1928 had been removed. Entrenchment at the bridge site is not necessarily typical of the entire arroyo, because the flow of the San Pedro River is artificially concentrated at this site where severe erosion would be anticipated. Nevertheless, the historical record provides a first-order estimate of the rate of arroyo incision. (Source: Cooke, R. U., and R. W. Reeves. 1976. San Pedro Arroyo at the Benson, Arizona railroad bridge. In *Arroyos and environmental change in the American south-west.* Fig. 11.7, p. 44. By permission of Oxford University Press.)

Attempts to control the damage from arroyo cutting began in the 1930s with the reduction of the number of grazing livestock and the construction of control structures. Ranchers, even though acutely aware of the erosion menace and loss of productive land, were generally unconvinced that grazing livestock was responsible for erosion. Studies of isolated tracts of land near arroyos that were fenced to keep out livestock to promote revegetation proved inconclusive—some areas experienced significant recovery as shown by stabilized arroyo walls, but recovery in other areas was insignificant. Experiments, including placing debris into arroyo channels to slow stream flow, planting trees along banks, and constructing silt-trapping structures, have been tried, but such controls are expensive. It appears there is no economically viable solution to the arroyo problem in this region.

Sources: Aby, S., A. Gellis, and M. Pavich. The Rio Puerco arroyo cycle and the history of landscape changes. Available at: http://geochange.er.usgs.gov/sw/impacts/geology/puercol; Bauer, R. W., R. R. Lozinsky, C. J. Condie, and L. G. Geer. 2003. *Albuquerque, a guide to its geology and culture.* Socorro, New Mexico Bureau of Geology and Mineral Resources; Cooke, R. U., and R. W. Reeves. 1976. *Arroyos and environmental change.* New York: Oxford University Press, 89 p.; Vogt, B. J. The arroyo problem in the southwestern United States. Available at: http://geochange.er.usgs.gov/sw/impacts/geology/arroyos.

Figure 3 (a) The Rio Puerco near Cabezon, New Mexico, in 1885 before incision of the arroyo. Notice the cottonwood trees along the river and the numerous small trees in the gullies on the distant mesa. (b) The Rio Puerco in 1977 from the same site as photo in (a), with several meters of incision. Note the high caving banks, which add to the sediment load of the river; no cottonwood trees remain, and there are fewer small trees in the distant gullies after 92 years of arroyo incision, lowering of the water table, and a warming climate.

(b)

(a)

PHOTO GALLERY

life and death in dry places

Desertification affects people, animals, and landscapes in a variety of ways, some of which are positive, but many of which cause severe hardship. Loss of water dooms plants and animals to a dusty death. Moving sand in deserts makes travel difficult, whereas the dust raised by windstorms makes life uncomfortable and can have significant health effects. In many arid and semiarid areas of our planet, people have adapted to changing and difficult environmental conditions, but in others, farms go bust and people starve when the rains fail to come.

1 In southern Sudan, women leave for their homes, carrying food aid, as a wind-driven dust storm sweeps through their town. The wind erodes soil that is no longer stabilized by vegetation—a grim cause and effect of desertification. ALESSANDRO ABBONIZIO/AFP/Getty Images

2 In Oman, desertification is a serious environmental issue. Here, a camel tries to find food in the form of the few remaining trees. In the distance, sand dunes advance. © 1987 STEVE MCCURRY/ National Geographic Image Collection

A Texas cotton farmer walking in a wind storm amid his desiccated, dead crops. Jim Brandenburg/ Minden Pictures

3 The course of ephemeral arid-region streams can change unpredictably, in part as shifting dunes block their course. Here, in the Namibian desert, a pan that once had enough water to support trees is now dry. Less than a kilometer away is another pan, now filled with water by the rains of 2011. For people and plants depending on the water, such shifts can be life-threatening. P. Bierman

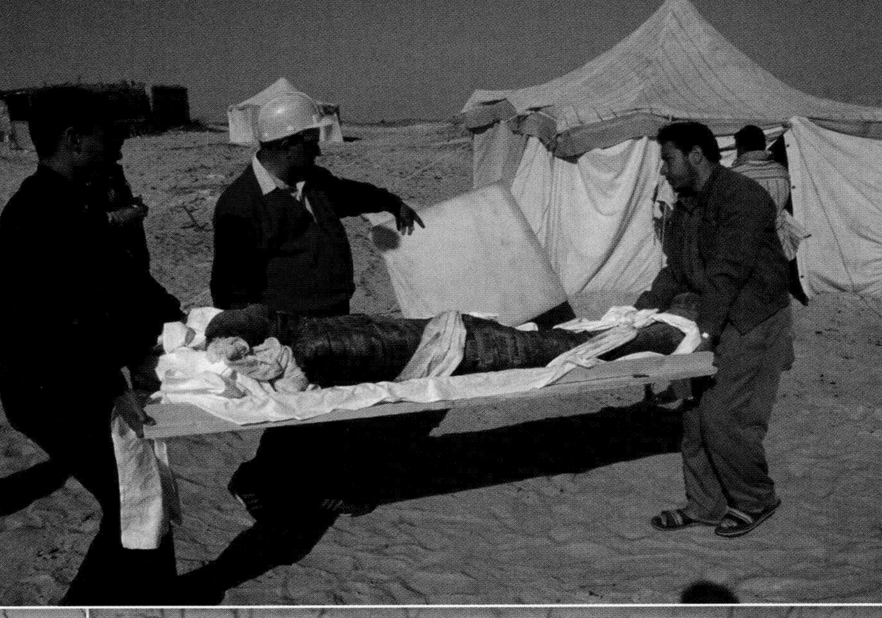

4 The dry of the desert is not all bad. The lack of water preserves many things that would otherwise rot away. Here, near an Egyptian oasis, a mummy, wrapped in a resin-soaked shroud, was excavated from beneath the shifting desert sands. Such finds are gold mines for archeologists seeking to understand how people lived and died thousands of years ago. © 1999/KENNETH GARRETT/National Geographic Image Collection

5 Shifting sands make travel slow, difficult, or impossible in many desert regions requiring specialized vehicles and a different approach to moving around the landscape. Here, tourists in a specialized four-wheel drive vehicle move over the coastal dunes in hyperarid Namibia. © 2010/PETE MCBRIDE/National Geographic Image Collection

SUMMARY Arid Lands, Winds, and Desertification

Deserts

Description
Regions where annual precipitation averages less than 25 centimeters (10 inches) and that are so lacking in vegetation as to be incapable of supporting abundant life.

Causes
High-pressure belts of subsiding, warming air that minimizes cloud formation; isolation from moist maritime air masses by position in deep continental interior; windward mountain barrier that blocks passage of maritime air.

Classification
May be classified into five categories determined by climate belts.

Types
Polar deserts, regions of perpetual cold and low precipitation; *mid-latitude deserts* within the interior of continents in the middle latitudes, distant from the moist influence of oceans; *subtropical deserts*, Earth's largest realm of arid regions, in and on the equatorial side of subtropical zones of subsiding high-pressure air masses in the central portions of continents; *coastal deserts* on the coastal side of a land or mountain barrier in subtropical latitudes—because they are bordered by ocean, they are cool, humid, and often foggy; *rain-shadow deserts* are leeward of a mountain barrier that traps moist ocean air.

Desertification
The expansion of deserts into formerly productive lands because of both natural and human-induced factors.

Wind as a Geological Agent

Erosion
Deflation—earth materials are lifted up and blown away.
Abrasion—mineral grains are blown against each other and into other objects.

Control of Migrating Sand
Accomplished with sand fences, paving, windbreaks, and stabilizing plants.

KEY TERMS

abrasion
alluvial fan
arroyo incision
coastal desert
deflation
desert
desertification
dune
dust storm
mid-latitude desert
polar desert
qanat
rain-shadow desert
Sahel
semiarid grassland
subtropical desert
upwelling

STUDY QUESTIONS

1. Explain why the world's largest deserts are located in the subtropics.
2. What factors other than drought are responsible for the desertification that has resulted in starvation in Africa and elsewhere? (Hint: Consider carrying capacity, as discussed in Chapter 1.)
3. List measures that can be taken to halt sand dune migration.
4. List agricultural practices in arid and semiarid regions of the United States (such as California's Coachella Valley and western Nebraska) that increase the likelihood of desertification.
5. Total rainfall from a desert cloudburst rarely exceeds 5 to 10 centimeters (2–4 inches) over a period of an hour or so. However, downpours lasting much longer are common in humid regions. Explain the cause of the spectacular erosion associated with short desert rainstorms.
6. How does the land–atmosphere interaction cause changes in rainfall?
7. Does "global dimming" from excess atmospheric dust produce a positive or negative feedback loop to global warming?

REFERENCES AND FURTHER READINGS

Ahrens, C. D. 2009. *Meteorology today: An introduction of weather, climate, and the environment*, 9th ed. Belmont, CA: Brooks/Cole, 537 p.

Cook, E. R., et al. 2004. Long-term aridity changes in the western United States. *Science* 306:1015–1018.

Cooke, R. U., and R. W. Reeves. 1976. *Arroyos and environmental change in the American south-west*. London: Oxford University Press, 213 p.

Cooke, R. U., A. Warren, and A. Goudie. 1993. *Desert geomorphology*. London: CRC Press, 534 p.

Ellis, W. S. 1987. Africa's Sahel: The stricken land. *National Geographic* 172 (2): 140–179.

Flannery, T. 2006. Liquid gold: Changes in rainfall. In *The weather makers* (pp. 123–134). New York: Atlantic Monthly Press.

Gianni, A., R. Saravanan, and P. Chang. 2003. Oceanic forcing of Sahel rainfall on interannual and interdecadal timescales. *Science* 302: 1027–1030.

Griffen, D. W, C. A. Kellog, V. H. Garrison, and E. A. Shin. 2002. The global transport of dust. *American Scientist* 90(3): 228–235.

Hoerling, M., and A. Kumar. 2003. The perfect ocean for drought. *Science* 299: 691–694.

Monroe, J. S., R. Wicander, and R. Hazlett. 2007. The work of wind and deserts. In *Physical geology, exploring the Earth* (pp. 566–595), 6th ed. Pacific Grove, CA: Brooks/Cole.

Prospero, J. M., and P. J. Lamb. 2003. African droughts and dust transport to the Caribbean: Climate change implications. *Science* 302: 236–237.

Reheis, M. C. 2006. A 16-year record of Aeolian dust in southern Nevada and California, USA: Controls on dust generation and accumulation. *Journal of Arid Environments* 67: 487–520.

Reisner, M. 1987. *Cadillac desert*. New York: Penguin Books, 582 p.

Walker, A. S. 1996. *Deserts: Geology and resources*. Denver, CO: U.S. Geological Survey.

Webster, D., and G. Steinmetz. 2002. China's unknown Gobi. *National Geographic* 201(1): 48–75.

Wilshire, H. G, J. K. Nakata, and B. Hallet. 1981. Field observations of the December 1977 wind storm, San Joaquin Valley, California. In Pewe, Troy L. (ed.), *Desert dust: Origin, characteristics, and effect on man* (pp. 233–251). Geological Society of America Special Paper 186. Geological Society of America, Boulder, CO.

Wilshire, H. G, J. E Nielson, and R W. Hazlett. 2008. *The American West at risk: Science, myths, and politics of land abuse and recovery*. New York: Oxford University Press, 619 p.

Zeng, N. 2003. Drought in the Sahel. *Science* 302: 999–1000.

Access an interactive eBook and chapter-specific interactive learning tools, including flashcards, quizzes, videos and more in your Geology CourseMate, accessed through CengageBrain.com.

Newmont's Gold Quarry Mine, Carlin, Nevada.
Calvin Larsen/Photo Researchers, Inc.

Have you ever
wondered...

13

Mineral Resources and Society

Entire society rests upon—and is dependent upon—our water, our land, our forests, and our minerals. How we use these resources influences our health, security, economy, and well-being.

—John F. Kennedy, 35th president of the United States, February 23, 1961

1. How are rich mineral deposits of precious or base metals discovered?

2. What is the difference between placer mining and hardrock mining?

3. Why are most of the U.S. metal mines in the western states?

4. Why do so many major copper deposits occur in a belt along the western parts of North and South America?

5. What happens to the ton of waste rock after the 0.03 ounce of gold per ton is removed?

Nevada's New Bonanza

Figure 1 Barrick Gold Corporation's Meikle Mine on the Carlin Trend near Elko, Nevada, is the largest underground gold mine in the United States. The mine is the largest in terms of production (413,186 troy ounces in 2007) and in estimated reserves (6 million ounces) in the ground.

We are currently in the midst of the biggest gold-mining boom in U.S. history. The bonanza is the Carlin trend, a 64-by-8 kilometer (40-by-5 mi) belt of enormous low-grade gold deposits in north-central Nevada. Current mining on the Trend accounts for about 79 percent of the gold production in the U.S. and about 7.2 percent of world production. By 2011 the Trend had poured over 70 million ounces of gold since its discovery in 1962, an amount worth over $112 billion at early 2012 price levels.

The Gold Quarry mine, the largest of Newmont Gold Company's several open-pit mines, is representative of the large mines on the Carlin Trend (Chapter Opener photo) and other Nevada open-pit mines. Newmont began operating on the Trend in 1965, with the combined production of its three open-pit mines exceeding about 2 million ounces per year. The cost of recovering gold from Newmont's mines runs about $300 per ounce with an additional cost of about $1 for each ounce of gold produced that is set aside to meet eventual costs of reclamation.

About 30 major mines and several smaller ones, most of them open-pit heap-leach operations, were mining the Trend in the first decade of the 21st century. The Trend, the most prolific gold producing mining district in the Western Hemisphere, is a valuable asset to Nevada's economy, contributing over 81.5 percent to the state's annual income.

The Barrick Goldstrike property on the Trend includes three mines: one open-pit and two underground mines (**Figure 1**). The three Goldstrike mines produce about 1.7 million ounces of gold each year with additional estimated reserves of 12 million ounces. It is estimated that the Battle Mountain-Eureka Trend (Nevada's second-most productive gold belt), the Getchell Trend, and other mineralized belts of the state (**Figure 2**) contain another 70 million ounces of gold reserves and resources, enough to maintain production at current levels for perhaps anther two decades.

The prolific gold belts result from the collision of two tectonic plates about 350 million years ago during the Mississippian Period. The collision caused increased temperatures and pressures that produced numerous hot springs along the collision zone, and subsequent crustal magmatism in the entire Basin and Range caused the hot springs to carry dissolved minerals toward the surface where they precipitated along faults and numerous fissures primarily in limey sedimentary rocks of Paleozoic age. The Carlin Trend, the Battle Mountain-Eureka Trend, and other mineralized belts in the region, are referred to as sediment-hosted gold deposits.

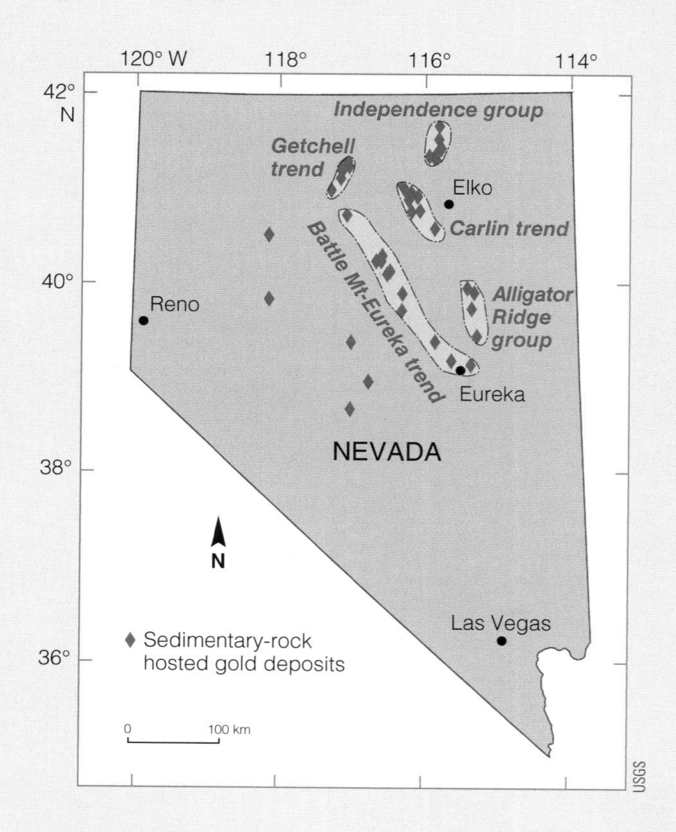

Figure 2 The Carlin Trend and other belts of sediment-hosted gold deposits in Nevada. (Source: USGS)

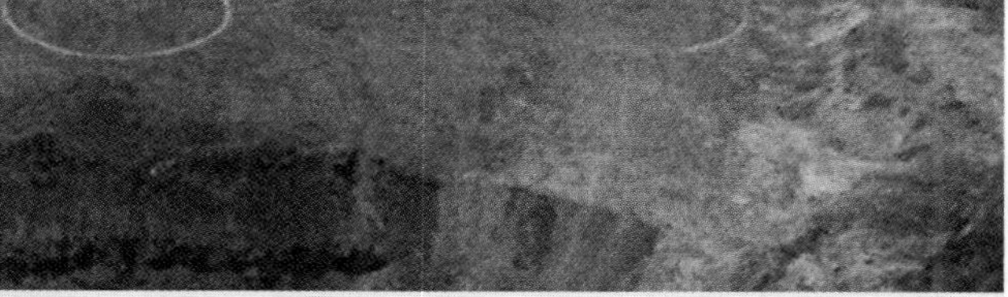

After mining ends, which is estimated to be about 2030, the pits will be closed off with berms and fences, and the waste-rock dumps will be encapsulated with a 4-foot cover of topsoil and planted with native plants in accordance with the mine's operating permits. A network of ditches is designed to collect and divert surface water flow from the waste piles, which is intended to keep them dry and free from potential seepage of toxic water. The protective berms will only be a few feet high, and the fences must be maintained forever for reasons of public safety, because the berms could create attractive jumps for ATV riders.

questions to ponder

1. How do gold and other important elements and minerals become concentrated in economic deposits?
2. How is gold, which is only present in rocks and only in small amounts, recovered?
3. What is the difference between the technologies of underground mining and open-pit mining?
4. Is there a difference between reserves and resources?
5. If the Goldstrike property has reserves of 12.8 million ounces, and the production is a steady 1.71 million ounces per year, how much longer will the mine continue to operate, assuming current economic conditions?

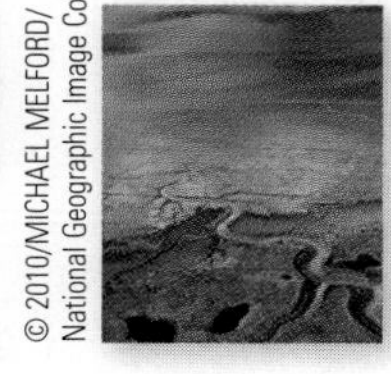

CourseMate Connection

Start off this chapter by listening to a podcast with author Edwin Dobb as he talks about the possibility of open-pit gold and copper mines in the Bristol Bay region of Alaska. **Log in to the CourseMate at www .cengagebrain.com to listen now.**

THE MINING INDUSTRY IS FOND of saying, "If it isn't grown, it's mined." And it's true: Every material used in modern industrial society is either grown in or derived from Earth's natural mineral resources, which are usually classified as either metallic or nonmetallic. Production and distribution of the thousands of manufactured products and the food we eat are dependent on the utilization of metallic and nonmetallic mineral resources. The per capita U.S. consumption of mineral resources used directly or indirectly in providing shelter, transportation, energy, and clothing is enormous (**Figures 13.1** and **13.2**), and the amount of minerals, metals, and fuels estimated by the Minerals Information Institute to be needed during the lifetime of a child born in 2012 is staggering (**Figure 13.3**). The availability and cost of mineral and rock products influence our nation's standard of living, domestic national product, and position in the world.

With the value of nonfuel minerals produced in the United States in 2010 running about $64 billion, and the value of recycled scrap (glass, aluminum, steel, etc.) adding another $15.2 billion, the value of materials produced from minerals in 2010 is estimated at $2.1 billion. Clearly, minerals and mining are a vital factor in the U.S. economy. It is amazing that the general public has relatively little knowledge of where these minerals occur naturally, the methods by which they are mined and processed, and the extent to which we depend on them. It is important that the public recognize that exploitable mineral resources occur only in particular places, having formed there because of unique geological conditions, and that all deposits are exhaustible. Furthermore, the development and exploitation of mineral resources have environmental consequences and are capital-intensive, requiring a substantial long-term investment.

Understanding the origins, economics, and methods of mining and processing of mineral resources enables an individual to understand many of the difficult resource-

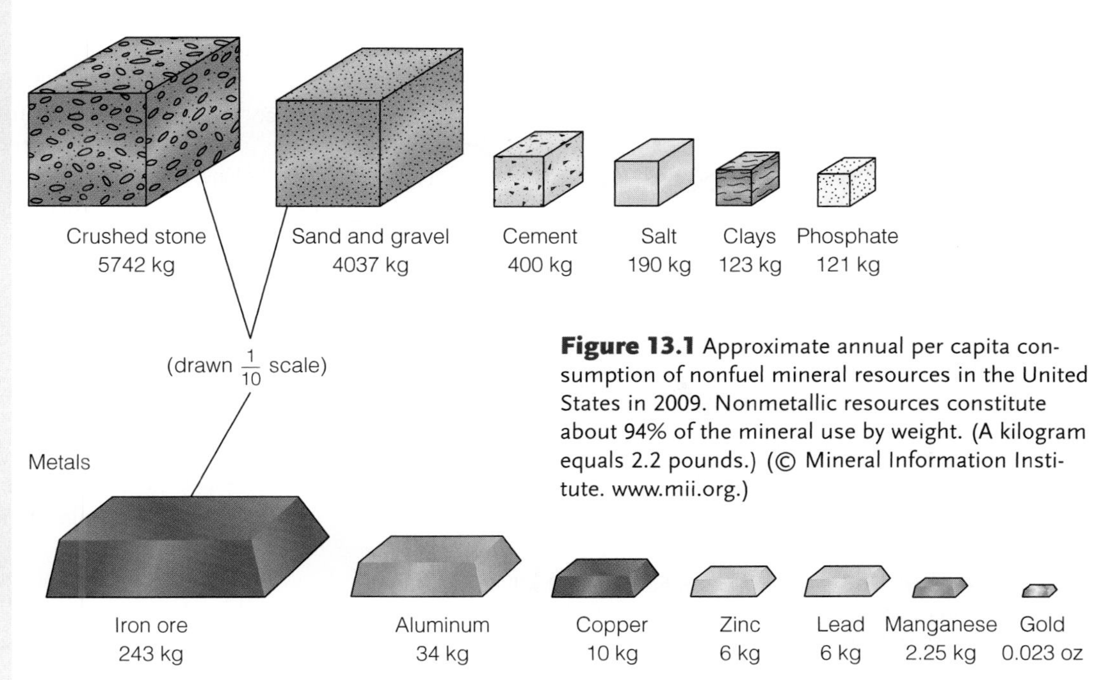

Figure 13.1 Approximate annual per capita consumption of nonfuel mineral resources in the United States in 2009. Nonmetallic resources constitute about 94% of the mineral use by weight. (A kilogram equals 2.2 pounds.) (© Mineral Information Institute. www.mii.org.)

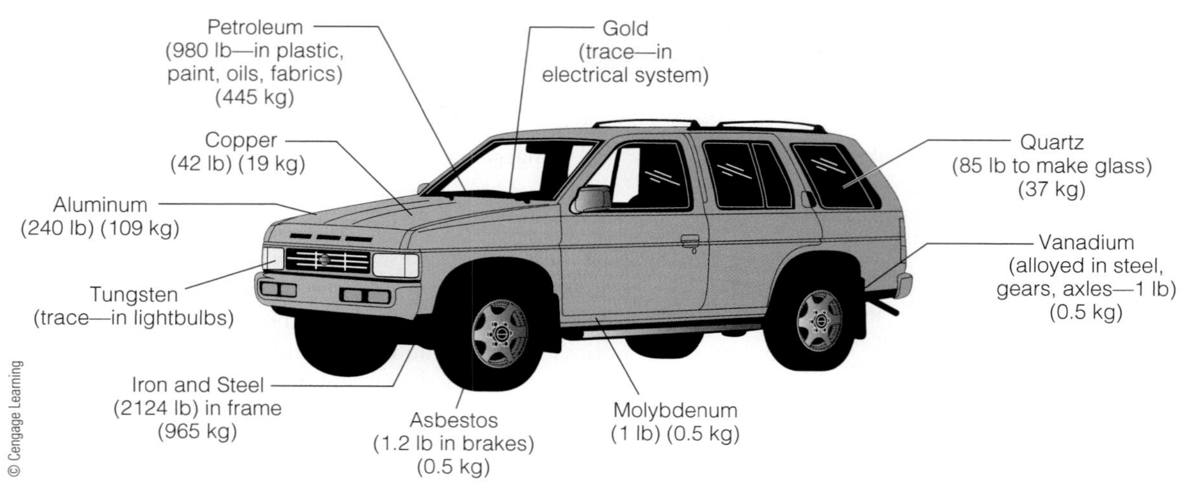

Figure 13.2 Estimated amounts of a few of the more than 30 mineral materials used in a midsize automobile.

related issues that face all levels of government. Increased land-use competition among the interests of mining, housing, wilderness preservation, agriculture, logging, recreation, wetlands preservation, and industrial development is forcing our government officials to make decisions that affect mineral-resource development and management and, thus, all of our lives.

Mineral Abundances and Distribution

Economic Mineral Concentration

The elements forming the minerals that make up Earth's crust exist in many forms in a great variety of rocks. The greater the quantity in place, the richer the deposit and the more economically feasible it is to extract and process the desired element. Locally rich concentrations of minerals are called **mineral deposits**. If they are sufficiently enriched, they may be **mineral reserves**, deposits of earth materials from which useful commodities are economically and legally recoverable with existing technology. Reserves of metallic minerals are called **ores**, and the minerals composing the deposits are referred to as **ore minerals**.

A particular ore deposit may be described in terms of its **concentration factor**, its enrichment expressed as the ratio of the element's abundance in the deposit to its average continental-crustal abundance. **Table 13.1** illustrates that the minimum concentration factors for profitable mining vary widely for eight metallic elements. Note in the table that aluminum, one of the most common elements in Earth's crust, has an average crustal abundance of 8%. A "good" aluminum-ore deposit contains about 35% aluminum; thus, the present concentration factor for profitable mining of aluminum ore is about 4. In contrast, deposits of some rare elements, such as uranium, lead, gold, and mercury, must have concentration factors in the thousands to be considered mineral *resources*.

A basic tenet of mineral commodity economics is that it is unlikely that we will ever run out of a useful substance, because there are always deposits of any substance that have lower concentrations than are currently economical to mine. If the supply of currently economic deposits is reduced, market forces will cause the price to increase and the concentration factor to increase. Some people see a problem with this rationale. This topic will be explored later in this chapter.

Some elements are extremely common. In fact, 97% of Earth's crust is composed of only eight elements: oxygen, silicon, aluminum, iron, sodium, calcium, potassium, and

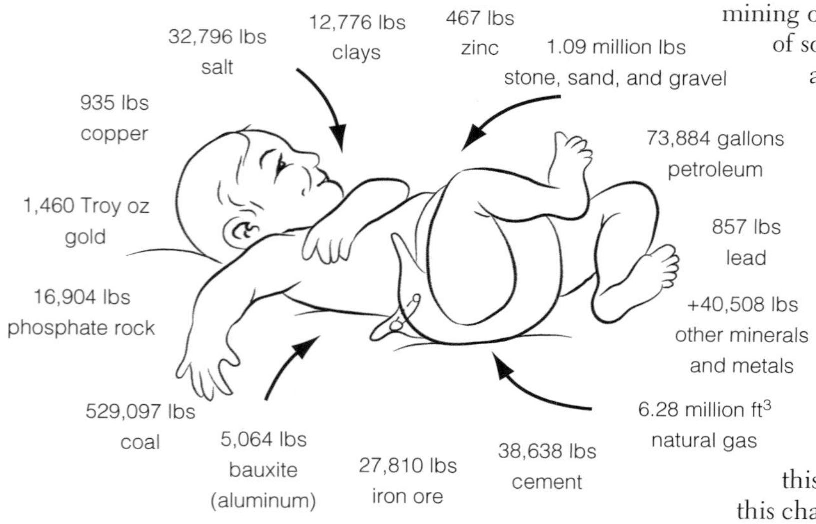

Figure 13.3 Every American born in 2012 will need 2.96 million pounds of metals, minerals, and fuels in his or her lifetime. (Data from Mineral Information Institute.)

Table 13.1 Concentration Factors for Profitable Mining of Selected Metals

	Percentage Abundance		
Metal	Average in Earth's Crust	In Ore Deposit	Concentration Factor
Aluminum	8.0	24–32	3–4
Iron	5.83	5–41	6–7
Titanium	0.86	22–86	25–100
Zinc	0.00829	2.5	300
Copper	0.0058	0.6–1.2	100–200
Silver	0.00000896	0.00896	1,000
Platinum	0.000000596	0.00036	600
Uranium	0.00016	0.08–0.16	500–1,000
Gold	0.000000296	0.0012–0.0015	4,000–5,000

Source: Earth Sciences Australia

magnesium. For most elements, the **average crustal abundance**, the amount of a particular element present in the continental crust, is only a fraction of a percent (see Table 13.1). Copper, for example, is rather rare; its average crustal abundance is only 0.0058%. In some localities, however, natural geochemical processes have concentrated it in mineral deposits that are 2% to 4% copper. A deposit with a high concentration of a desired element is called a **high-grade deposit**. A deposit in which the mineral content is minimal but still exploitable is called a **low-grade deposit**. For either grade of deposit, localities where desirable elements are concentrated sufficiently for economic extraction are relatively few.

Factors That Change Reserves

Reserves are not static, because they are defined by the current economics and technology, as well as by the amount of a mineral that exists. Reserves fluctuate because of several factors: changing demand, discoveries of new deposits, and changing technology. In a free economy, a mineral deposit will not be developed at an economic loss, and prices will rise with demand. Consequently, low-grade deposits that are marginal or submarginal in today's economic climate may, if demand and prices rise, eventually become profitable to exploit. Advancements in technology also may increase reserves by lowering the cost of development or processing.

For example, the exploitable reserves of gold increased dramatically in the United States in the late 1960s because of a combination of changes in government policies and advances in technology. In 1968, when the price of gold was $35 per troy ounce, the Treasury Department suspended gold purchases to back the dollar and began allowing the metal to be traded on the open market. (A troy ounce equals 31.103 grams, whereas the avoirdupois ounce of U.S. daily life equals 28.330 grams.) Then in the 1970s, the federal government removed restrictions prohibiting private ownership of gold bullion. In 1980, the gold price rose to more than $800 an ounce, and by 2006, it had settled at about $550 to $600 an ounce. Thus, in a period of about 38 years, the price of gold increased by about 1,700%, meaning that many submarginal mining claims became economically profitable mines in that time. (By February 2012, the gold price had climbed to more than $1,700 an ounce.) Furthermore, gold exploitation became more cost-efficient in the 1980s, with the development of the technology of cyanide heap-leaching, by which the disseminated gold is dissolved from low-grade deposits and then recovered from the solution. A combination of changing economics and the new heap-leaching technology initiated a new gold rush in the West beginning in the 1980s (see chapter opener).

Other factors may reduce reserves, because extracting Earth's riches almost always requires the trade-off of aesthetic and environmental consequences (**Table 13.2** provides some examples). For example, large, low-grade mineral deposits can be mined only by open-pit methods. Not only do these methods devastate the landscape, but their excavations also may reach below the water table, which can lead to groundwater contamination, and they produce enormous amounts of waste rock and tailings that can pollute surface-water runoff. **Smelters**, large industrial plants that process ore concentrates and extract the desired elements, produce more air pollution, in the form of flue dust, than any other single industrial activity. Smelters also produce slag, a solid residue that can contain thousands of times the natural levels of lead, zinc, arsenic, copper, and cadmium, but are locked up by chemical bonding and are not bioavailable.

The real or perceived negative environmental impacts that rule out the exploitation and development of known deposits also reduce reserves. In recent years, there have been several widely publicized instances of wildlife, wilderness, and other values being judged more important than the exploitation of mineral or petroleum deposits:

Location	Type of Operation	Consequences
Questa, New Mexico	Molybdenum mining	Drainage from waste rock from an abandoned molybdenum mine in upper Rio Grande Valley has contaminated watersheds, destroyed wildlife habitat, killed fish, polluted irrigation ditches and water wells, affected agricultural production, and caused illness to area residents. Estimated cleanup cost is $129–$368 million.
Baia Mare, Romania	Gold mining	A spill of about 1 million m^3 (3.5 million ft^3) in January 2000 of cyanide-laced tailings operations ravaged a 250-mile stretch of the Danube River system, killing all aquatic life, including thousands of fish.
North-central Montana	Gold mining	The Zortman and Landusky open-pit, large-scale cyanide heap-leach gold mines suffered cyanide leaks and spills, causing surface water and groundwater contamination and bird and wildlife fatalities. When faced with cleanup costs, the mining company declared bankruptcy in 1998, forfeiting its cleanup bond, which was estimated to be $8.5 million short of cleanup costs.
Summitville, San Juan Mountains, Colorado	Gold mining	A leaking leach pad, contaminated with cyanide and toxic metals, poisoned 27 km (17 mi) of the Alamosa River, on which the region's agriculture depends. When cleanup costs exceeded the posted bond, the Canadian mining company and its parent declared bankruptcy. Estimated cleanup will cost U.S. taxpayers $232 million.
Guyana, South America	Gold mining	A cyanide spill from the Omai Mine—backed by the same investors as Colorado's Summitville Mine—released 860 million gallons of cyanide-laced tailings into one of Guyana's largest rivers. The spill killed fish, causing a panic in Guyana's seafood markets.
Sudbury, Ontario, Canada	Nickel-copper mining and smelting	This was one of the world's best-known environmental "dead zones." Little vegetation survived in the 10,400-hectare (40 mi^2) area around the smelter. Acid fallout destroyed fish populations in lakes within 65 km (40 mi). Conditions improved after completion of a $530 million SO_2 smelter-abatement project and construction of the world's tallest smokestack; the city has a successful landscape reclamation program, but there is still significant SO_2 damage downwind of Sudbury.
Pará State, Brazil	Grande Carajas iron ore project	Wood requirements for smelting ore will require cutting 50,000 hectares (193 mi^2) of tropical forest annually during the 250-year life of the project.
Southeastern Peru, adjacent Bolivia and Brazil	Placer gold mining	Starting in 2007, tens of thousands of miners began rampant, illegal digging for gold, some in biological reserves, clearing forests, clogging rivers with sediment, and releasing many tons of mercury annually into the ecosystem.
Ilo-Locumbacoua area, Peru	Copper mining and smelting	Each year, the Ilo smelter emits 600,000 tons of sulfur compounds, and nearly 40 million cubic meters (523 million yd^3) of tailings containing lead, zinc, copper, aluminum, and traces of cyanide are dumped into the ocean, poisoning marine life in a 20,000-hectare (77 mi^2) area. Nearly 800,000 tons of slag are dumped in the sea yearly.

Table 13.2 Some Negative Environmental Consequences of Mineral Extraction and Processing Operations

- An extraordinary agreement was reached between President Clinton and Crown Butte Mines in 1996, which halted proposed mining activity on the doorstep of Yellowstone National Park.
- A 2-year moratorium was placed on hard-rock mining along a 100-mile-long region of Montana's Rocky Mountain Front, one of the nation's most important wildlife areas.
- U.S. federal government has imposed a 20-year ban on new mining claims of federal land near Grand Canyon National Park; the ban affects all new claims but is primarily aimed at uranium mining. The area is already home to about 3,000 mining claims, none of which will not be impacted by this new ban. The concern has been the potential for poisoning of surface and groundwater. Furthermore, the economic driver for the region is considered to be tourism, not mining, although the area contains about 40% of the nation's uranium, valued to be in the billions of dollars.
- A decision was made by the Office of Surface Mining and Reclamation not to allow surface mining of coal in the watershed of Fern Lake in Kentucky because it would be incompatible with Cumberland National Historic Park's land-use plan, would destroy historic and natural systems, and would damage the region's water quality.

We can expect that controversies over land use will become increasingly common in the future because larger and larger areas will need development as more and more lower grade deposits will need to be exploited when higher grade deposits become exhausted.

Political factors also are related to the economics of mineral resources. The United States imports substantial amounts of mineral resources from other countries. (This is discussed in some detail later in this chapter.) Some geopolitical authorities estimate that about a quarter of the roughly 50 wars and armed conflicts in 2001 had a strong "resource dimension." Exploitation of diamonds by Angolan rebels in 1992–2001 resulted in an estimated income of about $4 billion, and emeralds and lapis lazuli from the mid-1990s to 2001 netted Afghan groups (Taliban and the Northern Alliance) $90 to $100 million per year. The Democratic Republic of Congo's (DRC) vast deposits of cobalt, copper, tin, tungsten, diamonds, and gold have sparked a number of conflicts since the Belgian colonial period. The Second Congo War that raged from 1998 to the early 2000s and the continuing conflict into the 2010s have taken the lives of more than 2 million people and damaged the bodies of tens of thousands. Militias and a shadowy network of smugglers, and rebel groups have been fighting right up to the current time for control of the Congo's mineral wealth. A key player in one of the deadliest ongoing DRC conflicts is *col-tan*—the name for a mineral composed of columbium and tantalum, elements used in mobile phones, DVD players, hearing aids, digital cameras, computers, and other electronic devices. Purchases of col-tan by Western nations has helped finance the decade-long war in the DRC. Furthermore, resource availability may affect diplomatic policy. In spite of the general trade sanctions the United States enacted in the 1980s against South Africa to protest its apartheid policy, 10 minerals the United States imports from South Africa were exempted because of their importance to the United States.

Distribution of Mineral Resources

The global distribution of mineral resources has no relationship to the locations of political boundaries or technological capability. Some mineral deposits, such as those of iron, lead, zinc, and copper, originate by such a variety of igneous, sedimentary, and metamorphic processes that they are widely distributed, although in varying abundance. On the other hand, bauxite, essentially the only ore mineral of aluminum (one of the most common elements in Earth's crust), is concentrated by only one geochemical process: deep chemical weathering in a humid tropical climate. For this reason, economically exploitable deposits of bauxite and other such minerals are unequally distributed and restricted to a few highly localized geological and geographical sites. Such localization of mineral deposits means that desired mineral resources do not occur in all countries. When this unequal supply pattern results in the scarcity of particular mineral resources that are crucial to a nation's economy, those minerals become critical or even strategic in the event of a national emergency. What constitutes "critical" minerals changes from time to time, mainly because of technological advances in mining or processing technologies.

Throughout history, nations' domestic mineral-resource bases and their needs for minerals unavailable within their borders have dictated international relations. Empires and kingdoms have risen and fallen because of the availability or scarcity of critical mineral resources. Two classic examples are those of the Greek and Roman empires. Silver mined by the Greeks near Athens financed the building of their naval fleet, which defeated the Persians at the Bay of Salamis in 480 B.C.E. A century later, Greek gold supported the ambitious conquests of Alexander the Great. The Romans helped finance their vast empire by mining tin in Britain, mercury in Spain, and copper in Cyprus. In the 20th century, the efforts of Japan and Germany to secure mineral and fuel resources lacking within their borders were major factors leading to World War II, and the Soviet Union's annexation of the Karelian highlands of Finland following World War II provided the Soviets with new reserves of copper and nickel. More recently, Morocco acquired valuable phosphate deposits when it occupied part of the Spanish Sahara in 1975.

Origins of Mineral Deposits

Specific geological processes must occur to concentrate minerals or native elements in a mineral deposit. A simple classification of mineral deposits is based on their concentrating processes (**Table 13.3**). The geological processes responsible for many mineral deposits may be understood in terms of plate tectonics, as illustrated in **Figure 13.4**, with mineral concentrations occurring in distinct tectonic regimes. For example, the great porphyry copper deposits of the Western Hemisphere were formed by igneous processes at convergent plate margins at the sites of former volcanic island arcs. On the other hand, deposits of tungsten, tin,

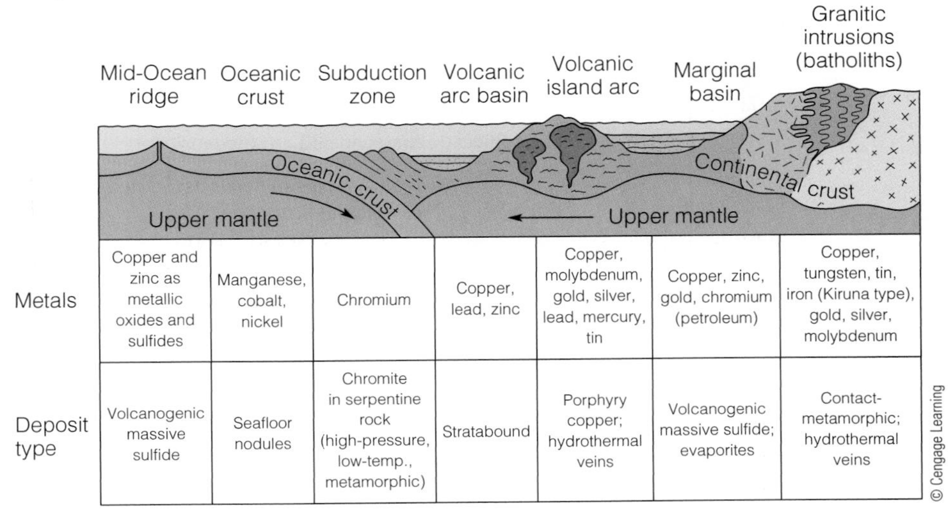

Figure 13.4 Relationships between metallic ore deposits and tectonic processes.

Table 13.3 Genetic Classification of Mineral Deposits

Geologic Setting	Genetic Type	Examples of Minerals Formed	Significant Locations
		WEATHERING	
SURFACE	Lateritic deposits	Bauxite, nickel, gold, clay, soil	Weipa, Australia; Jamaica; New Caledonia
	Secondary enrichment	Cuprite (copper oxide)	Globe-Miami District, Arizona; Ely, Nevada
		PHYSICAL SEDIMENTATION (Flowing water, both stream and beach)	
	Placer deposits	Gold, platinum group metals, heavy mineral sands (diamonds, titanium, zirconium, tungsten, rare earth minerals)	Yukon Territory, Canada; Platinum, Alaska; Queensland, Australia
		CHEMICAL SEDIMENTATION (Precipitation from or in water)	
	Marine evaporites	Gypsum, halite, anhydrite, calcite	Carlsbad, New Mexico; Great Salt Lake, Utah
	Non-marine evaporites	Borax, trona	Owens Lake, Searles Lake, and Death Valley, California
	Chemical precipitates in marine waters	banded iron formation	Lake Superior region; Hamersley Ranges, Australia
		Volcanic massive (copper) sulfides	Cypress; Rio Tinto, Spain; Jerome, Arizona
		Iron-manganese nodules	Ocean floor; worldwide
		ORGANIC SEDIMENTATION	
	Organic activity and/or accumulation	Hydrocarbons (coal, petroleum, natural gas)	Pennsylvania; Wyoming; Texas; Saudi Arabia
		Phosphate rock, sulfur	Florida; Texas
		IGNEOUS PROCESSES	
SUBSURFACE	Pegmatite	Tourmaline, tantalum and lithium-bearing minerals	Oxford County, Maine; San Diego County, California; Minas Gerais, Brazil
		Rare earths	Bokan Mountain, Alaska
	Crystal fractionation and settling	Chromite, magnetite, platinum group metals, vanadium minerals	Bushveld complex, So. Africa; Stillwater complex, Montana
	Disseminated	Copper-molybdenum minerals in granite porphyry, diamonds	Western North and South America; Kimberlite pipes in South Africa, Canada, and Australia
	Hydrothermal	Lead-zinc-silver deposits in carbonate host rocks	Mississippi River Valley; Leadville, Colorado
		Gold-silver quartz veins	San Juan County, Colorado
		METAMORPHIC PROCESSES	
	Contact-metamorphic	Tungsten and molybdenum minerals	Eastern Sierra Nevada, California
		Lead and zinc minerals associated with porphyritic granite	Ouray, Colorado
	Regional metamorphic	Asbestos	Quebec, Canada
	Metamorphic dehydration hydrothermal	Gold-quartz veins	Mother Lode gold belt, California
		OTHER	
	Meteor impact and mantle melt	Nickel and copper minerals, platinum, palladium, gold	Sudbury, Ontario, Canada

Figure 13.5 Granitic pegmatite, a very coarse-grained igneous rock. The grains in this sample have lengths up to 3 centimeters.

Figure 13.6 Tourmaline crystals from Dunton, Maine. Tourmaline is commonly found as an accessory mineral in granitic pegmatites.

some iron ore, gold, silver, and molybdenum originated by hydrothermal activity or by contact metamorphism accompanying the emplacement of granitic plutons (batholiths) within continental lithospheric plates near convergent boundaries.

Igneous Processes

Intrusive deposits—**Pegmatites** are small, tabular-shaped, very coarse-grained, intrusive igneous bodies that may be important sources of mica, quartz, feldspar, beryllium, lithium, and gemstones. Reserves of pegmatites in the Western Hemisphere occur in the Black Hills of South Dakota; at Dunton, Maine; at Minas Gerais, Brazil; and in the tin-spodumene deposits of North Carolina (**Figures 13.5** and **13.6**).

Crystal settling within a cooling magma chamber of ultramafic composition appears to be responsible for forming layers according to density, with the densest and earliest formed on the bottom and the less dense and later-to-crystallize above or toward the top (**Figure 13.7**), although alternate ideas for the layering have been proposed. Examples are the layered deposits of chromite at Red Mountain, Alaska, and the Bushveld Complex in the Republic of South Africa, the largest single igneous mass on Earth (roughly the size of Maine). Chromium, nickel, vanadium, and platinum are obtained from this source. Some layered nickel-copper-cobalt deposits also are important sources of platinum-group elements.

Disseminated deposits—**Disseminated** (scattered) **deposits** occur when a multitude of mineralized veinlets develop near or at the top of a large igneous intrusion. Probably the most common of these are the **porphyry copper** deposits (**Figure 13.8** and Case Study 13.1), in which the ore minerals are widely distributed throughout a large volume of granitic rocks that have been emplaced in regions of current or past plate convergence. America's largest, and the world's third largest, open-pit mine, at Bingham Canyon, Utah, has extracted more than $6 billion worth of copper and associated minerals from a porphyry copper-molybdenum body (**Figure 13.9**).

Hydrothermal deposits—**Hydrothermal** (hot-water) **deposits** originate from hot, mineral-rich fluids that are squeezed from cooling magma bodies during crystallization. Solidifying crystals force the liquids into cracks, fissures, and pores in both the magmatic rocks and the adjacent "country" rocks (see Figure 13.8). Hydrothermal fluids can also be of metamorphic origin, or they may form from subsurface waters that are heated when they circulate near a cooling magma. Regardless of their origin, the heated waters may

Figure 13.7 Layered mineral deposit; Red Mountain, Alaska. The black layers are chromite ($FeCr_2O_4$): the light layers are granitic rocks.

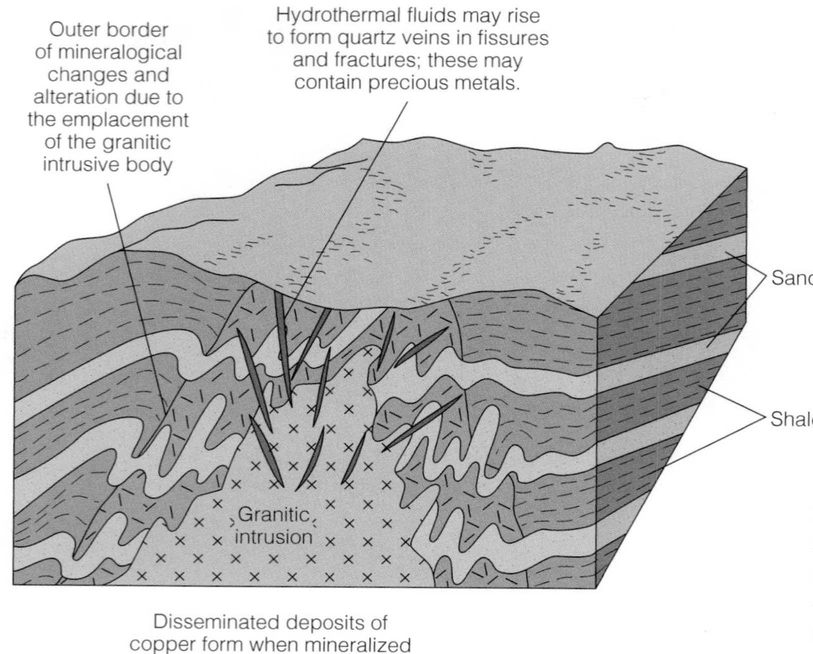

Figure 13.8 A disseminated porphyry copper deposit. (From Alaska's Oil/Gas & Mineral Industry, © 1982, with permission of Graphic Arts Books.)

dissolve, concentrate, and remove valuable minerals and redeposit them elsewhere. When the fluids cool, minerals crystallize and create hydrothermal **vein deposits**, which miners often refer to as **lodes**. The specific minerals formed vary with the composition and temperature of the hydrothermal fluids, but they commonly include such suites (associations) of elements as gold-quartz and lead-zinc-silver. The prolific gold and silver deposits of Nevada (see chapter opener) are sediment-hosted hydrothermal deposits.

Volcanogenic deposits—When volcanic activity vents fluids to the surface, sometimes associated with ocean-floor hot-spring **black smoker** activity, **volcanogenic deposits** are formed (**Figure 13.10**). These deposits are so named because they occur in marine sedimentary rocks that are associated with basalt flows or other volcanic rocks, and the ore bodies they produce are called **massive sulfides**. The rich copper deposits of the island of Cyprus, which have supplied all or part of the world's copper needs for more then 3,000 years, are this type. (The word *copper* is derived from Latin *Cyprium*, "Cyprian metal.") Cyprus's copper sulfide ore formed millions of years ago adjacent to hydrothermal vents near a seafloor spreading center. Warping of the copper-rich seafloor caused by convergence of the European and African tectonic plates brought the deposits to the surface when the island formed. The rich copper deposit mined at Jerome, Arizona, is of similar volcanogenic origin.

Weathering Processes

The deep chemical weathering of rock in hot, humid tropical climates promotes mineral enrichment, because the solution and removal of more soluble materials leave a residual soil of less soluble minerals. Because iron and aluminum are relatively insoluble under these conditions, they tend to

Figure 13.9 Kennecott Corporation's Bingham Canyon mine south of Salt Lake City, Utah. Although not the world's largest open-pit mine, it is the deepest at more than 800 meters (1/2 mi). The mine is 4 kilometers (2 ½ mi wide), with tailings ponds covering about 9,000 acres—more than half the area of New York's Manhattan Island—and waste rock dumps contain more than 3.5 billion tons of material. Since 1906, the mine has produced more than 16 million tons of copper, 21 million ounces of gold, 185 million ounces of silver plus molybdenum, lead, platinum, and palladium. Currently, about 500,000 tons of rock is excavated daily. New discoveries, expected to extend the mine's life to 2034, will require cutting back one wall about 300 meters (1,000 ft) to allow deepening the mine another 90 meters (300 ft) to reach an additional 700 million tons of copper ore.

Kennecott Corp.

Figure 13.10 A black smoker hydrothermal vent on the East Pacific rise at a depth of 2,800 meters (9,184 ft). The "smoke" is hot water saturated with dissolved metallic-sulfide minerals, which precipitate into black particles on contact with the cold seawater.

remain behind in *laterite*, a highly weathered red subsoil or material that is rich in oxides of iron and aluminum and lacking in silicates. When the iron content of the parent rock is low or absent, however, this lateritic weathering produces rich deposits of **bauxite** (**Figure 13.11**), the principal ore mineral of aluminum.

Groundwater moving downward through a disseminated sulfide deposit may dissolve the dispersed metals from above the water table to produce an enriched deposit below the water table by **secondary enrichment** (**Figure 13.12**). At Miami, Arizona, for example, the primary disseminated copper-ore body when first discovered in the 1870s was of marginal to submarginal grade, containing less than 1% copper, but secondary enrichment improved the grade to more than 3%, resulting in an economically profitable deposit.

Figure 13.11 Bauxite, the ore of aluminum, forms in horizon B of laterites derived from aluminum-rich parent materials.

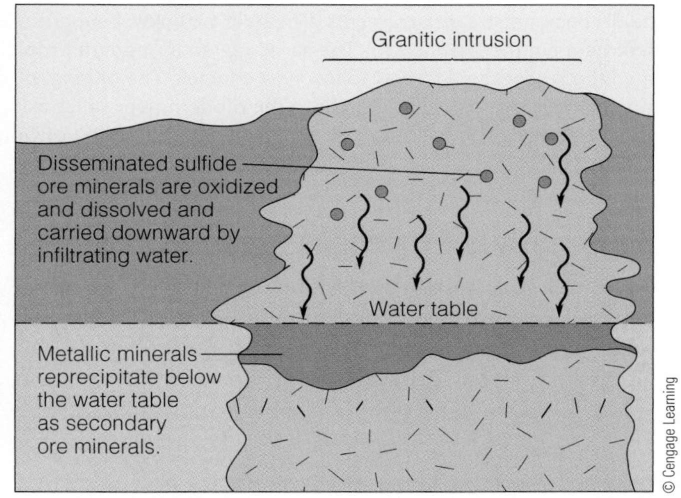

Figure 13.12 The development of ores by secondary enrichment. Descending groundwater oxidizes and dissolves soluble sulfide minerals, carrying them downward and leaving a brightly colored residue of limonite and hematite. Below the water table, new metallic minerals are precipitated as secondary ore minerals.

The rich silver deposits of Nevada, the self-proclaimed "Silver State," are due to secondary enrichment. The multitude of faults throughout the state provided the plumbing system for hot mineralized fluids to boil up through the fractured crust and emplace shallow, low-grade silver sulfide hydrothermal deposits. Then, for millions of years, rainwater and snow melt seeped downward into these low-grade deposits, gradually converting the silver sulfides into incredibly rich, heavily concentrated deposits of silver chloride.[1]

[1]By the 1860s, these rich deposits, called *surface bonanzas* by the miners, were discovered in range after range across the state. Intense mining began, especially in the 1870s, but by 1900, the deposits were virtually exhausted and the silver mining boom collapsed.

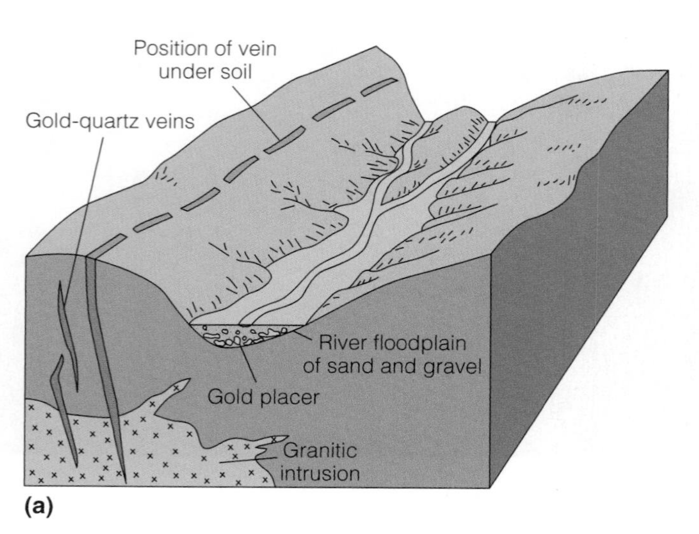

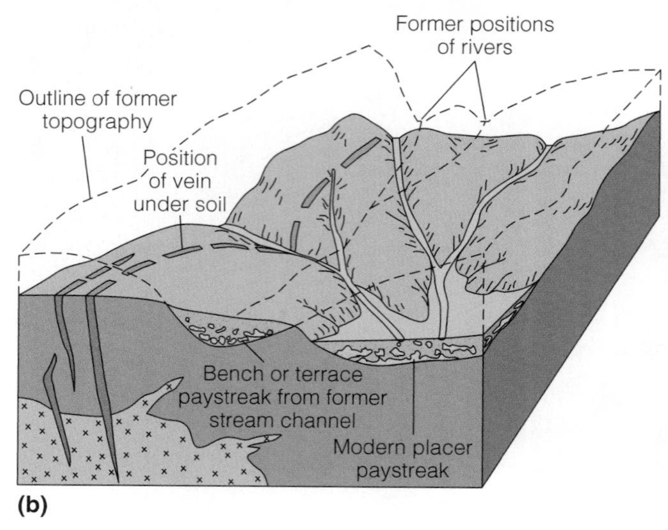

Figure 13.13 The origin of gold placer deposits. (a) An ancient landscape with a weathered and eroding gold-quartz vein shedding small amounts of gold and other mineral grains, which eventually become stream sediments. The gold particles, being heavier, settle to the bottom of the sediments in the channel. (b) The same region in modern time. Streams have eroded and changed the landscape and now follow new courses. The original placer deposits of the ancient stream channels are now elevated above the modern river valley as "bench" placers. (From Alaska's Oil/Gas & Mineral Industry, © 1982, with permission of Graphic Arts Books.)

Sedimentary Processes

Physical sedimentation—The simplest type of mineral deposits are formed by physical sedimentation by moving water in river gravels, and beach or even dune sands, and are called **placer deposits** (from Spanish, pronounced "plass-er"). Dense, erosion-resistant minerals, such as gold, platinum, diamonds, and tin, are readily concentrated in placers by the washing action of moving water. The less dense grains of sand and clay are carried away, leaving gold or other heavy minerals concentrated at the bottom of the stream channel. Such deposits formed by the action of rivers are referred to as **alluvial placers**; ancient river deposits that are now elevated as stream terraces above the modern channels are called **bench placers** (**Figures 13.13** and **13.14**). Considerable amounts of uranium are mined from Precambrian placer deposits in southern Africa. The deposits formed under the unique conditions existing on Earth's surface before free oxygen was present in the atmosphere. Some placer deposits occur on beaches, the classic example being the gold beach placers at Nome, Alaska.

Sand and gravel concentrated by rivers into alluvial deposits are important sources of aggregate for concrete. River deposits formed by **glacial outwash**—resulting from the runoff of glacial meltwater—are another type of placer deposit.

T. K. Buntzen, Silverado Mines U.S., Inc.

Figure 13.14 All that glitters *is* gold—a gold nugget weighing 41.3 troy ounces (1.284 kg; 2.82 lb), worth $71,139.25 at $1,722.50 per ounce, from a placer mine in the Brooks Range, Alaska.

Chemical sedimentation—Very large, rich mineral deposits may result from evaporation and direct **precipitation** (the process of a solid forming from a solution) of salts in ocean water, usually in shallow marine basins. Minerals formed this way are called **evaporites**. They can be grouped into two types: marine evaporites (which are primarily salts of sodium and potassium, gypsum, anhydrite, and bedded phosphates) and nonmarine evaporates (which are mainly calcium and sodium carbonate, nitrate, sulfate, and borate compounds). Other important sedimentary mineral deposits are the **banded-iron formation** ores (**Fig-**

Joseph Kirschvink

Figure 13.15 Western Australia has extensive areas of banded-iron formation, which is the iron ore resource exploited by large-scale mining. Some individual layers can be traced for more than 300 kilometers (186 mi). The dark layers are iron oxides; the red layers are chert containing fine-grained iron oxides.

ure 13.15). These deposits formed in Precambrian time, more than a billion years ago, when Earth's atmosphere lacked free oxygen. Without free oxygen, the iron that dissolved in surface water could be carried in solution by rivers from the continents to the oceans, where it precipitated with silica to form immense deposits of red chert and iron ore. Banded-iron deposits are found in the Great Lakes region, northwestern Australia, Brazil, and elsewhere, and they are enormous; they provide 200 years or more of reserves even without substantial conservation measures.

Manganese nodules (**Figure 13.16**) are formed by precipitation on the deep-ocean floor. The nodules are mixtures of manganese and iron oxides and hydroxides, with small amounts of cobalt, copper, nickel, and zinc, that grow in onion-like concentric layers by direct precipitation from ocean waters. Any commercial recovery of this resource appears to be many decades away because of technological, economic, international political, and environmental limitations, but eventually it may be necessary to exploit the deposits.

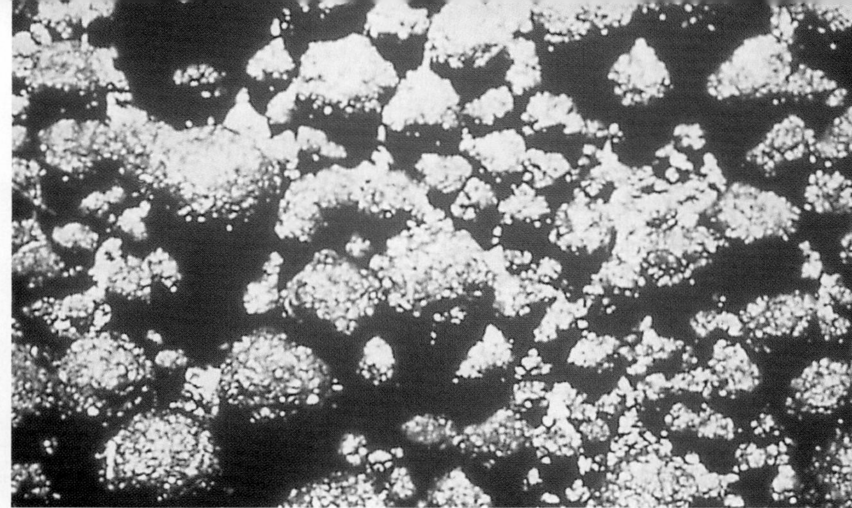

Figure 13.16 Manganese nodules, an example of deep-ocean precipitation. The nodules consist mainly of manganese and iron, with lower amounts of nickel, copper, cobalt, and zinc.
Bruce Heezen, Scripps Institute of Oceanography

consider this . . . How can a prospector use placer mineral deposits as a guide to finding ore deposits in bedrock?

Organic sedimentation—Biologic activity forming skeletal material in marine organisms, such as clams, oysters, corals, consists largely of calcium carbonate ($CaCO_3$) minerals (calcite and aragonite), and form coral reefs and limestone layers that are mined for limestone. Skeletal material of other marine organisms consists of apatite ($Ca_5(PO_4)_3(F,Cl,OH)$), which accumulates on the seafloor and with time forms beds of *phosphate rock*. Buried organic matter rich in carbon and hydrogen that becomes heated and chemically changed as it is buried beneath younger sediment may convert into *petroleum*, *coal*, and *natural gas* (methane, NH_3). When bacteria are involved in the alteration of the organic material, sulfate ions (SO_4^{-2}) are converted into hydrogen sulfide (H_2S) gas, which, in turn, may be reduced to elemental sulfur.

Metamorphic Processes

The high temperatures, high pressures, and ion-rich fluids that accompany the emplacement of intrusive igneous rocks produce a distinct metamorphic halo around the intrusive body. The result, in concert with the accompanying hydrothermal mineralization, is known as a **contact-metamorphic deposit**. If, for example, granite intrudes limestone, a diverse and colorful group of contact-metamorphic minerals may be produced, such as a tungsten-molybdenum deposit (**Figure 13.17**). Asbestos and talc originate by **regional metamorphism** that affects an entire region. Much of the nation's past production of lead and zinc came from contact-metamorphic deposits associated with granite porphyry intrusions.

Sufficient heating and widespread regional metamorphism of marine sediments may result in dehydration reactions, where water is driven off hydrous minerals (like the

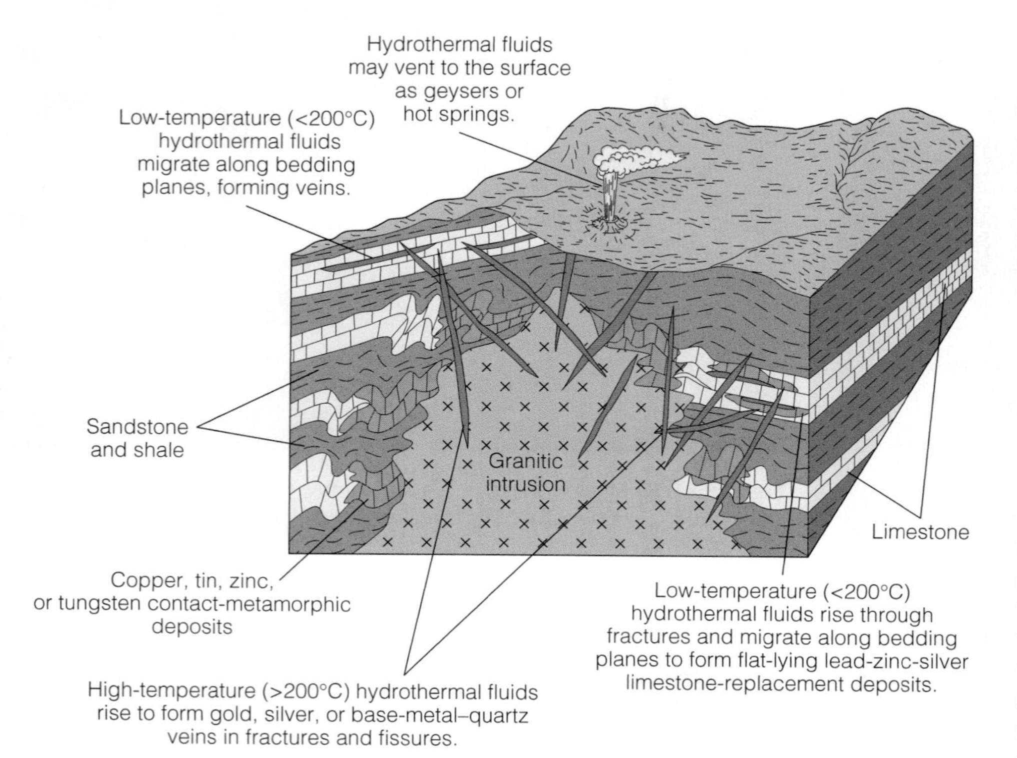

Figure 13.17 The relationship of limestone and a granitic intrusion in the origin of contact-metamorphic mineral deposits. (From Alaska's Oil/Gas & Mineral Industry. © 1982, with permission of Graphic Arts Books.)

micas and hornblende). As the hot water ascends to shallower levels, it may dissolve and become enriched in some minerals such as quartz, silver, and gold. When the metal-rich waters cool, they deposit their elemental content at shallow depths, forming gold- and silver-bearing veins (*lodes* in miner's jargon), such as those mined in the famous Mother Lode belt of California's western Sierra Nevada.

Other

Perhaps the world's most unusual metal mineral deposit is the Sudbury Nickel Irruptive, a major geologic structure in Ontario, Canada. The Sudbury deposit is one of the world's principal sources of nickel and copper ores, along with gold and platinum group elements. The deposit formed in a crater created by a meteor that impacted the area about 1.85 billion years ago.

Metallic Mineral Reserves

The metallic minerals are often grouped on the basis of their relative crustal abundance.

The Abundant Metals

The geochemically abundant metals—aluminum, iron, magnesium, manganese, and titanium—have abundances in excess of 0.1% by weight of the average continental crust. Economically valuable ore bodies of the abundant metals, such as iron and aluminum, need only comparatively small concentration factors for profitable mining (see Table 13.1) and are recovered mainly from ore minerals that are oxides and hydroxides. Even though they are abundant, these metals require large amounts of energy for production. Because of this, it is not surprising that the world's industrialized nations are the greatest consumers of these metals.

The scarce metals—The scarce metals have crustal abundances that are less than 0.1% by weight of average continental crust. Most of these metals are concentrated in sulfide deposits, but the rarest occur more commonly, or even solely, in rare rock types. Included within this category are copper, lead, zinc, and nickel—which are widely known and used, but nevertheless scarce—along with the more obvious gold and platinum-group minerals. U.S. deposits of chromium, manganese, nickel, tin, and platinum-group metals are scarce or of submarginal grade. Important, less well-known scarce metals include cobalt, columbium, tantalum, and cadmium, which are important in space-age ferroalloys; antimony, which is used as a pigment and as a fire retardant; and gallium, which is essential in the manufacture of solid-state electronic components. The United States has all of these critical metals in the National Defense Stockpile, a supply of about 100 critical mineral materials valued at about $4 billion; this stockpile could supply the nation's needs for 3 years in a national emergency.

Nonmetallic Mineral Reserves

The yearly per capita consumption of nonfuel mineral resources in North America exceeds 10 tons, 94% of which is nonmetallic, with the major bulk being construction materials (see Figure 13.1). The nonmetallic minerals are subdivided for convenience into three groups based on their use in industry, agriculture, and construction, although some minerals may not be easily classified as they may fit into two or more categories. For example, gypsum is an important player in industry (cement manufacturing), agriculture (soil amendment, soil conditioner, and fertilizer), and construction (making wallboard). The number and diversity of nonmetallic mineral resources are so great that it is impossible to treat the entire subject in a book of this sort. Thus, this subsection discusses only the nonmetallic resources generally considered most important.

Industrial minerals—The industrial minerals include those that contain specific elements or compounds that are used in the chemical industry—sulfur and halite, for example—and those that have important physical properties, such as the materials used in ceramics manufacture, and the abrasives, such as diamond and corundum. Sulfur, one of the most widely used and most important industrial chemicals, is obtained as a by-product of petroleum refining (**Figure 13.18**) and from the tops of salt domes in the central Gulf Coast states, where it is associated with anhydrite ($CaSO_4$) or gypsum ($CaSO_42H_2O$). Salt domes originate from evaporite beds that are compressed by overlying sedimentary rock layers, causing the salt to be mobilized into a dome, which rises, piercing the overlying rocks and concentrating sulfur, gypsum, and anhydrite at the top. Superheated water is pumped into the sulfur-bearing caprock to melt the sulfur, which is then piped to a processing plant. More than 80% of domestic U.S. sulfur is used to make sulfuric acid, the most important use for which is the manufacture of phosphate fertilizers. Salt domes are also important sources of nearly pure halite (NaCl)—common salt—and many domes have oil fields on their flanks (see Chapter 14).

D. D. Trent

Figure 13.18 Sulfur awaiting shipment to Asia at Vancouver, British Columbia. The sulfur originated as a contaminant in natural gas produced from gas fields in Alberta. Processing plants have separated the sulfur from the natural gas to produce a gas whose combustion products meet environmental air-quality standards. The sulfur is a useful by-product of the process.

In addition to being mined from salt domes, halite is mined from flat-lying evaporite beds in New Mexico, Kansas, and Michigan. The most important use of halite is for manufacturing sodium hydroxide and chlorine gas; about 15% of halite production is used for de-icing highways, and significant amounts are used in food products and in the manufacture of steel and aluminum.

Diamond is by far the most important industrial abrasive, although corundum, emery (an impure variety of corundum), and garnet are also used. Traditionally, natural diamonds have been important both as gemstones and as industrial abrasives, but artificial diamonds manufactured from graphite now hold about 70% of the market for diamond-grit abrasives. The major source of natural diamonds is kimberlite (named for Kimberley, South Africa), a rare and unique ultramafic rock that forms in pipelike bodies. Not only are kimberlite pipes rare, but most of them lack diamonds. The important diamond-bearing kimberlites are in South Africa, Ghana, the Republic of Zaire, Botswana, Australia, Belarus (formerly Belorussia of the USSR), Yakutia (the largest republic of the Russian Federation), and Canada's Northwest Territories, where North America's first diamond mine began operations in 1998 (**Figure 13.19**).

Jiri Hermann/BHP Billiton Diamonds, Inc.

Figure 13.19 The Ekati Diamond Mine™—Canada's first diamond mine, located in the remote Arctic tundra region known as the Barrenlands—is about 300 kilometers (188 mi) northeast of Yellowknife, Northwest Territories. The mine, with a potential life of about 25 years, is developed in kimberlite pipes similar to the famous diamond pipes of South Africa. Development of the property includes consulting with the leadership of the Aboriginal residents of the region and minimizing the impact on the environment. Remediation of disturbed lands is continually ongoing.

The metamorphic minerals corundum and emery, natural abrasives for which U.S. reserves are scarce or nonexistent, have largely been replaced by synthetic abrasives, such as carbides, manufactured alumina, and nitrides. The United States still imports limited amounts of corundum from South Africa.

Agricultural minerals—With the world's population heading toward 9 billion by 2050, it seems obvious that food production and the corresponding fertilizers and agricultural chemicals will also continue to expand. Therefore, nitrate, potassium, and phosphate compounds will continue to be in great demand. Whereas nearly all agricultural nitrate is derived from the atmosphere, phosphate and potassium come only from Earth's crust. Phosphate reserves occur in many parts of the world, but the major sources are in the United States, Morocco, Turkmenistan, and Buryat (the last two are republics of the former Soviet Union). The primary sources of phosphate in the United States are marine sedimentary rocks in North Carolina and Florida (see **Figure 13.20**), and there are other valuable deposits in Idaho, Montana, Wyoming, and Utah. The main U.S. supply of potassium comes from widespread nonmarine evaporite beds beneath New Mexico, Oklahoma, Kansas, and Texas, with the richest beds being in New Mexico. Canada also has large reserves of potassium salts.

Construction materials—About two thirds of the value of industrial minerals mined in the United States are construction materials, with aggregates (crushed stone, sand, and gravel) valued at $13.5 billion, and cement (made from limestone and shale) valued at $6.5 billion. Eighty percent of the aggregate is used in road building. Other major construction materials are clay, for bricks and tile, and gypsum, the primary component of plaster and wallboard. It is the mining of aggregate that is probably the most familiar to urban dwellers, because quarries customarily are sited near cities, the major market for the product, to minimize transportation costs. The quantity of aggregate needed to build a 1,500-square-foot house is impressive: 67 cubic yards are required, each cubic yard consisting of 1 ton of rock and gravel and 0.7 ton of sand. This amounts to 114 tons of rock, sand, and gravel per dwelling, including the garage, sidewalks, curbings, and gutters. Thus, it is no wonder that aggregate has the greatest commercial value of all the mineral products mined in most states, ranking second only in those states that produce natural gas and petroleum. The annual per capita production of sand, gravel, and crushed stone in the United States amounts to about 6 tons, the total value of which exceeds $13 billion. The principal sources of aggregate are open-pit quarries in modern and ancient floodplains, river channels, and alluvial fans. In areas of former glaciation, aggregate is mined from glacial outwash and other deposits of sand and gravel that remained after the retreat of the great Pleistocene ice sheets in the northern United States, Canada, and northern Europe (see Chapter 11).

How does the nation's need for mineral materials affect U.S. foreign policy?

Mineral Resources for the Future

Just as the world's fossil-fuel reserves are finite (see Chapter 14), so are its mineral reserves. Earth has been well explored, and fairly reliable estimates exist on the reserves of many important minerals. In the early 1970s, it was estimated that reserves of some minerals would be exhausted by early in the 21st century. Obviously, that is not happening, at least not yet, and part of the reason is that precise estimates are subject to the vacillations of several factors that are difficult to foresee: the rate of recycling; new discoveries in unexpected

IMC Fertilizer, Inc.

Figure 13.20 A rich phosphate bed in Florida is strip-mined by a mammoth dragline dredge. The bucket, swinging from a 100-meter boom (that's longer than a football field), can scoop up nearly 100 tons of phosphate at a time. The phosphate is then piped as a slurry to a plant for processing. Fine-grained gypsum, a by-product of processing, is pumped to a settling pond (upper left), where it becomes concentrated by evaporation.

Table 13.4 U.S. Consumption and Production of Selected Metallic Minerals

Mineral	Consumption, Thousands of Metric Tons	Primary Production, Thousands of Metric Tons*	Primary Production as Percentage of Consumption†
Aluminum	4,610	1,720	37.3
Chromium	400	0	0
Cobalt	8,500	0	0
Copper	1,730	1,120	64.7
Iron ore	47	49	104.3
Lead	1,400	400	28.6
Manganese	720	0	0
Nickel	229,000	0	0
Rare earths	2,800	0	0
Tin	58,000	Negligible	0
Zinc	901	720	79.9
Platinum-group metals	254,514	151,000	59.3

Source: *Mineral Commodity Summaries.* U.S. Geological Survey, 2011.

*Primary production is production from new ore.

†Primary production divided by consumption and rounded to the appropriate number of significant figures.

Table 13.5 U.S. Consumption and Production of Selected Nonmetallic Minerals

Mineral	Consumption, Thousands of Metric Tons	Primary Production, Thousands of Metric Tons*	Primary Production as Percentage of Consumption†
Clay	23,000	27,000	117.4
Gypsum	19,000	9,000	46.4
Potash	5,200	900	17.3
Salt	59,000	45,000	76.3
Sulfur	12,000	9,900	82.5
Sand and gravel	24,760	27,260	110.1
Crushed stone	1,190	1,150	96.6

Source: *Mineral Commodity Summaries.* U.S. Geological Survey, 2011.

*Primary production is production from new ore.

†Primary production divided by consumption and rounded to the appropriate number of significant figures.

places; new technological breakthroughs in processing; changing costs of energy for extracting, transporting, and processing; political factors; and changes in mineral economics because of increasing demands that may elevate submarginal grades to profitable grades. For example, the commodity values of copper, iron, and tungsten shot up significantly after 2004 because of the greater needs of China and India as their manufacturing capabilities increase. Consequently, the mining of previously uneconomic ores that provide those metals has become profitable.

U.S. consumption and production data for selected metallic and nonmetallic mineral resources appear in **Tables 13.4** and **13.5**, respectively. Notice in Table 13.4 that we consume greater amounts of many metallic mineral commodities than we produce. In the cases of chromium, cobalt, manganese, nickel, tin, and platinum-group metals, all or virtually all of those materials must be imported. The extent of U.S. import reliance for these and other mineral resources is shown in **Figure 13.21**. Thus, the United States is far from self-sustaining. By early in the 21st century, the United States imported an estimated $60 billion worth of raw and processed mineral materials annually. If nothing else, the data constitute a clear case against isolationism if we hope to sustain our present lifestyle.

A brighter forecast is shown by the data in Table 13.5 Note in the table that the United States was self-sustaining in its consumption of domestic nonmetallic and rock resources as of 2012, and that it was producing a surplus of one mineral commodity. There are some apparent disagreements between these tables and Figure 13.21. The lack of

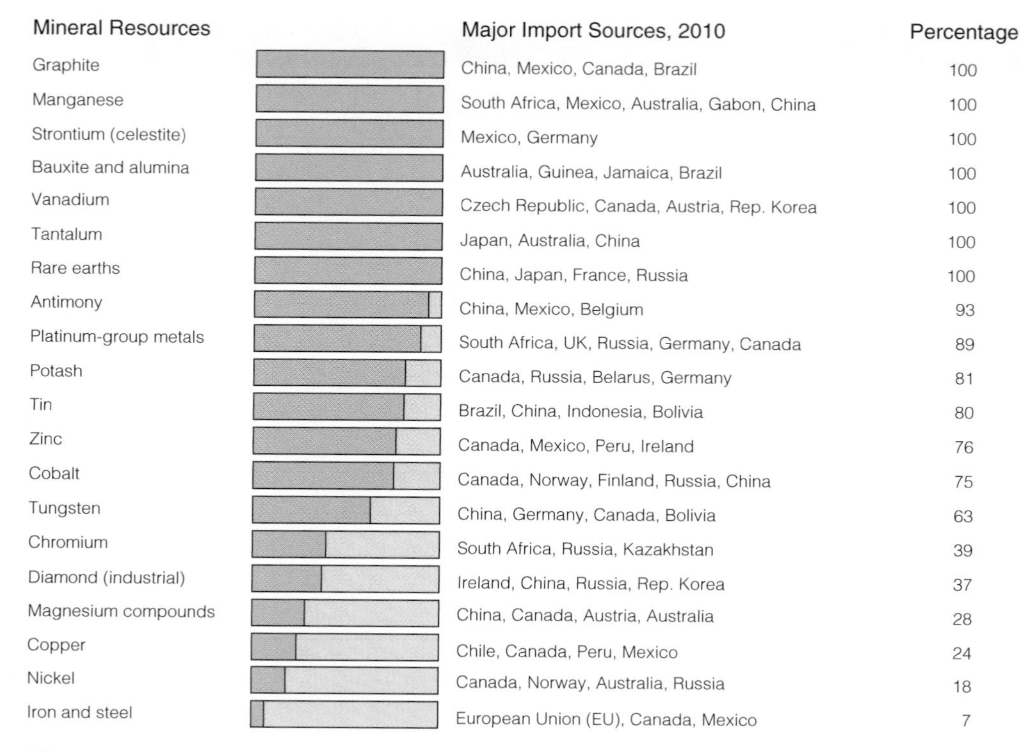

Figure 13.21 U.S. net import reliance for selected nonfuel materials. 2010. (Source: U.S. Geological Survey *Mineral Commodity Summaries, 2011.*)

imported lead shown in Figure 13.21, for example, does not seem to agree with the value shown in Table 13.4. The discrepancy results from extensive recycling of lead in the United States. Further, the high domestic production of aluminum reported in Table 13.4 results from a high reliance on imported bauxite (reported in Figure 13.21), because domestic sources of bauxite are insignificant.

As the richer deposits of some resources diminish in the coming decades, we can expect shortages of some minerals that presently are common. In many cases, technological advances will develop substitute materials that will suffice, but in other cases, we will have to do without. Some minerals will become so sufficiently scarce that deposits now classed as "submarginal" will be upgraded to "low-grade" and will be economically exploitable because of market forces.

The need for mineral resources may require exploration and exploitation in more remote regions and more hostile environments. Coal has been mined for more than a half-century well north of the Arctic Circle on Norway's island of Spitsbergen (Svalbard), for example, and sources of fuels and minerals are currently being sought and exploited in the harsh climates of Canada's Northwest Territories and Arctic Islands. Exploration by drilling cores has been done at a lead-zinc-silver prospect in western North Greenland. The Red Dog zinc-lead-silver mine, North America's largest zinc producer, on the lands of the Inupiat Eskimos north of the Arctic Circle in Alaska went into full production in 1991. Major Arctic gold mines already operate in Siberia and Greenland, and a huge iron mine on Baffin Island is expected to begin production in 2016. According to Barry Zellen, research director of the Arctic Security Project at the Naval Post-graduate School, "The estimated oil, natural gas, and strategic mineral reserves in the Arctic are staggering and of strategic importance."

For decades, Europe has been critically dependent on imports of metals, and needs to better know and use its own resources to decrease its dependence and cut trade deficits. European geologists report there is good evidence that continental Europe has hitherto undiscovered and underevaluated resources, specifically for deposits of high-tech metals (e.g., rare-earth elements, titanium, germanium, gadolinium, germanium, and platinum-group elements). In the United Kingdom, the prospects of exploiting copper, lead, and zinc from Parys Mountain in Wales are being assessed—at the site of the world's largest copper mine of the early 18th century. South Crofty, a Cornish tin mine, is set to begin full production in 2014, and Yorkshire is being explored for potash deposits. Cash is being raised to finance a tungsten mine in Devon at what is believed to be the world's fourth-largest deposit of the metal. Britain's stable politics and benign climate make it a miner's dream, and because it has gone unexplored for so many decades, new technologies may unlock resources hitherto overlooked. Further exploration and exploitation of these resources can be expected in the near future.

Even environmentally sensitive Antarctica, which is known to have extensive deposits of coal and iron, may become a target of future exploitation as more is learned about that continent's mineral wealth. In addition, the deeper parts of the continents and deep-sea mining of the ocean floor may also be targeted for exploration and exploitation as geophysical innovations and technological advances open up these presently inaccessible regions.

In the future, we will experience more recycling of critical metals, such as has been done with gold for thousands of years. The gold in the crown on your tooth, for example, may have been in jewelry during the age of Pericles in ancient Greece. Such metals as lead, aluminum, iron, and copper have been recycled for years. In many cities, the large-scale recycling of metals, especially lead and aluminum, as well as of paper, glass, and plastics, is being conducted successfully as part of municipal trash collection (see Chapter 15).

The recycling of metals in the United States amounts to about 150 metric tons per year, including 5 million tons of iron and steel, 2.7 million tons of aluminum, 1.8 million tons of copper, 2 million tons of stainless steel, 1.2 million tons of lead, and 85,000 tons of zinc. Other metals, such as brass, magnesium, and tin, are also recycled. Recycling 1 ton of steel conserves 2,500 pounds of iron ore, 1,400 pounds of coal, and 120 pounds of limestone. Recycling not only helps conserve mineral reserves but also saves energy. For example, recycling a ton of scrap aluminum saves up to 8 tons of bauxite ore and 14 MW-hr of electricity, and the recycling of all metals results in significant energy savings. Collectively, the value of recycled metals and mineral scrap in 2011 amounted to more than 32% of the nation's mineral production.

Despite recycling, mining will continue—it is essential for our way of living—but it cannot continue forever. As we explore and develop in more remote areas, more energy is required. We are caught in a twin dilemma: The declining richness and remoteness of ores require an increasingly greater input of energy, and the era of cheap fuels is also in decline.

Mining in the past was conducted largely by underground methods with a surface plant for milling and processing, and for hoisting workers, ore, and equipment (see chapter opener Figure 2). Surface mining methods, typified by vast open-pit excavations, such as the Bingham Canyon mine (see Figure 13.9), have now largely replaced underground methods. The surface plant remains much the same, however, although it usually operates on a much larger scale than in the past. Both surface and underground mining create significant environmental impacts on the land and air, and on biological and water resources. In addition, the needs for housing and services in mining areas have social impacts.

Mining and Its Environmental Impacts

When the ores are washed, the water which has been used poisons the brooks and streams, and either destroys the fish or drives them away.

Georgius Agricola, *De Re Metallica, 1550*

The General Mining Law of 1872 was important legislation; it helped establish the mining industry as a fundamental element in the U.S. economy. At the same time, the law was the ultimate in laissez-faire regulation, as it allowed miners to exploit, to take profits at public expense, to pay little in return for the privilege of mining on federal land, and to walk away from the scars and waste materials remaining when they abandoned their mines. The original 19th-century law allowed miners to stake a claim on potentially profitable public land and, should the claim prove to contain valuable minerals, to obtain a patent (legal title) for $2.50 an acre and reap the profits. This law helped "win" the West by enticing thousands of prospectors to seek their fortunes in gold, silver, and other valuable commodities.

Since 1872, the law has allowed approximately $250 billion worth of the public's hardrock minerals to be exploited by private corporations with no consideration for the environmental consequences of mining for about a century, until **National Environmental Policy Act (NEPA)**, **Surface Mining Control and Reclamation Act (SMCRA)**, and **Comprehensive Environmental Response, Compensation and Liability Act (CERCLA)** laws and regulations were enacted. The 1872 law is staunchly defended today by some public officials and the mining industry, who claim that it does a good job of providing the country with valuable minerals, employment, and economic development. The same law exists today, but it has been much modified over the years by more than 50 amendments, so that today's body of mining law fills six volumes. Two examples of the changes are: (1) the Mineral Lands Leasing Act of 1920, by which the government retains title to all federal lands possessing energy resources (oil, gas, and coal) and assesses royalties on the profits from the lands leased by developers; and (2) all mining claim, mill, and tunnel site locations must be filed with the Bureau of Land Management (BLM) and the County Recording Office within 90 days from the date of locating the claim or site. Initially, all new locations require a payment of a $15 service charge, a $34 location fee, and a $140 maintenance fee by September 1 of that year. Claims will expire if annual filings are not paid. On existing claims, an annual $140 maintenance payment must be paid to the BLM or, if a miner has fewer than 10 claims, a waiver filed with an assessment form including a $10 fee per claim. Maintenance fees for the many claims held by a major mining company may easily amount to $100,000 or more a year. Both existing and new claims must have $100 worth of labor or improvements performed annually per claim. (These fees are the only return taxpayers receive for hard rock minerals taken from public lands, which amounts to $25 to $30 billion per year, with the funds collected used to enforce surface mining regulations.) In addition, environmental restrictions added in the 1970s and 1980s dictate that modern mining must be conducted within a different policy framework than the simply drafted law of 1872. Although interest in environmental protection and reclamation in coal-mining districts began to develop in the 1930s, serious efforts at regulation and reclamation did not begin until the late 1960s because of the increasing amounts of unreclaimed land, as well as water pollution and problems arising from the lack of uniform standards among state mining programs. Consequently, the SMCRA was passed in 1977. This act established the coordination of federal and state efforts to regulate the coal industry to prevent the abuses that had prevailed in the past. SMCRA regulations apply to all lands—private, state, and federal. SMCRA established a fee that is charged on coal production that generates funds for reclaiming abandoned coal-mine lands. SMCRA also provides for state primacy once a state's coal-mine reclamation laws and policies have been approved by the Secretary of the Interior. This gives state regulators primary SMCRA-enforcement jurisdiction as long as the state programs meet federal standards.

Federal reclamation standards and other safeguards established by SMCRA for the coal-mining industry have no counterpart for noncoal-underground and surface mineral mining on state or private lands. Although there are federal regulations for noncoal mining on federal lands, it is left to the individual state governments to codify safeguards and enforce noncoal-mine-reclamation requirements.

Mining on federal lands must comply with the regulations of the NEPA of 1969. A mining company must file an operating plan and an *Environmental Impact Report* (EIR), and must conduct public hearings before the administrating agency can approve an operating permit. Even though an applicant may spend several million dollars complying with the NEPA permit application requirements, there is no assurance that an application will be approved.

Impacts of Coal Mining

Much of the coal now mined in the United States is extracted by surface operations, which involves removing *overburden*, rock and soil that overlies the coal beds. Compared with underground mining (**Figure 13.22**), surface mining is generally less expensive, is safer for miners, and facilitates more complete recovery of coal. Surface mining, however, causes more extensive disturbances to the land surface and has the potential for serious environmental consequences unless the land is carefully reclaimed. The three major methods of surface mining of coal are contour mining, area mining, and mountaintop-removal mining.

Contour mining is typical in the hilly areas of the eastern United States, where coal beds occur in outcrops along hillsides. Mining is accomplished by cutting into the hillside to expose the coal and then following the coal seam around the perimeter of the hill. Successive, roughly horizontal strips are cut, enlarging each strip around the hillside until the thickness of the overburden is so great that further exposure of the coal bed would not be cost-effective. At each level of mining is a *highwall*, a clifflike, excavated face of exposed overburden and coal that remains after mining is completed. Augers (giant drill bits) are used to extract coal from the areas beneath the highwall. The SMCRA requires that the sites be reclaimed—that is, returned to the original contour, and highwalls must be covered and stabilized after contour mining is completed. This is accomplished by backfilling **spoil**, the broken fragments of waste rock removed to mine the coal, against the highwall, and spreading and compacting as necessary to stabilize the reclaimed hillside. *Hydroseeding*, spraying seeds mixed with water, mulch, fertilizer, and lime onto regraded soil, is used on steep reclaimed slopes to aid in establishing vegetation that will help prevent soil erosion (see Case Study 13.3).

Area mining is commonly used to mine coal in flat and gently rolling terrain, principally in the midwestern and western states. Because the pits of active area mines may be several kilometers long, enormous equipment is used to remove the overburden, mine the coal, and reclaim the land. Topsoil is stockpiled in special areas and put back in place when the mining is completed. After replacing, it is tilled with traditional farming methods to reestablish it as cropland or pastureland. Sometimes, the land is more productive after reclamation than before mining (**Figures 13.23** and **13.24**).

Some states have enacted State Environmental Policy Acts (SEPAs), which are patterned after NEPA, whereas other states may have little regulation. The differences in the various states' permit procedures, their mining and reclamation standards, and their enforcement of underground and surface mining laws can be illustrated by comparing the laws of three states, each of which has more than a century of mining activity: Colorado, Arizona, and Michigan. Colorado has strong regulatory provisions relating to mine operation, environmental protection, reclamation bonding, inspections, emergency response by the state, and even the authority over mine operation. In contrast, Arizona, the only state in the country without a mine-reclamation law until 1994, allows companies to self-bond their reclamation fund, to wait as long as 17 years after mining has stopped to begin reclamation, and to set their own reclamation standards. Furthermore, there is an escape clause that may allow companies a variance from any of the law's provisions. Michigan, which ranks sixth among all states in the production value of nonfuel minerals, has no comprehensive mining law. No financial bond is required to ensure that a mine site will be properly closed and reclaimed if the company abandons the project or declares bankruptcy. No permit is needed to open a mine except a local zoning permit and a federal water-discharge permit—permits that are required of *every* industrial operation. Reportedly, there is no enforcement of the state's voluntary reclamation law.

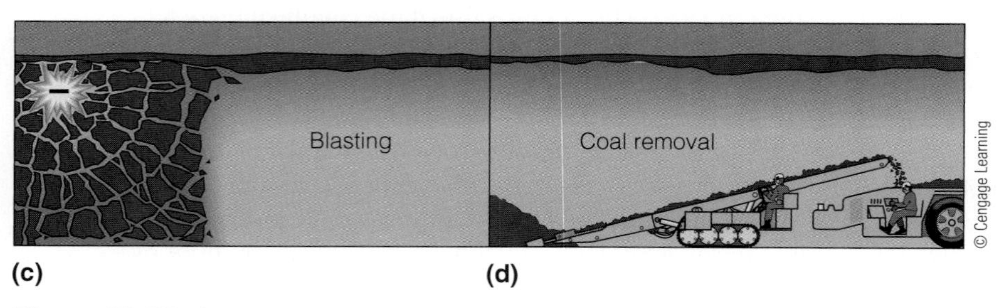

Figure 13.22 The steps in conventional large-scale underground bituminous-coal mining.

Alan Berger, *Reclaiming the American West*, Princeton Architectural Press, 2002

Figure 13.23 Colowyo surface (strip) coal mine near Meeker, Colorado. Reclamation follows the progress of the mining pit to minimize erosion. The reclaimed and revegetated land is to the right of the pit, which is advancing to the left.

high, the company may decide to go elsewhere where there are fewer or no bonding requirements.

After mining ceases, bonds are released to the mining company when it has met all of the requirements agreed on in the approved postmining reclamation plan. Each state determines its own bond release policy. Nevada, for instance, releases bonds when 25% of the required reclamation is complete. Other states' requirements range from 20% to 85% completion before the release of bonds.

There is no specific fund for cleaning up abandoned noncoal mines on state and private lands, nor are there uniform regulatory requirements for these lands. Action is left to the states and the mining companies. However, once any state with coal production has implemented safeguards and standards for coal mining, it may use the SMCRA abandoned-mine fund to address the reclamation of abandoned mines that exploited other commodities. For example, Utah's abandoned-mine reclamation program uses SMCRA funds to reclaim metals mines.

As we consider the environmental impacts of mining in the United States in this section, it is important to distinguish between: (1) the legacy of careless exploitation that is obvious at long-abandoned mines, with their residual scars and toxic wastes; and (2) the current mining scene, in which, depending on the state and whether the site is on state-owned land, land administered by the U.S. Forest Service or the BLM, more than 30 different permits may be needed to develop and operate a mine legally. Many mine owners consider the current permitting requirements to be excessive, and they question whether they will be able to continue to operate in the United States at all if the regulations become more restrictive.

Before a mining company can turn a spade full of dirt, the western states require miners to prove financial responsibility, called *bonding*, to pay reclamation costs in the event the mining company is unable (e.g., by bankruptcy) to do so. Bonding is a sort of catch-22: If the state bonding requirements are set too low and the miner declares bankruptcy, there will be insufficient funds to reclaim the site. A classic example of this occurred when Pegasus Gold Mining Company declared bankruptcy and left more than 30 unreclaimed mines in several western states, including the Zortman and Landusky mines in Montana (see Case Study 13.2). If the bond requirements are too

Environmental pollution abatement has become a major concern of domestic gold producers in some states, especially in Alaska, where there are many small placer mines. The reclamation of mined land has become an integral part of an increasing number of gold-mine operating plans. Abandoned mine sites, as well as other contaminated industrial sites that are deemed to pose serious threats to health or safety, may be cleaned up under the CERCLA of 1980, which is overseen by the EPA. For example, the Forest Service, a branch of the U.S. Department of Agriculture, has CERCLA authority to clean up contaminated mines on National Forest System lands. It was CERCLA that established the so-called Superfund to clean up the worst toxic or hazardous waste sites that have been assigned to the National Priorities List (see Chapter 15).

Mountaintop-removal mining (MTR) is used primarily in the eastern United States to recover coal that underlies

Figure 13.24 A reclaimed strip-mined land with sediment ponds; Rosebud Coal Mine, Colstrip, Montana. The freshly planted grasses and water source are ideal for cattle grazing. This mine leases reclaimed land to ranchers. The coal-fired power plant in the background supplies energy to three western states.
Alan Berger, *Reclaiming the American West*, Princeton Architectural Press, 2002

the tops of mountains. It is now the dominant force in altering land use in the central Appalachian coalfields. These coalfields cover about 48,000 square kilometers (12 million acres) in Kentucky, West Virginia, Tennessee, and southwestern Virginia. MTR mining involves clearing forests, stripping and disposal of overburden (soil and rock materials above the coal bed), the use of explosives to shatter rocks to expose and mine the coal beds, and it permits nearly complete removal of the coal. In some instances, whole communities and the people whose families lived for generations along the now-nonexistent stream banks have been displaced with minimal financial compensation. Not even cemeteries, some as much as 200 years old, are safe from the mining; they get shoved into the nearest hollow along with the other overburden. After the coal is removed, the mined area is returned to approximately its original configuration or is left as flat terrain (**Figure 13.25**) that historically has been revegetated by planting a few herb and grass species. To date, less than 5% of the flattened land has been developed, and many reclaimed areas exhibit little or no regrowth of woody vegetation and minimal carbon storage even after 15 years. Proponents of MTR point to its ability to provide jobs, greater safety for miners than underground mining, its efficiency, and the resulting increase of flat land in a region where otherwise there is little. The coal industry claims that MTR mining brings prosperity, yet regions where MTR mining is practiced have some of Appalachia's highest poverty rates.

A major environmental problem with MTR mining is disposal of excess overburden and waste rock, which includes toxic heavy metals found in coal and former rock layers—now pulverized—such as mercury and arsenic. The waste—millions of tons of it—is dumped into hollows or valleys, burying biologically crucial headwater streams (**Figure 13.26**). Some of the valley fills are miles long and hundreds of feet high. So far, 2,000 miles of the region's valleys have been filled in, ecosystems have been devastated, whole valleys have been eliminated, and residents have been subjected to health hazards. By 2020, it is expected that some 14 million acres of land will be affected. A study by the University of Maryland Center for Environmental Science, and the U.S. EPA's Programmatic Environmental Impact Statement released in 2005 reveals that after MTR mining, the levels of major chemical ions, such as iron, manganese, sulfate (SO_4), aluminum, and selenium, are persistently elevated downstream, with elevated selenium levels that bioaccumulate in organisms reaching concentrations that are toxic to birds and fish. The degraded water attains contamination levels that are lethal to laboratory test organisms, and macroinvertebrate and fish populations are consistently degraded. Appalachia's human population is known to suffer disproportionately higher instances of disease and death compared with the United States as a whole, and research shows that Appalachian health issues are worse in areas of coal mining even when accounting for the variable effects of poverty, race, health insurance, smoking, and education.

Following mine-site remediation attempts to return a site to premined conditions, SO_4 pollution persists long after mining ceases, and recovery of biodiversity in mining waste-impacted streams has not been observed. Furthermore, groundwater samples from domestic supply wells have higher levels of mine-derived contaminants than water from unmined areas. Despite regulatory requirements to minimize impacts, existing mitigation procedures meant to compensate for lost stream habitat and functions fail to do so. Consequently, legal action charged that waste-disposal practices in mountaintop-removal coal mining violates the Clean Water Act and various mining regulations that prohibit unnecessary and undue degradation of the public's land and water. To counter this charge, in 2002, the federal government issued a final clarification rule that a company can apply to the Corps of Engineers to put virtually anything labeled as fill into U.S. waters (streams, rivers, and lakes) as long as it either turns a portion or all of a water body into dry land or changes the bottom elevation of U.S. waters. This final rule is astounding, because it redefines fill and appears to violate both the stated goals of the Clean Water Act and mining regulations. Furthermore, the final rule applies broadly to all mining and industrial wastes, including those that are hazardous to the environment and human health, and it may apply to many sites that would qualify as Superfund sites.

Vivian Stockman

Figure 13.25 A massive dragline (on the right) looks like a child's toy when compared with the huge scale of a mountaintop-removal coal-mining operation near Kayford Mountain, West Virginia.

Vivian Stockman

Figure 13.26 Waste from mountaintop-removal coal mining forms an enormous valley fill in the Birchton Curve Valley, West Virginia. Such valley fills have buried many miles of streams, eliminating aquatic vegetation, fish, and other wildlife habitat, as well as entire communities and their residents, who lived for generations along now-buried valleys in West Virginia.

Despite many years of litigation supporting or opposing MTR mining, and numerous bills introduced in Congress to eliminate MTR mining or to at least strengthen MTR mining regulations, little has been decided and MTR mining continues much as it has in past decades.

Impacts of Underground Mining

The most far-reaching effects of underground mining are ground subsidence (see Chapter 7), the collapse of the overburden into mined-out areas (**Figure 13.27**), and **acid mine drainage (AMD)**, the drainage of acidic water from mine sites. AMD is generally considered to be the most serious environmental problem facing the mining industry today, as some acid drainage may continue for hundreds of years. AMD can result when high-sulfur coal or metallic-sulfide ore bodies are mined. In both cases, pyrite (FeS_2) and other metallic-sulfide minerals are prevalent in the walls of underground mines and open pits; in the **tailings**, finely ground, sand-sized waste material from the milling process that remains after the desired minerals have been extracted; and in other mine waste. The reaction of pyrite and other sulfide minerals with oxygen-rich water produces sulfur dioxide (SO_2), which with water forms sulfuric acid (H_2SO_4), which, in turn, liberates free metallic ions such as those of iron, lead, zinc, copper, and manganese into the environment (**Figure 13.28**). Such substances kill aquatic life and erode human-made structures, such as concrete drains, bridge piers, sewer pipes, and well casings. Certain bacteria thrive in this acidic environment, deriving energy from the conversion of one type of iron (ferrous, Fe^{+2}) to another (ferric, Fe^{+3}). The bacteria cause a "cascading" effect in the conversion of Fe^{+2} to Fe^{+3}, increasing the rate by as much as a million times what it would be if the bacteria were absent. The chemical and biological activity is exothermic and the heat produced provides positive feedback to the entire process, further speeding up the reactions. The acidic water produced by the reaction interacts with other components of the biosphere—aquatic life in streams and rivers, and land plants and animals, including humans, that use water—often with injurious effects. (Zinc concentrations as low as 0.06 mg/L and copper concentrations as low as 0.0015 mg/L are lethal to some species of fish.) Estimates are that between 8,000 and 16,000 kilometers (5,000 and 10,000 mi) of U.S. streams have been ruined by acid drainage. The oxidized iron in AMD colors some polluted streams rust-red (**Figure 13.29**). Where polluted water is used for livestock or irrigation, it may diminish the productive value of the affected land. Although expensive to implement, technologies do exist for preventing and controlling AMD (see Case Study 13.4). The subject of surface water and groundwater protection is addressed later in this chapter.

Ed Nuhfer

Figure 13.27 Store-building subsiding into an abandoned underground coal mine south of Pittsburgh, Pennsylvania.

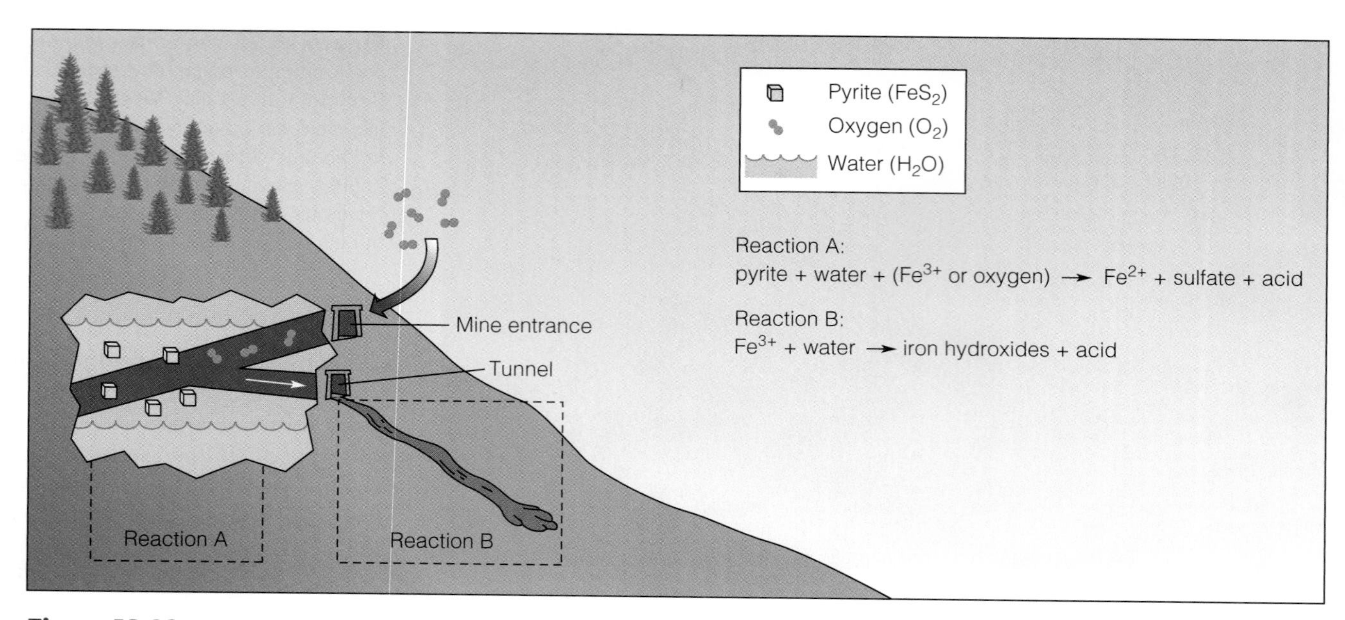

Figure 13.28 Pyrite reacts with air and water to produce sulfuric acid. (Based on USGS data.)

Pyrite (fool's gold) is common in deposits of gold, lead, zinc, silver, and copper. Why is a conspicuous rusty-colored outcropping often a clue to the existence of buried ore?

Impacts of Surface Mining

A variety of surface-mining operations contribute to environmental problems. Hydraulic mining is largely an activity of the past, but **dredging**—scooping up earth material below a body of water from a barge or raft equipped to process or transport materials—continues in the 21st century (**Figure 13.30**). Much of the current U.S. production of sand and gravel is accomplished by dredging rivers, and large dredges are used to mine placer tin deposits in Southeast Asia. Dredging causes significant disruption to the landscape; it washes away soil, leaves a trail of boulders, and severely damages biological systems. Scarification remains from former dredging in river bottoms in many western states and elsewhere in the world (**Figure 13.31**). Few gold-dredge areas have been reclaimed, but an outstanding example of what can be accomplished appears in **Figure 13.32**. Another example is a 73-square-kilometer (28 mi^2) dredge-scarified area near Folsom, California. The area was reclaimed by reshaping the waste piles, covering the surface with topsoil, and then building a subdivision of homes

Figure 13.29 Rust-red sludge of acid mine drainage from West Virginia coal mines that were abandoned in the 1960s. The acidic water has eaten away at the Portland-cement–based concrete retaining wall, and the bridge supports will soon share that fate. Problems such as this are rarely caused by modern surface mines, because current mining and reclamation practices eliminate or minimize acid mine drainage.

W. Virginia Geological and Economic Survey

Figure 13.30 Placer mining with a bucket-conveyer dredge near Platinum, Alaska, 1958. The dredge operated from the early 1930s until the late 1970s, during which time it was the major producer of platinum-group metals in North America.
D. D. Trent

Alan Berger, *Reclaiming the American West*, Princeton Architectural Press, 2002

Figure 13.31 Abandoned dredge tailings from placer mining near Fairplay, Colorado. These dredge tailings could be used as aggregate in construction. But the site is too far from a market to be economically feasible.

on it. The area is landscaped with trees and grasses, and no evidence remains of the once scarified surface.

In **hydraulic mining**, or *hydraulicking*, a high-pressure jet of water is blasted through a nozzle, called a *monitor*, against hillsides of ancient alluvial deposits (**Figure 13.33**). Hydraulicking requires the construction of ditches, reservoirs, a penstock (vertical pipe), and pipelines. Once constructed, the mine can wash thousands of cubic yards of gold-bearing gravel per day. The gold is recovered in sluice boxes (long, open-ended boxes with transverse slats on the bottom over which the slurry of muddy water flows), where mercury (the liquid metal that alloys with the gold and silver to form an amalgam) may be added to aid recovery at low cost. Hydraulicking is efficient but highly destructive to the land (**Figure 13.34**). Because 19th-century hydraulicking was found to create river sediment that increased downstream flooding, clogged irrigation systems, and ruined farmlands, court injunctions stopped most hydraulic mining in the United States before 1900 (although its use continued in California's Klamath-Trinity Mountains until the 1950s). Hydraulicking is still being used in Russia's Baltic region for mining amber, and hydraulicking and mercury amalgamation are used to extract placer gold in remote areas of Brazil's Amazon Basin, as well as parts of Chile and Peru. It is estimated that 100 tons of mercury is working its way into the ecosystem of the Amazon Basin each year. The danger of using mercury is that, when it escapes from mining activity, it takes on different geochemical forms, including a suite of organic compounds, the most important being methylmercury, which is readily incorporated into biological tissues and is the most toxic to humans. In California's Mother Lode region, where hundreds of hydraulic placer gold mines operated from the 1860s through the early 1900s, the mercury loss to the environment from these operations is estimated to

Figure 13.32 Reclaimed dredge-mined site, a former placer gold mine; Fox Creek, Fairbanks District, Alaska. The State of Alaska awarded a certificate of commendation for the reclamation work.

T. K. Buntzen

Figure 13.33 Hydraulicking a bench placer deposit.
Donald B. Sayner Collection

Marc Jay Plumley

Figure 13.34 A monument to the destructive force of hydraulic mining of bench placer deposits; Malakoff Diggins State Park, Nevada County, California. When the hydraulic mining operations here ended in 1884, Malakoff Diggins was the largest and richest hydraulic gold mine in the world, with a total output of about $3.5 million. More than 31 million meters^3 (41 million yds^3) of earth had been excavated to obtain gold. The site is now marked by colorful, eroded cliffs along the sides of an open pit that is some 11,200 meters (7,000 ft) long, 480 meters (300 ft) wide, and as much as 800 meters (500 ft) deep.

have been 3 to 8 million pounds, and it still remains in the ecosystem.

Strip mining is used most commonly when the resource lies parallel and close to the surface. The phosphate deposits of North Carolina and Florida are strip-mined by excavating the shallow, horizontal beds to a depth of about 8 meters (26 ft) (see Figure 13.20). After the phosphate beds are removed, the excavated area is backfilled, returning the surface to its original form and converted into parks and playgrounds.

Open-pit mining is the only practical way to extract many minerals when they occur in a very large low-grade-ore body near the surface. The technique requires processing enormous amounts of material and is devastating to the landscape. The epitome of open-pit mining is the Bingham Canyon copper mine in Utah, where about 3.4 billion tons of material—seven times the volume moved in constructing the Panama Canal—have been removed since 1906 (see Figure 13.9). Now a half-mile deep, and scheduled to become even deeper, the pit is one of the largest human excavations in the world, and the waste-rock piles literally form mountains.

There are several environmental consequences of open-pit mining. The mine itself disrupts the landscape, and the increased surface area of the broken and crushed rocks from mining and milling sets the stage for erosion and the leaching of toxic metals to the environment. This is especially true of sulfide-ore bodies. They produce AMD, because the waste rocks and tailings are highly susceptible to chemical weathering.

Impacts of Mineral Processing

Except for some industrial minerals, excavating and removing raw ore are only the first steps in producing a marketable product. Once metallic ores are removed from the ground, they are processed at a mill to produce an enriched ore, referred to as a *concentrate*. The concentrate is then sent to a smelter for refining into a valuable commodity.

Concentration and smelting are complex processes, and a thorough discussion of them is well beyond the scope of this book. In brief, the concentration process requires the following steps:

1. Crushing the ore to a fine powder
2. Classifying the crushed materials by particle size by passing them through various mechanical devices and passing on those particles of a certain size to the next step
3. Separating the desired mineral components from the noneconomic minerals by a flotation, gravity, magnetic, or chemical method.

The four means of separation, as well as the smelting process, are discussed individually in this section.

Flotation—The **flotation** separation process is widely used, especially for recovering sulfide-ore minerals, such as lead, zinc, and copper sulfides, from host rock. The process is based on the principles of the wettability of mineral particles and surface tension of fluids. After crushing and concentration, the wettability of the *undesired* mineral particles is increased by chemically treating the crushed ore—usually with liquid hydrocarbons—to ensure that the undesired minerals will sink. Air is then bubbled into the slurry of crushed ore and water, forming a froth that collects the *desired* mineral particles of low wettability. The froth, with the attached desirable mineral particles, is skimmed off the top of the flotation tank and dried; this is the concentrate. The undesired mineral particles, the tailings, sink to the bottom of the flotation tank. They are drawn off and piped to the tailings pond. Although they are usually environmentally undesirable, tailings are an unavoidable waste product of mining.

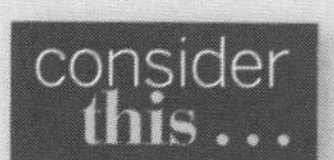

As the geological sources of raw materials are exhausted, the recycling of many metals will eventually be absolutely necessary. How would you educate the general public about the urgency of recycling?

Gravity separation—**Gravity separation** methods are used in recovering high-density ore minerals, such as gold, platinum-group metals, tungsten, and tin. By this process, mineral particles mixed with water are caused to flow across a series of riffles placed in a trough. The riffles trap the desired high-density particles, and water carries away the undesired low-density minerals, the tailings.

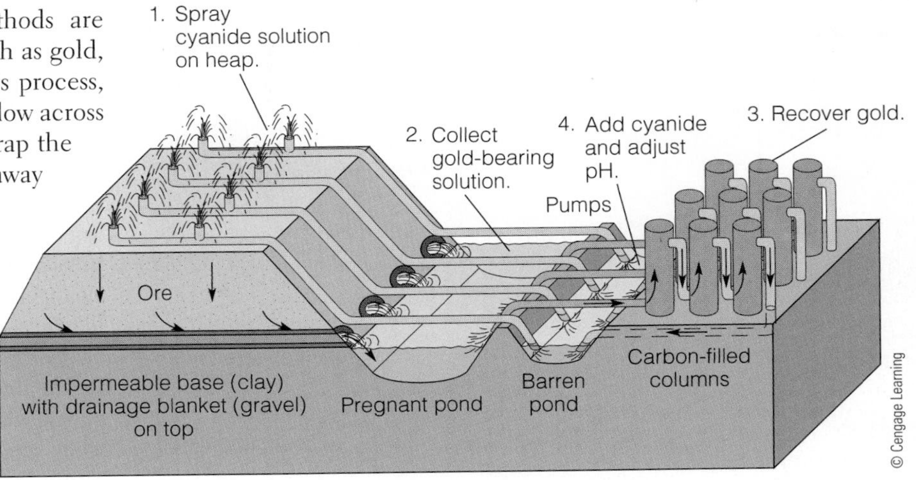

Figure 13.35 The major components of a cyanide heap-leach gold recovery.

Magnetic separation—Magnetic separation uses the magnetic, or nonmagnetic character of minerals. There are two classifications of magnetic minerals: (1) strongly magnetic minerals, such as iron and magnetite, which are easily separated by a magnetic separator having a low intensity magnetic field; and (2) weakly magnetic minerals that contain some iron, such as rutile, ilmenite, and chromite, which require a much higher intensity magnetic field for separation. *Diamagnetic* minerals are those that are repulsed by a magnetic field, such as copper, and are easily separated from the magnetic minerals. Minerals such as quartz, pyrite, galena, and silver simply fall off the end of the separator.

Chemical methods—For minerals whose physical properties make them unsuitable for separation by flotation or gravity methods, chemical processes are used, the major ones being leaching and cyanidation. *Leaching* is often used in treating copper-oxide ores. Sulfuric acid is added to crushed ore to dissolve the copper and produce a solution of copper sulfate. The dissolved copper is then recovered by placing scrap iron in the copper sulfate solution; the copper plates out onto the iron. The acidic waste materials are chemically neutralized by treating them with lime. Among the other potential toxic contaminants used in mineral processing are ammonia, benzene, bromine, chlorine, cyanide compounds, cyclohexane, ethylbenzene, glycol, ethers, hydrazine, hydrochloric acid, naphthalene, nitric acid, phenol, propylene, sulfuric acid, thiourea, toluene, and xylene.

The use of cyanide in gold and silver recovery is widespread and controversial. Since 1890, *cyanidation* has made use of the special property of cyanide to dissolve gold and silver. An innovation of cyanide recovery, **cyanide heap-leaching**, began to be used widely in the United States in the 1980s. By the turn of the 21st century, about 100 heap-leach operations were active in the western United States and South Carolina. Although heap-leaching is efficient, it is controversial in environmental circles, because the open-pit mining, waste-rock dumps, and tailings piles are destructive to the landscape. Furthermore, cyanide is perceived as a hazard to wildlife and a contaminant to ground and surface waters. In Montana alone, 51 cyanide releases were documented between 1982 and 1999. The use of cyanide is illustrated in **Figure 13.35**. Ore from an open-pit excavation is pulverized, spread out in piles over an impervious clay or high-density polyethylene liner, and sprayed with a dilute cyanide solution, commonly about 200 parts per million (0.02%) (**Figure 13.36**). The solution dissolves gold and silver (and several other metals) present in small amounts in

Figure 13.36 Cyanide drip lines on a leach heap. The cyanide concentration is 200 parts per million (0.02%), and the solution is kept highly alkaline to prohibit cyanide gas from forming.

D. D. Trent

Figure 13.37 A typical "pregnant pond" for collecting the cyanide solution containing dissolved gold from a leach heap. The face of the leach heap is at the upper right of the photograph. The pipes on the far side of the pond drain the "pregnant" solution from beneath the leach heap. The pond is lined with two layers of high-density polyethylene underlain by a layer of sand and a leakage-detection system of perforated PVC pipe. In the rare chance that leakage should occur, it is trapped in the pregnant pond, and appropriate repairs are made. The pond is covered with fine-mesh netting to keep birds from the toxic solution.

the ore as it works its way through the heap to the pregnant pond (**Figure 13.37**). The gold and silver are recovered from the resulting "pregnant" solution by adsorption on activated charcoal, and the barren cyanide solution is recycled to the leach heap. The precious metals are removed from the charcoal by chemical and electrical techniques, melted in a furnace, and poured into molds to form ingots.

Monitoring wells are placed downslope from leach pads and ponds containing cyanide for detecting possible leakage. Maintaining the low concentrations of cyanide required by state regulations requires regular monitoring of the solution. The cyanide solution is kept highly alkaline by additions of sodium hydroxide, a strong base, to inhibit the formation of lethal cyanide gas. No measurable cyanide gas has been detected above leach heaps where instrumental testing for escaping gas has been carried out. Unforeseen peculiarities of the ore chemistry and unusual weather have caused unexpected increases in cyanide values at some mines. This has required an occasional shutdown of operations until the chemistry of the solution could be corrected to the established standards. In a few cases, stiff fines have been assessed where cyanide levels exceeded the limits of the operator's permit. Upon abandonment of a leach-extraction operation, federal and state regulations require flushing and detoxification of any residual cyanide from the leach pile. Regular sampling and testing at groundwater-monitoring wells may also be required for several years following abandonment.

Securing the open cyanide ponds to keep wildlife from drinking the lethal poison is an environmental concern at several operations. After 900 birds died on the cyanide-tainted tailings pond at a mine in Nevada, the company pleaded guilty to misdemeanor charges, paid a $250,000 fine, and contributed an additional $250,000 to the Nature Conservancy for preservation of a migratory bird habitat.

Various techniques for discouraging wildlife from visiting cyanide-extraction operations have been tried: stringing lines of flags across leach ponds in an effort to frighten birds away, covering pregnant ponds with plastic sheeting, and blasting recorded heavy-metal rock music from loudspeakers. At one mine, the heavy-metal recordings were effective in keeping migratory waterfowl away, but resident birdlife adjusted to the din and remained in the area. One wonders what effect the blasting of recordings by Queens of the Stone Age, Nirvana, or Dead Weather might have. The current standard practice is to enclose the operations with chain-link fencing to keep out the larger animals and to stretch netting completely across the ponds to protect birds (see Figure 13.37). The fences and netting have been successful in some 20 active heap-leach operations in the western states: for example, bird losses have been reduced to only 50 to 60 a year in cyanide ponds. The number is less than trivial when compared with the thousands of birds killed in the western states each year when they fly into building windows, roof-mounted television antennas, and the paths of automobiles.

Smelting—Smelting is the process that produces a metal from its ore by using heat from a fuel, such as coke or charcoal, to remove the oxygen, and other elements, from the ore to leave the metal. Smelting is done in industrial plants called *smelters*. Historically, smelters have had a bad reputation for causing extensive damage to the environment. Sulfurous fumes emitted as a by-product of smelting processes have polluted the air, and toxic substances from smelting operations have contaminated soils and destroyed vegetation. Because water was necessary for operating early-day smelters, they were located near streams. The accepted practice at the time was to discharge mill wastes and tailings into streams and settling ponds, which when filled commonly spilled over into streams. Improved smelting technologies are now eliminating these problems. For example, Kennecott's former smelter at the Bingham Canyon, Utah, copper mine released 2,136 kilograms (4700 lb) of sulfur dioxide (SO_2) per hour to the atmosphere. Kennecott's new smelter, designed to meet or exceed all existing and anticipated future federal and state emission standards, was put into service in 1995. It is now considered the cleanest of its kind in the world, recovering 99.9% of all the sulfur dioxide emissions.

Mine-Land Reclamation

The volume of coal-mine spoil and hard-rock-mining waste rock is usually greater than the volume occupied by the rock before mining. SMCRA requires that coal-mine spoil be disposed of in fills, usually in the upper reaches of valleys near the mine site. Because some settlement of this material can be expected over time, construction on these reclaimed sites may be risky. Waste rock from hard-rock mining is usually deposited in large, mesa-like piles near the mines. In recent years, many of these piles have been contoured to break up their unnatural, flat-topped appearance. In both cases, the features are engineered and structured for stability with terraces and diversion ditches, for controlling surface-water flow and preventing erosion; then they are landscaped. Groundwater within the fills is commonly channeled through subsurface drains.

Surface-water and Groundwater Protection

High-quality, reliable water is critical for domestic, industrial, and agricultural activities, and it is especially precious in arid and semiarid regions. SMCRA requires that a surface coal-mining operation be conducted in a way that will maintain hydrologic balance and ensure the availability of an adequate water supply for postmining use. In the early 1990s, nearly 1,500 abandoned U.S. coal mines had been identified as having water problems. Of these, several hundred have been reclaimed or have been funded to begin reclamation.

During surface coal mining, all of the runoff water that collects in the pit is required to be collected and treated. Because surface mining destroys the original plant cover and exposes the soil to erosion, mitigation of erosion and sediment loss is needed during mining and reclamation. AMD problems have arisen from the hard-rock mining of sulfide ores in the western states, with more than 100,000 abandoned hard-rock mines in the West that may pose AMD problems (see Case Studies 13.1 and 13.4), in many areas of Canada, and from coal mining in West Virginia, Maryland, Ohio, and Pennsylvania.

Pennsylvania, recognizing the scars of 200 years of mining, has adopted the initiative *Reclaim PA*. Abandoned mine lands in the state—more than 250,000 acres of abandoned surface mines, 2,400 miles of streams polluted with mine drainage (see Figure 13.29), widespread subsidence problems (see Figure 13.27), more than 7,000 abandoned gas and oil wells, numerous hazardous mine openings, and affected water supplies—represent as much as one third of the total problem nationally. In the 1960s, Pennsylvania became a national leader in establishing regulations and laws to ensure that plugging and reclamation are completed after active operations close down. *Reclaim PA* includes policy and management objectives and legislative initiatives designed to enhance the reclamation efforts of the Pennsylvania Department of Environmental Protection, volunteers, and mine operators. The price tag for correcting Pennsylvania's problems is estimated at $15 billion.

At the federal level, SMCRA requires coal-mine operators to treat water from their mines before releasing it into streams and rivers. Control measures for AMD include:

- Holding mine drainage water in entrapment ponds and neutralizing it by adding alkaline materials from other industries, such as kiln dust, slag, alkaline fly ash, or limestone, before releasing it from the mine-permit area
- Grading and covering acid-forming materials to promote surface-water runoff and inhibit infiltration
- Backfilling underground mines with alkaline fly ash, which absorbs water and turns into weak concrete, to minimize the flow of water from the mine and reduce the amount of oxygen in the mine; filling is done by pumping a slurry down several boreholes to reduce AMD and stabilize the ground surface of many communities built over mined-out areas
- Using wetlands to treat AMD issuing from both operating and abandoned coal mines; cattails and other wetland plants have been found to be effective in removing some toxic metals and other substances from water; one such Pennsylvania wetland has saved an electric power company $50,000 in water-treatment costs each year since it was built in 1985

Wastewater from hard-rock mines, from processed ore tailings, and from waste rock dumps is another source of contamination. These waters are usually treated by chemical or physical processes before release to the environment. When those processes proved unsuitable at the Homestake Gold Mine at Lead, South Dakota,[2] the company developed a bacterial biooxidation process that broke down residual cyanide from the mill's wastewater into harmless components that are environmentally compatible with the disposal creek's ecosystem. The creek now supports a healthy trout fishery for the first time in more than a century.

Since 1990, new technologies have been perfected that produce faster chemical neutralization at less expense, and with less risk for water contamination. Nevertheless, federal and state regulators do not allow mining companies to post bonds for use of the new technologies, because they have not yet been proved over 20-year trial periods. Upon abandonment, heap-leach operations result in a potentially toxic threat to surface water and groundwater. Neutralizing is customarily accomplished by flushing the leach pad with freshwater (which is considerably more expensive and time-consuming than newer methods), and bonding is still required for the amount equivalent to the cost of flushing with freshwater. One could argue that federal or state funding to promote developing more effective reclamation technologies, especially for mine sites where freshwater is limited and acid rock and AMD are present, would benefit mine operators, agency regulators, and the public. An example of newer technology has been used at Coeur Rochester Mine in Nevada (**Figure 13.38**). Naturally occurring microbes within the heap nutrients (fatty acids, sugars, and alcohols) are placed on

[2]Although the Homestake Mine shut down operations in 2002, it still holds the record as the oldest, largest, and deepest mine in the Western Hemisphere, with workings more than 12,900 meters (8,000 ft) below the town of Lead.

Courtesy Coeur d'Alene Mines Corporation

Figure 13.38 Coeur Rochester Mine, Nevada, is one of America's largest silver mines. The leach pad is in the lower foreground, with open pits and waste rock repositories in the distance. Reclamation has been ongoing at the mine during its operation using naturally occurring microorganisms to metabolize the chemicals and trace metals in the leach pad. Since 2000, it has received the Habitat Restoration Award from the Nevada Division of Wildlife, the Corporate Conservationist of the Year Award, and the Nevada Excellence in Mining Reclamation–Wildlife Enhancement Award.

the top of the leach pad. Once the original nutrients are consumed, the microbes turn to the remaining chemicals and trace metals within the leach pad. During the metabolic process, the trace metals and chemicals are degraded or become insoluble in water. The result is a relatively clean effluent, yet no freshwater is used in the process.

At some unprotected mine properties, siltation clogs streams and increases the threat of flooding. This problem may be mitigated by constructing settling ponds for trapping the sediment downstream from the source (**Figure 13.39**).

Of all mining operations, sand and gravel mining are the most widespread and the most obvious to urban dwellers—especially when the operations are along or within rivers. Such modification to rivers results in a variety of consequences: increased erosion to channels and riverbanks, degradation of water quality, the disconnection of links between terrestrial and aquatic ecosystems, and the creation of unacceptable habitats for spawning fish, such as anadromous (migrating) salmonids (**Figure 13.40**). Fortunately, disrupted river channels can be restored, and more states are requiring restoration elements in mines' operating plans.

consider this . . . Why doesn't mining that exposes limestone or marble as it removes ore have AMD problems?

Mill and Smelter Waste Contamination

As ore is milled to produce concentrate, more than 95% of the mined material is discarded as tailings. Smelters that produce metallic-sulfide concentrates release waste products that can contain thousands of times the natural levels of heavy metals, such as lead, arsenic, zinc, cadmium, and beryllium. These contaminants remain in smelter slag (rendered inert by the smelting process, so that it is not bioavailable) and in flue dust that commonly remains around former smelter sites long after smelting has ended. The primary concerns with such contamination are the environmental impact of stormwater runoff into nearby streams or lakes and the risks to human health that result when people unwittingly inhale or ingest the heavy metals. If the site is on the EPA's Superfund list, the cleanup will take one of two directions—either removal or in-place remediation, such as enclosing or encapsulating the heavy-metal sources—determined by which one is considered appropriate for preventing human contact. At one Superfund site, the discovery of elevated levels of arsenic in the hair and blood of children in a small community built on contaminated soil resulted in the EPA's moving the entire commu-

John E. Gray, USGS Bulletin 2156

Figure 13.39 A placer gold-mining operation near Fairbanks, Alaska. A downstream settling pond traps placer tailings from the mine.

E. Frank Schnitzer, Oregon Dept. of Geology and Mineral Industries

E. Frank Schnitzer, Oregon Dept. of Geology and Mineral Industries

Figure 13.40 (a) Site of a sand- and gravel-mining operation along the Molalla River in Oregon before restoration and (b) after reclamation. Not only is the site more attractive, but also the pit has been modified to decrease erosion, shallow ponds have been created to provide habitat suitable for migrating and spawning salmon, and the overall setting is improved for the general benefit of wildlife.

nity of about 30 families and destroying the houses. If a contaminated Superfund area is populated, the EPA may require an ongoing blood-testing program to monitor residents' blood levels of heavy metals. Contaminated stormwater runoff may require the construction of a stormwater management system, with engineered drainage channels, ponds, and artificial wetlands.

The EPA procedure for remediating contaminated smelter slag, flue dust, and tailings that pose a threat to surface water or groundwater is to remove the material and deposit it in specially constructed repository sites (**Figure 13.41**). Commonly, these are pits lined with high-density polyethylene (**Figure 13.42**), into which the contaminated waste is placed and covered with crushed limestone (the most common method of neutralizing the acid rock waste), sealed with topsoil, and revegetated. To eliminate leaching of the material by rainwater or snowmelt, a subsoil drainage system collects excess water and disposes of it in such a way that it poses no threat to human health or the environment. (The disposal of toxic wastes is discussed thoroughly in Chapter 15.)

Revegetation and Wildlife Restoration

SMCRA regulations for coal mining and most state reclamation regulations for hard-rock mining require healthy vegetation to be reestablished once mining has ended. Permanent vegetation is the principal means of minimizing erosion and reducing stream siltation. The types of vegetation that are to be used in reclamation are stipulated in the original mine permit, based on premining vegetation and intended postmining uses. Commonly, a straw mulch or chemical soil stabilizer is applied after seeding to inhibit erosion and retain moisture. Mine operators are responsible for maintaining the new plant cover until it is successfully

ARCO

Figure 13.41 The removal of mill tailings at a Superfund site at Butte, Montana. Immense amounts of material must be removed from some Superfund sites.

ARCO

Figure 13.42 Preparation of a repository site for contaminated smelter-flue dust at Anaconda, Montana. The pit will have a high-density polyethylene liner. Once filled, it will be capped with soil and planted with native vegetation. The stack in the distance is all that remains of Kennecott's Anaconda smelter, built in 1919.

reestablished; the SMCRA minimum is 5 years in the East and Midwest and 10 years in the semiarid West.

Wildlife habitat is one of the most common uses of postmined land. Among the techniques the mining industry uses to meet the EPA, SMCRA, and state regulations for attracting and supporting desirable species of wildlife are contouring the land, introducing plant species that will support browsing and foraging, creating wetland habitats, and stocking ponds for sport fishing (**Figure 13.43**).

Courtesy Coeur d'Alene Mines Corporation

Figure 13.43 Deer on a restored waste-rock repository in 2009 at the Coeur Rochester mine, Nevada.

The Future of Mining

Concern about the future reserves of mineral resources has led to the obvious question: Are we running out? Numerous studies have been conducted over the years to determine reserves, and the answer to the question seems to be not yet. The question of availability is probably not the important one, however. Increasing needs for resources will require the mining of lower and lower grades of minerals. This will require improved technologies and larger, more powerful machines, which, in turn, will produce greater quantities of waste material. This will impose more stress on environmental systems. Because of these factors, the question that *should* be asked is, Can we afford the environmental and human costs required to satisfy our increasing need for minerals?

consider this . . . What is meant by the statement that the true cost of mining is externalized?

Today, most residents of industrialized countries enjoy living in comfortable homes, traveling by automobile, and having labor-saving appliances, all of which provide us with lifestyles that would have been considered luxurious only 100 years ago. These lifestyles are possible only because of the availability of inexpensive raw materials and inexpensive, easily obtained energy for mining and processing ores. Together, these inexpensive resources have allowed us to manufacture the products to which we have grown accustomed. Few of Earth's people realize that today's prices of mineral commodities typically reflect only the short-term, tangible costs of wages, equipment, fuel, financing, and transportation—just as they have since mining began in ancient times. Worldwide, much of the *real cost* of exploiting mineral and fossil-fuel resources is intangible, and it has been externalized. Historically, the environment has borne a great deal of the cost of extracting raw materials; the consumer has paid only part of the real cost. Even in the United States, where regulations designed to protect the environment are imposed, mining and processing still cause substantial damage.

Worldwide, poor mining and mineral-processing practices contribute significantly to soil erosion, water contamination, air pollution, and deforestation, especially in developing nations, as illustrated by the examples in Table 13.2. Fortunately, the negative environmental impacts of mining and mineral processing are gaining more attention in the developing nations. For example, Mexico adopted new mining and environmental laws in 1992 before encouraging mining by privatizing that industry. And in 1994, Tanzania and Guinea began requiring environmental-impact assessments to be submitted along with applications for mining licenses.

Mining generates twice as much solid waste as all other industries and cities combined, and most mining wastes in the United States are currently unregulated by federal law. In addition, the mineral industry is one of the greatest consumers of energy and a contributor to air pollution and global warming. It is estimated, for example, that the processing of bauxite into aluminum alone consumes about 1% of the world's total energy budget. Miners admit that mines leave holes in the ground, but they go on to point out that, when they abandon their currently active mines in the United States, they will be restoring the site to as near a natural appearance as is reasonable, except for the open pits.

In evaluating mining and its impact on the environment, we must recognize that the only way not to disturb the landscape is *not to mine*. This would mean not having the raw materials to build automobiles, houses, farm machinery, airplanes, computers, television sets—virtually all of the objects we have come to view as essential to a modern society. The question thus becomes: What must be done to minimize the impacts, but keeping in mind that if it isn't grown, it's mined?

Mining Legislation and Regulation

In the 1970s, some genuinely important concerns related to mining and processing mineral materials became obvious in the United States and many other countries, and laws began being written to address these concerns. Consequently, in the United States, the EPA now administers a number of acts regulating water quality and toxic wastes, and the Fish and Wildlife Service requires noninterference with rare and endangered species. Beginning in 2001, however, federal regulations dealing with some environmental laws and policies began changing, including regulations dealing with mining.

Various constraints are imposed by numerous state and local laws requiring that mining operations meet certain air-quality standards. Many states will not issue a permit for an operation to begin without prior approval of an adequately funded reclamation plan specifying that the waste and tailings piles will be restored to resemble the surrounding topography to the extent that it is possible, that the disturbed ground and the tailings will be replanted with native species, that all buildings and equipment will be removed, and that the open pits will be fenced and posted with warning signs. Most states require that mining companies be bonded to ensure that funds will be available for reclamation in the event of bankruptcy or abandonment of their operation before remediation is completed (see Case Study 13.2).

In 2003, California's State Mining and Geology Board approved permanent regulations requiring new open-pit metallic mines to backfill open pits and recontour their mine sites on completion of mining. Subsequent to the regulation, a Canadian mining company filed suit against the U.S. government under the provisions of the North American Free Trade Act (NAFTA). NAFTA allows foreign investors to sue signatory governments for the recovery of real or perceived financial losses caused by policies that foreign investors feel violate their rights. At issue were mining claims on California's Mojave Desert on land that is held sacred to Native Americans and that the Canadian company wished to develop. Because of the additional costs required by the new backfill regulation, the company realized that mining the property had become noneconomical and dropped the case. The new backfill regulation for metallic mines in California has essentially ended any new open-pit mining in the state.

It is claimed by some that the enforcement of federal regulations pertaining to mining has been weak. Initially, the EPA did little to regulate mining wastes. Further, Congress specifically exempted hard-rock–mining wastes and tailings from regulation as hazardous wastes in the Resource Conservation and Recovery Act (1976). Complicating the issues is the fact that, instead of the federal government, it is the individual states that play the role of regulator on state and private lands.

Further reform of the General Mining Law of 1872 is one of the eternal issues addressed by Congress. Beginning in the 1980s, public concern grew over the problems of protecting nonmineral values on public lands, the lack of federal reclamation of public lands, the lack of meaningful federal reclamation standards, limited environmental protection, and the lack of royalty collection for exploited public land (i.e., private profit from public land), all issues of public-resource management that are not addressed in the current mining law. Bills revamping the hard-rock–mining law have been introduced in nearly every session of Congress since the early 1990s.

Over the years, the House of Representatives has passed several reform bills, but they have all stalled out in the Senate, where they usually have been stopped by Harry Reid (D-NV), who has a record of protecting the mining industry and Nevada's gold mines. In mid-2009, U.S. Secretary of the Interior Ken Salazar informed the Senate that the time had come to make important changes to the law. Salazar said, "The mining industry must come to grips with meaningful reform of the 1872 mining law, patent reforms, and addressing environmental consequences of modern mining practices in meaningful and substantive ways for extraction of these valuable resources." For the first time, a revised bill emerged in the Senate that would establish a 2% to 5% royalty on all hard-rock–mining production and would end patentings—for several years, annual riders from Congress have prohibited patenting. A major point of contention in revising the mining law has been the imposition of royalties on mining income, with the funds generated earmarked for the reclamation of some 100,000 abandoned hard-rock mines in the western states, much as is done by SMCRA with abandoned coal mines using the Abandoned Mine Fund. Some critics of mining law reform believe that charging royalties for hard-rock operations on federal lands will close down U.S. mining and drive more mineral production to foreign countries. Senator Harry Reid (D-NV), who was thought to oppose the bill as written, announced that, because of other legislative priorities, the bill would not be acted on before Congress adjourned and the bill died. However, efforts at reform are expected to continue.

consider this . . . Growth in the consumption of natural (nonrenewable) resources increases exponentially, at a faster rate than the growth of Earth's population (which doubled twice during the 20th century). How long can this continue?

case study

13.1 Porphyry Copper and Plate Tectonics

Copper is a unique metal. It is the first metal to be mined and crafted by humans, which began in Mesopotamia about 9000 B.C.E., and in today's world it is a vital element in virtually all modern electronic devices. Copper deposits occur in several geologic environments, *porphyry copper deposits (PCDs)*, *volcanogenic massive sulfides*, *sedimentary and volcanic-hosted deposits*, *magmatic copper deposits*, and *supergene copper deposits*. PCDs, by far, are the world's most important sources of copper, accounting for more than 60% of annual global production and about 65% of the world's known copper reserves. In addition, PCDs may also be major sources of other metals, namely, molybdenum, lead, zinc, silver, gold, even arsenic. Despite the deposits being low to medium grade, usually less than 1%, PCDs are significant because of their enormous size (commonly hundreds of millions to billions of tons), the large scale of mining operations needed to exploit them, and long mine lives. PCDs, and other porphyry base-metal deposits, are unusual in their overall economic impact and the relatively recent recognition of their geologic, geochemical, mineralogic, and other characteristics of this class of deposits. The term *porphyry* refers to the igneous rock texture, large crystals that formed in the early history of crystallization of an igneous melt that later were surrounded by smaller crystals. *Porphyritic texture* may occur in both volcanic and plutonic igneous rocks. PCDs are of hydrothermal origin that formed in and around igneous intrusions in the crust that fed arc volcanoes (**Figure 1**). With the passage of time, the volcanoes are eroded and the landscape is eventually cut down to reveal the porphyry igneous body. The deposits are disseminated and consist of small, intersecting veins and veinlets called *stockworks* that contain quartz, chalcopyrite (copper iron sulfide), pyrite (iron sulfide), often molybdenite (molybdenum sulfide), and gold and silver. The porphyry "core" of the fabulously rich copper deposit at Butte, Montana, included major veins of sphalerite (zinc sulfide) and galena (lead sulfide), with the outer portions of the deposit enriched in rhodochrosite (manganese carbonate), making Butte the most important manganese district in the United States.

PCDs form when crystal-rich magma moves upward in the crust. As PCDs cool and harden, hydrothermal (very hot) fluids circulate through the magma and the surrounding rocks, shattering them to form a plumbing system of fissures and fractures into which the mineral-rich hydrothermal fluids can flow and deposit copper-bearing minerals to form veins and veinlets. A significant aspect of PCDs uniqueness is that they are found at the sites of former or modern-day convergent plate margins, such as in the western United States (mainly Arizona and New Mexico), Mexico, Canada, and in Chile, Indonesia, and the Philippines. The world's two largest porphyry copper mines, Chuquicamata and Escondida, are in Chile, followed as a close third by the Bingham Canyon Mine, Utah (see Figure 13.9).

The most common method of mining PCDs is by open-pit methods, such as those dotting the region of southern Arizona south of Tucson (**Figure 2**). Each day these operations move hundreds of thousands of tons of ore and waste rock and are perhaps the major earth moving efforts on the planet.

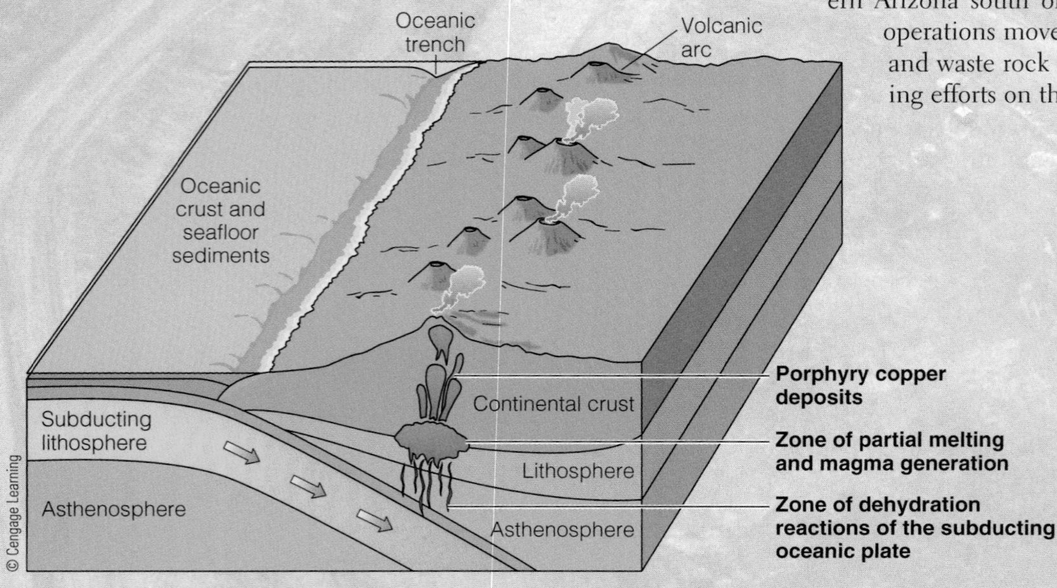

Figure 1 Schematic diagram illustrating the tectonic setting of porphyry copper deposits.

Figure 2 Satellite image of open-pit mines in southern Arizona south of Tucson. The Asarco-Mission Mine complex (upper left) produces copper and molybdenum, processing about 53,700 tons of ore per day. The Twin Buttes Mine (upper right), currently shut down, is owned by Freeport-McMoRan Copper & Gold Corporation. The Sierrita Mine (lower right), also owned by Freeport-McMoRan, produces molybdenum as well as copper, and the rare metal rhenium: This mine is the only U.S. source of rhenium. The mine processes over 115,000 tons of ore per day.

13.2 Cleaning Up a Mountain

The Little Rocky Mountains in north-central Montana on the southern edge of Fort Belknap Reservation, homeland to the Assiniboine and Gros Ventre Tribes, is the site of the Zortman and Landusky (Z-L) gold mines. The mines are in an area of the Little Rocky Mountains known to the tribes as the "Island Mountains." Historically, the tribes used the mountains for hunting and ceremonial purposes, and they also serve as the major source of the tribe's drinking water. In the 1880s, when gold was discovered in part of the reservation lands, federal commissioners from Washington negotiated with the tribes to return the mineralized land to the United States. In 1895, the commissioners acquired the land by threatening starvation by withholding food supplies, and the tribes signed over to the government 14,758 acres of land for $360,000 in an exchange known as the Grinnell Agreement. Most of the mining took place south of the tribal lands or south of the Grinnell treaty lands. A stampede of prospectors in the 1880s followed the discovery of placer gold in eight creeks draining the Little Rockies. Lode gold was discovered in 1893, and by 1917, at the peak of the large-scale mining operations, nine underground mines were active that supplied ore to several mills for processing. In addition to gold, the mines recovered silver, copper, lead, and zinc. Mining continued into the 1940s until 1942, when all nonessential mining was shut down for the duration of World War II by order of the War Production Board. Sporadic mining resumed after the war, but by 1959, large-scale underground mining was very limited, finally ending in the 1950s. Interestingly, very little gold was extracted from the actual tribal lands, and tribal members continue to use the mountains for hunting, burials, and vision quests.

Large-scale modern surface mining at the Zortman and Landusky mines began in 1979 by the Pegasus Gold Corporation, a Canadian company. The company's surface mining operations consisted of five open pits and six leach pads, the first major open-pit cyanide heap-leach gold mining operation in the United States. What followed was a multilayered environmental disaster involving bad science, some regulatory gaps, the bottom dropping out of gold's commodity price, and bankruptcy. The Z-L complex became the poster child for the inadequacy of the state and federal mining regulations of the time. The mine proved to be a failure because of a combination of inexperience with the new cyanide heap-leach process by the mining company and state regulators, a lack of adequate state regulations, an ambitious production schedule by Pegasus Gold Corporation, and some unexpected financial problems with Pegasus's mining operations in Australia, with all of these problems topped off by the collapse of gold's market value. The final result was Pegasus's filing for bankruptcy protection in 1998.

Before Pegasus could move a shovel full of dirt at the Z-L mines, the company had to submit an operating plan and post financial surety bonds to cover the expected costs of cleanup should the company have to walk away from the operation. Upon bankruptcy, these bonds were forfeited and the amount was adequate to cover 100% of the company's approved reclamation plan. However, over a period of time as cleanup proceeded, more items were added to the reclamation plan, which substantially increased the expense beyond the amount of the bonds. By 2012, the cleanup had cost at least $60 million, which includes an extra $6 million from the state of Montana, and about $12 million from the BLM. The estimated shortfall for cleanup by 2017 is about $10,500,000. Funding beyond 2017 will come from a state trust fund and will amount to an additional $34,000, although some authorities believe this is an inadequate amount given current interest rates and operating costs. Thus, in just two decades, Pegasus produced $300,000 in gold and, even after forfeiting its approved surety bond, has left Montana with a cleanup bill of tens of millions of dollars.

The science of what went wrong at the Z-L complex is the geology. Pegasus's initial exploration indicated they would be working an oxide deposit. However, as mining was undertaken, it proved to be oxide mineralization near the surface and above the water table, but primarily sulfide mineralization below the water table. Between the oxide and sulfide zones is a thick zone of partially oxidized and partially sulfide rock, with oxides dominating in the more fractured rock through which groundwater moves easily. The open-pit mining removed everything, and considerable sulfide-laden rock became exposed in areas that originally were thought to be oxide. In the 1970s, before open-pit mining, the drainage from the area's historic mines was of better quality than after excavating the open pits. Apparently, water infiltrated through the pit floors and sloping pit walls, generating acid drainage along the way that percolated into the older underground workings to produce the AMD that needed treatment. This plus **acid rock drainage (ARD)**—the natural background drainage from springs and seeps in the mining areas—ended up requiring four treatment plants to remove heavy metals and sulfides including arsenic, cadmium, copper, lead, and zinc (**Figure 1**). The treatment plants now process water from five open pits, seepage water from precipitation that infiltrates into four waste dumps, five leach pads, and several natural springs and seeps. The drainage water, collected in interception trenches, is pumped to the Z-L treatment plants where lime is added to raise the pH and precipitate out much of the heavy metals as sludge.

In addition to the drainage-water treatment plants, remediation focused on isolation, encapsulation, and keeping the sulfide-rich pits and waste dumps dry. Accomplishing this has been by partially backfilling open pits with materials from waste dumps and leach pads, and recontouring and grading the pit complexes. Backfill measures were used to cover most of the pit walls where sulfides were exposed. In most instances, a layer of limestone was placed along the bottom of pit walls. The backfilled pits have been graded to divert any runoff water, covered with 6 inches of compacted clay, this covered with geosynthetic clay liners (GCLs; a combination of high-density polyethylene and sodium bentonite, an expansive clay, that forms a highly impermeable barrier), then covered with 46 to 61 centimeters (18 to 24 inches) of non–acid-generating material, and topped with 61 centimeters (2 feet) of soil to limit stormwater and snowmelt infiltration. The covered pit floors and the benches were seeded with native grasses and plants. Those waste dumps and leach pads remaining after backfilling were reclaimed with soil cover. One dump was covered with a barrier cover to reduce infiltration, and lined water storage ponds were constructed on another leach pad that also reduces infiltration into the leach pad.

Dean Stiffarm, Fort Belknap Environmental Control Office

Figure 1 Swift Gulch Creek, one of the eight creeks draining the south slope of the Little Rocky Mountains, in October 2003, before water treatment. The creek is reddish brown from acid mine drainage (AMD) from the Zortman and Landusky mining complex. The water has a pH of 3.4 and contains high amounts of zinc, nickel, manganese, iron, and sulfate.

Calvin Larsen/Photo Researchers, Inc.

Dean Stiffarm, Fort Belknap Environmental Control Office

Figure 2 The Landusky Biological Water Treatment plant and holding pond that cleans up contaminated acid mine drainage at the abandoned Zortman and Landusky mine complex. The four water treatment plants will need to treat the contaminated drainage water for an indeterminate time, possibly 100 years or more.

Even though all of the leach pads were graded, covered with a soil layer, and revegetated, some rainwater and snowmelt falling on the reclaimed leach pads will work its way through the covers and percolate through the spent ore to become trapped in the lined leach pad basins. As the water percolates through the spent ore in the leach pad, it becomes more acidic, acquiring more heavy metals and sulfate, as well as some residual products from the cyanide leaching process. Consequently, all Z-L leach pad drainage is piped to the Biological Treatment Plant where bacteria reduce the metals and precipitate them from solution (**Figure 2**). This plant was designed especially to remove cyanide, selenium, and nitrates. In general, all of the treatment plants function well at eliminating acidity and removing most of the heavy metals, selenium, and nitrates.

At times in the past there was talk of heavy-metal–contaminated groundwater and polluted local drinking water. Each year, the state of Montana spends more than $40,000 testing the waters leaving the Z-L area to verify water quality. Testing has shown that no one living in the region drinks any water discharged from the four treatment plants, and there are no contaminated water wells. Nevertheless, the state has determined that pollution generated by the mines is severe enough that the water treatment plants will need to operate for possibly 100 years or more. This will be expensive. For example, in 2010, the total operating cost for reclamation maintenance and water treatment amounted to about $2,300,000, with bond funds only covering $731,321 of the expense.

The surety bond funding covers only anticipated expenses. An example of the unanticipated expenses that can arise in mine remediation occurred in May 2011, when Montana experienced a 500-year rainstorm. The storm triggered a debris flow that took out a portion of the Zortman waste dump, which then flowed down a small, steep canyon below the mine site (**Figure 3**). The failure probably resulted from a combination of rainwater infiltration through the dump augmented by discharge from old underground workings buried beneath the dump (which normally are above the water table). The water could not drain out of the dump fast enough, in part because the sulfide-rich waste rock had decomposed in the years since it was originally reclaimed into material that included much more sand and clay-sized material. It is estimated that it will cost more than $1 million to stabilize the failed waste rock dump. Because the event is considered an environmental accident, an event that was not reasonably foreseeable, it is

Bill Maehl

Figure 3 Waste dump failure at the Zortman Mine, May 2011. The creek bottom is buried by a debris flow from the failed waste dump. The failed slope of the waste dump and the head scarp of the debris flow is in the distance. It is estimated that it will cost more than $1 million to stabilize the waste dump.

Calvin Larsen/Photo Researchers, Inc.

not covered by the financial surety bond. Funds to cover the cleanup, should they ever become available, will come from government agencies, that is, the taxpayer.

questions to ponder

1. If you were an Assiniboine or a Gros Ventre living on the Fort Belknap Reservation in the 1890s, how would you have reacted to your treatment by the federal government? How would you feel about the situation now?
2. What changes in the General Mining Law of 1872 would you make in dealing with foreign mining companies that develop mines in the United States with inadequate bonding and that may leave American taxpayers with the cleanup costs if the miners go bankrupt?
3. When isolating and encapsulating sulfide waste rock or leach pads, a layer of crushed limestone is placed on the top before capping with soil and planting vegetation. What is the function of the limestone? Eventually, the limestone will be dissolved—what happens then?
4. Toxic heavy metals are naturally present in mineralized areas all over the West, but the runoff water is not contaminated. What is it about mining that can result in such a high amount of metals loading that the water is unfit for fish to survive or for human consumption?

13.3 Reclamation of Open-Pit Coal Mines

The five steps in reclaiming abandoned coal mine lands are: (1) drainage control to eliminate acid mine drainage; (2) the stabilization of landforms (**Figure 1**); (3) revegetation (**Figures 2** and **3**); (4) ongoing monitoring; and (5) return of the land to use (**Figure 4**). The reclamation work pictured in Figures 1, 2, and 3 was carried out near Beckley, West Virginia. The site in Figure 4 is near Harding, West Virginia.

Ed Nuhfer

Figure 2 Immediate seeding, done here with a hydroseeder, follows emplacement of topsoil to establish vegetation quickly. A biodegradable green dye is added to the seed to help the operator determine the coverage.

Figure 1 Drainage control and diversion at the disturbed area. Spreading and compacting waste rock, subsoil, and soil against the highwall stabilize slopes and bury reactive sulfide minerals.
Ed Nuhfer, The Citizens Guide to Geologic Hazards

Figure 3 A fast-growing, temporary cover crop, such as rye grass, prevents soil erosion and adds organic matter to the soil. The cover crop is selected carefully to avoid using a plant that may be undesirably dominant or persistent.

Ed Nuhfer

Chuck Meyers, Office of Surface Mining, Reclamation, and Enforcement

Figure 4 After the cover crop has died back, permanent legumes and grasses take over. They eventually restore the disturbed land to meadow or pasture, so that it can be used for wildlife habitat, recreation, or livestock grazing. Trees eventually may return the site to forest. Reclamation of this site eliminated more than a mile of highwall and sealed four hazardous abandoned mine openings. An underdrain was placed along the length of the highwall to collect drainage from the site. Today the site bears no resemblance to its appearance before reclamation.

question to ponder

1. The successful reclamation photos shown in this case study are for sites disturbed by underground and area coal mining (strip mining). Will reclamation of areas disturbed by MTR coal mining be equally successful?

13.4 Cleaning Up a Century of Hard-Rock Mining: The Abandoned Mine Lands Initiative

The challenges of cleaning up abandoned mining lands are well illustrated by ongoing activities in Colorado and Montana, where contaminated mine drainage is a serious environmental problem. Several thousand miles of these states' streams are contaminated with heavy metals from mine drainage. The water emitted from the mines, AMD, is commonly more acidic than lemon juice (citric acid) or vinegar (acetic acid)—often with a pH of less than 4—and contains concentrations of dissolved metals that often exceed the limits of state and federal standards. The high metal concentrations are due to acid dissolution of the country rock and to oxidation of metallic-sulfide minerals. These chemical reactions can produce harmful concentrations of such elements as iron, manganese, zinc, lead, copper, arsenic, and cadmium in the waters. Streams near abandoned mining lands can contain metal concentrations and (or) be so acidic that aquatic insects and fish cannot live in them.

From 1997 through 2001, the U.S. Geological Survey (USGS) conducted a program, the Abandoned Mine Lands (AML) Initiative, to provide technical assistance to support Federal Land Management Agency (FLMA) actions to remediate contamination associated with abandoned hard-rock mining. The strategy used a watershed approach in which contaminated sites were identified and remediated based on their impact on the surface waters and ecosystems within the watershed. The initiative began with two pilot projects in the Upper Animas River Basin in southwestern Colorado and the Boulder River Basin in southwestern Montana.

Upper Animas River Watershed, Colorado

In the area of Silverton, in the San Juan Mountains, many millions of dollars' worth of lead, zinc, silver, and gold were produced between 1874 and 1991. By the late 1990s, most of the 1,500 mines in the watershed had been abandoned (**Figure 1**), and metals-laden AMD issuing from the old mines affected the river's aquatic life and water quality for more than 100 miles downstream. Heavy metals, such as cadmium, copper, lead, and zinc, from both mining activity

case study

D. D. Trent

Figure 1 Longfellow Mine, at Red Mountain Pass, one of an estimated 1,500 abandoned mines in the Animas River watershed near Silverton, Colorado.

and natural sources are a threat to the environment and possibly human health. Contamination of the watershed is mainly from water discharged from some of the 1,500 inactive mines in the basin, although some of the contamination is from natural weathering of the region's highly sulfide-mineralized rock; such contaminated water is termed ARD. Part of the assessment of an area is to determine the environmental conditions and natural ARD that existed before mining to establish realistic cleanup goals for an area.

In response to the Colorado Water Control Division's upgrading of water-quality standards for the Upper Animas River Basin, the Animas River Stakeholders Group (ARSG) was formed in 1994. The ARSG is a coalition of elected officials; mining companies; environmental groups; local, state, and federal government agencies; and local interest groups. The ARSG has become part of the AML strategy of the U.S. Department of the Interior and the U.S. Department of Agriculture to clean up federal lands. The AML watershed approach to cleaning up the surface waters is a three-step process: monitoring, feasibility and site characterization, and implementation.

Monitoring consists of studying the mines in the areas, sampling and analyzing any waters draining from the mines. These studies showed that, about 350 days a year, zinc levels in the upper Animas River exceeded the standards proposed by the Colorado Water Standards Control Division. Feasibility and site characterization consists of determining remediation processes that work best in the area and prioritizing those sites for possible cleanup on the basis of:

- The contaminating site's level of environmental impact—that is, the relative "metal-loading" contributions to the watershed from a mine site's AMD, as well as ARD from natural sources
- Identification of the pathways of the toxic contaminants
- The feasibility of treatment
- Cost-effectiveness
- Preservation of the area's historic, natural, and cultural values

Of the 186 draining abandoned mines in the upper Animas watershed, the ARSG recognized that 33 are producing 90% of the metals loading. The implementation of remediation at those sites is the group's highest priority. The technology necessary for remediation, funding, and access to mine sites on private property were the critical factors in deciding which sites should be restored and in what order. Funding for the remediation comes from various state, federal, and local sources.

Implementation in the Animas watershed has included plugging and flooding AMD-producing mines, and capping or diverting drainage water around mine waste dumps. Perhaps the most complex and interesting remediation is that completed in 2005 to the Pride of the West Mine northeast of Silverton. The mine began working a polymineralic ore body in the late 1870s and continued to be worked sporadically until the 1970s. The ARSG recognized AMD from the mine as an important source of contamination to a major tributary of the Animas River.

The Pride of the West Mine is located on extremely steep terrain at an elevation of about 3354 meters (11,000 feet). Implementation consisted of covering two large, open stopes—large, vertical openings that were excavated upward from underground workings that broke through to the surface. For decades before remediation, meltwater streams coursed into these openings, dropped through approximately 213 meters (700 feet) of mine workings, and eventually discharged the toxic cocktail it had acquired into a creek and ultimately into the Animas River. Remediation included constructing a massive complex of steel beams supporting steel plates that were airlifted by helicopter to the site and fixed in place in the steep terrain by a construction crew. Working conditions on the precarious slopes required workers to use technical rock-climbing techniques, including wearing a climbing harness and being belayed with climbing ropes (**Figure 2**). The steel

Calvin Larsen/Photo Researchers, Inc.

Figure 2 (a) An early stage in the remedial work at the Pride of the West Mine in the Animas River Basin. Steel I-beams were lowered into place by a helicopter (out of view) over two open stopes. The motion and placement of the beams were controlled by workers using tag lines. (b) A late stage in the remediation of the mine. Steel plates were welded, torch cut, and pulled into place over the stopes. The finished project covered the stopes and diverted snow meltwater runoff from entering the stopes. Notice the steep terrain, requiring workers to be belayed from above for safety.

a and b: Daniel Hinds, Frontier Environmental Services

plates, welded in place, now covering the stope's openings, prevent snow, surface-water runoff, and avalanche deposits from entering the stopes, which previously caused much of the AMD metal loading emerging from the mine's entrance portal at lower elevation. The underground mine workings, which for decades have been largely filled with ice, will be entered once the ice melts, to plug any drill holes and fractures that are issuing water.

It was anticipated that these measures would greatly reduce the metals loading from the mine and improve the chemical quality of the creek. By the summer of 2009, however, the ARSG had found that flow measurements and metals loading had changed little. This case illustrates the difficulty of cleaning up abandoned mines. Nevertheless, it is anticipated that eventually the metals loading from the Pride of the West Mine will be reduced and the aquatic productivity of the Animas River watershed will be enhanced.

Boulder River Basin, Montana

Lands in the Boulder River watershed in Montana have been mined since the late 19th century. Metals extracted from the area include gold, lead, silver, and zinc. Environmental degradation resulting from a century of mining in the watershed drew together a cooperative partnership of more than 20 landowners, local communities, and several state and federal governmental agencies in an effort to clean up the messes left behind. A textbook example of the massive restoration effort in the mining area is represented by remediation efforts on High Ore Creek, an area with 26 inactive or abandoned mining sites.

The major contributor of contaminants to High Ore Creek has been the Comet Mine and Mill at Comet, Montana, about 5 miles northeast of the town of Basin. Mining at Comet began about 1860 (**Figure 3**). Several mines were active in the district, but the Comet Mine was the key

Figure 3 Comet Mine and Mill, 1918. Notice the huge pile of waste rock and tailings, to right of center.

Montana Historical Society, Helena

Figure 4 High Ore Creek Valley, Montana, in (a) 1997, when the creek had eroded through a century's accumulation of metals-loaded tailings from the Comet Mine and Mill, washing them downstream 6 miles to the Boulder River and enriching the water with heavy-metal concentrations, some of which were toxic and accumulating in the aquatic food chain, and (b) 2006, the valley restored by the removal of toxic mine waste, regrading, the addition of topsoil, revegetation, and the rebuilding of the channel of High Ore Creek. The pollutants have been substantially reduced, and trout are returning to the creek after an absence of many decades.

a: David Nimick/USGS; b: D. D. Trent

player and a major producer of gold, silver, lead, zinc, and copper between 1904 and 1950. By the 1950s, its operations had been shut down, and the mill dismantled, but it left behind an abandoned open pit and a legacy of toxic heavy metals contained in a vast tailings dump.

For nearly a century, High Ore Creek wound its way through the 3.7-mile floodplain composed of mining wastes and mill tailings from the Comet Mine (**Figure 4a**). In the process, the creek carried the heavy-metals–loaded waters downstream to the Boulder River at Basin. Sediment samples collected along High Ore Creek had metal concentrations greater than three times the natural background for arsenic, cadmium, copper, lead, manganese, mercury, silver, and zinc. Elevated levels of arsenic were detected in streambank tailings deposits as far as Basin, where soil samples collected at the school playground exceeded the cancer-risk level. At its best, fish could not survive in High Ore Creek's toxic cocktail of dissolved metals.

From 1997 to 2001, a cooperative effort of the BLM, the Montana Department of Environmental Quality (MDEQ), and the Montana Bureau of Mines and Geology restored the abandoned mine lands at Comet. The BLM took the cleanup effort an additional step by selecting it as one of two pilot projects of the AML Initiative. The remediation proved to be a classic example of what can be done to improve water quality and restore abandoned mines scattered across the West (see Figure 4b). Reclamation consisted of three basic steps: isolate, encapsulate, and keep dry.

The effort, focused on the extensive Comet Mine and Mill site, consisted of removing 229,500 cubic meters (300,00 yd^3) of tailings and waste rock that were disposed of in the abandoned open pit; constructing runoff diversion channels to keep the site dry and protect the excavated areas from erosion; removing 30,600 cubic meters (40,000 yd^3) of tailings and placing them in a nearby repository, where they were isolated from the environment; rebuilding large portions of High Ore Creek that flowed through the area of the former mine dump; removing 47,430 m^3 (62,400 yd^3) of streamside tailings and waste rock along a 4-mile length of the stream channel; grading waste-rock slopes to natural contours; amending the soil on the reclaimed area with a mix of compost and lime; seeding with native plants; covering the area with erosion-control nets; and constructing toxic sediment settling ponds (**Figure 5**). Cost for the project totaled \$3.5 million, which was funded by a 35-cent tax on each ton of coal mined in the West.

Postreclamation water analyses reveal that heavy-metal concentrations are substantially lower than before reclamation, and by 2006, the area was well on its way to recovery, with willows sprouting along the streambanks and grass and wildflowers growing on the reclaimed tailings site (see Figure 4b), where before restoration it had been a bare and sterile landscape; and a trout fishery has now returned to High Ore Creek after an absence of almost a century.

State officials have identified 270 other abandoned mine sites in southwestern Montana that are threatening the environment or human health. The successful restoration techniques used at Comet will be important in the remediation of other abandoned mining lands in the state.

D. D. Trent

Figure 5 High Ore Creek Valley in 2006, as viewed from the Rumley Mine at Comet, Montana. The grass-covered valley bottom is the area of the remediated tailings dump. In the distance are the toxic sediment settling ponds.

questions to ponder

1. The remediation of old historic mines is expensive and may come with mixed results—is it worth it?
2. When a present-day mine is developed in a sulfide ore body, what preventative measures might be taken to avoid the AMD problems such as those in the upper Animas River watershed?

case study

13.5 Rare Earths and Green Cars

Among the plans for reducing the world's emissions of CO_2 is dotting the landscape with wind farms and replacing gasoline-guzzling automobiles with electric cars that are charged up with electricity generated from renewable resources. A factor shared by hybrid and electric automobiles and wind turbines, however, is the need for the little-known elements *neodymium*, *dysprosium*, and *terbium* to make the magnets that are essential in their motors and generators. These three elements, members of a group called *rare-earth elements* (REEs), have unusual arrangements of electrons orbiting their nuclei, which give them significant powerful magnetic characteristics. As far as is known, no other materials can be substituted for these three elements.

REEs consist of 17 chemical elements: scandium, yttrium, and 15 lanthanides, of which lanthanum, cerium, praseodymium, neodymium, europium, and ytterbium are key elements used in these cars. For example, a Toyota Prius hybrid contains 10 to 15 kilograms (22–33 lb) of lanthanum and a small amount of cerium in its batteries, and 1 kilogram (2.2 lb) of neodymium, a major component in the alloy used for the 25 permanent magnets in the car (**Figure 1**). Furthermore, REEs are key components in TV screens, computers, cell phones, missiles, and numerous other electronic devices. Currently, nearly all hybrid components, such as motors and batteries, are manufactured in Asia.

REEs actually are not that rare, but high-grade deposits that are economic to exploit are. The manufacturing of REE hybrid cars, wind turbines, and electronic devices is vulnerable to a supply shortage, as the world's manufacturers of such products have been reminded recently that China has a chokehold on their production. China, the world's dominant REE producer, made headlines in 2010 when it abruptly curtailed exports to Japan. But it had been squeezing the global market for years. For example, China exported 47,000 tons of REEs in 2000, but by 2010, it had cut back exports to about 30,000 tons. The cutback seems to be the result of unfair quotas and export taxes. The United States, Japan, and Europe took the case to the World Trade Organization in March 2012, as it appears that China's export restrictions are a disguised restriction on international trade. Although the country has produced some 90% of REEs for several decades, it controls less than half of the world resource base. Higher prices and limited supply have led to the development of big mines in Australia and the United States where Molycorp's Mountain Pass mine, near the border of California and Nevada, reopened in 2011.

For many years, the Mountain Pass mine supplied all of the U.S. needs. It is the world's richest REE deposit and was shut down several years ago when REEs from China became available at a cheaper price. Beginning in 2009,

Calvin Larsen/Photo Researchers, Inc.

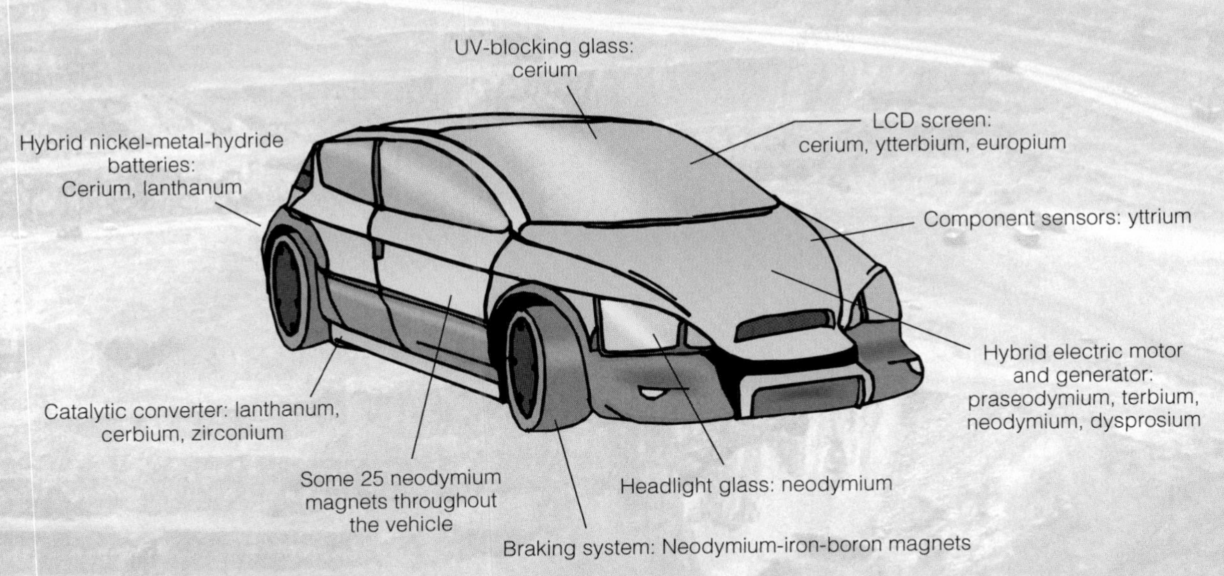

Figure 1 The major uses of rare-earth elements in hybrid and electric vehicles.

however, the Mountain Pass operation began producing about 1.8 kilotons (2,000 tons) per year of REEs from stockpiled ore. Because of increasing demand and China's export restrictions, it is planned to have the mine in full production by 2013, when it is anticipated it will produce about 19 kilotons (21,000 tons) per year.

questions to ponder

1. How might the need for REEs affect China's foreign policy?
2. Are there any substitutes for REEs in today's electronics?

PHOTO GALLERY the legacy of mines & mining

The collection of photographs in this gallery illustrates some aspects of past and present mining districts.

1 The Lucky Friday Mine in the Coeur d'Alene Mining District near Mullan, Idaho. The mine works a silver-lead-zinc vein. The mine is unique for two reasons: (1) It has been one of Hecla Mining Company's mainstay properties for more than 40 years, producing more than 130 million ounces of silver during its operation; and (2) once ore is removed from along the trend of the veins, the workings are backfilled with cemented tailings, thus reducing potential impacts on the environment. Hecla Mining Company and Idaho Geological Survey

2 Humans can be exposed to waters carrying high concentrations of toxic metals, even in undisturbed natural settings. Several thousand people living in and near the Brooks Range of northern Alaska rely on subsistence fishing and hunting for food. Massive metallic-sulfide deposits in the region pose a potential hazard to all life there if acid rock drainage (ARD) from these deposits enters streams and rivers. In this photo, a scientist is collecting water samples downstream from the Drenchwater massive sulfide deposit in the northwestern part of the range. Water bodies' pH and concentrations of various metals are monitored to evaluate the health hazard of this lead-zinc-silver deposit. U.S. Geological Survey

A massive tiered open-cut uranium mine within Kakadu National Park, Australia.
© 2009/JASON EDWARDS/National Geographic Image Collection

3 Even some national parks are not free of contamination from abandoned mining activity. An example is the Mariscal Mercury Mine in Big Bend National Park, Texas. Intermittently between 1900 and 1943, it produced mercury (quicksilver) from cinnabar ore. A detailed site investigation conducted in 1995 by the U.S. Bureau of Mines reported that minimal contamination remains in only a few isolated spots onsite, and there is minimal to no risk for short-term visitors and transient workers at the site. John Burghardt/National Park Service, U.S. Dept. of Interior

4 A creek near an open-pit copper mine at Morenci, Arizona. The water's distinct blue color reveals a high concentration of dissolved copper leached from mine wastes. D. D. Trent

5 The *entire* landscape visible in this photo is productive land reclaimed following extensive surface coal mining in the 1960s and 1970s; near Cologne, Germany. D. D. Trent

6 Designed by Jack Nicklaus, the Old Works Golf Course has been built on the Superfund site of Montana's historic, century-old, copper smelter at Anaconda, the first course ever built on an Environmental Protection Agency (EPA) Superfund site. Among the mining artifacts remaining at the site are huge piles of black slag (middle distance), by-products of the copper-smelting process. The slag is also used in the bunkers (foreground), which contrast with the lush green of the course. The fairway borders are planted with native vegetation, which blends with the area's natural plant community. The remediation approach here is to isolate and keep dry. The entire golf course is underlain by a complex subsoil underdrain network of perforated pipe to catch excess irrigation water and stormwater runoff, which is diverted to a lake for irrigation recycling. The underlying smelter waste is encapsulated with a soil capping, a water-tight liner, and a layer of crushed limestone to neutralize any possible acidic seepage. D. D. Trent

SUMMARY Mineral Resources and Society

Mineral Abundances, Definitions

Mineral Deposit
Locally rich concentration of minerals.

Ore
Metallic mineral resources that can be economically and legally extracted.

Mineral Reserves
Known deposits of earth materials from which useful commodities may be economically and legally recoverable with existing technology.

Factors That Change Reserves

Technological Changes
Affect costs of extraction and processing.

Economic Changes
Affect price of the commodity and costs of extraction.

Political Changes
Cut off or open up sources of important mineral materials.

Aesthetic and Environmental Factors
Public demands for wildlife, wilderness, and National Park protection and potential for surface- and groundwater pollution may reduce reserves.

Description of Mineral Resources

Description
Mineral deposits are highly localized; they are neither uniformly nor randomly distributed.

Cause
Concentrations of valuable mineral deposits are due to special, sometimes unique geochemical processes.

Origins of Mineral Deposits

Igneous Deposits
Intrusive—occur (a) as a pegmatite, an exceptionally coarse-grained igneous rock with interlocking crystals, usually found as irregular dikes, lenses, or veins, especially at the margins of batholiths; and (b) by crystal settling, the sinking of crystals in a magma because of their greater density, sometimes aided by magmatic convection.
Disseminated—a mineral deposit, especially of a metal, in which the minerals occur as scattered particles in the rock but in sufficient quantity to make the deposit a worthwhile ore.
Hydrothermal—a mineral deposit precipitated from a hot aqueous solution with or without evidence of igneous processes.
Volcanogenic—of volcanic origin (e.g., volcanogenic sediments).

Sedimentary Deposits
Mineral deposits resulting from the accumulation or precipitation of sediment: surficial marine and nonmarine precipitation, deep-ocean precipitation.

Weathering Deposits
Lateritic weathering—concentrations of minerals caused by the gradual chemical and physical breakdown of rocks in response to exposure at or near Earth's surface.
Secondary enrichment—mineral development that occurred later than that of the enclosing rock, usually at the expense of earlier primary minerals by chemical weathering.

Metamorphic Deposits
Concentrations of minerals, such as gem or ore minerals, in metamorphic rocks or from metamorphic processes.

Categories of Mineral Reserves

Abundant Metals
Average continental-crustal abundances are in excess of 0.1%.

Scarce Metals
Average continental-crustal abundances are less than 0.1%.

Nonmetallic Minerals
Minerals that do not have metallic properties include industrial minerals, agricultural minerals, and construction materials.

Environmental Impacts of Mining

1. Surface coal mining and land reclamation is regulated by federal law, the Surface Mining Control and Reclamation Act (SMCRA).
2. SMCRA requires surface coal-mine pits to be backfilled with the mine waste (spoil) and then covered with topsoil and landscaped. Some states in which hard-rock mining is conducted require similar handling of waste rock and tailings at currently operating mines.
3. Ground subsidence caused by underground coal mining is less common than in the past because of the required backfilling. Surface collapse from hard-rock (underground) mining is not common, and stabilization may be expensive.
4. Acid mine drainage (AMD) may be a problem at active and abandoned coal and sulfide-ore mines. Pit water in surface coal mines is required to be held and treated before it is released to streams or rivers. Sealing abandoned mines, covering waste rock with soil and landscaping, and installing drainage courses to direct surface drainage off tailings and waste-rock piles are some methods of controlling AMD formation. Wells are installed below waste-rock and tailings dumps to monitor water quality.
5. Scarified ground, despoiled landscapes, and disrupted drainage that remain after dredging and hydraulic mining can be reclaimed by reshaping, adding topsoil, and landscaping.

6. Abandoned open pits with oversteepened side walls despoil the landscape and may pollute hydrologic systems. They are difficult or impossible to reclaim. The public is protected by surrounding the pits with chain-link fencing and posted warning signs. The pit floor may be partially backfilled and sealed with impermeable clay to protect groundwater from pollution.
7. Cyanide heap-leach gold-extraction methods normally cause minimal environmental problems because they are strictly regulated by state and federal agencies. There are occasional leaks of cyanide into the environment, however. Wildlife losses have been mitigated by placing nets over the lethal ponds. Surface waters and groundwaters are monitored by sampling and monitoring wells, respectively.
8. The main concerns associated with heavy-metal contaminated mill and smelter waste are the environmental impacts of stormwater runoff into nearby surface waters and the human health problems caused by inhaling or ingesting the heavy metals in dust. Superfund cleanups require either removal or in-place remediation by encapsulating the heavy-metal sources. Wetlands with appropriate plantlife may be constructed, or water-treatment plants constructed, to clean toxins from surface runoff.

The Future of Mining

Worldwide

Mining and processing of mineral commodities are the basic economic foundation for many countries. But improper mining may contribute to soil erosion, water contamination, air pollution, and deforestation.

The United States

Minerals, mining, and processing of nonfuel mineral commodities are a vital factor in the United States, contributing nearly $70 billion dollars to the U.S. economy each year. Every material used in modern industrial societies is either grown or derived from mineral resources. Today's mining in the United States must meet standards of air and water quality set by the EPA and other state and federal agencies. The General Mining Law of 1872, although much amended, is praised by the mining industry, but criticized in light of current economics and conflicting land uses.

KEY TERMS

acid mine drainage (AMD)
acid rock drainage (ARD)
alluvial placer
area mining
average crustal abundance
banded-iron formation
bauxite
bench placer
black smoker
Comprehensive Environmental Response, Compensation and Liability Act (CERCLA)
concentration factor
contact-metamorphic deposit
contour mining
crystal settling
cyanide heap-leaching
disseminated deposit
dredging
evaporite
flotation
glacial outwash
gravity separation
high-grade deposit
hydraulic mining
hydrothermal deposit
lode
low-grade deposit
manganese nodule
massive sulfide
mineral deposit
mineral reserves
mountaintop-removal mining
National Environmental Policy Act (NEPA)
open-pit mining
ore
ore mineral
pegmatite
placer deposit
porphyry copper
precipitation
regional metamorphism
secondary enrichment
smelter
spoil
strip mining
Surface Mining Control and Reclamation Act (SMCRA)
tailings
vein deposit
volcanogenic deposit

STUDY QUESTIONS

1. Distinguish among mineral deposits, ores, and reserves.
2. Why do some metals, such as gold, require concentration factors in the thousands, whereas others, such as iron, require only single-digit concentration factors?
3. Of which metals does the United States have ample reserves? How do we obtain the scarce mineral resources we need but lack within our borders?
4. How do you expect domestic and foreign supplies of critical mineral resources to change in the next 25 years?
5. What would cause currently marginal or submarginal mineral deposits to become important mineral reserves?
6. Where are the major deposits of porphyry copper and porphyry copper–molybdenum deposits located? How do these deposits relate to plate tectonic theory?
7. What steps are involved in the origin of a placer gold deposit? Of a bench placer deposit?
8. What minerals may eventually be harvested from the deep-ocean floor? What might be the environmental consequences of such mining?
9. What is hydrothermal activity? What metallic mineral deposits are commonly associated with hydrothermal action?
10. Contrast the mining and extraction technology of the 19th century with those of today. What are the environmental legacies of 19th-century methods? What environmental concerns accompany modern mining technology?
11. Contrast the environmental hazards and mitigation procedures of underground, open-pit, and surface mining.
12. Why are mercury and cyanide, very dangerous poisons if mishandled, used in gold recovery?
13. What are the differences between metallic and nonmetallic minerals? Of what importance are the nonmetallic minerals?
14. It has been suggested that, as society exhausts Earth's reserves of critical minerals, lower-grade mineral deposits will be mined, and supply-and-demand economics will dictate prices and costs. This rationale could be extended to a scenario in which average rock would eventually be mined for critical materials. Is there a fallacy to such a rationale? Explain.
15. Discuss the ethics of changing the General Mining Law of 1872 to allow the federal government to collect royalties from current mine operations to be used for the cleanup of historic mines that were abandoned by companies that no longer exist.

REFERENCES AND FURTHER READINGS

Alpers, C. H., et al. 2005. *Mercury contamination from historical gold mining in California*. U.S.G.S. Fact Sheet, 2005–3014.

Barker, J. C., and B. S. Van Gosen. 2012. Alaska's rare earth deposits and resource potential. *Mining Engineering* 64(1): 20–32.

Berger, Alan. 2002. *Reclaiming the American West*. New York: Princeton Architectural Press, 233 p.

Clayton, Mark. 2008, August 20. Should uranium be mined here? *Christian Science Monitor* 13–16.

Drew, Lawrence, and Brian K. Fowler. 1999. The megaquarry: A conversation on the state of the aggregate industry. *Geotimes* 44(6): 17–22.

Dunn, James, Carol Russell, and Art Morrissey. 1999. Remediating historic mine sites in Colorado. *Mining Engineering* 51(8): 32–35.

Gray, J. E., and R. F. Sanzolone, eds. 1996. *Environmental studies on mineral deposits in Alaska*. U.S. Geological Survey Bulletin 2156. Washington, DC: U.S. Government Printing Office.

James, Patrick M. 1999. The miner and sustainable development. *Mining Engineering* 51(6): 89–92.

Kesler, S. E. 1994. *Mineral resources, economics and the environment*. New York: McMillan 391 p.

Lindberg, T. Ty, et al. 2011. Cumulative impacts of mountaintop mining on an Appalachian watershed. *Proceedings of the National Academy of Sciences* 108(52): 20924–20934.

Muller, N. Z., R. Mendelsohn, and W. Nordhaus. 2011. Environmental accounting for pollution in the United States economy. *American Economic Review* 101(5): 1645–1675.

Palmer, M. A., et al. 2010. Mountaintop mining consequences. *Science* 327(5962): 148.

Peters, S. G., T. V. V. King, T. J. Mack, M. P. Chornack, eds., and the U.S. Geological Survey Afghanistan Mineral Assessment Team. 2011. *Summaries of important areas for mineral investment and production opportunities of nonfuel minerals in Afghanistan: U.S. Geological Survey Open-File Report 2011–1204,* 1810 p. plus appendixes on DVD. (Also available at: http://pubs.usgs.gov/of/2011/1204/.)

Romero, Simon. 2009, February 3. In Bolivia, untapped bounty meets nationalism. *New York Times*. A1, A8.

Silva, Michael A. 1988. Cyanide heap leaching in California. *California Geology* 41(7): 147–156.

Spectrum Engineering. 2006. *Engineering evaluation/cost analysis (EE/CA) for water management at the Zortman and Landusky mines Phillips County, Montana*. Bureau of Land Management: Billings, MT, 186 p. (Also available at http://www.blm.gov/mt/st/en/fo/lewiston_field_office/zortman.html.)

Stewart, K. C., and R. C. Severson, eds. 1994. *Guidebook on the geology, history, and surface-water contamination and remediation in the area from Denver to Idaho Springs, Colorado*. U.S. Geological Survey circular 1097. Washington, DC: U.S. Government Printing Office.

U.S. Environmental Protection Agency. 2011. *The effects of mountaintop mines and valley fills on aquatic ecosystems of the central Appalachian coalfields*. Washington, DC: Office of Research and Development, National Center for Environmental Assessment, EPA/600/R-09/138F.

U.S. Geological Survey. 2011. Mining review. *Mining Engineering* 63(5): 45–75.

Wheeler, Gregory R. 1999. Teaching mineral resources as an aid to understanding international policy issues. *Journal of Geoscience Education* 47: 464–468.

Zullig, K. J., and M. Hendryx. 2010. Health-related quality of life among central Appalachian residents in mountaintop mining counties. *American Journal of Public Health* 101(5): 848–853.

Is the sun now setting on the future of oil production? If so, what energy source will replace oil?

Mike Theiss/National Geographic Image Collection

Have you ever

wondered...

14

Energy and the Environment

It is difficult for people living now, who have become accustomed to the steady exponential growth in the consumption of energy from the fossil fuels, to realize how transitory the fossil–fuel epoch will eventually prove to be when it is viewed over a longer span of human history.

—M. King Hubbert, 1971

1. What effect did the perfection of the steam engine have on the human scene?
2. What effect did the development of the modern automobile have on the human scene?
3. What is the origin of petroleum?
4. How is petroleum found?
5. Is carbon-based (fossil fuel) sustainable in the future as an energy source?

Energy and the Human Enterprise

Joshua Wolfe

Figure 1 The wide-open spaces near McCarney, Texas, have been the site of nodding oil pump jacks for many decades. The area is also highly suitable for wind farms, with Texas now having the greatest wind-energy-generating capacity in the United States. Will wind turbines eventually replace the pump jack as the icon of energy production?

The shift in use from one form of energy to another began long before today's potential transition from petroleum to renewables, such as wind turbines (**Figure 1**). Consider the energy source required by the Babylonians in the 18th century B.C.E. to build their massive ziggurats and the conventional energy available when the ancient Greeks were mining silver underground at Laurium. And where did the Romans get their energy to construct their marvelous buildings of heavy marble blocks?

For thousands of years, raw animal (and human) muscle, along with the burning of wood, were the major forms of conventional energy until alternative sources were discovered and tapped. Some alternative forms of energy appeared in ancient times: The earliest known sailboats, perhaps as early as 5000 B.C.E., were developed when the Egyptians discovered how wind can be used for travel, and about the same time, the Greeks were using water wheels for grinding grain. By 200 B.C.E., simple windmills were pumping water in China, and vertical-axis windmills were milling grain in the Middle East. Later, the Crusaders carried the knowledge of windmills back to Europe, where the new technology was adapted for draining swamps and lakes in Holland. Nevertheless, conventional energy for heavy work came largely from animal and human muscle.

The transition away from draft animals and human muscle, and the growth of food that was needed, began in the first century C.E., as coal began to be used for heat. But it took more than a thousand years before it became the conventional energy source for most heavy work. In the late 1600s, steam engines were being developed and, by the early 1800s, significant improvements by James Watt and George Stephenson had given rise to the wood- and coal-fired steam engine. Steam engines were recognized as more convenient than water or wind for generating power and less expensive than maintaining stables of draft animals. By the middle of the 19th century, steam engines were powering factories, farm machinery, and railroad locomotives and had replaced water-pumping windmills. The Industrial Revolution was well under way, with coal, the new conventional energy, largely displacing wood (**Figure 2**). The need for animal muscle, wind, and water energy was displaced by the concentrated energy found in a fossil fuel. In 1899, a coal-powered steam engine was attached to an electric generator and the modern world began.

In the 1890s, hydropower, essentially a water wheel connected to an electric generator, appeared on the scene, with the electricity being utilized for powering machinery, lighting cities, and operating streetcars. Hydropower, although important, had limitations. Also in the late 19th century, an even more concentrated form of fossil fuel was catching on: kerosene derived from petroleum. Initially, it was used to provide light by replacing whale oil, but by the turn of the 20th century, gasoline, a by-product of producing kerosene from oil, was being used to fuel internal combustion engines. Thus, two fossil fuels had become the conventional energy (see Figure 2). Low-cost automobiles, and the widespread use of electricity from coal-fired power plants, forever changed societies' energy use. Electric power plants became ever larger, and electricity reached into rural areas during the Great Depression of the 1930s. Energy use from fossil fuels grew quickly in the early and mid-20th century, doubling every 10 years, and automobiles made commuting from suburbs possible, leading to urban sprawl and the demand for even more gasoline.

Automobiles became heavier and larger, with gasoline use growing unchecked through the 1960s. In 1970, a gallon of gas in the United States cost about 25 cents, and American cars averaged about 13 miles per gallon. Electric power plants fueled by coal or natural gas were proliferating. About the same time, a new alternative energy source, the atom, appeared on the scene. Use of "the peaceful atom" led to the generation of electricity produced by several nuclear power plants, with some 200 planned for the United States. It is good to keep in mind that, by the 1970s, the average mechanical energy expended by each American every day was equivalent to the energy expended by roughly

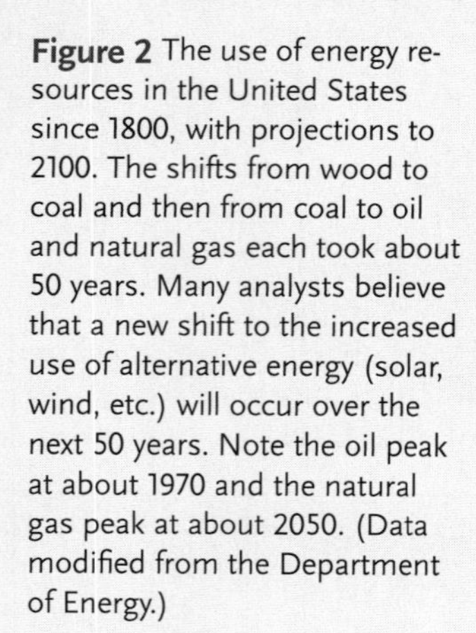

Figure 2 The use of energy resources in the United States since 1800, with projections to 2100. The shifts from wood to coal and then from coal to oil and natural gas each took about 50 years. Many analysts believe that a new shift to the increased use of alternative energy (solar, wind, etc.) will occur over the next 50 years. Note the oil peak at about 1970 and the natural gas peak at about 2050. (Data modified from the Department of Energy.)

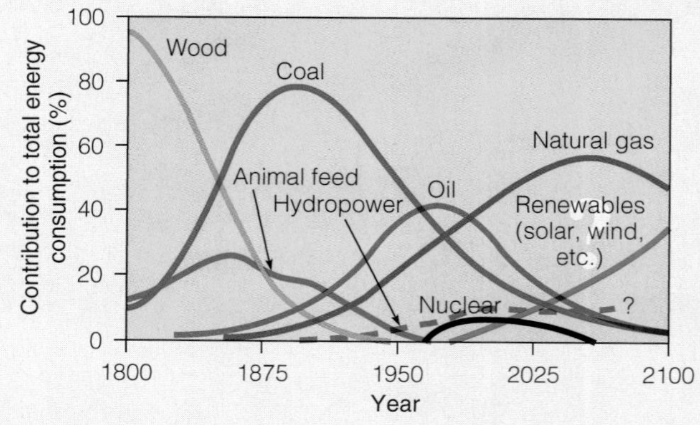

100 human slaves working full-time in the days of ancient Egypt and the Roman empire.

In 1973, U.S. support of Israel in the Arab-Israeli war resulted in an oil embargo by the Organization of the Petroleum Exporting Countries (OPEC). Within a few weeks, oil prices tripled, and in 1979, following the overthrow of the Shah of Iran, oil prices skyrocketed, increasing by 150% in a few weeks. By 1980, the average price of a barrel of oil was almost $45, and the resulting energy crisis caused an increased demand for U.S. coal.

Only 3 months after the fall of the shah, the United States got another wake-up call: The Three Mile Island nuclear power station in Pennsylvania suffered a partial meltdown following a series of operator mistakes and technical failures. The public, after years of being told that a nuclear accident could never happen, was shocked. The accident exacerbated the sense of crisis. All new nuclear plants on order were canceled, and no new plants were even planned after 1978.

Beginning in the 1990s, there were reports of dwindling energy reserves at the same time the global population had reached 6 billion. By the summer of 2012, the U.S. Census Bureau reported the world's population at 7 billion and on the way to 8 billion by about 2020. Such an increasing population demands more and more energy, which has initiated serious consideration of alternative energy sources: wind, geothermal, ocean waves, tidal currents, and biofuels (see Figure 2). Can a transition to such alternatives supersede the energy supplied by fossil fuels and keep providing us with inexpensive options to muscle power for the future of civilization? The reality of fossil fuels and alternative energy sources is the subject of this chapter.

questions to ponder

1. Is sustainable economic growth possible?
2. How much petroleum is left and how long will it last?
3. How do geologists go about finding sources of petroleum?
4. Are unconventional sources of fossil fuels (e.g., tar sands, oil shales) viable substitutes for conventional oil?
5. Are renewable alternative energy sources (e.g., wind, solar, ocean tides) adequate to substitute for carbon-based (fossil fuel) energy?

Michael T. Klare

CourseMate Connection

Start off this chapter by listening to a podcast with author Michael Klare as he discusses the planet's dwindling energy resources. **Log in to the CourseMate at www.cengagebrain.com to listen now.**

FROM THE SIMPLEST ALGAL SCUM to the most complex ecosystem, energy is essential to all life. Derived from the Greek word *energia*, meaning "in work," *energy* is defined as the capacity to do work. The units of energy are the same as those for work, and the energy of a system is diminished only by the amount of work it does.

Prosperity and quality of life in an industrialized society such as ours depend in large part on the society's energy resources and its ability to use them productively. We can illustrate this in a semiquantitative fashion with the following equation:

$$L = \frac{R + E + I}{\text{population}}$$

where L represents quality of life (or "standard of living"), R represents the raw materials that are consumed, E represents the energy that is consumed, and I represents an intangible we will call *ingenuity*. As the equation expresses, when high levels of raw materials, energy, and ingenuity are shared by a small population, a high material quality of life results. If, on the other hand, a large population must share low levels of resources and energy, a low standard of living is expected. Some highly ingenious societies with few natural resources and little energy can and do enjoy a high quality of life. Japan is a prime example. Some other countries that are self-sufficient in resources and energy, such as Argentina, are having difficult times. Thus, ingenuity, which is reflected in a country's political system, technologies, skills, and education, is heavily weighted in the equation and can cancel out a lack of resources and a large population.

To a physicist, there is no energy shortage, because he or she knows that energy is neither created nor destroyed; it is simply converted from one form to another, such as from nuclear energy to heat energy. Fuels of all kinds are warehouses of energy, which can be tapped by some means and applied in some way to do work. Coal and oil, for example, are fossil fuels that have been storing solar energy in the lithosphere for millions of years.

Some forms of energy are **renewable resources**—that is, they are replenished at a rate equal to or greater than the rate at which they are used. Examples include solar, water, wood, wind, ocean and lake thermal gradients, geothermal, and tidal energy. The energy in all of these resources—except for geothermal and tidal (gravitational) energy—was originally derived from the Sun. Renewable resources are dependable only if they are consumed at a rate less than or equal to their rate of renewal. If they are overexploited, some period of time will be required to replenish them. Peat, a fuel used extensively for space heating and cooking in Ireland and Russia, is estimated to accumulate at a remarkable 3 metric tons per hectare per year (1.3 tons/acre/year). Nonetheless, the conversion from plant litter to peat may take a hundred years. Wood energy may renew in a matter of a few decades, and water and wind are renewed continuously.

Nonrenewable resources are not replenished as fast as they are used and, once consumed, they are gone forever. Crude oil, oil shales, tar sands, coal, and fissionable elements are nonrenewable energy resources. These quantities are finite. Supplies of crude oil, for example, may be within a few decades of exhaustion. Oil underground was discovered almost 50 years before the first automobile was operational. Prior to then, gasoline was a minor by-product of refining oil for kerosene lamps. In 1885, Germans Gottlieb Daimler and Karl Benz independently developed gasoline engines, and in 1893, Massachusetts bicycle makers Charles and Frank Duryea built the first successful U.S. gasoline-powered automobile. By the turn of the century, production automobiles were hitting the roads, horses were being put out to pasture, and refineries were stepping up production to satisfy an increasing demand for gasoline. The transportation revolution to automobiles spawned the largest private enterprise on Earth: the exploration and production of petroleum. This has led to major geopolitical and economic problems, because the nations that consume the most do not produce in comparable amounts.

Petroleum

Although the carbon content of Earth's crust is less than 0.1% by weight, carbon is one of the most important elements to humankind. It is indispensable to life, and it is the principal source of energy and the raw material of many manufactured products. The concentrated energy available from petroleum and natural gas (and coal), and the development of technology to utilize this energy, powered the growth of industry, the expansion of massive agricultural production, and indeed the entire spectrum of economic growth of the United States and much of the rest of the world. An important factor affecting consideration of petroleum and other energy resources is the **energy return on investment** (EROI). The EROI is the ratio of the energy returned relative to the energy invested to discover and obtain that energy.[1] It can be expressed as

$$\text{EROI} = \frac{\text{Summation of energy content of fuel returned to society}}{\text{Energy required (invested) to produce, deliver, and use that energy}}$$

or

$$\text{EROI} = \frac{E_{out} - E_{in}}{E_{in}}$$

Fossil fuels are highly concentrated forms of energy and, when in great abundance, they have high EROIs. For example, in 1930, oil wells were relatively shallow and easily drilled, and the energy in one barrel of oil could be used to produce 100 barrels, thus having an EROI of 100:1. By 1970, the EROI of U.S. petroleum had dropped to 30:1, and since then the EROI of U.S. petroleum has fluctuated between 11:1 and 18:1. The EROI ratio for petroleum is likely to continue to decline with time as a consequence of decreasing energy returns as oil reservoirs are increasingly depleted and as energy costs increase as exploration and development shift to remote regions and deeper offshore sites (see Case Study 14.1). The EROIs of potential liquid alternatives, such as from oil shale and corn ethanol, are low, generally less than 5:1. **Figure 14.1** illustrates the EROIs for fossil fuels and several alternatives. The estimated EROIs of various energy sources will be included in the pages that follow.

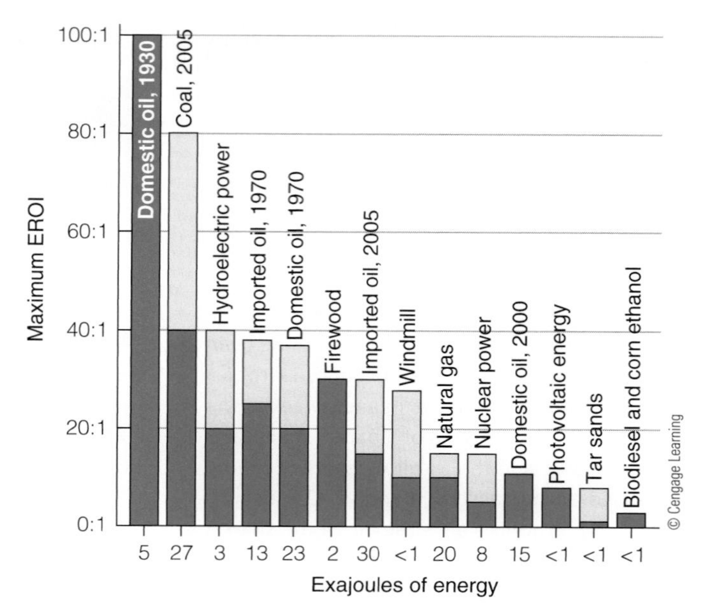

Figure 14.1 The energy return on investment (EROI) is the energy cost of acquiring an energy resource; one of the objectives is to get out far more than you put in. Domestic oil production's EROI has decreased from about 100:1 in 1930 to 40:1 in 1970 to about 14:1 today. The EROI of most "green" energy sources, such as photovoltaics, is presently low. (Lighter colors indicate a range of possible EROIs due to varying conditions and uncertain data.) EROI does not necessarily correspond to the total amount of energy in exajoules produced by each resource.

Crude oil, or petroleum (Latin *petra*, "rock," and *oleum*, "oil"), is composed of many **hydrocarbon compounds**, simple and complex combinations of hydrogen and carbon (**Table 14.1**). Petroleum occurs beneath Earth's surface in liquid and gaseous forms, and at the surface as oil seeps, tar sands, solid bitumen (gilsonite), and oil shales. In addition to their use as a fuel, hydrocarbon compounds derived from petroleum are used in producing paints, plastics, fertilizers, insecticides, soaps, synthetic fibers (nylon and acrylics, for example), and synthetic rubber. Carbon combines chemically with itself and with hydrogen in an infinite variety of bonding schemes; about 2 million hydrocarbon molecules have been identified. The manufacturing process of separating crude oil into its various components is known as *refining*, or *cracking*.

Origin and Accumulation of Hydrocarbon Deposits

Carbon and hydrogen did not combine directly to form petroleum; they were chemical components of living organisms before their transformation to complex hydrocarbons in crude oil. Porphyrin compounds found in petroleum are derived either from chlorophyll, the green coloring in plants, or from hemin, the red coloring matter in blood, and their presence is solid evidence for an organic origin for crude oil. The

[1] Hall, C. A. S., and C. J. Cleveland. 1981. Petroleum drilling and production in the United States: Yield per effort and net energy analysis. *Science* 211: 576-579.

Table 14.1 Hydrocarbon Compounds Typically Found in Crude Oil

Molecular Type				
Name	General Formula	Hydrocarbon Compound		Percentage of Weight in Medium-Grade Crude Oil
Paraffins	C_nH_{2n+2}	Methane, CH_4	Lighter ↑	25
		Ethane, C_2H_6		
		Propane, C_3H_8		
		Butane, C_4H_{10}*		
		Pentane, C_5H_{12}†		
Aromatics	C_nH_{2n-6}‡	Benzene, C_6H_6		17
Naphthenes	C_nH_{2n}	Asphalt		50
Asphaltenes	Solid hydrocarbons	Gilsonite	↓ Heavier	8
				100

*Butane gives gasoline quick-starting capability.
†Pentane gives smooth engine warm-up.
‡Aromatics improve mileage and "knock" resistance.

fact that large quantities of oil are not found in igneous or metamorphic rocks also rules out an inorganic source of oil.

Four conditions are necessary for the formation and accumulation of an exploitable petroleum deposit in nature:

- A source rock for oil
- A reservoir rock in which it can be stored
- A caprock to confine it
- A geological structure or favorable strata to "trap" the oil

Even where these geological conditions are met, the human elements of exploration, location, and discovery still remain. Oil companies utilize the skills of geologists to interpret surface and subsurface geology, to locate the potential oil-bearing structure or stratum, and to specify the optimal location for a discovery well. Until a well is drilled, the geologists' interpretation remains in doubt; much like a medical doctor's diagnosis of an ailment subject to surgery, the diagnosis is tentative until the patient is opened up.

A **source rock** is any volume of rock that is capable of generating and expelling commercial quantities of oil or gas. Source rocks are sedimentary rocks, mostly shales or limestones, usually of marine (ocean) origin and sometimes of *lacustrine* (lake) origin. The biological productivity (biomass) of surface waters must have been high enough to generate a settling "rain" of dead organisms, and the bottom waters must have been low enough in oxygen to prevent the deposited organic matter from being oxidized or consumed by scavengers. Almost all source rocks are dark-colored, indicative of high organic-matter content, and some of them carry a fetid or rotten-egg odor.

Favorable marine environments are rich in microscopic single-celled plants known as *diatoms* (phytoplankton), which form the largest biomass in the sea. Where there are diatoms, we also find zooplankton—tiny protozoans and larvae of large animals. These together with diatoms provide the molecules that make up crude oil. As the rocks are buried, they are heated. The conversion from organic matter to petroleum takes place mostly between 50°C and 200°C. Thus, a proper thermal history, ideally between 100°C and 120°C, is necessary to form liquid petroleum—too cool, and oil does not form; too hot, and the hydrocarbons "boil" away.

After petroleum has formed in a source bed, it is squeezed out and migrates through a simple or complex plumbing system into a **reservoir rock**. This migration is a critical element in the formation of an economically exploitable accumulation of oil. Reservoir rocks are porous and permeable (see Chapter 8). Commonly, they are sandstones, porous limestones, or fractured shales. Reservoir porosities range from 20% to 50%, meaning that for each cubic foot of reservoir rock, we will find 1 to 4 gallons of oil. The unit of oil volume is the barrel (equal to 42 gallons), and a so-called giant field, such as the Prudhoe Bay field on the north slope of Alaska, will yield a billion or more barrels of oil in its lifetime. Using standard recovery techniques, however, as much as 80% of the oil may be left in pore spaces and as films on mineral grains.

An impermeable **caprock** prevents oil from seeping upward to form tar pits at the surface or dissipate into other rocks. Such seals are analogous to aquicludes in groundwater systems and are most commonly clay shales or limestones of low permeability. Our discussion of the fourth requirement, a suitable geological structure for trapping the oil, requires a separate subsection.

Geological Traps: Oil and Gas Stop Here

Structural traps—An **anticline** is an ideal structure for trapping gas and oil. It is a convex-upward fold in stratified rock (**Figure 14.2a**). Analogous to a teacup inverted in a pan of water that traps a layer of air inside it, an anticline holds a reservoir of gas and oil. This occurs because crude oil floats on water and natural gas rises to the top of the reservoir. The anticlinal theory of oil accumulation was not developed until 1900, 41 years after oil was discovered. Strata dip away from the central axis of an anticline at the ground surface, and most anticlines with surface geological expression have been drilled. Today the search for oil is more difficult, because less obvious and geologically more complex traps need to be discovered. Many times, faults form impermeable barriers to hydrocarbon migration, and oil becomes trapped against them (**Figures 14.2a** and **14.3**). Faulted and folded stratified rocks in a single oil field may contain many isolated oil reservoirs.

Salt domes are of much interest to geologists (see Figure 14.2b). Not only do they create oil traps, but also they are valuable sources of salt and sulfur, and are potential underground storage sites for petroleum and hazardous waste. More than 500 salt domes have been located along the U.S. Gulf Coast, both offshore and on land (**Figure 14.4**). They rise as flowing fingers of salt, literally puncturing their way through the overlying rocks and buckling the overlying shallow strata into a dome (**Figure 14.5**). Some fingers rise as high as 13 kilometers (8 miles) above the "mother" salt bed, the Louann salt, and would be taller than Mount Everest if they were at the surface of Earth. Because the salt is considerably less dense than the overlying rock, buoyancy forces drive the salt upward—much as a blob of oil will rise through water. Brittle solids, such as salt or ice, will flow over long periods of time; many salt domes are still rising measurably. The "mother" salt was deposited in Jurassic time by the evaporation of seawater when the embryonic Gulf Coast was connected to the open ocean by a shallow opening. The opening allowed seawater to enter the basin but did not allow the denser salt brines at the bottom to escape.

Thus, the brine became concentrated to the point of saturation, and the great thickness of salt that now underlies the entire Gulf Coast was deposited. Oil accumulates against the salt column and in the dome overlying the salt.

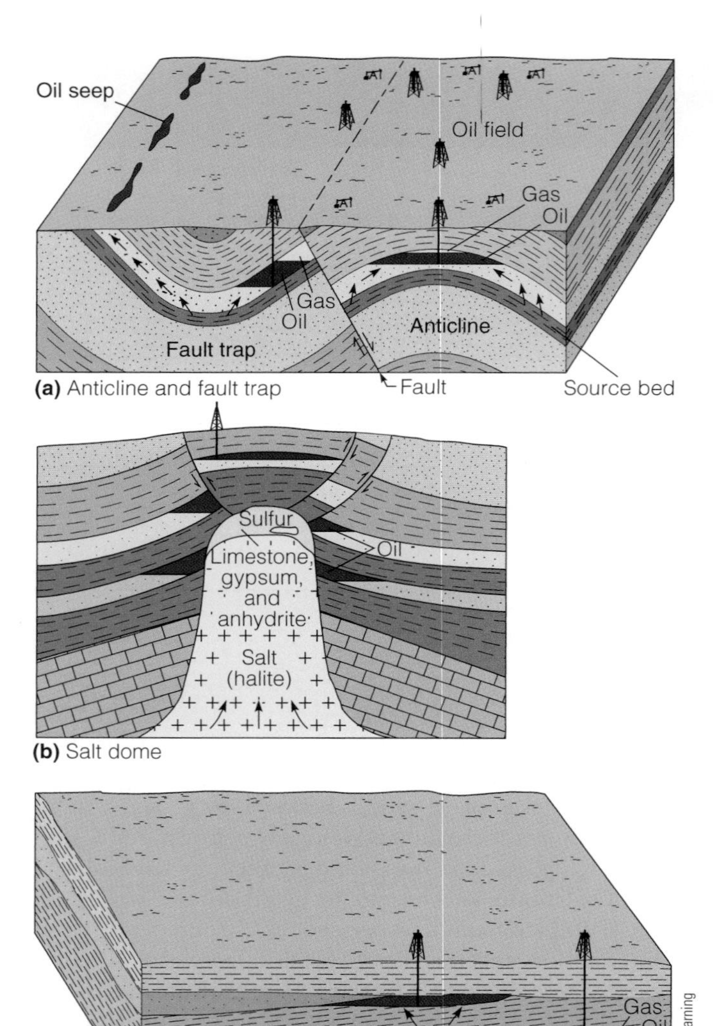

Figure 14.2 (a, b) Common structural oil traps; (c) stratigraphic oil traps.

Figure 14.3 Two wells in the Silver Thread oil field near Santa Paula, California. The wells are pumping oil from a 2.4-kilometer (7,874 ft)-deep fault trap oil deposit. The folded shale beds in the cliff are part of the drag folds in the uplifted block of the fault that formed the oil trap (see Figure 14.2a).

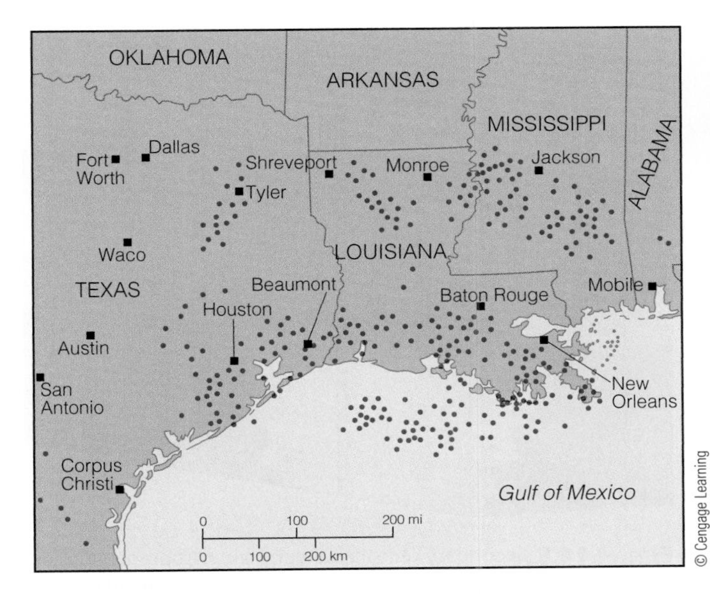

Figure 14.4 Locations of salt domes on the U.S. Gulf Coast. More than 500 domes have been discovered on land and in the shallow parts of the Gulf of Mexico, and more are known to be in deep water offshore.

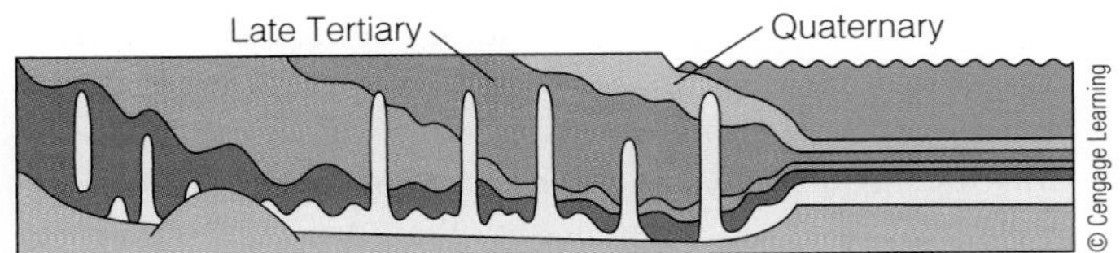

Figure 14.5 Ancient salt deposits are buried deeply and, because salt is less dense than the overlying sediments, it rises buoyantly as pillars of salt, creating salt domes.

"Old Spindletop," a salt dome near Beaumont, Texas, which was discovered in 1901, is perhaps the most famous U.S. oil field. Soon after its discovery, it was producing more oil than the rest of the world combined, and the price of oil plummeted to 2 cents a barrel.

Stratigraphic traps—Any change in sedimentary rock lithology (its physical character) that causes oil to accumulate is known as a **stratigraphic trap** (in contrast with anticlines and salt domes, which are structural traps). Thus, if a stratum changes laterally from a permeable sandstone to an impermeable shale or mudstone, oil may be trapped in the stratum (see Figure 14.2c).

Ancient coral reefs are ideal reservoirs, because they are porous and were biologically productive when they were living. Oil may be trapped in the porous, permeable debris on the flanks of the reef; production from such fields can be measured in thousands of barrels per day (bbl/d). Many oil fields of the Middle East are this type, and their production potential is tremendous. A comparison of reef production with that of sandstone reservoirs (such as those of California or Texas, which typically yield only a few hundred barrels per day), explains why the Middle East can control oil production and, therefore, price.

Oil Production

The first successful oil well in the United States was drilled in Titusville, Pennsylvania, in 1859. In modern jargon, this well would be called a **wildcat well**, because it was the discovery well of a new field. There is 1 chance in 50 of a wildcat well being successful—less favorable odds than those of winning at roulette. The probability improves when we consider all wells drilled, including those in known oil fields. Independent oil entrepreneurs take considerable risks; it is not a business for the faint of heart.

Most successful wells require pumping. If gas and water pressures are sufficient, however, oil may simply flow to the ground surface. High reservoir pressures develop from water pressing upward (buoyancy) beneath the oil and gas pressure pushing downward on the oil (**Figure 14.6**). In some cases, dissolved gases "drag" the oil along with them as they spew forth, as though from a bottle of champagne. If high pressures are not controlled, rocks, gas, oil, and even drill pipe may shoot into the air as a "gusher."

It is possible to drill a well so that the drill hole slants from the vertical to penetrate reservoir rocks far from the drilling site. This is desirable where the oil structure is offshore and must be slant drilled from land or from a drilling platform (see Case Study 14.1). **Slant drilling** is also used for tapping reservoirs beneath developed land, such as in Beverly Hills, California, where a large number of oil wells are slant drilled under fashionable commercial areas. Modern methods use remotely controlled "smart" drill bits. Producers can withdraw oil from below a large area more economically while minimizing the visual blight of drilling towers.

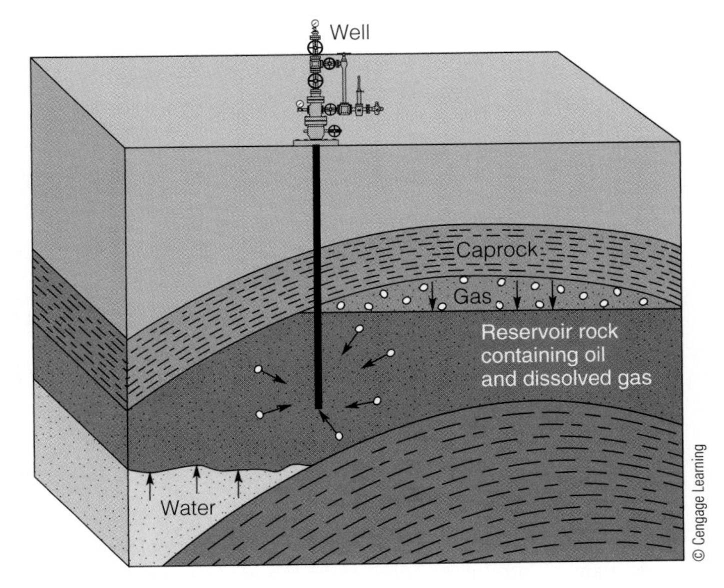

Figure 14.6 Oil in an anticline is driven by gas pressure from above and by buoyant water pressure from below.

The ultimate drilling technology of the 1990s was horizontal drilling, a technique by which the drill bit and pipe follow gently inclined reservoir sands and limestones. By this means, a single borehole can provide access to reservoirs beneath a much larger area than provided by vertical drilling or slant drilling. This method is now superseded by multilateral drilling, which enables operators to reach multiple oil-bearing reservoirs by means of lateral extensions from one well (**Figure 14.7**). It is worth noting that some offshore drilling platforms are in water more than a mile deep and have strings of drill pipe dangling below them 8.5 kilometers (5 mi), about the height of Mount Everest. Exploratory wells are being drilled in water depths greater than 3,100 meters (10,000 feet). (See Case Study 14.1.) The drilling of such exploratory wells comes at an enormous price, especially in remote regions. Drilling platforms may cost more than a billion dollars, and the chance of a successful discovery is not assured. For example, Cairn Energy, a Scottish oil and gas exploration company, spent $600 million to drill an exploratory well in Arctic waters off Greenland; it found no oil.

Secondary recovery—**Secondary recovery** methods extract oil that remains in the reservoir rock after normal withdrawal methods have ceased to be productive. As much as 75% of the total oil may remain. Secondary recovery methods can be grouped into three categories: thermal, chemical, and fluid-mixing (**miscible**) methods (**Figure 14.8**). All of these methods require *injection* wells for injecting a fluid or gas and *extraction* wells for removing the remobilized oil.

Thermal methods include steam injection, which makes the adhering oil less viscous and thus more free to flow, and fire flooding, in which air is injected into the reservoir to set fire to the oil, and thus produce gases and heat that will increase the flow of oil. Water injection is a chemical method that utilizes large-molecule compounds, which when added to water thicken it and increase its ability to wash or sweep the adhering oil films and globules migrating toward an extraction well. Fluids that mix with oil, that are miscible, are very effective in removing "stuck" oil from the reservoir. Miscible recovery methods use mixtures of water with propane or ethane extracted from natural gas or mixtures of CO_2 and water. Even after the use of these secondary recovery methods, as much as a fifth of the original oil may remain in the reservoir.

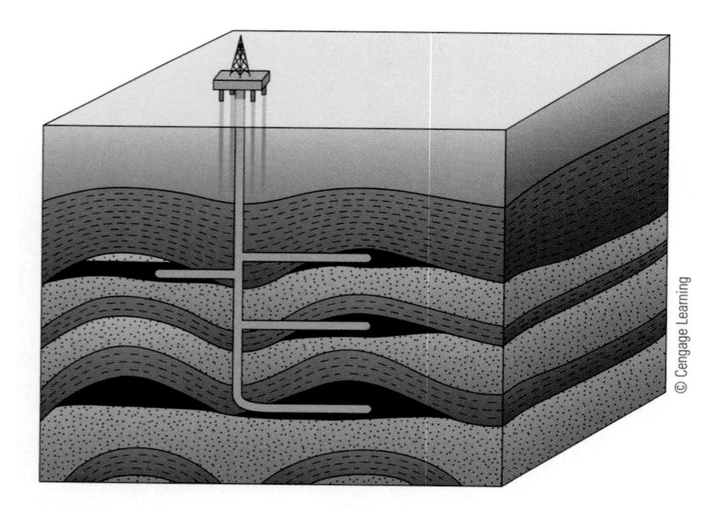

Figure 14.7 Multilateral drilling from an offshore platform allows many oil-producing zones to be tapped from one platform. In this illustration, four zones are tapped by lateral horizontal pipes.

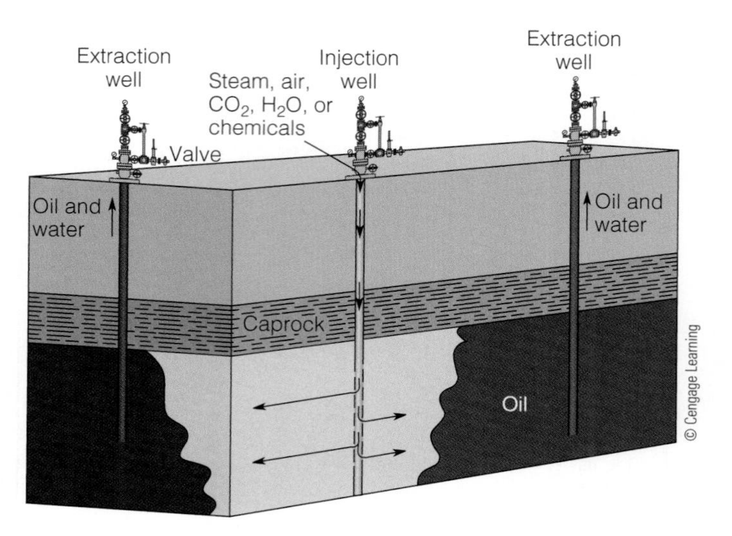

Figure 14.8 Secondary recovery. Steam, air, carbon dioxide, or chemicals dissolved in water are injected into a sluggishly producing formation to stimulate the flow of oil to extraction wells.

Quality and Price

The price of a barrel of crude (unrefined) oil varies with its grade—its quality—and with market demand at the time. Light oils bring higher prices, because they contain large proportions of paraffins and aromatic hydrocarbons, which are desirable for gasoline and bottled gases. Heavy crudes contain lower proportions of those components and greater proportions of the heavy, less valuable asphalts and tars (see Table 14.1). A scale of crude-oil quality has been established by the American Petroleum Institute (API) based on its weight. Light crude oil is very fluid and yields a high percentage of gasoline and diesel fuel. Heavy crude, on the other hand, is about the consistency of molasses. The percentages of fuels and lubricating oils yielded by a barrel of medium-weight crude oil are shown in **Table 14.2**.

U.S. Consumption, Production, and Imports of Petroleum

The United States has about 4.4% of the world's population, yet consumes 28% of the world's energy supplies (**Figure 14.9**). Each day, Americans use some 3,000 products, including such things as gasoline, lubricating oils, jet and other fuels, fertilizer, makeup, synthetic fabrics, and pharmaceuticals, which are derived from petroleum. Even with

Table 14.2 Typical Composition of an American Petroleum Institute Medium-Grade Crude Oil

Common Name	Number of Carbon Atoms*	Percentage of Crude-Oil Weight
Gasoline	5–10	27
Kerosene	11–13	13
Diesel fuel	14–18	12
Heavy gas oil	19–25	10
Lubricating oil	26–40	20
Heavy fractions (tars, asphalt)	>40	18
		100

*Volatility decreases as the number of carbon atoms increases, C_3 to C_{40}.

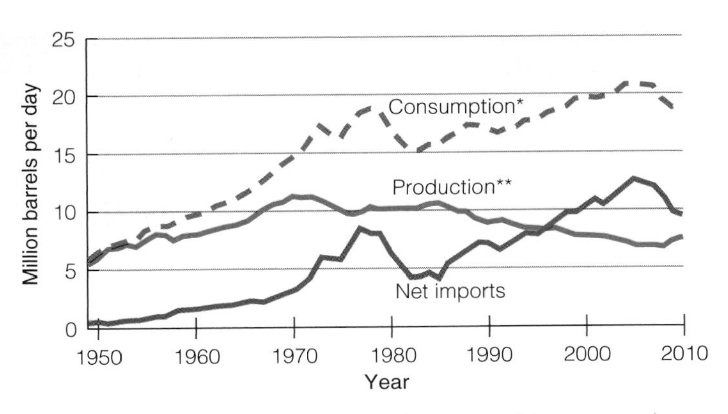

Figure 14.10 U.S. consumption, production, and import trends, 1950–2010. Notice the drop in consumption and imports beginning about 2006, but an increase in production beginning in 2008. (Energy Information Association, *Annual Energy Review 2010* (2011))

conservation efforts, our consumption of petroleum generally increases annually (although there has been a decrease since 2008 because of a worldwide economic downturn, policies that boost fuel efficiency, and an increase in renewable fuels). Examining **Figure 14.10** shows U.S. petroleum production peaked in 1970, but consumption increased as much as 40% between 1983 and 2006. By 2010, the United States was consuming more than 19 million barrels a day (down from 21 million in 2007) out of some 85 million barrels a day used worldwide (see **Figures 14.10** and **14.11**). To put U.S. oil consumption in a different context, of those 19 million barrels a day, about 10.4 million (down from about 12 million in 2007) were imported (see Figure 14.10). Using the Energy Information Administration (EIA) 2010 average price of $74 a barrel of imported oil, nearly $740 million *each day* in 2010 went to other countries to pay for our thirst for oil. Beginning about 2005, U.S. petroleum consumption and net imports began dropping, and in 2008, domestic production began to increase. Moreover, by 2011, U.S. domestic oil production reached an 8-year high, with more than 53% of all drilling rigs in the world here in the United States. Although U.S. oil production had increased greatly, the nation still remained a net importer of crude oil. Nevertheless, U.S. became an oil-product exporter for the first time since 1949, sending nearly 440,000 barrels a day overseas. Of all exports in 2011, refined petroleum products ranked second in value, up 60% from 2010—automotive vehicles ranked number one in export value.

Production, Reserves, and the Future for Oil

> **The time when we could count on cheap oil and even cheaper natural gas is clearly ending.**
>
> Dave O'Reilly, *chairman, Chevron-Texaco, February 2005*

On a global scale, proven petroleum reserves at the beginning of 2011 were estimated to be about 1,383.2 billion barrels of oil (**Table 14.3**). (**Reserves** are the amount of oil and natural gas that can be extracted at a profit. However, there are several kinds of reserves: proven, probable, possible developed, economically recoverable, technically recoverable, and undeveloped, and the use of these terms is not universally consistent.) The 1,383.2 billion barrels of global reserves may mean little to us until we realize that, in 2011, about 60% of the proven reserves were in predominantly Islamic states (see Table 14.3 and Figure 14.11). Of these, Saudi Arabia has about one third of the total, and Iran, Iraq, Kuwait, and the United Arab Emirates each claim to have about 100 billion barrels of proven reserves, according to the *BP Statistical Review of World Energy 2011*.

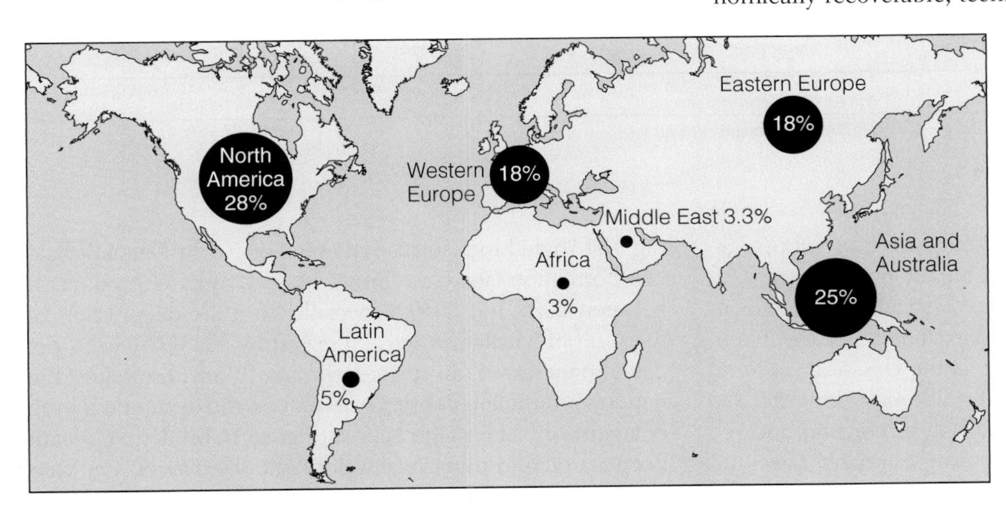

Figure 14.9 World energy consumption. (Colorado Geological Survey)

Table 14.3 Comparison of Proven Global Oil Reserves, 2008 and 2011

Country	2008 Proven Reserves (Billion Barrels)	2011 Proven Reserves (Billion Barrels)	2011 Percentage
Saudi Arabia†	264.0	264.5	19.1
Iran†	137.8	137.0	9.9
Iraq†	115.0	115.0	8.3
United Arab Emirates†	97.8	97.8	7.1
Kuwait†	101.5	101.5	7.3
Venezuela†	99.4	211.2	15.3
Russian Federation	79.0	77.4	5.6
Kazakhstan*	39.8	39.8	2.9
Libya†	43.7	46.4	3.4
Nigeria†	36.2	37.2	2.7
United States	30.5	30.9	2.2
Canada	28.6	32.1	2.3
China	15.5	14.8	1.1
Qatar†	27.3	5.5	1.9
Mexico	11.9	11.4	0.8
Algeria†	12.2	12.2	0.9
Brazil	12.6	14.2	1.0
Norway	7.5	6.7	0.5
Angola	13.5	13.5	1.0
Oman*	5.6	5.5	0.4
India	5.8	9.0	0.7
Malaysia*	5.5	5.8	0.4
Indonesia†	3.7	4.2	0.3
United Kingdom	3.4	2.8	0.2
Yemen*	2.7	2.7	0.2
Rest of world	52.5	84.1	4.5
Total world	1258.0	1383.2	100.0
Total OPEC	955.8	1032.5	76.2
Total for dominantly Islamic nations	856.6	833.9	58.8

Sources: British Petroleum statistical review of world energy, 2009; British Petroleum statistical review of world energy, 2011.

*Dominantly Islamic nation.

†OPEC nation.

In spite of the continuing production from those countries where for years, collectively, they annually produced tens of billions of barrels of oil, there has been virtually no change in their reported proven reserves for more than a decade. Even more thought-provoking is the fact that the U.S. reserves have been decreasing overall since 1970, although domestic reserves increased as Alaska's North Slope production and reserves became recognized, and reserve estimates have increased slightly in the last few years with the development of new drilling techniques and use of **hydraulic fracturing** ("fracking") of "tight" rock formations such as North Dakota's Bakken Formation (see Case Study 14.2). Crude oil production between 2005 and 2010, especially from the deep waters of the Gulf of Mexico, which increased by 334,000 barrels per day, also increased our reserve estimates. Notwithstanding, the nation's production continues to plot on the descending limb of a generally bell-shaped curve (**Figure 14.12**) despite recent deep-sea oil and other recent domestic discoveries. On average, since 1981, the world has discovered less than one barrel of new oil for every four consumed.

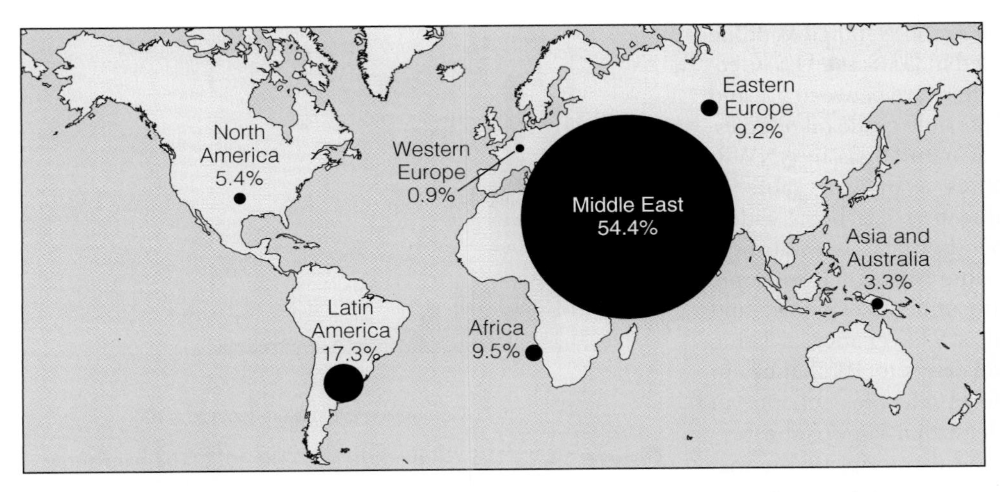

Figure 14.11 World proven petroleum reserves. Notice the scale change for North America. (Source: *BP Statistical Review of World Energy*, 2011, Colorado Geological Survey)

Let us look more closely at the significance of the reserve figures in Table 14.3. It is easy to overlook the margins of uncertainty concerning the actual size of proven reserves. For example, for apparent political reasons, some nations do not allow audits of their oil fields. This is especially true for nations that belonged to the former USSR and for the OPEC countries, a cartel of 12 oil-producing countries: Algeria, Angola, Ecuador, Iran, Iraq, Kuwait, Libya, Nigeria, Qatar, Saudi Arabia, the United Arab Emirates, and Venezuela. Reported reserves for the OPEC nations are especially suspect. In 1985, Kuwait increased its reported proven reserves by 50% when it was competing for its OPEC quota of exported oil, although nothing of significance had changed in its oil fields. Three years later, Abu Dhabi, one of the United Arab Emirates, tripled its reserves. Not to be outdone, Iran upped its reserves from 49 to 93 billion barrels, and Iraq jumped from 47 to 100 billion barrels. Saudi Arabia was already reporting more, but in the following year, to protect itself in this "OPEC quota war," the Saudis announced their reserves as 259 billion barrels (up from 170 billion barrels[2]), and from 2002 to 2011, Saudi Arabia reported reserves of 264 billion barrels (see Table 14.3). In those intervening 9 years, however, Arabia was producing about 10 million barrels a day, clearly drawing down its reserves, with no important new discoveries during that period. Consequently, all of the reserve estimates from the OPEC states are open to question, and the total world reserves shown in Table 14.3 are probably inflated.

consider this . . . California's Sierra Nevada is composed of granitic and metamorphic rocks. An acquaintance has offered you shares in a sure-shot wildcat oil-drilling venture in the Sierra Nevada. How should you respond to the offer?

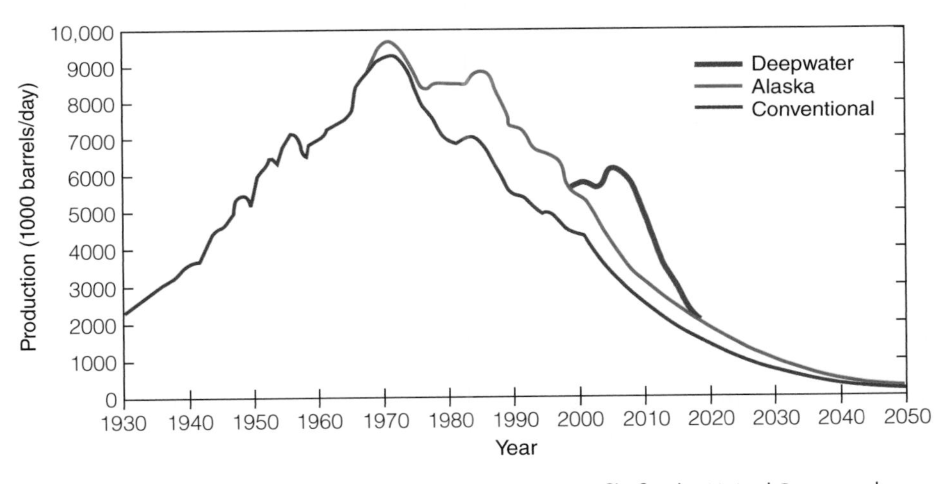

Figure 14.12 Cumulative and projected oil-production profile for the United States and deepwater areas currently being developed. Production peaked in the lower 48 states in about 1971 at 9.4 million barrels per day, as predicted by M. King Hubbert in 1956. The supergiant Prudhoe Bay, Alaska, field forms an upward bump on the descending limb of the production curve as does deepwater production from the Gulf of Mexico. Additional deepwater discoveries may add another small bump at lower production levels on the descending curve. New production from North Dakota's Bakken Formation essentially offsets the depletion rate in Prudhoe Bay production and will have virtually little impact on the overall slope of the descending limb. U.S. production in 2011 was 5.7 million barrels per day. (Source: Campbell, C.J. "The Assessment and Importance of Oil Depletion," in Proceedings of the First International Workshop on Oil Depletion, ed. K. Aleklett and C.J. Campbell (Uppsala, Sweden: 2002); Energy Information Administration, International Energy Statistics (April 9, 2012).)

Although the United States is the third largest producer of petroleum (Saudi Arabia and Russia rank first and second, respectively), it is the most energy deficient of the largest oil-producing countries. Since 1970, overall U.S. production (see Figure 14.12) has been dropping, and at the current rate of production, some authorities expect the U.S. reserves to last about 10 years, although other authorities, such as Daniel Yergin,[3] expect the reserves to last far longer.

[2]Data from *The Association for the Study of Peak Oil and Gas Newsletter*, January 2008.

[3]Yergin is a Pulitzer Prize historian and energy analyst, and one of the world's greatest optimists about oil reserves. His views are expressed in his book, *The Quest: the Global Race for Energy Security and Power* (Penquin, 2011).

What about "all the oil" in the Arctic National Wildlife Refuge (ANWR)? In 2002, upgraded in 2008, the U.S. Geological Survey (USGS) estimated there are between 5.7 and 16 billion barrels of probable and possible crude oil reserves, and 35 trillion cubic feet (Tcf) of natural gas in ANWR. The average of these values gives a median value of 10.4 billion barrels of oil, a value that has been widely quoted in the media.[4] The number, however, is a statistical fiction that represents a median value between a very large amount, with only a 5% probability of being extracted, and a more likely small amount, with a 95% chance of being found. (The 95% probability level refers to 19 chances in 20, whereas the 5% probability level refers to 1 chance in 20, that the amounts will be at least that large). The very high estimate (very low chance) unduly pumps up the median, a figure the uncritical public may not recognize.

Let's assume that the high value, 16 billion barrels of crude oil, is available. How much oil would that actually provide us? Our present thirst for oil is about 19 million barrels per day; thus 16 billion barrels divided by 19 million barrels per day gives an equivalent of 842 days, or only about 2.3 *years* use at our present rate of consumption, an impressively low volume. Even if the refuge were opened tomorrow, it would be some 10 years before a single barrel reached the lower 48 states, and it would take some 15 to 20 years to recover that amount.

Great effort has been expended in the search for discoveries of new conventional oil. Those discoveries are generally small, and located in regions that are climatically extreme or physically or politically hazardous, such as the offshore waters in the Arctic Ocean that Shell Oil is starting to drill. Another example, the recent deep-water oil and gas discoveries of the Jack Field in the Gulf of Mexico, came with warnings of the difficulty of producing oil from a 29,000-foot-deep well drilled in 7,000 feet of water in the hurricane alley of the Gulf (see Case Study 14.1). If successful, the field might increase the world's reserves by an additional 3 to 15 billion barrels. Fifteen billion barrels would increase the U.S. reserves by 50%. If true, the 15 billion barrels would amount to only about 175 days of global consumption at current levels of consumption.

The 20th century is sometimes referred to as the "Hydrocarbon Age" or the "Oil Age," and various authorities have speculated on when it might end. The era of the United States as the dominant player in oil production began to wane in 1930, the peak year for the discovery of new oil fields, with discoveries of new oil fields declining ever since (**Figure 14.13**). Using data from discovery and consumption rates, M. King Hubbert, a well-known geophysicist, made a startling prediction in 1949: The fossil-fuel era would be of very short duration. In 1956, while working for Shell Oil Company, he developed a formula for predicting when oil production would peak, the year when about 50%

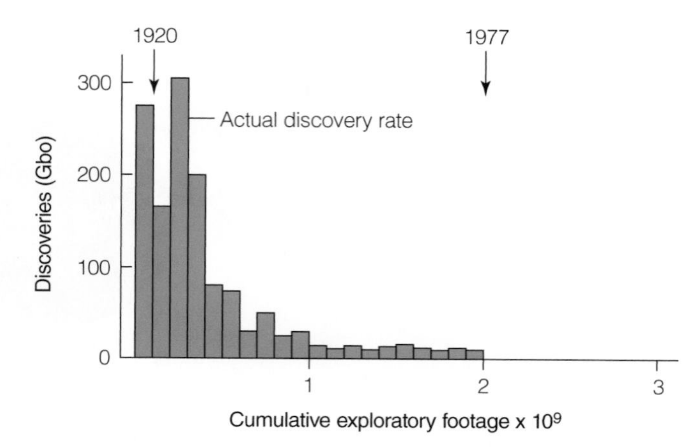

Figure 14.13 U.S. oil discoveries, 1910–1977. The bars show actual discovery rates in billions of barrels of oil (Gbo) plotted against total length of all exploratory wells. Total discoveries peaked about 1930. (U.S. Geological Survey Professional Paper 1193, 1982, 3–10.)

of the nation's oil endowment had been consumed and production would begin to decline. His prediction was scoffed at then, but his analysis has since proved remarkably accurate—the peak was reached about 1971 and is referred to as Hubbert's Peak (see Figure 14.12). New finds since 1971 have been of little consequence; even the opening of the Prudhoe Bay field, Alaska, North America's largest oil field, had little effect.

Although U.S. oil production began to plunge in the early 1970s, it seemed to be of no concern to most Americans. Beginning in the 1970s and continuing into the 21st century, Americans have driven more and more, pushing gasoline consumption and petroleum imports even higher. By 2006, each American, on average, was consuming his or her weight in petroleum every 7 days, and some 63% of that oil was imported. However, by 2010, U.S. dependence on imported oil fell to 49% for the first time since 1997, because of a combination of the weak economy, improved energy efficiency, more fuel-efficient vehicles, and crude oil production, especially in the deep waters of the Gulf of Mexico, that increased by more than 300,000 barrels a day between 2005 and 2010. The good news is that the EIA expects the moderating trend in U.S. oil imports to continue through the 2010s because of higher fuel economy standards and improvements in energy efficiency.

By the time Hubbert died in 1989, he had begun revising his prediction, pushing the peak into the future because of improvements in vehicle fuel efficiency that exceeded his expectations. The significance of his theory is not whether he got the exact year correct, but rather the overall accuracy of his theory and its implications.

Globally, new oil field discoveries in other parts of the world peaked about 1964, and world production began to level off in 2004 (**Figure 14.14**). Using Hubbert's methodology, the timing of the world peak was predicted to be about 2008 or 2009, to be followed by declining production (**Fig-**

[4]USGS. 2008. *Arctic National Wildlife Refuge, 1002 Area, Petroleum Assessment, 1998, Including Economic Analysis.* Fact Sheet 0028-01: Online Report.

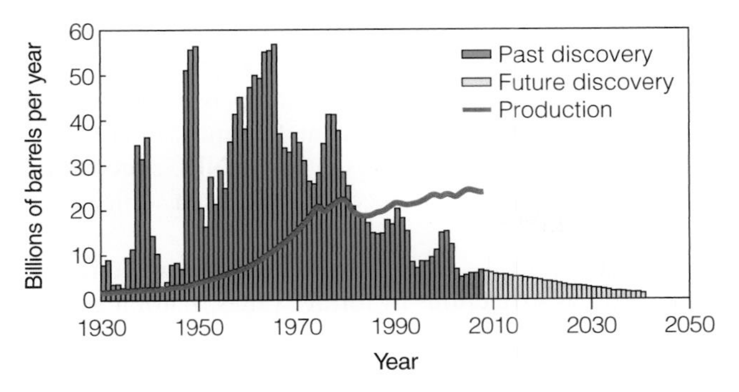

Figure 14.14 Worldwide oil discoveries, 1930–2009, with new discovery estimates to 2040. The rate of new oil discoveries dropped for decades (blue), with the bars showing actual discoveries in billion barrels of oil per year (Gb/a) plotted against time. World's discovery peak was 1964, with declining discoveries since then despite increased exploration. World production is projected to drop off even more in future years (green). The rate of worldwide consumption, however, is still climbing (red line). (Source: C. A. S. Hall, courtesy of American Scientist; data from ASPO, April 2009.)

ure 14.15). As global production drops, the oil exporting nations will be unable to meet world demands.

Some experts predict that the decline in the world's supply of affordable oil has already begun; others say the crunch will not start until 2040 or 2050. They refer to this time as *the end of cheap oil*. When it will happen, and the speculation about when the world will run out of oil, is not relevant. What does matter is when production will fall off as demand inexorably continues to rise. In other words, will the price adjust according to the well-known supply-and-demand curves of the economists? Figure 2 in the chapter opener shows the predicted decline in the use of oil in the 21st century and the increasing use of natural gas and projections for the future of alternative energy sources. Many energy economists report that natural gas is still reasonably plentiful—a view that is open to question—and it burns cleaner than coal, gasoline, or heating oil. Natural gas currently accounts for 75% of total U.S. energy use in space heating and cooking, and it is in wide use as fuel for electric power plants.

Energy Gases and the Future

Natural Gas

Natural gas is primarily methane, CH_4. It occurs with petroleum, in coal beds, as shale gas, and as methane clathrates. It is an important fuel, a major feedstock for manufacturing plastics and fertilizers, and a potent greenhouse gas, although it emits far less CO_2 than burning either petroleum or coal. The entire U.S. natural gas industry accounts for only about 3.2% of the nation's total greenhouse gas emissions. Before natural gas can be used as a fuel, it must undergo extensive processing to remove materials other than methane, such by-products as ethane, butane, pentane, propane, higher-molecular-weight hy-

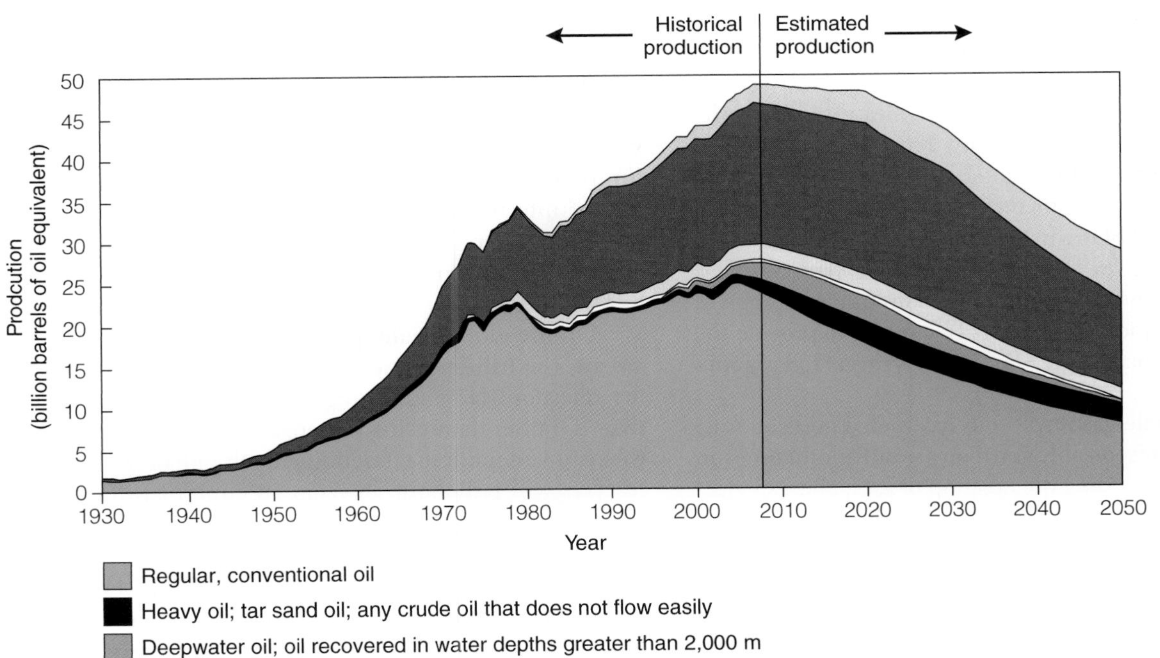

Figure 14.15 The global oil and natural gas production profiles for historical and projected production, 1930–2050, based on the best available production and reserve data as of 2008. (Source: Association for the Study of Peak Oil Newsletter, April 2009.)

drocarbons, and elemental sulfur. Natural gas is measured as cubic feet of gas at an atmospheric pressure of 101,353 pascals (14.7 pounds per square inch) at a temperature of 15.6°C (60°F). A trillion cubic feet (Tcf) of gas would fill a cube 3.2 kilometers (2 miles) on a side (33 cubic kilometers or 8 cubic miles). The EROI of natural gas is in the range of 10:1 to 15:1.

For many years, the United States supplemented its own natural gas output by importing 20% of its needs, primarily from Canada. Such imports of natural gas peaked in August 2007 at 10 billion cubic feet per day (Bcfpd), and it was thought that the United States would become increasingly dependent on imported liquefied natural gas (LNG) because of what appeared at the time to be a limited supply. Since then, however, this is no longer the case, and the imports of natural gas have been falling. By 2011, the daily natural gas average net imports—imports minus exports—was just over 5 Bcfpd, the lowest level since 1992. The drop in net import is due to both lower imports and higher exports. The National Petroleum Council (NPC) reports, "The natural gas resource base [in the U.S.] is enormous and its development—if carried out in acceptable ways—is potentially transformative for the American economy, energy security, and the environment, including reduction of air emissions."[5] The NPC glowingly estimates that gas reserves in North America range from "five or more decades" to "over 100 years demands at today's consumption rates." Shale gas and other unconventional gas resources are the new game changers, with significant conventional resources being produced both offshore and onshore. As a consequence, the United States is now the world's number-one natural gas producer, which, together with Canada, accounts for more than 25% of global natural gas production.

The U.S. natural gas reserve estimates vary widely depending on the reporting authority. For example, in 2011, the CIA and British Petroleum (BP) report natural gas reserves at 8 trillion cubic meters (272.5 Tcf), which would provide the United States with an 11.3-year supply at today's consumption rates. In contrast, the EIA's *Energy Outlook 2012* reports technically recoverable natural gas reserves at 63 trillion cubic meters (2,214 Tcf), enough to provide the United States with a supply of nearly 92 years at today's consumption rate.

The U.S. shale gas revolution has fueled an economic revolution. It is responsible for many positive outcomes in North America including the creation of jobs, cheaper electricity, and lower heating bills for many. It holds promise for reducing long-term dependence on foreign oil by replacing liquid fuel-powered transport with electricity from natural gas, or compressed natural gas. This, however, will take decades and great cost to accomplish. Just how it will work out depends on economics, as was well illustrated in 2012 when most shale-gas producers had scaled back their gas well drilling to more oil-prone prospects amid the rising price of crude oil and the falling price of gas. Despite these recent changes, many authorities expect shale gas to be a major factor, along with shale oil, Canadian tar sands, and recently discovered Brazilian deepwater oil, in the NPC's prediction that the United States will become almost self-sufficient in energy by 2030.

On the whole, the United States holds the world record for natural gas extraction. In former times, natural gas extracted with oil was considered a dangerous by-product and was burned off—called *flaring* in oil field jargon. Most flaring ended years ago, but even now the United States still flares off about 100 million cubic feet of gas each year by burning it at refineries and wellheads, an amount sufficient for a year's supply to about 1.5 million homes.

consider this . . . Many authorities have stated that natural gas, including coal-bed methane, will be the petroleum substitute that will provide the energy necessary for society to make the transition from a petroleum fuel-based economy to an alternative fuel-based economy. Are there adequate reserves and sufficient time to make this transition?

The United States faces an important decision with regard to natural gas: whether to export natural gas from LNG terminals in light of the expanded production of recent years. Applications for export are pending and, if approved, the United States could export some 20% of its production. Should this happen, the EIA estimates it could raise domestic gas prices by 24% or more, thus substantially increasing energy bills for consumers and have negative impacts on U.S. manufacturing. Keeping natural gas at home would permit greater amounts to be used in the domestic transportation and electric power sectors, and accelerate the transition away from burning coal and importing foreign oil. This would make U.S. energy consumption cheaper, cleaner, and more secure from foreign import interruptions. Such use could potentially reduce crude oil imports by 1 million barrels per day, reduce greenhouse gas emissions, and reduce coal consumption by 9% by 2035.

On the global scale, proven world reserves of natural gas are 187 trillion cubic meters (6,608.9 Tcf). This raises the question: How long might these reserves be productive? If, rather than using the present rate of consumption, we assume a modest extraction growth rate of 2.5%, global reserves will last about 38 years, in contrast with the estimates cited earlier (see Appendix 5 for the method of calculation). (As a precautionary note, it is important to recognize that the estimated expiration times of fossil fuels are not predictions or even forecasts. They are merely first-order estimates of the life expectancies based on the proven reserves consumed under conditions of steady growth, in this case, 2.5%, that our society and government consider necessary for the survival of our economic systems.) Distribution of the world's natural gas reserves are shown in **Figure 14.16**.

[5]National Petroleum Council. 2011. *Prudent development: Realizing the potential of North America's abundant natural gas and oil resources.*

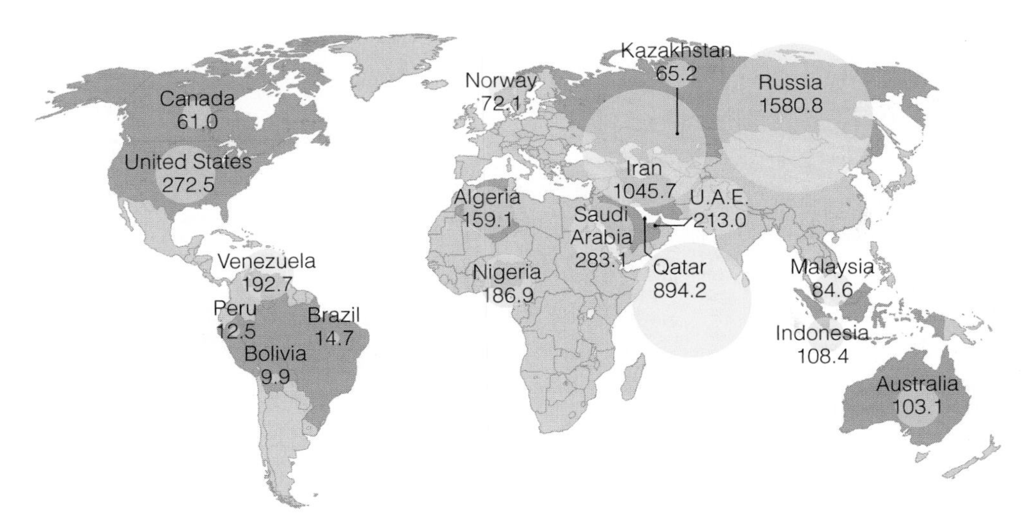

Figure 14.16 Major world sources of proven natural gas reserves in trillion cubic feet (Tcf). Canadian reserves include 21 Tcf from tar sands. (Source: Data from *BP Statistical Review of World Energy*, 2011.)

Unconventional Fossil Fuels

Unconventional oil and natural gas differs from conventional oil and gas by it being produced from reservoir rocks with very low porosity and low permeability—"tight" reservoirs in oil patch jargon. Examples are tight sands, relatively unfractured shales, methane hydrates, and coal seams (see Case Study 14.2). Unconventional petroleum is extracted using technology that is less efficient than conventional well-established oil well drilling and production techniques, and it comes with low EROI values. Some unconventional oil extraction processes have greater environmental impacts than conventional oil production. Unconventional oil includes oil shales, tar sands (or heavy-oil-sands)–based synthetic crude oil, liquids processed from natural gas, coal-based fuels, and biomass-based fuels. Some of these unconventional fuels may become sources of motor vehicle fuel for transportation; synthetic gasoline from tar sands and corn-based ethanol is already in limited use as a vehicle fuel.

Tight oil, sometimes called *light tight oil*, is a *light crude oil* that is difficult to extract because it is contained within "tight" oil-bearing formations such as shale that have low porosity and permeability. Tight oil should not be confused with shale oil or oil shale. In this case, the oil flows readily at room temperature because of its content of light hydrocarbon fractions. It is highly desirable because it produces a high percentage of gasoline and diesel fuel when processed at a refinery (see Table 14.2) and commands a better price than heavy crude on the commodities market. Extraction requires the same horizontal well drilling and hydraulic fracturing ("fracking") technology used in production of shale gas and coal-bed methane (CBM; see Case Study 14.3). Among the important tight oil deposits in the United States are the Bakken Shale in North Dakota and adjacent parts of Montana and Canada, the Barnett Shale, and Eagle Ford Shale of Texas (**Figure 14.17**). Tight oil also occurs elsewhere in the world, for example, in the Persian Gulf, West Siberia, Oman, and Mexico.

In 2009, the U.S. oil production began climbing after being in decline for 23 years. A major contributor to this increase in oil production was the development of the Bakken Shale in North Dakota and adjacent parts of Montana and Canada. The USGS estimates that the Bakken has about 3.65 billion barrels of **technically recoverable oil**. (It's most important to recognize that technically recoverable oil represents that portion of in-place petroleum that *may* be recoverable using current recovery technology, regardless of cost or effort; in contrast is **economically recoverable oil**, petroleum for which the costs of discovery, development, production, transport, and return of capital can be produced at a given market price.). The giant oil field at Prudhoe Bay, Alaska, originally had 25 billion barrels of oil, more than six times as much as the Bakken. Bakken production may prove to do little more than offset the drop in Prudhoe Bay production. A more thorough treatment of tight oil and the Bakken is presented in Case Study 14.2.

Coal-Bed Methane

Coal-bed methane (CBM) is natural gas trapped on the surface of, and in the fractures of, coal beds. It is a hazard to coal mining, and extraction was initiated to make coal mining safer. By the 1990s, several states had recognized CBM as a world-class commodity, and there was a 90% probability that 849 billion cubic meters (30 Tcf) could be found in the coal-mining regions of Wyoming, Colorado, Utah, Montana, and New Mexico. Commercial development of CBM in the United States has taken off and now accounts for approximately 10% of total natural gas production. Australia, Canada, and China also produce CBM. CBM is easy to extract by a drilled well (**Figure 14.18**), as opposed to extracting coal, which involves destructive surface mining. But gas production, as with all the nonconventional fuels, comes at a price, because it requires disruptive wells, a network of roads, disposal of salty water, pipelines, and methane leakage, all of which degrade the environment.

The completion of CBM wells requires technical enhancement by *hydraulic fracturing* (or hydrofracturing, or "fracking") which uses high-pressure fluids to increase porosity and permeability by shattering the rocks (see Case Study 14.3). Precisely what chemicals are used in the fracking fluids has long been a trade secret, but one is benzene,

Figure 14.17 Tight-oil and shale gas plays in the lower 48 states. (Source: Energy Information Administration, based on data from various published studies. Updated May 9, 2011. http://www.eia.gov/oil_gas/rpd/shale_gas.pdf)

a virulent carcinogen, which surfaces with the gas. The Bureau of Land Management (BLM) documented benzene contamination in Sublette County, Wyoming, groundwater where more than 3,000 coal-bed gas wells were drilled between 2000 and 2008. Sublette County is rural; there is no source of benzene other than from drilling and fracking. Nevertheless, the U.S. Environmental Protection Agency (EPA) and Congress have exempted hydrofracking fluids from the Safe Drinking Water Act that sets standards for underground injection of toxic chemicals. Wyoming, however, a state not known for its love of regulation, in September 2011, became the first state requiring gas drillers to disclose the chemicals used in fracking. Specific formulas are still proprietary, but the chemical ingredients are not.

The disposal of coal-bed wastewater presents other environmental problems. In the Powder River Basin, a large drainage basin straddling the Montana–Wyoming border, CBM producers pump salty groundwater out of the coal seams to release the gas. The water is discharged into streams, ponds, and dry washes, where the salt becomes a threat to irrigated crops and aquatic ecosystems. Ranchers worry that pumping will deplete the groundwater aquifers upon which

Todd Wilkinson

Figure 14.18 Drilling for coal-bed methane where the deer and the antelope play, near Pinedale, Wyoming.

they depend. In 2006, most of the 24,200 wells in the basin were on the Wyoming side of the border, but energy companies began moving into Montana. Montana government officials recognize the potential economic benefits to their state from the extraction of CBM, but they view Wyoming's development as uncontrolled and destructive. Thus, Montana will require producers to re-inject the water into the ground or remove the salt and other pollutants before dumping it into surface water bodies in accordance with the EPA's enforcement of the Clean Water Act. Re-injection and water treatment both require an expenditure of energy, which would cut into the profits of the developers, reduce the net amount of energy produced, and reduce the tax revenue to the state. Montana's tough water-quality standards would require rivers draining north from Wyoming into Montana to meet these standards. Wyoming politicians have accused Montana of "targeting" their state's thriving CBM industry. In addition to the two state governments as protagonists, two federal government agencies are also at odds. The EPA wants to enforce the Clean Water Act, but the Department of Energy (DOE) has focused on the potential impact on the national energy supply, stating that re-injection would cut the gas production significantly, resulting in losses of up to $21 billion worth of gas. Montana and the EPA are facing an uphill fight with Wyoming and the DOE; it may be that, in this case, there is no magic bullet to satisfy both environmental requirements and energy needs.

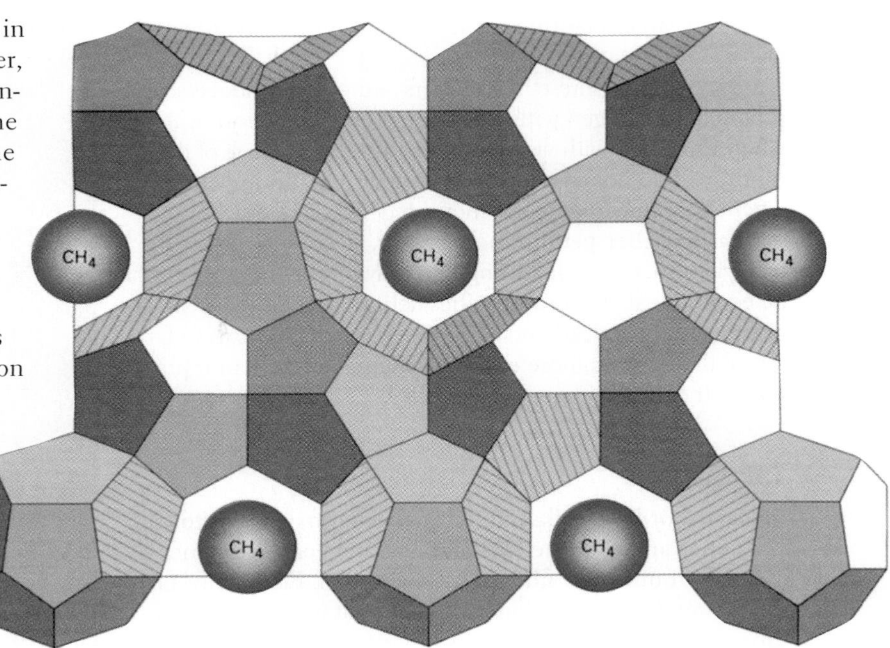

Figure 14.19 Crystalline structure of methane hydrate. Methane gas (CH_4) generated by bacterial digestion of organic matter in seafloor sediments is sometimes trapped in "cages" of frozen H_2O (the large polygons) to form methane hydrate. (Peter McCabe et al., 1993, USGS Circular 1115: 46.)

Methane Hydrates

A possible new source of gas is methane from **methane hydrate**. The gas is trapped in a cage of ice (**Figure 14.19**) and may eventually supply a share of the world's future fuel needs, *when and if* practical extraction techniques can be developed. Methane hydrates are widely distributed in deep-sea sediments in all oceans (**Figure 14.20**) and are present in permafrost in the Arctic and subarctic. India and Japan, at present energy-poor, are most interested in developing methane hydrates as an energy source. The exact origin of the gas is unknown, but it is strongly suspected that sea-bottom bacteria consume organic-rich detritus and generate methane as a waste product. Under appropriate conditions, the gas becomes trapped in ice within the sediment instead of being released to the overlying water column. The bad news is that these deep-sea hydrated molecular structures can release their caged methane into the atmosphere if there is a slight decrease in pressure (if sea level lowers, for example) or a slight warming of the surrounding seawater. Methane hydrates buried in permafrost can be released if sufficient warming of the atmosphere melts the permafrost.

Hydrogen

Another gas that has been touted as a potential fuel is **hydrogen**. Some authorities consider it to be the energy of the future, as it burns without pollutants, emitting only water vapor. Unfortunately, there are no large reservoirs of hydrogen, and it is not a primary fuel source. Hydrogen is an energy carrier and must be manufactured using fossil fuel, which then shifts the burden of pollution to a coal- or gas-fired power plant. It may offer promise as an effective way to store solar energy, as a fuel for airplanes and autos, and it might become economically viable if it can be pro-

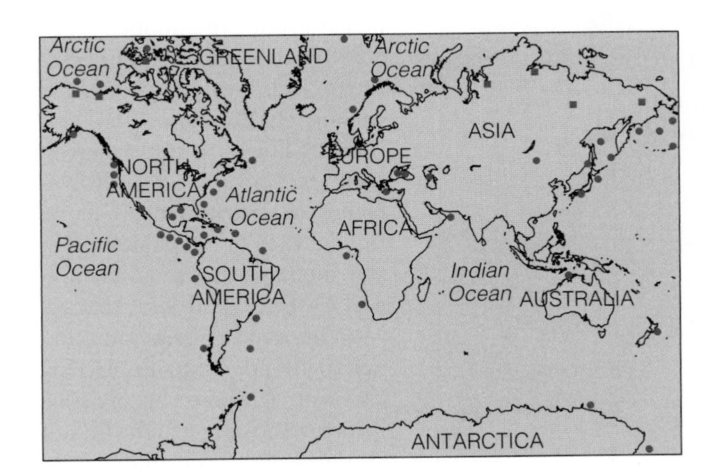

Figure 14.20 Location of the world's known methane hydrate deposits. The dots are marine deposits; the squares are permafrost deposits. (Source: Keith A. Kenvolden)

duced in large quantities using solar, wind, or hydroelectric energy.

In 2009, Chinese researchers announced success at producing hydrogen with exceptional efficiency from water when irradiated with visible light by using a catalyst of cadmium sulfide doped with palladium and platinum sulfide. Their announcement has motivated other researchers to search for better performing catalysts. A commercial-scale catalytic process is something that may or may not prove to be economical. Many authorities believe developing hydrogen as a fuel is a dead-end technology. Current efforts so far reveal that it takes more energy to manufacture than it produces. It comes with a negative EROI.

Tar Sands

Tar sands, or heavy-oil sands, consist of layers of sandstone, claystone, and siltstone containing water-coated sediment grains surrounded by films of tarry bitumen. Essentially, tar sands contain oil that is too thick to flow at normal temperatures. These deposits, found in Alberta (Canada), Venezuela, Madagascar, the United States, and elsewhere, are surface-mined like coal, using enormous diesel-powered shovels and haul trucks, or drilled using steam to mobilize and recover the oil. The size of Canadian and Venezuelan deposits is mind-boggling. Canada's Athabasca sands alone (**Figure 14.21**) constitute the largest oil field in the world, covering an area about the size of North Carolina. Canada claims that 1.73 trillion barrels of recoverable oil will come from the deposits, with proven oil reserves eight times that of the United States. Thanks to the prodigious Athabasca tar sands, Canada is second only to Saudi Arabia in proven oil reserves.

Figure 14.21 Locations of Alberta, Canada's tar sands.

We focus here on the Alberta, Canada, deposits, those that are most important to the United States. Alberta's tar sands underlie nearly 142,000 square kilometers (55,000 square miles) of land, but only about 3%, or 4,790 square kilometers (1,850 square miles) of that land could be impacted by surface tar sand mining. The remaining reserves, about 97% of the tar sands surface area, are too deep to be mined; they are exploited by in situ methods that require minimal surface land disturbance. Most of Alberta's heavy oil is obtained by open-pit mining. Excavation is done by massive Bucyrus shovels, which dig out 100 tons with each bite, which are then hauled to the extraction plant in three-story-high, 400-ton haul trucks (**Figure 14.22**).

Figure 14.22 Tar sands mining, Fort McMurray, Alberta, Canada. Tar sands open-pit mining is done in benches, each about 12 meters (40 ft) high. Massive Bucyrus shovels dig out 100 tons with each bite and dump it into three-story high haul trucks that have up to 400-ton capacities.

Processing of the tar sands requires 4 tons of ore to get one barrel of synthetic oil and requires enormous amounts of natural gas and hot water to produce a marketable product from the heavy oil. Two to four barrels of water are needed to produce the steam that separates the sand from the bitumen to make one barrel of synthetic crude oil.

Tar sands that are too deep (greater than 500 meters—1,640 feet) for surface exploitation underlie Alberta's Cold Lake area. Boreholes produce this oil by a process called *in situ extraction*. Steam at 300°C (575°F) is injected through boreholes at high pressure and allowed to "soak" for a week or more to melt the bitumen. The oil released from the sand is then brought to the land surface by "rocking horse" pump jacks. More than 2,000 steam-injection and oil-production wells deliver 150,000 barrels a day. The Cold Lake area is estimated to contain 220 billion barrels of recoverable oil.

At 10°C (50°F), bitumen is as hard as a hockey puck.

Development of synthetic gasoline from the tar sands is controversial. Claims and counterclaims flash back and forth, some claiming increased production and import of tar sand crude to the United States will lower U.S. gasoline prices, whereas others point out that the United States has a glut of refined petroleum, so much that refined petroleum products have become the nation's second ranking major economic export, yet as of mid-2012, the U.S. pump price keeps climbing.

Exploitation of the tar sands comes with a huge environmental footprint. By 2012, the tar sand mines had con-

Garth Lenz

Figure 14.23 Reclaimed disturbed land from tar sands mining. The Suncor refinery is in the upper left, and boreal forest in fall foliage is in the far distance. Reclamation is reported to have been ongoing since mining began in the 1960s, but only 0.15% of the mined land has been certified as reclaimed by the provincial government.

verted 673 square kilometers (260 square miles) of Canadian boreal forest into dirt, dust, and toxic tailings ponds, a major disturbance to the landscape, an area the size of Austin, Texas. Eventually, 4,790 square kilometers (1,850 square miles) could be impacted. The Canadian Association of Petroleum Producers reports that reclamation of mined land is an ongoing process since mining began in the 1960s and that 22% has been reclaimed. However, in the four decades since, the Pembina Institute claims only 0.15% of the active mining footprint is certified as reclaimed by the provincial government[6] (**Figure 14.23**). In cooperation with the Fort McKay First Nation, Syncrude, the tar sand mining company—a joint venture of several major oil companies—has successfully developed wood bison habitats on reclaimed lands, instead of restoring boreal forest and muskeg. More than 300 bison now graze on reclaimed lands once the sites of tar sand mining and tailings operations.

Another environmental concern is the large quantity of water used in processing. Although 80% to 95% is recycled, some ends up in highly contaminated tailings ponds. In April 2008, several hundred migrating waterfowl died after landing on one of the ponds. Several devices are used to deter birds from landing on ponds, including air cannons and radar/laser deterrent mechanisms. Toxic leakage from one huge tailings pond is seeping into the Athabasca River, causing First Nation (Indigenous Canadian) communities downstream to worry about toxins affecting their drinking water and the fish they eat. Greenhouse gas emission is another issue. The Syncrude refinery at Fort McMurray is one of Canada's major emitters of greenhouse gases, as it requires about 5,650,100 cubic meters (196 million cubic feet) of natural gas per day, around 40% of Alberta's total usage, to produce the steam required to make one barrel of heavy oil. In other words, the equivalent of one barrel of oil must be burned for every eight barrels manufactured. IEA figures show that in the "mine-to-wheels" lifetime of tar sand synthetic crude, carbon dioxide (CO_2) emissions from the tar sands are 20% higher than average emissions from conventional oil.

Clearly, substitutes for conventional oil do not come cheap. Some Canadians believe that using natural gas to produce oil from tar sand is analogous to turning gold into lead; they are asking if their finite natural gas resources are more valuable for their domestic use than for producing tar sand oil for export to the United States. Despite environmental issues and the relatively low EROI value of synthetic oil from tar sands, from about 2:1 to 6:1, synthetic oil from Alberta's tar sands is expected to grow, especially when the price of crude oil is $90 or more per barrel. Production is expected to reach 30 days of U.S. consumption by 2030. Currently, Canada's tar sands produce about 17% of what is consumed in the United States, and it is unlikely that it will ever be able to satisfy all of the U.S. demand.

Oil Shale

Oil shale is sedimentary rock that yields petroleum when heated. Found on all the continents, oil shales were originally sediments deposited in lakes, marshes, or the ocean. The original rock oil used in kerosene lamps was produced from black oil shale. Other nations with reserves of oil shales have been mining them for decades for power generation and other uses. American interest in oil shales has waxed and waned with fluctuating conventional oil prices. The most extensive U.S. deposit, the Green River Formation, was formed during Eocene time in huge freshwater lakes in the present-day states of Colorado, Utah, and Wyoming. Oil shale is not oily, like tar sands. Oil shale is not even shale; it is a clay-rich limestone containing **kerogen**, a solid, waxy hydrocarbon. When heated sufficiently, the rock yields oil (**Figure 14.24**). A ton of oil shale may yield 10 to 150 gallons of high-quality oil. Deposits of potential economic interest yield 25 to 50 gallons per ton, and the

[6]*About Canada's Oil Sands*. 2011. CAPP, 56 p.; http://www.pembina.org/oil-sands/os101/reclamation.

Figure 14.24 Oil shale from the Green River Formation in Wyoming and a beaker of the heavy oil that can be extracted from it.

USGS includes 3 trillion barrels of unconventional oil from shale in its optimistic petroleum reserve estimates. (Compare this figure with the proven oil reserves for the United States listed in Table 14.3.)

Where oil shales are shallow, they can be mined in open pits, but deeper deposits can be mined by underground methods. Converting solid kerogen into gasoline requires heating it to about 370°C (698°F) in extraction retorts. Adding hydrogen and other chemical treatment is needed to produce a finished product. There are environmental issues: Processing it is energy intensive, so extracting fuel from it is no solution to global warming, and it requires three to four barrels of water to produce one barrel of oil, conditions that do not bode well for oil shale development in the arid lands of the western United States. In addition, open-pit mining and the extraction process destroy the natural landscape, and some of the richest deposits are in scenic wilderness areas.

Experiments with in situ oil shale extraction have been done on a small scale. The process involves the underground heating of 1000-feet-thick columns of rock by electricity to 370°C (698°F) for 3 or 4 years, during which time the surrounding volume of rock is frozen to prevent moving groundwater from carrying away heat. The colossal amount of energy needed for this would require the construction of several coal-fueled power plants and the development of more coal mines to fuel the power needed, thus significantly increasing the carbon footprint of the operation. Consequently, it is unlikely that in situ production is economical.

A total of all the energy costs required for oil shale production (mining, crushing, hauling, heating, adding hydrogen, chemical treatment, and waste disposal) is probably a net energy loss. Even as we are forced to come to terms with the end of cheap oil, fuels derived from oil shale do not appear to be economical. At its very best, fuel extracted from oil shale has a low estimated EROI of perhaps 1.5:1 to 2:1.

Coal

Coal is the best of fuels; it is the worst of fuels.

Kenneth S. Deffeys, *2005, Beyond Oil, the view from Hubbert's Peak*

Coal is the carbonaceous residue of plant matter that has been preserved and altered by heat and pressure. Next to oil and oil shales, coal is Earth's most abundant reserve of stored energy. It is the best of fuels because it provides energy at a lower cost than other fuels. It is the worst because of atmospheric carbon dioxide, acid mine drainage (see Chapter 13), and acid rain, smog, and mercury pollution, which are explained later in this chapter. Coal deposits are known from every geological period since Devonian time and the appearance of widespread terrestrial plant life some 390 million years ago. Permian coal is found in Antarctica, Australia, and India—pre-continental drift Gondwanaland (see Chapter 2). The large fields of North America, England, and Europe were deposited during the Carboniferous period, so named because of the extensive coal deposits in rocks of that age. In the United States, the Carboniferous period is divided into the Mississippian and Pennsylvanian periods, named after the states where the deposits are found. Tertiary coals are found in such diverse locations as Spitsbergen Island in the Arctic Ocean, the western United States, Japan, India, Germany, and Russia.

Coalification and Rank

The first stage in the process of *coalification* is the accumulation of large amounts of plant debris under conditions that will preserve it. This requires high plant production in a low-oxygen depositional environment, the conditions usually found in nonmarine brackish water swamps. The accumulated plant matter must then be buried to a depth sufficient for the heat and pressure to expel water and volatile matter. The degree of metamorphism (conversion) of plant material to coal is denoted by its **rank**. From lowest to highest rank, the metamorphism of coal follows the sequence *peat, lignite, subbituminous, bituminous, anthracite*. Because of the various ranks of coal, the EROIs of coal range from about 20:1 (subbituminous) to 40:1 (highest anthracite rank). The sequence is accompanied by increasing amounts of fixed carbon and heat (British thermal units [Btu]) content and a decreasing amount of quickly burned volatile material. Once ignited, it is the carbon that burns (oxidizes) and gives off heat, just as wood charcoal does in a barbecue pit.

Bituminous, or soft, coal usually occurs in flat-lying beds at shallow depths that are amenable to surface-mining techniques. Anthracite, the highest rank of coal, is formed when coal-bearing rocks are subjected to intense heat and pressure—a situation that sometimes occurs in areas of plate convergence. In the Appalachian coal basin, high-volatile bituminous coals occur in the western part of the basin and increase in rank to low-volatile bituminous coals to the east. Anthracite formed close to the ancient plate-collision zone of North America and Africa (pre-Pangaea), where the coals were more intensely folded and subject to higher levels of heat. Traveling east from the Appalachian Plateau to the folded Ridge and Valley Province, one goes from flat-lying bituminous terrain to folded anthracite terrain. West Virginia, Kentucky, and Pennsylvania are the leading coal producers of the eastern states. The interior coal basins of the Midwest are characterized by high-sulfur bituminous coal, whereas the younger western deposits are low-sulfur lignite and subbituminous coal. Wyoming, with its low-sulfur coal, was the nation's leading producer in 2010 (442.5 million short tons), followed in order by West Virginia, Kentucky, Pennsylvania, and Illinois (**Figure 14.25**).

Reserves and Production

The future of coal-derived energy in the United States is more optimistic than the outlook for petroleum. Coal constitutes 80% of U.S. energy stores but only about 18% of the usage. The electric utilities today consume about 90% of U.S. coal production, compared with 19% 60 years ago, with residential and industrial use accounting for the remainder. It is estimated that there are 826 billion (826×10^9) recoverable tonnes[7] of coal in the world. The United States has 238 billion tonnes of recoverable coal reserves, which conventional wisdom suggests might last over 200 years at the current rate of production and consumption.

For a half-century, we have been given false hope by such statements as that in a report to the U.S. Congress: "At current levels of output and recovery these reserves [of U.S. coal] can be expected to last more than 500 years."[8] Such statements, and conventional wisdom, ignore a critical factor—*exponential growth*. This is dangerous—it is easy to miss the significance of the caveat "At *current* levels of output. . ." Coal production has generally increased on the average since 1890. (There were some years of declining coal production, but annual increases between 1950 and 2012 outnumber decreases by about two to one.) Since 1990, for example, the average annual increase in coal production in the United States has averaged 1.8%, and in 2010, it increased by 2.1% over 2009 (the most recent figure available).

[7]Data from BP Statistical Review of World Energy, 2011. One tonne = 1000 kilograms = 1.1023 short tons.

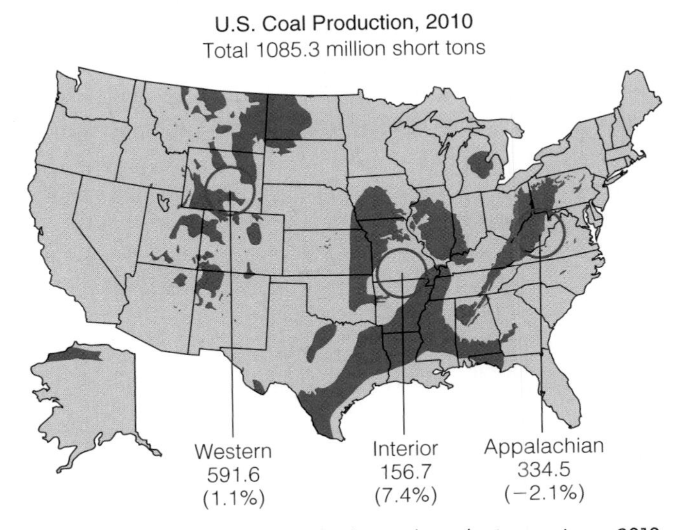

Figure 14.25 Coal production by coal-producing regions, 2010. Production in million short tons (and percentage change from 2009) for each coal producing region, and the total production for the United States. Note the major production is in the western coal states. (Source: Energy Information Administration)

Table 14.4 Energy (Power) Units

Unit	Explanation
British thermal unit (Btu)	The amount of heat required to raise the temperature of 1 pound of water 1°F; energy released by a burning match
Quad (quad)	10^{15} Btu = 172 million bbl/oil
Barrel of oil	5.8 million Btu = 42 gallons
Bituminous coal (average)	25 million Btu/ton
Natural gas	Variable Btu content, measured in cubic feet (cf)
Watt (W)	1 joule/sec
Exajoule (EJ)	0.948 quad = 948 X 10^{12} Btu
Kilowatt-hour (KWh)	3.6 megajoules (MJ)

Annual energy demands in the 21st century are expected to be between 100 and 150 quads (a quadrillion British thermal units, or 10^{15} Btu), compared with 80 quads just 35 years ago (**Table 14.4**). Coal production will need to double to meet this demand, but can it? What do some simple calculations show about the longevity of U.S. coal reserves? Assuming the annual exponential production growth rate of 2.1% between 2009 and 2010 continues, the U.S. recoverable coal reserves *may last only a mere 89 years,* not the 200 to 500 years reported by some energy analysts (see Appendix 5 for the method of calculation).

Should there be any increased production, it will be mainly in the western coal basins (**Figure 14.26**), where coal is more easily mined by relatively inexpensive, large-scale surface-mining methods rather than the underground methods still practiced in much of the eastern coal fields. Underground mining costs more, requires greater investment, takes more time to get into production, and miners working underground face greater risks than do those who work on the surface. (The various methods of mining coal are illustrated in **Figure 14.27**.) On a global scale, the EIA in 2011 estimated that coal reserves would last about 125 years, but with the usual caveat, "at the current extraction rate." However, for many years the world's annual rate of coal production has

[8]*National Fuels and Energy Policy Study Serial* No. 93-9 (92-44) 1973. Washington, DC: U.S. Government Printing Office, 15, 41-42. 267 pp.

John Carr

Figure 14.26 A 115-car train loaded with low-sulfur coal from the Powder River Basin, Wyoming, near Ellsworth, Nebraska, on its way to supply three coal-fired power plants in Missouri. Up to 100 such trains leave the Wyoming coal fields each day to supply coal to Midwestern and East Coast power plants. The same number of empty trains return for reloading each day.

increased between 2% and 3% per year, but the extraction rate increase jumped to 6.3% in 2009. Using a modest exponential growth rate of 2.5%, the calculated global depletion time is only 58 years. (Bear in mind that these expiration times are not predictions, merely first-order estimates determined from known quantities for the life expectancy of coal under the conditions of steady growth that many governments and societies believe is necessary to maintain strong economies.) In addition, California Institute of Technology Professor David Rutledge, using widely accepted reserve estimates and rigorous mathematics, forecasts 90% exhaustion for the world's oil, coal, and gas in only 60 years.

A potential key element in coal's economic future lies in its use in manufacturing a substitute for oil and gas. The technology is in place to make methane from coal; the product, called *synthesis gas*, or **syngas**, has been used on a large scale to manufacture petrochemicals, including methanol (methyl alcohol) and ammonia. Synthesis gas is a compound consisting of carbon monoxide, hydrogen, and commonly some carbon dioxide; it has less than half the energy density of natural gas. Direct liquefaction of coal also shows some promise. Using coal- or petroleum-based solvents, coal is dissolved and hydrogenated, resulting in about 75% gasoline and 25% propane or butane. About 5.5 barrels of liquid are derived from a ton of coal at a cost less than the most expensive oil during the oil shock of 2008 (which averaged about $92 a barrel). Some authorities predict that a major facet of our energy future will be converting coal to liquid or gases—*synthetic fuels*, or **synfuels**.

Clean Coal: IGCC and CCS

At least a quarter of the 30 billion tons of new CO_2 added each year to Earth's atmosphere comes from coal-fired power plants. Despite renewable technologies, such as wind and solar power, generating more electricity in the future, coal is likely to remain the dominant fuel for several decades. It is attractive because of its relatively stable price over the long term and because of its abundance in the United States and elsewhere in the world. Burning coal accounts for about 50% of the U.S. needs for electricity. Many new coal-fired power plants are being built in India and China, and the United States needs to take up the challenge of developing ways to reduce emissions from them. The most promising technology to meet the problems of CO_2, sulfur dioxide (SO_2), and mercury emissions is **integrated gasification combined cycle (IGCC)** technology coupled with **carbon capture and storage (CCS)**. The IGCC power plant uses a technology that turns coal into syngas, which is used to power a gas turbine that generates elec-

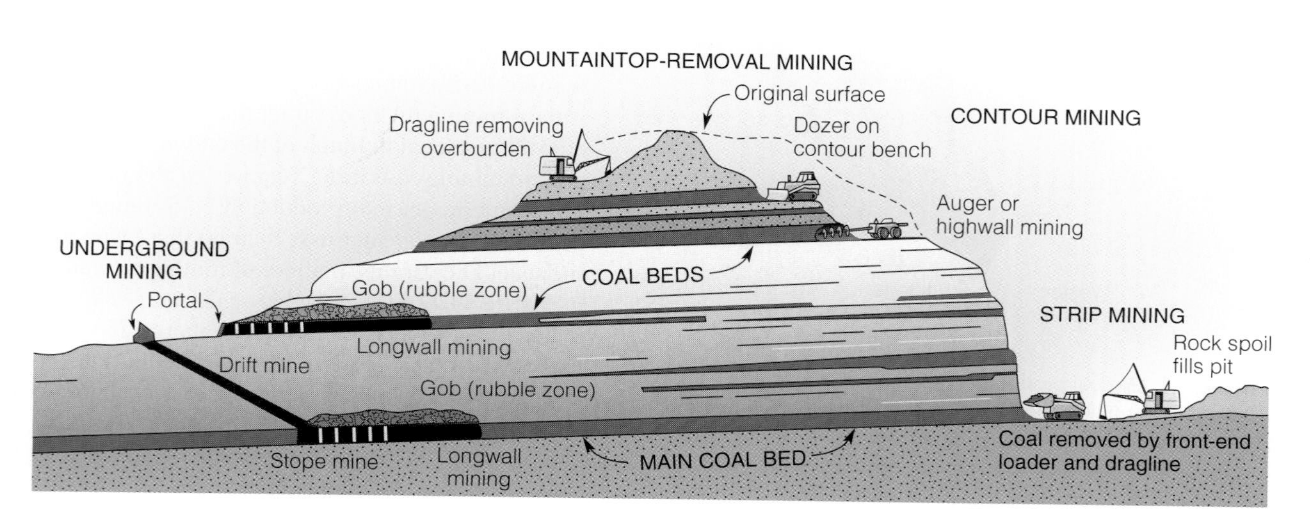

Figure 14.27 A schematic cross section illustrating conventional underground and surface coal-mining technology. (Source: Adapted from Colorado Geological Survey.)

tricity. The waste heat is then passed to a combined cycle gas turbine. The advantage of this technology is that the particulates, SO_2, and mercury emitted by burning coal can be removed easily. In addition to IGCC being cleaner, it offers greater efficiency over conventional pulverized coal–fired power plants. The technology can also remove CO_2 during the process, which can be used in oil fields for the secondary recovery of petroleum or stored underground in geological formations. IGCC technology has been widely heralded as the solution to the growing problem of pollution and CO_2 emissions to the atmosphere from burning coal.

The Department of Energy (DOE) has funded a number of projects across the United States to determine whether large quantities of CO_2 from fossil-fuel–fired power plants can be sequestered permanently deep underground. A \$61 million grant was awarded to a collaborating partnership of more than 35 members that includes many of the country's leading energy companies, eight states, universities, and state geological surveys. As a result, a clean coal IGCC electric power project is being constructed near Edwardsport, Indiana, to test the technology of underground carbon sequestration.

The plant will cost \$2.3 billion, which will be offset in part by \$460 million in federal, state, and local tax incentives. By 2013, the plant is expected to capture up to 1 million tons of its own CO_2 each year and store it underground. Nine other IGCC plants are at various stages of permitting and planning in the United States. In January 2009, Illinois passed a "clean coal" law requiring state utilities to obtain 5% of their electricity from power plants by CCS by 2015. The state also set a goal of 25% of its energy from "clean coal" by 2025.

Critics of IGCC and CCS raise important questions. First, CO_2 is one of the most active and dynamic of the common gases. Because of its chemical activity, it is questioned whether the gas can be sequestered securely in deep underground formations. The leakage of just 1% from storage would unravel any advantages of reducing greenhouse gas. With current technology, it would take a minimum of 20% of the electrical output from a standard fossil-fuel–fired power plant to capture and compress its CO_2 emissions for carbon sequestration. Does this seem sensible? Third, many people feel we should halt the construction of any new coal-fueled power plants altogether and invest in alternative energy sources instead. Even if cost concerns are ignored, there is still concern that coal mining in general, and mountaintop-removal mining specifically, is an inherently dirty process. Alternative energy technologies are presented in the following pages; mountaintop-removal mining is discussed in Chapter 13.

consider this . . . The United States would need to drill 67,000 new carbon sequestration wells over the next 20 years to maintain CO_2 emissions near the 1990 level.

Problems of Fossil-Fuel Combustion

Air Pollution

Obtaining energy by burning fossil fuels creates environmental problems of global proportions. It produces oxides of carbon, sulfur, and nitrogen and fine-particulate ash. Carbon monoxide (CO), an oxide produced by the combustion of all fossil and plant fuels, is converted to carbon dioxide (CO_2), which contributes to global warming. Burning coal also releases sulfur oxides (SO_x) to the atmosphere, where they form environmentally deleterious compounds. Nitrogen oxides (NO_x), mostly NO and NO_2, are products of combustion in auto engines and are the precursors of the photochemical oxidants ozone and peroxyacetyl nitrate (PAN), which we associate with smog. Water and oxygen in the atmosphere combine with SO_2 and NO_2 to form sulfuric acid (H_2SO_4) and nitric acid (HNO_3), the main components of **acid rain**—rain with increased acidity because of environmental factors, such as atmospheric pollutants.

The Clean Air Act of 1963 as amended in 1970 and 1990 specifies standards for pollutant oxides and hydrocarbon emissions. By 2009, new cars were 90% cleaner than their 1970 counterparts. The Clean Air Act Amendments of 1990 (CAAA) set goals and timetables that are affecting how and what we drive. Beginning in 1992, fuel suppliers were required to sell only reformulated gasoline (RFG) in 39 areas where winter air quality was a problem. Reformulations using methyl tertiary butyl ether (MTBE) add as much as 2.5% oxygen to the fuel, causing it to burn cleaner and create less ozone, at a cost of about 10 cents more per gallon. Unfortunately, RFG is stored in underground tanks, which sometimes leak and pollute groundwater. As a result, a Blue Ribbon Panel recommended the following:

- Remove the Clean Air Act requirement for 2% oxygen in RFG.
- Enhance underground storage tanks, thus improving protection of the nation's drinking water.
- Reduce the use of MTBE nationwide, maintaining current air-quality benefits.

So much for MTBE and reformulated fuels.

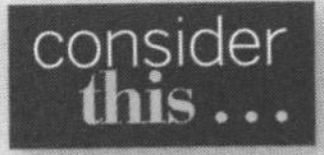

Historically, Pennsylvania crude oil (such brands as Quaker State and Pennzoil) has been known for its superior lubricating properties compared with, say, heavier California or Texas crude oils. Why do you suppose this is so? (Table 14.1 can help you deduce a reasonable answer.)

Sulfur emissions and acid rain—Sulfur occurs in coal as tiny particles of iron sulfide, most of which is the mineral pyrite, or "fool's gold" (**Figure 14.28**). Coal also contains organic sulfur originally contained in the coal-forming vegetation. Upon combustion, mostly in coal-

Hobart King, Mansfield Univ

Figure 14.28 Small pyrite masses in a bituminous coal matrix (gray). The light-gray material is the cell walls of the original plant material, and the darker gray is the solid material that was contained within the cell cavities. Earth pressures compacted the plant material, squeezed out any water, and collapsed the cavities. That pyrite formed within the coal before completion of compaction is evidenced by plant materials bent around solid pyrite masses. The pyrite "blebs" are about 50 microns across.

burning electrical generating plants, the sulfide is oxidized to sulfur dioxide (SO_2) and carried out through smokestacks into the environment. Emissions of SO_2 are known to have a negative impact on human health, particularly lung problems. Although most coal-burning in the United States occurs in the Midwest and the East, significant volumes of SO_2 can travel 1,000 kilometers (600 miles) or more from their source.

Sulfur dioxide emitted from a smokestack may be deposited dry near the facility, where it will damage plant life and pose a health threat to animals. If the gas plume from the smokestack is transported away from the facility, its SO_2 will react with the atmosphere and eventually fall to Earth as acid rain, called *wet deposition*. Acid rain can damage crops; acidify soils; corrode rocks, buildings, and monuments; and contaminate streams, lakes, and drinking water.

Nitrogen oxide emissions (NO_x) from both stationary (coal-burning) and mobile (vehicular) sources are also significant contributors to acid rain and to smog. To combat and reduce both SO_2 and NO_x, the Acid Rain Program was established as an amendment to the 1990 Clean Air Act. The Act requires the electric power industry to lower emissions of SO_2 and NO_x by about 50% from the levels of 1980 (**Figure 14.29**). The reduction is accomplished by using scrubbers, devices that selectively react with the undesired compounds and absorb or neutralize them. Many of the newer coal-burning facilities have switched to a process known as **fluidized-bed combustion** (FBC), which eliminates SO_2, produces less NO_x, but increases CO_2. It is thus a trade-off between health benefits of reduced atmospheric sulfur and the environmental effects of increased contributions to global warming. FBC plants

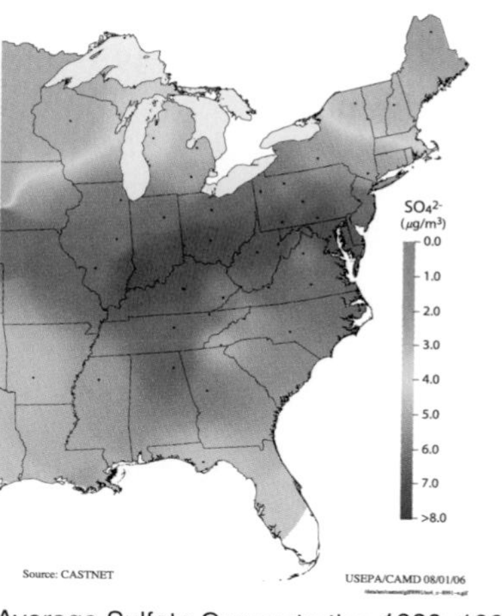

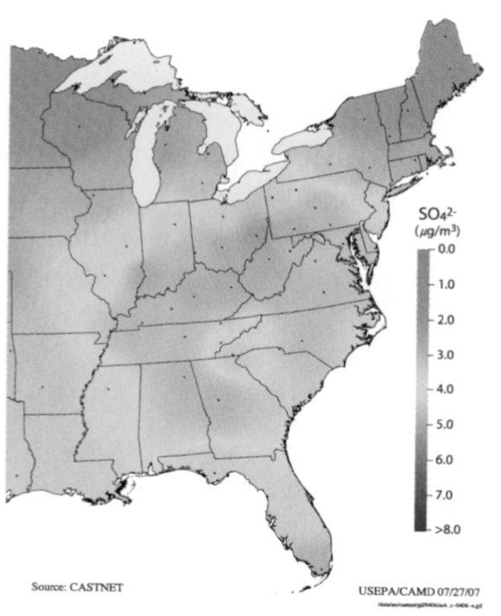

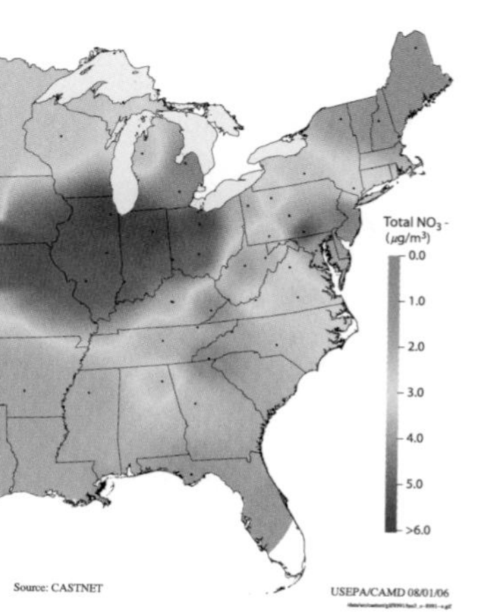

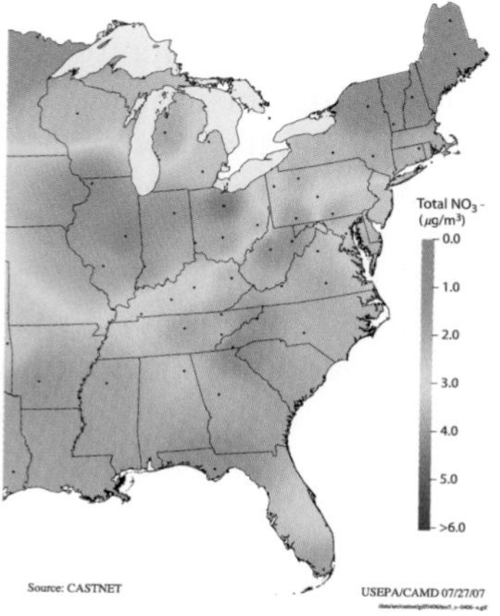

Figure 14.29 (a) Improvement in ambient sulfate (SO_4^{2-}) concentrations from 1989–1991 and 2004–2006. (b) Improvement in ambient nitrate (NO_3^-) concentrations from 1989–1991 and 2004–2006. $\mu g/m^3$ = millionths of a gram per cubic meter. Black dots mark CASTNET sampling sites. (Source: Clean Air Status and Trends Network [CASTNET] of the Environment Protection Agency [EPA])

are more flexible than conventional coal-fired plants because they can be fueled with coal, biomass, or other fuels. In addition, the FBC process results in greater combustion efficiency than conventional power plants.

Mercury emissions also are of concern. The source of about 98% of the United States mercury pollution is from the coal-fired electricity power sector. Because of this, many older, electric power plants are scheduled to be shut down as a result of new EPA regulations to curb mercury emissions. Power companies claim that retrofitting to meet the new emission standards would not be economical, and it is more cost-effective to shut down the plants. Consequently, 106 older coal-fired plants (= 43,000 megawatts (MW) of electricity-generating capacity) are slated for closure by about 2015, and coal industry representatives believe that enforcement of the new federal air quality rules could threaten the U.S. electricity supply.

The Acid Rain Program has been successful in achieving major reductions in SO_2 and NO_x (see Figure 14.29) because of the *cap and trade* program authorized by the amendment to the Clean Air Act. Cap and trade is a policy that sets an overall cap, or limit, on the amounts of emissions from a group of sources, which is important because it ensures that emissions of a pollutant are reduced. Then allowances for emissions can be traded between sources in order that economic market forces allow large emissions reductions to be cost-effective.

Mobile sources of NO_x are the most difficult to control, and their contribution to total NO_x increases as the number of automobiles increases. NO_x forms when engine combustion causes nitrogen in air to oxidize to NO and NO_2, which then react with oxygen, free radicals (incomplete molecules that are highly reactive), and unburned hydrocarbons to produce **photochemical smog**. Photochemical smog requires sunlight for its formation. Although the word *smog* is a contraction of the words smoke and fog, the air pollution in Los Angeles, Mexico City, and Denver has little direct relationship to either smoke or fog.

Both NO_2 and ozone (O_3) are reactive oxidants that cause respiratory problems. A level of only 6 parts per million (ppm) of ozone can kill laboratory animals by pulmonary edema (water in the lungs) and hemorrhage within 4 hours. In humans, alcohol consumption, exercise, and high temperatures have been found to increase adverse reactions to ozone, whereas vitamin C and previous exposure seem to lessen the reaction. One investigator noted that the person most likely to succumb to ozone would be "an alcoholic with a 'snootful' arriving at LAX for the first time on a hot, smoggy day and jogging all the way to Beverly Hills." Such a scenario may not be all that unlikely in Southern California. Energy-efficient cars and new blended fuels have gone a long way toward decreasing NO_x emissions. Exhaust-system catalytic conversion of NO_x, CO, and hydrocarbons into CO_2, nitrogen, and water is mandatory on new cars in the United States.

Domestic coal burning—China is the world's largest producer of coal, accounting for 50% of global output. It currently produces 24% of the global emissions of CO_2, and it is estimated that by 2025 it will emit more CO_2 than the United States, Canada, and Japan combined. Coal provides 76% of the huge country's commercial energy, and it is the only fuel millions of people use.

In a typical home, coal is burned in unvented stoves for cooking, space heating, and water heating. In many areas, foods are brought indoors in the fall for drying with coal fires, as the climate is too cool and damp to dry them outdoors. At least 3,000 people in Guizhou Province in southwest China suffer from severe arsenic poisoning caused by consuming chili peppers dried over high-arsenic coal fires (**Figure 14.30**). Although fresh chili peppers have less than

Robert Finkelman, USGS

Figure 14.30 Drying chili peppers over a coal fire that emits arsenic fumes increases the arsenic content of the chilies up to 500 times, and eating them causes arsenic poisoning. Also, as families gather around coal fires at night, they are exposed to polycyclic aromatic hydrocarbons from incomplete combustion of the coal. Breathing these substances can lead to lung and esophageal cancer.

1 ppm arsenic, chili peppers dried over high-arsenic coal have up to 500 ppm arsenic. More than 10 million people in the same province suffer dental and skeletal fluorosis caused by eating corn that has been cooked over coal containing large amounts of fluorine. About 3.5 billion people worldwide are exposed to toxic fumes indoors caused by coal burning. Each year, an estimated 178,000 people die prematurely because of indoor air pollution in China alone. To combat these health hazards, Chinese and USGS geologists are working together to identify coal deposits with high toxic-element concentrations, making them unsuitable for domestic use.

Mine Collapse

Whenever coal is extracted from shallow underground seams, there is a risk for collapse of the overburden into the mine opening. Lignite and bituminous (soft) coals are most susceptible to collapse, because they are exploited from near-horizontal beds at relatively shallow depths. The "room and pillar" method recovers about 50% of the coal, leaving the remainder as pillars for supporting the "roof" of the mine. As a mine is being abandoned, the pillars are sometimes removed or reduced in size, increasing the yield from the mine but also increasing the possibility of subsidence at sometime in the future. Even when the pillars are left, failure is possible. The weight of the overlying strata can cause the pillars to rupture, and sometimes it physically drives the pillars into the underlying shale beds. Subsidence has been documented in Pittsburgh, Scranton, Wilkes-Barre, and many other areas in Pennsylvania (see Figure 13.27). In 1982, an entire parking lot disappeared into a hole above a mined-out coal seam 88 meters (290 feet) below the surface. Wyoming, Montana, South Dakota, and other western "lignite" states are pockmarked with subsidence features, but human activities and human lives are less likely to be affected by subsidence, or collapse, in those areas.

Beyond Petroleum: Alternatives for the Future

> **The world faces a significant challenge to supply energy required for economic development and improved standards of living while managing greenhouse gas emissions and the risks of climate change. It's going to take integrated solutions and the development of all commercially viable energy sources, improved efficiency and effective steps to curb emissions. It's also going to include the development of new technology.**
>
> Emil Jacobs, *Vice President of Research and Development, Exxon Mobil Research and Engineering Co., June 2009*

The race is on to find a suitable alternative for gasoline for use in automobiles. The contenders are **ethanol, methanol, biodiesel**, electricity, hydrogen, **liquefied natural gas**, and **compressed natural gas**.

Leading the **biofuel** pack is ethanol (ethyl alcohol) derived from corn. Biofuels are any fuels derived from **biomass** (straw; agricultural waste; sugarcane; food leftovers; waste from the wood, paper, and forestry industrial sectors; corn; switchgrass; any organic waste; cow manure; etc.). Ethanol made from sugarcane has been used in Brazil for many years as an auto fuel, and at one time almost 90% of the cars there burned it.

As with petroleum and coal, biomass is a form of stored solar energy that has been captured by growing plants by the process of photosynthesis. Ethanol derived from corn has received the most attention, the costs supposedly being closely competitive with gasoline. A big plus is that, as a high-percentage blend in automobile fuel, it releases less CO_2 than gasoline, but only in high-percentage blends; when used as a low-percentage additive in a blend with gasoline, it worsens CO_2 emissions. Ethanol fuel does cut ozone-forming gases, and it is a safe substitute for the fuel oxygenate gasoline additive MTBE. A drawback is that ethanol has only 67% of the energy content of an equal volume of gasoline. One reason that it appears to be price competitive is that the corn-ethanol industry is heavily subsidized by the federal government as a result of policies promoted by the politically powerful corn-growing states. Corn ethanol has three major problems:

- It takes considerable energy to make corn ethanol, and that energy from coal-fired ethanol plants would completely undermine the reduced greenhouse gas emission of ethanol. Furthermore, it requires energy to plant, fertilize, harvest, and transport the corn, all of which produces considerable greenhouse gases that outweigh the advantages.
- There is a supply problem. If the entire U.S. corn production were devoted to producing E85 (a blend of 85% ethanol and 15% unleaded gasoline), it would meet only about 12% of the nation's current demands. Consequently, for ethanol to win the race, it will need feedstock other than corn.
- It is corrosive, and thus not suitable for sending through pipelines. Instead, it must be shipped in tank cars constructed of special materials.

The complex question of energy consumed in converting corn to ethanol compared with the energy content of the ethanol has created controversy, with estimates ranging widely from a net loss of energy of 20% to 29% to a net gain of 67%. The range of estimates is due to differences in energy input values, changes in ethanol production processes, and accreditation of energy value of production by-products (which are used mainly as animal feed). In 2012, the United States had more than 200 ethanol plants in operation or under construction. The newer plants are designed to use coal for power, because coal is much less expensive than natural gas. The feedstock for almost all of the plants is corn or corn and milo, although 10 ethanol plants use beer waste, wood waste, or food-processing waste as feedstock. It is now recognized that ethanol made in natural gas-powered plants increases greenhouse gas emissions by 5%, compared with petroleum-based gasoline. For ethanol from coal-powered plants, the gain in greenhouse gas (GHG) is a whopping 34%. Furthermore, corn-based ethanol may have a negative EROI, although some experts report a low EROI value of about 1.3:1. Sugar-cane ethanol's EROI ranks higher, about 5:1 or 6:1.

Most early studies of biofuel as a substitute for gasoline show a reduction of GHG, because biofuels sequester carbon through the growth of feedstock. But these studies fail to account for the carbon emissions that occur as farmers worldwide respond to higher prices by converting grassland and forest to new croplands, needed to replace the grain diverted to biofuels. Researchers at Iowa State University found that corn-based ethanol, instead of producing a 20% savings in GHG emission, as reported in earlier studies, nearly doubles the GHG emissions over 30 years and will increase GHG for 167 years. In recent years, farmers growing corn have benefitted from federal subsidies of about $6 billion a year, all in the name of ethanol as a gasoline additive. Those subsidies ended in 2012 when the companies producing ethanol lost a tax credit of 46 cents a gallon. What the ethanol industry hopes is that the tax credit will continue for ethanol made from switchgrass, wood chips, and corn stalks and leaves. Known as *cellulosic ethanol*, it is not yet on the market because of high research, development, and production costs.

Biodiesel—Biodiesel, an alternative fuel produced from animal fats, used cooking oils, plant oils, or algae, is cleaner than other fuels. Biodiesel contains no petroleum, but it can be blended at various levels with petroleum diesel. Biodiesel blends can be used as a fuel in most compression-ignition (diesel) engines with no modification. Pond scum (algae) may have a role to play in producing biofuels, especially biodiesel, and in fighting global warming.

At least 15 small start-up companies are working on the idea of algae cultivation of ethanol and biodiesel from CO_2 from coal-fired power plant emissions and for reducing CO_2 emissions. Algae cultivation looks attractive because it reproduces four times as much ethanol as the same weight of corn. Essentially, emissions from coal-fired power plants are routed to an algae bioreactor, where carbon dioxide and other pollutants are absorbed and utilized by the algae to grow at an exponential rate. Once harvested, the algae can be processed to produce gases, such as methane; biofuels, such as biodiesel and ethanol; and a variety of solids, such as bioplastics and proteins. Biodiesel has a low EROI, in the range of 1.5:1, but that is offset by its high marks in cutting GHG emissions: Carbon dioxide is cut by about 40%, and nitrogen oxides are reduced by up to 85%. One or more firms are experimenting with collecting coastal algae blooms—oxygen-sucking, fish-killing masses of algae in the ocean, whose growth is stimulated by fertilizer runoff from farm fields—to produce biofuels.

Methanol—Methanol (methyl alcohol) is a common industrial chemical that has been used as an alternative blended-liquid transportation fuel, and it is under consideration for wider use. Methanol uses natural gas (methane) as a feedstock, but the United States has reserves of coal that also can be used as feedstock. West Virginia and Montana already have systems in place for making coal-derived methanol. One company, Norcal Waste Systems in San Francisco, California, is even developing a process for making methane feedstock from dog droppings. (For many years, San Francisco dogs have been a major producer of the stuff, wastefully compelling naïve visitors to the city to carry it away, stuck to their shoes.) Methanol can be used directly as fuel; Indy race cars and model airplanes have used it for years. Methanol is an interesting fuel: Its combustion generates little nitric oxide (NO) and ozone (O_3), and it has a high octane number—about 110—which has made it attractive for race cars and high-compression engines. Although it has lower heat (Btu) content than gasoline, it has only about 50% of the energy content of an equal volume of gasoline in contrast with ethanol's 67%, and it burns more efficiently, which may even things out. On the negative side, methanol releases greater emissions of CO_2, and it has a very low EROI, with some authorities claiming a negative EROI. Methanol can be used in the process of making biodiesel, although biodiesel is currently made from soybeans and other feedstocks. Challenges facing all of the biofuel contenders in the race are the ability to cost-effectively collect, store, and transport biomass feedstocks and the development of a sustainable and safe supply system.

Kicking the Carbon Habit

We will never have a competitive alternative for carbon-based fuels—oil, coal, and natural gas—especially at the scale of today's consumption. Contrary to popular assertions among the "green" crowd, we simply do not have the technology in place for renewables to provide that much cost-effective, carbon-free power. Some people feel we have the technology and all we need are government-based policies to make the transformation to renewables. But it's not just a question of technology; it's a question of cost and the magnitude of the transition. So, what do we do? First, there's the matter of energy efficiency. As pointed out by Nathan Lewis:

> It's much cheaper to save a joule of energy than it is to make it, because of the losses all along the supply chain are such that saving a joule at the end means you save making, say, five joules at the source. So lowering demand with energy-efficient LED lighting, fuel cells, "green" buildings, and so on is going to pay off much sooner than clean energy supplies. On the other hand if we save as much energy as we currently use, combined, we will still need to make at least as much carbon-neutral energy by 2050 as we currently use [because of global population growth], combined, merely to hold

> CO_2 levels to double where they are now. That's the scale of the challenge.[9]

This section of the chapter will deal with carbon-neutral and carbon-free energy sources. We begin by examining nuclear energy.

Nuclear Energy

At one time, nuclear energy was considered *the* solution to the world's future energy needs—being clean; having no CO_2 emissions or blots on the landscape other than an occasional cooling tower and containment dome; and generating power "too cheap to meter." More than 400 nuclear power plants were built worldwide, 109 of them in the United States (5 U.S. nuclear plants are now shut down).

Globally, in January 2011, 31 countries operate 436 one-gigawatt (GW) nuclear power plants with a net installed capacity of about 377 GW, an amount that can provide electricity for approximately 1,700,000 average homes.[10] Another 63 plants, with an additional installed capacity of 61 GW, are under construction in 15 countries. Nuclear energy accounts for 14% of the planet's electrical power (**Figure 14.31**). The United States gets nearly 20% of its electrical energy from 104 operating nuclear plants. Some countries have gone heavily into nuclear. France, with 58 reactors, obtains more than 74% of its power from nuclear fission, and China, facing a shortage of electricity, is building new reactors at a rapid rate—one or two a year. It is estimated the world needs about 10 terrawatts (TW) of electricity to maintain business as usual; to get that means we would need to build 10,000 of our current 1-GW reactors. In other words, a new nuclear reactor would need to come online every other day for the next 50 years. In the 5 years since the publication of the previous edition of this textbook, we should have built some 1,000 new reactors, or twice as many as have ever been built.

However, the allure of nuclear power began to tarnish with the 1979 Three Mile Island event in the United States, along with the Chernobyl accident that followed in April 1986. The Fukushima Daiichi, Japan, nuclear power plant disaster, the largest nuclear disaster since Chernobyl, exacerbated concerns and further raised the specter of danger from nuclear electricity generation. The Fukushima disaster was the result of equipment failures, nuclear meltdowns, and the release of radioactive material following the four Tohoku earthquakes of M_w 6+ and 7.2, and the tsunami of 11 March 2011 (see Chapter 4). It so fueled apprehension and fear of nuclear power that within days of the catastrophe, Germany experienced large antinuclear protests with the result that in June 2011, Chancellor Angela Merkel shut down 7 of Germany's 17 reactors that had been online since 1981. She also imposed a 3-month moratorium on planned extensions of the country's existing nuclear plants. In addi-

[9]Lewis, Nathan S. 2007. Powering the planet. *Engineering and Science* 70 (2): 12–23.

[10]This value assumes 22 kW per household; 377 GW ÷ 22 kW/home = 17,000,000 homes.

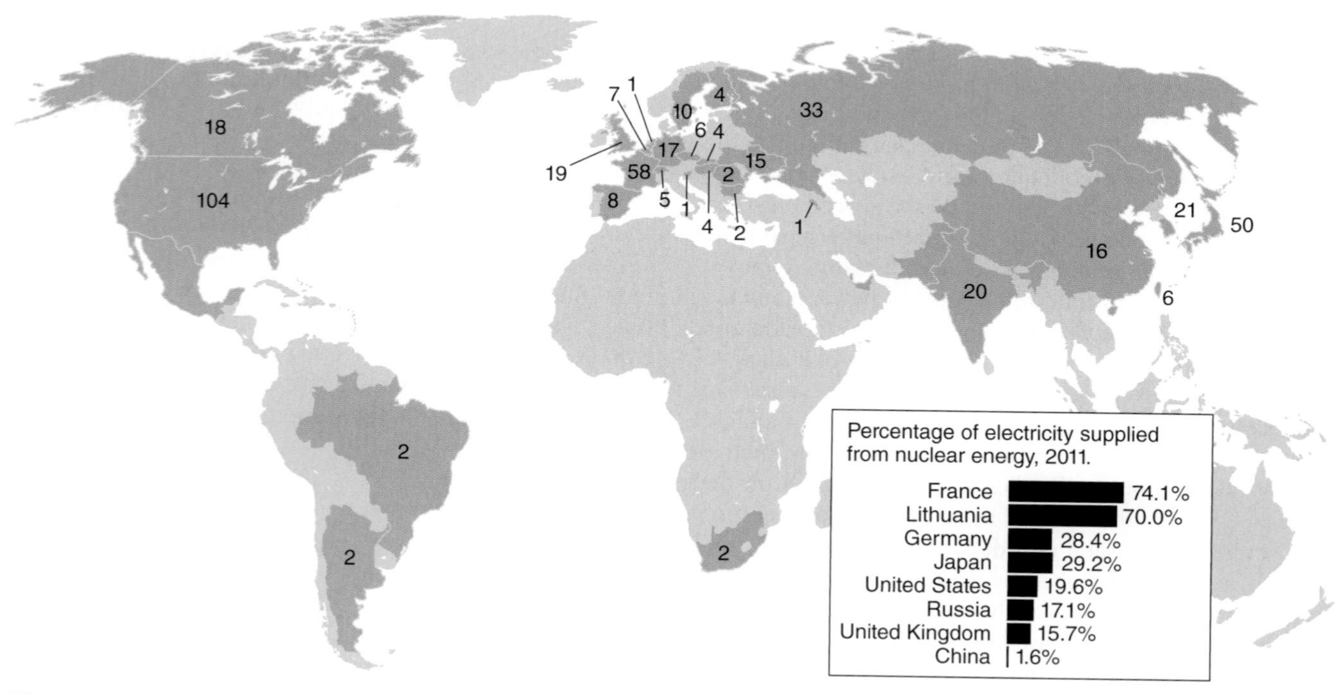

Figure 14.31 The worldwide number of operating nuclear plants, January 2011, and the percentage of electricity supplied by nuclear reactors for selected countries. Note France is the world's most nuclear-energy-dependent nation, with 58 reactors generating over 74% of its electricity. In 2012, 30 countries had 436 nuclear plants, with 65 new plants under construction in 14 countries. (Source: International Atomic Energy Agency)

tion to these nuclear disasters is the challenge of radioactive waste disposal, nuclear's poor economics compared with fossil-fuel–powered plants, and the ongoing threat of terrorist attacks. All of these concerns have reinforced people's anxieties about nuclear power. These concerns have led to cancellation of orders for new "nuke" plants in the United States, and in June 2011, Germany and Switzerland announced the eventual phase-out of nuclear power in their countries. Other countries—Australia, Austria, Denmark, Ireland, Liechtenstein, Portugal, Israel, Norway, and others—remain opposed to any nuclear power.

Nuclear reactors and fossil-fuel plants generate electricity by similar processes. Both heat a fluid, which then directly or indirectly makes steam, which spins turbine blades to drive an electrical generator. Uranium is the fuel of atomic reactors, because its nuclei are so packed with protons and neutrons that they are capable of sustained nuclear reactions. All uranium nuclei have 92 protons and between 142 and 146 neutrons. The common isotope uranium-238 (^{238}U) has 146 neutrons and is barely stable. The next most common isotope, ^{235}U, is so unstable that a stray so-called slow neutron penetrating its nucleus can cause it to split apart completely, a process known as **fission**—a term particle physicists borrowed from the biological sciences. Nuclei that are split easily, such as those of ^{235}U, ^{233}U, and plutonium-239 (^{239}Pu), are called **fissile isotopes**. Fission occurs when a fissile nucleus absorbs a neutron and splits into lighter elements, called *fission products*, while emitting several "fast" neutrons and energy, 90% of which is heat (**Figure 14.32**). The lighter fission products are elements that recoil from the split nucleus at high speeds. This energy of motion and subsequent collisions with other atoms and molecules creates heat, which raises the temperature of the surrounding medium.

Meanwhile, the stray fast neutrons collide with other nuclei, repeating the process in what is called a **chain reaction**. A controlled chain reaction occurs when one free neutron (on average) from each fission event goes on to split another nucleus. If more than one nucleus is split by each emitted neutron (on average), the rate of reaction increases rapidly, and the reaction eventually goes out of control. Because ^{235}U makes up only 0.7% of all naturally occurring uranium, the rest being ^{238}U, it must be enriched for use in commercial nuclear reactors.

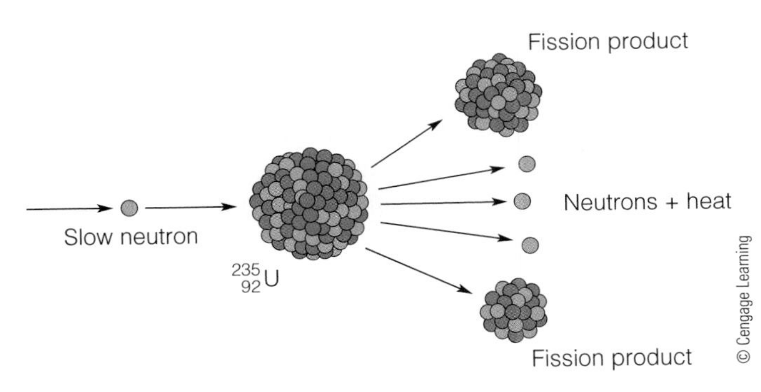

Figure 14.32 Nuclear fission. A slow neutron penetrates the nucleus of a U-235 atom, creating fission products (also called *daughter isotopes*), two or three fast neutrons, and heat.

The major concerns associated with nuclear energy are contamination during processing, transportation, and the disposal of high-level radioactive waste products. Other problems are the potential for nuclear weapons proliferation, terrorists, the decommissioning of antiquated reactors, radiation leaks from nuclear power plants that cannot be totally prevented, and nuclear fuel supply limitations. Notwithstanding, the Nuclear Regulatory Commission, the nuclear power industry, the media, and even some environmental groups promote nuclear energy as a clean and safe alternative to fossil fuels. Nuclear power proponents see it as "the one demonstrably practical technology that could decisively shift U.S. carbon emissions [away from burning fossil fuels] in the near term" and slow global warming. The EROI of nuclear power ranges from 5:1 to about 18:1.

It is important to recognize that the insufficiency of known uranium supplies to fuel the world's current nuclear reactors argues against business-as-usual nuclear plants. The average grade of uranium ore mined today is only 0.15%, and the average grade can only decrease through time. We could obtain uranium from seawater, but that would require the energy equivalent of 3,000 Niagara Falls for 24 hours, 7 days a week. In our current reactors, the fissionable material is run through only once. If more nuclear energy is on the horizon, we will need to use the only credible nuclear energy source, plutonium, which is generated in **breeder reactors**. Breeder reactors generate new fissionable material at a greater rate than they consume their initial fuel—a normal reactor consumes only about 1% of the initial fissionable material. A breeder reactor uses essentially the entire amount of initial fissionable material. Furthermore, breeder reactors can utilize thorium, which is more abundant than uranium, and because of greater efficiency, breeders generate much less radioactive waste. Because breeder reactors produce plutonium-239 (^{239}Pu), which is used to make nuclear bombs, there is fear that some of the ^{239}Pu could be stolen and end up in the hands of terrorists. Because of this, in 1977, President Jimmy Carter signed an order banning the reprocessing of nuclear fuel that would have required using breeder reactors. Nuclear scientists point out that it is nearly impossible to separate sufficiently pure ^{239}Pu from the other fissionable isotopes for use in making nuclear weapons and such concern is unwarranted. The best scientists in France, Russia, and the United States have tried to do so and have failed—anything short of pure ^{239}Pu in a nuclear weapon will fizzle. This fact is never discussed by politicians.

The Energy Policy Act of 2005 provided $6 billion in subsidies for incentives to build new nuclear power plants in the United States. As H. G. Wilshire and colleagues point out:

> Some environmentalists support the nuclear option to reduce greenhouse emissions and global warming. But producing nuclear power requires reactor building,

maintaining, securing, and decommissioning; waste repository construction, maintenance, and security; and shipping wastes long distances for disposal. All together, these processes and activities release enough greenhouse gas to cancel out the low CO_2 benefit of nuclear-generated electricity.[11]

Geological considerations—One of the many criteria in the siting of nuclear reactors is geological stability. The selected site must be free from landsliding, tsunamis, volcanic activity, flooding, and the like. The Nuclear Regulatory Commission (NRC), the federal government body that licenses nuclear reactors for public utilities, requires that all active and potentially active faults within a distance of 320 kilometers (200 miles) of a nuclear power plant be located and described. Reactors must be built at a distance from an active fault, determined by the fault's earthquake-generating potential. Further, the NRC defines an active fault as one that has moved once within the last 30,000 years or twice in 500,000 years. This is a very strict definition, and sites that meet these criteria are difficult to find along active continental margins. To provide additional safety and allow for geological uncertainty, reactors are programmed to shut down immediately at a seismic acceleration of 0.05 *g*.[12] This is conservative, indeed, and the shutdown would prevent overheating or meltdown, should the earthquake be a damaging one.

Finally, nuclear power is far from renewable, because the readily mineable uranium mineral deposits won't last much more than 50 years.

Hydroelectric Energy

Falling water, our largest renewable resource next to wood, has been used as an energy source for thousands of years. First used to generate electricity about 100 years ago on the Fox River near Appleton, Wisconsin, today it provides a fourth of the world's electricity. In Norway, 99% of the country's electricity and 50% of its total energy are produced by falling water. The principle is relatively simple: Impound water with a dam and then cause the water to fall through a system of turbines and generators to produce electricity. Because of this simplicity, electricity coming from existing facilities is also the cheapest source of this power. China's Three Gorges Dam on the Yangtze River is the world's largest producing electrical power station. It generates 18,300 megawatts (MW) and is expected eventually to produce 22,500 MW. (1 MW is enough electricity to power 250 to 300 homes.) The world's second largest source of hydropower, the Aitutu Dam on the Paraná River between Paraguay and Brazil, completed in 1982, produces 14,000 MW. It provides much more electricity than the present demand in this part of South America.

Because of land costs and environmental considerations, it is doubtful that any more large dams will be built in the United States. Therefore, dams that were built for other purposes have become attractive for retrofitting with generators. Hydroelectric facilities are particularly useful for providing power during times of peak demand in areas where coal, oil, or nuclear generation provides the base load. Hydroelectricity can be turned on or off at will to provide peak power, and many utilities have built pumped-water-storage facilities for just that reason. During off-peak hours, when plenty of power is available, water is pumped from an aqueduct or another source to a reservoir at a higher level. Then during peak demand, the water is allowed to fall to its original level, fulfilling the temporary need for added electricity.

Although hydroelectric energy is clean, dams and reservoirs change natural ecological systems into ones that require extensive management. One of the social consequences of dam building is displaced persons. This displacement is unacceptable in most societies. China's Three Gorges Dam set the record for number of people displaced (more than 1.2 million) and number of communities flooded (1,350 villages, 140 towns, and 13 cities). In addition, the project has been plagued by spiraling costs, corruption, technological problems, landslides triggered by the reservoir's rising waters, and resettlement difficulties. Also unacceptable are recurring attempts to dam areas of great aesthetic value, such as the Grand Canyon. Finally, dam failures—most often, geological failures of the dam's foundations—result in floods that can take enormous human and economic tolls. Nevertheless, hydropower is attractive, with EROI estimates ranging from 20:1 to 40:1, based on uncertain data. This places hydropower in a favorable position relative to conventional power-generation technologies, but it is doubtful that there is much opportunity for expanding hydropower, as the world's most favorable sites have already been developed.

Geothermal Energy

Earth's interior is an enormous reservoir of heat produced by the decay of small amounts of natural radioactive elements that occur in all rocks. When the heat rises to shallower depths, this **geothermal energy** can be obtained for human use. Geothermal energy can be used for electricity generation, space heating, and heat pumps. Unlike other renewable energies, such as solar or wind, geothermal can function nearly 24 hours a day, 7 days a week. Geothermal is ideal for providing nearly zero-emission renewable base-load electricity.

The 3,186 MW of geothermal capacity in the United States in 2012 (**Table 14.5**) amounted to about 8% of the nonhydropower renewable electric-generating capacity, but only about 0.2% of the nation's total electric-generating ca-

[11]Wilshire, H. G., J. E. Nielson, and R. W. Hazlett. 2008. *The American West at risk: Science, myths, and politics of land abuse and recovery*. New York. Oxford University Press.

[12]Seismic acceleration is the acceleration value of the ground or a structure during an earthquake. An acceleration of 1 g is equal to the acceleration of a falling object due to Earth's gravity.

Table 14.5 U.S. Geothermal Power Capacities Online, By State as of March 2012

State	Electrical Capacity (MW)
California	2615
Nevada	469
Utah	42
Hawaii	43
Idaho	16
Alaska	0.73
Wyoming	0.25
New Mexico	0.24
Total	3186.2

Source: *U.S. geothermal power production and development update.* March 2012. Geothermal Energy Association.

pacity. Geothermal operations contribute about 5% of California's electricity and 20% of Hawaii's. Globally, geothermal energy is used commercially in 70 countries, and geothermal capacity worldwide is about 11,225 MW.

Economists argue that technology will overcome depletion. Geologists argue that depletion will overwhelm technology.

Geothermal is cleaner than many other energy sources, and there are fewer negative environmental consequences of using it. Not surprisingly, the prospects for using geothermal energy are best at or near plate boundaries where active volcanoes and high heat flow are found. The Pacific Rim (Ring of Fire), Iceland on the Mid-Atlantic Ridge, and the Mediterranean belt offer the most promise. Today, the United States, Japan, New Zealand, Mexico, and countries of the former USSR are utilizing energy from Earth's heat. Iceland uses geothermal energy directly for space heating (one geothermal well there actually began spewing lava), and many other countries have geothermal potential. The versatility of Earth heat ranges from using it to grow mushrooms to driving steam-turbine generators (**Figure 14.33**).

Geothermal energy fields may be found and exploited where magma exists at shallow depths and where sufficient underground **hydrothermal** (hot-water) systems exist to form steam. These systems are classified as follows:

- *Hot-water hydrothermal systems*, with temperatures of about 200°C (390°F) or more, produce steam at sufficient pressure to drive turbine generators. A classic example is the four power plants operating at the Coso geothermal field in south-central California.
- *Vapor-dominated hydrothermal systems* are those in which the pore spaces in rocks of the high-temperature system are saturated with steam, rather than liquid water. In this instance, only steam is produced and is routed directly to turbine generators. Vapor-dominated systems are rare, the most efficient, the simplest, and the most desired for electrical power production. The Geysers, near Santa Rosa, California, with 21 operating power plants, supplies about 20% of California's renewable electrical energy and is the largest vapor-dominated system in the world (**Figure 14.34**).
- *Moderate-temperature hydrothermal systems* are incapable of producing steam at high enough pressure to drive a turbine generator directly. They utilize geothermal heat to vaporize a second, "working" fluid (e.g., isobutene) whose boiling temperature is lower than that of water. The vapor from the second fluid drives the turbine generator. Such a system is called a **binary-cycle system** or, simply, a binary system. A successful binary-cycle system operates near Mammoth Lakes, California.

Geothermal energy is clean. The amount of CO_2 emitted is about one tenth that emitted by coal-fueled plants per

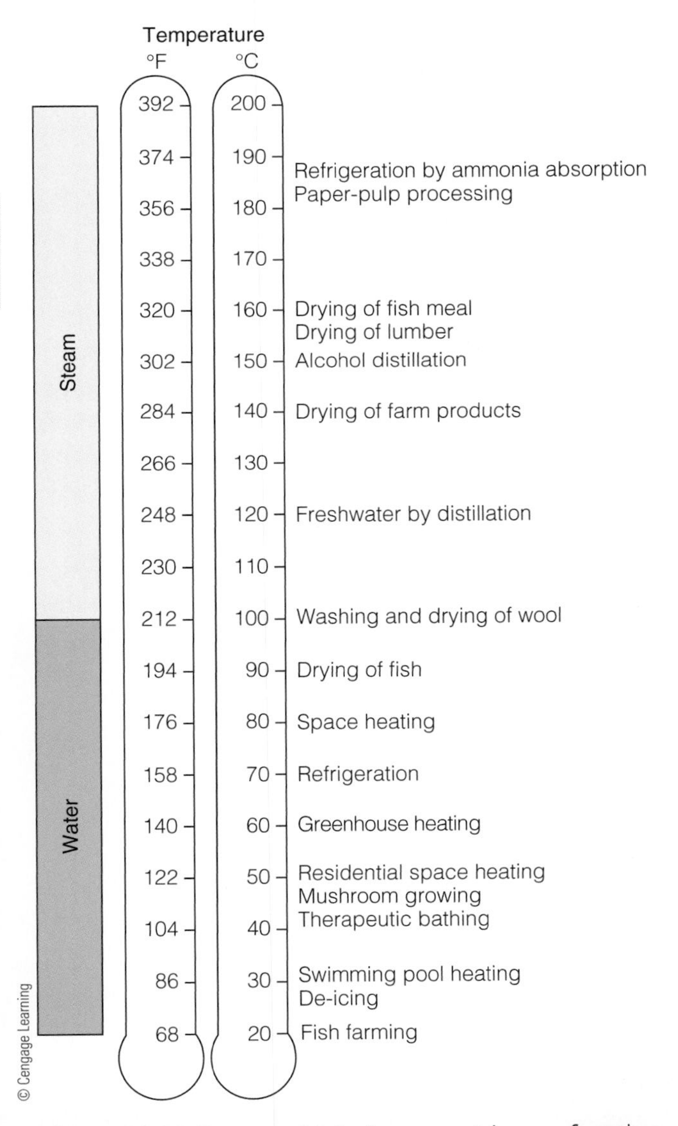

Figure 14.33 Some established commercial uses of geothermal energy.

explorer's perspective

Photograph by Sofia Ruzo/See http://www.nationalgeographic.com/explorers/bios/andres-ruzo/

ANDRÉS RUZO is seeking to reveal Peru's hidden energy resource to help solve global energy problems.

Andrés Ruzo grew up on a farm in Nicaragua that is on Casita Volcano. Later, as a student at Southern Methodist University (SMU), his childhood memories of living on a volcano inspired him to take a volcanology course. The course inspired him to become a geologist, especially to learn about Earth's heat, which eventually led him to pursue a Ph.D. at the SMU Geothermal Laboratory.

Andrés envisions using Earth's heat to generate electrical power to help solve energy issues. He states that "energy is a kingpin problem. By solving our energy issues, we simultaneously take care of other major world problems."

Andrés is now doing fieldwork and research to produce a detailed geothermal map of northern Peru. Fieldwork consists of lowering a thermometer down abandoned oil and gas wells to measure downhole temperatures, this in a desert region where day time temperatures can reach above 54°C (130°F) and vultures are their regular companions. He notes that energy and the environment seem to have become political issues that stir emotional responses that are rooted in political ideology. However, Andres perceives that "being 'green' is not an issue of 'left' or 'right.' It's a human issue." Furthermore, his view is that reusable energy lies in the hands of the petroleum industry. "Because the development of geothermal energy uses virtually the same technology and personnel as the oil and gas industry," it will be relatively easy to make the transition. He states that "for many, it may be surprising to learn that Chevron is the world's largest producer of geothermal energy."

By working to develop geothermal energy he is not only fighting for the environment itself, but addressing a problem with far reaching impacts, including the future security of nations and their international relations.

Source: http://newswatch.nationalgeographic.com/2012/05/31/revealing-perus-hidden-energy-source/?source=link_twex20120601ruzo-blog

A panoramic view of the Talara Desert near Los Organos, Piura, Peru.
Photo courtesy Sofia Ruzo

Figure 14.34 One of several geothermal power plants at The Geysers, Sonoma County, near Santa Rosa, California. The Geysers is a vapor-dominated system in which only steam is produced and is routed directly to turbine generators. It is the largest vapor-dominated system in the world and produces about 20% of California's renewable energy.

megawatt of electricity produced, and about one sixth that of relatively clean-burning natural gas. Similar contrasts exist for emissions of sulfur and sulfur oxides. The wastewater from these geothermal energy systems is injected back into the subsurface to help extend the useful life of the hydrothermal system.

The problems associated with geothermal energy production are related to groundwater withdrawal and water quality. Some geothermal waters contain toxic elements, such as arsenic, boron, and selenium, and heavy metals, such as gold, silver, and copper. In some geothermal areas, the removal of underground water can cause surface subsidence and perhaps even earthquakes. At The Geysers, subsidence of 13 centimeters (5 inches) has been measured without noticeable impact. Reinjecting the cooled wastewater into the geothermal reservoir decreases the subsidence risk and helps ensure a continued supply of steam to the electrical generators. All geothermal fields in California are monitored for possible seismic activity related to steam production. Small quakes have been reported from The Geysers, but none of significant magnitude.

The future of geothermal energy looks good, especially with an estimated EROI of perhaps 7:1; it is competitive with photovoltaic and sugarcane ethanol. About 20 geothermal fields in the United States generate electricity, and direct-use hydrothermal systems are being developed at more than 30 sites. Electricity was first produced from geothermal energy in 1904 in Larderello, Italy, where the initial output illuminated four lightbulbs. In just over 100 years, geothermal energy production grew to nearly 10,700 MW in 24 countries around the world. The United States alone produces over 3,100 MW of electricity from geothermal energy. This is equivalent to burning 68 million barrels of oil each year.

Geothermal energy can be used over a wide range of temperatures and volumes—in contrast with energies that can be utilized only when the resource exceeds some minimum size and grade for profitable operation. For example, at the low end of the range, geothermal energy can help cool or heat a single building. To do so requires only the burial of piping a few meters deep, where there is little seasonal temperature change through the year. By using a geothermal heat pump to circulate water or some other fluid through the piping, thermal energy can be deposited in the ground during the hottest season and extracted during the coldest season. A heat pump and Earth's thermal energy can effectively and economically provide a viable cooling and heating system. Such systems are currently operating in about 350,000 buildings in the United States.

Direct Solar Energy

Although the amount of solar energy that reaches Earth far exceeds all human energy needs, it is very diffuse. For example, the amount that strikes the atmosphere above the British Isles is 80 times Great Britain's energy needs; it averages about 1 kilowatt per square meter (10.8 square feet). Cloud cover reduces the amount by 80%, but there is still more than enough, so why aren't there solar-electrical plants in every city and hamlet? The reasons are cost and space. There are essentially three ways to put the Sun to work for us:

1. **Solar-thermal** methods, which use either power towers or parabolic-trough systems, both of which heat a fluid, which is then used for generating electricity
2. A combination dish/engine concentrator that converts the Sun's energy to generate electricity
3. **Photovoltaic** (PV) cells, which use various semiconductor materials to convert sunlight directly into electricity

Power towers use thousands of sun-tracking mirrors (or heliostats) to focus the Sun's heat on a central receiver mounted atop a tower. The receiver collects the reflected heat in a working fluid—either using water for high-temperature steam or molten salt—which then powers a turbine generator to produce solar electricity. The feasibility of solar power towers was demonstrated in the 1980s and 1990s in the Mojave Desert near Barstow, California. In August 2009, a 5-MW power tower system began operating in

Lancaster, California (**Figure 14.35**) and supplies power to as many as 4,000 homes during peak generation. This is the only commercial power tower system operating in North America to feed electricity into the power grid.

Another solar-thermal method uses **parabolic troughs**. Parabolic mirrors focus the Sun's rays on a fluid-filled heat receiver pipe (**Figure 14.36**). The heated fluid is routed to heat exchangers, where water vaporized into steam powers a conventional turbine-powered generator. Nine such trough power plants now operate in the Mojave Desert, with a combined capacity of 354 MW, making this system the world's largest solar-power operation. These nine plants are hybrids, designed to operate on natural gas after dark or on overcast days, with gas generating about 25% of the total output. A competing technology has been developed that replaces the parabolic mirrors with an inexpensive, thin, reflective low-cost film that is more durable than standard glass.

A third type of solar-thermal system consists of a **dish/engine** concentrator. Thousands of small, computer-controlled mirrors reflect the sunlight to a solar tower, which transfers the Sun's energy to a **Stirling cycle heat engine**, which converts the energy into electricity. Of all the practical solar technologies, the dish/engine system has the highest efficiency (about 31%), in contrast with the trough solar-thermal system (between 15% and 19%) and PV (between 8% and 15%). The Stirling engine system has another advantage: The engines are air-cooled, not water-cooled, as are the other solar-thermal systems. A 1.5-MW demonstration dish/engine solar plant began operating at Scottsdale, Arizona, in January 2010, but could not compete with the falling costs of solar PV and filed for bankruptcy the following September.

Power generated by all of the large solar-thermal systems has an EROI of about 4:1. Power tower electricity costs about 14 cents per kilowatt-hour (kWh), which is close to the cost of natural gas plants (about 12 cents per kWh). Currently, pulverized-coal plants generate electricity for about 6 cents per kWh. If future coal-fired power plants are required to capture and sequester carbon, however, the cost will increase to about 10 cents per kWh, which is still cheaper than power tower electricity.

As potentially attractive as these large solar-thermal systems are, they come with environmental issues. They take up large amounts of land and require total removal of the native vegetation, which disrupts or kills wildlife. Water usage is probably the largest issue. Solar-thermal energy plants are similar to coal- and natural-gas–fueled power plants in that they need massive amounts of water to convert to steam to power the turbine generators and to cool water. In the arid lands of California, Arizona, and Nevada, water is already in short supply, with its use already overappropriated in some areas. These solar-thermal installations in remote areas of the desert will require new long-distance transmission lines, and these corridors will impact desert wildlife—especially desert tortoises, which are on the federal endangered list. Promoters contend that the best use of desert land is for human activity. Environmental opponents maintain that changing desert lands is an irreversible commitment of land—the "barren desert" may seem barren, but it supports an ecosystem. Instead, they assert that PV panels installed on the roofs of houses or other structures would satisfy both energy needs and environmental concerns. Additionally,

D. D. Trent

Figure 14.36 A parabolic trough solar energy system in the Mojave Desert, California. The parabolic reflectors are computer-controlled to track the sun, and the reflected sunlight is focused on heat-receiver tubes to heat a fluid. The hot fluid is routed to heat exchangers where water is vaporized to power turbine-powered electric generators.

Figure 14.35 The Sierra SunTower power tower system operates in Lancaster, California, and produces 5 MW, enough electricity to supply as many as 4,000 homes during peak generation.

eSolar

they would eliminate the need for the long, public environmental review process for power plant siting and location of new transmission line corridors. Financing PV on private homes could be accomplished with a rooftop solar loan program and **feed-in tariffs** (an incentive system to promote solar, wind, and other renewable energy systems), such as already exist in parts of Germany; Vermont; Gainsville, Florida; Sacramento and Palm Desert, California; and elsewhere (see Case Study 14.5).

PV cells absorb pulses of light energy on semiconductor materials that turn light energy into an electric current. They are used widely as the power source for satellites, calculators, watches, and remote communications and instrumentation systems. The early PV cells utilized silicon as the energy converter, but cadmium telluride (CdTe), copper indium gallium selenide (CIGS), and other, newer types of PV cells are more efficient and have an EROI of about 7:1.

Nationwide, there is increasing interest in *distributed PV programs*, in which utility companies invest in company-owned PV programs. One major utility is installing up to 10 MW of roof- and ground-mounted PV systems on space leased from businesses, homes, and utility-owned property. The PV system routes all of the solar-generated electricity into the regional power grid, and the property owners receive financial compensation. For example, a PV array installed at the California Institute of Technology in Pasadena, California, generates 320,000 kWh per year and earns a $0.632 per kWh rebate from the local electric utility (**Figure 14.37**). One of the largest cost-effective, city-owned PV systems in the United States is installed at San Francisco's Moscone Convention Center. The system, combined with savings from various energy-efficiency measures, delivers the equivalent energy to power approximately 8,500 homes.

In 2012, the electricity produced from PV cells cost an average of about 16 cents per kWh, more than that produced by fossil fuels. Natural gas, for example, costs about 12 cents per kWh, depending on location. According to the EIA, the average retail rate of conventional electrical energy per kWh varies from 6.1 cents in Wyoming to 18.1 cents in Connecticut. Moreover, the costs of installed PV systems and solar panels have been decreasing at about 15% per year, with the price of solar panels dropping 30% since 2010 because of increasing efficiency of solar cells and major improvements in manufacturing technology. Competition between manufacturers also, has driven down prices with the PV solar industry, growing at a rate of about 25% per year, even during the recession years since 2008. It is estimated that PV electricity costs will continue to decline with technological improvements and within 3 to 5 years will be cheaper than electricity generated by fossil fuels and nuclear reactors (**Table 14.6**). In 2010, the PV industry was the fastest growing segment of the nation's economy, with more than 100,000 Americans employed in the industry, twice the number in 2009.

Figure 14.37 A photovoltaic array atop a parking structure at the California Institute of Technology in Pasadena, California. The array produces 320,000 kWh per year, earning a $0.623 per kWh rebate from the local electric utility.

California Institute of Technology

Table 14.6 Cost of Electricity by Energy Source

Energy Source	Approximate Cost Cents/KWH
Sun	12–17
Wind	5–8*
Nuclear power	15
Conventional coal	6
Natural gas	12
Gasified coal with CCS†	16–18
Photovoltaic cells‡	10–22

Note: The price of electricity varies greatly across the United States. Demand moves from east to west with the Sun, following human activity and afternoon peaks in air-conditioning loads. On many days, some utilities are burning expensive natural gas as they strain to meet peak demand, whereas others have cheap capacity standing idle. Often, wholesale electricity is sold in one region for 20% to 50% less than others are paying elsewhere.

*Wind electricity cost varies with location of the wind farm.

†Carbon capture and sequestration (CCS) is a technology that has yet to be tested on a large scale.

‡Costs depend on the size of the system, and the geographical locality (because the amount of solar radiation received varies with the average cloudiness, Sun angle, and the path of the Sun relative to the orientation of the panel). Cost at a Phoenix, Arizona, 10-MW PV installation, for example, ranges from 15 to 22¢/kWh.

PV cells have a major environmental issue. They use toxic chemicals, such as gallium arsenide, CdTe, CIGS, and cadmium sulfide, in their manufacture. The disposal of used cells could become a significant environmental problem, because the chemicals persist in the environment for hundreds of years. Although not as efficient, low-cost silicon cells are environmentally less toxic.

Indirect Solar Energy

Wind energy—Wind power is the biggest success story in renewable energy. Wind power generates electricity when the wind turns rotors (essentially, windmills) attached to a turbine. Globally, it is the fastest-growing energy segment, increasing installed capacity by a factor of 32 from 6,100 MW in 1996 to 194.4 gigawatts (GW) in 2010 (see

Figure 14.41). (1 GW = 1 billion watts.) During that time, the cost of wind power dropped by over 90%, and present production pales in comparison with what electrical wind generation will be in 20 years. The DOE envisions that, by 2030, wind could satisfy 20% of the nation's energy needs.

The costs of wind generation are a function of wind velocity, turbine design, and the diameter of the rotor. First, the energy that can be tapped from the wind is proportional to the cube of the wind velocity. For this reason, a slight increase in wind speed yields a large increase in electricity produced. For instance, an increase of wind speed from 14 to 16 miles per hour increases electrical generation nearly 50%. Second, longer blades requiring a higher tower yield greater production. Increasing the blade diameter from 10 meters (32.8 feet) (1980s technology) to 50 meters (164 feet) (2000 technology) gives a 55-fold increase in yearly electrical production. The state-of-the-art "smart" wind turbines (2011 technology) may stand more than 120 meters (394 feet), are rated at 2.3 MW, and have rotor diameters that exceed the length of a football field (**Figure 14.38**). GE's giant 3.6 MW series, designed for offshore installations, has a rotor diameter of 111 meters (341 feet). The latest model turbines are programmed to adjust automatically for varying wind conditions, and they produce electricity at prices competitive with other sources (see Table 14.6). Depending on location and other factors, the estimate EROI for wind power may range from 10:1 to 30:1, and EROIs for windy regions, such as Denmark's islands, may be even higher.

In the past, California has led the nation in wind power, and the world's first large wind farms were constructed in the state at three locations: Tehachapi Pass, southeast of Bakersfield; San Gorgonio Pass near Palm Springs; and Altamont Pass east of San Jose (**Figure 14.39**). California is no longer the leading state for wind power—both Texas (10,223 MW) and Iowa (4,495 MW) now surpass California (3,927 MW). Since 2006, wind power in California increased by only 8%, whereas wind power in Texas and Iowa increased by approximately 160% and 200%, respectively. California's negligible wind-power growth is attributed to the high cost of obtaining state permits, as well as regulators who are concerned about the deaths of birds and bats that fly through the windy passes where wind farms are located. Nevertheless, wind provides about 5% of California's energy needs.

Wind accounts for about 4% of all energy used in Texas, and the state even exports wind to other states. In 2011, Texas produced 10,223 MW from more than 40 wind farms. One of these, the Roscoe Wind Farm (781 MW), is the world's largest wind farm. Among other large Texas wind farms are the Horse Hollow Wind Energy Center, Buffalo Gap Wind Farm, and Wildorado Wind Ranch.

The wind potential of the lower 48 states is shown in **Figure 14.40**. One of the most attractive features of wind is that it can produce electricity during peak demand periods, on hot summer days, and early in the evening in winter, thus avoiding the construction of additional fossil-fuel plants to meet the need. However, there are environmental problems that accompany wind generation, such as noise, land acquisition, TV interference, and, most important, visual blight. Modern generators have largely solved the problem of noise and TV reception, but a couple of hundred windmills are hard to ignore. The feeling is that they destroy the scenic values on public lands. Other environmental issues are that large hillside wind farms, such as those in California, also become major road-building projects that exacerbate erosional processes. Constructing access roads and turbine pads disturbs large acreages. Wildlife, too, is affected.

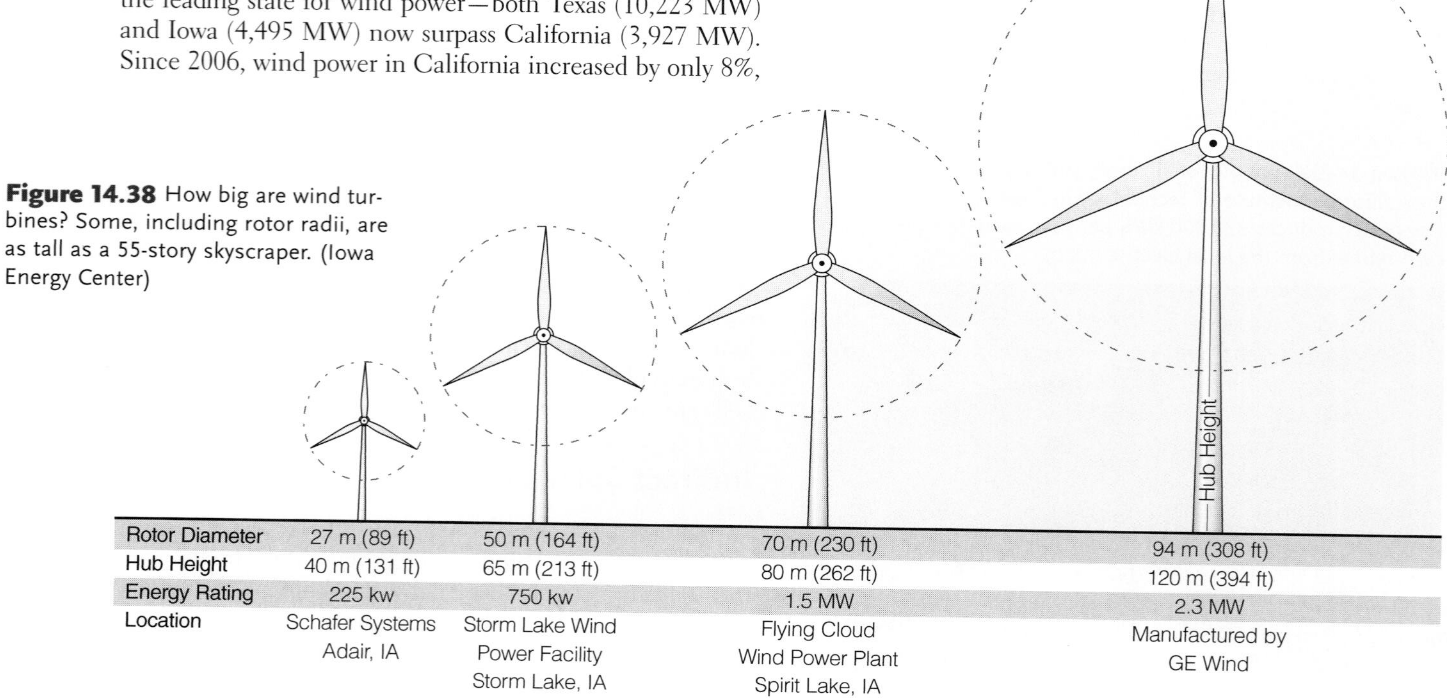

Rotor Diameter	27 m (89 ft)	50 m (164 ft)	70 m (230 ft)	94 m (308 ft)
Hub Height	40 m (131 ft)	65 m (213 ft)	80 m (262 ft)	120 m (394 ft)
Energy Rating	225 kw	750 kw	1.5 MW	2.3 MW
Location	Schafer Systems Adair, IA	Storm Lake Wind Power Facility Storm Lake, IA	Flying Cloud Wind Power Plant Spirit Lake, IA	Manufactured by GE Wind

Figure 14.38 How big are wind turbines? Some, including rotor radii, are as tall as a 55-story skyscraper. (Iowa Energy Center)

California Energy Commission

Figure 14.39 Wind farm at Altamont Pass, California—one of the world's largest wind farms. Its capacity averages about 125 MW and 1.1 terrawatt-hours (TWh) a year. Some observers feel it is the worst possible place to put a wind farm, as it lies in an important bird migration route and has been responsible for an enormous number of bird kills. The older, smaller turbines are especially dangerous to raptors that hunt for ground squirrels in the area, with about 4,700 birds killed annually, including the federally protected golden eagle. These older, less efficient wind turbines are being replaced with larger, more cost-effective units whose rotors are elevated higher and, turning more slowly, are less hazardous to birds.

Fences may block game migration routes, and bird deaths are frequent at wind farms located in mountain passes, which are common routes for migrating birds.

Environmental groups have raised the issue of bird fatalities by wind turbines, but a study by the American Birding Association shows that at least 500 million birds a year are killed by cats in comparison with 440,000 reportedly killed by collision with wind turbines. Ranchers are delighted with wind turbines. They can graze their cattle while receiving land-use royalties. The only solution to bird kills seems to place the wind farms in areas away from bird habitats. Bats, too, succumb to wind turbines, especially those species that migrate seasonally. Because bats are nocturnal, the easy solution is to feather the rotor blades and shut down the turbines at night during migration periods. Energy demands are lower at night, so the economic loss is not great.

Europe is the global leader in wind-energy development, with a wind power capability in the European Union (EU) of nearly 100,000 MW, or 6.3% of EU's electricity needs. By 2011, the growth of wind-energy development in

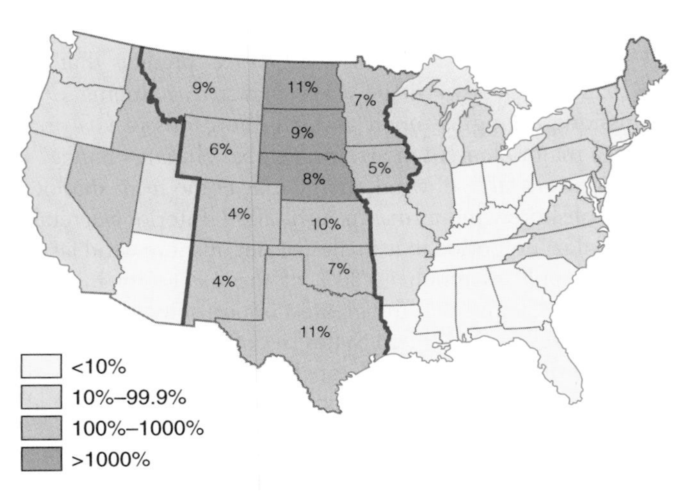

Figure 14.40 Wind-electric potential of the 48 contiguous states as a percentage of each state's energy need. Twelve states in the central part of the country could provide about 90% of the total U.S. need. (From Weinberg, C. J., and R. Williams. 1990, Sept. Energy from the Sun. *Scientific American*.)

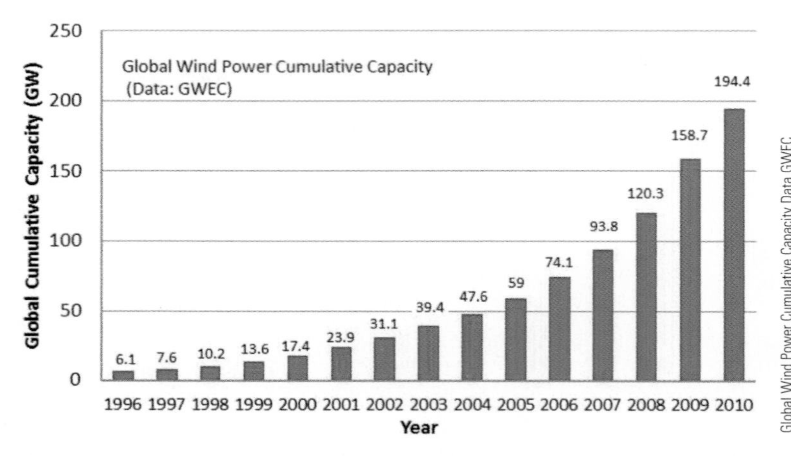

Global Wind Power Cumulative Capacity Data GWEC

Figure 14.41 Growth of global cumulative installed wind-power electricity capacity, 1996–2010. Globally, wind power is the fastest-growing alternative energy system, with generating capacity increasing nearly 32 times from 1996 to 2010. (Global Wind Energy Council)

the EU averaged 15.6% over the previous 17 years. In 2011, the United States had 46,919 MW of installed capacity that provided almost 3% of the nation's electrical energy, with the newer state-of-the-art turbines producing electricity in the 5 to 8 cents/kWh range. This puts the price of wind energy in close competition with conventional power generation. China ranks second to the United States with an installed wind-energy capacity of 44,700 MW. By 2011, Germany had 29,075 MW of installed wind-energy capacity, which provided 8% of the country's electricity. (It's worth noting that by 2011, all renewables in Germany accounted for a full 20.8% of the country's energy production, well on track to meet its goal of at least 35% of production from renewables by 2020.) The Global Wind Energy Council (GWEC) and the American Wind Energy Association (AWEA) anticipate that the pace of global wind energy growth experienced since 1996, with wind-generating capacity increasing nearly 32 times by 2010, will continue (**Figure 14.41**). Furthermore, the GWEC predicts that, by 2020, wind energy will be a key player in reducing CO_2 emissions, saving as much as 1.5 billion tons of CO_2 per year, a major element in arresting global climate change.

A new source of wind energy may come from shallow-water offshore wind farms. In April 2009, Interior Secretary Ken Salazar reported that shallow-water offshore wind farms could supply as much as 20% of the electricity in most coastal states, and that the greatest opportunity for offshore wind energy in the United States lies off the Atlantic Coast, which holds the potential for 1,000 gigawatts (GW) of electricity—one quarter of national demand. He stated, "More than three-fourths of the nation's electricity demand comes from coastal states and the wind potential off the coasts of the lower 48 states actually exceeds our entire U.S. energy demand."

In North Carolina, where high winds in the Outer Banks are legendary, some coastal county officials and state legislators are preparing standards for siting wind farms. State and local governments from Florida to Maine have begun drafting plans and regulations for offshore siting and permitting within 3 miles of their coastlines. (Beyond 3 miles, the federal government holds jurisdiction.)

In September 2009, the first deepwater, full-scale floating wind turbine, Hywind, went into operation off the southwest coast of Norway (**Figure 14.42**). Hywind is financed by the state-owned oil company Statoil, and it incorporates engineering technology drawn from Statoil's long-term expertise in offshore oil and gas exploitation. Hywind lies some 7 miles from the nearest land and generates 2.3 MW, with the tower height 65 meters (213 feet) above the waterline and a rotor diameter of more than 82 meters (269 feet). Hywind delivered 7.3 gigawatt-hours (GWh) of electricity in 2010, and has survived stormy seas with 11-meter (36 foot)-high waves with seemingly no wear. A key fac-

Hild Bjelland Vik/Statoil

Figure 14.42 Hywind—the world's first deepwater, full-scale floating 2.3 MW wind turbine, Norway. The tower stands 65 meters (213 ft) above the waterline, supported by a 100-meter (328-ft) submerged body anchored to the seafloor, with an 82-meter (269-ft) rotor diameter. It can be installed in water depths up to 700 meters (2,300 ft) and has withstood stormy seas with wave heights up to 11 meters (36 ft) high.

tor of Hywind is that it can be anchored in water depths of 122 to 701 meters (400–2,300 feet), allowing it to be positioned sufficiently far from the shore that it eliminates the common complaint that ocean-based wind farms are scenically objectionable. Statoil plans additional pilot Hywind installations in the North Sea off Scotland and offshore the United States.

Energy from the Sea

As the need for alternative energy sources and solutions for global warming becomes more urgent, scientists, engineers, and entrepreneurs are searching for new technologies to achieve clean, renewable energy. Perhaps the most promising source of renewable energy is one of Earth's largest: the ocean. There may be as much as 500 times the global demand for electrical power available in the raw energy of the ocean, a potential for as much energy as 2 to 4 trillion kilowatt-hours (GWh) per year. The World Energy Council estimates that 10% of the global energy demand could be met by harvesting ocean energy. Three promising areas of research and development for utilizing the ocean's inexhaustible energy are **ocean thermal energy conversion, wave energy**, and **tidal energy**. Because these new technologies are in the research and early developmental stages, EROIs cannot be estimated yet.

Ocean thermal energy conversion—Ocean thermal energy conversion (OTEC) utilizes the temperature difference between warm surface water and colder deep waters. The ideal OTEC location is a coastal area where water depth increases rapidly enough to achieve the appropriate temperature differential and close enough to shore for power transmission. The Hawaiian Islands have the population, need, and oceanographic conditions for this form of energy conversion. The temperature differential between the surface water and the deep ocean is 4.4°C (40°F), ideal conditions for OTEC. In 1979, scientists at the National Energy Laboratory of Hawaii Authority (NELHA), located at Keahole Point near Kona on the Big Island (Hawaii), successfully produced 10 to 15 kW of net electrical power by OTEC. Since then, the NELHA has experimented with a so-called open-cycle OTEC system, which not only can produce electricity but also could produce 600,000 gallons of pure freshwater per day as a by-product of the process. The freshwater would be a valuable commodity on the arid western side of Hawaii that could be used by several commercial enterprises that are important in the Hawaiian economy.

In the 1980s, another OTEC system operated in Hawaii, using a low-boiling-point fluid to run a generator. A mini-OTEC of about 50 kW powered a continuously burning lightbulb on a float near the village of Mokapu, demonstrating the capability of OTEC technology.

Wave energy—The U.S. Electric Power Research Institute (EPRI) estimates that wave energy in the United States offers an energy-generation potential equal to that currently generated in the United States by hydropower. The EPRI projects that the first 100-MW **wave energy conversion** (WEC) to be deployed will provide electricity at a nearly competitive cost of 9 cents per kWh, compared with new wind projects at 7 cents per kWh and new coal-fired power plants at 6 to 8 cents per kWh (assuming the plants will not be charged penalty costs for CO_2 emissions). Capturing this wave energy and converting it to electricity is not a new idea, but until recently its practicality has been little improved since 1900. Various technologies are currently being tested and developed. Research in wave energy conversion technology by a renewable-energy company based in New Jersey has resulted in its PowerBuoy system. The rising and falling of waves causes a floating buoy to move freely up and down. The resultant mechanical motion is converted to electricity by a power take-off device that drives an electrical generator; the power is transmitted ashore by an underwater electric cable. The first commercial wave park on the West Coast of the United States uses this PowerBuoy system. The wave park, 2.5 miles offshore Reedsport, Oregon, consists of 10 PowerBuoys and the cable and equipment to deliver the renewable power to the Pacific Northwest electric grid. Each PowerBuoy has a maximum generating capacity of 150 kW, with their combined power sufficient to supply up to 375 homes. PowerBuoy systems are also in service off Atlantic City, New Jersey, and at the U.S. Marine Corps Base at Kaneohe Bay, Hawaii (an animated video of the PowerBuoy system can be accessed at http://www.oceanpowertechnologies.com/technology.htm). Other installations in various stages of deployment are in waters off the northern coast of Spain; the Orkney Islands, Scotland; and Cornwall, Great Britain.

A second type of WEC is a permanent magnet linear generator device developed by the Oregon State University School of Electrical Engineering (OSU) in collaboration with the U.S. Navy and a private company. The core of this WEC is a floating spar, tethered to the seafloor, that remains relatively motionless as waves pass by. Wave motion causes a buoyant float containing permanent magnets to move freely up and down relative to the spar. The rise and fall of the magnet generates a current in an electric coil mounted to the spar (**Figure 14.43**).

A very different WEC is the Pelamis Wave Energy Converter, a long, semisubmerged, articulated device composed of cylindrical sections linked by hinged joints, which is oriented perpendicular to the direction of wave propagation. Wave motion causes flexing at the joints between segments, which drives internal pumps to power electrical generators (**Figure 14.44**). Developed by a Scottish company, Pelamis Wave Power began operating three of their first-generation converters at the world's first commercial wave farm off Póvoa Varzim, Portugal, in 2008. They have an installed capacity of 2.25 MW, enough to meet the average electricity demands of 1,500 homes. Testing of a second-generation Pelamis converter began in 2010 off the Orkney Islands.

Wave energy technology faces some challenges. The marine environment is aggressive—marine organisms cause

Annette von Jouanne, OSU School of Electrical Engineering and Computer Science

Figure 14.43 A permanent magnet linear generator wave energy convertor developed through a collaboration of OSU, the U.S. Department of Energy, and a private corporation, being deployed for testing. Wave motion causes a buoyant float containing permanent magnets to move up and down relative to a fixed spar containing the generator coils. The rise and fall of the magnets generates a current in the generator coils.

fouling, intense storms can impart large loads to WECs, and seawater is corrosive. Furthermore, WEC projects experience environmental, policy, and permitting issues. Nevertheless, wave energy is a viable source of renewable energy and it deserves serious attention.

Tidal energy—Coastal dwellers have used tidal currents to power mills for at least a thousand years. Restored working tidal mills in New England and Europe are popular tourist attractions. Both flood (landward-moving) and ebb (seaward-moving) tidal currents are used. The EPRI estimates that as much as 10% of U.S. electricity could be supplied eventually by tidal power, a potential equaled in the United Kingdom and surpassed at significant coastal sites, such as Canada's Bay of Fundy. Compared with solar and wind energy, tidal-energy generation has the advantage of predictability. Tides are reliable, being driven by the gravitational pull of the Sun and Moon, as opposed to the weather, and their timing and strength can be determined in advance with a high level of certainty. Cosmetically, the underwater machinery is less obtrusive than wind turbines.

There are two basic types of tidal-energy–generating technology. The first type uses a barrage (a damlike barrier) on a bay with a large tidal range—the height difference between high and low tides—with a large area behind the bay for storing the elevated water. Power is usually generated during the ebb tide, as the barrage creates a substantial head of water, much as a hydroelectric dam does.

The first and best-known large-scale tidal power plant is on the Rance River near St. Malo on the English Channel in northwestern France. The tidal range there varies from 9 to 14 meters (30–40 feet), and the plant generates power at both the flood and ebb tides. Because the hydrostatic head (water drop) is only a few meters, 24 small 10-MW generators with a total capacity of 240 MW are used. The plant has been in operation since 1966 and, on balance, has been successful.

Annapolis Royal in Nova Scotia, Canada, is the only modern tidal-generating plant in North America (**Figure 14.45**). It generates more than 30 million kilowatts per year, enough electricity to support 4,500 homes. An island in the mouth of the Annapolis River was selected as the site for the powerhouse, which opened in 1984 after 4 years of construction (**Figure 14.46**). It is an ideal location, because the Bay of Fundy has the highest tides in the world, and there was an existing causeway on the river with sluice gates that dammed a pond in the river. On the incoming tide, the sluice gates are opened and the incoming seawater fills the pond. The sluice gates are then closed, which traps the seawater in the pond upstream of the generating turbine. As the tide recedes, the level of the water below the pond drops and a hydraulic head develops. When the difference is

Figure 14.44 A Pelamis Wave Energy Converter during sea trials. Three of these convertors constitute the world's first commercial wave energy conversion farm off Portugal. Collectively, the converters generate 2.25 MW, enough to meet the needs of 1,500 homes.

Pelamis Wave Power

Figure 14.45 Annapolis Royal tidal electricity-generating station in Nova Scotia, Canada—the only modern tidal generator in North America. The plant generates more than 30 million kilowatts per year, enough electricity to support 4,500 homes.
D. D. Trent

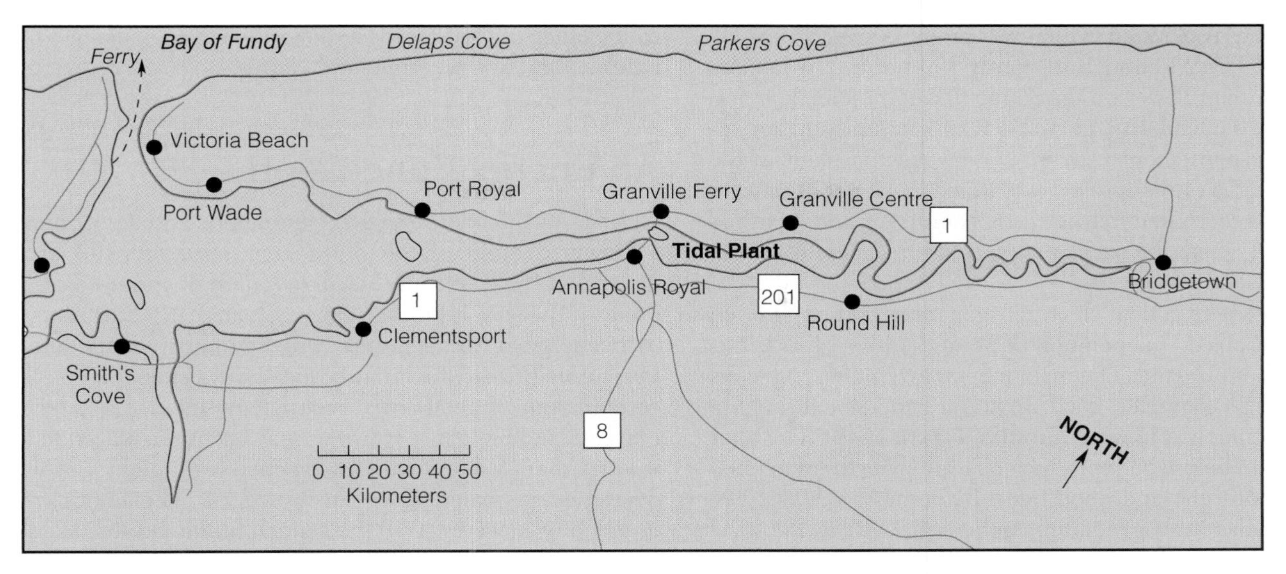

Figure 14.46 Map of the site of the Annapolis Royal, Canada, tidal-generating station. The Bay of Fundy has the highest tides in the world, which are put to work at the station to generate electricity. (Reprinted by permission of Nova Scotia Power, Inc.)

1.6 meters (5 feet) or more, the gates are opened and water flows through the turbine to generate electricity.

The second type of tidal-energy generation utilizes fast-moving marine currents caused by tidal action. Ocean tides cause the water to flow inward from the ocean during flood tides and outward during ebb tides. Shallow or narrow constrictions that restrict the tidal flow produce the fastest and most powerful tidal currents, whose energy can be harnessed using submerged turbine generators. The submerged turbines are very much like windmills, but because water is more than 800 times denser than air, the blades can be quite small and more compact, so that a relatively small device can create a relatively large amount of energy. Because tides vary in time and intensity from day to day, but do so in a predictable manner, tidal power can be integrated easily into the local power grid. This is a new and evolving technology, with only two prototypes tested and two grid-connected operations installed, one in Ireland and one in Norway.

Currently, there are two advanced demonstrations of ocean-current power generation. One, the Free Flow Kinetic Hydropower System, is at the bottom of New York City's East River, which is actually a tidal channel. In 2006, turbines were moored to the riverbed under 9 meters (30 feet) of water. Each turbine swings a 5-meter (16-foot) diameter rotor and turns at up to 32 rpm. The project has operated more than 9,000 hours and generated more than 70 MWh of energy from the natural tidal currents, making it the industry leader (**Figure 14.47**). The placement of

Kris Unger/Verdant Power, Inc.

Figure 14.47 Installing Verdant Power's first tidal turbine, the Free Flow Kinetic Hydropower System, in New York City's East River. The project has clocked over 9,000 turbine hours and generated over 70 MWh of energy from tidal currents, making it the industry leader.

30 turbines is planned, which will produce a combined capacity of 1 MW, enough to supply 300 homes. In January 2012, Verdant Power was given the first commercial license from the Federal Energy Regulatory Commission for installing a permanent Free Flow tidal electric plant in the East River. A similar project is planned in Massachusetts for the Muskeget Channel between Nantucket and Martha's Vineyard where the project is expected to generate up to 5 MW of electricity by 2013.

Elsewhere, a tidal electricity generator named SeaGen has been placed in operation in Strangford Lough Narrows in Northern Ireland. The machine has two rotors, 16 meters (52 feet) in diameter, each driving a generator through a gearbox, much as in wind turbines (**Figure 14.48**). The rotor blades can be pitched through 180°, allowing them to operate on both ebb and flood tides. Beginning in May 2008, SeaGen has been generating sufficient power to the local electricity grid to meet the needs of about 1,000 homes. Environmental monitoring began in 2005, before SeaGen was deployed, and concluded that the machine had no major impacts. Other than avoiding the machine, porpoises and sea lions seemed unconcerned with its presence. The study also showed that the seafloor life surrounding SeaGen's foundations has recovered, and there has been no significant change in tidal speeds or flow directions, or changes in marine traffic within Strangford Narrows.

An Energy Conclusion

Clearly, our petroleum-fueled, automobile-based economy is "mature," and certainly our energy sources for the later decades of the 21st century will be different and not based on fossil fuels, as it has been since about 1910. Earth's primary energy source, of course, will continue to be the Sun. Our future primary fuel may be electricity generated from a combination of geothermal technology, solar, wind, tidal currents, and biofuels. Hydrogen may become a player, too, if—and that's a big *if*—a cost-effective technology can be developed to manufacture the hydrogen. The important question is, however, Will the general public be able to adjust to a new energy paradigm?

consider this . . . Assume you are an investor looking for attractive venture capital projects for investment. Which of the potential nonfossil-fuel energy sources seem to be the best investment? Why?

Figure 14.48 The SeaGen twin-turbine tidal energy system in the narrows of Northern Ireland's Strangford Lough. It is the world's first tidal turbine to generate 1.2 MW and is the most powerful marine energy device to date that is connected to an electrical grid. It generates sufficient power to supply about 1,000 homes.

Marine Current Turbines

Concern about a shift to new energy sources is well expressed by David MacKay, a physicist at the Cavendish Laboratory, University of Cambridge:

> Given the general tendency of the public to say "no" to wind farms, "no" to nuclear power, "no" to tidal barges—"no" to anything other than fossil fuel power systems—I am worried that we won't actually get off fossil fuels when we need to. Instead, we'll settle for half-measures: slightly-more-efficient fossil-fuel power stations, cars, and home heating systems; a fig leaf of a carbon trading system; a sprinkling of wind turbines; an inadequate number of nuclear power stations. . . .
>
> We need to stop saying no and start saying yes. We need to stop the Punch and Judy show and get building.[13]

[13]David J. C. MacKay. 2009. *Sustainable energy: Without the hot air.* Cambridge, England: UTT Cambridge Ltd.

case study

14.1 Industry Cracks Open a New Oil Source in the Gulf of Mexico

In what could become the biggest new source of domestic oil in the United States since the discovery of Alaska's North Slope in the 1970s is the oil production from the ultra-deep waters of the Gulf of Mexico. The Jack field, discovered in 2006, about 435 kilometers (270 miles) southwest of New Orleans, and a dozen comparable discoveries in the same region, may hold as much as 15 billion barrels of oil and gas reserves (**Figure 1**). If successfully exploited, these discoveries will boost the U.S. reserves by about 50%, to more than 29 billion barrels. Production from these ultra-deepwater discoveries may well exceed the output of Alaska's giant Prudhoe Bay field, the largest of all U.S. oil fields. The new discoveries are sizable, but the Gulf's ultra-deepwater oil doesn't come close to that of Mexico's huge Cantarell field or the enormous fields in the Middle East.

The Jack well, one of the world's deepest producing wells at a total depth of 5 miles, taps a reservoir in rocks of Early Tertiary age. The well was drilled at an estimated cost of more than $100 million. Further development will cost several billion dollars for building platforms (**Figure 2**), drilling, and laying pipelines to the mainland. Such deepwater discoveries are now possible by advances in seismic exploration and by today's large floating drilling platforms. These technological innovations have made it possible for geologists to search out and drill below the massive salt beds that typically overlie and obscure petroleum reservoirs in these deepwater Early Tertiary sedimentary rocks.

In 2009, a consortium of three oil companies announced the discovery of a new oil pool in the Gulf of Mexico, the Tiber prospect, after drilling the deepest oil well in the world (see Figure 2). At more than 35,000 feet,

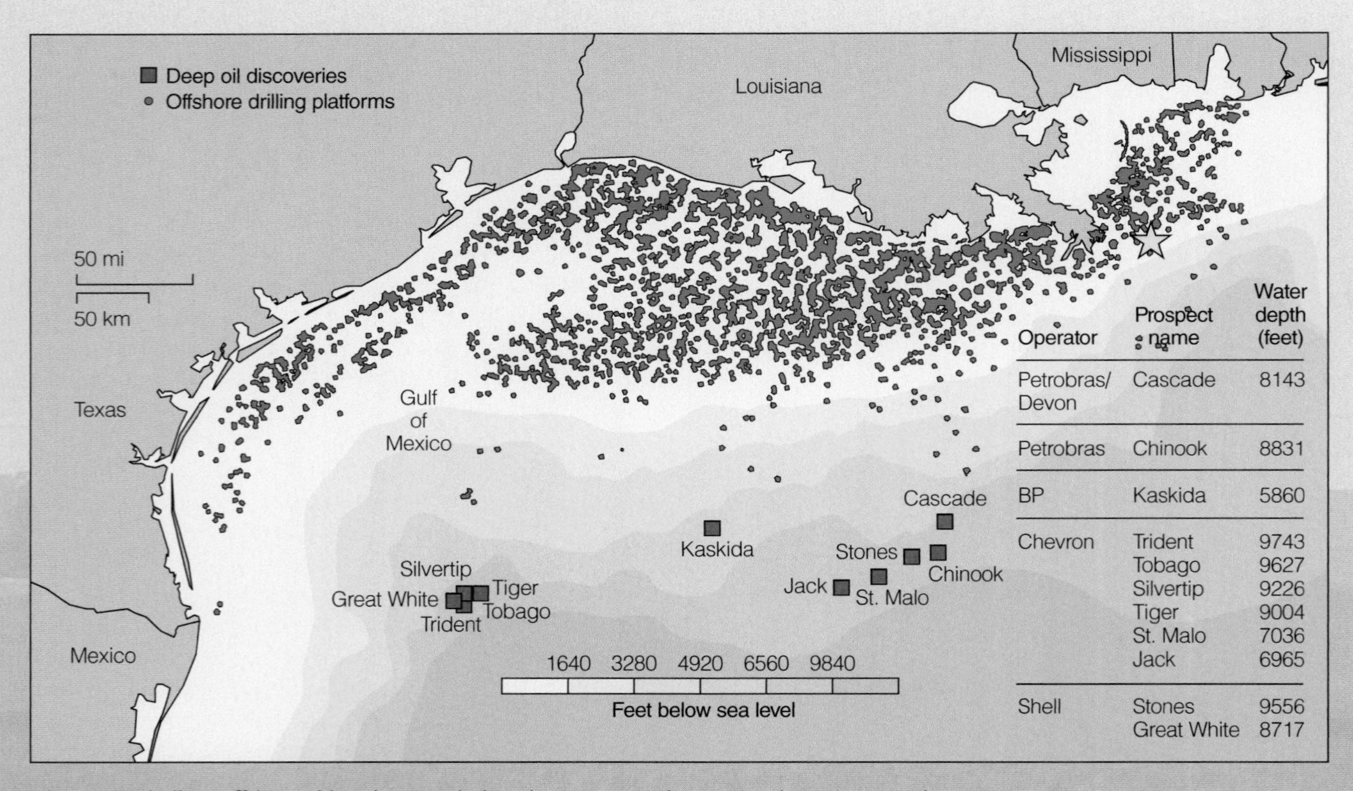

Operator	Prospect name	Water depth (feet)
Petrobras/ Devon	Cascade	8143
Petrobras	Chinook	8831
BP	Kaskida	5860
Chevron	Trident	9743
	Tobago	9627
	Silvertip	9226
	Tiger	9004
	St. Malo	7036
	Jack	6965
Shell	Stones	9556
	Great White	8717

Figure 1 Shallow offshore (blue dots) and ultra-deepwater (red squares) discoveries in the Gulf of Mexico. Shown are more than 4,000 oil well drilling platforms. Oil companies in waters up to 3 kilometers (2 mi) deep (red squares) have made some significant discoveries. The first well of Chevron's Jack field, which is under 2.1 kilometers (7,000 ft) of water, was completed in 2006 at a total depth of 8.7 kilometers (28,543 ft, or 5.4 mi) and was successful. Ten other ultra-deep prospects in the Gulf offer promise. The yellow star marks the site of BP's Macondo prospect and the *Deepwater Horizon* drilling platform disaster of April 22, 2010. (Source: Department of the Interior, Minerals Management Service/BOEM)

Mike Theiss/National Geographic Image Collection

Transocean photographer Ken Childress

Figure 2 The *Deepwater Horizon* in better days. The rig drilled the deepest oil well in history in the Tiber field in September 2009. It reached a total depth of 10,685 meters (35,055 ft) while working for BP and its co-owners in the U.S. Gulf of Mexico. It was an ultra deepwater semisubmersible oil well drilling rig that was destroyed in an explosion and fire in April, 2010.

it is as deep as Mount Everest is above sea level, and it is in a water depth of more than 4,000 feet. It will require the investment of more money and years of work before the oil comes online; such delays are common for deepwater wells. The Gulf's Kaskida play, for example, which may have more than 3 billion barrels of oil in place, was discovered in 2006 but has yet to produce any oil.

The reserves of the Tiber discovery are believed to be 4 billion barrels. With a potential of 35% recovery, it would produce 868 million barrels. This may seem like a lot of oil, but in 2008 the United States imported 12 million barrels a day. At this rate, those 868 million barrels are the equivalent of 72 days of imported oil, and it might take 20 years or more to obtain it. Another recent deepwater development in the Gulf, the Thunder Horse field, extracts about 300,000 barrels of oil equivalent per day, about half as much per day as Alaska's North Slope. As much as we need to extract this Gulf oil, it is still a trickle compared with Saudi Arabia's supergiant Ghawar field, which yields 5 *million* barrels per day. (Supergiant fields are those with proven reserves exceeding 5 or even 10 billion barrels.) The oil produced from these discoveries is light and sweet (i.e., low in sulfur), the kind of petroleum that is in demand and realizes the best prices. Nevertheless, the discoveries are somewhat provocative. As stated by international oil expert Colin Campbell, these discoveries "represent a remarkable technological achievement but at the same time [reflect] the extreme desperation of the industry, evidently having little easier [remaining] to test."

The risks of offshore and deepwater drilling were exemplified on 20 April 2010, when the *Deepwater Horizon*, drilling British Petroleum's (BP's) Macondo Prospect in 1,524 meters (5,000 feet) of water and 3,962 meters (13,000 feet) of rock, suffered a blowout, explosion and fire, the tragic loss of 11 lives, the destruction of the entire drilling platform, and the uncontrolled flow of more than 750 million liters (200 million gallons) of oil into the Gulf of Mexico. Flames from the fire roared up to almost 100 meters (300 feet) and could be seen from 90 kilometers (56 miles) away. Two days after the blowout, the *Deepwater Horizon* sank, extinguishing the fire, but leaving the well gushing oil on the seafloor and causing the largest offshore oil spill in history. A relief well was drilled that took 83 days to successfully seal off the flow. An investigation by BP and government authorities revealed that the immediate cause of the explosion was a bubble of gas shooting up the drill column, bursting through the blowout preventer, various safety barriers and seals, and exploding at the drilling floor. It is worth noting that U.S.'s OSHA regulations had no jurisdiction on the safety measures practiced on the *Deepwater Horizon* because it, as with many other offshore drilling rigs, was registered in the Marshall Islands. This process, called a "flag of convenience," is widely practiced by the maritime industry because their vessels are subject to less stringent safety regulations and lower taxes when so registered. Details of the factors and underlying causes of the blowout are contained in the *Final Report on the Investigation of the Macondo Well Blowout*, the official report by the *Deepwater Horizon* Study Group. A key conclusion of the report is that, "At the time of the Macondo blowout, BP's corporate culture remained one that was embedded in risk-taking and cost-cutting . . ."

The losses from this event are hard to imagine. The value of the drilling platform exceeded a billion dollars, and the operation of such a deepwater operation, with support boats, numerous services, and helicopter support, amounts to about $1 million per day. Furthermore, the loss of the Gulf's bird and marine life was immense, with oceanographers recording evidence of long-range damage from the oil spill to marine life. Dead and dying deep-sea corals coated in "brown gunk" were found as far away as 11 kilometers (7 miles) from the well at a water depth of 1,311 meters (4,300 feet). A study of bottlenose dolphins in Barataria Bay, Louisiana, revealed that many of the animals were underweight and suffering from liver and lung disease attributed to seawater contamination. The use of chemical dispersants in the clean-up effort, meant to separate the films of oil into tiny droplets, was easily spread by wind and waves, also had adverse biological impacts.

On a human scale, the financial losses experienced by Gulf Coast businesses and commercial fishermen were catastrophic. The scope of final insurance settlements and the ultimate roll call of damage are still unknown. Because of the unprecedented scale and complexity of the disaster, the litigation may continue for 20 years.

questions to ponder

1. What effect will the estimated reserves of the Gulf of Mexico's deepwater wells have on our future energy supply? (Hint: See Figures 14.12, 14.13, and 14.14.)
2. The Thunder Horse field extracts 300,000 barrels of oil per day. How many Thunder Horse fields would it take to equal one day's production of Saudi Arabia's Ghawar oil field?
3. As oil companies make new discoveries in remote or hostile environments, like the deepwater oil in the Gulf of Mexico, the costs of discovery, production, and transportation will increase, eventually causing an increase in the price of gasoline at the pump. Hypothetically, when the price of oil exceeds its market value, oil will no longer be extracted. Will this scenario ever occur? If so, what will it mean for society?

14.2 Tight Gas and Oil

Greater demand, sustained higher prices, and improved technology are turning unconventional tight gas and oil resources into a major element of the U.S. energy picture. Unconventional oil and gas are produced from "tight" reservoir rocks with very low porosity and permeability, such rocks as tight sands, relatively unfractured shales, and coal seams. Currently, unconventional tight gas accounts for one third of the annual domestic production of the slightly more than 18 trillion cubic feet (Tcf) of natural gas from the lower 48 states.

The character of tight sands is well illustrated by the Bakken Formation. Interest in the tight oil and gas of the Bakken was renewed in 2008 with the release of generous reserve estimates by the USGS, which received widespread enthusiasm among the investment sector. (This is not proven conventional oil but unconventional oil, such as tar sands and oil from oil shales.) Oil was first produced from the Bakken some 50 years ago, but until the 1980s that production was mainly from a few conventional vertical wells. The Bakken is only one of several hydrocarbon-producing horizons in the Williston Basin, a large sedimentary basin underlying parts of North Dakota, Montana, and Saskatchewan. The Bakken is generally thin, from 2 to 6 meters (6.5–20 feet) thick, and the depth to the top of the formation can vary from a few thousand feet in Canada to more than 10,000 feet in North Dakota. It is characterized by very low porosity (as low as 10%–15%, in contrast with good producing formations with porosities of 30% or more), low permeability (the ability of a rock to transmit fluids to the well head), and 15% to 25% water cut. The opportunity for "new" oil from the Bakken requires drilling vertically around 10,000 feet, then drilling out horizontally with long laterals as much as 9,000 feet through the productive layer to maximize contact with the pay zone. The oil recovered is good-quality light oil, but the pay zones require technical enhancement, called *hydrofracturing*, or "fracking" (see Case Study 14.3). The process is expensive and the bottom line is that it requires investing lots of energy in to get any energy out of the Bakken.

Estimates of recoverable oil from the Bakken run the gamut from 3.65 to 200 billion barrels. It is difficult enough to estimate oil reserves from conventional fields, but even more difficult from tight fields. The 2008 USGS report estimates that the Bakken contains 3.65 billion barrels of technically recoverable oil, 1.8 Tcf of natural gas, and more than 100 million barrels of natural gas liquids, with roughly 90% of the energy in the oil. Let's take a closer look at the USGS average figure. The 3.65 billion figure is the average of two guesses: 3.05 billion at a 95% chance and 4.32 bil-

Mike Theiss/National Geographic Image Collection

lion at a 5% chance. This means that the average is really a mathematical fiction, and this estimate is for technically recoverable oil, which in all likelihood is not the same as economically recoverable oil. In contrast, some North Dakota oil industry professionals claimed the Bakken has 24 billion barrels of oil, but they do not specify whether it is technically or economically recoverable. Production from individual Bakken wells is not overly impressive; of production records from nearly 9,000 wells, only 24 produced more than 800 barrels a day, with the average well producing a mere 52 barrels a day. It is expected that the field will be producing for another 15 to 20 years, but a concern expressed by some oil authorities is what kind of long-term production rates can be expected from horizontal fracked wells. Any increased production from the Bakken will depend on rapid replacement of wells that have undergone the steep initial decline shown in **Figure 1**. Overall production can increase only until they run out of new drill sites.

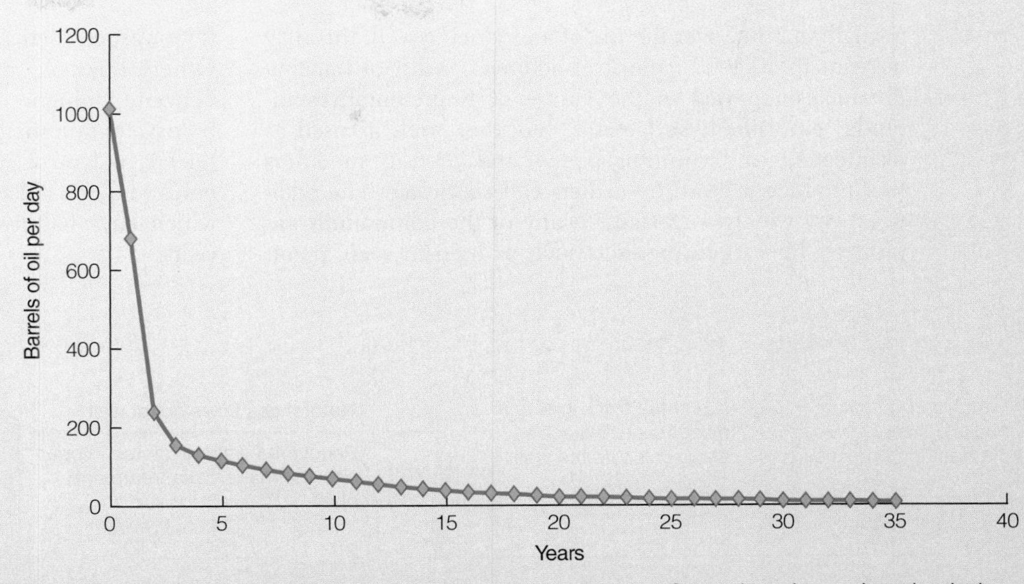

Figure 1 Decline rate of a Bakken well, with initial production of 1,000 barrels per day. The decline rate is just as rapid for wells coming in at less than 1,000 barrels per day. (North Dakota Department of Mineral Resources)

How significant is the USGS estimate? In 2010, the United States imported 10.4 million barrels per day (the most recent figure available) and there are 365 days in a year. Thus, 10.4 million barrels per day × 365 days per year = 3.796 billion barrels per year of imported oil. Consequently, the USGS estimate of 3.65 billion barrels is less than 1 year's supply of oil imported to the United States, and this is only if we can recover all that oil. Because the oil is in a tight formation with very low porosity and permeability, we could never extract it all even using the very latest fracturing and drilling technology. (Modest recoveries from conventional petroleum reservoirs are usually only about 15% to 35%; good reservoirs yield about 40% to 50%.)

At this writing, thousands of wells are being drilled in the United States, most notably in the tight oil regions of the Bakken, North Dakota, and Eagle Ford, Texas, plays. These tight-formation wells each average less than 100 barrels a day. The *OPEC Annual Statistical Review 2010/2011* reports that worldwide there are roughly a million producing wells averaging about 2,957 barrels per well per day. Drilling the tight oil plays in the United States does not appear to be the solution to solving our current dependency on imported oil.

14.3 Fracking

Hydraulic fracturing, also known as hydrofracturing, and "fracking," is the method of improving the production of oil and gas wells by injecting a high-pressure fluid, infused with chemicals and sand, deep into "tight" formations like shale, to fracture the formation to increase its porosity and permeability, and promote the release of oil and/or natural gas into the well bore (**Figure 1**). Fracking is truly an impressive technology that, in concert with improvements in horizontal drilling technology, is one of the most significant energy production innovations of the 20th century. Nevertheless, its use may bring with it large costs to the public, as it is clear that it produces toxic, and sometimes radioactive, wastewater that, despite emphatic industry denials, has the potential to contaminate surface water and water wells.

Fracking treatments may require up to 6 million gallons of fluid and 100,000 pounds of proppants (sand that holds open the fractures). The fluid includes less than 2% of chemical additives, rust inhibitors, microorganism-killing agents, and friction reducers. Although only 2% may seem

Mike Theiss/National Geographic Image Collection

a small amount, over the life of a producing well, this may amount to 100,000 gallons of additives. A list of fracking chemicals reported to the House of Representatives includes emulsifiers, surfactants, biocides (such as used as disinfectants in swimming pools), and viscosity modifiers that produce a fluid the drillers call *slickwater*. The additives vary widely in toxicity. Many of the compounds are ordinary household products such as used in soap, paint, floor wax, cosmetics, with some even used in food products. Others, however, are toxic: carcinogens; neurotoxins (e.g., benzene, toluene, xylene, ethylene glycol [antifreeze]); butoxyethanol; methanol; and radioactive isotopes, used as tracers, a few of which are iodine-127, lanthanum-140, potassium-40, cobalt-58, rubidium-86, and strontium-85, which have half-lives ranging from a few hours to a few years.

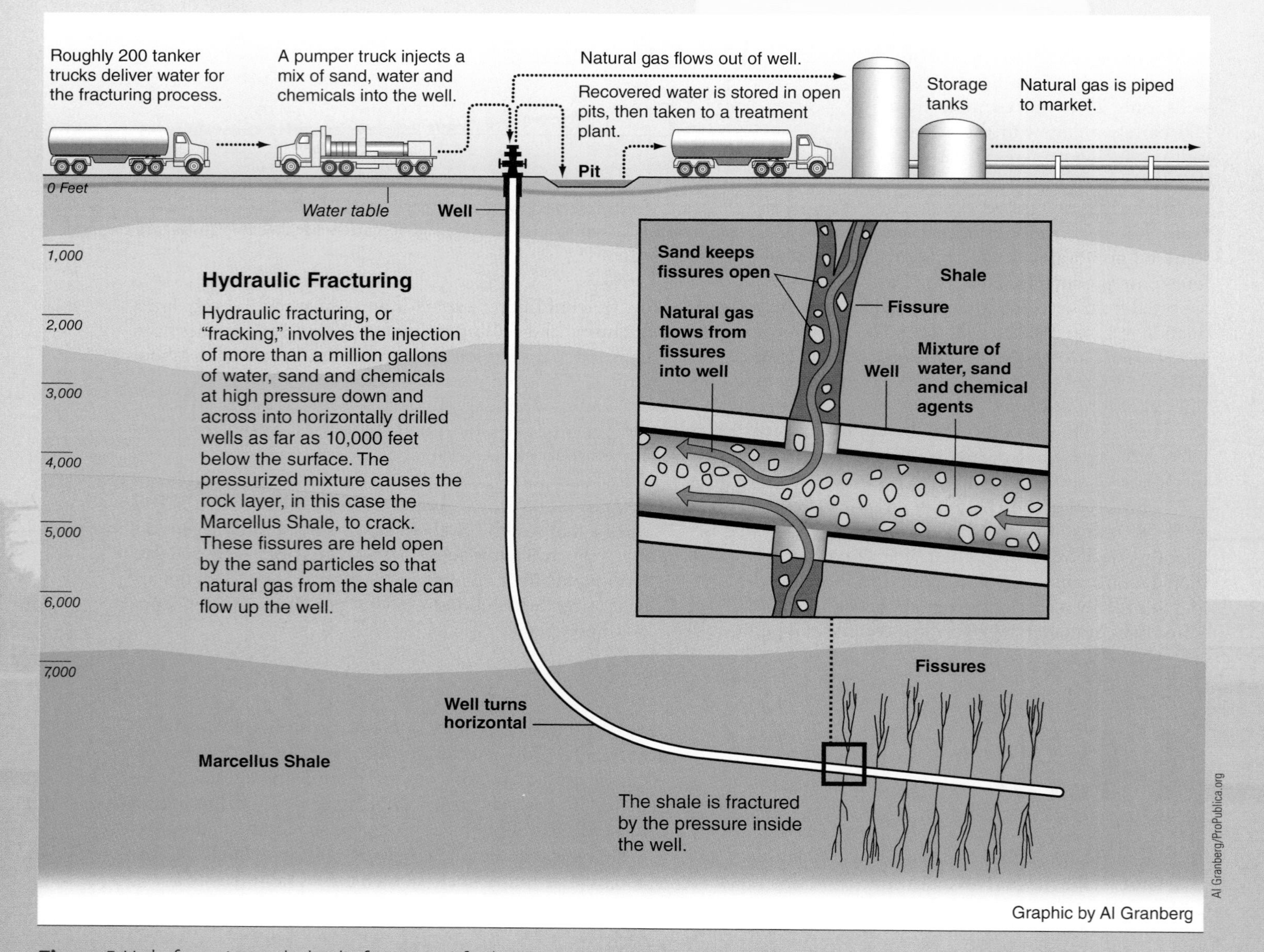

Figure 1 Hydrofracturing, or hydraulic fracturing ("fracking"), requires injecting over a million gallons of water, chemicals, and sand at high pressure to crack the rock and release oil and/or gas into the well. (Source: http://www.propublica.org/special/hydraulic-fracturing-national)

Frequent claims are voiced by activist groups that the components of the fracking fluid are secret, protected by legislation. Yet, in Pennsylvania, Wyoming, and probably other states, lists of the chemical compounds are readily available to first responders in emergency situations.

Halliburton, a major oil field services provider, acquired patent rights for the fracking process in the 1940s, and the practice is now used worldwide. Of the 75,000 wells drilled across 30 states since 2006, most were fracked because it is the only technology that can be used to extract oil or natural gas trapped in so-called tight rock formations such as underlie the entire San Juan Basin, New Mexico, the Sprayberry trend in west Texas, the Denver Julesburg Basin, Colorado, and the Marcellus formation of West Virginia, Pennsylvania, eastern Ohio, and southern New York.

The fracking technology has raised contamination issues, especially in Wyoming and Pennsylvania where it has initiated studies, reports, claims, and counterclaims that have resulted in a media frenzy. In 2011, after a 3-year study, the EPA confirmed the presence of toxic chemicals often associated with fracking in contaminated well water in Pavillion, Wyoming, an area that is home to nearly 170 natural gas wells. Consequently, in 2010, Wyoming became the first state to require a public listing of the chemicals used in fracking. The formulae for fracking fluids still remain proprietary secrets, but the list of chemicals must be disclosed.

Pennsylvania has experienced a phenomenal increase in drilling since 2000. By 2011, the state had more than 71,000 active gas wells, and accompanying the drilling has come numerous complaints of contamination to drinking water. For that reason, the EPA has been investigating contamination reports around Dimock, Pennsylvania, where at least 20 families claim their wells have been tainted with methane and chemicals after fracking began nearby. The EPA studies for the first group of homes tested showed that the water meets federal standards for drinking water. Traces of arsenic, which is not used in fracking fluids, was found in two homes, and sodium, chromium, methane, or bacteria were found in samples from six homes, but overall, no concentrations above safe levels were found. (In some cases, perhaps all, contamination may be caused by poor well construction [e.g., bad casing jobs] not the fracking itself). In 2009 and 2010, the Pennsylvania Department of Environmental Protection (DEP) found no frack fluids, chemicals, or other frack contaminants in any of 18 water wells tested in Dimock, although methane was detected that had migrated from gas wells and contaminated the wells. The DEP ordered corrective measures, including repairing gas wells and installation of methane treatment systems to address the problems in these 18 wells. By 2011, after remedies had been taken, methane levels had been reduced to safe levels in most of the 18 wells and the water was safe to drink.

The 2010 documentary film *Gasland* focused on the impact of fracking, and includes a key scene in southern New York showing burning gas from a kitchen water tap. Unmentioned in the film is that "burning water" is a common phenomenon in the region, having been observed in that region since the 17th century. Data from more than 7,000 wells drilled in Pennsylvania revealed that 22% of the well water sampled *prior to drilling operations* already had measureable methane. Well-water samples taken before and after drilling at one highly publicized site revealed virtually no changes in gas content. The homeowners knew their well contained gas and had it equipped with a methane gas-venting cap *long before the onset of nearby drilling*.

Disposal of the toxic fracking flowback wastewater is perhaps the most significant fracking problem. Somewhere between 10% and 40% of the fracking fluid pumped down a well returns to the surface as flowback, a murky, highly salty water enriched in drilling fluids and other contaminants dissolved from the producing formation. The flowback is recovered and stored in aboveground ponds until removed by tanker truck for disposal. The remaining fracking fluid remains deep underground where a possibility exists for contaminating groundwater, although the industry claims this "very unlikely." However, contamination by wastewater leaking through a cracked pond liner is a very real possibility.

Pennsylvania is the only state allowing drillers to send wastewater to sewage treatment plants, which, after treatment, is discharged into rivers. In 2008, some treatment plants were processing and discharging such enormous amounts of salty wastewater that downstream utilities complained the river water was corroding their equipment. John H. Quigley, former secretary of Pennsylvania's Department of Conservation and Natural Resources, stated, "In shifting away from coal and toward natural gas, we're trying for cleaner air, but we're producing massive amounts of toxic wastewater with salts and naturally occurring radioactive materials, and it's not clear we have a plan for properly handling this waste." Consequently, Pennsylvania has developed new fracking regulations that include "presumptive responsibility" for any water-well contamination observed within 760 meters (2,500 feet) of a drill site as opposed to 305 meters (1,000 feet) earlier.

Air pollution from natural gas drilling is another growing problem. For example, in 2009, Wyoming failed to meet federal air quality standards for the first time in history, in part because of fumes containing toluene and benzene from about 27,000 gas wells, most drilled since 2004.

case study

An effort by the 111th Congress to define hydraulic fracturing as a federally mandated technology under the Safe Drinking Water Act required disclosure by the energy industry of the chemical additives in the fracking fluid, information that has largely been proprietary. The bill, the Fracturing Responsibility and Awareness of Chemicals Act, known as the FRAC Act, was introduced in 2009. The gas industry opposed the potential regulation claiming the information was already available through the Frac Focus Web site, although the list is incomplete as there is no mention of benzene, toluene, xylene, or some other additives. The 111th Congress took no action on the measure, and the FRAC Act was reintroduced again in both houses of the 112th Congress. As of April 2012, the bill had not been brought up for vote by either house. It comes as no surprise that the natural gas industry is a heavy contributor to Congressional campaign funds.

questions to ponder

1. Oil companies are working hard at increasing production from older, nearly exhausted oil fields by using specialized recovery technologies and discovering new "plays" (potential oil fields) that contain unconventional "tight" oil that requires "fracking." Would investing in new recovery techniques and in "tight" plays be good short-term investments? Would they be good long-term investments?

14.4 Geothermal Energy, Volcanoes, and Magma

Geologists have long known that high-temperature geothermal systems are associated with molten rock (magma) in Earth's crust, in areas of current or past volcanic activity. Because of the connection between magma and geothermal systems, geologists of the USGS developed a method to relate the temperature and approximate size of a magma body in the crust from the age and volume of magma recently erupted from the magma reservoir. Another feature of the method is that it is possible to determine how long it takes for the crystal magma body to cool. These determinations have resulted in a three-part classification of volcanic areas: those with high geothermal potential, those with low geothermal potential, and a transition zone where the potential is uncertain. In simpler terms, a young, large magma body is more likely to be a good source of geothermal heat than an older, smaller magma body (**Figure 1**).

Research in the 1970s revealed that a very large magma body, with a temperature of at least 650°C (1202°F), lay beneath the Coso Hot Springs area of California. The age/volume relation placed the Coso area in the transition zone of geothermal potential (see Figure 1). Subsequent exploration and drilling confirmed the Coso area as a substantial

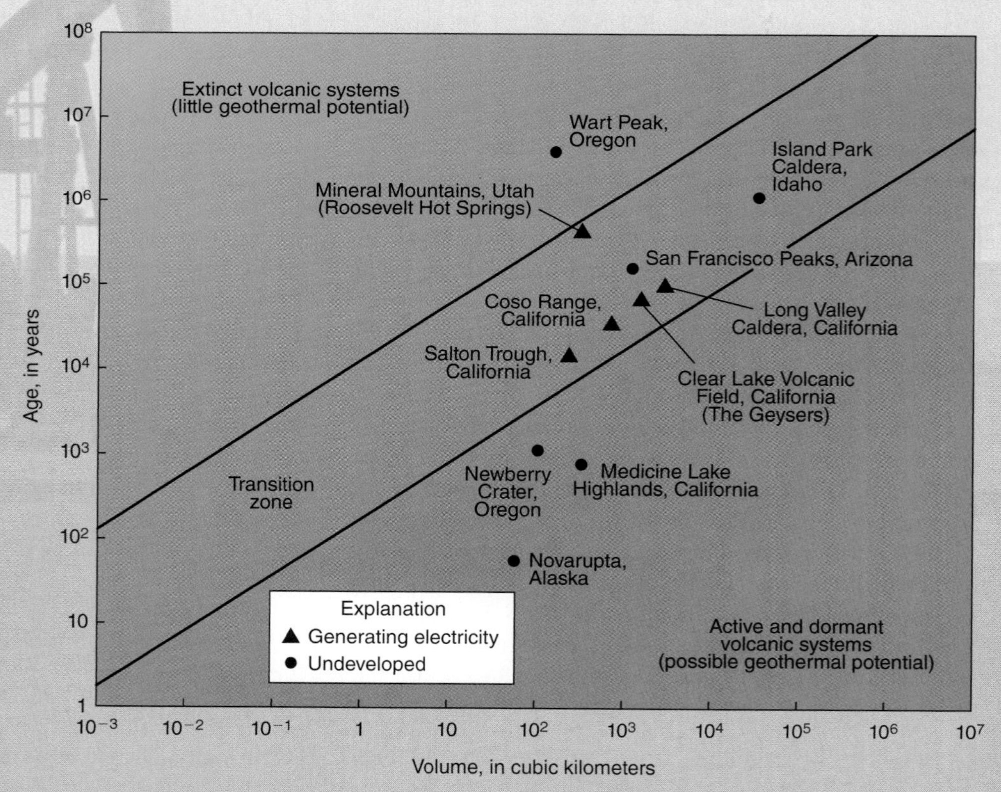

Figure 1 Ages of selected volcanic areas versus volumes of their associated magma bodies for several areas in the United States. (Source: USGS)

resource, and today it provides 270 MW of electricity, enough to accommodate a city of about 270,000 people.

Similar studies in the 1970s and 1980s suggested that a significant geothermal resource exists in the Island Park Caldera, Idaho; the Medicine Lake Highlands volcanic area in northern California; and the San Francisco Peaks volcanic area near Flagstaff, Arizona. The Novarupta area in Alaska's Lake Clark National Park and Preserve is underlain by a magma body of about 100 cubic kilometers (24 cubic miles) at 650°C (1202°F), and an even larger hot magma body underlies Yellowstone National Park. Because of their National Park status, the last two areas cannot be developed.

Source: Duffield, W. A., and J. H. Sass. 2004, *Geothermal energy—Clean power from the Earth's heat.* USGS Circular 1249.

questions to ponder

1. Upon examining the plot of age/volume relations shown in Figure 1, what regions of the United States offer the most promise for developing geothermal power?
2. Will geothermal power ever become a major source of electricity for the United States?
3. Dutch Harbor, Alaska, located near Makushin volcano, is the busiest and most important commercial fishing port in the entire Western Hemisphere. Currently, diesel-fueled generators produce electric power for the community. What studies might be undertaken to evaluate the area's potential for developing geothermal power?

14.5 Feed-in Tariffs

Freiburg, Germany, a town of 200,000 people in the Black Forest, has almost as much installed solar photovoltaic (PV) power as the entire United Kingdom. There is such competition among German towns that Ulm has recently overtaken Freiburg as solar capital of the world. Virtually all of the PV cells are roof-mounted installations.

By 2011, Germany generated nearly 21% of its electricity from various renewables, whereas the United States produced about 11% of its energy from renewables. The secret of German success is the *Stromeinspeisungsgesetz* (feed-in tariff, or FIT). Anyone generating electricity from wind, water, or PV cells receives a guaranteed payment for 20 years—four times the market rate—currently about 56 cents a kWh. This reduces the payback time on renewable technologies to less than 10 years and offers an 8% to 9% return on investment. The utility companies distribute the extra cost among all users, adding about $1.65 to the monthly bill. FITs have now been adopted in 19 European Union countries, 47 countries worldwide, and are being introduced in several states and cities in the United States. In the United States, the FITs differ in details from those in Europe. FITs are most commonly associated with PV solar panels, but some utilities in Vermont and California also pay a set price for energy generated from other sources, such as wind. Representative of various state's FITs are those adopted by California, Florida and Hawaii.

In 2008, the California Public Utilities Commission approved an FIT to homeowners who produce more electricity than they use and sell the excess back to their local utility, but with some restrictions. City commissioners in Gainesville, Florida, approved an FIT that is open to small-, medium-, and large-scale projects but is capped at a maximum of 4 MW per year. In 2011, Gainesville had increased its solar-generating electricity from 328 to 7,391 KW, which makes up about 1.2% of peak load energy demands. In Hawaii, the Hawaiian Electric Company pays above-market prices on a 20-year FIT contract for excess electricity generated by renewables and fed into the electric grid.

questions to ponder

1. Does spending $12,000 to $18,000 (not counting potential subsidies) for a roof-mounted PV system on your house, with an 8% return over 10 years, seem like a good investment?
2. Would an investment such as the one in question 1 result in a significant reduction of CO_2 emissions, compared with the amount of CO_2 released by conventional gas- or coal-fired power plants for the same amount of electricity?

PHOTO GALLERY energy is where you find it

Humans devote a great deal of mental and physical energy to harnessing Earth's energy. Energy, like freshwater, is invaluable and, when it is in short supply, a family, tribe, or country will do almost anything to obtain it. The five photos in this gallery illustrate some of Earth's nonconventional energy sources and some relics of fossil energy. Keep in mind that all these energy resources derive from our Sun.

1 A community gathers around its photovoltaic-powered television set; the Republic of Niger. More than likely, they are watching a cricket or soccer match, not MTV or *Seinfeld* reruns. John Chiasson/Gamma Liaison

2 This could be a scene in West Texas, Oklahoma, or Tampico, Mexico, in the 1920s and 1930s. Little regulation of the impact of oil production on the environment existed at the time. In reality, this is the Balakhany oil field in Northern Baku, Azerbaijan, in 2003. Peter Bentham, BP Exploration

The Navajo Generating Plant, near Page, Arizona, provides electricity to millions of people in Arizona, Nevada, and California. As a coal-fired plant, it is highly polluting and causes much of the haze in the Grand Canyon.
© 2009/SCOTT S. WARREN/National Geographic Image Collection

3 Three versions of a new vertical-axis wind turbine (VAWT) that went into commercial deployment in 2006. The inner blades, each shaped as a half-cylinder, rotate around a central vertical axis, while the outer blades, shaped like airplane wings, are fixed. The result is that the VAWTs are 43%–45% efficient, in contrast to conventional propeller-type turbines, with only 25%–40% efficiencies. The VAWTs can function at wind speeds up to 110 kilometers per hour (68 mph), which would shut down conventional turbines. Furthermore, in 8 years of experience, the VAWTs have proved to be bird- and bat-friendly. Terra Moya Aqua, Inc.

4 The Blue Lagoon, a hot-water pool 40 kilometers (25 mi) from Reykjavik, Iceland. It uses geothermal power plant outlet water, which is salty and grows algae, giving it its blue color. Open every day, the lagoon attracts about 100,000 visitors annually. D. D. Trent

5 Ten offshore wind turbines, part of renewable-energy development on the small Danish island of Samsø. In just 10 years, beginning in 1997, the island residents invested $70 million in solar panels, wind turbines, woodchip-fired district heating plants, and biofuels. Except for some necessary gasoline, they have totally liberated themselves from fossil fuels. By using only renewables, the island cut its carbon footprint by 140%. Producing more energy than they can use, they sell the surplus of "clean" power to the mainland. On average, wind power in Denmark provides nearly 20% of the entire country's electricity. Samsø Energy Academy

SUMMARY Energy and the Environment

Energy

Defined

Capacity or ability to do work. A society's quality of life depends in large part on the availability of energy.

Types

Renewable energy (e.g., solar, wind, geothermal, and hydrologic energy) is replaced at least as fast as it is consumed. Nonrenewable energy (e.g., energy derived from coal, oil, and natural gas [fossil fuels]) is replenished much more slowly than it is utilized.

Petroleum

Defined

Volatile hydrocarbons composed mostly of hydrogen and carbon. Crude oil contains many different hydrocarbon compounds.

Origin and Accumulation

The necessary conditions for an oil field are:

Source rock—origin as organic matter
Reservoir rock—porous and permeable
Caprock—impermeable overlying stratum
Geological trap—such as an anticline, a faulted anticline, reef, or stratigraphic trap

Production

Drilled wells are pumped, or they flow if under gas and water pressure. Even with the best drilling and producing techniques, primary production may recover only half the oil.

Secondary Recovery

Oil remaining in pore spaces underground is stimulated to flow to a recovery well by injecting water, chemicals, steam, or CO_2. Much of the "stuck" oil can be stimulated to flow to an extraction well by these methods.

Value

Varies with weight. Light crude oil is the most valuable. U.S. oil production peaked about 1970. The world's oil production may peak between 2010 and 2030—the exact date is unknown. Alternate energy sources will become increasingly important.

Energy Gases

May be the major fuels of the next 50 years: natural gas from unconventional sources, such as coal-bed methane; methane hydrates, ice balls containing natural gas; and hydrogen gas.

Coal

Defined

Carbonaceous residue of plants that has been preserved and altered by heat and pressure.

Rank

As coal matures, it increases in heat content and decreases in the amount of water and volatile matter it contains. The sequence is plant matter to peat to lignite to subbituminous to bituminous to anthracite.

Reserves

The United States may only have enough coal to last about 90 years at the same growth rate of production as in 2010. Worldwide, coal reserves may last about 60 years, assuming no growth in demand.

Synfuels

The manufacture of combustible gas, methanol, and gasoline from coal may have a promising but short-lived future because of the estimated exponential expiration date of coal.

Other Fossil Fuels

Tar Sands

Sands containing oil that is too thick to flow and that can be surface mined. Oil is then washed from the sand. One place the sands are mined is the Athabasca field in Alberta, Canada.

Oil Shale

Eocene lake deposits containing light kerogen oil that can be removed by heating. Deposits are found in Wyoming, Utah, and Colorado.

Problems with Fossil-Fuel Combustion

Air Pollution

Products of combustion contribute to global warming, smog, and acid rain. Acid rain is a hazard in the U.S. East and Southeast and in eastern Canada, which receives some of the U.S. emissions. This has been a major international air-pollution problem, but the situation is improving.

Mitigation

Sulfur content must be reduced below 1%. Methods include fluidized-bed combustion, neutralizing sulfur dioxide in smokestacks with limestone scrubbers, cleaning coals by gravity separation, and chemical leaching methods.

Subsidence and Collapse

Areas underlain by shallow lignite are subject to subsidence.

Alternative Energy Sources

Direct Solar Energy

Solar-thermal collectors warm a fluid, which powers a turbine generator. Photovoltaic cells directly convert sunlight to electricity.

Indirect Solar Energy
Wind farms, both onshore and offshore, are increasingly important, especially in Europe.

Synfuel
A fuel manufactured from organic materials, corn, farm waste, or coal.

Geothermal Energy
Facilities for generating electricity with geothermal energy currently are in place and may supply as much as 2% of U.S. electrical energy needs. Geothermal heat is also used worldwide for heating interior spaces, keeping sidewalks free of winter snow and ice, etc.

Hydroelectric Energy
Hydroelectric dams harness the power of water distributed over Earth by the hydrologic cycle.

Oceans
The energy of waves, currents, and tides can be converted to do work for humans. Tidal-energy, wave-energy, and thermal ocean energy conversion are in various stages of development. Thermal energy conversion is still in the early stages of development. At present, tide-generated electricity is being produced on the Rance River, France, in Northern Ireland, and Nova Scotia, Canada. Electricity on a small scale is being generated from wave motion in Portugal.

Nuclear Energy
Provides 16% of world and 20% of U.S. energy needs. Heat generated by fission of uranium isotopes and their daughter isotopes, such as plutonium, is converted to electricity. There are 104 commercial nuclear power plants operating in the United States. Mining, transportation, and disposal of high-level nuclear materials are the most dangerous aspects of the nuclear energy cycle. If the chain reaction runs out of control, the heat generated may lead to a central core meltdown, such as occurred in nuclear power plants at Chernobyl in 1986 and Fukushima in 2010. Very careful geological site studies and investigations are performed to assess hazards from earthquakes, mass wasting, subsidence, and coastal and riverine flooding.

KEY TERMS

acid rain
anticline
binary-cycle system
biodiesel
biofuel
biomass
breeder reactor
caprock
carbon capture and storage (CCS)
chain reaction
coal-bed methane
compressed natural gas
dish/engine
economically recoverable oil
energy return on investment (EROI)
ethanol
feed-in tariffs
fissile isotope
fission
fluidized-bed combustion
geothermal energy
hydraulic fracturing
hydrocarbon compound
hydrogen
hydrothermal
integrated gasification combined cycle (IGCC)
kerogen
liquefied natural gas
methane hydrate
methanol
miscible
natural gas
nonrenewable resource
ocean thermal energy conversion
oil shale
parabolic trough
photochemical smog
photovoltaic (PV)
power tower
rank
renewable resource
reserves
reservoir rock
salt dome
secondary recovery
slant drilling
solar-thermal
source rock
Stirling cycle heat engine
stratigraphic trap
synfuels
syngas
tar sands
technically recoverable oil
tidal energy
tight oil
unconventional oil and gas
wave energy
wave energy conversion (WEC)
wildcat well

STUDY QUESTIONS

1. What four conditions are required for an oil deposit to form?
2. What oceanographic conditions favor the formation of a good source rock for oil?
3. Draw a geological cross section that shows an anticline and a faulted anticline. Assume that your cross section is looking north. What are the directions of the dips and strikes of the two limbs of the anticline? Dip and strike determination is explained in Appendix 3.
4. What are the similarities and differences of conventional and unconventional oil and gas deposits?
5. Is hydraulic fracturing, or "fracking," safe? Explain.
6. How does M. King Hubbert's 1956 prediction of petroleum decline apply today?
7. Which would be the best renewable energy development for investment and why?
8. Describe atomic fission. How does it create electrical energy?
9. Should the world build more nuclear reactors to meet the energy needs of the 21st century? Why or why not?
10. What are the environmental consequences of burning fossil fuels? What will be the impact upon the global environment of continued dependence on fossil fuels?
11. How can solar energy be collected and used for heating, generating electricity, and producing fuels?
12. What sector of the U.S. economy uses the most fossil fuel, and what international problem is connected with this use?

REFERENCES AND FURTHER READINGS

Fact sheets may be obtained from the USGS free of charge. Write to U.S. Geological Survey (Information Services), P.O. Box 25286, Denver Federal Center, Denver, CO 80225.

Bartlett, Albert A. 2004. *The essential exponential! For the future of our planet*. Lincoln, NE: Center for Science and Mathematics, 291 p.

Beaumont, Peter. 2012, March 31. Gulf's dolphins pay heavy price for Deepwater oil spill. *The Guardian*. Available at: http://guardian.co.uk/

Canadian Association of Petroleum Geologists. 2011. *About Canada's oil sands*. CAPP, 56 p.

Deffeyes, Kenneth S. 2003. *Hubbert's Peak: The impending world oil shortage*. Princeton, NJ: Princeton University Press, 208 p.

Deffeyes, Kenneth S. 2005. *Beyond oil: The view from Hubbert's Peak*. New York: Hill and Wang, 200 p.

Duffield, W. A., and J. H. Sass. 2003. *Geothermal energy—Clean power from the Earth's heat*. U.S. Geological Survey Circular 1249, 36 p.

Finch, Warren I. 1997. *Uranium, its impact on the national and global energy mix*. U.S. Geological Survey circular 1141, 24 p.

Fisher, William L. 2002. The coming methane economy. *Geotimes* 47 (11): 20–22.

Goldstein, David. 2004. *Out of gas: The end of the age of oil*. New York: W. W. Norton, 140 p.

Hall, Charles A. S., and John W. Day. 2009. Revisiting the limits to growth after peak oil. *American Scientist* 97 (3): 230-237.

Hall, Charles A. S., Stephen Balogh, and D. J. R. Murphy. 2009. What is a minimum EROI that a sustainable society must have? *Energies* 2: 25–47.

Heinberg, Richard. 2004. *Powerdown: Options and actions for a post-carbon world*. Gabriola Island, British Columbia: New Society Publishers, 209 p.

Hubbard, N. King. 1971. The energy resources of the Earth. *Scientific American* 225 (3): 60–70.

Kolbert, Elizabeth. 2012, April 3. Bill McKibben on Keystone XL and the power of fossil fuel industry. *The Guardian*. Available at: http://www.guardian.co.uk/environment/2012/apr/03/keystone-xl-fossil-fuel-industry.

Kunzig, Robert. 2009. The Canadian oil boom. *National Geographic* 215 (3): 34–59.

Lewis, Nathan S. 2007. Powering the planet. *Engineering and Science* 70 (2): 12–23.

Nuccitelli, Dana. 2011, August 23. What tar sands and the Keystone XL pipeline mean for climate change. *The Guardian*. Available at: http://www.guardian.co.uk/environment/2011/aug/23/tar-sands-keystone-xl-climate

Rutledge, David. 2011. Estimating long-term world coal production with logit and probit transforms. *International Journal of Coal Geology* 85: 23–33.

U.S. Geological Survey. 2000. *USGS world petroleum assessment 2000*. Fact Sheet FS-070-00, April.

U.S. Geological Survey. 2001. *Arctic National Wildlife Refuge, 1002 area petroleum assessment, 1998*. Fact Sheet FS-040-98, April.

U.S. Geological Survey. 2001. *Coal-bed gas resources of the Rocky Mountain region*. Fact Sheet FS-110-01, November.

U.S. Geological Survey. 2002. *Natural gas production in the United States*. Fact Sheet FS-113-01, January.

Wilshire, H. G., J. E. Nielson, and R. W. Hazlett. 2008. *The American West at risk: Science, myths, and politics of land abuse and recovery*. New York: Oxford University Press, 619 p.

Access an interactive eBook and chapter-specific interactive learning tools, including flashcards, quizzes, videos and more in your Geology CourseMate, accessed through CengageBrain.com.

An active sanitary landfill.

Have you ever
wondered...

We aren't going to solve world peace if we can't figure out what to do with garbage.

—Willie Brown, Mayor of San Francisco, 1996–2004

15

Waste Management and Geology

1. How much trash does the average American generate each year?

2. How much of that trash is recycled? Composted?

3. Are Americans improving in their effort to compost and recycle?

4. What happens to municipal solid waste (MSW) when a community does not have a suitable landfill site?

5. Is there really an "away" when you throw something away?

The Plastic Pacific

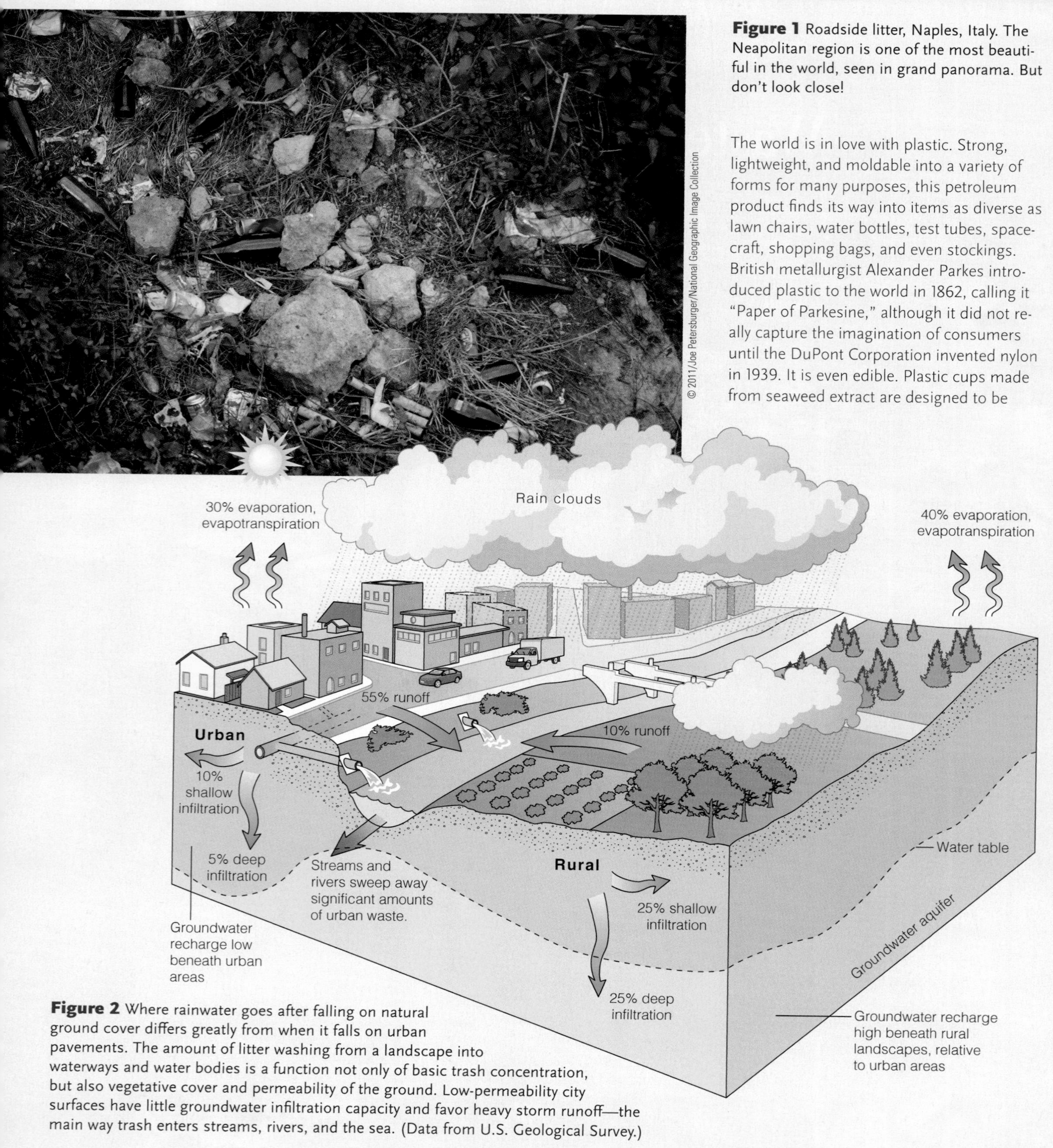

Figure 1 Roadside litter, Naples, Italy. The Neapolitan region is one of the most beautiful in the world, seen in grand panorama. But don't look close!

The world is in love with plastic. Strong, lightweight, and moldable into a variety of forms for many purposes, this petroleum product finds its way into items as diverse as lawn chairs, water bottles, test tubes, spacecraft, shopping bags, and even stockings. British metallurgist Alexander Parkes introduced plastic to the world in 1862, calling it "Paper of Parkesine," although it did not really capture the imagination of consumers until the DuPont Corporation invented nylon in 1939. It is even edible. Plastic cups made from seaweed extract are designed to be

Figure 2 Where rainwater goes after falling on natural ground cover differs greatly from when it falls on urban pavements. The amount of litter washing from a landscape into waterways and water bodies is a function not only of basic trash concentration, but also vegetative cover and permeability of the ground. Low-permeability city surfaces have little groundwater infiltration capacity and favor heavy storm runoff—the main way trash enters streams, rivers, and the sea. (Data from U.S. Geological Survey.)

Figure 3 Laysan albatrosses live amidst litter washed up from the great North Pacific Garbage Patch in the remote Hawaiian Leeward Islands, one of the most geographically isolated wildlife refuges in the world.

Table 1 Time For Products to Break Down at Sea

Product	Total Decay Interval
Paper towel	2–4 weeks
Newspaper	6 weeks
Cardboard box	2 months
Apple core	2 months
Waxed milk carton	3 months
Plywood	1–3 years
Plastic bags	10–20 years
Plastic bottle	100 years
Aluminum cans	200 years
Plastic beverage holders	400 years
Monofilament fishing line	600 years

Source: Mote Marine Laboratory, Florida.

eaten as one drinks their contents. And they come in a variety of delicious flavors—from lemon to rosemary-beet. But many plastics also contain potentially harmful additives, including *bisphenol A* (BPA) and *phthalates,* that interfere with the normal estrogen-producing functions of the human body—and those of wild animals too. Problems with brain development, reproduction, and other organ functions have been attributed to chronic exposure to these substances. This would not be such a serious environmental concern, however, if plastics did not enter the natural world so easily.

Plastics are the principal form of long-lasting litter in the world today. Being lightweight and easily disposable, much of this waste simply comes from casual discards or articles falling off of vehicles along roads and highways. Who litters the most? Circumstances vary greatly from region to region and over time. Geographer Cory Rhodes undertook a 2006 statistical study of littering in the Dog River watershed around Mobile, Alabama, and found no direct correlation between age, race, sex, and littering there, but he did find significant correlation with educational attainment, labor force status, and income. Wealthier, well-educated people tend to litter less; but they can also afford to clean up their neighborhoods more easily too. The "cleanest" states in the United States also have among the highest per capita incomes, including Vermont, Connecticut, Minnesota, Iowa, and New Hampshire, according to a 2008 study by the American Society for Public Administration. The same study showed that the "dirtiest" states are located in the southern and southwestern United States, where per capita income levels are notably low relative to the national average.

Why do people litter? An Ohio Department of Natural Resources Division of Recycling and Litter Prevention study in 2003–2004 showed that the greatest litter concentrations in Ohio occurred along national interstate freeways and highways, with 25% less litter along state and county roads. Rural littering is significantly greater than urban littering. These trends and others suggest the following factors are important for motivating litterers: lax law enforcement and lack of any personal connection with a particular space. Also important are simple convenience ("throw it out of the car so that we don't have to deal with the mess") and a disconnection from reality all too common in the origin of environmental problems ("don't worry; someone else will clean it up"). Moreover, as well-known in studies of urban graffiti, litter simply encourages more litter (**Figure 1**).

What usually "cleans it up" however, is Nature in the form of heavy rains. Runoff swiftly conveys rubbish to streams, rivers, and storm drains, and washes it into wetlands, lakes, and ultimately the sea (**Figure 2**). Most types of storm debris break down rapidly, but not plastics (**Table 1**). Since the end of the 1960s, substantial amounts of floating trash have accumulated in the central gyres of the world's oceans, especially in the northern Pacific and Atlantic—largely the product of storm runoff from the surrounding continents. Research undertaken by the U.S. Geological Survey (USGS) in 2003 suggested that 60% to 80% of this maritime debris derives from urban settings, where runoff during storms is especially intense because of pavements, and where would-be human litterers are concentrated (**Figure 3**). The rest comes from ships, and falls into three general categories: *flotsam* (material washed off of boats accidentally or uncontrollably, usually during storms), *jetsam* (material deliberately discarded by ships dumping their waste at sea), and *ligan* (flotsam that is tagged by owners who seek to recover whatever is lost when it washes ashore or is otherwise fished out of the water). Severe storms and high waves toss 2,000 to 10,000 cargo containers from the decks of ships each year.

The most common types of *visible* waste reported by the USGS in the "garbage patch" of the Pacific gyre include plastic water bottles, cigarette lighters, plastic bags, tampon applicators, and toy soldiers and other figurines. Traffic cones, plastic caps, fishing floats, tennis shoes, and many other items also bob about (see Figure 3). Not immediately visible floating within a meter or so of the surface is an even greater volume of litter—tiny bits of broken-up plastic resembling natural food, which are readily consumed by fish and other marine animals. In a 2001 central Pacific study, Captain Charles Moore (who discovered the garbage patch there in 1997) found plastic particles in concentrations averaging more than 330,000 pieces per square kilometer of ocean surface—some 5,100 grams (11.3 pounds) of debris. The mean diameter of particles was 5 millimeters (0.2 inch). The actual extent of impact on natural biota is very uncertain—perhaps hundreds of thousands of seabirds alone die each year. Swallowed by a gull or turtle, the plastics and other waste items can block the digestive track, perforate the gut, and cause malnutrition because of displacement of ordinary food. Laws now exist for prevention of disposal of plastics from seagoing vessels, including the International Convention for the Prevention of Pollution from Ships (MARPOL Annex V), and the Marine Plastic Pollution Research and Control Act. Many ships now bring plastic waste back to shore each year where it is recycled. The many tiny bits of shredded plastic that might otherwise end up in the oceans now go to fill pillows and fleece blankets, and to make flower pots and artificial lumber. But the scale of the problem remains staggering, given the long residence times of unrecovered floating plastics. Barring extraordinary technology and regulations, or massive change in ordinary human behavior, nothing can prevent massive fresh inputs from land. The March 2011 tsunami in Japan alone swept an estimated 20 to 40 million tons of plastic waste into the Pacific.

Figure 4 Nature meets trash in a Tahsis, British Columbia, dump. At the southern edge of their ranges, polar bears are rapidly becoming synanthropes—wild creatures that thrive within human habitats, in many cases, lured by litter. Better known examples of synanthropes include pigeons, squirrels, crows, gulls, and rats.

The plastic water bottles that make up so much of our litter today only entered the market around 1970. They represent a surprisingly wasteful use of natural resources, all for the sake of convenience. According to a 2009 Pacific Institute study, 1 million tons of polyethylene water bottles were produced in the United States in 2006. This manufacture used the equivalent of enough oil to run a million cars per year with average mileage and distances driven. Each ton of plastics that is recycled embodies the equivalent of two Americans' average annual use, or 2 months of water consumption per person. The Clean Air Council points out that the State of California spends $25 million each year sending plastic bags to landfills, and another $8.5 million to remove littered bags from streets. The energy use expended to transport a typical water bottle for sale in Los Angeles is 100 to 2,000 times the energy cost of providing an equivalent amount of tap water, perfectly safe to drink in that city. None of this is necessary. Reusable water bottles are generally easy to purchase, even in lesser developed countries, and textile shopping bags could well replace those derived from plastics.

What do we do with all of this waste? How do we protect the environment, and ourselves, from the harm it can bring? Getting rid of trash is a major environmental problem, especially in industrialized nations, which produce so much highly toxic material (**Figure 4**). This is the subject of this chapter.

questions to ponder

1. How can we isolate trash on an ordinary basis so that it doesn't escape into the environment in harmful ways?
2. How do we deal with the most dangerous forms of waste, such as toxic chemicals and radioactive materials?
3. How do liquid wastes spread if they are not effectively contained?
4. How is a typical residential septic system designed?

Figure 15.1 Wrangell's open trash dump before it was closed and sealed in 1999.

Coursemate Connection

Start off this chapter by listening to a podcast with National Geographic Emerging Explorer Ashley Murray as she talks about her work in Ghana, transforming a crude human waste treatment plant into a profitable fish farming business. **Log in to the CourseMate at www.cengagebrain.com to listen now.**

WRANGELL, ALASKA, is a town of about 2,000 people on Wrangell Island in the state's beautiful southern panhandle. In 1999, the town's open trash dump was deemed unacceptable and was sealed (**Figure 15.1**). It was determined that the island has no geologically suitable landfill site, and that the town's solid waste must be transferred by ocean barge to acceptable sites in the lower 48 states, hundreds of kilometers away. Much of the town's small budget goes to pay for this means of waste disposal.

In sharp contrast with the picture at Wrangell is an engineered sanitary landfill in Southern California (**Figure 15.2**). Graded slopes are designed to provide stability, and a plastic liner has been placed beneath a compacted, impermeable clay layer. The clay-layer "seal" prevents liquids from migrating out of the fill from the decaying locally derived waste into the groundwater environment. Geology is very important in the world of trash.

Municipal solid waste (MSW) generated in the United States in 2010 amounted to about 250 million tons. Of that amount, 85 million tons were composted and recycled, equivalent to a 34% recycling rate (**Figures 15.3** and **15.4**). On average, Americans generated 4.4 pounds per person per day, of which 1.5 pounds was composted or recycled. This category of waste includes yard cuttings, garbage, construction materials, paper products, metal cans, plastics, and glass. It does not include wastes from agriculture, industry, utilities, or mining, which account for more than 96% of the waste generated in the United States (**Figure 15.5**). After incineration and recovery for recycling and composting, 57% went to a landfill. Just one day's total U.S. waste would cover 15 square kilometers (almost 6 square miles) to a depth of 3 meters (10 feet). If it were loaded into 10-ton trucks lined up bumper to bumper, the trucks would stretch around the world 20 times. Those trucks containing only the day's *municipal solid waste* would circle Earth almost three times. The spectrum of solid-waste disposal problems ranges from the need to isolate highly dangerous nuclear waste to the challenge of dismantling and scrapping an astounding number of cars every year, es-

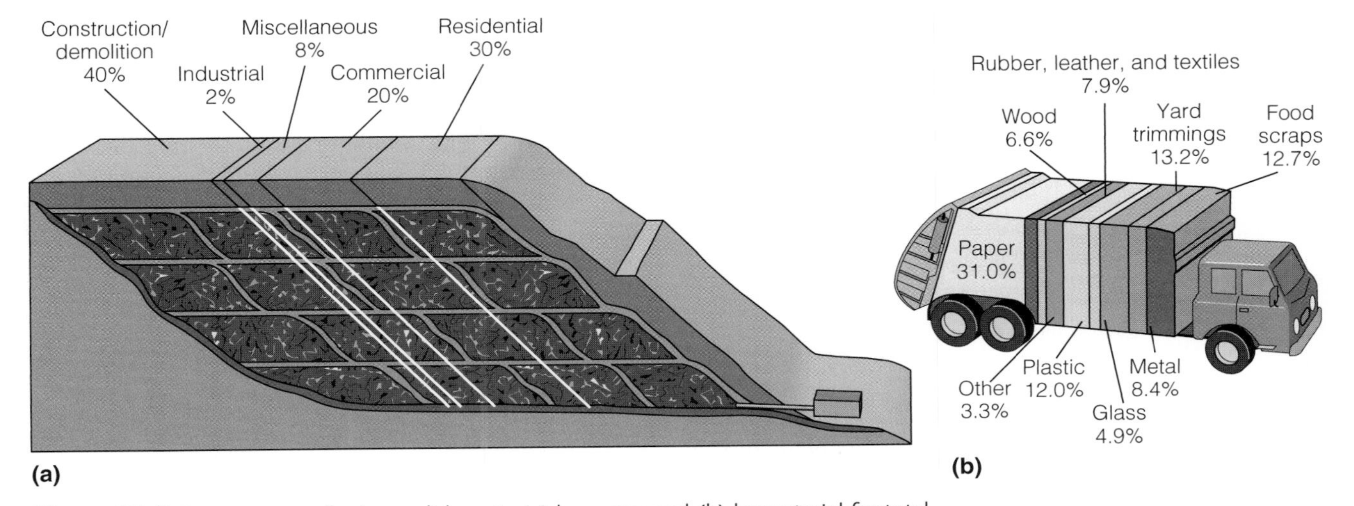

Figure 15.2 Components of urban solid waste (a) by sector and (b) by material for total waste generated in 2008 (249.6 million tons before recycling). (EPA)

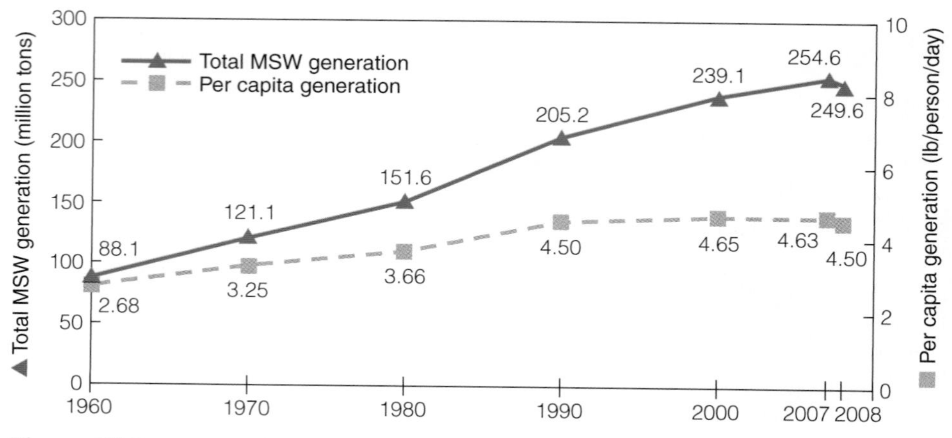

Figure 15.3 U.S. municipal solid waste (MSW) rates, 1960–2008. (EPA)

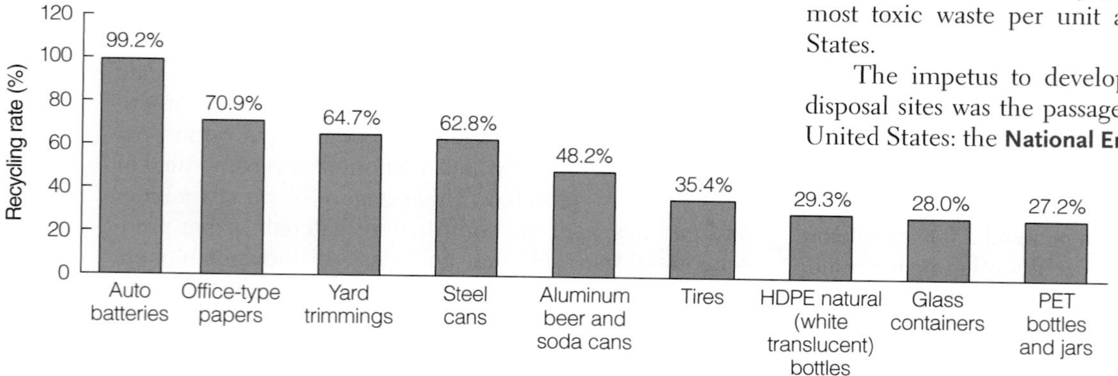

Figure 15.4 Recycling rates of selected materials recovered from the waste stream for 2008 as a percentage of the total weight recovered, not including combustion (with energy recovery). HDPE = high-density polyethylene; PET = polyethylene terephthalate. (Data from EPA.)

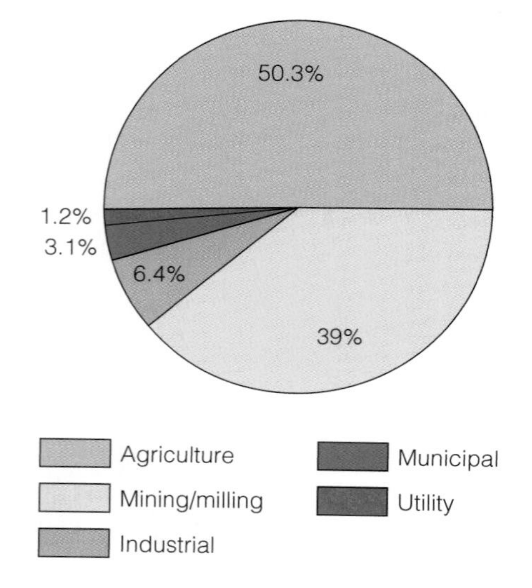

Figure 15.5 Estimated relative contributions to total U.S. solid waste, 1992. The ratios had not changed appreciably by 2010. (EPA)

timated at 15 million in the United States alone. (One year, 60,000 cars were abandoned on the streets of New York City.)

The amount of trash a country or governmental entity generates *per unit of land area* is just as important as the *total* amount it generates. If small countries with limited disposal sites generate large quantities of trash, major problems are created (**Figure 15.6**). Although the United States generates the highest total amount of waste in the world, the tiny city-state of Singapore produces the most trash per unit area of land, almost 2,500 tons per square kilometer (6,500 tons per square mile) each year. This is in comparison with less than 2 tons per square kilometer (5.2 tons per square mile) for Canada and 20 tons per square kilometer (52 tons per square mile) for the United States. Poland generates the most *industrial* waste per unit area, followed by Japan; Hungary produces the most toxic waste per unit area, followed by the United States.

The impetus to develop environmentally safe trash-disposal sites was the passage of two important laws in the United States: the **National Environmental Policy Act (NEPA)** of 1969 and the **Resource Conservation and Recovery Act (RCRA)** of 1976. The NEPA requires an in-depth field study and the issuance of an environmental impact statement on the consequences of all projects on federal land. The RCRA mandates federal regulation of waste products and encourages solid-waste planning by the states. The RCRA addresses the problems of hazardous waste and its impact on surface-water and groundwater quality, and it created the framework for regulating hazardous wastes. These laws sent the message that the then-prevailing "out of sight, out of mind" approach to disposing of wastes of all kinds, hazardous and benign, was no longer acceptable. The laws mandated "cradle to the grave" systems for impounding wastes and monitoring them to ensure their low potential for migrating into freshwater supplies. The **Environmental Protection Agency (EPA)** is charged with monitoring and policing the provisions of these and subsequent laws and amendments intended to prevent pollution of natural systems (see Case Study 15.1).

Principle Waste Management Approaches

Disposing of solid and liquid wastes, regardless of how they are generated, falls into three main categories:

- **Isolation**—encapsulating, burying, or in some other way removing waste from the environment.

© 2010/Ibrahim Muneez/National Geographic Image Collection

Figure 15.6 Smoke rises from the island of Thilafushi, which is a wholly dedicated trash dump for the Maldive Republic in the Indian Ocean, a string of coral atolls and home to 330,000 people crowded onto less than 300 square kilometers (115 square miles) of land.

- **Incineration**—the burning of trash leaving behind a much smaller volume residue that then can be attenuated or isolated
- **Attenuation**—diluting or spreading trash or a pollutant so thinly that it has little impact (*"Dilution is the solution."*)

Let's consider each of these approaches in turn.

Isolation

In many less developed rural areas around the world, and in the early history of the United States, people disposed of much of their household waste by simply feeding it to livestock, in particular pigs. With increasing development, the need for more organized, centralized concentration of wastes arose, leading to the establishment of open air dumps. Many consisted of trash thrown freely into small excavations or gullies at the edges of towns. Such garbage dumps are no longer acceptable where people can afford alternatives because of their attendant insect, vermin, odor, and air-pollution problems.

Most wastes in the developed world today are sorted, concentrated, and buried at carefully designed disposal sites called **sanitary landfills,** or simply **landfills.** The method involves spreading trash in thin layers, compacting it to the smallest practical volume with heavy machinery, and then covering the day's accumulation with at least 15 centimeters (6 inches) of soil (**Figure 15.7**). Compartments of waste handled in this way are isolated from the environment in rhomb-shaped compartments, or cells. When finished, the landfill is sealed by 50 additional centimeters (20 inches) of compacted soil and is graded, so that water will drain off the finished surface. This prevents water infiltration and the production of potentially toxic fluids within the fill (**Figure 15.8**). A well-operated sanitary landfill is relatively free of odors, blowing dust, and debris (see Case Study 15.2).

Leachate is the water from rainfall, snowmelt, decay, or spraying for dust control that filters down through a landfill, acquiring (leaching out) dissolved chemical compounds and fine-grained solid and microbial contaminants as it

© Cengage Learning

B. Pipkin

Figure 15.7 (a) The method of dump and cover used in a sanitary landfill. (b) An urban sanitary landfill in action in Palos Verdes, California.

Table 15.1 Classifications of Disposal Sites and Waste Groups	
Geology of Disposal Sites	
Class 1	No possibility of discharge of leachate to usable waters. Inundation and washout must not occur. The underlying lining material, whether soil or synthetic, must be essentially impermeable (i.e., must have a permeability less than 0.3 centimeters or a tenth of an inch/year). All waste groups may be received (see Figure 15.9a).
Class II	Site overlies or is adjacent to usable groundwater. Artificial barriers may be used for both vertical and lateral leachate migration. Geological formation or artificially constructed liners or barriers should have a permeability of less than 30 centimeters or 11 inches/year. Groups 2 and 3 waste may be accepted (see Figure 15.9b).
Class III	Inadequate protection of underground- or surface-water quality. Includes filling of areas that contain water, such as marshy areas, pits, and quarries. Only inert Group 3 wastes may be accepted (see Figure 15.9c).
Constituents of Waste Groups	
Group 1	Consists of but not limited to toxic substances that could impair water quality; examples are saline fluids, toxic chemicals, toilet wastes, brines from food processing, pesticides, chemical fertilizers, toxic compounds of arsenic, and chemical-warfare agents
Group 2	Household and commercial garbage, tin cans, metals, paper products, glass, cloth, wood, yard clippings, small dead animals, and hair, hide, and bones
Group 3	Non–water-soluble, nondecomposable inert solids, such as concrete, asphalt, plasterboard, rubber products, steel-mill slag, clay products, glass, and asbestos shingles

goes. Protecting the local environment from groundwater leachate is a major consideration in designing and operating waste-disposal sites, as we discuss later. Landfills and individual disposal sites within a given landfill are classified on the basis of their underlying geology and the potential impact of leachate on local groundwater or surface water. Each of the three classes is certified to accept specific wastes of differing degrees of reactivity (**Table 15.1** and **Figure 15.9**). Some landfill sites encompass all three classifications within their boundaries.

The world's first solid-waste sanitary landfill was established in Great Britain in 1912. Known as "controlled tipping," this method of isolating wastes expanded rapidly and had been adopted by thousands of municipalities by the 1940s. The city of Fresno, California, claims to be the first city in the United States to use a sanitary landfill for waste disposal. Perhaps counterintuitively, the number of U.S. landfills has declined in the last few decades, from more than 7,900 in 1988 to just around 1,900 in 2009. Despite the drop, total national landfill capacity has increased as individual landfills have grown larger. Most landfill closures have occurred because they had reached their capacity, but many others could not meet the stringent EPA standards for protecting groundwater.

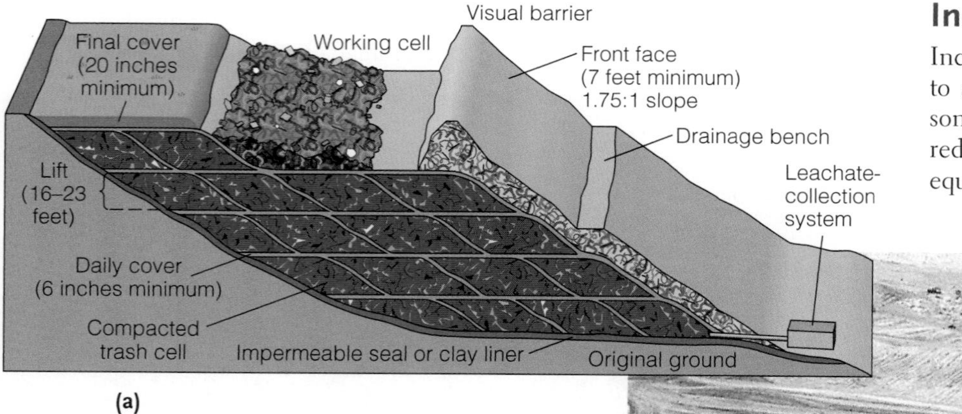

(a)

(b)

Figure 15.8 (a) Local government landfill specifications for cells, final cover, and leachate-collection system. Note the barrier for reducing the visual blight of the landfill. (b) A full, graded landfill before planting in Puente Hills, Los Angeles County, California. Final grades and slope benches for intercepting runoff are visible.

Incineration

Incineration, or burning, is the only proven way to significantly reduce the volume of garbage, sometimes by as much as 90%. In addition to reducing volume, combustors, when properly equipped, can convert water into steam to gen-

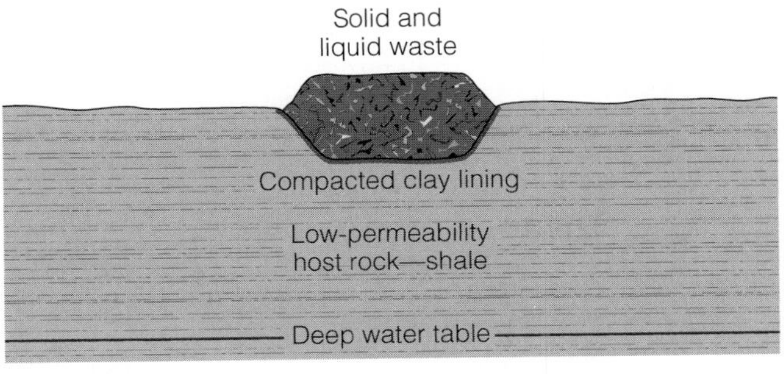

(a) Class I landfill

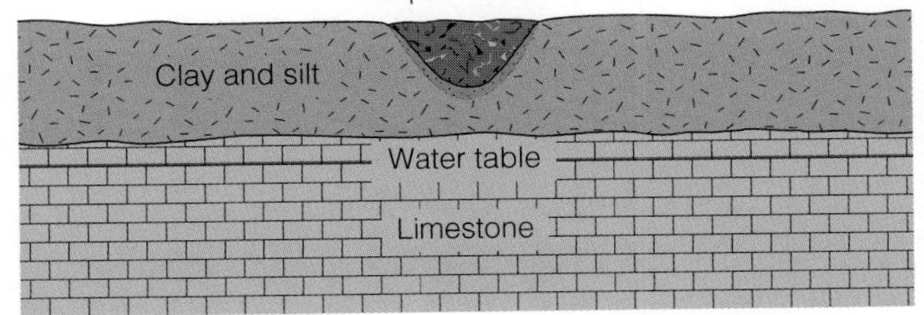

(b) Class II landfill

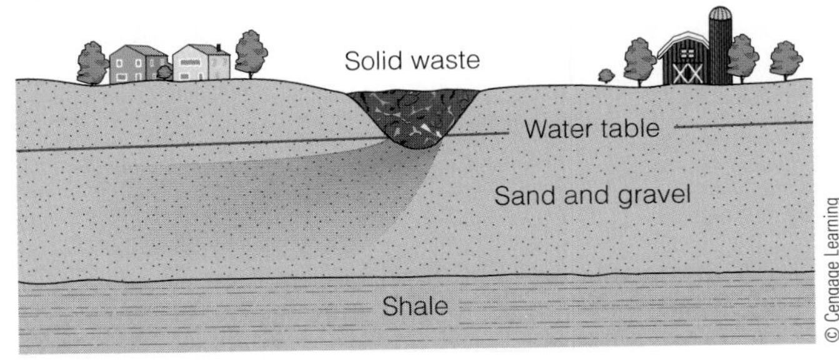

(c) Class III landfill

Figure 15.9 The geology of the three classes of landfills (a–c). Note that a Class I fill offers maximum protection to underground water and a Class III fill offers no protection.

erate electricity or fuel heating systems. Recycling can also be accommodated at incineration facilities. More than one fifth of the U.S. municipal waste incinerators use refuse-derived fuel, which is the combustible fraction remaining after the recovery of recyclables (metals, glass, cans, etc.). Scrubbers (devices that neutralize acid gases) and filters (devices that remove tiny ash particles) significantly reduce gases and other pollutants emitted to the atmosphere. All combustion ash and air emissions must meet the applicable federal and state regulations. The use of incineration is highest in areas with high population density, little open land, or high water tables. Massachusetts, New Jersey, Connecticut, Maine, Delaware, and Maryland, for instance, burn more than 20% of their waste.

New York City is a leader in trash incineration, burning about 720,000 tons per year, or 11% of all its waste. Wastes must be burned at very high temperatures, and incinerator exhausts are fitted with sophisticated scrubbers that remove dioxins and other toxic air pollutants. Incinerator ash presents another problem, because it is itself a hazardous waste, containing high ratios of heavy metals that are chemically active (**Figure 15.10**; also see Case Study 15.2).

The City of Philadelphia operates trash incinerators but lacks landfills for isolating the ash. In desperation, in the mid-1980s, 14,500 metric tons (16,000 tons) of the ash from two of its incinerators were loaded onto the steamship *Khian Sea* and sent to sea (**Figure 15.11**). Like the sailor in Coleridge's *Rime of the Ancient Mariner*, the *Khian Sea* sailed "alone, alone on a wide, wide sea" for 2 years, seeking a place to offload its unwanted cargo, with port after port refusing. In 1988, a miracle must have occurred, as the ship sailed into Singapore empty. No explanation has ever been given.

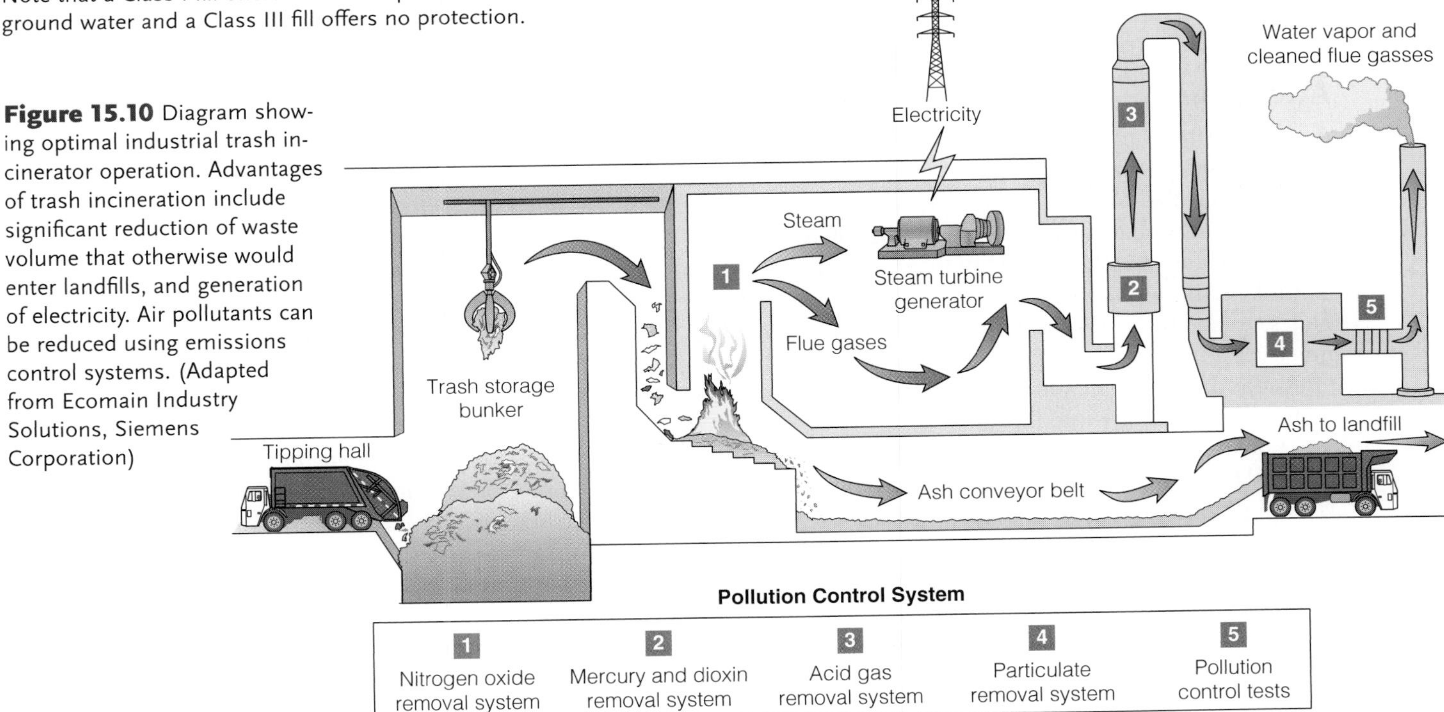

Figure 15.10 Diagram showing optimal industrial trash incinerator operation. Advantages of trash incineration include significant reduction of waste volume that otherwise would enter landfills, and generation of electricity. Air pollutants can be reduced using emissions control systems. (Adapted from Ecomain Industry Solutions, Siemens Corporation)

Figure 15.11 Greenpeace hung a protest banner on this loaded garbage barge in New York Harbor. In the past, trash such as this was destined for sea burial. Disposal of solid waste is a problem in urban areas in all parts of the world.
Dennis Capolongo, Greenpeace

More recently, incineration has fallen into disfavor, with only 89 plants remaining in 2007 from the 186 operating in 1990. The reasons for the reduction include more stringent federal and state air pollution emission requirements, public opposition, and expense. The disposal of municipal solid waste in landfills is less expensive than disposal of the residual ash.

Attenuation

A common form of attenuation waste management is the emission of wastes in factory or utility smokestacks. Though we now know that much of the gas and fine particulate matter coming out of a smokestack cannot be safely dispersed downwind, the principle of dilution as a means of waste control is still applied widely. Federal laws in the United States, guided by NEPA and EEPA regulations, seek to minimize or eliminate harmful constituents of smokestack emissions through mandating technologies to scrub or filter them out before entry into the air.

Looking seaward, coastal populations have long been tempted by the ultimate attenuation opportunity. The ocean is so big, nothing, it seems, could possibly be wrong with simply dumping community wastes out beyond the watery horizon. Various currents could sweep the waste away, and the simple vastness of the water could dilute its toxins to safe levels in terms of human exposure. Offshore waste disposal became an especially attractive means of getting rid of sewage sludge, the gray–brown solid effluent flushed into sewage treatment plants. (The liquid component of sewage is generally treated separately, and after advanced processing returned to waterways and aquifers). For decades, cities hauled raw garbage out to sea and dumped it rather than take it to landfills, although that practice ceased when New York City banned it in 1934 when too much unwelcome debris began washing back onto local swimming beaches.

An extreme example of ocean dumping came during the Cold War. Early during this era, radioactive materials accumulated faster than people could establish a sensible means of dealing with all the waste. So, why not tow it out to sea? Between 1946 and 1970, the U.S. government sank nearly fifty thousand 210-liter (55-gallon) drums of radioactive waste, mostly from laboratories, across 1,400 square kilometers (530 square miles) of ocean floor in the Farallon National Wildlife Refuge, a sanctuary famous for its abundance of marine mammals and great white sharks, 50 kilometers (30 miles) west of San Francisco (**Figure 15.12**). Laboratory wastes weren't the only radiation sources deposited there. The Navy also scuttled a 10,000-ton aircraft carrier, the *U.S.S. Independence*—irradiated during a 1946 South Pacific nuclear test—here in 1951. News of that sinking did not become public until 2001. Most radioactivity had dissipated by the 1980s, although a 1996 biological study revealed an order of magnitude greater concentration of plutonium isotopes 239 and 240 in Refuge fish relative to average values measured elsewhere in California waters—a sure indicator that some contamination had taken place.

Figure 15.12 The Farallon Islands off San Francisco; an unlikely place for a nuclear waste dump.

The United States was, of course, not alone in dumping radioactive wastes at sea during the Cold War. The Soviet Union sank a large number of decommissioned nuclear submarines and reactor parts in the seas surrounding its national territory. In the Barents Sea alone, 8 scuttled submarine hulls and 16 nuclear reactors (6 still containing fuel), plus 9,000 tons of discarded fuel assemblies lie on the shallow seabed (**Figure 15.13**). Large amounts of low- to intermediate-level radioactive waste were routinely dumped at more than 50 sites in the North Pacific and Atlantic oceans before an international moratorium in 1982. Combined with this were liquid circuit effluents and many other toxic military-industrial pollutants. Throughout, people placed faith in the *assimilative capacity* approach to waste disposal—that is, the assumption that a body of seawater can hold a certain amount of material without adverse biological impact. And so far, no "Godzilla's" have emerged from the ocean. But the legacy of possible Cold War contamination will likely continue in unpredictable, often undetected ways for millennia to come.

Today, ocean pollution is governed by the Marine Protection, Research and Sanctuaries Act of 1972, better known as the *Ocean Dumping Act.* This act requires anyone dumping waste into the ocean to have an EPA permit and to provide proof that the dumped material will not degrade the marine environment or endanger human health. In U.S. waters, the only legal dumping now consists of relatively harmless sediment dredged from coastal shipping channels and ports. The only potential risk this poses is the possible re-introduction of contaminants into the water column stirred up by dredging and settling. According to the 1988 Ocean Dumping Ban Act, even municipal U.S. sludge has to be disposed of on land (**Figure 15.14**).

consider this . . . Incinerating all our waste would go a long way toward solving the landfill capacity problem, and the heat that would be generated could be put to use. Some negative aspects of burning trash must be considered, however. What are they?

Some Special Situations

Sewage Sludge

Greek mythology relates how Herakles (Latin, *Hercules*) was given the task of cleaning King Augeas' stables, which contained 30 years' accumulation of filth from 3,000 head of cattle. Inasmuch as one cow produces roughly 18 wet

Figure 15.13 Highly radioactive waste from dumped nuclear reactors and scuttled or sunken submarines is a major concern in some waters of the world. Former Norwegian Minister of Foreign Affairs Johan Jorgen Holst called pollution from Russia "the biggest security problem Norway faces."

© 2011/Dennis Finley/National Geographic Image Collection

Figure 15.14 Archaeological researchers from the University of Arizona sampled the composition of our landfill trash, both biological and inorganic, as part of the Garbage Project (1973–2005). Professor William Rathje (aka "Captain Planet") set up the study to learn how Americans were being wasteful.

tons of manure per year, Herakles was up to his ears in 1.6 million tons of you-know-what, about the amount of sewage sludge New York City generates in 4 months. Because Herakles was to perform the cleanup in a day, he ingeniously diverted the courses of two rivers to make them flow through the stables and wash the filth into an estuary. Had an environmental impact statement been required, it would have noted that a huge mass of solid sludge would cover a large area of the wetlands and bury bottom-dwelling organisms. In addition, phosphates and nitrates in the effluent would promote explosive algal blooms at the expense of other organisms, and dissolved oxygen in the water would decrease to the point of mass mortality of swimming and bottom-dwelling organisms.

Coastal cities that once ocean-dumped were forced to begin dewatering sludge from sewage treatment plants and exporting it as a soil additive or depositing it in landfills. From 1992 to 2001, at the west Texas town of Sierra Blanca, trains from New York City, more than 3,200 kilometers (2,000 miles) away, arrived daily to dump 250 tons per week of wet sewage sludge, where it was spread like peanut butter on 31,800 hectares (78,500 acres) of a "ranch" owned by a Long Island, New York, company. The contract was not renewed in 2001, and the sludge ranch, probably the world's largest, now lies idle. As EPA employee and sludge critic Hugh Kaufman said, *"The fish in New York are being protected. The people in Texas are being poisoned."* Analyses show that there is 1 pound of lead in every ton of New York City sludge. There are also other heavy metals, polychlorinated biphenyls (PCBs), dioxin, banned pesticides, and various pathogens with which the residents of Sierra Blanca will have to live. Bill Addington, a Sierra Blanca businessman, described the noxious odors as an *"ammonia odor mixed with fecal smell"* and reported that the residents experience strange rashes, a higher than average incidence of flu, and other health issues.

Similarly, waste managers began spreading sludge near the city of Tucson, Arizona, on eroded, formerly grazed land with the hope of helping reduce erosion and fertilize vegetative recovery. Unfortunately, the sludge represented contributions from 1,500 industries containing more than 80 pollutants ranked as "priority" concerns by the EPA. These included such toxic substances as toluene, phenols, dioxin, cadmium, and arsenic. Spread human waste in "pure" form can be an effective fertilizer like that of many livestock animals, and for centuries has been applied by subsistence rice farmers in East Asia where it is called *night soil.* The European Union, in fact, endorses sludge applications on farms and degraded landscapes. As the world's geological reserve of phosphorus for agriculture becomes exhausted (in roughly 100–250 years), "recycled" phosphorus in the form of applied sludge may become increasingly important for maintaining human-carrying capacity, as well as dealing with urban waste issues. But when people mix industrial wastes with "natural" ones passing into a city's sewer lines, the combination quickly becomes far more problematic. Nevertheless, efforts continue to find a practical way of disposing of sludge in its current form. New York City imaginatively processes sludge into pellets at an organic fertilizer facility in the Bronx (creating an acrid stench across the borough on hot summer days that has caused quite a lot of howling). About 60% of the pelletized sludge ends up in Florida to be used as fertilizer on citrus groves—an uncomfortable thought for orange juice lovers. In commercial form, sludge sold for crops and treatment of degraded land comes in various names, such as *recyclite, nutricake,* and *powergro.* Critics have suggested alternatives such as *bioslurp, geoslime,* and *hu-doo.*

One costly solution to the issue of sewage sludge would be to re-sewer cities so that the most dangerous industrial chemicals were isolated from more "benign" purely biological components of the waste stream, and transported separately to sanitary landfills, deep injection wells, or other yet-to-be-developed treatment technologies. But basic human behavior also needs changing. The temptation to flush *anything* that will make it down a toilet or wash it off a driveway simply to get it "away" is too great for many, who in any event may simply be unaware of options. Cleaning solvents, soluble medications, paint thinners, greases, oils, and various other products often find their way into sewers and septic systems from ordinary households. Many municipalities in developed countries now provide "hazardous waste" delivery stations for residents to safely dispose of their unwanted substances. As ever, it is mostly an issue of personal convenience.

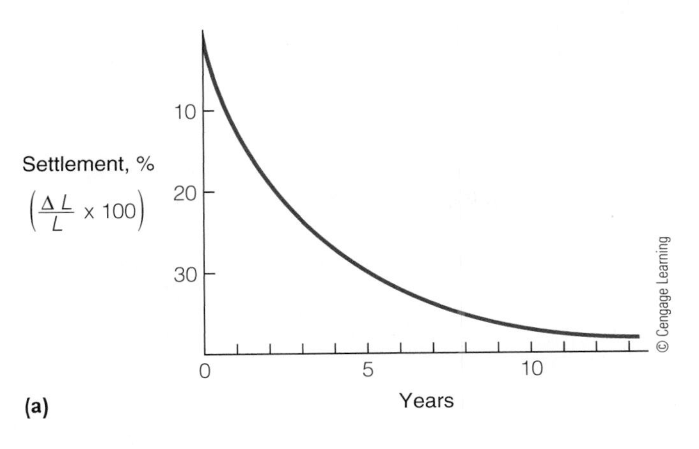

Figure 15.15 (a) Landfill settlement over time. The percentage of settlement is calculated by dividing the amount of settlement (ΔL) by the original thickness of the landfill (L) and multiplying by 100. The settlement values here are for illustration only and are not real values. (b) A landfill weigh station on the Palos Verdes Peninsula, California, built on trash-filled land. The station eventually collapsed, requiring that a new structure and scale be established at a different location.

Landfill Instability

Organic refuse, including sludge in landfills, ultimately decomposes. As it does, the ground surface settles; methane (CH_4), carbon dioxide (CO_2), and other gases are generated; and noxious or even toxic leachate from it may seep into the water table or bordering streams. Thus, a landfill's *pollution potential* depends on the change in volume that accompanies decomposition, the waste's reactivity (which determines the chemistry of the leachate), the local geology, and the climate. Organic matter decomposes more slowly in cold, dry climates than in warm, wet ones.

Landfills continue to settle as decomposition goes on for many decades after burial. Typically, settlement is rapid at first and diminishes to a much slower rate after about 10 years (**Figure 15.15**). The rate of subsidence depends on many factors, including the climate, the kind of waste, and the amount of compaction during filling. Thirty-ton tractors left on a fill over a weekend have been known to settle as much as a meter (3.3 feet) into the trash, and total settlement of as much as 30% has been measured in landfills. Variation is great. A 6-meter-(20 foot) thick landfill in Seattle settled 23% in its first year, whereas a 23-meter-(76-foot) thick fill in Los Angeles settled only 1% per year in its first 3 years. Settlement is a problem, because it results in cracks, fissures, and infestation by vermin. It is monitored to ensure that water does not pond and percolate downward, creating a high volume of polluting leachate.

consider this... Each ton of recycled mixed paper amounts to an energy savings equivalent to 700 liters (185 gallons) of gasoline.

As microorganisms break down organic material (*biodegradation*), they release heat, methane (natural gas), and other gases, including hydrogen sulfide, carbon dioxide, and oxygen. Research at the University of Arizona and elsewhere has shown that biodegradation can be surprisingly slow, as illustrated by exhumed garbage containing virtually unblemished 10- to 20-year-old newspapers (**Figure 15.16**), a 14-year-old mound of recognizable guacamole, and "ancient" hot dogs (testimony to the effectiveness of sausage preservatives). The most efficient decomposers are the voracious oxygen-using aerobic bacteria, such as those that quickly turn grass cuttings into compost for our gardens. Most landfills are sealed off from the atmosphere, however, giving rise to anaerobic (no oxygen) microorganisms that manufacture methane and hydrogen sulfide in the slow process of decomposing cellulose. Pumping air into a fill, a procedure known as **composting**, stimulates aerobes, accelerates decomposition and settlement, and creates more space for trash.

Progress-Bulletin
5¢
Pomona Temperature Hits 82; 20 Below in S. Dakota
16 Deaths Attributed To Weather
Virginia Warren To Wed John Daly
U. S. Tanker

Figure 15.16 A newspaper found underground. It is still readable after at least 10 years of burial in a Pomona, California, landfill.

Figure 15.17 Methane-extraction well and drill rig. The extracted gas is conveyed to a central location for conversion to electricity.

Figure 15.18 One of 660 methane-extraction wells on the Staten Island Fresh Kills landfill site in New York. The materials in the landfill decompose, releasing methane gas from the interior, which is funneled to the extraction wells. It is then piped to a power-generation plant on the site. The system not only provides electrical power, it also reduces the danger of landfill explosions and caps greenhouse gas emissions.

Landfill Gas Generation

Methane, a gaseous hydrocarbon, is the principal component of natural gas. Also known as *marsh gas,* it bubbles forth from stagnant ponds and swamps, the result of complex chemical fermentation of plant material by bacteria. It is the same gas that causes the explosions in coal mines that have taken innumerable lives. Methane is explosive when present in air at concentrations between 5% and 15%. Although enormous volumes of methane are generated by anaerobic microorganisms in a landfill, there is little danger of the fill's exploding, because no oxygen is present. Being lighter than air, methane migrates upward in a fill. Methane leaking through the cover of a landfill converted to a golf course near Bel Air, California, was known to "pop" from ignition when cigarettes were dropped on its greens. Methane can also migrate laterally, and it did so several hundred meters (yards) into the wall spaces of structures near a landfill south of Los Angeles. Fortunately, the potentially explosive situation was discovered, and wells were drilled to intercept and extract the gas (**Figure 15.17**). One strategy for handling the large volumes of methane generated at large landfills is to extract it through perforated plastic pipes, remove the impurities (mostly carbon dioxide), and use it as fuel—the so-called refuse-derived fuel, or *RDF.* This can be accomplished either by piping the gas away via pipelines or by using it onsite to generate electricity (**Figure 15.18**). Because the gas may be as much as 50% carbon dioxide, it may be more expedient to use the fuel directly to generate electricity than to clean out the impurities and put it into pipelines. Enough electricity is generated from 660 methane extraction wells at the Staten Island, New York, Fresh Kills landfill to provide electricity to about 2,200 homes (see Case Study 15.2), and the Los Angeles County (California) Sanitation District extracts methane from several landfills and generates sufficient electricity to supply up to 45,000 homes.

Leachate Pollution Patterns

The composition of leachate ("garbage juice") in and leaving landfills is highly variable, but it should be considered dangerous to the environment until established otherwise. A leachate may be such that a receiving body of water can assimilate it without any impairment in water quality. Some natural purification of a leachate occurs as it filters through clayey soils, and some purification occurs when contained organic pollutants are oxidized. Groundwater-quality measurements taken in the vicinity of a landfill, at the fill, and downslope from the fill are listed in **Table 15.2**. Note that the leachate was causing the deterioration of groundwater quality in the vicinity of the monitoring well. That is a red flag signaling need to take action.

Where leachates infiltrate a poorly operated landfill for a number of years, an underground area of polluted groundwater called a **contaminant plume** will develop (**Figure 15.19**). Individual plumes can extend up to kilometers (miles) from their sources in the direction of groundwater flow, and may be hundreds of meters (up to a thousand feet or more) thick. Where groundwater flows rapidly (meters a day) the plume outline is typically long and narrow in the direction of flow. A length-to-width ratio, or *aspect ratio,* of 10:1 is not uncommon. Slower developing plumes (centimeters or inches a day) tend to be stubby in general outline

Table 15.2 Groundwater Quality Measured Near One Landfill

	Measurement Site		
Water Characteristics	Local Groundwater	Leachate	Monitoring Well
Total dissolved solids, parts per million (ppm)	636	6,712	1,506
BOD,* milligram/liter	20	1,863	71
Hardness, ppm	570	4,960	820
Sodium content, ppm	30	806	316
Chloride content, ppm	18	1,710	248

Source: Brunner, D. R., and D. J. Keller. 1972. *Sanitary landfill design and operations.* Washington, DC: Environmental Protection Agency.

*Biological oxygen demand (BOD): the amount of oxygen per unit volume of water required for total aerobic decomposition of organic matter by microorganisms.

and are confined to the earth directly beneath a landfill as they begin forming. In addition to spreading leachates underground, landfill-related plumes generally release volatile organic compounds (VOC), gases from organic contaminants that they are carrying along. VOC gas detection is an effective way of identifying the presence of a plume that otherwise is hidden from the surface. More often, however, plumes are first detected through a physically notable change in the water quality of some local residents' or a town's well (see Table 15.2).

Leachates passing underground may be classified several ways. *Conservative contaminants* do not react with solid material as they pass underground and spread through an aquifer. They spread quite rapidly, as fast as the groundwater itself. *Reactive contaminants,* however, do not move as fast and often quite as far as conservative contaminants. They interact chemically and physically with the soil, mineral, and rocky material through which they move, in many instances, absorbed by clays, organic substances, and various oxides.

Underground leachates may also be categorized based on their relative density with respect to water: floating, soluble, and sinking. *Floating contaminants,* including lighter weight oils and greases and other organic compounds, will infiltrate down to the water table but then float along its surface (see Chapter 8); they are less dense than water, in other words. *Soluble contaminants* dissolve or mix well with groundwater. These include metals under the right conditions of acidity, nitrates from farm fertilizers, and miscible pesticides and herbicides such as atrazine, which has become a major concern with regard to possible infertility issues in males. *Sinking contaminants* sink all the way through an aquifer and may pool at the base, where the rock is no longer permeable. These include many solvents, such as vinyl chloride, benzene, and toluene. Common sources of sinking contaminants other than badly managed landfills include dry cleaners, printers, gas stations, septic tanks undergoing chemical cleaning, and even people who are washing their cars in their driveways. Many commercial car washes are required by law to have special bays that collect contaminated water to keep it from infiltrating the underground environment; this is not so for private homeowners.

Uncontrolled, a contaminant plume can devastate a community by destroying its supply of drinking water. In the 1970s, southern California introduced methyl butyl tertiary ether (MTBE) to its gasoline stations. The MTBE reduced the amount of carbon monoxide released by automobiles using the fuel—a major bonus for the greater Los Angeles area, which suffers from air pollution. But spills of MTBE from gas tanks and nozzles made their way into the water supply. In 1996, the first southern California city, Santa Monica, discovered that about 50% of its groundwater was heavily polluted with this toxic substance. Santa Monica had to divert precious city funds to buy drinking water from elsewhere to sustain its population. Cleanup efforts continue to the present.

Unfortunately, groundwater cleanup is not cheap. Various strategies have been used, depending on the scale of the problem. Probably the most effective is the construction of wells to pump out and clean up groundwater, which then can be re-injected underground or, if beyond recovery, stored elsewhere. Rows of *extraction wells* must be constructed at a right angle to the direction of plume movement to maximize impact (see Figure 15.19). In some instances, waste managers have also proposed injecting key bacterial species into contaminated aquifers to consume and break down leachates. For instance, grease-eating bacteria have proved effective in cleaning up clogged drain and sewer pipes. Could or *should* we bioengineer bacterial ecologies at a much larger scale to assure that our groundwater aquifers remain clean and healthful?

Perhaps the best way to deal with leachate toxins, however, is to be better about not letting them escape underground in the first place. A landfill that is in a geological setting where leachate threatens local water bodies may accept only inert (Group 3) wastes. And well-monitored landfills will not leak at all.

A well-constructed state-of-the-art landfill is illustrated in **Figure 15.20**. It is designed to collect leachate above the bottom liner and to recover methane for use in power generation. Note the inclusion of leachate- and gas-monitoring wells in the design. Double-lined landfills have been required in the United States since 1996 (most older landfills, though, are still unlined). These landfills are designed to accept waste for 10 to 40 years, and owners are required to monitor the fills for at least 30 years after closure. If they begin to leak after that period, the contamination problem belongs to whoever is living in the area at the time.

Recycling

Resource recovery—the removal of certain materials from the waste stream for the purpose of recycling or composting them—not only saves materials but also saves energy. Recy-

SPOTLIGHT FIGURE 15.19 Waste Management and Geology

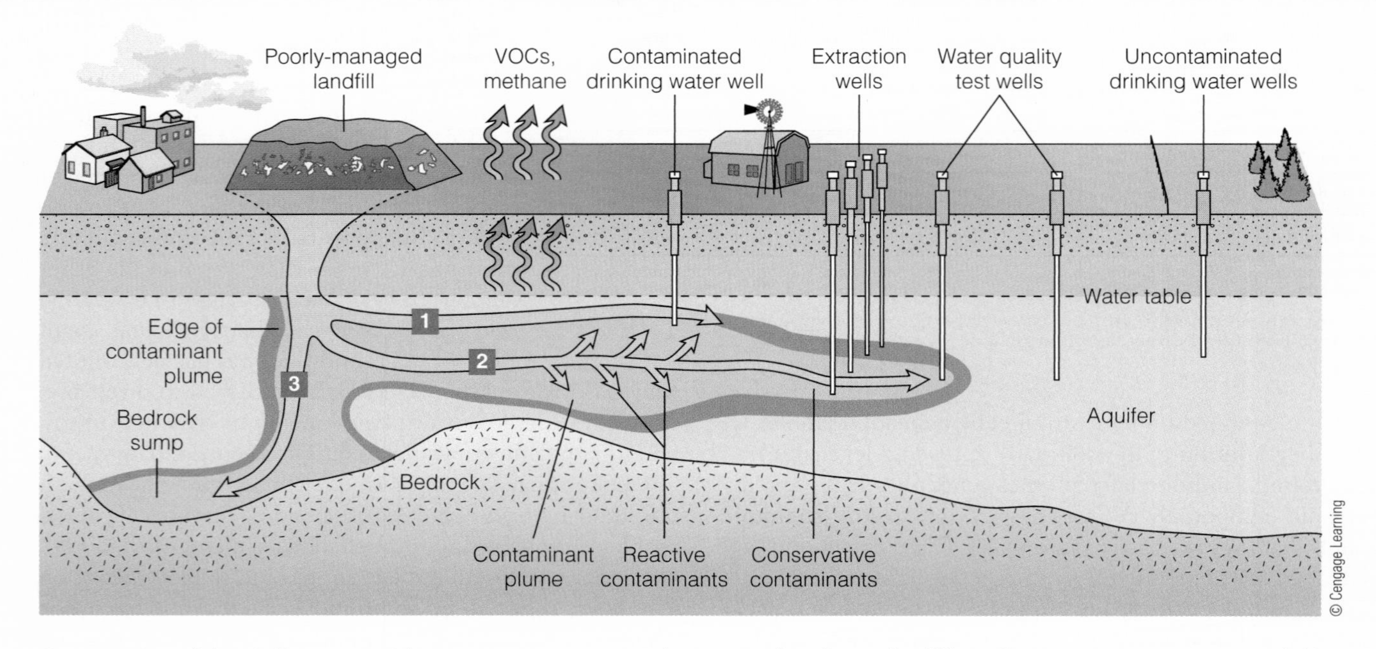

Cross section of the shallow crust with an active contaminant plume extending from a landfill; (1) floating contaminants; (2) soluble contaminants; (3) sinking contaminants.

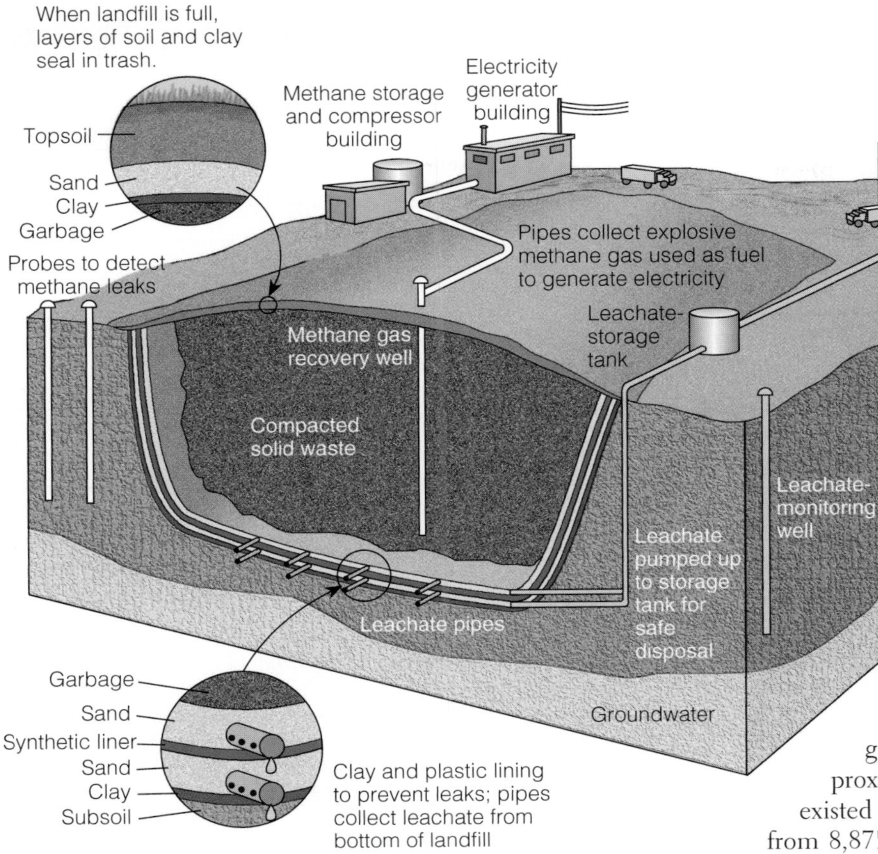

Figure 15.20 State-of-the-art landfills are designed to eliminate the problems of toxic leachate and methane escaping into the environment. Note the double lining (clay and plastic layers) to prevent leakage. Unfortunately, 85% of U.S. sanitary landfills are unlined.

cled aluminum cuts energy use 96% over processing aluminum ore. For recycled steel, the savings is 74%; for copper, 87%. In 2008, 83 million tons of material was recycled in the United States, which saved an amount of energy equal to 3.5 months' electrical demand. Remanufacturing—the process of disassembling a worn-out product, cleaning and replacing its own parts, and then reassembling it—is estimated to save 85% of the energy required to make the product new from raw materials. Remanufacturing is estimated worldwide to be equal to the electricity generated by five nuclear power plants and saves enough raw materials to fill 155,000 railroad cars, forming a train 1,100 miles long. The industrial nations are the "Saudi Arabias" of trash, and all have the same problems—dwindling resources and dwindling space for solid waste.

Until recently, recycling held little appeal; that is, it cost more to separate, sort, and recycle trash than to landfill it. Recycling is now mandated in many states, however, because existing landfills are filling, and acceptable sites for new ones are difficult to find. An effective way to extend the lifetimes of landfills and natural resources is to recycle municipal wastes. The consuming public is "thinking green" these days, and manufacturers are realizing favorable marketing results by using recycled materials in packaging their products. Because of a growing awareness that mineral and other resources are finite, recycling increased from 10% to 33% of municipal waste between 1985 and 2008 (**Figure 15.21**). One indicator of the anticipated value of trash is that glass and high-grade plastics have joined corn and pork bellies as commodities bought and sold on the Chicago Board of Trade.

Curbside recycling programs have grown enormously since 1980. In 2008, approximately 8,660 curbside recycling programs existed nationwide, although it was down slightly from 8,875 in 2002. Since 1980, the national recycling rate has increased from less than 10% to more than 33%, with hundreds of communities reporting levels greater than 50% and several exceeding 60%. Many communities use a "pay as you throw" program, which bases collection charges on the amount of true waste that is added to a landfill; materials sorted out for recycling are collected free of charge. Many communities require separation of glass, metal, paper products, and plastics from other trash materials. Plastic recycling is difficult to handle and expensive because there are seven kinds of plastic that must be separated by hand (**Table 15.3**). Only 2% of waste plastic was recycled in 1995; however, by 2008, 51.1% was being recycled. When it comes to recycling plastics, ingenuity pays; a small Massachusetts company grew into a $1.3 million business by recycling spent plastic ink-jet printer cartridges. It sells refurbished cartridges to consumers, and the worn plastic and metal components to recyclers.

The newest thing entering the waste stream is electronic waste, or "e-waste." Computers, printers, fax machines, and cell phones have short lives. The National Safety Council estimates that the life span of a cell phone is about 18 months, with PCs being replaced every 2 years, on average. It is estimated that, between 2000 and 2010, a half billion computers became obsolete—e-waste. And where do the outmoded electronic gadgets and discarded computers end up? Stored by their owners in garages, basements, storerooms, or left on the curb to end up as toxic waste in landfills? Because older computer monitors may contain more than 100 chemicals, including toxic metals (such as

Table 15.3 The Seven Types of Plastics

Plastics are categorized by number corresponding to general chemical compositions and physical properties. The numbers are labeled in triangular recycling logos on many products, in some instances, on their bases. Some types can be recycled and some cannot, as indicated below:

Number of Plastic	Composition	Notes	Recyclable?
1	Polyethylene terephthalate (PET or PETE)	Clear in color; most soda and water bottles; safe, but should not be reused because bacteria and flavoring will accumulate	Most curbside recycling programs will pick up
2	High-density polyethylene (HDPE)	Most milk jugs, detergent and juice bottles, containers for butter and toilet products; often opaque	The most often recycled plastic
3	Polyvinyl chloride (PVC)	Food wrap, plumbing pipes, cooking oil bottles; contains potentially harmful phthalates (don't cook with food wrap, especially in microwave ovens)	Generally not accepted by recycling programs
4	Low-density polyethylene (LDPE)	LDPE goes into grocery bags, some food wraps (notably breads and baked goods), squeezable food bottles	Not harmful to human health, but still not accepted by most curbside recycling programs; some markets accept in specialized recycling collectors
5	Polypropylene	Goes into medicine bottles, ketchup and syrup bottles, straws, yogurt cups, and many water bottles (with cloudy finishes)	Not harmful to human health but just beginning to be accepted by many curbside recycling programs
6	Polystyrene, or Styrofoam	Disposable cups, plates, picnic coolers; will release potentially harmful chemicals, especially when heated	Not accepted by most curbside recycling programs
7	All other types of plastics	Includes some potentially harmful substances, including bisphenol A in polycarbonate products; includes DVDs and CDs, baby and many water bottles, medical and dental devices	Not accepted by most curbside recycling programs

Source: Green Guide Network, Green Business Directory, 2012.

cadmium, mercury, and lead), these devices need to be removed from the waste stream.

Because more than 100 million cell phones and 250 million personal computers are thrown away each year, several computer manufacturers, such as Dell, HP, Intel, and Apple, offer "e-cycling" programs to their customers. Most manufacturers' recycling programs require their customers to buy a new computer to recycle the old one for free. Most large cities have commercial recycling firms that accept e-waste.[1]

In 2002, the European Union began requiring European manufacturers to pay for the entire cost of e-cycling their products—everything from laptops to stereos to computers. In the United States, grassroots computer-recycling

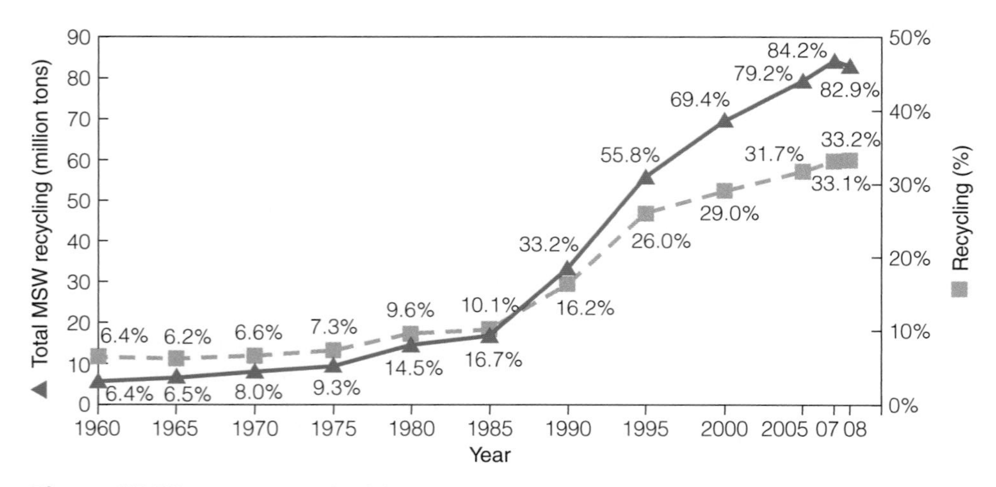

Figure 15.21 U.S. municipal solid waste (MSW) recycling rates, 1960–2008.

[1] Visit Earth911.com for recycling e-waste (e-cycling) information. Earth 911 has consolidated nationwide environmental Web sites, hot lines, and other information sources into one network. Once you contact Earth 911, you will find community-specific information on e-cycling and much more. My Green Electronics and TechSoup are organizations for consumers wishing to purchase green products and/or who are searching for local opportunities to recycle, donate, or refurbish used electronic hardware. For e-cycling portable rechargeable batteries—commonly found in cordless power tools, cell and cordless phones, laptop computers, camcorders, digital cameras, and so on—the Rechargeable Battery Recycling Corporation can help. Access its Web site and search for collection sites by zip code.

nonprofits, such as Free Geek in Portland, Oregon, are springing up. E-cycling is catching on, with some more aggressive states, including Maine, California, Minnesota, and Massachusetts, banning such electronics as cathode-ray tubes, the bulky picture tubes in older computer monitors and TVs, from landfills for fear that toxic materials will leak into groundwater. Washington State requires manufacturers to be responsible for both collecting and recycling their products.

Not easily solvable is the problem of what to do with old automobile tires. At least 3 billion used tires are stockpiled in the United States, a number that is growing at a rate of 250 million per year. When placed in landfills, they tend to add bulk very rapidly, a fact that has caused 33 states to ban them there. In 2000, the nation used 32 million scrap tires (California alone contributes this many) in civil engineering projects, an increase of 23% over 1999. The incidental uses of old tires are the construction of small dams, planting stacks for potatoes, erosion control, houses (New Mexico is the leader), fencing, rifle ranges, playgrounds (**Figure 15.22**), and grain-storage structures.

Figure 15.22 Three thousand tires were recycled as children's playground equipment and landscaping material at this park in Japan.

Figure 15.23 Worker at the Kirby Tire Dump in Sycamore, Ohio, rakes shredded tires for recycling into landfill liners, an environmentally friendly way of dealing with tire waste, provided the liners themselves don't significantly degrade.

Used tires are being used experimentally as bedding for livestock corrals, as a replacement for stone in septic systems, as reinforcement in flood-control structures, even (ironically) as crumb rubber modifier to make asphalt for new roads. Such roads may contain as much as 20% by weight tire scraps, and provide a generally quieter ride. Unfortunately, tires contain a number of potentially toxic substances that can be released to the environment through burning or ordinary weathering, including phenols, butylated hydroxyanisole, selenium, zinc, lead, and cadmium. Because old tires are such an enormous environmental problem, many engineers and scientists are working on a technological "fix" that will rid us of this blight (**Figures 15.23** and **15.24**).

As more curbside separation of garbage for recycling is mandated, and as more states enact container-bill legislation encouraging the return of beverage containers, the flow of recyclables increases. The "champion" states (percentage recycled) in the United States are Minnesota (41%), New Jersey (39%), and Washington (35%).

consider this . . . Recycling and composting 83 million tons of MSW saves about 1.3 quadrillion Btu of energy per year—the equivalent of more than 38.7 billion liters (10.2 billion gallons) of gasoline. This gives a benefit of eliminating 182 million tons of carbon dioxide per year, comparable with the annual greenhouse gas emissions from more than 33 million passenger cars. Taiwan is the world champion paper recycler; 98% of the paper used in that timber-poor country is recycled. But for all categories of waste, Switzerland takes the prize, with a national recycling rate of more than 50%. Recycling is provided for free in that alpine country, while homeowners are charged to throw out their trash according to the quantity produced. (This illustrates the value of well-conceived public policy as a tool for tackling environmental problems.)

Figure 15.24 Discarded tires are used to weigh down a plastic covering over a landfill in Northeast Kingdom, Vermont.

Composting

When decomposed by bacteria, biodegradable wastes, the so-called green garbage, can be used as a fertilizer and soil amendment. Yard cuttings, most kitchen wastes, and organic wastes from commercial operations are acceptable compost materials. Because yard wastes are second only to paper in landfill tonnage, many states and municipalities have mandated separate collection for them. As a result, the number of composting facilities in the United States grew dramatically from 651 facilities in 1988 to 3,510 in 2009.

Multiple Land-Use Strategies

Land, like trash, can be recycled. Where the local geology is amenable to it, abandoned rock quarries, as well as sand and gravel pits, have been converted to sanitary landfills. Population pressure and the need to conserve resources will require that we not only dispose of our waste efficiently, but also benefit from the stored trash and the disposal site. The average new home requires about 65 cubic meters (85 cubic yards) of concrete in its construction, and its inhabitants will generate about a ton of trash a year. Thus, it *makes sense* to place the household's trash in an opening created to build the house. After a landfill is completed, methane gas can be recovered from it and converted to energy, and the land may be reshaped into a golf course, an athletic field, a public garden, or the like. An ideal, beneficial cycle of multiple land uses is shown in **Figure 15.25**.

As explained earlier, structures placed directly on a landfill are subject to settlement and to methane emanations. A restaurant and shops built on a thin landfill (<8 meters; 26.5 feet thick) south of Los Angeles were placed on piles driven through the fill to keep them from settling. The walkways around the buildings were not, however, and after a few years they settled and had to be built up to the base of the structures (**Figure 15.26**). In the same commercial development, methane is extracted and used to illuminate exterior gas lanterns.

As cities expand and coalesce—such as in the New York to Boston corridor and the strip between Los Angeles and San Diego—waste-disposal site selection must become very creative.

consider this . . . During the past few decades in the United States, recycling waste has evolved from a concept to a major component of waste management for many communities. Why do some environmental groups object to waste-problem solutions that emphasize recycling?

Hazardous-Waste Disposal

Hazardous wastes are primarily such industrial products as sludges, solvents, acids, pesticides, PCBs, highly toxic organic fluids used in plastics and electrical insulation, and nuclear waste. The EPA estimates that U.S. industries generate 260 million metric tons of hazardous waste per year. A typical rate at which liquid hazardous waste was generated in the early 1980s was 150 million metric tons per year, which can be thought of as 40 billion gallons. If that amount could be placed in standard 210-liter (55-gallon) drums end to end, they would encircle Earth 16 times (**Figure 15.27**).

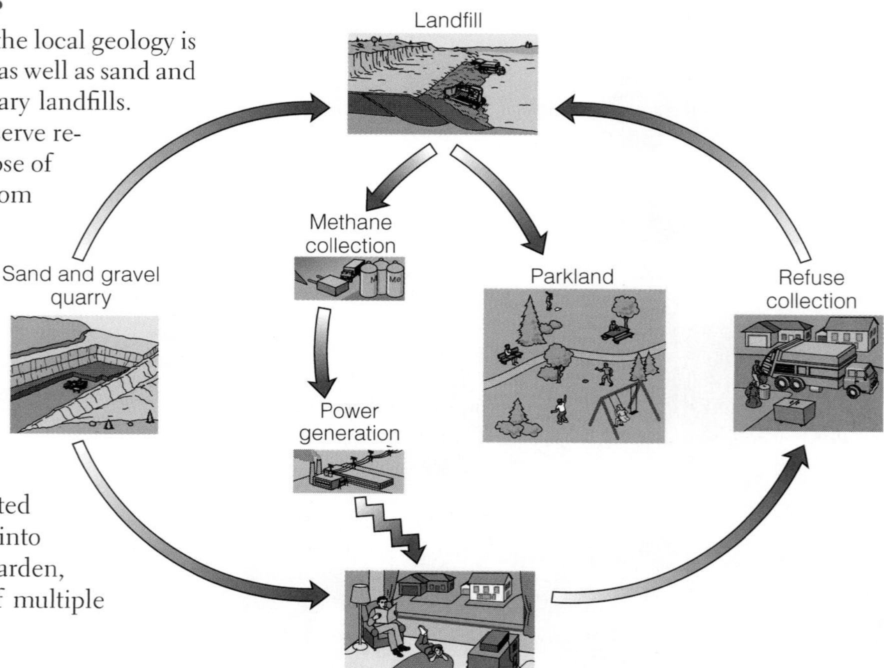

Figure 15.25 Multiple land use, a cycle that can benefit everyone.

B. Pipkin

Figure 15.26 Commercial development built on a landfill. Whereas the buildings are founded on piles through the fill to bedrock, the pavement was placed directly on the fill. The lanterns are fueled by methane from the fill.

Figure 15.28 Drums and containers of dangerous chemicals await storage at a 50-hectare (120-acre) landfill in southwest Idaho. The crates contain discarded transformers full of toxic polychlorinated biphenyls (PCBs).

By 2005, this amount had increased almost 30 times. Examples of hazardous wastes that individuals introduce into the environment include drain cleaners, paint products, used crankcase oil, and discarded fingernail polish (not trivial).

Secure landfills are designed to isolate the received waste from the environment (**Figure 15.28**). Wastes are packaged and placed in an underground vault surrounded by a clay barrier, thick HDPE plastic liners ("geomembranes"), and a leachate-removal system. Unfortunately, few of the nation's existing 75,000 industrial landfills have geomembranes, clay barriers, or leachate drains. Many older fills will either have to be retrofitted with modern environmental-protection systems or shut down. Additional protection is provided by existing regulations that mandate groundwater monitoring at disposal sites. An ideal secure landfill for hazardous waste would be a site in a dry climate with a very low water table. In humid climates where water tables are high, alternative means of safe disposal must be found.

Deep-well injection into permeable rocks far below freshwater aquifers is another method of isolating hazardous liquid substances (**Figure 15.29**). A *hazardous* liquid is one that may exhibit one or more of the following four properties: ignitability, corrosiveness, reactivity, or chemical harmfulness to human health (**Figure 15.30**). Before such a well is approved, thorough studies of subsurface geology must be made to determine the location of faults, the state of rock stresses at depth, and the impact of fluid pressures on the receiving formations (see Case Study 15.3). The injection boreholes are lined with steel casing to prevent the hazardous fluids from leaking into freshwater zones above the injection depth. Multiple classes of injection wells are legally permitted in the United States by the EPA's Underground Well Injection Program; each class is based on the types and depths of waste disposed (**Table 15.4**). The mandate for legal well operation is that underground sources of drinking water (USDW) will not be impacted—or *if* potentially impacted, not for at least 10,000 years. As many as 750,000 injection wells for both hazardous and nonhazardous

Figure 15.27 A breakdown by category of the 40 billion gallons of toxic waste U.S. industries generated in 1981; by 2005, this amount had increased almost 30 times. Note: The total of all categories is greater than 40 billion because of overlap; for example, a fluid may be both corrosive and reactive.

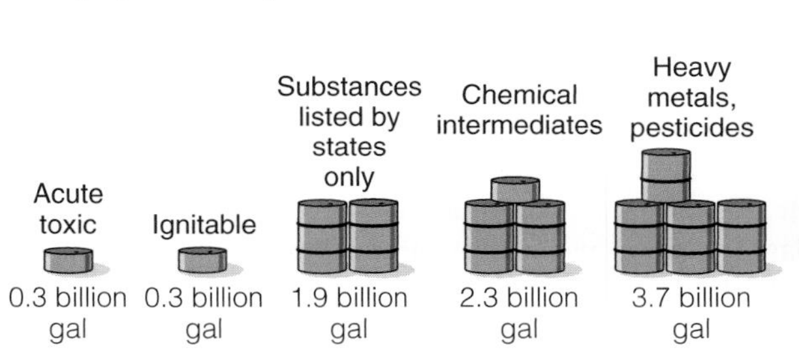

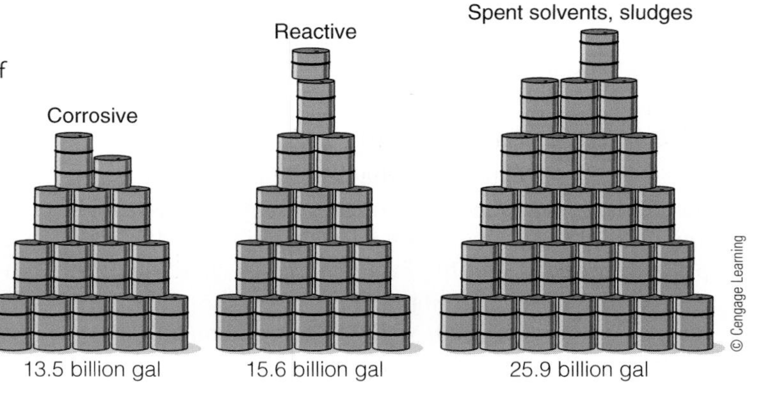

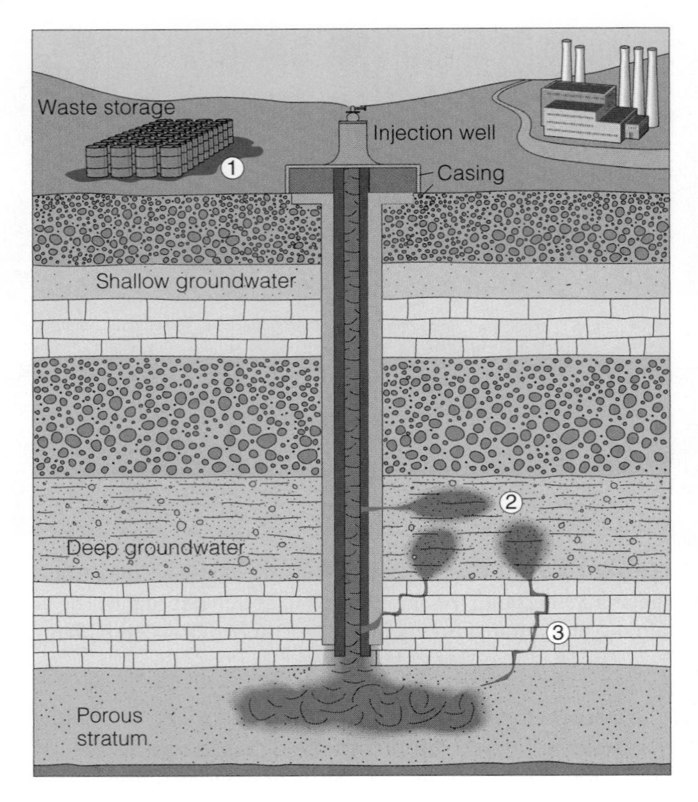

Figure 15.29 Deep-well injection of hazardous waste. This method presumes that wastes injected into strata of porous rock deep within the ground are isolated from the environment forever. What could go wrong? (1) Toxic spills may occur at the ground surface. (2) Corrosion of the casing may allow injected waste to leak into an aquifer. (3) Waste may migrate upward through rock fractures into the groundwater. (Data from American Institute of Professional Geologists [AIPG].)

Figure 15.30 A drum corroded from the hazardous liquid waste it once contained lies at a dump in Natchez, Mississippi.

substances operate annually across the United States. On the order of 34 billion liters (9 billion gallons) of hazardous wastes disappears into the nation's crust through many of these wells. To treat and render harmless these substances otherwise would be extremely expensive.

The disposal of all U.S. hazardous wastes, both liquid and solid, is regulated by the RCRA of 1976. Exempt from the requirements of this act are businesses that generate less than a metric ton of waste per month. According to EPA estimates, about 700,000 of these small firms generate 90% of hazardous waste in the United States. Consequently, most of these wastes end up in sanitary landfills.

The Problem with Coal

Discharge into sealed pits is the least expensive way to dispose of large amounts of water containing relatively small amounts of hazardous substances (see Table 15.4). If the pit is well sealed and the evaporation of water equals or exceeds the input of contaminated water, the pit may receive and hold hazardous waste almost indefinitely. Problems can arise from leaky seals and overflow of holding ponds during heavy storms or floods. A noteworthy case is that of the coal-mining industry and its surface wastewater storage.

Coal combustion in power plants produces around half of the world's energy, and a major by-product of this, besides air pollution, is solid combustion residue—hydrocarbon ash. About 80% of the ash remains in the plant combustion chamber that provides the heat for producing power. The rest flies out of the smokestack—so-called *fly ash*—which is a serious form of air pollution in many regions (e.g., eastern China; **Figure 15.31**). Irrespective of its fate, the ash contains not only organic residues but radioactive particles (uranium, thorium) and toxic metals including arsenic, cadmium, and lead, which contribute to organ and nerve damage. As much as a third of the residue may be mixed into cement to dispose of combustion chamber ash. The ash imparts strength to this common construction material and helps it resist weathering. It also makes it easier to pour concrete. Cement is used in making dams all around the world in large part because of these additive properties. The toxicity of ash in cements, however, is so low as to be inconsequential because of attenuation and mixing.

Table 15.4 Classes of Waste Injection Wells in the United States

These classes are divided into two groups here; shallow injection (above the water table or USDW supplies), and deep injection (in rock layers below or remote from USDW supplies—typically 1,700 to 10,000 feet deep).

Category	Number of Wells in United States	Notes
Shallow-Injection Wells (above the water table or USDW supplies)		
Class V	500,000–650,000	Injection of nonhazardous fluids only. Includes cesspools and septic tanks. Fluids tend to lose their harmfulness by natural breakdown during percolation.
Class IV	40	Hazardous and radioactive wastes injected. These wells are no longer legal but are still used to return treated contaminated water back into original aquifers as part of cleanup efforts.
Deep-Injection Wells (in rock layers below or remote from USDW supplies—typically 1,700–10,000 feet deep)		
Class III	17,000	Used by mining industry to inject water into ground to dissolve valuable minerals that are extracted when water with solutes is returned to surface. 50% of all salt and 80% of all uranium is produced this way in the United States. After minerals are extracted, the remaining fluid is then re-injected.
Class II	147,000	Used by oil and gas industry to extract fossil fuels. Brine (highly saline water) brought up with the oil and gas is separated and re-injected into oil and gas reservoirs through these wells.
Class I Nonhazardous	366	Used for injection of municipal liquid wastes (sewage, etc.); mostly in Florida and Texas.
Hazardous	163	Used for deep injection of toxic liquids, which may be radioactive.

USDW = underground sources of drinking water.

Getting rid of the rest of the ash is a huge problem, however. It cannot simply pile up in combustion chambers, or no room for burning fresh supplies of coal would be available. Of the 7 billion tons of coal burned worldwide each year, according to Duke University, about 12% ends up as coal ash—that's a billion tons. A typical 500-megawatt coal-fired power plant will produce 125,000 tons of solid ash and 193,000 tons of ash sludge (mud) per year. The sludge comes from retrieval of fly ash from smokestacks by wet scrubbers, which are units that spray mist across escaping smoke to capture particles that otherwise might exit into the atmosphere. The total solid plant waste is transported to coal-waste landfills, of which 21 are presently designated in the United States. Meanwhile, liquid waste from scrubbing is stored in impoundment ponds. If one includes the inactive depots, 600 such sites presently exist in the United States. To date, 67 cases of serious coal waste contamination from these sites have been reported. The waste cannot be regarded as harmless. EPA studies indicate that the general level of cancer risk associated with smoking a pack of cigarettes every day is 1,000 cases of cancer per million people. For continuous exposure to coal ash, it is around 9,000 (**Figure 15.32**). Yet, because of political lobbying on the part of industries, coal waste is not defined as "toxic," and the monitoring of individual coal impoundment sites and landfills is left to the individual states, which often have little incentive or funding to put into expensive coal waste management. The result, unfortunately, is a

Figure 15.31 Truck disposes of power plant coal ash in waste storage lot in Fairbanks, Alaska. Under dry, windy conditions, rising coal dust can spread for great distances.

general environmental mess, and one of the worst failings of the federal Clean Water Act.

One recent case of serious contamination occurred on 22 December 2008, when a 21-hectare (84-acre) ash sludge impoundment reservoir near Tennessee's Fossil coal-fired power plant burst open and sent a slurry flood down slope into the nearby Emory River, destroying houses and covering more than 120 hectares (300 acres) of land. The slurry poured into two watersheds: that of the Emory and the Clinch rivers, both major tributaries of the Tennessee—a total of 4.2 billion liters (1.1 billion gallons) of contaminated mud.

A few months after this disaster, a scientific team from Duke University traveled to these rivers and took 200 water samples, collecting them in open-flow conditions and from sediment pores at depths of 10 to 50 cm (4–20 inches) beneath each river bed. They searched for high arsenic concentrations that might remain from the slurry. On a positive note, they were pleased to discover that just a few months of precipitation and snowmelt were sufficient to cleanse the rivers where they were *actually flowing*—no problem for the fish swimming in them. But levels of arsenic measured as high as 2,000 parts per billion (ppb) in the associated pore waters. Lack of flow in the spaces between the silt, sand, and gravel particles meant that the arsenic simply stayed put at immediate post-slurry levels. The EPA safe level drinking standard is only 20 ppb (which had been doubled from an earlier level a few years previously because of intensive industry lobbying). The level of arsenic regarded as "safe" for aquatic life is 150 ppb. River bottoms are where the food chain begins in waterways. Worms and other invertebrates burrow into the bottom sediments scavenging for algae, microbes, and one another. Plants send their roots into this substrate, too, and they incorporate arsenic and other metals, which, in turn, bioaccumulate within plant-eating invertebrates and fish—which may well end up on a dinner plate somewhere.

Superfund

What about the thousands of old and mostly unrecorded sites where hazardous waste has been dumped improperly in the past? Hazardous substances have been disposed of indiscriminately in the United States since the 18th century. **Superfund**, the Comprehensive Environmental Response, Compensation and Liability Act (CERCLA), was passed in 1980 to rectify past and present abuses from toxic-waste dumping. It authorizes the EPA to clean up a spill and send the bill to the responsible party. CERCLA also empowered the EPA to establish a National Priorities List (NPL) for cleaning up the worst of the 40,000 identified abandoned toxic- and hazardous-waste sites now known as *Superfund sites*, some of which date back many decades (see Case Study 15.4). By 2006, 1,305 sites had been designated or were pending. New Jersey has the dubious distinction of harboring 6 of the worst 15 Superfund sites as ranked by impact on the environment—5 landfills and 1 industrial site. **Table 15.5** lists the top 10 states with Superfund sites.

Although the cost of cleaning them up is staggering, toxic dump sites do not just "go away" on their own. They need to be cleaned up. Some problems with Superfund also need to be cleaned up. The Superfund concept was well intended, but the program has been poorly executed. As a result, 40% of its funding has gone to litigation expenses. Superfund has become a "lawyers' money pit," as one journalist put it. And many potential sources of harmful pollution remain waiting for attention for lack of public funding. (Many of the original industries contributing to the pollution have long gone out of business.)

Figure 15.32 Two young coal miners in Beckley, West Virginia, deal continuously with pollutants as part of their livelihoods—and provide fuel for half of America's electricity.

Table 15.5 Top 10 States on the National Priorities List (Superfund), 2007	
State	**Number of Sites**
New Jersey	116
Pennsylvania	96
California	96
New York	87
Michigan	67
Florida	49
Illinois	49
Washington	48
Texas	47
Ohio	38

Source: *EPA hazardous waste sites on the National Priority List by state*, Table 367, 2007; see http://www.epa.gov/superfund/sites/npl/.

The polluter pay principle states that whoever causes pollution should clean it up, but quite often the cost of cleaning up many forms of pollution exceeds the ability of the polluter to pay. Yet, some would argue that the social and economic benefits of the goods produced (e.g., medicines, automobiles, CDs) are arguably "well worth the cost," and justify public expense for cleaning up the environmental problems created. Do you agree?

Nuclear-Waste Disposal

There is probably no more sensitive and emotional issue in geology today than the need to provide for the safe disposal of the tons upon tons of radioactive materials that have accumulated over many decades. These wastes are the largest deterrent to further development of nuclear energy, because they present a legitimate hazard to humans, their offspring, and generations yet to come. Nuclear-waste products must be isolated because they emit high-energy radiation that kills cells, causes cancer and genetic mutations, and causes death to individuals exposed to large doses. The need to develop the means for safe, permanent disposal led Congress to pass the Nuclear Waste Policy Act (NWPA) in 1982. This law gave the Department of Energy (DOE) responsibility for locating a fail-safe underground disposal facility, a **geological repository**, for high-level nuclear waste and designated nine locations in six states as potential sites. Based on preliminary studies, President Reagan approved three of the sites for intensive scientific study (known as *site characterization*): Hanford, Washington; Deaf Smith County, Texas; and Yucca Mountain, Nevada. In 1987, Congress passed the Nuclear Waste Policy Amendments Act, which directed the DOE to study only the Yucca Mountain site (**Figure 15.33**). This site is discussed further later in the chapter.

Figure 15.33 (a) Location of the proposed Yucca Mountain nuclear repository. (b) Aerial view of the Yucca Mountain study area. The structures are at the site of the proposed nuclear-waste repository. The site was shelved in 2009. (Data from the Department of Energy.)

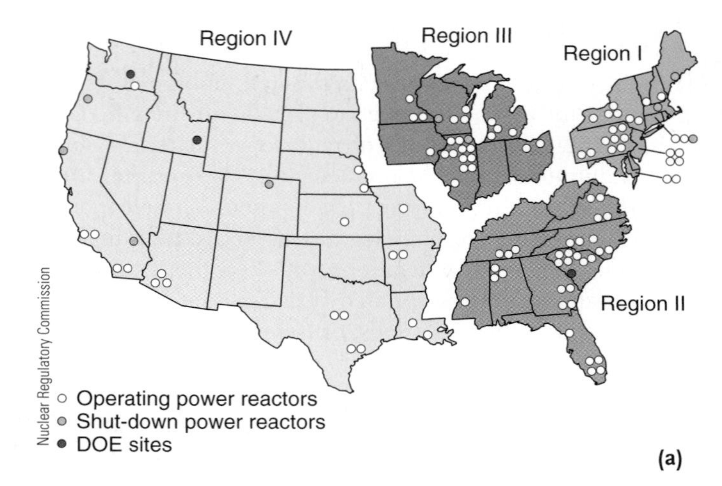

Figure 15.34 (a) Location of temporarily stored spent (no longer usable) nuclear fuel and/or high-level nuclear waste awaiting permanent storage, 2008. DOE = Department of Energy. (Sources: Nuclear Regulatory Commission: http://www.nrc.gov/info-finder/reactor/#USMap; National Research Council. Going the distance? The safe transport of spent nuclear fuel and high-level radioactive waste in the United States. Washington, DC: National Academies Press, Table 2.) (b) At certain areas on the Nevada test site, soils contaminated from nuclear testing are quarantined and kept strictly off limits—the best approach that can be taken toward "storage" of this hazardous material at present.

High level wastes—**High-level wastes**, by-products of nuclear-power generation and military uses, are the most intensely radioactive and dangerous wastes. About once a year, a third of a reactor's fuel rods are removed and replaced with fresh rods. Replacement is necessary because, as the fissionable uranium in a reactor is consumed, the fission products capture more neutrons than the remaining uranium produces. This causes the chain reaction to slow, and eventually to stop. The radioactive rods, called *spent fuel*, are the major form of high-level nuclear waste. Currently, they are stored in water-filled pools at the individual reactor sites, supposedly a temporary measure until a final grave is prepared for them. More than 63,000 metric tons of heavy metal of spent fuel, mostly uranium, is stored across the country and, at many of the reactors, pool storage capacity is nearly filled (**Figure 15.34**). The number does not include other materials, such as the tubes that contain the fuel and structural materials. (One metric ton is 1,000 kilograms, or 2,200 pounds). Two major concerns about pool storage are that an unintended nuclear reaction might start in the pool, and that the rods might deteriorate and release fuel pellets. Although the fuel removed from a reactor every year weighs about 31,000 kilograms (65,000 pounds), because of its high density, it is only about the size of an automobile.

Alternatively, the rods are chopped up and *reprocessed* to recover unused uranium-235 and plutonium-239. However, the liquid waste remaining after reprocessing contains more than 50 radioactive isotopes, such as strontium (^{90}Sr) and iodine (^{131}I), that pose significant health hazards. In 1992, the DOE decided to phase out reprocessing of spent fuel. Unfortunately, the volume of existing reprocessed waste in temporary storage would cover a football field to a depth of 60 meters (200 feet). These wastes are currently stored in metal tanks at four U.S. sites. The oldest site, the one at Hanford, Washington, originally stored wastes in single-walled steel tanks. In 1956, a tank leak was detected, and since then, 2,840,000 liters (750,000 gallons) of high-level waste have leaked underground from 60 of 149 storage tanks. It is said that "radioactive waste was invented here." Cleanup at Hanford may be the costliest in world history, at least $57 billion.

Low-level wastes—**Low-level wastes**, by Nuclear Regulatory Commission (NCR) definition, are anything that is not high-level waste or transuranic waste. Transuranic waste includes tools, residues, clothing, and other such items contaminated with small amounts of radioactive elements, mainly plutonium. These radioactive elements have an atomic number greater than uranium's—hence *transuranic* ("beyond uranium")—and are human-made. Low-level wastes are created in research laboratories, hospitals, and industry, as well as in reactors. Ocean dumping of these wastes in 208 liter (55-gallon) drums was a common practice between 1946 and 1970, as noted earlier. Since then, shallow land burial has been the preferred disposal method, but many sites have been found to leak.

Mill tailings—Uranium-mill tailings, the finely ground residue that remains after uranium ore is processed, are typically found in mountainous piles outside mill facilities. Large volumes of tailings exist that still contain low concentrations of radioactive materials, including thorium (^{230}Th) and radium (^{226}Ra), both of which produce radioactive radon gas (^{222}Rn), the so-called hidden killer (see Case Study 15.5). Most of the abandoned tailings piles are in sparsely settled areas of the West. A former Atomic Energy Commission policy permitted the use of uranium-mill tailings in the manufacture of building materials, such as concrete blocks

Figure 15.35 Radioactive waste is hauled on a train through the town of Patrick, South Carolina, on its way to some distant, temporary repository.

and cement. For almost 20 years, 30,000 citizens of Grand Junction, Colorado, lived in homes whose radon levels were up to 7 times the maximum allowed for uranium miners. Uranium tailings were used in building materials in Durango, Rifle, and Riverton, Colorado; Lowman, Idaho; Shiprock, New Mexico; and Salt Lake City, Utah.

The Uranium Mill Tailings Radiation Control Act (1978) makes the DOE responsible for 24 inactive sites, in 10 states, left from uranium operations. The act specifies that the federal government will pay 90% of the cleanup cost, with each state to pay the remainder. Uranium-mill tailings could become a serious problem because they contain the largest volume of radioactive waste in the country.

Isolation of nuclear waste—Scientists agree that nuclear waste must be isolated from contact with all biological systems for at least 250,000 years for the radioactive materials—particularly, plutonium-239, whose half-life is 24,000 years—to decay to harmless levels. Should the ban on plutonium recovery be rescinded, the conditions for high-level waste disposal may change. Several methods of storing wastes have been proposed, but any method must meet these conditions:

- Safe isolation for at least 250,000 years
- Safety from terrorists or accidental entry
- Safety from natural disasters (hurricanes, landslides, floods, etc.)
- Noncontamination to any nearby natural resources
- Geologically stable site (no volcanic activity, earthquakes, etc.)
- Fail-safe handling and transport mechanisms (**Figure 15.35**)

A number of disposal methods have been proposed. The more serious ones are discussed here.

1. *Continue the present tank storage.* This is not a permanent solution because the tanks have limited lifetimes and leakage has already occurred.

2. *Dispose of waste in a convergent plate boundary (subduction zone).* The feasibility of this very imaginative proposal has not been demonstrated because the rate of plate movement is very slow.
3. *Place containerized waste on the ice sheets of Antarctica or Greenland.* Radioactive heat would cause the containers to melt downward to the ice-bedrock contact surface, which could place the radioactive waste 2 miles beneath ice in either Greenland or Antarctica. This proposal assumes that long-term global warming will not occur in the 250,000-year period. This is unsound, however, because the last ice sheet of the Pleistocene melted back to Greenland in less than 10,000 years.
4. *Place waste in a geological repository: salt mines or domes.* Salt is dry, flows readily, and is self-healing. Because it is a good conductor of heat, temperatures in the repository rocks would not become excessive. Salt is subject to underground solution, however, and some hydrated minerals in salt beds give up their water when heated. Salt beds and salt domes occur in geologically stable regions, such as Texas, Louisiana, and Mississippi.
5. *Place waste in a geological repository: deep chambers in granite, volcanic rocks, or some other very competent, relatively dry rock formation.* This would require detailed site analysis to ensure that all geological criteria are met.

Nuclear waste and Yucca Mountain—starting over—The DOE's Civilian Radioactive Waste Management Program was established in the early 1980s, and the site characterization of Yucca Mountain, Nevada, as a high-level nuclear-waste repository began in 1987. In the early stages of characterization, the site seemed ideal. It is isolated, it is located within the area of the Nevada Test Site (where atmospheric and underground testing of nuclear weapons had been carried out for more than 40 years), the area is dry desert, and the geochemical nature of the volcanic tuff comprising the site was thought capable of containing any potential leakage of radioactive materials. However, the rocks proved to be not as dry as originally thought. Nevada does have earthquakes, which sometimes cause ground breakage, and there are geologically young volcanoes nearby. By 2009, the costs of development of the site had reached $13 billion, with a projected final cost estimated at an astounding $76 billion.

In May 2009, U.S. Secretary of Energy Steven Chu stated:

> *Yucca Mountain as a repository is off the table [as the potential repository for U.S. spent nuclear fuel (SNF) and high-level nuclear waste (HLW).] What we're going to be doing is saying, let's step back. We realize that we know a lot more today than we did 25 or 30 years ago. The NRC [Nuclear Regulatory Commission] is saying that the dry cask storage [of nuclear waste] at current sites would be safe for many decades, so that gives us time to figure out what we should do for a long-term strategy. We will be assembling a blue-ribbon panel to look at the issue. We're looking at reactors that have a high-energy neutron spectrum that can actually allow you to burn down the long-lived actinide[2] waste. So the real thing is, let's get some really wise heads together and figure out how you want to deal with the interim and long-term storage. Yucca was supposed to be everything to everybody, and I think, knowing what we know today, there's going to have to be several regional areas.*

Thus, the decision to close Yucca Mountain ends a 30-year effort that began in 1987, when Congress selected Yucca as the only site to be investigated, with no backup option if Yucca failed technically or politically.

What went wrong at Yucca Mountain? The site is geologically complex. A number of unresolved technical and geological issues exist (e.g., seismicity climate change and volcanism). The EPA's original 10,000-year radiation dosage standard for the repository—an arbitrarily selected figure—was belatedly upgraded in September 2008 to 1 million years; that is, the new radiation standard sets tighter limits on radiation doses for up to 1,000,000 years after it closes. Because the youngest volcanic activity in the area, only a few kilometers (miles) from Yucca Mountain, erupted only 80,000 years ago, the new ruling throws into serious doubt that Yucca Mountain can safely contain nuclear waste. Contributing factors were local opposition and unreliable funding, as expenditures for development were subject to yearly congressional appropriations. It was the appropriation process that ultimately put the Yucca Mountain repository on hold.

So, what now? Where and how is the United States to securely store spent nuclear fuel and high-level nuclear waste? There are four basic options: (1) restart the search and develop one or more new geological repositories; (2) consolidate waste from decommissioned sites at one or more central storage sites; (3) store the nuclear waste indefinitely in 35 states, and more than 70 reactor sites; or (4) implement a combination of these options.

At present, the accumulated high-level nuclear waste intended for places like Yucca Mountain is being warehoused in dry casks consisting of steel encased in concrete. Each cylindrical cask stands roughly 6 meters (20 feet tall) and is 3 meters (10 feet) wide. The casks began to be introduced as a stopgap measure in the 1970s, when the swimming pools at nuclear power plants designed to store spent reactor fuel rods temporarily began to fill up. (All remaining pool space will likely be completely filled by 2015.) Each cask is meant to retain the overflow waste for only a period of 20 years, with possible renewal authorized by the NRC for up to 40 years. But now it looks like dry cask storage will grow and continue indefinitely in the United States until the country decides what to do on a more permanent basis. As of November 2010, 1,400 high-level nuclear waste casks were in existence at 63 sites called *Independent Spent Fuel*

[2]The actinide waste consists of 15 metallic radioactive elements, 13 of which, including plutonium, are generated in nuclear reactions.

Storage Installations (ISFSIs). Each resembles a larger, well-guarded parking lot. ISFSIs may continue to grow in number over the next few centuries even if a new geological repository comes into operation, given the tremendous amount of nuclear waste in need of storage.

How is nuclear waste stored elsewhere in the world? Sweden and Finland have successfully sited repositories in stable granite permeated with oxygen-depleted water.[3] The Scandinavians store their nuclear wastes in copper canisters embedded in protective bentonite clay. Switzerland, France, and Belgium are looking at potential nuclear-waste repositories in clay. The stability of granite and clay host rocks, their great age, and their lack of tectonism or volcanism for eons increase the confidence in their stability and long-term repository performance. U.S. volcanologist Wendell Duffield states, *"There are vast, sparsely populated tracts of the United States that have no volcanic eruptions (and accompanying groundbreaking earthquakes) for tens of millions of years and longer"* that would be suitable repository sites. Such sites exist along the Texas–New Mexico border where the **Waste Isolation Pilot Plant (WIPP)** has been operating since 1999, and a new low-level radioactive waste facility was licensed in 2008.[4]

Waste Isolation Pilot Plant—In 1999, the DOE opened the Waste Isolation Pilot Plant (WIPP), the nation's first operating underground repository for defense-generated transuranic waste.

Located in southeastern New Mexico, near Carlsbad, the site covers 41 square kilometers (16 square miles). WIPP is a project designed to demonstrate safe, cost-effective, and environmentally sound storage of defense-related radioactive waste. Elsewhere, transuranic wastes are stored at 5 major DOE sites and 18 other locations across the United States.

The concerns about WIPP are comprehensive, including the safe transportation in containers certified by the NRC in special WIPP trucks driven by highly trained drivers, notification of states and tribal authorities when materials will be passing through their jurisdictions, and monitoring the location of each shipment by a satellite tracking system while in transit. Oversight of the WIPP program is provided by numerous state and federal agencies and the Environmental Evaluation Group, an independent group that participates in, and comments on, various WIPP activities.

Storage of the radioactive materials is in excavated rooms 655 meters (2,150 feet—nearly ½ mile) underground in the Salado Formation, an ancient, stable salt bed (**Figures 15.36** and **15.37**). The 610-meter (2,000-ft)-thick salt deposit was chosen as the site because there is very little earthquake activity, there is no flowing water that could move waste to the surface, and because salt readily heals its own fractures because of its plastic character. Furthermore, the salt will slowly flow into and fill the mined areas, further isolating the waste from the environment.

A system of permanent markers was designed to tell our descendants that the area of the WIPP is not in its natural state, and that it has been marked for good reason. The site must remain inviolate for 10,000 years, and the warning components include a 9-meter (30 foot)-high berm, an information center, monuments along the perimeter, and a summary in six languages printed on archival paper with warnings that it must be preserved for the 10,000-year regulatory period. Yet, a serious, largely academic question still remains about just how one constructs a warning system that will both *endure* and *remain intelligible* to our descendants through such a long span of time—essentially a geological interval. To study this basic question, the DOE set up a group called the Expert Judgment Panel, in connection with its earlier Yucca Mountain planning. After a decade of study, the group found that virtually nothing we can use as construction material will survive in satisfactory condition for so long, save, perhaps, a new synthetic product called *synroc*, made up of the rare minerals hollandite, zirconolite, and perovskite. If we leave a message rather than some sort of warning symbols, then what should it say? Thomas Sebeok, lead author of the report *Communication Measures to Bridge Ten Millennia*, proposed that we initiate a transgenerational Atomic Priesthood:

> *Each successive generation of the Atomic Priesthood would be charged with the responsibility of seeing to it that our behest is heeded by the general public—if not for legal reasons, than perhaps for moral ones—with the veiled threat that to ignore our mandate [not to drill wells or dig in this area] would be tantamount to inviting some sort of supernatural retribution.*

Adapting this authoritative overtone, the Expert Judgment Panel ultimately settled on the following message to be inscribed monumentally above the Yucca Mountain site, should it ever be developed: *"This is not a place of honor. No esteemed deed is commemorated here. Nothing of value is buried here. This place is a message, and part of a system of messages. Sending this message was significant for us. Ours was considered an important culture."* How would you interpret such a sign if you were to stumble across it in a desert wilderness? Furthermore, after 10,000 years, would the language even be recognizable? University of Iowa Professor

[3]The waste, which contains uranium dioxide (UO_2), is unstable in a wet oxidizing environment, converting quickly into other, more readily soluble oxides, which could contaminate groundwater. Toxic loads of radioactive groundwater from 41 years of detonating underground nuclear warheads at the Nevada Test Site already move to the Southwest. Fortunately, the tainted water migrates slowly, from 7.6 centimeters to 5.5 meters (3 inches to 18 feet) a year, and it will not reach Beatty, the nearest town, for about 6,000 years.

[4]Approval of this site has met with controversy. All of the scientists on the Texas Commission on Environmental Quality (TCEQ) studying the site ruled against it because of radioactive contamination to groundwater and a variety of technical and other reasons. Their scientific judgment was overruled by the Texas Compact Commission, which approved the site despite concerns over radioactive dose rate assessments, engineering design, financial assurance, and other concerns. Four TCEQ employees resigned over the decision to issue the license, and the matter has gone to the state district court.

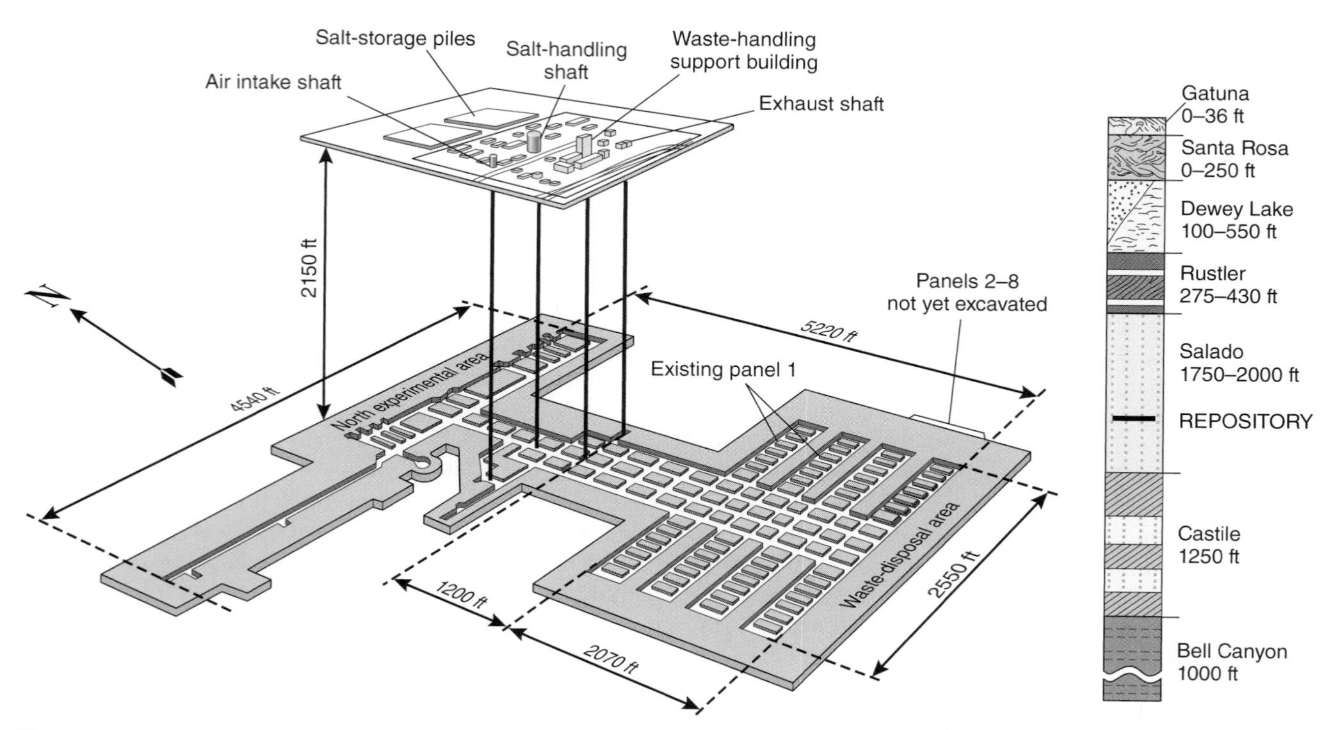

Figure 15.36 Schematic representation of the Waste Isolation Pilot Plant underground facility for storage of transuranic waste and the sedimentary rock sequence near Carlsbad, New Mexico. Only two of the storage panels have been mined. (Department of Energy)

Figure 15.37 Underground storage of military-generated transuranic waste in Waste Isolation Pilot Plant excavated salt caverns.
Department of Energy

explorer's perspective

Photograph by Matthew Muspratt/See: http://www.nationalgeographic.com/explorers/bios/ashley-murray

ASHLEY MURRAY, engineer and entrepreneur, has found a way of making waste generate income sustainably.

Ashley Murray is a Ph.D. graduate of the Energy and Resources Group, University of California, Berkeley, with a background in civil and environmental engineering. She is a model of a citizen putting her knowledge to constructive purpose by addressing environmental issues and actually building sustainability. Well aware that 850 million people worldwide are too poor to afford safe sanitation and waste management services, and that 350 million people in Asia and Africa alone can't even access standard hygienic toilets, she founded and now directs Waste Enterprisers in Ghana to tackle the problem head-on. Rather than let human waste in the impoverished world spread perilously across the landscape, her company captures and diverts it to useful purposes, including catfish aquaculture and the making of various industrial fuels. The goal is to generate enough revenue to reinvest in pit latrine/septic tank emptying. The state does not have to risk taxing the poor beyond their means to finance basic public sanitation. Recent collaboration with researchers at Columbia and Kwame Nkrumah universities seeks to perfect the manufacture of biodiesel fuel from sewer sludge. The Bill and Melinda Gates Foundation provided $1.5 million to fund pilot studies. Ashley comments, "What's novel about our approach is that we show how waste can bring money back into the sanitation sector ... I wanted to highlight that sanitation is a business and that it can be profitable."

Source: http://www.nationalgeographic.com/explorers/bios/ashley-murray

Release of treated wastewaters, west Africa.
Photo from http://www.waste-enterprisers.com/about-us/

John D'Agata provided the following translation using this very wording in Old English—our commonly spoken tongue until 800 to 1,100 years ago. How much of it do you understand?

> *Nis weorðful stow. Nis last mære dæde na. Her nis naht geoweorðes bebyrged. Þeos stow bið ærend woruldes. Giemþ wel! We sindon eornost! Þeos ærengiefu wæs niedmicel us. Hit þuhte us þuhte ys þæt we wæron formicel cynn.*

consider this... What message of "nuclear warning" would *you* leave for the future?

Health and safety standards for nuclear waste—Nuclear waste emits three types of radiation:

1. Alpha (α) particles: the most energetic but least penetrating type of radiation (they can be stopped by a piece of paper). If *a* emitters are inhaled or ingested, alpha radiation can attack lung and body tissue.
2. Beta (β^-) radiation: penetrating but most harmful when a β^- emitter is inhaled or ingested. For example, strontium-90 is a β^- emitter that behaves exactly like calcium (builds bone) in the food chain.
3. Gamma (γ) rays: nuclear X-rays that can penetrate deeply into tissue and damage human organs (**Figure 15.38**).

Nuclear-waste policy limits the radioactivity that may be emitted from a repository, as measured by cancer deaths directly attributable to leaking radiation. The *rem* is the unit used for measuring the amounts of ionizing radiation that affect human tissue, and a millirem (mrem) is 0.001 rem. Doses on the order of 10 rems can cause weakness, redden the skin, and reduce blood-cell counts. A dose of 500 rems would kill half of the people exposed. The mortality limit is 10 deaths per 100 years, and for this reason, the maximum allowable radiation dose per year per individual in the proximity of a repository is set at 25 mrem. Rem units take into account the type of radiation—alpha, beta, or gamma—and the amount of radioactivity deposited in body tissue. **Figure 15.39** shows how the average annual dose of 208 mrem is partitioned for U.S. citizens and common sources of radiation. (Case Study 15.5 explains the hazards of high levels of radon gas in the home and methods for testing for and mitigating them.) It is estimated that a person living within 5 kilometers (3 miles) of the proposed Yucca Mountain repository would have received less than 1 mrem per year from the nuclear waste stored there. The 25 mrem standard for maximum annual dosage is equal to the radiation a person would receive from two or three medical X-rays.

Is there a future for nuclear energy?—Nuclear reactors are very complex; a typical U.S. reactor may have 40,000 valves, whereas a coal-fired plant of the same size has only 4,000. This fact alone makes reactors relatively unforgiving of errors in operation, construction, or design. A newer generation of "standard" reactors is now under consideration. These use natural forces, such as gravity and convection, instead of the network of pumps and valves currently required. For example, if the core of one of these new reactors should overheat, a deluge of water stored in tanks above the reactor would inundate the core. In addition, design would be standardized in the new reactors, which would greatly simplify construction, maintenance, and repair. Standardization would also facilitate better operator training and would eliminate the possibility of some human errors.

In the final analysis, it seems likely that nuclear power production will be required for at least some decades as the human race comes to grips with the double-edged sword of possible climate destabilization and peaking fossil-fuel resources. But in the end, conventional nuclear power production is not sustainable, and this creates an egregious waste situation. It can only serve as a measure to buy us time.

The ultimate clean nuclear-energy source is **fusion**, whereby two nuclei merge to form a heavier nucleus and in the process give off tremendous amounts of heat. Deuterium ($^2_1$H) is a heavy isotope of hydrogen that at very high temperatures fuses to form helium ($^{-4}_{2}$He). This is the same nuclear process that energizes our Sun. Ultimately derived from water, the fuel is essentially inexhaustible. The reaction cannot run out of control in an escalating fashion, and

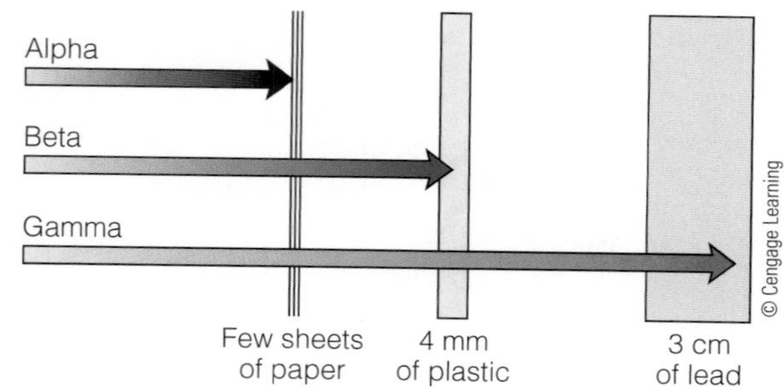

Figure 15.38 The penetrating abilities of alpha, beta, and gamma radiation, the three types of radiant energy emitted by nuclear wastes.

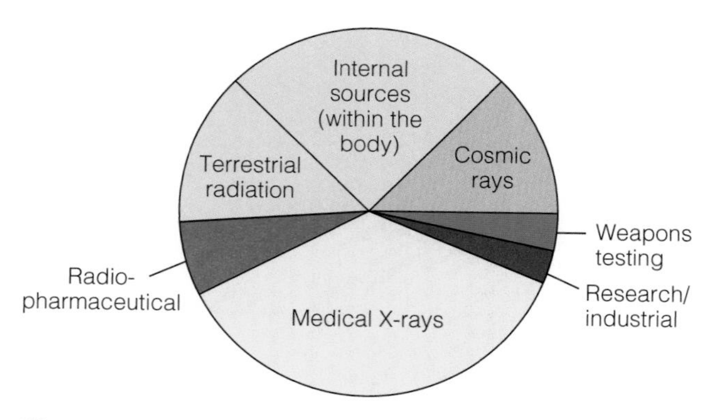

Figure 15.39 The typical American receives 208 millirems of radiation per year from various sources. Some medical, research, and industrial workers' exposures may exceed the 208-millirem average. (National Academy of Sciences, Nuclear Regulatory Commission)

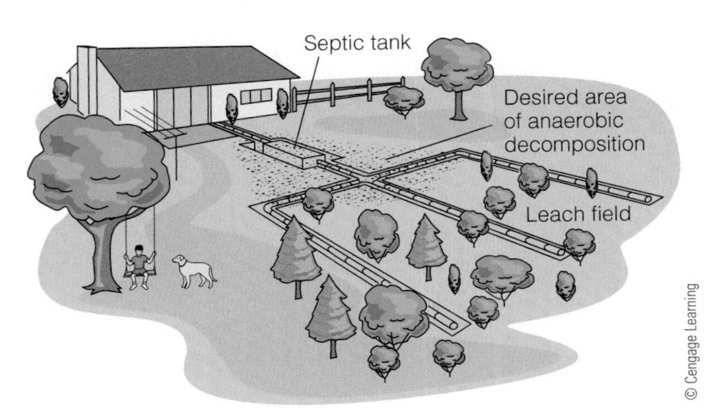

Figure 15.40 A septic tank near a house drains into a leach field of drain tiles that are laid in trenches and covered with a filter gravel and soil.

it is for all practical purposes environmentally nonpolluting. But getting a reaction to go in a controlled way has vexed some of the world's top physical engineers for the past 60 years, with no commercial breakthrough in sight.

Private Sewage Disposal

Modern household sewage-disposal systems use a septic tank, which separates solids from the effluent, and leach lines, which disseminate the liquids into the surrounding soil or rock (**Figure 15.40**). A septic tank is constructed in such a way that all solid material stays within the tank and only liquid passes into the leach field. A leach field consists of rows of tile pipe surrounded by a gravel filter, through which the effluent percolates before seeping into the soil or underlying rock. Sewage is broken down by *anaerobic* bacteria in the septic tank and the first few feet of the leach lines. As the effluent percolates into the soil or bedrock, it is filtered and oxidized, and aerobic conditions are established. Studies have shown that the movement of effluent through a few feet of unsaturated, fine-grained soil under aerobic conditions reduces bacterial counts to almost nil. Fine-grained soils have been found to be the best filters of harmful pathogens, whereas coarse-grained soils allow pathogenic organisms and viruses to disperse over greater distances.

Under the best of conditions, leach lines have a life expectancy of 10 to 12 years. The factors that reduce infiltration rates, and thus limit the life of a septic system, are dispersion of clay particles by transfer of sodium ions (Na^+) in sewage to clay particles in the soil, plugging by solids, physical breakdown of the soil by wetting and drying, and biological "plugging." This plugging occurs when the leach system becomes overloaded with sewage and anaerobic conditions invade the leach lines, resulting in a growth of organisms that plug the filter materials.

After passing through the leach field, the effluent must be kept underground until oxidation and bacteria can purify it. The geological condition that most commonly causes a private sewage-disposal system to fail is a high water table that intersects or is immediately beneath the leach lines (**Figure 15.41**). This condition initiates the anaerobic growth that plugs filter materials and tile drains. Water tables may rise because of prolonged heavy precipitation or the addition of a number of septic tanks in a small area. One way to avoid exceeding the soil's ability to accept effluent is to design leach fields on the basis of percolation tests performed during wet periods. If necessary, a system's lifetime can be extended by constructing multiple leach fields whose use can be rotated regularly by switching a valve. Another problem is the improper placement of leach lines. This results in effluent seepage on hill slopes or stream-valley walls (see Figure 15.41b). Although typical plumbing codes specify a minimum of 5 meters (16 feet) of horizontal distance from a hill face, this may not be sufficient if the vertical permeability is low and the hill slope is steep.

Nitrates in sewage effluent pose a **eutrophication** problem if households are built near a lake or pond (see Chapter 8). Cultural (human-caused) eutrophication accelerates the process and can convert an attractive, "real-estate" lake into a weed-choked bog in a short time. Nutrients from sewage systems and fertilized lawns must not be allowed to flow unchecked into bodies of water. Solving this problem requires the help of such experts as aquatic biologists and hydrogeologists.

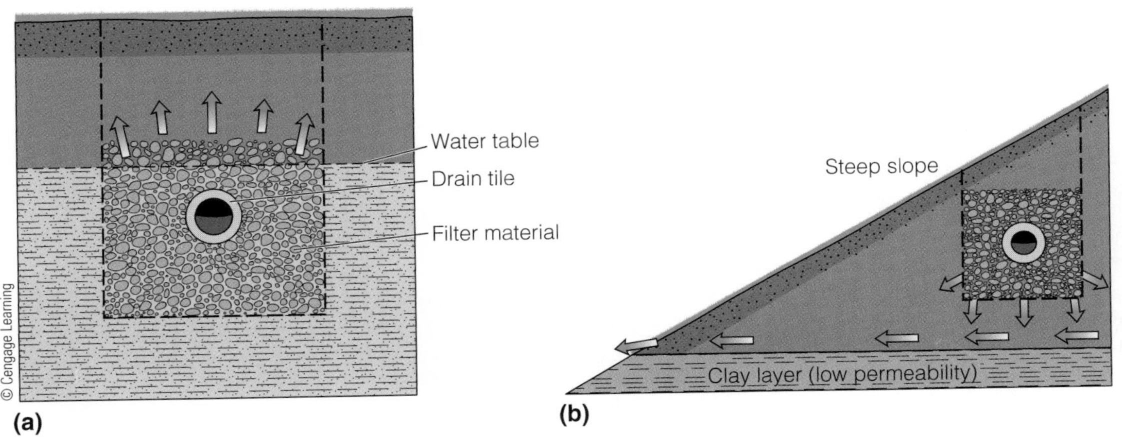

Figure 15.41 Conditions under which raw effluent can seep into the ground. (a) A drain tile and filter material intersect a high water table. When this occurs, sewage rises, and the growth of anaerobic bacteria is promoted in the leach field. This leads to biological plugging of the filter material. (b) A leach line is underlain by low-permeability layers adjacent to a steep slope.

15.1 The Sociology of Waste Disposal

The EPA is charged with the Herculean tasks of monitoring existing land and ocean dump sites, approving new hazardous-waste disposal sites, and supervising the cleanup of historic toxic sites and accidental spills and leaks. In carrying out these responsibilities, EPA personnel have encountered the spectrum of political and public attitudes about waste disposal. A wall poster at EPA headquarters in Washington, DC, provides the following acronyms for those attitudes:

NIMBY	Not In My Back Yard
NIMFYE	Not In My Front Yard Either
PIITBY	Put It in Their Back Yard
NIMEY	Not In My Election Year
NIMTOO	Not In My Term Of Office
LULU	Locally Unavailable Land Use
NOPE	Not On Planet Earth

questions to ponder

1. Which of the acronyms in the list apply to the public's attitudes concerning environmental issues where you live?
2. How might it be possible to overcome some of these attitudes?

15.2 The Highest Point between Maine and Florida

Fresh Kills landfill on Staten Island, New York, can claim fame for two reasons: It stands 165 meters (500 feet) above sea level, the highest elevation on the eastern shore of the United States, and before closure it was the world's largest landfill, with 25 times the volume of the Great Pyramid of Giza, Egypt (**Figure 1**). Fresh Kills (Dutch *kil,* "stream") served the five boroughs of New York City, its suburbs, and parts of New Jersey (**Figure 2**). The fill was established in 1948 on a salt marsh, with no provisions for constraining leachate. For more than five decades, greater than 175,000 tons of municipal waste per week ended up at the site, and up to 4.2 million liters (1 million gallons) of leachate leaked into the marsh and nearby waters each day.

Included in the waste stream to the fill was ash residue from the three active trash incinerators operated by New York City. Because the ash contained toxic metals, specially designed ash-only fill areas were constructed, each with a double liner and a double leachate-collection system to keep toxic substances within the fill. A clay slurry "wall" was excavated around the fill to further confine leachate if there were leaks. Methane was extracted and used for heating and cooking in nearby homes, and construction debris was recycled.

The landfill was scheduled for closure in March 2001, because of public pressure and EPA support. However, following the attacks on the World Trade Center on September 11, 2001, the landfill was reopened to accommodate the disposal of debris from the destruction.

Most of the debris ended up being sold for scrap and recycled, and the site was officially closed in 2003. The

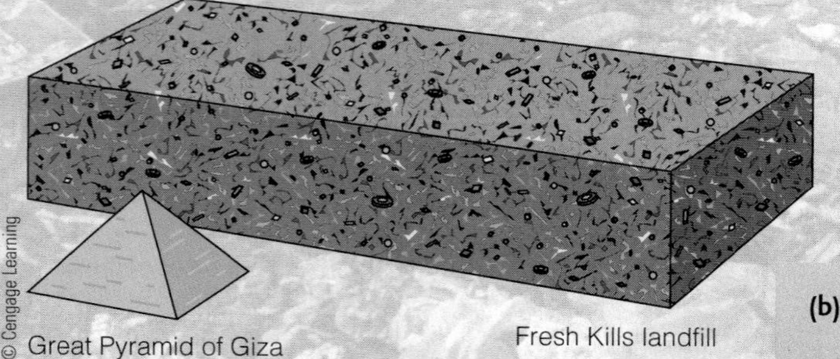

(a)

(b)

Figure 1 (a) Comparison of the volumes of the Great Pyramid of Giza and the Fresh Kills landfill. (b) The Great Pyramid (right) built by King Cheops at Giza on the west bank of the Nile River about 2500 B.C.E. The length of each side at the base averages 756 feet, and its present height is 451 feet (original height, 481 feet). The smaller pyramid was built by Chephren, a later king. The automobiles give an idea of the immensity of these structures.

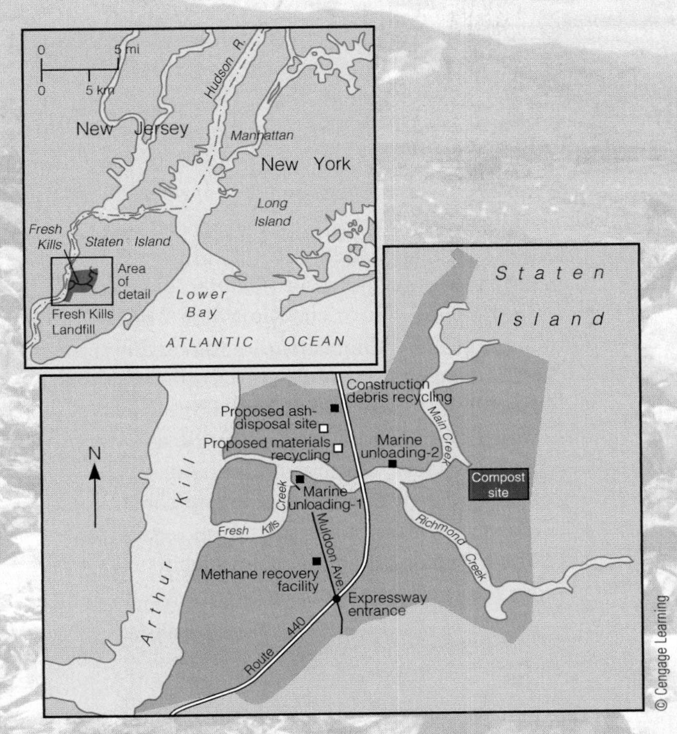

Figure 2 Fresh Kills landfill on Staten Island.

New York City Department of Parks and Recreation soon proposed a thirty-year plan for developing the site as a public park, and construction plans are presently on the drafting board. The new green space will be three times larger than Central Park in Manhattan.

Closure of the Fresh Kills landfill and the other five landfills that had reached their capacity in 2001 solved one problem for the city but created another: how to dispose of thousands of tons of waste per day (TPD) without landfilling and incineration. The city was left with one option: exportation.

The 8 million residents and thousands of businesses in the city's five boroughs generate 26,000 TPD. The New York City Department of Sanitation is responsible for the disposal of about 13,000 TPD; the remainder is handled by commercial waste-handling companies. The 13,000 TPD is shipped to facilities out of the city, mostly in Virginia, Pennsylvania, and Ohio. The commercial waste-handling companies also export to facilities outside the city. As a consequence of recent state controls on accepting waste from other states, exporting cannot last forever. Two new technologies for waste disposal that are being investigated by New York City are plasma pyrolysis, which not only destroys the waste but generates energy and by-products that can be sold, and anaerobic digestion (see the Solid Waste Alternative Technology Team reference in For Further Information section).

questions to ponder

1. State-of-the-art sanitary landfills have a double lining of a clay sealer and two layers of high-density polyethylene, with subsurface drains to catch any leachate. Will these last forever?
2. What action, if any, should be undertaken to treat the 85% of U.S. sanitary landfills that are unlined?
3. Does shipping sewage waste from large cities, such as New York, to repositories in other states make sense? How might this practice be overcome?
4. What temperatures are reached in plasma pyrolysis, and what waste products are produced?

15.3 Deep-Well Injection and Human-Caused Earthquakes

The U.S. Army's Rocky Mountain Arsenal in suburban Denver drilled a 3,764-meter (2.5-mile)-deep well for disposing of chemical-warfare wastes in 1961. The Tertiary and Cretaceous sedimentary rocks into which the well was drilled were cased off, and the lower 23 meters (75 feet) were open in Precambrian granite. About 40 kilometers (25 miles) to the west of the arsenal is the frontal fault system of the Rocky Mountains (**Figure 1**).

Injection began in 1962 at a rate of about 700 cubic meters (915 cubic yards) per day (equivalent to 176,000 grams or 388 pounds/day). Soon after injection began, numerous earthquakes were felt in the Denver area. The earthquake epicenters lay within 8 kilometers (5 miles) of the arsenal and plotted along a line northwest of the well. A local geologist attributed the quakes to the fluid injection at the arsenal. Although the army denied any cause-and-effect relationship, a graph of injection rates versus the number of earthquakes showed an almost direct correlation—especially during the quiet period, when fluid injection was stopped (**Figure 2**).

The fractured granite into which the fluid was being injected was in a stressed condition, and the high fluid pres-

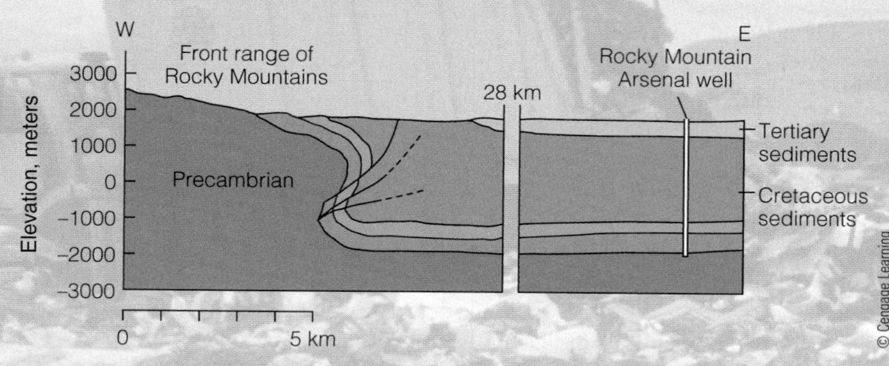

Figure 1 Relationship between the Rocky Mountain Arsenal and the Rocky Mountain frontal fault system.

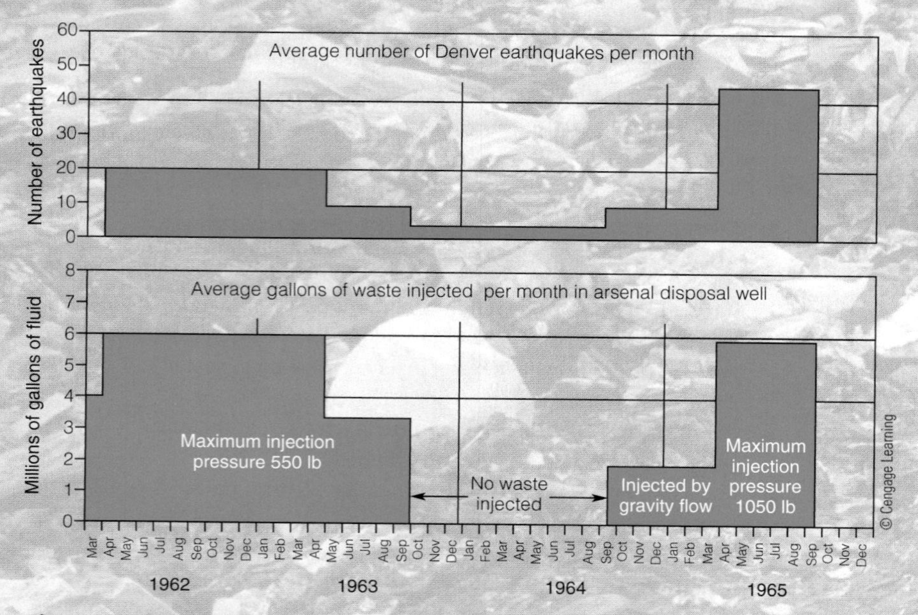

Figure 2 The correlation between fluid-waste injection and number of earthquakes per month in the Denver area, 1962–1965.

sures at the bottom of the hole decreased the frictional resistance on the fracture surfaces. Earthquakes were generated along a fracture propagating northwest of the well as the rocks adjusted to relieve internal stresses.

Thus, earthquakes *can* be "triggered" by human activities, and the Denver earthquakes initiated research into the possibility of modifying fault behavior by fluid injection—that is, relieving fault strain by producing a large number of small earthquakes. Seismic studies conducted during a water-injection (see Chapter 14) program at the nearby Rangely Oil Field confirmed that earthquakes *can* be turned on or off at will, so to speak, in areas of geological strain. To date, no group or government entity has assumed responsibility for triggering earthquakes by fluid injection, for obvious reasons.

questions to ponder

1. Does injecting fluids into active fault zones to gradually release accumulated strain (such as that along the active fault that caused the disastrous Haiti earthquake of January 2010) seem like a good idea?
2. What are the potential benefits of such a procedure?
3. What are the potential risks with such a procedure?

15.4 Love Canal: A Precedent in Human-Caused Environmental Pollution

Love Canal, in the honeymoon city of Niagara Falls, New York, may sound very romantic, but it's not. Excavated by William T. Love in the 1890s but never finished, the canal is 1,000 meters long and 25 meters wide (3,300 × 80 feet). Unused and drained, Love Canal was bought by Hooker Chemical and Plastics Corporation in 1942 for use as a waste dump. In the next 11 years, Hooker dumped 21,800 tons of toxic waste into the canal, which was believed to be impermeable, and thus an ideal "grave" for hazardous substances.

After the canal was filled and covered with a clay cap, Hooker sold it to the Niagara Falls School Board for $1. The company inserted in the deed a disclaimer denying legal liability for any injury caused by the wastes. In 1957, Hooker warned the school board not to disturb the clay cap because of the possible danger from leakage of toxic chemicals. A school was built on the landfill, and hundreds of homes and all the infrastructure needed to support a suburban community were built nearby.

In the early 1970s, after several years of heavy rainfall, water leaked through the "impermeable" clay cap and into basements and yards in the Love Canal area, as well as into the local sewer system. Soon the new community was experiencing high rates of miscarriages, birth defects, liver cancer, and seizure-inducing diseases among children. Residents complained that toxic chemicals in the water were causing these adverse health effects.

In 1978, the New York State Health Commissioner requested the EPA's assistance in investigating the chemistry of fluids that were leaking into a few houses around the canal. The study revealed the presence of 82 toxic chemicals, including benzene, chlorinated hydrocarbons, and dioxin. Five days later, then President Carter declared it a federal disaster area, the first human-caused environmental problem to be so designated in the United States. The recommendation that pregnant women and children younger than 2 be evacuated from the area followed, and New York State appropriated $22 million to buy the homes and repair the leaks. In all, a thousand families were relocated (**Figure 1**).

Love Canal was a precedent-setting case, because it spurred Congress to enact CERCLA, which provides federal money for toxic cleanup without long appeals. It also authorizes the EPA to sue polluters for the costs of the cleanup and relocation of victims. Although Love Canal has been essentially contained, some chemicals still infect a nearby stream and schoolyard, and maintenance costs amount to a half-million dollars annually. The EPA no longer ranks Love Canal high among the nation's most dangerous waste sites, and the New York State Health Commissioner has announced that recent federal-state studies found four of the seven polluted Love Canal areas to be habitable. In spite of the commissioner's statement, there has been no stampede to reinhabit the area.

After 16 years of litigation, the legal battles over Love Canal ended in 1995. Occidental Chemical Corporation, which purchased Hooker Chemical in 1968, agreed to pay Superfund $102 million, and the Federal Emergency Management Agency agreed to pay $27 million. The State of New York made a $98 million settlement with the EPA in 1994. The cleanup is estimated to have cost $275 million.

Michael J. Phillippot/Sygma/CORBIS

Figure 1 Where the honeymoon ended and the nightmares began: Love Canal, Niagara Falls, New York.

questions to ponder

1. The Love Canal incident was a tragic and precedent-setting situation. What good resulted from the incident?
2. What regulations are in effect where you live that ensure that similar toxic incidents do not happen?
3. Now that the New York State Health Department has announced that several areas of the Love Canal district are habitable, what measures might you take to confirm the safety of a house and property you might be interested in purchasing?

15.5 Radon and Indoor Air Pollution

Radon (^{222}Rn) is a radioactive gas formed by the natural disintegration of uranium as it transmutes step by radioactive step to form stable lead. The immediate parent element of radon is radium (^{226}Ra), which emits an alpha particle (^{4}He) to form ^{222}Rn. ^{222}Rn has a half-life of only 3.8 days, but it, in turn, breaks down into other elements that emit dangerous radionuclides. *Emanation* is sometimes used to describe the behavior of this element, and concern has been growing about the health hazard posed by radon emanating from rocks and soils, and seeping into homes. The EPA estimates that between 5,000 and 20,000 people die every year of lung cancer because they have inhaled radon and its decay products.

Radon is a problem only in areas underlain by rocks whose uranium concentrations are greater than 10 ppm—three times the average amount found in granite. For comparison, uranium ore contains more than 1,000 ppm of uranium. As radon disintegrates in the top few meters of rock and soil, it seeps upward into the atmosphere or into homes through cracks in concrete slabs or around openings in pipes. The measure of radon activity is *picocuries per liter* (pCi/L)—the number of nuclear decays per minute per liter of air. One picocurie per liter represents 2.2 potentially cell-damaging disintegrations per minute. Inasmuch as the average human inhales between 7,000 and 12,000 liters (2,000–3,000 gallons) of air per day, high radon concentrations in household air can be a significant health hazard.

Cancer caused by the bombardment of lung tissue by breathing ^{222}Rn or its decay products is a major concern of health scientists. The EPA has established 4 pCi/L as the maximum allowable indoor radon level. This is equivalent to smoking a half pack of cigarettes per day or having 300 chest X-rays in a year. **Figure 1** vividly illustrates the dangers of inhaling indoor radon (and of smoking). Such cancer is rarely apparent before 5 to 7 years of exposure, and its incidence is rare in individuals younger than 40.

Widespread interest in radon pollution first appeared in the 1980s with the discovery of high radon levels in houses in Pennsylvania, New Jersey, and New York. The discovery was made accidentally when a nuclear-plant worker set off the radiation-monitoring alarm when he arrived at work one day. Because his radioactivity was at a safe level, he did not worry about it. Returning to work after a weekend at home, he once again triggered the alarm. An investigation of his home revealed very high levels of radon. This was the first time radon was recognized as a public health hazard.

There are currently two types of radon "detectors." One is a charcoal canister that is placed in the household air for a week and then returned to the manufacturer for analysis. The other is a plastic-film alpha-track detector that is placed in the home for 2 to 4 weeks. It is recommended that the less expensive canister type be used first, and if a high radon level is indicated, a follow-up measurement with a plastic-film detector be made. Any reading greater than 4 pCi/L requires follow-up measurements. If the reading is between 20 and 200 pCi/L, one should consider taking measures to reduce radon levels in the home. Common methods are increasing the natural ventilation below the house in the basement or crawlspace, using forced air to circulate subfloor gases, and intercepting radon before it enters the house by installing gravel-packed plastic pipes below the floor slabs. For houses with marginal radon pollution, sealing all openings from beneath the house that go through the floor or the walls is usually effective.

pCi/L	Comparable Exposure Level	Comparable Risk*
200	1000 times average outdoor level	More than 75 times nonsmoker risk of dying from lung cancer
		4-pack/day smoker
100	100 times average indoor level	10,000 chest X-rays per year
40		30 times nonsmoker risk of dying from lung cancer
20	100 times average outdoor level	2-pack/day smoker
10	10 times average indoor level	1-pack/day smoker
4		3 times nonsmoker risk of dying from lung cancer
2	10 times average outdoor level	200 chest X-rays per year
		Nonsmoker risk of dying from lung cancer
1	Average indoor level	
0.2	Average outdoor level	

*Based on lifetime exposure

Figure 1 Lung cancer risk because of radon exposure, smoking, and X-ray exposure. A few houses have been found with radon radiation levels greater than 100 picocuries/liter, and one in Pennsylvania had a level of 2,000 picocuries/liter.

questions to ponder

1. Do you live in a region that has had reports of excess radon gas seeping from the ground? Have any incidents of radon poisoning been reported?
2. If you live in a region that may have radon poisoning, have you had your home checked out with a radon detector?
3. Consider that you have used a radon detector and discovered that your bedroom (where you spend 6–8 hours each night) has 3 picocuries per liter (pCi/L) radon. How does this threaten your health? What action would you take in light of this information?

15.6 Waste and the Birth of the Environmental Justice Movement

The term *environmental justice* refers to the equitable spatial distribution of the *burdens* of environmental problems, and the *benefits* of the natural world on groups all across the social spectrum. Although any group may be vulnerable, "unjust burdens" have generally fallen upon racial minorities in lower income brackets. Historically, these populations have concentrated in areas with the lowest comparative residential costs and property values because of proximity to sources of pollution, such as incinerators, chemical plants, and waste repositories. In many cases, facilities such as these have been sited close to already well-emplaced low-income populations, nominally on economic grounds (**Figure 1**).

Environmental justice as a political movement began with a particularly wicked incident in North Carolina in the summer of 1978. At that time, the U.S. government was considering a ban on PCBs, which were used in a wide variety of industrial applications, including circuit board and transformer manufacture. Chronic exposure to PCBs in the environment can be lethal to people, and has been implicated in a number of illnesses, including Japan's infamous Yusho disease. Seeking to rid itself of an expensive PCB waste disposal process as mandated by the new Toxic Substance Control Act regulations, the Ward PCB Transformer Company secretly dumped 120,400 liters (31,800 gallons) of PCB-contaminated oil along 380 kilometers (240 miles) of highway shoulders in 14 counties and at the Fort Bragg Army Base. The crime took place over a period of 2 weeks as drivers of a black-painted truck drove along rural roads spraying the liquid toxin first along one shoulder one night, then the other shoulder the next. The press called the act "The Midnight Dumpings." In response, the governor had the Highway Department erect large yellow warning signs along the roads saying, "Caution PCB Chemical Spills along Highway Shoulders."

The perpetrators were caught but served minimal time in jail. In the meantime, the state proposed to skim off the 60,000 tons of contaminated soil and put it a new 7.7 hectare (19 acre) managed landfill in Warren County, then a poor, primarily African American agricultural backwater. The local residents set up a fierce protest. The proposed landfill would be built on low ground, with the water table only a few meters down. Neighbors were concerned that any leakages could readily contaminate their groundwater supplies. Protests held off landfill construction for 4 years, and culminated in a 6-week-long nonviolent direct action campaign that was the largest civil protest in the American South since Martin Luther King marched through Alabama. The *environmental justice movement* was born.

The residents were wise to be concerned. Despite their opposition, the need to sequester the poisoned soils *somewhere* proved overwhelming, and Warren County would just have to do, according to the state. But almost as soon as the landfill went up, it began leaking PCBs, initially polluting the air as an odorless gas within a kilometer (0.6 miles) of the site. Shockingly, local residents were not informed of this situation for 15 years. The landfill operators eventually put a plastic liner atop the fill to seal in the gas. Finally, in 1993, North Carolina decided to detoxify the landfill. Too much water was building up within it to retain a poisonous overflow any longer. The wastewater would soon leak out above the base liner and infiltrate contaminants into community streams and wells. Detoxification continued until the end of 2003. It cost taxpayers $17 million.

questions to ponder

1. How was the cause of environmental justice served, or not served, in the Ward Transformer/Warren County case study?
2. Do you know of any potential environmental justice issues in your own community or home region? What is the potentially impacted population, and how is it impacted?
3. How can we build for the future so that we minimize incidents of environmental injustice?

Figure 1 A shantytown built atop a landfill near Port-au-Prince, Haiti. Residents have no electricity, plumbing, or convenient freshwater supply.

PHOTO GALLERY Waste disposal then and now

All humans generate solid waste, and disposing of it has always been a societal challenge. Safely disposing of containers of all kinds, old tires, large and small appliances, and just plain trash presents major problems today. As individuals, we must recycle, demand recycled and recyclable containers, and actively assist in reducing the volume of waste (**Figure 1**). Historians point out that ancient Rome was a filthy city; its cobblestones were noisy, its air was polluted from thousands of wood fires, and municipal sewage disposal was nonexistent. Medieval castles, monasteries, and convents had primitive waste-disposal systems, most of which were unhealthy and emanated noxious odors (**Figure 2**). Conditions are not much better in some parts of rural America today (**Figure 3**). And it should be pointed out that sanitary landfills are not all work and no play (**Figure 4**).

1 Norris McDonald organized a major cleanup of the Anacostia River in Washington, DC, and adjacent Maryland.
Robert Rathe Photography

Old Volkswagens in a junkyard; Monterrey, Mexico.
© 2009/RAUL TOUZON/National Geographic Image Collection

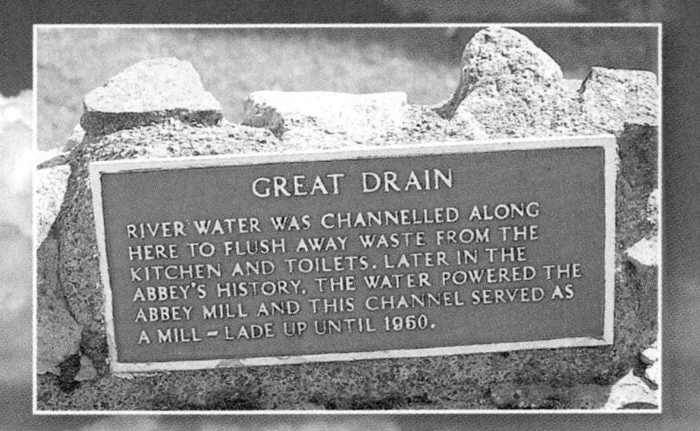

2 An adjacent stream was diverted to this medieval abbey in Scotland to create what was called a "great drain." All of the abbey's wastes went into the drain, which carried them to the stream, thus polluting its water for downstream users. Little wonder that life spans were so short in those days; diseases, such as cholera, typhoid fever, and dysentery, and pestilence caused by the lack of sanitation, took their toll. The correlation between stream pollution and disease was not widely recognized until the 1800s. B. Pipkin

3 An all-too-common method of private sewage disposal in the rural United States in times past. Joe Halbig, University of Hawaii-Hilo

4 Caterpillar tractor operators play "king of the mountain" at a landfill. The 25-ton "Cats" are used to compact and flatten the debris to reduce its volume. B. Pipkin

SUMMARY Waste Management and Geology

Waste Management

Management Components

Geology must be considered in all components:

1. Waste disposal
2. Source reduction
3. Waste recovery
4. Recycling

Rationale

Extending the lifetimes of landfills and natural resources.

Categories of Waste-Disposal Methods

Isolation and attenuation.

Municipal Waste Disposal

Sanitary Landfill

Trash is covered with clean soil each day.

Ocean Dumping

Favored in the past for disposal of sewage, sludge, dredgings, and canistered nuclear wastes. Now regulated by Marine Protection, Research and Sanctuaries Act (Ocean Dumping Act).

Landfill Problems

Leachate, settlement, gas generation, visual blight.

Resource Recovery

Amount of waste delivered to landfills can be reduced through recycling programs, composting, and waste-reduction strategies.

Multiple Land-Use Concept

A mine or quarry becomes a landfill, which when full may be converted to a park or other recreational area. Although some shallow landfills have been built on, settlement is always a problem.

Hazardous-Waste Disposal

Secure Landfill

Lined with plastic and impermeable clay seal. Built with leachate drains.

Deep-Well Injection

Into dry permeable rock body sealed from aquifers above and below. Has triggered earthquakes in Denver area.

Superfund

Comprehensive Environmental Response, Compensation and Liability Act (CERCLA) created in 1980 to correct and stop abuses from irresponsible toxic-waste dumping.

Private Sewage Disposal

Requires permeability and porosity, and can lead to local pollution.

Radioactive-Waste Disposal

High-Level Wastes

High-level nuclear waste from reactors (spent fuel rods), and 63,000 metric tons of heavy metal, is currently kept at individual reactors in pools of water. This is not a good situation. Yucca Mountain, Nevada, was for 30 years considered the most likely site for the final geological repository of civilian nuclear waste. But in 2009, the site was shelved as the potential repository because of the recognition of unsuitable complex geological conditions and excessive costs.

Low-Level Wastes

These are disposed of by shallow burial.

Mill Tailings

There are 24 inactive sites in 10 states, many of which are being cleaned up, with the government paying 90% of the costs.

Health and Safety Standards

The standard for radiation that a person may receive is 25 mrem per year, about equal to the dosage that a person would receive from two or three medical X-rays.

The Future

There is a future for nuclear energy and it lies in new technology and training.

KEY TERMS

attenuation
composting
contaminant plume
deep-well injection
Environmental Protection Agency (EPA)
eutrophication
fusion
geological repository
high-level wastes
incineration
isolation
leachate
low-level wastes
National Environmental Policy Act (NEPA)
radon
Resource Conservation and Recovery Act (RCRA)
resource recovery
sanitary landfill
secure landfill
Superfund
Waste Isolation Pilot Plant (WIPP)
Yucca Mountain

STUDY QUESTIONS

1. Explain the federal legislation known as NEPA and RCRA, and describe their impacts on the environmental safety of landfills.
2. What is leachate, and what methods are used to prevent landfill leachates from entering bodies of water?
3. Explain the "cell" method of placing trash in a landfill.
4. How much solid waste does the average American generate each day in kilograms? In pounds?
5. What problems are associated with operating a sanitary landfill?
6. What resources are found in a landfill, and how can they be recovered? What can individuals do to assist in recovering and extending the lifetimes of certain limited natural resources?
7. Describe two common methods of isolating hazardous wastes from the environment. Explain the steps that are taken to keep these wastes from polluting underground water.
8. What constitutes a good geological repository?
9. What are some of the health hazards connected with exposure to radiation, and how are dosages measured?
10. Transuranic waste must be contained and remain inviolate for 10,000 years. Is this possible?

REFERENCES AND FURTHER READINGS

Alexander, J. H. 1993. *In defense of garbage*. New York: Praeger Publishers, 239 p.

Barber, D. A. 2002. *Building with tires*. University of Arizona, Report on Research. Tucson, AZ: University Publications.

Brunner, D. R., and D. K. Keller. 1972. *Sanitary landfill design and operations*. Washington, DC: U.S. Environmental Protection Agency, 59 p.

Chester, M., and E. Martin. 2009. Cellulosic ethanol from municipal solid waste: A case study of economic, energy, and greenhouse gas impacts in California. *Journal of Environmental Science and Technology* 43: 5183–5189.

Crowley, K. D., and J. Ahearne. 2002. Managing the environmental legacy of U.S. nuclear weapons production. *American Scientist* 90: 514–520.

D'Agata, J. 2010. *About a mountain*. New York: Norton Publishing, 236 p.

DiGregorio, B. E. 2012. Tracking plastic in the ocean. *Earth* 57(2): 28–35.

Duffield, W. A. 2009. Yucca Mountain's volcanic risk. *Earth* 54(2): 5.

Environmental Protection Agency. 2010. *Municipal solid waste in the United States: 2009 facts and figures* [EPA530-R-10-012], 189 p.

Ewing, R. C., and E. R. von Hippel. 2009. Nuclear waste management in the United States. *Science* 325: 151–152.

George, R. 2008. *The big necessity: The unmentionable world of human waste and why it matters*. New York: Metropolitan Books, 283 p.

Grossman, E. 2006. *High tech trash: Digital devices, hidden toxics, and human health*. Washington, DC: Island Press, 331 p.

Miller, G. T., and S. Spoolsman. 2011. *Environmental science*, 13th ed. Belmont, CA: Cengage-Brooks/Cole, 522 p.

Rathje, W. L. 1991, May. Once and future landfills. *National Geographic* 114–134.

Rathje, W. L., and C. Murphy. 2001. *Rubbish! The archeology of garbage*. Tucson, AZ: University of Arizona Press, 263 p.

Sebeok, T. 1984. *Communication measures to bridge ten millennia*. Columbus, OH: Battelle Memorial Institute, Office of Nuclear Waste Isolation.

Solid Waste Alternative Technology (SWAT) Team. 2004. *Solid waste alternative technologies for New York City's waste disposal*. New York: New York School of International and Public Affairs, Columbia University.

White, P. T. 1992, April. The wonderful world of trash. *National Geographic* 424–457.

APPENDIX 1

Periodic Table of the Elements

Key:
- 1 — Atomic Number
- H — Symbol of Element
- 1.008 — Atomic Mass (rounded to four significant figures)

Period	IA	IIA	IIIB	IVB	VB	VIB	VIIB	VIII	VIII	VIII	IB	IIB	IIA	IVA	VA	VIA	VIIA	Noble Gases
1	1 H 1.008																	2 He 4.003
2	3 Li 6.941	4 Be 9.012											5 B 10.81	6 C 12.01	7 N 14.01	8 O 16.00	9 F 19.00	10 Ne 20.18
3	11 Na 22.99	12 Mg 24.31											13 Al 26.98	14 Si 28.09	15 P 30.97	16 S 32.06	17 Cl 35.45	18 Ar 39.95
4	19 K 39.10	20 Ca 40.08	21 Sc 44.96	22 Ti 47.90	23 V 50.94	24 Cr 52.00	25 Mn 54.94	26 Fe 55.85	27 Co 58.93	28 Ni 58.71	29 Cu 63.85	30 Zn 65.37	31 Ga 69.72	32 Ge 72.59	33 As 74.92	34 Se 78.96	35 Br 79.90	36 Kr 83.80
5	37 Rb 85.47	38 Sr 87.62	39 Y 88.91	40 Zr 91.22	41 Nb 92.91	42 Mo 95.94	43 Tc[x] 98.91	44 Ru 101.1	45 Rh 102.9	46 Pd 106.4	47 Ag 107.9	48 Cd 112.4	49 In 114.8	50 Sn 118.7	51 Sb 121.8	52 Te 127.6	53 I 126.9	54 Xe 131.3
6	55 Cs 132.9	56 Ba 137.3	57 ● La 138.9	72 Hf 178.5	73 Ta 180.9	74 W 183.9	75 Re 186.2	76 Os 190.2	77 Ir 192.2	78 Pt 195.1	79 Au 197.0	80 Hg 200.6	81 Tl 204.4	82 Pb 207.2	83 Bi 209.0	84 Po[x] (210)	85 At[x] (210)	86 Rn[x] (222)
7	87 Fr[x] (223)	88 Ra[x] 226.0	89 ■ Ac[x] (227)	104 Unq[x] (261)	105 Unp[x] (262)	106 Unh[x] (263)	107 Uns[x] (262)	108 Uno[x] (265)	109 Une[x] (266)									

Series														
● Lanthanide series	58 Ce 140.1	59 Pr 140.9	60 Nd 144.2	61 Pm[x] (147)	62 Sm 150.4	63 Eu 152.0	64 Gd 157.3	65 Tb 158.9	66 Dy 162.5	67 Ho 164.9	68 Er 167.3	69 Tm 168.9	70 Yb 173.0	71 Lu 175.0
■ Actinide series	90 Th[x] 232.0	91 Pa[x] 231.0	92 U[x] 238.0	93 Np[x] 237.0	94 Pu[x] (244)	95 Am[x] (243)	96 Cm[x] (247)	97 Bk[x] (247)	98 Cf[x] (251)	99 Es[x] (254)	100 Fm[x] (257)	101 Md[x] (258)	102 No[x] (255)	103 Lr[x] (256)

x: All isotopes are radioactive.

() indicates mass number of isotope with longest known half-life.

Some Important Minerals

Mineral	Chemical Composition	Color	Hardness	Other Characteristics	Primary Rock Occurrence
1. amphibole (e.g., hornblende) (hydrous)	Na, Ca, Mg, Fe, Al silicate	green, blue brown, black	5–6	often forms needle-like crystals; two good cleavages forming 60°–120° angle crystals hexagonal in cross section	igneous and metamorphic, minor amounts, all types
2. apatite	$Ca_5(PO_4)_3$ (F, Cl, OH)	blue-green	5	crystals hexagonal in cross section	minor amounts, all types
3. biotite (a mica)	hydrous K, Mg, Fe silicate	brown to black	2.5–3	excellent cleavage into thin sheets	all types
4. calcite	$CaCO_3$	variable; colorless if pure	3	effervesces in weak acid	sedimentary and metamorphic
5. dolomite	$CaMg(CO_3)_2$	white or pink	3.5–4	powdered mineral effervesces in acid	sedimentary
6. galena	PbS	silver-gray	2.5	metallic luster; cubic cleavage; heavy	sulfide veins
7. garnet	Ca, Mg, Fe, Al silicate	variable; often dark red	7	glassy luster	metamorphic
8. graphite	C	dark gray	1–2	streaks like pencil lead	metamorphic
9. gypsum	$CaSO_4 \cdot 2H_2O$	colorless	2	two directions of cleavage	sedimentary
10. halite	NaCl	colorless	2.5	salty taste; cleaves into cubes	sedimentary
11. hematite	Fe_2O_3	red or dark gray	5.5–6.5	red-brown streak regardless of color	oxidation of iron-bearing minerals
12. limonite	$Fe_2O_3 \cdot 3H_2O$	yellow-brown	2–3	earthy luster; yellow-brown streak	hydration of iron oxides
13. magnetite	Fe_3O_4	black	6	strongly magnetic	minor amounts, all types
14. muscovite (mica)	K, Al silicate	colorless to light brown	2–2.5	excellent basal cleavage	all types
15. olivine	Fe, Mg silicate	yellow-green	6.5–7	glassy luster	igneous and metamorphic
16. plagioclase feldspar	Na, Ca, Al silicate	white to gray	6	two cleavages at 90°, striations on cleavage surfaces	all types
17. orthoclase feldspar	K, Al silicate	white; or colored pink or aqua	6	two good cleavages forming 90° angle; no striations	all types
18. pyrite	FeS_2	brassy yellow	6–6.5	metallic luster; black streak	all types
19. pyroxene (e.g., augite)	Na, Ca, Mg, Fe, Al silicate	usually green or black	5–7	two good cleavages forming 90° angle	igneous and metamorphic
20. quartz	SiO_2	variable; commonly colorless or white	7	glassy luster; conchoidal fracture; hexagonal crystals	all types

The Elements

Element	Symbol	Atomic Number	Element	Symbol	Atomic Number
Actinium	Ac	89	Mercury	Hg	80
Aluminum	Al	13	Molybdenum	Mo	42
Americium	Am	95	Neodymium	Nd	60
Antimony	Sb	51	Neon	Ne	10
Argon	Ar	18	Neptunium	Np	93
Arsenic	As	33	Nickel	Ni	28
Astatine	At	85	Niobium	Nb	41
Barium	Ba	56	Nitrogen	N	7
Berkelium	Bk	97	Nobelium	No	102
Beryllium	Be	4	Osmium	Os	76
Bismuth	Bi	83	Oxygen	O	8
Boron	B	5	Palladium	Pd	46
Bromine	Br	35	Phosphorus	P	15
Cadmium	Cd	48	Platinum	Pt	78
Calcium	Ca	20	Plutonium	Pu	94
Californium	Cf	98	Polonium	Po	84
Carbon	C	6	Potassium	K	19
Cerium	Ce	58	Praseodymium	Pr	59
Cesium	Cs	55	Promethium	Pm	61
Chlorine	Cl	17	Protactinium	Pa	91
Chromium	Cr	24	Radium	Ra	88
Cobalt	Co	27	Radon	Rn	86
Copper	Cu	29	Rhenium	Re	75
Curium	Cm	96	Rhodium	Rh	45
Dysprosium	Dy	66	Rubidium	Rb	37
Einsteinium	Es	99	Ruthenium	Ru	44
Erbium	Er	68	Samarium	Sm	62
Europium	Eu	63	Scandium	Sc	21
Fermium	Fm	100	Selenium	Se	34
Fluorine	F	9	Silicon	Si	14
Francium	Fr	87	Silver	Ag	47
Gadolinium	Gd	64	Sodium	Na	11
Gallium	Ga	31	Strontium	Sr	38
Germanium	Ge	32	Sulfur	S	16
Gold	Au	79	Tantalum	Ta	73
Hafnium	Hf	72	Technetium	Tc	43
Helium	He	2	Tellurium	Te	52
Holmium	Ho	67	Terbium	Tb	65
Hydrogen	H	1	Thallium	Tl	81
Indium	In	49	Thorium	Th	90
Iodine	I	53	Thulium	Tm	69
Iridium	Ir	77	Tin	Sn	50
Iron	Fe	26	Titanium	Ti	22
Krypton	Kr	36	Tungsten	W	74
Lanthanum	La	57	Uranium	U	92
Lawrencium	Lr	103	Vanadium	V	23
Lead	Pb	82	Xenon	Xe	54
Lithium	Li	3	Ytterbium	Yb	70
Lutetium	Lu	71	Yttrium	Y	39
Magnesium	Mg	12	Zinc	Zn	30
Manganese	Mn	25	Zirconium	Zr	40
Mendelevium	Md	101			

Planes in Space: Dip and Strike

Geologists describe the orientation of a plane surface by the plane's "strike" and "dip." Geologists measure such tangible planes as those made by sedimentary strata, foliation, and faults, because it allows them to anticipate hazards during excavations for such things as tunnels, dam foundations, and roadcuts. Accurate surface measurements of planes also enable geologists to predict the underground locations of valuable resources such as oil reservoirs, ore deposits, and water-bearing strata.

Strike is the direction of the trace of an inclined plane on the horizontal plane, and the direction is referenced to north. In the illustrated example, the strike is 30° west of north. In geologic notation, this is written N30° W.

Dip is the vertical angle that the same inclined plane makes with the horizontal plane. Dip is assessed perpendicular to the strike, because any other angle will give an "apparent dip" that is less than the true dip. Because a plane that strikes northwest may be inclined to either the southwest or the northeast, the general direction of dip must be indicated. In our example, the dip is 60° to the northeast, which is notated 60° NE.

Thus, a geologist's notes for the bedding planes in this example would describe their orientation as "strike N30° W, dip 60° NE." All other geologists would understand that the plane has a strike that is 30° west of north and a dip that is 60° to the northeast.

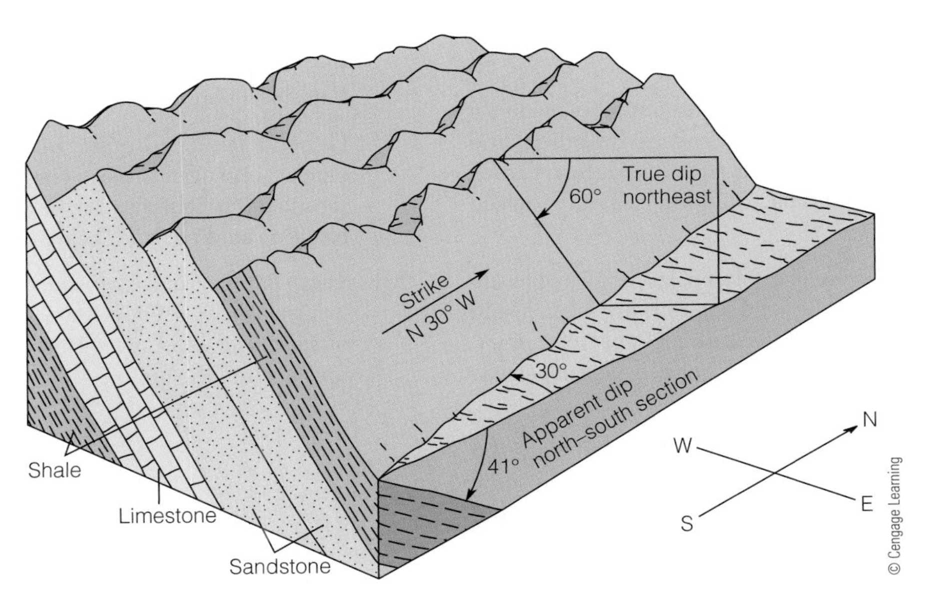

Block diagram showing sedimentary rock that is striking N30° W and dipping 60° northeast. The apparent dip is shown on the north–south cross section at the lower front of the diagram.

Modified Mercalli Scale

The Rossi–Forel intensity scale was the first scale developed for evaluating the effects of an earthquake on structures and humans. It was developed in Europe in 1878 and assigned Roman numerals from *I* (for *barely felt*) to X (for *total destruction*) for varying levels of intensity. This scale was first modified by Father Giuseppe Mercalli in Italy and later in 1931 by Frank Neumann and Harry O. Wood in the United States. The Modified Mercalli scale is the scale most widely used today for evaluating the effects of earthquakes in the field. An abbreviated version is given here.

I. Not felt except by a very few persons under especially favorable circumstances.

II. Felt by a few persons at rest, especially by persons on upper floors of multistory buildings, and by nervous or sensitive persons on lower floors.

III. Felt quite noticeably indoors, especially on upper floors, but many people do not recognize it as an earthquake; vibrations resemble those made by a passing truck.

IV. Felt indoors by many persons and outdoors by only a few. Dishes, windows, and doors are disturbed, and walls make creaking sounds as though a heavy truck had struck the building. Standing cars are rocked noticeably.

V. Felt by nearly everyone. Many sleeping people are awakened. Some dishes and windows are broken, and some plaster is cracked. Disturbances of trees, poles, and other tall objects may be noticed. Some persons run outdoors.

VI. Felt by all. Many people are frightened and run outdoors. Some heavy furniture is moved, and some plaster and chimneys fall. Damage is slight, but humans are disturbed.

VII. General fright and alarm. Everyone runs outdoors. Damage is negligible in buildings of good design and construction, considerable in those that are poorly built. Noticeable in moving cars.

VIII. General fright approaching panic. Damage is slight in specially designed structures; considerable in ordinary buildings, with partial collapse; great in older or poorly built structures. Chimneys, smokestacks, columns, and walls fall. Sand and mud are ejected from ground openings (liquefaction).

IX. General panic. Damage is considerable in specially designed structures, well-designed frame structures being deformed; great in substantial buildings, with partial collapse. Ground is cracked; underground pipes are broken.

X. General panic. Some well-built wooden structures and most masonry and frame structures are destroyed. Ground is badly cracked; railway rails are bent; underground pipes are torn apart. Considerable landsliding on riverbanks and steep slopes.

XI. General panic. Ground is greatly disturbed, with cracks and landslides common. Sea waves of significant height may be seen. Few, if any, structures remain standing; bridges are destroyed.

XII. Total panic. Total damage to human engineering works. Waves seen on the ground surface. Lines of sight and level are distorted. Objects are thrown upward into the air.

Calculating the Expiration Time for U.S. Coal Reserves

When a quantity such as the rate of consumption of a resource grows at a fixed percentage rate per year, the growth is exponential. The equation for determining the estimated exponential lifetime of coal, or any other finite resource, is

$$T_e = (1/k)\ \ln\left([kR/r_o] + 1\right)$$

where T_e is the exponential expiration time (i.e., how long it will last), k is the annual rate of growth in production or consumption, R is the size of today's resource (i.e., the estimated reserve amount), r_o is the production in a given year (in this case, for 2010), and ln is the natural logarithm.[1] The raw data used to calculate the estimated expiration time, T_e, for United States coal reserves is from authoritative published sources:[2]

R = 260,000 million tonnes
= 286,600 million short tons;
k = 2.1%, or 0.021, the production increase in 2010 in comparison with 2009;
r_o = 1,085 million short tons[3]

Substituting these values in the above equation and rounding to three significant figures:

$$T_e = \left(\frac{1}{0.021}\right) \ln\left[\frac{(0.021)(286{,}600 \times 10^6 \text{ tons})}{1{,}085 \times 10^6 \text{ tons}} + 1\right] \text{ years}$$

$$T_e = 89.5 \text{ years}$$

A word of caution: The above calculation for the expiration time of U.S. coal reserves is not a prediction, merely a first-order estimate determined from known quantities for the life expectancy under the condition of steady growth that our government and our society believe is necessary to maintain our economy.

[1] The method of calculation is from Albert A. Bartlett. 1978. The forgotten fundamentals of the energy crisis. *American Journal of Physics* 46. (9): 876–888.
[2] Data for k and r_o are from the Energy Information Agency, 2011; R is from the *British Petroleum Statistical Review of World Energy*, June 2011.
[3] One tonne = 1,000 kilograms = 1.1023 short tons.

a'ā A basaltic lava that has a rough, blocky surface.

ablation All processes by which snow or ice is lost from a glacier, floating ice, or snow cover. These processes include melting, evaporation, and calving.

ablation zone The part of a glacier or a snowfield in which ablation exceeds accumulation over a year's time; the region below the firn line.

abrasion The mechanical grinding, scraping, wearing, or rubbing away of rock surfaces by friction or impact. Abrasion agents include wind, running water, waves, and ice.

absolute-age dating Assigning units of time, usually years, to fossils, rocks, and geological features or events. The method uses mostly the decay of radioactive elements, but ages may also be obtained by counting tree rings, measuring the build-up of isotopes formed by cosmic rays, or annual sedimentary layers.

accretion The mechanism by which the Earth is believed to have formed from a small nucleus by additions of solid bodies, such as meteorites, asteroids, or planetesimals. Also used in plate tectonics to indicate the addition of terranes to a continent.

accumulation zone The part of a glacier or snowfield in which build-up of snow and ice exceeds ablation over a year's time; the region above the firn line.

acid mine drainage (AMD) The acid runoff from a mine or mine waste.

acid rain Corrosive precipitation caused by the solution of atmospheric sulfuric acid in rain, which makes it acidic. Although it is generally considered human caused, it can also be natural, as from volcanic activity.

acid rock drainage The natural background of spring or stream runoff from a mineralized rock mass.

active layer The zone in permafrost that is subject to seasonal freezing and thawing.

active sea cliff A cliff whose formation is dominated by wave erosion.

active volcano A volcano with a historical record of eruptions, or one in a current state of eruption.

adiabatic heating The heating that occurs in air that is compressed, as individual gas molecules are driven closer together and increase their collisions.

adsorption The attraction of ions or molecules to the surface of particles, in contrast with *absorption,* the process by which a substance is taken into a clay-mineral structure.

aerosol A suspension of very fine, liquid and solid particles in the atmosphere.

A horizon The upper soil horizon (topsoil) capped by a layer of organic matter. It is the zone of leaching of soluble soil materials.

albedo The reflectivity of surfaces, measured as a decimal fraction of incident sunlight reflected. (Snow and ice have an albedo >0.9, meaning >90% reflectivity.)

alluvial fan A gently sloping, fan-shaped mass of alluvium deposited in arid or semiarid regions where a stream flows from a narrow canyon onto a plain or valley floor.

alluvial placer A placer deposit that is concentrated in a streambed.

alluvium Stream-deposited sediment.

alpha decay The transformation that occurs when the nucleus of an atom spontaneously emits a helium atom, reducing its atomic mass by 4 units and its atomic number by 2; for example, $^{238}{}_{92}$ UÆ $^{234}{}_{90}$ Th + $_{2}{}^{4}$a.

amorphous Descriptive term applied to glasses and other substances that lack an orderly internal crystal structure.

andesite line The petrographic boundary between basalts and andesites. It runs roughly around the Pacific basin margin. It also marks the boundary between continental (explosive) and oceanic (Hawaiian-type) volcanoes in the Pacific Ocean.

angle of repose The maximum stable slope angle that granular, cohesionless material can assume. For dry sand, this is about 34°.

anomaly A deviation from the average or expected value. In paleomagnetism, a "positive" anomaly is a stronger-than-average magnetic field at the Earth's surface.

antecedence The precipitation history in the days, weeks and months before a flood event. A wet period before a rainstorm leads to greater flooding.

Anthropocene The proposed new epoch in the geological time scale defined by a world in which human activity creates great physical, biological, and chemical change; change sufficient to leave a long-standing trace in the geological record.

anthropogenic Relating to or resulting from the influence of humans on nature.

anticline A generally convex-upward fold that has older rocks in its core. (Contrast with *syncline.*)

aphanitic texture A fine-grained, igneous-rock texture resulting from rapid cooling. Aphanites are fine-grained igneous rocks.

A.P.I. gravity (American Petroleum Institute) The weight of crude oil arbitrarily compared with that of water at 60°F, which is assigned a value of 10°. Most crude oils range from 5° to 30° A.P.I.

aquiclude A layer of low-permeability rock or sediment that hampers groundwater movement. Often applied to the confining bed in a confined aquifer.

aquifer A water-bearing body of rock or sediment that will yield water to a well or spring in usable quantities.

arc volcano One of a set of volcanoes that lie above a subduction zone, which line up in a broadly arcuate line typically stretching for many hundreds of kilometers. Examples of arc volcanoes include Mt. Rainier (Cascade arc), and Fujiyajma (southwest Honshu arc, Japan).

area mining The mining of coal in flat terrain (strip mining).

arroyo incision Erosional deepening of arroyos, desert gullies that may contain water only a few hours or days a year.

artesian pressure surface An imaginary surface defined by the level to which water will rise in an artesian well.

artificial levee An embankment constructed to contain a stream at flood stage. See also *levee.*

ash Very fine, gritty volcanic dust; particles are less than 4 mm in diameter.

ashfall A "rain" of airborne volcanic ash.

asthenosphere The *plastic* layer of the Earth, at whose upper boundary the brittle overlying lithosphere is detached from the underlying mantle.

atmosphere The envelope of gases enclosing a planet.

atom The smallest particle of an element that can exist either alone or in combination with similar particles of the same or a different element.

atomic mass The sum of the number of protons and neutrons (uncharged particles) in the nucleus of an atom.

atomic number The number of protons (positively charged particles) in an atomic nucleus. Atoms with the same atomic number belong to the same chemical element.

attenuation (waste disposal) To dilute or spread out the waste material.

average crustal abundance The percentage of a particular element in the composition of the Earth's crust.

back slope The gentler slope of a dune, scarp, or fault block.

ball mill A large, rotating drum containing steel balls that grinds ore to a fine powder. Batteries of ball mills are used in modern milling operations.

banded-iron formation An iron deposit consisting essentially of iron oxides and chert occurring in prominent layers or bands of brown or red and black.

barchan dune An isolated, moving, crescent-shaped sand dune lying transverse to the direction of the prevailing wind. A gently sloping, convex side faces the wind, and the horns of the crescent point downwind.

barrier island (barrier beach) A narrow sand ridge rising slightly above high tide level that is oriented generally parallel with the coast and separated from the coast by a lagoon or a tidal marsh.

basalt A dark-colored, fine-grained volcanic rock that is commonly called *lava*. Basaltic lavas are quite fluid.

base isolation Building design option using a movable base under the building that allows the ground to move but minimizes building vibration and sway.

base level The level below which a stream cannot erode its bed. The ultimate base level is sea level.

base shear Caused by a rapid horizontal ground acceleration during an earthquake. It can cause a building to part from its foundation, thus the term.

batholith A large, discordant, intrusive body greater than 100 square kilometers (40 square miles) in surface area with no known bottom. Batholiths are granitic in composition.

bauxite An off-white, grayish, brown, or reddish brown rock composed of a mixture of various hydrous aluminum oxides and aluminum hydroxides.

beach cusps A series of regularly spaced points of beach sand or gravel separated by crescent-shaped troughs.

beach nourishment The addition of sand to a beach by trucking or pumping in order to build up the beach after it has eroded.

bedding See *stratification.*

bedding planes The divisions or surfaces that separate successive layers of sedimentary rocks from the layers above and below them.

bench placer Remnant of an ancient alluvial placer deposit that was concentrated beneath a former stream and that is now preserved as a bench on a hillside.

bentonite A soft, plastic, light-colored clay that swells to many times its volume when placed in water. Named for outcrops near Fort Benton, Wyoming, bentonite is formed by chemical alteration of volcanic ash.

beta decay The radioactive emission of a nuclear electron (beta ray, β^-) that converts a neutron to a proton. The element is changed to the element of next-highest atomic number; for example, ${}_{6}^{14}C$ ${}_{7}^{14}N + \beta^-$.

binary-cycle system A geothermal system that uses geothermal heat to vaporize a second fluid (e.g., isobutane) with a lower boiling temperature than water, to power a turbine generator.

biodiesel A diesel-fuel equivalent derived from plant or animal sources.

biofuel Any fuel derived from biomass, e.g., sugarcane, food waste, straw, corn, or switchgrass.

biogenic sedimentary rock Rocks resulting directly from the activities of living organisms, such as a coral reef, shell limestone, or coal.

biological oxygen demand (B.O.D.) Indicator of organic pollutants (microbial) in an effluent measured as the amount of oxygen required to support the organisms. The greater the B.O.D. (parts per million or milligrams/liter), the greater the pollution and less oxygen is available for higher aquatic organisms.

biomass fuels Fuels that are manufactured from plant materials; for example, grain alcohol (ethanol) is made from corn.

bioremediation Use of bacteria to degrade petroleum-based pollutants. Used primarily for remediating soil contamination.

biosphere The sum total of life inhabiting the Earth.

black smoker A submarine hydrothermal vent associated with a mid-oceanic spreading center that has emitted or is emitting a "smoke" of metallic sulfide particles.

block (volcanology) A mass of cold country rock greater than 32 millimeters (1¼ inch) in size that is ripped from the vent and ejected from a volcano.

block glide (translational slide) A landslide in which movement occurs along a well-defined plane surface or surfaces, such as bedding planes, foliation planes, or faults.

body wave Earthquake wave that travels through the body of the Earth.

bolide A large meteor that explodes as it passes through the Earth's atmosphere.

bomb (volcanology) A molten mass greater than 32 millimeters (1¼ inches) in diameter that is ejected from a volcano and cools sufficiently before striking the ground to retain a rounded, streamlined shape.

boreal The northern and mountainous regions of the Northern Hemisphere in which the average temperature of the hottest season does not exceed 18°C (64.4°F).

breakwater An offshore rock or concrete structure, attached at one end or parallel to the shoreline, intended to create quiet water for boat anchorage purposes.

breeder reactor A reactor that produces more fissile (splittable) nuclei than it consumes.

brine exclusion effect The generation of saltier, denser water at depth as freezing to ice occurs at the surface of the sea.

British thermal unit (Btu) The amount of heat, in calories, needed to raise the temperature of 1 pound of water 1°F.

caldera A large crater caused by explosive eruption and/or subsidence of the cone into the magma chamber, usually more than 1.5 kilometers (1 mile) in diameter.

caliche In arid regions, sediment or soil that is cemented by calcium carbonate ($CaCO_3$).

calving The breaking away of a block of ice, usually from the front of a tidewater glacier, that produces an iceberg.

cap-and-trade A form of emissions trading; an administrative approach to controlling pollution that provides economic incentives for achieving reductions in the emissions of pollutants. A limit (cap) is set on the amount of pollution. Those companies that require greater emissions allowances must buy credits from those that pollute less. The transfer of permitted emissions is called a *trade.*

capillary fringe The moist zone in an aquifer above the water table. The water is held in it by capillary action.

capillary water Water that is held in tiny openings in rock or soil by capillary forces.

caprock An impermeable stratum that caps an oil reservoir and prevents oil and gas from escaping to the ground surface.

carbon capture and sequestration (or storage) (CCS) The process of removing carbon from a fossil-fuel–burning power plant or other industry that is stored underground or in the deep ocean.

carrying capacity The number of creatures a given tract of land can adequately support with food, water, and other necessities of life.

CCS See *carbon capture and sequestration.*

chain reaction A self-sustaining nuclear reaction that occurs when atomic nuclei undergo fission. Free neutrons are released that split other nuclei, causing more fissions and the release of more neutrons, which split other nuclei, and so on.

chemical sedimentary rock Rock resulting from precipitation of chemical compounds from a water solution. See also *evaporite*.

chlorofluorocarbons (CFCs) A group of compounds that, on escape to the atmosphere, break down, releasing chlorine atoms that destroy ozone molecules; for example, Freon, which is widely used in refrigeration and air-conditioning, and formerly was a propellant in aerosol-spray cans.

cinder cone A small, straight-sided volcanic cone consisting mostly of pyroclastic material (usually basaltic).

cinders Glassy and vesicular (containing gas holes) volcanic fragments, only a few centimeters across, that are thrown from a volcano.

clastic Describing rock or sediment composed primarily of detritus of preexisting rocks or minerals.

clastic sedimentary rock Rock composed of fragments of minerals or other rock materials.

clay minerals Hydrous aluminum silicates that have a layered atomic structure. They are very fine grained and become plastic (moldable) when wet. Most belong to one of three clay groups: kaolinite, illite, and smectite, the last of which are expansive when they absorb water between layers.

claypan A dense, relatively impervious subsoil layer that owes its character to a high clay-mineral content because of concentration by downward-percolating waters.

cleavage The breaking of minerals along certain crystallographic planes of weakness that reflects their internal structure.

climate controls The relatively constant factors that dictate a region's climate.

climax avalanche A deep, thick avalanche that tends to occur in late spring when the snowpack has warmed and weakened.

cloud-albedo effect The process by which clouds reflect sunlight or trap incident solar radiation to change atmospheric temperature beneath them, thereby influencing any further cloud development.

cloud reflectivity enhancement A proposal for creating cloud condensation nuclei that would lighten the cloud's albedo and reflect more sunlight away from Earth.

coal-bed methane A form of natural gas, methane (NH_3), extracted from coal beds and seams. In the early 21st century, it has become an important source of energy in the United States, Canada, and other countries.

coastal desert A desert on the western edge of continents in tropical latitudes; for example, near the Tropic of Capricorn or Cancer. The daily and annual temperature fluctuations are much less than in inland tropical deserts.

cohesion The tendency of a material to stick to itself; the strength a material has when no normal force is applied.

columnar jointing Polygonal columns in a solid lava flow or shallow intrusion formed by contraction on cooling and oriented perpendicular to the cooling surface. They are usually formed in basaltic lavas.

compaction The reduction in bulk volume of fine-grained sediments caused by increased overburden weight or by tighter packing of grains. Clay is more compressible than sand and, therefore, compacts more.

complex landslide When both rotational and translational sliding are found within one landslide mass.

composite volcano A volcano built of mixed lava flows, pyroclastic debris, and breccias (rubble). Also known as a stratovolcano.

composting Bacterial recycling of organic material.

compound growth equation An equation quantifying how rapidly a certain quantity will grow at a fixed rate of increase (given, for example, as a percentage per year).

Comprehensive Environmental Response, Compensation and Liability Act (CERCLA) Comprehensive Environmental Response, Compensation and Liability Act of 1980. See also *Superfund*.

compressed natural gas A fossil fuel substitute for gasoline, diesel, or propane/LPG used in vehicles and compressed to pressures above 3,100 pounds per square inch for storage in tanks.

concentrate Enriched ore, often obtained by flotation, produced at a mill.

concentration factor The enrichment of a deposit of an element, expressed as the ratio of the element's abundance in the deposit to its average crustal abundance.

cone of depression The cone shape formed by the water table around a pumping well.

confined aquifer A water-bearing formation bounded above and below by impermeable beds or beds of distinctly lower permeability.

contact metamorphism A change in the mineral composition of rocks in contact with an invading magma from which fluid constituents are carried out to combine with some of the country-rock constituents to form a new suite of minerals.

contaminant plume A volume of contaminated groundwater extending away from a surface pollution source, such as a poorly sealed landfill or waste disposal site. The plume extends in the direction that the groundwater is flowing.

continental drift The term applied by American geologists to Wegener's hypothesis of long-distance horizontal movement of continents he called "continental displacement."

contour mining The mining of coal by cutting into and following the coal seam around the perimeter of the hill.

convection The circulatory movement within a body of nonuniformly heated fluid. Warmer material rises in the center, and cooler material sinks at the outer boundaries. It has been proposed as the mechanism of plate motion in plate tectonic theory.

convergent boundary Tectonic boundary where two plates collide. Where a tectonic plate sinks beneath another plate and is destroyed, this boundary is known as a *subduction zone*. Subduction zones are marked by deep-focus seismic activity, strong earthquakes, and violent volcanic eruptions.

Cordillera The mountain system along the western parts of North and South America that includes the Sierra Nevada, Cascade Range, Rocky Mountains, Sierra Oriental, Sierra Occidental, and Andes.

core Center of the Earth, consisting of dense, metallic, iron-rich material.

Coriolis effect The deflection of wind and ocean currents by rotation of the Earth. Deflection is toward the right in the Northern Hemisphere and toward the left in the Southern Hemisphere.

cosmogenic isotopes Rare isotopes or types of an element, such as ^{10}Be or ^{36}Cl, formed as cosmic rays, such as neutrons interact with minerals at and near Earth's surface.

craton A part of the Earth's continental crust that has attained stability—that has not been deformed for a long period of time. The term is restricted to continents and includes their most stable areas, the continental shields.

creep (soil) The imperceptibly slow downslope movement of rock and soil particles by gravity.

cross-bedding Arrangement of strata, greater than 1 centimeter thick, inclined at an angle to the main stratification.

crust A thin "skin" of aluminum and alkali-rich silicate rocks that surrounds the Earth's mantle; it is still evolving.

crystal settling In a magma, the sinking of crystals because of their greater density, sometimes aided by magmatic convection. See also *gravity separation.*

cubic feet per second (cfs) Unit of water flow or stream discharge. One cubic foot per second is 7.48 gallons passing a given point in 1 second.

cutoff A new channel that is formed when a stream cuts through a very tight meander. See also *oxbow lake.*

cyanide heap leaching A process of using cyanide to dissolve and recover gold and silver from ore.

cyclone A storm characterized by a circular twisting of winds around a central core, or "eye." Cyclones include hurricanes and typhoons, and are much larger than tornadoes.

c-zone Soil zone of weathered parent material leading downward to fresh bedrock.

Darcy's law A derived formula for the flow of fluids in a porous medium.

debris avalanche A sudden, rapid movement of a water–soil–rock mixture down a steep slope.

debris flow A moving mass of a water, soil, and rock intimately mixed. More than half of the soil and rock particles are coarser than sand, and the mass has the consistency of wet concrete.

deep-well injection A liquid waste-disposal process that pumps liquids hundreds of feet underground into aquifers that have no potential to contaminate potential or existing potable water aquifers.

deflation The lifting and removal of loose, dry, fine-grained particles by the action of wind.

degassing Loss of volcanic gases from a volcano during an eruption or more gradually via fumarolic activity.

delta The nearly flat land formed where a stream empties into a body of standing water, such as a lake or the ocean.

depth hoar Ice crystals formed in the snowpack by migration of water vapor. Forms a weak layer upon which avalanches can run.

desert An arid region of low rainfall, usually less than 25 centimeters (10 inches) annually, and of high evaporation or extreme cold. Deserts are generally unsuited for human occupation under natural conditions.

desertification The process by which semi-arid grasslands are converted to desert.

desert pavement An interlocking cobble mosaic that remains on the earth's surface in a dry region.

detrital See *clastic.*

dip The angle that an inclined geological planar surface (sedimentary bedding plane, fault, joint) makes with the horizontal plane.

discharge (Q) The volume of water, usually expressed in cubic feet per second (cfs) or cubic meters per second (cms), that passes a given point within a given period of time.

dish/engine A solar power technology that uses thousands of small mirrors to focus reflected sunlight on a tower. The Sun's energy is then transferred to a Sterling cycle heat engine that runs a turbine which generates electricity.

disseminated deposit A mineral deposit, especially of a metal, in which the minerals occur as scattered particles in the rock, but in sufficient quantity to make the deposit a commercially worthwhile ore.

divergent boundary Found at mid-ocean ridge or spreading center where new crust is created as plates move apart. For this reason, it is sometimes referred to as a *constructive* boundary and is the site of weak, shallow-focus earthquakes and volcanic action.

doldrums The area of warm rising air, calm winds, and heavy precipitation between 5° north latitude and 5° south latitude. It is where the northeast and southeast trade winds meet.

dormant volcano A volcano that has no recent historical record of volcanic activity (if any record at all) but which may appear to have been active at some time in the recent geological past, including fresh appearing volcanic deposits, areas lacking vegetative cover, etc. A dormant volcano could erupt again, though implicitly—and perhaps deceptively—no time soon.

doubling time The number of years required for a population to double in size.

drainage basin The tract of land that contributes water to a particular stream, lake, reservoir, or other body of surface water.

drainage divide The boundary between two drainage basins.

drawdown The lowering of the water table immediately adjacent to a pumping well.

dredging The excavation of earth material from the bottom of a body of water by a floating barge or raft equipped to scoop up, discharge by conveyors, and process or transport materials.

drift A general term for all rock material that is transported and deposited by a glacier or by running water emanating from a glacier.

driving force (landslide) Gravity.

dry snow zone The area of a snow field that is rarely warm enough for snow to melt.

dune A mound, bank, ridge, or hill of loose, windblown granular material (generally sand), either bare or covered with vegetation. It is capable of movement but maintains its characteristic shape.

dust storm Area of the atmosphere where high winds are moving dust eroded from Earth's surface. Visibility can be very low in dust storms and such storms can sandblast or etch glass and damage paint.

eccentricity The amount by which Earth's orbit deviates from a perfect circle around the Sun. Earth's orbit varies from a slight ellipse to a strong ellipse with regularity over periods of tens of thousands of years.

ecological overshoot The condition in which a population of organisms reaches a level above and beyond the capacity of needed natural resources to support it.

economically recoverable oil Petroleum for which the costs of discovery, development, production, transport, and return of capital can be produced at a given market price.

effective stress The average normal force per unit area (stress) that is transmitted directly across grain-to-grain boundaries in a sediment or rock mass.

effusive eruptions Eruptions predominantly of low-viscosity molten lava, generally initiated from fissures.

El Niño A phenomena of the equatorial Pacific Ocean every 3 to 12 years when the warming of the eastern Pacific alters the normal jet stream, affects storm tracks, and creates unusual global weather patterns. Typically, El Niños bring excessive dry periods to the North American Midwest and western Pacific, and excessive rain to the eastern Pacific. See also *La Niña* and *ENSO.*

El Niño–Southern Oscillation See *ENSO.*

elastic Said of a body in which strains are totally recoverable, as in a rubber band. Contrast with *plastic.*

elastic rebound theory The theory that movement along a fault and the resulting seismicity are the result of an abrupt release of stored elastic strain energy between two rock masses on either side of the fault.

element A substance that cannot be separated into different substances by usual chemical means.

end moraine (terminal moraine) A moraine that is produced at the front of an actively flowing glacier; a moraine that has been deposited at the lower end of a glacier.

energy return on investment (EROI) The ratio of energy returned relative to the energy invested to discover and obtain that energy.

engineering geology The application of the science of geology to human engineering works.

ENSO Abbreviation for the periodic meteorological event known as El Niño–Southern Oscillation. See also *El Niño* and *La Niña*.

environment Literally "surroundings," more specifically the natural world or world that functions without direct human intervention and manipulation.

environmental geology The relationship between humans and their geological environment.

eolian Of, produced by, or carried by wind.

EPA Environmental Protection Agency.

epicenter The point at the surface of Earth directly above the focus of an earthquake.

EROI See *energy return on investment.*

erosion The weathering and transportation of the materials of the Earth's surface.

erratic A rock fragment that has been carried by glacial ice or floating ice and deposited when the ice melted at some distance from the outcrop from which it was derived.

esker A long, sinuous ridge winding across a landscape that has been formed by sediment deposited on the bed of a subglacial or englacial river that flowed in ice tunnels in or beneath a melting glacier.

estuary A semi-enclosed coastal body where outflowing river water meets tidal seawater.

ethanol A colorless liquid chemical compound, commonly called *grain alcohol,* with the formula CH_3CH_2OH.

eutrophication The increase in nitrogen, phosphorus, and other plant nutrients in the aging of an aquatic system. Blooms of algae develop, preventing light penetration and causing reduction of oxygen needed in a healthy system.

evaporite A nonclastic sedimentary rock composed primarily of minerals produced when a saline solution becomes concentrated by evaporation of the water; especially a deposit of salt that precipitated from a restricted or enclosed body of seawater or from the water of a salt lake.

evapotranspiration That portion of precipitation that is returned to the air through evaporation and transpiration, the latter being the escape of water from the leaves of plants.

exfoliation The process whereby slabs of rock bounded by sheet joints peel off the host rock, usually a granite or sandstone.

exfoliation dome A large, rounded dome resulting from exfoliation; for example, Half Dome in Yosemite National Park.

exotic terranes Fault-bounded bodies of rock that have been transported some distance from their place of origin, and that are unrelated to adjacent rock bodies or terranes.

expansive soils Clayey soils that expand when they absorb water and shrink when they dry out.

extinct volcano A volcano that has not erupted in such a long time that, typically, it has become significantly eroded. Extinct volcanoes will not ever erupt again—though eruptions could recur in the same area or region.

extrusive rocks Igneous rocks that formed at or near the surface of the Earth. Because they cooled rapidly, they generally have an *aphanitic* texture.

factor of safety The balance between resisting and driving forces in a landslide. Failure occurs when the factor of safety is less than 1—that is, when resisting forces are no longer greater than driving forces.

fan Triangular or fan-shaped deposit of sediment where a stream or gully deposits sediment on a less-sloped plain. There the sediment cannot be transported and so accumulates.

fault A major fracture in Earth's crust along which there has been displacement.

fault creep (tectonic) The gradual slip or motion along a fault without an earthquake.

feedback A spontaneous, natural response to some process occurring in nature. See also *positive feedback* and *negative feedback.*

feed-in tariffs Payments by utility companies to customers that generate electricity by renewable power sources.

felsic Descriptive of magma or rock with abundant light-colored minerals and a high silica content. The mnemonic term is derived by combining the words *fel*dspar and *si*lica.

fetch The unobstructed stretch of sea over which the wind blows to create wind waves.

fiamme Shred-like lenses of obsidian representing clots of pumice that have been flattened, compressed, and remelted in a thick PDC deposit generated by a Plinian eruption.

firn A transitional material between snow and glacial ice, being older and denser than snow, but not yet transformed into glacial ice. Snow becomes firn after surviving one summer melt season; firn becomes ice when its permeability to liquid water becomes zero.

fissile isotopes Isotopes with nuclei that are capable of being split into other elements. See also *fission.*

fission The rupture of the nucleus of an element into lighter elements (fission products) and free neutrons spontaneously or by absorption of a neutron.

fjord A narrow, steep-walled inlet of the sea between cliffs or steep slopes, excavated or at least shaped by the passage of a glacier. A drowned glacial valley.

flood basalt A lava flow resulting from the rapid outpouring of an exceptionally large volume of lava. Flood basalt flows stack up to develop plains, plateaus, and submarine platforms. They originate from long fissure systems, and represent a type of effusive eruption.

flood frequency The average time interval between floods that are equal to or greater than a specified discharge.

floodplain The portion of a river valley adjacent to the channel that is built of sediments deposited during times when the river overflows its banks at flood stage.

floodwall A heavily reinforced wall designed to contain a stream at flood stage.

floodway Floodplain area under federal regulation for flood insurance purposes.

flotation The process of concentrating minerals with distinct nonwettable properties by floating them in liquids that contain soapy frothing agents, such as pine oil.

fluidization The process whereby granular solids (corn, wheat, volcanic ash, and lapilli) under high gas pressures become fluid-like and flow downslope or can be pumped.

fluidized-bed combustion A process of burning powdered coal more cleanly and efficiently, when suspended in upward-moving jets of air or gas during the combustion process.

fluvial Pertaining to rivers.

focus (hypocenter) Point within Earth where an earthquake originates.

foliation The planar or wavy structure that results from the flattened growth of minerals in a metamorphic rock.

forecast In science, a forecast is a description of what may happen in the future, usually stated in terms of probability.

foreshore slope The zone of a beach that is regularly covered and uncovered by the tide.

fossil fuels Coal, oil, natural gas, and all other solid or liquid hydrocarbon fuels.

fracking Oil company jargon for improving oil or gas production by hydraulic fracturing.

fracture Break in a rock caused by tensional, compressional, or shearing forces. See also *joint.*

frictional strength The strength of a material that is proportional to the normal force applied.

frost wedging The opening of joints and cracks by the freezing and thawing of water.

fully developed sea The largest possible waves that form in a wind of a certain strength blowing over a set length of water.

fumaroles Vents in a volcanic area that issue volcanic gases.

fusion The combination of two nuclei to form a single, heavier nucleus accompanied by a loss of some mass that is converted to heat.

Gaia The hypothesis that proposes life has a controlling influence on the oceans and atmosphere. It states that the Earth is a giant, self-regulating body with close connections between living and nonliving components.

gaining stream A stream that receives water from the zone of saturation.

garbage patches Large fields of floating trash in certain parts of the oceans; primarily small bits of plastic, but also including much larger debris.

geoengineering Technological options that involve large-scale engineering of our environment in order to combat or counteract the effects of changes in atmospheric chemistry.

Geographic Information System (GIS) A way of presenting data by overlaying different "layers" of information on a single map. The information may come from a variety of sources, from census tracks to satellite measurements or data acquired via geological field work.

geological agents All geological processes—for example, wind, running water, glaciers, waves, mass wasting—that erode, move, and deposit earth materials.

geological repository (nuclear waste) An underground vault in a terrane area free from geological hazards. Repositories in salt mines, granite, and welded tuff have been studied.

geothermal energy Energy derived from circulating hot water and steam; usually associated with cooling magma or hot rocks.

Geze classification The classification of eruptions into explosive, effusive, and extrusive categories. Useful for generally characterizing the eruptions of a volcano.

GHG See *greenhouse gases.*

gigawatt (GW) A thousand million (a billion) watts; 1,000 megawatts.

glacial outwash Deposits of stratified sand, gravel, and silt that have been removed from a glacier by meltwater streams.

glacial till See *till.*

glacier A large mass of ice formed, at least in part, on land by the compaction and recrystallization of snow. It moves slowly downslope by creep or outward in all directions because of the stress of its weight, and it survives from year to year.

global conveyor The integrated system of marine circulation initiated by the descent of large masses of cold water into the deep sea from the edges of Antarctica and Greenland. This water upwells elsewhere in the world ocean as it loses density because of warming and mixing, eventually flowing back to replace fresh masses of sinking water. (In this way, the sea surface remains nearly level as it circulates.)

Global Positioning System (GPS) A navigational system of coded satellite signals that can be processed in a GPS electronic receiver, enabling the receiver to compute position, elevation, and velocity.

global warming The general term for the steady rise in average global temperatures over the last 100 years.

Gondwana succession The fossil assemblage used by Alfred Wegener and others to discern the past existence of southern Pangaea.

GPS See *Global Positioning System.*

graben An elongate, flat-floored valley bounded by faults on each side.

graded stream A stream that is in equilibrium, showing a balance of erosion, transporting capacity, and material supplied to it. Graded streams have a smooth profile.

gradient (stream) The slope of a streambed usually expressed as the amount of drop per horizontal distance in meters per kilometer (m/km) or in feet per mile (ft/mi).

granular material A material with no cohesion between grains.

gravity separation See *crystal settling.*

greenhouse effect The added atmospheric temperature imparted by the trapping and reradiation of heat energy by certain gases—notably, carbon dioxide, methane, and water vapor in Earth's atmosphere.

greenhouse gases (GHG) Those gases responsible for the greenhouse effect. They absorb infrared radiation entering the atmosphere from space, from the surface below, and from neighboring gas molecules and reradiate it.

greenhouse warming Rise in average temperature of a body of air caused by an increase in the concentration of heat trapping gas—notably, carbon dioxide, methane, and water vapor in Earth's atmosphere.

groin A structure of rock, wood, or concrete built roughly perpendicular to a beach to trap sand.

groundwater That part of subsurface water that is in the zone of saturation (below the water table).

gullying The cutting of channels into the landscape by running water. When extreme, it renders farmland useless.

Gutenberg–Richter law (GRL) The observation that numbers of earthquakes in a region increase exponentially with decrease in magnitude. The GRL also provides a forecast of the numbers and sizes of aftershocks following large earthquakes.

Hadley cells Large, vertical, circulatory currents of air in Earth's atmosphere caused as air currents move from the equator toward the poles and vice versa. Particular cells are restricted to certain latitudinal zones.

half-life The period of time during which half of a given number of atoms of a radioactive element or isotope will disintegrate.

halocarbons Hydrogen/carbon-bearing compounds primarily of industrial origin that also contain atoms of halogen elements, particularly chlorine and fluorine.

hardpan Impervious layer just below the land surface produced by calcium carbonate ($CaCO_3$) in the B horizon. See also *claypan.*

hard rock (a) A term loosely used for an igneous or metamorphic rock, as distinguished from a sedimentary rock. (b) In mining, a rock that requires drilling and blasting for economical removal.

Hawaiian eruption The rapid outpouring of great volumes of lava, initially from a fissure. Little or no pyroclastic material erupts with the lava. A shield volcano or lava shield may grow in time from sustained Hawaiian eruptions.

high-grade deposit A mineral deposit with a high concentration of a desired element.

high-level nuclear waste Mostly radioactive elements in spent fuel with long half-lives.

homeostasis A self-regulatory property of many both biological and physical systems in which processes occur to minimize long-term drastic change.

horse latitudes Latitudes between 30° and 35° north and south of the equator, at sea, in which winds tend to be mild or nonexistent and the air dry and warm for long periods of time. Sailing vessels have trouble moving in such conditions.

hot spot A point (or area) on the lithosphere over a plume of magma rising from the mantle.

human carrying capacity The size of human population that can be indefinitely supported by essential natural resources.

hurricanes The name for cyclones originating in tropical waters of the eastern Pacific and Atlantic oceans.

hydration A process whereby anhydrous minerals combine with water.

hydraulic conductivity A groundwater unit expressed as the volume of water that will move in a unit of time through a unit of area measured perpendicular to the flow direction. Commonly called *permeability*, it is expressed as the particular aquifer's cubic meters/(day/square meters) (or meters/day) or cubic feet/(day/square feet) (or feet/day).

hydraulic fracturing A method for improving oil and gas production by injecting high-pressure fluid infused with chemicals and sand into "tight" formations like shale to increase porosity and permeability.

hydraulic gradient The slope or vertical change (feet/feet or meter/meter) in water-pressure head with horizontal distance in an aquifer.

hydraulic mining, hydraulicking A mining technique by which high-pressure jets of water are used to dislodge unconsolidated rock or sediment so that it can be processed.

hydrocarbon One of the many chemical compounds solely of hydrogen and carbon atoms; may be solid, liquid, or gaseous.

hydrocompaction The compaction of dry, low-density soils due to the heavy application of water.

hydrogen A colorless, odorless, highly flammable gaseous element, the lightest of all gases and the universe's most abundant element. It is used in the production of synthetic methanol, in petroleum refining, and in many other uses.

hydrogeologist A person who studies the geology and management of underground water and related aspects of surface water and groundwater.

hydrograph A graph of the stage (height) or discharge of a body of water over time.

hydrologic cycle (water cycle) The constant circulation of water from the sea to the atmosphere, to the land, and eventually back to the sea. The cycle is driven by solar energy.

hydrology The study of liquid and solid water on, under, and above the Earth's surface, including economic and environmental aspects.

hydrolysis The chemical reaction between hydrogen ions in water with a mineral, commonly a silicate, usually forming clay minerals.

hydrosphere The sum total of water present at the Earth's surface (or in shallow ground).

hydrothermal Of or pertaining to heated water, the action of heated water, or a product of the action of heated water—such as a mineral deposit that precipitated from a hot aqueous solution.

hydrothermal alteration The transformation of pre-existing rock by highly acidic or corrosive vapors or fluids and mineralizing solutions derived from magma or magmatic heating of groundwater. Hydrothermally altered rocks have distinctive coloration (often bright yellows and reds) relative to unaffected adjacent rocks, and can be quite soft and pliable. Some contain valuable ores, such as gold and cinnabar (mercury).

hypocenter See *focus*.

hypothesis An idea that can be rigorously tested and, if need be, falsified and rejected by field or laboratory observations to verify as a theory or fact.

ice crusts Ice layer on top or within a snowpack.

IGCC See *integrated gasification combined cycle*.

igneous rocks Rocks that crystallize from molten material at the surface of Earth (volcanic) or within Earth (plutonic).

illuviation Movement of clay through a soil profile by the action of water.

inactive sea cliff A cliff whose erosion is dominated by subaerial processes rather than wave action.

incineration A waste-treatment process that involves the combustion of waste at high temperatures and converts the waste into heat (that can be used to generate electricity), emits gases to the atmosphere, and makes residual ash.

indicated reserves The hydrocarbons that can only be recovered by secondary recovery techniques, such as fluid injection. These reserves are in addition to the reserves directly recoverable from a known reservoir.

infiltration The movement of water, including rain and snowmelt, into the soil surface.

integrated gasification combined cycle (IGCC) A technology that fuels an electric power plant with syngas.

intensity scale An earthquake rating scale (I–XII) based on subjective reports of human reactions to ground shaking and on the damage caused by an earthquake.

Intergovernmental Panel on Climate Change (IPCC) A scientific intergovernmental body tasked with evaluating the risk and impacts of climate change.

Intertropical Convergence Zone (ITCZ) See *doldrums*.

intrusive rocks Igneous rocks that have "intruded" into the crust; hence they are slowly cooled and generally have phaneritic texture.

inundation Flooding by water.

ion An atom with a positive or negative electrical charge because it has gained or lost one or more electrons.

IPAT equation A relationship describing how human environmental impact is a result of population, affluence, and choice of technology.

IPCC See *Intergovernmental Panel on Climate Change*.

island arcs Curving archipelagos of volcanic islands which overlay oceanic subduction zones.

isolation (waste disposal) Buried or otherwise sequestered waste.

isoseismals Lines on a map that enclose areas of equal earthquake shaking based on an intensity scale.

isostatic rebound Uplift of the crust of Earth that results from unloading, as results from the melting of ice sheets.

isotope One of two or more forms of an element that have the same atomic number but different atomic masses because they have different numbers of neutrons.

J-curve The shape of geometric (exponential) growth. Growth is slow at first, and the rate of growth increases with time.

jetty A structure built perpendicular to the shoreline to improve harbor inlets or river outlets.

joint Separation or parting in a rock that has not been displaced. Joints usually occur in groups ("sets"), the members of a set having a common orientation.

karst terrain, karst topography Area underlain by soluble limestone or dolomite and riddled with caves, caverns, sinkholes, lakes, and disappearing streams.

kerogen A carbonaceous residue in sediments that has survived bacterial metabolism. It consists of large molecules from which hydrocarbons and other compounds are released on heating.

kettle A bowl-shaped depression without surface drainage in glacial-drift deposits, often containing a lake, believed to have formed by the melting of a large, detached block of stagnant ice left behind by a retreating glacier.

Kīpuka Hawaiian word referring to a patch of older land surrounded by younger lava flows. "Steptoe" is a European equivalent term.

Kyoto Protocol 1997 agreement by which participating countries would reduce CO_2 emissions.

Lacroix classification System of classifying volcanic eruptions based on observational characteristics, such as relative explosivity, type of material issued, and associated vent formation. Eruption categories are named for particular volcanoes or regions where particular eruptions are common. For example, Hawaiian and Strombolian are two categories in the Lacroix classification.

lahar A debris flow or mudflow consisting of volcanic material.

land conversion The transformation of land from natural conditions to human use, primarily for agricultural development.

landslide The downslope movement of rock and/or soil as a semicoherent mass on a discrete slide surface or plane. See also *mass wasting*.

La Niña A phenomenon of the equatorial Pacific bringing cooler than normal waters to the eastern Pacific, causing wetter than normal conditions in the Pacific Northwest, warmer than normal conditions in the North Central states, and cooler than normal in the Southwest and Southeast. See also *El Niño* and *ENSO*.

lapilli (cinders) Fragments only a few centimeters in size that are ejected from a volcano.

lateral spreading The horizontal movement on nearly level slopes of soil and mineral particles caused by liquefaction of quick clays.

laterite A highly weathered brick-red soil characteristic of tropical and subtropical rainy climates. Laterites are rich in oxides or iron and aluminum, and have some clay minerals and silica. Bauxite is an aluminum-rich deposit of a similar origin.

lava Molten igneous material at the surface of the Earth.

lava dome (volcanic dome) Steep-sided protrusion of viscous, glassy lava, sometimes within the crater of a larger volcano (e.g., Mount St. Helens) and sometimes free-standing (e.g., Mono Craters).

lava plateau A broad, elevated tableland, thousands of square kilometers in extent, underlain by a thick sequence of lava flows.

leachate The water that percolates through landfills. If released to the surface or reaches the underground environment, leachate can pollute surface and underground water.

levees (natural) Embankments of sand or silt built by a stream along both banks of its channel. They are deposited during floods when waters overflowing the stream banks deposit sediment as they slow down.

limnic eruption Sudden eruption of gas, typically carbon dioxide, from a lake.

liquefaction See *spontaneous liquefaction*.

liquified natural gas (LNG) Gas, primarily methane (CH_4), that is cooled to approximately -160°C (-256°F) so that it can be transported in a small volume as a liquid to regions that use more natural gas than they produce.

lithification The conversion of sediment into solid rock through processes of compaction, cementation, and crystallization.

lithified Changed into stone, as in the transformation of loose sand to sandstone.

lithosphere The solid, rocky outer shell of Earth.

littoral drift Sediment (sand, gravel, silt) that is moved parallel to an ocean or lake shore by longshore currents.

lode A mineral deposit consisting of a zone of veins in consolidated rock, as opposed to a *placer deposit*.

loess A blanket deposit of buff-colored silt that shows little or no stratification. It covers wide areas in Europe, eastern China, and the Mississippi Valley. It is generally windblown dust of Pleistocene age.

longitudinal wave (P-wave) A type of seismic wave involving particle motion that alternates in expansion and compression in the direction of wave propagation. Also known as *compressional waves*, they resemble sound waves in their motion.

long-period seismic waves Seismic waves that arrive infrequently at any given seismic station. They characteristically have longer wavelengths than short-period seismic waves and impact larger buildings as a result.

longshore current (littoral current) The current adjacent and parallel to a shoreline that is generated by waves striking the shoreline at an angle.

longshore drift The movement of sand along a beach caused by longshore currents.

losing stream A stream or reach of a stream (typically in arid regions) that contributes water to the zone of saturation.

low-grade deposit A mineral deposit in which the mineral content is minimal but still exploitable at a profit.

low-level nuclear wastes Wastes generated in research laboratories, hospitals, and industry. Shallow burial in containers is sufficient to sequester this waste.

luster The manner in which a mineral reflects light, described as metallic, resinous, silky, or glassy.

maar A low-relief, broad volcanic crater formed by multiple shallow explosive eruptions. Maars commonly contain a lake and are surrounded by a low rampart or ring of ejected material. Typically they form by *phreatic eruptions*.

mafic Descriptive of a magma or rock rich in iron and magnesium. The mnemonic term is derived from *ma*gnesium and *f*erric.

magma Molten material within the earth that is capable of intrusion or extrusion and from which igneous rocks form.

magnetic polarity A property indicating the position of Earth's magnetic poles relative to a rock during its formation, as expressed by the alignment of tiny iron-bearing magnetic grains contained within the rock.

magnitude (earthquake) A measure of the strength or the strain energy released by an earthquake at its source. See also *moment magnitude* and *Richter magnitude scale*.

manganese nodule A small, irregular, black to brown, laminated concretionary mass consisting primarily of manganese minerals with some iron oxides and traces of other metallic minerals, abundant on the floors of the world's oceans.

mantle A thick shell of magnesium-silicate matter that surrounds Earth's core.

mantle lid Part of Earth's rigid outer shell that also includes the crust.

marine trenches Steep-walled abyssal valleys that fringe some ocean floors.

massive sulfide deposit Usually volcanogenic, often rich in zinc and sometimes in lead.

mass wasting A general term for all downslope movements of soil and rock material under the direct influence of gravity.

maximum contaminant level (MCL) The highest level of a contaminant that is allowed in drinking water. MCLs are set as close to the MCL goals as feasible using the best available treatment technology.

maximum contaminant level goal (MCLG) The level of a contaminant in drinking water below which there is no known or expected risk to health. MCLGs allow for a margin of safety.

M-discontinuity (Moho) The contact between Earth's crust and mantle. There is a sharp increase in earthquake P-wave velocity across this boundary.

meander One of a series of sinuous curves in the course of a stream.

megawatt (MW) A million watts; 1,000 kilowatts.

metamorphic rocks Preexisting rocks that have been altered by heat, pressure, or chemically active fluids.

methane clathrate or **methane hydrate** A solid compound in which a large amount of methane gas is trapped within a cage of ice, and is widely distributed in deep-sea sediments and in permafrost in the Arctic and subarctic.

methanol A colorless liquid chemical compound, commonly called *wood alcohol*, that is volatile, flammable, and poisonous, with the chemical formula CH_3OH.

mid-latitude desert A desert area occurring within latitudes 30° to 40° north and south of the equator in the deep interior of a continent, usually on the lee side of a mountain range that blocks the path of prevailing winds, and commonly characterized by a highly seasonal and dry climate. See also *rain-shadow desert.*

mid-ocean ridges Belts of broad volcanic highlands that bisect the seafloor worldwide.

millirem A measure of radioactive emissions in the environment.

mineral Naturally occurring crystalline substance with well-defined physical properties and a definite range of chemical composition.

mineral deposit A localized concentration of naturally occurring mineral material (e.g., a metallic ore or a nonmetallic mineral), usually of economic value, without regard to its mode of origin.

mineral reserves Known mineral deposits that are recoverable under present conditions but as yet undeveloped. The term excludes *potential ore*.

mineral resources The valuable minerals of an area that are presently legally recoverable or that may be so in the future; include both the known ore bodies (*mineral reserves*) and the *potential ores* of a region.

miscible Soluble; capable of mixing.

modified Mercalli scale (*MM*) Earthquake intensity scale from I (not felt) to XII (total destruction) based on damage and reports of human reactions.

Mohs hardness scale A scale that indicates the relative hardness of minerals on a scale of 1 (talc, very soft) to 10 (diamond, very hard).

moment magnitude (M_w or M) A scale of seismic energy released by an earthquake based on the product of the rock rigidity along the fault, the area of rupture on the fault plane, and the amount of slip.

monogenetic Referring to volcanoes whose full development takes place during a single eruption. Monogenetic volcanoes erupt only once in their lifetimes.

moraine A mound, ridge, hill, or other distinct accumulation of unsorted, unstratified glacial sediment, predominantly till, deposited chiefly by direct action of glacial ice in a variety of landforms.

mother lode (a) A main mineralized unit that may not be economically valuable in itself but to which workable deposits are related; for example, the Mother Lode of California. (b) An ore deposit from which a placer is derived; the *mother rock* of a placer.

mountaintop-removal mining The mining of coal that underlies the tops of mountains.

natural gas Gas, primarily methane (CH_4), that occurs naturally with liquid petroleum, in coal beds, and as methane clathrates. It is an important fuel, a feedstock for manufacturing fertilizers and plastics, and a potent greenhouse gas.

natural levee See *levee*.

negative anomaly Magnetization in ancient rocks caused by a magnetic field oriented opposite to the one existing at present; that is, the northern magnetic pole existed in the south, and the southern pole existed in the north.

negative feedback A natural response to a process occurring in nature that tends to lessen the effects of that process.

NEPA National Environmental Policy Act of 1969.

neutral buoyancy zone The level in the crust at which the density of magma matches that of the surrounding hardened crust, causing the magma to stall out as it ascends, and accumulate to form a magma chamber or reservoir. Neutral buoyancy can also refer to the atmosphere, where a growing cloud spreads out in the thinner air above.

neutron An electronically neutral particle (zero charge) of an atomic nucleus with an atomic mass of approximately 1.

nonrenewable resource An energy or material resource that once used is not available for reuse in human time spans. Coal, oil, and metallic minerals are examples.

normal force The force applied perpendicular to the failure plane of a landslide.

nucleus The positively charged central core of an atom containing protons and neutrons that provide its mass.

nuée ardenté (pyroclastic flow) French for "glowing cloud," it is a highly heated, almost incandescent cloud of volcanic gases and pyroclastic material that travels with great velocity down the slopes of a volcano. Produced by the explosive disintegration of viscous lava in a vent.

NWPA Nuclear Waste Policy Act of 1982 mandating selection of a repository for nuclear waste.

obliquity The tilt of Earth's axis.

oceanic gyres Great looping currents in shallow seawater. Each hemisphere has its own set of gyres, the sizes and shapes of which are determined by the adjoining continental coastlines. The directions of circulation reflect the Coriolis effect.

ocean thermal energy conversion (OTEC) A technology that utilizes the temperature difference between warm surface ocean water and colder deep water to generate electricity.

oil shale A group of fine-grained sedimentary rocks rich enough in organic material (kerogen) that will yield petroleum on distillation.

open-pit mining Mining from open excavations, most commonly for low-grade copper and iron deposits, as well as coal.

ore A volume of rock containing useful minerals in concentrations that can be profitably mined, transported, and processed.

ore deposit The same as a *mineral reserve* except that it refers only to a metal-bearing deposit.

ore mineral The part of an ore, usually metallic, that is economically desirable.

outflow Water moving away from the beach.

outgassing The release of gases and water vapor from molten rocks, leading to the formation of Earth's atmosphere and oceans.

outlet glacier A tongue-shaped glacier that originates from an ice sheet or mountain icefield.

overturn Process by which stagnant bottom waters in a lake come to the surface and are refreshed with oxygen.

oxbow lake A crescent-shaped lake formed along a stream course when a tight meander is cut off and abandoned.

ozone hole An area in the atmospheric ozone layer that is thin or absent.

pāhoehoe Basaltic lava typified by smooth, billowy, or ropy surfaces.

paleoseismicity The rock record of past earthquake events in displaced beds and liquefaction features in trenches or natural outcrops.

Pangaea A supercontinent that existed from about 300 to 200 million years ago, which included all the continents we know today.

parabolic trough Parabolic mirrors that focus sunlight on a fluid-filled reservoir pipe to generate electricity.

passive margins Coastlines, such as the Eastern Seaboard of the United States, that are no longer active in terms of volcanism and mountain building.

pedalfer Soil of humid regions characterized by an organic-rich A horizon, and clays and iron oxides in the B horizon.

pedocal Soil of arid or semiarid regions that is rich in calcium carbonate.

pedologist (Greek: *pedo,* "soil," and *logos,* "knowledge") A person who studies soils.

pedology The study of soils.

pegmatite An exceptionally coarse-grained igneous rock with interlocking crystals, usually found as irregular dikes, lenses, or veins, especially at the margins of batholiths.

perched water table The upper surface of a body of groundwater held up by a discontinuous impermeable layer above the static water table.

permafrost Permanently frozen ground, with or without water, occurring in arctic, subarctic, and alpine regions.

permafrost table The upper limit of permafrost.

permeability The degree of ease with which fluids flow through a porous medium. See also *hydraulic conductivity.*

phaneritic texture A coarse-grained, igneous-rock texture resulting from slow cooling. Phanerites are coarse-grained igneous rocks.

photochemical smog Atmospheric haze that forms when automobile-exhaust emissions are activated by ultraviolet radiation from the Sun to produce highly reactive oxidants known to be health hazards.

photovoltaics (PV) A technology for generating electricity using various semiconductor materials, such as silicon or germanium arsenide, that converts sunlight directly into direct current (DC) electricity.

phreatic eruption Volcanic eruption, mostly steam, caused by the interaction of hot magma with underground water, lakes, or seawater. Where significant amounts of new (magmatic) material are ejected in addition to steam, the eruptions are described as *phreatomagmatic.*

phytoremediation The use of plants to remove soil contaminants, particularly excess salt.

pillow lava Accumulations of basaltic lava resembling piled sacks that form during many underwater effusive eruptions.

pingo An ice-cored, round hill in permafrost formed as groundwater migrates toward the surface and freezes.

piping Subsurface erosion in sandy materials caused by the percolation of water under pressure or the influence of gravity.

placer deposit (pronounced "plasser") A surficial mineral deposit formed by settling from streams of mineral particles from weathered debris. See also *lode.*

plastic Capable of being deformed continuously and permanently in any direction without rupture.

plate Broad region of Earth's crust that is structurally distinctive.

plate tectonics A scientific theory that describes the large-scale motions of curved plates of Earth's lithosphere.

Pleistocene The geological epoch of the Quaternary period during which glaciations were common, falling by definition between 2.0 million and 10,000 years ago.

Plinian column A steady, turbulent, nearly vertical column of ash and steam released at high velocity from a volcanic vent in an explosive eruption.

plutonic Pertaining to rock formed by any process at depth, usually with a phaneritic texture.

point bar A deposit of sand and gravel found on the inside curve of a *meander.*

polar air The masses of frigid, dry air residing over Earth's poles.

polar desert A bitterly cold, arid region at high latitudes that receives very little precipitation.

polar front The zones of collision between warm, moist air heading poleward and polar air occurring at high latitudes encircling both of Earth's poles.

polar high The region of cold, high-density air that exists in the polar regions.

polar wandering The apparent shifting of Earth's magnetic poles over geologic time, which turns out to be illusory, because it is the continents, and not the magnetic poles, that have actually shifted.

polygenetic Referring to volcanoes whose full development takes place from multiple eruptions, generally occurring over a period of many centuries or millennia.

population dynamics The study of how populations grow.

population growth rate The number of live births less deaths per 1,000 people, usually expressed as a percentage; a growth rate of 40 would represent 4.0% growth.

pore pressure The stress exerted by the fluids that fill the voids between particles of rock or soil.

porosity A material's ability to contain fluid. The ratio of the volume of pore spaces in a rock or sediment to its total volume, usually expressed as a percentage.

porphyry copper A copper deposit, usually of low grade, in which the copper-bearing minerals occur in disseminated grains and/or in veinlets through a large volume of rock.

positive anomaly Magnetization in ancient rocks caused by a magnetic field oriented much as at present ("positive" pole in the north and "negative" pole in the south).

positive feedback A natural response to a process occurring in nature that tends to enhance the effects of that process.

potable water Water that is suitable for drinking.

potential ore As yet undiscovered mineral deposits and known mineral deposits for which recovery is not yet economically feasible.

potentiometric surface The hydrostatic head of groundwater and the level to which the water will rise in a well. The water table is the potentiometric surface in an unconfined aquifer. Also called *piezometric surface.*

powder-snow avalanche Avalanche of dry, recently fallen snow that has little cohesion.

power tower A tall tower with a fluid-filled central receiver that is heated by Sun-tracking mirrors to generate electricity.

precession The slow gyration of Earth's axis, analogous to that of a spinning top, which causes the slow change in the orientation of Earth's axis relative to the Sun that accounts for the reversal of the seasons of winter and summer in the Northern and Southern hemispheres every 11,500 years.

precipitation (a) The separation of a solid substance from a solution by a chemical reaction. (b) rain, sleet or snow falling to the ground from clouds.

precursors (of earthquakes) Observable phenomena that occur before an earthquake and indicate that an event is soon to occur.

prediction In science, it is a statement that under specific circumstances something will occur, leading to recognition of cause and effect.

pregnant pond A catchment basin at a mine that holds a gold- and silver-bearing cyanide solution.

proglacial lake An ice-marginal lake formed just beyond the frontal moraine of a retreating glacier and generally in direct contact with the ice.

projection In science, it is a "what if" forecast; there is no assurance that a particular scenario will actually take place.

proton A particle in an atomic nucleus with a positive electrical charge and an atomic mass of approximately 1.

protostar A star in the process of formation that has entered the slow gravitational contraction phase.

proxy A substitute.

pumice Frothy-appearing rock composed of natural glass ejected from high-silica, gas-charged magmas. It is the only rock that floats.

P-wave See *longitudinal wave.*

pyroclastic ("fire broken") Descriptive of the fragmental material, ash, cinders, blocks, and bombs ejected from a volcano. See also *tephra.*

pyroclastic density current (PDC) A deadly, ground-hugging cloud of hot gases and pyroclastic debris.

qanats Subsurface tunnels excavated in alluvial fans in arid climates (e.g., Iraq and Iran) to extract water. An ancient technology that effectively transports water by gravity without the need for pumping.

quad A quadrillion British thermal units (10^{15} Btu).

quick clay (sensitive clay) A clay possessing a "house-of-cards" sedimentary structure that collapses when it is disturbed by an earthquake or other shock.

radioactivity The property possessed by certain elements that spontaneously emit energy as the nucleus of the atom changes its proton-to-neutron ratio.

radiometric dating See *absolute age dating.*

radon A radioactive gas emitted in the breakdown of uranium.

rain-shadow desert A desert in the middle latitudes on the lee side of a mountain or a mountain range. See also *mid-latitude desert.*

rank (coal) A coal's carbon content depending on its degree of metamorphism.

RCRA Resource Conservation and Recovery Act.

recessional moraine One of a series of nested end moraines that record the stepwise retreat of a glacier at the end of an ice age.

recrystallization The formation in the solid state of new mineral grains in a metamorphic rock. The new grains are generally larger than the original mineral grains.

recurrence interval The return period of an event, such as a flood or an earthquake, of a given magnitude. For flooding, it is the average interval of time within which a given flood will be equaled or exceeded by the annual maximum discharge.

recycling Reprocessing of materials into new products.

regional metamorphism A general term for metamorphism that affects an extensive area.

regolith Unconsolidated rock and mineral fragments at the surface of Earth.

regulatory floodplain The part of a floodplain that is subject to federal government regulation for insurance purposes.

relative-age dating The chronological ordering of geological strata, features, fossils, or events without reference to their absolute age.

renewable resource An energy or material resource that continually renews itself as it is being consumed. Timber, solar energy, wind, and water are examples.

replacement rate The birth rate at which just enough offspring are born to replace their parents.

reserves Those portions of an identified resource that can be recovered economically.

reservoir rock A permeable, porous geological formation that will yield oil or natural gas.

residence time The average length of time a substance (atom, ion, molecule) remains in a given reservoir.

residual soil Soil formed in place by decomposition of the rocks on which it lies.

resisting force (landslide) Friction along the slide plane and normal forces across the plane.

resonance The tendency of a system (a structure) to vibrate with maximum amplitude when the frequency of the applied force (seismic waves) is the same as the vibrating body's natural frequency.

resource recovery The removal of certain materials from the waste stream for the purpose of recycling or composting.

revetment A rock or concrete structure built landward of a beach to protect coastal property.

Richter magnitude scale (earthquakes) The logarithm of the maximum trace amplitude of a particular seismic wave on a seismogram, corrected for distance to epicenter and type of seismometer. A measure of the energy released by an earthquake.

rift zone (volcanology) A linear zone of weakness on the flank of a shield volcano that is the site of frequent flank eruptions.

rill erosion The carving of small channels, up to 25 centimeters (10 inches) deep, in soil by running water.

riparian Pertaining to or situated on the bank of a body of water, especially of a river.

rip current A local, focused current that returns water from the beach zone to deeper water.

rock cycle The cycle or sequence of events involving the formation, alteration, destruction, and re-formation of rocks as a result of processes such as erosion, transportation, lithification, magnetism, and metamorphism.

rocks Aggregates of minerals or rock fragments.

rotational landslide (slump) A slope failure in which sliding occurs on a well-defined, concave-upward, curved surface, producing a backward rotation of the slide mass.

runaway feedback A process that once set in motion accelerates by setting into action secondary processes that promote it further.

runoff That part of precipitation falling on land that runs into surface streams.

run-up The elevation to which a wind wave or tsunami advances onto the land.

Sahel Belt of populated semiarid lands along the southern edge of North Africa's Sahara Desert.

salt-crystal growth The growth of salt crystals from solution, such as those of sodium chloride, which can generate large pressures sufficient to break rock.

salt dome A column or plug of rock salt that rises from depth because of its low density and pierces overlying sediments.

saltwater intrusion The movement of denser saltwater into the pore space of an aquifer as less dense freshwater is removed by groundwater pumping and not replaced.

sanitary landfill A method of burying waste where each day's dumping is covered with soil to protect the surrounding area. This is also monitored to ensure protection of local underground water.

seafloor spreading A hypothesis that oceanic crust forms by convective upwelling of lava at mid-ocean ridges and moves laterally to trenches where it is destroyed.

seawall A rock or concrete structure built landward of a beach to protect the land from wave action. See also *revetment*.

secondary enrichment Processes of near-surface mineral deposition, in which oxidation produces acidic solutions that leach metals, carry them downward, and reprecipitate them, thus enriching sulfide minerals already present.

secondary recovery Production of oil or gas by artificially stimulating a depleted reservoir with water, steam, or other fluids to mobilize the remaining hydrocarbons.

secure landfill A landfill that totally isolates the received waste from the environment.

sedimentary rock A layer of rock deposited from water, ice, or air and subsequently lithified to form a coherent rock.

seif (longitudinal) dune A very large, sharp-crested, tapering chain of sand dunes; commonly found in the Arabian Desert.

seismogram The recording from a seismograph.

seismograph An instrument used to measure and record seismic waves from earthquakes.

seismology The study of earthquakes.

semiarid grasslands Extensive treeless grassland regions that are marginal to many mid-latitude deserts.

sensitive clay See *quick clay*.

settlement This occurs when the applied load of a structure's foundation is greater than the bearing strength of the foundation material (soil or rock).

sheet erosion The removal of thin layers of surface rock or soil from an area of gently sloping land by broad, continuous sheets of running water (sheet flow), rather than by channelized streams.

sheet joints Cracks more or less parallel to the ground surface that result from expansion because of deep erosion and the unloading of overburden pressure. See also *joint*.

shield volcano (cone) The largest type of volcano, composed of piles of lava in a convex-upward slope. Found mainly in Hawaii and Iceland.

shoaling The steepening of waves as they approach the shore and "feel" the bottom; eventually causes the wave to break.

short-period seismic waves Seismic waves that arrive frequently at any given seismic station.

sinkhole Circular depression formed by the collapse of a shallow cavern in limestone.

slab avalanche A coherent block of snow that fails as a slide and then moves downslope in chunks.

slant drilling (whipstocking) Purposely deflecting a drill hole from the vertical to tap a reservoir not directly below the drill site. The deflecting tool is a whipstock.

slide plane (slip plane) A curved or planar surface along which a landslide moves.

slip face The steeper, lee side of a dune, standing at or near the angle of repose of loose sand and advancing downwind by a succession of slides whenever that angle is exceeded.

slump See *rotational landslide*.

SMCRA See *Surface Mining Control and Reclamation Act*.

smelter An industrial plant that mechanically and chemically produces metals from their ores.

snow-albedo positive feedback The impact of high snow albedo (solar reflectivity) by cooling the surrounding atmosphere. This may encourage the accumulation of additional snowfall.

soil (a) Loose material at the surface of Earth that supports the growth of plants (pedological definition). (b) All loose surficial earth material resting on coherent bedrock (engineering definition).

soil horizon A layer of soil that is distinguishable from adjacent layers by properties such as color, texture, structure, and chemical composition.

soil order A means by which major soil types can be classified.

soil profile A vertical section of soil that exposes all of its horizons.

solar-thermal A technology for converting solar energy into thermal energy (heat) which heats a fluid, which is then used to run a turbine and generate electricity.

solution (weathering) The dissolving of rocks and minerals by natural acids or water.

source rock The geological formation in which oil and/or gas originates.

specific yield The ratio of the volume of water drained from an aquifer by gravity to the total volume considered.

speleology The study of caves and caverns.

spheroidal weathering Weathering of rock surfaces that creates a rounded or spherical shape as the corners of the rock mass are weathered faster than its flat faces.

spit A narrow stretch of sand that extends out from the coastline, often formed by longshore drift.

spoil Refuse rock material that results from mining, excavation, or dredging.

spontaneous liquefaction Process whereby water-saturated sands, clays, or artificial fill suddenly becomes fluid upon shaking, as in an earthquake.

stage The height of floodwaters in feet or meters above an established datum plane.

stalactites Deposits of calcium carbonate precipitated from solution and building downward from the top of a cave.

stalagmites Deposits of calcium carbonate precipitated from solution and building upward from the bottom of a cave.

steppe An extensive, treeless grassland in the semiarid mid-latitudes of Asia and southeastern Europe. It is drier than the prairie that develops in the mid-latitudes of the United States.

Sterling cycle heat engine A closed-cycle hot air engine that operates by compression and expansion of a working fluid, commonly air. The working fluid is in a closed system permanently contained within the engine. Sterling cycle engines are more efficient than steam engines, and heat can be supplied by any fuel, even the Sun.

storm surge The sudden rise of sea level on an open coast because of a storm. Storm surge is caused by strong onshore winds and results in water being piled up against the shore. It may be enhanced by high tides and low atmospheric pressure.

strain Deformation resulting from an applied stress (force per unit area). It may be elastic (recoverable) or ductile (nonreversible).

stratification (bedding) The arrangement of sedimentary rocks in strata, or layers.

stratigraphic trap An accumulation of oil that results from a change in the character (permeability) of the reservoir rock, rather than from structural deformation.

stratovolcano A large volcano that is shaped slightly concave upward and is composed of layers of ash and lava.

stress The force applied on an object per unit area, expressed in pounds per square foot (lb/ft^2) or kilograms per square meter (kg/m^2).

strike The direction west or east of north of the trace of the inclined plane (sedimentary bedding plane, fault, or joint surface) on the horizontal plane.

strip mining Surficial mining, in which the resource is exposed by removing the overburden.

Strombolian eruption Characterized by pulsating bursts or explosions from a volcanic vent that throw out large quantities of cinders and volcanic bombs (blobs of lava).

subduction The sinking of one lithospheric plate beneath another at a convergent plate margin. See also *convergent boundary*.

subpolar low A region of low atmospheric pressure located at about latitudes 60° north and 60° south.

subsidence Sinking or downward settling of Earth's surface caused by solution, compaction, withdrawal of underground fluids, cooling of the hot lithosphere, or loading with sediment or ice.

subtropical desert A hot, dry desert occurring between latitudes 15° to 30° north and south of the equator where subtropical air masses prevail.

subtropical highs The climatic belts at latitudes of approximately 30° to 35° north and south of the equator characterized by semipermanent high-pressure air masses, moderate seasonal rainfall, summer heat, dryness, and generally clear skies.

Suess effect The build-up in Earth's atmosphere of isotope ^{12}C, from the burning of fossil fuels through transportation emissions and industrial activity.

Superfund A spinoff of CERCLA insuring financial accountability for environmental polluters.

surcharge Weight added to the top of a natural or artificial slope by sedimentation or by artificial fill.

surface hoar Deposit of ice directly from water vapor onto the top of the snowpack. Forms a very weak layer on which avalanches run.

SMCRA Surface Mining Control and Reclamation Act of 1977.

surface wave Seismic waves that are trapped and spread along Earth's surface. Two types exist: Love and Rayleigh waves. They are the most damaging of seismic waves.

surf zone The near-shore area where waves are breaking.

surging glacier A glacier that alternates periodically between brief periods (usually 1–4 years) of very rapid flow, called *surges*, and longer periods (usually 10–100 years) of near stagnation.

suspension (a) A mode of sediment transport in which the upward currents in eddies of turbulent flow are capable of supporting the weight of sediment particles and keeping them held indefinitely in the surrounding fluid (such as silt in water or dust in air). (b) The state of a substance in such a mode of transport.

sustainable society A population capable of maintaining itself while leaving sufficient resources for future generations.

sustained yield (hydrology) The amount of water an aquifer can yield on a daily basis over a long period of time.

S-wave See *transverse wave*.

syncline A concave-upward fold that contains younger rocks in its core. Contrast with *anticline*.

synfuels (synthetic fuels) Manufactured fuels; for example, fuel derived by liquefaction or gasification of coal.

syngas Synthesis gas, a mixture of hydrogen and carbon monoxide manufactured by the gasification of coal or other organic materials. It has less than half the energy density of natural gas.

tailings The worthless rock material discarded from mining operations.

talus Coarse, angular rock fragments lying at the base of the cliff or steep slope from which they were derived.

tar sand A sand body that is large enough to hold a commercial reserve of asphalt or other thick oil. It is usually surface excavated to remove the thick hydrocarbons.

technically recoverable oil Petroleum that may be recoverable using current technology, regardless of cost.

tectonism (adj. tectonic) Deformation of the Earth's crust by natural processes leading to the formation of ocean basins, continents, mountain systems, and other Earth features.

teleseisms The records on seismographs of the arrival of very long period seismic waves from large, distant earthquakes.

temperate climate The mild seasonality of cool winters and warm summers characteristic of many regions of the world between the tropics and polar latitudes.

tephra A collective term for all *pyroclastic* material ejected from a volcano.

terrain anchors Surface roughness, such as trees or rocks, which prevent avalanches from releasing by holding back snow.

terranes See *exotic terranes*.

texture The size, shape, and arrangement of the component particles or crystals in a rock.

thermal advection The conveyance of heat from one place to another by a moving body, such as magma ascending in the crust, or a plume of hot air rising from a desert floor into the cooler atmosphere above.

tidal energy A technology for harnessing the natural energy of oceanic flood and ebb tides to generate electricity. Tidal energy has been used in the past to power grain mill water wheels that produced mechanical power.

tight oil Crude oil contained in petroleum-bearing formations of low permeability and porosity, such as shale.

till Unsorted and unstratified drift glacial sediment, generally firm and compacted by the weight of overlying ice.

tiltmeter A device designed to measure change in the angle of a slope.

tipping point A critical threshold in a feedback process (e.g., the snow-albedo effect), beyond which that feedback accelerates to overwhelm any counteractive feedbacks.

total fertility rate Average number of children born to women of childbearing age.

trace element An element that occurs in minute quantities in rocks or plant or animal tissue. Some are essential for human health.

trade wind desert See *subtropical desert*.

trade winds Steady winds blowing from areas of high pressure at 30° north and south latitudes toward the area of lower pressure at the equator. This pressure differential produces the northeast and southeast trade winds in the Northern and Southern hemispheres, respectively.

transform boundaries Plate boundaries with mostly horizontal (lateral) movement that connect spreading centers to each other or to subduction zones.

translational slide See *block glide*.

transported soils Soils developed on *regolith* transported and deposited by a geological agent, hence the terms *glacial* soils, *alluvial* soils, *eolian* soils, *volcanic* soils, and so forth.

transuranic elements Radioactive elements that have an atomic number greater than uranium and are human-made.

transverse dune A strongly asymmetrical sand dune that is elongated perpendicular to the direction of the prevailing winds. It has a gentle windward slope and a steep leeward slope that stands at or near the angle of repose of sand.

transverse wave (S-wave) A seismic wave propagated by a shearing motion that involves oscillation perpendicular to the direction of travel. It travels only in solids.

tropical climate The warm to hot, humid, and often rainy conditions characteristic of tropical latitudes, at and around the equator.

troposphere That part of the atmosphere next to Earth's surface in which temperature generally decreases rapidly with altitude, and clouds and convection are active. Where the weather occurs.

tsunami Long-wavelength sea wave produced by a submarine earthquake, a volcanic eruption, or a landslide. Incorrectly known as a tidal wave.

tuff A coherent rock formed from volcanic ash, cinders, or other *pyroclastic* material.

tundra (soil) A treeless plain characteristic of Arctic regions with organic rich, poorly drained tundra soils and permanently frozen ground.

typhoons The name for cyclones occurring in the western Pacific region.

umbrella (cloud) Refers to the spreading out of an ash-laden eruption cloud at the level of neutral buoyancy in the atmosphere.

unconfined aquifer Underground body of water that has a free (static) water table—that is, water that is not confined under pressure beneath an aquiclude.

unconventional oil and gas Petroleum and natural gas that is obtained using different techniques than production of conventional oil by oil wells. Unconventional oil and gas extraction is less efficient and has a greater environmental impact than producing conventional oil.

upwelling A process whereby cold, nutrient-rich water is brought to the surface. Coastal upwelling occurs mostly in trade wind belts.

vapor extraction The use of vacuum pumps placed in shallow bore holes to remove vaporized organic contaminants in the soil, such as gasoline or TCE.

vein deposit A thin, sheetlike igneous intrusion into a crevice.

ventifact Any stone or pebble that has been shaped, worn, faceted, cut, or polished by the abrasive or sandblast action of windblown sand, generally under arid conditions.

ventifaction The process of making ventifacts, or wind eroded stones, by abrasion from windblown particles such as sand and silt.

volatiles Substances that readily liquefy or become gaseous upon heating, including water vapor and the sulfurous fumes of some hot springs and virtually all volcanic vents.

volcanic conduit An opening or channel in Earth's crust through which magma and related gases ascend to the surface. Volcanic conduits may be pipe or sheet-like.

Volcanic Explosivity Index (VEI) A scale for rating volcanic eruptions according to the volume of the ejecta, the height to which the ejecta rise, and the duration of the eruption.

volcanogenic deposit Of volcanic origin; for example, volcanogenic sediments or ore deposits.

Waste Isolation Pilot Plant (WIPP) Sequesters defense-related transuranic waste.

waste management A complex of processes including waste disposal, reducing the amount of waste at its sources, waste recovery, and recycling.

water cycle See *hydrologic cycle.*

water table The contact between the zone of aeration and the zone of saturation.

wave base The depth at which water waves no longer move sediment, equal to approximately half the wavelength.

wave crest The top or highest point of a wave.

wave energy Energy produced by ocean surface waves, from which energy may be obtained to do work, such as generating electricity. The technology for capturing of wave energy is accomplished by a wave energy converter (WEC).

wave energy conversion A mechanical process of capturing wave energy by offshore wave energy conversion (OWEC) devices.

wave frequency The frequency with which individual seismic waves arrive at any given location.

wavelength The distance between two equivalent points on two consecutive waves; for example, crest to crest or trough to trough.

wave period (*T*) The time (in seconds) between passage of equivalent points on two consecutive waveforms.

wave refraction The bending of wave crests as they move into shallow water.

weathering The physical and chemical breakdown of materials of the Earth's crust by interaction with the atmosphere and biosphere.

welded tuff A glassy pyroclastic rock that has been made hard by the welding together of its particles under the action of retained heat.

wetted perimeter (stream) The length of streambed and bank that are in contact with running water in the stream in a given cross section.

wildcat well An oil well drilled in a region not known to be an oil field.

Wilson cycle An evolutionary sequence for oceans that identifies five stages in their birth and death; named for J. Tuzo Wilson.

wind The movement of air from one place to another.

WIPP See *Waste Isolation Pilot Plant.*

wobble The regular shift in orientation of Earth's rotational axis over a period of thousands of years.

Yucca Mountain The proposed repository in southern Nevada for the storage of high-level nuclear wastes. The repository proposal has recently been withdrawn because of concerns about geological stability and integrity of the site.

zero population growth A population growth rate of zero; that is, the number of deaths in a population is equal to the number of live births on average.

zone of aeration (vadose zone) The zone of soil or rock through which water infiltrates by gravity to the water table. Pore spaces between rock and mineral grains in this zone are filled with air, hence the name.

zone of flowage The lower part of a glacier that deforms because of the glacier's own weight.

zone of saturation The groundwater zone where voids in rock or sediment are filled with water.

INDEX